BIOCHEMICAL INTERACTIONS

Biochemistry is bundled with *Biochemical Interactions,* a CD-ROM that expands on the information presented in the textbook through the use of a variety of interactive three-dimensional molecular graphics displays and animations. These take the following forms, and they are all keyed to the text by a mouse icon ( ).

Interactive Exercises, 56 Chime™-based molecular graphics displays of proteins and nucleic acids that can be interactively rotated and otherwise manipulated.

Kinemages, alternative types of molecular graphics displays. These are presented in the form of 21 Exercises comprising 54 kinemages that amplify specific aspects of protein and nucleic acid structures.

Guided Explorations, 31 more complex interactive computer graphics displays and computerized animations, dealing with specific subjects in the textbook.

Animated Figures, text figures that have been animated for better understanding.

The Interactive Exercises and Guided Explorations were produced by ScienceMedia Inc. in collaboration with Donald Voet and Judith G. Voet. The Kinemages were produced by Donald Voet and Judith G. Voet. The Animated Figures were created by Super Nova in collaboration with Steven Vik and Judith G. Voet.

For the student, the CD-ROM extends the learning process from the textbook to the multimedia environment by drawing on motion, color, and three-dimensionality to illustrate aspects of molecular form and function that would otherwise be difficult to envision.

For the instructor, the CD-ROM is designed to be used as a teaching tool in computer presentation–equipped classrooms.

The Tables of Contents for these exercises (with text references in parentheses) *may be found on the last few pages of the text.*

One- and Three-Letter Symbols for the Amino Acids[a]

A	Ala	Alanine
B	Asx	Asparagine or aspartic acid
C	Cys	Cysteine
D	Asp	Aspartic acid
E	Glu	Glutamic acid
F	Phe	Phenylalanine
G	Gly	Glycine
H	His	Histidine
I	Ile	Isoleucine
K	Lys	Lysine
L	Leu	Leucine
M	Met	Methionine
N	Asn	Asparagine
P	Pro	Proline
Q	Gln	Glutamine
R	Arg	Arginine
S	Ser	Serine
T	Thr	Threonine
V	Val	Valine
W	Trp	Tryptophan
Y	Tyr	Tyrosine
Z	Glx	Glutamine or glutamic acid

[a]The one-letter symbol for an undetermined or nonstandard amino acid is X.

Thermodynamic Constants and Conversion Factors

Joule (J)
$1\ \text{J} = 1\ \text{kg} \cdot \text{m}^2 \cdot \text{s}^{-2}$ $\quad$ $1\ \text{J} = 1\ \text{C} \cdot \text{V}$ (coulomb volt)
$1\ \text{J} = 1\ \text{N} \cdot \text{m}$ (newton meter)

Calorie (cal)
1 cal heats 1 g of H_2O from 14.5 to 15.5°C
1 cal = 4.184 J

Large calorie (Cal)
1 Cal = 1 kcal $\quad$ 1 Cal = 4184 J

Avogadro's number (N)
$N = 6.0221 \times 10^{23}$ molecules · mol^{-1}

Coulomb (C)
$1\ \text{C} = 6.241 \times 10^{18}$ electron charges

Faraday ($\mathcal{F}$)
$1\ \mathcal{F} = N$ electron charges
$1\ \mathcal{F} = 96{,}485\ \text{C} \cdot \text{mol}^{-1} = 96{,}485\ \text{J} \cdot \text{V}^{-1} \cdot \text{mol}^{-1}$

Kelvin temperature scale (K)
0 K = absolute zero $\quad$ 273.15 K = 0°C

Boltzmann constant (k_B)
$k_B = 1.3807 \times 10^{-23}\ \text{J} \cdot \text{K}^{-1}$

Gas constant (R)
$R = Nk_B$ $\quad$ $R = 1.9872\ \text{cal} \cdot \text{K}^{-1} \cdot \text{mol}^{-1}$
$R = 8.3145\ \text{J} \cdot \text{K}^{-1} \cdot \text{mol}^{-1}$ $\quad$ $R = 0.08206\ \text{L} \cdot \text{atm} \cdot \text{K}^{-1} \cdot \text{mol}^{-1}$

The Standard Genetic Code

First Position (5′ end)	*Second Position* U	C	A	G	*Third Position (3′ end)*
U	UUU Phe	UCU Ser	UAU Tyr	UGU Cys	U
	UUC Phe	UCC Ser	UAC Tyr	UGC Cys	C
	UUA Leu	UCA Ser	UAA Stop	UGA Stop	A
	UUG Leu	UCG Ser	UAG Stop	UGG Trp	G
C	CUU Leu	CCU Pro	CAU His	CGU Arg	U
	CUC Leu	CCC Pro	CAC His	CGC Arg	C
	CUA Leu	CCA Pro	CAA Gln	CGA Arg	A
	CUG Leu	CCG Pro	CAG Gln	CGG Arg	G
A	AUU Ile	ACU Thr	AAU Asn	AGU Ser	U
	AUC Ile	ACC Thr	AAC Asn	AGC Ser	C
	AUA Ile	ACA Thr	AAA Lys	AGA Arg	A
	AUG Met[a]	ACG Thr	AAG Lys	AGG Arg	G
G	GUU Val	GCU Ala	GAU Asp	GGU Gly	U
	GUC Val	GCC Ala	GAC Asp	GGC Gly	C
	GUA Val	GCA Ala	GAA Glu	GGA Gly	A
	GUG Val	GCG Ala	GAG Glu	GGG Gly	G

[a]AUG forms part of the initiation signal as well as coding for internal Met residues.

BIOCHEMISTRY

BIOCHEMISTRY

VOLUME TWO

The Expression and Transmission of Genetic Information

DONALD VOET
University of Pennsylvania

JUDITH G. VOET
Swarthmore College

WILEY JOHN WILEY & SONS, INC.

About the Cover: The cover is an illustration of horse heart cytochrome *c* designed to show the influence of amino acid side chains on the protein's three-dimensional folding pattern. It was drawn by Irving Geis, in collaboration with Richard Dickerson, in 1972, the same year that the Protein Data Bank was established and had one structure deposited. As of the beginning of 2003, there were 20,000 structures available for download and visualization on desktop and laptop computers that did not even exist when this drawing was created. It reminds us that biochemistry is a process that is driven by the creativity of the human mind. Our visualization tools have developed from pen, ink, and colored pencils to sophisticated computer software available to all. Without creativity, however, these tools have little use.

Executive Editor *David Harris/Patrick Fitzgerald*
Senior Marketing Manager *Robert Smith*
Developmental Editor *Barbara Heaney*
Production Editor *Sandra Dumas*
Photo Editor *Hilary Newman*
Photo Researcher *Elyse Rieder*
Cover/Text Designer *Madelyn Lesure*
Production Management Services *Suzanne Ingrao*
Illustration Editor *Sigmund Malinowski*
Cover illustration and Part Openers, Irving Geis. Image from the Irving Geis Collection/ Howard Hughes Medical Institute. Rights owned by HHMI. Reproduction by permission only.

This book was typeset in 9.5/11.5 Times Ten Roman by TechBooks and printed and bound by Von Hoffmann Corporation. The cover was printed by Von Hoffmann Corporation.

The paper in this book was manufactured by a mill whose forest management programs include sustained yield harvesting of its timberlands. Sustained yield harvesting principles ensure that the number of trees cut each year does not exceed the amount of new growth.

This book is printed on acid-free paper. ∞

Voet, Donald
Biochemistry, Third Edition Volume Two/Donald Voet, Judith G. Voet

ISBN 0-471-25089-9 (cloth)

Printed in the United States of America.

10 9 8 7 6 5 4 3 2 1

To
Our parents, who encouraged us,
Our teachers, who enabled us, and
Our children, who put up with us.

PREFACE

Biochemistry is a field of enormous fascination and utility, arising, no doubt, from our own self-interest. Human welfare, particularly its medical and nutritional aspects, has been vastly improved by our rapidly growing understanding of biochemistry. Indeed, scarcely a day passes without the report of a biomedical discovery that benefits a significant portion of humanity. Further advances in this rapidly expanding field of knowledge will no doubt lead to even more spectacular gains in our ability to understand nature and to control our destinies. It is therefore essential that individuals embarking on a career in biomedical sciences be well versed in biochemistry.

This textbook is a distillation of our experiences in teaching undergraduate and graduate students at the University of Pennsylvania and Swarthmore College and is intended to provide such students with a thorough grounding in biochemistry. We assume that students who use this textbook have had the equivalent of one year of college chemistry and at least one semester of organic chemistry so that they are familiar with both general chemistry and the basic principles and nomenclature of organic chemistry. We also assume that students have taken a one-year college course in general biology in which elementary biochemical concepts were discussed. Students who lack these prerequisites are advised to consult the appropriate introductory textbooks in these subjects.

In the eight years since the second edition of *Biochemistry* was published, the field of biochemistry has continued its phenomenal and rapidly accelerating growth. This remarkable expansion of our knowledge, the work of thousands of talented and dedicated scientists, has been characterized by numerous new paradigms, as well as an enormous enrichment of almost every aspect of the field. For example, the number of known protein and nucleic acid structures as determined by X-ray and NMR techniques has increased by over fourfold. Moreover, the quality and complexity of these structures have significantly improved, thereby providing enormous advances in our understanding of structural biochemistry. Bioinformatics, an only recently coined word, has come to dominate the way that many aspects of biochemistry are conceived and practiced. When the second edition of *Biochemistry* was published, no genome had yet been sequenced. Now over 100 genome sequences, including that from humans, have been determined with a new one being reported almost weekly. Likewise, the state of knowledge has exploded in such subdisciplines as eukaryotic and prokaryotic molecular biology, metabolic control, protein folding, electron transport, membrane transport, immunology, signal transduction, etc. Indeed, these advances have affected our everyday lives in that they have changed the way that medicine is practiced, the way that we protect our own health, and the way in which food is produced.

THEMES

In writing this textbook we have emphasized several themes. First, biochemistry is a body of knowledge compiled by people through experimentation. In presenting what is known, we therefore stress how we have come to know it. The extra effort the student must make in following such a treatment, we believe, is handsomely repaid since it engenders the critical attitudes required for success in any scientific endeavor. Although science is widely portrayed as an impersonal subject, it is, in fact, a discipline shaped through the often idiosyncratic efforts of individual scientists. We therefore identify some of the major contributors to biochemistry (many of whom are still professionally active) and, in many cases, consider the approaches they have taken to solve particular biochemical puzzles. The student should realize, however, that most of the work described could not have been done without the dedicated and often indispensable efforts of numerous co-workers.

The unity of life and its variation through evolution is a second dominant theme running through the text. Certainly one of the most striking characteristics of life on earth is its enormous variety and adaptability. Yet, biochemical research has amply demonstrated that all living things are closely related at the molecular level. As a consequence, the molecular differences among the various species have provided intriguing insights into how organisms have evolved from one another and have helped delineate the functionally significant portions of their molecular machinery.

A third major theme is that biological processes are organized into elaborate and interdependent control networks. Such systems permit organisms to maintain relatively constant internal environments, to respond rapidly to external stimuli, and to grow and differentiate.

A fourth theme is that biochemistry has important medical consequences. We therefore frequently illustrate biochemical principles by examples of normal and abnormal human physiology and discuss the mechanisms of action of a variety drugs.

ORGANIZATION AND COVERAGE

As the information explosion in biochemistry has been occurring, teachers have been exploring more active learning methods such as problem-based learning, discovery-based learning, and cooperative learning. These new teaching and learning techniques involve more interaction among students and teachers and, most importantly, require more in-class time. In writing the third edition of this textbook, we have therefore been faced with the dual pressures of increased content and pedagogical innovation.

We have responded to this challenge by presenting the subject matter of biochemistry as thoroughly and accurately as we can so as to provide students and instructors alike with this information as they explore various innovative learning strategies. In this way we deal with the widespread concern that these novel methods of stimulating student learning tend to significantly diminish course content. We have thus written a textbook that permits teachers to direct their students to areas of content that can be explored outside of class as well as providing material for in-class discussion.

We have reported many of the advances that have occurred in the last eight years in the third edition of *Biochemistry* and have thereby substantially enriched nearly all of its sections. Nevertheless, with the several exceptions noted below, the basic organization of the third edition remains the same as those of the first and second editions.

The text is organized into five parts:

I. Introduction and Background: An introductory chapter followed by chapters that review the properties of aqueous solutions and the elements of thermodynamics.

II. Biomolecules: A description of the structures and functions of proteins, nucleic acids, carbohydrates, and lipids.

III. Mechanisms of Enzyme Action: An introduction to the properties, reaction kinetics, and catalytic mechanisms of enzymes.

IV. Metabolism: A discussion of how living things synthesize and degrade carbohydrates, lipids, amino acids, and nucleotides with emphasis on energy generation and consumption.

V. Expression and Transmission of Genetic Information: An expansion of the discussion of nucleic acid structure that is given in Part II followed by an exposition of both prokaryotic and eukaryotic molecular biology.

This organization permits us to cover the major areas of biochemistry in a logical and coherent fashion. Yet, modern biochemistry is a subject of such enormous scope that to maintain a relatively even depth of coverage throughout the text, we include more material than most one-year biochemistry courses will cover in detail. This depth of coverage, we feel, is one of the strengths of this book; it permits the instructor to teach a course of his/her own design and yet provide the student with a resource on biochemical subjects not emphasized in the course.

The order in which the subject matter of the text is presented more or less parallels that of most biochemistry courses. However, several aspects of the textbook's organization deserve comment:

1. Chapter 5 (Nucleic Acids, Gene Expression, and Recombinant DNA Technology) now introduces molecular biology early in the narrative in response to the central role that recombinant DNA technology has come to play in modern biochemistry. For the same reason, the chapter that contained the review of genetics and the discussion of how we came to know the role of DNA has been subsumed into Chapters 1 (Life) and 5 and the sections on nucleic acid sequencing and the synthesis of oligonucleotides now appear in Chapter 7 (Covalent Structures of Proteins and Nucleic Acids). Likewise, the burgeoning field of bioinformatics is discussed in a separate section of Chapter 7.

2. We have split our presentation of thermodynamics between two chapters. Basic thermodynamic principles—enthalpy, entropy, free energy, and equilibrium—are discussed in Chapter 3 because these subjects are prerequisites for understanding structural biochemistry, enzyme mechanisms, and kinetics. Metabolic aspects of thermodynamics—the thermodynamics of phosphate compounds and oxidation–reduction reactions—are presented in Chapter 16 since knowledge of these subjects is not required until the chapters that follow.

3. Techniques of protein purification are described in a separate chapter (Chapter 6) that precedes the discussion of protein structure and function. We have chosen this order so that students will not feel that proteins are somehow "pulled out of a hat." Nevertheless, Chapter 6 has been written as a resource chapter to be consulted repeatedly as the need arises. Techniques of nucleic acid purification are now also discussed in this chapter for the above-described reasons.

4. Chapter 10 describes the properties of hemoglobin in detail so as to illustrate concretely the preceding discussions of protein structure and function. This chapter introduces allosteric theory to explain the cooperative nature of hemoglobin oxygen binding. The subsequent extension of allosteric theory to enzymology in Chapter 13 is a relatively simple matter.

5. Concepts of metabolic control are presented in the chapters on glycolysis (Chapter 17) and glycogen metabolism (Chapter 18) through the consideration of flux generation, allosteric regulation, substrate cycles, covalent enzyme modification, cyclic cascades, and a newly added discussion of metabolic control analysis. We feel that these concepts are best understood when studied in metabolic context rather than as independent topics.

6. The rapid growth in our knowledge of biological signal transduction necessitates that this important subject now have its own chapter, Chapter 19.

7. There is no separate chapter on coenzymes. These substances, we feel, are more logically studied in the context of the enzymatic reactions in which they participate.

8. Glycolysis (Chapter 17), glycogen metabolism (Chapter 18), the citric acid cycle (Chapter 21), and electron transport and oxidative phosphorylation (Chapter 22) are detailed as models of general metabolic pathways with emphasis placed on many of the catalytic and control mechanisms of the enzymes involved. The principles illustrated in these chapters are reiterated in somewhat less detail in the other chapters of Part IV.

9. Consideration of membrane transport (Chapter 20) precedes that of mitochondrially based metabolic pathways

such as the citric acid cycle, electron transport, and oxidative phosphorylation. In this manner, the idea of the compartmentalization of biological processes can be easily assimilated. We have moved the discussion of neurotransmission to this chapter because it is intimately involved with membrane transport.

10. Discussions of both the synthesis and the degradation of lipids have been placed in a single chapter (Chapter 25), as have the analogous discussions of amino acids (Chapter 26) and nucleotides (Chapter 28).

11. Energy metabolism is summarized and integrated in terms of organ specialization in Chapter 27, following the descriptions of carbohydrate, lipid, and amino acid metabolism.

12. The principles of both prokaryotic and eukaryotic molecular biology are expanded from their introduction in Chapter 5 in sequential chapters on DNA replication, repair and recombination (Chapter 30), transcription (Chapter 31), and translation (Chapter 32). Viruses (Chapter 33) are then considered as paradigms of more complex cellular functions, followed by discussions of eukaryotic gene expression (Chapter 34).

13. Chapter 35, the final chapter, is a series of minichapters that describe the biochemistry of a variety of well-characterized human physiological processes: blood clotting, the immune response, and muscle contraction.

The old adage that you learn a subject best by teaching it simply indicates that learning is an active rather than a passive process. The problems we provide at the end of each chapter are therefore designed to make students think rather than to merely regurgitate poorly assimilated and rapidly forgotten information. Few of the problems are trivial and some of them (particularly those marked with an asterisk) are quite difficult. Yet, successfully working out such problems can be one of the most rewarding aspects of the learning process. Only by thinking long and hard for themselves can students make a body of knowledge truly their own. The answers to the problems are worked out in detail in the solutions manual that accompanies this text. However, this manual can only be an effective learning tool if the student makes a serious effort to solve a problem before looking up its answer.

We have included lists of references at the end of every chapter to provide students with starting points for independent biochemical explorations. The enormity of the biochemical research literature prevents us from giving all but a few of the most seminal research reports. Rather, we list what we have found to be the most useful reviews and monographs on the various subjects covered in each chapter.

Finally, although we have made every effort to make this text error free, we are under no illusions that we have done so. Thus, we are particularly grateful to the many readers of the first and second editions, students and faculty alike, who have taken the trouble to write us with suggestions on how to improve the textbook and to point out errors they have found. We earnestly hope that the readers of the third edition will continue this practice.

Donald Voet
Judith G. Voet

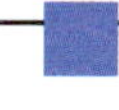

ANCILLARY MATERIALS

The third edition of *Biochemistry* is accompanied by the following ancillary materials:

FOR THE STUDENT

- A CD-ROM that accompanies this textbook, which was produced by ScienceMedia, Inc. in collaboration with the authors. It contains an extensive series of computer-animated Interactive Exercises and Guided Explorations. The CD also contains a series of Kinemages by Donald Voet and Judith G. Voet. These are computer-animated color images of selected proteins and nucleic acids that are discussed in the text and which students can manipulate. Finally, the CD contains a series of animations of figures in the textbook. All of these items are keyed to the textbook as indicated by a mouse icon ( ).

- A *Solutions Manual* containing detailed solutions for all of the textbook's end-of-chapter problems.

FOR THE INSTRUCTOR

- A CD-ROM containing nearly all of the illustrations in the textbook. With computerized projection equipment, these full-color images can be shown in any prearranged order to provide "slide shows" to accompany lectures.

- A set of transparencies for overhead projection containing a selection of illustrations from the textbook.

ACKNOWLEDGMENTS

This textbook is the result of the dedicated effort of many individuals, several of whom deserve special mention:

David Harris, our Executive Editor, adroitly directed the entire project. Patrick Fitzgerald, our new Editor, helped us bring this edition successfully to market.

Barbara Heaney, our Developmental Editor, deftly coordinated both the art and the writing programs and kept our noses to the grindstone.

Suzanne Ingrao, our Production Editor, skillfully and patiently managed the production of the textbook.

Connie Parks, our Copy Editor, put the final polish on the manuscript and eliminated an enormous number of stylistic and typographical errors.

Laura Ierardi combined text, figures, and tables in designing each of the textbook's pages.

Madelyn Lesure designed the textbook's typography and its covers.

Hilary Newman and Elyse Reider acquired many of the photographs in this textbook and kept track of all of them.

Edward Starr and Sigmund Malinowski coordinated the illustration program, with the able help of Ken Liao.

Much of the art in this third edition of *Biochemistry* is the creative legacy of the drawings made for its first and second editions by John and Bette Woolsey and Patrick Lane of J/B Woolsey Associates.

Linda Muriello oversaw the development of the CD-ROM that accompanies this textbook.

The late Irving Geis provided us with his extraordinary molecular art and gave freely of his wise counsel.

The atomic coordinates we have used to draw many of the proteins and nucleic acids that appear in this textbook were obtained from the Protein Data Bank, which is managed by the Research Collaboratory for Structural Bioinformatics (RCSB). The drawings were created using the molecular graphics programs RIBBONS by Mike Carson; GRASP by Anthony Nicholls, Kim Sharp, and Barry Honig; INSIGHT II from BIOSYM Technologies; and RasMol by Roger Sayle. Many of the drawings generously contributed by others were made using these programs or MOLSCRIPT by Per Kraulis.

The interactive computer graphics diagrams that are presented in the CD-ROM that accompanies this textbook are either Chime images or Kinemages. Chime, which is based on the program RosMol, was developed and generously made publicly available by MDL Information Systems, Inc. Kinemages are displayed by the program MAGE, which was written and generously provided by David C. Richardson, who also wrote and provided the program PREKIN, which we used to help generate the Kinemages.

We wish especially to thank those colleagues who reviewed this textbook, in both its current and earlier editions, and provided us with their prudent advice:

Joseph Babitch, *Texas Christian University*

E.J. Berhman, *Ohio State University*

Karl D. Bishop, *Bucknell University*

Robert Blankenshop, *Arizona State University*

Charles L. Borders, Jr., *The College of Wooster*

Kenneth Brown, *University of Texas at Arlington*

Larry G. Butler, *Purdue University*

Carol Caparelli, *Fox Chase Cancer Center*

W. Scott Champney, *East Tennessee Stage University*

Paul F. Cook, *The University of Oklahoma*

Glenn Cunningham, *University of Central Florida*

Eugene Davidson, *Georgetown University*

Don Dennis, *University of Delaware*

Walter A. Deutsch, *Louisiana State University*

Kelsey R. Downum, *Florida International University*

William A. Eaton, *National Institutes of Health*

David Eisenberg, *University of California at Los Angeles*

Jeffrey Evans, *University of Southern Mississippi*

David Fahrney, *Colorado State University*

Paul Fitzpatrick, *Texas A&M University*

Robert Fletterick, *University of California at San Francisco*

Norbert C. Furumo, *Eastern Illinois University*

Scott Gilbert, *Swarthmore College*

Guido Guidotti, *Harvard University*

James H. Hageman, *New Mexico State University*

Lowell Hager, *University of Illinois at Urbana–Champaign*

James H. Hammons, *Swarthmore College*

Edward Harris, *Texas A&M University*

Angela Hoffman, *University of Portland*

Ralph A. Jacobson, *California Polytechnic State University*

Eileen Jaffe, *Fox Chase Cancer Center*

Jan G. Jaworski, *Miami University*

William P. Jencks, *Brandeis University*

Mary Ellen Jones, *University of North Carolina*

Jason D. Kahn, *University of Maryland*

Tokuji Kimura, *Wayne State University*

Barrie Kitto, *University of Texas at Austin*
Daniel J. Kosman, *State University of New York at Buffalo*
Robert D. Kuchta, *University of Colorado, Boulder*
Thomas Laue, *University of New Hampshire*
Albert Light, *Purdue University*
Dennis Lohr, *Arizona State University*
Larry Louters, *Calvin College*
Robert D. Lynch, *University of Lowell*
Harold G. Martinson, *University of California at Los Angeles*
Michael Mendenhall, *University of Kentucky*
Sabeeha Merchant, *University of California at Los Angeles*
Christopher R. Meyer, *California State University at Fullerton*
Ronald Montelaro, *Lòuisiana State University*
Scott Moore, *Boston University*
Harry F. Noller, *University of California at Santa Cruz*
John Ohlsson, *University of Colorado*
Gary L. Powell, *Clemson University*
Alan R. Price, *University of Michigan*
Paul Price, *University of California at San Diego*
Thomas I. Pynadath, *Kent State University*
Frank M. Raushel, *Texas A&M University*
Ivan Rayment, *University of Wisconsin*
Frederick Rudolph, *Rice University*
Raghupathy Sarma, *State University of New York at Stony Brook*
Paul R. Schimmel, *The Scripps Research Institute*
Thomas Schleich, *University of California at Santa Cruz*
Allen Scism, *Central Missouri State University*
Charles Shopsis, *Adelphi University*
Marvin A. Smith, *Brigham Young University*
Thomas Sneider, *Colorado State University*
Jochanan Stenish, *Western Michigan University*
Phyllis Strauss, *Northeastern University*
JoAnne Stubbe, *Massachusetts Institute of Technology*
William Sweeney, *Hunter College*
John Tooze, *European Molecular Biology Organization*
Mary Lynn Trawick, *Baylor University*
Francis Vella, *University of Saskatchewan*
Harold White, *University of Delaware*
William Widger, *University of Houston*
Ken Willeford, *Mississippi State University*
Lauren Williams, *Georgia Institute of Technology*
Jeffery T. Wong, *University of Toronto*
Beulah M. Woodfin, *The University of New Mexico*
James Zimmerman, *Clemson University*

D.V.
J.G.V.

BRIEF CONTENTS

(Parts I through IV are also available as a separate volume.)

(*Chapter 35 is available as a web-based chapter. Go to* www.wiley.com/college/voet)

CONTENTS

(Parts I through IV are also available as a separate volume.)

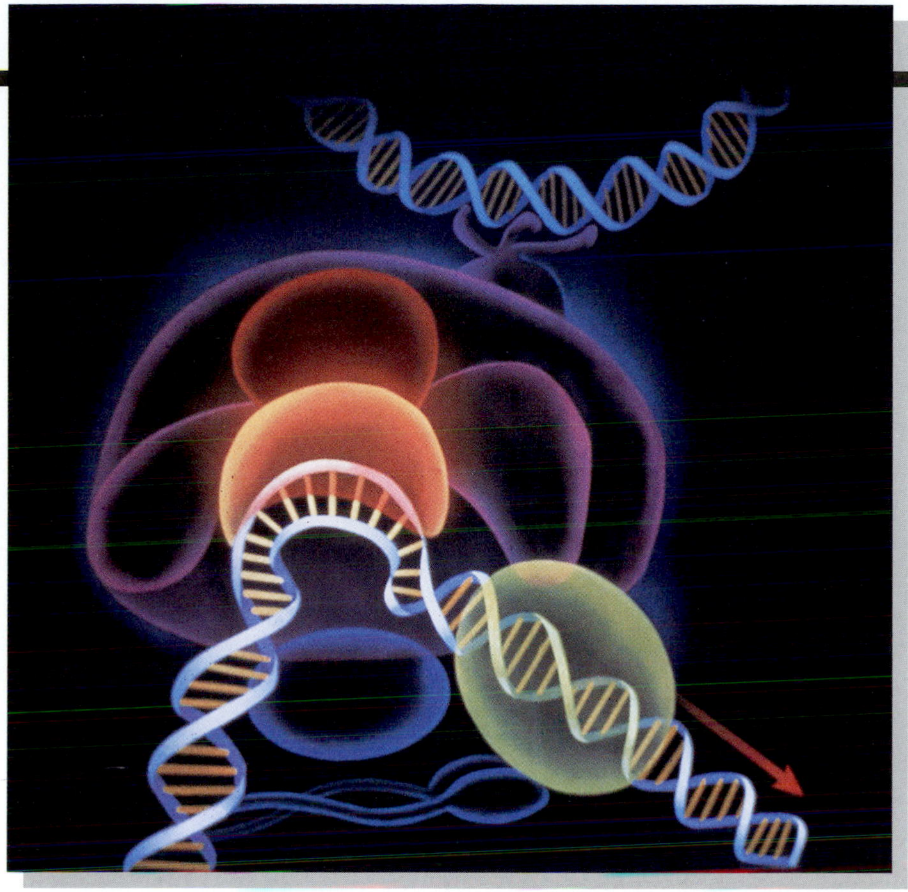

Schematic diagram of the eukaryotic preinitiation complex that is required for the transcription of DNA to messenger RNA. The TATA-box binding protein is shown in orange.

PART

V

EXPRESSION AND TRANSMISSION OF GENETIC INFORMATION

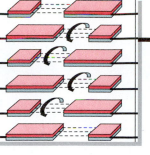

Chapter 29 Nucleic Acid Structures

There are two classes of nucleic acids, ***deoxyribonucleic acid (DNA)*** *and* ***ribonucleic acid (RNA).*** *DNA is the hereditary molecule in all cellular life-forms, as well as in many viruses.* It has but two functions:

1. To direct its own **replication** during cell division.

2. To direct the **transcription** of complementary molecules of RNA.

RNA, in contrast, has more varied biological functions:

1. The RNA transcripts of DNA sequences that specify polypeptides, **messenger RNAs (mRNAs),** direct the ribosomal synthesis of these polypeptides in a process known as **translation.**

2. The RNAs of ribosomes, which are about two-thirds RNA and one-third protein, have functional as well as structural roles.

3. During protein synthesis, amino acids are delivered to the ribosome by molecules of **transfer RNA (tRNA).**

4. Certain RNAs are associated with specific proteins to form **ribonucleoproteins** that participate in the post-transcriptional processing of other RNAs.

5. In many viruses, RNA, not DNA, is the carrier of hereditary information.

The structure and properties of DNA are introduced in Section 5-5. In this chapter we extend this discussion with emphasis on DNA; the structures of RNAs are detailed in Sections 31-4A and 32-2B. Methods of purifying, sequencing, and chemically synthesizing nucleic acids are discussed in Sections 6-6, 7-2, and 7-5, and recombinant DNA techniques are discussed in Section 5-5. Bioinformatics, as it concerns nucleic acids, is outlined in Section 7-4, and the Nucleic Acid Database is described in Section 8-3C.

1 ■ DOUBLE HELICAL STRUCTURES

See Guided Exploration 23. DNA Structures Double helical DNA has three major helical forms, B-DNA, A-DNA, and Z-DNA, whose structures are depicted in Figs. 29-1, 29-2, and 29-3. In this section we discuss the major characteristics of each of these helical forms as well as those of double helical RNA and DNA–RNA hybrid helices.

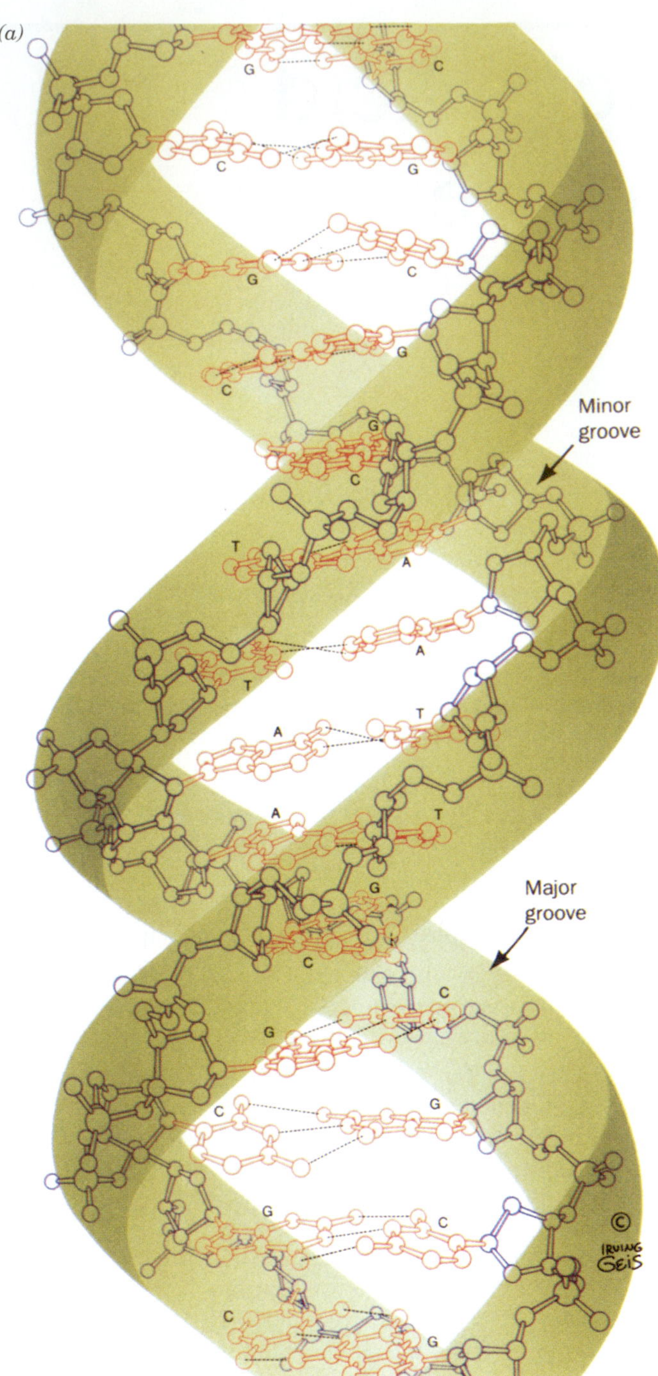

FIGURE 29-1 Structure of B-DNA. The structure is represented by ball-and-stick drawings and the corresponding computer-generated space-filling models. The repeating helix is based on the X-ray structure of the self-complementary dodecamer d(CGCGAATTCGCG) determined by Richard Dickerson and Horace Drew, California Institute of Technology (PDBid 1BNA). (*a*) View perpendicular to the helix axis. In the drawing, the sugar–phosphate backbones, which wind about the periphery of the molecule, are blue, and the bases, which occupy its core, are red. In the space-filling model, C, N, O, and P atoms are white, blue, red, and green, respectively. H atoms have been omitted for clarity in both drawings. Note that the two sugar–phosphate chains run in opposite directions. (*b*) (*Opposite*) View down the helix axis. In the drawing, the ribose ring O atoms are red and the nearest base pair is white. Note that the helix axis passes through the base pairs so that the helix has a solid core. [Illustration, Irving Geis/Geis Archives Trust. Copyright Howard Hughes Medical Institute. Reproduced with permission.] **See Kinemage Exercises 17-1 and 17-4**

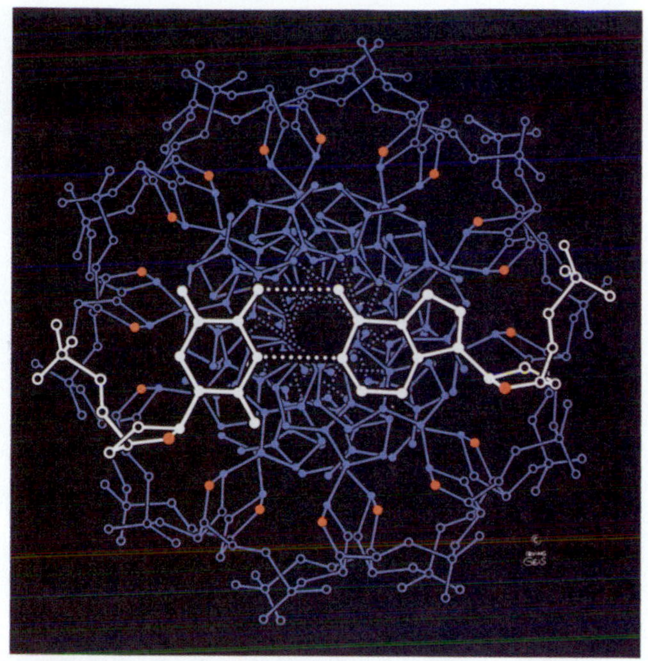

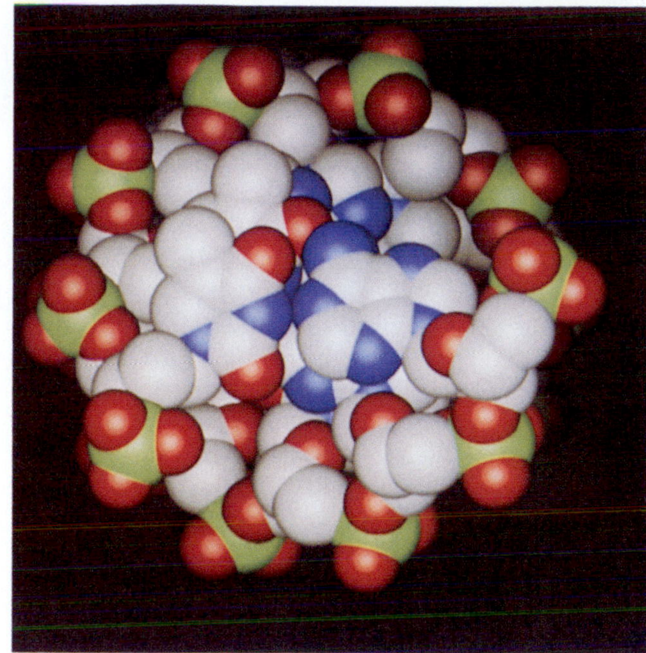

FIGURE 29-1 *(b)*

A. *B-DNA*

The structure of **B-DNA** (Fig. 29-1), the biologically predominant form of DNA, is described in Section 5-3A. To recapitulate (Table 29-1), B-DNA consists of a right-handed double helix whose two antiparallel sugar–phosphate chains wrap around the periphery of the helix. Its aromatic bases (A, T, G, and C), which occupy the core of the helix, form complementary A · T and G · C Watson–Crick base pairs (Fig. 5-12), whose planes are nearly perpendicular to the axis of the double helix. Neighboring base pairs, whose aromatic rings are 3.4 Å thick, are stacked in van der Waals contact, with the helix axis passing through the middle of each base pair. B-DNA is ~20 Å in diameter and has two deep grooves between its sugar–phosphate chains: the relatively narrow **minor groove,** which exposes that edge of the base pairs from which the glycosidic bonds (the bonds from the base N to the ribose

TABLE 29-1 Structural Features of Ideal A-, B-, and Z-DNA

	A-DNA	B-DNA	Z-DNA
Helical sense	Right-handed	Right-handed	Left-handed
Diameter	~26 Å	~20 Å	~18 Å
Base pairs per helical turn	11.6	10	12 (6 dimers)
Helical twist per base pair	31°	36°	9° for pyrimidine–purine steps; 51° for purine–pyrimidine steps
Helix pitch (rise per turn)	34 Å	34 Å	44 Å
Helix rise per base pair	2.9 Å	3.4 Å	7.4 Å per dimer
Base tilt normal to the helix axis	20°	6°	7°
Major groove	Narrow and deep	Wide and deep	Flat
Minor groove	Wide and shallow	Narrow and deep	Narrow and deep
Sugar pucker	C3′-*endo*	C2′-*endo*	C2′-*endo* for pyrimidines; C3′-*endo* for purines
Glycosidic bond	Anti	Anti	Anti for pyrimidines; syn for purines

Source: Mainly Arnott, S., *in* Neidle, S. (Ed.), *Oxford Handbook of Nucleic Acid Structure, p.* 35, Oxford University Press (1999).

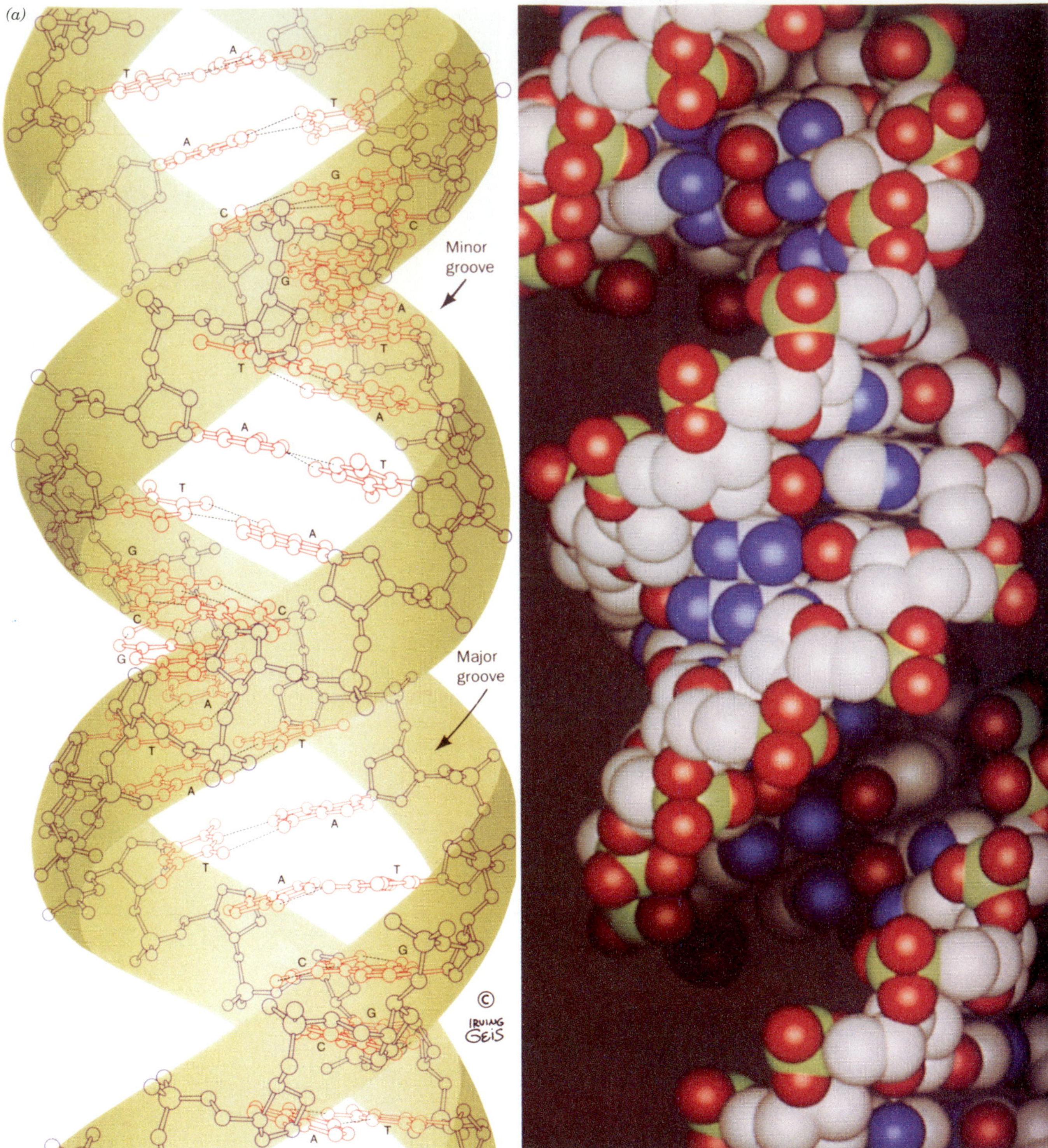

FIGURE 29-2 Structure of A-DNA. Ball-and-stick drawings and the corresponding space-filling models of A-DNA are viewed (*a*) perpendicular to the helix axis and (*b*) (*Opposite*) down the helix axis. The color codes are given in the legend to Fig. 29-1. The repeating helix was generated by Richard Dickerson based on the X-ray structure of the self-complementary octamer d(GGTATACC) determined by Olga Kennard, Dov Rabinovitch, Zippora Shakked, and Mysore Viswamitra, Cambridge University, U.K. (Nucleic Acid Data Base ID ADH010). Note that the base pairs are inclined to the helix axis and that the helix has a hollow core. Compare this figure with Fig. 29-1. [Illustration, Irving Geis/Geis Archives Trust. Copyright Howard Hughes Medical Institute. Reproduced with permission.] **See Kinemage Exercises 17-1 and 17-5**

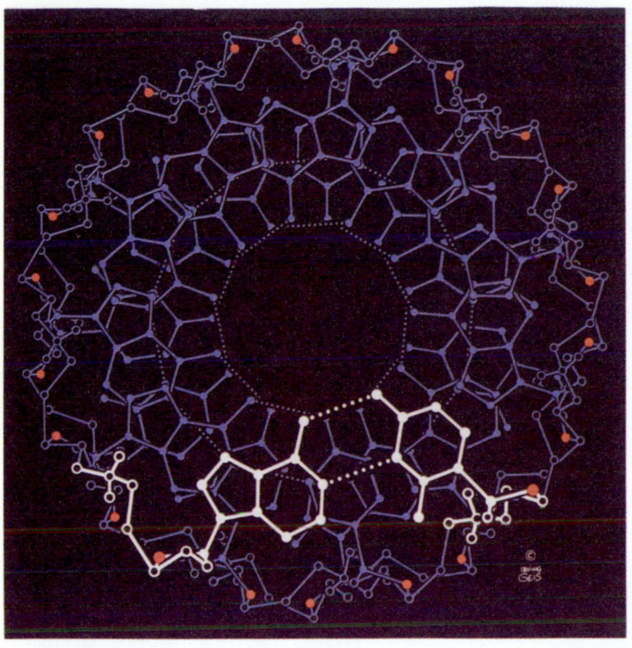

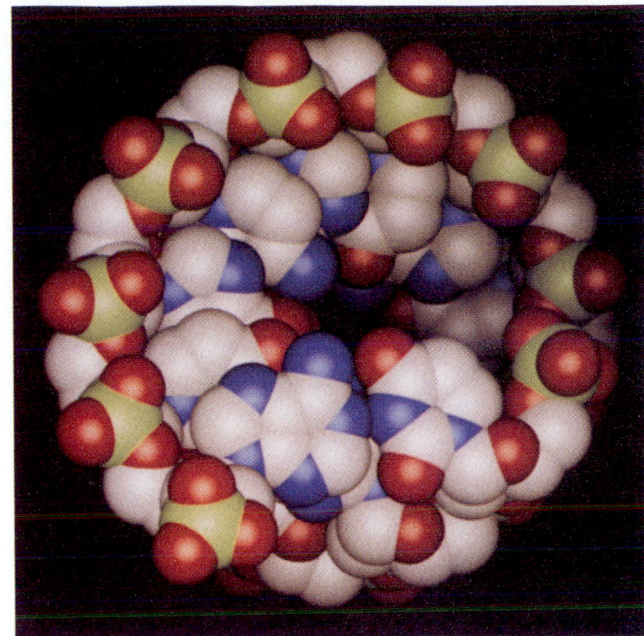

FIGURE 29-2 *(b)*

C1′) extend (toward the bottom of Fig. 5-12), and the relatively wide **major groove,** which exposes the opposite edge of each base pair (toward the top of Fig. 5-12). Canonical (ideal) DNA has a helical twist of 10 base pairs (bp) per turn and hence a pitch (rise per turn) of 34 Å.

The Watson–Crick base pairs in either orientation are structurally interchangeable, that is, A · T, T · A, G · C, and C · G can replace each other in the double helix without altering the positions of the sugar–phosphate backbones' C1 atoms. In contrast, any other combination of bases would significantly distort the double helix since the formation of a non-Watson–Crick base pair would require considerable reorientation of the DNA's sugar–phosphate backbones.

a. Real DNA Deviates from the Ideal Watson–Crick Structure

The DNA samples that were available when James Watson and Francis Crick formulated the Watson–Crick structure in 1953 were extracted from cells and hence consisted of molecules of heterogeneous lengths and base sequences. Such elongated molecules do not crystallize, but can be drawn into threadlike fibers in which the helix axes of the DNA molecules are all approximately parallel to the fiber axis but are poorly aligned, if at all, in any other way. The X-ray diffraction patterns of such fibers provide only crude, low-resolution images in which the base pair electron density is the average electron density of all the base pairs in the fiber. The Watson–Crick structure was based, in part, on the X-ray fiber diffraction pattern of B-DNA (Fig. 5-10).

By the late 1970s, advances in nucleic acid chemistry permitted the synthesis and crystallization of ever longer oligonucleotides of defined sequences (Section 7-6A), many of which could be crystallized. Consequently, some 25 years after the Watson–Crick structure was formulated, its X-ray crystal structure was clearly visualized for the first time when Richard Dickerson and Horace Drew determined the first X-ray crystal structure of a B-DNA, that of the self-complementary dodecamer d(CGCGAATTCGCG), at near-atomic (1.9 Å) resolution. This molecule, whose structure was subsequently determined at significantly higher (1.4 Å) resolution by Loren Williams, has an average rise per residue of 3.3 Å and has 10.1 bp per turn (a helical twist of 35.5° per bp), values that are nearly equal to those of canonical B-DNA. However, individual residues depart significantly from this average conformation (Fig. 29-1). For example, the helical twist per base pair in this dodecamer ranges from 26° to 43°. Each base pair further deviates from its ideal conformation by such distortions as propeller twisting (the opposite rotation of paired bases about the base pair's long axis; in the 1.4-Å resolution structure, this quantity ranges from −23° to −7°) and base pair roll (the tilting of a base pair as a whole about its long axis; this quantity ranges from −14° to 17°).

X-Ray and NMR studies of numerous other double helical DNA oligomers have amply demonstrated that *the conformation of DNA, particularly B-DNA, is irregular in a sequence-specific manner,* although the rules specifying how sequence governs conformation have proved to be surprisingly elusive. This is because *base sequence does not so much confer a fixed conformation on a double helix as it establishes the deformability of the helix.* Thus, 5′-R–Y-3′ steps (where R and Y are the abbreviations for purines and pyrimidines, respectively) in B-DNA are easily bent because they exhibit relatively little ring–ring overlap between adjacent base pairs. In contrast, both Y–R steps and R–R steps (the latter, due to base pairing, are equiv-

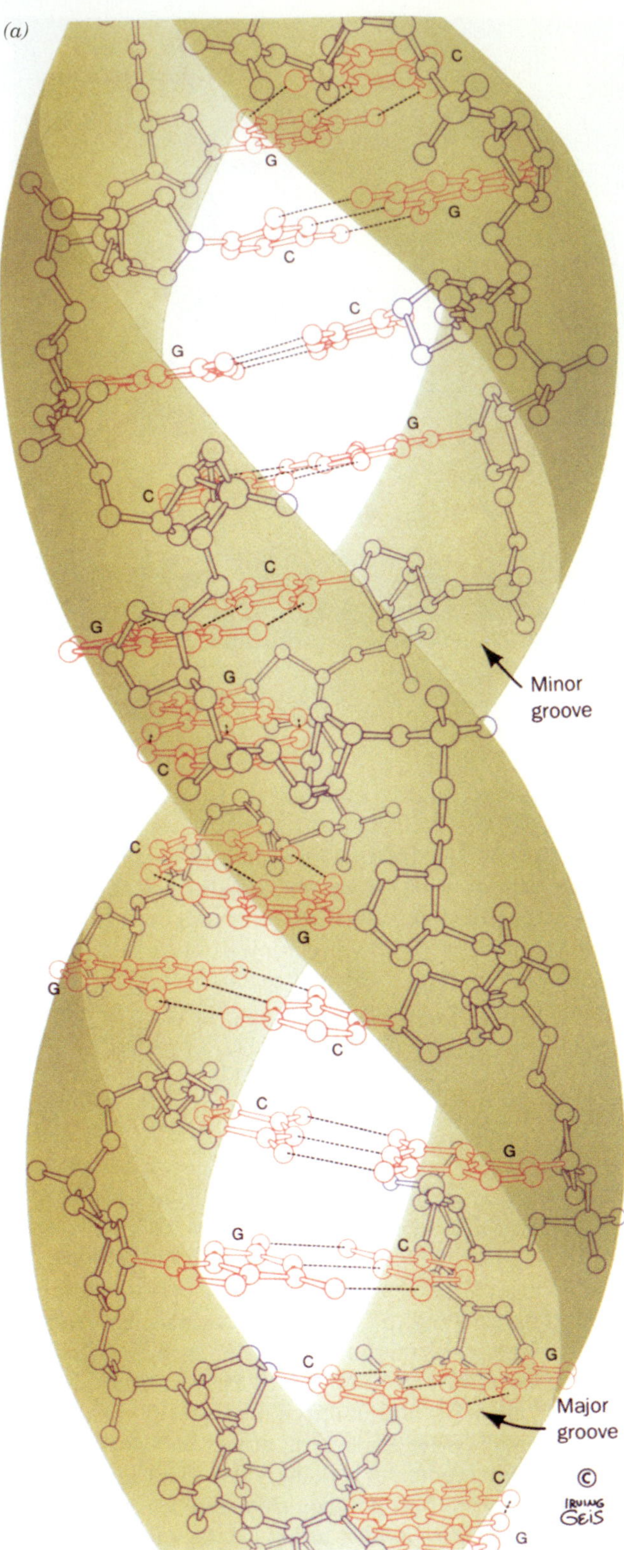

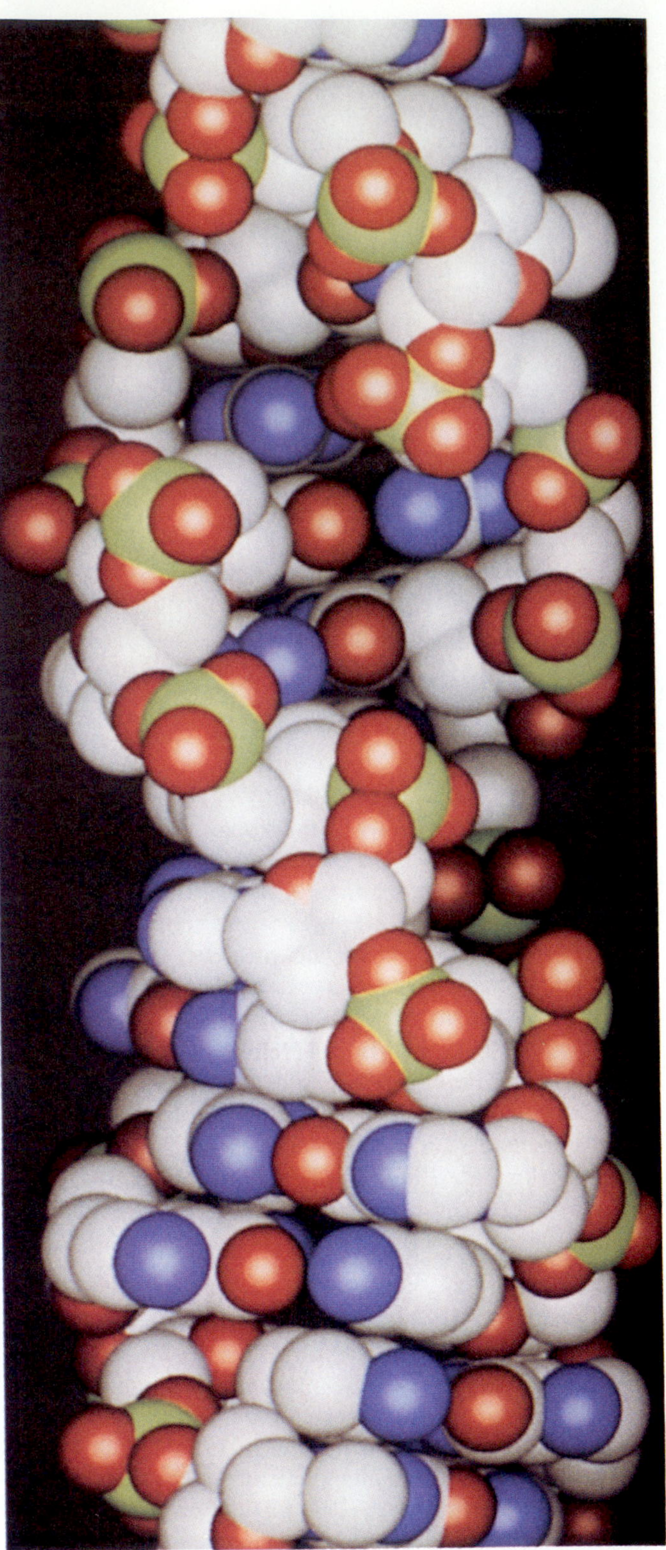

FIGURE 29-3 Structure of Z-DNA. Ball-and-stick drawings and the corresponding space-filling models of Z-DNA are viewed (*a*) perpendicular to the helix axis and (*b*) (*Opposite*) down the helix axis. The color codes are given in the legend to Fig. 29-1. The repeating helix was generated by Richard Dickerson based on the X-ray structure of the self-complementary hexamer d(CGCGCG) determined by Andrew Wang and Alexander Rich, MIT (PDBid 2DCG). Note that the helix is left handed and that the sugar–phosphate chains follow a zigzag course (alternate ribose residues lie at different radii in Part *b*), indicating that the Z-DNA's repeating motif is a dinucleotide. Compare this figure with Figs. 29-1 and 29-2. [Illustration, Irving Geis/Geis Archives Trust. Copyright Howard Hughes Medical Institute. Reproduced with permission.] **See Kinemage Exercises 17-1 and 17-6**

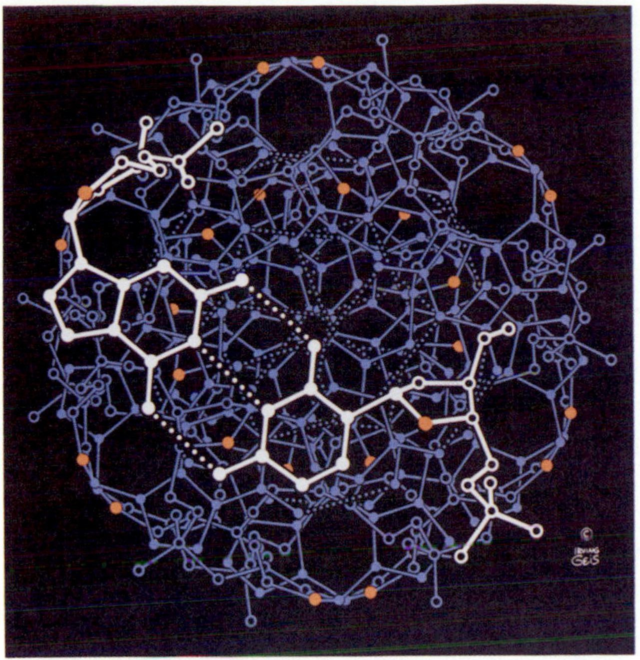

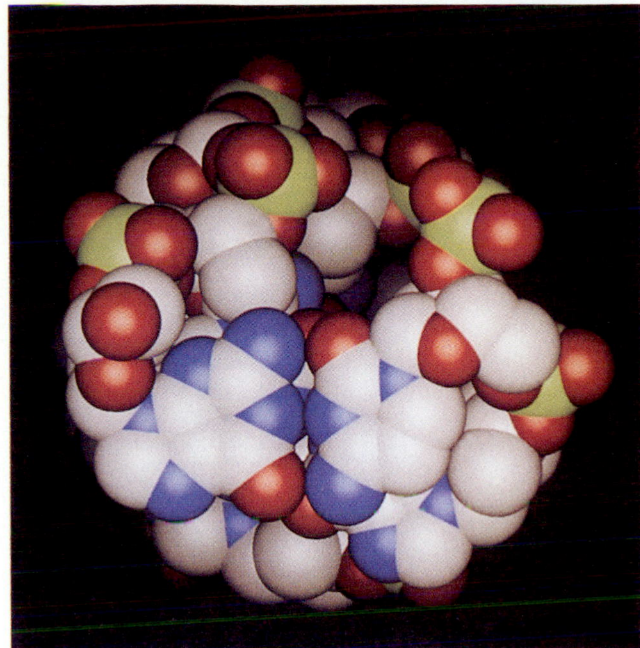

FIGURE 29-3 *(b)*

alent to Y–Y steps), and most notably A–A steps, are more rigid because the extensive ring–ring overlap between their adjacent base pairs tends to keep these base pairs parallel. *This phenomenon, as we shall see, is important for the sequence-specific binding of DNA to proteins that process genetic information.* This is because many of these proteins wrap their target DNAs around them, in many cases by bending them by well over 90°. DNAs with different sequences than the target DNA would not bind so readily to the protein because they would resist deformation to the required conformation more than the target DNA.

B. *Other Nucleic Acid Helices*

X-Ray fiber diffraction studies, starting in the mid-1940s, revealed that *nucleic acids are conformationally variable molecules.* Indeed, double helical DNA and RNA can assume several distinct structures that vary with such factors as the humidity and the identities of the cations present, as well as with base sequence. For example, fibers of B-DNA form in the presence of alkali metal ions such as Na^+ when the relative humidity is 92%. In this subsection, we describe the other major conformational states of double-stranded DNA as well as those of double-stranded RNA and RNA–DNA hybrid helices.

a. A-DNA's Base Pairs Are Inclined to the Helix Axis

When the relative humidity is reduced to 75%, B-DNA undergoes a reversible conformational change to the so-called A form. Fiber X-ray studies indicate that ***A-DNA** forms a wider and flatter right-handed helix than does B-DNA* (Fig. 29-2; Table 29-1). A-DNA has 11.6 bp per turn and a pitch of 34 Å, which gives A-DNA an axial hole (Fig. 29-2*b*). A-DNA's most striking feature, however, is that the planes of its base pairs are tilted 20° with respect to the helix axis. Since its helix axis passes "above" the major groove side of the base pairs (Fig. 29-2*b*) rather than through them as in B-DNA, A-DNA has a deep major groove and a very shallow minor groove; it can be described as a flat ribbon wound around a 6-Å-diameter cylindrical hole. Most self-complementary oligonucleotides of <10 base pairs, for example, d(GGCCGGCC) and d(GGTATACC), crystallize in the A-DNA conformation. Like B-DNA, these molecules exhibit considerable sequence-specific conformational variation although the degree of variation is less than that in B-DNA.

A-DNA has, so far, been observed in only two biological contexts. A ~3-bp segment of A-DNA is present at the active site of DNA polymerase (Section 30-2A). In addition, Gram-positive bacteria undergoing **sporulation** (the formation, under environmental stress, of resistant although dormant cell types known as **spores;** a sort of biological lifeboat) contain a high proportion (20%) of **small acid-soluble spore proteins (SASPs).** Some of these SASPs induce B-DNA to assume the A form, at least *in vitro.* The DNA in bacterial spores exhibits a resistance to UV-induced damage that is abolished in mutants that lack these SASPs. This occurs because the B→A conformation change inhibits the UV-induced covalent cross-linking of pyrimidine bases (Section 30-5A), in part by increasing the distance between successive pyrimidines.

b. Z-DNA Forms a Left-Handed Helix

Occasionally, a seemingly well-understood or at least familiar system exhibits quite unexpected properties. Over 25 years after the discovery of the Watson–Crick

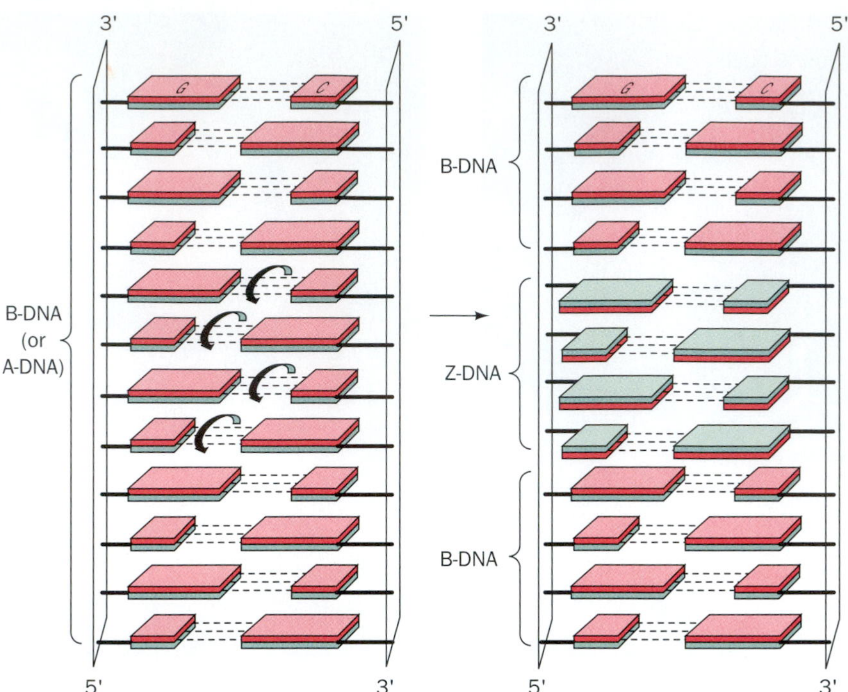

FIGURE 29-4 Conversion of B-DNA to Z-DNA. The conversion, here represented by a 4-bp DNA segment, involves a 180° flip of each base pair (*curved arrows*) relative to the sugar–phosphate chains. Here, the different faces of the base pairs are colored red and green. Note that if the drawing on the left is taken as looking into the minor groove of unwound A- or B-DNA, then in the drawing on the right, we are looking into the major groove of the unwound Z-DNA segment. [After Rich, A., Nordheim, A., and Wang, A.H.-J., *Annu. Rev. Biochem.* **53,** 799 (1984).]

structure, the crystal structure determination of the self-complementary hexanucleotide d(CGCGCG) by Andrew Wang and Alexander Rich revealed, quite surprisingly, *a left-handed double helix (Fig. 29-3; Table 29-1).* A similar helix is formed by d(CGCATGCG). *This helix, which has been dubbed* ***Z-DNA,*** *has 12 Watson–Crick base pairs per turn, a pitch of 44 Å, and, in contrast to A-DNA, a deep minor groove and no discernible major groove* (its helix axis passes "below" the minor groove side of its base pairs; Fig. 29-3*b*). Z-DNA therefore resembles a left-handed drill bit in appearance. The base pairs in Z-DNA are flipped 180° relative to those in B-DNA (Fig. 29-4) through conformational changes discussed in Section 29-2A. As a consequence, the repeating unit of Z-DNA is a dinucleotide, d(XpYp), rather than a single nucleotide as it is in the other DNA helices. The line joining successive phosphate groups on a polynucleotide strand of Z-DNA therefore follows a zigzag path around the helix (Fig. 29-3*a*; hence the name Z-DNA) rather than a smooth curve as it does in A- and B-DNAs (Figs. 29-1*a* and 29-2*a*).

Fiber diffraction and NMR studies have shown that complementary polynucleotides with alternating purines and pyrimidines, such as poly d(GC) · poly d(GC) and poly d(AC) · poly d(GT), take up the Z-DNA conformation at high salt concentrations. Evidently, *the Z-DNA conformation is most readily assumed by DNA segments with alternating purine–pyrimidine base sequences (for structural reasons explained in Section 29-2A).* A high salt concentration stabilizes Z-DNA relative to B-DNA by reducing the otherwise increased electrostatic repulsions between closest approaching phosphate groups on opposite strands (8 Å in Z-DNA vs 12 Å in B-DNA). The methylation of cytosine residues at C5, a common biological modification (Section 30-7), also promotes Z-DNA formation since a hydrophobic methyl group in this position is less exposed to solvent in Z-DNA than it is in B-DNA.

Does Z-DNA have any biological function? Rich has proposed that the reversible conversion of specific segments of B-DNA to Z-DNA under appropriate circumstances acts as a kind of switch in regulating genetic expression, and there are indications that it transiently forms behind actively transcribing RNA polymerase (Section 31-4B). It has nevertheless been surprisingly difficult to prove the *in vivo* existence of Z-DNA. A major difficulty is demonstrating that a particular probe for detecting Z-DNA, for example, a Z-DNA-specific antibody, does not in itself cause what would otherwise be B-DNA to assume the Z conformation—a kind of biological uncertainty principle (the act of measurement inevitably disturbs the system being measured). Recently, however, Rich has discovered a family of Z-DNA-binding protein domains named **Zα,** whose existence strongly suggests that Z-DNA does, in fact, exist *in vivo.* The X-ray structure of the 81-residue Zα domain from the RNA editing enzyme **ADAR1** (Section 31-4A) in complex with d(TCGCGCG) has been determined (Fig. 29-5). The CGCGCG segment of this heptanucleotide is self-complementary, and therefore forms a twofold symmetric, 6-bp segment of Z-DNA with an overhanging dT at the 5′ end of each strand (although these dT's are disordered in the X-ray structure). A monomeric unit of Zα binds to each strand of the Z-DNA, out of contact with the Zα that binds to the opposite strand. The protein primarily interacts with Z-DNA via hydrogen bonds and salt bridges between polar and basic protein side chains and the Z-DNA's sugar–phosphate backbone. Note that none of the DNA's bases participate in these associations. The protein's DNA-binding surface, which is complementary in shape to the Z-DNA, is positively charged, as is expected for a protein that interacts with several closely spaced, anionic phosphate groups. It is postulated that ADAR1's Zα domain targets it to the Z-DNA upstream of actively transcribing genes (for reasons discussed in Section 31-4A).

c. RNA-11 and RNA–DNA Hybrids Have an A-DNA-Like Conformation

Double helical RNA is unable to assume a B-DNA-like conformation because of steric clashes involving its 2′-OH groups. Rather, it usually assumes a conformation resembling A-DNA (Fig. 29-2), known as **A-RNA** or **RNA-11,** which ideally has 11.0 bp per helical turn, a pitch of 30.9 Å, and its base pairs inclined to the helix axis by 16.7°. Many RNAs, for example, transfer and ribosomal RNAs (whose structures are detailed in Sections 32-2A and 32-3A), contain complementary sequences that form double helical stems.

Hybrid double helices, which consist of one strand each of DNA and RNA, are also predicted to have A-RNA-like conformations. In fact, the X-ray structure, by Barry Finzel, of a 10-bp complex of the DNA oligonucleotide d(GGCGCCCGAA) with the complementary RNA oligonucleotide r(UUCGGGCGCC) reveals (Fig. 29-6) that it forms a double helix with A-RNA-like character (Table 29-1) in that it has 10.9 bp per turn, a pitch of 31.3 Å, and its base pairs are, on average, inclined to the helix axis by 13.9°. Nevertheless, this hybrid helix also has B-DNA-like qualities in that the width of its minor groove (9.5 Å) is intermediate between those for canonical B-DNA (7.4 Å) and A-DNA (11 Å) and in that some of the ribose rings of its DNA strand have conformations characteristic of B-DNA (Section 29-2A), whereas others have conformations characteristic of A-RNA. Note that this structure is of biological significance because short segments of RNA · DNA hybrid helices occur in both the transcription of RNA on DNA templates (Section 31-2C) and in the initiation of DNA replication by short lengths of RNA (Section 30-1D). The RNA component of this helix is a substrate for **RNase H,** which specifically hydrolyzes the RNA strands of RNA · DNA hybrid helices *in vivo* (Section 30-4C).

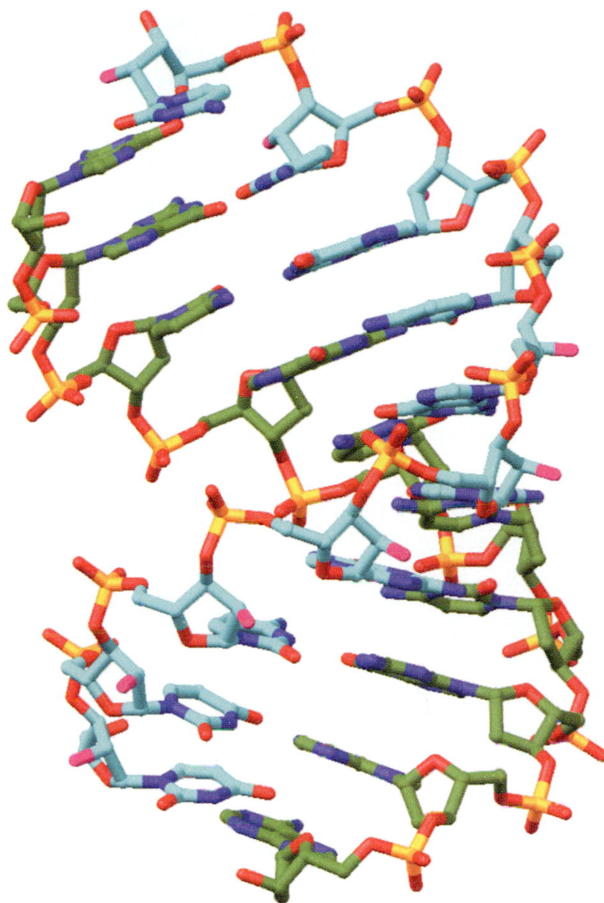

FIGURE 29-6 X-Ray structure of a 10-bp RNA–DNA hybrid helix consisting of d(GGCGCCCGAA) in complex with r(UUCGGGCGCC). The structure is shown in stick form with RNA C atoms cyan, DNA C atoms green, N blue, O red except for RNA O2′ atoms, which are magenta, and P gold. [Based on an X-ray structure by Barry Finzel, Pharmacia & Upjohn, Inc., Kalamazoo, Michigan. PDBid 1FIX.] See the Interactive Exercises

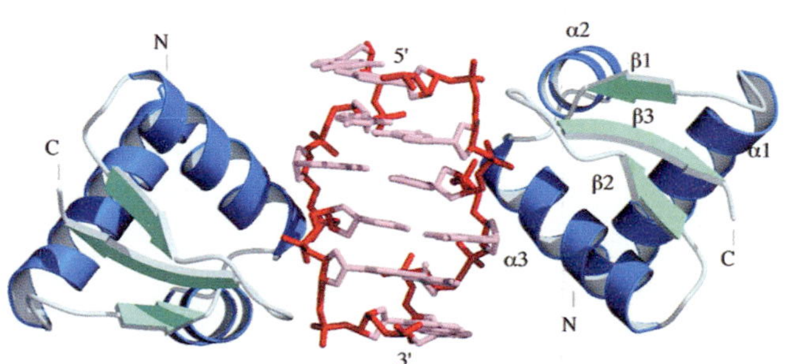

FIGURE 29-5 X-Ray structure of two ADAR1 Zα domains in complex with Z-DNA. The complex is viewed along its 2-fold axis of symmetry. The duplex of self-complementary d(CGCGCG) hexamers is shown in stick form with its backbones red and its remaining portions pink. The Zα domains are drawn in ribbon form with helices blue and sheets cyan. Note that each Zα domain contacts only one strand of the Z-DNA. [Courtesy of Alexander Rich, MIT. PDBid 1QBJ.]

2 ■ FORCES STABILIZING NUCLEIC ACID STRUCTURES

Double-stranded DNA does not exhibit the structural complexity of proteins because it has only a limited repertoire of secondary structures and no comparable tertiary or quaternary structures (although see Section 29-3). This is perhaps to be expected since there is a far greater range of chemical and physical properties among the 20 amino acid residues of proteins than there is among the four DNA bases. However, many RNAs have well-defined tertiary structures (Sections 31-4A, 32-2B, and 32-3A).

In this section we examine the forces that give rise to the structures of nucleic acids. These forces are, of course, much the same as those that are responsible for the structures of proteins (Section 8-4) but, as we shall see, the way they combine gives nucleic acids properties that are quite different from those of proteins.

A. *Sugar–Phosphate Chain Conformations*

The conformation of a nucleotide unit, as Fig. 29-7 indicates, is specified by the six torsion angles of the sugar–phosphate backbone and the torsion angle describing the orientation of the base about the glycosidic bond. It would seem that these seven degrees of freedom per nucleotide would render polynucleotides highly flexible. Yet, as we shall see, these torsion angles are subject to a variety of internal constraints that greatly restrict their conformational freedom.

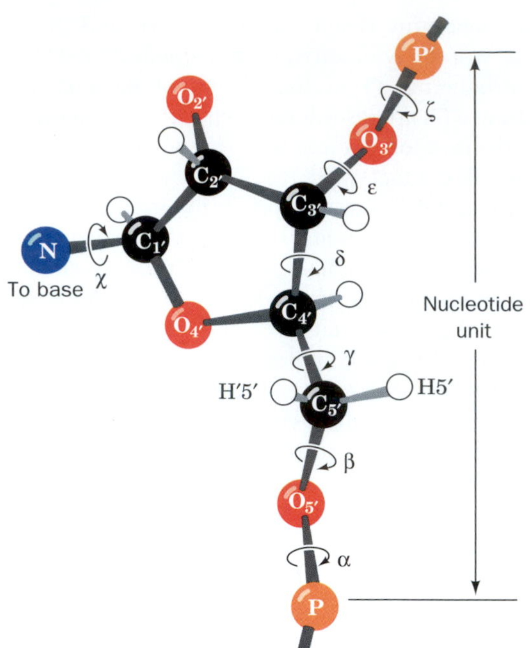

FIGURE 29-7 The conformation of a nucleotide unit is determined by the seven indicated torsion angles.

a. Torsion Angles about Glycosidic Bonds Have Only One or Two Stable Positions

The rotation of a base about its glycosidic bond is greatly hindered, as is best seen by the manipulation of a space-filling molecular model. Purine residues have two sterically permissible orientations relative to the sugar known as the **syn** (Greek: with) and **anti** (Greek: against) conformations (Fig. 29-8). For pyrimidines, only the anti conformation is easily formed because, in the syn conformation, the sugar residue sterically interferes with the pyrimidine's C2 substituent. In most double helical nucleic acids, all bases are in the anti conformation (e.g., Figs. 29-1*b* and 29-2*b*). The exception is Z-DNA (Section 29-1B), in which the alternating pyrimidine and purine residues are anti and syn (Fig. 29-3*b*). *This explains Z-DNA's pyrimidine–purine alternation.* Indeed, the base pair flips that convert B-DNA to Z-DNA (Fig. 29-4) are brought about by rotating each purine base about its glycosidic bond from the anti to syn conformation, whereas it is the sugars that rotate in the pyrimidine nucleotides, thereby maintaining them in their anti conformations.

b. Sugar Ring Pucker Is Largely Limited to Only a Few of Its Possible Arrangements

The ribose ring has a certain amount of flexibility that significantly affects the conformation of the sugar–phosphate backbone. The vertex angles of a regular pentagon are 108°, a value quite close to the tetrahedral angle (109.5°), so that one might expect the ribofuranose ring to be nearly flat. However, the ring substituents are eclipsed when the ring is planar. To relieve the resultant crowding, which even occurs between hydrogen atoms, the ring **puckers;** that is, it becomes slightly nonplanar, so as to reorient the ring substituents (Fig. 29-9; this is readily observed by the manipulation of a skeletal molecular model).

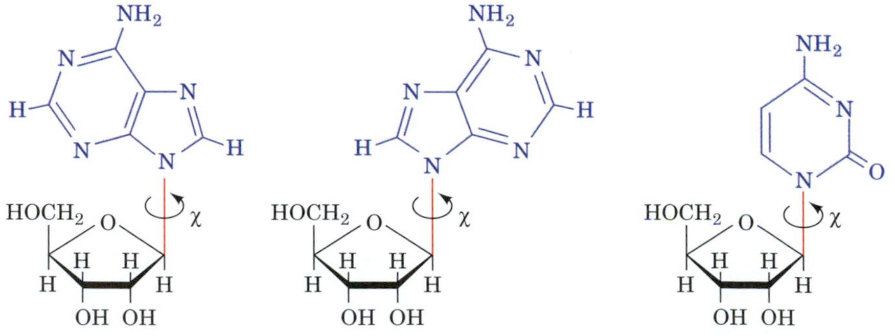

syn-Adenosine **anti-Adenosine** **anti-Cytidine**

FIGURE 29-8 The sterically allowed orientations of purine and pyrimidine bases with respect to their attached ribose units.

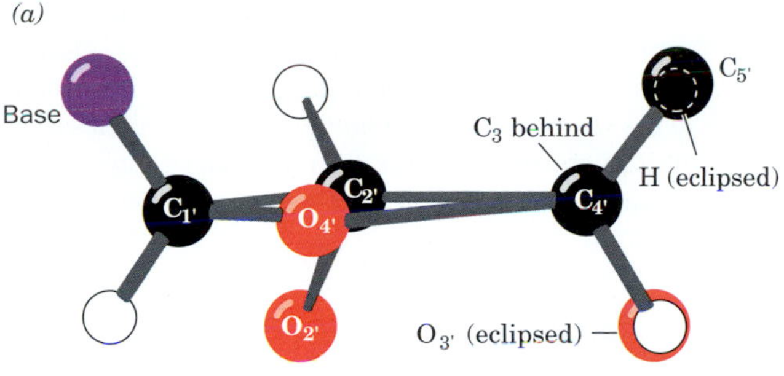

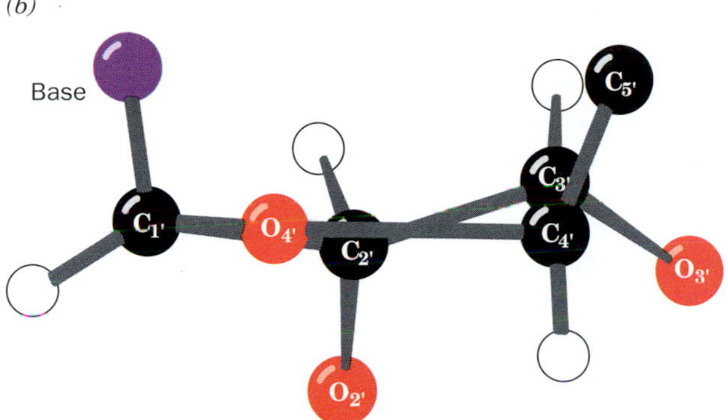

FIGURE 29-9 Sugar ring pucker. The substituents to (*a*) a planar ribose ring (here viewed down the C3′—C4′ bond) are all eclipsed. The resulting steric strain is partially relieved by ring puckering such as in (*b*), a half-chair conformation in which C3′ is the out-of-plane atom.

One would, in general, expect only three of a ribose ring's five atoms to be coplanar since three points define a plane. Nevertheless, in the great majority of the >50 nucleoside and nucleotide crystal structures that have been reported, four of the ring atoms are coplanar to within a few hundreths of an angstrom and the remaining atom is out of this plane by several tenths of an angstrom (the **half-chair** conformation). If the out-of-plane atom is displaced to the same side of the ring as atom C5′, it is said to have the **endo** conformation (Greek: *endon,* within), whereas displacement to the opposite side of the ring from C5′ is known as the **exo** conformation (Greek: *exo,* out of). In the great majority of known nucleoside and nucleotide structures (molecules that are subject to few of the conformational constraints of double helices), the out-of-plane atom is either C2′ or C3′ (Fig. 29-10). C2′-*endo* is the most frequently occurring ribose pucker with C3′-*endo* and C3′-*exo* also being common. Other ribose conformations are rare.

The ribose pucker is conformationally important in nucleic acids because it governs the relative orientations of the phosphate substituents to each ribose residue (Fig. 29-10). For instance, it is difficult to build a regularly repeating model of a double helical nucleic acid unless the sugars are either C2′-*endo* or C3′-*endo*. In fact, canonical B-DNA has the C2′-*endo* conformation, whereas canonical A-DNA and RNA-11 are C3′-*endo*. In canonical Z-DNA, the purine nucleotides are all C3′-*endo* and the pyrimidine nucleotides are C2′-*endo,* which is another reason that the repeating unit of Z-DNA is a dinucleotide. The sugar puckers observed in the X-ray structures of A-DNA are, in fact, almost entirely C3′-*endo*. However, those of B-DNAs, although predominantly C2′-*endo,* exhibit significant variation including C4′-*exo,* O4′-*endo,* C1′-*exo,* and C3′-*exo*. This variation in B-DNA's sugar pucker is probably indicative of its greater flexibility relative to other types of DNA helices.

c. The Sugar–Phosphate Backbone Is Conformationally Constrained

If the torsion angles of the sugar–phosphate chain (Fig. 29-7) were completely free to rotate, there could probably be no stable nucleic acid structure. However, the comparison, by Muttaiya Sundaralingam, of some 40 nucleoside and nucleotide crystal structures has revealed that these angles are really quite restricted. For example, the torsion

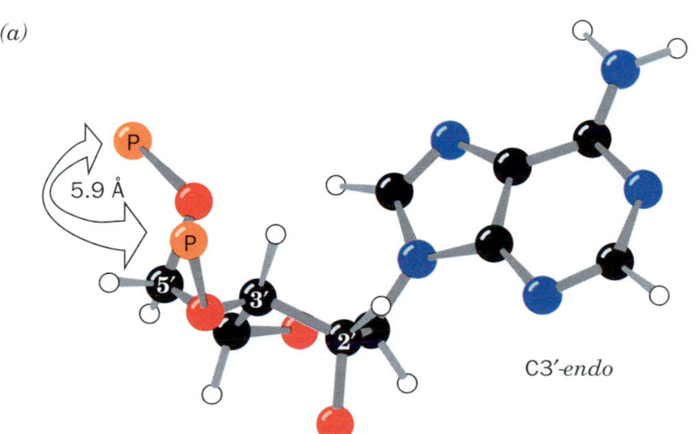

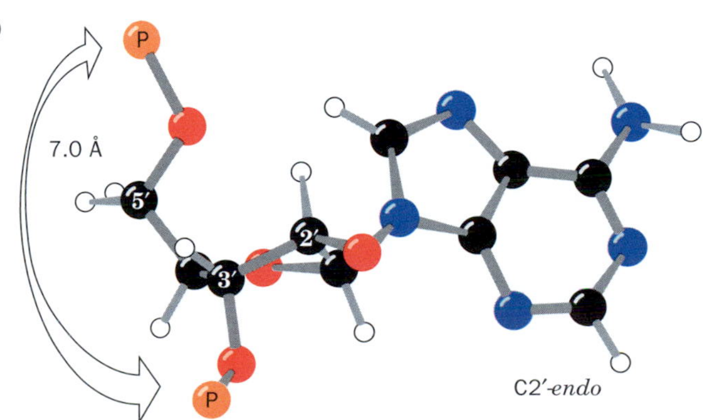

FIGURE 29-10 Nucleotide sugar conformations. (*a*) The C3′-*endo* conformation (on the same side of the sugar ring as C5′), which occurs in A-RNA and RNA-11. (*b*) The C2′-*endo* conformation, which occurs in B-DNA. The distances between adjacent P atoms in the sugar–phosphate backbone are indicated. [After Saenger, W., *Principles of Nucleic Acid Structure,* p. 237, Springer-Verlag (1983).] **See Kinemage Exercises 17-3**

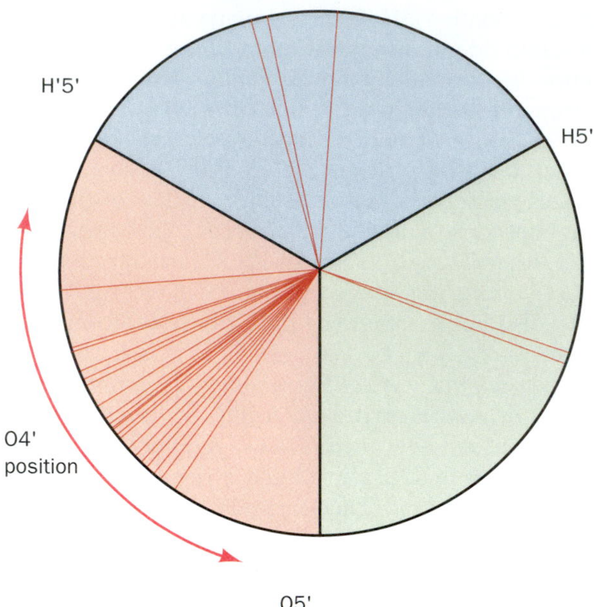

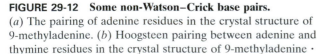

FIGURE 29-11 Conformational wheel showing the distribution of the torsion angle about the C4′—C5′ bond. The torsion angle (γ in Fig. 29-7) was measured in 33 X-ray structures of nucleosides, nucleotides, and polynucleotides. Each radial line represents the position of the C4′—O4′ bond in a single structure relative to the substituents of C5′ as viewed from C5′ to C4′. Note that most of the observed torsion angles fall within a relatively narrow range. [After Sundaralingam, M., *Biopolymers* **7,** 838 (1969).]

angle about the C4′—C5′ bond (γ in Fig. 29-7) is rather narrowly distributed such that O4′ usually has a gauche conformation with respect to O5′ (Fig. 29-11). This is because the presence of the ribose ring together with certain noncovalent interactions of the phosphate group stiffens the sugar–phosphate chain by restricting its range of torsion angles. These restrictions are even greater in polynucleotides because of steric interference between residues.

The sugar–phosphate conformational angles of the various double helices are all reasonably strain free. *Double helices are therefore conformationally relaxed arrangements of the sugar–phosphate backbone.* Nevertheless, the sugar–phosphate backbone is by no means a rigid structure, so, on strand separation, it assumes a random coil conformation.

B. *Base Pairing*

Base pairing is apparently a "glue" that holds together double-stranded nucleic acids. Only Watson–Crick pairs occur in the crystal structures of self-complementary oligonucleotides. It is therefore important to understand how Watson–Crick base pairs differ from other doubly hydrogen bonded arrangements of the bases that have reasonable geometries (e.g., Fig. 29-12).

a. Unconstrained A · T Base Pairs Assume Hoogsteen Geometry

When monomeric adenine and thymine derivatives are cocrystallized, the A · T base pairs that form invariably have adenine N7 as the hydrogen bonding acceptor (**Hoogsteen geometry;** Fig. 29-12*b*) rather than N1 (Watson–Crick geometry; Fig. 5-12). This suggests that Hoogsteen geometry is inherently more stable for A · T pairs than is Watson–Crick geometry. Apparently steric and other environmental influences make Watson–Crick geometry the preferred mode of base pairing in double helices. A · T pairs with Hoogsteen geometry are nevertheless of biological importance; for example, they help stabilize the tertiary structures of tRNAs (Section 32-2B). In contrast, monomeric G · C pairs always cocrystallize with Watson–Crick geometry as a consequence of their triply hydrogen bonded structures.

b. Non-Watson–Crick Base Pairs Are of Low Stability

The bases of a double helix, as we have seen (Section 5-3A), associate such that any base pair position may interchangeably be A · T, T · A, G · C, or C · G without

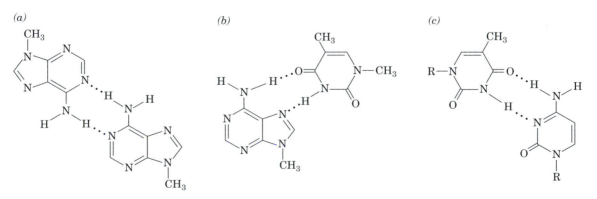

FIGURE 29-12 Some non-Watson–Crick base pairs. (*a*) The pairing of adenine residues in the crystal structure of 9-methyladenine. (*b*) Hoogsteen pairing between adenine and thymine residues in the crystal structure of 9-methyladenine · 1-methylthymine. (*c*) A hypothetical pairing between cytosine and thymine residues. Compare these base pairs with the Watson–Crick base pairs in Fig. 5-12.

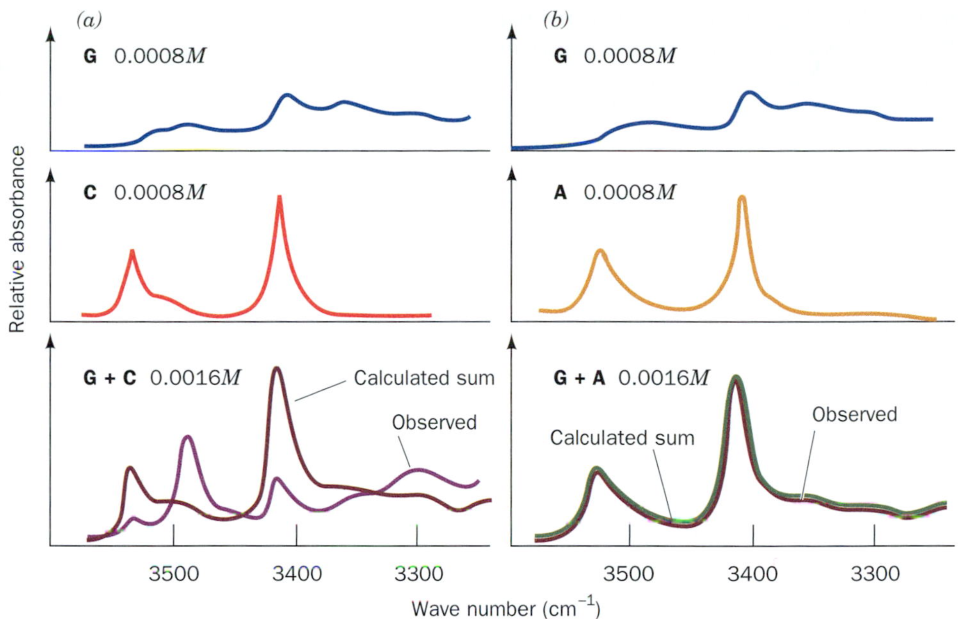

FIGURE 29-13 The IR spectra, in the N—H stretch region, of guanine, cytosine, and adenine derivatives. The derivatives were analyzed both separately and in the indicated mixtures. The solvent, $CDCl_3$, does not hydrogen bond with the bases and is relatively transparent in the frequency range of interest. (*a*) G + C. The brown curve in the lower panel, which is the sum of the spectra in the two upper panels, is the calculated spectrum of G + C for noninteracting molecules. The band near 3500 cm^{-1} in the observed G + C spectrum (*purple*) is indicative of a specific hydrogen bonding association between G and C. (*b*) G + A. The close match between the calculated and observed spectra of the G + A mixture indicates that G and A do not significantly interact. [After Kyogoku, Y., Lord, R.C., and Rich, A., *Science* **154,** 5109 (1966).]

affecting the conformations of the sugar–phosphate chains. One might reasonably suppose that this requirement of **geometric complementarity** of the Watson–Crick base pairs, A with T and G with C, is the only reason that other base pairs do not occur in a double helical environment. In fact, this was precisely what was believed for many years after the DNA double helix was discovered.

Eventually, the failure to detect pairs of different bases in nonhelical environments other than A with T (or U) and G with C led Richard Lord and Rich to demonstrate, through spectroscopic studies, that *only the bases of Watson–Crick pairs have a high mutual affinity.* Figure 29-13*a* shows the infrared (IR) spectrum in the N—H stretch region of guanine and cytosine derivatives, both separately and in a mixture. The band in the spectrum of the G + C mixture that is not present in the spectra of either of its components is indicative of a specific hydrogen bonding interaction between G and C. Such an association, which can occur between like as well as unlike molecules, may be described by ordinary mass action equations.

$$B_1 + B_2 \rightleftharpoons B_1 \cdot B_2 \qquad K = \frac{[B_1 \cdot B_2]}{[B_1][B_2]} \qquad [29.1]$$

From analyses of IR spectra such as Fig. 29-13, the values of K for the various base pairs have been determined. The self-association constants of the Watson–Crick bases are given in the top of Table 29-2 (the hydrogen bonded association of like molecules is indicated by the appearance of new IR bands as the concentration of the molecule is increased). The bottom of Table 29-2 lists the association constants of the Watson–Crick pairs. Note that each of these latter quantities is larger than the self-association constants of either of their component bases, so that Watson–Crick base pairs preferentially form from their constituents. In contrast, the non-Watson–Crick base pairs, A · C, A · G, C · U, and G · U, whatever their geometries, have association constants that are negligible compared with the self-pairing association constants of

TABLE 29-2 Association Constants for Base Pair Formation

Base Pair	K (M^{-1})[a]
Self-association	
A · A	3.1
U · U	6.1
C · C	28
G · G	10^3–10^4
Watson–Crick Base Pairs	
A · U	100
G · C	10^4–10^5

[a]Data measured in deuterochloroform at 25°C.

Source: Kyogoku, Y., Lord, R.C., and Rich, A., *Biochim. Biophys. Acta* **179,** 10 (1969).

their constituents (e.g., Fig. 29-13*b*). *Evidently, a second reason that non-Watson–Crick base pairs do not occur in DNA double helices is that they have relatively little stability.* Conversely, the exclusive presence of Watson–Crick base pairs in DNA results, in part, from an **electronic complementarity** matching A to T and G to C. The theoretical basis of this electronic complementarity, which is an experimental observation, is obscure. This is because the approximations inherent in theoretical treatments make them unable to accurately account for the minor (few kJ · mol^{-1}) energy differences between specific and nonspecific hydrogen bonding associations. The double helical segments of many RNAs, however, contain occasional non-Watson–Crick base pairs, most often G · U, which have functional as well as structural significance (e.g., Sections 32-2B and 32-2D).

c. Hydrogen Bonds Only Weakly Stabilize DNA

It is clear that hydrogen bonding is required for the specificity of base pairing in DNA that is ultimately responsible for the enormous fidelity required to replicate DNA with almost no error (Section 30-3D). Yet, as is also true for proteins (Section 8-4B), *hydrogen bonding contributes little to the stability of the double helix.* For instance, adding the relatively nonpolar ethanol to an aqueous DNA solution, which strengthens hydrogen bonds, destabilizes the double helix, as is indicated by its decreased melting temperature (T_m; Section 5-3C). This is because hydrophobic forces, which are largely responsible for DNA's stability (Section 29-2C), are disrupted by nonpolar solvents. In contrast, *the hydrogen bonds between the base pairs of native DNA are replaced in denatured DNA by energetically nearly equivalent hydrogen bonds between the bases and water.* This accounts for the thermodynamic observation that hydrogen bonding contributes only 2 to 8 kJ/mol of hydrogen bonds to base pairing stability.

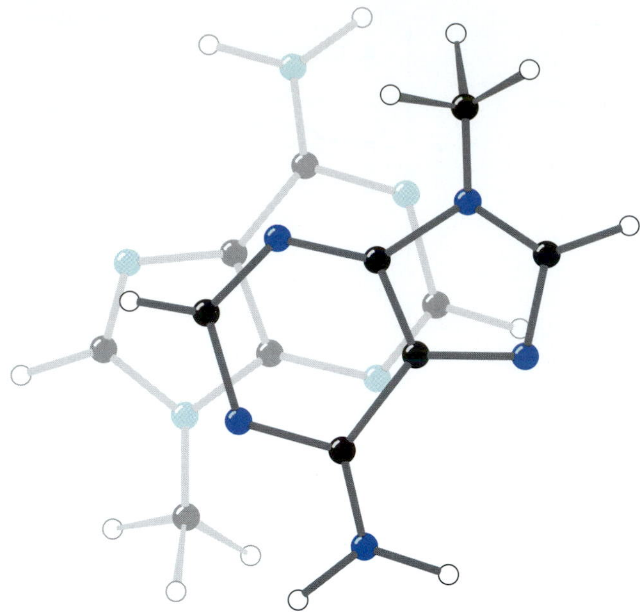

FIGURE 29-14 Stacking of adenine rings in the crystal structure of 9-methyladenine. The partial overlap of the rings is typical of the association between bases in crystal structures and in double helical nucleic acids. [After Stewart, R.F. and Jensen, L.H., *J. Chem. Phys.* **40,** 2071 (1964).]

C. *Base Stacking and Hydrophobic Interactions*

Purines and pyrimidines tend to form extended stacks of planar parallel molecules. This has been observed in the structures of nucleic acids (Figs. 29-1, 29-2, and 29-3) and in the several hundred reported X-ray crystal structures that contain nucleic acid bases. The bases in these structures are usually partially overlapped (e.g., Fig. 29-14). In fact, crystal structures of chemically related bases often exhibit similar stacking patterns. Apparently stacking interactions, which in the solid state are a form of van der Waals interaction (Section 8-4A), have some specificity, although certainly not as much as base pairing.

a. Nucleic Acid Bases Stack in Aqueous Solution

Bases aggregate in aqueous solution, as has been demonstrated by the variation of osmotic pressure with concentration. The van't Hoff law of osmotic pressure is

$$\pi = RTm \qquad [29.2]$$

where π is the osmotic pressure, m is the molality of the solute (mol solute/kg solvent), R is the gas constant, and T is the temperature. The molecular mass, M, of an ideal solute can be determined from its osmotic pressure since $M = c/m$, where $c = g$ solute/kg solvent.

If the species under investigation is of known molecular mass but aggregates in solution, Eq. [29.2] must be rewritten:

$$\pi = \phi RTm \qquad [29.3]$$

where ϕ, the **osmotic coefficient,** indicates the solute's degree of association. ϕ varies from 1 (no association) to 0 (infinite association). The variation of ϕ with m for nucleic acid bases in aqueous solution (e.g., Fig. 29-15) is consistent with a model in which the bases aggregate in successive steps:

$$A + A \rightleftharpoons A_2 + A \rightleftharpoons A_3 + A \rightleftharpoons \cdots \rightleftharpoons A_n$$

where n is at least 5 (if the reaction goes to completion, $\phi = 1/n$). This association cannot be a result of hydrogen bonding since $\boldsymbol{N^6,N^6}$**-dimethyladenosine,**

H_3C — N — CH_3; N; N; N; N; Ribose

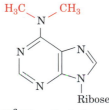

$\boldsymbol{N^6,N^6}$**-Dimethyladenosine**

which cannot form interbase hydrogen bonds, has a greater degree of association than does adenosine (Fig. 29-15). Apparently *the aggregation arises from the formation of*

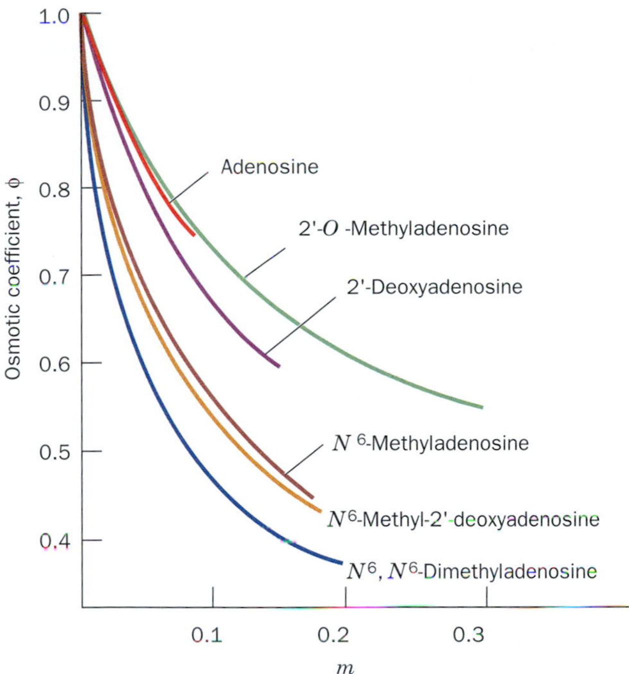

FIGURE 29-15 Variation of the osmotic coefficient ϕ with the molal concentrations *m* of adenosine derivatives in H_2O. The decrease of ϕ with increasing *m* indicates that these derivatives aggregate in solution. [After Broom, A.D., Schweizer, M.P., and Ts'o, P.O.P., *J. Am. Chem. Soc.* **89,** 3613 (1967).]

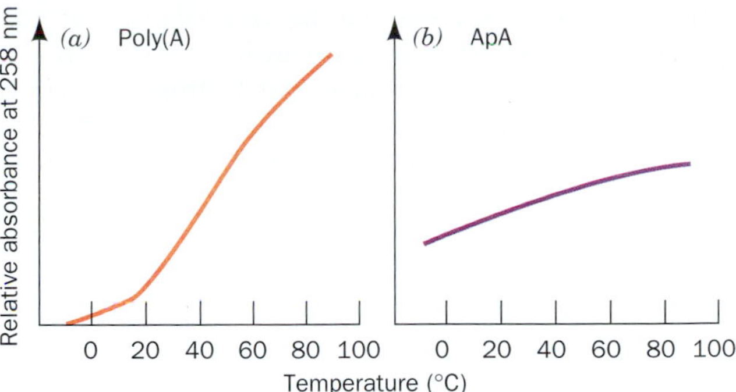

FIGURE 29-16 Melting curves for poly(A) and ApA. The broad temperature range of hyperchromic shifts at 258 nm of (*a*) poly(A) and (*b*) ApA is indicative of noncooperative conformational changes in these substances. Compare this figure with Fig. 5-16. [After Leng, M. and Felsenfeld, G., *J. Mol. Biol.* **15,** 457 (1966).]

stacks of planar molecules. This model is corroborated by proton NMR studies: The directions of the aggregates' chemical shifts are compatible with a stacked but not a hydrogen bonded model. The stacking associations of monomeric bases are not observed in nonaqueous solutions.

Single-stranded polynucleotides also exhibit stacking interactions. For example, poly(A) shows a broad increase of UV absorbance with temperature (Fig. 29-16*a*). This hyperchromism (which is indicative of nucleic acid denaturation; Section 5-3C) is independent of poly(A) concentration, so that it cannot be a consequence of intermolecular disaggregation. Likewise, it is not due to a reduction in intramolecular hydrogen bonding because poly(N^6,N^6-dimethyl A) exhibits a greater degree of hyperchromism than does poly(A). The hyperchromism must therefore arise from some sort of stacking associations within a single strand that melt out with increasing temperature. This is not a very cooperative process, as is indicated by the broadness of the melting curve and the observation that short polynucleotides, including dinucleoside phosphates such as ApA, exhibit similar melting curves (Fig. 29-16*b*).

b. Nucleic Acid Structures Are Stabilized by Hydrophobic Forces

Stacking associations in aqueous solutions are largely stabilized by hydrophobic forces. One might reasonably suppose that hydrophobic interactions in nucleic acids are similar in character to those that stabilize protein structures. However, closer examination reveals that these two types of interactions are qualitatively different in character. Thermodynamic analysis of dinucleoside phosphate melting curves in terms of the reaction

$$\text{Dinucleoside phosphate } (\textit{unstacked}) \rightleftharpoons \text{dinucleoside phosphate } (\textit{stacked})$$

(Table 29-3) indicates that *base stacking is enthalpically driven and entropically opposed. Thus the hydrophobic interactions responsible for the stability of base stacking associations in nucleic acids are diametrically opposite in character to those that stabilize protein structures* (which are enthalpically opposed and entropically driven; Section 8-4C). This is reflected in the differing structural properties of these interactions. For example, the aromatic side chains of proteins are almost never stacked and the crystal structures of aromatic hydrocarbons such as benzene, which resemble these side chains, are characteristically devoid of stacking interactions.

Hydrophobic forces in nucleic acids are but poorly understood. The observation that they are different in character from the hydrophobic forces that stabilize proteins is nevertheless not surprising because the nitrogenous bases are considerably more polar than the hydrocarbon residues of proteins that participate in hydrophobic inter-

TABLE 29-3 Thermodynamic Parameters for the Reaction

$$\underset{(\textit{unstacked})}{\text{Dinucleoside phosphate}} \rightleftharpoons \underset{(\textit{stacked})}{\text{dinucleoside phosphate}}$$

Dinucleoside Phosphate	$\Delta H_{stacking}$ (kJ · mol^{-1})	$-T\Delta S_{stacking}$ (kJ · mol^{-1} at 25°C)
ApA	−22.2	24.9
ApU	−35.1	39.9
GpC	−32.6	34.9
CpG	−20.1	21.2
UpU	−32.6	36.2

Source: Davis, R.C. and Tinoco, I., Jr., *Biopolymers* **6,** 230 (1968).

actions. There is, however, no theory available that adequately explains the nature of hydrophobic forces in nucleic acids (our understanding of hydrophobic forces in proteins, it will be recalled, is similarly incomplete). They are complex interactions of which base stacking is probably a significant component. Whatever their origins, hydrophobic forces are of central importance in determining nucleic acid structures.

D. *Ionic Interactions*

Any theory of the stability of nucleic acid structures must take into account the electrostatic interactions of their charged phosphate groups. Polyelectrolyte theory approximates the electrostatic interactions of DNA by considering the anionic double helix to be a homogeneously charged line or cylinder. We shall not discuss the details of this theory here, but note that it is often in reasonable agreement with experimental observations.

The melting temperature of duplex DNA increases with the cation concentration because these ions bind more tightly to duplex DNA than to single-stranded DNA due to the duplex DNA's higher anionic charge density. An increased salt concentration therefore shifts the equilibrium toward the duplex form, thus increasing the DNA's T_m. The observed relationship for Na^+ is

$$T_m = 41.1X_{G+C} + 16.6\log[Na^+] + 81.5 \qquad [29.4]$$

where X_{G+C} is the mole fraction of G · C base pairs (recall that T_m increases with the G + C content; Fig. 5-17); the equation is valid in the ranges $0.3 < X_{G+C} < 0.7$ and $10^{-3}M < [Na^+] < 1.0M$. Other monovalent cations such as Li^+ and K^+ have similar nonspecific interactions with phosphate groups. Divalent cations, such as Mg^{2+}, Mn^{2+}, and Co^{2+}, in contrast, specifically bind to phosphate groups, so that *divalent cations are far more effective shielding agents for nucleic acids than are monovalent cations.* For example, an Mg^{2+} ion has an influence on the DNA double helix comparable to that of 100 to 1000 Na^+ ions. Indeed, enzymes that mediate reactions with nucleic acids or just nucleotides (e.g., ATP) usually require Mg^{2+} for activity. Moreover, Mg^{2+} ions play an essential role in stabilizing the complex structures assumed by many RNAs such as transfer RNAs (tRNAs; Section 31-2B) and ribosomal RNAs (Section 31-3A).

3 ■ SUPERCOILED DNA

See Guided Exploration 24. DNA Supercoiling Genetic analyses indicate that numerous viruses and bacteria have circular genetic maps, which implies that their chromosomes are likewise circular. This conclusion has been confirmed by electron micrographs in which circular DNAs are seen (Fig. 29-17). Some of these circular DNAs have a peculiar twisted appearance, a phenomenon that is known equivalently as **supercoiling, supertwisting,** and **superhelicity.** Supercoiling arises from a biologically important topological property of covalently closed circular duplex DNA that is the subject of this section. It is occasionally referred to as DNA's tertiary structure.

A. *Superhelix Topology*

Consider a double helical DNA molecule in which both strands are covalently joined to form a circular duplex molecule as is diagrammed in Fig. 29-18 (each strand can be joined only to itself because the strands are antiparallel). *A geometric property of such an assembly is that the number of times one strand wraps about the other cannot be altered without first cleaving at least one of its polynucleotide strands.* You can easily demonstrate this to yourself with a buckled belt in which each edge of the belt represents a strand of DNA. The number of times the belt is twisted before it is buckled cannot be changed without unbuckling or cutting the belt (cutting a polynucleotide strand).

This phenomenon is mathematically expressed

$$L = T + W \qquad [29.5]$$

in which:

1. L, the **linking number** (also symbolized Lk), is the number of times that one DNA strand winds about the other. This integer quantity is most easily counted when the molecule's duplex axis is constrained to lie in a plane (see below).

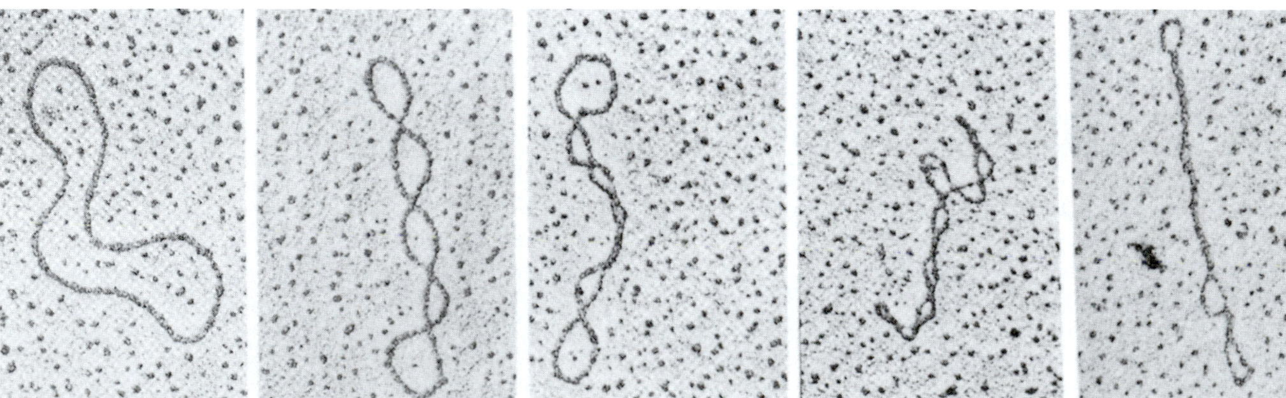

FIGURE 29-17 Electron micrographs of circular duplex DNAs. Their conformations vary from no supercoiling (*left*) to tightly supercoiled (*right*). [Electron micrographs by Laurien Polder. From Kornberg, A. and Baker, T.A., *DNA Replication* (2nd ed.), *p.* 36, W.H. Freeman (1992). Used with permission.]

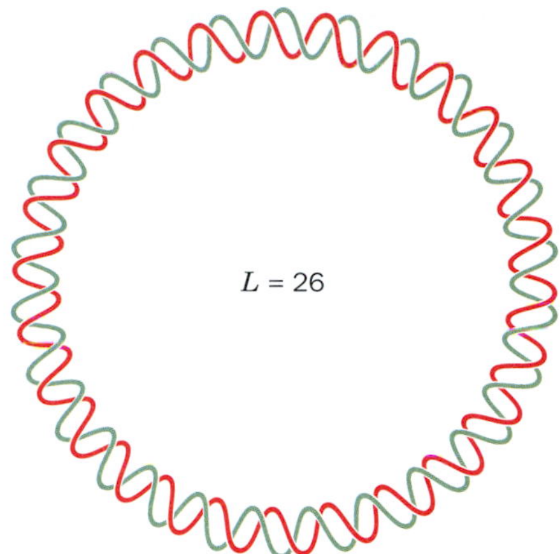

FIGURE 29-18 Schematic diagram of covalently closed circular duplex DNA that has 26 double helical turns. Its two polynucleotide strands are said to be **topologically bonded** to each other because, although they are not covalently linked, they cannot be separated without breaking covalent bonds.

However, *the linking number is invariant no matter how the circular molecule is twisted or distorted so long as both its polynucleotide strands remain covalently intact; the linking number is therefore a topological property of the molecule.*

2. *T,* the **twist** (also symbolized *Tw*), is the number of complete revolutions that one polynucleotide strand makes about the duplex axis in the particular conformation under consideration. By convention, *T* is positive for right-handed duplex turns, so that, for B-DNA in solution, the twist is normally the number of base pairs divided by 10.4 (the number of base pairs per turn of the B-DNA double helix under physiological conditions; see Section 29-3B).

3. *W,* the **writhing number** (also symbolized *Wr*), is the number of turns that the duplex axis makes about the superhelix axis in the conformation of interest. *It is a measure of the DNA's superhelicity.* The difference between writhing and twisting is illustrated by the familiar example in Fig. 29-19. $W = 0$ when the DNA's duplex axis is constrained to lie in a plane (e.g., Fig. 29-18); then $L = T$, so L may be evaluated by counting the DNA's duplex turns.

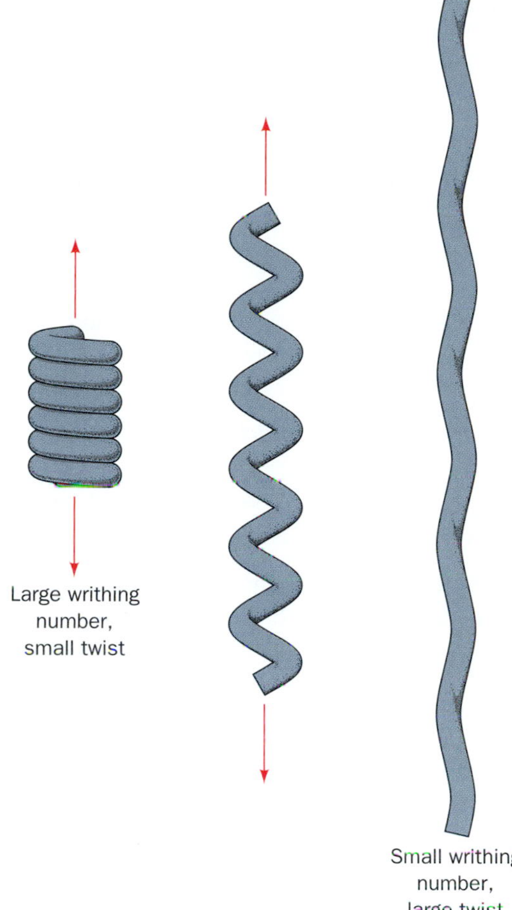

FIGURE 29-19 The difference between writhing and twist as demonstrated by a coiled telephone cord. In its relaxed state (*left*), the cord is in a helical form that has a large writhing number and a small twist. As the coil is pulled out (*middle*) until it is nearly straight (*right*), its writhing number becomes small as its twist becomes large.

The two DNA conformations diagrammed on the right of Fig. 29-20 are topologically equivalent; that is, they have the same linking number, *L,* but differ in their twists and

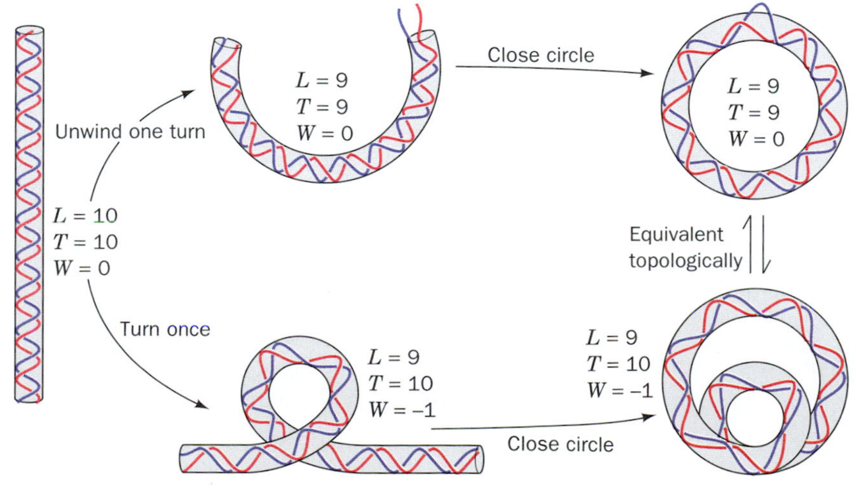

FIGURE 29-20 Two ways of introducing one supercoil into a DNA with 10 duplex turns. The two closed circular forms shown (*right*) are topologically equivalent; that is, they are interconvertible without breaking any covalent bonds. The linking number *L,* twist *T,* and writhing number *W* are indicated for each form. Strictly speaking, the linking number is only defined for a covalently closed circle.

writhing numbers. Note that T and W need not be integers, only L.

Since L is constant in an intact duplex DNA circle, for every new double helical twist, ΔT, there must be an equal and opposite superhelical twist, that is, $\Delta W = -\Delta T$. For example, a closed circular DNA without supercoils (Fig. 29-20, *upper right*) can be converted to a negatively supercoiled conformation (Fig. 29-20, *lower right*) by winding the duplex helix the same number of positive (right-handed) turns.

a. Supercoils May Be Toroidal or Interwound

A supercoiled duplex may assume two topologically equivalent forms:

1. **A toroidal helix** in which the duplex axis is wound as if about a cylinder (Fig. 29-21*a*).

2. An **interwound helix** in which the duplex axis is twisted around itself (Fig. 29-21*b*).

Note that these two interconvertible superhelical forms have opposite handedness. Since left-handed toroidal turns may be converted to left-handed duplex turns (see Fig. 29-19), left-handed toroidal turns and right-handed interwound turns both have negative writhing numbers. Thus an underwound duplex ($T <$ number of bp/10.4), for example, will tend to develop right-handed interwound or left-handed toroidal superhelical turns when the constraints causing it to be underwound are released (the molecular forces in a DNA double helix promote its winding to its normal number of helical turns).

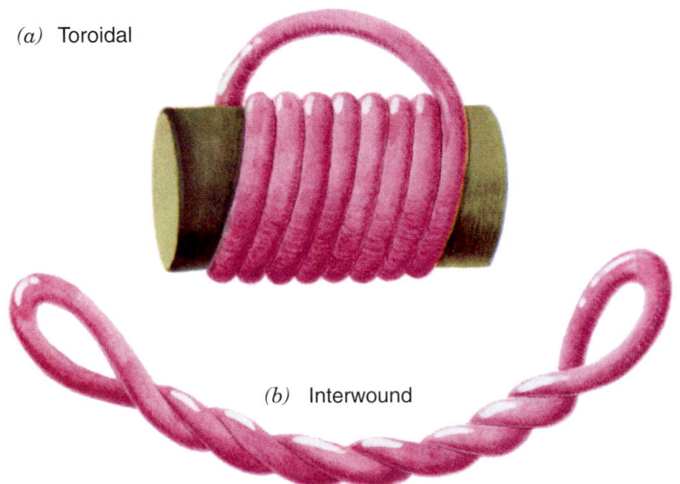

FIGURE 29-21 Toroidal and interwound supercoils. A rubber tube that has been (*a*) toroidally coiled in a left-handed helix around a cylinder with its ends joined such that it has no twist jumps to (*b*) an interwound helix with the opposite handedness when the cylinder is removed. Neither the linking number, the twist, nor the writhing number are changed in this transformation.

b. Supercoiled DNA Is Relaxed by Nicking One Strand

Supercoiled DNA may be converted to **relaxed circles** (as appears in the leftmost panel of Fig. 29-17) by treatment with **pancreatic DNase I,** an **endonuclease** (an enzyme that cleaves phosphodiester bonds within a polynucleotide strand) that cleaves only one strand of a duplex DNA. *One single-strand nick is sufficient to relax a supercoiled DNA.* This is because the sugar–phosphate chain opposite the nick is free to swivel about its backbone bonds (Fig. 29-7) so as to change the molecule's linking number and thereby alter its superhelicity. Supercoiling builds up elastic strain in a DNA circle, much as it does in a rubber band. This is why the relaxed state of a DNA circle is not supercoiled.

B. *Measurements of Supercoiling*

Supercoiled DNA, far from being just a mathematical curiosity, has been widely observed in nature. In fact, its discovery in polyoma virus DNA by Jerome Vinograd stimulated the elucidation of the topological properties of superhelices rather than *vice versa.*

a. Intercalating Agents Control Supercoiling by Unwinding DNA

All naturally occurring DNA circles are underwound; that is, their linking numbers are less than those of their corresponding relaxed circles. This phenomenon has been established by observing the effect of ethidium ion binding on the sedimentation rate of circular DNA (Fig. 29-22). Intercalating agents such as ethidium (a planar aromatic cation; Section 6-6C) alter a circular DNA's degree of superhelicity because they cause the DNA double helix to unwind (untwist) by ~26° at the site of the intercalated molecule (Fig. 29-23). $W < 0$ in an unconstrained underwound circle because of the tendency of a duplex DNA to maintain its normal twist of 1 turn per 10.4 bp. The titration of a DNA circle by ethidium unwinds the duplex (decreases T), which must be accompanied by a compensating increase in W. This, at first, lessens the superhelicity of an underwound circle. However, as the circle binds more and more ethidium, its value of W passes through zero (relaxed circles) and then becomes positive, so that the circle again becomes superhelical. Thus the sedimentation rate of underwound DNAs, which is a measure of their compactness and therefore their superhelicity, passes through a minimum as the ethidium concentration increases. This is what is observed with native DNAs (Fig. 29-22). In contrast, the sedimentation rate of an overwound circle would only increase with increasing ethidium concentration.

b. DNAs Are Separated According to Their Linking Number by Gel Electrophoresis

Gel electrophoresis (Sections 6-4 and 6-6C) also separates similar molecules on the basis of their compactness, so that the rate of migration of a circular duplex DNA increases with its degree of superhelicity. The agarose gel electrophoresis pattern of a population of chemically iden-

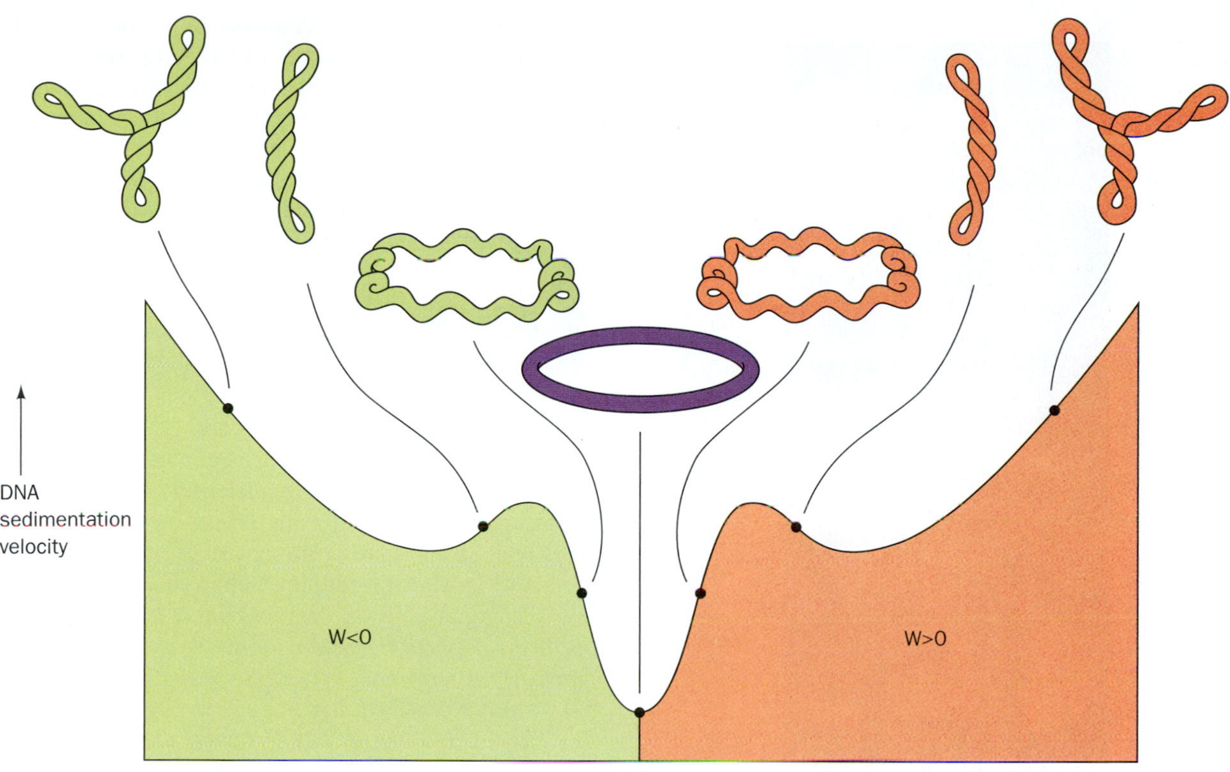

FIGURE 29-22 Sedimentation rate of underwound closed circular duplex DNA as a function of ethidium bromide concentration. The intercalation of ethidium between the base pairs locally untwists the double helix (Fig. 29-23) which, since the linking number of the circle is constant, is accompanied by an equivalent increase in the writhing number. As the negatively coiled superhelix untwists, it becomes less compact and sediments more slowly. At the low point on the curve, the DNA circles have bound sufficient ethidium to become fully relaxed. As the ethidium concentration is further increased, the DNA supercoils in the opposite direction, yielding a positively coiled superhelix. The supertwisted appearances of the depicted DNAs have been verified by electron microscopy. [After Bauer, W.R., Crick, F.H.C., and White, J.H., *Sci. Am.* **243**(1): 129 (1980). Copyright © 1981 by Scientific American, Inc.]

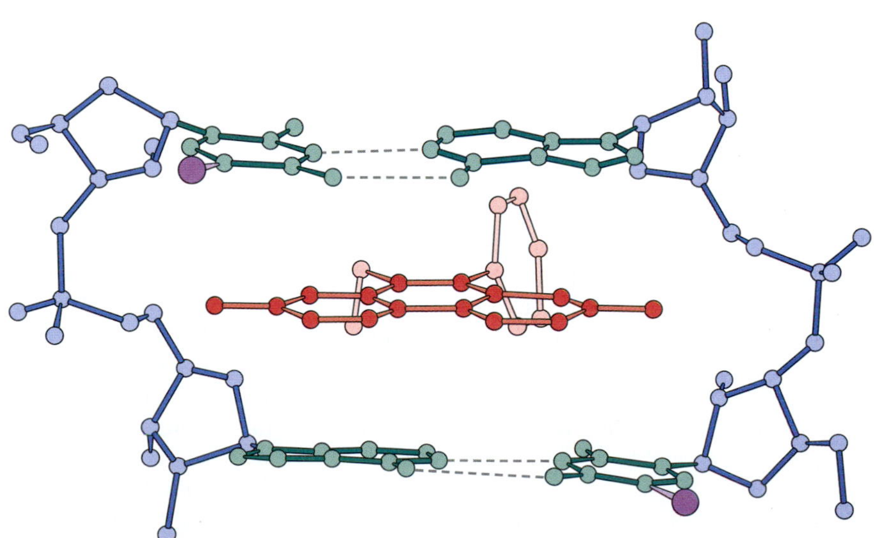

FIGURE 29-23 X-Ray structure of a complex of ethidium with 5-iodo-UpA. Ethidium (*red*) intercalates between the base pairs of the double helically paired dinucleoside phosphate and thereby provides a model for the binding of ethidium to duplex DNA. [After Tsai, C.-C., Jain, S.C., and Sobell, H.M., *Proc. Natl. Acad. Sci.* **72,** 629 (1975).]

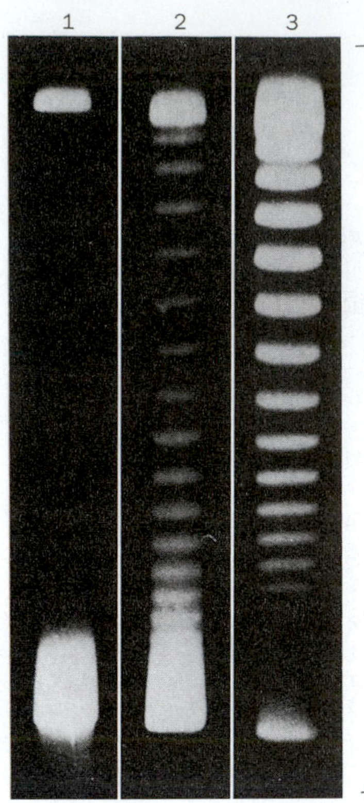

FIGURE 29-24 Agarose gel electrophoresis pattern of SV40 DNA. Lane 1 contains the negatively supercoiled native DNA (*lower band;* the DNA was applied to the top of the gel). In lanes 2 and 3, the DNA has been exposed for 5 and 30 min, respectively, to an enzyme, known as a type IA topoisomerase (Section 29-3C), that relaxes negative supercoils one at a time by increasing the DNA's linking number (L). The DNAs in consecutively higher bands of a given gel have successively increasing linking numbers ($\Delta L = +1$). [From Keller, W., *Proc. Natl. Acad. Sci.* **72,** 2553 (1975).]

tical DNA molecules with different linking numbers therefore consists of a series of discrete bands (Fig. 29-24). The molecules in a given band all have the same linking number and differ from those in adjacent bands by $\Delta L \pm 1$.

Comparison of the electrophoretic band patterns of **simian virus 40 (SV40)** DNA that had been enzymatically relaxed to varying degrees and then resealed (Fig. 29-24) reveals that 26 bands separate native from fully relaxed SV40 DNAs. Native SV40 DNA therefore has $W = -26$ (although it is somewhat heterogeneous in this quantity). Since SV40 DNA consists of 5243 bp, it has 1 negative superhelical turn per ~19 duplex turns. Such a **superhelix density** (W/T) is typical of circular DNAs from various biological sources.

c. DNA in Physiological Solution Has 10.4 Base Pairs per Turn

The insertion, using genetic engineering techniques (Section 5-5C), of an additional x base pairs into a superhelical DNA with a given linking number will increase the DNA's twist and hence decrease its writhing number by $x/h°$, where $h°$ is the number of base pairs per duplex turn. Such an insertion shifts the position of each band in the DNA's gel electrophoretic pattern by $x/h°$ of the spacing between bands. By measuring the effects of several such insertions, James Wang established that $h° = 10.4 \pm 0.1$ bp for B-DNA in solution under physiological conditions.

C. *Topoisomerases*

The normal biological functioning of DNA occurs only if it is in the proper topological state. In such basic biological processes as RNA transcription and DNA replication, the recognition of a base sequence requires the local separation of complementary polynucleotide strands. The negative supercoiling of naturally occurring DNAs results in a torsional strain that promotes such separations since it tends to unwind the duplex helix (an increase in T must be accompanied by a decrease in W). *If DNA lacks the proper superhelical tension, the above vital processes (which themselves supercoil DNA; Sections 30-2C and 31-2C) occur quite slowly, if at all.*

The supercoiling of DNA is controlled by a remarkable group of enzymes known as ***topoisomerases.*** They are so named because they alter the topological state (linking number) of circular DNA but not its covalent structure. There are two classes of topoisomerases:

1. **Type I topoisomerases** act by creating transient single strand breaks in DNA. Type I enzymes are further classified into **type IA** and **type IB topoisomerases** on the basis of their amino acid sequences and reaction mechanisms (see below).

2. **Type II topoisomerases** act by making transient double strand breaks in DNA.

a. Type I Topoisomerases Incrementally Relax Supercoiled DNA

Type I topoisomerases *catalyze the relaxation of supercoils in DNA by changing their linking number in increments of one turn until the supercoil is entirely relaxed.* Type IA enzymes, which are present in all cells, relax only negatively supercoiled DNA, whereas type IB enzymes, which are widely present in prokaryotes (but not *E. coli*) and eukaryotes, relax both negatively and positively coiled DNA. Although types IA and IB topoisomerases are both monomeric, ~100-kD enzymes, they share no apparent sequence or structural similities and function, as we shall see, via different enzymatic mechanisms.

A clue to the mechanism of type IA topoisomerase was provided by the observation that it reversibly **catenates** (interlinks) single-stranded circles (Fig. 29-25*a*). Apparently the enzyme operates by cutting a single strand, passing a single-strand loop through the resulting gap, and then resealing the break (Fig. 29-25*b*), thereby twisting double helical DNA by one turn. In support of this **strand passage** mechanism, the denaturation of type IA enzyme that has been incubated with single-stranded circular DNA yields a linear DNA that has its 5′-terminal phosphoryl group linked to the enzyme via a phosphoTyr diester linkage.

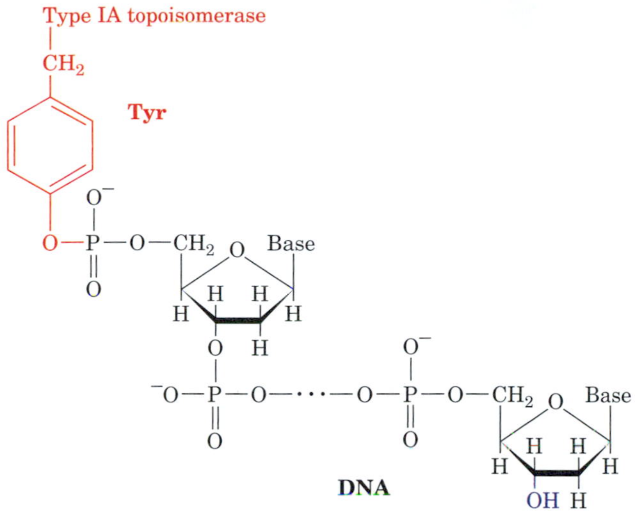

Type IA topoisomerase

Tyr

DNA

In contrast, denatured type IB enzyme is linked to the 3′ end of DNA via a phosphoTyr linkage. *By forming such covalent enzyme–DNA intermediates, the free energy of the cleaved phosphodiester bond is preserved, so that no energy input is required to reseal the nick.*

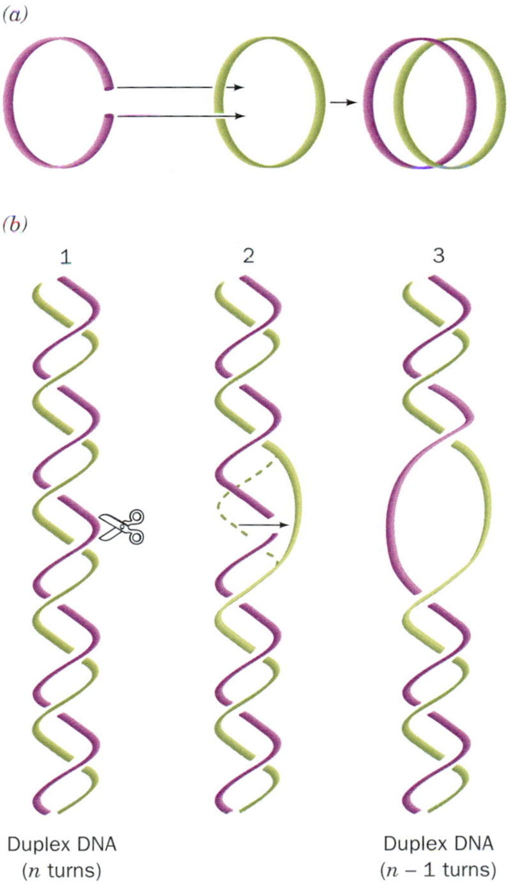

(a)

(b)

1 2 3

Duplex DNA (n turns)

Duplex DNA (n – 1 turns)

FIGURE 29-25 Type IA topoisomerase action. By cutting a single-stranded DNA, passing a loop of a second strand through the break, and then resealing the break, a type IA topoisomerase can (*a*) catenate two single-stranded circles or (*b*) unwind duplex DNA by one turn.

b. Type IA Topoisomerase Probably Functions via a Strand Passage Mechanism

Cells of *E. coli* contain two type IA topoisomerases named **topoisomerase I** and **topoisomerase III.** Topoisomerase III's Tyr 328 is the active site residue that forms a 5′-phosphoTyr linkage with the cleaved DNA. The X-ray structure of the inactive Y328F mutant of topoisomerase III in complex with the single-stranded octanucleotide d(CGCAACTT), determined by Alfonso Mondragón (Fig. 29-26), reveals that this 659-residue monomer folds into four domains which enclose an ~20 by 28 Å hole that is large enough to contain a duplex DNA and which is lined with numerous Arg and Lys side chains. The octanucleotide binds in a groove that is also lined with Arg and Lys side chains with its sugar–phosphate backbone in contact with the protein and with most of its bases exposed for possible base pairing. Curiously, this single-stranded DNA assumes a B-DNA-like conformation even though its complementary strand would be sterically excluded from the groove. The DNA strand is oriented with its 3′ end near the active site, where, if the mutant Phe 328 were the wild-type Tyr, its side chain would be properly positioned to nucleophilically attack the phosphate group bridging the DNA's C6 and T7 to form a 5′-phosphoTyr linkage with T7 and release C6 with a free 3′-OH. This structure and that of the homologous and structurally sim-

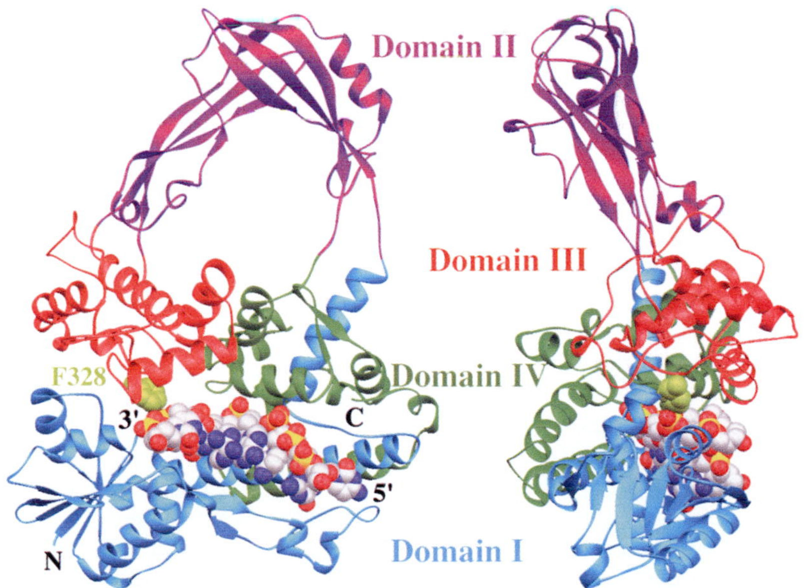

FIGURE 29-26 X-Ray structure of the Y328F mutant of *E. coli* topoisomerase III, a type IA topoisomerase, in complex with the single-stranded octanucleotide d(CGCAACTT). The two views shown are related by a 90° rotation about a vertical axis. The DNA is drawn in space-filling form with C gray, N blue, O red, and P yellow. The enzyme's active site is marked by the side chain of Phe 328, which is shown in space-filling form in yellow-green. [Based on an X-ray structure by Alfonso Mondragón, Northwestern University. PDBid 1I7D.]

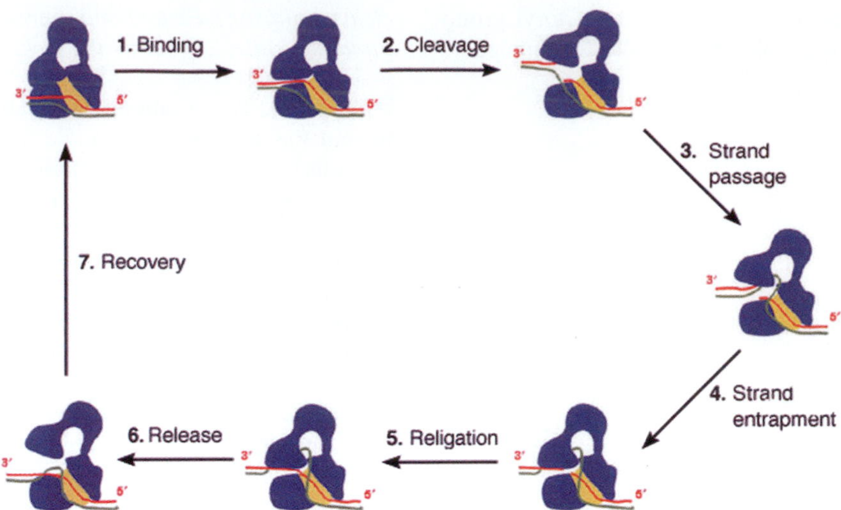

FIGURE 29-27 Proposed mechanism for the strand passage reaction catalyzed by type IA topoisomerases. The enzyme is shown in blue with the yellow patch representing the binding groove for single-stranded (ss) DNA. The two DNA strands, which are drawn in red and green, could represent the two strands of a covalently closed circular duplex or two ss circles. **(1)** The protein recognizes a ss region of the DNA, here the red strand, and binds it in its binding groove. This is followed by or occurs simultaneously with the opening of a gap between domains I and III. **(2)** The DNA is cleaved with the newly formed 5′ end, becoming covalently linked to the active site Tyr and the segment with the newly formed 3′ end remaining tightly but noncovalently bound in the binding groove. **(3)** The unbroken (*green*) strand is passed through the opening or gate formed by the cleaved (*red*) strand to enter the protein's central hole. **(4)** The unbroken strand is trapped by the partial closing of the gap. **(5)** The two cleaved ends of the red strand are rejoined in what is probably a reversal of the cleavage reaction. **(6)** The gap between domains I and III reopens to permit the escape of the red strand, yielding the reaction product in which the green strand has been passed through a transient break in the red strand. **(7)** The enzyme returns to its initial state. If the two strands form a negatively supercoiled duplex DNA, its linking number, L, has increased by 1; if they are separate ss circles, they have been catenated or decatenated. For duplex DNA, this process can be repeated until all of its supercoils have been removed ($W = 0$). [After a drawing by Alfonso Mondragón, Northwestern University.]

ilar *E. coli* topoisomerase I suggest the mechanism for the type IA topoisomerase-catalyzed strand passage reaction that is diagrammed in Fig. 29-27.

c. Type IB Topoisomerase Appears to Function via a Controlled Rotation Mechanism

Human **topoisomerase I (topo I)** is a 765-residue type IB topoisomerase (and hence is unrelated to *E. coli* topoisomerase I). It mediates the transient cleavage of one strand of a duplex DNA through the nucleophilic attack of Tyr 723 on a DNA P atom to yield a 3′-linked phosphoTyr diester bond and a free 5′-OH group on the succeeding nucleotide. Limited proteolysis studies revealed that topo I consists of four major regions: its N-terminal, core, linker, and C-terminal domains. The ~210-residue, highly polar, N-terminal domain, which is poorly conserved, contains several nuclear targeting signals and is dispensable for enzymatic activity.

The X-ray structure of the catalytically inactive Y723F mutant of topo I lacking its N-terminal 214 residues and in complex with a 22-bp palindromic duplex DNA was determined by Wim Hol (Fig. 29-28). The core domain of this bilobal protein is wrapped around the DNA in a tight embrace. If the mutant Phe 723 were the wild-type Tyr, its OH group would be colinear with the scissile P—O5′ bond and hence ideally positioned to nucleophilically attack this P atom so as to form a covalent linkage with the 3′ end of the cleaved strand. As expected, the protein interacts with the DNA in a largely sequence independent manner: Of the 41 direct contacts that the protein makes to the DNA, 37 are protein–phosphate interactions and only one is base-specific. The protein interacts to a much greater extent with the five base pairs of the DNA's downstream segment (which would contain the cleaved strand's newly formed 5′ end; 29 of the 41 contacts) than it does with the base pairs of the DNA's upstream segment (to which Tyr 723 would be covalently linked; 12 of the 41 contacts).

Topo I does not seem sterically capable of unwinding supercoiled DNA via the strand passage mechanism that type IA topoisomerases appear to follow (Fig. 29-27). Rather, as is diagrammed in Fig. 29-29, it is likely that topo I relaxes DNA supercoils by permitting the cleaved duplex DNA's loosely held downstream segment to rotate relative to the tightly held upstream segment. This rotation can only occur about the sugar–phosphate bonds in the uncleaved strand (α, β, γ, ε, and ζ in Fig. 29-7) that are opposite the cleavage site because the cleavage frees these bonds to rotate. In support of this mechanism, the protein region surrounding the downstream segment contains 16 conserved, positively charged residues that form a ring about this duplex DNA, which would presumably hold the DNA in the ring but not in any specific orientation. Nevertheless, the downstream segment is unlikely to rotate freely because the cavity containing it is shaped so as to interact with the

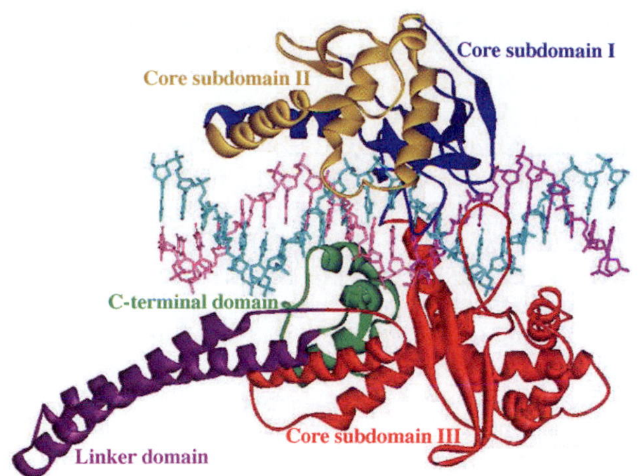

FIGURE 29-28 X-Ray structure of the N-terminally truncated, Y723F mutant of human topoisomerase I in complex with a 22-bp duplex DNA. The protein's various domains and subdomains are drawn in different colors. The DNA's uncleaved strand is cyan, and the upstream and downstream portions of the scissile strand are magenta and pink, respectively. [Courtesy of Wim Hol, University of Washington. PDBid 1A36.]

downstream segment during some portions of its rotation. Hence, type IB topoisomerases are said to mediate a **controlled rotation** mechanism in relaxing supercoiled DNA. This unwinding is driven by the superhelical tension in the DNA and hence requires no other energy input. Eventually, the DNA is religated by a reversal of the cleavage reaction and the now less supercoiled DNA is released.

d. Type II Topoisomerases Function via a Strand Passage Mechanism

The prokaryotic type II topoisomerase known as **DNA gyrase** is an ~375-kD A_2B_2 heterotetramer in which the A and B subunits are named **GyrA** and **GyrB.** *This enzyme catalyzes the stepwise negative supercoiling of DNA with the concomitant hydrolysis of an ATP to ADP + P_i.* It can also catenate and decatenate double-stranded circles as well as tie knots in them. All other type II topoisomerases, both eukaryotic and prokaryotic, only relax supercoils, although they hydrolyze ATP in doing so (DNA supercoiling in eukaryotes is generated differently from that in prokaryotes; Section 34-1B).

DNA gyrases are inhibited by a variety of substances including **novobiocin,** a member of the *Streptomyces*-

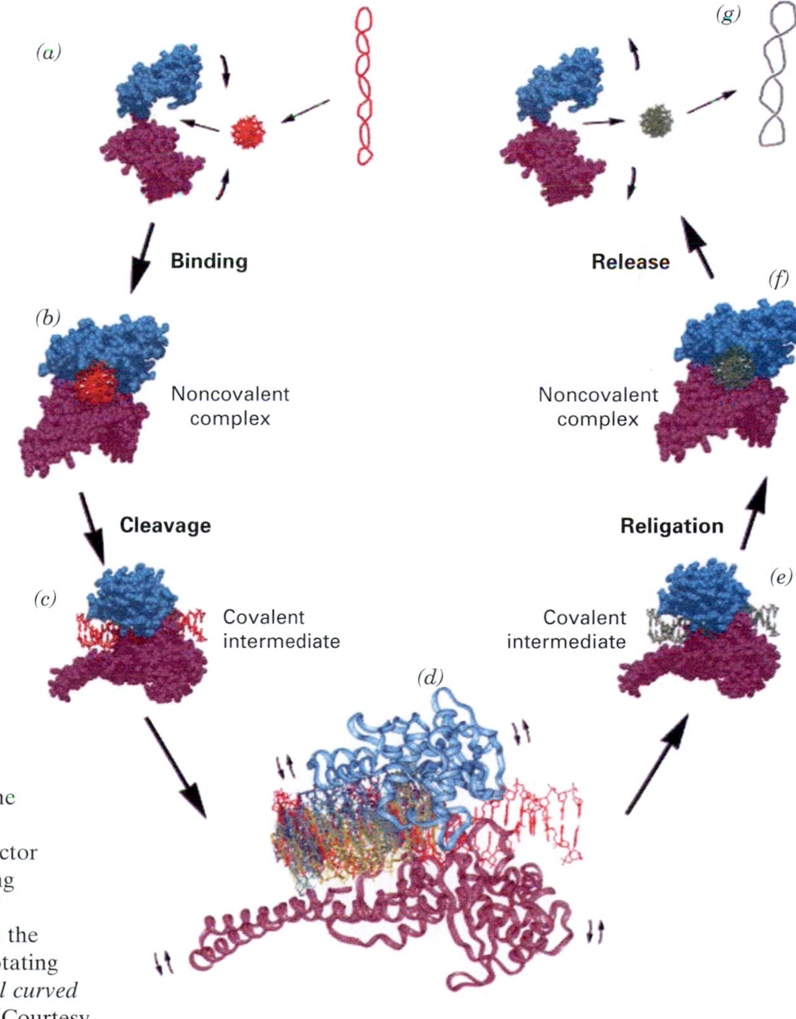

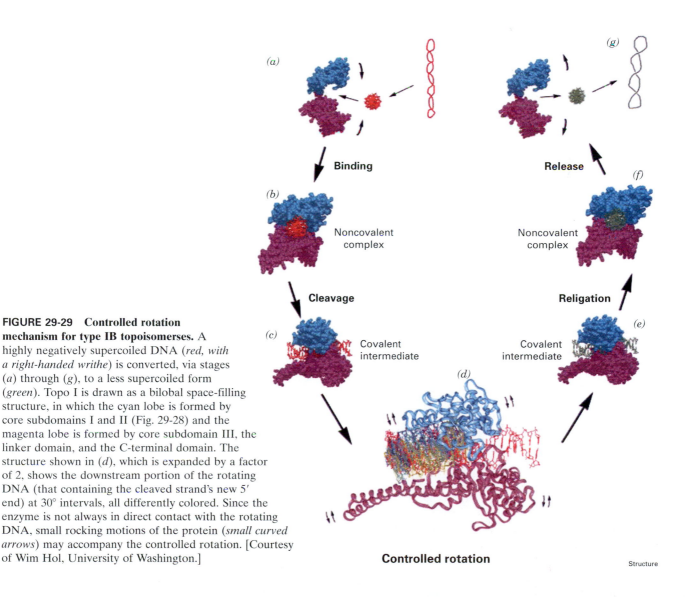

FIGURE 29-29 Controlled rotation mechanism for type IB topoisomerses. A highly negatively supercoiled DNA (*red, with a right-handed writhe*) is converted, via stages (*a*) through (*g*), to a less supercoiled form (*green*). Topo I is drawn as a bilobal space-filling structure, in which the cyan lobe is formed by core subdomains I and II (Fig. 29-28) and the magenta lobe is formed by core subdomain III, the linker domain, and the C-terminal domain. The structure shown in (*d*), which is expanded by a factor of 2, shows the downstream portion of the rotating DNA (that containing the cleaved strand's new 5′ end) at 30° intervals, all differently colored. Since the enzyme is not always in direct contact with the rotating DNA, small rocking motions of the protein (*small curved arrows*) may accompany the controlled rotation. [Courtesy of Wim Hol, University of Washington.]

Structure

derived **coumarin** family of antibiotics, and **ciprofloxacin** (trade name **Cipro**), a member of the synthetically generated **quinolone** family of antibiotics (their coumarin and quinolone groups are drawn in red):

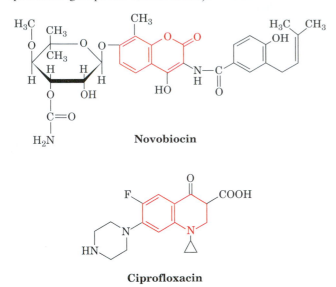

Novobiocin

Ciprofloxacin

These agents profoundly inhibit bacterial DNA replication and RNA transcription, thereby demonstrating the importance of properly supercoiled DNA in these processes. Studies using *E. coli* DNA gyrase mutants resistant to these substances have demonstrated that ciprofloxacin associates with GyrA and novobiocin binds to GyrB.

The gel electrophoretic pattern of duplex circles that have been exposed to DNA gyrase shows a band pattern in which the linking numbers differ by increments of 2 rather than 1, as occurs with type I topoisomerases. *Evidently, DNA gyrase acts by cutting both strands of a duplex, passing the duplex through the break, and resealing it* (Fig. 29-30). This hypothesis is corroborated by the observation that when DNA gyrase is incubated with DNA and ciprofloxacin, and subsequently denatured with guanidinium chloride, a GyrA subunit remains covalently linked to the 5′ end of each of the two cut strands through a phosphoTyr linkage. These cleavage sites are staggered by 4 bp, thereby yielding sticky ends.

Saccharomyces cerevisiae (baker's yeast) **topoisomerase II (topo II),** a type II topoisomerase, is a homodimer of subunits whose N- and C-terminal segments are homologous to DNA gyrase's B and A subunits, respectively. Hence these subfragments are designated B′ and A′. The 92-kD segment encompassing residues 410 to 1202 of this 1429-residue protein can cleave duplex DNA but cannot transport it through the break because it lacks the enzyme's ATPase domain (residues 1–409). However, the C-terminal fragment (residues 1203–1429) appears to be dispensable.

The X-ray structure of the 92-kD segment (Fig. 29-31*a*), determined by James Berger, Stephen Harrison, and Wang, reveals that its two crescent-shaped monomers associate to form a heart-shaped dimer with its two B′ subfragments (residues 410–633) associating at the top of the heart and its two A′ subfragments (residues 683–1202) coming together at its base (point). The 49-residue segment between these two subfragments is disordered. The dimer encloses a large triangular central hole (55 Å wide and 60 Å in height). Tyr 783, the residue that forms a transient phosphoTyr covalent link with the 5′ end of a cleaved DNA strand, is located at the interface between the A′ and B′ subfragments of the same subunit, at the end of a narrow tunnel that opens up into the central hole. Here, the A′ subfragment forms a positively charged semicircular groove that funnels into this active site tunnel. B-DNA can be modeled into this groove with a 4-nt overhang of its 5′-ending strand extending into the active site tunnel. The dimer's two active site Tyr residues are located 27 Å apart such that they must move 35 to 40 Å toward and past each other to achieve positions that are properly staggered to link to the 5′ ends of a cleaved duplex DNA.

The X-ray structure of an *E. coli* GyrB fragment comprising residues 2 to 393 of the 804-residue subunit in complex with the nonhydrolyzable ATP analog ADPNP was determined by Guy Dodson and Eleanor Dodson (Fig. 29-31*b*). This protein fragment, which dimerizes in solution in the presence of ADPNP, consists of two domains. The N-terminal domain, which has been implicated in ATP hydrolysis, binds Mg^{2+}–ADPNP. The C-terminal domains form the walls of a 20-Å-diameter hole through the dimer, the same diameter as that of the B-DNA double helix. All

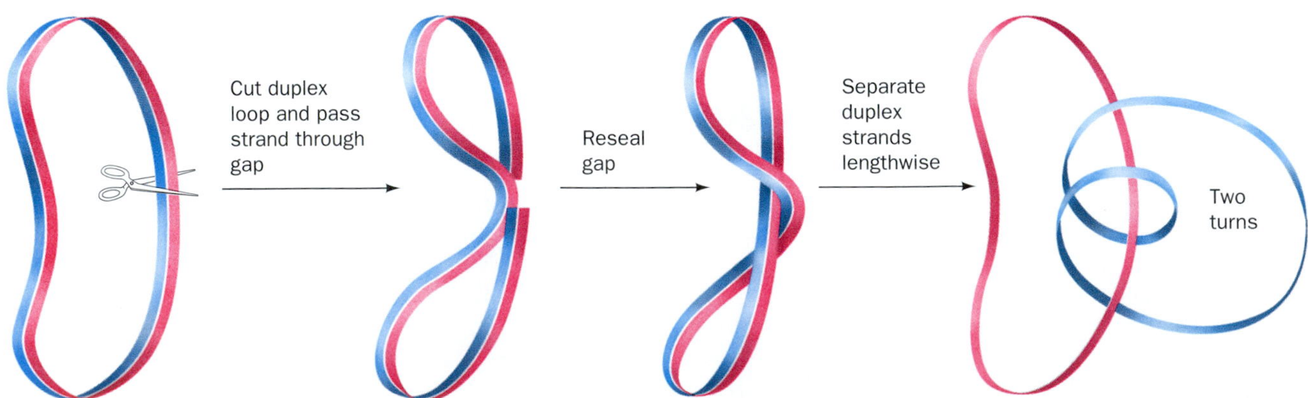

FIGURE 29-30 A demonstration, in which DNA is represented by a ribbon, that cutting a duplex circle, passing the double helix through the resulting gap, and then resealing the break changes the linking number by 2. Separating the resulting single strands (slitting the ribbon along its length; *right*) indicates that one single strand makes two complete revolutions about the other.

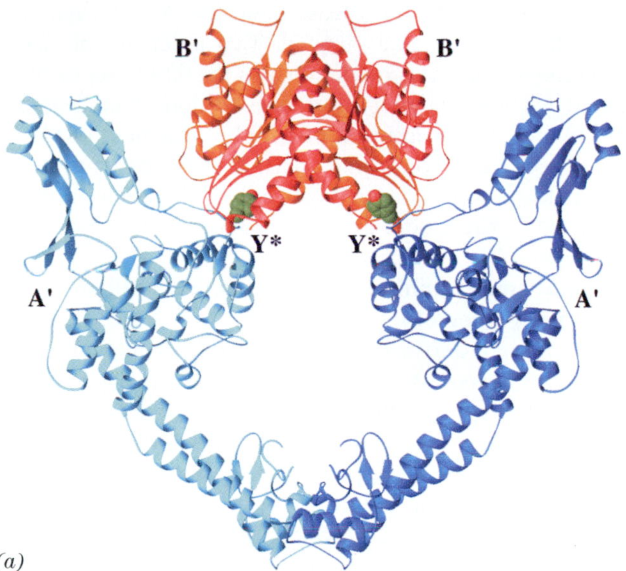

(a)

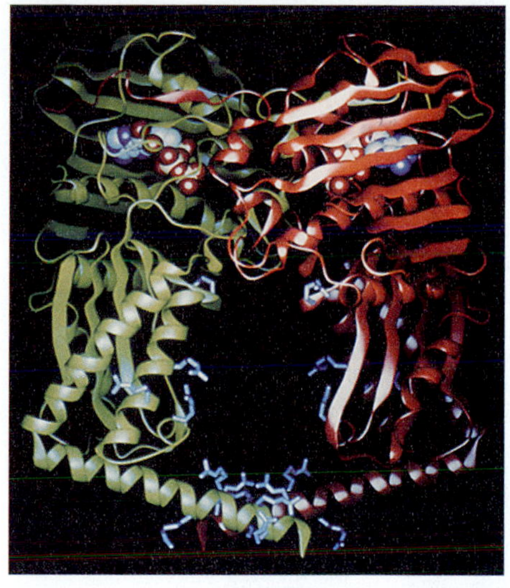

(b)

FIGURE 29-31 Structures of topoisomerase II. (*a*) X-Ray structure of the 92-kD segment of the yeast topoisomerase II (residues 410–1202) dimer as viewed with its twofold axis vertical. The A′ and B′ subfragments of one subunit are blue and red and those of the other subunit are cyan and orange. The active site Tyr 783 side chains are shown in space-filling form (C green and O red) and labeled Y*. [Based on an X-ray structure by James Berger, Stephen Harrison, and James Wang, Harvard University. PDBid 1BGW.] (*b*) X-Ray structure of a dimer of the N-terminal fragment of *E. coli* GyrB (residues 2–393) in complex with ADPNP as viewed with its twofold axis vertical. The two identical subunits, which are colored red and green, each fold into two domains, which are represented by lighter and darker shades of color. The side chains of the Arg residues lining the 20-Å-diameter hole through the protein are shown in stick form (*blue*) and the bound ADPNP molecules are shown in space-filling form. [Courtesy of Eleanor Dodson and Guy Dodson, University of York, U.K.] **See the Interactive Exercises**

of this domain's numerous Arg residues line the walls of the cavity, as might be expected for a DNA-binding surface.

Consideration of the foregoing two structures led to a type of strand passage model for the mechanism of type II topoisomerases (Fig. 29-32) in which the DNA duplex to be cleaved binds in the above-described groove across the top of the heart. ATP binding to the ATP-binding domain (which is absent in the 92-kD fragment) then induces a series of conformational changes in which the DNA's so-called G-segment (G for gate) is cleaved and the resulting two frag-

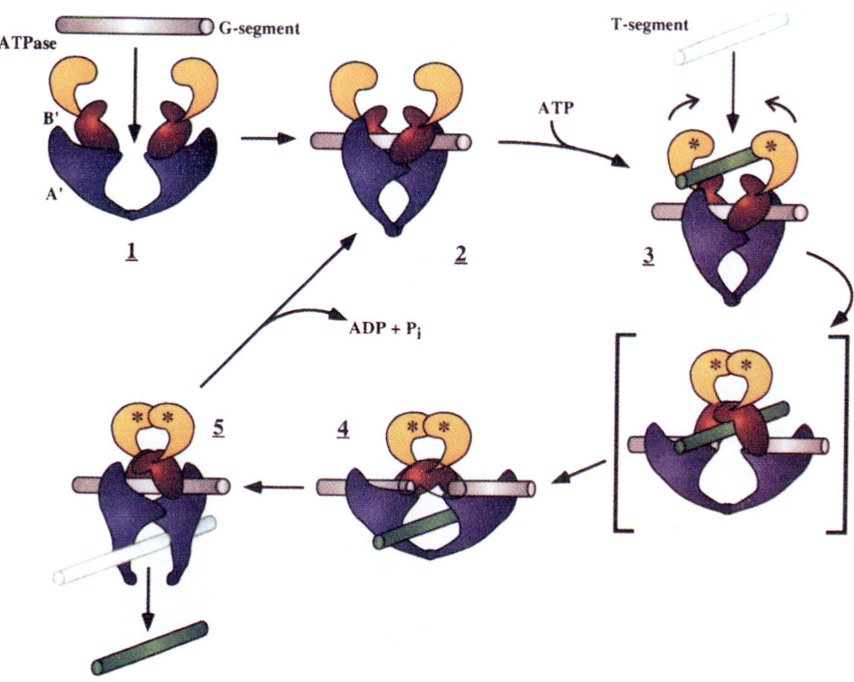

FIGURE 29-32 Model for the enzymatic mechanism of type II topoisomerases. The protein's ATPase, B′, and A′ domains are colored yellow, red, and purple, respectively, and the DNA's G- and T-segments are colored gray and green, respectively. In **1**, the G-segment binds to the enzyme, thereby inducing the conformational change drawn in **2.** The binding of ATP (represented by asterisks) and a T-segment (**3**) induces a series of conformational changes in which the G-segment is cleaved by the A′ subfragments as they separate from one another. The ATPase domains simultaneously dimerize, and the T-segment is transported through the break into the central hole (**4;** here the B′ subfragment in front is transparent for clarity). The DNA transport step is shown as proceeding through the hypothetical intermediate shown in square brackets. The G-segments are then resealed and the T-segment is released through the separation of the A′ subfragments at their dimer interface (**5**). This interface then reforms and the ATP is hydrolyzed and released to yield the enzyme in its starting state (**2**). [Courtesy of James Wang, Harvard University.]

ments are spread apart by at least 20 Å through the action of the protein. This permits the passage of the DNA's so-called T-segment (T for transported) from the top of the heart through the break and into the central hole, thereby incrementing the DNA's linking number by 2. Then, in a process that is accompanied by ATP hydrolysis, the two B′ subfragments come together to reseal the cleaved DNA, and the DNA occupying the central hole is released from the bottom of the heart by the spreading apart of the two contacting A′ subfragments (or GyrA subunits). Finally, the resulting ADP and P_i are released and the A′ subfragments rejoin to yield recycled enzyme. Two independent X-ray structures support this model: that of the 92-kD segment of topo II crystallized under different conditions than the structure in Fig. 29-31*a* and that of a 59-kD fragment of GyrA (Fig. 29-33). The conformations that these latter proteins appear to be representative of some of the conformations predicted by the model depicted in Fig. 29-32.

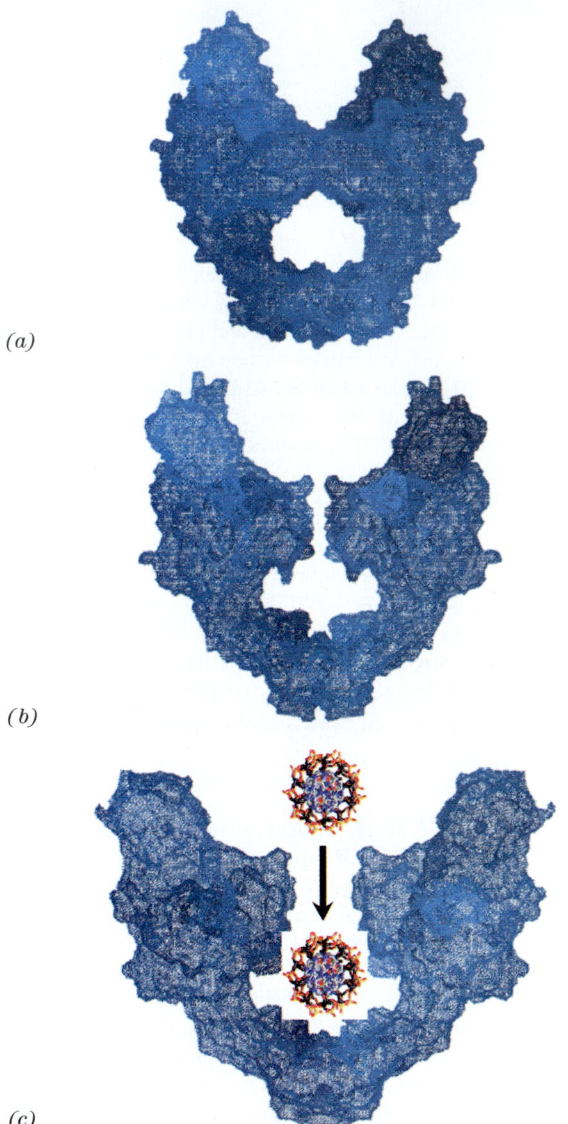

FIGURE 29-33 Surface representations of X-ray structures of type II topoisomerase A′ subfragment dimers. The proteins are viewed with their 2-fold axes vertical. (*a*) *E. coli* GyrA subunits (residues 2–523), the minimal fragments which, when complexed to Gyr B, have DNA cleavage activity. (*b*) The A′ subfragments in the X-ray structure of the 92-kD segment of topo II crystallized under different conditions from that in Fig. 29-31*a*. (*c*) The A′ subfragments in the X-ray structure of topo II (the blue and cyan portions of Fig. 29-31*a*) shown with the modeled passage of DNA into the enzyme's central hole. [Courtesy of James Berger, University of California at Berkeley PDBids (*a*) 1AB4, (*b*) 1BJT, and (*c*) 1BJW.]

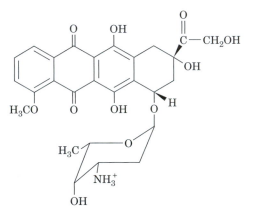

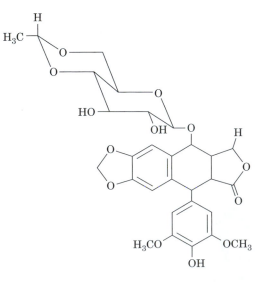

e. Topoisomerse Inhibitors Are Effective Antibiotics and Cancer Chemotherapy Agents

Coumarin derivatives such as novobiocin, and quinolone derivatives such as ciprofloxacin, specifically inhibit DNA gyrases and are therefore antibiotics. In fact, ciprofloxacin is the most efficacious oral antibiotic against gram-negative bacteria presently in clinical use (novobiocin's adverse side effects and the rapid generation of bacterial resistance to it have resulted in the discontinuation of its use in the treatment of human infections). A number of substances, including **doxorubicin** (also called **adriamycin;** a product of *Streptomyces peucetius*) and **etoposide** (a synthetic derivative),

Doxorubicin (Adriamycin)

Etoposide

inhibit eukaryotic type II topoisomerases and are therefore widely used in cancer chemotherapy.

Type II topoisomerase inhibitors act in either of two ways. Many of them, including novobiocin, inhibit their target enzyme's ATPase activity (novobiocin is a competitive inhibitor of ATP because it tightly binds to GyrB in a way that prevents the binding of ATP's adenine ring). They therefore kill cells by blocking topoisomerase activity, which results in the arrest of DNA replication and RNA transcription. However, other substances, including ciprofloxacin, doxorubicin, and etoposide, enhance the rate at which their target type II topoisomerases cleave double-stranded DNA and/or reduce the rate at which these enzymes reseal these breaks. Consequently, these agents induce higher than normal levels of transient protein-bridged breaks in the DNA of treated cells. These protein bridges are easily ruptured by the passage of the replication and transcription machinery, thereby rendering the breaks permanent. Although all cells have extensive enzymatic machinery to repair damaged DNA (Section 30-5), a sufficiently high level of DNA damage results in cell death. Consequently, since rapidly replicating cells such as cancer cells have elevated levels of type II topoisomerases, they are far more likely to incur lethal DNA damage through the inhibition of their type II topoisomerases than are slow-growing or quiescent cells.

Type IB topoisomerases are specifically inhibited by the quinoline-based alkaloid **camptothecin**

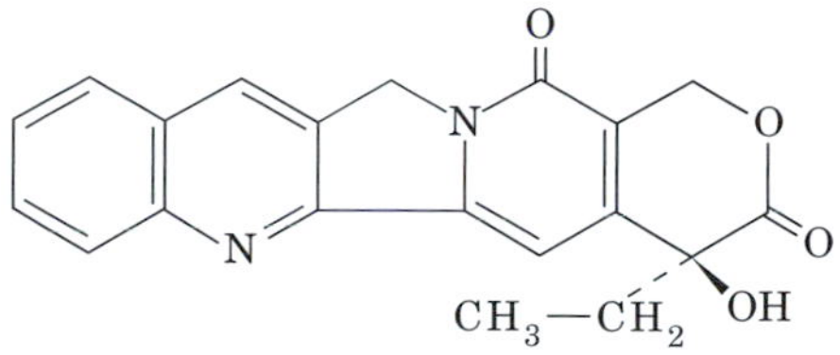

$CH_3—CH_2$ OH

Camptothecin

(a product of the Chinese tree *Camptotheca acuminata*) and its derivatives, which act by stabilizing the covalent topoisomerase I–DNA complex. These compounds, the only known naturally occurring topoisomerase IB inhibitors, are therefore potent anticancer agents.

CHAPTER SUMMARY

1 ■ Double Helical Structures B-DNA consists of a right-handed double helix of antiparallel sugar–phosphate chains with ~10 bp per turn of 34 Å and with its bases nearly perpendicular to the helix axis. Bases on opposite strands hydrogen bond in a geometrically complementary manner to form A · T and G · C Watson–Crick base pairs. At low humidity, B-DNA undergoes a reversible transformation to a wider, flatter right-handed double helix known as A-DNA. Z-DNA, which is formed at high salt concentrations by polynucleotides of alternating purine and pyrimidine base sequences, is a left-handed double helix. Double helical RNA and RNA · DNA hybrids have A-DNA-like structures. The conformation of DNA, particularly that of B-DNA, varies with its base sequence largely because DNA's deformability varies with its base sequence.

2 ■ Forces Stabilizing Nucleic Acid Structures The orientations about the glycosidic bond and the various torsion angles in the sugar–phosphate chain are sterically constrained in nucleic acids. Likewise, only a few of the possible sugar pucker conformations are commonly observed. Watson–Crick base pairing is both geometrically and electronically complementary. Yet hydrogen bonding interactions do not greatly stabilize nucleic acid structures. Rather, the structures are largely stabilized by hydrophobic interactions. Nevertheless, the hydrophobic forces in nucleic acids are qualitatively different in character from those that stabilize proteins. Electrostatic interactions between charged phosphate groups are also important structural determinants of nucleic acids.

3 ■ Supercoiled DNA The linking number (L) of a covalently closed circular DNA is topologically invariant. Consequently, any change in the twist (T) of a circular duplex must be balanced by an equal and opposite change in its writhing number (W), which indicates its degree of supercoiling. Supercoiling can be induced by intercalation agents. The gel electrophoretic mobility of DNA increases with its degree of superhelicity. Naturally occurring DNAs are all negatively supercoiled and must be so in order to participate in DNA replication and RNA transcription. Type IA topoisomerases relax negatively supercoiled DNAs via a strand passage mechanism in which they cleave a single-strand of DNA to form a 5′-phosphoTyr bond, pass a single-strand DNA segment through the gap, and then reseal the gap. Type IB topoisomerases relax both negatively and positively supercoiled DNAs via a controlled rotation mechanism involving a single-strand cleavage in which a transient phosphoTyr bond is formed with the newly generated 3′ end. Type II topoisomerases relax duplex DNA in increments of two supertwists at the expense of ATP hydrolysis by making a double-strand scission in the DNA so as to form two transient 5′-phosphoTyr linkages, passing the duplex through the break, and resealing it. DNA gyrase also generates negative supertwists in an ATP-dependent manner. Topoisomerases are the targets of various antibiotics and chemotherapeutic agents.

REFERENCES

GENERAL

Bloomfield, V.A., Crothers, D.M., and Tinoco, I., Jr., *Nucleic Acids: Structures, Properties, and Functions,* University Science Books (2000).

Calladine, C.R. and Drew, H.R., *Understanding DNA,* Academic Press (1992). [The molecule and how it works.]

Neidle, S. (Ed.), *Oxford Handbook of Nucleic Acid Structure,* Oxford University Press (1999).

Saenger, W., *Principles of Nucleic Acid Structure,* Springer-Verlag (1984). [A detailed and authoritative exposition.]

Sinden, R.R., *DNA Structure and Function,* Academic Press (1994).

The double helix–50 years, *Nature* **421,** 395–453 (2003). [A supplement containing a series of articles on the historical, cultural, and scientific influences of the DNA double helix celebrating the fiftieth anniversary of its discovery.]

Travers, A. and Buckle, M. (Eds.) *DNA–Protein Interactions. A Practical Approach,* Oxford University Press (2000). [A laboratory manual for numerous physicochemical methods that are used to probe the interactions of DNA and proteins.]

STRUCTURES AND STABILITIES OF NUCLEIC ACIDS

Dickerson, R.E., Sequence-dependent B-DNA conformation in crystals and in protein complexes, *in* Sarma, R.H. and Sarma, M.H. (Eds.), *Structure, Motion, Interaction and Expression in Biological Molecules, pp.* 17–35, Adenine Press (1998); *and* DNA bending: the prevalence of kinkiness and the virtues of normality, *Nucleic Acids Res.* **26,** 1906–1926 (1998).

Fairhead, H., Setlow, B., and Setlow, P., Prevention of DNA damage in spores and in vitro by small, acid-soluble proteins from *Bacillus* species, *J. Bacteriol.* **175,** 1367–1374 (1993).

Joshua-Tor, L. and Sussman, J.L., The coming of age of DNA crystallography, *Curr. Opin. Struct. Biol.* **3,** 323–335 (1993).

Rich, A., Nordheim, A., and Wang, A.H.-J., The chemistry and biology of left-handed Z-DNA, *Annu. Rev. Biochem.* **53,** 791–846 (1984).

Schwartz, T., Rould, M.A., Lowenhaupt, K., Herbert, A., and Rich, A., Crystal structure of the Zα domain of the human editing enzyme ADAR1 bound to left-handed Z-DNA, *Science* **284,** 1841–1845 (1999).

Sundaralingam, M., Stereochemistry of nucleic acids and their constituents. IV. Allowed and preferred conformations of nucleosides, nucleoside mono-, di-, tri-, and tetraphosphates, nucleic acids and polynucleotides, *Biopolymers* **7,** 821–860 (1969).

Voet, D. and Rich, A., The crystal structures of purines, pyrimidines and their intermolecular structures, *Prog. Nucleic Acid Res. Mol. Biol.* **10,** 183–265 (1970).

Wing, R., Drew, H., Takano, T., Broka, C., Tanaka, S., Itakura, K., and Dickerson, R.E., Crystal structure analysis of a complete turn of B-DNA, *Nature* **287,** 755–758 (1980); and Shui, X., McFail-Isom, L., Hu, G.G., and Williams, L.D., The B-DNA decamer at high resolution reveals a spine of sodium, *Biochemistry* **37,** 8341–8355 (1998). [The Dickerson dodecamer at its original 2.5 Å resolution and at its later-determined 1.4 Å resolution.]

SUPERCOILED DNA

Bates, A.D. and Maxwell, A. *DNA Topology,* IRL Press (1993). [A monograph.]

Berger, J.M., Type II DNA topoisomerases, *Curr. Opin. Struct. Biol.* **8,** 26–32 (1998).

Berger, J.M., Gamblin, S.J., Harrison, S.C., and Wang, J.C., Structure and mechanism of DNA topoisomerase II, *Nature* **379,** 225–232 (1996); Morais Cabral, J.H., Jackson, A.P., Smith, C.V., Shikotra, N., Maxwell, A., and Liddington, R.C., Crystal structure of the breakage–reunion domain of DNA gyrase, *Nature* **388,** 903–906 (1997); *and* Fass, D., Bogden, C.E., and Berger, J.M., Quaternary changes in topoisomerase II may direct orthogonal movement of two DNA strands, *Nature Struct. Biol.* **6,** 322–326 (1999).

Champoux, J.J., DNA topoisomerases: Structure, function, and mechanism, *Annu. Rev. Biochem.* **70,** 369–413 (2001).

Changela, A., DiGate, R., and Mondragón, A., Crystal structure of a complex of a type IA DNA topoisomerase with a single-stranded DNA, *Nature* **411,** 1077–1081 (2001); Mondragón, A. and DiGate, R., The structure of *Escherichia coli* DNA topoisomerase III, *Structure* **7,** 1373–1383 (1999); *and* Lima, C.D., Wang, J.C., and Mondragón, A., Three-dimensional structure of the 67K N-terminal fragment of *E. coli* DNA topoisomerase I, *Nature* **367,** 138–146 (1994).

Froelich-Ammon, S.J. and Osheroff, N., Topoisomerase poisons: Harnessing the dark side of enzyme mechanism, *J. Biol. Chem.* **270,** 21429–21432 (1995).

Horton, N.C. and Finzel, B.C., The structure of an RNA/DNA hybrid: A substrate of the ribonuclease activity of HIV-1 reverse transcriptase, *J. Mol. Biol.* **264,** 521–533 (1996).

Kanaar, R. and Cozarelli, N.R., Roles of supercoiled DNA structure in DNA transactions, *Curr. Opin. Struct. Biol.* **2,** 369–379 (1992).

Lebowitz, J., Through the looking glass: The discovery of supercoiled DNA, *Trends Biochem. Sci.* **15,** 202–207 (1990). [An informative eyewitness account of how DNA supercoiling was discovered.]

Li, T.-K. and Liu, L.F., Tumor cell death induced by topoisomerase-targeting drugs, *Annu. Rev. Pharmacol. Toxicol.* **41,** 53–77 (2001).

Maxwell, A., DNA gyrase as a drug target, *Biochem. Soc. Trans.* **27,** 48–53 (1999).

Redinbo, M.R., Stewart, L., Kuhn, P., Champoux, J.J., and Hol, W.G.J., Crystal structures of human topoisomerase I in covalent and noncovalent complexes with DNA, *Science* **279,** 1504–1513 (1998); Stewart, L., Redinbo, M.R., Qiu, X., Hol, W.G.J., and Champoux, J.J., A model for the mechanism of human topoisomerase I, *Science* **279,** 1534–1541 (1998); *and* Redinbo, M.R., Champoux, J.J., and Hol, W.G.J., Structural insights into the function of type IB topoisomerases, *Curr. Opin. Struct. Biol.* **9,** 29–36 (1999).

Wang, J.C., DNA topoisomerases, *Annu. Rev. Biochem.* **65,** 635–692 (1996).

Wang, J.C., Moving one DNA double helix through another by a type II DNA topoisomerase: The story of a simple molecular machine, *Q. Rev. Biophys.* **31,** 107–144 (1998).

Wang, J.C., Cellular roles of DNA topoisomerases: a molecular perspective, *Nature Rev. Mol. Cell Biol.* **3,** 430–440 (2002).

Wigley, D.B., Davies, G.J., Dodson, E.J., Maxwell, A., and Dodson, G., Crystal structure of an N-terminal fragment of DNA gyrase B, *Nature* **351,** 624–629 (1991).

PROBLEMS

1. A · T base pairs in DNA exhibit greater variability in their propeller twisting than do G · C base pairs. Suggest the structural basis of this phenomenon.

***2.** At Na^+ concentrations $>5M$, the T_m of DNA decreases with increasing [Na^+]. Explain this behavior. (*Hint:* Consider the solvation requirements of Na^+.)

***3.** Why are the most commonly observed conformations of the ribose ring those in which either atom C2′ or atom C3′ is out of the plane of the other four ring atoms? (*Hint:* In puckering a planar ring such that one atom is out of the plane of the other four, the substituents about the bond opposite the out-of-plane atom remain eclipsed. This is best observed with a ball-and-stick model.)

4. Polyoma virus DNA can be separated by sedimentation at neutral pH into three components that have sedimentation coefficients of 20, 16, and 14.5S and that are known as Types I, II, and III DNAs, respectively. These DNAs all have identical base compositions and molecular masses. In 0.15*M* NaCl, both Types II and III DNA have melting curves of normal cooperativity and a T_m of 88°C. Type I DNA, however, exhibits a very broad melting curve and a T_m of 107°C. At pH 13, Types I and III DNAs have sedimentation coefficients of 53 and 16S, respectively, and Type II separates into two components with sedimentation coefficients of 16 and 18S. How do Types I, II, and III DNAs differ from one another? Explain their different physical properties.

5. When the helix axis of a closed circular duplex DNA of 2340 bp is constrained to lie in a plane, the DNA has a twist (T) of 212. When released, the DNA takes up its normal twist of 10.4 bp per turn. Indicate the values of the linking number (L), writhing number (W), and twist for both the constrained and unconstrained conformational states of this DNA circle. What is the superhelix density, σ, of both the constrained and unconstrained DNA circles?

6. A closed circular duplex DNA has a 100-bp segment of alternating C and G residues. On transfer to a solution containing a high salt concentration, this segment undergoes a transition from the B conformation to the Z conformation. What is the accompanying change in its linking number, writhing number, and twist?

7. You have discovered an enzyme secreted by a particularly virulent bacterium that cleaves the C2′—C3′ bond in the deoxyribose residues of duplex DNA. What is the effect of this enzyme on supercoiled DNA?

8. A bacterial chromosome consists of a protein–DNA complex in which its single DNA molecule appears to be supercoiled, as demonstrated by ethidium bromide titration. However, in contrast to the case with naked circular duplex DNA, the light single-strand nicking of chromosomal DNA does not abolish this supercoiling. What does this indicate about the structure of the bacterial chromosome, that is, how do its proteins constrain its DNA?

9. Although types IA and II topoisomerases exhibit no significant sequence similarity, it has been suggested that they are distantly related based on the similarites of certain aspects of their enzymatic mechanisms. What are these similarities?

10. Draw the mechanism of DNA strand cleavage and rejoining mediated by topoisomerase IA.

Chapter 29 also appeared in Volume 1 of *Biochemistry,* 3rd edition, by Donald Voet and Judith G. Voet. It is included in this Volume 2 as well, as part of the Expression and Transmission of Genetic Information.

Chapter **30**

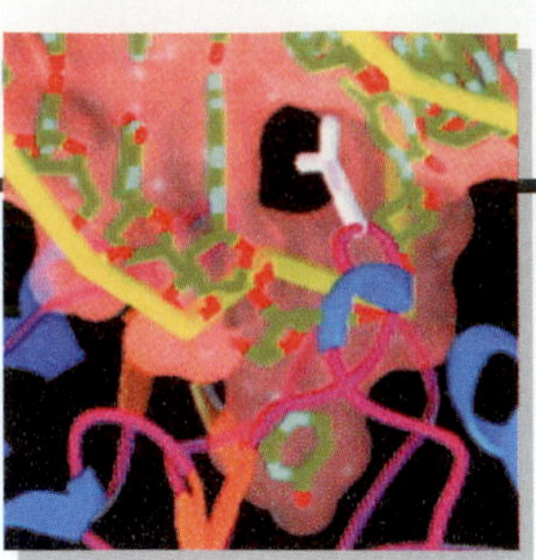

DNA Replication, Repair, and Recombination

People are DNA's way of making more DNA.

Anon.

Here we begin a three-chapter series on the basic processes of gene expression: DNA replication (this chapter), transcription (Chapter 31), and translation (Chapter 32). These processes have been outlined in Section 5-4. We shall now discuss them in greater depth with an emphasis on how we have come to know what we know.

1 ■ DNA REPLICATION: AN OVERVIEW

Watson and Crick's seminal paper describing the DNA double helix ended with the statement: "It has not escaped our notice that the specific pairing we have postulated immediately suggests a possible copying mechanism for the genetic material." In a succeeding paper they expanded on this rather cryptic remark by pointing out that a DNA strand could act as a template to direct the synthesis of its complementary strand. Although Meselson and Stahl demonstrated, in 1958, that DNA is, in fact, semiconservatively replicated (Section 5-3B), it was not until some 20 years later that the mechanism of DNA replication in prokaryotes was understood in reasonable detail. This is because, as we shall see in this chapter, the DNA replication process rivals translation in its complexity but is mediated by often loosely associated protein assemblies that are present in only a few copies per cell. *The surprising intricacy of DNA replication compared to the chemically similar transcription process (Section 31-2) arises from the need for extreme accuracy in DNA replication so as to preserve the integrity of the genome from generation to generation.*

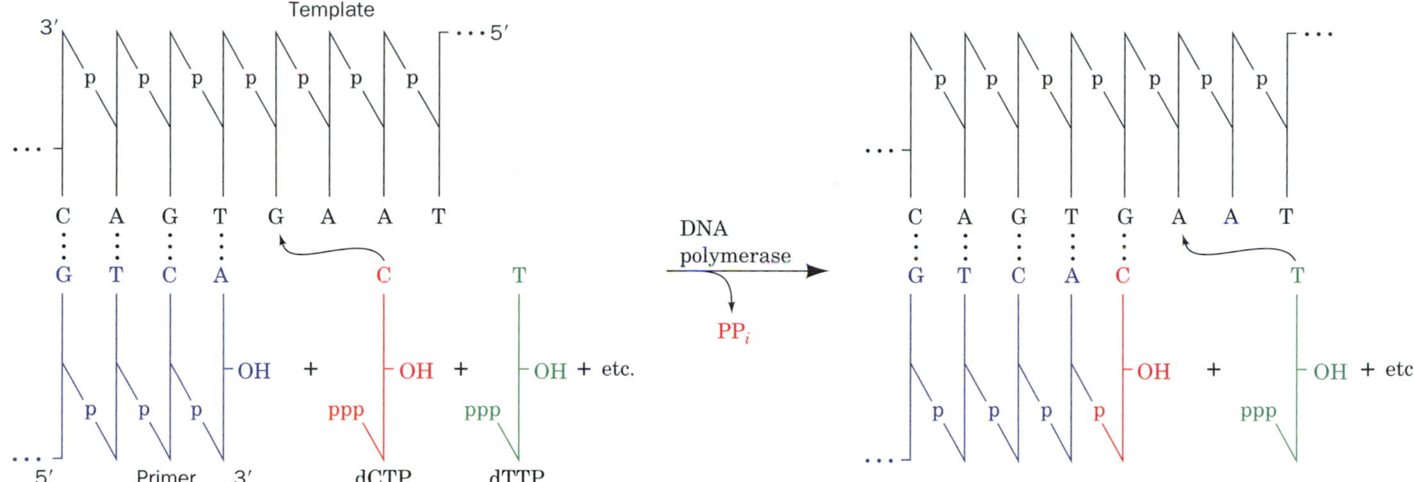

FIGURE 30-1 Action of DNA polymerase. DNA polymerases assemble incoming deoxynucleoside triphosphates on single-stranded DNA templates such that the growing strand is elongated in its 5′ → 3′ direction.

A. Replication Forks

DNA is replicated by enzymes known as ***DNA-directed DNA polymerases*** *or simply* ***DNA polymerases.*** These enzymes utilize single-stranded DNA as templates on which to catalyze the synthesis of the complementary strand from the appropriate deoxynucleoside triphosphates (Fig. 30-1). The incoming nucleotides are selected by their ability to form Watson–Crick base pairs with the template DNA so that the newly synthesized DNA strand forms a double helix with the template strand. *Nearly all known DNA polymerases can only add a nucleotide donated by a nucleoside triphosphate to the free 3′-OH group of a base paired polynucleotide so that DNA chains are extended only in the 5′ → 3′ direction.* DNA polymerases are discussed further in Sections 30-2A, 30-2B, and 30-4B.

a. Duplex DNA Replicates Semiconservatively at Replication Forks

John Cairns obtained the earliest indications of how chromosomes replicate through the autoradiography of replicating DNA. Autoradiograms of circular chromosomes grown in a medium containing [^{3}H]thymidine show the presence of replication "eyes" or "bubbles" (Fig. 30-2). These so-called **θ structures** (after their resemblance to the Greek letter theta) indicate that *double-stranded DNA* ***(dsDNA)*** *replicates by the progressive separation of its two parental strands accompanied by the synthesis of their complementary strands to yield two semiconservatively replicated duplex daughter strands (Fig. 30-3).* DNA replication involving θ structures is known as **θ replication.**

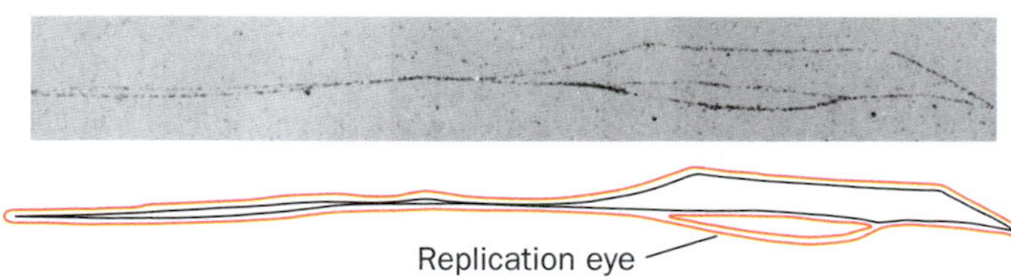

FIGURE 30-2 Autoradiogram and its interpretive drawing of a replicating *E. coli* chromosome. The bacterium had been grown for somewhat more than one generation in a medium containing [^{3}H]thymidine, thereby labeling the subsequently synthesized DNA so that it appears as a line of dark grains in the photographic emulsion (*red lines in the interpretive drawing*). The size of the replication eye indicates that the circular chromosome is about one-sixth duplicated in the present round of replication. [Courtesy of John Cairns, Cold Spring Harbor Laboratory.]

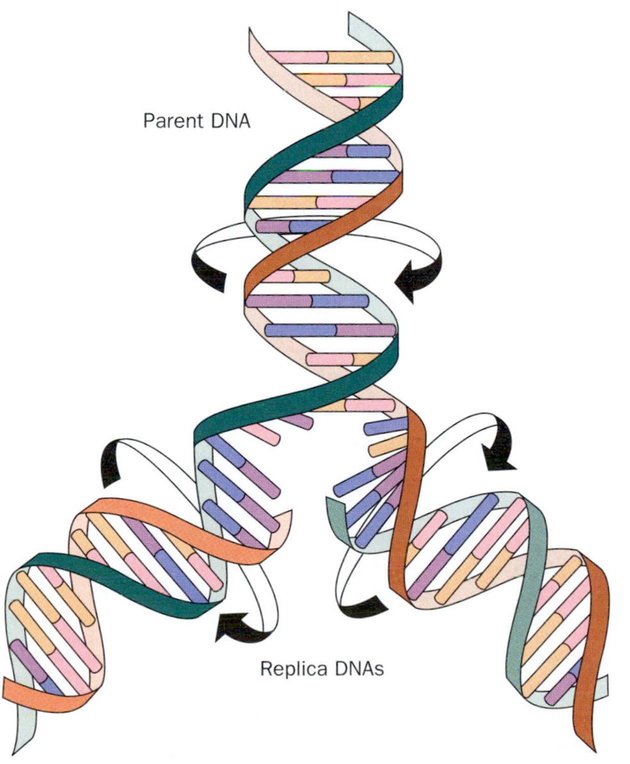

FIGURE 30-3 Replication of DNA.

A branch point in a replication eye at which DNA synthesis occurs is called a **replication fork.** A replication bubble may contain one or two replication forks **(unidirectional** or **bidirectional replication).** Autoradiographic studies have demonstrated that θ replication is almost always bidirectional (Fig. 30-4). Moreover, such experiments, together with genetic evidence, have established that prokaryotic and bacteriophage DNAs have but one **replication origin** (point where DNA synthesis is initiated).

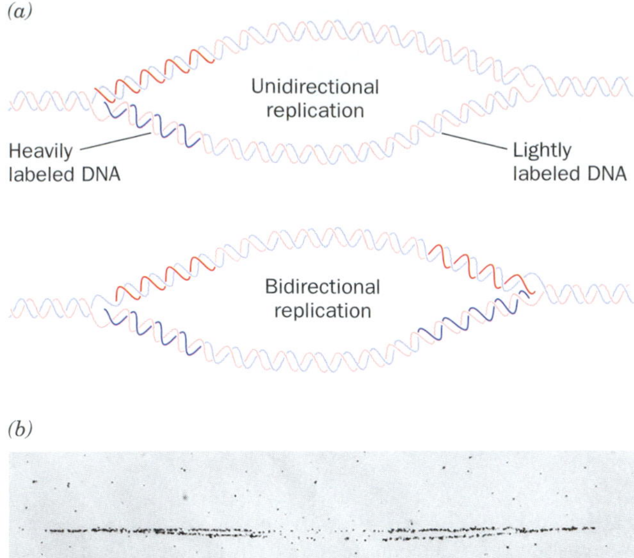

FIGURE 30-4 Autoradiographic differentiation of unidirectional and bidirectional θ replication of DNA. (*a*) An organism is grown for several generations in a medium that is lightly labeled with [^{3}H]thymidine so that all of its DNA will be visible in an autoradiogram. A large amount of [^{3}H]thymidine is then added to the medium for a few seconds before the DNA is isolated **(pulse labeling)** in order to label only those bases near the replication fork(s). Unidirectional DNA replication will exhibit only one heavily labeled branch point (*above*), whereas bidirectional DNA replication will exhibit two such branch points (*below*). (*b*) An autoradiogram of *E. coli* DNA so treated, demonstrating that it is bidirectionally replicated. [Courtesy of David M. Prescott, University of Colorado.]

B. *Role of DNA Gyrase*

The requirement that the parent DNA unwind at the replication fork (Fig. 30-3) presents a formidable topological obstacle. For instance, *E. coli* DNA is replicated at a rate of ~1000 nucleotides/s. If its 1300-μm-long chromosome were linear, it would have to flail around within the confines of a 3-μm-long *E. coli* cell at ~100 revolutions/s (recall that B-DNA has ~10 bp per turn). But since the *E. coli* chromosome is, in fact, circular, even this could not occur. Rather, the DNA molecule would accumulate +100 supercoils/s (see Section 29-3A for a discussion of supercoiling) until it became too tightly coiled to permit further unwinding. Naturally occurring DNA's negative supercoiling promotes DNA unwinding but only to the extent of ~5% of its duplex turns (recall that naturally occurring DNAs are typically underwound by one supercoil per ~20 duplex turns; Section 29-3B). In prokaryotes, however, negative supercoils may be introduced into DNA through the action of a Type II topoisomerase (DNA gyrase; Section 29-3C) at the expense of ATP hydrolysis. This process is essential for prokaryotic DNA replication as is demonstrated by the observation that DNA gyrase inhibitors, such as novobiocin, arrest DNA replication except in mutants whose DNA gyrase does not bind these antibiotics.

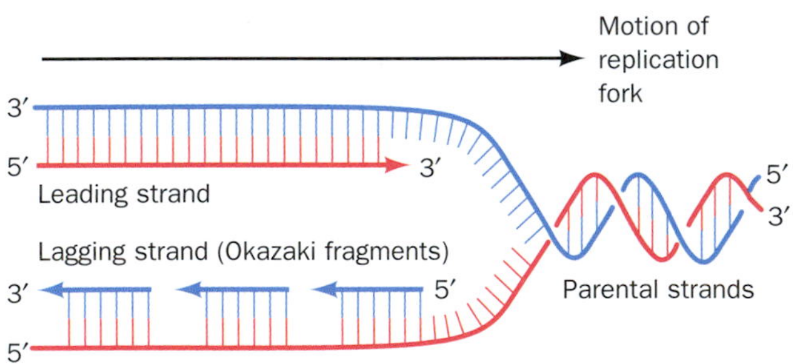

FIGURE 30-5 Semidiscontinuous DNA replication. In DNA replication, both daughter strands (*leading strand red, lagging strand blue*) are synthesized in their 5′ → 3′ directions. The leading strand is synthesized continuously, whereas the lagging strand is synthesized discontinuously.

C. *Semidiscontinuous Replication*

The low-resolution images provided by autoradiograms such as Figs. 30-2 and 30-4*b* suggest that dsDNA's two antiparallel strands are simultaneously replicated at an advancing replication fork. Yet, all known DNA polymerases can only extend DNA strands in the 5′ → 3′ direction. How, then, does DNA polymerase copy the parent strand that extends in the 5′ → 3′ direction past the replication fork? This question was answered in 1968 by Reiji Okazaki through the following experiments. If a growing *E. coli* culture is pulse-labeled for 30 s with [^{3}H]thymidine, much of the radioactive and hence newly synthesized DNA has a sedimentation coefficient in alkali of 7S to 11S. These so-called **Okazaki fragments** evidently consist of only 1000 to 2000 nucleotides (**nt;** 100–200 nt in eukaryotes). If, however, following the 30 s [^{3}H]thymidine pulse, the *E. coli* are transferred to an unlabeled medium (a **pulse–chase** experiment), the resulting radioactively labeled DNA sediments at a rate that increases with the time that the cells had grown in the unlabeled medium. The Okazaki fragments must therefore become covalently incorporated into larger DNA molecules.

Okazaki interpreted his experimental results in terms of the **semidiscontinuous replication** model (Fig. 30-5). The

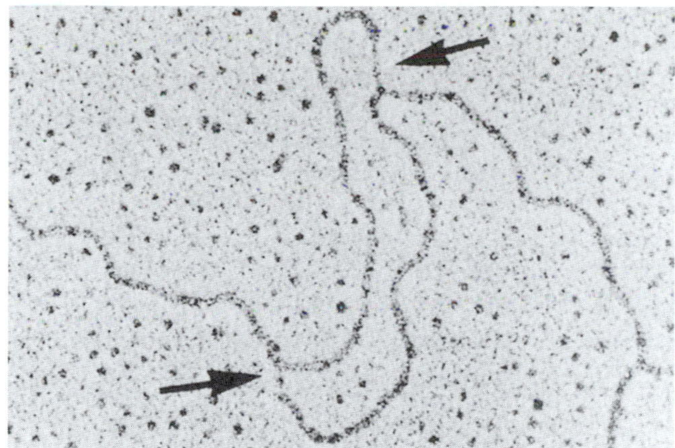

FIGURE 30-6 Electron micrograph of a replication eye in *Drosophila melanogaster* DNA. Note that the single-stranded regions (*arrows*) near the replication forks have the trans configuration consistent with the semidiscontinuous model of DNA replication. [From Kreigstein, H.J. and Hogness, D.S., *Proc. Natl. Acad. Sci.* **71,** 173 (1974).]

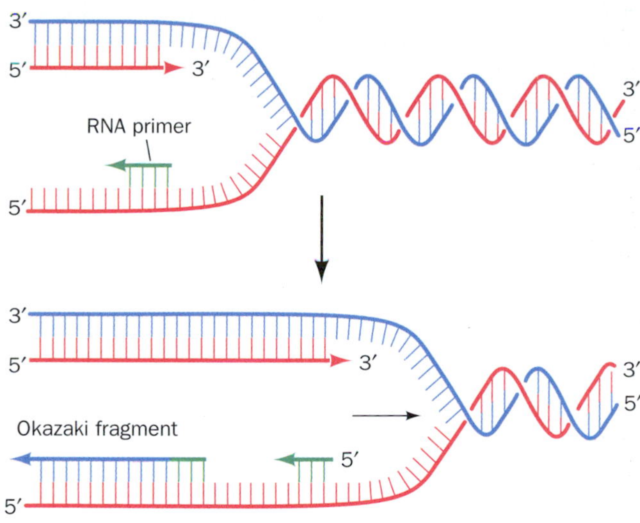

FIGURE 30-7 Priming of DNA synthesis by short RNA segments.

two parent strands are replicated in different ways. *The newly synthesized DNA strand that extends 5′ → 3′ in the direction of replication fork movement, the so-called* ***leading strand,*** *is essentially continuously synthesized in its 5′ → 3′ direction as the replication fork advances. The other newly synthesized strand, the* ***lagging strand,*** *is also synthesized in its 5′ → 3′ direction but discontinuously as Okazaki fragments. The Okazaki fragments are only covalently joined together sometime after their synthesis in a reaction catalyzed by the enzyme* ***DNA ligase*** *(Section 30-2D).*

The semidiscontinuous model of DNA replication is corroborated by electron micrographs of replicating DNA showing single-stranded regions on one side of the replication fork (Fig. 30-6). In bidirectionally replicating DNA, moreover, the two single-stranded regions occur, as expected, on diagonally opposite sides of the replication bubble.

D. *RNA Primers*

DNA polymerases' all but universal requirement for a free 3′-OH group to extend a DNA chain poses a question that was emphasized by the establishment of the semidiscontinuous model of DNA replication: How is DNA synthesis initiated? Careful analysis of Okazaki fragments revealed that *their 5′ ends consist of RNA segments of 1 to 60 nt (a length that is species dependent) that are complementary to the template DNA chain (Fig. 30-7). E. coli* has two enzymes that can catalyze the formation of these **RNA primers: RNA polymerase,** the ~459-kD multisubunit enzyme that mediates transcription (Section 31-2), and the much smaller **primase** (60 kD), the monomeric product of the *dnaG* gene.

Primase is insensitive to the RNA polymerase inhibitor **rifampicin** (Section 31-2C). The observation that rifampicin inhibits only leading strand synthesis therefore indicates that *primase initiates the Okazaki fragment primers.* The initiation of leading strand synthesis in *E. coli,* a much rarer event than that of Okazaki fragments, can be mediated *in vitro* by either RNA polymerase or primase alone but is greatly stimulated when both enzymes are present. It is therefore thought that these enzymes act synergistically *in vivo* to prime leading strand synthesis.

Mature DNA does not contain RNA. The RNA primers are eventually removed and the resulting single-strand gaps are filled in with DNA by a mechanism described in Section 30-2A.

2 ■ ENZYMES OF REPLICATION

DNA replication is a complex process involving a great variety of enzymes. It requires, to list only its major actors in their order of appearance: (1) DNA topoisomerases, (2) enzymes known as helicases that separate the DNA strands at the replication fork, (3) proteins that prevent them from reannealing before they are replicated, (4) enzymes that synthesize RNA primers, (5) a DNA polymerase, (6) an enzyme to remove the RNA primers, and (7) an enzyme to covalently link successive Okazaki fragments. In this section, we describe the properties and functions of many of these proteins.

A. *DNA Polymerase I*

In 1957, Arthur Kornberg reported that he had discovered an enzyme that catalyzes the synthesis of DNA in extracts of *E. coli* through its ability to incorporate the radioactive label from [^{14}C]thymidine triphosphate into DNA. This enzyme, which has since become known as **DNA polymerase I** or **Pol I,** consists of a monomeric 928-residue polypeptide.

Pol I couples deoxynucleoside triphosphates on DNA templates (Fig. 30-1) in a reaction that occurs through the nucleophilic attack of the growing DNA chain's 3′-OH group on the α-phosphoryl of an incoming nucleoside triphosphate. The reaction is driven by the resulting elimination of PP_i and its subsequent hydrolysis by inorganic pyrophosphatase. The overall reaction resembles that catalyzed by RNA polymerase (Fig. 5-23) but differs from it by the strict requirement that the incoming nucleoside be linked to a free 3′-OH group of a polynucleoside that is base paired to the template (RNA polymerase initiates transcription by linking together two ribonucleoside triphosphates on a DNA template; Section 31-2C). The complementarity between the product DNA and the template was at first inferred through base composition and hybridization studies but was eventually directly established by base sequence determinations. The error rate of Pol I in copying the template is extremely low, as was first demonstrated by its *in vitro* replication of the 5386-nt DNA from bacteriophage **φX174** to yield fully infective phage DNA. In fact, its measured error rate is around one wrong base per 10 million.

Pol I is said to be **processive** in that it catalyzes a series of successive polymerization steps, typically 20 or more, without releasing the template. Pol I can, of course, work in reverse by degrading DNA through pyrophosphorolysis. This reverse reaction, however, probably has no physiological significance because of the low *in vivo* concentration of PP_i resulting from the action of inorganic pyrophosphatase.

a. Pol I Recognizes the Incoming dNTP According to the Shape of the Base Pair It Forms with the Template DNA

The specificity of Pol I for an incoming base arises from the requirement that it form a Watson–Crick base pair with the template rather than direct recognition of the incoming base (recall that the four base pairs, A · T, T · A, G · C, and C · G, have nearly identical shapes; Fig. 5-12). Thus, as Eric Kool demonstrated, when the "base" **2,4-difluorotoluene (F),**

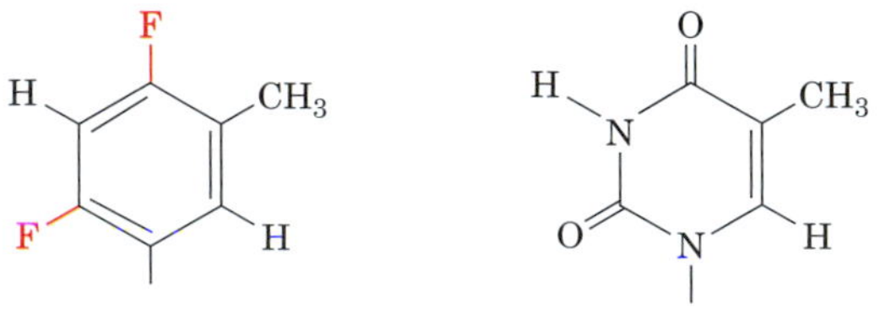

2,4-Difluorotoluene base (F) **Thymine (T)**

which is isosteric with (has the same shape as) thymine but does not accept hydrogen bonds, is synthetically inserted into a template DNA, Pol I incorporates A opposite the F with a similar rate of mismatches as it incorporates A opposite T. Likewise, dFTP is incorporated opposite template A with a similar fidelity as is dTTP. Yet the incorporation of F opposite an A in DNA destabilizes the double helix by 15 kJ/mol relative to T opposite this A. Evidently, *Pol I selects an incoming dNTP largely according to its ability to form a Watson–Crick-shaped pair with the template base but with little regard for its hydrogen bonding properties.* Indeed, the NMR structure of a 12-bp DNA containing a centrally located F opposite an A reveals that it assumes a B-DNA conformation in which the F–A pair closely resembles a T · A base pair in the same position of an otherwise identical DNA.

b. Pol I Can Edit Its Mistakes

In addition to its polymerase activity, Pol I has two independent hydrolytic activities:

1. It can act as a 3′ → 5′ exonuclease.
2. It can act as a 5′ → 3′ exonuclease.

The 3′ → 5′ exonuclease reaction differs chemically from the pyrophosphorolysis reaction (the reverse of the polymerase reaction) only in that H_2O rather than PP_i is the nucleotide acceptor. Kinetic and crystallographic studies, however, indicate that these two catalytic activities occupy separate active sites (see below). The 3′ → 5′ exonuclease function is activated by an unpaired 3′-terminal nucleotide with a free OH group. If Pol I erroneously incorporates a wrong (unpaired) nucleotide at the end of a growing DNA chain, the polymerase activity is inhibited and the 3′ → 5′ exonuclease excises the offending nucleotide (Fig. 5-36). The polymerase activity then resumes DNA replication. *Pol I therefore has the ability to* ***proofread*** *or* ***edit*** *a DNA chain as it is synthesized so as to correct its mistakes.* This explains the great fidelity of DNA replication by Pol I: The overall fraction of bases that the enzyme misincorporates, $\sim 10^{-7}$, is the product of the fraction of bases that its polymerase activity misincorporates and the fraction of misincorporated bases that its 3′ → 5′ exonuclease activity fails to excise. The price of this high fidelity is that ~3% of correctly incorporated nucleotides are also excised.

The Pol I 5′ → 3′ exonuclease binds to dsDNA at single-strand nicks with little regard to the character of the 5′ nucleotide (5′-OH or phosphate group; base paired or not). It cleaves the DNA in a base paired region beyond the nick such that the DNA is excised as either mononucleotides or oligonucleotides of up to 10 residues (Fig. 5-33). In contrast, the 3′ → 5′ exonuclease removes only unpaired mononucleotides with 3′-OH groups.

c. Pol I's Polymerase and Two Exonuclease Functions Each Occupy Separate Active Sites

The 5′ → 3′ exonuclease activity of Pol I is independent of both its 3′ → 5′ exonuclease and its polymerase activi-

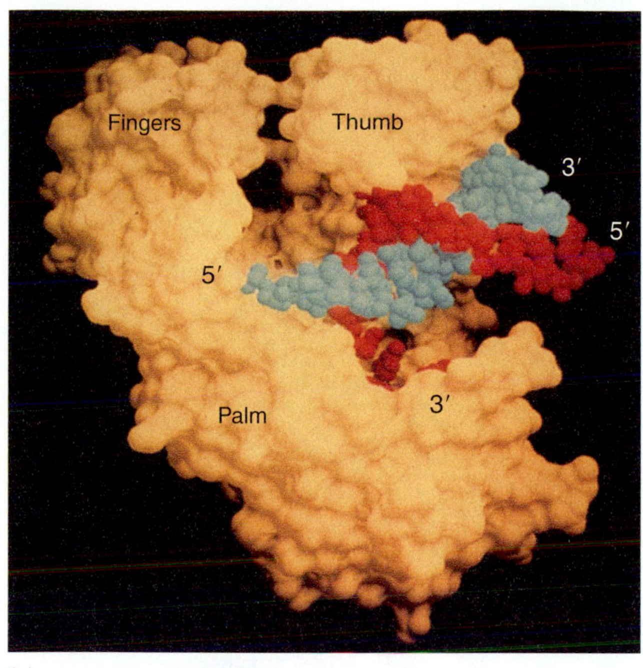

(a)

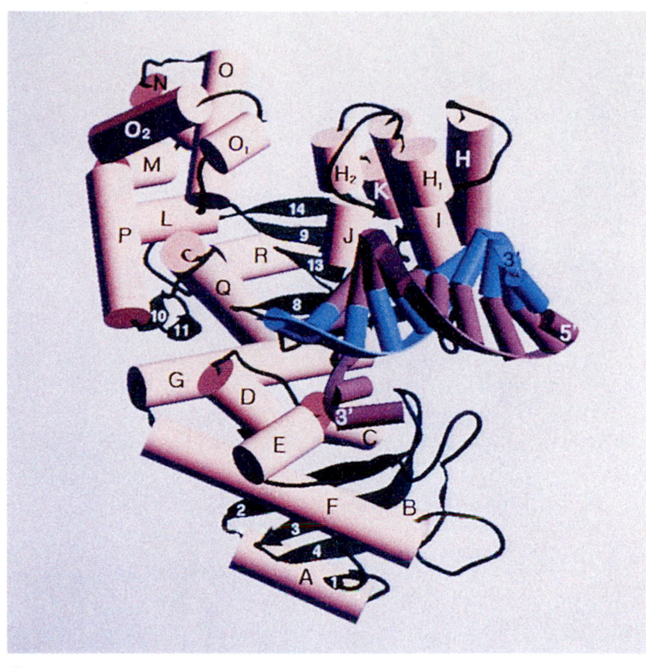

(b)

FIGURE 30-8 X-Ray structure of *E. coli* DNA polymerase I Klenow fragment (KF) in complex with a dsDNA. (*a*) The solvent-accessible surface of KF (*yellow*) with the 12-nt template strand in cyan and the 14-nt primer strand in red. (*b*) A tube-and-arrow representation of the complex in the same orientation as Part *a* in which the template strand is blue and the primer strand is purple. [Courtesy of Thomas Steitz, Yale University. PDBid 1KLN.] **See the Interactive Exercises.**

ties. In fact, as we saw in Section 7-2A, proteases such as subtilisin or trypsin cleave Pol I into two fragments: a larger C-terminal or **Klenow fragment (KF;** residues 324–928), which contains both the polymerase and the 3′ → 5′ exonuclease activities; and a smaller N-terminal fragment (residues 1–323), which contains the 5′ → 3′ exonuclease activity. Thus Pol I contains three active sites on a single polypeptide chain.

d. The X-Ray Structure of Klenow Fragment Indicates How It Binds DNA

The X-ray structure of KF, determined by Thomas Steitz, reveals that this protein consists of two domains (Fig. 30-8). The smaller domain (residues 324–517) contains the 3′ → 5′ exonuclease site, as was demonstrated by the absence of this function but not polymerase activity in a genetically engineered Klenow fragment mutant that lacks the divalent metal ion–binding sites known to be essential for 3′ → 5′ exonuclease activity but which otherwise has a normal structure. The larger domain (residues 521–928; helix G and beyond in Fig. 30-8*b*) contains the polymerase active site at the bottom of a prominent cleft, a surprisingly large distance (~25 Å) from the 3′ → 5′ exonuclease site. The cleft, which is lined with positively charged residues, has the appropriate size (~22 Å wide by ~30 Å deep) and shape to bind a B-DNA molecule in a manner resembling a right hand grasping a rod (in which the "thumb" consists of helices H–I, the "fingers" consist of helices L–P, and the remainder of the larger domain, the "palm," includes a 6-stranded antiparallel β sheet that forms the floor of the cleft and contains the polymerase function's active site residues). Indeed, the active sites of all DNA and RNA polymerases of known structure are located at the bottoms of similarly shaped clefts (Sections 30-4B, 30-4C, and 31-2A).

e. DNA Polymerase Distinguishes Watson–Crick Base Pairs via Sequence-Independent Interactions That Induce Domain Movements

The C-terminal domain of the thermostable *Thermus aquaticus (Taq)* DNA polymerase I **(Klentaq1)** is 50% identical in sequence and closely similar in structure to the large domain of Klenow fragment, although Klentaq1 lacks a functional 3′ → 5′ exonuclease site. Gabriel Waksman crystallized Klentaq1 in complex with an 11-bp DNA that has a GGAAA-5′ overhang at the 5′ end of its template strand, and the crystals were incubated with 2′,3′-dideoxy-CTP (ddCTP; which lacks a 3′-OH group). The X-ray

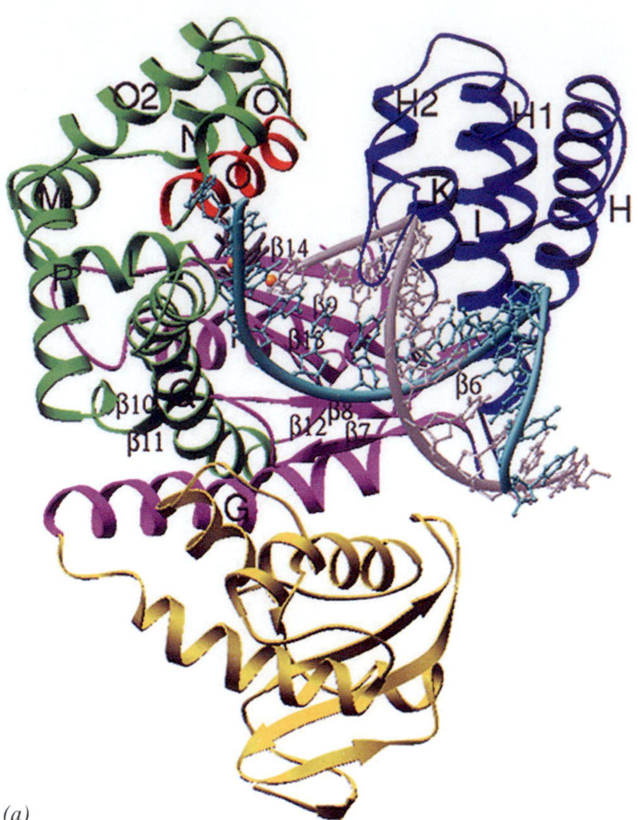

(*a*)

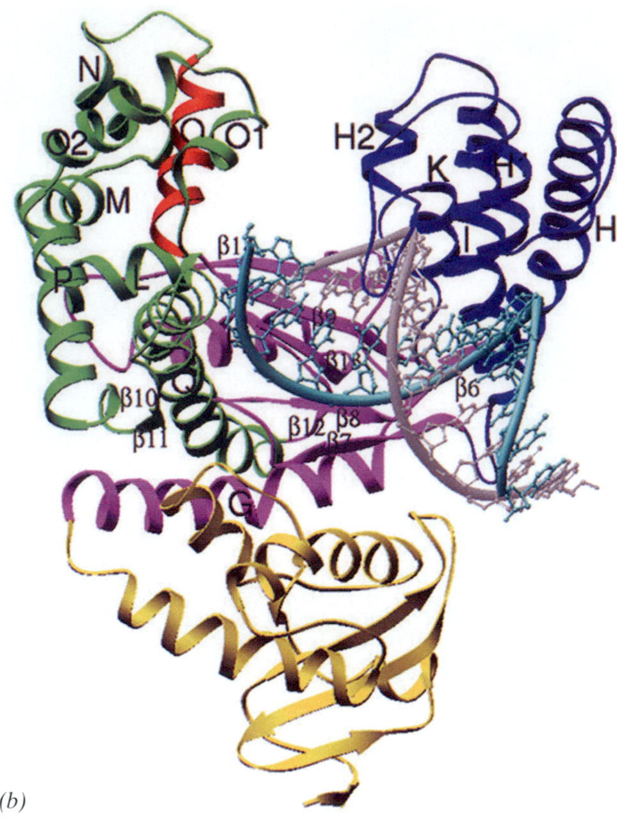

(*b*)

FIGURE 30-9 X-Ray structure of Klentaq1 in complex with DNA and ddCTP. (*a*) The closed conformation. (*b*) The open conformation. The protein, which is viewed similarly to that in Fig. 30-8, is represented in ribbon form with its N-terminal, palm, fingers, and thumb domains colored yellow, magenta, green, and dark blue, respectively, and with the O helix in the fingers domain red. The DNA is shown in stick form and its sugar–phosphate backbone is also represented in tube form with the template strand cyan and the primer strand silver. In Part *a,* the bound ddCTP is shown in stick form in black, and its two bound metal ions are represented by orange spheres. [Courtesy of Gabriel Waksman, Washington University School of Medicine. PDBids 3KTQ and 2KTQ.]

structure of these crystals (Fig. 30-9*a*) reveals that a ddC residue had been covalently linked to the 3′ end of the primer and formed a Watson–Crick pair with the template overhang's 3′ G. Moreover, a ddCTP molecule (to which the primer's new 3′-terminal ddC residue is incapable of forming a covalent bond) occupies the enzyme's active site where it forms a Watson–Crick pair with the templates next G. Clearly, Klentaq1 retains its catalytic activity in this crystal.

A DNA polymerase must distinguish correctly paired bases from mismatches and yet do so via sequence-independent interactions with the incoming dNTP. The foregoing X-ray structure reveals that this occurs through an active site pocket that is complementary in shape to Watson–Crick base pairs. This pocket is formed by the stacking of a conserved Tyr side chain on the template base, as well as by van der Waals interactions with the protein and with the preceding base pair. In addition, although the dsDNA is mainly in the B conformation, the 3 base pairs nearest the active site assume the A conformation, as has also been observed in the X-ray structures of several other DNA polymerases in their complexes with DNA. The resulting wider and shallower minor groove (Section 29-1B) permits protein side chains to form hydrogen bonds with the otherwise inaccessible N3 atoms of the purine bases and O2 atoms of the pyrimidine bases. The positions of these hydrogen bond acceptors are sequence-independent as can be seen from an inspection of Fig. 5-12 [in contrast, the positions of the hydrogen bonding acceptors in the major groove vary with both the identity (A · T vs G · C) and the orientation (e.g., A · T vs T · A) of the base pair]. However, with a non-Watson–Crick pairing, these hydrogen bonds would be greatly distorted if not completely disrupted. The protein also makes extensive sequence-independent hydrogen bonding and van der Waals interactions with the DNA's sugar–phosphate backbone.

The above Klentaq1 · DNA · ddCTP crystals were partially depleted of ddCTP by soaking them in a stabilizing solution that lacks ddCTP. The X-ray structure of the ddCTP-depleted crystals (Fig. 30-9*b*) revealed that Klentaq1's fingers domain assumed a so-called open conformation, which differs significantly from that in the so-called closed conformation described above. In particular, the O, O_1, and O_2 helices in the closed conformation have moved via a hingelike motion in the direction of the active site relative to their positions in the open complex (Fig. 30-9*a*) so as to bury the bound ddCTP, thereby assembling the productive ternary complex. These observations are consistent with kinetic measurements on Pol I, indicating that the binding of the correct dNTP to the enzyme induces a rate-limiting conformational change that yields a tight ternary complex. It therefore appears that the enzyme rapidly samples the available dNTPs in its open conformation but only when it binds the correct dNTP in a Watson–Crick pairing with the template base does it form the catalytically competent closed conformation. The subsequent reaction steps then rapidly yield the product complex which, following a second conformational change, releases the product PP_i. Finally, the DNA is translocated in the active site, probably via a linear diffusion mechanism, so as to position it for the next reaction cycle.

The comparison of the above X-ray structures with that of Klentaq1 alone indicates that on binding DNA, the thumb domain moves to wrap around the DNA. It is likely that this conformational change is largely responsible for Pol I's processivity. In both Klentaq1 · DNA structures, neither the dsDNA nor the single-stranded DNA **(ssDNA)** passes through the cleft between the thumb and fingers domain as the shape and position of the cleft suggest. Rather, the template strand makes a sharp bend at the first unpaired base, thereby unstacking this base and positioning this ssDNA on the same side of the cleft as the dsDNA. Similar arrangements have been observed in X-ray structures of other DNA polymerases in their complexes with DNA.

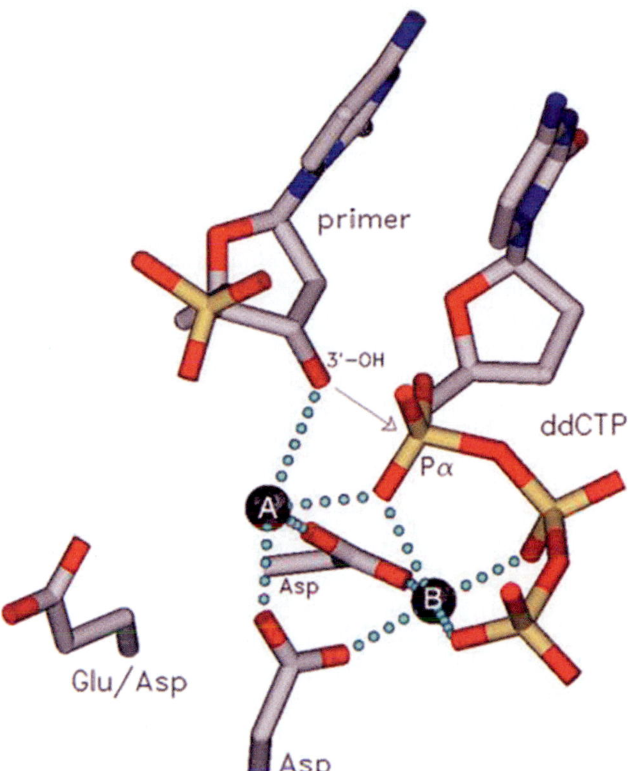

FIGURE 30-10 **Schematic diagram for the nucleotidyl transferase mechanism of DNA polymerases.** A and B represent enzyme-bound metal ions that usually are Mg^{2+}. Atoms are colored according to atom type (C gray, N blue, O red, and P yellow) and metal ion coordination is represented by green dotted lines. Metal ion A activates the primer's 3′-OH group for in-line nucleophilic attack on the incoming dNTP's α-phosphate group (*arrow*), whereas metal ion B acts to orient and electrostatically stabilize the negatively charged triphosphate group. [Courtesy of Tom Ellenberger, Harvard Medical School.]

f. The DNA Polymerase Catalytic Mechanism Involves Two Metal Ions

The X-ray structures of a variety of DNA polymerases suggest that they share a common catalytic mechanism for nucleotidyl transfer (Fig. 30-10). Their active sites all contain two metal ions, usually Mg^{2+}, that are liganded by two invariant Asp side chains in the palm domain. Metal ion B in Fig. 30-10 is liganded by all three phosphate groups of the bound dNTP, whereas metal ion A bridges the α-phosphate group of this dNTP and the primer's 3′-OH group. Metal ion A presumably activates the primer's 3′-OH group for an in-line nucleophilic attack on the α-phosphate group (Fig. 16-6*b*), whereas metal ion B functions to orient its bound triphosphate group and to electrostatically shield their negative charges as well as the additional negative charge on the transition state leading to the release of the PP_i ion (Section 16-2B).

g. Editing Complexes Contain the Primer Strand in the 3′ → 5′ Exonuclease Site

Steitz cocrystallized KF with a 12-nt DNA "template" strand (5′-TGCCTCGCGGCC-3′), a 7-nt "primer" strand (3′-GCGCCGG-5′) that is complementary to the 3′ end of the template strand, and **2′,3′-epoxy-ATP,**

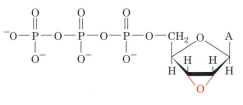

2',3'-Epoxy-ATP

which promotes the tight binding of DNA to the polymerase site. The X-ray structure of the resulting complex (Fig. 30-8) shows that the primer strand base pairs, as expected, to the 3′ end of the template strand to form a distorted segment of B-DNA, and that the polymerase has apparently appended an epoxy-A residue to the primer's 3′ end (where it base pairs to a T on the template strand;

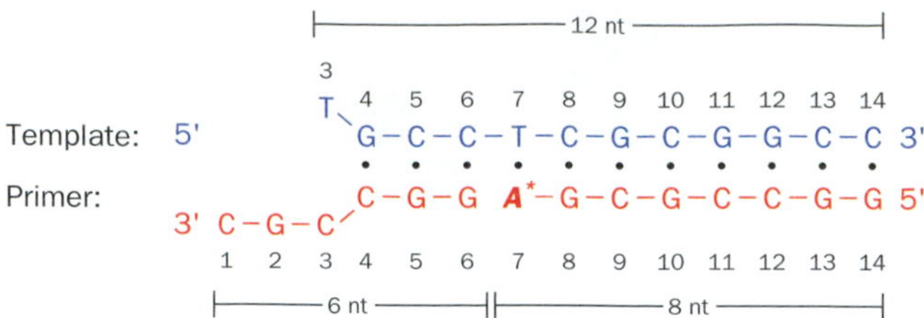

FIGURE 30-11 Probable sequence of the double-stranded DNA seen in the X-ray structure of KF. *A** represents a 2′,3′-epoxy-A residue that KF has appended to the 3′ end of the 7-nt DNA (*red*) with which the KF crystal (Fig. 30-8) was incubated (together with the 12-nt DNA; *blue*). The two-part primer strand's 3′ segment appears to be a second copy of the 7-nt DNA with its 3′ nucleotide removed.

Fig. 30-11). In addition, a second primer strand, whose 3′ G residue has apparently been removed, continues the 3′ end of the primer (after a break in the sugar–phosphate chain) by base pairing, via its 5′-terminal three nucleotides, to the template strand. Thus, this complex contains an 11-bp dsDNA, which has a single nucleotide overhang at the 5′ end of its template strand and a 3-nt overhang at the 3′ end of its primer strand. The 3′-terminal nucleotide of the primer strand (the last one that an active polymerase would have added) is bound at the 3′ → 5′ exonuclease active site, whereas the 5′-terminal nucleotide of the template strand is bound at the entrance of the polymerase active site. Evidently, the KF has bound the DNA in an "editing" complex rather than in the polymerase cleft.

In *E. coli* Pol I, how does the 3′ end of the primer strand transfer between the polymerase active site and the 3′ → 5′ exonuclease active site? This appears to occur through the competition of these sites for the 3′ end of the primer strand, which base pairs to form dsDNA in the polymerase site and binds as a single strand to the exonuclease site. Thus, the formation of a Watson–Crick base pair would tend to bind the primer strand in the polymerase site preparatory for the next round of chain extension, whereas a mismatched base pair would promote the binding of the primer strand as a single strand to the exonuclease site for the subsequent excision of the offending nucleotide. Comparison of the editing complex with those of the Klentaq1 · DNA complexes suggests that the transfer of the primer strand from the polymerase to the editing sites of KF requires that the dsDNA translocate backward (toward the 3′ end of the template strand) by several angstroms along the helix axis.

h. Pol I Functions Physiologically to Repair DNA

For some 13 years after Pol I's discovery, it was generally assumed that this enzyme was *E. coli*'s DNA replicase because no other DNA polymerase activity had been detected in *E. coli.* This assumption was made untenable by Cairns and Paula DeLucia's isolation, in 1969, of a mutant *E. coli* whose extracts exhibit <1% of the normal Pol I activity (although it has nearly normal levels of the 5′ → 3′ exonuclease activity) but which nevertheless reproduce at the normal rate. This mutant strain, however, is highly susceptible to the damaging effects of UV radiation and **chemical mutagens** (substances that chemically induce mutations; Section 32-1A). *Pol I evidently plays a central role in the repair of damaged (chemically altered) DNA.*

Damaged DNA, as we discuss in Section 30-5, is detected by a variety of DNA repair systems. Many of them endonucleolytically cleave the damaged DNA on the 5′ side of the lesion, thereby activating Pol I's 5′ → 3′ exonuclease. While excising this damaged DNA, Pol I simultaneously fills in the resulting single-strand gap through its polymerase activity. In fact, its 5′ → 3′ exonuclease activity increases 10-fold when the polymerase function is active. Perhaps the simultaneous excision and polymerization activities of Pol I protect DNA from the action of cellular nucleases that would further damage the otherwise gapped DNA.

i. Pol I Catalyzes Nick Translation

Pol I's combined 5′ → 3′ exonuclease and polymerase activities can replace the nucleotides on the 5′ side of a single-strand nick on otherwise undamaged DNA. These reactions, in effect, translate (move) the nick toward the DNA strand's 3′ end without otherwise changing the molecule (Fig. 30-12). This **nick translation** process, in the presence of labeled deoxynucleoside triphosphates, is synthetically employed to prepare highly radioactive DNA (the required nicks may be generated by treating the DNA with a small amount of pancreatic **DNase I**).

j. Pol I's 5′ → 3′ Exonuclease Functions Physiologically to Excise RNA Primers

Pol I's 5′ → 3′ exonuclease also removes the RNA primers at the 5′ ends of newly synthesized DNA while its DNA polymerase activity fills in the resulting gaps (Fig. 5-34). The importance of this function was demonstrated by the isolation of temperature-sensitive *E. coli* mutants that neither are viable nor exhibit any 5′ → 3′ exonuclease activity at the restrictive temperature of ~43°C (the low level of polymerase activity in the Pol I mutant isolated by Cairns and DeLucia is apparently sufficient to carry out this essential gap-filling process during chromosome replication). Thus Pol I has an indispensable role in *E. coli* DNA replication although a different one than was first supposed.

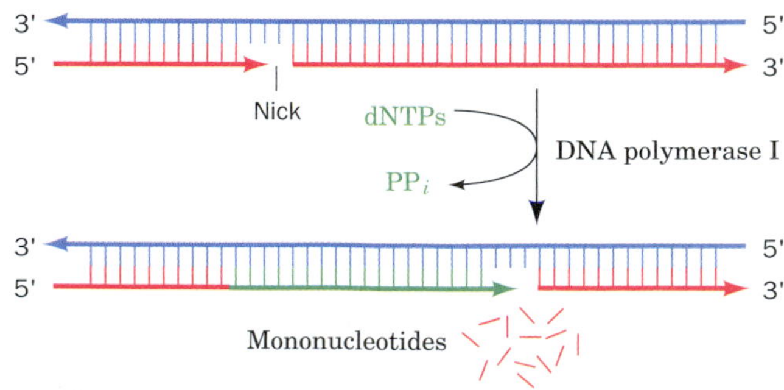

FIGURE 30-12 Nick translation as catalyzed by Pol I.

B. *DNA Polymerase III*

The discovery of normally growing *E. coli* mutants that have very little Pol I activity stimulated the search for an additional DNA polymerizing activity. This effort was rewarded by the discovery of two more enzymes, designated, in the order they were discovered, **DNA polymerase II (Pol II)** and **DNA polymerase III (Pol III).** The properties of these enzymes are compared with that of Pol I in Table 30-1. Pol II and Pol III had not previously been detected because their combined activities in the assays used are normally <5% that of Pol I.

A mutant *E. coli* lacking measurable Pol II activity grows normally. However, Pol II has been implicated as a participant in repairing DNA damage via the **SOS response** (Section 30-5D), as have two additional *E. coli* enzymes that were recently discovered: **DNA polymerase IV (Pol IV)** and **DNA polymerase V (Pol V)** (Section 30-5D).

a. Pol III Is *E. coli's* DNA Replicase

The cessation of DNA replication in temperature-sensitive ***polC*** mutants above the restrictive (high) temperature demonstrates that *Pol III is E. coli's DNA replicase.* Its **Pol III core** has the subunit composition $\alpha\varepsilon\theta$ where α, the *polC* gene product (Table 30-2), contains the polymerase function. The catalytic properties of Pol III core resemble those of Pol I (Table 30-1) except for Pol III core's inability to replicate primed ssDNA or nicked dsDNA. Rather, Pol III core acts *in vitro* at single-strand gaps of <100 nucleotides, a situation that probably resembles the state of DNA at the replication fork. The Pol III $3' \rightarrow 5'$ exonuclease function, which resides on the enzyme's ε subunit, is DNA's primary editor during replication; it enhances the enzyme's replication fidelity by up to 200-fold. However, the Pol III $5' \rightarrow 3'$ exonuclease acts only on single-stranded DNA, so it cannot catalyze nick translation. θ is an accessory protein that stimulates the editing function of ε.

Pol III core functions in vivo as part of a complicated and labile multisubunit enzyme, the ***Pol III holoenzyme,*** *which consists of at least 10 types of subunits (Table 30-2).* The latter 7 subunits in Table 30-2 act to modulate Pol III core's activity. For example, Pol III core has a processivity of 10 to 15 residues; it can only fill in short single-stranded regions of DNA. However, Pol III core is rendered processive by association with the **β subunit** in the presence of the 7-subunit **γ complex** ($\gamma\tau_2\delta\delta'\chi\psi$). Assembly of the processive enzyme is a two-stage process in which the γ complex transfers the β subunit to the primed template in an ATP-dependent reaction followed by the assembly of Pol III core with the β subunit on the DNA (Section 30-3C). The β subunit confers essentially unlimited processivity (>5000 residues) on the core enzyme even if the γ complex is subsequently removed. In fact, the β subunit is very strongly bound to the DNA, although it can freely slide along it.

b. The β Subunit Forms a Ringlike Sliding Clamp

The observation that a β subunit clamped to a cut circular DNA slides to the break and falls off suggests that the β subunit forms a closed ring around the DNA, thereby preventing its escape. The X-ray structure of the β subunit (Fig. 30-13*a*), determined by John Kuriyan, reveals that it forms a dimer of C-shaped, 366-residue monomer units which associate to form an ~80-Å-diameter doughnut-shaped structure that is equivalently known as the **sliding clamp** and the **β clamp.** The sliding clamp's central hole is ~35 Å in diameter, which is larger than the 20- and 26-Å diameters of B- and A-DNAs (recall that the hybrid helices which RNA primers make with DNA have A-DNA-like conformations; Section 29-2B). Each β subunit consists of six $\beta\alpha\beta\beta\beta$ motifs of identical topology, which associate in pairs to form three pseudo-2-fold symmetric domains of very similar structures (although with <20% sequence identity). The dimeric ring therefore has the shape of a

TABLE 30-1 Properties of *E. coli* DNA Polymerases

	Pol I	Pol II	Pol III
Mass (kD)	103	90	130
Molecules/cell	400	?	10–20
Turnover number[a]	600	30	9000
Structural gene	*polA*	*polB*	*polC*
Conditionally lethal mutant	+	−	+
Polymerization: $5' \rightarrow 3'$	+	+	+
Exonuclease: $3' \rightarrow 5'$	+	+	+
Exonuclease: $5' \rightarrow 3'$	+	−	−

[a]Nucleotides polymerized $\text{min}^{-1} \cdot \text{molecule}^{-1}$ at 37°C.

Source: Kornberg, A. and Baker, T.A., *DNA Replication* (2nd ed.), p. 167, Freeman (1992).

TABLE 30-2 Components of *E. coli* DNA Polymerase III Holoenzyme

Subunit	Mass (kD)	Structural Gene
α[a]	130	*polC (dnaE)*
ε[a]	27.5	*dnaQ*
θ[a]	10	*holE*
τ[b]	71	*dnaX*[c]
γ[b]	45.5	*dnaX*[c]
δ[b]	35	*holA*
δ'[b]	33	*holB*
χ[b]	15	*holC*
ψ[b]	12	*holD*
β	40.6	*dnaN*

[a]Components of the Pol III core.
[b]Components of the γ complex.
[c]The γ and τ subunits are encoded by the same gene sequence; the γ subunit comprises the N-terminal end of the τ subunit.

Sources: Kornberg, A. and Baker, T.A., *DNA Replication* (2nd ed.), p. 169, Freeman (1992); *and* Baker, T.A. and Wickner, S.H., *Annu. Rev. Genet.* **26,** 450 (1992).

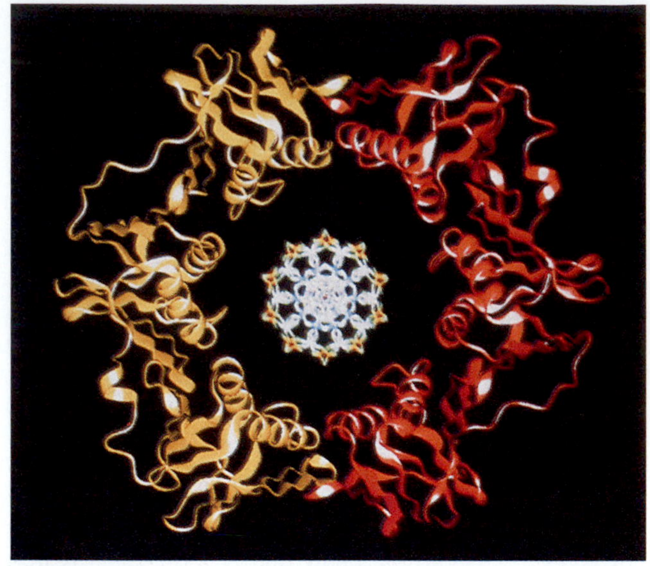

(a)

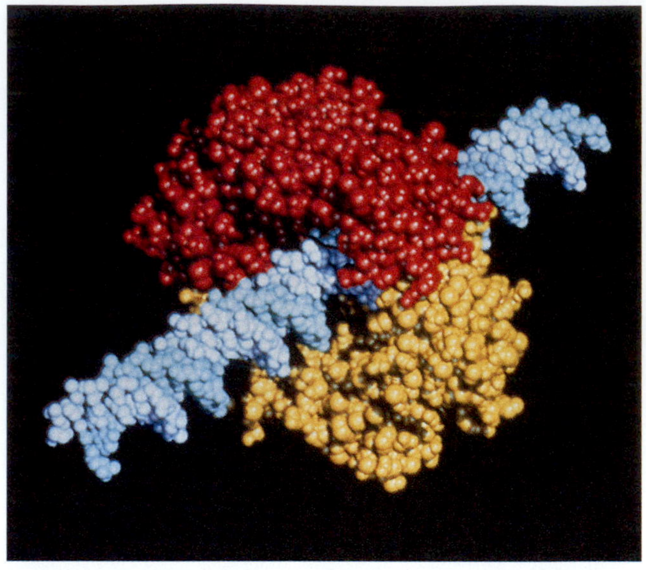

(b)

FIGURE 30-13 X-Ray structure of the β subunit of *E. coli* Pol III holoenzyme. (*a*) A ribbon drawing showing the two monomeric units of the dimeric sliding clamp in yellow and red as viewed along the homodimer's 2-fold axis. A stick model of B-DNA is placed with its helix axis coincident with the sliding clamp's twofold axis. (*b*) A space-filling model of the sliding clamp in the hypothetical complex with the B-DNA shown in Part *a*. The protein is colored as in Part *a* and the DNA is cyan. [Courtesy of John Kuriyan, The Rockefeller University. PDBid 2POL.]

6-pointed star in which the 12 helices line the central hole and the β strands associate in six β sheets that form the protein's outer surface. Electrostatic calculations indicate that the interior surface of the ring is positively charged, whereas its outer surface is negatively charged.

Model building studies in which a B-DNA helix is threaded through the sliding clamp's central hole (Fig. 30-13) indicate that the helices are all oriented such that they are perpendicular to their radially adjacent segments of sugar–phosphate backbone. These helices therefore span the major and minor grooves of the DNA rather than entering into them as do many helices that make sequence-specific interactions with dsDNA (Section 31-3C). Since A- and B-DNAs have 11 and 10 bp per turn, whereas the sliding clamp has a pseudo-12-fold symmetry, it appears that the sliding clamp is designed to minimize its associations with its threaded DNA. This presumably permits the sliding clamp to freely slide along the DNA helix. Indeed, the radius of the sliding clamp's central hole is at least 3.5 Å larger than that of DNA, so any interactions between them are likely to be attenuated by a sheath of intervening water molecules.

C. *Unwinding DNA: Helicases and Single-Strand Binding Protein*

Pol III holoenzyme, unlike Pol I, cannot unwind dsDNA. Rather, *three proteins,* DnaB *protein (the product of the dnaB gene; proteins may be assigned the name of the gene specifying them but in roman letters with the first letter capitalized), Rep helicase, and single-strand binding protein (SSB) (Table 30-3), work in concert to unwind the DNA before an advancing replication fork (Fig. 30-14) in a process that is driven by ATP hydrolysis.*

a. Hexameric Helicases Mechanically Separate the Strands of dsDNA by Climbing Up One Strand

Access to the genetic information encoded in a double-helical nucleic acid requires that the double helix be unwound. The proteins that do so, which are known as **helicases,** form a diverse group of enzymes that facilitate a variety of functions including DNA replication, recombination, and repair, as well as transcription termination (Section 31-2A), RNA splicing, and RNA editing (Section 31-4A). Indeed, all forms of life contain helicases, 12 varieties of which occur in *E. coli.* Helicases function by translocating along one strand of a double-helical nucleic acid so as to unwind the double helix in their path. This, of course, requires free energy, and hence helicases are driven by the hydrolysis of NTPs. Helicases have been classified in several ways: whether they translocate along their bound single strand in the $5' \rightarrow 3'$ direction or the $3' \rightarrow 5'$

TABLE 30-3 Unwinding and Binding Proteins of *E. coli* DNA Replication

Protein	Subunit Structure	Subunit Mass (kD)
DnaB protein	hexamer	50
SSB	tetramer	19
Rep protein	monomer	68
PriA protein	monomer	76

Source: Kornberg, A. and Baker, T.A., *DNA Replication* (2nd ed.), p. 366, Freeman (1992).

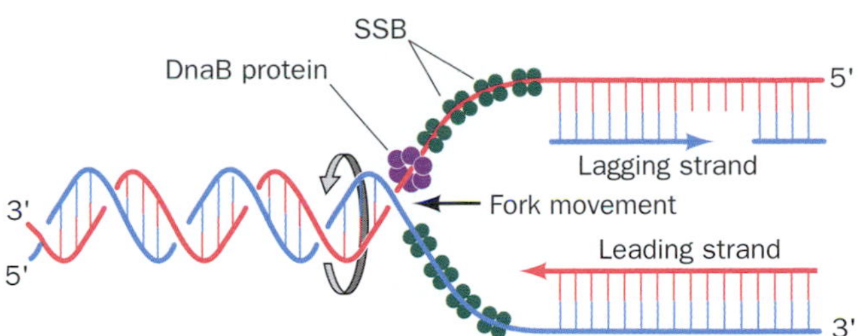

FIGURE 30-14 Unwinding of DNA by the combined action of DnaB and SSB proteins. The hexameric DnaB protein moves along the lagging strand template in the $5' \rightarrow 3'$ direction. The resulting separated DNA strands are prevented from reannealing by SSB binding.

direction, whether they function as hexameric rings or as dimers, and whether their sequences contain certain signature motifs.

E. coli **DnaB** protein, a hexameric helicase of identical 471-residue subunits, separates the strands of dsDNA by translocating along the lagging strand template in the $5' \rightarrow 3'$ direction while hydrolyzing ATP (it can also use GTP and CTP but not UTP). Electron microscopy studies reveal that DnaB forms a hexameric ring that, depending on conditions, exhibits C_3 or C_6 symmetry and which encloses an ~30-Å-diameter central channel. Similarly, the bacteriophage **T7 gene 4 helicase/primase** (bacteriophage T7 infects *E. coli*) forms a two-tiered hexagonal ring (Fig. 30-15) whose smaller N-terminal domains (residues 1–271) contain its primase activity and whose larger C-terminal domains (residues 272–566) carry out its helicase function. T7 gene 4 helicase/primase (also called **T7 gp4;** gp for *g*ene *p*roduct) preferentially hydrolyzes dTTP but also hydrolyzes dATP and ADP.

The X-ray structure of DnaB has not been determined. However, Dale Wigley has determined that of the mainly C-terminal domain (residues 241–566) of T7 gene 4 helicase/primase in complex with ADPNP. This helicase, as expected, forms a hexagonal ring (Fig. 30-16), which appears to be largely held together by each subunit's N-terminal arm binding to an adjacent subunit. Two loops from each subunit that extend into the hexamer's central channel and which contain several conserved basic residues presumably form the hexamer's DNA-binding surface.

The hexagonal ring exhibits only 2-fold (C_2) rotational symmetry. If the adjacent subunits of the asymmetric half of this ring are labeled A, B, and C, subunit B is related to subunit A by a 15° rotation about an axis lying in the plane of the ring (after a 60° rotation about the 6-fold axis) and subunit C is similarly related to subunit B (a 30° rotation relative to subunit A). The DNA-binding loops thereby form a helical ramp that is approximately complementary in shape to the sugar–phosphate backbone of single-stranded DNA **(ssDNA)** when it is in the B-DNA conformation (although keep in mind that the helical ramp is discontinuous at the interface between adjacent C and A subunits). ADPNP is bound to the A and B subunits but not to the C subunits.

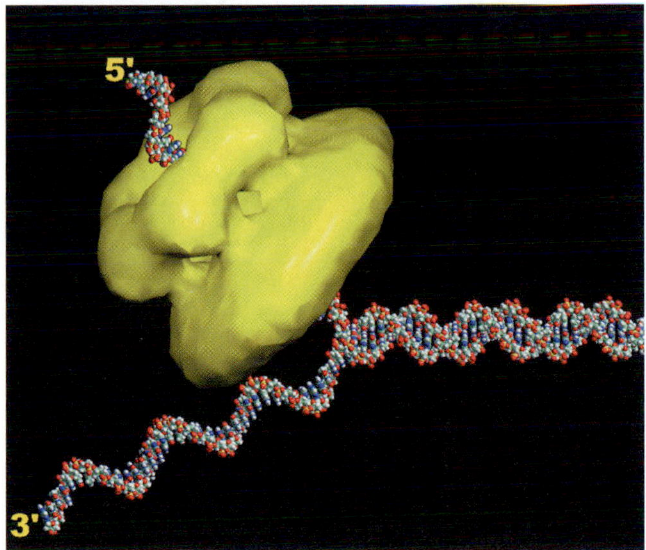

FIGURE 30-15 Electron microscopy–based image reconstruction of T7 gene 4 helicase/primase. In this two-tiered hexameric ring (*yellow*), the smaller lobe of each subunit forms the N-terminal primase domain and the larger lobe forms the C-terminal helicase domain. The protein is postulated to interact with DNA as is depicted by this model of a DNA fork consisting of a 30-bp duplex segment and two 25-nt single-stranded segments with the 5′ tail threaded through the hexameric ring. The way in which the 3′ tail interacts with the protein, if at all, is unknown. [Courtesy of S.S. Patel and K.M. Picha, University of Medicine and Dentistry of New Jersey.]

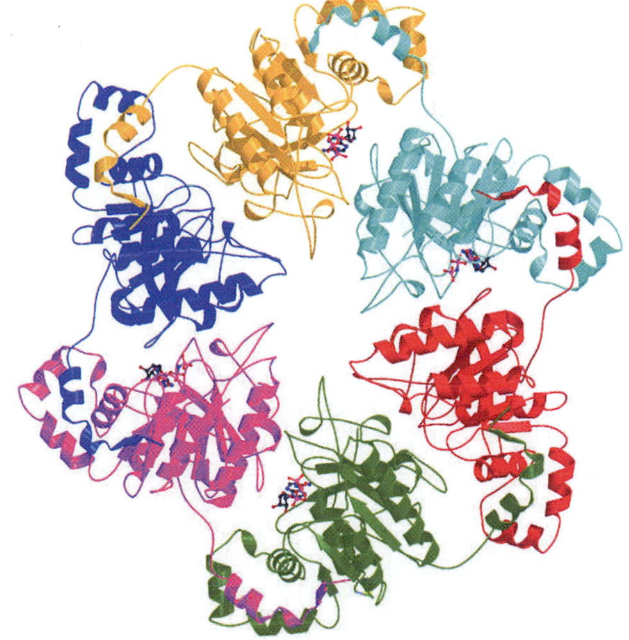

FIGURE 30-16 X-Ray structure of the helicase domain of T7 gene 4 helicase/primase. Each subunit of this cyclic hexamer is drawn in a different color. The four bound ADPNP molecules are represented in ball-and-stick form. Note that the conformations of adjacent subunits are not identical. [Courtesy of Dale Wigley, Cancer Research U.K. London Research Institute. PDBid 1E0J.]

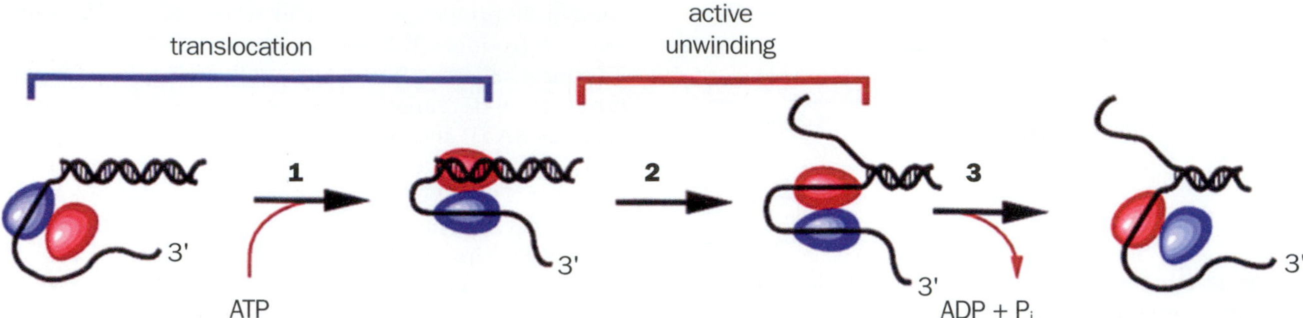

FIGURE 30-17 Active, rolling mechanism for DNA unwinding by Rep helicase. (1) The subunit of dimeric Rep helicase that is not bound to ssDNA binds to dsDNA accompanied by ATP binding. **(2)** The subunit bound to dsDNA unwinds the double strand and remains bound to the 3′-ending strand. **(3)** In a process that is accompanied by the release of the ATP hydrolysis products, the subunit closer to the 3′ end of the bound ssDNA releases it preparatory for a new cycle of dsDNA unwinding. [Courtesy of Gabriel Waksman, Washington University School of Medicine.]

The different conformations and ADPNP-binding properties of the hexamer's chemically identical subunits are reminiscent of the binding change mechanism of ATP synthesis by the F_1F_0-ATPase (Section 22-3C). Wigley has therefore suggested that when a helicase binds and hydrolyzes NTP, it undergoes a conformational change that pulls the ssDNA through the center of the hexameric ring by the lever action of its DNA-binding loops. The conformational states, A, B, and C, are presumed to arise through the sequential binding of an NTP, its hydrolysis, and the release of the product NDP and P_i as similarly occurs in the binding change mechanism (Fig. 22-42). Hence the translocation process is postulated to be motivated by a wave of coupled subunit rotations that ripple around the hexameric ring. The helicase would thereby mechanically separate the strands of dsDNA by essentially pulling itself along the groove of one strand in the $5' \rightarrow 3'$ direction but without turning relative to the DNA.

b. Rep Helicase Dimers Separate the Strands of dsDNA via an "Active, Rolling" Mechanism

Two other helicases, **Rep helicase** and **PriA protein,** have been implicated in the replication of various *E. coli* phage DNAs (Section 30-3B) and also participate in certain aspects of *E. coli* DNA replication (Section 30-3C). Both proteins translocate along DNA in the $3' \rightarrow 5'$ direction (and hence along the opposite strand from DnaB) while hydrolyzing ATP. Rep helicase is not essential for *E. coli* DNA replication but the rate at which *E. coli* replication forks propagate is reduced ~2-fold in *rep*⁻ mutants.

Rep helicase is a 673-residue monomer in solution but dimerizes on binding to DNA. Both subunits of the Rep dimer bind to ssDNA or dsDNA such that DNA binding to one subunit strongly inhibits DNA binding to the other (negative cooperativity). This observation led Timothy Lohman to propose the "active, rolling" mechanism for Rep-mediated DNA unwinding in which the two subunits of the dimer alternate in binding dsDNA and the 3′ end of the ssDNA at the ssDNA/dsDNA junction (Fig. 30-17). The two subunits then "walk" up the DNA while unwinding it in an ATP-dependent manner via a subunit switching mechanism in which the helicase subunit that is bound to the dsDNA displaces its 5′-starting strand while remaining bound to its 3′-starting strand. Release of the other subunit from the 3′-starting ssDNA then permits this subunit to bind to and unwind the new end of the dsDNA, thereby continuing the cycle.

The X-ray structure of *E. coli* Rep helicase in complex with the short ssDNA $dT(pT)_{15}$ and ADP (Fig. 30-18), determined by Lohman and Waksman, reveals that the rel-

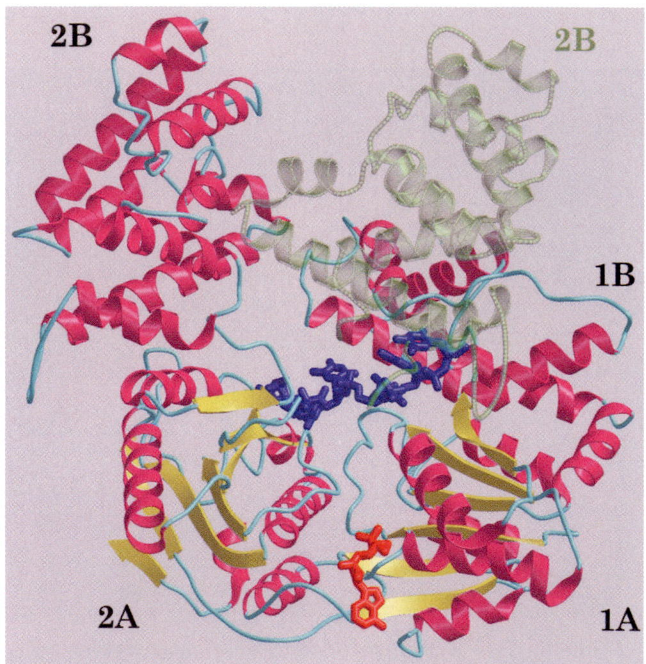

FIGURE 30-18 X-Ray structure of Rep helicase in complex with $dT(pT)_{15}$ and ADP. The monomer in the open conformation is drawn in ribbon form colored according to secondary structure (helices magenta, β sheets yellow, and coil cyan) with its bound ssDNA segment and ADP drawn in stick form in blue and in red. In the closed conformation, subdomain 2B (transparent green ribbon) has rotated via a 130° hinge motion so as to close over the ssDNA. [Courtesy of Gabriel Waksman, Washington University School of Medicine. PDBid 1UAA.]

atively straight ssDNA molecule binds two contacting Rep monomers. A Rep monomer consists of two domains, 1 and 2, each of which is formed by two subdomains, A and B, with the two N-terminal subdomains (1A and 2A) homologous to each other. In the two Rep monomers that are bound to the same ssDNA, subdomain 2B exhibits strikingly different orientations with respect to the other three subdomains (Fig. 30-18). The Rep monomer that is bound to the 5′ end of the ssDNA (which it contacts between bases 1 and 8) assumes the "open" conformation in which the four subdomains form an assembly that is reminiscent of a crab claw with one pincer (subdomain 2B) larger than the other (subdomain 1B). The DNA is bound at the bottom of the resulting cleft, whose floor is formed by subdomains 1A and 2A. In the Rep monomer that binds to the 3′ end of the ssDNA (which it contacts between bases 9 and 16), subdomain 2B has reoriented relative to the other subdomains via a 130° rotation about a hinge region between subdomains 2A and 2B, thereby closing the cleft about the DNA to form the "closed" conformation. This conformation change is consistent with the active, rolling mechanism even though the way in which two Rep monomers form the dimer observed in solution remains unknown. The ADP binds to Rep between its subdomains 1A and 2A in close proximity to the DNA, suggesting that conformation changes at the ATP-binding site arising from ATP hydrolysis are transmitted to the DNA-binding site via the secondary structural elements that contact both sites. The way in which Rep separates the two strands of dsDNA is, as yet, unknown.

c. Single-Strand Binding Protein Prevents ssDNA from Reannealing

If left to their own devices, the separated DNA strands behind an advancing helicase would rapidly reanneal to reform dsDNA. What prevents them from doing so is the binding of **single-strand binding protein (SSB).** It also prevents ssDNA from forming fortuitous intramolecular secondary structures (helical stems) and protects it from nucleases. Numerous copies of SSB cooperatively coat ssDNA, thereby maintaining it in an unpaired state. Note, however, that ssDNA must be stripped of SSB before it can be replicated by Pol III holoenzyme.

E. coli SSB is a homotetramer of 177-residue subunits. SSB binds ssDNA in several distinct modes referred to as $(SSB)_n$, which differ by the number of nucleotides (n) bound to each tetramer. The two major modes are $(SSB)_{35}$, in which only two of the tetramer's subunits strongly interact with the ssDNA, and $(SSB)_{65}$, in which all four subunits interact with the ssDNA. The $(SSB)_{35}$ mode displays unlimited cooperativity in that it forms extended strings of contacting tetramers along the length of a bound ssDNA, whereas the $(SSB)_{65}$ mode has limited cooperativity in that it forms beaded clusters on ssDNA that consist of only a few contacting tetramers.

Proteolysis studies have shown that SSB's ssDNA-binding site is contained within its 115 N-terminal residues. The X-ray structure of *E. coli* SSB's chymotryptic fragment (residues 1–135) in complex with $dC(pC)_{34}$, determined by

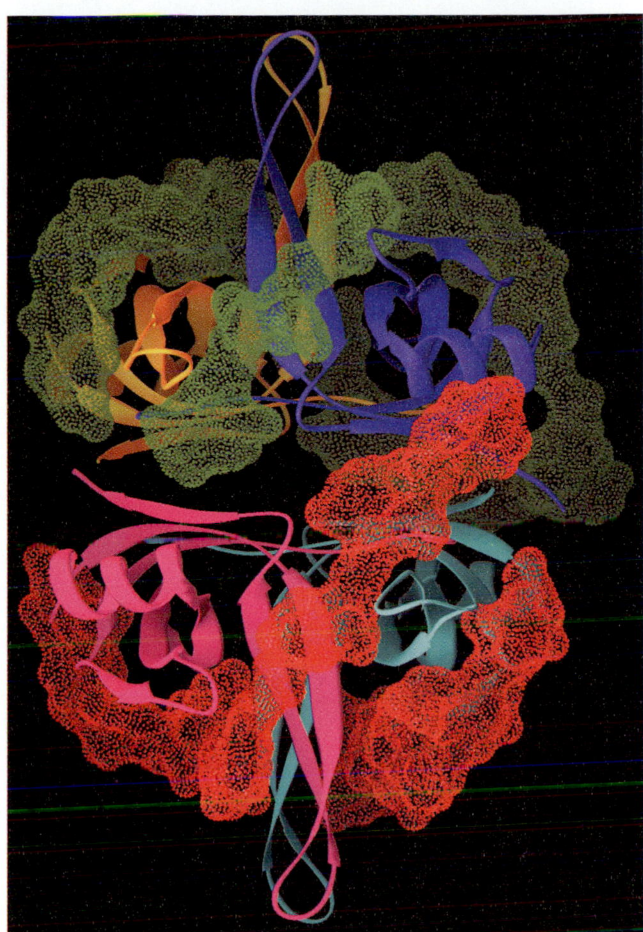

FIGURE 30-19 X-Ray structure of the N-terminal 135 residues of *E. coli* SSB in complex with $dC(pC)_{34}$. The homotetrameric protein is viewed along one of its 2-fold axes with its other 2-fold axes horizontal and vertical. Each of its subunits is differently colored. The two bound ssDNA molecules are represented by dot surfaces with the observed 28-residue segment of one of the ssDNAs green and the observed 14- and 9-residue segments of the other ssDNA red. [Based on an X-ray structure by Timothy Lohman and Gabriel Waksman, Washington University School of Medicine. PDBid 1EYG.]

Lohman and Waksman, reveals that the tetrameric protein has D_2 symmetry and binds two molecules of $dC(pC)_{34}$ (Fig. 30-19). For one of these 35-mers, 28 nucleotides (residues 3–30) were visible and these assumed the shape of an elongated horseshoe that wrapped around two SSB subunits with approximate 2-fold symmetry and with its apex contacting a third subunit. The other bound ssDNA was partially disordered such that only two segments were visible, one with 14 nt (residues 3–16) and the other with 9 nt (residues 19–27). The paths of the ssDNA segments along the surface of the SSB suggested models that rationalize the different properties of $(SSB)_{35}$ and $(SSB)_{65}$. In the $(SSB)_{65}$ model, the two ends of a 65-nt segment emerge from the same side of the tetramer, which would limit the number of SSB tetramers that can bind to contiguous 65-nt segments of ssDNA. However, in the $(SSB)_{35}$ model, the two ends of a 35-nt segment emerge from opposite ends of the tetramer, thereby permitting an unlimited series of SSB tetramers to interact end-to-end along the length of an ssDNA.

D. DNA Ligase

Pol I, as we saw in Section 30-2A, replaces the Okazaki fragments' RNA primers with DNA through nick translation. *The resulting single-strand nicks between adjacent Okazaki fragments, as well as the nick on circular DNA after leading strand synthesis, are sealed in a reaction catalyzed by* ***DNA ligase.*** The free energy required by this reaction is obtained, in a species-dependent manner, through the coupled hydrolysis of either NAD^+ to NMN^+ + AMP or ATP to PP_i + AMP. The *E. coli* enzyme, a 671-residue monomer that utilizes NAD^+, catalyzes a three-step reaction (Fig. 30-20):

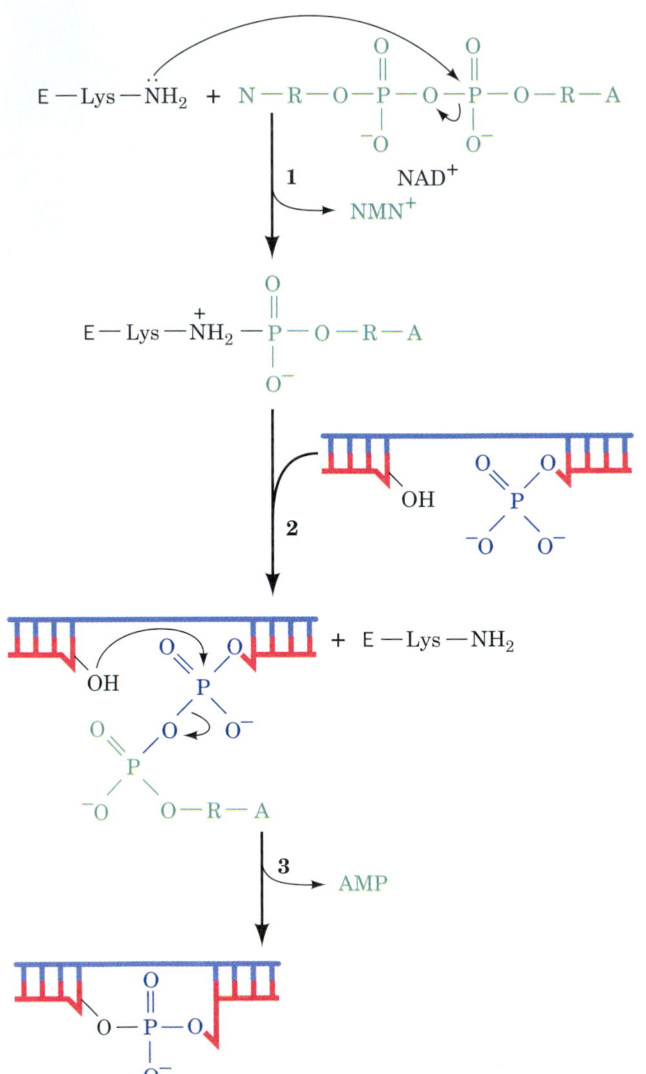

FIGURE 30-20 The reactions catalyzed by *E. coli* DNA ligase. In eukaryotic and T4 DNA ligases, NAD^+ is replaced by ATP so that PP_i rather than NMN^+ is eliminated in the first reaction step. The numbered steps are described in the text.

1. The adenylyl group of NAD^+ is transferred to the ε-amino group of an enzyme Lys residue to form an unusual phosphoamide adduct that is, nevertheless, readily isolated.

2. The adenylyl group of this activated enzyme is transferred to the 5′-phosphoryl terminus of the nick to form an adenylylated DNA. Here, AMP is linked to the 5′-nucleotide via a pyrophosphate rather than the usual phosphodiester bond.

3. DNA ligase catalyzes the formation of a phosphodiester bond by attack of the 3′-OH on the 5′-phosphoryl group, thereby sealing the nick and releasing AMP.

ATP-requiring DNA ligases, such as those of all eukaryotes and bacteriophage T4, release PP_i in the first step of the reaction rather than NMN^+. T4 ligase is also noteworthy in that, at high DNA concentrations, it can link together two duplex DNAs **(blunt end ligation)** in a reaction that is a boon to genetic engineering (Section 5-5C).

The X-ray structure of the 667-residue NAD^+-dependent DNA ligase from *Thermus filiformis,* determined by Se Won Suh, reveals a deeply clefted monomeric protein that consists of four domains (Fig. 30-21): Domain 1 (residues 1–317), which contains the adenylylated Lys residue (Lys 116); Domain 2 (residues 318–403); Domain 3 (residues 404–581), which contains a Zn^{2+} ion that is tetrahedrally liganded by four Cys residues and appears to have a structural but not a catalytic role; and Domain 4 (residues 582–660), which is poorly ordered (its side chains are not visible) and hence appears to be highly mobile. The shape and predominantly positive surface charge of the cleft strongly suggest that it forms the enzyme's DNA-binding site. Nevertheless, this putative DNA-binding cleft contains a highly negative patch in the vicinity of Lys 116 that is formed by the highly conserved side chains of Asp 118, Glu 281, and Asp 283. This suggests that these side chains help form the active site for the ligation reaction, which is chemically similar to the polymerization reaction catalyzed by DNA polymerases and hence is likely to have a similar divalent metal ion mechanism involving acidic side chains (Fig. 30-10). Indeed, several DNA ligases have been shown to require divalent metal ion for activity. The apparent high mobility of Domain 4 suggests that it folds out via a hinge-like mechanism to allow the enzyme's nicked dsDNA substrate to bind to the active site and then folds back over to immobilize the DNA as is drawn in Fig. 30-21.

E. Primase

The primases from bacteria and several bacteriophages track the moving replication fork in close association with its DNA helicase. Thus as we have seen (Section 30-2C), the N-terminal domain of T7 gene 4 helicase/primase forms its primase function, whereas *E. coli* primase (DnaG) forms a noncovalent complex with DnaB. Since this DNA helicase translocates along the lagging strand template DNA in its 5′ → 3′ direction (Fig. 30-14), the primase must reverse its direction of travel in order to synthesize an

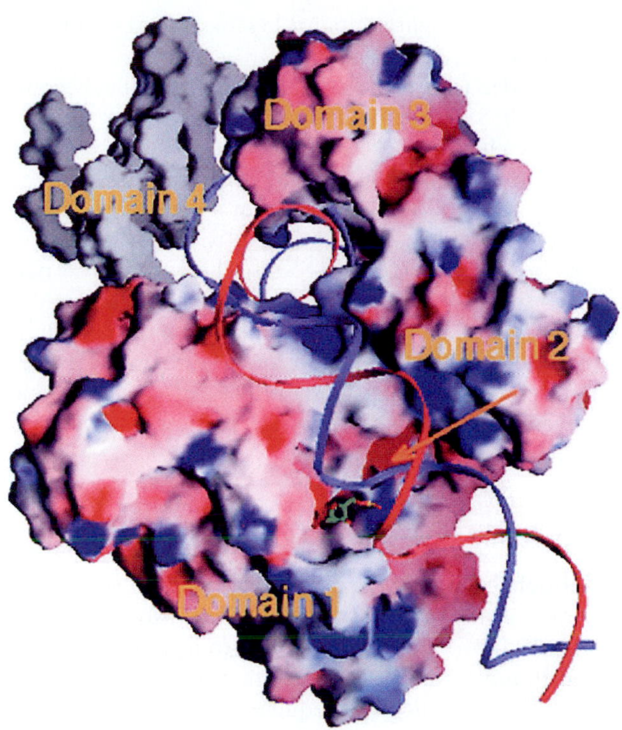

FIGURE 30-21 X-Ray structure of DNA ligase from *Thermus filiformis.* The monomeric protein is represented by its surface diagram, which is colored according to its electrostatic potential with blue, white, and red indicating positive, near neutral, and negative potentials, respectively (Domain 4, which is partially behind Domain 1 in this view, is gray because its side chains were not observed). The exposed ribose–phosphate moiety of the AMP that is covalently bound to the side chain of Lys 116 is drawn in ball-and-stick form with C green, O red, and P yellow. dsDNA has been modeled into the putative DNA-binding cleft (*blue and red ribbons*). The orange arrow points to a highly negative surface patch in the vicinity of the AMP residue that consists of the side chains of the conserved Asp 118, Glu 281, and Asp 283 and that is postulated to form the active site for the divalent metal–catalyzed ligase reaction. [Courtesy of Se Won Suh, Seoul National University, Korea. PDBid 1DGT.]

RNA primer in its $5' \rightarrow 3'$ direction. *E. coli* primase can synthesize up to 60-nt primers *in vitro,* although *in vivo,* primers have the length of 11 ± 1 nt. Since the replication fork in *E. coli* moves at ~1000 nt per second and Okazaki fragments are ~1000 nt in length, primase must synthesize about one RNA primer per second.

E. coli primase is a 581-residue monomeric protein. Proteolysis studies have shown that it consists of three domains: an N-terminal Zn^{2+}-binding domain (residues 1–110), which tetrahedrally ligands a Zn^{2+} ion via three Cys residues and a His residue and is implicated in recognizing ssDNA; a central catalytic domain (residues 111–433) that catalyzes RNA synthesis; and a C-terminal domain (residues 434–581) that interacts with DnaB. The X-ray structure of the catalytic domain (Fig. 30-22), which was independently determined by James Berger and by Kuriyan, reveals a cashew-shaped protein whose fold is unrelated to those of any other DNA or RNA polymerases. However, it contains an ~100-residue segment that is similar in both sequence and structure to segments in types IA and II topoisomerases (Section 29-3C) and has therefore been named the **Toprim fold** (for topoisomerase and primase). The Toprim fold consists of a 4-stranded parallel β sheet flanked by three helices that resembles the nucleotide-binding (Rossmann) fold (Section 8-3B).

The primase catalytic domain contains a groove at the center of its concave surface that is surrounded by residues that are highly conserved in DnaG-type primases. Among them are a Glu and two Asp residues, which are invariant in all known Toprim folds and which in the X-ray structure of a type II topoisomerase coordinate an Mg^{2+} ion. This suggests that these three acidic residues are located at the primase active site, which is known to require Mg^{2+} for activity. Model building by Kuriyan suggests that an RNA–DNA hybrid helix (which has an A-DNA-like conformation; Section 29-1B) binds in the groove with one of its phosphate groups adjacent to the putative Mg^{2+}-binding site (Fig. 30-22). Extension of this double helix into the upper portion of the protein is prevented by a narrowing of the groove. Consequently, it appears that the primase can accommodate only an ~10-bp segment of an RNA–DNA helix, thereby accounting for this enzyme's limited processivity.

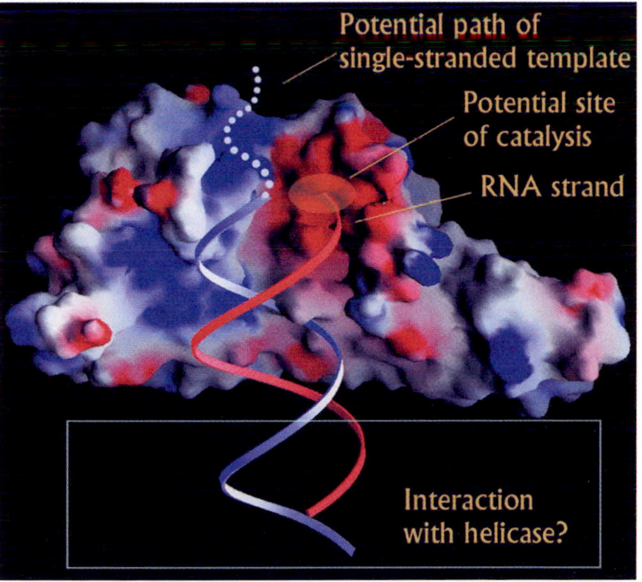

FIGURE 30-22 X-Ray structure of *E. coli* primase. The protein is represented by its molecular surface colored according to its electrostatic potential with red negative, white nearly neutral, and blue positive. The yellow patch marks the enzyme's putative Mg^{2+}-binding site. A segment of an RNA–DNA hybrid double helix has been modeled into the protein so as to place the end of the RNA strand (*red*) near the Mg^{2+}-binding site. Note that the template DNA strand (*blue*) runs along a region of mainly positive charge. [Courtesy of John Kuriyan, The Rockefeller University. PDBid 1EQN.]

3 ■ PROKARYOTIC REPLICATION

Bacteriophages are among the simplest biological entities and their DNA replication mechanisms reflect this fact. Much of what we know about how DNA is replicated therefore stems from the study of this process in various phages. In this section we examine DNA replication in the **coliphages** (bacteriophages that infect *E. coli*) **M13** and ϕX174 and then consider DNA replication in *E. coli* itself. Eukaryotic DNA replication is discussed in Section 30-4.

A. *Bacteriophage M13*

Bacteriophage M13 carries a 6408-nt single-stranded circular DNA known as its **viral** or (+) strand. On infecting an *E. coli* cell, this strand directs the synthesis of its complementary or (−) strand to form the circular duplex **replicative form (RF),** which may be either nicked **(RF II)** or supercoiled **(RF I).** This replication process (Fig. 30-23) may be taken as a paradigm for leading strand synthesis in duplex DNA.

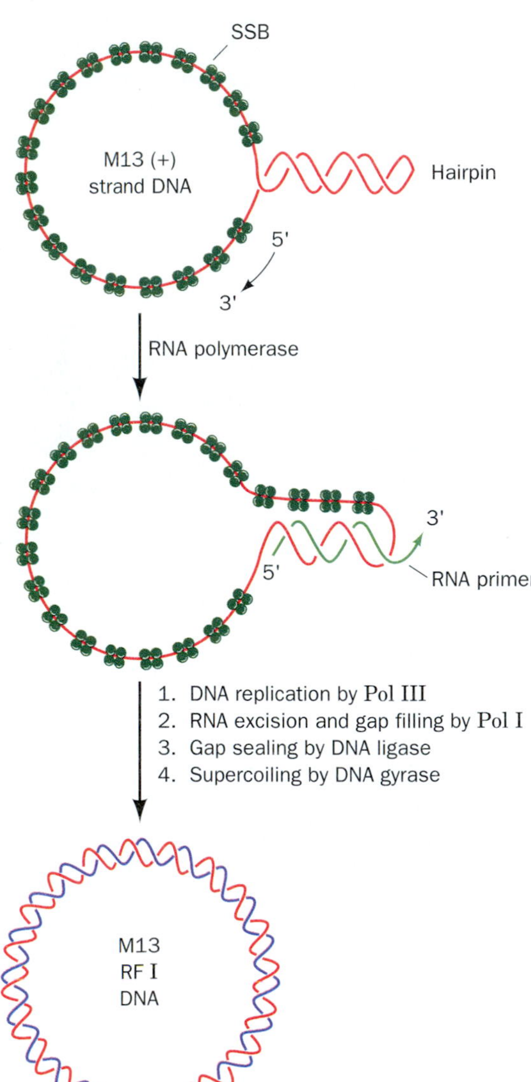

FIGURE 30-23 The synthesis of the M13 (−) strand DNA on a (+) strand template to form M13 RF I DNA.

As the M13 (+) strand enters the *E. coli* cell, it becomes coated with SSB except at a palindromic 57-nt segment that forms a hairpin. RNA polymerase commences primer synthesis 6 nt before the start of the hairpin and extends the RNA 20 to 30 residues to form a segment of RNA–DNA hybrid duplex. The DNA that is displaced from the hairpin becomes coated with SSB so that when RNA polymerase reaches it, primer synthesis stops. Pol III holoenzyme then extends the RNA primer around the circle to form the (−) strand. The primer is removed by Pol I-catalyzed nick translation, thereby forming RF II, which is converted to RF I by the sequential actions of DNA ligase and DNA gyrase.

B. *Bacteriophage ϕX174*

Bacteriophage ϕX174, as does M13, carries a small (5386 nt) single-stranded circular DNA. Curiously, the *in vivo* conversion of the ϕX174 viral DNA to its replicative form is a much more complex process than that for M13 DNA in that ϕX174 replication requires the participation of a nearly 600-kD protein assembly known as a **primosome** (Table 30-4).

a. ϕX174 (−) Strand Replication Is a Paradigm for Lagging Strand Synthesis

ϕX174 (−) strand synthesis occurs in a six-step process (Fig. 30-24):

1. The reaction sequence begins in the same way as that for M13: The (+) strand is coated with SSB except for a 44-nt hairpin. A 70-nt sequence containing this hairpin, known as ***pas*** (for *p*rimosome *a*ssembly *s*ite), is then recognized and bound by the PriA, **PriB,** and **PriC** proteins.

2. DnaB and **DnaC** proteins in the form of a $DnaB_6 \cdot DnaC_6$ complex add to the DNA with the help of **DnaT protein** in an ATP-requiring process. DnaC protein is then released yielding the **preprimosome.** The preprimosome, in turn, binds primase yielding the primosome.

TABLE 30-4 Proteins of the Primosome[a]

Protein	Subunit Structure	Subunit Mass (kD)
PriA	monomer	76
PriB	dimer	11.5
PriC	monomer	23
DnaT	trimer	22
DnaB	hexamer	50
DnaC[b]	monomer	29
Primase (DnaG)	monomer	60

[a]The complex of all primosome proteins but primase is known as the preprimosome.

[b]Not part of the preprimosome or the primosome.

Source: Kornberg, A. and Baker, T.A., *DNA Replication* (2nd ed.), pp. 286–288, Freeman (1992).

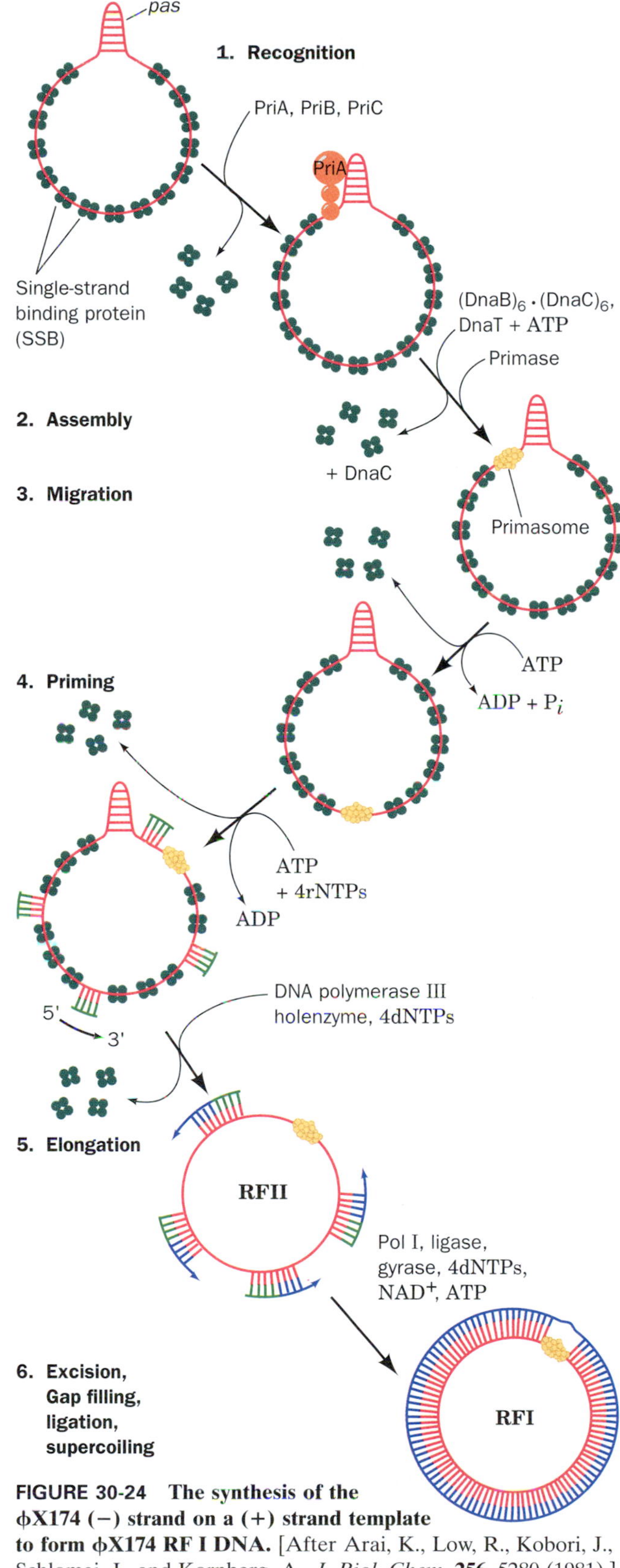

FIGURE 30-24 The synthesis of the φX174 (−) strand on a (+) strand template to form φX174 RF I DNA. [After Arai, K., Low, R., Kobori, J., Schlomai, J., and Kornberg, A., *J. Biol. Chem.* **256,** 5280 (1981).]

3. The primosome is propelled in the $5' \rightarrow 3'$ direction along the (+) strand by the PriA and DnaB helicases at the expense of ATP hydrolysis. This motion, which displaces the SSB in its path, is opposite in direction to that of template reading during DNA chain propagation.

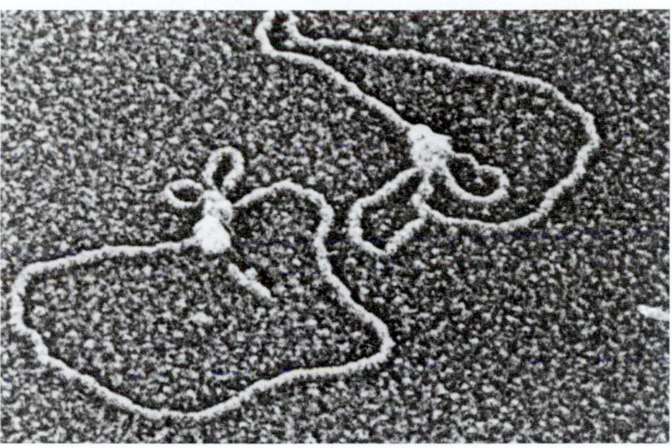

FIGURE 30-25 Electron micrograph of a primosome bound to a φX174 RF I DNA. Such complexes always contain a single primosome with one or two associated small DNA loops. [Courtesy of Jack Griffith, Lineberger Cancer Research Center, University of North Carolina.]

4. At randomly selected sites, the primosome reverses its migration while primase synthesizes an RNA primer. The initiation of primer synthesis requires the participation of DnaB protein which, through concomitant ATP hydrolysis, is thought to alter template DNA conformation in a manner required by primase.

5. Pol III holoenzyme extends the primers to form Okazaki fragments.

6. Pol I excises the primers and replaces them by DNA. The fragments are then joined by DNA ligase and supercoiled by DNA gyrase to form the φX174 RF I.

The primosome remains complexed with the DNA (Fig. 30-25) where it participates in (+) strand synthesis (see below).

b. φX174 (+) Strand Replication Serves as a Model for Leading Strand Synthesis

One strand of a circular duplex DNA may be synthesized via the **rolling circle** or **σ-replication** mode (so called because of the resemblance of the replicating structure to the Greek letter sigma; Fig. 30-26). *The φX174 (+) strand is synthesized on an RF I template by a variation on this process, the* ***looped rolling circle mode*** *(Fig. 30-27):*

1. (+) strand synthesis begins with the primosome-aided binding of the phage-encoded 513-residue enzyme **gene A protein** to its ~30-bp recognition site. There, gene A protein cleaves a specific phosphodiester bond on the (+) strand nucleotide (near the beginning of gene A) by forming a covalent bond between a Tyr residue and the DNA's 5′-phosphoryl group, thereby conserving the cleaved bond's energy.

2. Rep helicase (Section 30-2C) subsequently attaches to the (−) strand at the gene A protein and, with the aid

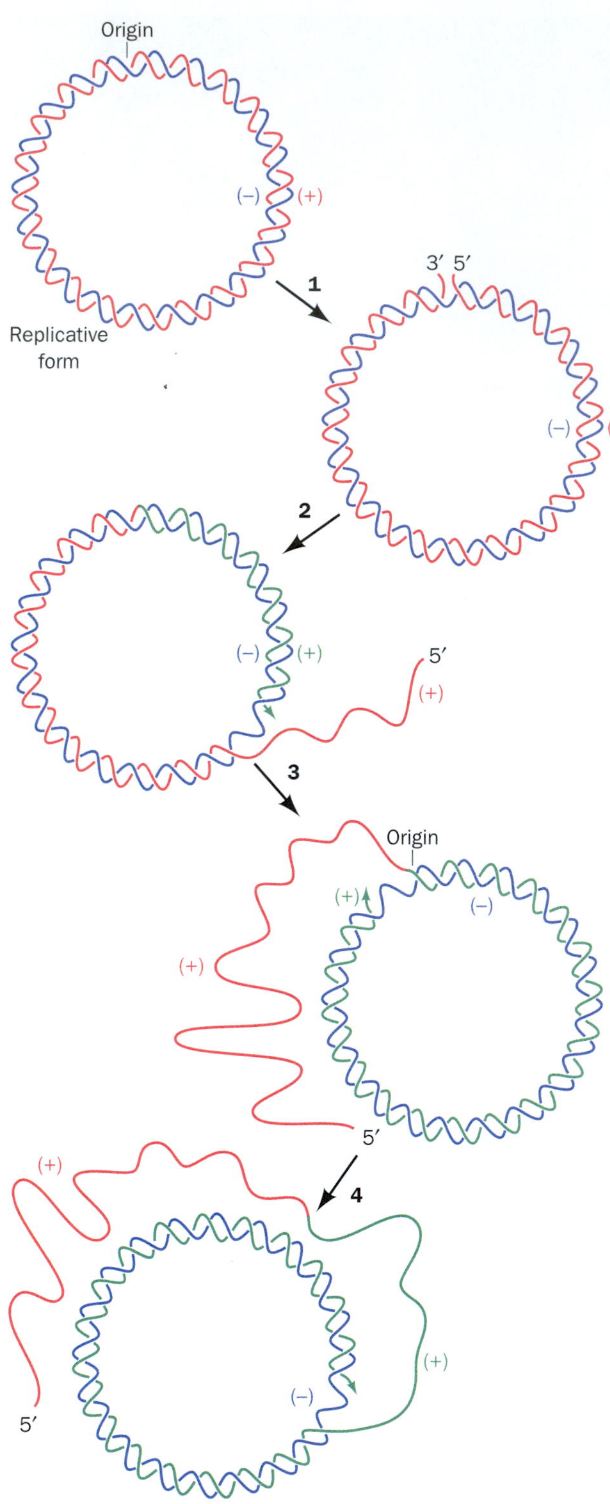

FIGURE 30-26 The rolling circle mode of DNA replication. The (+) strand being synthesized is extended from a specific cut made at the replication origin **(1)** so as to strip away the old (+) strand **(2 and 3).** The continuous synthesis of the (+) strand on a circular (−) strand template produces a series of tandemly linked (+) strands **(4),** which may later be separated by a specific endonuclease.

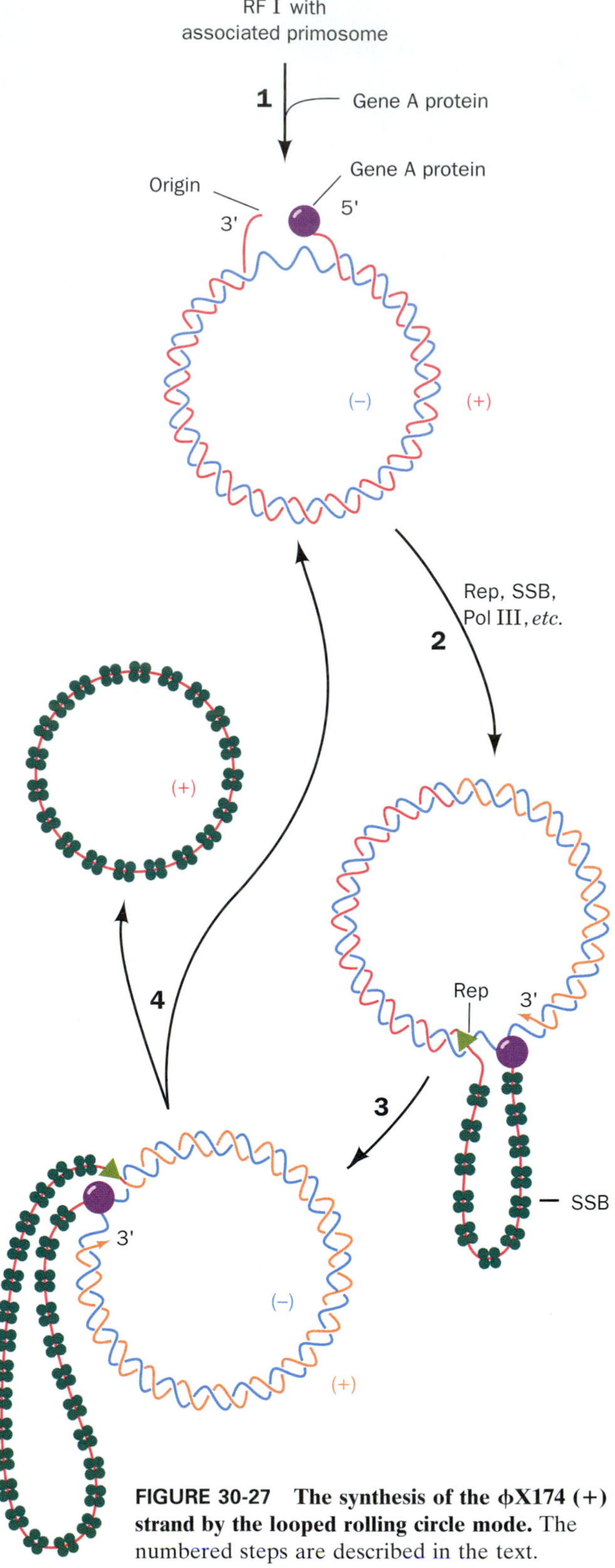

FIGURE 30-27 The synthesis of the ϕX174 (+) strand by the looped rolling circle mode. The numbered steps are described in the text.

of the primosome still associated with the (+) strand, commences unwinding the duplex DNA from the (+) strand's 5′ end. The displaced (+) strand is coated with SSB, which prevents it from reannealing to the (−) strand. Rep helicase is essential for the replication of ϕX174 DNA, but not for the *E. coli* chromosome, as is demonstrated by the inability of ϕX174 to multiply in *rep*⁻ *E. coli.* Pol III holoenzyme extends the (+) strand from its free 3′-OH group.

3. The extension process generates a **looped rolling circle** structure in which the 5′ end of the old (+) strand remains linked to the gene A protein at the replication fork. It is thought that as the old (+) strand is peeled off the RF, the primosome synthesizes the primers required for the later generation of a new (−) strand.

4. When it has come full circle around the (−) strand, the gene A protein again makes a specific cut at the replication origin so as to form a covalent linkage with the new (+) strand's 5′ end. Simultaneously, the newly formed 3′-terminal OH group of the old, looped-out (+) strand nucleophilically attacks its 5′-phosphoryl attachment to the gene A protein, thereby liberating a covalently closed (+) strand. This is possible because the gene A protein has two closely spaced Tyr residues that alternate in their attachment to the 5′ ends of successively synthesized (+) strands. The replication fork continues its progress about the duplex circle, producing new (+) strands in a manner reminiscent of linked sausages being pulled off a reel.

In the intermediate stages of a ϕX174 infection, each newly synthesized (+) strand directs the synthesis of the (−) strand to form RF I as described above. In the later stages of infection, however, the newly formed (+) strands are packaged into phage particles.

C. *E. coli*

See Guided Exploration 25. The Replication of DNA in *E. coli*. *The E. coli chromosome replicates by the bidirectional θ mode from a single replication origin (Section 30-1A).* The most plausible model for events at the *E. coli* replication fork (Fig. 30-28) is largely derived from studies on the

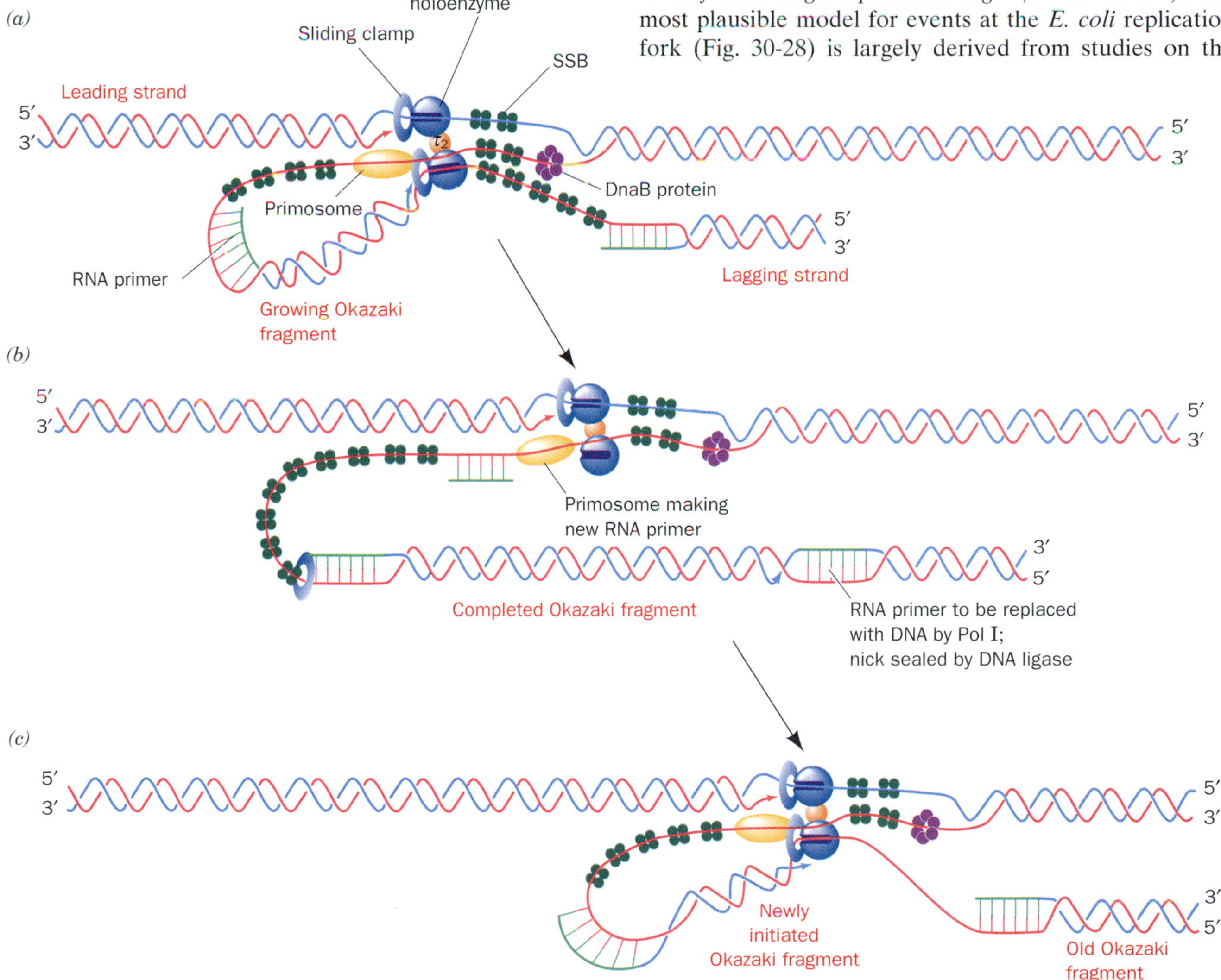

FIGURE 30-28 The replication of *E. coli* DNA. (*a*) The *E. coli* DNA replisome, which contains two DNA polymerase III holoenzyme complexes, synthesizes both the leading and the lagging strands. The lagging strand template must loop around to permit the holoenzyme to extend the primosome-primed lagging strand. (*b*) The holoenzyme releases the lagging strand template when it encounters the previously synthesized Okazaki fragment. This possibly signals the primosome to initiate the synthesis of lagging strand RNA primer. (*c*) The holoenzyme rebinds the lagging strand template and extends the RNA primer to form a new Okazaki fragment. Note that in this model, leading strand synthesis is always ahead of lagging strand synthesis.

simpler and more experimentally accessible DNA replication mechanisms of coliphages such as M13 and φX174. Duplex DNA is unwound by DnaB helicase on the lagging strand template, where it is joined by the primosome. The separated single strands are immediately coated by SSB. Leading strand synthesis is catalyzed by Pol III holoenzyme, as is that of the lagging strand after priming by primosome-associated primase. Both leading and lagging strand syntheses occur on a single ~900-kD multisubunit particle, the **replisome,** which contains two Pol III cores (αεθ) that are joined together by a dimer of τ subunits that bridges the α subunits. Hence, the lagging strand template must be looped around (Fig. 30-28). The τ_2 dimer also binds the DnaB helicase (not shown in Fig. 30-28), thereby stimulating its helicase action while holding it to the replication fork. After completing the synthesis of an Okazaki fragment, the lagging strand holoenzyme relocates to a new primer near the replication fork, the primer heading the previously synthesized Okazaki fragment is excised by Pol I-catalyzed nick translation, and the nick is sealed by DNA ligase.

a. *E. coli* DNA Replication Is Initiated at *oriC* in a Process Mediated by DnaA Protein

The replication origin of the *E. coli* chromosome consists of a unique 245-bp segment known as the ***oriC*** locus. This sequence, segments of which are highly conserved among gram-negative bacteria, supports the bidirectional replication of the various plasmids into which it has been inserted. Experiments with such plasmids pioneered by Kornberg indicate that replication initiation in *E. coli* occurs via the following multistep process (Fig. 30-29):

1. DnaA protein (467 residues) recognizes and binds *oriC*'s five so-called **DnaA boxes** (which each contain a highly conserved 9-bp segment of consensus sequence 5′-TTATCCACA-3′) to form a complex of negatively supercoiled *oriC* DNA wrapped around a central core of five DnaA protein monomers. This process is facilitated by two related DNA-binding proteins named **HU** and **integration host factor (IHF)** that induce DNA bending (IHF and HU are discussed in Section 33-3C).

2. The DnaA protein subunits then successively melt three tandemly repeated, 13-bp, AT-rich segments (consensus sequence 5′-GATCTNTTNTTTT-3′ where N marks nonspecific positions) located near *oriC*'s "left"

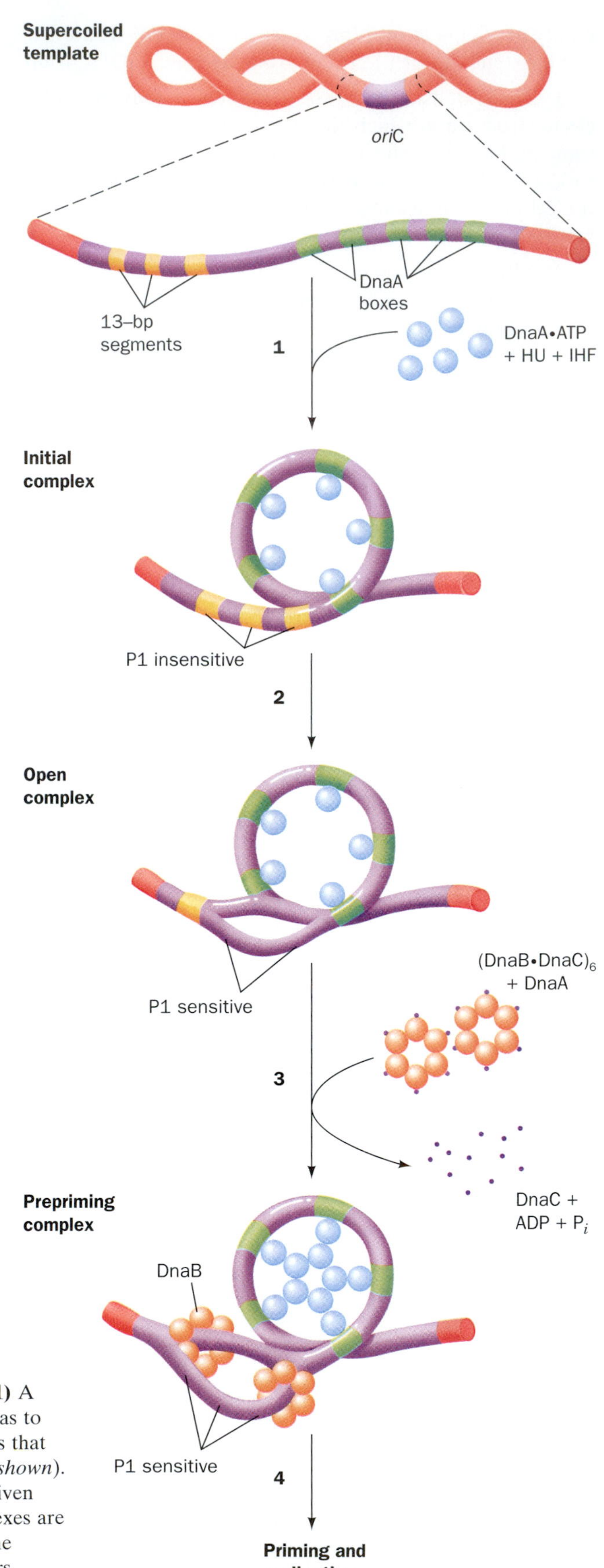

FIGURE 30-29 A model for DNA replication initiation at *oriC*. (1) A DnaA protein subunit binds to each of *oriC*'s five DnaA boxes so as to wrap the suitably supercoiled *oriC* around the proteins in a process that is aided by the binding of HU and IHF proteins to the DNA (*not shown*). **(2)** The three AT-rich 13-bp repeats are then melted in an ATP-driven reaction to form an open complex. **(3)** Two $DnaB_6 \cdot DnaC_6$ complexes are recruited to opposite ends of the open complex accompanied by the binding of five additional DnaA subunits to form five DnaA dimers. **(4)** The open complex is further unwound through the helicase action of DnaB protein, thereby preparing the complex for priming and bidirectional replication.

boundary. The existence of the resulting ~45-bp open complex was established through its sensitivity to **P1 nuclease,** an endonuclease produced by *Penicillium citrinum* that is specific for single strands. The formation of the open complex requires the presence of DnaA protein and ATP (which DnaA protein tightly binds and hydrolyzes in a DNA-dependent manner). The AT-rich nature of the 13-bp repeats, no doubt, facilitates the melting process.

3. The DnaA complex then recruits two $DnaB_6 \cdot DnaC_6$ complexes to opposite ends of the melted region to form the **prepriming complex** in a process that is accompanied by the binding of five additional DnaA subunits to form five DnaA dimers bound at the DnaA boxes. DnaC, an ATPase that facilitates DnaB loading, is subsequently released.

4. In the presence of SSB and gyrase, DnaB helicase further unwinds the DNA in the prepriming complex in both directions so as to permit the entry of primase and RNA polymerase. The participation of both these enzymes in leading strand primer synthesis (Section 30-1D), together with limitation of this process to the *oriC* site, suggests that the RNA polymerase activates primase to synthesize the primer. This perhaps explains the similarity of *oriC*'s AT-rich 13-mers to RNA polymerase's transcriptional promoters (Section 31-2B).

The stage is thereby set for bidirectional DNA replication by Pol III holoenzyme as described above.

b. The Initiation of *E. coli* DNA Replication Is Strictly Regulated

Chromosome replication in E. coli occurs only once per cell division, so this process must be tightly controlled. The doubling (cell generation) time of *E. coli* at 37°C varies with growth conditions from <20 min to ~10 h. Yet the constant ~1000 nt/s rate of movement of each replication fork fixes the 4.6×10^6-bp *E. coli* chromosome's replication time, *C*, at ~40 min. Moreover, the segregation of cellular components and the formation of a septum between them, which must precede cell division, requires a constant time, $D = 20$ min, after the completion of the corresponding round of chromosome replication. *Cells with doubling times $< C + D = 60$ min must consequently initiate chromosome replication before the end of the preceding cell division cycle.* This results in the formation of **multiforked chromosomes** as is diagrammed in Fig. 30-30 for a cell division time of 35 min.

Even in cells that contain multiple *oriC* sites, DNA replication is initiated at each such site once and only once per cell generation. However, after initiation has occurred, chain elongation proceeds at a uniform, largely uncontrolled rate. This suggests that a post-initiation *oriC* site is somehow sequestered from (prevented from interacting with) the replication initiation machinery, a phenomenon called **sequestration.** There is extensive morphological evidence, such as shown in Fig. 30-31, that the *E. coli* chromosome is associated with the cell membrane. This attachment would help explain how replicated chromosomes are segregated into different cells during cell division. But what is the mechanism of sequestration?

The sequence most commonly methylated in *E. coli* is the palindrome GATC, which is methylated at N6 of both its A bases by **Dam methyltransferase** (Section 30-7). GATC occurs 11 times in *oriC*, including at the beginning of all four of its 13-bp repeats (see above). Newly replicated GATC segments are hemimethylated, that is, the GATC sequences on the newly synthesized strand are unmethylated. Although Dam methyltransferase begins methylating most hemimethylated GATC segments immediately after their synthesis (within <1.5 min), those on *oriC* remain hemimethylated for around one-third of a cell

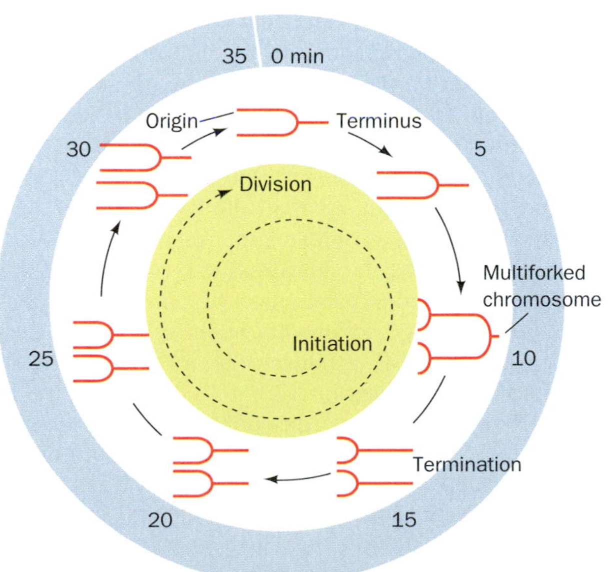

FIGURE 30-30 Multiforked chromosomes in *E. coli*. In cells that are dividing every 35 min, the fixed 60-min interval between the initiation of replication and cell division results in the production of multiforked chromosomes. [After Lewin, B., *Genes VII*, p. 370, Oxford Univ. Press (2000).]

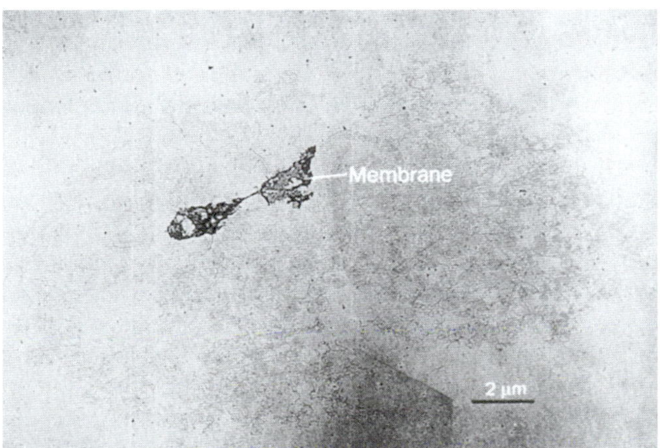

FIGURE 30-31 Electron micrograph of an intact and supercoiled *E. coli* chromosome attached to two fragments of the cell membrane. [From Delius, H. and Worcel, A., *J. Mol. Biol.* **82,** 108 (1974).]

generation. Consequently, the observation that membranes bind hemimethylated *oriC,* but not unmethylated or fully methylated *oriC,* suggests that hemimethylated *oriC* is bound to the membrane in a way that makes it inaccessible to both the initiation machinery and Dam methyltransferase.

The association of hemimethylated *oriC* with membrane requires the presence of the 181-residue **SeqA** protein, the product of *seqA* gene. Thus in *seqA*$^-$ cells: (1) the time to fully methylate hemimethylated GATC sites in *oriC* is reduced to 5 min, whereas the time to do so for other GATC sites is unaffected; (2) the synchrony of initiation of multiple *oriC* sites is lost; and (3) in the absence of functional Dam methyltransferase, fully methylated *oriC*-containing plasmids are replicated numerous times per cell generation, whereas in the presence of SeqA they are replicated only once. Evidently, sequestration occurs via the SeqA-mediated binding of hemimethylated *oriC* to the membrane. The hemimethylated promoter of the *dnaA* gene is similarly sequestered so as to repress its transcription, thereby providing an additional mechanism for preventing promiscuous initiation of DNA replication.

c. The Clamp Loader Loads the Sliding Clamp onto the DNA

The sliding clamp, which is responsible for Pol III's high processivity, is a ring-shaped dimer of β subunits through which the DNA strand being replicated is threaded (Section 30-2B). The two tightly associated β subunits ($K_D < 50$ nM) that form the sliding clamp dissociate with a half-life of ~100 min at 37°C. Yet, since each replisome synthesizes around one Okazaki fragment per second, a sliding clamp must be loaded onto the lagging strand template at this frequency. This loading function is carried out in an ATP-dependent process by the γ complex ($\gamma\tau_2\delta\delta'\chi\psi$). The τ and γ subunits are both encoded by the *dnaX* gene with τ (643 residues) the full-length product and γ (431 residues) its C-terminally truncated form. The γ complex bridges the replisome's two Pol III cores via the C-terminal segments of its two τ subunits, which also bind the DnaB helicase. However, since the χ and ψ subunits are not essential participants in the clamp loading process (their roles are poorly understood), we shall refer to the $\gamma\tau_2\delta\delta'$ complex as the **clamp loader.** How does the clamp loader do its job?

Of the clamp loader's five subunits, only δ is capable of binding to and opening up the sliding clamp on its own. Kuriyan and Mike O'Donnell determined the X-ray structure of the δ subunit in 1:1 complex with a β subunit that had two residues in its dimerization interface mutated so as to prevent its dimerization. The structure reveals (Fig. 30-32) that δ, which consists of three domains, inserts its β interaction element, a hydrophobic plug that forms the tip of its N-terminal domain, into a hydrophobic pocket on the surface of β. Comparison of δ in this structure with that in the $\gamma_3\delta\delta'$ complex (see below) reveals that the β interaction element undergoes a dramatic conformational change on binding to β in which its α4 helix rotates by 45° and translates by 5.5 Å. Moreover, in forming the β–δ complex, the β subunit increases its radius of curvature relative to that in the β dimer (Fig. 30-13) such that the β–δ interaction would induce the opening of one of the sliding clamp's β–β interfaces by ~15 Å. Such a gap is large enough to permit the passage of ssDNA but not dsDNA. Apparently, the clamp loader functions by trapping one β subunit of the sliding clamp in a conformation that prevents ring closure rather than actively pulling apart the two halves of the ring. This is corroborated by molecular dynamics simulations (Section 9-4) suggesting that a β_2 dimer has a stable conformation but that an isolated β subunit with the conformation it has in the β_2 dimer rapidly (in ~1.5 ns) converts to a conformation resembling that in the β–δ complex. Thus, the conformational change of the δ subunit's β interaction element on binding to a β subunit is reminiscent of the action of a plumber's wrench in unlatching the nearby β–β interface so as to allow the sliding clamp to spring open.

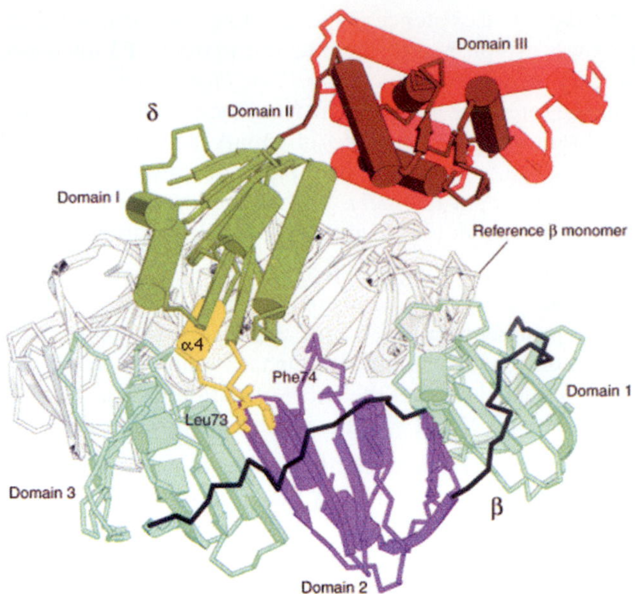

FIGURE 30-32 X-Ray structure of the β–δ complex. A second β subunit taken from the X-ray structure of the sliding clamp (Fig. 30-13), the "Reference β monomer," is drawn in gray. The view is along the edge of the β ring. The δ subunit's β interaction element (*yellow*) consists largely of the α4 helix and two hydrophobic residues, Leu 73 and Phe 74, whose side chains are drawn in stick form. [Courtesy of John Kuriyan, The Rockefeller University. PDB 1JQJ.]

The X-ray structure of the $\gamma_3\delta\delta'$ complex (the clamp loader with its two τ subunits lacking their C-terminal 212 residues; γ and τ are interchangeable in terms of their clamp loading functions), also determined by Kuriyan and O'Donnell, suggests how the clamp loader functions. The γ, δ, and δ′ subunits all have similar folds; they are all members of the widely distributed **AAA$^+$** family (for *A*TPases *a*ssociated with a variety of cellular *a*ctivities; DnaA and DnaC proteins are members of this family) even though only the γ (and τ) subunits bind and hydrolyze ATP. The conserved regions of AAA$^+$ proteins consist of two do-

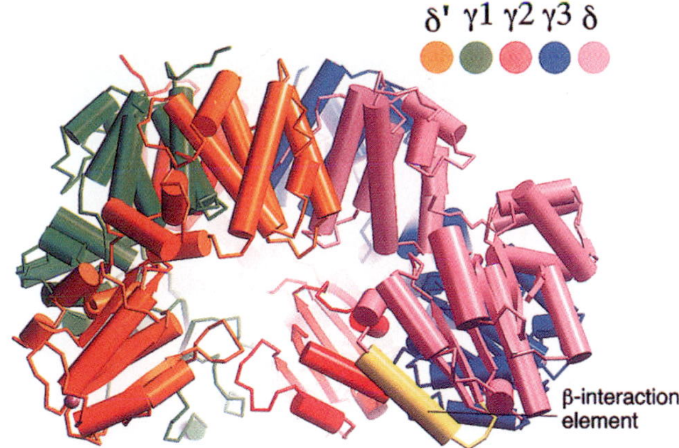
δ' γ1 γ2 γ3 δ
β-interaction element

FIGURE 30-33 X-Ray structure of the $\gamma_3\delta\delta'$ clamp loading complex. The subunits are colored as indicated with the β interaction element yellow. [Courtesy of John Kuriyan, The Rockefeller University. PDB 1JR3.]

mains, an N-terminal ATP-binding domain and a smaller domain composed of a 3-helix bundle, whose relative orientations vary with ATP binding. The $\gamma_3\delta\delta'$ complex's C-terminal domains form a ring-shaped collar in which the subunits are arranged in clockwise order δ′-γ1-γ2-γ3-δ (Fig. 30-33). The relative orientations of the three domains differ in each of the five subunits, thereby forming a highly asymmetric structure, particularly in its more N-terminal regions. Even though the X-ray structure of the $\gamma_3\delta\delta'$ complex is devoid of bound nucleotides, its γ subunits' ATP-binding sites have been identified by analogy with the known structures of the nucleotide complexes of other AAA^+ proteins, such as NSF protein (Section 12-4D; a supposition that was later confirmed by the X-ray structure of the first two domains of the γ subunit in complex with ATPγS). The γ subunits' ATP-binding sites, which are formed by their N-terminal domains, are all located on the inner surface of the $\gamma_3\delta\delta'$ complex and hence the clamp loader.

The clamp loader must tightly bind the sliding clamp prior to its loading on/unloading off the template DNA but must subsequently release the clamp to avoid interfering with its binding to the Pol III core (αεθ). The structures of the clamp loader and the β–δ complex, together with a variety of biochemical evidence, suggest a model of how this might occur (Fig. 30-34): The binding of ATP to γ1 (the γ subunit that contacts δ′) results in a conformational change that exposes the otherwise occluded ATP-binding site of γ2; ATP binding to γ2 likewise exposes γ3; and ATP binding to γ3 exposes the δ subunit's β interaction element, thereby permitting it to bind to a β subunit so as to spring open the sliding clamp. The eventual β- and DNA-stimulated hydrolysis of these bound ATPs reverses this process.

Once the clamp loader has loaded the sliding clamp onto the template DNA, it must dissociate from the clamp to allow the binding of the Pol III core. However, when the synthesis of an Okazaki fragment has been completed, the Pol III core must dissociate from the sliding clamp so that it can initiate the synthesis of the next Okazaki fragment. How does this occur?

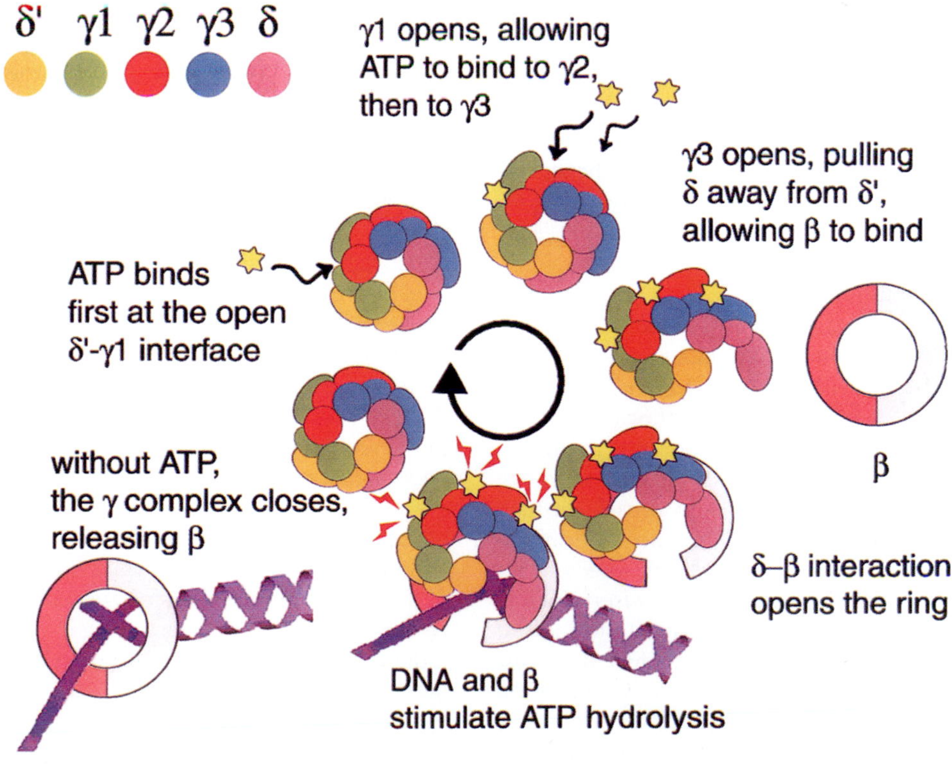
δ' γ1 γ2 γ3 δ
γ1 opens, allowing ATP to bind to γ2, then to γ3
γ3 opens, pulling δ away from δ', allowing β to bind
ATP binds first at the open δ'-γ1 interface
β
without ATP, the γ complex closes, releasing β
δ–β interaction opens the ring
DNA and β stimulate ATP hydrolysis

FIGURE 30-34 Schematic diagram of the clamp loading cycle. This speculative model is based on a combination of structural and biochemical information. [Courtesy of John Kuriyan, The Rockefeller University.]

The α and δ subunits bind to overlapping sites on one face of the β_2 sliding clamp. This was shown by the observations that the phosphorylation of a kinase recognition sequence that had been engineered into the C-terminal segment of β is inhibited by both α and δ. The β subunit has an ~30-fold greater affinity for the γ complex in the presence of ATP than it has for the Pol III core. However, when primed ssDNA is also present, this order of affinity is reversed with β preferring to bind to the Pol III core (possibly due to the additional contacts between the core and the DNA). Thus once the sliding clamp has been loaded onto the primed lagging strand template, the clamp loader is displaced by the Pol III core, which then blocks the clamp loader from unloading the clamp. Instead, the clamp loader loads a new clamp onto the lagging strand template in association with the primer that the primosome had synthesized in preparation for the next round of Okazaki fragment synthesis (Fig. 30-28*b*). When the Pol III core has completed its synthesis of the Okazaki fragment, that is, when the gap between the two successively synthesized Okazaki fragments has been reduced to a nick, it releases the DNA and the clamp. The Pol III core then binds to the newly primed template and its associated clamp (displacing the clamp loader), where it commences the synthesis of the next Okazaki fragment. Throughout this process, the Pol III holoenzyme is held at the replication fork by the leading strand Pol III core, which remains tethered to the DNA by its associated sliding clamp.

The sliding clamp that remains around the completed Okazaki fragment probably functions to recruit Pol I and DNA ligase so as to replace the RNA primer on the previously synthesized Okazaki fragment with DNA and seal the remaining nick. However, the sliding clamp must eventually be recycled. It was initially assumed that this was the job of the clamp loader. However, it is now clear that the release of the sliding clamp from its associated DNA is carried out by free δ subunit (the "wrench" in the clamp loader that cracks apart the β subunits forming the sliding clamp), which is synthesized in 5-fold excess over that required to populate the cell's few clamp loaders.

d. Replication Termination

The *E. coli* replication terminus is a large (350 kb) region flanked by seven nearly identical nonpalindromic ~23-bp terminator sites, ***TerE, TerD,*** and ***TerA*** on one side and ***TerG, TerF, TerB,*** and ***TerC*** on the other (Fig. 30-35; note that *oriC* is directly opposite the terminus region on the *E. coli* chromosome). A replication fork traveling counterclockwise as drawn in Fig. 30-35 passes through *TerG, TerF, TerB,* and *TerC* but stops on encountering either *TerA, TerD,* or *TerE* (*TerD* and *TerE* are presumably backup sites for *TerA*). Similarly, a clockwise-traveling replication fork transits *TerE, TerD,* and *TerA* but halts at *TerC* or, failing that, *TerB* or *TerF* or *TerG*. Thus, these termination sites act as one-way valves that allow replication forks to enter the termination region but not to leave it. This arrangement guarantees that the two replication forks generated by bidirectional initiation at *oriC* will meet in the replication terminus even if one of them arrives there well ahead of its counterpart.

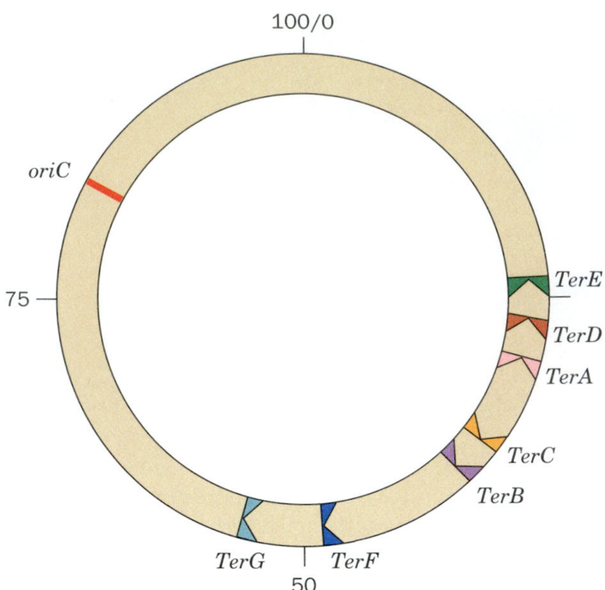

FIGURE 30-35 Map of the *E. coli* chromosome showing the positions of the *Ter* sites and the *oriC* site. The *TerG, TerB, TerF,* and *TerC* sites, in combination with Tus protein, allow a counterclockwise-moving replisome to pass but not a clockwise-moving replisome. The opposite is true of the *TerE, TerD,* and *TerA* sites. Consequently, two replication forks that initiate bidirectional DNA replication at *oriC* will meet between the oppositely facing *Ter* sites.

The arrest of replication fork motion at *Ter* sites requires the action of **Tus** protein, a 309-residue monomer that is the product of the ***tus*** gene (for *t*erminator *u*tilization *s*ubstance). Tus specifically binds to a *Ter* site, where it prevents strand displacement by DnaB helicase, thereby arresting replication fork motion. The X-ray structure of Tus in complex with a 15-bp *Ter* sequence-containing DNA with a single T overhang at each 5′ end, determined by Kosuke Morikawa, reveals that Tus consists of two domains that form a deep positively charged cleft that largely envelops the bound DNA (Fig. 30-36). A 5-bp segment of the DNA near the side of Tus that permits the passage of the replication fork (the lower side of Fig. 30-36) is deformed and underwound relative to canonical (ideal) B-DNA such that its major groove becomes deeper and its minor groove is significantly expanded. The protein makes polar contacts with more than two-thirds of the phosphate groups in a 13-bp region and its interdomain β sheet penetrates the deepened major groove to make sequence-specific contacts with the exposed bases. The importance of this interdomain region for Tus function is demonstrated by the observation that most single residue mutations that reduce the ability of Tus to arrest replication occur in this interdomain region.

When Tus is fused to another DNA-binding protein, replication is inhibited at the other protein's binding site. This suggests that Tus does not act as a simple DNA-

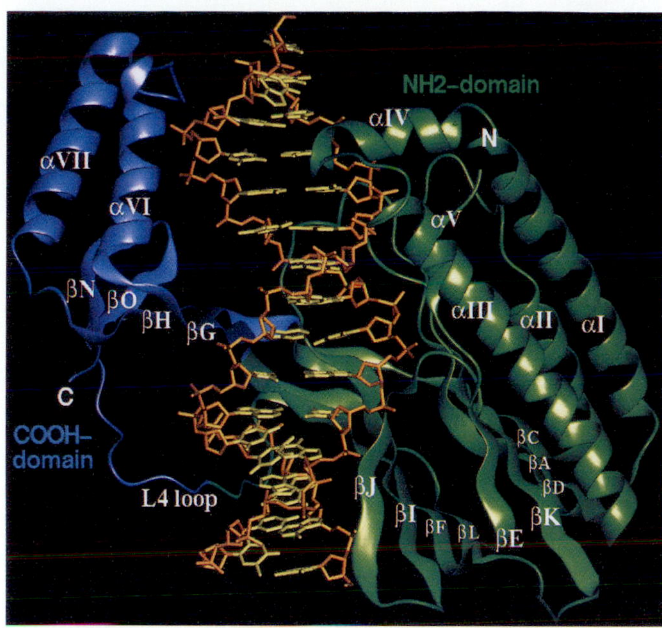

FIGURE 30-36 X-Ray structure of *E. coli* Tus protein in complex with a 15-bp *Ter*-containing DNA. The protein's N- and C-terminal domains are green and blue and its bound DNA is shown in stick form with its bases yellow and its sugar–phosphate backbone gold. [From Kamada, K., Horiuchi, T., Ohsumi, K., Simamato, M., and Morikawa, K., *Nature* **383,** 599 (1996). Used with permission. PDBid 1ECR.] **See the Interactive Exercises**

binding clamp, but interacts with DnaB to inhibit its helicase action. Apparently, Tus interferes with the progress of DnaB in unwinding DNA from one side of Tus but not the other, although how Tus and DnaB interact is unknown. Curiously, however, this termination system is not essential for termination. When the replication terminus is deleted, replication simply stops, apparently through the collision of opposing replication forks. Nevertheless, this termination system is highly conserved in gram-negative bacteria.

D. *Fidelity of Replication*

Since a single polypeptide as small as the Pol I Klenow fragment can replicate DNA by itself, why does *E. coli* maintain a battery of >20 intricately coordinated proteins to replicate its chromosome? The answer apparently is *to ensure the nearly perfect fidelity of DNA replication required to preserve the genetic message's integrity from generation to generation.*

The rates of reversion of mutant *E. coli* or T4 phage to the wild type indicates that only one mispairing occurs per 10^8 to 10^{10} base pairs replicated. This corresponds to ~1 error per 1000 bacteria per generation. Such high replication accuracy arises from four sources:

1. Cells maintain balanced levels of dNTPs through the mechanism discussed in Section 28-3A. This is an important aspect of replication fidelity because a dNTP present at aberrantly high levels is more likely to be misincorporated and, conversely, one present at low levels is more likely to be replaced by the dNTPs present at higher levels.

2. The polymerase reaction itself has extraordinary fidelity. This is because, as we have seen (Section 30-2A), the polymerase reaction occurs in two stages: (1) a binding step in which the incoming dNTP base pairs with the template while the enzyme is in an open conformation that cannot catalyze the polymerase reaction; and (2) a catalysis step in which the polymerase forms a closed conformation about the newly formed base pair, which properly positions its catalytic residues (induced fit). Since the formation of the closed conformation requires that the incoming dNTP form a Watson–Crick-shaped base pair with the template, the conformation change constitutes a double check for correct base pairing.

3. The $3' \rightarrow 5'$ exonuclease functions of Pol I and Pol III detect and eliminate the occasional errors made by their polymerase functions. In fact, mutations that increase a DNA polymerase's proofreading exonuclease activity decrease the rates of mutation of other genes.

4. A remarkable battery of enzyme systems, contained in all cells, function to repair residual errors in the newly synthesized DNA as well as any damage that it may incur after its synthesis through chemical and/or physical insults. We discuss these DNA repair systems in Section 30-5.

In addition, *the inability of a DNA polymerase to initiate chain elongation without a primer is a feature that increases DNA replication fidelity.* The first few nucleotides of a chain to be coupled together are those most likely to be mispaired because of the cooperative nature of base pairing interactions (Section 29-2). The editing of a short duplex oligonucleotide is similarly an error-prone process. The use of RNA primers eliminates this source of error since the RNA is eventually replaced by DNA under conditions that permit accurate base pairing to be achieved.

One might wonder why cells have evolved the complex system of discontinuous lagging strand synthesis rather than a DNA polymerase that could simply extend DNA chains in their $3' \rightarrow 5'$ direction. Consideration of the chemistry of DNA chain extension also leads to the con-

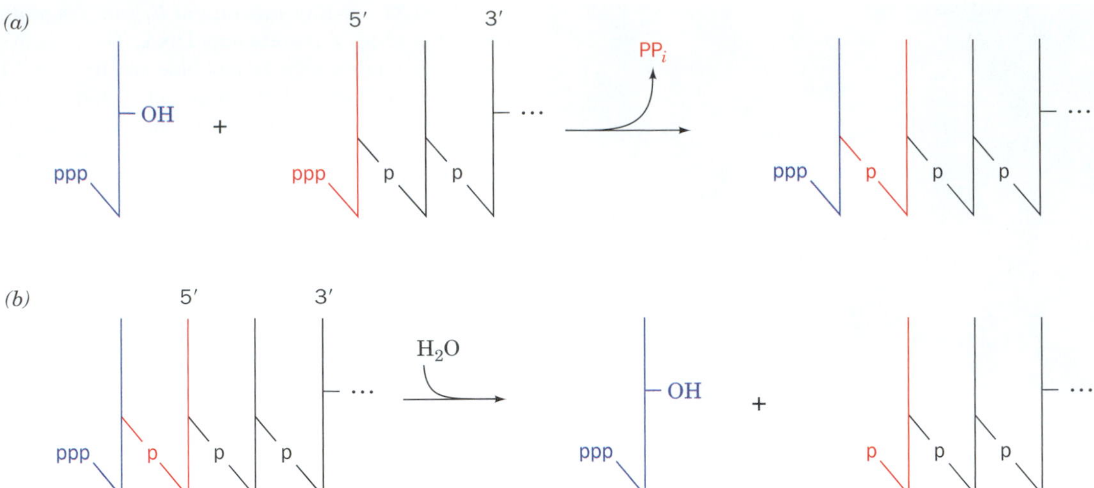

FIGURE 30-37 Chemical consequences if a DNA polymerase could synthesize DNA in its 3′ → 5′ direction. (*a*) The coupling of each nucleoside triphosphate to the growing chain would be driven by the hydrolysis of the previously appended nucleoside triphosphate. (*b*) The editorial removal of an incorrect 5′-terminal nucleoside triphosphate would render the DNA chain incapable of further extension.

clusion that this system promotes high-fidelity replication. The linking of 5′-deoxynucleotide triphosphates in the 3′ → 5′ direction would require the retention of the growing chain's 5′-terminal triphosphate group to drive the next coupling step (Fig. 30-37*a*). On editing a mispaired 5′-terminal nucleotide (Fig. 30-37*b*), this putative polymerase would—in analogy with Pol I, for example—excise the offending nucleotide, leaving either a 5′-OH or a 5′-phosphate group. Neither of these terminal groups is capable of energizing further chain extension. A proofreading 3′ → 5′ DNA polymerase would therefore have to be capable of reactivating its edited product. The inherent complexity of such a system has presumably selected against its evolution.

4 ■ EUKARYOTIC REPLICATION

There is a remarkable degree of similarity between eukaryotic and prokaryotic DNA replication mechanisms. Nevertheless, there are important differences between these two replication systems as a consequence of the vastly greater complexity of eukaryotes in comparison to prokaryotes. For example, eukaryotic chromosomes are structurally complicated and dynamic complexes of DNA and protein (Section 34-1) with which the replication machinery must interact in carrying out its function. Consequently, as is true of most aspects of biochemistry, our knowledge of how DNA is replicated in eukaryotes has lagged well behind that for prokaryotes, although in recent years there has been significant progress in our understanding of this essential process. In this section, we outline what is known about DNA replication in eukaryotes. We also discuss two DNA polymerases that are peculiar to eukaryotic systems: reverse transcriptase and telomerase.

A. *The Cell Cycle*

The **cell cycle,** the general sequence of events that occur during the lifetime of a eukaryotic cell, is divided into four distinct phases (Fig. 30-38):

1. Mitosis and cell division occur during the relatively brief **M phase** (for *m*itosis).

2. This is followed by the **G_1 phase** (for *g*ap), which covers the longest part of the cell cycle. This is the main period of cell growth.

3. G_1 gives way to the **S phase** (for *s*ynthesis) which, in contrast to events in prokaryotes, *is the only period in the cell cycle when DNA is synthesized.*

4. During the relatively short **G_2 phase,** the now tetraploid cell prepares for mitosis. It then enters M phase once again and thereby commences a new round of the cell cycle.

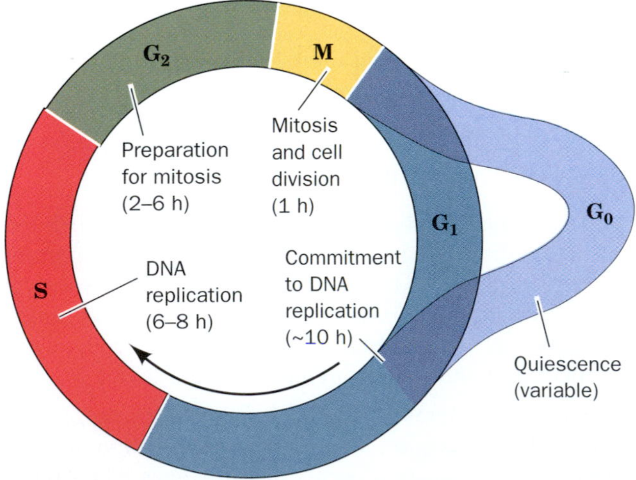

FIGURE 30-38 The eukaryotic cell cycle. Cells in G_1 may enter a quiescent phase (G_0) rather than continuing about the cycle.

The cell cycle for cells in culture typically occupies a 16- to 24-h period. In contrast, cell cycle times for the different types of cells of a multicellular organism may vary from as little as 8 h to >100 days. Most of this variation occurs in the G_1 phase. Moreover, many terminally differentiated cells, such as neurons or muscle cells, never divide; they assume a quiescent state known as the **G_0 phase.**

A cell's irreversible "decision" to proliferate is made during G_1. Quiescence is maintained if, for example, nutrients are in short supply or the cell is in contact with other cells **(contact inhibition).** Conversely, DNA synthesis may be induced by various agents such as carcinogens or tumor viruses, which trigger uncontrolled cell proliferation (cancer; Sections 19-3B and 34-4C); by the surgical removal of a tissue, which results in its rapid regeneration; or by proteins known as **mitogens,** which bind to cell surface receptors and induce cell division (Section 34-4D).

a. The Cell Cycle Is Controlled by Cyclins and Cyclin-Dependent Protein Kinases

The progression of a cell through the cell cycle is regulated by proteins known as **cyclins** and **cyclin-dependent protein kinases (Cdks).** Cyclins are so named because they are synthesized during one phase of the cell cycle and completely degraded during a succeeding phase (protein degradation is discussed in Section 32-6). A particular cyclin specifically binds to and thereby activates its corresponding Cdk(s) to phosphorylate its target proteins, thus activating these proteins to carry out the processes comprising that phase of the cell cycle. In order to enter a new phase in the cell cycle, a cell must satisfy a corresponding **checkpoint,** which monitors whether the cell has satisfactorily completed the preceding phase [e.g., the attachment of all chromosomes to the mitotic spindle must precede mitosis (Section 1-4A); if this were not the case for even one chromosome, one daughter cell would lack this chromosome and the other would have two, both deleterious if not lethal conditions]. If the cell has not met the criteria of the checkpoint, the cell cycle is slowed or even arrested until it does so. We further discuss cell cycle control in Section 34-4C.

B. *Eukaryotic Replication Mechanisms*

Much of what we know about eukaryotic DNA replication has been learned from studies on budding yeast (*Saccharomyces cerevisiae*) and fission yeast (*Schizosaccharomyces pombe*), the simplest eukaryotes, and on simian virus 40 (SV40), which has a 5243-bp circular DNA chromosome (Fig. 5-40) that has only one replication origin. However, studies of DNA replication in the cells of **metazoa** (multicellular animals), particularly *Drosophila, Xenopus laevis* (an African clawed toad, whose eggs are easily studied), and humans, have also led to important advances in our knowledge.

a. Eukaryotic Cells Contain Numerous DNA Polymerases

The many known DNA polymerases can be classified into six families based on phylogenetic relationships: family A (e.g., *E. coli* Pol I), family B (e.g., *E. coli* Pol II), family C (e.g., *E. coli* Pol III), and families D, X, and Y. Animal cells express at least four distinct types of DNA polymerases that are implicated in DNA replication. They are designated, in the order of their discovery, DNA polymerases (pols) α, γ, δ, and ε (alternatively, POLA, POLG, POLD1, and POLE). Their functions were, in part, elucidated by their different responses to various inhibitors (Table 30-5).

Pol α, a B-family enzyme, occurs only in the cell nucleus and participates in the replication of chromosomal DNA. This function was largely established through the use of its specific inhibitor **aphidicolin**

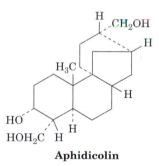

Aphidicolin

TABLE 30-5 Properties of Some Animal DNA Polymerases

	α	β	γ	δ
Location	nucleus	nucleus	mitochondrion	nucleus
Subunit masses (kD)[a]	167, ~83, 58, 48	68	143	125, 55, 40, 22
	(165, 67, 58, 48)	(39)	(125, 43)	(125, 66, 50)
Family	B	X	A	B
Inhibitors:				
Aphidicolin	strong	none	none	strong
Dideoxy NTPs	none	strong	strong	weak
N-Ethylmaleimide (NEM)[b]	strong	none	strong	strong

[a]Yeast *S. cerevisiae* (mammalian cells).
[b]A cysteine-alkylating agent (Section 12-4D).

Source: Kornberg, A. and Baker, T.A., *DNA Replication* (2nd ed.), p. 199, Freeman (1992); *and* Hübscher, U., Nasheuer, H.-P., and Syväoja, J.E., *Trends Biochem. Sci.* **25,** 143 (2000).

and by the observation that pol α activity varies with the rate of cellular proliferation. Pol α, as do all DNA polymerases, replicates DNA by extending a primer in the 5′ → 3′ direction under the direction of a single-stranded DNA template. This enzyme lacks exonuclease activity but tightly associates with a primase (which consists of a 48-kD subunit that contains the primase catalytic site and a 58-kD subunit that is required for full primase activity) and an ~83-kD subunit that is implicated in the regulation of initiation, yielding a protein named **pol α/primase.**

The X-ray structure of the B-family DNA polymerase encoded by bacteriophage RB69 **(RB69 pol),** determined by Steitz, reveals that this enzyme consists of five domains arranged around a central hole that contains its polymerase active site (Fig. 30-39). RB69 pol has the right-hand-like architecture first seen in A-family DNA polymerases (Figs. 30-8 and 30-9), and its palm domain has a structurally similar core that contains the two invariant Asp residues implicated in the nucleotidyl transfer mechanism (Fig. 30-10). However, in comparison with A-family enzymes, the fingers domain of RB69 pol exhibits a 60° rotation and its editing domain lies on the opposite side of its palm domain.

Pol δ, a B-family enzyme, is a nuclear enzyme with inhibitor sensitivities similar to those of pol α (Table 30-5). It lacks an associated primase but exhibits a proofreading 3′ → 5′ exonuclease activity. Moreover, whereas pol α exhibits only moderate processivity (~100 nucleotides), that of pol δ is essentially unlimited (replicates the entire length of a template), but only when it is in complex with a protein named **proliferating cell nuclear antigen (PCNA;** so named because it occurs only in the nuclei of proliferating cells and reacts with antibodies produced by a subset of patients with the autoimmune disease systemic lupus erythematosus). The X-ray structure of PCNA (Fig. 30-40), determined by Kuriyan, reveals that it forms a trimeric ring with almost identical structure (and presumably function) as the *E. coli* β_2 sliding clamp (Fig. 30-13). Thus, each PCNA subunit consists of four rather than six of the structurally similar βαβββ motifs from which the *E. coli* β subunit is constructed. Intriguingly, PCNA and the β subunit exhibit no significant sequence identity, even when their structurally similar portions are aligned.

Pol δ in complex with PCNA is required for both leading and lagging strand synthesis. In contrast, pol α/primase functions to synthesize 7- to 10-nt RNA primers, which it extends by an additional ~15 nt of DNA. Then, in a process called **polymerase switching,** the eukaryotic counterpart of the *E. coli* γ complex (the clamp loader), **replication factor C (RFC),** displaces the pol α and loads PCNA on the template DNA near the primer strand, following which pol δ binds to the PCNA and processively extends the DNA strand.

Pol ε, a B-family nuclear enzyme which superficially resembles pol δ, differs from it in that pol ε is highly processive in the absence of PCNA and has a 3′ → 5′ exonuclease activity that degrades single-stranded DNA to 6- or 7-residue oligonucleotides rather than to mononucleotides, as does that of pol δ. Although pol ε is required for the viability of yeast, its essential function can be carried out by only the noncatalytic C-terminal half of its 256-kD catalytic

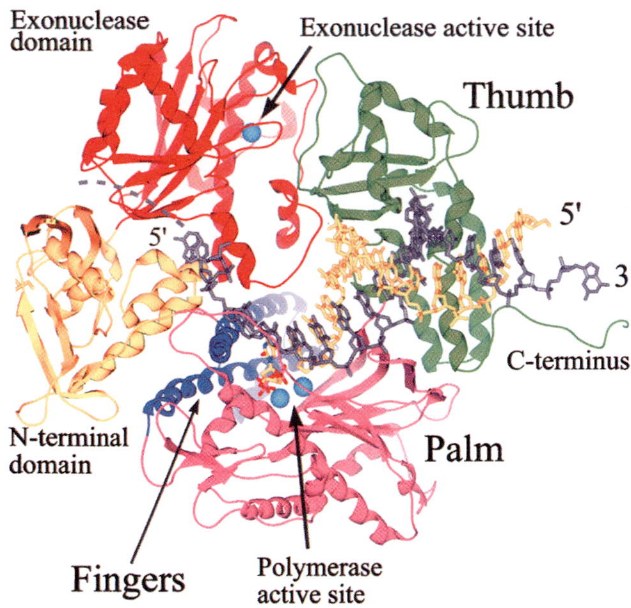

FIGURE 30-39 X-Ray structure of RB69 DNA polymerase (RB69 pol) in complex with primer–template DNA and dTTP. The protein is drawn in ribbon form with its various domains differently colored. The DNA is drawn in stick form with its primer strand gold and its template strand gray. The incoming dTTP, also in stick form, is colored according to atom type (C gold, N blue, O red, and P magenta). The two Ca^{2+} ions at the polymerase site are represented by cyan balls as is the Ca^{2+} ion at the exonuclease site. The probable extension of the path taken by the single-stranded template as it enters the polymerase active site is shown as a dashed gray line. [Courtesy of Thomas Steitz, Yale University. PDBid 1IG9.]

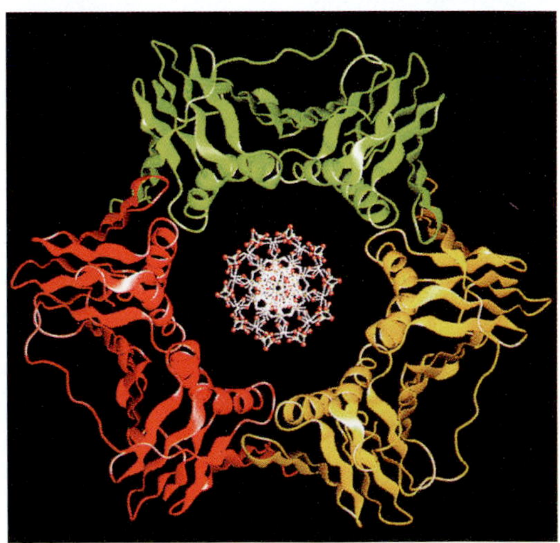

FIGURE 30-40 X-Ray structure of PCNA. Its three subunits (*red, green, and yellow*) form a 3-fold symmetric ring. A model of duplex DNA viewed along its helix axis has been drawn in the center of the PCNA ring. Compare this structure with that of the β subunit dimer of *E. coli* Pol III holoenzyme (Fig. 30-13). [Courtesy of John Kuriyan, The Rockefeller University. PDBid 1PLQ.] **See the Interactive Exercises**

subunit, which is unique among B-family DNA polymerases. Moreover, the only DNA polymerases required to replicate SV40 DNA are pol α and pol δ. It therefore appears that, at least in yeast, pol ε has an essential control function but not a catalytic function.

Pol γ, an A-family enzyme, occurs exclusively in the mitochondrion, where it presumably replicates the mitochondrial DNA. Chloroplasts contain a similar enzyme.

Eukaryotic cells contain batteries of DNA polymerases. These include the DNA polymerases that participate in chromosomal DNA replication (pols α, δ, and ε) and several that take part in DNA repair processes (Section 30-5) including **pols β, η, ι, κ,** and **ζ** (alternatively, POLB, POLH, POLI, POLK, and POLZ). Pol β, an X-family enzyme, is remarkable for its small size (a 335-residue monomer in rat). The X-ray structure of a stable proteolytic fragment (residues 85–335) of rat pol β (Fig. 30-41), independently determined by Zdenek Hostomsky and Joseph Kraut, reveals that this protein has the right-hand-like shape of other polymerases of known structure (e.g., Figs. 30-8, 30-9, and 30-39). However, its folding topology is unique, which suggests that it does not share a common ancestor with these other polymerases.

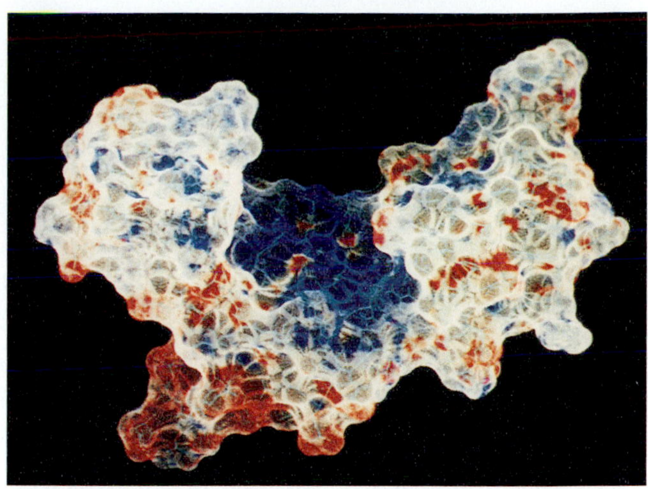

FIGURE 30-41 X-Ray structure of the catalytic domain of rat DNA polymerase β. The protein, which is oriented with its N-terminal fingers subdomain at the top left, is represented by its solvent-accessible surface colored according to charge, with red negative, blue positive, and white neutral. The strong positive charge of the putative DNA binding cleft no doubt facilitates the binding of the polyanionic DNA. [Courtesy of Zdenek Hostomsky, Agouron Pharmaceuticals, San Diego, California. PDBid 1RPL.]

b. Eukaryotic Chromosomes Consist of Numerous Replicons

Eukaryotic and prokaryotic DNA replication systems differ most obviously in that eukaryotic chromosomes have multiple replication origins in contrast to the single replication origin of prokaryotic chromosomes. Eukaryotic cells replicate DNA at the rate of ~50 nt/s (~20 times slower than does *E. coli*) as was determined by autoradiographically measuring the lengths of pulse-labeled sections of eukaryotic chromosomes. Since a eukaryotic chromosome typically contains 60 times more DNA than those of prokaryotes, its bidirectional replication from a single origin would require ~1 month to complete. Electron micrographs such as Fig. 30-42, however, reveal that eukaryotic chromosomes contain multiple origins, one every 3 to 300 kb depending on both the species and the tissue, so that S phase usually occupies only a few hours.

Cytological observations indicate that the various chromosomal regions are not all replicated simultaneously; rather, clusters of 20 to 80 adjacent **replicons** (replication units; DNA segments that are each served by a replication origin) are activated simultaneously. New replicons are activated throughout S phase until the entire chromosome has been replicated. During this process, replicons that have already been replicated are distinguished from those that have not; that is, *a cell's chromosomal DNA is replicated once and only once per cell cycle.*

c. The Assembly of the Eukaryotic Initiation Complex Occurs in Two Stages

The once-and-only-once replication of eukaryotic DNA per cell cycle is conferred by a type of binary switch. A **pre-replicative complex (pre-RC)** is assembled at each replication origin during the G_1 phase of the cell cycle. This is the only period of the cell cycle during which the pre-RC can form and hence this process is known as **licensing.** However, a licensed pre-RC cannot initiate DNA replication. Rather, it must be activated to do so, a process that occurs only during S phase. *This temporal separation of pre-RC assembly and origin activation ensures that a new pre-RC cannot assemble on an origin that has already "fired" (commenced replication) so that an origin can only fire once per cell cycle.* How does this occur?

The elucidation of the licensing process and how the pre-RC is activated to form an initiation complex is still in its early stages. Thus, although it appears that most of the proteins forming these complexes have been identified, their structures, interactions, and, in many cases, their functions are largely unknown. Keeping this in mind, let us consider what is known about these processes.

Replication origins are surprisingly variable among species, often within the same organism, and even vary with a given organism's developmental stage. Thus, whereas *S. cerevisiae* origins, which are known as

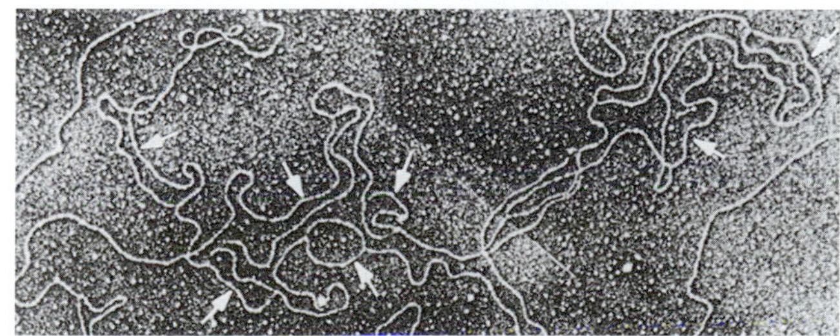

FIGURE 30-42 Electron micrograph of a fragment of replicating *Drosophila* DNA. The arrows indicate its multiple replication eyes. [From Kreigstein, H.J. and Hogness, D.S., *Proc. Natl. Acad. Sci.* **71,** 136 (1974).]

autonomously replicating sequences (ARS), contain a highly conserved 11-bp AT-rich sequence within a less well defined ~125-bp region, some metazoan origins are dispersed over 10 to 50 kb "initiation zones" that contain multiple origins and, in some cases, require no specific DNA sequence at all. Despite this disparity, the proteins that participate in eukaryotic DNA replication are highly conserved from yeast to humans.

The assembly of the pre-RC (Fig. 30-43) begins late in M phase or early in G_1 phase with the binding of the **origin recognition complex (ORC),** a hexamer of related proteins **(Orc1** through **Orc6),** to the origin, where it remains bound during most or all of the cell cycle. ORC, the functional analog of DnaA protein in *E. coli* replication initiation (Section 30-3C), then recruits two proteins, **Cdc6** in *S. cerevisiae* (**Cdc18** in *S. pombe*; Cdc for *c*ell *d*ivision *c*ycle) and **Cdt1.** These proteins then cooperate with the ORC to load the **MCM complex** [named for its *m*inichromosome (plasmid) *m*aintenance functions], a hexamer of related subunits (**Mcm2** through **Mcm7**), onto the DNA to yield the licensed pre-RC. The MCM complex, a ring-shaped ATP-driven helicase, is the analog of *E. coli* DnaB helicase, whereas Cdc6/Cdc18 together with Cdt1 appears to be an analog of *E. coli* DnaC (which facilitates DnaB loading). With the exception of Cdt1, all of these proteins, Orc1 through Orc6, Cdc6/Cdc18, Mcm2 through Mcm7, as well as *E. coli* DnaA, DnaB, and DnaC, are AAA$^+$ ATPases.

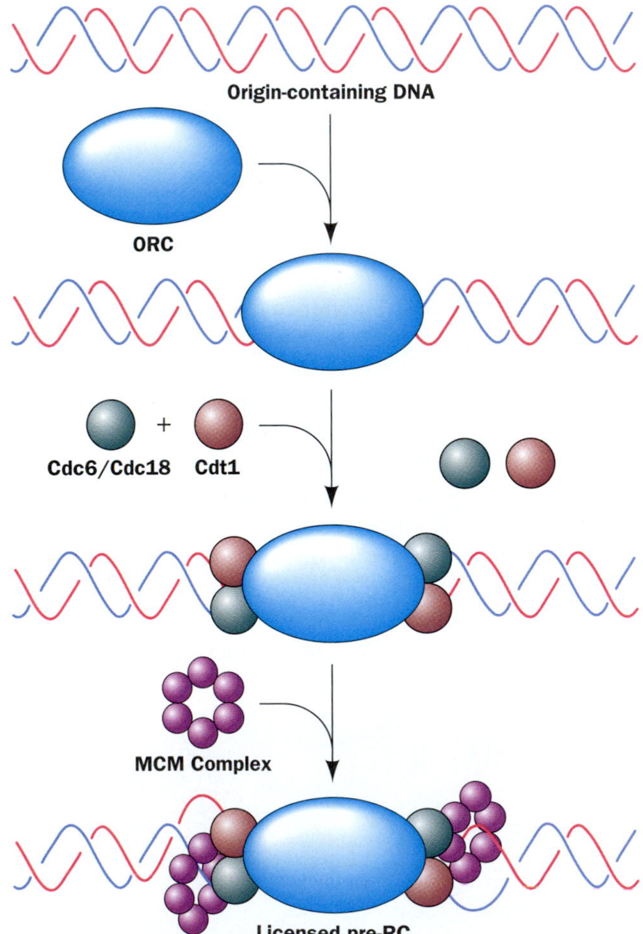

FIGURE 30-43 Schematic diagram for the assembly of the eukaryotic pre-replicative complex (pre-RC). The actual stoichiometries, positions, and interactions of its various components are largely unknown. The pre-RC only forms during the G_1 phase of the cell cycle.

The conversion of a licensed pre-RC to an active initiation complex requires the addition of pol α/primase, pol ε, and several accessory proteins, which only occurs at the onset of S phase. This process begins with addition of **Mcm10** protein (which shares no sequence similarity with any of the subunits of the MCM complex) to the pre-RC, which probably displaces Cdt1. This is followed by the addition of at least two protein kinases, a Cdk and **Ddk,** the latter being a heterodimer of the protein kinase **Cdc7** with its activating subunit **Dbf4** (Ddk stands for *D*bf4-*d*ependent *k*inase). Ddk acts to phosphorylate five of the six MCM subunits (all but Mcm2) so as to activate the MCM complex as a helicase. In contrast, the way in which Cdks activate the pre-RC is poorly understood although several ORC and MCM proteins as well as Cdc6/Cdc18 are phosphorylated by Cdks. Ddk together with a Cdk also recruits **Cdc45** to the growing initiation complex. Cdc45, in turn, is required for the assembly of the initiating synthetic machinery at the replication fork, including pol α/primase, pol ε, PCNA, and **replication protein A (RPA),** the heterotrimeric eukaryotic counterpart of SSB, thereby forming an active initiation complex.

d. Re-Replication Is Prevented through the Actions of Cdks and Geminin

Once initiation (priming) has occurred, the initiation complex is joined by RFC and pol δ and, as is described above, is converted to an active replicative complex by polymerase switching. DNA replication then proceeds bidirectionally until each replication fork has collided with an oppositely traveling replication fork, thereby completing the replication of the replicon. An active replication fork will destroy any licensed pre-RCs and unfired initiation complexes in its path, thereby preventing the DNA at such sites from being replicated twice. Eukaryotes appear to lack termination sequences and proteins analogous to the *Ter* sites and Tus protein in *E. coli.*

Several redundant mechanisms ensure that a pre-RC can initiate DNA synthesis only once. Cdks are active from late G_1 phase through late M phase. These elevated Cdk levels, which are required to activate initiation, also prevent reinitiation. The Cdk-mediated phosphorylation of Cdc6/Cdc18, which occurs late in G_1 after the pre-RCs have formed, causes Cdc6/Cdc18 to be proteolytically degraded in yeast and exported from the nucleus in mammalian cells. Evidently, Cdc6/Cdc18 is only required for the assembly of the pre-RC, not its activation. The helicase activity of the MCM complex is inhibited by phosphorylation, at least *in vitro.* Moreover, MCM proteins are exported from the nucleus in G_2 and M phases, a process that is interrupted by Cdk inactivation. However, the function of Cdk-mediated phosphorylation of ORC proteins is unclear.

Metazoan cells have yet another mechanism to prevent the assembly of a licensed pre-RC on already replicated DNA. High levels of a protein named **geminin** appear in S phase and continue to accumulate until late M phase, when geminin is degraded. Geminin associates with Cdt1 (which together with Cdc6/Cdc18 loads the MCM complex onto the ORC) so as to inhibit the assembly of the pre-RC. This inhibition can be reversed by the addition of excess Cdt1. It therefore seems likely that the presence of geminin provides protection against DNA re-replication under conditions when Cdks are inhibited by checkpoint activation.

Finally, cells that have shifted to the G_0 (quiescent) phase of the cell cycle (Fig. 30-38)—the majority of cells in the human body—cease making DNA. Such cells are characterized by the absence of Cdk activity. In proliferating cells, this would permit the re-replication of DNA. However, cells in G_0 also lack the proteins of the MCM complex and are therefore incapable of assembling licensed pre-RCs. Since cancerous cells are characterized by being in a state of rapid proliferation (Section 19-3B), the presence of MCM complex proteins in what should be quiescent cells is a promising diagnostic marker for cancer.

e. Primers Are Removed by RNase H1 and Flap Endonuclease-1

The RNA primers of eukaryotic Okazaki fragments are removed through the actions of two enzymes: **RNase H1** removes most of the RNA leaving only a 5′ ribonucleotide adjacent to the DNA, which is then removed through the action of **flap endonuclease-1 (FEN1).** However, as we have seen, pol α/primase extends the RNA primers it has made by ~15 nt of DNA before it is displaced by pol δ. Since pol α lacks proofreading ability, this primer extension is more likely to contain errors than the DNA synthesized by pol δ. However, FEN1 provides what is, in effect, pol α's proofreading function: It is also an endonuclease that excises mismatch-containing oligonucleotides up to 15 nt long from the 5′ end of an annealed DNA strand. Moreover, FEN1 can make several such excisions in succession to remove more distant mismatches. The excised segment is later replaced by pol δ as it synthesizes the succeeding Okazaki fragment.

f. Mitochondrial DNA Is Replicated in D-Loops

Mitochondrial DNA is replicated by a process in which leading strand synthesis precedes lagging strand synthesis (Fig. 30-44). The leading strand therefore displaces the lagging strand template to form a **displacement** or **D-loop.** The 15-kb circular mitochondrial chromosome of mammals normally contains a single 500- to 600-nt D-loop that undergoes frequent cycles of degradation and resynthesis. During replication, the D-loop is extended. When it has reached a point approximately two-thirds of the way around the chromosome, the lagging strand origin is exposed and its synthesis proceeds in the opposite direction around the chromosome. Lagging strand synthesis is therefore only about one-third complete when leading strand synthesis terminates.

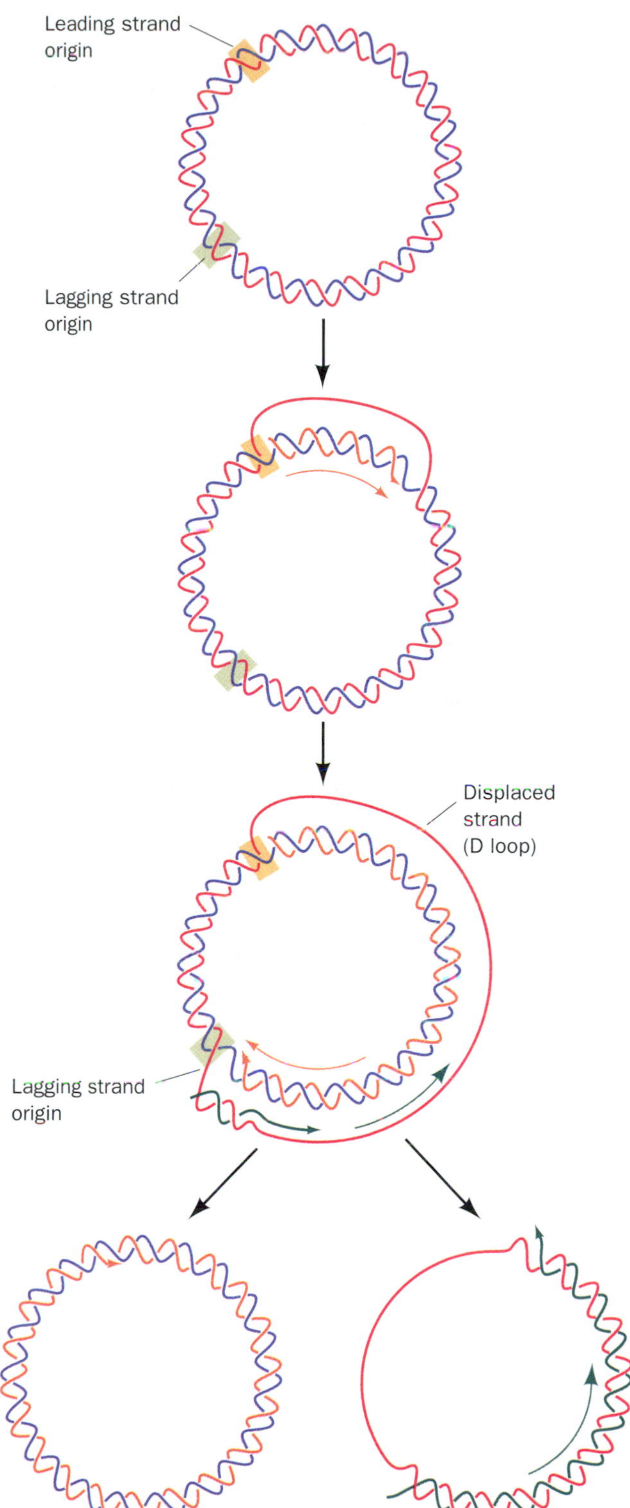

FIGURE 30-44 **The D-loop mode of DNA replication.**

C. *Reverse Transcriptase*

The **retroviruses,** which are RNA-containing eukaryotic viruses such as certain tumor viruses and human immunodeficiency virus (HIV), contain an **RNA-directed DNA polymerase (reverse transcriptase).** This enzyme, which was independently discovered in 1970 by Howard Temin and David Baltimore, acts much like Pol I in that it synthesizes DNA in the $5' \rightarrow 3'$ direction from primed templates. In the case of reverse transcriptase, however, RNA is the template.

The discovery of reverse transcriptase caused a mild sensation in the biochemical community because it was perceived by some as being heretical to the central dogma of molecular biology (Section 5-4). There is, however, no thermodynamic prohibition to the reverse transcriptase reaction; in fact, under certain conditions, Pol I can likewise copy RNA templates.

Reverse transcriptase transcribes the retrovirus's single-stranded RNA genome to a double-stranded DNA as follows (Fig. 30-45):

1. The retroviral RNA acts as a template for the synthesis of its complementary DNA (RNA-directed DNA polymerase activity), yielding an RNA–DNA hybrid helix. The DNA synthesis is primed by a host cell tRNA whose 3′ end partially unfolds to base pair with a complementary segment of the viral RNA.

2. The RNA strand is then nucleolytically degraded (**RNase H** activity; H for hybrid).

3. Finally, the DNA strand acts as a template for the synthesis of its complementary DNA (DNA-directed DNA polymerase activity), yielding double-stranded DNA.

The DNA is then integrated into a host cell chromosome.

Reverse transcriptase has been a particularly useful tool in genetic engineering because of its ability to transcribe mRNAs to complementary strands of DNA (cDNA). In transcribing eukaryotic mRNAs, which have poly(A) tails (Section 31-4A), the primer can be oligo(dT). cDNAs have been used, for example, as probes in Southern blotting (Section 5-5D) to identify the genes coding for their corresponding mRNAs. An RNA's base sequence can be easily determined by sequencing its cDNA (Section 7-2A).

a. X-Ray Structure of HIV-1 Reverse Transcriptase

HIV-1 reverse transcriptase (RT) is a dimeric protein whose subunits are synthesized as identical 66-kD polypeptides, known as **p66** (p for *p*rotein), that each contain a polymerase domain and an RNase H domain. However, the RNase H domain of one of the two subunits is proteolytically excised, thereby yielding a 51-kD polypeptide named **p51.** Thus, RT is dimer of p66 and p51.

The first drugs to be clinically approved to treat AIDS, **3′-azido-3′-deoxythymidine (AZT; zidovudine), 2′,3′-dideoxyinosine (ddI; didanosine), 2′,3′-dideoxycytidine (ddC; zalcitabine),** and **2′,3′-didehydro-3′-deoxythymidine (stavudine),**

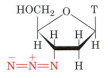

HOCH$_2$ O T; H H H H; $\overset{-}{N}=\overset{+}{N}=N$ H

3′-Azido-3′-deoxythymidine (AZT; zidovudine)

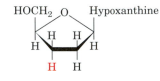

HOCH$_2$ O Hypoxanthine; H H H H; H H

2′,3′-Dideoxyinosine, (ddI; didanosine)

HOCH$_2$ O C; H H H H; H H

2′,3′-Dideoxycytidine (ddC; zalcitabine)

HOCH$_2$ O T; H H; H H

2′,3′-Didehydro-3′-deoxythymidine (stavudine)

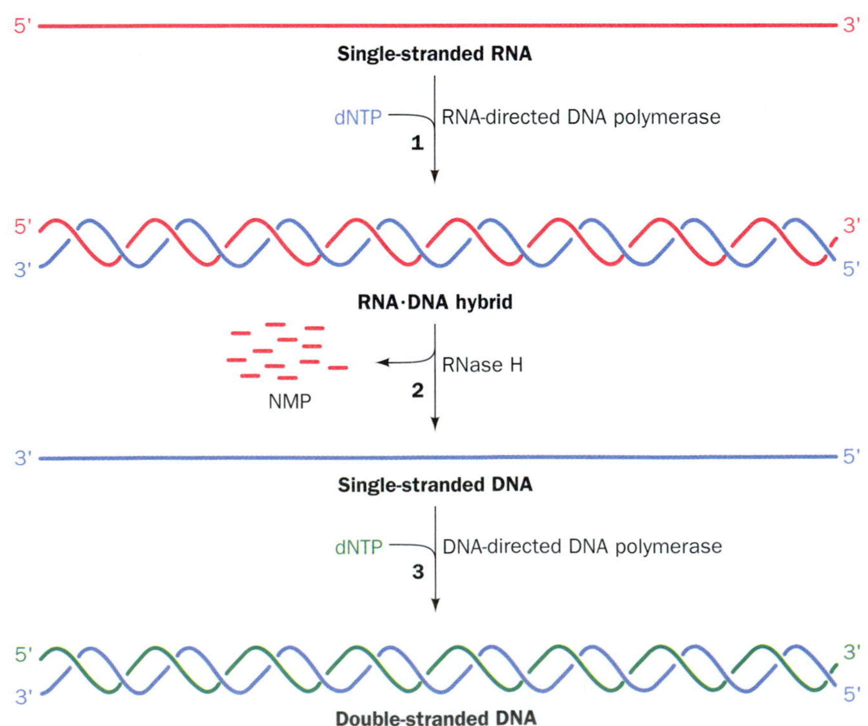

FIGURE 30-45 **The reactions catalyzed by reverse transcriptase.**

are RT inhibitors. Unfortunately, resistant strains of HIV-1 arise quite rapidly because RT lacks a proofreading exonuclease function and hence is highly error prone. Thus, as we have seen (Section 15-4C), effective long-term anti-HIV therapy requires the concurrent administration of at least one RT inhibitor and an HIV protease inhibitor.

The X-ray structure of RT complexed to an 18-bp DNA with a 1-nt overhang at the 5′ end of one strand was determined by Edward Arnold (Fig. 30-46). This complex also contains a monoclonal **Fab fragment** (the antigen-binding segment of an immunoglobulin; Section 35-2B) that specifically binds to RT and presumably facilitated the crystallization of the complex. Steitz independently determined the X-ray structure of RT in the absence of DNA.

The two RT structures are closely similar, although there appear to be shifts in some secondary structural elements, particularly those that contact the DNA and the Fab fragment. The polymerase domains of p66 and p51 each contain four subdomains, which, because of their collective resemblance in p66 to DNA polymerases, are named, from N- to C-terminus, "fingers," "palm," "thumb," and "connection." In p66, the RNase H domain follows the connection.

p51 has undergone a remarkable conformational change relative to p66: The connection has rotated by 155° and translated by 17 Å to bring it from a position in p66 in which it contacts the RNase H domain (Fig. 30-46*a*) to one in p51 in which it contacts all three other polymerase subdomains (Fig. 30-46*b*). This permits p66 and p51 to bring different surfaces of their connections into juxtaposition to form, in part, RT's DNA-binding groove. Thus, the chemically identical polymerase domains of p66 and p51 are not related by 2-fold molecular symmetry (a rare but not unprecedented phenomenon), but, rather, associ-

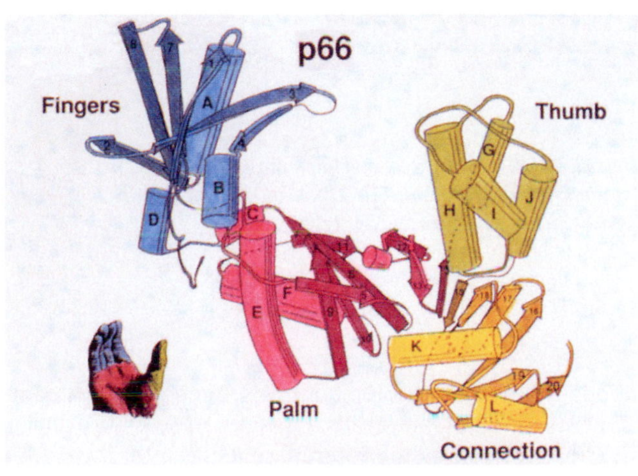

(*a*)

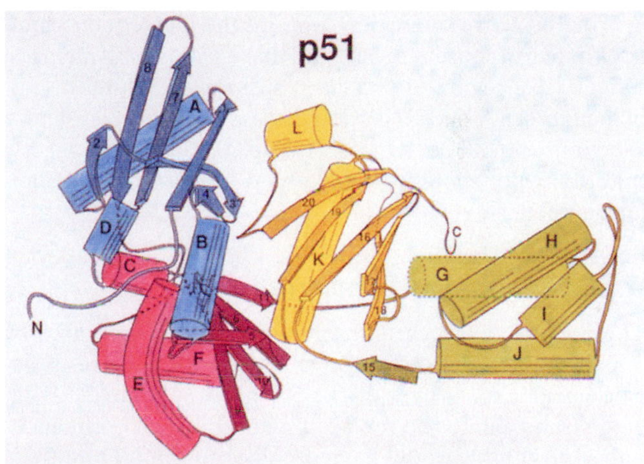

(*b*)

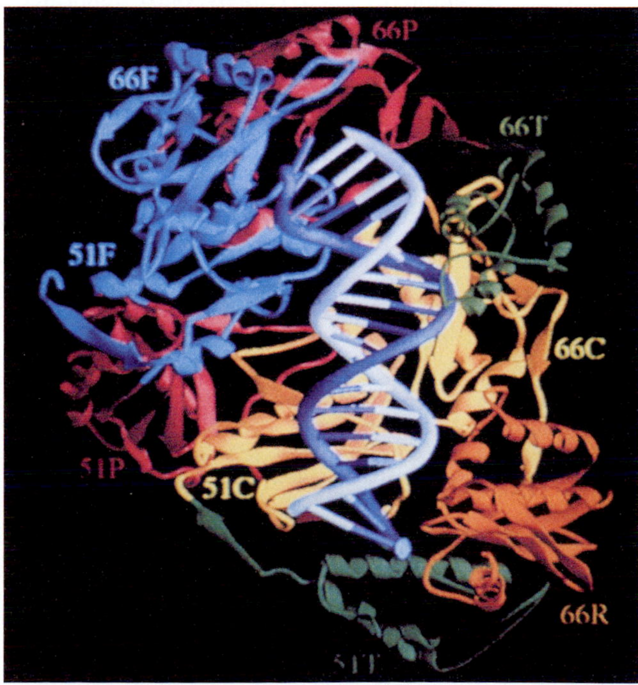

(*c*)

FIGURE 30-46 X-Ray structure of HIV-1 reverse transcriptase. (*a*) A tube-and-arrow representation of the p66 subunit's polymerase domain in which the N-terminal finger subdomain is blue, the palm is pink, the thumb is green, and the connection is yellow. The RNase H domain (not shown) follows the connection. (*b*) The p51 subunit with its pink palm subunit oriented identically to that in p66. Note the different relative orientations of the four subdomains in the two subunits. The G helix is shown in dashed outline because its electron density is weak and ambiguous. (*c*) A ribbon diagram of the HIV-1 RT p66/p51 heterodimer in complex with DNA. The subdomains of p66 and p51 are colored as in Parts *a* and *b* and the RNase H subdomain of p66 is orange [the labels indicate subunit and (sub)domain; e.g., 51F and 66R denote the p51 finger subdomain and the p66 RNase H domain]. The DNA is shown in ladder representation with the 18-nt primer strand white and the 19-nt template strand blue. The complex is oriented with its p66 polymerase domain toward the top of the figure and viewed from above the protein's template–primer binding cleft (whose floor is largely composed of the connection subdomains of p66 and p51) so as to show the bend in the DNA. [Parts *a* and *b* courtesy of Thomas Steitz, Yale University. PDBid 3HVT. Part *c* courtesy of Edward Arnold, Rutgers University. PDBid 2HMI.] **See the Interactive Exercises**

ate in a sort of head-to-tail arrangement. Consequently, RT has only one polymerase active site and one RNase H active site. This is an example of viral genetic economy: HIV-1, with its limited genome size, has succeeded in using a single polypeptide for what are essentially two different functions.

The DNA assumes a conformation that, near the polymerase active site, resembles A-DNA but, near the RNase H domain, more closely resembles B-DNA (Fig. 30-46*c*), a phenomenon that also has been observed in several structures of DNA polymerases in their complexes with DNA (Section 30-2A). The 3′-OH group at the end of the 18-nt DNA strand, the so-called primer strand, is near p66's three catalytically essential Asp side chains, where it is properly positioned to nucleophilically attack the α phosphate of an incoming dNTP that had been shown to bind near this site. Most of the protein–DNA interactions involve the DNA's sugar–phosphate backbone and the residues of p66's palm, thumb, and fingers.

The RT active site region contains the few sequence motifs that are conserved among the various polymerases. Indeed, this region of p66 has a striking structural resemblance to DNA and RNA polymerases of known structure (Sections 30-2A, 30-4B, 31-2A, and 31-2F). This suggests that other polymerases are likely to bind DNA in a similar manner.

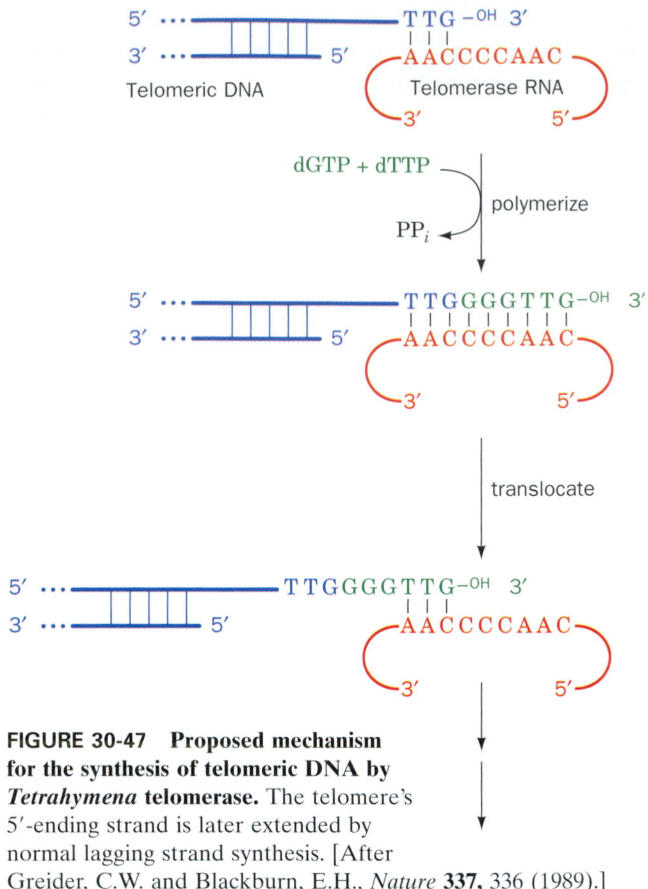

FIGURE 30-47 Proposed mechanism for the synthesis of telomeric DNA by *Tetrahymena* telomerase. The telomere's 5′-ending strand is later extended by normal lagging strand synthesis. [After Greider, C.W. and Blackburn, E.H., *Nature* **337,** 336 (1989).]

D. *Telomeres and Telomerase*

The ends of linear chromosomes cannot be replicated by any of the mechanisms we have yet considered. This is because the RNA primer at the 5′ end of a completed lagging strand cannot be replaced with DNA; the primer required to do this would have no place to bind. How, then, are the DNA sequences at the ends of eukaryotic chromosomes, the **telomeres** (Greek: *telos,* end), replicated?

Telomeric DNA has an unusual sequence: It consists of up to several thousand tandem repeats of a simple, species-dependent, G-rich sequence concluding the 3′-ending strand of each chromosomal terminus. For example, the ciliated protozoan *Tetrahymena* has the repeating telomeric sequence TTGGGG, whereas in all vertebrates it is TTAGGG. Moreover, this strand ends with an overhang that varies from ~20 nt in yeast to ~200 bp in humans.

Elizabeth Blackburn has shown that telomeric DNA is synthesized by a novel mechanism. The enzyme that synthesizes the G-rich strand of telomeric DNA is named **telomerase.** *Tetrahymena* telomerase, for example, adds tandem repeats of the telomeric sequence TTGGGG to the 3′ end of any G-rich telomeric oligonucleotide independently of any exogenously added template. A clue as to how this occurs came from the discovery that telomerases are ribonucleoproteins whose RNA components contain a segment that is complementary to the repeating telomeric sequence. This sequence apparently acts as a template in a kind of reverse transcriptase reaction that synthesizes the telomeric sequence, translocates to the DNA's new 3′ end, and repeats the process (Fig. 30-47). This hypothesis is confirmed by the observation that mutationally altering the telomerase RNA gene segment complementary to telomere DNA results in telomere DNA with the corresponding altered sequence. In fact, telomerase's protein component is homologous to known reverse transcriptases. The DNA strand complementary to the telomere's G-rich strand is apparently synthesized by the normal cellular machinery for lagging strand synthesis, thereby accounting for the 3′ overhang of the G-rich strand.

a. Telomeres Must Be Capped

Without the action of telomerase, a chromosome would be shortened at both ends by 50 to 100 nt with every cycle of DNA replication and cell division. It was therefore initially assumed that, in the absence of active telomerase, essential genes located near the ends of chromosomes would eventually be lost, thereby killing the descendents of the originally affected cells. However, it is now evident that telomeres serve a vital chromosomal function that is compromised before this can happen. Free DNA ends, which are subject to nuclease degradation, trigger DNA damage repair systems that normally function to rejoin the ends of broken chromosomes (as well as cell cycle arrest until this has happened). Thus exposed telomeric DNA would result in the end-to-end fusion of chromosomes, a process that leads to chromosomal instability and eventual cell death [fused chromosomes often break in mitosis (their two centromeres may cause them to be pulled in opposite directions), activating DNA damage checkpoints]. However, in a process known as **capping,** telomeric DNA is specifically bound by proteins that sequester the DNA ends. There is mounting evidence that capping is a dynamic process in which the probability of a telomere spontaneously upcapping increases as telomere length decreases. Since most somatic cells in multicellular organisms lack telomerase activity, this explains why such cells in culture can only undergo a limited number of doublings (20–60) before they reach senescence (a stage in which they cease dividing) and eventually die (Section 19-3B). Indeed, otherwise immortal *Tetrahymena* cultures with mutationally impaired telomerases exhibit characteristics reminiscent of senescent mammalian cells before dying off. Apparently, *the loss of telomerase function in somatic cells is a basis for aging in multicellular organisms.*

b. Telomere Length Correlates with Aging

There is strong experimental evidence in support of this theory of aging. The analysis of cultured human fibroblasts from a number of donors between 0 and 93 years old indicates that there is only a weak correlation between the proliferative capacity of a cell culture and the age of its donor. There is, however, a strong correlation, valid over the entire donor age range, between the initial telomere length in a cell culture and its proliferative capacity. Thus, cells that initially have relatively short telomeres undergo significantly fewer doublings than cells with longer telomeres. Moreover, fibroblasts from individuals with **progeria** (a rare disease characterized by rapid and premature aging resulting in childhood death) have short telomeres, an observation that is consistent with their known reduced proliferative capacity in culture. In contrast, sperm (which, being germ cells, are in effect immortal) from donors ranging in age from 19 to 68 years had telomeres that did not vary in length with donor age, which indicates that telomerase is active at some stage of germ cell growth. Likewise, those few cells in a culture that become immortal (capable of unlimited proliferation) exhibit an active telomerase and a telomere of stable length, as do the cells of unicellular eukaryotes (which are also immortal). It therefore appears that telomere erosion is a significant cause of cellular senescence and hence aging.

c. Cancer Cells Have Active Telomerases

What advantage might multicellular organisms gain by eliminating telomerase activity in their somatic cells? An intriguing possibility is that cellular senescence is a mechanism that protects multicellular organisms from cancer. The two defining characteristics of cancer cells are that they are immortal and grow uncontrollably (Sections 19-3B and 34-4C). If mammalian cells were normally immortal, the incidence of cancer would probably be far greater than it is since immortalization, which requires an active telomerase, is a major step toward **malignant transformation** (cancer formation), which requires several independent genetic changes (Section 19-3B). Indeed, nearly all human cancers exhibit high telomerase activity. Moreover, as Robert Weinberg demonstrated, human fibroblasts in culture can be malignantly transformed by the acquisition of only three genes, those encoding: (1) **TERT,** the protein subunit of telomerase (its 451-nt RNA subunit, **TR,** is normally expressed in somatic cells), (2) an oncogenic variant of H-Ras (an essential participant in intracellular signal transduction pathways; Section 19-3C), and (3) the SV40 **large-T antigen** [SV40 is a tumor virus whose large-T antigen binds and functionally inactivates the tumor suppressor proteins known as **Rb** and **p53** (Section 34-4C; it also functions as a helicase in viral DNA replication)]. This suggests that telomerase inhibitors may be effective antitumor agents.

d. Telomeric DNA Can Dimerize via G-Quartets

It has long been known that guanine forms strong Hoogsteen-type base pairs (Table 29-2) that can further associate to form cyclic tetramers known as **G-quartets** (Fig. 30-48*a*). Indeed, G-rich polynucleotides are notoriously difficult to work with because of their propensity to ag-

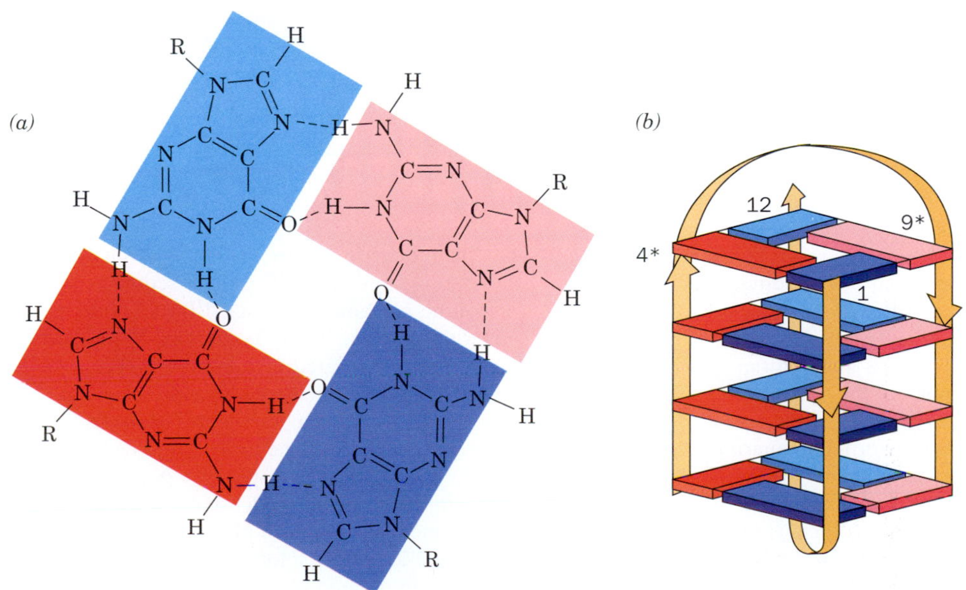

FIGURE 30-48 NMR structure of the telomeric oligonucleotide d(GGGGTTTTGGGG). (*a*) The base pairing interactions in the G-quartet at the end of the quadruplex in solution. (*b*) Schematic diagram of the NMR solution structure, in which the strand directions are indicated by arrows. The nucleotides are numbered 1 to 12 in one strand and 1* to 12* in the symmetry-related strand. Guanine residues G1 to G4 are represented by dark blue rectangles, G8 to G12 are light blue, G1* to G4* are red, and G9* to G12* are pink. [After Schultze, P., Smith, F.W., and Feigon, J., *Structure* **2,** 227 (1994). PDBid 156D.]

gregate. The G-rich overhanging strands of telomeres dimerize to form stable complexes in solution, presumably via the formation of G-quartet-containing structures.

The 3′-terminal telomeric overhang of the ciliated protozoan *Oxytricha nova* has the sequence $d(T_4G_4)_2$, which resembles the repeating telomeric sequences of other organisms. The NMR structure of the dodecamer $d(G_4T_4G_4)$, determined by Juli Feigon (Fig. 30-48*b*), reveals that each oligonucleotide folds back on itself to form a hairpin, two of which associate in an antiparallel fashion to form a structure that contains four stacked G-quartets, with the T_4 sequences forming the loops at the ends of each stack.

The **telomere end binding protein (TEBP)** of *O. nova* is a heterodimeric capping protein that binds to and protects the foregoing 3′ overhang. The X-ray structure of TEBP in complex with $d(G_4T_4G_4)$, determined by Steve Schultz, reveals that the DNA binds in a deep cleft between the protein's α and β subunits, where it adopts an irregular nonhelical conformation (Fig. 30-49). In addition, two other $d(G_4T_4G_4)$ molecules form a G-quartet–linked dimer with the same conformation they adopt in solution (Fig. 30-48). The G-quartet assembly fits snugly into a small positively charged cavity formed by the N-terminal domains of three symmetry-related (in the crystal) α subunits at sites distinct from their ssDNA binding sites. The presence of both the ssDNA and the G-quartet assembly in the X-ray structure supports the hypothesis that multiple DNA structures and, in particular, G-quartets, play a role in telomere biology. Nevertheless, no obvious TEBP homologs occur in yeast or vertebrates. However, both humans and fission yeast express a telomere end-binding protein named **Pot1** (for *p*rotection *o*f *t*elomeres), whose deletion causes rapid loss of telomeric DNA and chromosomal end joining.

e. Telomeres Form T-Loops

Mammalian telomeric DNA is also capped by two related proteins, **TRF1** and **TRF2** (TRF for *t*elomere

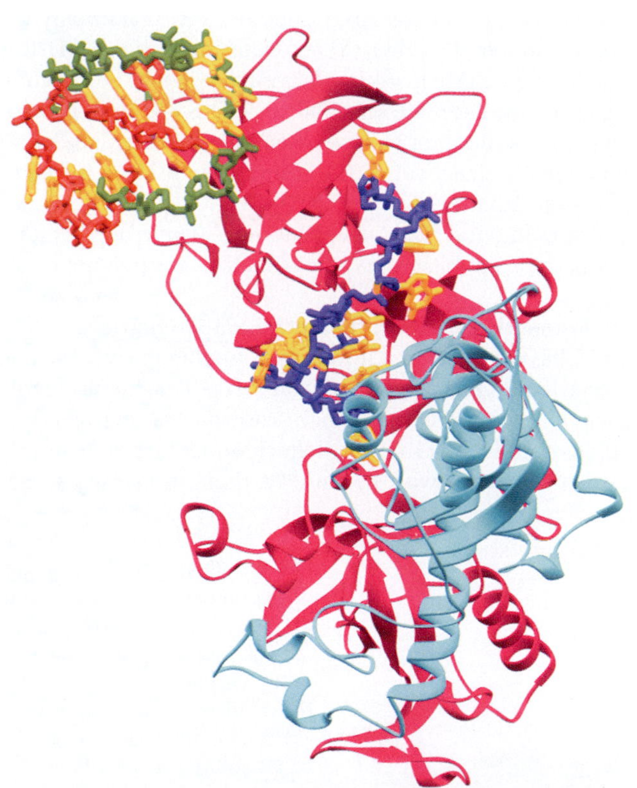

FIGURE 30-49 X-Ray structure of *Oxytricha nova* telomere end binding protein (TEBP) in complex with $d(G_4T_4G_4)$. The TEBP is drawn in ribbon form with its α and β subunits magenta and cyan. The DNA is drawn in stick form with its bases gold, the sugar–phosphate backbone of the single strand that binds in a cleft between the protein's α and β subunits blue, and the backbones of two strands that form a G-quartet–linked dimer red and green. The G-quartet–linked dimer binds in a cavity formed by the N-terminal domains of three symmetry related α chains, although only one of them is shown here. [Based on an X-ray structure by Steve Schultz, University of Colorado. PDBid 1JB7.]

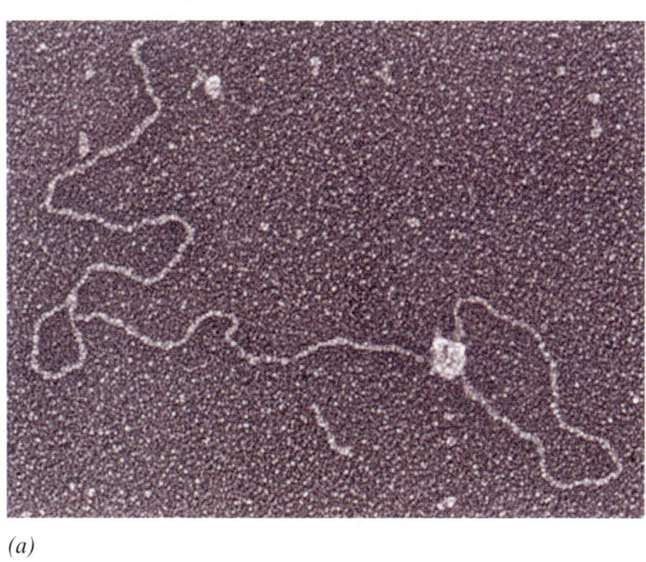

(*a*)

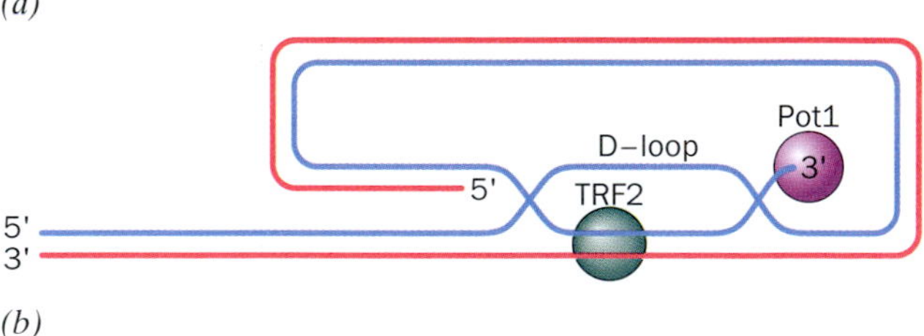

(*b*)

FIGURE 30-50 The telomeric T-loop. (*a*) An electron micrograph of a dsDNA consisting of a 3-kb unique sequence followed by ~2 kb of the repeating sequence TTAGGG on the strand that ends with a 150- to 200-nt 3′ overhang. This model telomeric DNA was then incubated with human TRF2. [Courtesy of Jack Griffith, University of North Carolina at Chapel Hill.] (*b*) The proposed structure of a T-loop. In a process that is mediated by TRF2, the repeating TTAGGG sequence in the DNA's 3′ overhang displaces a portion of the same strand (*blue*) in the double-stranded region of the DNA to form a duplex segment with the complementary strand (*red*), thereby generating a D-loop. The telomere end-binding protein Pot1 specifically binds to the end of the 3′ overhang.

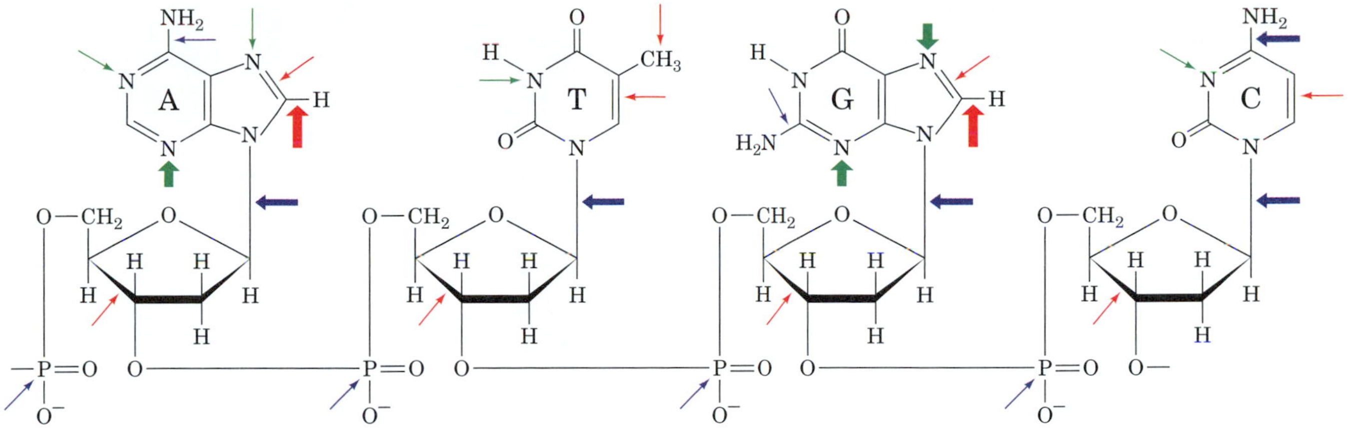

FIGURE 30-51 The types and sites of chemical damage to which DNA is normally susceptible *in vivo*. Red arrows indicate sites subject to oxidative attack, blue arrows indicate sites subject to spontaneous hydrolysis, and green arrows indicate sites subject to nonenzymatic methylation by *S*-adenosylmethionine. The width of an arrow is indicative of the relative frequency of the reaction. [After Lindahl, T., *Nature* **362,** 709 (1993).]

*r*epeat-binding *f*actor). Jack Griffith and Titia de Lange have shown through electron microscopy (EM) studies that, in the presence of TRF2, otherwise linear telomeric DNA forms large duplex end loops named **T-loops** (Fig. 30-50*a*). Moreover, the EM of DNA from mammalian telomeres, whose strands had been chemically cross-linked to preserve their structural relationships on deproteination, likewise revealed the presence of abundant T-loops of varying sizes. These observations suggest that T-loops are formed by the TRF2-induced invasion of the 3′ telomeric overhang into the repeating telomeric dsDNA (Fig. 30-50*b*) to form a D-loop (Section 30-4B). T-loops have also been observed in protozoa, suggesting that T-loops are a conserved feature of eukaryotic telomeres. TRF1 is implicated in controlling telomeric length, presumably by somehow limiting the number of TRF1 molecules that can bind to a telomere.

5 ■ REPAIR OF DNA

DNA is by no means the inert substance that might be supposed from naive consideration of genome stability. Rather, the reactive environment of the cell, the presence of a variety of toxic substances, and exposure to UV or ionizing radiation subjects it to numerous chemical insults that excise or modify bases and alter sugar–phosphate groups (Fig. 30-51). Indeed, some of these reactions occur at surprisingly high rates. For example, under normal physiological conditions, the glycosidic bonds of ~10,000 of the 3.2 billion purine nucleotides in each human cell hydrolyze spontaneously each day.

Any DNA damage must be repaired if the genetic message is to maintain its integrity. Such repair is possible because of duplex DNA's inherent information redundancy. The biological importance of DNA repair is indicated by the identification of at least 130 genes in the human genome that participate in DNA repair and by the great variety of DNA repair pathways possessed by even relatively simple organisms such as *E. coli*. In fact, *the major DNA repair processes in eukaryotic cells and E. coli are chemically quite similar.* These processes are outlined in this section.

A. *Direct Reversal of Damage*

a. Pyrimidine Dimers Are Split by Photolyase

UV radiation of 200 to 300 nm promotes the formation of a cyclobutyl ring between adjacent thymine residues on the same DNA strand to form an intrastrand **thymine dimer** (Fig. 30-52). Similar cytosine and thymine–cytosine

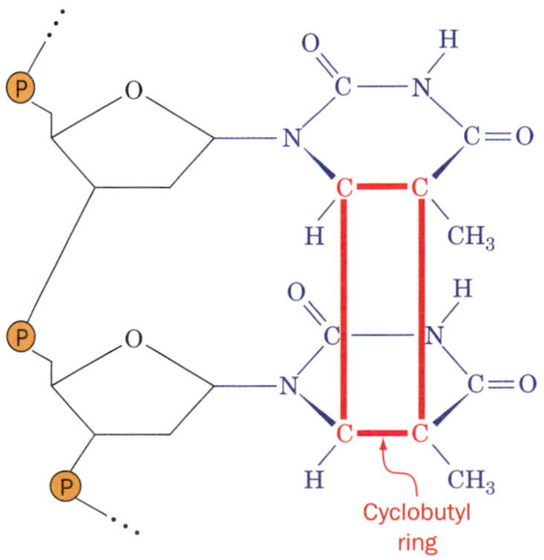

FIGURE 30-52 The cyclobutylthymine dimer that forms on UV irradiation of two adjacent thymine residues on a DNA strand. The ~1.6-Å-long covalent bonds joining the thymine rings (*red*) are much shorter than the normal 3.4-Å spacing between stacked rings in B-DNA, thereby locally distorting the DNA.

dimers are likewise formed but at lesser rates. Such **pyrimidine dimers** locally distort DNA's base paired structure such that it can be neither transcribed nor replicated. Indeed, a single thymine dimer, if unrepaired, is sufficient to kill an *E. coli.*

Pyrimidine dimers may be restored to their monomeric forms through the action of light-absorbing enzymes named **photoreactivating enzymes** or **DNA photolyases** that are present in many prokaryotes and eukaryotes (including goldfish, rattlesnakes, and marsupials, but not placental mammals). These enzymes are 55- to 65-kD monomers that bind to a pyrimidine dimer in DNA, a process that can occur in the dark. A noncovalently bound chromophore, in some species an N^5,N^{10}-methenyltetrahydrofolate (**MTHF;** Fig. 26-49) and in others a **5-deazaflavin,**

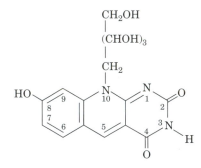

8-Hydroxy-7,8-didemethyl-5-deazariboflavin

then absorbs 300- to 500-nm light and transfers the excitation energy to a noncovalently bound $FADH^-$, which in turn transfers an electron to the pyrimidine dimer, thereby splitting it. Finally, the resulting pyrimidine anion rereduces the FADH· and the now unblemished DNA is released, thereby completing the catalytic cycle. DNA photolyases bind either dsDNA or ssDNA with high affinity but without regard to base sequence.

The X-ray structure of the 471-residue *E. coli* DNA photolyase, determined by Johann Deisenhofer, reveals that its MTHF and flavin rings are ~17 Å apart, which permits efficient energy transfer between them. The enzyme's putative DNA binding site (Fig. 30-53) is a positively charged flat surface that is penetrated by a hole that has the size and polarity complementary to that of a pyrimidine dimer–containing dinucleotide. A pyrimidine dimer bound in the hole would be in contact with the flavin ring and hence properly situated for efficient electron transfer from this ring system. This implies that a pyrimidine dimer in a double helix must flip (swing) out of the helix in order to interact with the enzyme. This flip-out is probably facilitated by the relatively weak base pairing interactions of the pyrimidine dimer and the distortions it imposes on the double helix. In the following discussions we shall see that **base flipping** is by no means an unusual process for enzymes that perform chemistry on the bases in dsDNA.

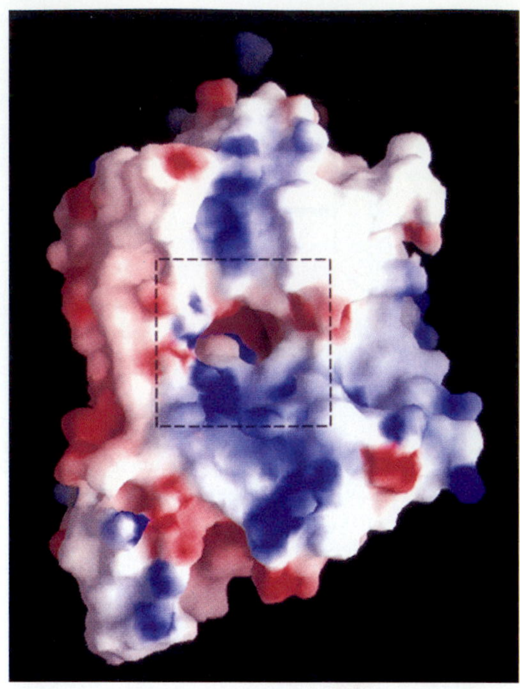

FIGURE 30-53 X-Ray structure of *E. coli* DNA photolyase showing its putative DNA binding surface. The enzyme is represented by its solvent-accessible surface, which is colored according to its electrostatic potential with blue most positive, red most negative, and white nearly neutral. The dashed box encloses the hole in the protein surface that is the presumed pyrimidine dimer binding site. [Courtesy of Johann Diesenhofer, Texas Southwestern Medical Center, Dallas, Texas. PDBid 1DNP.]

b. Alkyltransferases Dealkylate Alkylated Nucleotides

The exposure of DNA to alkylating agents such as ***N*-methyl-*N*′-nitro-*N*-nitrosoguanidine (MNNG)**

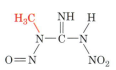

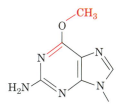

***N*-Methyl-*N*′-nitro-*N*-nitrosoguanidine (MNNG)**

***O*⁶-Methylguanine residue**

yields, among other products, ***O*⁶-alkylguanine** residues. The formation of these derivatives is highly mutagenic because on replication, they frequently cause the incorporation of thymine instead of cytosine.

***O*⁶-Methylguanine** and ***O*⁶-ethylguanine** lesions of DNA in all species tested are repaired by ***O*⁶-alkylguanine–DNA alkyltransferase,** which directly transfers the offending alkyl group to one of its own Cys residues. The reaction inactivates this protein, which therefore cannot be strictly

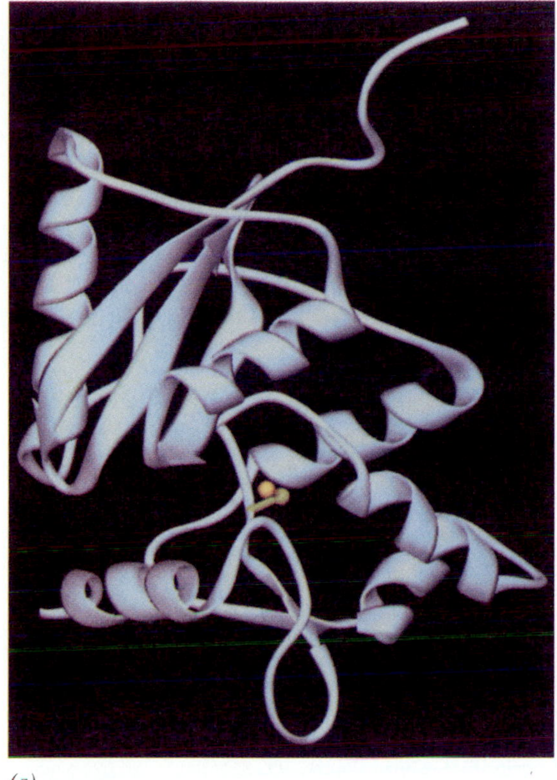

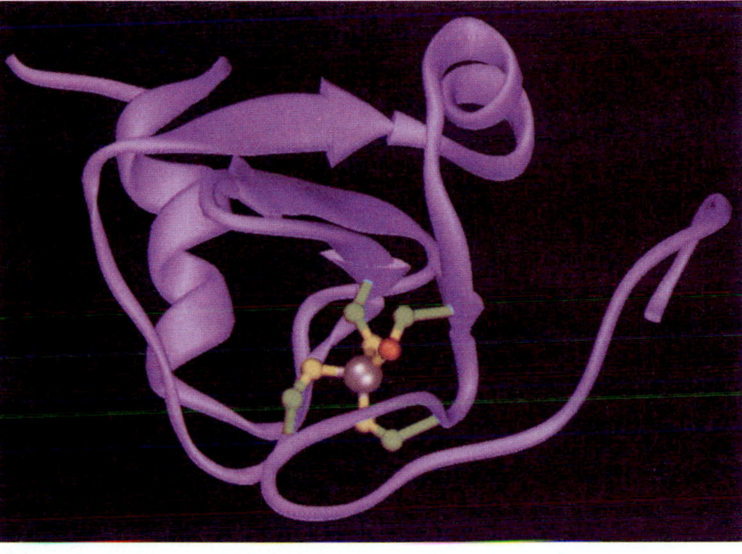

(a) (b)

FIGURE 30-54 The structure of *E. coli* Ada protein. (*a*) The X-ray structure of Ada's 178-residue C-terminal segment, which contains its O^6-alkylguanine–DNA alkyltransferase function. The side chain of Cys 146 (Cys 321 in the intact protein), to which the methyl group is irreversibly transferred, is shown in ball-and-stick form with C green and S yellow. Note that this residue is almost entirely buried within the protein. [Based on an X-ray structure determined by Eleanor Dodson and Peter Moody, University of York, U.K. PDBid 1SFE.] (*b*) The NMR structure of Ada's 92-residue, N-terminal segment, which mediates its methyl phosphotriester repair function. The protein's bound Zn^{2+} ion is represented by a silver sphere and its four tetrahedrally coordinating Cys side chains are shown in ball-and-stick form, with C green and S yellow except for the orange S atom of Cys 69, which becomes irreversibly methylated when the protein encounters a methylated phosphate group on DNA. [Based on an NMR structure determined by Gregory Verdine and Gerhard Wagner, Harvard University. PDBid 1ADN.]

classified as an enzyme. The alkyltransferase reaction has elicited considerable attention because carcinogenesis induced by methylating and ethylating agents is correlated with deficient repair of O^6-alkylguanine lesions.

The *E. coli* O^6-alkylguanine–DNA alkyltransferase activity occurs on the 178-residue C-terminal segment of the 354-residue **Ada protein** (the product of the ***ada*** gene). Its X-ray structure (Fig. 30-54*a*), determined by Eleanor Dodson and Peter Moody, reveals, unexpectedly, that its active site Cys residue, Cys 321, is buried inside the protein. Apparently, the protein must undergo a significant conformation change on DNA binding in order to effect the methyl transfer reaction.

Ada protein's 92-residue N-terminal segment has an independent function: It repairs methyl phosphotriesters in DNA (methylated phosphate groups) by irreversibly transferring the offending methyl group to its Cys 69. The NMR structure of Ada's N-terminal domain (Fig. 30-54*b*), determined by Gregory Verdine and Gerhard Wagner, reveals that Cys 69, together with three other Cys residues, tetrahedrally coordinates a Zn^{2+} ion. This presumably stabilizes the thiolate form of Cys 69 over its thiol form, thereby facilitating its nucleophilic attack on the methyl group.

Intact Ada protein that is methylated at its Cys 69 binds to a specific DNA sequence, which is located upstream of the *ada* gene and several other genes encoding DNA repair proteins, thereby inducing their transcription. Evidently, Ada also functions as a chemosensor of methylation damage.

B. *Excision Repair*

Cells employ two types of excision repair mechanisms: (1) **nucleotide excision repair (NER),** which functions to repair relatively bulky DNA lesions; and (2) **base excision repair (BER),** which repairs nonbulky lesions involving a single base.

a. Nucleotide Excision Repair

NER is a DNA repair mechanism found in all cells that eliminates damage to dsDNA by excising an oligonucleotide containing the lesion and filling in the resulting single-strand gap. NER repairs lesions that are characterized by the displacement of bases from their normal positions, such as pyrimidine dimers, or by the addition of a bulky substituent to a base. This system appears to be activated by a helix distortion rather than by the recognition of any particular group. In humans, NER is the major defense against two important carcinogens, sunlight and cigarette smoke. The mechanism of NER in prokaryotes is similar to that in eukaryotes. However, prokaryotic NER employs 3 subunits, whereas eukaryotic NER involves the actions of 16 subunits. The eukaryotic proteins are conserved from yeast to humans but none of them exhibit any sequence similarity to the prokaryotic proteins, suggesting that the two NER systems arose by convergent evolution.

In *E. coli,* NER is carried out in an ATP-dependent process through the actions of the **UvrA, UvrB,** and **UvrC** proteins (the products of the *uvrA, uvrB,* and *uvrC* genes). This system, which is often referred to as the **UvrABC endonuclease** (although, as we shall see, there is no complex that contains all three subunits), cleaves the damaged DNA strand at the seventh and at the third or fourth phosphodiester bonds from the lesion's 5′ and 3′ sides, respectively (Fig. 30-55). The excised 11- or 12-nt oligonucleotide is displaced by the binding of **UvrD** (also called **helicase II**) and replaced through the actions of Pol I and DNA ligase.

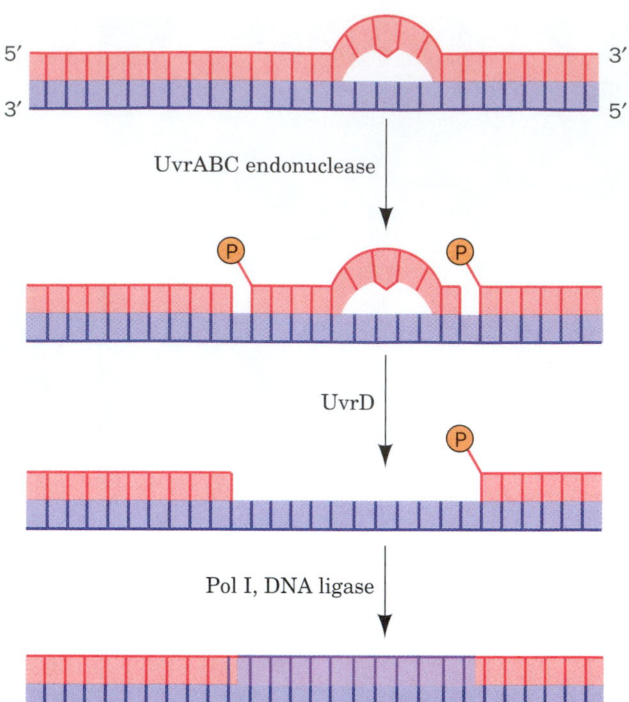

FIGURE 30-55 The mechanism of nucleotide excision repair (NER) of pyrimidine photodimers.

The mechanism of prokaryotic NER was elucidated mainly by Aziz Sancar. It begins with the damage recognition step in which a $(UvrA)_2UvrB$ heterotrimer binds tightly although nonspecifically to dsDNA, which it probes for damage according to its local propensity for bending and unwinding. The presence of a lesion activates the helicase function of UvrB to unwind 5 bp around the lesion in an ATP-driven process. This conformation change induces the dissociation of the UvrA from the complex, which allows the binding of UvrC. UvrB then makes the incision on the 3′ side of the lesion following which UvrC makes the incision on its 5′ side. UvrD binds to the resulting nicks in the DNA, which displaces UvrC and the lesion-containing oligomer. This makes the 5′ incision site accessible to Pol I, which fills in the gap and displaces UvrB. Finally, DNA ligase seals the remaining nick yielding refurbished DNA.

b. Xeroderma Pigmentosum and Cockayne Syndrome Are Caused by Genetically Defective NER

In humans, the rare inherited disease **xeroderma pigmentosum (XP;** Greek: *xeros,* dry + *derma,* skin) is mainly characterized by the inability of skin cells to repair UV-induced DNA lesions. Individuals suffering from this autosomal recessive condition are extremely sensitive to sunlight. During infancy they develop marked skin changes such as dryness, excessive freckling, and keratoses (a type of skin tumor; the skin of these children is described as resembling that of farmers with many years of sun exposure), together with eye damage, such as opacification and ulceration of the cornea. Moreover, they develop often fatal skin cancers at a 2000-fold greater rate than normal and internal cancers at a 10 to 20-fold increased rate. Curiously, many individuals with XP also have a bewildering variety of seemingly unrelated symptoms including progressive neurological degeneration and developmental deficits.

Cultured skin fibroblasts from individuals with xeroderma pigmentosum are defective in the NER of pyrimidine dimers. Cell-fusion experiments with cultured cells taken from various patients have demonstrated that this disease results from defects in any of 8 complementation groups (Section 1-4C), indicating that there must be at least 8 gene products, XPA through XPG and XPV, involved in this clearly important UV damage repair pathway. **Cockayne syndrome (CS),** an inherited disease which is also associated with defective NER, arises from defects in XPB, XPD, and XPG as well as in two additional complementation groups, CSA and CSB. Individuals with CS are hypersensitive to UV radiation and exhibit stunted growth as well as neurological dysfunction due to neuron demyelination, but, intriguingly, have a normal incidence of skin cancer. What is the biochemical basis for the diverse group of symptoms associated with impaired NER?

The free radicals produced by oxidative metabolism can damage DNA. Some of these oxidative lesions are repaired via NER. Since neurons have high rates of respiration and are long-lived nondividing cells, it seems likely that they would be particularly susceptible to oxidative damage in the absence of NER. This would explain the progressive neurological deterioration in XP.

The retarded development typical of XPB defects and perhaps the demyelination that occurs in CS appear to be due more to impaired transcription than to defective NER. Moreover, pyrimidine dimers are more efficiently removed from transcribed portions of DNA than from unexpressed sequences. These observations are explained by the discovery that some or all of the subunits of the eukaryotic transcription factor **TFIIH,** a helicase that participates in the initiation of mRNA transcription by RNA polymerase II (Section 34-3B), are required for NER. The coupling of NER and transcription also occurs in *E. coli* in which **Mfd** protein displaces RNA polymerase that has stalled on a damaged template strand (which it cannot transcribe), following which the Mfd recruits the proteins of the UvrABC system to the damage site.

c. Base Excision Repair

DNA bases are modified by reactions that occur under normal physiological conditions as well as through the action of environmental agents. For example, adenine and cytosine residues spontaneously deaminate at finite rates to yield hypoxanthine and uracil residues, respectively. *S*-Adenosylmethionine (SAM), a common metabolic methylating agent (Section 26-3E), occasionally nonenzymatically methylates a base to form derivatives such as 3-methyladenine and 7-methylguanine residues (Fig. 30-51). Ionizing radiation can promote ring opening reactions in bases. Such changes modify or eliminate base pairing properties.

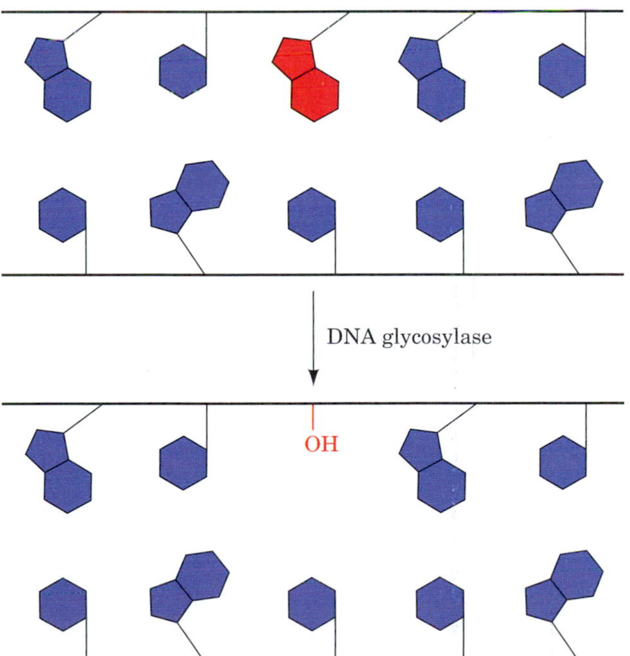

FIGURE 30-56 Action of DNA glycosylases. These enzymes hydrolyze the glycosidic bond of their corresponding altered base (*red*) to yield an AP site.

DNA containing a damaged base may be restored to its native state through base excision repair (BER). Cells contain a variety of **DNA glycosylases** that each cleave the glycosidic bond of a corresponding specific type of altered nucleotide (Fig. 30-56), thereby leaving a deoxyribose residue in the backbone. Such **apurinic** or **apyrimidinic (AP) sites** (also called **abasic sites**) are also generated under normal physiological conditions by the spontaneous hydrolysis of a glycosidic bond. The deoxyribose residue is then cleaved on one side by an **AP endonuclease,** the deoxyribose and several adjacent residues are removed by the action of a cellular exonuclease (possibly associated with a DNA polymerase), and the gap is filled in and sealed by a DNA polymerase and DNA ligase.

d. Uracil in DNA Would Be Highly Mutagenic

For some time after the essential functions of nucleic acids had been elucidated, there seemed no apparent reason for nature to go to the considerable metabolic effort of using thymine in DNA and uracil in RNA when these substances have virtually identical base pairing properties. This enigma was solved by the discovery of cytosine's penchant for conversion to uracil by deamination, either via spontaneous hydrolysis (Fig. 30-51), which is estimated to occur ~120 times per day in each human cell, or by reaction with nitrites (Section 32-1A). If U were the normal DNA base, the deamination of C would be highly mutagenic because there would be no indication of whether the resulting mismatched G · U base pair had initially been G · C or A · U. *Since T is DNA's normal base, however, any U in DNA is almost certainly a deaminated C.* U's that occur in DNA are efficiently excised by **uracil–DNA glycosylase [UDG;** also called **uracil *N*-glycosylase (UNG)]** and then replaced by C through BER.

UDG also has an important function in DNA replication. dUTP, an intermediate in dTTP synthesis, is present in all cells in small amounts (Section 28-3B). DNA polymerases do not discriminate well between dUTP and dTTP (recall that DNA polymerases select a base for incorporation into DNA according to its ability to base pair with the template; Section 30-2A) so that, despite the low dUTP level that cells maintain, newly synthesized DNA contains an occasional U. These U's are rapidly replaced by T through BER. However, since excision occurs more rapidly than repair, all newly synthesized DNA is fragmented. When Okazaki fragments were first discovered (Section 30-1C), it therefore seemed that all DNA was synthesized discontinuously. This ambiguity was resolved with the discovery of *E. coli* defective in UDG. In these ung^- mutants, only about half of the newly synthesized DNA is fragmented, strongly suggesting that DNA's leading strand is synthesized continuously.

e. Uracil–DNA Glycosylase Induces Uridine Nucleotides to Flip Out

The X-ray structure of human UDG in complex with a 10-bp DNA containing a U · G mismatch (which can form a doubly hydrogen bonded base pair whose shape differs from that of Watson–Crick base pairs; Section 32-2D), determined by John Tainer, reveals that the UDG has bound the DNA with its uridine nucleotide flipped out of the dsDNA (Fig. 30-57). Moreover, the enzyme has hydrolyzed uridine's glycosidic bond yielding the free uracil base and an AP site on the DNA, although both remain bound to the enzyme. The cavity in the DNA's base stack that would otherwise be occupied by the flipped out uracil is filled by the side chain of Leu 272, which intercalates into the DNA from its minor groove side. The X-ray structure of a similar complex in which the U · G mismatch was replaced by a U · A base pair contained essentially identical features. However, when the U in the U · A-containing complex was replaced by **pseudouridine** (in which the "glycosidic" bond is made to uracil's C5 atom rather than to N1),

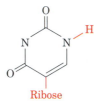

O, N, N—H, O, Ribose

Pseudouridine

the uracil remained covalently linked to the DNA because the UDG could not hydrolyze its now C—C "glycosidic" bond.

How does UDG detect a base paired uracil in the center of DNA and how does it discriminate so acutely between uracil and other bases, particularly the closely similar thymine? The above X-ray structures indicate that the phosphate groups flanking the flipped out nucleotide are 4 Å closer together than they are in B-DNA (8 Å vs 12 Å), which causes the DNA to kink by ~45° in the direction parallel to the view in Fig. 30-57. These distortions arise from the binding of three rigid protein loops to the DNA, which would be unable to simultaneously bind to undistorted B-DNA. This led Tainer to formulate the "pinch–push–pull" mechanism for uracil detection in which he postulated that UDG rapidly scans a DNA for uracil by periodically binding to it so as to compress and thereby slightly bend the DNA's backbone (pinch). The DNA's presumed low resistance to bending at a uracil-containing site (a U · G base pair is smaller than C · G and hence leaves a space in the base stack, whereas a U · A base pair is even weaker than T · A) permits the enzyme to flip out the uracil by intercalating Leu 272 into the minor groove (push), thereby fully bending and kinking the DNA. This process is aided by the tight binding of the flipped out uracil to the enzyme (pull). The exquisite specificity of this binding pocket for uracil prevents the binding and hence hydrolysis of any other base that the enzyme may have induced to flip out. Thus the overall shapes of adenine and guanine exclude them from this pocket, whereas thymine's 5-methyl group is sterically blocked by the rigidly held side chain of Tyr 147. Cytosine, which has approximately the same shape as uracil, is excluded through a set of hydrogen bonds emanating from the protein that mimic those made by adenine in a Watson–Crick A · U base pair.

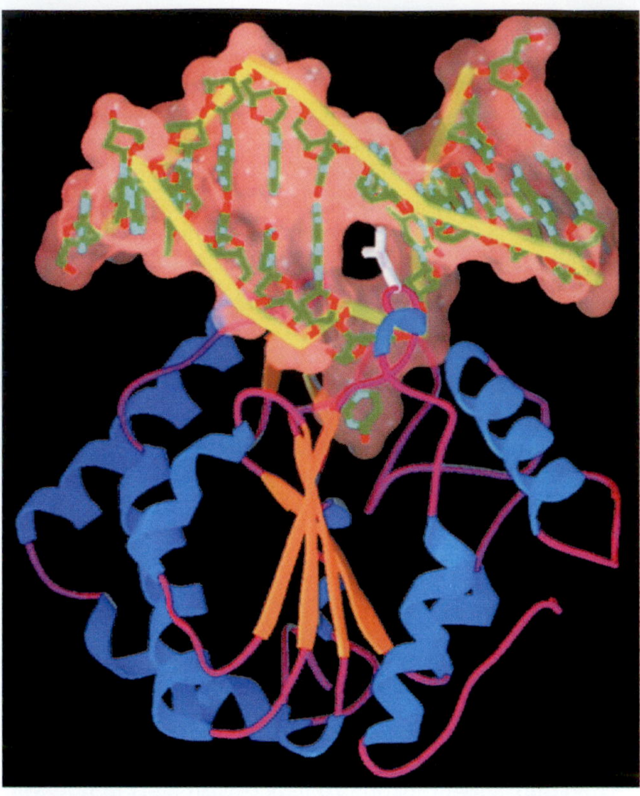

FIGURE 30-57 X-Ray structure of human uracil–DNA glycosylase (UDG) in complex with a 10-bp DNA containing a U · G base pair. The protein (the C-terminal 223 residues of the 304-residue monomer) is colored according to its secondary structure (helices blue, β strands orange, and other segments magenta). The DNA, viewed looking into its minor groove, is drawn in stick form colored according to atom type (C green, N cyan, O red) and with its phosphate backbone traced by yellow tubes (phosphate O atoms have been omitted for clarity). The DNA's transparent solvent-accessible surface is pink and the side chain of Leu 272 is white. The uridine nucleotide can be seen to have flipped out of the double helix (below the DNA) and to have been hydrolyzed to yield an AP nucleotide and uracil, which remains bound in UDG's binding pocket. The side chain of Leu 272 has intercalated into the DNA base stack to fill the space vacated by the flipped out uracil base. [Courtesy of John Tainer, The Scripps Research Institute, La Jolla, California. PDBid 4SKN.]

AP sites in DNA are highly cytotoxic because they irreversibly trap mammalian topoisomerase I in its covalent complex with DNA (Section 29-3C). Moreover, since the ribose at the AP site lacks a glycosidic bond, it can readily convert to its linear form (Section 11-1B), whose reactive aldehyde group can cross-link to other cell components. This rationalizes why AP sites remain tightly bound to UDG in solution as well as in crystals. UDG activity is enhanced by AP endonuclease, the next enzyme in the BER pathway, but the two enzymes do not interact in the absence of DNA. This suggests that UDG remains bound to an AP site it generated until it is displaced by the more tightly binding AP endonuclease, thereby protecting the cell from the AP site's cytotoxic effects. It seems likely that other damage-specific DNA glycosylases function similarly.

C. *Mismatch Repair*

Any replicational mispairing that has eluded the editing functions of the various participating DNA polymerases may still be corrected by a process known as **mismatch repair (MMR).** For example, *E. coli* Pol I and Pol III have error rates of 10^{-6} to 10^{-7} per base pair replicated but the observed mutational rates in *E. coli* are 10^{-9} to 10^{-10} per base pair replicated. In addition, the MMR system can correct insertions or deletions of up to 4 nt (which arise from the slippage of one strand relative to the other in the active site of DNA polymerase). The importance of MMR is indicated by the fact that defects in the human MMR system result in a high incidence of cancer, most notably **hereditary nonpolyposis colorectal cancer (HNPCC;** which affects several organs and may be the most common inherited predisposition to cancer).

If an MMR system is to correct errors in replication rather than perpetuate them, it must distinguish the parental DNA, which has the correct base, from the daughter strand, which has an incorrect although normal base. In *E. coli,* as we have seen (Section 30-3C), this is possible because newly replicated GATC palindromes remain hemimethylated until the Dam methyltransferase has had sufficient time to methylate the daughter strand.

E. coli mismatch repair, which was elucidated in large part by Paul Modrich, requires the participation of three proteins and occurs as follows (Fig. 30-58):

1. **MutS** (853 residues) binds to a mismatched base pair or unpaired bases as a homodimer.

2. The MutS–DNA complex binds **MutL** (615 residues), also as a homodimer.

3. The MutS–MutL complex translocates along the DNA in both directions, thereby forming a loop in the DNA. The translocation appears to be driven by the ATPase function of MutS.

4. On encountering a hemimethylated GATC palindrome, the MutS–MutL complex recruits **MutH** (228 residues) and activates this single strand endonuclease to make a nick on the 5′ side of the unmethylated GATC. This GATC may be located on either side of the mismatch and over 1000 bp distant from it.

5. MutS–MutL recruits UvrD helicase, which in concert with an exonuclease, separates the strands and degrades the nicked strand from the nick to beyond the mismatch. If the nick is on the 3′ side of the mismatch as shown, the exonuclease is **exonuclease I** (a 3′ → 5′ exonuclease), whereas if the nick is on the 5′ side of the mismatch, the exonuclease can be either **RecJ** or **exonuclease VII** (both 5′ → 3′ exonucleases).

The resulting gap is filled in by Pol III and sealed by DNA ligase, thereby correcting the mismatch. MutL is also an ATPase, which, it is postulated, functions to coordinate the various steps of mismatch repair.

Eukaryotic MMR systems are, not surprisingly, more complicated than those of *E. coli.* Eukaryotes express six homologs of MutS and five homologs of MutL that form heterodimers on mismatched DNA. However, homologs of MutH only occur in gram-negative bacteria. Eukaryotes must have some other way of differentiating the parental and daughter DNA strands. Perhaps a newly synthesized daughter strand is identified by its as-yet unsealed nicks.

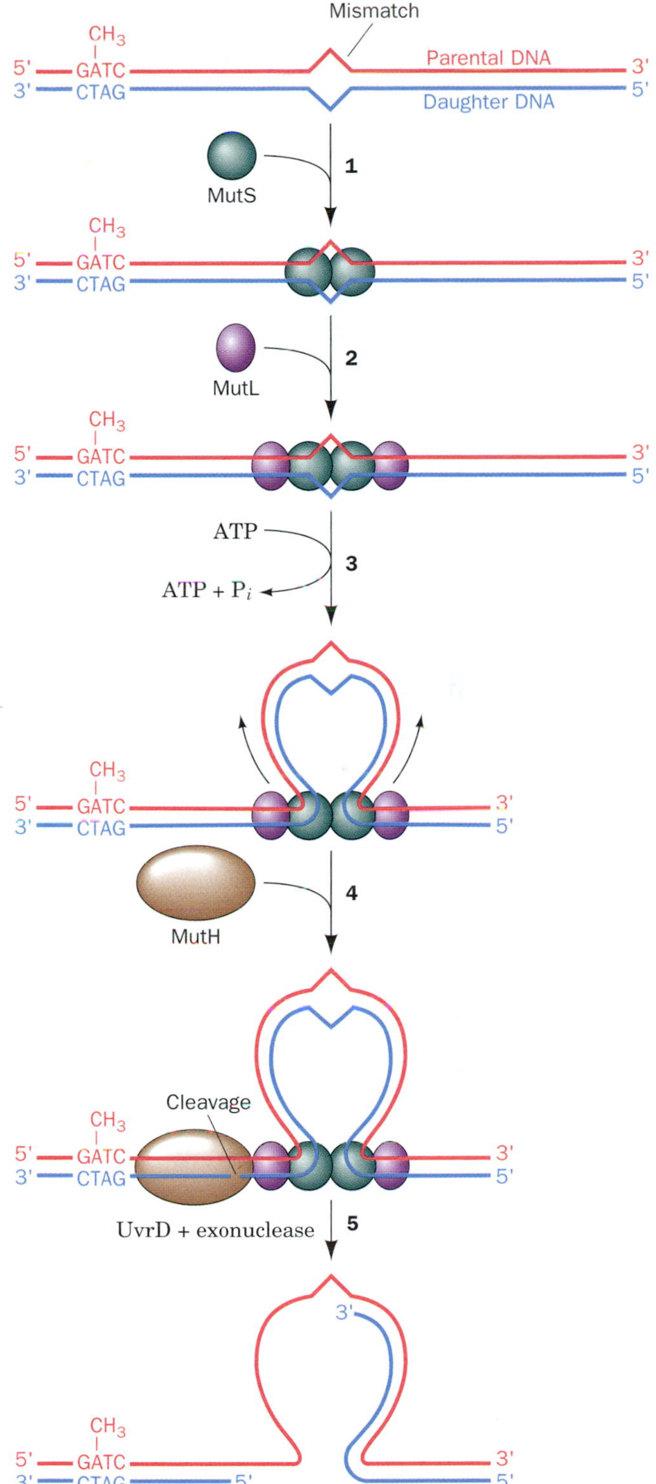

FIGURE 30-58 **The mechanism of mismatch repair in *E. coli*.**

D. *The SOS Response*

Agents that damage DNA, such as UV radiation, alkylating agents, and cross-linking agents, induce a complex system of cellular changes in *E. coli* known as the **SOS response.** *E. coli so treated cease dividing and increase their capacity to repair damaged DNA.*

a. LexA Protein Represses the SOS Response

Clues as to the nature of the SOS response were provided by the observations that *E. coli* with mutant *recA* or *lexA* genes have their SOS response permanently switched on. **RecA,** a 352-residue protein that coats DNA as a multimeric helical filament, plays a central role, as we shall see, in homologous recombination (Section 30-6A). When *E. coli* are exposed to agents that damage DNA or inhibit DNA replication, their RecA specifically mediates the proteolytic cleavage of **LexA** (202 residues) between its Asp 84 and Gly 85. RecA is activated to do so on binding to ssDNA (it was initially assumed that RecA catalyzes the proteolysis of LexA but subsequent experiments by John Little indicate that activated RecA stimulates LexA to cleave itself). Further investigations indicated that LexA functions as a repressor of 43 genes that participate in DNA repair and the control of cell division, including *recA, lexA, uvrA,* and *uvrB.* DNA sequence analyses of the LexA-repressible genes revealed that they are all preceded by a homologous 20-nt sequence, the so-called **SOS box,** that has the palindromic symmetry characteristic of operators (control sites to which repressors bind so as to interfere with transcriptional initiation by RNA polymerase; Section 5-4A). Indeed, LexA has been shown to specifically bind the SOS boxes of *recA* and *lexA.*

The preceding observations suggest a model for the regulation of the SOS response (Fig. 30-59). During normal growth, LexA largely represses the expression of the SOS

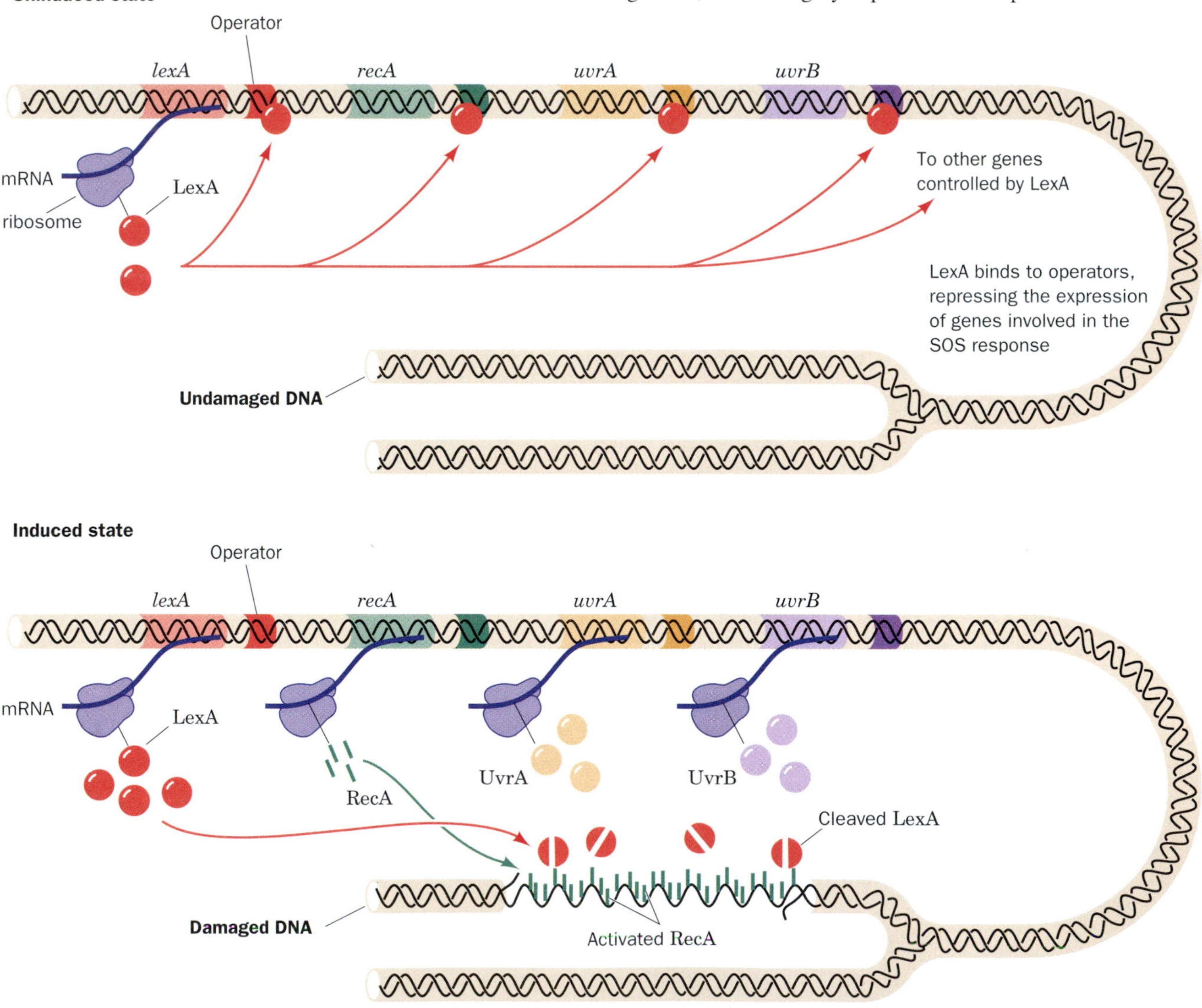

FIGURE 30-59 Regulation of the SOS response in *E. coli.* In a cell with undamaged DNA (*above*), LexA largely represses the synthesis of LexA, RecA, UvrA, UvrB, and other proteins involved in the SOS response. When there has been extensive DNA damage (*below*), RecA is activated by binding to the resulting single-stranded DNA to stimulate LexA self-cleavage. The consequent synthesis of the SOS proteins results in the repair of the DNA damage.

genes, including the *lexA* gene, by binding to their SOS boxes so as to inhibit RNA polymerase from initiating the transcription of these genes. When DNA damage has been sufficient to produce postreplication gaps, however, this ssDNA binds to RecA so as to stimulate LexA cleavage. The LexA-repressible genes are consequently released from repression and direct the synthesis of SOS proteins including that of LexA (although this repressor continues to be cleaved through the influence of RecA). When the DNA lesions have been eliminated, RecA ceases stimulating LexA's autoproteolysis. The newly synthesized LexA can then function as a repressor, which permits the cell to return to normality.

b. SOS Repair Is Error Prone

The *E. coli* Pol III holoenzyme is unable to replicate through a variety of lesions such as AP sites and thymine dimers. On encountering such lesions, the replisome stalls and disassembles by releasing its Pol III cores, a process that is called replication fork "collapse." Cells have two general modes for restoring collapsed replication forks, **recombination repair** and **SOS repair.** Recombination repair circumvents the damaged template by using a homologous chromosome as its template DNA in a process known as **homologous recombination,** which also functions to generate genetic diversity. Hence we shall postpone our discussion of recombination repair until after our consideration of homologous recombination in Section 30-6A. In the following paragraphs we discuss SOS repair.

In SOS repair, the Pol III core lost from the collapsed replication fork is replaced by one of two so-called **bypass DNA polymerases,** whose synthesis is induced by the SOS response: **DNA polymerase IV** (**Pol IV,** the 336-residue product of the *dinB* gene) or **DNA polymerase V** [**Pol V;** the heterotrimeric product of the *umuD* and *umuC* genes, **UmuD′$_2$C** (umu for *U*V *mu*tagenesis), where UmuD′ is produced by the RecA-assisted self-cleavage of the 139-residue **UmuD** to remove its N-terminal 24 residues, and UmuC consists of 422 residues]. Both of these enzymes are Y-family DNA polymerases, all of whose members lack $3' \rightarrow 5'$ proofreading exonuclease activity and replicate undamaged DNA with poor fidelity and low processivity and hence are also known as **error-prone DNA polymerases.**

Translesion synthesis (TLS) by Pol V, which was characterized in large part by O'Donnell and Myron Goodman, requires the simultaneous presence of the β_2 sliding clamp, the γ complex (clamp loader), and SSB, together with a RecA filament in complex with the ssDNA arising from the action of helicase on the dsDNA ahead of the stalled replication fork. This so-called **Pol V mutasome** tends to incorporate G about half as often as A opposite thymine dimers and AP sites, with pyrimidines being installed infrequently. This process is, of course, highly mutagenic. But even in replicating undamaged DNA, Pol V is at least 1000-fold more error prone than is Pol I or Pol III holoenzyme. However, after synthesizing ~7 nt, the Pol V mutasome is replaced by Pol III holoenzyme, which commences normal DNA replication after the now bypassed lesion. Pol II, a TLS participant that accurately replicates DNA, is also induced by the SOS response but it is synthesized well before Pol V appears (see below). The role of Pol II appears to be the mediation of error-free TLS, and only if this process fails, is it replaced by Pol V to carry out error-prone TLS.

There are numerous types of DNA lesions besides AP sites and thymine dimers that interfere with normal DNA replication. Depending on the type of lesion, Pol IV, which is also error prone, may instead be recruited to carry out TLS. With many lesions, TLS may skip over the altered nucleotide, resulting in deletion of one or two bases in the daughter strand opposite the lesion (yielding a **frameshift mutation,** so called because it would change a structural gene's reading frame from that point onward; Section 5-4B). Moreover, Pol IV is prone to generating frameshift mutations even when replicating undamaged DNA.

The Y-family DNA polymerase **Dpo4** from the archaebacterium *Sulfolobus solfataricus* P2, a homolog of *E. coli* Pol IV and Pol V, misincorporates ~1 base per 500 replicated nucleotides. The X-ray structure of a complex of Dpo4 with a primer–template DNA that had been incubated with ddATP (which is complementary to the template base), determined by Wei Yang, reveals the structural basis for this low fidelity (Fig. 30-60). The 352-residue protein contains the fingers, palm, and thumb domains common to all known DNA polymerases (although their orders differ in the sequences of the different families of DNA polymerases) and, in addition, has a C-terminal domain unique to Y-family DNA polymerases that has been

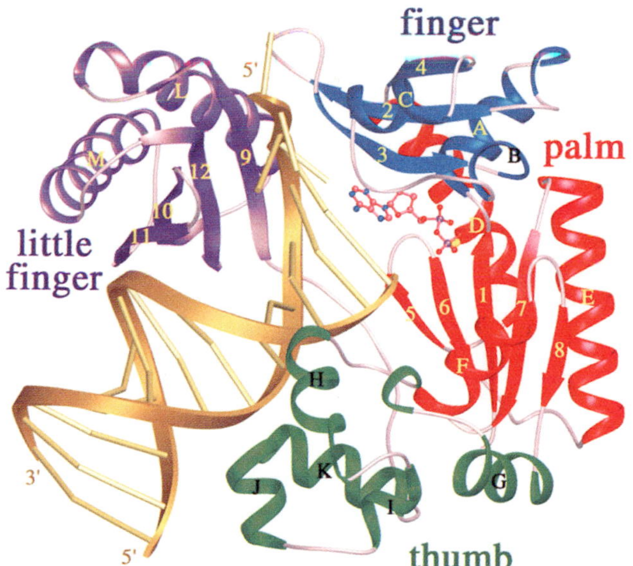

FIGURE 30-60 X-Ray structure of the bypass DNA polymerase Dpo4 from *Sulfolobus solfataricus* P2 in complex with a primer–template DNA and ddADP. The protein is drawn in ribbon form with its fingers, palm, thumb, and little finger domains blue, red, green, and purple, respectively. The DNA is gold with its backbones drawn as ribbons and its bases represented by rods. The ddADP, which is base paired to a template T in the enzyme's active site, is shown in ball-and-stick form colored according to atom type (C pink, N blue, O red, and P magenta). [Courtesy of Wei Yang, NIH, Bethesda, Maryland. PDBid 1JX4.]

dubbed the "little finger" domain. The enzyme, as expected, has incorporated a ddA residue at the 3′ end of the primer and, in addition, binds a ddADP in base paired complex to the new template T. The little finger domain binds in the major groove of the DNA. However, the fingers and thumb domains are small and stubby compared to those of replicative DNA polymerases such as Klentaq1 (Fig. 30-9) and RB69 pol (Fig. 30-39), and the residues that contact the base pair in the active site are all Gly and Ala rather than the Phe, Tyr, and Arg that mainly do so in the replicative DNA polymerases. Moreover, the bound DNA is entirely in the B form rather than in the A form at the active site as occurs in replicative DNA polymerases. Since the minor groove is more accessible in A-DNA than in B-DNA (Section 29-1B), this suggests that error-prone DNA polymerases have relatively little facility to monitor the base pairing fidelity of the incoming nucleotide. This accounts for the ability of error-prone DNA polymerases to accommodate distorted template DNA as well as non-Watson–Crick base pairs at their active sites.

SOS repair is an error-prone and hence mutagenic process. It is therefore a process of last resort that is only initiated ~50 min after SOS induction if the DNA has not already been repaired by other means. Yet, DNA damage that normally activates the SOS response is nonmutagenic in the *recA*$^-$ *E. coli* that survive. This is, as we saw, because bypass DNA polymerases will replicate over a DNA lesion even when there is no information as to which bases were originally present. Indeed, *most mutations in E. coli arise from the actions of the SOS repair system,* which is therefore a testimonial to the proposition that survival with a chance of loss of function (and the possible gain of new ones) is advantageous, in the Darwinian sense, over death, although only a small fraction of cells actually survive this process. It has therefore been suggested that, under conditions of environmental stress, the SOS system functions to increase the rate of mutation so as to increase the rate at which the *E. coli* adapt to the new conditions. Finally, it should be noted that the eukaryotic pols η, ι, and κ, all Y-family members, and pol ζ, an X-family member, are implicated in TLS and that pol η, the product of the *XPV* gene, is defective in the XPV form of xeroderma pigmentosum (Section 30-5B).

E. *Double-Strand Break Repair*

Double-strand breaks **(DSBs)** in DNA are produced by ionizing radiation and the free radical by-products of oxidative metabolism. Moreover, DSBs are normal intermediates in certain specialized cellular processes such as meiosis (Section 1-4A) and **V(D)J recombination** in lymphoid cells, which helps generate the vast diversity of antigen-binding sites in antibodies and *T*-cell receptors (Section 35-2C). Unrepaired or misrepaired DSBs can be lethal to cells or cause chromosomal aberrations that may lead to cancer. Hence the efficient repair of DSBs is essential for cell viability and genomic integrity.

Cells have two general modes to repair DSBs, recombination repair and **nonhomologous end-joining (NHEJ).**

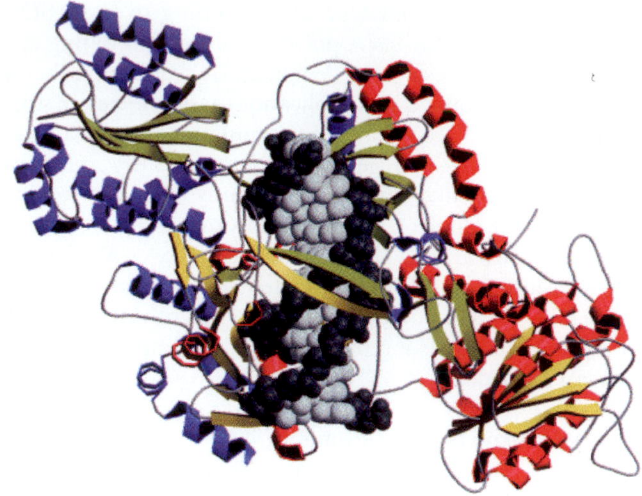

FIGURE 30-61 X-Ray structure of human Ku protein in complex with DNA containing 14 bp. The subunits of Ku70 (*red helices and yellow strands*) and Ku80 (*blue helices and green strands*) are viewed along the pseudo-2-fold axis relating them. The DNA, viewed with its DSB pointing upward, is drawn in space-filling form with its sugar–phosphate backbone dark gray and its base pairs light gray. Note that the DNA is surrounded by a ring of protein. [Courtesy of John Tainer, The Scripps Research Institute, La Jolla, California. Based on an X-ray Structure by Jonathan Goldberg, Memorial Sloan-Kettering Cancer Center, New York, New York. PDBid 1JEY.]

Here we discuss NHEJ, a process which, as its name implies, directly rejoins DSBs. The recombination repair of DSBs is discussed in Section 30-6A.

In NHEJ, the broken ends of the DSB must be aligned, its frayed ends trimmed and/or filled in, and their strands ligated. The core NHEJ machinery in eukaryotes includes the DNA end-binding protein **Ku** (a heterodimer of homologous 70- and 83-kD subunits, **Ku70** and **Ku80**), **DNA ligase IV,** and the accessory protein **Xrcc4.** Ku, an abundant nuclear protein, binds to a DSB, whether blunt or with an overhang, and hence appears to be the cell's primary DSB sensor. The X-ray structure of Ku in complex with a 14-bp DNA, determined by Jonathan Goldberg, reveals that the protein cradles the dsDNA segment along its entire length and encircles its central ~3 bp segment (Fig. 30-61). The protein ring is also present in the closely similar X-ray structure of Ku alone, thereby explaining why Ku that is bound to a dsDNA which is then circularized becomes permanently associated with it. Ku makes no specific contacts with the DNA's bases and few with its sugar–phosphate backbone, but instead fits snugly into the DNA's major and minor grooves so as to precisely orient it.

Ku–DNA complexes have been shown to dimerize so as to align the members of a DSB, both blunt ended and with short (1–4 bp) complementary single strands, for ligation as is diagrammed in Fig. 30-62. The DNA ends are exposed along one face of each Ku–DNA complex, presumably making them accessible to polymerases that fill in gaps and to nucleases that trim excess and inappropriate ends preparatory for ligation by DNA ligase IV in complex with Xrcc4. Nucleotide trimming, which of course generates mutations, appears to be carried out in an ATP-dependent manner by the evolutionarily conserved **Mre11 complex,** which consists of two **Mre11** nuclease subunits and two **Rad50** ATPase subunits. Ku is eventually released from the rejoined DNA, perhaps by proteolytic cleavage.

F. *Identification of Carcinogens*

Many forms of cancer are known to be caused by exposure to certain chemical agents that are therefore known as ***carcinogens.*** It has been estimated that as much as 80% of human cancer arises in this fashion. There is considerable evidence that the primary event in carcinogenesis is often damage to DNA (carcinogenesis is discussed in Section 34-4C). Carcinogens are consequently also likely to induce the SOS response in bacteria and thus act as indirect mutagenic agents. In fact, there is a high correlation between carcinogenesis and mutagenesis (recall, e.g., the progress of xeroderma pigmentosum; Section 30-5B).

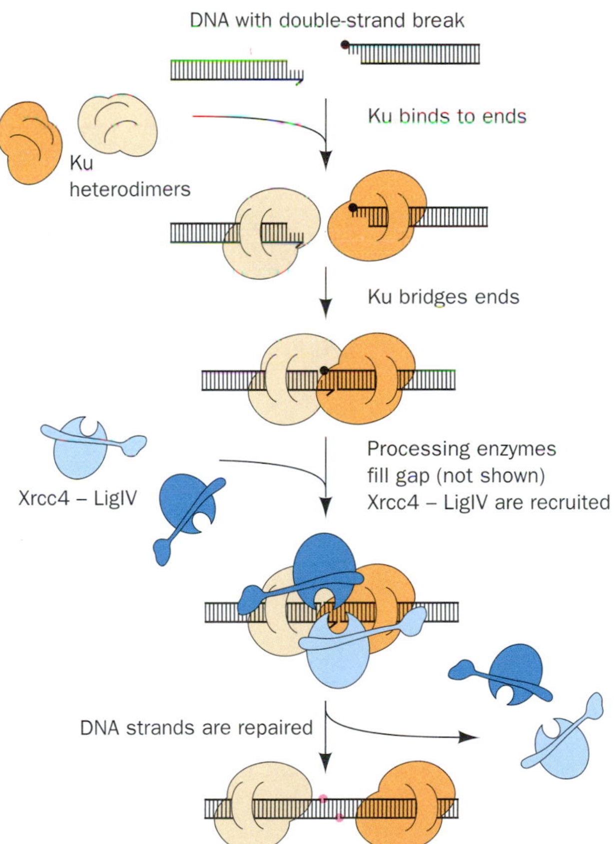
DNA with double-strand break
Ku binds to ends
Ku
heterodimers
Ku bridges ends
Processing enzymes
fill gap (not shown)
Xrcc4 – LigIV are recruited
Xrcc4 – LigIV
DNA strands are repaired

FIGURE 30-62 Schematic diagram of nonhomologous end-joining (NHEJ). The left dsDNA fragment is missing a base and the right fragment is blocked by a nonligatable group (*filled black circle*). The two Ku heterodimers are drawn in two shades of yellow and the Xrcc4–DNA ligase IV complexes are drawn in two shades of blue. The newly repaired links in the DNA are represented by pink circles. [After Jones, J.M., Gellert, M., and Yang, W., *Structure* **9,** 881 (2001).]

There are presently over 60,000 man-made chemicals of commercial importance and ~1000 new ones are introduced each year. The standard animal tests for carcinogenesis, exposing rats or mice to high levels of the suspected carcinogen and checking for cancer, are expensive and require ~3 years to complete. Thus relatively few substances have been tested in this manner.

a. The Ames Test Assays for Probable Carcinogenicity

Bruce Ames devised a rapid and effective bacterial assay for carcinogenicity that is based on the high correlation between carcinogenesis and mutagenesis. He constructed special tester strains of *Salmonella typhimurium* that are *his*$^-$ (cannot synthesize histidine so that they are unable to grow in its absence), have cell envelopes that lack the lipopolysaccharide coating that renders normal *Salmonella* impermeable to many substances (Section 11-3B), and have inactivated excision repair systems. Mutagenesis in these tester strains is indicated by their reversion to the *his*$^+$ phenotype.

In the **Ames test,** ~10^9 tester strain bacteria are spread on a culture plate that lacks histidine. Usually a mixture of several *his*$^-$ strains is used so that mutations due to both base changes and nucleotide insertions or deletions can be detected. A mutagen placed in the culture medium causes some of these *his*$^-$ bacteria to revert to the *his*$^+$ phenotype, which is detected by their growth into visible colonies after 2 days at 37°C (Fig. 30-63). The mutagenicity of a substance is scored as the number of such colonies less the few spontaneously revertant colonies that occur in the absence of the mutagen.

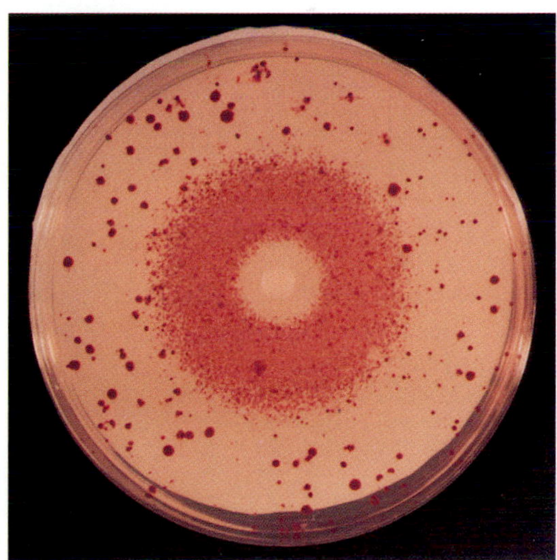

FIGURE 30-63 The Ames test for mutagenesis. A filter paper disk containing a mutagen, in this case the alkylating agent ethyl methanesulfonate, is centered on a culture plate containing *his*$^-$ tester strains of *Salmonella typhimurium* in a medium that lacks histidine. A dense halo of revertant bacterial colonies appears around the disk from which the mutagen diffused. The larger colonies distributed about the culture plate are spontaneous revertants. The bacteria near the disk have been killed by the toxic mutagen's high concentration. [Courtesy of Raymond Devoret, Institut Curie, Orsay, France.]

Many noncarcinogens are converted to carcinogens in the liver or in other tissues via a variety of detoxification reactions (e.g., those catalyzed by the cytochromes P450; Section 15-4B). A small amount of rat liver homogenate is therefore included in the Ames test medium in an effort to approximate the effects of mammalian metabolism.

b. Both Man-Made and Naturally Occurring Substances Can Be Carcinogenic

There is an ~80% correspondence between the compounds determined to be carcinogenic by animal tests and those found to be mutagenic by the Ames test. Dose–response curves, which are generated by testing a given compound at a number of concentrations, are almost always linear and extrapolate back to zero, indicating that *there is no threshold concentration for mutagenesis.* Several compounds to which humans have been extensively exposed that were found to be mutagenic by the Ames test were later found to be carcinogenic in animal tests. These include tris(2,3-dibromopropyl)phosphate, which was used as a flame retardant on children's sleepwear in the mid-1970s and can be absorbed through the skin; and furylfuramide, which was used in Japan in the 1960s and 1970s as an antibacterial additive in many prepared foods (and had passed two animal tests before it was found to be mutagenic). Carcinogens are not confined to man-made compounds but also occur in nature. For example, carcinogens are contained in many plants that are common in the human diet, including alfalfa sprouts. **Aflatoxin B_1,**

O O O O O OCH_3

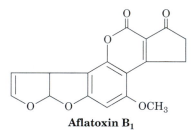

Aflatoxin B_1

one of the most potent carcinogens known, is produced by fungi that grow on peanuts and corn. Charred or browned food, such as occurs on broiled meats and toasted bread, contains a variety of DNA-damaging agents. Thus, with respect to carcinogenesis, as Ames has written, "Nature is not benign."

6 ■ RECOMBINATION AND MOBILE GENETIC ELEMENTS

The chromosome is not just a simple repository of genetic information. If this were so, the unit of mutation would have to be an entire chromosome rather than a gene because there would be no means of separating a mutated gene from the other genes of the same chromosome. Chromosomes would therefore accumulate deleterious mutations until they became nonviable.

It has been known from some of the earliest genetic studies that pairs of allelic genes may exchange chromo-

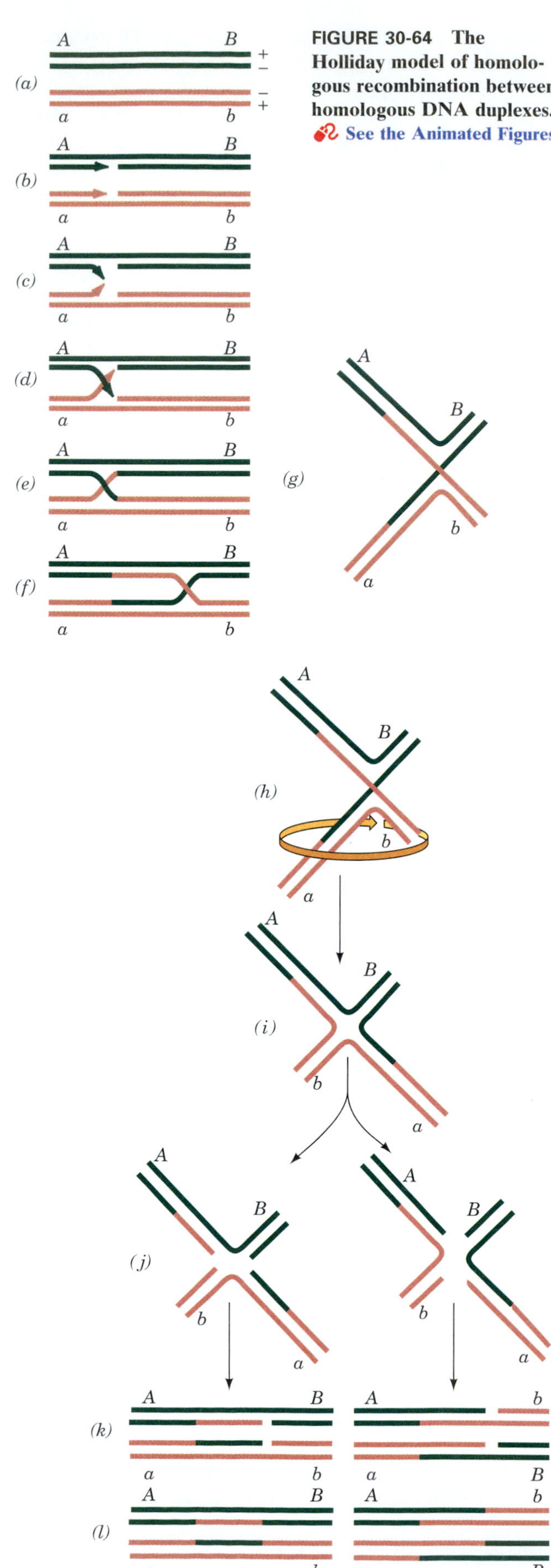

FIGURE 30-64 **The Holliday model of homologous recombination between homologous DNA duplexes.** See the Animated Figures

somal locations by a process known as **genetic recombination** (Section 1-4B). Mutated genes can thereby be individually tested, since their propagation is then not absolutely dependent on the propagation of the genes with which they had been previously associated. In this section, we consider the mechanisms by which genetic elements can move, both between chromosomes and within them.

A. *Homologous Recombination*

Homologous recombination *(also called* ***general recombination****) is defined as the exchange of homologous segments between two DNA molecules.* Both genetic and cytological studies have long indicated that such a crossing-over process occurs in higher organisms during meiosis (Fig. 1-27). Bacteria, which are normally haploid, likewise have elaborate mechanisms for the interchange of genetic information. They can acquire foreign DNA through transformation (Section 5-2A), through a process called **conjugation** (mating) in which DNA is directly transferred from one cell to another via a cytoplasmic bridge (Section 31-1A), and via **transduction** in which a defective bacteriophage that has erroneously acquired a segment of bacterial DNA rather than the viral chromosome transfers this DNA to another bacterial cell. In all of these processes, the foreign DNA is installed in the recipient's chromosome or plasmid through homologous recombination (to be propagated, a DNA segment must be part of a replicon; that is, be associated with a replication origin such as occurs in a chromosome, a plasmid, or a virus).

a. Recombination Occurs via a Crossed-Over Intermediate

The prototypical model for homologous recombination (Fig. 30-64) was proposed by Robin Holliday in 1964 on the basis of genetic studies on fungi. The corresponding strands of two aligned homologous DNA duplexes are nicked, and the nicked strands cross over to pair with the nearly complementary strands of the homologous duplex after which the nicks are sealed (Fig. 30-64*a–e*), thereby yielding a four-way junction known as a **Holliday junction** (alternatively, **Holliday intermediate;** Fig. 30-64*e*). A Holliday junction has, in fact, been observed in the X-ray structure of d(CCGGTACCGG), determined Shing Ho (Fig. 30-65), in which, perhaps unexpectedly, all the bases form normal Watson–Crick base pairs without any apparent strain. The crossover point can move in either direction, often thousands of nucleotides, in a process known

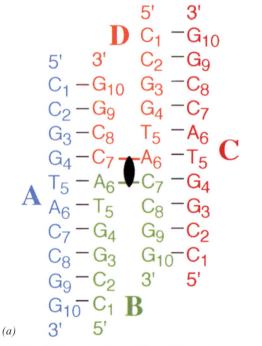

D
5' 3'
$C_1 - G_{10}$
$C_2 - G_9$
$G_3 - C_8$
$G_4 - C_7$
$T_5 - A_6$
$A_6 - T_5$
$C_7 - G_4$
$C_8 - G_3$
$G_9 - C_2$
$G_{10} - C_1$
3' 5'
C

A
5' 3'
$C_1 - G_{10}$
$C_2 - G_9$
$G_3 - C_8$
$G_4 - C_7$
$T_5 - A_6$
$A_6 - T_5$
$C_7 - G_4$
$C_8 - G_3$
$G_9 - C_2$
$G_{10} - C_1$
3' 5'
B

(*a*)

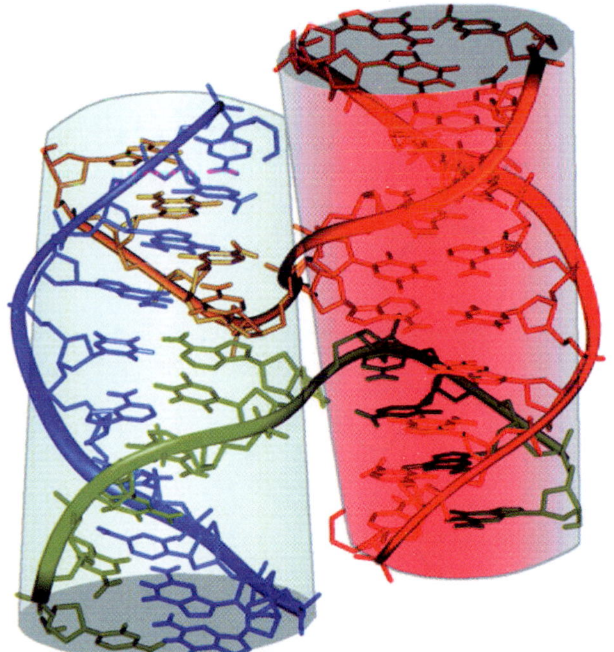

(*b*)

FIGURE 30-65 X-Ray structure of the self-complementary decameric DNA d(CCGGTACCGG). (*a*) The secondary structure of the four-stranded Holliday junction formed by this sequence in which the four strands, A, B, C, and D, are individually colored, their nucleotides are numbered 1 to 10 from their 5′ to 3′ termini, and Watson–Crick base pairing interactions are represented by black dashes. The 2-fold axis relating the two helices of this so-called **stacked-X conformation** is represented by the black lenticular symbol. (*b*) The observed three-dimensional structure of the Holliday junction, as viewed along its 2-fold axis, in which the oligonucleotides are represented in stick form with their backbones traced by ribbons, all colored as in Part *a*. With the exception of the backbones of strands B and D at the crossovers, the two arms of this structure each form an undistorted B-DNA helix, including the stacking of the base pairs flanking the crossovers. The two helices are inclined to each other by 41°. Note that Fig. 30-64*g* is a schematic representation of the stacked-X conformation as viewed perpendicular to both helices (from the side in this drawing and hence having the projected appearance of the letter X). A Holliday junction can also assume a so-called **open-X conformation,** which is represented by Fig. 30-64*i*. [Courtesy of Shing Ho, Oregon State University. PDBid 1DCW.]

as **branch migration** (Fig. 30-64*e, f*) in which the four strands exchange base pairing partners.

A Holliday junction can be resolved into two duplex DNAs in two equally probable ways (Fig. 30-64*g–l*):

1. The cleavage of the strands that did not cross over (right branch of Fig. 30-64*j–l*) exchanges the ends of the original duplexes to form, after nick sealing, the traditional recombinant DNA (Fig. 1-27*b*).

2. The cleavage of the strands that crossed over (left branch of Fig. 30-64*j–l*) exchanges a pair of homologous single-stranded segments.

The recombination of circular duplex DNAs results in the types of structures diagrammed in Fig. 30-66. Electron microscopic evidence for the existence of the postulated "figure-8" structures is shown in Fig. 30-67*a*. These figure-8 structures were shown not to be just twisted circles by cutting them with a restriction endonuclease to yield **chi structures** (after their resemblance to the Greek letter χ) such as that pictured in Fig. 30-67*b*.

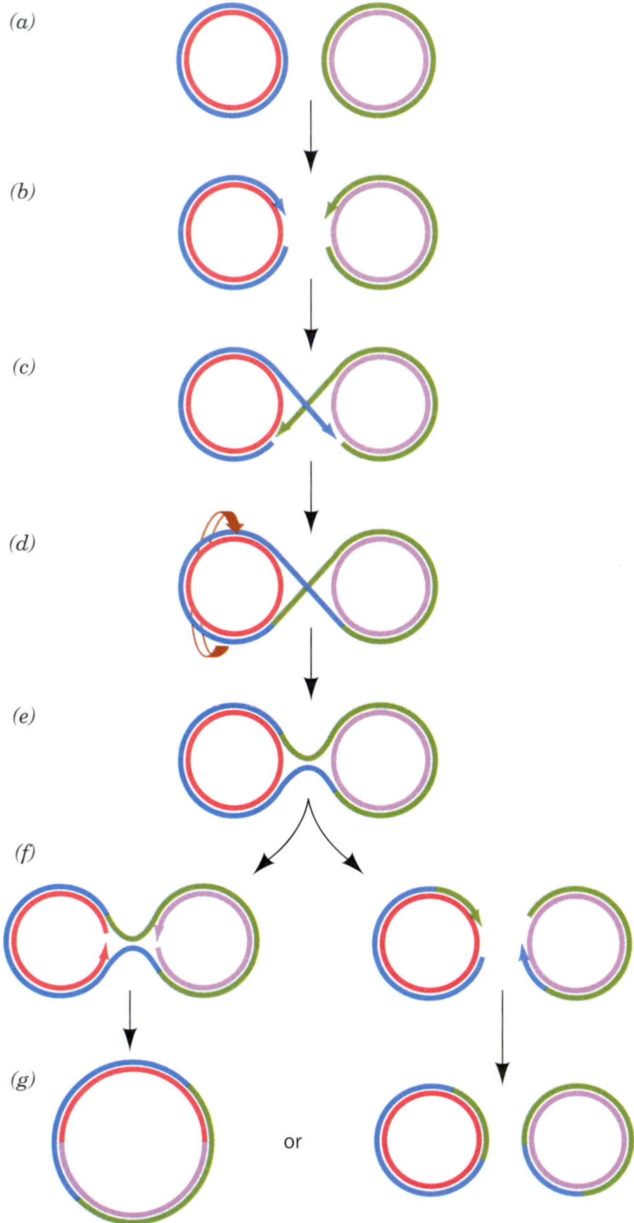

FIGURE 30-66 Homologous recombination between two circular DNA duplexes. This process can result either in two circles of the original sizes or in a single composite circle.

b. Homologous Recombination in *E. coli* Is Catalyzed by RecA

The observation that *recA*$^-$ *E. coli* have a 10^4-fold lower recombination rate than the wild-type indicates that *RecA protein has an important function in recombination.* Indeed, RecA greatly increases the rate at which complementary strands renature *in vitro.* This versatile protein (recall it also stimulates the autoproteolysis of LexA to trigger the SOS response and is an essential participant in the translesion synthesis of DNA; Section 30-5D) polymerizes cooperatively without regard to base sequence on ssDNA or on dsDNA that has a single-stranded gap. The resulting filaments, which may contain up to several thousand RecA monomers, specifically bind the homologous dsDNA, and, in an ATP-dependent reaction, catalyze strand exchange. EM studies by Edward Egelman (Fig. 30-68) reveal that RecA filaments bound to ssDNA or dsDNA form a right-handed helix with ~6.2 RecA monomers per turn and a

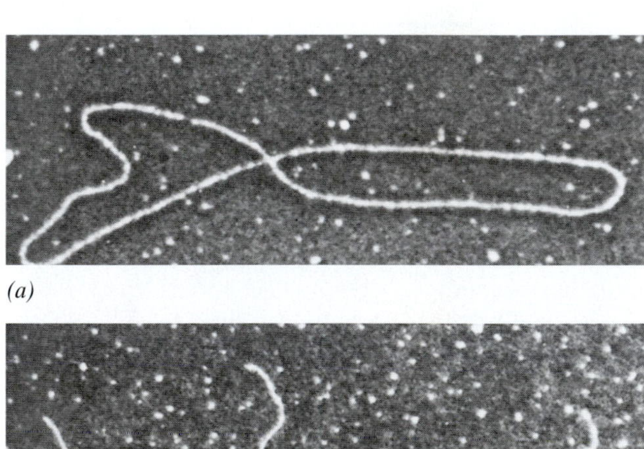

(*a*)

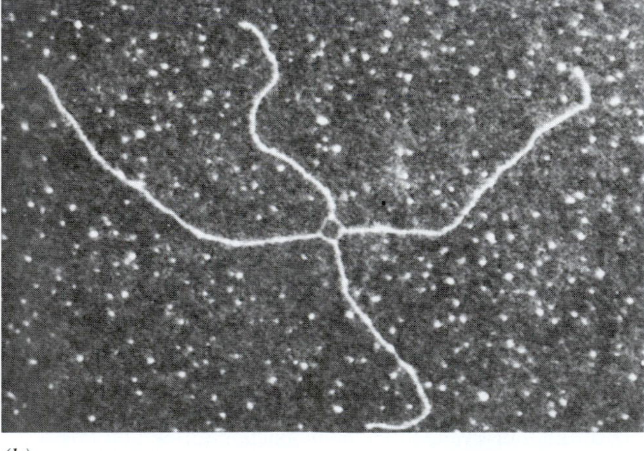

(*b*)

FIGURE 30-67 Electron micrographs of intermediates in the homologous recombination of two plasmids. (*a*) A figure-8 structure. This corresponds to Fig. 30-66*d*. (*b*) A chi structure that results from the treatment of a figure-8 structure with a restriction endonuclease. Note the thinner single-stranded connections in the crossover region. [Courtesy of Huntington Potter, University of South Florida, and David Dressler, Oxford University, U.K.]

pitch (rise per turn) of 95 Å. The DNA in these filaments, which binds to the protein with 3 nt (or bp) per RecA monomer unit and hence has ~18.6 nt (or bp) per turn, is so extended (having a rise of 5.1 Å/bp vs 3.4 Å/bp in B-DNA) that it must lie near the center of the helical filament as Fig. 30-68 indicates.

The X-ray structure of RecA (Fig. 30-69), determined by Steitz, reveals that the protein consists of a major central domain that is flanked by smaller N- and C-terminal domains. The monomers associate to form an ~120-Å-wide helical filament with six monomer units per turn and a pitch of 82.7 Å. The helical filament is remarkably open, so much so that there are gaps between the monomer units in successive turns. This arrangement results in a large he-

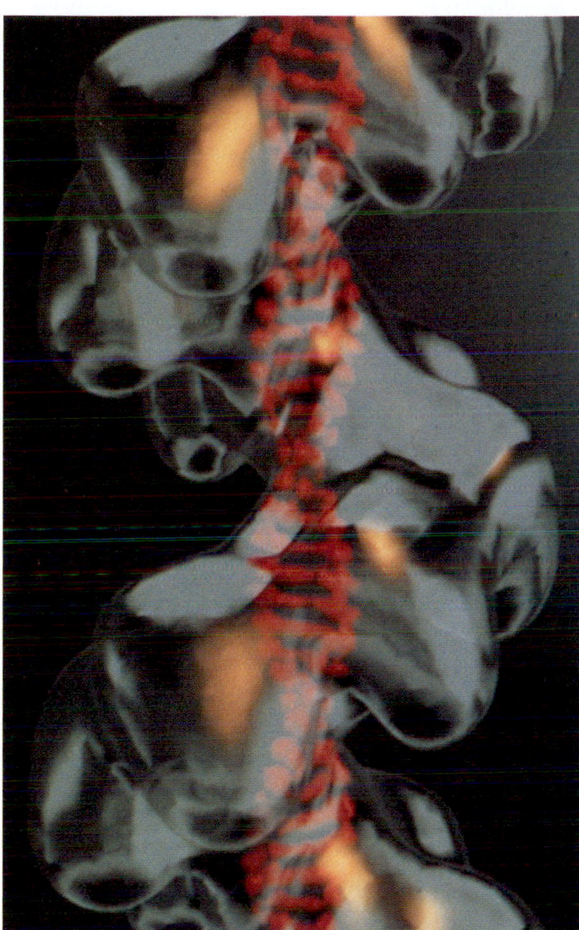

FIGURE 30-68 An electron microscopy–based image (*transparent surface*) of an *E. coli* RecA–dsDNA–ATP filament. The extended and untwisted dsDNA (*red*) has been modeled into this image. [Courtesy of Edward Egelman, University of Minnesota Medical School.]

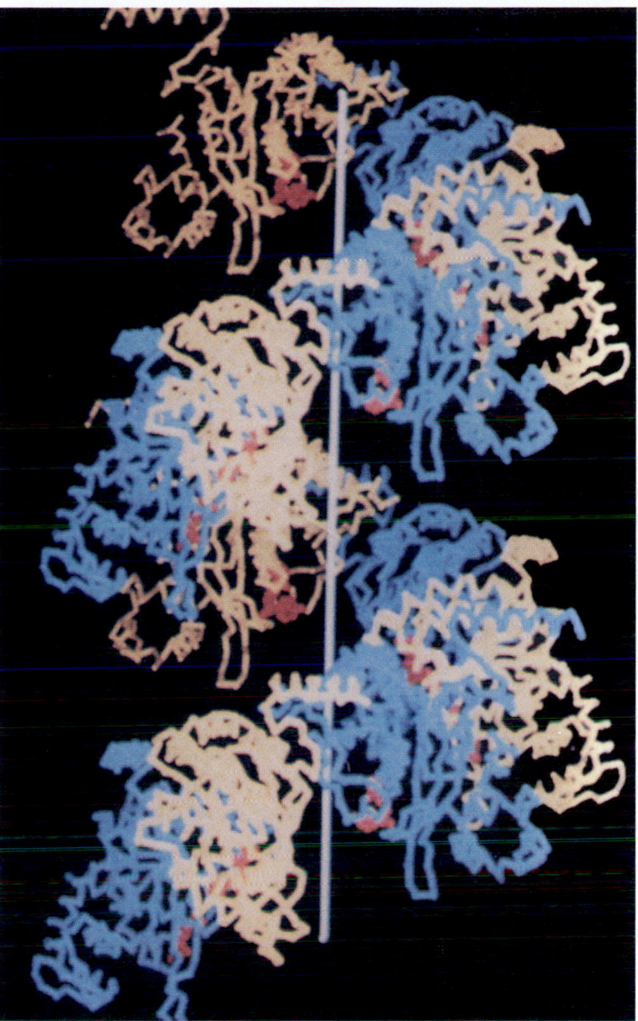

(*a*)

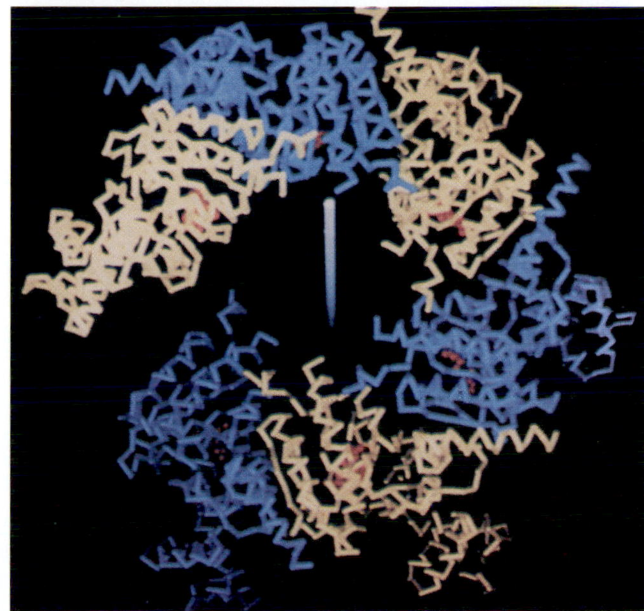

(*b*)

FIGURE 30-69 X-Ray structure of *E. coli* RecA protein. The RecA monomers are represented by their C_α chains. Alternate RecA monomers are yellow and blue, and their bound ADPs are red. (*a*) View perpendicular to the protein filament's helix axis (*light blue rod*) showing 12 monomers constituting two turns of the helix in the same orientation as in Fig. 30-68. (*b*) View nearly parallel to the helix axis showing one turn of the helix. [Courtesy of Thomas Steitz, Yale University. PDBid 1REA.]

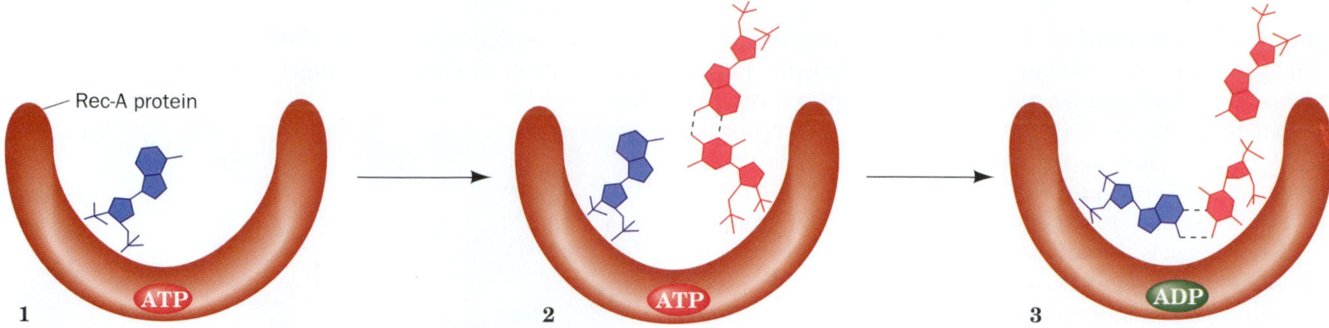

FIGURE 30-70 A model for RecA-mediated pairing and strand exchange between a single-stranded and a duplex DNA. **(1)** The ssDNA binds to RecA to form an initiation complex. **(2)** The dsDNA binds to the initiation complex so as to transiently form a three-stranded helix that mediates the correct pairing of the homologous strands. **(3)** RecA rotates the bases of the aligned homologous strands to effect strand exchange in an ATP-driven process. [After West, S.C., *Annu. Rev. Biochem.* **61,** 618 (1992).]

lical groove running the length of the filament that, when viewed down the helix axis, forms a 25-Å-wide central hole. This helical filament is strikingly similar to that of a RecA–duplex DNA filament as visualized at lower resolution by electron microscopy (Fig. 30-68; EM studies have also shown that LexA protein binds within this helical groove such that it spans the two RecA subunits on successive turns of the helix).

How does RecA mediate DNA strand exchange between single-stranded and duplex DNAs? On encountering a dsDNA with a strand that is complementary to its bound ssDNA, RecA partially unwinds the duplex and, in a reaction driven by RecA-catalyzed ATP hydrolysis, exchanges the ssDNA with the corresponding strand on the duplex. A model of how RecA might do so is diagrammed in Fig. 30-70. *This process tolerates only a limited degree of mispairing and requires that one of the participating DNA strands have a free end.* The assimilation (exchange) of a single-stranded circle with a strand on a linear duplex (Fig. 30-71) cannot proceed past the 3′ end of a highly mismatched segment in the complementary strand. *The invasion of the single strand must therefore begin with its 5′ end.*

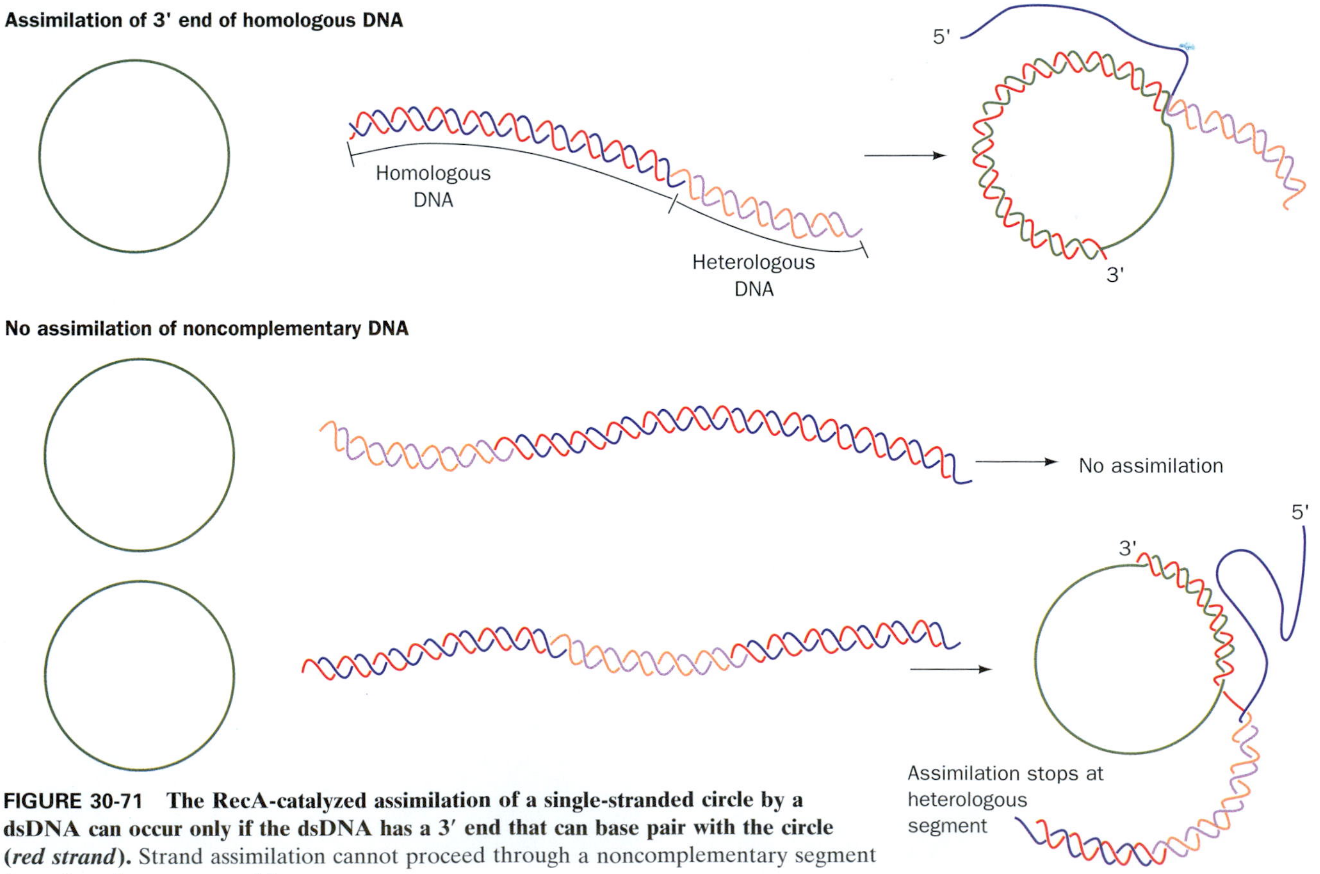

FIGURE 30-71 The RecA-catalyzed assimilation of a single-stranded circle by a dsDNA can occur only if the dsDNA has a 3′ end that can base pair with the circle (*red strand*). Strand assimilation cannot proceed through a noncomplementary segment (*purple and orange strands*).

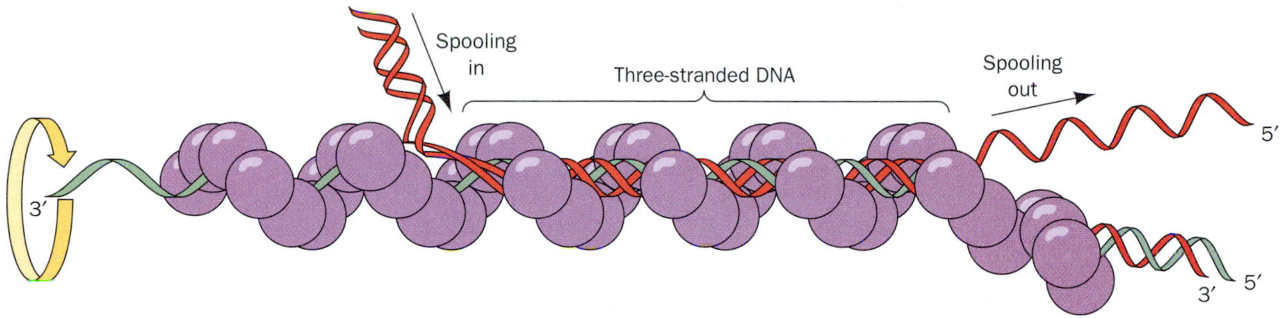

FIGURE 30-72 A hypothetical model for the RecA-mediated strand exchange reaction. Homologous DNA molecules are paired in advance of strand exchange in a three-stranded helix. The ATP-driven rotation of the RecA filament about its helix axis would cause duplex DNA to be "spooled in" to the filament, right to left as drawn. [After West, S.C., *Annu. Rev. Biochem.* **61,** 617 (1992).]

A model for the consequent branch migration process is diagrammed in Fig. 30-72. Of course, two such strand exchange processes must occur simultaneously in a Holliday junction (Figs. 30-64 and 30-66).

c. Eukaryotes Have RecA-Like Proteins

Yeast **RAD51** (339 residues) functions in the ATP-dependent repair and recombination of DNA in much the same way as does the 30% homologous *E. coli* RecA protein. The electron micrograph–based image reconstruction of RAD51 in complex with double-stranded DNA is nearly identical to that of RecA at low resolution: Both complexes form helical filaments in which the DNA has an ~5.1-Å rise per bp and 18.6 bp per turn. Since RAD51 homologs occur in chickens, mice, and humans, it is very likely that such filaments universally mediate DNA repair and recombination.

d. RecBCD Initiates Recombination by Making Single-Strand Nicks

The single-strand nicks to which RecA binds are made by the **RecBCD** protein, the 330-kD heterotrimeric product of the SOS genes *recB, recC,* and *recD,* which has both helicase and nuclease activities (Fig. 30-73). The process begins with RecBCD binding to the end of a dsDNA and then unwinding it via its ATP-driven helicase function. As it does so, it nucleolytically degrades the unwound single strands behind it, with the 3′-ending strand being cleaved more often and hence broken down to smaller fragments than the 5′-ending strand. However, on encountering the sequence GCTGGTGG from its 3′ end (the so-called **Chi sequence,** which occurs about every ~5 kb in the *E. coli* genome), it stops cleaving the 3′-ending strand and increases the rate at which it cleaves the 5′-ending strand, thereby yielding the 3′-ending single-strand segment to which RecA binds. This explains the observation that regions containing Chi sequences have elevated rates of recombination.

RecBCD can only commence unwinding DNA at a free duplex end. Such ends are not normally present in *E. coli,* which has a circular genome, but become available during such recombinational processes as bacterial transformation, conjugation, and viral transduction, as well as at collapsed replication forks.

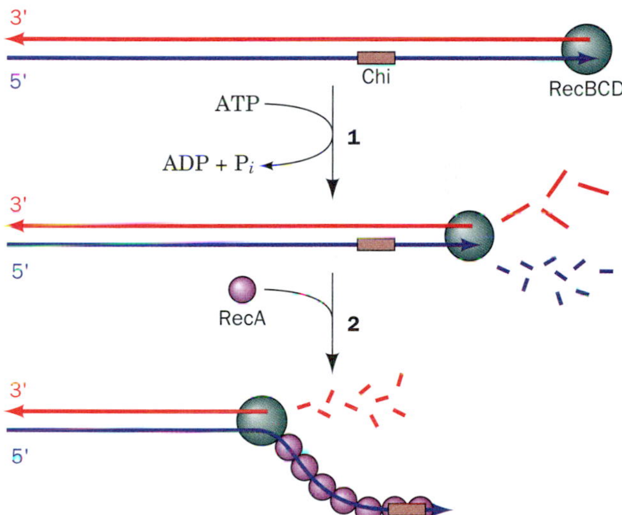

FIGURE 30-73 The generation of a 3′-ending single-strand DNA segment by RecBCD to initiate recombination. **(1)** RecBCD binds to a free end of a dsDNA and, in an ATP-driven process, advances along the helix, unwinding the DNA and degrading the resulting single strands behind it, with the 3′-ending strand cleaved more often than the 5′-ending strand. **(2)** When RecBCD encounters a properly oriented Chi sequence, it increases the frequency at which it cleaves the 5′-ending strand but stops cleaving the 3′-ending strand, thereby generating the potentially invasive 3′-ending strand segment to which RecA binds.

e. RuvABC Mediates the Branch Migration and the Resolution of the Holliday Junction

The branch migration of the RecA-generated Holliday junction (Fig. 30-64*e, f*) requires the breaking and reforming of base pairs as the bases exchange partners in passing from one double helical stem to the other. Since $\Delta G = 0$ for this process, it was initially assumed that it occurs spontaneously. However, such a process moves forward and backward at random and, moreover, is blocked by as little as a single mismatched base pair. In *E. coli,* and most other bacteria, branch migration is an ATP-dependent unidirectional process that is mediated by two proteins whose synthesis is induced by the SOS response (Section 30-5D): **RuvB** (336 residues), an ATP-powered pump that drives branch migration but binds only weakly to DNA; and **RuvA** (203 residues), which binds to both a Holliday junction and to RuvB, thereby targeting RuvB to the DNA.

The X-ray structure of *Mycobacterium leprae* (the cause of leprosy) RuvA in complex with a synthetic and immobile Holliday junction (Fig. 30-74*a*), determined by Morikawa, reveals that RuvA forms a homotetramer to which the Holliday junction binds in its open-X conformation (Fig. 30-74*b*). The RuvA tetramer, which has the appearance of a four-petaled flower (it has C_4 symmetry rather than the D_2 symmetry of the vast majority of homotetramers), is relatively flat ($80 \times 80 \times 45$ Å) with one square face concave and the other convex. The concave face (that facing the viewer in Fig. 30-74*b*), which is highly positively charged and is studded with numerous conserved residues, has four symmetry-related grooves that bind the Holliday junction's four arms. This face's four centrally located projections or "pins" are negatively charged, and hence the repulsive forces between them and the Holliday junction's anionic phosphate groups probably facilitate the separation of the single-stranded DNA segments and guide them from one double helix to another.

RuvB is a member of the AAA^+ family of ATPases (Section 30-3C). The X-ray structure of *Thermus thermophilus* RuvB crystallized in the presence of both ADP and ADPNP, determined by Morikawa, reveals two molecules of RuvB with somewhat different conformations: one binding ADP and the other binding ADPNP. Each RuvB molecule consists of three consecutive domains arranged in a crescentlike configuration with the adenine nucleotides binding at the interface between its N-terminal and middle domains. EM studies indicate that, in the presence of dsDNA, RuvB oligomerizes to form a hexamer (Fig. 30-75*a*), as do most other AAA^+ family members, including the D2 domain of NSF (Fig. 12-66). A hexameric model of RuvB (Fig. 30-75*b*), constructed by superimposing the N-terminal domain of the RuvB monomer on the ATPase domains of the NSF-D2 hexamer, agrees well with the EM-based image and contains no serious steric clashes. This 130-Å-diameter hexameric model contains a 30-Å-diameter hole through which a single dsDNA can readily be threaded (see below). Moreover the six β hairpins, one per monomer, that have been implicated in binding to RuvA are located on the top face of the hexamer (that pictured in Fig. 30-75*b*).

The EM images of the RuvAB–Holliday junction complex indicate that RuvA binds two oppositely located RuvB hexamers. This has led to the model of their interaction depicted in Fig. 30-76 in which RuvA binds the Holliday junction and helps load the RuvB hexameric rings onto two opposing arms of the Holliday junction. The two hexameric rings are postulated to counter-rotate, each in the anticlockwise direction looking toward the center of the junction, so as to screw the horizontal DNA strands

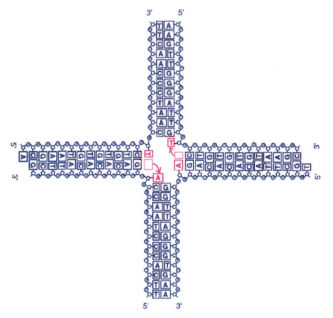

(*a*)

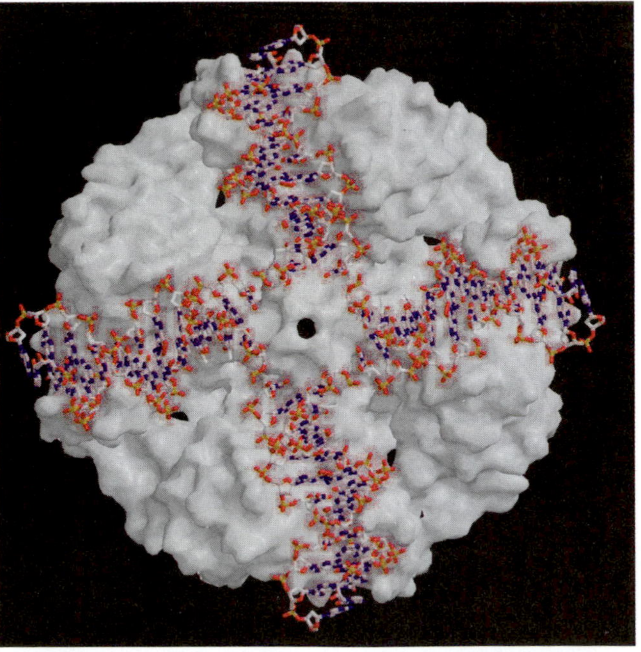

(*b*)

FIGURE 30-74 X-Ray structure of a RuvA tetramer in complex with a Holliday junction. (*a*) A schematic drawing of the synthetic and immobile Holliday junction in this structure showing its base sequence. The two A · T base pairs that are disrupted at the crossover (and which, if the Holliday junction consisted of two homologous dsDNAs, as it normally does, would exchange base pairing partners) are magenta. (*b*) The RuvA–Holliday junction complex as viewed along the protein tetramer's 4-fold axis. The protein is represented as its molecular surface (*gray*) and the DNA is drawn in stick form colored according to atom type (C white, N blue, O red, and P yellow). [Courtesy of Kosuke Morikawa, Biomolecular Engineering Research Institute, Osaka, Japan. PDBid 1C7Y.]

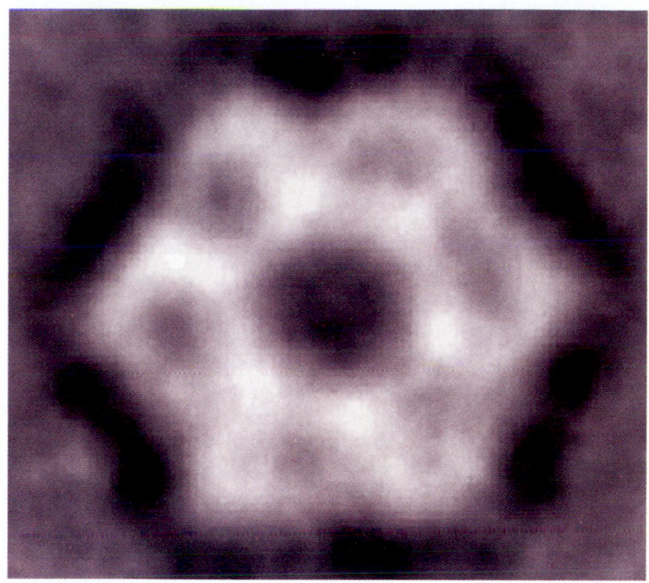

(a)

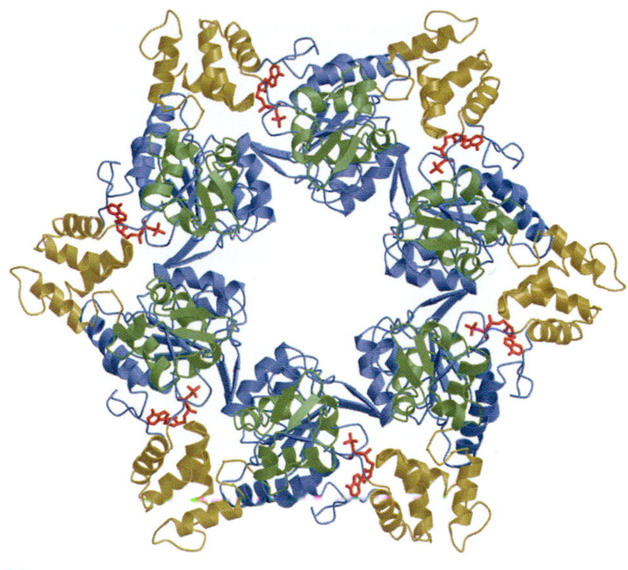

(b)

FIGURE 30-75 Proposed structure of the *T. thermophilus* RuvB hexamer. *(a)* An EM-based image reconstruction of RuvB complexed with a 30-bp DNA (not visible) as viewed along its 6-fold axis. The image resolution is 30 Å. *(b)* A model of the RuvB hexamer that was constructed from the X-ray structure of RuvB monomers by superimposing their N-terminal domains on the homologous ATPase domains of the NSF-D2 homohexamer (Fig. 12-66). The N-terminal, middle, and C-terminal domains are blue, yellow, and green, respectively, and its bound ADPNP is drawn in stick form in red. [Courtesy of Kosuke Morikawa, Biomolecular Engineering Research Institute, Osaka, Japan. PDBid 1HQC.]

through the center of the junction and into the top and bottom double helices, thereby effecting branch migration (although rather than actually rotating relative to RuvA,

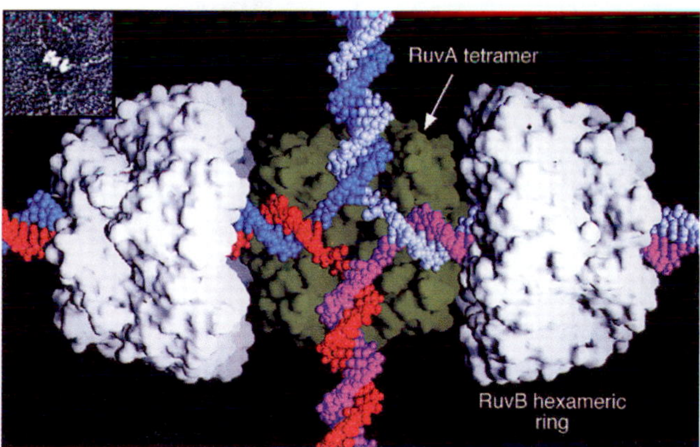

FIGURE 30-76 Model of the RuvAB–Holliday junction complex. The model is based on electron micrographs such as that in the inset. The proteins are represented by their surface diagrams with the RuvA tetramer, as seen in its X-ray structure, green and the two oppositely oriented RuvB hexamers gray. The DNA of the Holliday junction is drawn in space-filling form with its homologous blue and magenta strands complementary to its red and white strands. The complex is postulated to drive branch migration via the ATP-driven counter-rotation of the RuvB hexamers relative to the RuvA tetramer. This pumps (screws) the horizontal dsDNAs through the RuvB hexamers to the center of the Holliday junction, where their strands separate and then base pair with their homologs to form new dsDNAs, which are pumped out vertically. [Courtesy of Peter Artymiuk, University of Sheffield, U.K.]

a RuvB hexamer might pull the dsDNA through its central hole by "walking" up its grooves in a manner resembling that postulated for hexagonal helicases; Section 30-2C). The direction of branch migration depends on which pair of arms the RuvB hexamers are loaded.

The final stage in homologous recombination is the resolution of the Holliday junction into its two homologous dsDNAs. This process is carried out by **RuvC,** a 173-residue homodimeric exonuclease whose active sites are located ~30 Å apart on the same face of the protein. This suggests that RuvC sits down on the open face of the RuvAB–Holliday junction complex, that facing the viewer in Fig. 30-76, to cleave oppositely located strands at the Holliday junction. The resulting single-strand nicks in the now resolved dsDNAs are sealed by DNA ligase.

This so-called RuvABC resolvosome provides a satisfying mechanism for branch migration and Holliday junction resolution. However, there is a fly in this particular ointment. The X-ray structure of an *M. leprae* RuvA–Holliday junction complex crystallized under conditions different from that in Fig. 30-74, determined by Laurence Pearl, resembles the complex in Fig. 30-74*b* but with a second RuvA tetramer in face-to-face contact with the first. Hence the Holliday junction is contained in two intersecting tunnels running through the resulting RuvA octamer. Are both RuvA–Holliday junction structures biologically relevant, or is one an artifact of crystallization? Pearl argues that the extensive complementary contacts between the two RuvA tetramers, which are strongly conserved, are unlikely to be artifactual and that a single RuvA tetramer is unlikely to be able to withstand the torque exerted by the two (in effect) counter-rotating RuvB hexamers. However, if the RuvA octamer is biologically relevant, one of its tetramers would at some point have to dissociate in order to allow RuvC access to the Holliday junction. But modeling studies indicate that the RuvC dimer observed in its X-ray structure cannot contact the DNA strands bound to a RuvB tetramer without a significant conformational change. Further investigations are necessary to resolve these inconsistencies.

f. Recombination Repair Reconstitutes Damaged Replication Forks

Transformation, transduction, and conjugation are such rare events that the vast majority of bacterial cells never participate in these processes. Similarly, the only place in the metazoan life cycle at which gene shuffling through homologous recombination occurs is in meiosis (Section 1-4A). Why then do nearly all cells have elaborate systems for mediating homologous recombination? This is because damaged replication forks occur at a frequency of at least once per bacterial cell generation and perhaps 10 times per eukaryotic cell cycle. The DNA lesions that damage the replication forks can be circumvented via homologous recombination in a process named **recombination repair** [translesion synthesis, which is highly mutagenic, is a process of last resort (Section 30-5D)]. Indeed, the rates of synthesis of RuvA and RuvB are greatly enhanced by the SOS response. Thus, as Michael Cox pointed out, *the primary function of homologous recombination is to repair damaged replication forks.* In what follows, we describe recombination repair as it occurs in *E. coli.*

Recombination repair is called into play when a replication fork encounters an unrepaired single-strand lesion (Fig. 30-77):

1. DNA replication is arrested at the lesion but continues on the opposing undamaged strand for some distance before the replisome fully collapses (Section 30-5D).

2. The replication fork regresses to form a type of Holliday junction dubbed a "chicken foot." This process may occur spontaneously as driven by the positive supercoiling that has built up ahead of the replication fork, it may be mediated by RecA, or it may be promoted by **RecG,** an ATP-driven helicase that catalyzes branch migration at DNA junctions with three or four branches.

3. The single-strand gap at the collapsed replication fork, now an overhang, is filled in by Pol I.

4. Reverse branch migration mediated by RuvAB or RecG yields a reconstituted replication fork, which supports replication restart (see below).

Note that this process does not actually repair the single-strand lesion that has caused the problem but instead reconstructs the replication fork in a way that permits the previously discussed DNA repair systems (Section 30-5) to eventually eliminate the lesion.

A second situation that requires recombination repair is the encounter of a replication fork with an unrepaired single-strand nick (Fig. 30-78):

1. When a single-strand nick is encountered, the replication fork collapses.

2. The repair process begins via the RecBCD plus RecA-mediated invasion of the newly synthesized and undamaged 3′-ending strand into the homologous dsDNA starting at its broken end.

3. Branch migration, as mediated by RuvAB, then yields a Holliday junction, which exchanges the replication fork's 3′-ending strands.

4. RuvC then resolves the Holliday junction yielding a reconstituted replication fork ready for replication restart.

Thus, the 5′-ending strand of the nick has, in effect, become the 5′ end of an Okazaki fragment.

The final step in the recombination repair process is the restart of DNA replication. This process is, of necessity,

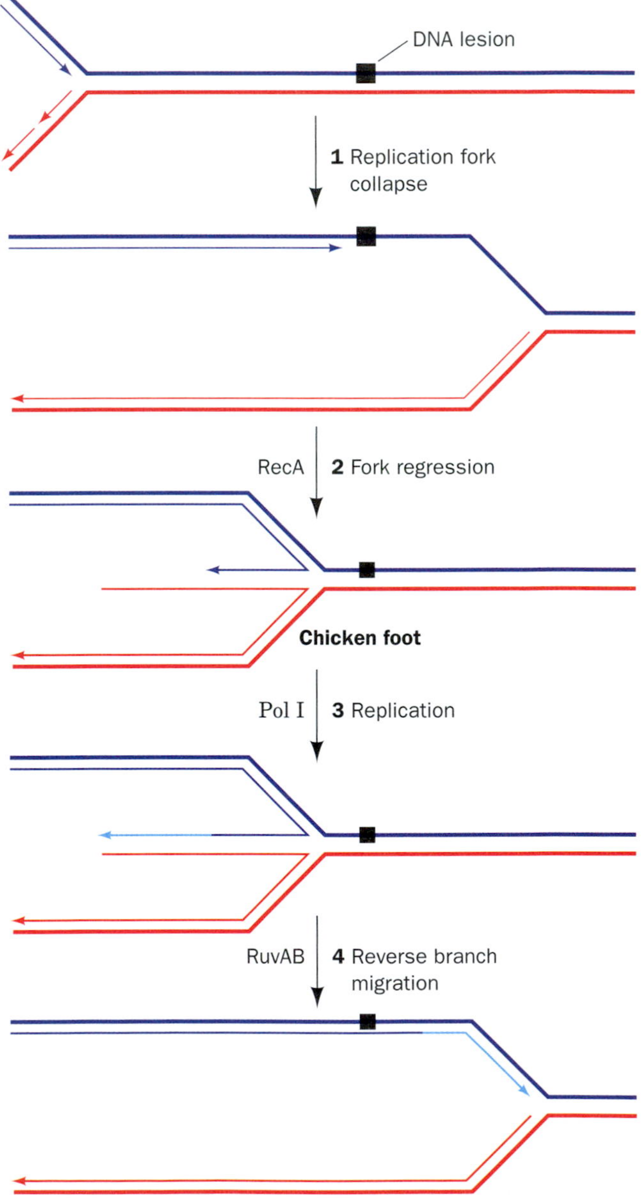

FIGURE 30-77 The recombination repair of a replication fork that has encountered a single-strand lesion. Thick lines indicate parental DNA, thin lines indicate newly synthesized DNA, the cyan lines indicate DNA that was synthesized by Pol I, and the arrows point in the $5' \rightarrow 3'$ direction. [After Cox, M.M., *Annu. Rev. Genet.* **35,** 53 (2000).]

distinct from the replication initiation that occurs at *oriC* (Section 30-3C). **Origin-independent replication restart** is mediated by the same seven-protein primosome that initiates the minus strand replication of bacteriophage φX174 (Table 30-4), which has therefore been named the **restart primosome.**

g. Recombination Repair Reconstitutes Double-Strand Breaks

We have seen that double-strand breaks (DSBs) in DNA can be rejoined, often mutagenically, by nonhomologous end-joining (NHEJ; Section 30-5E). DSBs may also be nonmutagenically repaired through a recombination repair process known as **homologous end-joining,** which occurs via two Holliday junctions (Fig. 30-79):

1. The DSB's double-stranded ends are resected to produce single-stranded ends. One of the 3′-ending strands invades the corresponding sequence of a homologous chromosome to form a Holliday junction, a process that, in eukaryotes, is mediated by the RecA homolog RAD51. The other 3′-ending strand pairs with the displaced strand segment on the homologous chromosome to form a second Holliday junction.

2. DNA synthesis and ligation fills in the gaps and seals the joints.

3. Both Holliday junctions are resolved to yield two intact double strands.

Thus, the sequences that may have been expunged in the formation of the DSB are copied from the homologous chromosome. Of course, a limitation of homologous end-joining, particularly in haploid cells, is that a homologous chromosomal segment may not be available.

The importance of recombination repair in humans is demonstrated by the observation that defects in the pro-

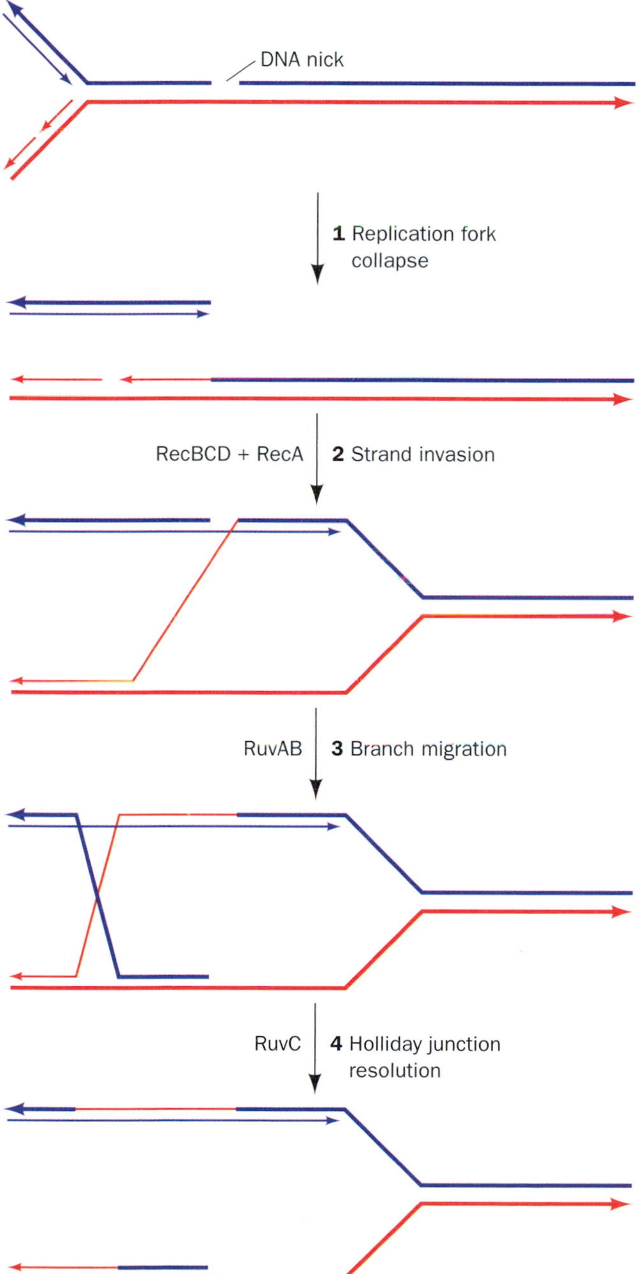

FIGURE 30-78 The recombination repair of a replication fork that has encountered a single-strand nick. Thick lines indicate parental DNA, thin lines indicate newly synthesized DNA, and the arrows point in the 5′ → 3′ direction. [After Cox, M.M., *Annu. Rev. Genet.* **35,** 53 (2000).]

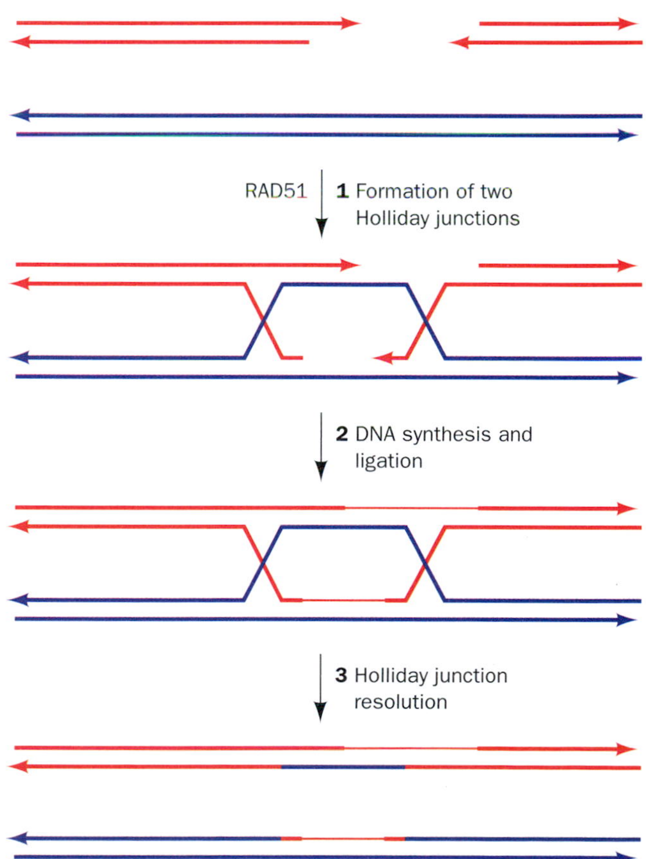

FIGURE 30-79 The repair of a double-strand break in DNA by homologous end-joining. Thick lines indicate parental DNA, thin lines indicate newly synthesized DNA, and the arrows point in the 5′ → 3′ direction. [After Haber, J.E., *Trends Genet.* **16,** 259 (2000).]

teins **BRCA1** (1863 residues) and **BRCA2** (3418 residues), both of which interact with RAD51, are associated with a greatly increased incidence of breast, ovarian, prostate, and pancreatic cancers. Indeed, individuals with mutant *BRCA1* or *BRCA2* genes have up to an 80% lifetime risk of developing cancer.

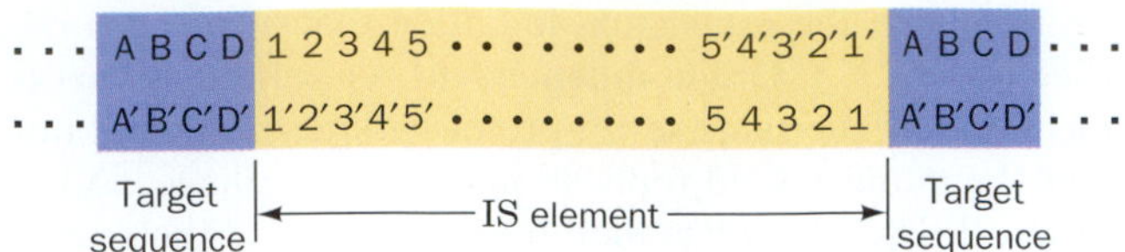

FIGURE 30-80 Structure of IS elements. These and other transposons have inverted terminal repeats (*numerals*) and are flanked by direct repeats of host DNA target sequences (*letters*).

B. *Transposition and Site-Specific Recombination*

In the early 1950s, on the basis of genetic analysis, Barbara McClintock reported that the variegated pigmentation pattern of maize (Indian corn) kernels results from the action of genetic elements that can move about the maize genome. This proposal was resoundingly ignored because it was contrary to the then held genetic orthodoxy that chromosomes consist of genes linked in fixed order. Another 20 years were to pass before evidence of mobile genetic elements was found in another organism, *E. coli.*

*It is now known that **transposable elements** or **transposons** are common in both prokaryotes and eukaryotes, where they influence the variation of phenotypic expression over the short term and evolutionary development over the long term. Each transposon codes for the enzymes that specifically insert it into the recipient DNA.* This process has been described as **illegitimate recombination** because it requires no homology between donor and recipient DNAs. Since the insertion site is chosen largely at random, transposition is a potentially dangerous process; the insertion of a transposon into an essential gene will kill a cell together with its resident transposons. Hence transposition is tightly regulated; it occurs at a rate of only 10^{-5} to 10^{-7} events per element per generation. The conditions that trigger transposition are, for the most part, unknown.

a. Prokaryotic Transposons

Prokaryotic transposons with three levels of complexity have been characterized:

1. The simplest transposons, and the first to be characterized, are named **insertion sequences** or **IS elements.** They are designated by "IS" followed by an identifying number. IS elements are normal constituents of bacterial chromosomes and plasmids. For example, a common *E. coli* strain has eight copies of **IS1** and five copies of **IS2.** IS elements generally consist of <2000 bp. These comprise a so-called **transposase** gene, and in some cases a regulatory gene, flanked by short inverted (having opposite orientation) terminal repeats (Fig. 30-80 and Table 30-6). The inverted repeats are essential for transposition; their genetic alteration invariably prevents this process. An inserted IS element is flanked by a directly (having the same orientation) repeated segment of host DNA (Fig. 30-80). This suggests that an IS element is inserted in the host DNA at a staggered cut that is later filled in (Fig. 30-81). The length of this target sequence (most commonly 5 to 9 bp), but not its sequence, is characteristic of the IS element.

2. *More complex transposons carry genes not involved in the transposition process, for example, antibiotic resistance genes.* Such transposons are designated "Tn" followed by an identifying number. For example, **Tn3** (Fig. 30-82) consists of 4957 bp and has inverted terminal repeats of 38 bp each. The central region of Tn3 codes for three proteins: (1) a 1015-residue transposase named **TnpA;** (2) a 185-residue protein known as **TnpR,** which mediates the **site-specific recombination** reaction necessary to complete the transposition process (see below) and also functions as a repressor for the expression of both *tnpA* and *tnpR;* and (3) a **β-lactamase** that inactivates ampicillin (Section 11-3B). The site-specific recombination occurs in an AT-rich region known as the **internal resolution site** that is located between *tnpA* and *tnpR.*

3. The so-called **composite transposons** (Fig. 30-83) consist of a gene-containing central region flanked by two identical or nearly identical IS-like modules that have either the same or an inverted relative orientation. It therefore seems that composite transposons arose by the association of two originally independent IS elements. Since the IS-like modules are themselves flanked by inverted repeats, the ends of either type of composite transposon must also be inverted repeats. Experiments demonstrate that composite transposons can transpose any sequence of DNA in their central region.

TABLE 30-6 Properties of Some Insertion Elements

Insertion Element	Length (bp)	Inverted Terminal Repeat (bp)	Direct Repeat at Target (bp)	Number of Copies in *E. coli* Chromosome
IS1	768	23	9	5–8
IS2	1327	41	5	5
IS4	1428	18	11–13	5
IS5	1195	16	4	1–2

Source: Mainly Lewin, B., *Genes VII,* p. 459, Oxford University Press (2000).

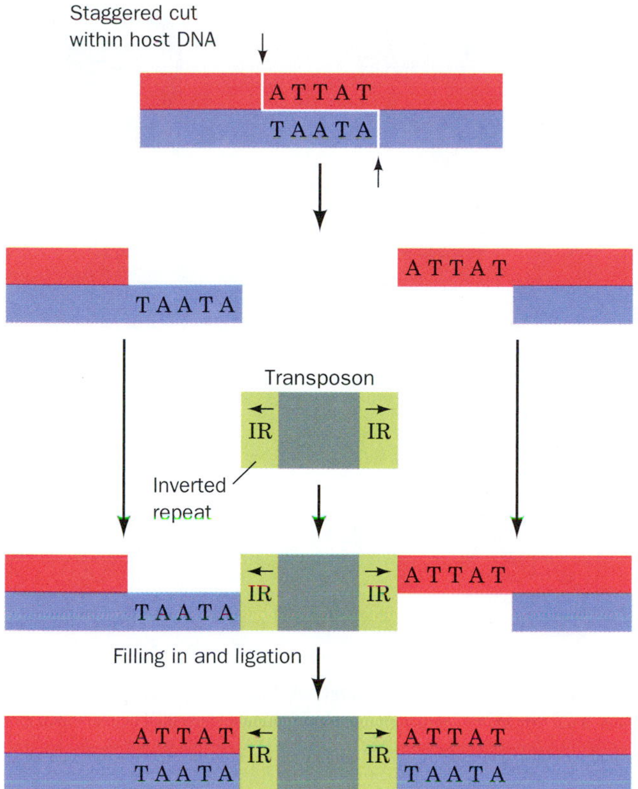

FIGURE 30-81 A model for the generation of direct repeats of the target sequence by transposon insertion.

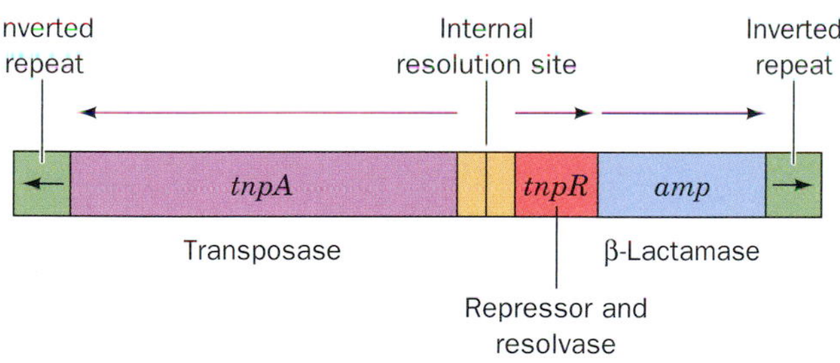

FIGURE 30-82 A map of transposon Tn3.

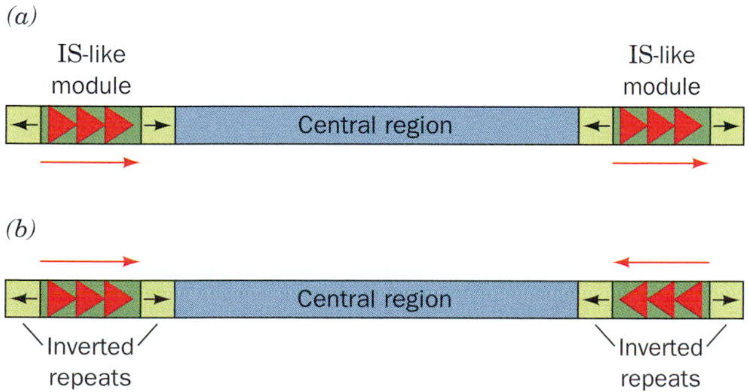

FIGURE 30-83 A composite transposon. This element consists of two identical or nearly identical IS-like modules (*green*) flanking a central region carrying various genes. The IS-like modules may have either (*a*) direct or (*b*) inverted relative orientations.

There are two modes of transposition: (1) **direct** or **simple transposition,** in which the transposon, as the name implies, physically moves from one DNA site to another; and (2) **replicative transposition,** in which the transposon remains at its original site and a copy of it is inserted at a target site. The two modes, as we shall see, have similar mechanistic features and, indeed, some transposons can move by either mode.

b. Direct Transposition of Tn5 Occurs by a Cut-and-Paste Mechanism

Tn5 is a 5.8-kb composite transposon that contains the gene encoding the 476-residue **Tn5 transposase** together with three antibiotic resistance genes. It is flanked by inverted IS-like modules ending in 19-bp sequences called outside end (OE) sequences. Tn5 undergoes direct transposition via a cut-and-paste mechanism that was elucidated in large part by William Reznikoff (Fig. 30-84):

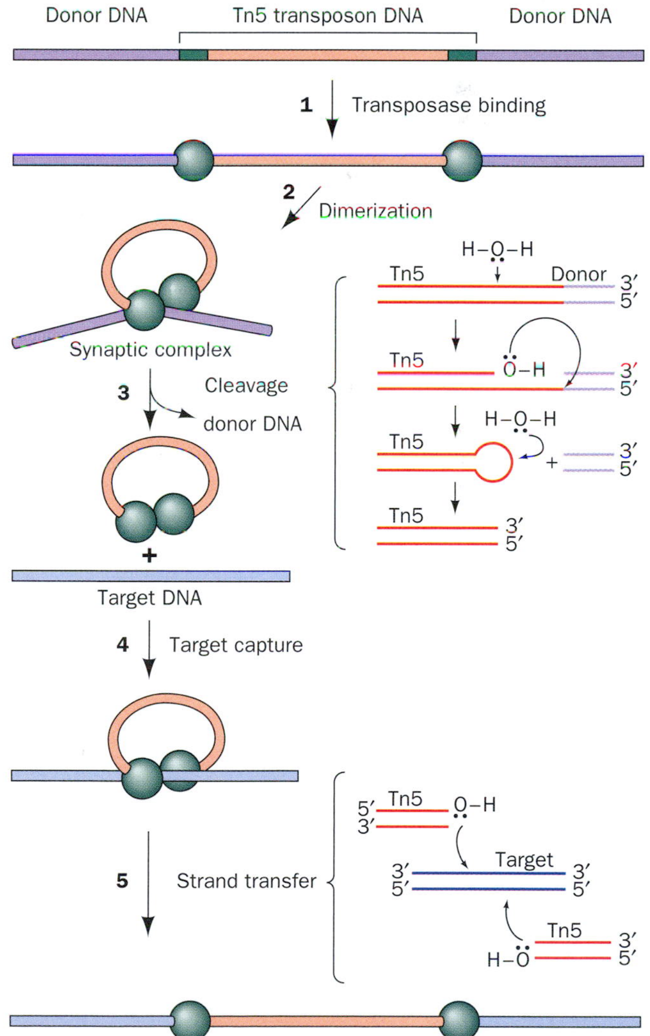

FIGURE 30-84 The cut-and-paste transposition mechanism catalyzed by Tn5 transposase. The reactions comprising Steps 3 and 5 are indicated beside the braces to the right of these steps. [After Davies, D.R., Goryshin, I.Y., Reznikoff, W.S., and Rayment, I., *Science* **289,** 77 (2000).]

1. Each of Tn5's two OE sequences on the donor DNA is bound by a monomer of Tn5 transposase.

2. The transposase dimerizes to form a catalytically active **synaptic complex** in which the transposon is held between the two transposase subunits.

3. Each transposase subunit activates a water molecule to nucleophilically attack the outermost nucleotide of its bound OE sequence, yielding a free 3′-OH group. This 3′-OH group is then activated to attack the opposite strand on the DNA to form a hairpin structure, thereby excising the transposon from the DNA. The hairpin is then hydrolyzed to yield a blunt-ended dsDNA at each end of the transposon, thus completing the "cut" portion of the transposition mechanism.

4. The synaptic complex binds to the target DNA.

5. The transposon's 3′-OH groups nucleophilically attack the target DNA on opposite strands spaced 9 bp apart, thereby installing the transposon at the target site. Remarkably, this reaction and the three preceding lytic reactions are all mediated by the same catalytic site. The repair of the oppositely located single-strand gaps (Fig. 30-81) completes the "paste" portion of the mechanism.

Although, strictly speaking, not part of the transposition process, the double-strand break in the donor DNA left by the excision of the transposon must be repaired if the donor DNA is to be propagated (in bacteria, the donor DNA is often a plasmid so that its loss has little effect on the cell since plasmids are generally present in multiple copies).

The X-ray structure of a Tn5 synaptic complex (Fig. 30-85), determined by Reznikoff and Ivan Rayment, provides a model of the synaptic complex at the stage following its cleavage from the donor DNA (the product of Step 3 in Fig. 30-84). This 2-fold symmetric complex consists of a dimer of Tn5 transposase subunits binding two 20-bp DNA segments containing the Tn5 transposon's 19-bp OE sequence with the outer end of each OE sequence bound to the protein (and whose opposite ends would, *in vivo,* be connected by the looped around transposon; Fig. 30-84). Both transposase subunits extensively participate in binding each DNA segment, thereby explaining why the individual subunits cannot cleave their bound DNA segments before forming the synaptic complex. The protein holds the DNA in a distorted B-DNA conformation with its two end pairs of nucleotides no longer base paired. Indeed, the penultimate base on the nontransferred strand is flipped out of the double helix and binds in a hydrophobic pocket. The transferred strand's free 3′-OH group, which occupies the active site, is bound in the vicinity of a cluster of three catalytically essential acidic residues, the so-called **DDE motif,** which is shared with other transposases. In the X-ray structure the DDE motif binds one Mn^{2+} ion, but physiologically it probably binds two Mg^{2+} ions. This suggests that transposases employ a metal-activated catalytic mechanism similar to that of the DNA polymerases (Section 30-2A). The facing surface of the protein in Fig. 30-85 is positively charged with a prominent groove running from upper left to lower right that forms the apparent binding site for the target DNA.

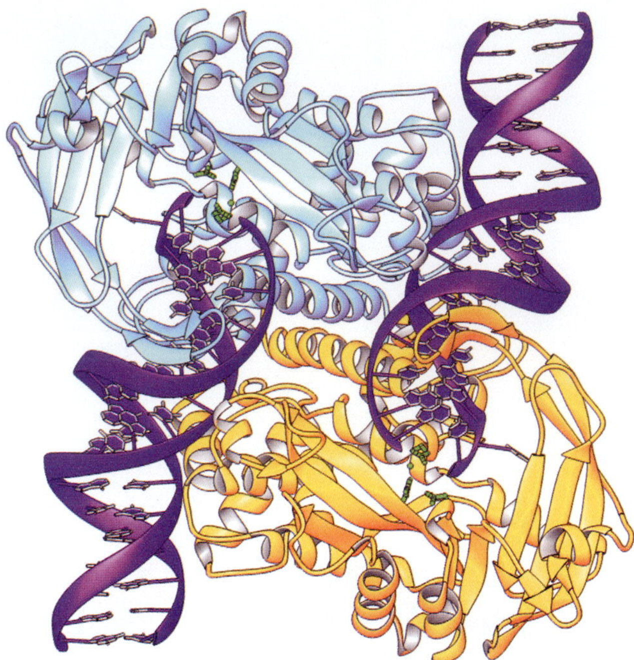

FIGURE 30-85 X-Ray structure of Tn5 transposase in complex with a 20-bp DNA containing the OE sequence. The complex, which represents the product of Step 3 in Fig. 30-84, is viewed along its 2-fold axis with its two identical subunits cyan and yellow. The three acidic residues of each DDE motif are drawn as green ball-and-stick structures with their bound Mn^{2+} ions represented by green balls. The DNAs' sugar–phosphate backbones are represented by purple ribbons and their bases are drawn as purple-filled gray stick figures. The DNAs' reactive 3′-OH groups are located at the ends of the inner strands where they contact the DDE motifs. [Courtesy of Ivan Rayment, University of Wisconsin. PDBid 1F3I.]

Wild-type Tn5 transposase has such low catalytic activity that it is undetectable *in vitro*. However, that in the X-ray structure is a hyperactive mutant form that contains the mutations E54K and L372P (an unusual circumstance in that it is far more common to mutationally inhibit an enzyme under crystallographic study so as to trap it at some specific stage along its reaction pathway). Lys 54 is hydrogen bonded to O4 of a thymine base on the transferred strand. In the wild-type transposase, Glu 54 would probably have an unfavorable charge–charge repulsion with a nearby phosphate group, thus providing a structural basis for the increased activity of the E54K mutant. The L372P mutation disorders the peptide segment between residues 373 and 391 (it is ordered in the X-ray structure of wild-type Tn5 transposase lacking its N-terminal 55 residues), thereby suggesting that this mutation facilitates a conformation change required for substrate binding.

c. Replicative Transposition

If a plasmid carrying a transposon resembling Tn3 is introduced into a bacterial cell carrying a plasmid that lacks

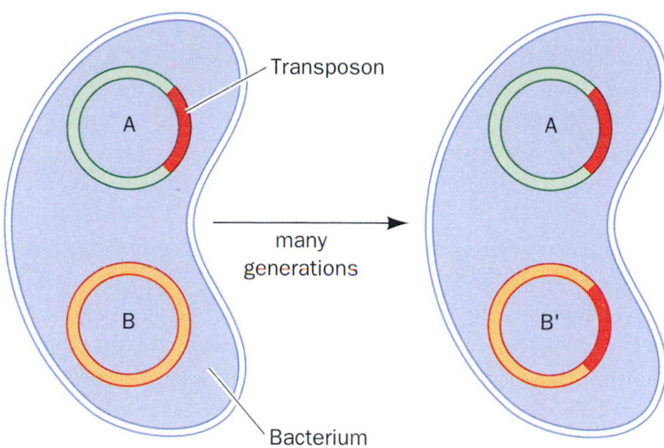

FIGURE 30-86 Replicative transposition. This type of transposition inserts a copy of the transposon at the target site while another copy remains at the donor site.

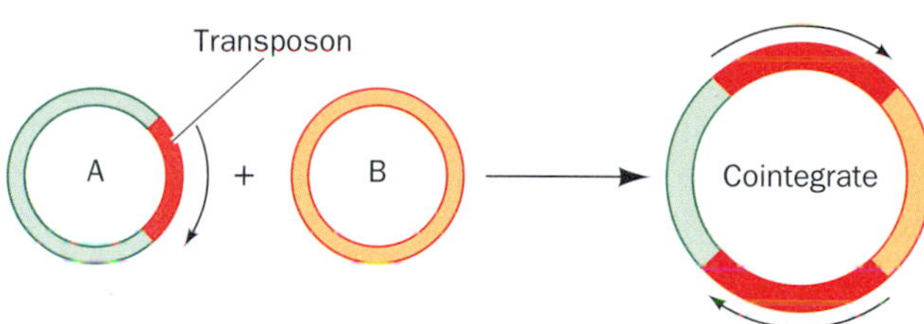

FIGURE 30-87 A cointegrate. This structure forms by the fusion of two plasmids, one carrying a transposon, such that both junctions of the original plasmid are spanned by transposons with the same orientation (*arrows*).

the transposon, in some of the progeny cells both types of plasmid will contain the transposon (Fig. 30-86). Evidently, *such transposition involves the replication of the transposon into the recipient plasmid rather than its transfer from donor to recipient.*

Two plasmids, one containing a replicative transposon, will occasionally fuse to form a so-called **cointegrate** containing like-oriented copies of the transposon at both junctions of the original plasmids (Fig. 30-87). Yet, some of the progeny of a cointegrate-containing cell lack the cointegrate and instead contain both original plasmids, each with one copy of the transposon (Fig. 30-86). The cointegrate must therefore be an intermediate in the transposition process.

Although the mechanism of replicative transposition has not been fully elucidated, a plausible model for this process (and there are several) that accounts for the foregoing observations consists of the following steps (Fig. 30-88):

1. A pair of staggered single-strand cuts, such as is diagrammed in Fig. 30-81, is made by the transposon-encoded transposase at the target sequence of the recipient plasmid so as to liberate 3′-OH ends. Similarly, single-strand cuts are made on opposite strands to either side of the transposon. Note that these reactions resemble those catalyzed by Tn5 transposase (Fig. 30-84).

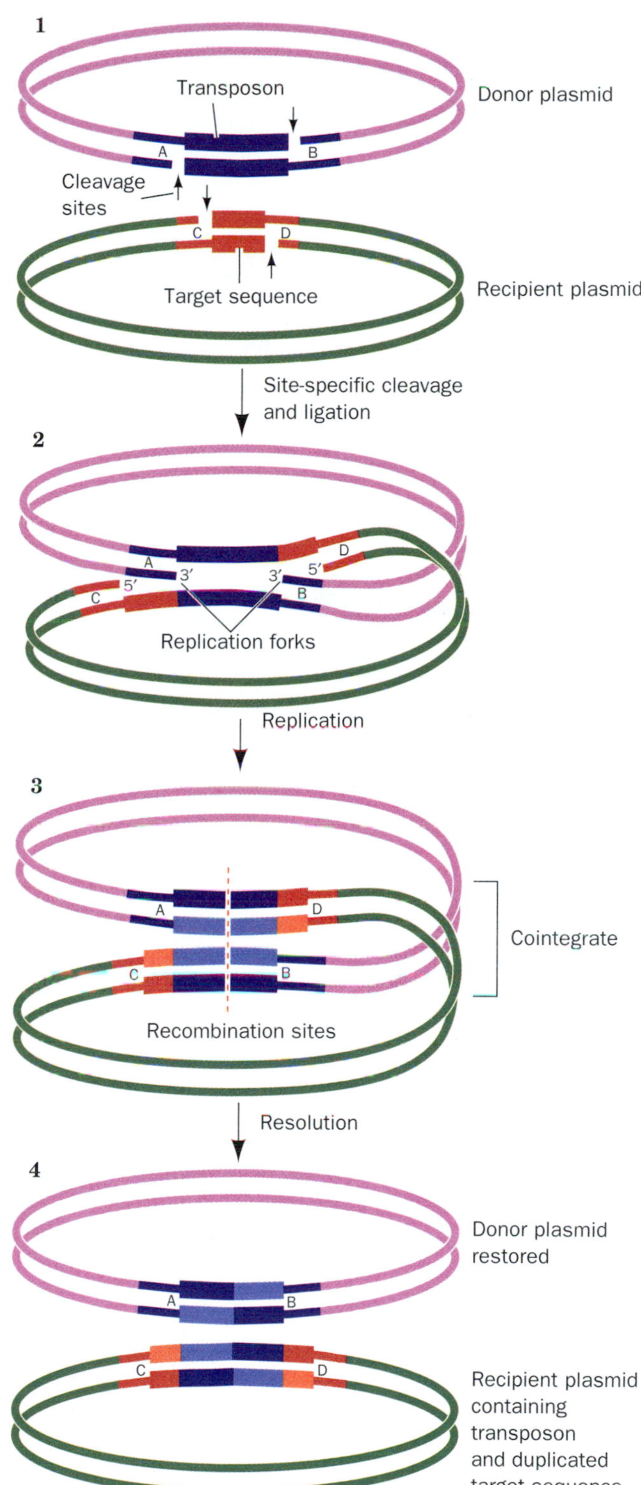

FIGURE 30-88 A model for transposition involving the intermediacy of a cointegrate. Here more lightly shaded bars represent newly synthesized DNA. [After Shapiro, J.A., *Proc. Natl. Acad. Sci.* **76,** 1934 (1979).]

2. Each of the transposon's free ends is ligated to a protruding single strand at the insertion site. This forms a replication fork at each end of the transposon.

3. The transposon is replicated, thereby yielding a cointegrate.

4. Through a site-specific recombination between the internal resolution sites of the two transposons, the cointegrate is resolved into the two original plasmids, each of which contains a transposon. This crossover process is catalyzed by a transposon-encoded **resolvase** (TnpR in Tn3) rather than RecA; transposition proceeds normally in *recA*$^-$ cells (although RecA will resolve a cointegrate containing a transposon with a mutant resolvase and/or an altered internal resolution site, albeit at a much reduced rate).

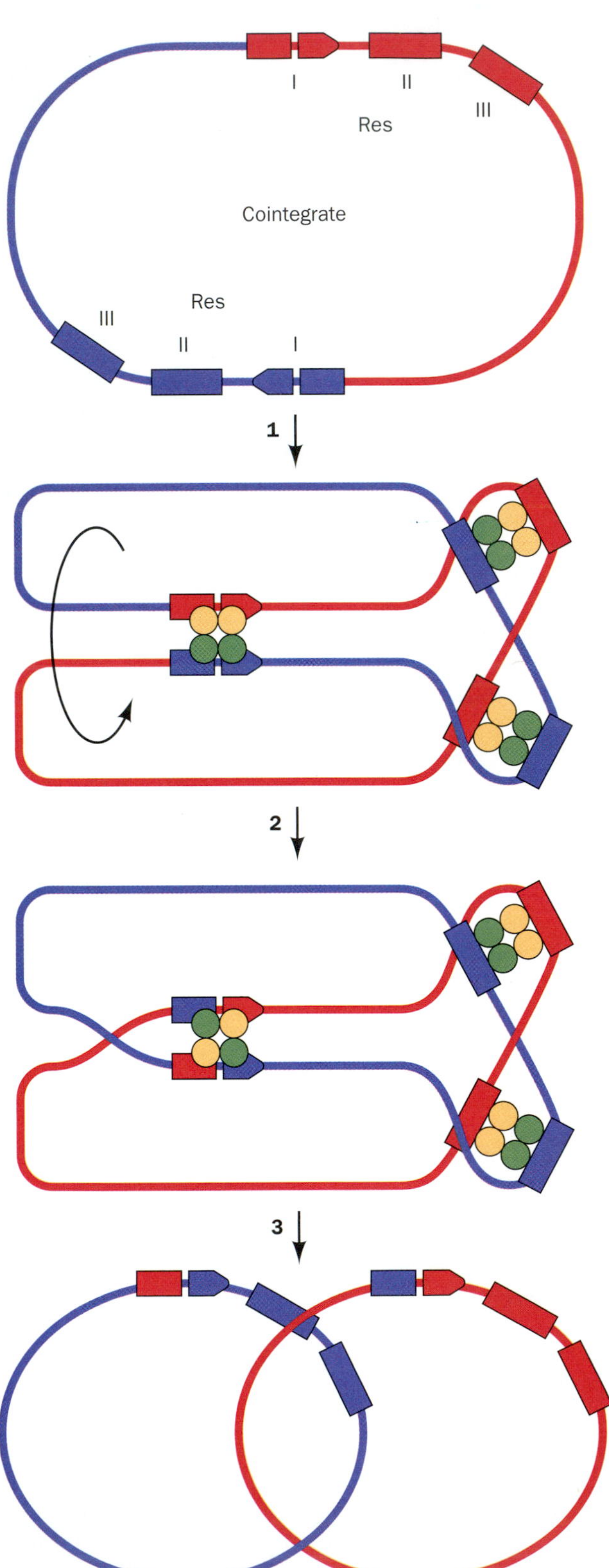

FIGURE 30-89 A model for the resolution of a cointegrate containing two γδ transposons to form two catenated dsDNA circles. (1) The γδ resolvase binds as six homodimers to its binding sites, I, II, and III, in each of the cointegrate's two *res* sites (yellow and green circles represent the γδ resolvase monomers initially bound to the red and blue *res* sites, respectively). Although not shown as such, the pair of dimers bound to sites I associate with the pairs of dimers bound at sites II and III to form, as seen in the electron microscope, a compact globule of unknown structure known as a **synaptosome. (2)** The dsDNA at sites I both undergo staggered (by 2 bp) double strand scissions via the transient formation of phosphoSer bonds between Ser 10 and the 5′ phosphates at the cleavage sites. The cleaved strands then exchange places (cross over) in a process that apparently requires the rotation of one of the pairs of resolvase monomers with respect to the other and are then ligated. **(3)** The dissociation of the synaptosome yields the catenated dsDNA circles. [Courtesy of Gregory Mullen, University of Connecticut Health Center.]

d. γδ Resolvase Catalyzes Site-Specific Recombination

The **γδ resolvase** is a TnpR homolog that is encoded by the **γδ transposon** (a member of the Tn3 family of replicative transposons; Fig. 30-82). It catalyzes a site-specific recombination event in which a cointegrate containing two copies of the γδ transposon is resolved, via double-strand DNA cleavage, strand exchange, and religation (the last step in Fig. 30-88), into two catenated (linked) dsDNA circles that each contain one copy of the γδ transposon (it also serves as its own transcriptional repressor as does TnpR). The γδ transposon contains a 114-bp *res* site that includes three binding sites for γδ resolvase dimers, each of which contains an inverted repeat of the γδ resolvase's 12-bp recognition sequence. The resolution of the cointegrate involves the binding of a γδ resolvase homodimer to all six of these binding sites in the cointegrate (three from each of its two transposons) as is diagrammed in Fig. 30-89. The reaction proceeds via the formation of a transient phosphoSer bond between Ser 10 and the 5′-phosphate at each cleavage site.

The X-ray structure of the γδ resolvase homodimer in complex with a 34-bp palindromic DNA segment containing an inverted repeat of the 12-bp recognition sequence separated by an 8-bp spacer (Fig. 30-90) was determined by Steitz. Each 183-residue resolvase monomer consists of a catalytic domain (residues 1–120) whose structure closely resembles that observed in the absence of DNA, a C-terminal DNA-binding domain (residues 148–183), and an extended arm (residues 121–147) that connects the N- and C-terminal domains.

The centrally located N-terminal domain dimer approaches the DNA from its minor groove side along the

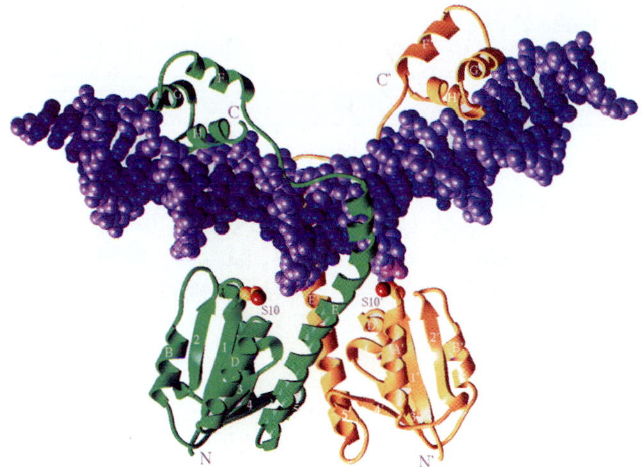

FIGURE 30-90 X-Ray structure of a γδ resolvase homodimer in complex with a 34-bp palindromic DNA containing its binding site. The DNA is drawn in space-filling form with its backbone purple, its bases blue, and its scissile phosphate groups highlighted in magenta. The protein subunits are green and gold and its Ser 10 residues are shown in space-filling form with C yellow and O red. The complex is viewed with its approximate 2-fold axis vertical. [Courtesy of Thomas Steitz, Yale University. PDBid 1GDT.]

structure's local 2-fold axis; the C-terminal domains each bind in the major groove of their target sequence on the opposite side of the DNA from the N-terminal domain dimer such that the two C-terminal domains are separated by two helical turns; and the extended arms which connect the two domains more or less run along the DNA's minor groove. The DNA, which otherwise closely assumes the B-DNA conformation, is centrally kinked by ~60° such that it bends toward its major groove, away from the N-terminal domain dimer. The C-terminal helix of the N-terminal domain (helix E) binds over the DNA's minor groove such that the dimer's two E helices grip the DNA like a pair of chopsticks (the segment of the E helix that contacts the DNA is disordered in the absence of the DNA). The C-terminal helix (helix H) binds in the major groove and, together with its preceding helix (helix G), forms a **helix–turn–helix (HTH) motif,** a common sequence-specific DNA-binding motif that occurs mainly in prokaryotic transcriptional repressors and activators (Section 31-3D). The structure is asymmetric with the active site Ser 10 in Fig. 30-90's yellow monomer much closer to the DNA than that in the green monomer, although both are quite distant from the scissile bonds in the DNA. This suggests that the two single-strand cleavage reactions catalyzed by the dimer may occur sequentially and, in any case, require significant conformational changes. Of course, a detailed understanding of the mechanism of the γδ resolvase reaction will require the knowledge of how all six γδ resolvase dimers that form the synaptosome participate in the reaction (Fig. 30-89).

e. Replicative Transposons Are Responsible for Much Genetic Remodeling in Prokaryotes

In addition to mediating their own insertion into DNA, *replicative transposons promote inversions, deletions, and rearrangements of the host DNA.* Inversions can occur when the host DNA contains two copies of a transposon in inverted orientation. The recombination of these transposons inverts the region between them (Fig. 30-91*a*). If, instead, the two transposons have the same orientation, the deletion of a DNA segment between two identical transposons with the same orientation. This process parcels one transposon each to the resulting two DNA segments.

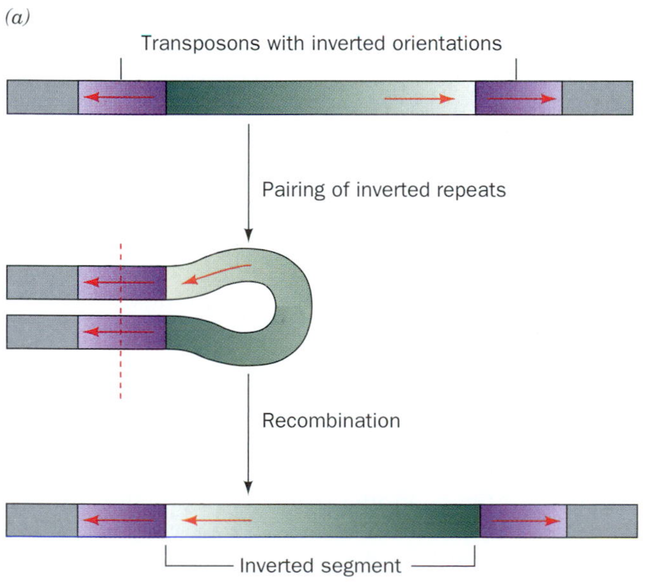

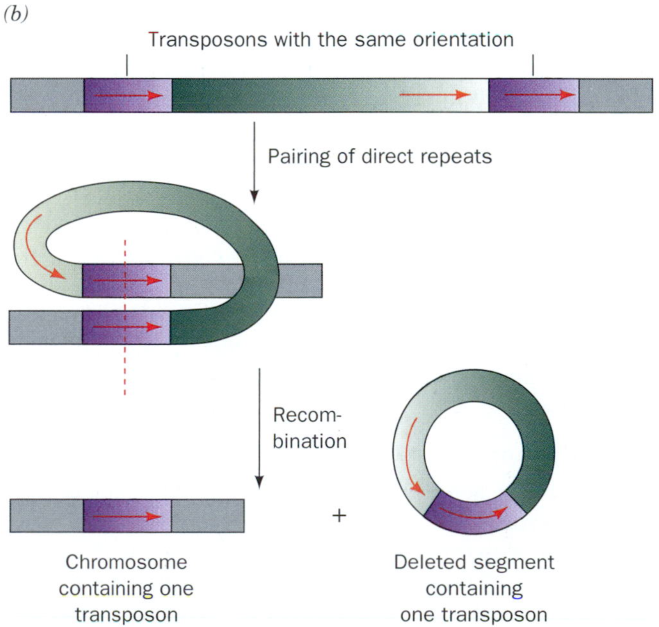

FIGURE 30-91 Chromosomal rearrangement via recombination. (*a*) The inversion of a DNA segment between two identical transposons with inverted orientations. (*b*) The

resolution of this cointegrate-like structure deletes the segment between the two transposons (Fig. 30-91*b*; if the deleted segment lacks a replication origin, it will not be propagated). The deletion of a chromosomal segment in this manner, followed by its integration into the chromosome at a different site by a separate recombinational event, results in chromosomal rearrangement.

Transposition appears to be important in chromosomal and plasmid evolution. Indeed, it has been suggested that transposons are nature's genetic engineering "tools." For example, the rapid evolution, since antibiotics came into common use, of plasmids that confer resistance to several antibiotics (Section 5-5B) has resulted from the accumulation of the corresponding antibiotic-resistance transposons in these plasmids. Transposon-mediated rearrangements may well have been responsible for organizing originally distant genes into coordinately regulated operons (Section 5-4A) as well as for forming new proteins by linking two formerly independent gene segments. Moreover, *the occurrence of identical transposons in unrelated bacteria indicates that the transposon-mediated transfer of genetic information between organisms is not limited to related species, in contrast to genetic transfers mediated by homologous recombination.*

f. Phase Variation Is Mediated by Site-Specific Recombination

Phenotypic expression in bacteria can be regulated by site-specific recombination. For example, certain strains of *Salmonella typhimurium* make two antigenically distinct versions of the protein **flagellin** (the major component of the whiplike flagella with which bacteria propel themselves; Section 35-3G) that are designated **H1** and **H2.** Only one of these proteins is expressed by any given cell but about once every 1000 cell divisions, in a process known as **phase variation,** a cell switches the type of flagellin it synthesizes. It is thought that phase variation helps *Salmonella* evade its host's immunological defenses.

What is the mechanism of phase variation? The two flagellin genes reside on different parts of the bacterial chromosome. *H2* is linked to the *rh1* gene that encodes a repressor of H1 expression (Fig. 30-92; *rh1*, *H2*, and *H1* are also known as *fljA*, *fljB*, and *fljC*, respectively). Hence, when the *H2–rh1* transcription unit is expressed, H1 synthesis is repressed; otherwise H1 is synthesized. Melvin Simon has shown that the expression of the *H2–rh1* unit is controlled by the orientation of a 995-bp segment that lies upstream of *H2* (Fig. 30-92) and that contains the following elements:

1. A promoter for *H2–rh1* expression.

2. The *hin* gene, which encodes the 190-residue **Hin DNA invertase.** Hin mediates the inversion of the DNA segment in a manner similar to that diagrammed in Fig. 30-91*a*. In fact, Hin is ~40% identical in sequence with the γδ resolvase, which strongly suggests that these proteins have similar structures.

3. Two closely related 26-bp sites, *hixL* and *hixR*, that form the boundaries of the segment and hence contain its cleavage sites. They each consist of two imperfect 12-bp inverted repeats separated by 2 nt.

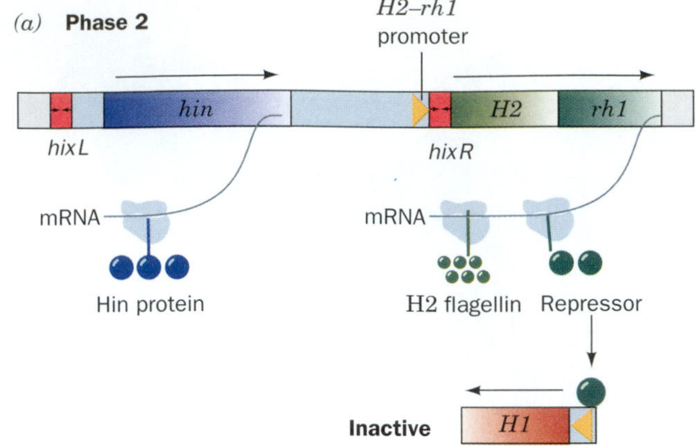

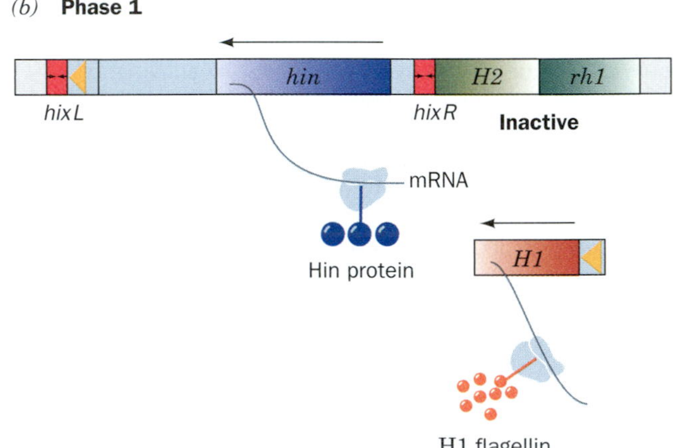

FIGURE 30-92 The mechanism of phase variation in *Salmonella.* (*a*) In Phase 2 bacteria, the *H2–rh1* promoter is oriented so that H2 flagellin and repressor are synthesized. Repressor binds to the *H1* gene, thereby preventing its expression. (*b*) In Phase 1 bacteria, the segment preceding the *H2–rh1* transcription unit has been inverted relative to its orientation in Phase 2 bacteria. Hence this transcription unit cannot be expressed because it lacks a promoter. This releases *H1* from repression and results in the synthesis of H1 flagellin. The inversion of the segment preceding the *H2–rh1* transcription unit is mediated by the Hin protein, which is expressed in either orientation by the *hin* gene.

In the Phase 2 orientation (Fig. 30-92*a*), the properly oriented promoter is just upstream of *H2,* so this gene and *rh1* are coordinately expressed, thereby repressing H1 synthesis. In Phase 1 bacteria (Fig. 30-92*b*), however, this segment has the opposite orientation. Consequently, neither *H2* nor *rh1*, which then lacks a promoter, is expressed so that H1 is synthesized.

g. Cre-Mediated Site-Specific Recombination Occurs via 3′-PhosphoTyr Intermediates

Bacteriophages, as we have seen (Fig. 1-31), replicate themselves within their host bacterial cells which, in most cases, they then lyse to release the progeny phage, a lifestyle that is therefore known as the **lytic** mode. However, certain bacteriophages can assume an alternative, nondestructive lifestyle, the **lysogenic** mode, in which they install their DNA, usually in the host chromosome via site-specific recombination, so that the phage DNA is passively replicated with the host DNA. However, if the bacterial host encounters conditions in which it is unlikely to survive, the phage DNA is excised from the bacterial chromosome via a reversal of the site-specific recombination reaction and it reenters the lytic mode so as to escape the doomed host. We discuss the genetic factors that maintain the balance between the lytic and lysogenic lifestyles in **bacteriophage λ** in Section 33-3.

The enzymes that mediate the foregoing site-specific recombination reactions are members of the **λ integrase (λ Int;** alternatively, **tyrosine recombinase)** family, whose >100 known members also occur in prokaryotes and eukaryotes. These include the **XerC** and **XerD** proteins of *E. coli* which, operating in concert, function to decatenate the two linked circular dsDNA products of homologous recombination (Fig. 30-66*g*, *left*), as well as type 1B topoisomerases (Section 29-3C).

The structurally best characterized member of the λ integrase family is the **Cre recombinase** of *E. coli* **bacteriophage P1.** In its lysogenic state, bacteriophage P1 is a single-copy circular plasmid (rather than being inserted in the host chromosome as is bacteriophage λ), but in the phage head (the lytic mode), P1 DNA is a linear dsDNA that has a 34-bp *loxP* site at each end. The main function of Cre, which is encoded by bacteriophage P1, is to mediate the site-specific recombination between these two *loxP* sites so as to circularize the linear DNA (Fig. 30-93).

The *loxP* site is palindromic except for its central 8-bp crossover region, which confers directionality on the site. In carrying out the recombination reaction, the 343-residue Cre subunits form a homotetramer that binds two *loxP* sites in an antiparallel orientation, with each Cre subunit binding half of a *loxP* site. Then, as is diagrammed in Fig. 30-94, oppositely located Cre subunits catalyze single-strand scissions on the 5′ side of the crossover region on one strand of each of the two dsDNAs. This occurs through the nucleophilic attack of each of these active Cre subunit's conserved Tyr 324 residues on the DNA's scissile phosphoester bond to yield a 3′-phosphoTyr intermediate on one side of the cleaved bond and a free 5′-OH group on the other side (as similarly occurs in the reactions catalyzed by type IB topoisomerases; Section 29-3C). Each of the liberated 5′-OH groups then nucleophilically attacks the 3′-phosphoTyr group on the opposite duplex to form a Holliday junction, thereby releasing the Tyr residues. The Holliday junction is resolved into two recombined dsDNAs when the two Cre subunits that had not yet participated in the reaction mediate the same cleavage and strand exchange reactions on the two heretofore unreacted single strands. This latter process must be preceded by a structural rearrangement (isomerization) of the Cre tetramer that positions the catalytic Tyr residues in the latter pair of subunits to participate in the reaction while those in the former pair of subunits are similarly removed from the scene of the action.

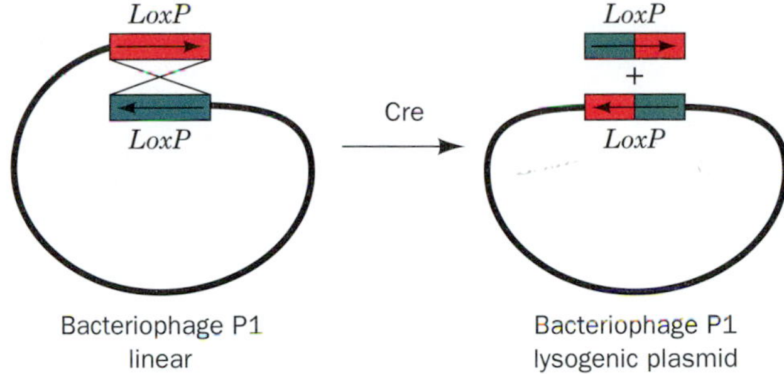

FIGURE 30-93 The circularization of linear bacteriophage P1 DNA. This occurs through the Cre-mediated site-specific recombination between its two terminally located *loxP* sites (*red and green*) to yield its lysogenic plasmid.

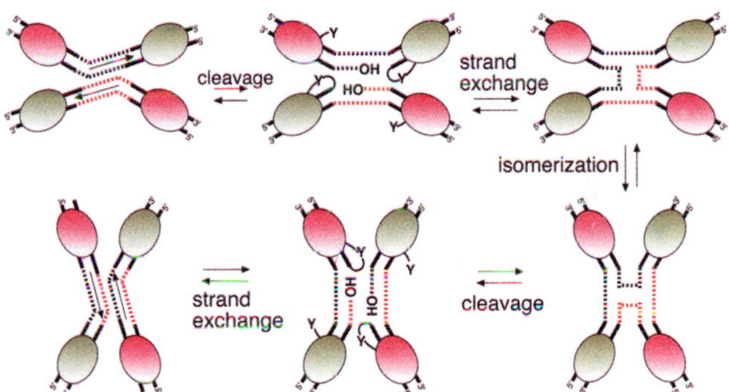

FIGURE 30-94 The mechanism of Cre–*loxP* site-specific recombination. The dashed lines represent the nonpalindromic crossover regions of the *loxP* sites. The green and magenta Cre subunits are active for cleavage in the top and bottom parts of the diagram, respectively, with their roles being switched by the isomerization step. Note that the mechanism does not require branch migration of the Holliday junction intermediate. [Courtesy of Gregory Van Duyne, University of Pennsylvania School of Medicine.]

The X-ray structures of Cre tetramers in their complexes with several *loxP* model DNAs, determined by Gregory Van Duyne, have helped elucidate the foregoing mechanism. When the DNA had a single-strand nick past the second nucleotide from the 5′ end of the crossover region, Cre-catalyzed strand scission yielded a free nucleotide (a CMP) that diffused away. Since this nucleotide contained the otherwise reactive 5′-OH group, the 3′-phosphoTyr intermediate was irreversibly trapped, that is, Cre could not carry out the strand exchange reaction in Fig. 30-94 (this nicked DNA is a suicide substrate for Cre; Section 28-3B). The X-ray structure of the Cre complex of this nicked DNA confirmed the presence of the 3′-phosphoTyr intermediate and indicated, through model building, that the 5′-OH group on the missing CMP residue would be well positioned to nucleophilically attack the

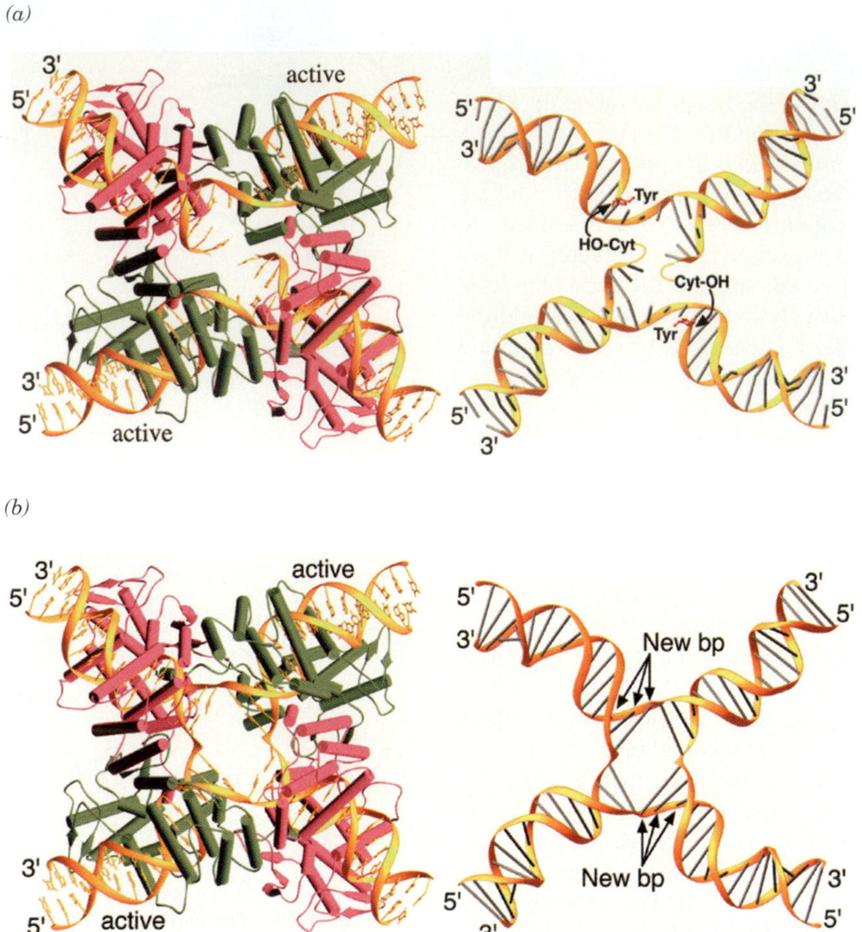

FIGURE 30-95 X-Ray structures of the Cre homotetramer in its complexes with model *loxP* DNAs. (*a*) Two identical dsDNAs that were nicked past the second nucleotide from the 5′ end of their crossover regions; and (*b*) an immobile Holliday junction. The left panels show the Cre–DNA complexes as viewed along their exact 2-fold and pseudo-4-fold axes, with the active and inactive subunits green and magenta, respectively (as in Fig. 30-94), and with the DNA gold. The right panels show only the DNAs in the X-ray structures as viewed from below the left panels. In the right panel of Part *a*, the active site Tyr that is covalently linked to the 3′-OH group of the cleaved DNA strand is shown in stick form (*red*) and the modeled-in position of the cleaved CMP's 5′-OH group is shown positioned to nucleophilically attack the 3′-phosphoTyr group on the opposite dsDNA (*curved arrows*). In the right panel of Part *b*, the three base pairs that form as a consequence of strand exchange are indicated. Note that the vertical strands in the crossovers but not the horizontal strands are distinctly kinked at their centers. [Courtesy of Gregory Van Duyne, University of Pennsylvania School of Medicine. PDBids 2CRX, 3CRX, 4CRX, and 5CRX.]

3′-phosphoTyr bond on the opposite strand (Fig. 30-95*a*). Note that this complex is only 2-fold symmetric although its four Cre subunits and much of the DNA are related by pseudo-4-fold symmetry. When the DNA was, instead, an immobile Holiday junction (Fig. 30-95*b*), the complex was also pseudo-4-fold symmetric with the single strands that had crossed over noticeably kinked at their centers. These structures revealed that the conformational changes necessary to carry out the strand exchange and isomerization reactions (Fig. 30-94) required surprisingly small movements on the part of the Cre subunits and that only the sugar–phosphate backbones of the strand-exchanged nucleotides needed to move in order to form the Holliday junction.

h. Most Transpositions in Eukaryotes Involve RNA Intermediates

Transposons similar to those in prokaryotes also occur in eukaryotes, including yeast, maize, *Drosophila,* and humans. In fact, ~3% of the human genome consists of DNA-based transposons although, in most cases, their sequences have mutated so as to render them inactive, that is, these transposons are evolutionary fossils. However, many eukaryotic transposons exhibit little similarity to those of prokaryotes. Rather, their base sequences resemble those of retroviruses (see below), which suggests that these transposons are degenerate retroviruses. The transposition of these so-called **retrotransposons** occurs via a pathway that resembles the replication of retroviral DNA

(Section 15-4C): (1) their transcription to RNA, (2) the reverse transcriptase–mediated copying of this RNA to cDNA (Section 30-4C), and (3) the largely random insertion of this DNA into the host organism's genome as mediated by enzymes known as **integrases** (which catalyze reactions similar to and structurally resemble cut-and-paste DNA transposases).

The involvement of RNA in retrotransposon-mediated transposition was ingeniously shown by Gerald Fink through his remodeling of **Ty1,** the most common transposable element in budding yeast (which has ~35 copies of this 6.3-kb element comprising ~13% of its 1700 kb genome; Ty stands for *T*ransposon *y*east), so that it contained a yeast intron (a sequence that is excised from an RNA transcript and hence is absent in the mature RNA; Section 5-4A) and was preceded by a galactose-sensitive yeast promoter. The transposition rate of this remodeled Ty1 element varied with the galactose concentration in the medium and the transposed elements all lacked the intron, thereby demonstrating the participation of an RNA intermediate.

A retroviral genome (Fig. 30-96*a*) is flanked by direct long terminal repeats **(LTRs)** of 250 to 600 bp and typically contains the genes encoding three polyproteins: **gag,** which is cleaved to the proteins comprising the viral core (Fig. 15-34); **pol,** which is cleaved to the above-mentioned reverse transcriptase and integrase, as well as the protease that catalyzes these cleavages; and **env,** which is cleaved to viral outer envelope proteins. Ty1 (Fig. 30-96*b*) is likewise flanked by LTRs (of 330 bp) but expresses only two polyproteins: **TYA** and **TYB,** the counterparts of gag and pol. Moreover, TYA and TYB, together with Ty1 RNA, form viruslike particles in the yeast cytoplasm. However, Ty1 lacks a counterpart of the retroviral *env* gene. Hence Ty1 is an "internal virus" that can only replicate within a genome, albeit at an extremely low rate compared to that of real retroviral infections. ***Copia*** (Latin for abundance), the most abundant retrotransposon in the *Drosophila* genome (which contains 20–60 copies of copia), resembles Ty1.

The LTRs in retroviruses and retrotransposons such as Ty1 and *copia* are essential elements for their transcription and hence for their transposition. Yet, vertebrate genomes also contain retrotransposons that lack LTRs and hence cannot be transcribed analogously to retroviruses. A common family of these **nonviral retrotransposons,** the 1- to 7-kb **long interspersed nuclear elements (LINEs),** each contain two open reading frames: *ORF1,* which contains sequences similar to those in *gag;* and *ORF2,* which contains sequences similar to those encoding reverse transcriptase. A proposed mechanism for the transposition of LINEs is diagrammed in Fig. 30-97.

Different types of transposons, DNA-only, retroviral, and nonviral, predominate in different organisms. Thus

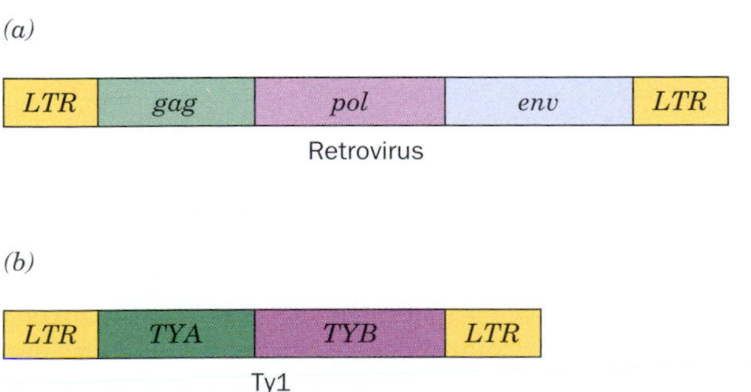

FIGURE 30-96 Gene sequences of (*a*) retroviruses and (*b*) the Ty1 retrotransposon.

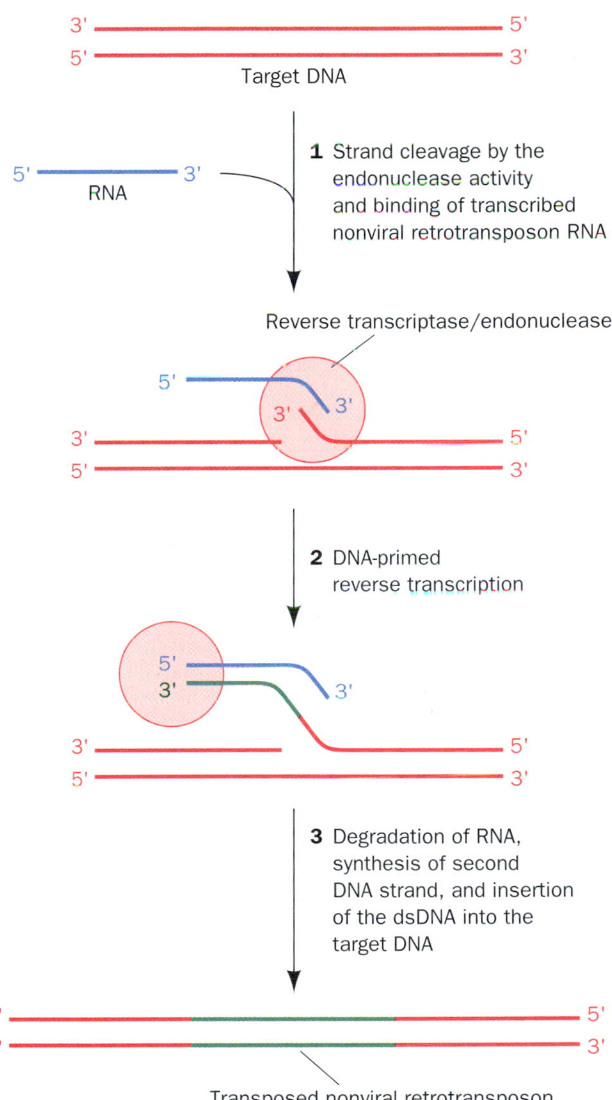

FIGURE 30-97 Proposed mechanism for the transposition of nonviral retrotransposons. (1) The retrotransposon encoded reverse transcriptase/endonuclease nicks one strand of the target DNA and then recruits the RNA transcript of the retrotransposon to this site. **(2)** The DNA-primed reverse transcription of the retrotransposon RNA. **(3)** The RNA is degraded, and the second DNA strand is synthesized using the first strand as its template (normal reverse transcriptase reactions; Section 30-4C), followed by the insertion of the resulting nonviral retrotransposon into the target DNA via a poorly understood process.

bacteria, as we have seen, contain nearly exclusively DNA-only transposons, yeast have mainly retroviral retrotransposons, *Drosophila* have all three types, and in humans LINES predominate. In fact, the human genome contains an estimated 1.4 million LINEs or LINE fragments that comprise ~20% of the 3.2-billion-bp human genome (genomic organization is discussed in Section 34-2). The great majority of these molecular parasites have mutated to the point of inactivity but a few still appear capable of further transposition. Indeed, several hereditary diseases are caused by the insertion of a LINE into a gene. Several other types of retrotransposons also comprise significant fractions of the human genome as we shall see in Section 34-2.

7 ■ DNA METHYLATION AND TRINUCLEOTIDE REPEAT EXPANSIONS

The A and C residues of DNA may be methylated, in a species-specific pattern, to form **N^6-methyladenine (m^6A)**, **N^4-methylcytosine (m^4C)**, and **5-methylcytosine (m^5C)** residues, respectively.

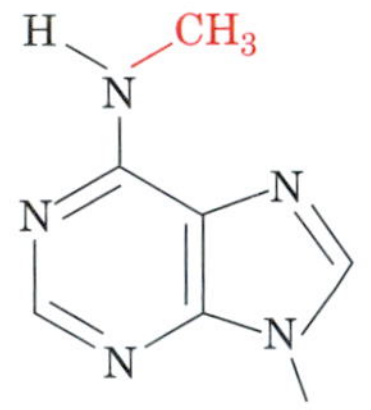

N^6-Methyladenine (m^6A) residue

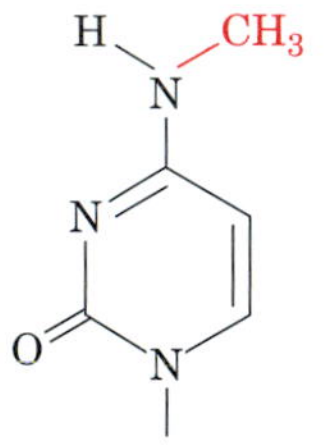

N^4-Methylcytosine (m^4C) residue

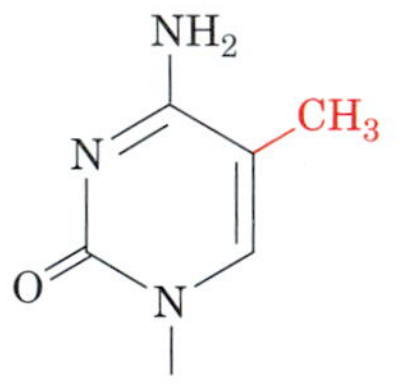

5-Methylcytosine (m^5C) residue

These are the only types of modifications to which DNA is subjected in cellular organisms (although all the C residues of T-even phage DNAs are converted to **5-hydroxymethylcytosine** residues,

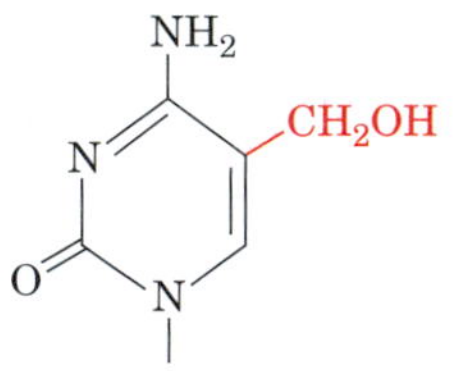

5-Hydroxymethylcytosine residue

which may, in turn, be glycosylated). These methyl groups project into B-DNA's major groove, where they can interact with DNA-binding proteins. In most cells, only a few percent of the susceptible bases are methylated, although this figure rises to >30% of the C residues in some plants.

Bacterial DNAs are methylated at their own particular restriction sites, thereby preventing the corresponding restriction endonucleases from degrading the DNA (Section 5-5A). These restriction–modification systems, however, account for only part of the methylation of bacterial DNAs. In *E. coli,* most DNA methylation is catalyzed by the products of the *dam* and *dcm* genes. The **Dam methyltransferase (Dam MTase)** methylates the A residue in all GATC sequences, whereas the **Dcm MTase** methylates both C residues in CC$^{A}_{T}$GG at their C5 positions. Note that both of these sequences are palindromic. We have seen that *E. coli* uses Dam MTase-mediated methylation to differentiate parental from newly synthesized DNA in mismatch repair (Section 30-5C) and in limiting *oriC*-based DNA replication initiation to once per cell generation via sequestration (Section 30-3C).

a. The MTase Reaction Occurs via a Covalent Intermediate in Which the Target Base Is Flipped Out

The Dam and Dcm MTases, as do all known DNA MTases, use *S*-adenosylmethionine (SAM) as their methyl donor. Indeed, all m^5C-MTases share a set of conserved sequence motifs. Daniel Santi has proposed that the catalytic mechanism of these m^5C-MTases (Fig. 30-98) is similar to that of thymidylate synthase (Fig. 28-19) in that both types of enzymes transfer methyl groups to pyrimidine C5 atoms via a reaction that is initiated by the nucleophilic attack of a Cys thiolate group on the pyrimidine's C6 position. The pyrimidine's C5 atom is thereby activated as a resonance-stabilized carbanion that nucleophilically attacks the methyl donor's methyl group (which in thymidylate synthase is donated by N^5,N^{10}-methylene-THF rather than SAM) to yield a covalent intermediate. This intermediate subsequently decomposes to products through the enzymatic abstraction of the proton substituent to C5 and elimination of the enzyme. The Cys thiolate nucleophile is a component of a Pro-Cys dipeptide that is invariant in all known m^5C-MTases and thymidylate synthases.

This mechanism is supported by the observation that the action of m^5C-MTases on a **5-fluorocytosine (f^5C)** residue

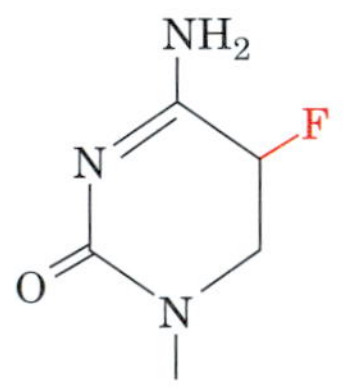

5-Fluorocytosine (f^5C) residue

irreversibly traps the covalent intermediate (and hence inactivates the enzyme) because the enzyme cannot abstract fluorine, the most electronegative element, as an F^+ ion (5-fluorodeoxyuridylate is likewise a suicide substrate for thymidylate synthase; Section 28-3B). Stereochemical principles dictate that the enzyme's Cys thiolate group can

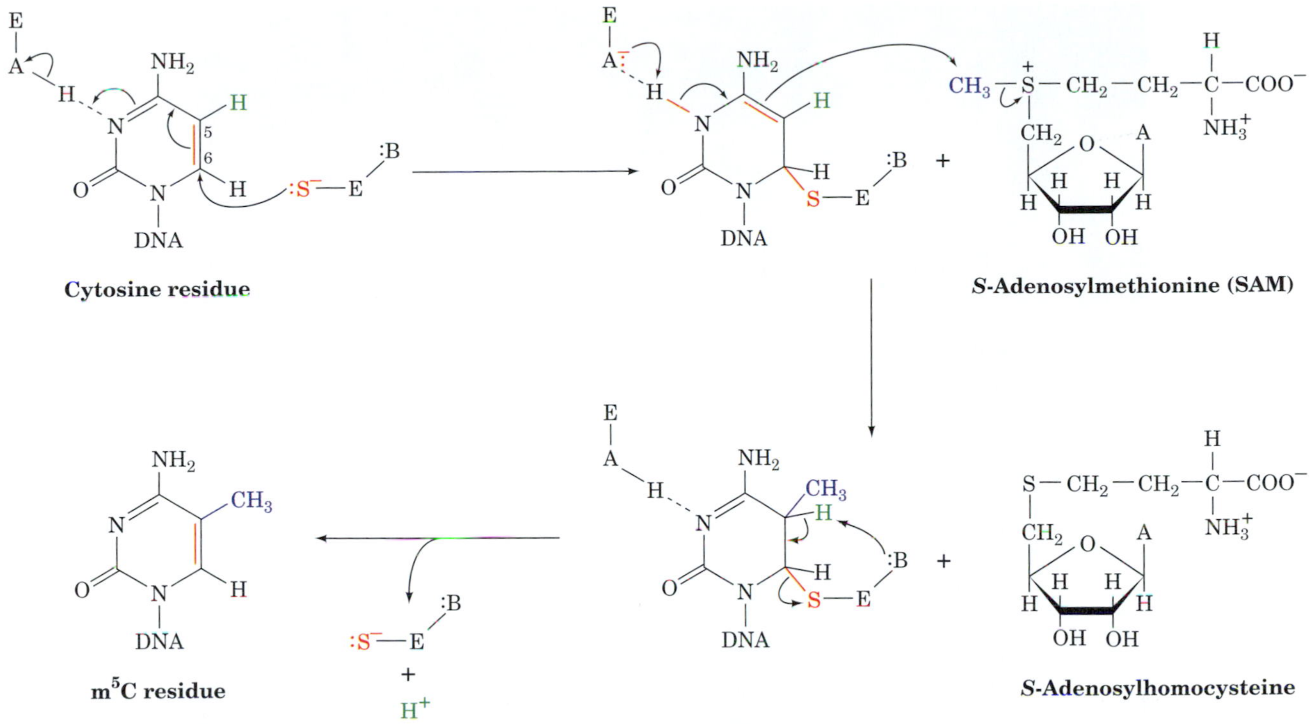

Cytosine residue

S-Adenosylmethionine (SAM)

m^5C residue

S-Adenosylhomocysteine

FIGURE 30-98 The catalytic mechanism of 5-methylcytosine methyltransferases (m^5C-MTases). The methyl group is supplied by SAM, which thereby becomes *S*-adenosylhomocysteine. In M.HhaI, the DNA MTase from *Haemophilus haemolyticus,* the active site thiolate group, $^-$S—E, is on Cys 81, the enzyme general acid, E—A, is Glu 119, and the enzyme general base, E—B, has not been identified. [After Verdine, G.L., *Cell* **76,** 198 (1994).]

nucleophilically attack cytosine's C5 position only from above or below the ring. This is possible because, as we shall see below, the enzyme induces its cytosine target to flip out of the DNA double helix.

The DNA MTase from *Haemophilus haemolyticus* **(M.HhaI),** a 327-residue monomer, is a component of this bacterium's restriction–modification system. M.HhaI methylates its recognition sequence, 5′-GCGC-3′ in double-stranded DNA, to yield 5′-G-m^5C-GC-3′. Richard Roberts and Xiaodong Cheng determined the X-ray structure of the inactivated M.HhaI–DNA complex formed by incubating the enzyme with the self-complementary sequence d(TGATA**G-f^{5}C-GC**TATC) (in which the enzyme's recognition sequence is in bold) in the presence of SAM. The DNA binds to the enzyme in a large cleft between its two unequally sized domains (Fig. 30-99). The structure's most striking feature is that the f^5C nucleotide has flipped out of the minor groove in the otherwise largely undistorted B-DNA helix and has inserted into the enzyme's active site. There, the f^5C has reacted with SAM so as to yield adenosylhomocysteine (SAM without its methyl group) and the methylated intermediate covalently linked to Cys 81. The side chain of Gln 237 fills the cavity in the DNA double helix left by the departure of the f^5C by hydrogen bonding to the opposing G base. Comparison of this structure with that of M.HhaI in complex only with SAM indicates that on binding the DNA the protein's so-called active site loop (residues 80–99) swings around to contact the DNA, a movement of up to 25 Å. Nearly all base-specific interactions are made in the major groove by two Gly-rich loops (residues 233–240 and 250–257), the so-called recognition loops. The protein also makes extensive sequence-nonspecific contacts with DNA phosphate groups.

Base flipping was first observed in the above X-ray structure. However, as is now clear from the structures of two other MTases as well as those of variety of DNA repair enzymes (e.g., Sections 30-5A and 30-5B), *base flipping is a common mechanism through which enzymes gain access to the bases in dsDNA on which they perform chemistry.*

b. DNA Methylation in Eukaryotes Functions in Gene Regulation

5-Methylcytosine is the only methylated base in most eukaryotic DNAs, including those of vertebrates. This modification occurs largely in the CG dinucleotide of various palindromic sequences. CG is present in the vertebrate genome at only about one-fifth its randomly expected frequency. The upstream regions of many genes, however, have normal CG frequencies and are therefore known as **CpG islands.**

The degree of eukaryotic DNA methylation and its pattern are conveniently assessed by comparing the Southern blots (Section 5-5D) of DNA cleaved by the restriction endonucleases *Hpa*II (which cleaves CCGG but not C-m^5C-GG) and *Msp*I (which cleaves both). Such

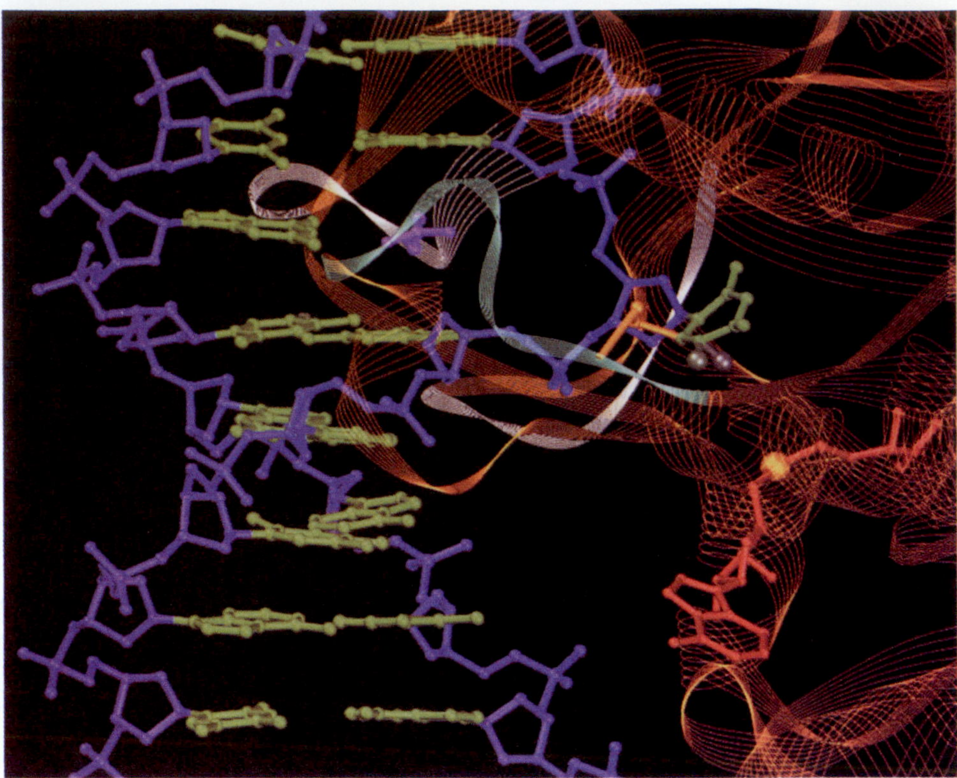

FIGURE 30-99 X-Ray structure of M.HhaI in complex with *S*-adenosylhomocysteine and a duplex 13-mer DNA containing a methylated f^5C residue at the enzyme's target site. The DNA is shown in ball-and-stick form with its bases green and its sugar–phosphate backbone purple. The protein backbone is represented by a multiline orange ribbon with its active site loop (residues 80–89) cyan and its two recognition loops (residues 233–240 and 250–257) white. The latter interact with the DNA's target sequence in its major groove from the back of the drawing. The methylated f^5C residue has swung out of the DNA into the enzyme's active site pocket, where its C6 forms a covalent bond with the S atom of Cys 81 (*yellow*). Both the methyl group and the F atom substituent to C5 of the f^5C base are represented by silver spheres because the X-ray structure cannot confidently differentiate them. The flipped out f^5C base is replaced in the DNA double helix by the side chain of Gln 237 (*magenta*), which hydrogen bonds to the "orphaned" guanine base. The adenosylhomocysteine (*red*) is shown in ball-and-stick form with its S atom, the methyl donor in the SAM that methylated the f^5C, represented by a yellow sphere. [Based on an X-ray structure determined by Richard Roberts, New England Biolabs, Beverly, Massachusetts, and Xiaodong Cheng, Cold Spring Harbor Laboratory, Cold Spring Harbor, New York. PDBid 1MHT.]

studies indicate that eukaryotic DNA methylation varies with the species, the tissue, and the position along a chromosome. The m^5C residues in a given DNA segment can be identified through **bisulfite sequencing,** in which the DNA is reacted with **bisulfite ion** (HSO_3^-), which selectively deaminates C (but not m^5C) residues to U, followed by PCR amplification (Section 5-5F), which copies these U's to T's and the m^5C's to C's. Comparison of the sequences of the amplified DNA with that of untreated DNA (as determined by the chain-terminator method; Section 7-2A) reveals which C's in the untreated DNA are methylated.

There is clear evidence that *DNA methylation switches off eukaryotic gene expression, particularly when it occurs in the promoter regions upstream of a gene's transcribed sequences.* For example, globin genes are less methylated in erythroid cells than they are in nonerythroid cells and, in fact, the specific methylation of the control region in a recombinant globin gene inhibits its transcription in transfected cells. In further support of the inhibitory effect of DNA methylation is the observation that **5-azacytosine (5-azaC),**

NH$_2$ / N / N / O / N

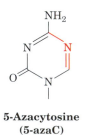

5-Azacytosine (5-azaC)

a base analog that cannot be methylated at its N5 position and that inhibits DNA MTases, stimulates the synthesis of several proteins and changes the cellular differentiation patterns of cultured eukaryotic cells. The observation that repetitive intragenic parasites such as LINEs are highly methylated in somatic tissues has led to the hypothesis that CpG methylation in mammals arose to prevent the spurious transcriptional initiation of these retrotransposons.

The way in which DNA methylation prevents gene expression is poorly understood. However, in many cases, DNA methylation is recognized by a family of proteins that contain a conserved **methyl-CpG binding domain (MBD).** Since the methyl groups of m^5C residues extend into dsDNA's major groove, MBDs can bind to them without perturbing DNA's double helical structure. MBD-containing proteins inhibit the transcription of their bound promoter-methylated genes by recruiting protein complexes that induce the alteration of the local chromosome structure in a way that prevents the transcription of the associated gene (eukaryotic chromosome structure is discussed in Section 34-1). Another possibility has been raised by the observation that the methylation of synthetic poly(GC) stabilizes its Z-DNA conformation. Perhaps the formation of Z-DNA, which has been detected *in vivo* (Section 29-1B), acts as a conformational switch to turn off local gene expression.

c. DNA Methylation in Eukaryotes Is Self-Perpetuating

The palindromic nature of DNA methylation sites in eukaryotes permits the methylation pattern on a parental DNA strand to direct the generation of the same pattern in its daughter strand (Fig. 30-100). This **maintenance methylation** would result in the stable "inheritance" of a methylation pattern in a cell line and hence cause these cells to all have the same differentiated phenotype. Such changes to the genome are described as being **epigenetic** (Greek: *epi,* upon or beside) because they provide an additional layer of information that specifies when and where specific portions of the otherwise fixed genome are expressed (an epigenetic change that we have already encountered is the lengthening of telomeres in germ cells; Section 30-4A). Epigenetic characteristics, as we shall see, are not bound by the laws of Mendelian inheritance.

There is considerable experimental evidence favoring the existence of maintenance methylation, including the observation that artificially methylated viral DNA, on transfection into eukaryotic cells, maintains its methylation pattern for at least 30 cell generations. Maintenance methylation in mammals appears to be mediated mainly by the **DNMT1** protein, which has a strong preference for methylating hemimethylated substrate DNAs. In contrast, prokaryotic DNA MTases such as M.HhaI do not differentiate between hemimethylated and fully methylated substrate DNAs. The importance of maintenance methylation is demonstrated by the observation that mice that are homozygous for deletion of the *DNMT1* gene die early in embryonic development.

The pattern of DNA methylation in mammals varies in early embryological development. DNA methylation levels are high in mature gametes (sperm and ova) but are nearly eliminated by the time a fertilized ovum has become a **blastocyst** (a hollow ball of cells, the stage at which the embryo implants into the uterine wall; embryonic development is discussed in Section 34-4A). After this stage, however, the embryo's DNA methylation levels globally rise until, by the time the embryo has reached the developmental stage known as a **gastrula,** its DNA methylation

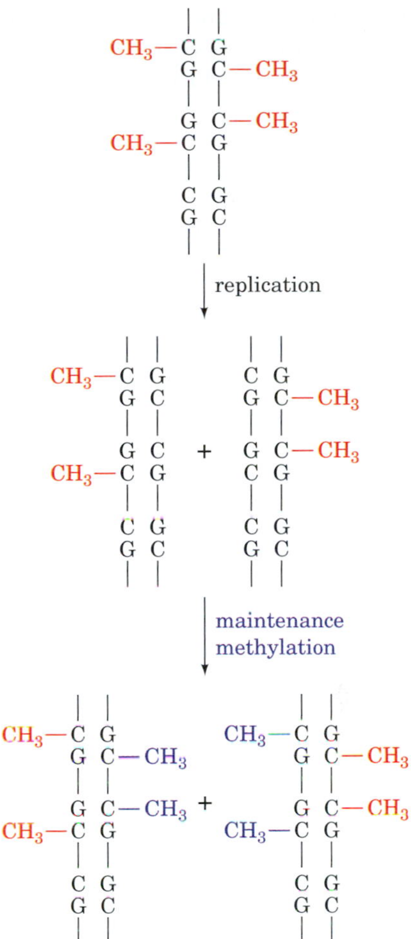

FIGURE 30-100 Maintenance methylation. The pattern of methylation on a parental DNA strand induces the corresponding methylation pattern in the complementary strand. In this way, a stable methylation pattern may be maintained in a cell line.

levels have risen to adult levels, where they remain for the lifetime of the animal. This *de novo* (new) methylation appears to be mediated by two DNA MTases distinct from DNMT1 named **DNMT3a** and **DNMT3b.** An important exception to this remethylation process is that the CpG islands of germline cells (cells that give rise to sperm or ova) remain unmethylated. This ensures the faithful transmission of the CpG islands to the succeeding generation in the face of the strong mutagenic pressure of m^5C deamination (which yields T, a mutation that mismatch repair occasionally fails to correct).

The change in DNA methylation levels (epigenetic reprogramming) during embryonic development suggests that the pattern of genetic expression differs in embryonic and somatic cells. This explains the observed high failure rate in cloning mammals (sheep, mice, cattle, etc.) by transferring the nucleus of an adult cell into an enucleated oocyte (immature ovum). Few of these animals survive to birth, many of those that do so die shortly thereafter, and most of the ~1% that do survive have a variety of abnormalities, most prominently an unusually large size.

However, the survival of any embryos at all is indicative that the oocyte has the remarkable capacity to epigenetically reprogram somatic chromosomes (although it is rarely entirely successful in doing so) and that mammalian embryos are relatively tolerant of epigenetic abnormalities. Presumably, the reproductive cloning of humans from adult nuclei would result in similar abnormalities and for this reason (in addition to social and ethical prohibitions) should not be attempted.

d. Genomic Imprinting Results from Differential DNA Methylation

It has been known for thousands of years that maternal and paternal inheritance can differ. For example, a mule (the offspring of a mare and a male donkey) and a hinny (the offspring of a stallion and a female donkey) have obviously different physical characteristics, a hinny having shorter ears, a thicker mane and tail, and stronger legs than a mule. This is because, in mammals only, certain maternally and paternally supplied genes are differentially expressed, a phenomenon termed **genomic imprinting.** The genes that are subject to genomic imprinting are, as Rudolph Jaenisch has shown, differentially methylated in the two parents during gametogenesis and the resulting different methylation patterns are resistant to the wave of demethylation that occurs during the formation of the blastocyst and to the wave of *de novo* methylation that occurs thereafter.

The importance of genomic imprinting is demonstrated by the observation that an embryo derived from the transplantation of two male or two female pronuclei into an ovum fails to develop (pronuclei are the nuclei of mature sperm and ova before they fuse during fertilization). Inappropriate imprinting is also associated with certain diseases. For example, **Prader-Willi syndrome (PWS),** which is characterized by the failure to thrive in infancy, small hands and feet, marked obesity, and variable mental retardation, is caused by a >5000-kb deletion in a specific region of the paternally inherited chromosome 15. In contrast, **Angelman syndrome (AS),** which is manifested by severe mental retardation, a puppetlike ataxic (uncoordinated) gait, and bouts of inappropriate laughter, is caused by a deletion of the same region from the maternally inherited chromosome 15. These syndromes are also exhibited by those rare individuals who inherit both their chromosomes 15 from their mothers for PWS and from their fathers for AS. Evidently, certain genes on the deleted chromosomal region must be paternally inherited to avoid PWS and others must be maternally inherited to avoid AS. Several other human diseases are also associated with either maternal or paternal inheritance or lack thereof.

e. DNA Methylation Is Associated with Cancer

The mutation of an m^5C residue to T (with its associated G to A mutation on the complementary strand) is, by far, the most prevalent mutational change in human cancers. Such mutations usually convert proto-oncogenes to oncogenes (Section 19-3B) or inactivate tumor suppressors (Fig. 34-4C). In addition, the hypomethylation of proto-oncogenes and the hypermethylation of genes encoding tumor suppressors are associated with cancers, although it is unclear whether these are initiating or consolidating events for malignancies.

f. Several Neurological Diseases Are Associated with Trinucleotide Repeat Expansions

Fragile X syndrome, whose major symptoms include mental retardation and a characteristic long, narrow face with large ears, afflicts 1 in 4500 males and 1 in 9000 females. Fragile X syndrome is so named because, in affected individuals, the tip of the X chromosome's long arm is connected to the rest of the chromosome by a slender thread that is easily broken. The genetics of this condition are bizarre. The maternal grandfathers of individuals having fragile X syndrome may be asymptomatic, both clinically and cytogenetically. Their daughters are likewise asymptomatic, but these daughters' children of either sex may have the syndrome. Evidently, the fragile X defect is activated by passage through a female. Moreover, the probability of a child having fragile X syndrome and the severity of the disease increase with each succeeding generation, a phenomenon termed **genetic anticipation.**

The affected gene in fragile X syndrome, *FMR1* (for *f*ragile X *m*ental *r*etardation *1*), encodes a 632-residue RNA-binding protein named **FMRP** (for *FMR* protein), which apparently functions in the transport of certain mRNAs from the nucleus to the cytoplasm (Section 34-3C), where it probably regulates their translation. FMRP, which is highly conserved in vertebrates, is expressed in most tissues but most heavily in brain neurons, where a variety of evidence indicates that its participation is required for the proper formation and/or function of synapses.

In the general population, the 5′ untranslated region of *FMR1* contains a polymorphic $(CGG)_n$ sequence with n ranging from 6 to 60 and often punctuated by one or two AGG interruptions. However, in certain asymptomatic individuals, n has increased from 60 to 200, a so-called premutation that males transmit in unchanged form to their daughters (they transmit a Y rather than an X chromosome to their sons). In the daughters' children, however, ~80% of the individuals inheriting a premutant *FMR1* gene exhibit an astonishing expansion (amplification) of the triplet repeat with n ranging from >200 to several thousand, as well as the symptoms of the disease, a so-called full mutation. These triplet repeats differ in size among siblings and often exhibit heterogeneity within an individual, suggesting that they are somatically generated.

These **dynamic mutations,** which expand more often than they contract, probably arise through slippage of the template DNA during replication. Slippage is thought to occur through the formation of loop-outs (Fig. 30-101) on either the newly synthesized strand (causing expansions) or on the template strand (causing contractions). Since the lagging strand has more single-stranded character than the leading strand, most slippage probably occurs during lagging strand synthesis. As expected, the frequency of slippage increases with the number of repeats.

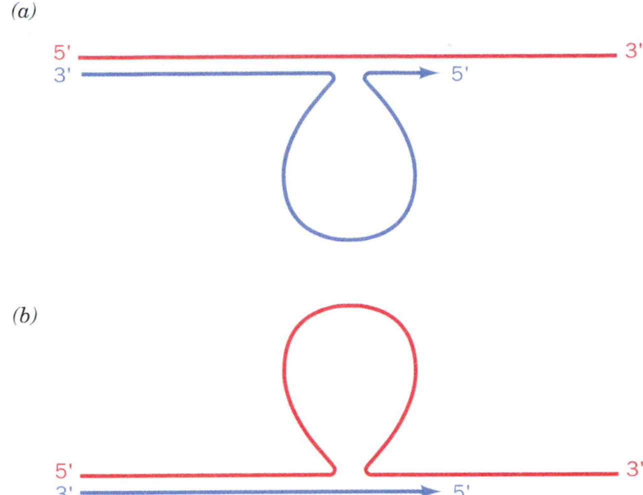

FIGURE 30-101 The loop-out mechanism for the alteration of the number of consecutive triplet repeats in DNA through its replication. Here the template strand is red and its nascent (newly synthesized) daughter strand is blue. With long tracts of repeating sequences, the probability of a loop-out occurring increases because its neighboring sequences will remain base paired. (*a*) If the daughter strand loops out, the number of repeats increases. (*b*) If the template strand loops out, the number of repeats decreases.

The peculiar genetics of fragile X syndrome is a consequence of genomic imprinting through methylation. The *FMR1* gene is unmethylated in normal individuals. However, it is hypermethylated in individuals with a maternally transmitted full mutation. This maintains the *FMR1* gene in a transcriptionally silent (inactive) state, thereby accounting for the symptoms of the disease. The lesser frequency and severity of fragile X syndrome in females are accounted for by the fact that females have two X chromosomes, one of which is unlikely to be mutated.

Thirteen other pathological instances of the expansion of a GC-rich trinucleotide repeat, all also neurological diseases, are known including the following:

1. **Myotonic dystrophy (DM),** the most common adult form of **muscular dystrophy** (its estimated incidence is 1 in 8000). It is a multisystem autosomal dominant disorder that is mainly characterized by progressive muscle weakness and wasting, the severity of which increases with successive generations while the age of onset decreases (genetic anticipation). Its most severe form, congenital DM, is exclusively maternally transmitted. DM arises from a trinucleotide expansion in the 3′ untranslated region of the gene encoding **myotonic dystrophy protein kinase (MDPK),** which is expressed in the neurons affected by DM. The repeating triplet, $(CAG)_n$, is present in between 5 and 30 copies in the MDPK gene of normal individuals but expands from at least 50 repeats in minimally affected individuals to ~2000 repeats in severely affected individuals.

2. **Huntington's disease (HD;** previously called **Huntington's chorea),** a devastating neurodegenerative disorder characterized by progressively choreic (disordered) movements, cognitive decline, and emotional disturbances over an average 18-year course that is inevitably fatal. This dominant autosomal disorder, which affects ~1 in 10,000 individuals and has an average age of onset of ~40 years, is a consequence of the selective loss of certain groups of neurons in the brain. The *HD* gene, which encodes a widely expressed 3145-residue polypeptide named **huntingtin,** contains a polymorphic trinucleotide repeat, $(CAG)_n$, within its polypeptide coding sequence. The *HD* genes from 150 independent families with HD all contained between 37 and 86 repeat units, whereas those from normal individuals had 11 to 34 repeats. Moreover, the *HD* repeat length is unstable: >80% of meiotic transmissions show either increases or decreases, with the largest increases occurring in paternal transmissions (genomic imprinting). The number of repeats in afflicted individuals is inversely correlated with the age of onset of HD.

CAG is the codon for Gln (Table 5-3) and hence mutant huntingtin contains a long poly(Gln) tract. Synthetic poly(Gln) aggregates as β sheets that are linked by hydrogen bonds involving both their main chain and side chain amide groups. Indeed, the nuclei of HD-affected neurons contain inclusions that presumably consist of aggregates of huntingtin or its proteolytic products. It is these inclusions, as Max Perutz pointed out, that apparently kill the neurons in which they are contained, although the mechanism of how they do so is unknown. The long incubation period before the symptoms of HD become evident is attributed to the lengthy nucleation time for aggregate formation, much like what we have seen occurs in the formation of amyloid fibrils (Section 9-5A).

3. **Spinocerebellar ataxia (SCA) type 1,** a progressive neurodegenerative disease whose age of onset is typically in the third or fourth decade, although it exhibits genetic anticipation. Like HD, it is caused by selective neuronal loss and is associated with an expansion of a CAG repeat in a coding region, in this case of a neuronal protein named **ataxin-1.** There it expands from ~28 to between 43 and 81 copies, thereby yielding a poly(Gln) tract of increased length (and tendency to aggregate). Four similar diseases, **SCA types 2, 3, 6,** and **7,** are caused by CAG expansions in different neuronal proteins.

CHAPTER SUMMARY

1 and 2 ■ DNA Replication DNA is replicated in the 5′ → 3′ direction by the assembly of deoxynucleoside triphosphates on complementary DNA templates. Replication is initiated by the generation of short RNA primers, as mediated in *E. coli* by primase and RNA polymerase. The DNA is then extended from the 3′ ends of the primers through the action of DNA polymerase (Pol III in *E. coli*). The leading strand at a replication fork is synthesized continuously, whereas the lagging strand is synthesized discontinuously by the formation of Okazaki fragments. RNA primers on newly synthesized DNA are excised and replaced by DNA through Pol I-catalyzed (in *E. coli*) nick translation. The single-strand nicks are then sealed by DNA ligase. Mispairing errors during DNA synthesis are corrected by the $3' \rightarrow 5'$ exonuclease functions of both Pol I and Pol III. The Klenow fragment of Pol I and other DNA polymerases of known structure have a right-hand-like structure with the active site located in the palm domain. Pol I recognizes the incoming nucleotide according to the shape of the base pair it forms with the template base and catalyzes the formation of a phosphodiester bond via a mechanism involving two metal ions. DNA synthesis in *E. coli* requires the participation of many auxiliary proteins including DNA gyrase, DnaB helicase, single-strand binding protein (SSB), primase, the β_2 sliding clamp, and DNA ligase.

3 ■ Prokaryotic Replication DNA synthesis commences from specific sites known as replication origins. In the synthesis of the bacteriophage M13 (−) strand on the (+) strand template, the origin is recognized and primer synthesis is initiated by RNA polymerase. The analogous process in bacteriophage ϕX174, as well as in *E. coli,* is mediated by a complex primase-containing particle known as the primosome. ϕX174 (+) strands are synthesized according to the looped rolling circle mode of DNA replication on (−) strand templates of the replicative form in a process that is directed by the virus-specific gene *A* protein.

The *E. coli* chromosome is bidirectionally replicated in the θ mode from a single origin, *oriC,* which is recognized by DnaA protein. Leading strand synthesis is probably primed by RNA polymerase and primase working together, whereas Okazaki fragments are primed by primase in the primosome. The uncontrolled initiation of DNA replication is prevented by the sequestration of newly synthesized and hence hemimethylated *oriC* by membrane-associated SeqA protein, which prevents the *oriC* from becoming fully methylated at its multiple GATC sites. The β_2 sliding clamp, which is responsible for Pol III's processivity, is loaded onto the DNA by the $\gamma_3\delta\delta'$ clamp loader in an ATP-driven process. The δ subunit, when unmasked by ATP binding to the γ subunits, acts as a molecular "wrench" to spring open the sliding clamp, thereby permitting the entry of a single-stranded template DNA. Replication termination is facilitated by Tus protein which, on binding to an appropriately oriented *Ter* site, arrests the motion of a replication fork by binding to DnaB helicase. The great complexity of the DNA replication process functions to ensure the enormous fidelity necessary to maintain genome integrity.

4 ■ Eukaryotic Replication Progression through the eukaryotic cell cycle is mediated by cyclins complexed to their cognate cyclin-dependent protein kinases (Cdks). Chromosomal DNA replication is initiated by pol α/primase, which synthesizes a primer followed by a short length of DNA. Then, via polymerase switching mediated by replication factor C (RFC), the eukaryotic clamp loader, pol δ processively synthesizes both the lagging and leading strands in complex with PCNA, the eukaryotic sliding clamp.

Eukaryotic chromosomal DNA is synthesized in multiple origin-containing segments known as replicons. Nevertheless, chromosomal DNA is synthesized once and only once per cell cycle. The re-replication of DNA is prevented because replication initiation is licensed only in the G_1 phase of the cell cycle by the formation of the pre-replicative complex (pre-RC) but DNA is synthesized only in S phase by the activation of the pre-RC. The pre-RC is assembled in G_1 phase by the binding of the origin recognition complex (ORC) to an origin, which recruits Cdc6/Cdc18 and Cdt1 followed by the MCM complex, the replicative helicase. The activation of the pre-RC begins in S phase with the addition of Mcm10 followed by the phosphorylation of many of the pre-RC's subunits by Cdks and Ddk. Cdc45 then binds followed by pol α/primase, pol ε, PCNA, and the replication protein A (RPA), the SSB counterpart, to yield the active initiation complex. Re-replication is prevented through the actions of Cdks, which cause the elimination of Cdc6/Cdc18 and inhibit the helicase activity of the MCM complex. In metazoan cells, re-replication is also prevented by the binding of geminin to Cdt1.

Mitochondrial DNA is replicated in the D-loop mode by DNA polymerase γ. Retroviruses produce DNA on RNA templates in a reaction sequence catalyzed by reverse transcriptase. Telomeric DNA, a G-rich repeating octamer on the 3′-ending strand, is synthesized by the RNA-containing enzyme telomerase. Telomerase is active in germ cells but not somatic cells, a phenomenon that is at least in part responsible for cellular senescence and aging. The observation that telomerase is active in nearly all cancer cells suggests that telomerase inactivation is a defense against the development of cancer. The free DNA ends of telomeres are capped to prevent them from triggering DNA damage checkpoints. The *O. nova* telomere end binding protein (TEBP) binds both single strands of telomere DNA and a G-quartet-containing dimer, whereas humans and yeast have an unrelated telomere end-binding protein named Pot1. Telomeric DNA forms T-loops which are formed by the TRF2-mediated invasion of the 3′ telomeric overhang into repeating telomeric dsDNA to form a D-loop.

5 ■ DNA Repair Cells have a great variety of DNA repair mechanisms. DNA damage may be directly reversed such as in the photoreactivation of UV-induced pyrimidine dimers or in the repair of O^6-alkylguanine lesions by the transfer of the offending alkyl group to a repair protein. Pyrimidine dimers, as well as many other types of DNA lesions, may also be removed by nucleotide excision repair (NER), which in *E. coli* involves the UvrABC system. Xeroderma pigmentosum, an inherited human disease characterized by marked UV-induced skin changes and a greatly increased incidence of cancer, is caused by defects in any of seven complementation groups that participate in NER. In base excision repair (BER), DNA glycosylases specifically remove the corresponding chemically altered bases, including uracil, through mechanisms that involve base flip-outs to form AP sites. The AP sites are cleaved on one side by an AP endonuclease, removed together with adjacent residues by an exonuclease, and replaced through the actions of a DNA polymerase and a DNA ligase. In mismatch repair (MMR), base pairing mismatches arising from replication errors are corrected. In *E. coli* MMR, MutS and MutL bind to the mismatch and then identify the daughter strand, which contains the error, according to which strand of the nearest hemimethylated GATC palindrome is unmethylated. MutH then cleaves this strand, which is excised past the mismatch and replaced.

DNA damage in *E. coli* induces the SOS response, a LexA- and RecA-mediated process in which the error-prone bypass DNA polymerases Pol IV and Pol V replicate a damaged template DNA even if it provides no information as to which base to incorporate. Double-strand break (DSB) repair by nonhomologous end-joining (NHEJ) is facilitated by Ku protein, which holds two dsDNA ends together for ligation by DNA ligase IV in complex with Xrcc4. The high correlation between mutagenesis and carcinogenesis permits the detection of carcinogens by the Ames test.

6 ■ Recombination and Mobile Genetic Elements Genetic information may be exchanged between homologous DNA sequences through homologous recombination, a process that occurs according to the Holliday model. In *E. coli,* strand invasion to form Holliday junctions is mediated by RecA after the RecBCD-mediated generation of the single-strand nicks to which RecA binds. Branch migration is mediated by RuvAB, which consists of a homotetramer (or a homooctamer) of RuvA, which binds both a Holliday junction and two oppositely located RuvB hexamers. In an ATP-driven process, the RuvB hexamers (effectively) counter-rotate to pump the dsDNA stems into the center of the RuvA-bound Holliday junction, where each of its single strands exchange base pairing partners to form new dsDNA stems, which are translocated toward the periphery of the complex. The Holliday junction is eventually resolved to its component dsDNAs by RuvC.

The primary function of homologous recombination is to repair damaged replication forks resulting from the encounters of replisomes with unrepaired single-strand lesions or breaks. DSBs may be rejoined via a recombination repair process called homologous end-joining.

Chromosomes and plasmids may be rearranged through the action of transposons. These DNA segments carry the genes that encode the proteins that mediate the transposition process as well as other genes. Tn5 transposase catalyzes the cut-and-paste transposition of the Tn5 transposon. Replicative transposition proceeds via the intermediacy of cointegrates, which are resolved through the action of enzymes such as the $\gamma\delta$ resolvase. Transposition may be important in chromosomal and plasmid evolution and has been implicated in the control of phenotypic expression such as phase alternation in *Salmonella,* a process that is catalyzed by the Hin DNA invertase, a homolog of the $\gamma\delta$ resolvase. Members of the λ integrase family of proteins, such as Cre recombinase, insert dsDNA segments into their target sites via a Holliday junction intermediate in which transient covalent bonds are formed between active site Tyr side chains and the 3′-OH groups at the cleavage sites. Retrotransposons undergo transposition through an RNA intermediate. Many retrotransposons, such as yeast Ty1, are "internal" retroviruses that can only replicate within a genome. Nonviral retrotransposons, such as LINEs, the dominant transposons in the human genome, have a different transpositional mechanism.

7 ■ DNA Methylation and Trinucleotide Repeat Expansions Prokaryotic DNA may be methylated at its A or C bases. This prevents the action of restriction endonucleases and permits the correct mismatch repair of newly replicated DNA. In most eukaryotes, DNA methylation, which occurs, mainly at CpG islands, through the formation of m^5C, has been implicated in the control of gene expression and, via maintenance methylation, in genomic imprinting.

Several inherited neurological diseases, including fragile X syndrome, myotonic dystrophy, and Huntington's disease, are characterized by the genetically bizarre expansion of segments of repeating GC-rich triplets. If an expanded triplet repeat occurs in an upstream noncoding region of a gene, its aberrant methylation, perhaps through genomic imprinting, may lead to the gene's transcriptional silencing, and if the expanded repeat is instead manifested as a poly(Gln) tract in a protein, the resulting protein aggregates may kill the neurons in which it occurs.

REFERENCES

GENERAL

Adams, R.L.P., Knowler, J.T., and Leader, D.P., *The Biochemistry of the Nucleic Acids* (11th ed.), Chapters 6 and 7, Chapman & Hall (1992).

Kornberg, A., *For Love of Enzymes: The Odyssey of a Biochemist,* Harvard University Press (1989). [A scientific autobiography.]

Kornberg, A. and Baker, T.A., *DNA Replication* (2nd ed.), Freeman (1992). [A compendium of information about DNA replication whose first author is the founder of the field.]

Lewin, B., *Genes VII,* Chapters 12–17 and 33–36, Oxford University Press (2000).

PROKARYOTIC DNA REPLICATION

Baker, T.A. and Wickner, S.H., Genetics and enzymology of DNA replication in *Escherichia coli, Annu. Rev. Genet.* **26,** 447–477 (1992).

Beese, L.S., Derbyshire, V., and Steitz, T.A., Structure of DNA polymerase I Klenow fragment bound to duplex DNA, *Science* **260,** 352–355 (1993).

Benkovic, J.J., Valentine, A.M., and Salinas, F., Replisome-mediated DNA replication, *Annu. Rev. Biochem.* **70,** 181–208 (2001).

Carr, K.M. and Kaguni, J.M., Stoichiometry of DnaA and DnaB protein in initiation at the *Escherichia coli* chromosomal origin, *J. Biol. Chem.* **276,** 44919–44925 (2001).

Caruthers, J.M. and McKay, D.B., Helicase structure and mechanism, *Curr. Opin. Struct. Biol.* **12,** 123–133 (2002).

Crooke, E., Regulation of chromosomal replication in *E. coli:* sequestration and beyond, *Cell* **82,** 877–880 (1995).

Davey, M.J., Jeruzalmi, D., Kuriyan, J., and O'Donnell, M., Motors and switches: AAA$^+$ machines within the replisome, *Nature Rev. Mol. Cell Biol.* **3,** 1–10 (2002).

Doublié, S., Sawaya, M.R., and Ellenberger, T., An open and closed case for all polymerases, *Structure* **7,** R31–R35 (1999). [Reviews the mechanisms of DNA polymerases.]

Frick, D.N. and Richardson, C.C., DNA primases, *Annu. Rev. Biochem.* **70,** 39–80 (2001).

Guckian, K.M., Krugh, T.R., and Kool, E.T., Solution structure of a DNA duplex containing a replicable difluorotoluene–adenine pair, *Nature Struct. Biol.* **5,** 954–959 (1998); *and* Moran, S., Ren, R.X.-F., and Kool, E.T., A thymine triphosphate shape analog lacking Watson–Crick pairing ability is replicated with high sequence specificity, *Proc. Natl. Acad. Sci.* **94,** 10506–10511 (1997).

Jeruzalmi, D., O'Donnell, M., and Kuriyan, J., Clamp loaders and sliding clamps, *Curr. Opin. Struct. Biol.* **12,** 217–224 (2002); Jeruzalmi, D., Yurieva, O., Zhao, Y., Young, M., Stewart, J., Hingorani, M., O'Donnell, M., and Kuriyan, J., Mechanism of processivity clamp opening by the delta subunit wrench of the clamp loader complex of *E. coli* DNA polymerase III, *Cell* **106,** 417–428 (2001); Jeruzalmi, D., O'Donnell, M., and Kuriyan, J., Crystal structure of the processivity clamp loader gamma (γ) complex of *E. coli* DNA polymerase III, *Cell* **106,** 429–441 (2001); *and* Podobnik, M., Weitze, T.F., O'Donnell, M., and Kuriyan, J., Nucleotide-induced conformational change in an isolated *Escherchia coli* DNA polymerase III clamp loader subunit, *Structure* **11,** 253–263 (2003).

Johnson, K.A., Conformational coupling in DNA polymerase fidelity, *Annu. Rev. Biochem.* **62,** 685–713 (1993).

Kamada, K., Horiuchi, T., Ohsumi, K., Shimamoto, N., and Morikawa, K., Structure of a replication–terminator protein complexed with DNA, *Nature* **383,** 598–603 (1996).

Keck, J.L., Roche, D.D., Lynch, A.S., and Berger, J.M., Structure of the RNA polymerase domain of *E. coli* primase, *Science* **287,** 2482–2486 (2000); *and* Podobnik, M., McInerney, P., O'Donnell, M., and Kuriyan, J., A TOPRIM domain in the crystal structure of the catalytic core of *Escherichia coli* primase confirms a structural link to DNA topoisomerases, *J. Mol. Biol.* **300,** 353–362 (2000).

Kelman, Z. and O'Donnell, M., DNA polymerase III holoenzyme: Structure and function of a chromosomal replicating machine, *Annu. Rev. Biochem.* **64,** 171–200 (1995).

Kiefer, J.R., Mao, C., Braman, J.C., and Beese, L.S., Visualizing DNA replication in a catalytically active *Bacillus* DNA polymerase crystal, *Nature* **391,** 304–307 (1998).

Kong, X.-P., Onrust, R., O'Donnell, M., and Kuriyan, J., Three-dimensional structure of the β subunit of *E. coli* DNA polymerase III holoenzyme: A sliding DNA clamp, *Cell* **69,** 425–437 (1992).

Kool, E.T., Active site tightness and substrate fit in DNA replication, *Annu. Rev. Biochem.* **71,** 191–219 (2002); *and* Hydrogen-bonding, base stacking, and steric effects in DNA replication, *Annu. Rev. Biophys. Biomol. Struct.* **30,** 1–2 (2001).

Korolev, S., Hsieh, J., Gauss, G.H., Lohman, T.M., and Waksman, G., Major domain swiveling revealed by the crystal structures of complexes of *E. coli* Rep helices bound to single-stranded DNA and ADP, *Cell* **90,** 635–647 (1997).

Kunkel, T.A. and Bebenek, K., DNA replication fidelity, *Annu. Rev. Biochem.* **69,** 497–529 (2000).

Lee, J.Y., Chang, C., Song, H.K., Moon, J., Yang, J.K., Kim, H.-K., Kwon, S.-T., and Suh, S.W., Crystal structure of NAD^+-dependent DNA ligase: modular architecture and functional implications, *EMBO J.* **19,** 1119–1129 (2000).

Li, Y., Korolev, S., and Waksman, G., Crystal structures of open and closed forms of binary and ternary complexes of the large fragment of *Thermus aquaticus* DNA polymerase I: structural basis for nucleotide incorporation, *EMBO J.* **17,** 7514–7525 (1998).

Naktinis, V., Turner, J., and O'Donnell, M., A molecular switch in the replication machine defined by internal competition for protein rings, *Cell* **84,** 137–145 (1996).

Patel, S.S. and Picha, K.M., Structure and function of hexameric helicases, *Annu. Rev. Biochem.* **69,** 651–697 (2000).

Raghunathan, S., Kozlov, A.G., Lohman, T.M., and Waksman, G., Structure of the DNA binding domain of *E. coli* SSB bound to ssDNA, *Nature Struct. Biol.* **7,** 648–652 (2000).

Singleton, M.R., Sawaya, M.R., Ellenberger, T., and Wigley, D.B., Crystal structure of T7 gene 4 ring helicase indicates a mechanism for sequential hydrolysis of nucleotides, *Cell* **101,** 589–600 (2000).

Soultanas, P. and Wigley, D.B., Unwinding the 'Gordian knot' of helicase action, *Trends Biochem. Sci.* **26,** 47–54 (2001).

Steitz, T.A., DNA polymerases: structural diversity and common mechanisms, *J. Biol. Chem.* **274,** 17395–17398 (1999).

Watson, J.D. and Crick, F.H.C., Genetical implications of the structure of deoxyribonucleic acid, *Nature* **171,** 964–967 (1953). [The paper in which semiconservative DNA replication was first postulated.]

EUKARYOTIC DNA REPLICATION

Allsopp, R.C., Vaziri, H., Patterson, C., Goldstein, S., Younglai, E.V., Futcher, A.B., Greider, C.W., and Harley, C.B., Telomere length predicts replicative capacity of human fibroblasts, *Proc. Natl. Acad. Sci.* **89,** 10114–10118 (1992).

Arezi, B. and Kuchta, R.D., Eukaryotic DNA primase, *Trends Biochem. Sci.* **25,** 572–576 (2000).

Bell, S.P. and Dutta, A., DNA replication in eukaryotic cells, *Annu. Rev. Biochem.* **71,** 333–374 (2002).

Blackburn, E.H., Telomerases, *Annu. Rev. Biochem.* **61,** 113–129 (1992).

Blackburn E.H., Switching and signaling at the telomere, *Cell* **106,** 661–673 (2001); *and* Telomere states and cell fates, *Nature* **408,** 53–56 (2000).

Blow, J.J. and Hodgson, B., Replication licensing—defining the proliferative state? *Trends Cell Biol.* **12,** 72–78 (2002).

Cech, T.R., Life at the end of the chromosome: Telomeres and telomerase, *Angew. Chemie* **39,** 34–43 (2000).

Clayton, D.A., Replication and transcription of vertebrate mitochondrial DNA, *Annu. Rev. Cell Biol.* **7,** 453–478 (1991).

Davies, J.F., II, Almassey, R.J., Hostomska, Z., Ferre, R.A., and Hostomsky, Z., 2.3 Å crystal structure of the catalytic domain of DNA polymerase β, *Cell* **76,** 1123–1133 (1994).

DePamphilis, M.L., Replication origins in metazoan chromosomes: fact or fiction, *BioEssays* **21,** 5–16 (1999).

Diffley, J.F.X., DNA replication: Building the perfect switch, *Curr. Biol.* **11,** R367–R370 (2001).

Ding, J., Das, K., Hsiou, Y., Sarafianos, S.G., Clark, A.D., Jr., Jacobo-Molina, A., Tantillo, C., Hughes, S.H., and Arnold, E., Structure and functional implications of the polymerase active site region in a complex of HIV-1 RT with a double-stranded DNA template-primer and an antibody Fab fragment at 2.8 Å resolution, *J. Mol. Biol.* **284,** 1095–1111 (1998); *and* Jacobo-Molina, A., Ding, J., Nanni, R.G., Clark, A.D., Jr., Lu, X., Tantillo, C., Williams, R.L., Kamer, G., Ferris, A.L., Clark, P., Hizi, A., Hughes, S.H., and Arnold, E., Crystal structure of human immunodeficiency virus type 1 reverse transcriptase complexed with double-stranded DNA at 3.0 Å resolution shows bent DNA, *Proc. Natl. Acad. Sci.* **90,** 6320–6324 (1993).

Franklin, M.C., Wang, J., and Steitz, T.A., Structure of the replicating complex of a pol α family DNA polymerase, *Cell* **105,** 657–667 (2001).

Gilbert, D.M., Making sense out of eukaryotic DNA replication origins, *Science* **294,** 96–100 (2001).

Griffith, J.D., Comeau, L., Rosenfield, S., Stansel, R.M., Bianchi, A., Moss, H., and de Lange, T., Mammalian telomeres end in a large duplex loop, *Cell* **97,** 503–514 (1999).

Hahn, W.C., Counter, C.M., Lundberg, A.S., Beijersbergen, R.L., Brooks, M.W., and Weinberg, R.A., Creation of human tumour cells with defined genetic elements, *Nature* **400,** 464–468 (1999).

Horvath, M.P. and Schultz, S.C., DNA G-quartets in a 1.86 Å resolution structure of an *Oxytricha nova* telomeric protein–DNA complex, *J. Mol. Biol.* **310,** 367–377 (2001).

Hübscher, U., Maga, G., and Spadari, S., Eukaryotic DNA polymerases, *Annu. Rev. Biochem.* **71,** 133–163 (2002); *and* Hübscher, U., Nasheuer, H.-P., and Syväoja, J.E., Eukaryotic DNA polymerases. A growing family, *Trends Biochem. Sci.* **25,** 143–147 (2000).

Jäger, J. and Pata, J.D., Getting a grip: polymerases and their substrate complexes, *Curr. Opin. Struct. Biol.* **9,** 21–28 (1999).

Kelleher, C., Teixeira, M.T., Förstemann, K., and Lingner, J., Telomerase: Biochemical considerations for enzyme and substrate, *Trends Biochem. Sci.* **27,** 572–579 (2002).

Kelly, T.J. and Brown, G.W., Regulation of chromosome replication, *Annu. Rev. Biochem.* **69,** 829–880 (2000).

Kohlstaedt, L.A., Wang, J., Friedman, J.M., Rice, P.A., and Steitz, T.A., Crystal structure at 3.5 Å resolution of HIV-1 reverse transcriptase complexed with an inhibitor, *Science* **256,** 1783–1790 (1992).

McEachern, M.J., Krauskopf, A., and Blackburn, E.H., Telomeres and their control, *Annu. Rev. Genet.* **34,** 331–358 (2000).

Neidle, S. and Parkinson, G., Telomere maintenance as a target for anticancer drug discovery, *Nature Rev. Drug Discov.* **1,** 383–393 (2002); *and* The structure of telomeric DNA, *Curr. Opin. Struct. Biol.* **13,** 275 (2003).

Schultze, P., Smith, F.W., and Feigon, J., Refined solution structure of the dimeric quadruplex formed from the *Oxytricha* telomeric oligonucleotide d(GGGGTTTTGGGG), *Structure* **2,** 221–233 (1994).

Takisawa, H., Mimura, S., and Kubota, Y., Eukaryotic DNA replication: from pre-replication complex to initiation complex, *Curr. Opin. Cell Biol.* **12,** 690–696 (2000).

Tye, B.K. and Sawyer, S., The hexameric eukaryotic MCM helicase: building symmetry from nonidentical parts, *J. Biol. Chem.* **275,** 34833–34836 (2000); *and* Tye, B.K., MCM proteins in DNA replication, *Annu. Rev. Biochem.* **68,** 649–686 (1999).

Urquidi, V., Tarin, D., and Goddison, S., Role of telomerase in cell senescence and oncogenesis, *Annu. Rev. Med.* **51,** 65–79 (2000).

Waga, S. and Stillman, B., The DNA replication fork in eukaryotic cells, *Annu. Rev. Biochem.* **67,** 721–751 (1998).

REPAIR OF DNA

Ames, B.N., Identifying environmental chemicals causing mutations and cancer, *Science* **204,** 587–593 (1979).

Beckman, K.B. and Ames, B.N., Oxidative decay of DNA, *J. Biol. Chem.* **272,** 19633–19636 (1997).

Devoret, R., Bacterial tests for potential carcinogens, *Sci. Am.* **241**(2), 40–49 (1979).

Friedberg, E.C., Wagner, R., and Radman, M., Specialized DNA polymerases, cellular survival, and the genesis of mutations, *Science* **296,** 1627–1630 (2002).

Friedberg, E.C., Walker, G.C., and Siede, W., *DNA Repair and Mutagenesis,* ASM Press (1995).

Goodman, M.F., Error-prone repair DNA polymerases in prokaryotes and eukaryotes, *Annu. Rev. Biochem.* **71,** 17–50 (2002).

Hall, J.G., Genomic imprinting: Nature and clinical relevance, *Annu. Rev. Med.* **48,** 35–44 (1997).

Harfe, B.D. and Jinks-Robertson, S., DNA mismatch repair and genetic instability, *Annu. Rev. Genet.* **34,** 359–399 (2000).

Hopfner, K.-P., Putnam, C.D., and Tainer, J.A., DNA double-strand break repair from head to tail, *Curr. Opin. Struct. Biol.* **12,** 115–122 (2002).

Jaenisch, R., DNA methylation and imprinting: why bother? *Trends Genet.* **13,** 322–329 (1997).

Jiricny, J., Replication errors: cha(lle)nging the genome, *EMBO J.* **17,** 6427–6436 (1998). [A review of mismatch repair.]

Kenyon, C.J., The bacterial response to DNA damage, *Trends Biochem. Sci.* **8,** 84–87 (1983).

Lalande, M., Parental imprinting and human disease, *Annu. Rev. Genet.* **30,** 173–195 (1997).

Lindahl, T., Instability and decay of the primary structure of DNA, *Nature* **363,** 709–715 (1993).

Lindahl, T. and Wood, R.D., Quality control by DNA repair, *Science* **286,** 1897–1905 (1999). [A review.]

Ling, H., Boudsocq, F., Woogate, R., and Yang, W., Crystal structure of a Y-family DNA polymerase in action: A mechanism for error-prone and lesion-bypass replication, *Cell* **107,** 91–102 (2001).

Marra, G. and Schär, P., Recognition of DNA alterations by the mismatch repair system, *Biochem. J.* **338,** 1–13 (1999).

McCullough, A.K., Dodson, M.L., and Lloyd, R.S., Initiation of base excision repair: glycosylase mechanism and structures, *Annu. Rev. Biochem.* **68,** 255–285 (1999).

Mitra, S. and Kaina, B., Regulation of repair of alkylation damage in mammalian genomes, *Prog. Nucleic Acid Res. Mol. Biol.* **44,** 109–142 (1993).

Modrich, P., Mismatch repair in replication fidelity, genetic recombination, and cancer biology, *Annu. Rev. Biochem.* **65,** 101–133 (1996).

Mol, C.D., Parikh, S.S., Putnam, C.D., Lo, T.P., and Tainer, J.A., DNA repair mechanism for the recognition and removal of damaged DNA bases, *Annu. Rev. Biophys. Biomol. Struct.* **28,** 101–128 (1999).

Moore, M.H., Gulbis, J.M., Dodson, E.J., Demple, B., and Moody, P.C.E., Crystal structure of a suicidal DNA repair protein: Ada O^6-methylguanine-DNA methyltransferase from *E. coli, EMBO J.* **13,** 1495–1501 (1994).

Myers, L.C., Verdine, G.L., and Wagner, G., Solution structure of the DNA methyl triester repair domain of *Escherichia coli* Ada, *Biochemistry* **32,** 14089–14094 (1993).

Parikh, S.S., Mol, C.D., Slupphaug, G., Bharati, S., Krokan, H.E., and Tainer, J.A., Base excision repair initiation revealed by crystal structures and binding kinetics of human uracil–DNA glycosylase with DNA, *EMBO J.* **17,** 5214–5226 (1998).

Park, H.-W., Kim, S.-T., Sancar, A., and Diesenhofer, J., Crystal structure of DNA photolyase from *Escherichia coli, Science* **268,** 1866–1872 (1995).

Pegg, A.E., Dolan, M.E., and Moschel, R.C., Structure, function, and inhibition of O^6-alkylguanine–DNA alkyltransferase, *Prog. Nucleic Acid Res. Mol. Biol.* **51,** 167–223 (1995).

Pham, P., Rangarajan, S., Woodgate, R., and Goodman, M.F., Roles of DNA polymerases V and II in SOS-induced error-prone and error-free repair in *Escherichia coli, Proc. Natl. Acad. Sci.* **98,** 8350–8354 (2001); *and* Goodman, M.F., Coping with replication 'train wrecks' in *Escherichia coli* using Pol V, Pol II, and RecA proteins, *Trends Biochem. Sci.* **25,** 189–195 (2000).

Sancar, A., DNA excision repair, *Annu. Rev. Biochem.* **65,** 43–81 (1996).

Scriver, C.R., Beaudet, A.L., Sly, W.S., and Valle, D. (Eds.), *The Metabolic & Molecular Bases of Inherited Disease* (8th ed.), Chaps. 28 and 32, McGraw-Hill (2001). [Discussions of xeroderma pigmentosum, Cockayne syndrome, and hereditary nonpolyposis colorectal cancer.]

Sutton, M.D., Smith, B.T., Godoy, V.G., and Walker, G.C., The SOS response: recent insights into *umuDC*-dependent mutagenesis and DNA damage tolerance, *Annu. Rev. Genet.* **34,** 479–497 (2000).

Tainer, J.A. and Friedberg, E.C. (Eds.), *Biological Implications from Structures of DNA Repair Proteins, Mutation Research* **460,** 139–335 (2000). [A series of authoritative reviews.]

Walker, J.R., Corpina, R.A., and Goldberg, J., Structure of the Ku heterodimer bound to DNA and its implications for double-strand break repair, *Nature* **412,** 607–614 (2001).

Wood, R.D., Nucleotide excision repair in mammalian cells, *J. Biol. Chem.* **272,** 23465–23468 (1997); *and* DNA repair in eukaryotes, *Annu. Rev. Biochem.* **65,** 135–167 (1996).

Yang, W., Damage repair DNA polymerases Y, *Curr. Opin. Struct. Biol.* **13,** 23–30 (2003).

RECOMBINATION AND MOBILE GENETIC ELEMENTS

Ariyoshi, M., Nishino, T., Iwasaki, H., Shinagawa, H., and Morikawa, K., Crystal structure of the Holliday junction DNA in complex with a single RuvA tetramer, *Proc. Natl. Acad. Sci.* **97,** 8257–8262 (2000).

Changela, A., Perry, K., Taneja, B., and Mondragón, A., DNA manipulators: caught in the act, *Curr. Opin. Struct. Biol.* **13,** 15–22 (2003).

Cox, M.M., Recombinational DNA repair of damaged replication forks in *Escherichia coli:* questions, *Annu. Rev. Genet.* **35,** 53–82 (2001); *and* Recombinational DNA repair in bacteria and the RecA protein, *Prog. Nucleic Acid Res. Mol. Biol.* **63,** 311–366 (2000).

Cox, M.M., Goodman, M.F., Kreuzer, K.N., Sherratt, D.J., Sandler, S.J., and Marians, K.J., The importance of repairing stalled replication forks, *Nature* **404,** 37–41 (2000).

Craig, N.L., Target site selection in transposition, *Annu. Rev. Biochem.* **66,** 437–474 (1997).

Craig, N.L., Craigie, R., Gellert, M., and Lambowitz, A.M. (Eds.), *Mobile DNA II,* ASM Press (2002). [A compendium of authoritative articles.]

Davies, D.R., Gorshin, I.Y., Reznikoff, W.S., and Rayment, I., Three-dimensional structure of the Tn5 synaptic complex transposition intermediate, *Science* **289,** 77–85 (2000); *and* Reznikoff, W.S., Bhasin, A., Davies, D.R., Gorshin, I.Y., Mahnke, L.A., Naumann, T., Rayment, I., Steiniger-White, M., and Twining, S.S., Tn5: a molecular window on transposition, *Biochem. Biophys. Res. Commun.* **266,** 729–734 (1999).

Egelman, E.H., What do X-ray crystallographic and electron microscopic structural studies of RecA protein tell us about recombination? *Curr. Opin. Struct. Biol.* **3,** 189–197 (1993).

Eichman, B.F., Vargason, J.M., Mooers, B.H.M., and Ho, P.S., The Holliday junction in an inverted repeat DNA sequence: sequence effects on the structure of four-way junctions, *Proc. Natl. Acad. Sci.* **97,** 3971–3976 (2000).

Feng, J.-A., Dickerson, R.E., and Johnson, R.C., Proteins that promote DNA inversion and deletion, *Curr. Opin. Struct. Biol.* **4,** 60–66 (1994).

Haber, J.E., Partners and pathways. Repairing a double-strand break, *Trends Genet.* **16**, 259–264 (2000); *and* DNA recombination: the replication connection, *Trends Biochem. Sci.* **24,** 271–275 (1999).

Haren, L., Ton-Hoang, B., and Chandler, M., Integrating DNA: transposases and retroviral integrases, *Annu. Rev. Microbiol.* **53,** 245–281 (1999).

Ho, P.S. and Eichman, B.F., The crystal structures of Holliday junctions, *Curr. Opin. Struct. Biol.* **11,** 302–308 (2001).

Kuzminov, A., Recombinational repair of DNA damage in *Escherichia coli* and bacteriophage λ, *Microbiol. Mol. Biol. Rev.* **63,** 751–813 (1999).

Lusetti, S.L. and Cox, M.M., The bacterial RecA protein and the recombinational DNA repair of stalled replication forks, *Annu. Rev. Biochem.* **71,** 71–100 (2002).

Marians, K.J., PriA-directed replication fork restart in *Escherichia coli, Trends Biochem. Sci.* **25,** 185–189 (2000).

Rice, P.A. and Baker, T.A., Comparative architecture of transposase and integrase complexes, *Nature Struct. Biol.* **8,** 302–307 (2001).

Roe, S.M., Barlow, T., Brown, T., Oram, M., Keeley, A., Tsaneva, I.R., and Pearl, L.H., Crystal structure of an octameric RuvA–Holliday junction complex, *Molecular Cell* **2,** 361–372 (1998).

Simon, M., Zieg, J., Silverman, M., Mandel, G., and Doolittle, R., Phase variation: evolution of a controlling element, *Science* **209,** 1370–1374 (1980).

Story, R.M., Weber, I.T., and Steitz, T.A., The structure of the *E. coli recA* protein monomer and polymer, *Nature* **355,** 318–325 (1992); *and* the erratum for this paper, *Nature* **355,** 367 (1992). [These two papers should be read together.]

Van Duyne, G.D., A structural view of Cre–*loxP* site-specific recombination, *Annu. Rev. Biophys. Biomol. Struct.* **30,** 87–104 (2001).

Yamada, K., Kunishima, N., Mayanagi, K., Ohnishi, T., Nishino, T., Iwasaki, H., Shinagawa, H., and Morikawa, K., Crystal structure of the Holliday junction migration motor protein RuvB from *Thermus thermophilus* HB8, *Proc. Natl. Acad. Sci.* **98,** 1442–1447 (2001).

West, S.C., Processing of recombination intermediates by the RuvABC proteins, *Annu. Rev. Genet.* **31,** 213–244 (1997).

Yang, W. and Steitz, T.A., Crystal structure of the site-specific recombinase γδ resolvase complexed with a 34 bp cleavage site, *Cell* **82,** 193–207 (1995).

DNA METHYLATION AND TRINUCLEOTIDE REPEAT EXPANSIONS

Bowater, R.P. and Wells, R.D., The intrinsically unstable life of DNA repeats associated with human hereditary disorders, *Prog. Nucleic Acid Res. Mol. Biol.* **66,** 159–202 (2001).

Cheng, X., Structure and function of DNA methyltransferases, *Annu. Rev. Biophys. Biomol. Struct.* **24,** 293–318 (1995).

Cummings, C.J. and Zoghbi, H.Y., Trinucleotide repeats: mechanisms and pathophysiology, *Annu. Rev. Genomics Hum. Genet.* **1,** 281–328 (2002); *and* Zoghbi, H.Y. and Orr, H.T., Glutamine repeats and neurodegeneration, *Annu. Rev. Neurosci.* **23,** 217–247 (2000).

Goodman, J. and Watson, R.E., Altered DNA methylation: a secondary mechanism involved in carcinogenesis, *Annu. Rev. Pharmacol. Toxicol.* **42,** 501–525 (2002).

Jones, P.A. and Baylin, S.B., The fundamental role of epigenetic events in cancer, *Nature Rev. Genet.* **3,** 415–428 (2002); *and* Jones, P.A. and Takai, D., The role of DNA methylation in mammalian epigenetics, *Science* **293,** 1068–1070 (2001).

Klimasauskas, S., Kumar, S., Roberts, R.J., and Cheng, X., HhaI methyltransferase flips its target base out of the DNA helix, *Cell* **76,** 357–369 (1994).

Marinus, M.G., DNA methylation in *Escherichia coli, Annu. Rev. Genet.* **21,** 113–131 (1987).

O'Donnell, W.T. and Warren, S.T., A decade of molecular studies of fragile X syndrome, *Annu. Rev. Neurosci.* **25,** 315–338 (2002).

Perutz, M.F. and Windle, A.H., Causes of neural death in neurodegenerative diseases attributable to expansion of glutamine repeats, *Nature* **12,** 143–144 (2001); *and* Perutz, M.F., Glutamine repeats and neurodegenerative diseases: molecular aspects, *Trends Biochem. Sci.* **24,** 58–63 (1999).

Reik, W., Dean, W., and Walter, J., Epigenetic reprogramming in mammalian development, *Science* **293,** 1089–1093 (2001).

Rideout, W.M., III, Eggan, K., and Jaenisch, R., Nuclear cloning and epigenetic reprogramming of the genome, *Science* **293,** 1093–1098 (2001).

Roberts, R.J. and Cheng, X., Base flipping, *Annu. Rev. Biochem.* **67,** 181–198 (1998).

Scriver, C.R., Beaudet, A.L., Sly, W.S., and Valle, D. (Eds.), *The Metabolic & Molecular Bases of Inherited Disease* (8th ed.),

Chaps. 64, 223, and 226, McGraw-Hill (2001). [Discussions of fragile X syndrome, Huntington's disease, and the spinocerebellar ataxias.]

Szyf, M. and Detich, N., Regulation of the DNA methylation machinery and its role in cellular transformation, *Prog. Nucleic Acid Res. Mol. Biol.* **69,** 47–79 (2001).

PROBLEMS

1. Explain how certain mutant varieties of Pol I can be nearly devoid of DNA polymerase activity but retain almost normal levels of $5' \rightarrow 3'$ exonuclease activity.

2. Why haven't Pol I mutants been found that completely lack $5' \rightarrow 3'$ activity at all temperatures?

3. Why aren't type I topoisomerases necessary in *E. coli* DNA replication?

***4.** The $3' \rightarrow 5'$ exonuclease activity of Pol I excises only unpaired 3′-terminal nucleotides from DNA, whereas this enzyme's pyrophosphorolysis activity removes only properly paired 3′-terminal nucleotides. Discuss the mechanistic significance of this phenomenon in terms of the polymerase reaction.

5. You have isolated *E. coli* with temperature-sensitive mutations in the following genes. What are their phenotypes above their restrictive temperatures? Be specific. (a) *dnaB,* (b) *dnaE,* (c) *dnaG,* (d) *lig,* (e) *polA,* (f) *rep,* (g) *ssb,* and (h) *recA.*

6. About how many Okazaki fragments are synthesized in the replication of an *E. coli* chromosome?

***7.** What are the minimum and maximum number of replication forks that occur in a contiguous chromosome of an *E. coli* that is dividing every 25 min; every 80 min?

8. To put the *E. coli* replication system on a human scale, let us imagine that the 20-Å-diameter B-DNA was expanded to 1 m in diameter. If everything were proportionally expanded, then each DNA polymerase III holoenzyme would be about the size of a medium-sized truck. In such an expanded system: (a) How fast would each replisome be moving? (b) How far would each replisome travel during a complete replication cycle? (c) What would be the length of an Okazaki fragment? (d) What would be the average distance a replisome would travel between each error it made? Provide your answers in km/hr and km.

9. Why can't linear duplex DNAs, such as occur in bacteriophage T7, be fully replicated by only *E. coli*-encoded proteins?

***10.** What is the half-life of a particular purine base in the human genome assuming that it is subject only to spontaneous depurination? What fraction of the purine bases in a human genome will have depurinated in the course of a single generation (assume 25 years)? The DNAs of ~4000-year-old Egyptian mummies have been sequenced. Assuming that mummification did not slow the rate of DNA depurination, what fraction of the purine bases originally present in the mummy would still be intact today.

11. Why is the methylation of DNA to form O^6-methylguanine mutagenic?

12. A replication fork encountering a single-strand lesion may either dissociate or leave a single-strand gap. The latter process is more likely to occur during lagging strand synthesis than during leading strand synthesis. Explain.

13. The *E. coli* genome contains 1009 Chi sequences. Do these sequences occur at random, and, if not, how much more or less frequently than random do they occur?

14. *Deinococcus radiodurans,* which the *Guiness Book of World Records* has dubbed the world's toughest bacterium, can tolerate doses of ionizing radiation ~3000-fold greater than those that are lethal to humans (it was first discovered growing in a can of ground meat that had been "sterilized" by radiation). It appears to have several strategies to repair radiation damage to its DNA (which large doses of ionizing radiation fragment to many pieces) including a particularly large number of genes encoding proteins involved in DNA repair and 4 to 10 copies per cell of its genome, which consists of two circular chromosomes and two circular plasmids. Yet, these strategies, alone, do not account for *D. radiodurans'* enormously high radiation resistance. However, in an additional strategy, it organizes its multiple identical dsDNA circles into stacks in which, it is thought, the identical genes in the neighboring circles are aligned side by side. How would this latter strategy help *D. radiodurans* efficiently repair its fragmented DNA?

15. CpG islands occur in eukaryotic genomes at about one-fifth their expected random frequency. Suggest an evolutionary (mutational) process that eliminates CpG islands.

16. Explain why the brief exposure of a cultured eukaryotic cell line to 5-azacytosine results in permanent phenotypic changes to these cells.

17. Explain why chi structures, such as that shown in Fig. 30-67*b*, have two pairs of equal length arms.

***18.** Single-stranded circular DNAs containing a transposon have a characteristic stem-and-double-loop structure such as that shown in Fig. 30-102. What is the physical basis of this structure?

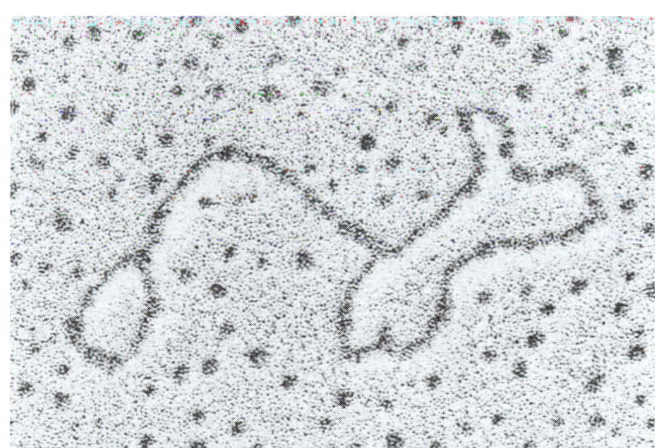

FIGURE 30-102 Electron micrograph of a single-stranded circular DNA containing a transposon. [Courtesy of Stanley N. Cohen, Stanford University School of Medicine.]

19. A composite transposon integrated in a circular plasmid occasionally transposes the DNA comprising the original plasmid rather than the transposon's central region. Explain how this is possible.

***20.** Cre recombinase has an additional function to that of circularizing the linear P1 dsDNA (Fig. 30-93). It is also required to resolve the circular dimers of P1 plasmids that result from their recombinational repair during replication, thereby permitting both daughter cells to receive a copy of the P1 plasmid. Using simple line diagrams, outline how these plasmids become dimerized and how Cre resolves them to circular monomers.

Chapter

31 Transcription

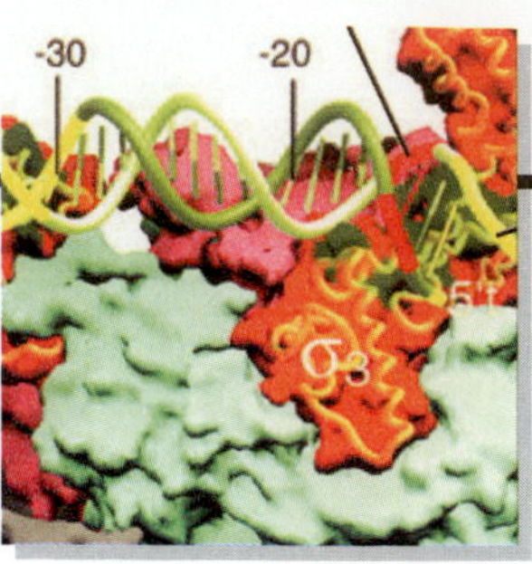

There are three major classes of RNA, all of which participate in protein synthesis: **ribosomal RNA (rRNA), transfer RNA (tRNA),** and **messenger RNA (mRNA).** All of these RNAs are synthesized under the direction of DNA templates, a process known as **transcription.**

RNA's involvement in protein synthesis became evident in the late 1930s through investigations by Torbjörn Caspersson and Jean Brachet. Caspersson, using microscopic techniques, found that DNA is confined almost exclusively to the eukaryotic cell nucleus, whereas RNA occurs largely in the cytosol. Brachet, who had devised methods for fractionating cellular organelles, came to similar conclusions based on direct chemical analyses. He found, in addition, that the cytosolic RNA-containing particles are also protein rich. Both investigators noted that the concentration of these RNA–protein particles (which were later named ribosomes) is correlated with the rate at which a cell synthesizes protein, implying a relationship between RNA and protein synthesis. Indeed, Brachet even suggested that *the RNA–protein particles are the site of protein synthesis.*

Brachet's suggestion was shown to be valid when radioactively labeled amino acids became available in the 1950s. A short time after injection of a rat with a labeled amino acid, most of the label that had been incorporated in proteins was associated with ribosomes. This experiment also established that *protein synthesis is not immediately directed by DNA because, at least in eukaryotes, DNA and ribosomes are never in contact.*

In 1958, Francis Crick summarized the then dimly perceived relationships among DNA, RNA, and protein by what he called the **central dogma** of molecular biology: *DNA directs its own replication and its transcription to RNA which, in turn, directs its translation to proteins (Fig. 5-21).*

The peculiar use of the word "dogma," one definition of which is a religious doctrine that the true believer cannot doubt, stemmed from a misunderstanding. When Crick formulated the central dogma, he was under the impression that dogma meant "an idea for which there was no reasonable evidence."

We begin this chapter by discussing experiments that led to the elucidation of mRNA's central role in protein synthesis. We then study the mechanism of transcription and its control in prokaryotes. Finally, in the last section, we consider posttranscriptional processing of RNA in both prokaryotes and eukaryotes. Translation is the subject of Chapter 32. Note that these subjects were outlined in Section 5-4. Here we shall delve into much greater detail.

1 ■ THE ROLE OF RNA IN PROTEIN SYNTHESIS

The idea that proteins are specified by mRNA and synthesized on ribosomes arose from the study of **enzyme induction,** a phenomenon in which bacteria vary the synthesis rates of specific enzymes in response to environmental changes. In this section, we discuss the classic experiments that explained the basis of enzyme induction and revealed the existence of mRNA. We shall see that *enzyme induction occurs as a consequence of the regulation* *of mRNA synthesis by proteins that specifically bind to the mRNA's DNA templates.*

A. *Enzyme Induction*

E. coli can synthesize an estimated ~4300 different polypeptides. There is, however, enormous variation in the amounts of these different polypeptides that are produced. For instance, the various ribosomal proteins may each be present in over 10,000 copies per cell, whereas certain regulatory proteins (see below) normally occur in <10 copies per cell. Many enzymes, particularly those involved in basic cellular "housekeeping" functions, are synthesized at a more or less constant rate; they are called **constitutive enzymes.** Other enzymes, termed **adaptive** or **inducible enzymes,** are synthesized at rates that vary with the cell's circumstances.

a. Lactose-Metabolizing Enzymes Are Inducible

Bacteria, as has been recognized since 1900, adapt to their environments by producing enzymes that metabolize certain nutrients, for example, lactose, only when those substances are available. *E. coli* grown in the absence of lactose are initially unable to metabolize this disaccharide. To do so they require the presence of two proteins: **β-galactosidase,** which catalyzes the hydrolysis of lactose to its component monosaccharides,

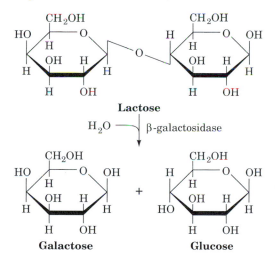

Lactose

H_2O → β-galactosidase

Galactose + **Glucose**

and **galactoside permease** (also known as **lactose permease;** Section 20-4B), which transports lactose into the cell. *E. coli* grown in the absence of lactose contain only a few (<5) molecules of these proteins. Yet, a few minutes after lactose is introduced into their medium, *E. coli* increase the rate at which they synthesize these proteins by ~1000-fold (such that β-galactosidase can account for up to 10% of their soluble protein) and maintain this pace until lactose is no longer available. The synthesis rate then returns to its miniscule **basal level** (Fig. 31-1). *This ability to produce a series of proteins only when the substances they metabolize are present permits bacteria to adapt to their environment without the debilitating need to continuously synthesize large quantities of otherwise unnecessary substances.*

Lactose or one of its metabolic products must somehow trigger the synthesis of the above proteins. Such a substance is known as an **inducer.** The physiological inducer of the lactose system, the lactose isomer **1,6-allolactose,**

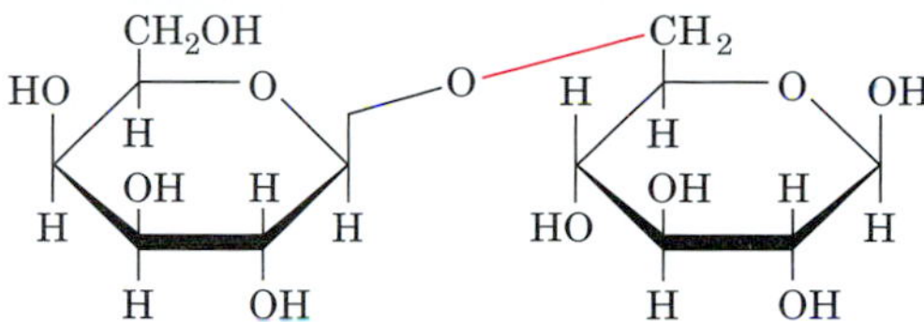

1,6-Allolactose

arises from lactose's occasional transglycosylation by β-galactosidase. Most studies of the lactose system use **isopropylthiogalactoside (IPTG),**

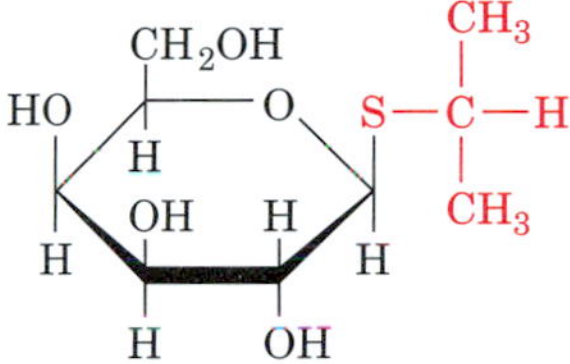

Isopropylthiogalactoside (IPTG)

a potent inducer that structurally resembles allolactose but that is not degraded by β-galactosidase.

Lactose system inducers also stimulate the synthesis of **thiogalactoside transacetylase,** an enzyme that, *in vitro,* transfers an acetyl group from acetyl-CoA to the C6-OH group of a β-thiogalactoside such as IPTG. Since lactose fermentation proceeds normally in the absence of thiogalactoside transacetylase, however, this enzyme's physiological role is unknown.

b. *lac* System Genes Form an Operon

The genes specifying wild-type β-galactosidase, galactoside permease, and thiogalactoside transacetylase are designated Z^+, Y^+, and A^+, respectively. Genetic mapping of the defective mutants Z^-, Y^-, and A^- indicated that

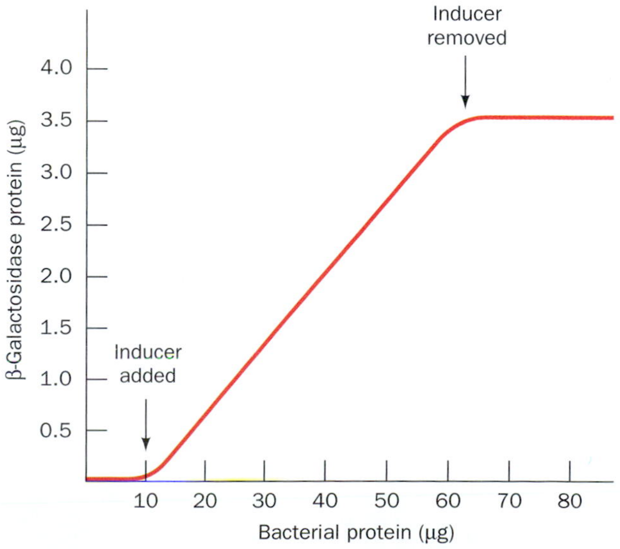

FIGURE 31-1 The induction kinetics of β-galactosidase in *E. coli.* [After Cohn, M., *Bacteriol. Rev.* **21,** 156 (1957).]

these *lac* **structural genes** (genes that specify polypeptides) are contiguously arranged on the *E. coli* chromosome (Fig. 31-2). *These genes, together with the control elements P and O, form a genetic unit called an* ***operon,*** *specifically the* ***lac operon.*** The nature of the control elements is discussed below. The role of operons in prokaryotic gene expression is examined in Section 31-3.

c. Bacteria Can Transmit Genes via Conjugation

An important clue as to how *E. coli* synthesizes protein was provided by a mutation that causes the proteins of the *lac* operon to be synthesized in large amounts in the absence of inducer. This so-called **constitutive mutation** occurs in a gene, designated *I*, that is distinct from although closely linked to the genes specifying the *lac* enzymes (Fig. 31-2). What is the nature of the *I* gene product? This riddle was solved in 1959 by Arthur Pardee, Francois Jacob, and Jacques Monod through an ingenious experiment that is known as the **PaJaMo experiment.** To understand this experiment, however, we must first consider **bacterial conjugation.**

Bacterial conjugation is a process, discovered in 1946 by Joshua Lederberg and Edward Tatum, through which some bacteria can transfer genetic information to others. The ability to conjugate ("mate") is conferred on an otherwise indifferent bacterium by a **plasmid** named **F factor** (for *f*ertility). Bacteria that possess an F factor (designated F^+ or male) are covered by hairlike projections known as **F pili.** These bind to cell-surface receptors on bacteria that lack the F factor (F^- or female), which leads to the formation of a cytoplasmic bridge between these cells (Fig. 31-3). The F factor then replicates and, as the newly replicated single strand is formed, it passes through the cytoplasmic bridge to the F^- cell where the complementary strand is synthesized (Fig. 31-4). This converts the F^- cell to F^+ so that the F factor is an infectious agent (a bacterial venereal disease?).

On very rare occasions, the F factor spontaneously integrates into the chromosome of the F^+ cell. In the result-

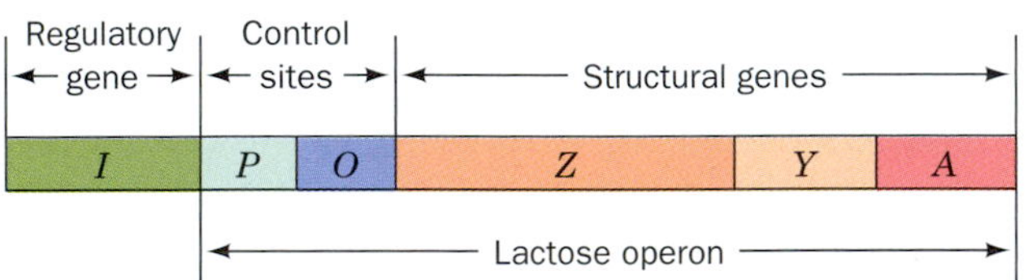

FIGURE 31-2 Genetic map of the *E. coli lac* operon. The map shows the genes encoding the proteins mediating lactose metabolism and the genetic sites that control their expression. The *Z*, *Y*, and *A* genes, respectively, specify β-galactosidase, galactoside permease, and thiogalactoside transacetylase.

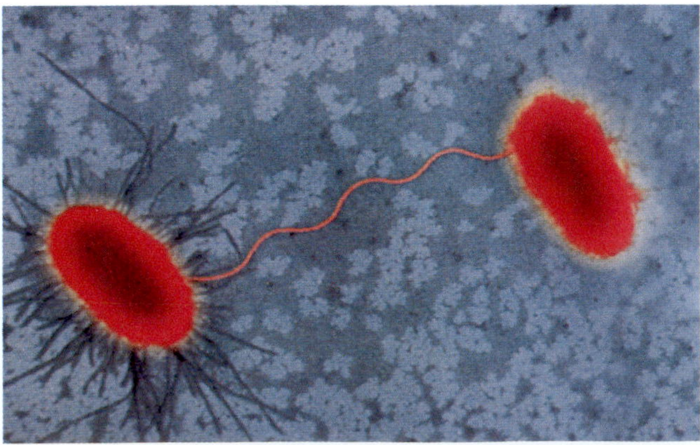

FIGURE 31-3 Bacterial conjugation. An electron micrograph shows an F^+ (*left*) and an F^- (*right*) *E. coli* engaged in sexual conjugation. [Dennis Kunkel/Phototake.]

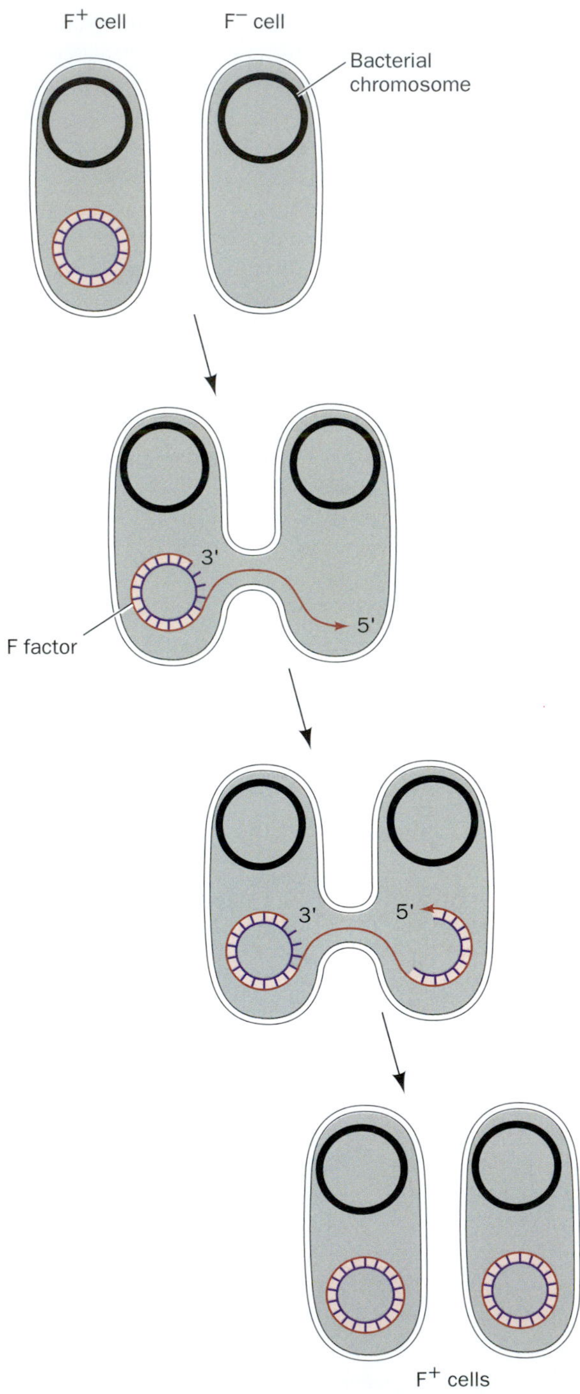

FIGURE 31-4 Diagram showing how an F^- cell acquires an F factor from an F^+ cell. A single strand of the F factor is replicated, via the rolling circle mode (Section 30-3B), and is transferred to the F^- cell where its complementary strand is synthesized to form a new F factor.

ing **Hfr** (for *h*igh *f*requency of *r*ecombination) cells, the F factor behaves much as it does in the autonomous state. Its replication commences at a specific internal point in the F factor, and the replicated section passes through a cytoplasmic bridge to the F^- cell, where its complementary strand is synthesized. In this case, however, the replicated chromosome of the Hfr cell is also transmitted to the F^- cell (Fig. 31-5). *Bacterial genes are transferred from the Hfr cell to the F^- cell in fixed order.* This is because the F factor in a given Hfr strain is integrated into the bacterial chromosome at a specific site and because only a particular strand of the Hfr chromosomal DNA is replicated and transferred to the F^- cell. Usually, only part of the Hfr bacterial chromosome is transferred during sexual conjugation because the cytoplasmic bridge almost always breaks off sometime during the ~90 min required to complete the transfer process. In the resulting **merozygote** (a partially diploid bacterium), the chromosomal fragment, which lacks a complete F factor, neither transforms the F^- cell to Hfr nor is subsequently replicated. However, the transferred chromosomal fragment recombines with the chromosome of the F^- cell (Section 30-6A), thereby permanently endowing the F^- cell with some of the traits of the Hfr strain.

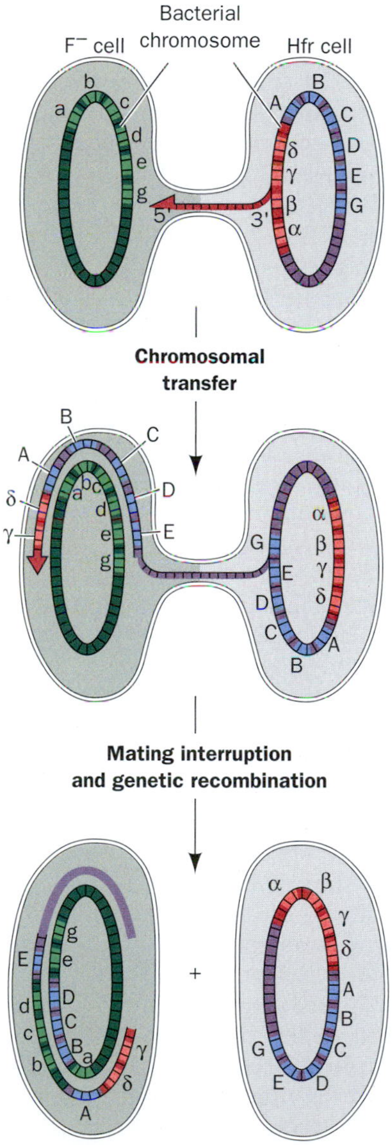

FIGURE 31-5 Transfer of the bacterial chromosome from an Hfr cell to an F^- cell and its subsequent recombination with the F^- chromosome. Here, Greek letters represent F factor genes, uppercase Roman letters represent bacterial genes from the Hfr cell, and lowercase Roman letters represent the corresponding alleles in the F^- cell. Since chromosomal transfer, which begins within the F factor, is rarely complete, the entire F factor is seldom transferred. Hence the recipient cell usually remains F^-.

The integrated F factor in an Hfr cell occasionally undergoes spontaneous excision to yield an F^+ cell. In rare instances, the F factor is aberrantly excised such that a portion of the adjacent bacterial chromosome is incorporated in the subsequently autonomously replicating F factor. Bacteria carrying such a so-called **F′ factor** are permanently diploid for its bacterial genes.

d. *lac* Repressor Inhibits the Synthesis of *lac* Operon Proteins

In the PaJaMo experiment, Hfr bacteria of genotype I^+Z^+ were mated to an F^- strain of genotype I^-Z^- in the absence of inducer while the β-galactosidase activity of the culture was monitored (Fig. 31-6). At first, as expected,

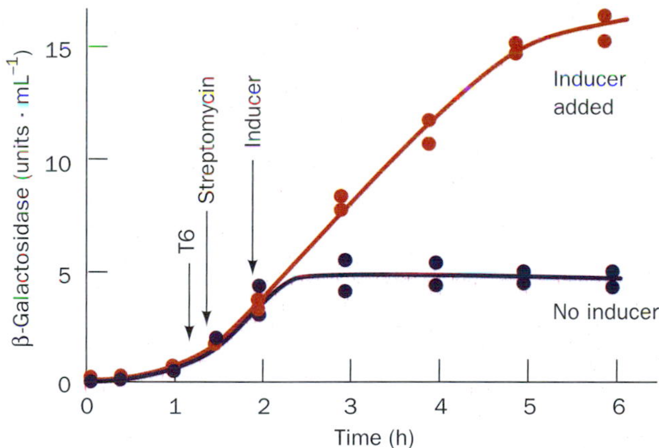

FIGURE 31-6 The PaJaMo experiment. This experiment demonstrated the existence of the *lac* repressor through the appearance of β-galactosidase in the transient merozygotes (partial diploids) formed by mating I^+Z^+ Hfr donors with I^-Z^- F^- recipients. The F^- strain was also resistant to both **bacteriophage T6** and **streptomycin,** whereas the Hfr strain was sensitive to these agents. Both types of cells were grown and mated in the absence of inducer. After sufficient time had passed for the transfer of the *lac* genes, the Hfr cells were selectively killed by the addition of T6 phage and streptomycin. In the absence of inducer (*lower curve*), β-galactosidase synthesis commenced at around the time at which the *lac* genes had entered the F^- cells but stopped after ~1 h. If inducer was added shortly after the Hfr donors had been killed (*upper curve*), enzyme synthesis continued unabated. This demonstrates that the cessation of β-galactosidase synthesis in uninduced cells is not due to the intrinsic loss of the ability to synthesize this enzyme but to the production of a repressor specified by the I^+ gene. [After Pardee, A.B., Jacob, F., and Monod, J., *J. Mol. Biol.* **1,** 173 (1959).]

there was no β-galactosidase activity because the Hfr donors lacked inducer and the F^- recipients were unable to produce active enzyme (only DNA passes through the cytoplasmic bridge connecting mating bacteria). About 1 h after conjugation began, however, when the I^+Z^+ genes had just entered the F^- cells, β-galactosidase synthesis began and only ceased after about another hour. The explanation for these observations is that the donated Z^+ gene, on entering the cytoplasm of the I^- cell, directs the synthesis of β-galactosidase in a constitutive manner. Only after the donated I^+ gene has had sufficient time to be expressed is it able to repress β-galactosidase synthesis. *The I^+ gene must therefore give rise to a diffusible product, the* ***lac repressor,*** *which inhibits the synthesis of β-galactosidase (and the other lac proteins).* Inducers such as IPTG temporarily inactivate *lac* repressor, whereas I^- cells constitutively synthesize *lac* enzymes because they lack a functional repressor. *Lac* repressor, as we shall see in Section 31-3B, is a protein.

B. *Messenger RNA*

The nature of the *lac* repressor's target molecule was deduced in 1961 through a penetrating genetic analysis by Jacob and Monod. A second type of constitutive mutation in the lactose system, designated O^c (for **operator constitutive**), which complementation analysis (Section 1-4C) has shown to be independent of the *I* gene, maps between the *I* and *Z* genes (Fig. 31-2). In the partially diploid F′ strain $O^cZ^-/F\ O^+Z^+$, β-galactosidase activity is inducible by IPTG, whereas the strain $O^cZ^+/F\ O^+Z^-$ constitutively synthesizes this enzyme. *An O^+ gene can therefore only control the expression of a Z gene on the same chromosome.* The same is true with the Y^+ and A^+ genes.

Jacob and Monod's observations led them to conclude that the proteins are synthesized in a two-stage process:

1. The structural genes on DNA are transcribed onto complementary strands of **messenger RNA (mRNA).**

2. The mRNAs transiently associate with ribosomes, which they direct in polypeptide synthesis.

This hypothesis explains the behavior of the *lac* system that we previously outlined in Section 5-4A (Fig. 5-25; **See Guided Exploration 2: Regulation of Gene Expression by the *lac* Repressor System**). *In the absence of inducer, the lac repressor specifically binds to the O gene (the* ***operator****) so as to prevent the enzymatic transcription of mRNA. On binding inducer, the repressor dissociates from the operator, thereby permitting the transcription and subsequent translation of the lac enzymes.* The operator–repressor–inducer system thereby acts as a molecular switch so that the *lac* operator can only control the expression of *lac* enzymes on the same chromosome. The O^c mutants constitutively synthesize *lac* enzymes because they are unable to bind repressor. The **coordinate** (simultaneous) expression of all three *lac* enzymes under the control of a single operator site arises, as Jacob and Monod theorized, from the transcription of the *lac* operon as a single **polycistronic mRNA** which directs the ribosomal synthesis of each of these proteins (the term **cistron** is a somewhat archaic synonym for gene). This transcriptional control mechanism is further discussed in Section 31-3. [DNA sequences that are on the same DNA molecule are said to be "in cis" (Latin: on this side), whereas those on different DNA molecules are said to be "in trans" (Latin: across). Control sequences such as the *O* gene, which are only active on the same DNA molecule as the genes they control, are called **cis-acting elements.** Genes such as *lacI,* which specify the synthesis of diffusible products and can therefore be located on a different DNA molecule from the genes they control, are said to direct the synthesis of **trans-acting factors.**]

a. mRNAs Have Their Predicted Properties

The kinetics of enzyme induction, as indicated, for example, in Figs. 31-1 and 31-6, requires that the postulated mRNA be both rapidly synthesized and rapidly degraded. An RNA with such quick turnover had, in fact, been observed in T2-infected *E. coli.* Moreover, the base composition of this RNA fraction resembles that of the viral DNA rather than that of the bacterial RNA (keep in mind that base sequencing techniques would not be formulated for another ~15 years). Ribosomal RNA, which comprises up to 90% of a cell's RNA, turns over much more slowly than mRNA. Ribosomes are therefore not permanently committed to the synthesis of a particular protein (a once popular hypothesis). Rather, *ribosomes are nonspecific protein synthesizers that produce the polypeptide specified by the mRNA with which they are transiently associated.* A bacterium can therefore respond within a few minutes to changes in its environment.

Evidence favoring the Jacob and Monod model rapidly accumulated. Sydney Brenner, Jacob, and Matthew Meselson carried out experiments designed to characterize the RNA that *E. coli* synthesized after T4 phage infection. *E. coli* were grown in a medium containing ^{15}N and ^{13}C so as to label all cell constituents with these heavy isotopes. The cells were then infected with T4 phages and immediately transferred to an unlabeled medium (which contained only the light isotopes ^{14}N and ^{12}C) so that cell components synthesized before and after phage infection could be separated by equilibrium density gradient ultracentrifugation in CsCl solution (Section 6-5B). No "light" ribosomes were observed, which indicates, in agreement with the above-mentioned T2 phage results, that no new ribosomes are synthesized after phage infection.

The growth medium also contained either ^{32}P or ^{35}S so as to radioactively label the newly synthesized and presumably phage-specific RNA and protein, respectively. Much of the ^{32}P-labeled RNA was associated, as was postulated for mRNA, with the preexisting "heavy" ribosomes (Fig. 31-7). Likewise, the ^{35}S-labeled proteins were transiently associated with, and therefore synthesized by, these ribosomes.

Sol Spiegelman developed the RNA–DNA hybridization technique (Section 5-3C) in 1961 to characterize the RNA synthesized by T2-infected *E. coli.* He found that this

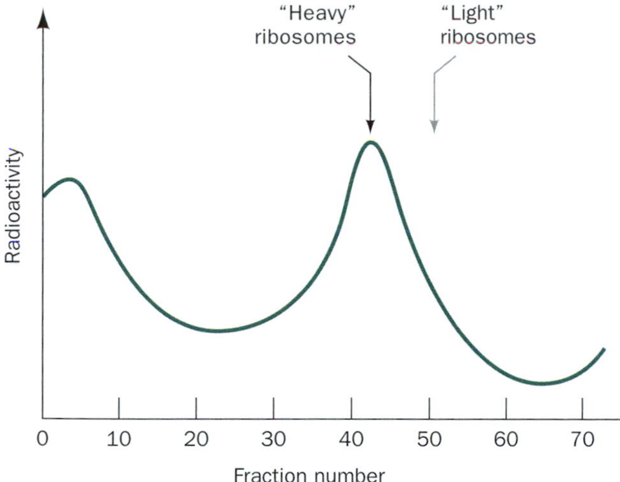

FIGURE 31-7 The distribution, in a CsCl density gradient, of ^{32}P-labeled RNA that had been synthesized by *E. coli* after T4 phage infection. Free RNA, being relatively dense, bands at the bottom of the centrifugation cell (*left*). Much of the RNA, however, is associated with the ^{15}N- and ^{13}C-labeled "heavy" ribosomes that had been synthesized before the phage infection. The predicted position of unlabeled "light" ribosomes, which are not synthesized by phage-infected cells, is also indicated. [After Brenner, S., Jacob, F., and Meselson, M., *Nature* **190,** 579 (1961).]

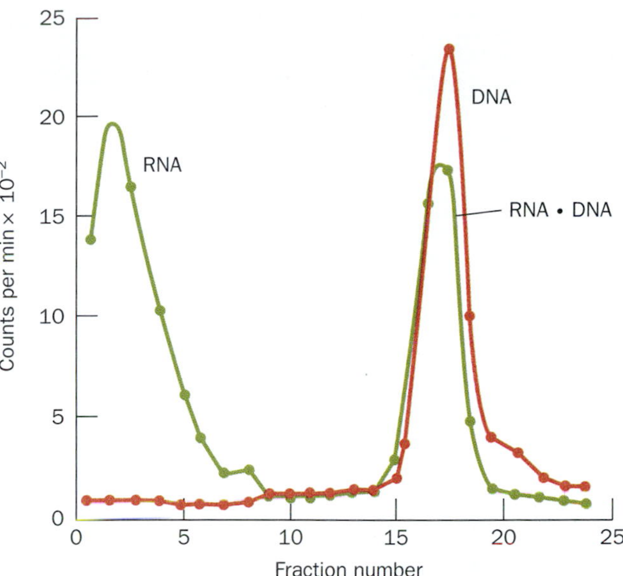

FIGURE 31-8 The hybridization of ^{32}P-labeled RNA produced by T2-infected *E. coli* with ^{3}H-labeled T2 DNA. On radioactive decay, ^{32}P and ^{3}H emit β particles (electrons) with characteristically different energies so that these isotopes can be independently detected. Although free RNA (*left*) in a CsCl density gradient is denser than DNA, much of the RNA bands with the DNA (*right*). This indicates that the two polynucleotides have hybridized and are therefore complementary in sequence. [After Hall, B.D. and Spiegelman, S., *Proc. Natl. Acad. Sci.* **47,** 141 (1961).]

phage-derived RNA hybridizes with T2 DNA (Fig. 31-8) but does not hybridize with DNAs from unrelated phage nor with the DNA from uninfected *E. coli.* This RNA must therefore be complementary to T2 DNA in agreement with Jacob and Monod's prediction; that is, the phage-specific RNA is a messenger RNA. Hybridization studies have likewise shown that mRNAs from uninfected *E. coli* are complementary to portions of *E. coli* DNA. In fact, other RNAs, such as transfer RNA and ribosomal RNA, have corresponding complementary sequences on DNA from the same organism. Thus, *all cellular RNAs are transcribed from DNA templates.*

2 ■ RNA POLYMERASE

RNA polymerase (RNAP), *the enzyme responsible for the DNA-directed synthesis of RNA, was discovered independently in 1960 by Samuel Weiss and Jerard Hurwitz. The enzyme couples together the ribonucleoside triphosphates ATP, CTP, GTP, and UTP on DNA templates in a reaction that is driven by the release and subsequent hydrolysis of* PP_i*:*

$$(\text{RNA})_{n\text{ residues}} + \text{NTP} \rightleftharpoons (\text{RNA})_{n+1\text{ residues}} + \text{PP}_i$$

All cells contain RNAP. In bacteria, one species of this enzyme synthesizes all of the cell's RNA except the RNA primers employed in DNA replication (Section 30-1D). Various bacteriophages encode RNAPs that synthesize only phage-specific RNAs. Eukaryotic cells contain four or five RNAPs that each synthesize a different class of RNA. In this section we first consider the properties of the bacterial RNAPs and then consider the eukaryotic enzymes.

E. coli RNAP's so-called **holoenzyme** is an ~459-kD protein with subunit composition $\alpha_2\beta\beta'\omega\sigma$ (Table 31-1) in which the β and β′ subunits contain several colinearly arranged homologous segments. Once RNA synthesis has been initiated, however, the σ subunit (also called **σ factor** or **σ^{70}** since its molecular mass is 70 kD) dissociates from the **core enzyme,** $\alpha_2\beta\beta'\omega$, which carries out the actual polymerization process (see below).

TABLE 31-1. Components of *E. coli* RNA Polymerase Holoenzyme

Subunit	Number of Residues	Structural Gene
α	329	*rpoA*
β	1342	*rpoB*
β′	1407	*rpoC*
ω	91	*rpoZ*
σ^{70}	613	*rpsD*

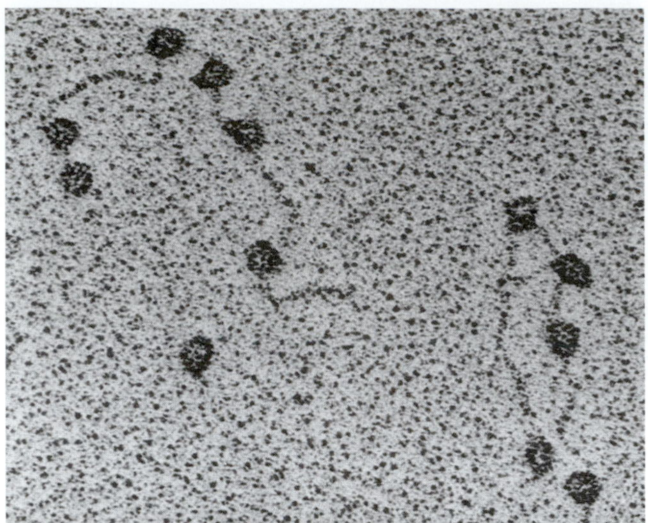

FIGURE 31-9 An electron micrograph of *E. coli* RNA polymerase (RNAP) holoenzyme attached to various promoter sites on bacteriophage T7 DNA. RNAP is one of the largest known soluble enzymes. [From Williams, R.C., *Proc. Natl. Acad. Sci.* **74,** 2313 (1977).]

Electron micrographs (Fig. 31-9) clearly indicate that RNAP, which has a characteristic large size, binds to DNA as a protomer. This large size is presumably a consequence of the holoenzyme's several complex functions including (1) template binding, (2) RNA chain initiation, (3) chain elongation, and (4) chain termination. We discuss these various functions below.

A. *Template Binding*

RNA synthesis is normally initiated only at specific sites on the DNA template. This was first demonstrated through hybridization studies of bacteriophage φX174 DNA with the RNA produced by φX174-infected *E. coli.* Bacteriophage φX174 carries a single strand of DNA known as the (+) strand. On its injection into *E. coli,* the (+) strand directs the synthesis of the complementary (−) strand with which it combines to form a circular duplex DNA known as the replicative form (Section 30-3B). The RNA produced by φX174-infected *E. coli* does not hybridize with DNA from intact phages but does so with the replicative form. Thus only the (−) strand of φX174 DNA, the so-called **antisense strand,** is transcribed, that is, acts as a template; the (+) strand, the **sense strand** (or **coding strand;** so called because it has the same sequence as the transcribed RNA), does not do so. Similar studies indicate that in larger phages, such as T4 and λ, the two viral DNA strands are the antisense (template) strands for different sets of genes. The same is true of cellular organisms.

a. Holoenzyme Specifically Binds to Promoters

RNA polymerase binds to its initiation sites through base sequences known as ***promoters*** *that are recognized by the corresponding σ factor.* The existence of promoters was first recognized through mutations that enhance or diminish the transcription rates of certain genes, including those of the *lac* operon. *Genetic mapping of such mutations indicated that the promoter consists of an ~40-bp sequence that is located on the 5′ side of the transcription start site.* [By convention, the sequence of template DNA is represented by its sense (nontemplate) strand so that it will have the same directionality as the transcribed RNA. A base pair in a promoter region is assigned a negative or positive number that indicates its position, upstream or downstream in the direction of RNAP travel, from the first nucleotide that is transcribed to RNA; this start site is +1 and there is no 0.] RNA, as we shall see, is synthesized in the 5′ → 3′ direction (Section 31-2C). Consequently, the promoter lies on the "upstream" side of the RNA's starting nucleotide. Sequencing studies indicate that the *lac* promoter (*lacP*) overlaps the *lac* operator (Fig. 31-2).

The holoenzyme forms tight complexes with promoters (dissociation constant $K \approx 10^{-14} M$) and thereby protects the bound DNA segments from digestion by DNase I. The region from about −20 to +20 is protected against exhaustive DNase I degradation. The region extending upstream to about −60 is also protected but to a lesser extent, presumably because it binds holoenzyme less tightly.

Sequence determinations of the protected regions from numerous *E. coli* and phage genes have revealed the "consensus" sequence of *E. coli* promoters (Fig. 31-10). *Their most conserved sequence is a hexamer centered at about the −10 position, the so-called* **Pribnow box** (named after David Pribnow, who pointed out its existence in 1975). It has a consensus sequence of TATAAT in which the leading TA and final T are highly conserved. *Upstream sequences around position −35 also have a region of sequence similarity,* TTGACA, which is most evident in efficient promoters. The sequence of the segment between the −10 and the −35 sites is unimportant but its length is critical; it ranges from 16 and 19 bp in the great majority of promoters. The initiating (+1) nucleotide, which is nearly always A or G, is centered in a poorly conserved CAT or CGT sequence. Most promoter sequences vary considerably from the consensus sequence (Fig. 31-10). Nevertheless, a mutation in one of the partially conserved regions can greatly increase or decrease a promoter's initiation efficiency. In addition, Richard Gourse discovered that certain highly expressed genes contain an A + T-rich segment between positions −40 and −60, the **upstream promoter (UP) element,** which binds to the C-terminal domain of RNAP's α subunits (αCTD; Section 31-3C). The UP element-containing genes include those encoding the ribosomal RNAs, the *rrn* genes (e.g., Fig. 31-10), which collectively account for 60% of the RNA synthesized by *E. coli. The rates at which genes are transcribed, which span a range of at least 1000, vary directly with the rate at which their promoters form stable initiation complexes with the holoenzyme.* Promoter mutations that increase or decrease the rate at which the associated gene is transcribed are known as **up mutations** and **down mutations.**

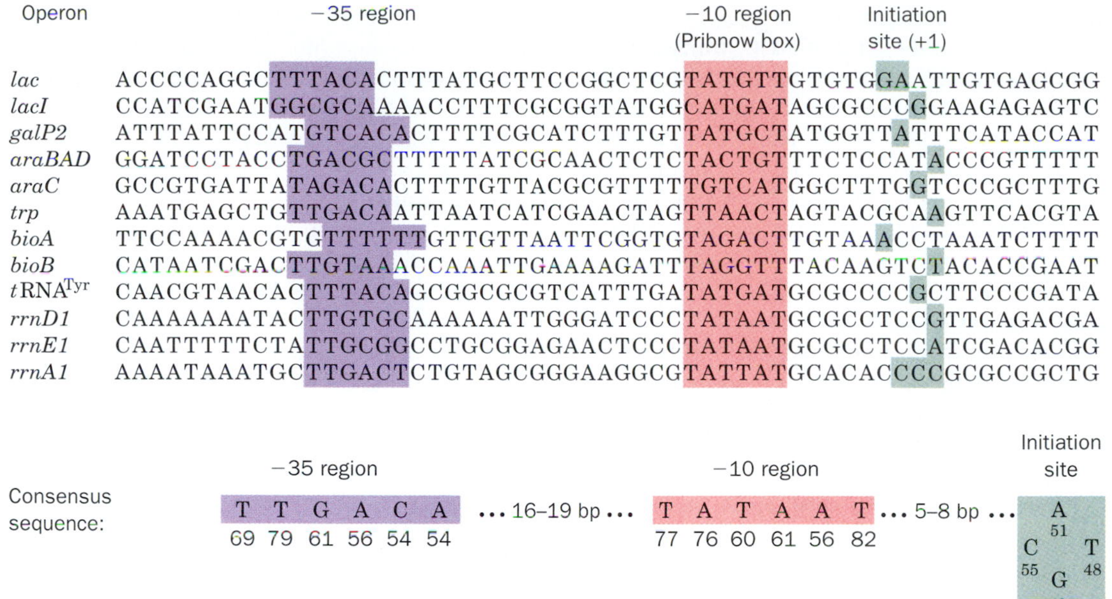

FIGURE 31-10 The sense (nontemplate) strand sequences of selected *E. coli* promoters. A 6-bp region centered around the −10 position (*red shading*) and a 6-bp sequence around the −35 region (*blue shading*) are both conserved. The transcription initiation sites (+1), which in most promoters occurs at a single purine nucleotide, are shaded in green. The bottom row shows the consensus sequence of 298 *E. coli* promoters with the number below each base indicating its percentage occurrence. The downstream portions of the *rrn* genes' UP elements can be seen. [After Rosenberg, M. and Court, D., *Annu. Rev. Genet.* **13,** 321–323 (1979). Consensus sequence from Lisser, S. and Margalit, H., *Nucleic Acids Res.* **21,** 1512 (1993).]

b. Initiation Requires the Formation of an Open Complex

The promoter regions in contact with the holoenzyme were identified by determining where the enzyme alters the susceptibility of the DNA to alkylation by agents such as dimethyl sulfate (DMS), a procedure named **DMS footprinting** (Section 34-3B). These experiments demonstrated that the holoenzyme contacts the promoter mainly around its −10 and −35 regions. These protected sites are both on the same side of the B-DNA double helix as the initiation site, which suggests that holoenzyme binds to only one face of the promoter.

DMS methylates G residues at N7, A residues at N1 and N3, and C residues at N3. Since N1 on A and N3 on C participate in base pairing interactions, however, they can only react with DMS in single-stranded DNA. This differential methylation of single- and double-stranded DNAs provides a sensitive test for DNA strand separation or "melting." Such chemical footprinting studies indicate that the binding of holoenzyme "melts out" the promoter in a region of ~14 bp extending from the middle of the −10 region to just past the initiation site. The need to form this "open complex" explains why promoter efficiency tends to decrease with the number of G · C base pairs in the −10 region; this presumably increases the difficulty in opening the double helix as is required for chain initiation (recall that G · C pairs are more stable than A · T pairs).

Core enzyme, which does not specifically bind promoter (except when it has an UP element), tightly binds duplex DNA (the complex's dissociation constant is $K \approx 5 \times 10^{-12} M$ and its half-life is ~60 min). Holoenzyme, in contrast, binds to nonpromoter DNA comparatively loosely ($K \approx 10^{-7} M$ and a half-life >1 s). Evidently, the σ subunit allows holoenzyme to move rapidly along a DNA strand in search of the σ subunit's corresponding promoter. Once transcription has been initiated and the σ subunit jettisoned, the tight binding of core enzyme to DNA apparently stabilizes the ternary enzyme–DNA–RNA complex.

B. *Chain Initiation*

The 5′-terminal base of prokaryotic RNAs is almost always a purine with A occurring more often than G. The initiating reaction of transcription is simply the coupling of two nucleoside triphosphates in the reaction

$$\text{pppA} + \text{pppN} \rightleftharpoons \text{pppApN} + \text{PP}_i$$

and hence, unlike DNA replication, does not require a primer. Bacterial RNAs therefore have 5′-triphosphate groups as was demonstrated by the incorporation of radioactive label into RNA when it was synthesized with [γ-^{32}P]ATP. Only the 5′ terminus of the RNA can retain the label because the internal phosphodiester groups of RNA are derived from the α-phosphate groups of nucleoside triphosphates.

The difficulty in forming an open complex is reflected in the observation that RNA synthesis is frequently aborted after usually 2 or 3 but up to 12 nucleotides have been joined. However, the holoenzyme does not release the promoter but, rather, reinitiates transcription. Eventually, the open complex forms and processive (continuous) RNA synthesis commences. At this point, σ factor dissociates from the core–DNA–RNA complex and can join with another core to form a new initiation complex. This is demonstrated by a burst of RNA synthesis on addition of core enzyme to a transcribing reaction mixture that initially contained holoenzyme.

a. RNAP Has a Highly Complex Structure

The X-ray structure of *E. coli* RNAP has not been determined. However, Seth Darst has elucidated the X-ray structures of the closely related *Thermus aquaticus* (*Taq*) RNAP core enzyme and holoenzyme. The X-ray structure of *Taq* core enzyme, in agreement with EM studies of *E. coli* RNAP, has the overall shape of a crab claw whose two "pincers" are formed by the β and β′ subunits (Fig. 31-11*a*). The protein is ~150 Å long (parallel to the pincers), ~115 Å high, and ~110 Å deep with the channel (really a cavern) between the two pincers ~27 Å high. A large internal segment of the β′ subunit as well as the C-terminal domains of both α subunits are disordered. The β and β′ subunits extensively interact with one another, particularly at the base of the channel where the active site Mg^{2+} ion is located, which is also where their homologous segments converge. The β′ subunit binds a Zn^{2+} ion via four Cys residues that are invariant in prokaryotes but not in eukaryotes.

The X-ray structure of the *Taq* holoenzyme indicates that its σ subunit (σ^A) consists of three flexibly linked domains, σ_2, σ_3, and σ_4, that extend across the top of the holoenzyme (Fig. 31-11*b*). The binding of σ^A causes the core enzyme's pincers to come together so as to narrow the channel between them by ~10 Å. The outer surface of the holoenzyme is almost uniformly negatively charged, whereas those surfaces presumed to interact with nucleic acids, particularly the inner walls of the main channel, are positively charged.

The X-ray structure of *Taq* holoenzyme in complex with a so-called fork-junction promoter DNA fragment (Fig. 31-12*a*) reveals that the DNA lies across one face of the holoenzyme, completely outside of the active site channel (Fig. 31-12*b*). All sequence-specific contacts that the holoenzyme makes with the DNA (with the −10 and −35 regions as well as the so-called extended −10 region just upstream of the −10 region) are mediated by the σ^A subunit via conserved residues. This structure presumably resembles the closed complex.

The foregoing X-ray structures, together with footprinting data, have led Darst to construct models for the closed (RP_c) and open (Rp_o) complexes (Fig. 31-13). RP_c (Fig. 31-13*a*) resembles the holoenzyme–DNA complex (Fig. 31-12*b*) but with an extended length of dsDNA (from −60 to +25) whose upstream end is bent around the enzyme such that the DNA's UP element contacts the 80-residue C-terminal domains of the α subunits (which are disordered in the above X-ray structures). In Rp_o (Fig. 31-13*b*), the template strand of the transcription bubble has slipped into a tunnel formed by the σ^A, β, and β′ subunits that is lined with universally conserved basic residues. This tunnel directs the template strand to the active site channel, where it base pairs with the initiating ribonucleotides at the *i* and *i* + 1 sites near the Mg^{2+} ion.

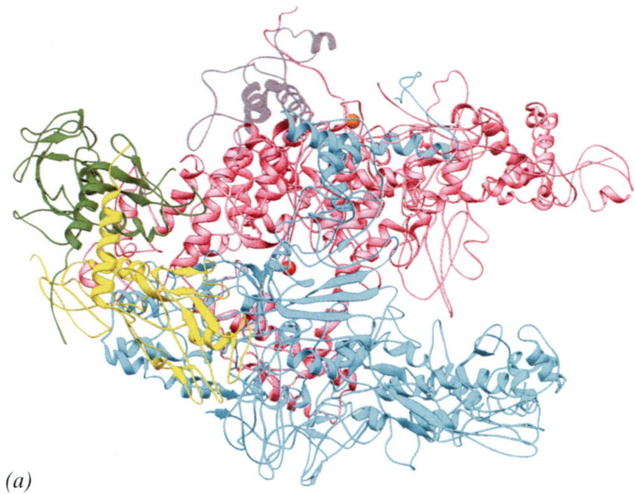

(*a*)

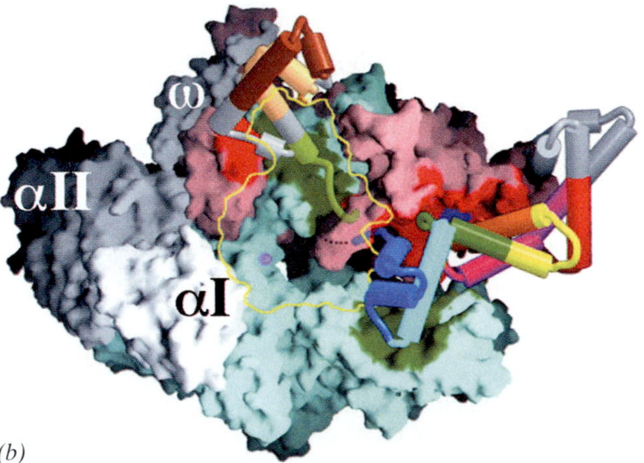

(*b*)

FIGURE 31-11 X-Ray structure of *Taq* RNAP. (*a*) The core enzyme in which the two α subunits are yellow and green, the β subunit is cyan, the β′ subunit is pink, and the ω subunit is gray. The bound Mg^{2+} and Zn^{2+} ions are represented by red and orange spheres, respectively. Note that residues 156 through 452 of the 1524-residue β′ subunit, which extend from the tip of its "pincer," are disordered and hence not visible. (*b*) The holoenzyme viewed as in Part *a*. Its core enzyme component is represented by its molecular surface with its α and ω subunits gray, its β subunit blue-green, and its β′ subunit pink. The σ^A subunit is represented by its C_α backbone with its helices drawn as cylinders, its various conserved segments in different colors, and its N-terminal end at the right. A portion of the β subunit, which is outlined in yellow, has been deleted to expose the connectivity of the σ^A subunit. Those portions of the β and β′ subunits within 4 Å of the σ^A subunit are colored green and red, respectively. The active site Mg^{2+} ion is represented by a magenta sphere. [Part *a* based on an X-ray structure by and Part *b* courtesy of Seth Darst, The Rockefeller University.]

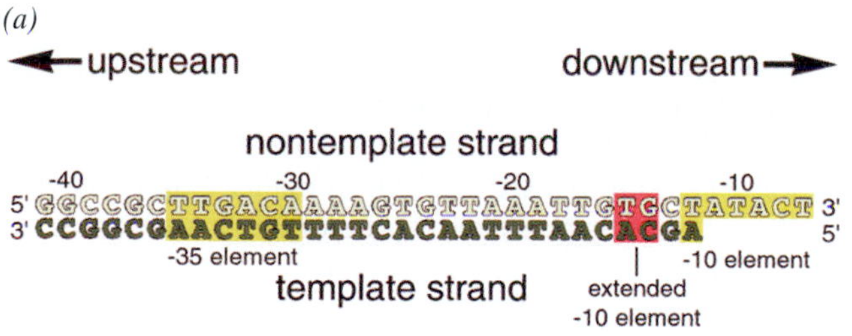

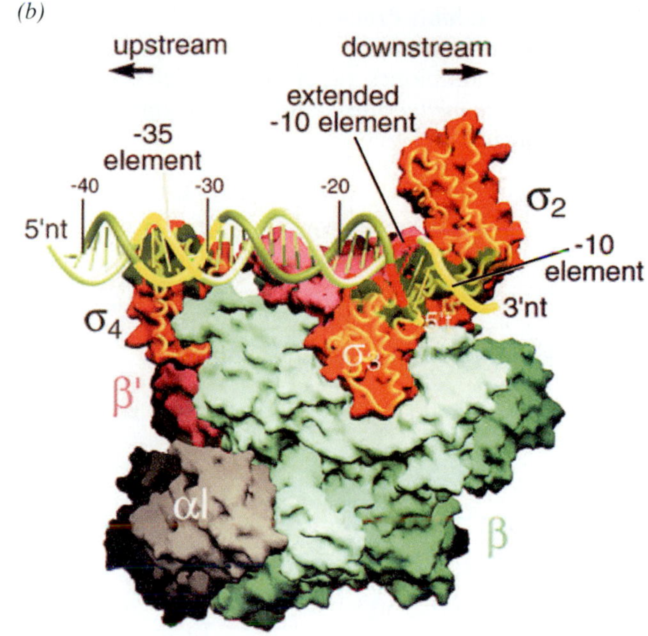

FIGURE 31-12 X-Ray structure of *Taq* holoenzyme in complex with a fork-junction promoter DNA fragment. (*a*) The sequence of the fork-junction DNA with the numbers indicating the base position relative to the transcription start site, +1. (*b*) The structure of the holoenzyme–fork-junction DNA complex viewed as in Fig. 31-11 but rotated counterclockwise by 45° about an axis perpendicular to the page. The DNA is drawn in ladder form with its template strand green and its nontemplate (sense) strand yellow-green, except that its −35 and −10 elements are yellow and its extended −10 element is red. The holoenzyme is represented by its molecular surface colored according to subunit with αI, αII, and ω gray, β light green, β′ pink, and σ^A orange and partially transparent to reveal its C_α backbone. Those portions of the holoenzyme <4 Å from the DNA (which occurs only on σ^A) are dark green. [Courtesy of Seth Darst, The Rockefeller University. PDBid 1L9Z.]

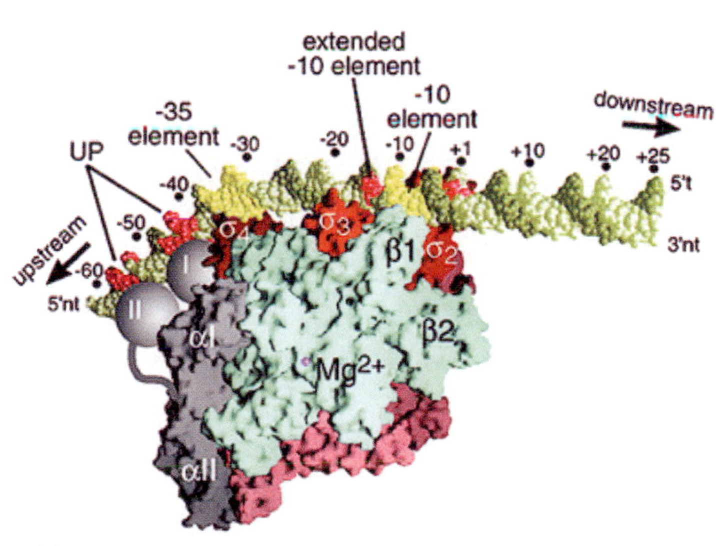

(*a*)

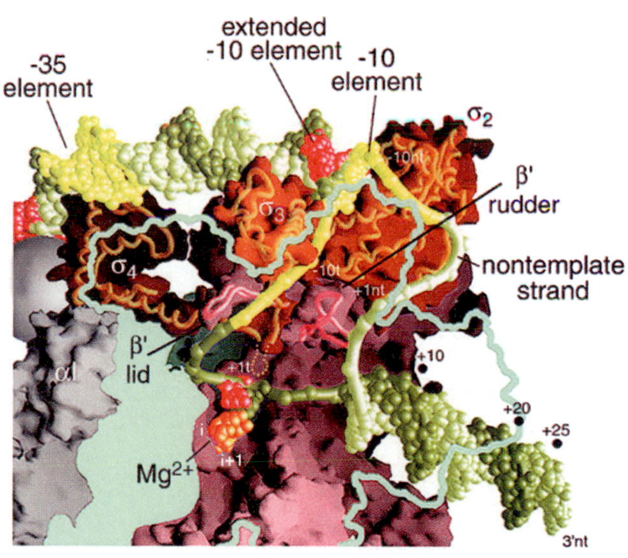

(*b*)

FIGURE 31-13 Models of the closed (RP_c) and open (Rp_o) complexes of *Taq* RNAP with promoter-containing DNA extending between positions −60 and +25. (*a*) A model of RP_c in which the protein is represented similarly to that in Fig. 31-12*b* but rotated by 65° about the horizontal axis and the β subunit is partially transparent to show the position of the active site Mg^{2+} ion (*magenta sphere*). The dsDNA is drawn in space-filling form with its template (t) strand green and its nontemplate (nt) strand light green except for the −35 and −10 elements, which are yellow, and the UP element, the extended −10 element, and the transcription start site (+1), which are red. The UP element bends around the enzyme to interact with the C-terminal domains of the α subunits (*gray spheres labeled I and II*). However, there is no evidence that the downstream segment of the dsDNA interacts with the protein, and hence this DNA is represented as extending away from the protein. (*b*) A model of Rp_o shown in magnified view relative to Part *a* and with obscuring portions of the β subunit removed (its outline is shown as a light green line) to reveal the transcription bubble and its interactions with the active site. Single-stranded portions of the DNA are represented by their linked P atoms. The initiating ribonucleotides at positions *i* and *i* + 1 are drawn in space-filling form (*red and orange*). Note how the downstream segment of the dsDNA emanates from the end of the active site channel. [Courtesy of Seth Darst, The Rockefeller University.]

b. Rifamycins Inhibit Prokaryotic Transcription Initiation

Two related antibiotics, **rifamycin B,** which is produced by *Streptomyces mediterranei,* and its semisynthetic derivative **rifampicin**

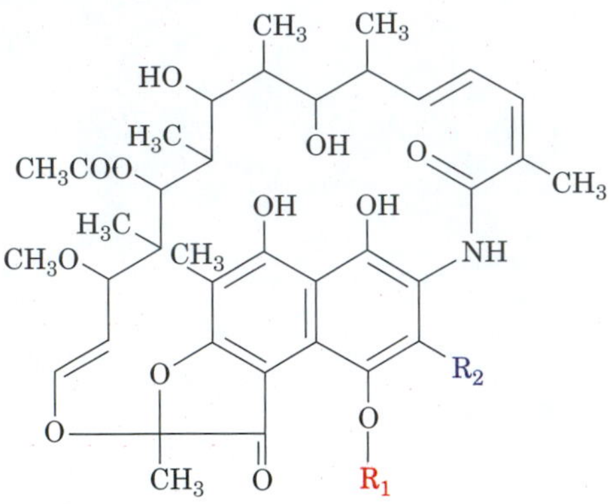

Rifamycin B $R_1 = CH_2COO^-$; $R_2 = H$

Rifampicin $R_1 = H$; $R_2 = CH{=}\overset{+}{N}$ (piperazine ring) $N{-}CH_3$

specifically inhibit transcription by prokaryotic, but not eukaryotic, RNAPs. This selectivity and their high potency (bacterial RNAP is 50% inhibited by $2 \times 10^{-8}M$ rifampicin) has made them medically useful bacteriocidal agents against gram-positive bacteria and tuberculosis. Indeed, few other antibiotics are effective against tuberculosis, which is reaching epidemic levels in some parts of the world.

The finding that the β subunits of rifamycin-resistant mutants have altered electrophoretic mobilities first demonstrated that this subunit contains the rifamycin-binding site. Rifamycins inhibit neither the binding of RNAP to the promoter nor the formation of the first phosphodiester bond, but they prevent further chain elongation. The inactivated RNAP remains bound to the promoter, thereby blocking its initiation by uninhibited enzymes. Once RNA chain initiation has occurred, however, rifamycins have no effect on the subsequent elongation process. The rifamycins are therefore useful research tools because they permit the transcription process to be dissected into its initiation and its elongation phases.

The X-ray structure of *Taq* core enzyme in complex with rifampicin reveals how this antibiotic inhibits RNAP. Rifampicin binds with close complementary fit but little conformational change in a pocket in the β subunit that is located within the main DNA–RNA channel, ~12 Å distant from the active site Mg^{2+} ion. Model building indicates that the bound rifampicin would sterically interfere with the RNA transcript at positions −2 to −5 in the transcription bubble. Thus, as is observed, rifampicin would not interfere with the initiation of transcription but would mechanically block the extension of the RNA transcript. The residues lining the pocket in which rifampicin binds are highly conserved among prokaryotes but not in eukaryotes, thereby explaining why rifamycins inhibit only prokaryotic RNAPs.

C. Chain Elongation

What is the direction of RNA chain elongation; that is, does it occur by the addition of incoming nucleotides to the 3′ end of the nascent (growing) RNA chain (5′ → 3′ growth; Fig. 31-14*a*) or by their addition to its 5′ terminus

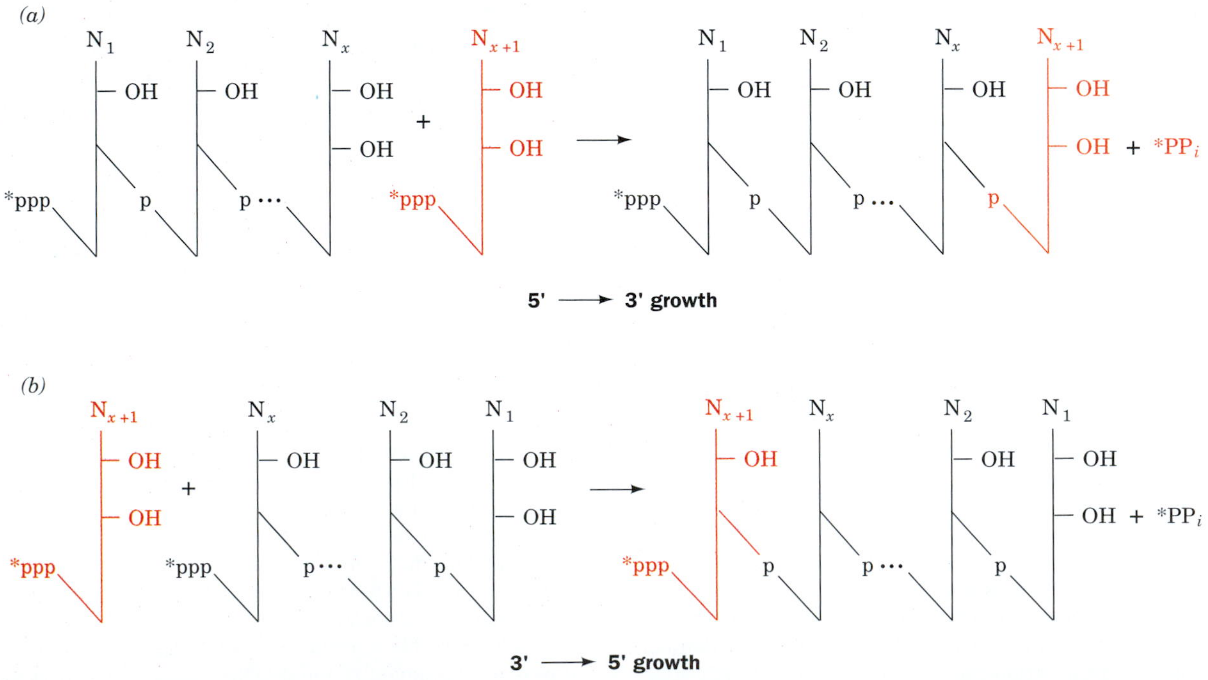

(a) N_1, N_2, N_x + N_{x+1} → N_1, N_2, N_x, N_{x+1} + $*PP_i$

5′ → 3′ growth

(b) N_{x+1} + N_x, N_2, N_1 → N_{x+1}, N_x, N_2, N_1 + $*PP_i$

3′ → 5′ growth

FIGURE 31-14 The two possible modes of RNA chain growth. Growth may occur (*a*) by the addition of nucleotides to the 3′ end and (*b*) by the addition of nucleotides to the 5′ end. RNA polymerase catalyzes the former reaction.

($3' \rightarrow 5'$ growth; Fig. 31-14*b*)? This question was answered by determining the rate at which the radioactive label from [γ-^{32}P]GTP is incorporated into RNA. For $5' \rightarrow 3'$ elongation, the $5'$ γ-P is permanently labeled and, hence, the chain's level of radioactivity would not change on replacement of the labeled GTP with unlabeled GTP. However, for $3' \rightarrow 5'$ elongation, the $5'$ γ-P is replaced with the addition of every new nucleotide so that, on replacement of labeled with unlabeled GTP, the nascent RNA chains would lose their radioactivity. The former was observed. *Chain growth must therefore occur in the $5' \rightarrow 3'$ direction (Fig. 31-14a),* the same direction as DNA is synthesized. This conclusion is corroborated by the observation that the antibiotic **cordycepin,**

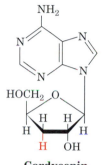

Cordycepin
(3'-deoxyadenosine)

an adenosine analog that lacks a $3'$-OH group, inhibits bacterial RNA synthesis. Its addition to the $3'$ end of RNA, as is expected for $5' \rightarrow 3'$ growth, prevents the RNA chain's further elongation. Cordycepin would not have this effect if chain growth occurred in the opposite direction because it could not be appended to an RNA's $5'$ end.

a. Transcription Supercoils DNA

RNA chain elongation requires that the double-stranded DNA template be opened up at the point of RNA synthesis so that the template strand can be transcribed to its complementary RNA strand. In doing so, the RNA chain only transiently forms a short length of RNA–DNA hybrid duplex, as is indicated by the observation that transcription leaves the template duplex intact and yields single-stranded RNA. The unpaired "bubble" of DNA in the open initiation complex apparently travels along the DNA with the RNAP. There are two ways this might occur (Fig. 31-15):

1. If the RNAP followed the template strand in its helical path around the DNA, the DNA would build up little supercoiling because the DNA duplex would never be unwound by more than about a turn. However, the RNA transcript would wrap around the DNA, once per duplex turn. This model is implausible since it is unlikely that its DNA and RNA could be readily untangled: The RNA would not spontaneously unwind from the long and often circular DNA in any reasonable time, and no known topoisomerase can accelerate this process.

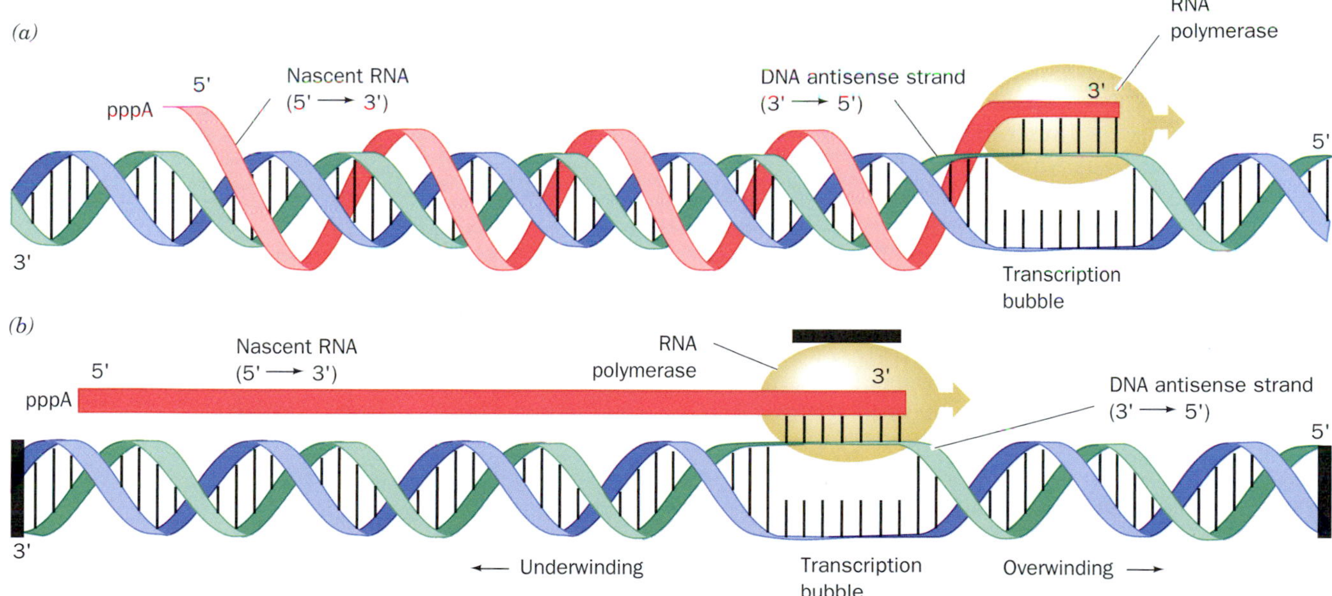

FIGURE 31-15 RNA chain elongation by RNA polymerase. In the region being transcribed, the DNA double helix is unwound by about a turn to permit the DNA's sense strand to form a short segment of DNA–RNA hybrid double helix with the RNA's $3'$ end. As the RNAP advances along the DNA template (here to the right), the DNA unwinds ahead of the RNA's growing $3'$ end and rewinds behind it, thereby stripping the newly synthesized RNA from the template (antisense) strand. (*a*) One way this might occur is by the RNAP following the path of the template strand about the DNA double helix, in which case the transcript would become wrapped about the DNA once per duplex turn. (*b*) A second and more plausible possibility is that the RNA moves in a straight line while the DNA rotates beneath it. In this case the RNA would not wrap around the DNA but the DNA would become overwound ahead of the advancing transcription bubble and unwound behind it (consider the consequences of placing your finger between the twisted DNA strands in this model and pushing toward the right). The model presumes that the ends of the DNA, as well as the RNAP, are prevented from rotating by attachments within the cell (*black bars*). [After Futcher, B., *Trends Genet.* **4,** 271, 272 (1988).]

2. If the RNAP moves in a straight line while the DNA rotates, the RNA and DNA will not become entangled. Rather, the DNA's helical turns are pushed ahead of the advancing transcription bubble so as to more tightly wind the DNA ahead of the bubble (which promotes positive supercoiling), and the DNA behind the bubble becomes equivalently unwound (which promotes negative supercoiling, although note that the linking number of the entire DNA remains unchanged). This model is supported by the observations that the transcription of plasmids in *E. coli* causes their positive supercoiling in gyrase mutants (which cannot relax positive supercoils; Section 29-3C) and their negative supercoiling in topoisomerase I mutants (which cannot relax negative supercoils). In fact, by tethering RNAP to a glass surface and allowing it to transcribe DNA that had been fluorescently labeled at one end, Kazuhiko Kinosita demonstrated, through fluorescence microscopy (using techniques similar to those showing that the F_1F_0-ATPase is a rotary engine; Section 22-3C), that single DNA molecules rotated in the expected direction during transcription.

Inappropriate superhelicity in the DNA being transcribed halts transcription (Section 29-3C). Quite possibly the torsional tension in the DNA generated by negative superhelicity behind the transcription bubble is required to help drive the transcriptional process, whereas too much such tension prevents the opening and maintenance of the transcription bubble.

b. Transcription Occurs Rapidly and Accurately

The *in vivo* rate of transcription is 20 to 50 nucleotides per second at 37°C as indicated by the rate at which *E. coli* incorporate ^{3}H-labeled nucleosides into RNA (cells cannot take up nucleoside triphosphates from the medium). Once an RNAP molecule has initiated transcription and moved away from the promoter, another RNAP can follow suit. The synthesis of RNAs that are needed in large quantities, ribosomal RNAs, for example, is initiated as often as is sterically possible, about once per second (Fig. 31-16).

RNA polymerase, unlike DNA polymerase, cannot rebind a polynucleotide that it has released. Hence, RNA synthesis must be entirely processive, even for the largest eukaryotic genes (which are longer than 2000 kb). Thus, RNAP does not exonucleolytically correct its mistakes as do many DNA polymerases. This accounts for the observations that the error frequency of RNA synthesis, as estimated from the analysis of transcripts of simple templates such as poly[d(AT)] · poly[d(AT)], is one wrong base incorporated for every ~10^4 transcribed, whereas, for example, *E. coli* Pol I incorporates one incorrect base in 10^7 (Section 30-2A). The former rate is tolerable because (1) most genes are repeatedly transcribed, (2) the genetic code contains numerous synonyms (Table 5-3), (3) amino acid substitutions in proteins are often functionally innocuous, and (4) large portions of many eukaryotic transcripts are excised in forming the mature mRNAs (intron excision; Section 31-4A).

c. Intercalating Agents Inhibit Both RNA and DNA Polymerases

Actinomycin D,

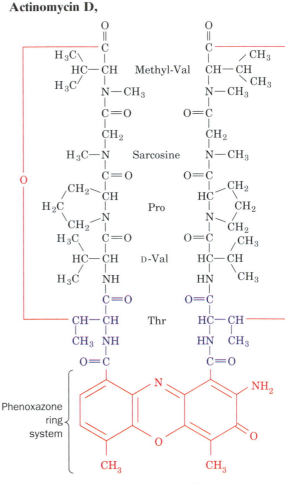

Actinomycin D

a useful antineoplastic (anticancer) agent produced by

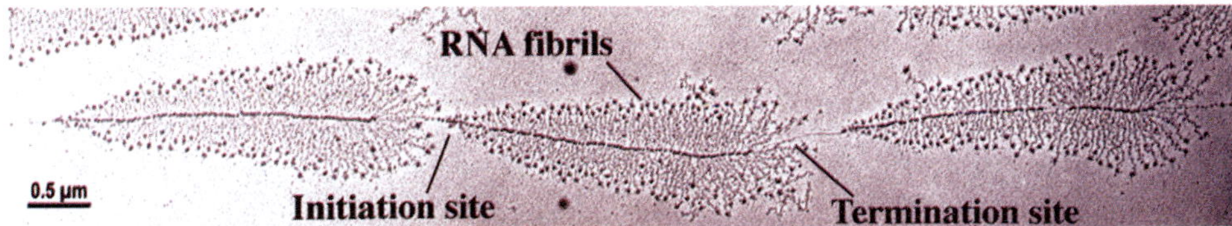

FIGURE 31-16 An electron micrograph of three contiguous ribosomal genes from oocytes of the salamander *Pleurodeles waltl* undergoing transcription. The "arrowhead" structures result from the increasing lengths of the nascent RNA chains as the RNAP molecules synthesizing them move from the initiation site on the DNA to the termination site. [Courtesy of Ulrich Scheer, University of Würzburg, Germany.]

FIGURE 31-17 X-Ray structure of actinomycin D in complex with a duplex DNA of self-complementary sequence d(GAAGCTTC). The complex is shown in space-filling form in which the DNA's sugar–phosphate backbone is yellow, its bases are white, and the actinomycin D is colored according to atom type with C green, N blue, and O red. The two symmetry-related DNA molecules that are shown stack vertically to form a pseudocontinuous helix. The upper DNA is viewed toward its minor groove into which its bound actinomycin D's two cyclic depsipeptides are tightly wedged. The lower DNA (which is turned 180° about its helix axis relative to the upper DNA) is viewed toward its major groove, into which its bound actinomycin D's intercalated phenoxazone ring system projects from the minor groove side. [Based on an X-ray structure by Fusao Takusagawa, University of Kansas. PDBid 172D.]

Streptomyces antibioticus, tightly binds to duplex DNA and, in doing so, strongly inhibits both transcription and DNA replication, presumably by interfering with the passage of RNA and DNA polymerases. The X-ray structure of actinomycin D in complex with a duplex DNA composed of two strands of the self-complementary octamer d(GAAGCTTC) reveals that the DNA assumes a B-like conformation in which the actinomycin's **phenoxazone ring system,** as had previously been shown, is intercalated between the DNA's central G · C base pairs (Fig. 31-17). Consequently, the DNA helix is unwound by 23° at the intercalation site and the central G · C base pairs are separated by 7.0 Å. The DNA helix is severely distorted from the normal B-DNA conformation such that its minor groove is wide and shallow in a manner resembling that of A-DNA. Actinomycin D's two chemically identical cyclic **depsipeptides** (having both peptide bonds and ester linkages), which assume different conformations, extend in opposite directions from the intercalation site along the minor groove of the DNA. The complex is stabilized through the formation of base–peptide and phenoxazone–sugar-phosphate backbone hydrogen bonds, as well as by hydrophobic interactions, in a way that explains the preference of actinomycin D to bind to DNA with its phenoxazone ring intercalated between the base pairs of a 5′-GC-3′ sequence. Several other intercalation agents, including ethidium and proflavin (Sections 6-6C and 29-3A), also inhibit nucleic acid synthesis, presumably by similar mechanisms.

D. *Chain Termination*

Electron micrographs such as Fig. 31-16 suggest that DNA contains specific sites at which transcription is terminated. The transcriptional termination sequences of many *E. coli* genes share two common features (Fig. 31-18*a*):

1. A series of 4 to 10 consecutive A · T's with the A's on the template strand. The transcribed RNA is terminated in or just past this sequence.

2. A G + C-rich region with a palindromic (2-fold symmetric) sequence that is immediately upstream of the series of A · T's.

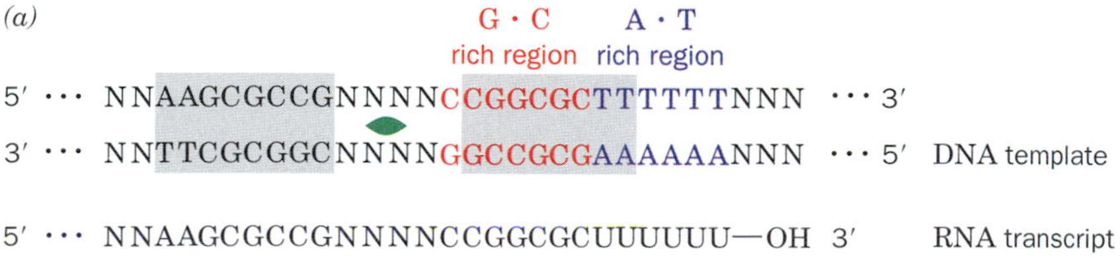

(a)

G · C rich region   A · T rich region

5′ ··· NNAAGCGCCGNNNNCCGGCGCTTTTTTNNN ··· 3′

3′ ··· NNTTCGCGGCNNNNGGCCGCGAAAAAANNN ··· 5′ DNA template

5′ ··· NNAAGCGCCGNNNNCCGGCGCUUUUUU—OH 3′ RNA transcript

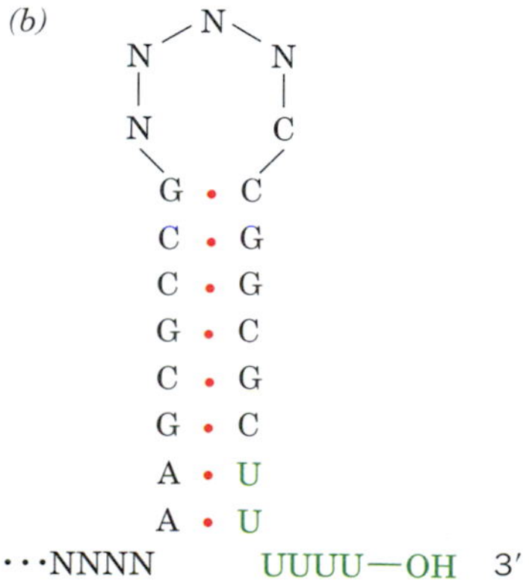

(b)

N N N N C (loop)
G · C
C · G
C · G
G · C
C · G
G · C
A · U
A · U
···NNNN UUUU—OH 3′

FIGURE 31-18 A hypothetical strong (efficient) *E. coli* terminator. The base sequence was deduced from the sequences of several transcripts. (*a*) The DNA sequence together with its corresponding RNA. The A · T-rich and G · C-rich sequences are shown in blue and red, respectively. The 2-fold symmetry axis (*green lenticular symbol*) relates the flanking shaded segments that form an inverted repeat. (*b*) The RNA hairpin structure and poly(U) tail that trigger transcription termination. [After Pribnow, D., *in* Goldberger, R.F. (Ed.), *Biological Regulation and Development,* Vol. 1, p. 253, Plenum Press (1979).]

The RNA transcript of this region can therefore form a self-complementary "hairpin" structure that is terminated by several U residues (Fig. 31-18*b*).

The stability of a terminator's G + C-rich hairpin and the weak base pairing of its oligo(U) tail to template DNA appear to be important factors in ensuring proper chain termination. In fact, model studies have shown that oligo(dA · rU) forms a particularly unstable hybrid helix although oligo(dA · dT) forms a helix of normal stability. The formation of the G + C-rich hairpin causes RNAP to pause for several seconds at the termination site. This, it has been proposed, induces a conformational change in the RNAP, which permits the nontemplate DNA strand to displace the weakly bound oligo(U) tail from the template strand, thereby terminating transcription. Consistent with this notion is the observation that mutations in the termination site that alter the strengths of these associations reduce the efficiency of chain termination and often eliminate it. Termination is similarly diminished when *in vitro* transcription is carried out with GTP replaced by **inosine triphosphate (ITP):**

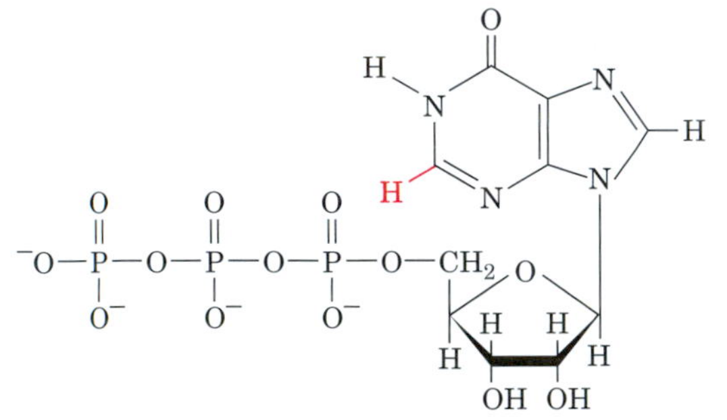

Inosine triphosphate (ITP)

I · C pairs are weaker than those of G · C because the hypoxanthine base of I, which lacks the 2-amino group of G, can only make two hydrogen bonds to C, thereby decreasing the hairpin's stability.

Despite the foregoing, experiments by Michael Chamberlin in which segments of highly efficient terminators were swapped via recombinant DNA techniques indicate that the RNA terminator hairpin and U-rich 3′ tail do not function independently of their upstream and downstream flanking regions. Indeed, terminators that lack a U-rich segment can be highly efficient when joined to the appropriate sequence immediately downstream from the termination site. Moreover, mutations in the β subunit of RNAP can both increase and decrease termination efficiency. This has led to an alternative model for termination in which the RNA transcript is stably bound to the RNAP via interactions with two single strand-specific RNA binding sites. Termination then occurs through the formation of a stable RNA hairpin that reduces the RNA's binding to one RNA-binding site. In this latter model, termination can occur on both single-stranded and double-stranded templates.

To distinguish between the above two models, Chamberlin constructed a system in which termination took place on a single-stranded template. Because RNA polymerase can only initiate transcription on double-stranded DNA, this was done by allowing RNAP to initiate transcription on dsDNA containing an initiation and a termination site, "walking" the RNAP past the start site to a specific site upstream of the terminator by adding and removing appropriate subsets of NTPs, and then adding **exonuclease III** (a processive 3′ → 5′ double strand-specific exonuclease), which digested both the template and the nontemplate strands from their 3′ ends to the stalled RNAP. The exonuclease III was then removed and NTPs were added, thereby permitting the RNA polymerase to resume transcription on the now single-stranded template. Since the terminator preceded the 5′ end of the template, transcripts that had normally terminated would be shorter than unterminated transcripts that had simply run off the end of the template. The results of these experiments, using three different terminators, revealed that termination is essentially as efficient on single-stranded templates as it is on double-stranded templates. Evidently, neither the nontemplate strand nor the transcription bubble is required for termination by *E. coli* RNAP.

a. Termination Often Requires the Assistance of Rho Factor

The above termination sequences induce the spontaneous termination of transcription. Around half the termination sites, however, lack any obvious similarities and are unable to form strong hairpins; *they require the participation of a protein known as* ***Rho factor*** *to terminate transcription.* The existence of Rho factor was suggested by the observation that *in vivo* transcripts are often shorter than the corresponding *in vitro* transcripts. Rho factor, a hexameric protein of identical 419-residue subunits, enhances the termination efficiency of spontaneously terminating transcripts as well as inducing the termination of nonspontaneously terminating transcripts.

Several key observations have led to a model of Rho-dependent termination:

1. Rho factor is a helicase that unwinds RNA–DNA and RNA–RNA double helices. This process is powered by the hydrolysis of NTPs to NDPs + P_i with little preference for the identity of the base. NTPase activity is required for Rho-dependent termination as is demonstrated by its *in vitro* inhibition when the NTPs are replaced by their β,γ-imido analogs,

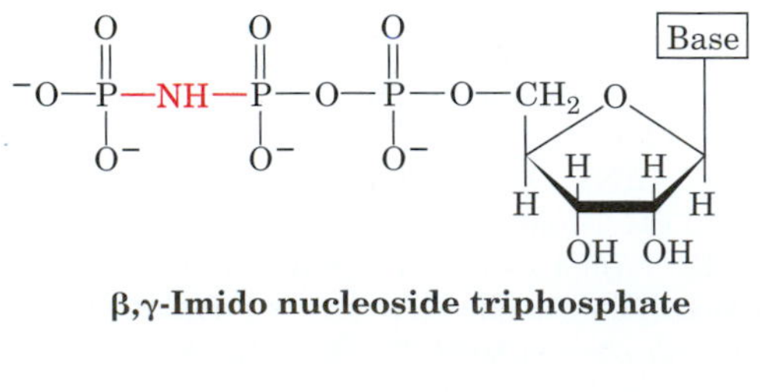

β,γ-Imido nucleoside triphosphate

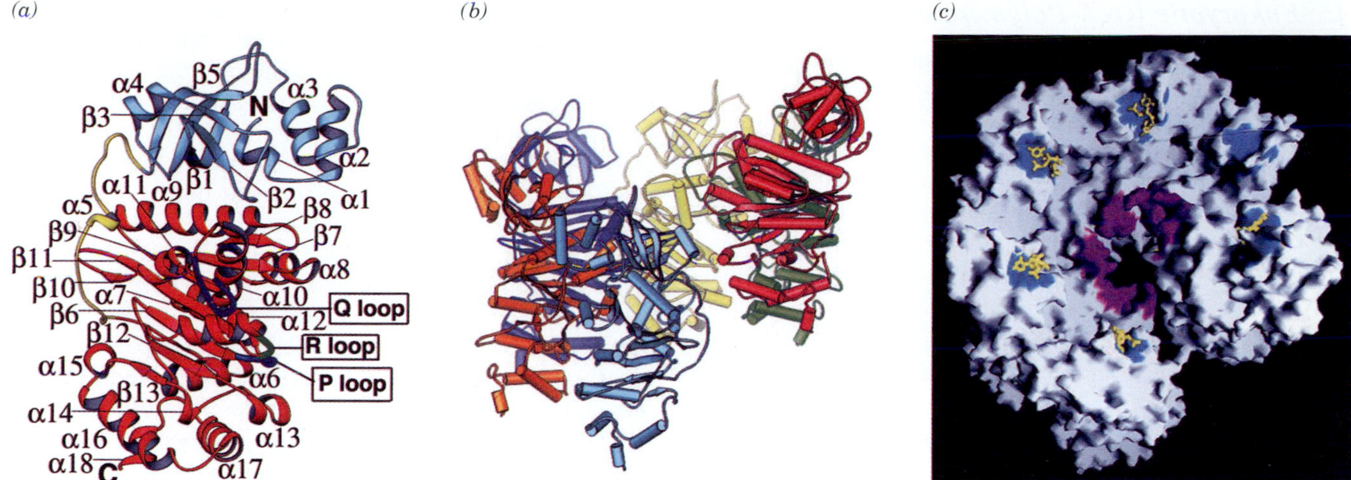

FIGURE 31-19 X-Ray structure of Rho factor in complex with RNA. (*a*) The Rho protomer with its N-terminal domain cyan, its C-terminal domain red, and their connecting linker yellow. The P loop (*blue*), Q loop (*magenta*), and R loop (*green*) have been implicated in mRNA binding and translocation as well as NTP binding. (*b*) The Rho hexamer. Its six subunits, each of which are drawn in a different color, form an open lock washer-shaped hexagonal ring. The yellow subunit is viewed similarly to that in Part *a*. (*c*) The solvent-accessible surface of the Rho hexamer as viewed from the top of Part *b*. The primary RNA binding sites, which occur on the N-terminal domain are cyan and the secondary RNA binding sites, which occur on the C-terminal domain, are magenta. The RNA, which is drawn in yellow stick form, is bound to the primary RNA binding sites. It is only partially visible in the X-ray structure. [Courtesy of James Berger, University of California at Berkeley. PDBid 1PVO.]

substances that are RNAP substrates but cannot be hydrolyzed by Rho factor.

2. Genetic manipulations indicate that Rho-dependent termination requires the presence of a specific recognition sequence on the newly transcribed RNA upstream of the termination site. The recognition sequence must be on the nascent RNA rather than the DNA as is demonstrated by Rho's inability to terminate transcription in the presence of pancreatic RNase A. The essential features of this termination site have not been fully elucidated; the construction of synthetic termination sites indicates that it consists of 80 to 100 nucleotides that lack a stable secondary structure and contain multiple regions that are rich in C and poor in G.

These observations suggest that Rho factor attaches to nascent RNA at its recognition sequence and then migrates along the RNA in the $5' \rightarrow 3'$ direction until it encounters an RNAP paused at the termination site (without the pause, Rho might not be able to overtake the RNA polymerase). There, Rho unwinds the RNA–DNA duplex forming the transcription bubble, thereby releasing the RNA transcript. Rho-terminated transcripts have 3′ ends that typically vary over a range of ~50 nucleotides. This suggests that Rho somehow pries the RNA away from its template DNA rather than "pushing" an RNA release "button."

Each Rho subunit consists of two domains that can be separated by proteolysis: Its N-terminal domain binds single-stranded polynucleotides and its C-terminal domain, which is homologous to the α and β subunits of the F_1-ATPase (Section 22-3C), binds an NTP. The X-ray structure of Rho in complex with ADPNP and an 8-nt RNA, $(UC)_4$, determined by James Berger, reveals that the N-terminal domain consists of a 3-helix bundle mounted on a 5-stranded β barrel and that the C-terminal domain consists of seven parallel β strands sandwiched between several helices (Fig. 31-19*a*). Rho forms the expected hexamer but, unlike the ring-shaped F_1-ATPase, forms a lock washer-shaped helix that is 120 Å in diameter with an ~30-Å diameter central hole and whose first and sixth subunits are separated by a 12-Å gap and rise of 45 Å along the helix axis (Fig. 31-19*b*). The RNA, which is only partially visible in the structure, binds to the so-called primary RNA binding sites on the N-terminal domains that face the interior of the helix, as do the so-called secondary RNA binding sites on the C-terminal domain that have been implicated in mRNA translocation and unwinding (Fig. 31-19*c*). Since electron microscopic images of Rho show both closed as well as notched hexameric rings, the X-ray structure probably represents an open state that is poised to bind mRNA that has entered its central cavity through the notch. Presumably, mRNA binding would cause the hexameric ring to close. Ensuing cycles of ATP hydrolysis, with their attendant conformational changes, would then propel Rho along the mRNA in the $5' \rightarrow 3'$ direction.

E. *Eukaryotic RNA Polymerases*

Eukaryotic nuclei, as Robert Roeder and William Rutter discovered, contain three distinct types of RNAPs that differ in the RNAs they synthesize:

1. **RNA polymerase I (RNAP I;** also called **Pol I** and **RNAP A),** which is located in the nucleoli (dense granular bodies in the nuclei that contain the ribosomal genes; Section 31-4B), synthesizes precursors of most ribosomal RNAs (rRNAs).

2. **RNA polymerase II (RNAP II;** also called **Pol II** and **RNAP B),** which occurs in the nucleoplasm, synthesizes mRNA precursors.

3. **RNA polymerase III (RNAP III;** also called **Pol III** and **RNAP C),** which also occurs in the nucleoplasm, synthesizes the precursors of 5S ribosomal RNA, the tRNAs, and a variety of other small nuclear and cytosolic RNAs.

Eukaryotic nuclear RNAPs have considerably greater subunit complexity than those of prokaryotes. These enzymes have molecular masses of up to 600 kD and, as is indicated in Table 31-2, each contains two nonidentical "large" (>120 kD) subunits comprising ~65% of its mass that are homologs of the prokaryotic RNAP β′ and β subunits, and up to 12 additional "small" (<50 kD) subunits, two of which are homologs of prokaryotic RNAP α, and one of which is a homolog of prokaryotic RNAP ω. Of these small subunits, five are identical in all three eukaryotic RNAPs and two others (the RNAP α homologs) are identical in RNAPs I and III. Two of the RNAP II subunits, Rbp4 and Rbp7, are not essential for activity and, in fact, are present in RNAP II in less than stoichiometric amounts. (Curiously, Rbp7 has a 102-residue segment that is 30% identical to a portion of σ^{70}, the predominant *E. coli* σ factor.) Thus 10 of the 12 RNAP II subunits are either identical or closely similar to subunits of RNAPs I and III (Table 31-2). Moreover, the sequences of these subunits are highly conserved (~50% identical) across species from yeast to humans (and to a lesser extent between eukaryotes and bacteria). In fact, in all ten cases tested, a human RNAP II subunit could replace its counterpart in yeast without loss of cell viability.

Rpb1, the β′ homolog in RNAP II, has an extraordinary C-terminal domain **(CTD).** In mammals, it contains 52 highly conserved repeats of the heptad PTSPSYS (26 repeats in yeast with other eukaryotes having intermediate values). Five of the seven residues in these particularly hydrophilic repeats bear hydroxyl groups and at least 50 of them, predominantly those on Ser residues, are subject to reversible phosphorylation by **CTD kinases** and **CTD phosphatases.** RNAP II initiates transcription only when the CTD is unphosphorylated but commences elongation only after the CTD has been phosphorylated, which sug-

TABLE 31-2. RNA Polymerase Subunits[a]

S. cerevisiae RNAP I (14 subunits)	*S. cerevisiae* RNAP II (12 subunits)	*S. cerevisiae* RNAP III (15 subunits)	*E. coli* RNAP Core (5 subunits)	Class[b]
Rpa1 (A190)	Rbp1 (B220)	Rpc1 (C160)	β′	Core
Rpa2 (A135)	Rbp2 (B150)	Rpc2 (C128)	β	Core
Rpc5 (AC40)	Rpb3 (B44.5)	Rpc5 (AC40)	α	Core
Rpc9 (AC19)	Rpb11 (B13.6)	Rpc9 (AC19)	α	Core
Rbp6 (ABC23)	Rbp6 (ABC23)	Rpb6 (ABC23)	ω	Core/common
Rpb5 (ABC27)	Rpb5 (ABC27)	Rpb5 (ABC27)		Common
Rpb8 (ABC14.4)	Rpb8 (ABC14.4)	Rpb8 (ABC14.4)		Common
Rbp10 (ABC10β)	Rpb10 (ABC10β)	Rpb10 (ABC10β)		Common
Rbp12 (ABC10α)	Rpb12 (ABC10α)	Rpb12 (ABC10α)		Common
Rpa9 (A12.2)	Rpb9 (B12.6)	Rpc12 (C11)		
Rpa8 (A14)[c]	Rpb4 (B32)	—		
Rpa4 (A43)[c]	Rpb7 (B16)	Rpc11 (C25)		
+2 others[d]		+4 others[d]		

[a]Homologous subunits occupy the same row. In the alternative subunit names in parentheses, the letter(s) indicates the RNAPs in which the subunit is a component (A, B, and C for RNAPs I, II, and III) and the numbers indicate its approximate molecular mass in kD.

[b]Core: sequence partially homologous in all RNAPs; common: shared by all eukaryotic RNAPs.

[c]Potential homologs of Rbp4 and Rbp7.

[d]Rpa3 (A49) and Rpa5 (A34.5) in RNAP I and Rpc3 (C74), Rpc4 (C53), Rpc6 (C34), and Rpc8 (C31) in RNAP III.

Source: Mainly Cramer, P., *Curr. Opin. Struct. Biol.* **12,** 89 (2002).

gests that this process triggers the conversion of RNAP II's initiation complex to its elongation complex. Charge–charge repulsions between nearby phosphate groups probably cause a highly phosphorylated CTD to project as far as 500 Å from the globular portion of RNAP II. Indeed, as we shall see, the phosphorylated CTD provides the binding sites for numerous auxiliary factors that have essential roles in the transcription process.

In contrast to the somewhat smaller prokaryotic RNAP holoenzymes, eukaryotic RNAPs do not independently bind their target DNAs. Rather, as we shall see in Section 34-3B, they are recruited to their target promoters through the mediation of complexes of transcription factors and their ancillary proteins that, in the case of RNAP II-transcribed genes, are so large and complicated that they collectively dwarf RNAP II.

In addition to the foregoing nuclear enzymes, eukaryotic cells contain separate mitochondrial and (in plants) chloroplast RNAPs. These small (~100 kD) single-subunit RNAPs, which resemble those encoded by certain bacteriophages, are much simpler than the nuclear RNAPs although they catalyze the same reaction.

a. X-Ray Structures of Yeast RNAP II Reveal a Transcribing Complex

In a crystallographic tour de force, Roger Kornberg determined the X-ray structure of yeast (*S. cerevisiae*) RNAP II that lacks its nonessential Rpb4 and Rpb7 subunits (Fig. 31-20). This enzyme, as expected, resembles *Taq* RNAP (Fig. 31-11) in its overall crab claw-like shape and in the positions and core folds of their homologous subunits although, of course, RNAP II is somewhat larger than and has several subunits that have no counterpart in *Taq* RNAP. RNAP II binds two Mg^{2+} ions at its active site (although one appears to be weakly bound and hence is only faintly visible in the X-ray structure; perhaps it accompanies the incoming NTP) in the vicinity of five conserved acidic residues, which suggests that RNAPs catalyze RNA elongation via a 2-metal ion mechanism similar to that employed by DNA polymerases (Section 30-2A). As is the case with *Taq* RNAP, the surface of RNAP II is almost entirely negatively charged except for the DNA-binding cleft and the region about the active site, which are positively charged.

Although, as mentioned above, RNAP II does not normally initiate transcription by itself, Kornberg found that it will do so on a dsDNA bearing a 3′ single-stranded tail at one end. Consequently, incubating yeast RNAP II with

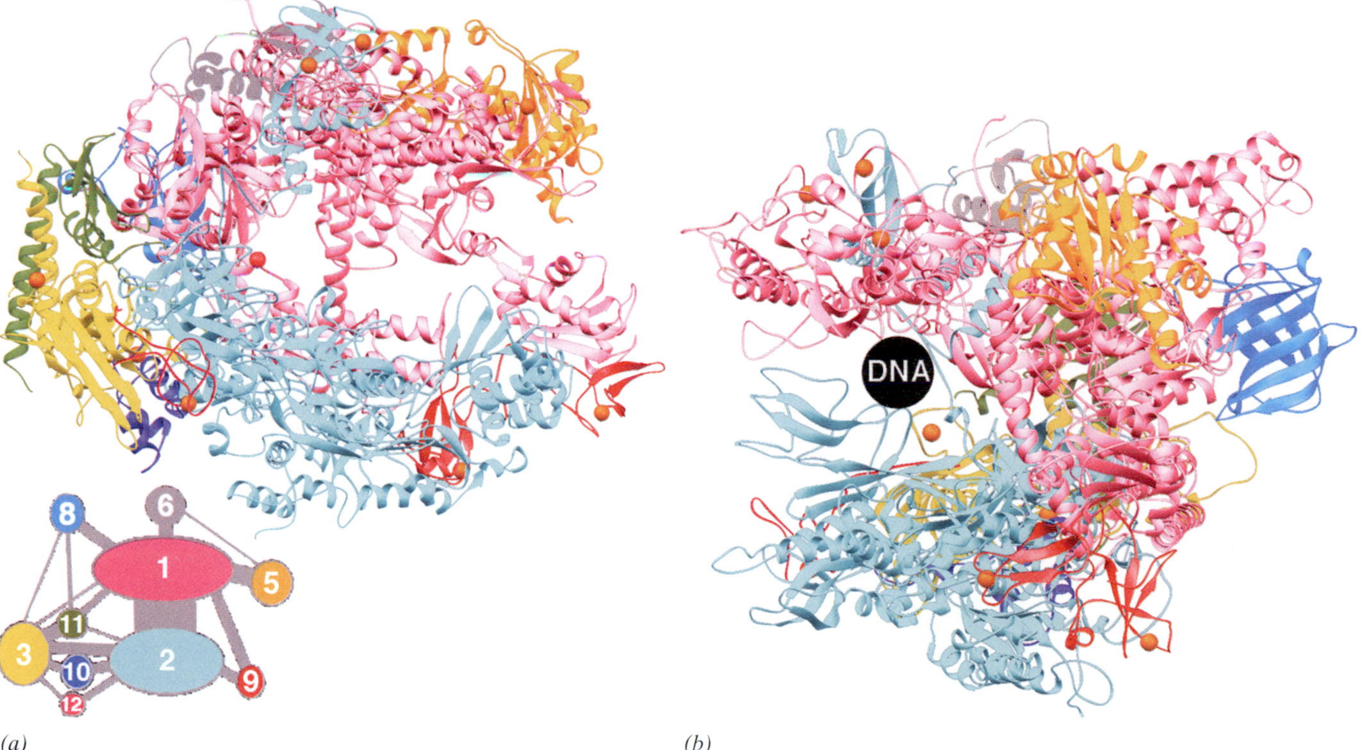

(*a*) (*b*)

FIGURE 31-20 The X-Ray structure of yeast RNAP II that lacks its Rpb4 and Rpb7 subunits. (*a*) The enzyme is oriented similarly to *Taq* RNAP in Fig. 31-11 and its subunits are colored as is indicated in the accompanying diagram, with the subunits homologous to those of *Taq* RNAP given the same colors. The strongly bound Mn^{2+} ion (physiologically Mg^{2+}) that marks the active site is shown as a red sphere and the enzyme's 8 bound Zn^{2+} ions are shown as orange spheres. The Rpb1 C-terminal domain (CTD) is not visible due to disorder. In the accompanying diagram, the area of each numbered ellipsoid is proportional to the corresponding subunit's size and the width of each gray line connecting a pair of subunits is proportional to the surface area of their interface. (*b*) View of the enzyme from the right in Part *a* showing its DNA binding cleft. The black circle has the approximate diameter of B-DNA. [Based on an X-ray structure by Roger Kornberg. PDBid 1I50.]

the DNA shown in Fig. 31-21*a* and all NTPs but UTP yielded the DNA · RNA hybrid helix diagrammed in Fig. 31-21*a* bound to RNAP II. The X-ray structure of this paused transcribing complex revealed, as expected, that the dsDNA had bound in the enzyme's cleft (Fig. 31-21*b, c;* transcription resumed on soaking the crystals in UTP, thereby demonstrating that the crystalline complex was active). In comparison with the X-ray structure of RNAP II alone, a massive (~50 kD) portion of Rpb1 and Rpb2 named the "clamp" has swung down over the DNA to trap it in the cleft, in large part accounting for the enzyme's essentially infinite processivity. The mainly rigid motion of the clamp is mediated by conformational changes at five so-called switch regions at the base of the clamp in which three of these switches, which are disordered in the structure of RNAP II alone, become ordered in the transcribing complex.

The DNA unwinds by three bases before entering the active site (which is contained on Rpb1). Past this point, however, a portion of Rpb2 dubbed the "wall" directs the

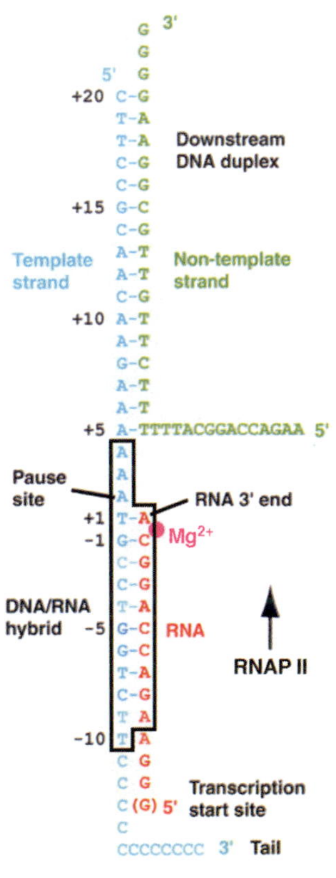

(*a*)

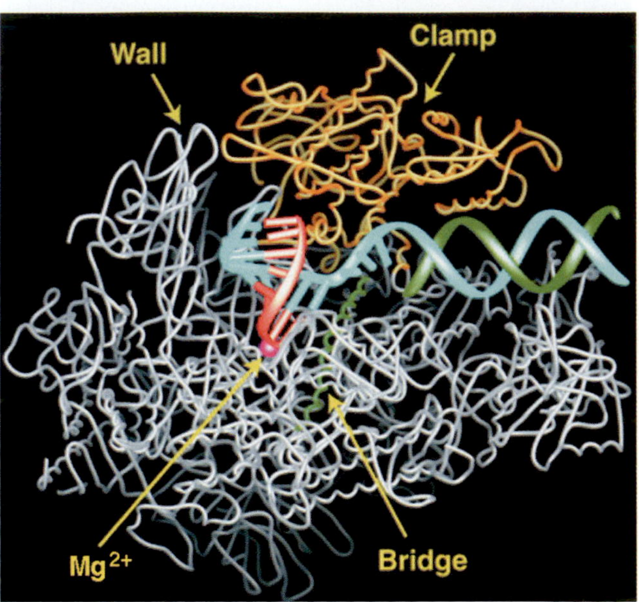

(*b*)

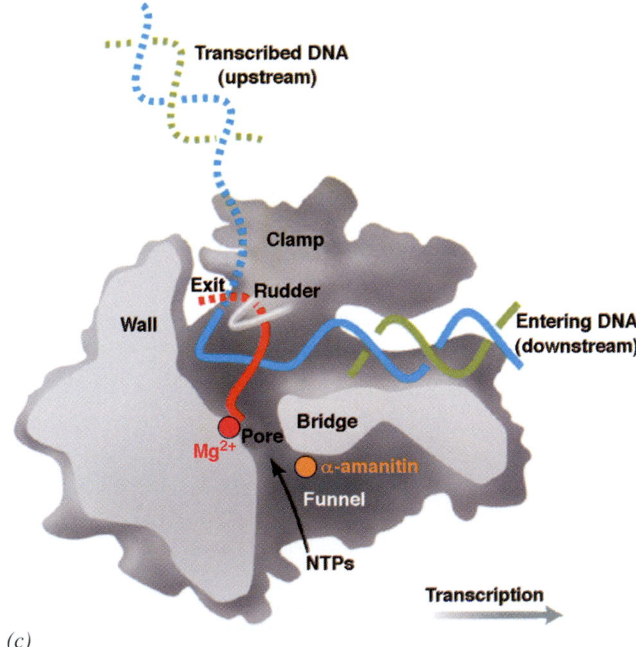

(*c*)

FIGURE 31-21 X-Ray structure of an RNAP II elongation complex. (*a*) The RNA · DNA complex in the structure with the template DNA cyan, the nontemplate DNA green, and the newly synthesized RNA red. The magenta dot marked Mg^{2+} represents the strongly bound active site metal ion. The black box encloses those portions of the complex that are clearly visible in the structure; the double-stranded portion of the DNA marked "Downstream DNA duplex" is poorly ordered, and the remaining portions of the complex are disordered. (*b*) View of the transcribing complex from the bottom of Fig. 31-20*a* in which portions of Rpb2 that form the near side of the cleft have been removed to expose the bound RNA · DNA complex. The protein is represented by its backbone in which the clamp, which is closed over the downstream DNA duplex, is yellow, the bridge helix is green, and the remaining portions of the protein are gray. The DNA and RNA are colored as in Part *a* with their well-ordered portions drawn in ladder form and their less ordered portions drawn in backbone form. The active site Mg^{2+} ion is represented by a magenta sphere. (*c*) Cutaway schematic diagram of the transcribing complex in Part *b* in which the cut surfaces of the protein are light gray, its remaining surfaces are darker gray, and several of its functionally important structural features are labeled. The DNA, RNA, and active site Mg^{2+} ion are colored as in Part *a* with portions of the DNA and RNA that are not visible in the X-ray structure represented by dashed lines. The α-amanitin binding site is marked by an orange dot. [Modified from diagrams by Roger Kornberg, Stanford University. PDBid 1I6H.]

template strand out of the cleft in an ~90° turn. As a consequence, the template base at the active site (+1) points toward the floor of the cleft where it can be read out by the active site. This base is paired with the ribonucleotide at the 3′ end of the RNA, which is positioned above a "pore" at the end of a "funnel" to the protein exterior through which NTPs presumably gain access to the otherwise sealed off active site. The RNA · DNA hybrid helix adopts a nonstandard conformation intermediate between those of A- and B-DNAs, which is underwound relative to that in the X-ray structure of an RNA · DNA hybrid helix alone (Fig. 29-6). Nearly all contacts that the protein makes with the RNA and DNA are to their sugar–phosphate backbones; none are with the edges of their bases. The specificity of the enzyme for a ribonucleotide rather than a deoxyribonucleotide is attributed to the enzyme's recognition of both the incoming ribose sugar and the RNA · DNA hybrid helix. After about one turn of hybrid helix, a loop extending from the clamp called the "rudder" separates the RNA and template DNA strands, thereby permitting the DNA double helix to reform as it exits the enzyme (although the unpaired 5′ tail of the nontemplate strand and the 3′ tail of the template strand are disordered in the X-ray structure). The models of the RNAP II transcribing complex and the *Taq* RNAP holoenzyme open complex differ mainly in the placement of the downstream segment of dsDNA (compare Figs. 31-13*b* and 31-21*c*). Of course, to become a transcribing complex, the *Taq* open complex must first jettison its σ subunit, to which the downstream dsDNA segment is bound.

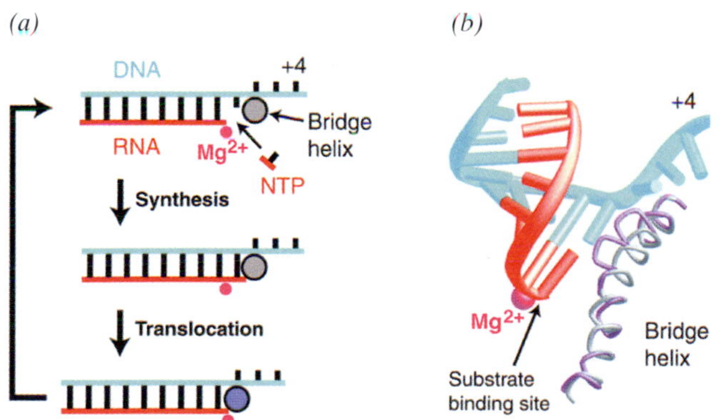

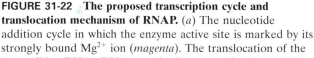

FIGURE 31-22 The proposed transcription cycle and translocation mechanism of RNAP. (*a*) The nucleotide addition cycle in which the enzyme active site is marked by its strongly bound Mg^{2+} ion (*magenta*). The translocation of the transcribing RNA · DNA complex is proposed to be motivated by a conformational change of the bridge helix from straight (*gray circle*) to bent (*violet circle*). The relaxation of the bridge helix back to its straight form would complete the cycle by yielding an empty NTP binding site at the active site. (*b*) The RNA · DNA complex in RNAP II viewed and colored as in Fig. 31-21*b*. The RNAP II bridge helix is gray and the superimposed (and bent) *Taq* polymerase bridge helix is violet. The side chains extending from the bent helix would sterically clash with the hybrid base pair at position +1. [Courtesy of Roger Kornberg, Stanford University.]

How does RNAP translocate its bound RNA–DNA assembly in preparation for a new round of synthesis? The highly conserved helical segment of Rpb1, dubbed the "bridge" because it bridges the two pincers forming the enzyme's cleft (Figs. 31-20 and 31-21), nonspecifically contacts the template DNA base at the +1 position. Although this helix is straight in all X-ray structures of RNAP II yet determined, it is bent in that of *Taq* RNAP (Fig. 31-11*a*). If the bridge helix, in fact, alternates between its straight and bent conformations, it would move by 3 to 4 Å. Kornberg has therefore speculated that translocation occurs through the bending of the bridge helix so as to push the paired nucleotides at position +1 to position −1 (Fig. 31-22). The recovery of the bridge helix to its straight conformation would then yield an empty site at position +1 for entry of the next NTP, thereby preparing the enzyme for a new round of nucleotide addition.

b. Amatoxins Specifically Inhibit RNA Polymerases II and III

The poisonous mushroom ***Amanita phalloides*** **(death cap),** which is responsible for the majority of fatal mushroom poisonings, contains several types of toxic substances, including a series of unusual bicyclic octapeptides known as **amatoxins. α-Amanitin,**

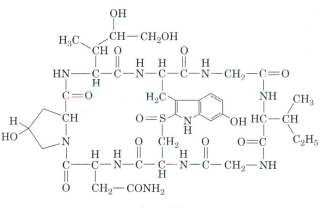

α-Amanitin

which is representative of the amatoxins, forms a tight 1:1 complex with RNAP II ($K = 10^{-8}M$) and a looser one with RNAP III ($K = 10^{-6}M$). Its binding slows an RNAP's rate of RNA synthesisis from several thousand to only a few nucleotides per minute. α-Amanitin is therefore a useful tool for mechanistic studies of these enzymes. RNAP I as well as mitochondrial, chloroplast, and prokaryotic RNAPs are insensitive to α-amanitin.

The X-ray structure of RNAP II in complex with α-amanitin, also determined by Kornberg, reveals that α-amanitin binds in the funnel beneath the protein's bridge helix (Fig. 31-21*c*) such that it interacts almost exclusively with the residues of the bridge helix and the adjacent part of Rpb1. The observation that RNAP II mutations that affect α-amanitin inhibition also map to this site indicates that this binding mode is not just an artifact of crystallization. The α-amanitin binding site is too far away from the enzyme active site to directly interfere with NTP entry or

RNA synthesis, consistent with the observation that α-amanitin does not influence the affinity of RNAP II for NTPs. Most probably, α-amanitin binding impedes the conformational change of the bridge helix postulated to motivate the RNAP translocation step (Fig. 31-22), which further supports this mechanism.

Despite the amatoxins' high toxicity (5–6 mg, which occurs in ~40 g of fresh mushrooms, is sufficient to kill a human adult), they act slowly. Death, usually from liver dysfunction, occurs no earlier than several days after mushroom ingestion (and after recovery from the effects of other mushroom toxins). This, in part, reflects the slow turnover of eukaryotic mRNAs and proteins.

c. Mammalian RNA Polymerase I Has a Bipartite Promoter

Since, as we shall see in Section 31-4B, the numerous rRNA genes in a given eukaryotic cell have essentially identical sequences, its RNAP I only recognizes one promoter. Yet, in contrast to the case for RNAPs II and III, RNAP I promoters are species specific, that is, an RNAP I only recognizes its own promoter and those of closely related species. This is because only closely related species exhibit recognizable sequence identities near the transcriptional start sites of their rRNA genes. RNAP I promoters were therefore identified by determining how the transcription rate of an rRNA gene is affected by a series of increasingly longer deletions approaching its start site from either its upstream or its downstream sides. Such studies have indicated, for example, that mammalian RNAPs I require the presence of a so-called **core promoter element,** which spans positions −31 to +6 and hence overlaps the transcribed region. However, efficient transcription additionally requires an **upstream promoter element,** which is located between residues −187 and −107. These elements, which are G + C-rich and ~85% identical, are bound by specific transcription factors which then recruit RNAP I to the transcription start site.

d. RNA Polymerase II Promoters Are Complex and Diverse

The promoters recognized by RNAP II are considerably longer and more diverse than those of prokaryotic genes but have not yet been fully described. The structural genes expressed in all tissues, the so-called housekeeping genes, which are thought to be constituitively transcribed, have one or more copies of the sequence GGGCGG or its complement (the **GC box**) located upstream from their transcription start sites. The analysis of deletion and point mutations in eukaryotic viruses such as SV40 indicates that GC boxes function analogously to prokaryotic promoters. On the other hand, structural genes that are selectively expressed in one or a few types of cells often lack these GC-rich sequences. Rather, *many contain a conserved AT-rich sequence located 25 to 30 bp upstream from their transcription start sites (Fig. 31-23).* Note that this so-called **TATA box** resembles the −10 region of prokaryotic promoters (TATAAT), although they differ in their locations relative to the transcription start site (−27 vs −10). The functions of these two promoter elements are not strictly analogous, however, since the deletion of the TATA box does not necessarily eliminate transcription. Rather, TATA box deletion or mutation generates heterogeneities in the transcriptional start site, thereby indicating that the TATA box participates in selecting this site.

The gene region extending between about −50 and −110 also contains promoter elements. For instance, many eukaryotic structural genes, including those encoding the various globins, have a conserved sequence of consensus CCAAT (the **CCAAT box**) located between about −70 and −90 whose alteration greatly reduces the gene's transcription rate. Globin genes have, in addition, a conserved **CACCC box** upstream from the CCAAT box that has also been implicated in transcriptional initiation. Evidently, the promoter sequences upstream of the TATA box form the initial DNA-binding sites for RNA polymerase II and the other proteins involved in transcriptional initiation (see below).

e. Enhancers Are Transcriptional Activators That Can Have Variable Positions and Orientations

Perhaps the most surprising aspect of eukaryotic transcriptional control elements is that some of them need not have fixed positions and orientations relative to their corresponding transcribed sequences. For example, the SV40 genome, in which such elements were first discovered, contains two repeated sequences of 72 bp each that are located upstream from the promoter for early gene expres-

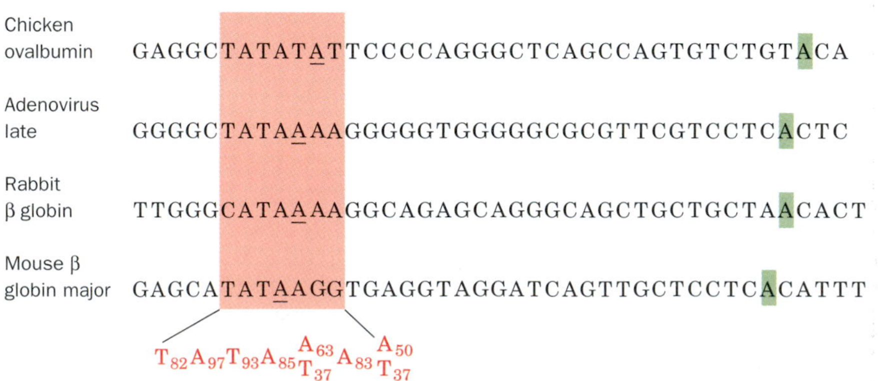

FIGURE 31-23 The promoter sequences of selected eukaryotic structural genes. The homologous segment, the TATA box, is shaded in red with the base at position −27 underlined and the initial nucleotide to be transcribed (+1) shaded in green. The bottom row indicates the consensus sequence of several such promoters with the subscripts indicating the percent occurrence of the corresponding base. [After Gannon, F., et al., *Nature* **278,** 433 (1978).]

sion. Transcription is unaffected if one of these repeats is deleted but is nearly eliminated when both are absent. The analysis of a series of SV40 mutants containing only one of these repeats demonstrated that its ability to stimulate transcription from its corresponding promoter is all but independent of its position and orientation. Indeed, transcription is unimpaired when this segment is several thousand base pairs upstream or downstream from the transcription start site. Gene segments with such properties are named **enhancers** to indicate that they differ from promoters, with which they must be associated in order to trigger site-specific and strand-specific transcription initiation (although the characterization of numerous promoters and enhancers indicates that their functional properties are similar). Enhancers occur in both eukaryotic viruses and cellular genes.

Enhancers are required for the full activities of their cognate promoters. It was originally thought that enhancers somehow acted as entry points on DNA for RNAP II (perhaps by altering DNA's local conformation or through a lack of binding affinity for the histones that normally coat eukaryotic DNA; Section 34-1A). However, it is now clear that *enhancers are recognized by specific transcription factors that stimulate RNA polymerase II to bind to the corresponding but distant promoter.* This requires that the DNA between the enhancer and promoter loop around so that the transcription factor can simultaneously contact the enhancer and the RNAP II and/or its associated proteins at the promoter. Most cellular enhancers are associated with genes that are selectively expressed in specific tissues. It therefore seems, as we discuss in Section 34-3B, that *enhancers mediate much of the selective gene expression in eukaryotes.*

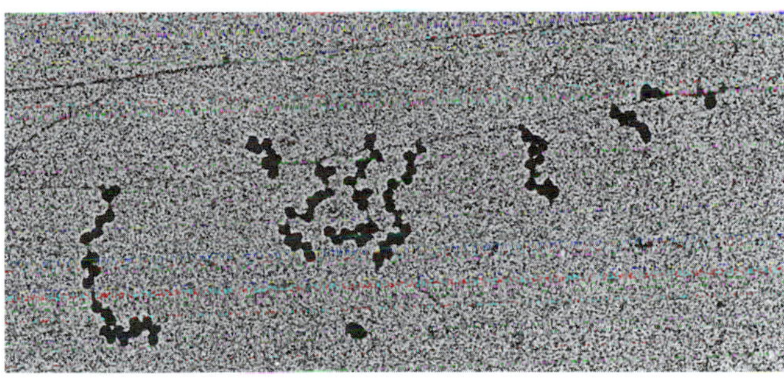

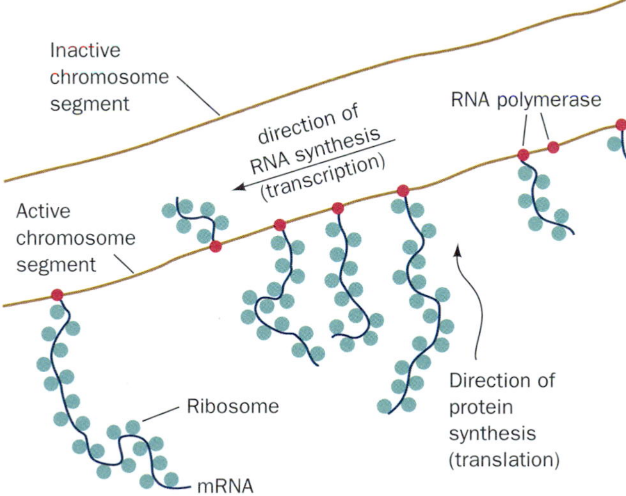

FIGURE 31-24 An electron micrograph and its interpretive drawing showing the simultaneous transcription and translation of an *E. coli* gene. RNA polymerase molecules are transcribing the DNA from right to left while ribosomes are translating the nascent RNAs (mostly from bottom to top). [Courtesy of Oscar L. Miller, Jr. and Barbara Hamkalo, University of Virginia.]

f. RNA Polymerase III Promoters Can Be Located Downstream from Their Transcription Start Sites

The promoters of genes transcribed by RNAP III can be located entirely within the genes' transcribed regions. Donald Brown established this through the construction of a series of deletion mutants of a *Xenopus borealis* 5S RNA gene. Deletions of base sequences that start from outside one or the other end of the transcribed portion of the 5S gene only prevent transcription if they extend into the segment between nucleotides +40 and +80. Indeed, a fragment of the 5S RNA gene consisting of only nucleotides 41 to 87, when cloned in a bacterial plasmid, is sufficient to direct specific initiation by RNAP III at an upstream site. This is because, as was subsequently demonstrated, the sequence contains the binding site for transcription factors that stimulate the upstream binding of RNAP III. Further studies have shown, however, that the promoters of other RNAP III-transcribed genes lie entirely upstream of their start sites. These upstream sites also bind transcription factors that recruit RNAP III.

3 ■ CONTROL OF TRANSCRIPTION IN PROKARYOTES

Prokaryotes respond to sudden environmental changes, such as the influx of nutrients, by inducing the synthesis of the appropriate proteins. This process takes only minutes because transcription and translation in prokaryotes are closely coupled: *Ribosomes commence translation near the 5′ end of a nascent mRNA soon after it is extruded from RNA polymerase (Fig. 31-24).* Moreover, *most prokaryotic mRNAs are enzymatically degraded within 1 to 3 min of their synthesis,* thereby eliminating the wasteful synthesis of unneeded proteins after a change in conditions (protein degradation is discussed in Section 32-6). In fact, the 5′ ends of some mRNAs are degraded before their 3′ ends have been synthesized.

In contrast, the induction of new proteins in eukaryotic cells frequently takes hours or days, in part because transcription takes place in the nucleus and the resulting mRNAs must be transported to the cytoplasm, where translation occurs. However, eukaryotic cells, particularly those of multicellular organisms, have relatively stable environments; major changes in their transcription patterns usually occur only during cell differentiation.

In this section we examine some of the ways in which prokaryotic gene expression is regulated through transcriptional control. Eukaryotes, being vastly more complex creatures than are prokaryotes, have a correspondingly more complicated transcriptional control system whose general outlines are beginning to come into focus. We therefore defer discussion of eukaryotic transcriptional control until Section 34-3B, where it can be considered in light of what we know about the structure and organization of the eukaryotic chromosome.

A. *Promoters*

In the presence of high concentrations of inducer, the *lac* operon is rapidly transcribed. In contrast, the *lacI* gene is transcribed at such a low rate that a typical *E. coli* cell contains <10 molecules of the *lac* repressor. Yet, the *I* gene has no repressor. Rather, it has such an inefficient promoter (Fig. 31-10) that it is transcribed an average of about once per bacterial generation. *Genes that are transcribed at high rates have efficient promoters.* In general, the more efficient a promoter, the more closely its sequence resembles that of the corresponding consensus sequence.

a. Gene Expression Can Be Controlled by a Succession of σ Factors

The processes of development and differentiation involve the temporally ordered expression of sets of genes according to genetically specified programs. Phage infections are among the simplest examples of developmental processes. Typically, only a subset of the phage genome, often referred to as *early* genes, are expressed in the host immediately after phage infection. As time passes, *middle* genes start to be expressed, and the *early* genes as well as the bacterial genes are turned off. In the final stages of phage infection, the *middle* genes give way to the *late* genes. Of course some phage types express more than three sets of genes and some genes may be expressed in more than one stage of an infection.

One way in which families of genes are sequentially expressed is through "cascades" of σ factors. In the infection of *Bacillus subtilis* by bacteriophage SP01, for example, the *early* gene promoters are recognized by the bacterial RNAP holoenzyme. Among these *early* genes is gene 28, whose gene product is a new σ subunit, designated σ^{gp28}, that displaces the bacterial σ subunit from the core enzyme. This reconstituted holoenzyme recognizes only the phage *middle* gene promoters, which all have similar −35 and −10 regions, but bear little resemblance to the corresponding regions of bacterial and phage *early* genes. The *early* genes therefore become inactive once their corresponding mRNAs have been degraded. The phage *middle* genes include genes 33 and 34, which together specify yet another σ factor, $\sigma^{gp33/34}$, which, in turn, permits the transcription of only *late* phage genes.

Several bacteria, including *E. coli* and *B. subtilis,* likewise have several different σ factors. These are not necessarily utilized in a sequential manner. Rather, those that differ from the predominant or primary σ factor (σ^{70} in *E. coli*) control the transcription of coordinately expressed groups of special purpose genes, whose promoters are quite different from those recognized by the primary σ factor. For example, sporulation in *B. subtilis,* a process in which the bacterial cell is asymmetrically partitioned into two compartments, the **forespore** (which becomes the **spore,** a germline cell from which subsequent progeny arise) and the **mother cell** (which synthesizes the spore's protective cell wall and is eventually discarded), is governed by five σ factors in addition to that of the **vegetative** (nonsporulating) cell: one that is active before cell partition occurs, two that are sequentially active in the forespore, and two that are sequentially active in the mother cell. Cross-regulation of the compartmentalized σ factors permits the forespore and mother cell to tightly coordinate this differentiation process.

B. *lac Repressor I: Binding*

In 1966, Beno Müller-Hill and Walter Gilbert isolated *lac* repressor on the basis of its ability to bind ^{14}C-labeled IPTG (Section 31-1A) and demonstrated that it is a protein. This was an exceedingly difficult task because *lac* repressor comprises only ~0.002% of the protein in wild-type *E. coli.* Now, however, *lac* repressor is available in quantity via molecular cloning techniques (Section 5-5G).

a. *lac* Repressor Finds Its Operator by Sliding along DNA

The *lac* repressor is a tetramer of identical 360-residue subunits, each of which binds one IPTG molecule with a dissociation constant of $K = 10^{-6}M$. In the absence of inducer, the repressor tetramer nonspecifically binds duplex DNA with a dissociation constant of $K \approx 10^{-4}M$. However, it specifically binds to the *lac* operator with far greater affinity: $K \approx 10^{-13}M$. Limited proteolysis of *lac* repressor with trypsin reveals that each subunit consists of two functional domains: Its 58-residue N-terminal peptide binds DNA but not IPTG, whereas the remaining "core tetramer" binds only IPTG.

The observed rate constant for the binding of *lac* repressor to *lac* operator is $k_f \approx 10^{10}M^{-1}\ s^{-1}$. This "on" rate is much greater than that calculated for the diffusion-controlled process in solution: $k_f = 10^7M^{-1}\ s^{-1}$ for molecules the size of *lac* repressor. Since it is impossible for a reaction to proceed faster than its diffusion-controlled rate, the *lac* repressor must not encounter operator from solution in a random three-dimensional search. Rather, *it appears that lac repressor finds operator by nonspecifically binding to DNA and diffusing along it in a far more efficient one-dimensional search.*

b. *lac* Operator Has a Nearly Palindromic Sequence

The availability of large quantities of *lac* repressor made it possible to characterize the *lac* operator. *E. coli* DNA that had been sonicated to small fragments was mixed with *lac* repressor and passed through a nitrocellulose filter.

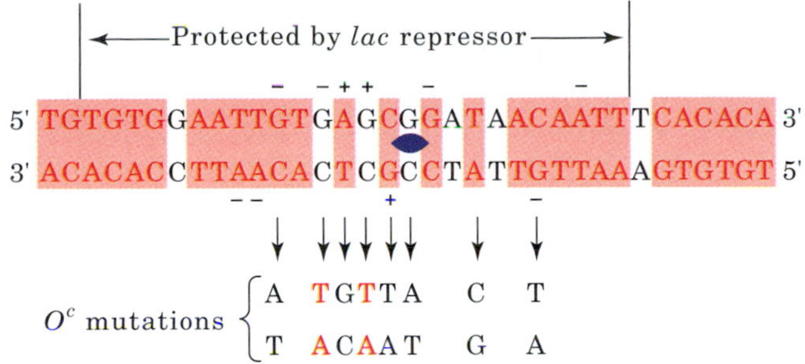

FIGURE 31-25 The base sequence of the *lac* operator. The symmetry related regions (*red*) comprise 28 of its 35 bp. A "+" denotes positions at which repressor binding enhances methylation by dimethyl sulfate (which methylates G at N7 and A at N3) and a "−" indicates where this footprinting reaction is inhibited. The bottom row indicates the positions and identities of different point mutations that prevent *lac* repressor binding (O^c mutants). Those in red increase the operator's symmetry. [After Sobell, H.M., *in* Goldberger, R.F. (Ed.), *Biological Regulation and Development,* Vol. 1, *p.* 193, Plenum Press (1979).]

Protein, with or without bound DNA, sticks to nitrocellulose, whereas duplex DNA, by itself, does not. The DNA was released from the filter-bound protein by washing it with IPTG solution, recombined with *lac* repressor, and the resulting complex treated with DNase I. The DNA fragment that *lac* repressor protects from nuclease degradation consists of a run of 26 bp that is embedded in a nearly 2-fold symmetric sequence of 35 bp (Fig. 31-25, *top*). *Such palindromic symmetry is a common feature of DNA segments that are specifically bound by proteins* (recall, for example, that restriction endonuclease recognition sites are also palindromic; Section 5-5A).

Palindromic DNA sequences, as we have seen, bind to proteins that have matching 2-fold symmetry. However, methylation protection experiments on the *lac* repressor–operator system do not fully support this model: There is an asymmetric pattern of differences between free and repressor-bound operator in the susceptibility of its bases to reaction with DMS (Fig. 31-25). Furthermore, point mutations in the operator that render it operator constitutive (O^c), and that invariably weaken the binding of repressor to operator, may increase as well as decrease the operator's 2-fold symmetry (Fig. 31-25).

c. *lac* Repressor Prevents RNA Polymerase from Forming a Productive Initiation Complex

Operator occupies positions −7 through +28 of the *lac* operon relative to the transcription start site (Fig. 31-26). Nuclease protection studies, it will be recalled, indicate that, in the initiation complex, RNA polymerase tightly binds to the DNA between positions −20 and +20 (Section 31-2B). Thus, *the lac operator and promoter sites overlap.* It was therefore widely assumed for many years that *lac* repressor simply physically obstructs the binding of RNA polymerase to the *lac* promoter. However, the observation that *lac* repressor and RNA polymerase can simultaneously bind to the *lac* operon indicates that *lac* repressor must act by somehow interfering with the initiation process. Closer investigation of this phenomenon revealed that, in the presence of bound *lac* repressor, RNA polymerase holoenzyme still abortively synthesizes oligonucleotides, although they tend to be shorter than those made in the absence of repressor. Evidently, *lac repressor acts by somehow increasing the already high kinetic barrier for RNA polymerase to generate the open complex and commence processive elongation.*

We discuss the *lac* repressor structure and further aspects of *lac* operator organization in Section 31-3F.

C. *Catabolite Repression: An Example of Gene Activation*

Glucose is E. coli's metabolite of choice; the availability of adequate amounts of glucose prevents the full expression of >100 genes that encode proteins involved in the fermentation of numerous other catabolites, including lactose (Fig.

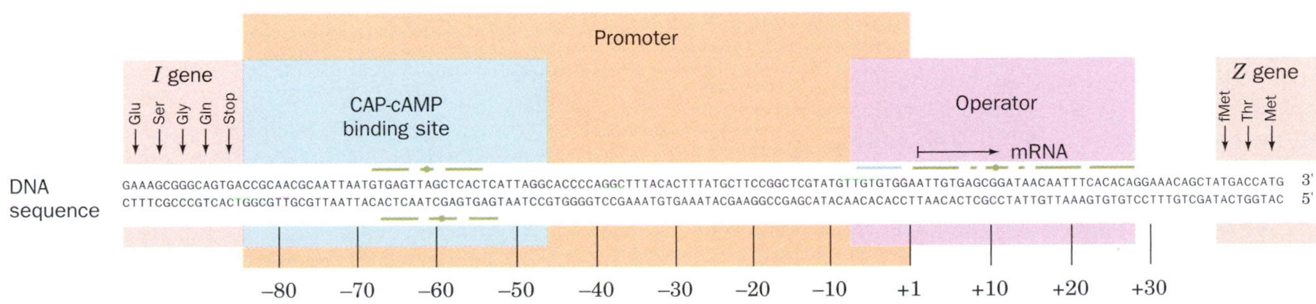

FIGURE 31-26 The nucleotide sequence of the *E. coli lac* promoter–operator region. The region extends from the C-terminal portion of *lacI* (*left*) to the N-terminal portion of *lacZ* (*right*). The palindromic sequences of the operator and the CAP-binding site (Section 31-3C) are overscored or underscored. [After Dickson, R.C., Abelson, J., Barnes, W.M., and Reznikoff, W.A., *Science* **187,** 32 (1975).]

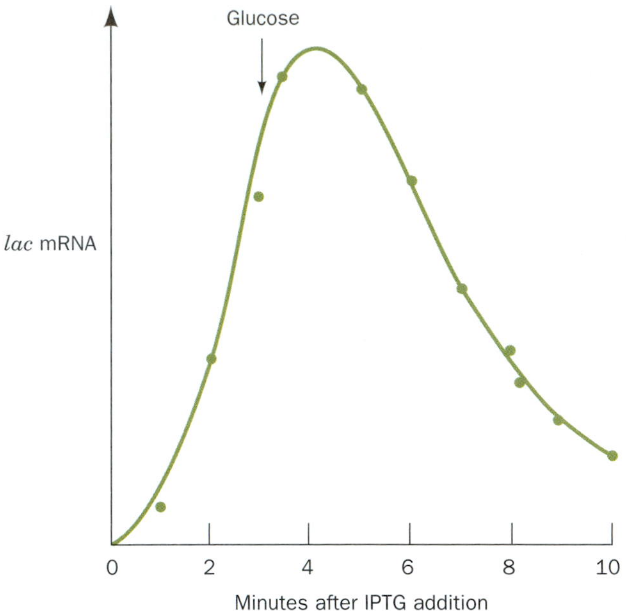

FIGURE 31-27 The kinetics of *lac* operon mRNA synthesis following its induction with IPTG, and of its degradation after glucose addition. *E. coli* were grown on a medium containing glycerol as their only carbon-energy source and ^{3}H-labeled uridine. IPTG was added to the medium at the beginning of the experiment to induce the synthesis of the *lac* enzymes. After 3 min, glucose was added to stop the synthesis. The amount of ^{3}H-labeled *lac* RNA was determined by hybridization with DNA containing the *lacZ* and *lacY* genes. [After Adesnik, M. and Levinthal, C., *Cold Spring Harbor Symp. Quant. Biol.* **35,** 457 (1970).]

31-27), arabinose, and galactose, even when these metabolites are present in high concentrations. This phenomenon, which is known as **catabolite repression,** prevents the wasteful duplication of energy-producing enzyme systems.

a. cAMP Signals the Lack of Glucose

The first indication of the mechanism of catabolite repression was the observation that, in *E. coli,* the level of cAMP, which was known to be a second messenger in animal cells (Section 18-3E), is greatly diminished in the presence of glucose. This observation led to the finding that the addition of cAMP to *E. coli* cultures overcomes catabolite repression by glucose. Recall that, in *E. coli,* adenylate cyclase is activated by the phosphorylated enzyme $EIIA^{glc}$ (or possibly inactivated by dephospho-$EIIA^{glc}$), which is dephosphorylated on the transport of glucose across the cell membrane (Section 20-3D). *The presence of glucose, therefore, normally lowers the cAMP level in E. coli.*

b. CAP–cAMP Complex Stimulates the Transcription of Catabolite Repressed Operons

Certain *E. coli* mutants, in which the absence of glucose does not relieve catabolite repression, are missing a cAMP-binding protein that is synonymously named **catabolite gene activator protein (CAP)** and **cAMP receptor protein (CRP).** CAP is a homodimer of 210-residue subunits that undergoes a large conformational change on binding cAMP. Its function was elucidated by Ira Pastan, who showed that *CAP–cAMP complex, but not CAP itself, binds to the lac operon (among others) and stimulates transcription from its otherwise low-efficiency promoter in the absence of lac repressor.* CAP is therefore a **positive regulator** (turns on transcription), in contrast to *lac* repressor, which is a **negative regulator** (turns off transcription).

The X-ray structure, by Thomas Steitz, of CAP–cAMP in complex with a palindromic 30-bp segment of duplex DNA whose sequence resembles that of the CAP binding sequence (Fig. 31-26) reveals that the DNA is bent by ~90° around the protein (Fig. 31-28*a*). The bend arises from two ~45° kinks in the DNA between the fifth and sixth bases out from the complex's 2-fold axis in both directions. This distortion results in the closing of the major groove and an enormous widening of the minor groove at each kink.

Why is the CAP–cAMP complex necessary to stimulate the transcription of its target operons? And how does it do so? The *lac* operon has a weak (low-efficiency) promoter; its −10 and −35 sequences (TATGTT and TTTACA; Fig. 31-10) differ significantly from the corresponding consensus sequences of strong (high-efficiency) promoters (TATAAT and TTGACA; Fig. 31-10). Such weak promoters evidently require some sort of help for efficient transcriptional initiation.

Richard Ebright has shown that CAP interacts directly with RNAP via the C-terminal domain of its 85-residue α subunit (αCTD) in a way that stimulates RNAP to initiate transcription from a nearby promoter. The αCTD also binds dsDNA nonspecifically but does so with higher affinity at A + T-rich sites such as those of UP elements (Section 31-2A). It is flexibly linked to the rest of the α subunit and hence is not seen in the X-ray structure of *Taq* RNAP (Fig. 31-11) due to disorder.

Three classes of CAP-dependent promoters have been characterized:

1. Class I promoters, such as that of the *lac* operon, require only CAP–cAMP for transcriptional activation. The CAP binding site on the DNA can be located at various distances from the promoter provided that CAP and RNAP bind to the same face of the DNA helix. Thus, CAP–cAMP activates the transcription of the *lac* operon if its DNA binding site is centered near positions −62 (its wild-type position; Fig. 31-26), −72, −83, −93, or −103, all of which are one helical turn apart. For the latter sites, this requires that the DNA loop around to permit CAP–cAMP to contact the αCTD. Such looping is likely to be facilitated by the bending of the DNA around CAP–cAMP.

2. Class II promoters also require only CAP–cAMP for transcriptional activation. However, in class II promoters, the CAP binding site only occupies a fixed position that overlaps the RNAP binding site, apparently by replacing the promoter's −35 promoter region. CAP then interacts with RNAP via interactions with both the αCTD and the α subunit's N-terminal domain.

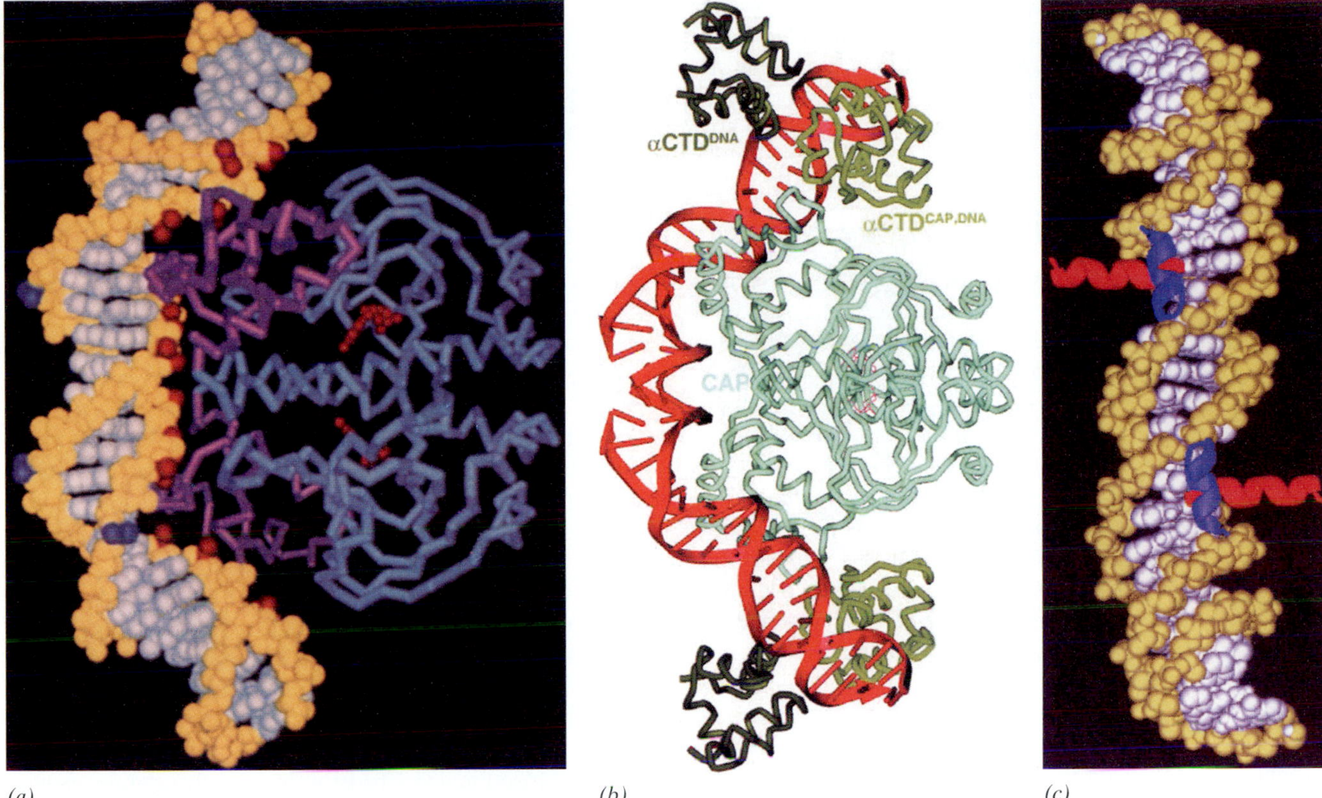

FIGURE 31-28 X-Ray structures of CAP–cAMP complexes. (*a*) CAP–cAMP in complex with a palindromic 30-bp duplex DNA. The complex is viewed with its molecular 2-fold axis horizontal. The protein is represented by its C_α backbone with its N-terminal cAMP-binding domain blue and its C-terminal DNA-binding domain purple. The DNA is shown in space-filling form with its sugar–phosphate backbone yellow and its bases white (the atoms are drawn with slightly less than van der Waals radii). The DNA phosphates whose ethylation interferes with CAP binding are red. Those in the complex that are hypersensitive to DNase I are blue (these latter phosphates bridge the CAP-induced kinks and hence occur where the minor groove has been dramatically widened, which apparently increases their susceptibility to DNase I digestion). The bound cAMPs are shown in ball-and-stick form in red. (*b*) CAP–cAMP in complex with a 44-bp palindromic DNA and the αCTD oriented similarly to Part *a*. The DNA is shown in ladder form in red (it contains two symmetrically related single-phosphate gaps that are separated by 4 bp); the proteins are represented by their backbones with the CAP dimer blue-green, each $\alpha CTD^{CAP,DNA}$ light green, and each αCTD^{CAP} dark green; and cAMP is drawn in wireframe form in red. (*c*) The same structure as in Part *a* showing the binding of the CAP dimer's two helix–turn–helix (HTH) motifs in successive major grooves of the DNA. The HTH motif's N-terminal helix is blue and its C-terminal recognition helix is red. The view is along the molecular 2-fold axis and is related to that in Part *a* by a 90° rotation about its vertical axis. [Part *a* courtesy of and Part *c* based on an X-ray structure by Thomas Steitz, Yale University. PDBid 1CGP. Part *b* courtesy of Helen Berman and Richard Ebright, Rutgers University. PDBid 1LB2.] **See the Interactive Exercises**

3. Class III promoters require multiple activators to maximally stimulate transcription. These may be two or more CAP–cAMP complexes or a CAP–cAMP complex acting in concert with promoter-specific activators as occurs in the *araBAD* operon (Section 31-3E).

The X-ray structure of CAP–cAMP in complex with the *E. coli* αCTD and a 44-bp palindromic DNA containing the 22-bp CAP–cAMP binding site and 5′-AAAAAA-3′ at each end, determined by Helen Berman and Ebright, reveals how these components interact (Fig. 31-28*b*). The 2-fold symmetric CAP–cAMP–αCTD complex contains two differently located pairs of αCTDs. Each member of the pair designated $\alpha CTD^{CAP,DNA}$ binds to both CAP and to the DNA. CAP and $\alpha CTD^{CAP,DNA}$ interact over a surprisingly small surface area involving only six residues on each protein that mutagenesis experiments had previously implicated. $\alpha CTD^{CAP,DNA}$ also interacts with the minor groove of a 6-bp segment of the DNA (5′-AAAAAG-3′) centered 19 bp from the center of the DNA. Each member of the other pair of αCTDs, designated αCTD^{DNA}, interacts with the minor groove of an UP element-like sequence (5′-GAAAAA-3′) that is fortuitously present in the DNA but it makes no contacts with other protein molecules. The common portions of the two CAP complexes pictured in Fig. 31-28*a*, *b* are closely superimposable, thereby indicating that the conformation of CAP and its interaction with DNA are not significantly altered by its association with the αCTD. Evidently, CAP–cAMP transcriptionally activates RNAP via a simple "adhesive" mechanism that facilitates and/or stabilizes its interaction with the promoter DNA. The structures of αCTD[CAP,DNA] and αCTD[CAP] and their interactions with DNA are nearly identical, thereby suggesting that they are representive of the interaction of an αCTD with an UP element.

D. *Sequence-Specific Protein–DNA Interactions*

Since genetic expression is controlled by proteins such as CAP and *lac* repressor, an important issue in the study of gene regulation is how these proteins recognize their target base sequences on DNA. Sequence-specific DNA-binding proteins generally do not disrupt the base pairs of the duplex DNA to which they bind. Consequently, these proteins can only discriminate among the four base pairs (A · T, T · A, G · C, and C · G) according to the functional groups of these base pairs that project into DNA's major and minor grooves. An inspection of Fig. 5-12 reveals that the groups exposed in the major groove have a greater variation in their types and arrangements than do those that are exposed in the minor groove. Indeed, the positions of the hydrogen bonding acceptors in the major groove vary with both the identity and orientation of the base pair, whereas in the minor groove they are largely sequence independent. Moreover, the ~5-Å-wide and ~8-Å-deep minor groove of canonical (ideal) B-DNA is too narrow to admit protein structural elements such as an α helix, whereas its ~12-Å-wide and ~8-Å-deep major groove can do so. Thus, in the absence of major conformational changes to B-DNA, it would be expected that proteins could more readily differentiate base sequences from its major groove than from its minor groove. We shall see below that this is, in fact, the case.

a. The Helix–Turn–Helix Motif Is a Common DNA Recognition Element in Prokaryotes

See Guided Exploration 30. Transcription factor-DNA Interactions The CAP dimer's two symmetrically disposed F helices protrude from the protein surface in such a way that they fit into successive major grooves of B-DNA (Fig. 31-28). *CAP's E and F helices form a* ***helix–turn–helix (HTH) motif*** *(supersecondary structure) that conformationally resembles analogous HTH motifs in numerous other prokaryotic repressors of known X-ray and NMR structure,* including the *lac* repressor, the *E. coli* ***trp*** **repressor** (Section 31-3F), and the **cI repressors** and **Cro proteins** from **bacteriophages λ** and **434** (Section 33-3D). HTH motifs are ~20-residue polypeptide segments that form two α helices which cross at ~120° (Fig. 31-28*c*). They occur as components of domains that otherwise have widely varying structures, although all of them bind DNA. Note that HTH motifs are structurally stable only when they are components of larger proteins.

The X-ray and NMR structures of a number of protein–DNA complexes (see below) indicate that *DNA-binding proteins containing an HTH motif associate with their target base pairs mainly via the side chains extending from the second helix of the HTH motif, the so-called* ***recognition helix*** (helix F in CAP, E in *trp* repressor, and α3 in the phage proteins). Indeed, replacing the outward-facing residues of the 434 repressor's recognition helix with the corresponding residues of the related **bacteriophage P22** yields a hybrid repressor that binds to P22 operators but not to those of 434. Moreover, the HTH motifs in all these proteins have amino acid sequences that are similar to each other and to polypeptide segments in numerous other prokaryotic DNA-binding proteins, including *lac* repressor. Evidently, *these proteins are evolutionarily related and bind their target DNAs in a similar manner.*

How does the recognition helix recognize its target sequence? Since each base pair presents a different and presumably readily differentiated constellation of hydrogen bonding groups in DNA's major groove, it seemed likely that there would be a simple correspondence, analogous to Watson–Crick base pairing, between the amino acid residues of the recognition helix and the bases they contact in forming sequence-specific associations. The above X-ray structures, however, indicate this idea to be incorrect. Rather, base sequence recognition arises from complex structural interactions. For instance:

1. The X-ray structures of the closely similar N-terminal domain of 434 repressor (residues 1–69) and the entire 71-residue 434 Cro protein in their complexes with the identical 20-bp target DNA (434 phage expression is regulated through the differential binding of these proteins to the same DNA segments; Section 33-3D) were both determined by Stephen Harrison. Both dimeric proteins, as seen for CAP (Fig. 31-28), associate with the DNA in a 2-fold symmetric manner with their recognition helices bound in successive turns of the DNA's major groove (Figs. 31-29 and 31-30). In both complexes, the protein closely conforms to the DNA surface and interacts with its paired bases and sugar–phosphate chains through elaborate networks of hydrogen bonds, salt bridges, and van der Waals contacts. Nevertheless, the detailed geometries of these associations are significantly different. In the repressor–DNA complex (Fig. 31-29), the DNA bends around the protein in an arc of radius ~65 Å which compresses the minor groove by ~2.5 Å near its center (between the two protein monomers) and widens it by ~2.5 Å toward its ends. In contrast, the DNA in complex with Cro (Fig. 31-30), although also bent, is nearly straight at its center and has a less compressed minor groove (compare Figs. 31-29*a* and 31-30*a*). This explains why the simultaneous replacement of three residues in the repressor's recognition helix with those occurring in Cro does not cause the resulting hybrid protein to bind DNA with Cro-like affinity: *The different conformations of the DNA in the repressor and Cro complexes prevents any particular side chain from interacting identically with the DNA in the two complexes.*

2. Paul Sigler determined the X-ray structure of *E. coli trp* repressor in complex with a DNA containing an 18-bp palindrome (TGTACTAGTTAACTAGTAC, where the *trp* repressor's target sequence is underlined) that closely resembles the *trp* operator (Section 31-3F). The dimeric

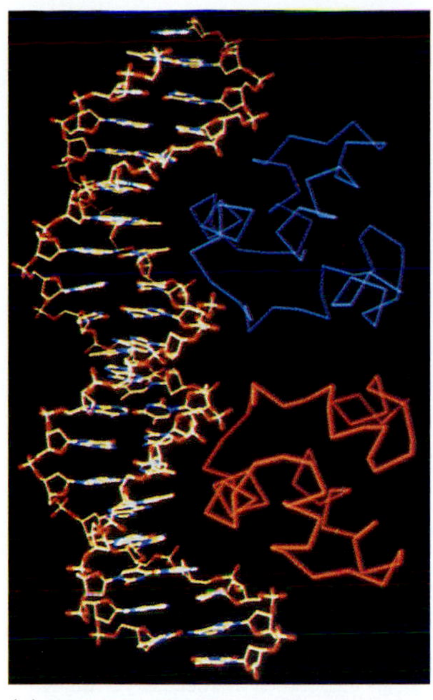

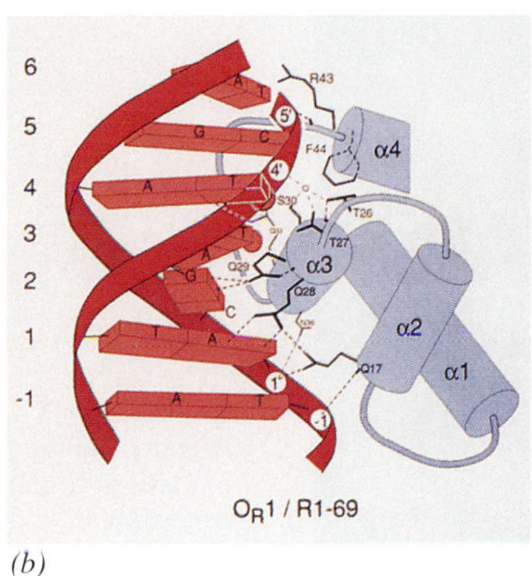

(*b*)

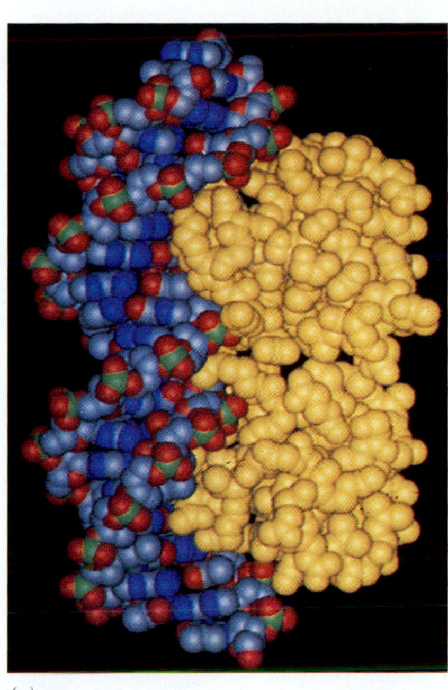

(*a*) (*c*)

FIGURE 31-29 X-Ray structure of the 69-residue N-terminal domain of 434 phage repressor in complex with a 20-bp fragment of its target sequence. One strand of the DNA (*left*) has the sequence d(TATACAAGAAAGTTTGTACT). The complex is viewed perpendicular to its 2-fold axis of symmetry. (*a*) A skeletal model with the protein's two identical subunits (*blue and red;* C_α backbone only). Only the first 63 residues of the protein are visible. (*b*) A schematic drawing indicating how the helix–turn–helix motif, which encompasses helices α2 and α3, interacts with its target DNA. Short bars emanating from the polypeptide chain represent peptide NH groups, hydrogen bonds are represented by dashed lines, and DNA phosphates are represented by numbered circles. The small circle is a water molecule. (*c*) A space-filling model corresponding to Part *a*. The DNA is colored according to atom type (C gray, N blue, O red, and P green) and the protein's non-H atoms are all yellow. [Courtesy of Aneel Aggarwal, John Anderson, and Stephen Harrison, Harvard University. PDBid 2OR1.] **See the Interactive Exercises and Kinemage Exercise 19-1**

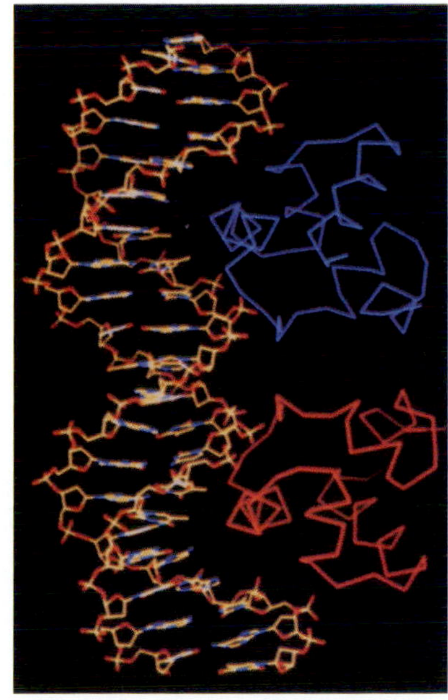

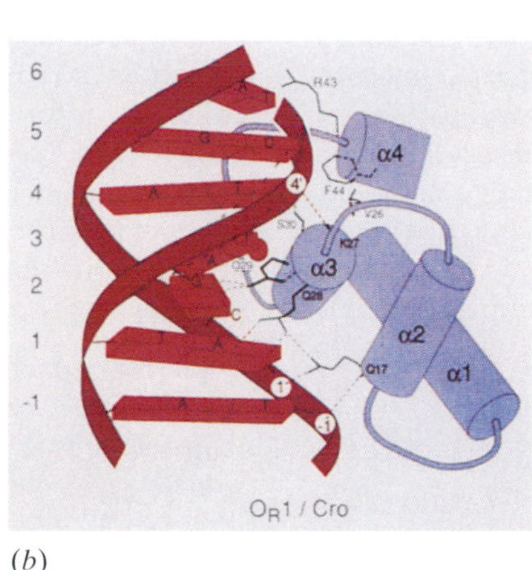

(*b*)

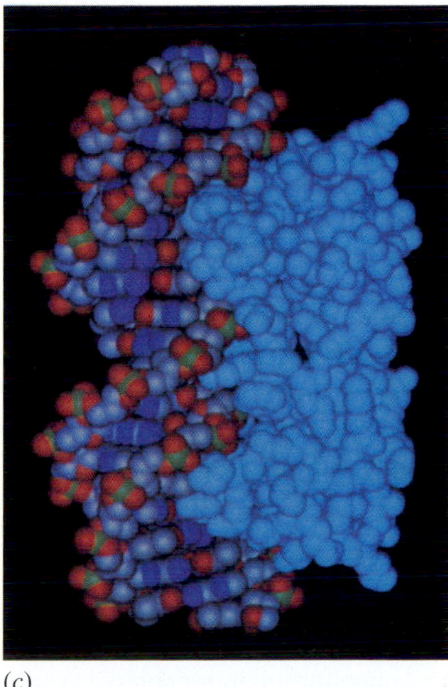

(*a*) (*c*)

FIGURE 31-30 X-Ray structure of the 72-residue 434 Cro protein in complex with the same 20-bp DNA shown in Fig. 31-29. The complex is viewed perpendicular to its 2-fold axis of symmetry. Only the first 64 residues of the protein are visible. Parts *a*, *b*, and *c* correspond to those in Fig. 31-29 with the protein in Part *c* shown in cyan. Note the close but not identical correspondence between the two structures. [Courtesy of Alfonso Mondragón, Cynthia Wolberger, and Stephen Harrison, Harvard University. PDBid 3CRO.]

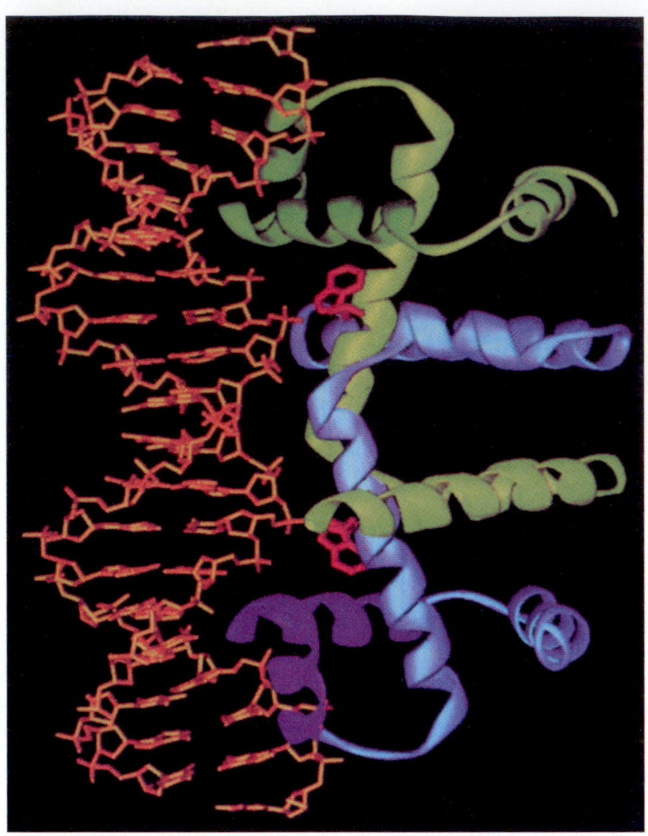

FIGURE 31-31 X-Ray structure of an *E. coli trp* repressor–operator complex. The complex is viewed with its molecular 2-fold axis horizontal. The protein's two identical subunits are green and blue with their HTH motifs (helices D and E) more deeply colored. The 18-bp-containing self-complementary DNA is gold. *trp* repressor binds its operator only when L-tryptophan (*red*) is simultaneously bound. Note that the protein's recognition helices (E) bind, as expected, in successive major grooves of the DNA but extend approximately perpendicular to the DNA duplex axis. In contrast, the recognition helices of 434 repressor and Cro proteins are nearly parallel to the major grooves of their bound DNAs (Figs. 31-29 and 31-30), whereas those of CAP assume an intermediate orientation (Fig. 31-28). [Based on an X-ray structure by Paul Sigler, Yale University. PDBid 1TRO.] **See the Interactive Exercises**

protein's recognition helices bind, as expected, in successive major grooves of the DNA, each in contact with an operator half-site (ACTAGT; Fig. 31-31). There are numerous hydrogen bonding contacts between the *trp* repressor and its bound DNA's nonesterified phosphate oxygens. Astoundingly, however, *there are no direct hydrogen bonds or nonpolar contacts that can explain the repressor's specificity for its operator. Rather, all but one of the side chain–base hydrogen bonding interactions are mediated by bridging water molecules* (the one direct interaction involves a base that can be mutated without greatly affecting repressor binding affinity). Such buried water molecules have therefore been described as "honorary" protein side chains. In addition, the operator contains several base pairs that are not in contact with the repressor but whose mutation nevertheless greatly decreases repressor binding affinity. This suggests that the operator assumes a sequence-specific conformation that makes favorable contacts with the repressor. Indeed, comparison of the X-ray structure of an uncomplexed 10-bp self-complementary DNA containing the *trp* operator's half-site (CCACTAGTGG) with that of the DNA in the *trp* repressor–operator complex reveals that the ACTAGT half-site assumes nearly identical idiosyncratic conformations and patterns of hydration in both structures. However, the B-DNA helix, which is straight in the DNA 10-mer, is bent by 15° toward the major groove in each operator half-site of the repressor–operator complex. Other DNA sequences could conceivably assume the repressor-bound operator's conformation but at too high an energy cost to form a stable complex with repressor (*trp* repressor's measured 10^4-fold preference for its operator over other DNAs implies an ~23 kJ · mol^{-1} difference in their binding free energies). This phenomenon, in which a protein senses the base sequence of DNA through the DNA's backbone conformation and/or flexibility, is referred to as **indirect readout.** 434 repressor apparently also employs indirect readout: Replacing the central A · T base pair of the operator shown in Fig. 31-29 with G · C reduces repressor binding affinity by 50-fold even though 434 repressor does not contact this region of the DNA.

It therefore appears that *there are no simple rules governing how particular amino acid residues interact with bases. Rather, sequence specificity results from an ensemble of mutually favorable interactions between a protein and its target DNA.*

b. *met* Repressor Contains a Two-Stranded Antiparallel β Sheet That Binds in Its Target DNA's Major Groove

The *E. coli* ***met*** **repressor (MetJ),** when complexed with *S*-adenosylmethionine (SAM; Fig. 26-18), represses the transcription of its own gene and those encoding enzymes involved in the synthesis of methionine (Fig. 26-60) and SAM. The X-ray structure of the *met* repressor–SAM–operator complex (Fig. 31-32), determined by Simon

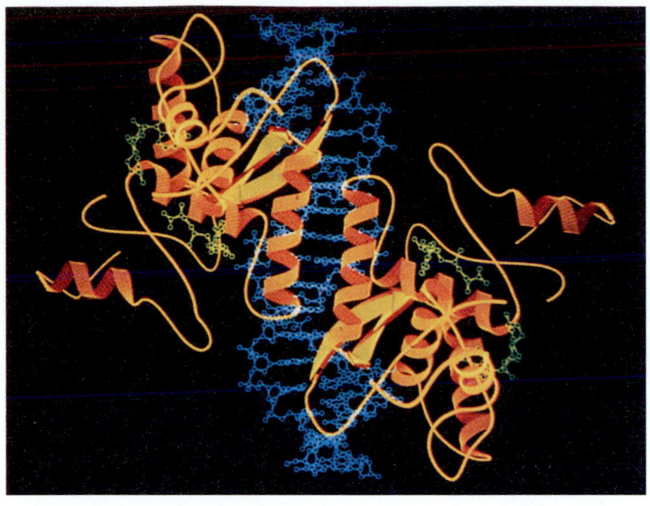

(a)

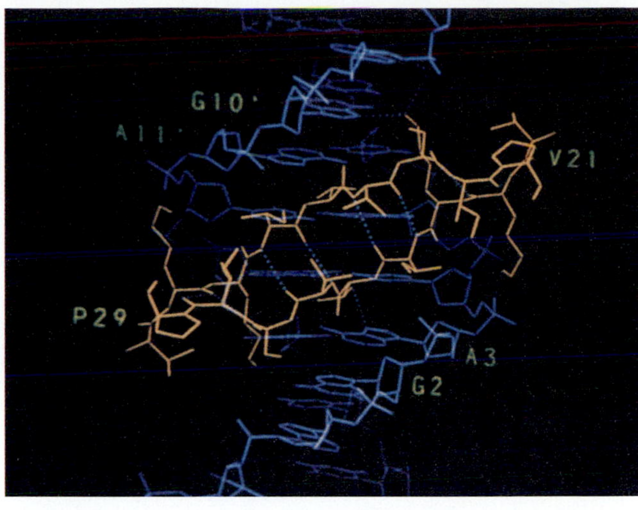

(b)

FIGURE 31-32 X-Ray structure of the *E. coli met* repressor–SAM–operator complex. (*a*) The overall structure of the complex as viewed along its 2-fold axis of symmetry. The 104-residue repressor subunits are shown in gold. The self-complementary 19-bp DNA and SAM, which must be bound to the repressor for it to also bind DNA, are shown in ball-and-stick form with the DNA blue and SAM green. Note that the DNA has four bound repressor subunits: Pairs of subunits form symmetric dimers in which each subunit donates one strand of the 2-stranded antiparallel β ribbon that is inserted in the DNA's major groove (*upper left and lower right*). Two such dimers pair across the complex's 2-fold axis via their antiparallel N-terminal helices which contact one another over the DNA's minor groove. (*b*) Detailed view of the 2-stranded antiparallel β ribbon (*yellow*, residues 21–29) inserted into the DNA's major groove (*blue*). Hydrogen bonds are indicated by dashed lines. [Courtesy of Simon Phillips, University of Leeds, U.K. PDBid 1CMA.] **See the Interactive Exercises**

Phillips, reveals a symmetric homodimer of intertwined monomers that lacks an HTH motif. Rather, *met* repressor binds to its palindromic target DNA sequence through a symmetry-related pair of symmetrical two-stranded antiparallel β sheets (called **β ribbons**) that are inserted in successive major grooves of the DNA. Each β ribbon makes sequence-specific contacts with its target DNA sequence via hydrogen bonding and, probably, indirect readout.

Phillips first determined the X-ray structure of *met* repressor in the absence of DNA. Model building studies aimed at elucidating how *met* repressor binds to its palindromic target DNA assumed that the 2-fold rotation axes of both molecules would be coincident, as they are in all prokaryotic protein–DNA complexes of known structure. There were, consequently, two reasonable choices: (1) The protein could dock to the DNA with the above pairs of β ribbons entering successive major grooves; or (2) a symmetry-related pair of protruding α helices on the opposite face of the protein could do so in a manner resembling the way in which the recognition helices of HTH motifs interact with DNA. A variety of structural criteria suggested that the α helices make significantly better contacts with the DNA than do the β ribbons. Thus, the observation that it is, in fact, the β ribbons that bind to the DNA provides an important lesson: *The results of model building studies must be treated with utmost caution.* This is because our imprecise understanding of the energetics of intermolecular interactions (Sections 8-4 and 29-2) prevents us from reliably predicting how associating macromolecules conform to one another. In the case of the Met repressor, unpredicted mutual structural accommodations of the protein and DNA yielded a significantly more extensive interface than had been predicted by simply docking the uncomplexed Met repressor to canonical B-DNA.

The numerous prokaryotic transcriptional regulators of known structure either contain an HTH motif or pairs of β ribbons like the *met* repressor (although numerous prokaryotic DNA-binding proteins, including CAP, contain an elaboration of the HTH motif known as the **winged helix** motif in which two protein loops, one of which contacts the DNA's minor groove, flank the HTH recognition helix like the wings of a butterfly). Moreover, most of these proteins are homodimers that bind to palindromic or pseudopalindromic DNA target sequences. However, eukaryotic transcription factors, as we shall see in Section 34-3B, employ a much wider variety of structural motifs to bind their target DNAs, many of which lack symmetry.

E. *araBAD* Operon: Positive and Negative Control by the Same Protein

Humans neither metabolize nor intestinally absorb the plant sugar L-arabinose. Hence, the *E. coli* that normally inhabit the human gut are periodically presented with a banquet of this pentose. Three of the five *E. coli* enzymes

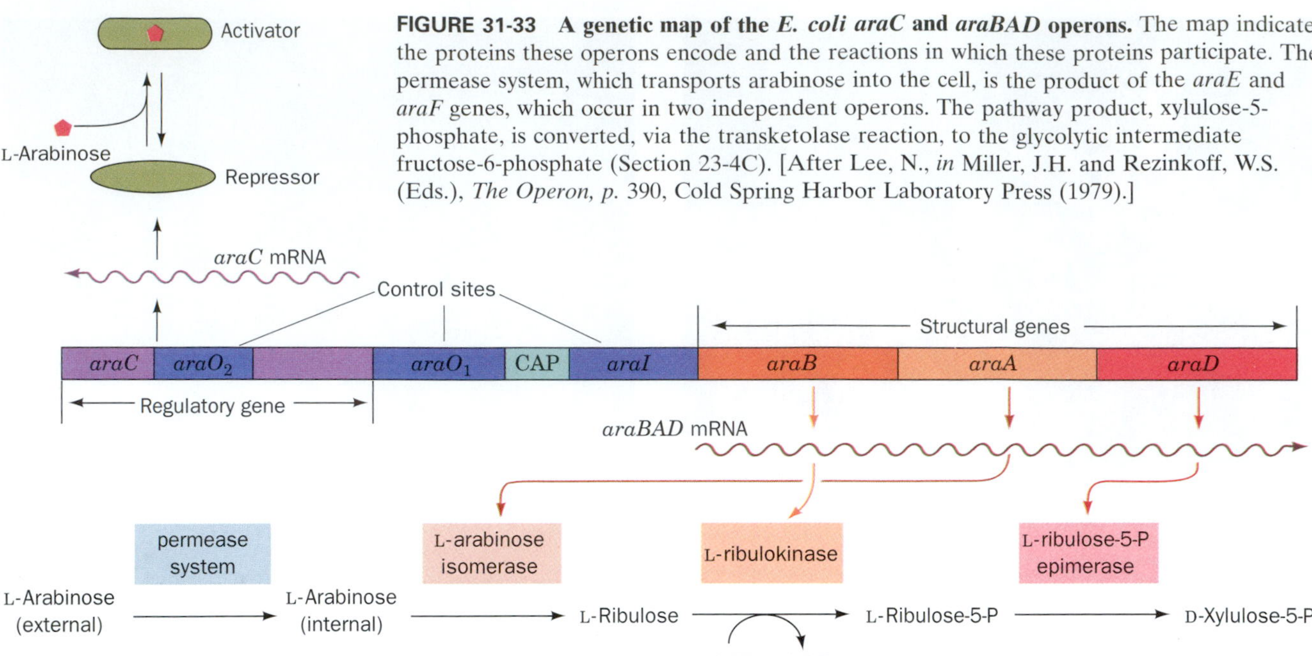

FIGURE 31-33 A genetic map of the *E. coli araC* and *araBAD* operons. The map indicates the proteins these operons encode and the reactions in which these proteins participate. The permease system, which transports arabinose into the cell, is the product of the *araE* and *araF* genes, which occur in two independent operons. The pathway product, xylulose-5-phosphate, is converted, via the transketolase reaction, to the glycolytic intermediate fructose-6-phosphate (Section 23-4C). [After Lee, N., *in* Miller, J.H. and Rezinkoff, W.S. (Eds.), *The Operon, p.* 390, Cold Spring Harbor Laboratory Press (1979).]

that metabolize arabinose are products of the catabolite repressible ***araBAD* operon** (Fig. 31-33).

The *araBAD* operon, as Robert Schleif has shown, contains, moving upstream from its transcriptional start site, the *araI*, *araO*$_1$, and *araO*$_2$ control sites (Fig. 31-34*a*). The *araI* site (*I* for inducer) consists of two closely similar 17-bp half-sites, *araI*$_1$ and *araI*$_2$, that are direct repeats separated by 4 bp and are oriented such that *araI*$_2$, which overlaps the −35 region of the *araBAD* promoter, is downstream of *araI*$_1$. Likewise, *araO*$_1$ consists of two directly repeating half-sites, O_{1L} and O_{1R}. Intriguingly, however, *araO*$_2$ consists of a single half-site that is located in a noncoding upstream region of the *araC* gene (see below), at position −270 relative to the *araBAD* start site.

The transcription of the *araBAD* operon is regulated by both CAP–cAMP and the arabinose-binding protein **AraC.** Each 292-subunit of the homodimeric AraC consists of an N-terminal, arabinose-binding, dimerization domain (residues 1–170) connected via a flexible linker to a C-terminal DNA-binding domain (residues 178–292). Regulation of the *araBAD* operon occurs as follows (Fig. 31-34):

1. In the absence of AraC, RNA polymerase initiates transcription of the *araC* gene in the direction away from its upstream neighbor, *araBAD*. The *araBAD* operon is expressed at a low basal level.

2. When AraC is present, but neither arabinose nor CAP–cAMP (high glucose), AraC binds to *araO*$_2$ and *araI*$_1$. The binding of AraC to *araI*$_1$ prevents RNAP from initiating transcription of the *araBAD* operon (negative control). A series of deletion mutations indicate that the presence of *araO*$_2$ is also required for the repression of *araBAD*. The remarkably large 211-bp separation between *araO*$_2$ and *araI*$_1$ therefore strongly suggests that the DNA between them is looped such that a dimeric molecule of AraC protein simultaneously binds to both *araO*$_2$ and *araI*$_1$. This is corroborated by the observation that the level of repression is greatly diminished by the insertion of 5 bp (half a turn) of DNA between these two sites, thereby transferring *araO*$_2$ to the opposite face of the DNA helix relative to *araI*$_1$ in the putative loop. Yet, the insertion of 11 bp (one turn) of DNA has no such effect. Moreover, looping does not readily occur unless the DNA is supercoiled, which presumably drives the looping process. The AraC dimer also binds to *araO*$_1$, the operator of the *araC* gene, so as to block the transcription of *araC* but only at high concentrations. Thus, it is likely that DNA looping itself represses the transcription of *araC*. In either case, the expression of *araC* is autoregulatory.

3. When arabinose is present, it allosterically induces the AraC subunit bound to *araO*$_2$ to instead bind to *araI*$_2$. This activates RNAP to transcribe the *araBAD* genes (positive control). When the cAMP level is high (low glucose), CAP–cAMP, whose presence is required to achieve the maximum level of transcriptional activation, binds to a site between *araO*$_1$ and *araI*$_1$, where it functions to help break the loop between *araO*$_2$ and *araI*$_1$ and hence to increase the affinity of AraC for *araI*$_2$. The orientation of *araO*$_1$ with respect to *araC* is opposite to that of *araI* with respect to *araBAD*, and hence the binding of AraC–arabinose at *araO*$_1$ blocks RNAP binding at the *araC* promoter, that is, it represses the expression of AraC.

If the *araI*$_2$ subsite is mutated so as to increase AraC's affinity for it, arabinose is no longer required for transcriptional activation. This suggests that arabinose does not conformationally transform AraC to an activator but, rather, weakens its binding affinity for *araO*$_2$. If the *araI* site is turned around or if it is moved upstream so that *araI*$_2$

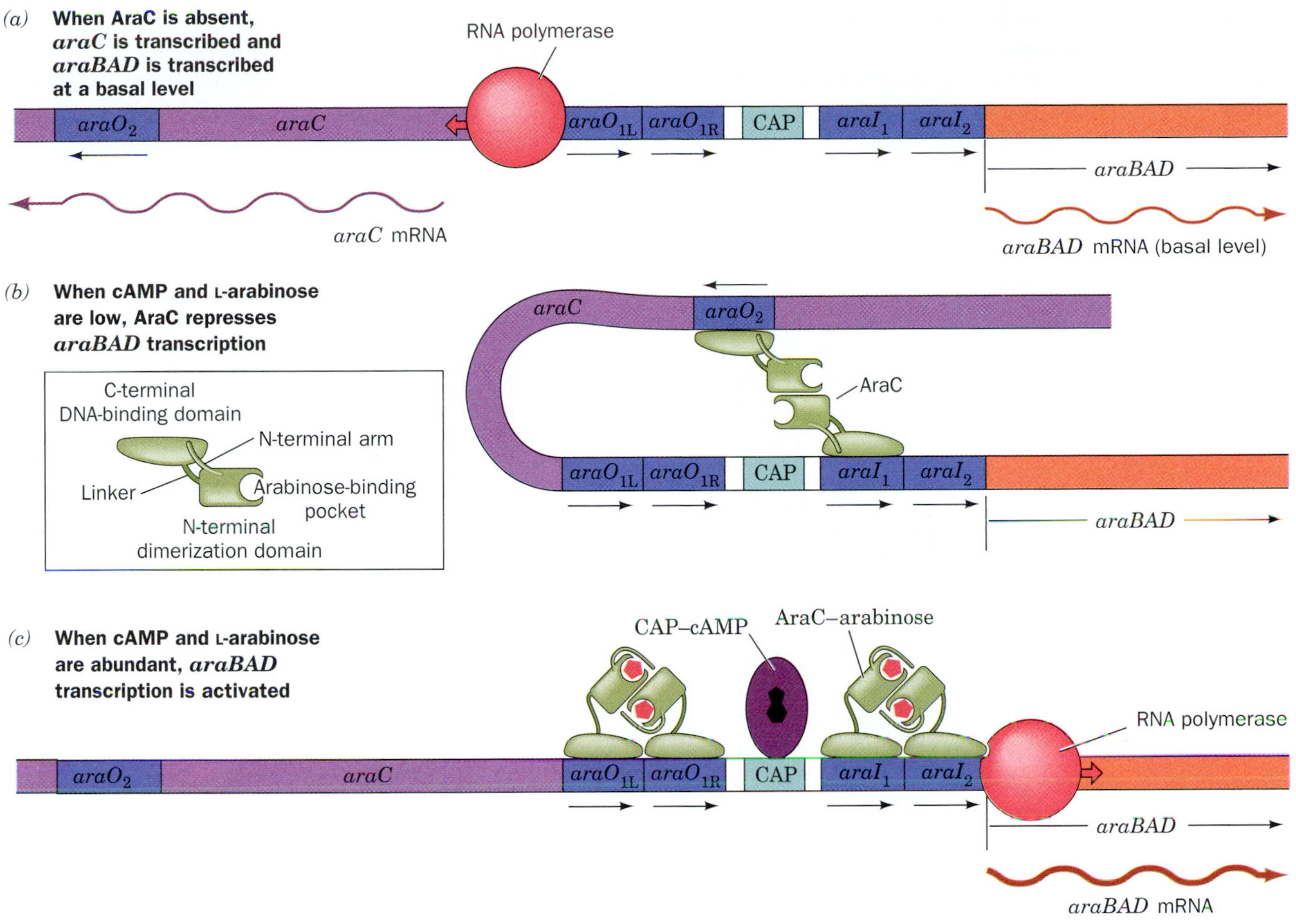

(*a*) **When AraC is absent, *araC* is transcribed and *araBAD* is transcribed at a basal level**

RNA polymerase

$araO_2$ *araC* $araO_{1L}$ $araO_{1R}$ CAP $araI_1$ $araI_2$

araBAD

araC mRNA

araBAD mRNA (basal level)

(*b*) **When cAMP and L-arabinose are low, AraC represses *araBAD* transcription**

C-terminal DNA-binding domain

N-terminal arm

Linker

Arabinose-binding pocket

N-terminal dimerization domain

araC $araO_2$ AraC $araO_{1L}$ $araO_{1R}$ CAP $araI_1$ $araI_2$ *araBAD*

(*c*) **When cAMP and L-arabinose are abundant, *araBAD* transcription is activated**

CAP–cAMP

AraC–arabinose

RNA polymerase

$araO_2$ *araC* $araO_{1L}$ $araO_{1R}$ CAP $araI_1$ $araI_2$ *araBAD*

araBAD mRNA

FIGURE 31-34 The mechanism of *araBAD* regulation. (*a*) In the absence of AraC, RNAP initiates the transcription of *araC*. *araBAD* is also expressed but at a low basal level. (*b*) When AraC is present, but not L-arabinose or cAMP, AraC links together $araO_2$ and $araI_1$ to form a DNA loop, thereby repressing both *araC* and *araBAD*. (*c*) When AraC and L-arabinose are both present and cAMP is abundant, the resulting AraC–arabinose complex releases $araO_2$ and instead binds $araI_2$, thereby activating *araBAD* transcription. This process is facilitated by the binding of CAP–cAMP. *araC* is repressed by

does not overlap the *araBAD* promoter, AraC cannot stimulate transcription. Evidently, *AraC activates RNAP through specific and relatively inflexible protein–protein interactions.*

The X-ray structures of the N-terminal domain of AraC (residues 2–178), in both the presence and the absence of arabinose, were determined by Schleif and Cynthia Wolberger. In the presence of arabinose, this domain consists of an 8-stranded β barrel followed by two antiparallel α helices (Fig. 31-35). Two such domains associate via an antiparallel coiled coil between each of their C-terminal helices to form the protein's dimerization interface. An arabinose molecule binds in a pocket of each β barrel via a network of direct and water-mediated hydrogen bonds with side chains that line the pocket. Residues 7 to 18 of the N-terminal arm lie across the mouth of the sugar-binding pocket (residues 2–6 are disordered), thereby fully enclosing the arabinose. The structure of the N-terminal domain in the absence of arabinose is largely superimposable on that in the complex with arabinose, with the

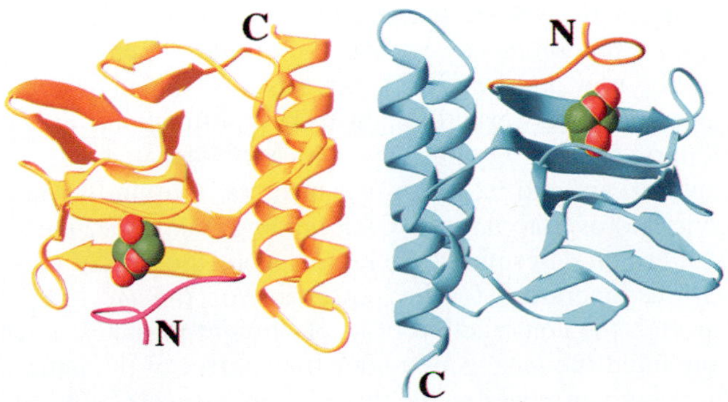

FIGURE 31-35 X-Ray structure of *E. coli* AraC in complex with L-arabinose. The homodimeric protein is viewed along its 2-fold axis with its subunits cyan and gold except for their N-terminal arms, which are orange and magenta. The arabinose is drawn in space-filling form with C green and O red. [Based on an X-ray structure by Robert Schleif and Cynthia Wolberger, Johns Hopkins University. PDBid 2ARC.]

exception that the N-terminal arm is disordered, a not unexpected observation considering that it interacts with bound arabinose via a series of hydrogen bonds.

How does arabinose binding induce the AraC subunit bound at *araO*$_2$ to instead bind to *araI*$_2$? Several lines of evidence indicate that, in the absence of arabinose, AraC's N-terminal arm binds to its DNA-binding domain in a way that favors loop formation: (1) the deletion of the N-terminal arm beyond its sixth residue makes AraC act as if arabinose is present; (2) mutations to surface residues on the DNA-binding domain that presumably eliminate its binding of the N-terminal arm also constitutively activate AraC; and (3) mutations in the DNA-binding domain that weaken the binding of arabinose to the protein, presumably by strengthening the binding of the N-terminal arm, can be suppressed by a second mutation in the N-terminal arm or by the deletion of its five N-terminal residues. Evidently, *the binding of the N-terminal arms to the DNA-binding domains in the absence of arabinose rigidifies the AraC dimer such that it cannot simultaneously bind to the directly repeated araI*$_1$ *and araI*$_2$ *and hence induce the transcription of araBAD.* This is corroborated by the observations that (1) joining two AraC DNA-binding domains by flexible polypeptide linkers yields proteins that behave like AraC in the presence of arabinose, and (2) a construct consisting of two double-stranded *araI*$_1$ half-sites flexibly connected by a 24-nt segment of ssDNA binds wild-type AraC with an affinity that is unaffected by arabinose.

F. *lac Repressor II: Structure*

Here we continue our discussions of the *lac* repressor, but now in terms of the concepts learned in Sections 31-3CDE.

a. Loop Formation Is Important in the Expression of the *lac* Operon

DNA loop formation, which is now known to occur in numerous bacterial and eukaryotic systems, apparently permits several regulatory proteins and/or regulatory sites on one protein to simultaneously influence transcription initiation by RNAP. In fact, *the lac repressor has three binding sites on the lac operon:* the primary operator (Fig. 31-25), now known as O_1, and two so-called pseudo-operators (previously thought to be nonfunctional evolutionary fossils), O_2 and O_3, which are located 401 bp downstream and 92 bp upstream of O_1 (within the *lacZ* gene and overlapping the CAP binding site, respectively). Müller-Hill determined the relative contributions of these various operators to the repression of the *lac* operon through the construction of a set of eight plasmids: Each contained the *lacZ* gene under the control of the natural *lac* promoter as well as the three *lac* operators (O_1, O_2, and O_3), which were either active or mutagenically inactive in all possible combinations. When all three operators are active, *lacZ* expression is repressed 1300-fold relative to when all three operators are inactive. The inactivation of only O_1 results in almost complete loss of repression whereas the inactivation of only O_2 or O_3 causes only a ~2-fold loss in repression. However, when O_2 and O_3 are both inactive, repression is decreased ~70-fold. These results suggest that efficient repression requires the formation of a DNA loop between O_1 and either O_2 or O_3. Indeed, such loop formation, and/or the cooperativity of repressor binding arising from it, appears to be a greater contributor to repression than repressor binding to O_1 alone, which provides only 19-fold repression.

b. The *lac* Repressor Is a Dimer of Dimers

Ponzy Lu and Mitchell Lewis determined the X-ray structures of the *lac* repressor alone, in its complex with IPTG, and in its complex with a 21-bp duplex DNA segment whose sequence is a palindrome of the left half of O_1 (Fig. 31-25). Each repressor subunit consists of five functional units (Fig. 31-36): (1) an N-terminal DNA-binding domain (residues 1–49) which is known as the "headpiece" because it is readily proteolytically cleaved away from the remaining still tetrameric "core" protein; (2) a hinge helix (residues 50–58) that also binds to the DNA; (3 & 4) a sugar-binding domain (residues 62–333) that is divided into an N-subdomain and a C-subdomain; and (5) a C-terminal tetramerization helix (residues 340–360).

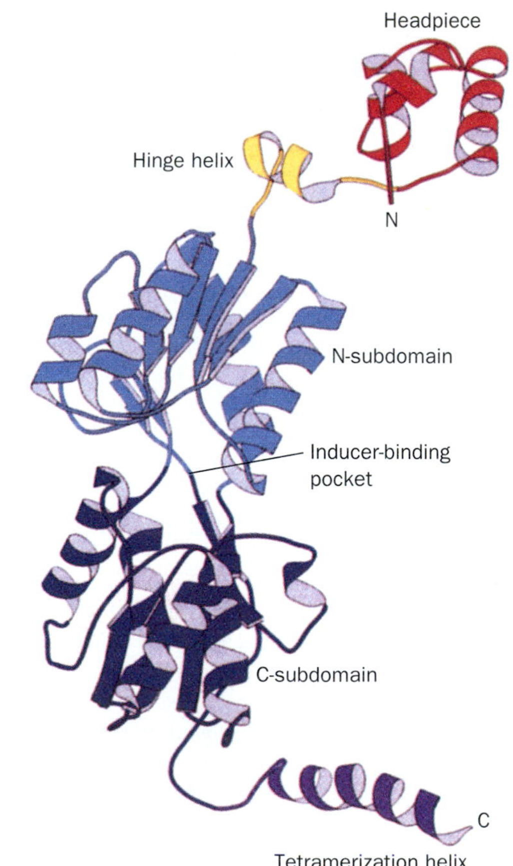

FIGURE 31-36 X-Ray structure of the *lac* repressor subunit. The DNA-binding domain (the headpiece), which contains an HTH motif, is red, the DNA-binding hinge helix is yellow, the N-subdomain of the sugar-binding domain is light blue, its C-subdomain is dark blue, and the tetramerization helix is purple. [Courtesy of Ponzy Lu and Mitchell Lewis, University of Pennsylvania. PDBid 1LBI.]

The *lac* repressor has an unusual quaternary structure (Fig. 31-37*a*). Whereas nearly all homotetrameric nonmembrane proteins of known structure have D_2 symmetry (three mutually perpendicular 2-fold axes; Fig. 8-64*b*), *lac* repressor is a V-shaped protein that has only 2-fold symmetry. Each leg of the V consists of a locally symmetric dimer of closely associated repressor subunits. Two such dimers associate rather tenuously, but with 2-fold symmetry, at the base (point) of the V to form a dimer of dimers.

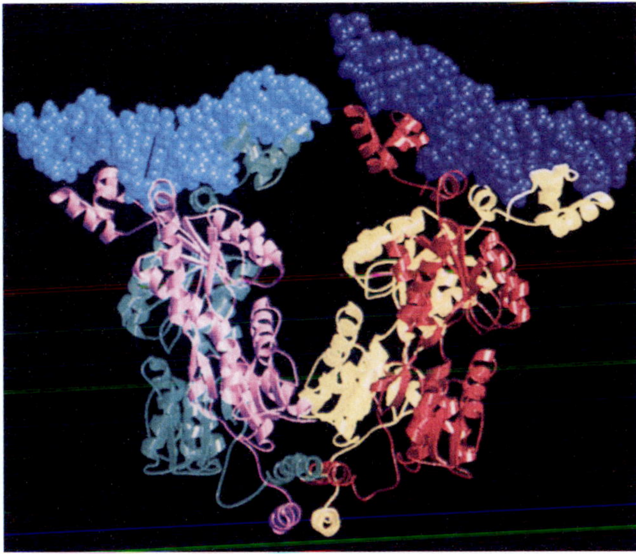

(*a*)

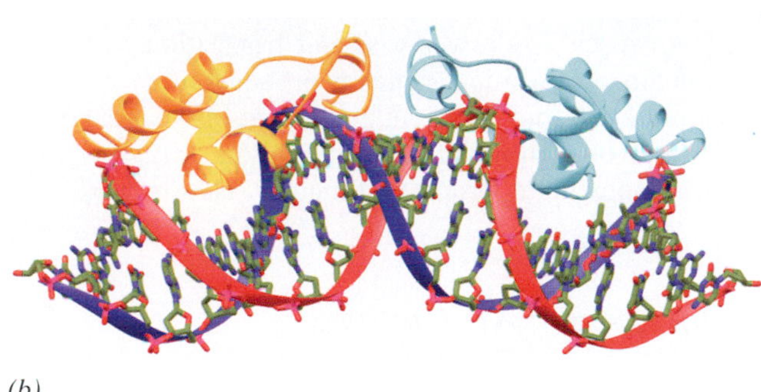

(*b*)

FIGURE 31-37 The structure of the *lac* repressor in complex with DNA. (*a*) The X-ray structure of the *lac* repressor tetramer bound to two 21-bp segments of symmetric *lac* operator DNA. The protein monomers are green, pink, yellow, and red and the DNA segments, drawn in space-filling form, are cyan and blue. [Courtesy of Ponzy Lu and Mitchell Lewis, University of Pennsylvania. PDBid 1LBG.] (*b*) The NMR structure of a 22-bp symmetric *lac* operator DNA in complex with two segments of the *lac* repressor consisting of its DNA-binding domain and its hinge helix. The two protein subunits are cyan and gold. The DNA is represented in stick form colored according to atom type (C green, N blue, O red, and P magenta) with its sugar–phosphate backbones traced by red and blue ribbons. The complex is viewed with its 2-fold axis vertical. Note that the protein dimer's two HTH motifs are inserted in successive major grooves at the periphery of the complex and that the insertion of the two centrally located hinge helices into the DNA's minor groove greatly widens and flattens the minor groove at this point and kinks the DNA in a downward bend. [Based on an NMR structure by Robert Kaptein, Utrecht University, The Netherlands. PDBid 1CJG.]

In the structures of *lac* repressor alone and that of its IPTG complex, the DNA-binding domain is not visible, apparently because the hinge region that loosely tethers it to the rest of the protein is disordered. However, in the DNA complex, in which one DNA duplex binds to each of the two dimers forming the repressor tetramer, the DNA domain forms a compact globule containing three helices, the first two of which form a helix–turn–helix (HTH) motif. The two DNA-binding domains extending from each repressor dimer (at the top of each leg of the V) bind in successive major grooves of a DNA molecule via their HTH motifs, much as is seen, for example, in the complexes of 434 phage repressor and *trp* repressor with their target DNAs (Figs. 31-29 and 31-31). The binding of the *lac* repressor distorts the operator DNA such that it bends away from the DNA-binding domain with an ~60 Å radius of curvature due to an ~45° kink at the center of the operator that widens the DNA's minor groove to over 11 Å and reduces its depth to less than 1 Å. These distortions permit the now ordered hinge helix to bind in the minor groove so as to contact the identically bound hinge helix from the other subunit of the same dimer. NMR structures by Robert Kaptein reveal that the DNA-binding domain, when cleaved from the repressor, binds to the *lac* operator without distorting the DNA. However, the DNA-binding domain together with the hinge helix forms a complex with the *lac* operator in which the hinge helix binds in the DNA's distorted minor groove (Fig. 31-37*b*) as in the X-ray structure. Thus, the binding of the two hinge helices to the *lac* operator appears necessary for DNA distortion. The two DNA duplexes that are bound to each repressor tetramer are ~25 Å apart and do not interact.

The sugar-binding domain consists of two topologically similar subdomains that are bridged by three polypeptide segments (Fig. 31-36). The two sugar-binding domains of a dimer make extensive contacts (Fig. 31-37*a*). IPTG binds to each sugar-binding domain between its subdomains. This does not significantly change the conformations of these subdomains, but it changes the angle between them. Although the hinge helix is not visible in the IPTG complex, model building indicates that, since the dimer's two hinge helices extend from its sugar-binding domains, this conformation change levers apart these hinge helices by 3.5 Å such that they and their attached HTH motifs can no longer simultaneously bind to their operator half-sites. Thus, inducer binding, which is allosteric within the dimer (has a positive homotropic effect; Section 10-4B), greatly loosens the repressor's grip on the operator.

The C-terminal helices from each subunit, which are located on the opposite end of each subunit from the DNA-binding portion (at the point of the V), associate to form a 4-helix bundle that holds together the two repressor dimers, thereby forming the tetramer (Fig. 31-37*a*). The allosteric effects of inducer binding within each dimer are apparently not transmitted between dimers. Moreover, the *E. coli* **purine repressor (PurR),** which is homologous to the *lac* repressor but lacks its C-terminal helix, crystallizes

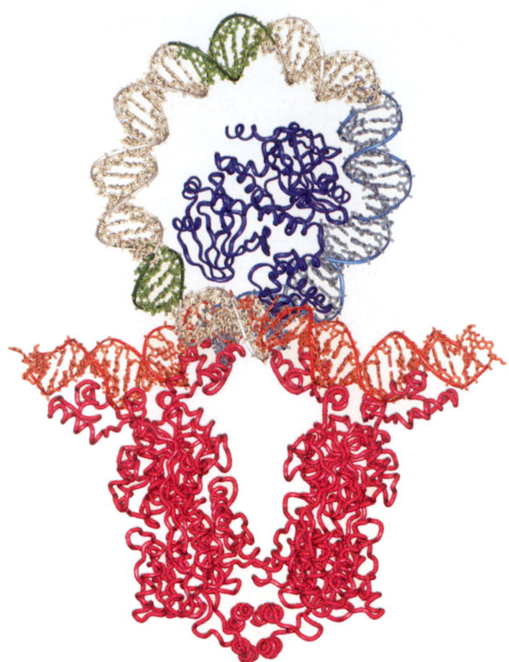

FIGURE 31-38 Model of the 93-bp DNA loop formed when *lac* repressor binds to O_1 and O_3. The proteins are represented by their C_α backbones and the DNA is drawn in stick form with its sugar–phosphate backbones traced by helical ribbons. The model was constructed from the X-ray structure of the *lac* repressor (*magenta*) in complex with two 21-bp operator DNA segments (*red*) and the X-ray structure of CAP–cAMP (*blue*) in complex with its 30-bp target DNA (*cyan;* Fig. 31-28). The remainder of the DNA loop was generated by applying a smooth curvature to canonical B-DNA (*white*) with the −10 and −35 regions of the *lac* promoter highlighted in green. [Courtesy of Ponzy Lu and Mitchell Lewis, University of Pennsylvania.]

as a dimer whose X-ray structure closely resembles that of the *lac* repressor dimer. What then is the function of *lac* repressor tetramerization?

Model building suggests that when the *lac* repressor tetramer simultaneously binds to both the O_1 and O_3 operators, the 93-bp DNA segment containing them forms a loop ~80 Å in diameter (Fig. 31-38). Furthermore, the CAP–cAMP binding site is exposed on the inner surface of the loop. Adding the CAP–cAMP at its proper position to this model reveals that the ~90° curvature which CAP–cAMP binding imposes on DNA (Fig. 31-28) has the correct direction and magnitude to stabilize the DNA loop, thereby stabilizing this putative CAP–cAMP–*lac* repressor–DNA complex. It may seem paradoxical that the binding of CAP–cAMP, a transcriptional activator, stabilizes the repressor–DNA complex. However, when both glucose and lactose are in short supply, it is important that the bacterium lower its basal rate of *lac* operon expression in order to conserve energy. The binding site (promoter) for RNAP is also located on the inner surface of the loop. Thus, the large size of the RNAP molecule would prevent it from fully engaging the promoter in this looped complex, thereby maximizing repression.

c. Combining Genetic and Structural Studies of the *lac* Repressor Reveals Its Allosterically Important Residues

The phenotypes of 4042 point mutations of the *lac* repressor, which encompass nearly all of its 360 residues (making the *lac* repressor the most exhaustively mutationally characterized protein known) have been mapped onto its X-ray structure. Mutations with an I^- phenotype (*lac* repressors that fail to bind to the *lac* operator, so that β-galactosidase is constitutively synthesized) are located at the *lac* repressor's DNA-binding interface, at its dimer interface, or at internal residues of its inducer-binding core domain. Residues whose mutations result in the I^S phenotype (S for super-repressed; *lac* repressors that, in the presence of inducer, continue to repress the synthesis of β-galactosidase) appear to be of two types: (1) residues that are in direct contact with the inducer, whose alteration therefore interferes with inducer binding; and (2) residues at the dimer interface that are >8 Å from (not in direct contact with) the inducer-binding site. These latter observations reveal which residues mediate the *lac* repressor's allosteric mechanism rather than directly binding the inducer or the DNA. Most of the allosterically important residues are located at the dimer interface and are members of the N-subdomain of the core domain, which links the inducer-binding sites to the operator DNA-binding sites. This is consistent with the observation that inducer binding causes a relative twist and translation of the N-subdomain, a movement which is propagated to the hinge helix and DNA-binding domain. This study demonstrates the power of combining genetic analysis with structural studies to elucidate structure–function relationships.

G. *trp* Operon: Attenuation

We now discuss a sophisticated transcriptional control mechanism named **attenuation** through which bacteria regulate the expression of certain operons involved in amino acid biosynthesis. This mechanism was discovered through the study of the *E. coli* ***trp* operon** (Fig. 31-39), which encodes five polypeptides comprising three enzymes that mediate the synthesis of tryptophan from chorismate (Section 26-5B). Charles Yanofsky established that the *trp* operon genes are coordinately expressed under the control of the *trp* repressor, a dimeric protein of identical 107-residue subunits that is the product of the *trpR* gene (which forms an independent operon). *The trp repressor binds* L-*tryptophan, the pathway's end product, to form a complex that specifically binds to trp operator (trpO, Fig. 31-40) so as to reduce the rate of trp operon transcription 70-fold.* The X-ray structure of the *trp* repressor–operator complex (Section 31-3D) indicates that tryptophan binding allosterically orients *trp* repressor's two symmetry related helix–turn–helix "DNA reading heads" so that they can simultaneously bind to *trpO* (Fig. 31-31). Moreover, the bound tryptophan forms a hydrogen bond to a DNA

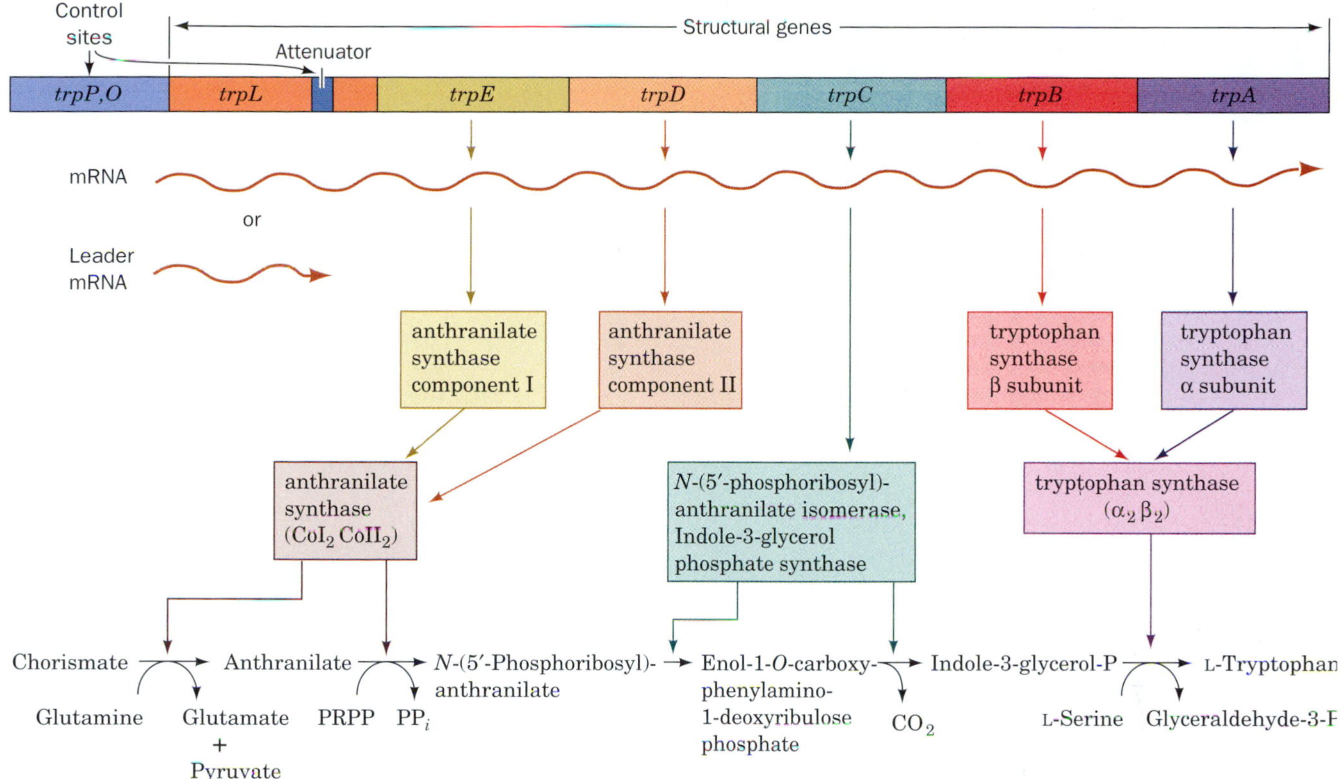

FIGURE 31-39 A genetic map of the *E. coli trp* operon indicating the enzymes it specifies and the reactions they catalyze. The gene product of *trpC* catalyzes two sequential reactions in the synthesis of tryptophan. [After Yanofsky, C., *J. Am. Med. Assoc.* **218,** 1027 (1971).]

phosphate group, thereby strengthening the repressor–operator association. Tryptophan therefore acts as a **corepressor;** its presence prevents what is then superfluous tryptophan biosynthesis (SAM similarly functions as a corepressor with the *met* repressor; Fig. 31-32*a*). The *trp* repressor also controls the synthesis of at least two other operons: the ***trpR* operon** and the ***aro*H operon** (which encodes one of three isozymes that catalyze the initial reaction of chorismate biosynthesis; Section 26-5B).

a. Tryptophan Biosynthesis Is Also Regulated by Attenuation

The *trp* repressor–operator system was at first thought to fully account for the regulation of tryptophan biosynthesis in *E. coli.* However, the discovery of *trp* deletion mutants located downstream from *trpO* that increase *trp* operon expression 6-fold indicated the existence of an additional transcriptional control element. Sequence analysis established that *trpE,* the *trp* operon's leading structural gene, is preceded by a 162-nucleotide **leader sequence** (*trpL*). Genetic analysis indicated that the new control element is located in *trpL,* ~30 to 60 nucleotides upstream of *trpE* (Fig. 31-39).

When tryptophan is scarce, the entire 6720-nucleotide polycistronic *trp* mRNA, including the *trpL* sequence, is synthesized. As the tryptophan concentration increases, the rate of *trp* transcription decreases as a result of the *trp* repressor–corepressor complex's consequent greater abundance. Of the *trp* mRNA that is transcribed, however, an increasing proportion consists of only a 140-nucleotide segment corresponding to the 5′ end of *trpL. The availability of tryptophan therefore results in the premature termination of trp operon transcription.* The control element responsible for this effect is consequently termed an **attenuator.**

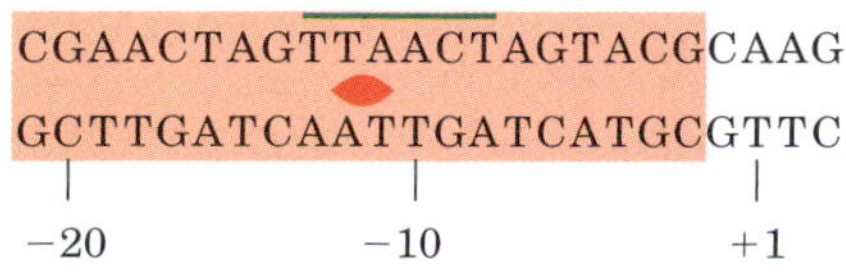

FIGURE 31-40 The base sequence of the *trp* operator. The nearly palindromic sequence is boxed and its −10 region is overscored.

b. The *trp* Attenuator's Transcription Terminator Is Masked when Tryptophan Is Scarce

What is the mechanism of attenuation? The attenuator transcript contains four complementary segments that can form one of two sets of mutually exclusive base paired hair-

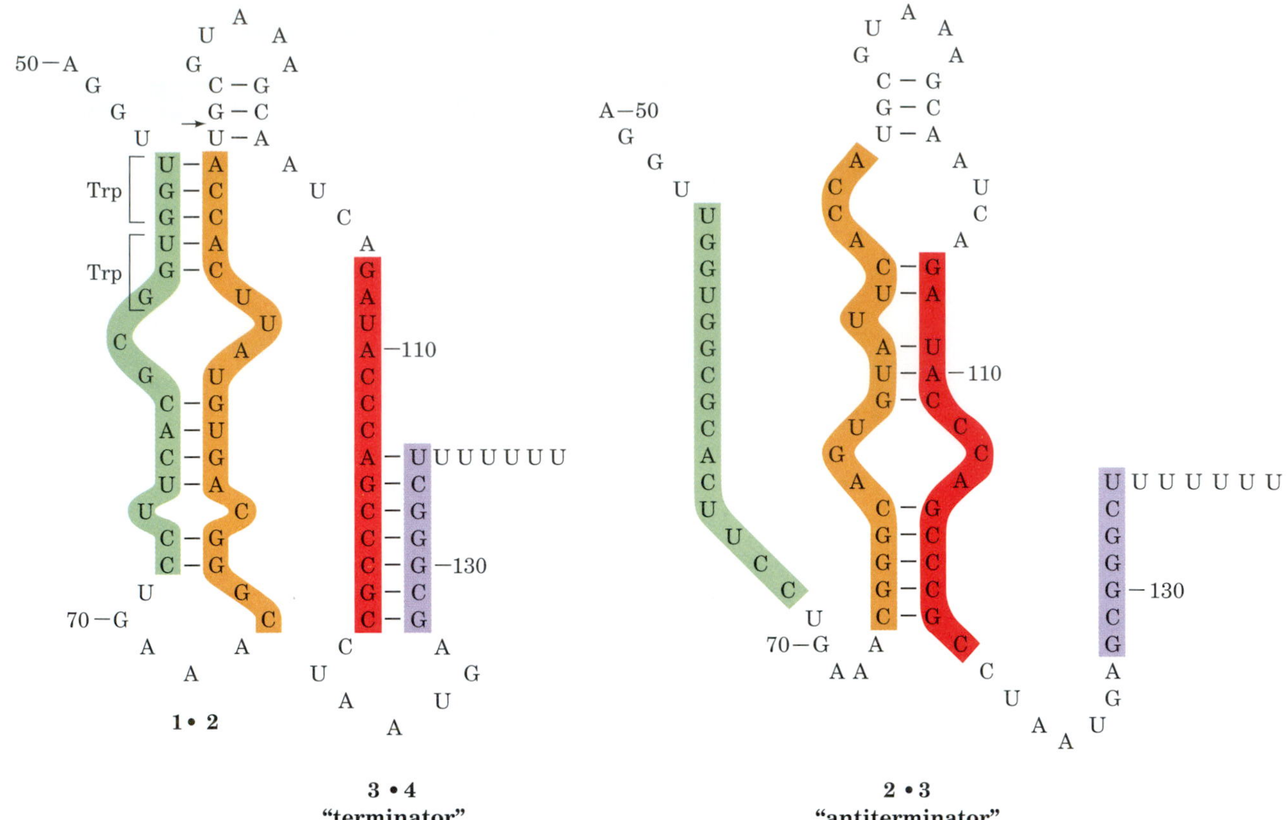

FIGURE 31-41 The alternative secondary structures of *trpL* mRNA. The formation of the base paired 2 · 3 (antiterminator) hairpin (*right*) precludes the formation of the 1 · 2 and 3 · 4 (terminator) hairpins (*left*) and vice versa. Attenuation results in the premature termination of transcription immediately after nucleotide 140 when the 3 · 4 hairpin is present. The arrow indicates the mRNA site past which RNA polymerase pauses until approached by an active ribosome. [After Fisher, R.F. and Yanofsky, C., *J. Biol. Chem.* **258,** 8147 (1983).]

pins (Fig. 31-41). *Segments 3 and 4 together with the succeeding residues comprise a normal rho-independent transcription terminator (Section 31-2D):* a G + C-rich sequence that can form a self-complementary hairpin structure followed by several sequential U's (compare with Fig. 31-18). *Transcription rarely proceeds beyond this termination site unless tryptophan is in short supply.*

A section of the leader sequence, which includes segment 1 of the attenuator, is translated to form a 14-residue polypeptide that contains two consecutive Trp residues (Fig. 31-41, *left*). The position of this particularly rare dipeptide segment (~1% of the residues in *E. coli* proteins are Trp) provided an important clue to the mechanism of attenuation. An additional essential aspect of this mechanism is that ribosomes commence the translation of a prokaryotic mRNA shortly after its 5′ end has been synthesized.

The above considerations led Yanofsky to propose the following model of attenuation (Fig. 31-42). An RNA polymerase that has escaped repression initiates *trp* operon transcription. Soon after the ribosomal initiation site of the *trpL* gene has been transcribed, a ribosome attaches to it and begins translation of the leader peptide. When tryptophan is abundant, so that there is a plentiful supply of **tryptophanyl–tRNA^{Trp}** (the transfer RNA specific for Trp with an attached Trp residue; Section 32-2C), the ribosome follows closely behind the transcribing RNA polymerase so as to sterically block the formation of the 2 · 3 hairpin. Indeed, RNA polymerase pauses past position 92 of the transcript and only continues transcription on the approach of a ribosome, thereby ensuring the proximity of these two entities at this critical position. The prevention of 2 · 3 hairpin formation permits the formation of the 3 · 4 hairpin, the transcription terminator pause site, which results in the termination of transcription (Fig. 31-42*a*). When tryptophan is scarce, however, the ribosome stalls at the tandem UGG codons (which specify Trp; Table 5-3) because of the lack of tryptophanyl–tRNA^{Trp}. As transcription continues, the newly synthesized segments 2 and 3 form a hairpin because the stalled ribosome prevents the otherwise competitive formation of the 1 · 2 hairpin (Fig. 31-42*b*). The formation of the transcriptional terminator's 3 · 4 hairpin is thereby preempted for sufficient time for RNA polymerase to transcribe through it and consequently through the remainder of the *trp* operon. The cell is thus provided with a regulatory mechanism that is responsive to the tryptophanyl–tRNA^{Trp} level, which, in turn, depends on the protein synthesis rate as well as on the tryptophan supply.

There is considerable evidence supporting this model of attenuation. The *trpL* transcript is resistant to limited

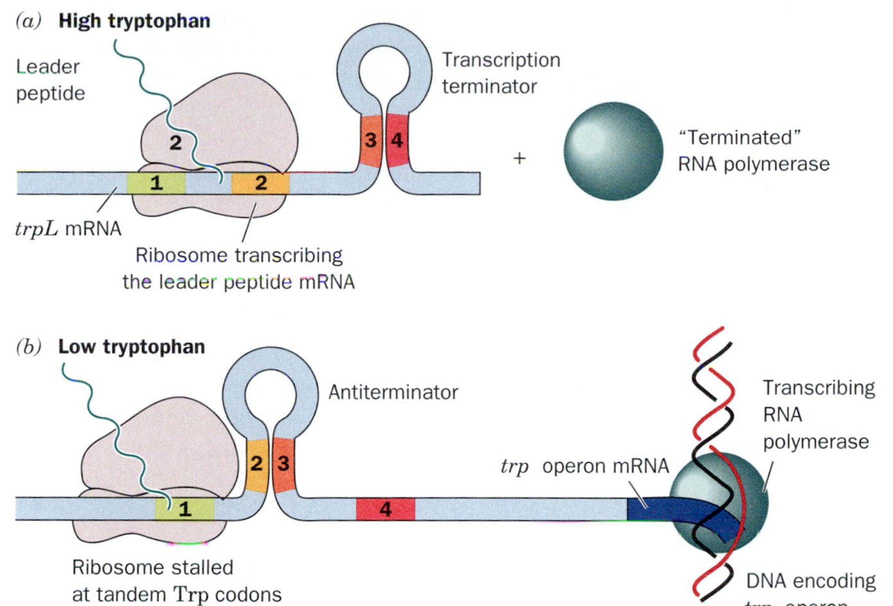

FIGURE 31-42 Attenuation in the *trp* operon. (*a*) When tryptophanyl–tRNA^{Trp} is abundant, the ribosome translates *trpL* mRNA. The presence of the ribosome on segment 2 prevents the formation of the base paired 2 · 3 hairpin. The 3 · 4 hairpin, an essential component of the transcriptional terminator, can thereby form, thus aborting transcription. (*b*) When tryptophanyl–tRNA^{Trp} is scarce, the ribosome stalls on the tandem Trp codons of segment 1. This situation permits the formation of the 2 · 3 hairpin which, in turn, precludes the formation of the 3 · 4 hairpin. RNA polymerase therefore transcribes through this unformed terminator and continues *trp* operon transcription.

RNase T1 digestion, indicating that it has extensive secondary structure. The significance of the tandem Trp codons in the *trpL* transcript is corroborated by their presence in *trp* leader regions of several other bacterial species. Moreover, the leader peptides of the five other amino acid–biosynthesizing operons known to be regulated by attenuation (most exclusively so) are all rich in their corresponding amino acid residues (Table 31-3). For example, the *E. coli* ***his*** **operon,** which specifies enzymes synthesizing histidine (Fig. 26-65), has seven tandem His residues in its leader peptide whereas the ***ilv*** **operon,** which specifies enzymes participating in isoleucine, leucine, and valine biosynthesis (Fig. 26-61), has five Ile's, three Leu's, and six Val's in its leader peptide. Finally, the leader transcripts of these operons resemble that of the *trp* operon in their capacity to form two alternative secondary structures, one of which contains a trailing termination structure.

H. *Regulation of Ribosomal RNA Synthesis: The Stringent Response*

E. coli cells growing under optimal conditions divide every 20 min. Such cells contain up to 70,000 ribosomes and hence must synthesize ~35,000 ribosomes per cell division cycle. Yet RNAP can initiate the transcription of an rRNA

TABLE 31-3. Amino Acid Sequences of Some Leader Peptides in Operons Subject to Attenuation

Operon	Amino Acid Sequence[a]
trp	Met-Lys-Ala-Ile-Phe-Val-Leu-Lys-Gly-TRP-TRP-Arg-Thr-Ser
pheA	Met-Lys-His-Ile-Pro-PHE-PHE-PHE-Ala-PHE-PHE-PHE-Thr-PHE-Pro
his	Met-Thr-Arg-Val-Gln-Phe-Lys-HIS-HIS-HIS-HIS-HIS-HIS-HIS-Pro-Asp
leu	Met-Ser-His-Ile-Val-Arg-Phe-Thr-Gly-LEU-LEU-LEU-LEU-Asn-Ala-Phe-Ile-Val-Arg-Gly-Arg-Pro-Val-Gly-Gly-Ile-Gln-His
thr	Met-Lys-Arg-ILE-Ser-THR-THR-ILE-THR-THR-THR-ILE-THR-ILE-THR-THR-Gln-Asn-Gly-Ala-Gly
ilv	Met-Thr-Ala-LEU-LEU-Arg-VAL-ILE-Ser-LEU-VAL-VAL-ILE-Ser-VAL-VAL-VAL-ILE-ILE-ILE-Pro-Pro-Cys-Gly-Ala-Ala-Leu-Gly-Arg-Gly-Lys-Ala

[a]Residues in uppercase are synthesized in the pathway catalyzed by the operon's gene products.

Source: Yanofsky, C., *Nature* **289,** 753 (1981).

gene no faster than about once per second. If *E. coli* contained only one copy of each of the three types of rRNA genes (those specifying the so-called 23S, 16S, and 5S rRNAs; Section 32-3A), fast-growing cells could synthesize no more than ~1200 ribosomes during their cell division cycle. However, *the E. coli genome contains seven separately located rRNA operons, all of which contain one nearly identical copy of each type of rRNA gene.* Moreover, rapidly growing cells contain multiple copies of their replicating chromosomes (Section 30-3C), thereby accounting for the observed rRNA synthesis rate.

Cells have the remarkable ability to coordinate the rates at which their thousands of components are synthesized. For example, *E. coli* adjust their ribosome content to match the rate at which they can synthesize proteins under the prevailing growth conditions. The rate of rRNA synthesis is therefore proportional to the rate of protein synthesis. One mechanism by which this occurs is known as the **stringent response:** *A shortage of any species of amino acid-charged tRNA (usually a result of "stringent" or poor growth conditions) that limits the rate of protein synthesis triggers a sweeping metabolic readjustment.* A major facet of this change is an abrupt 10- to 20-fold reduction in the rate of rRNA and tRNA synthesis. This **stringent control,** moreover, depresses numerous metabolic processes (including DNA replication and the biosynthesis of carbohydrates, lipids, nucleotides, proteoglycans, and glycolytic intermediates) while stimulating others (such as amino acid biosynthesis). The cell is thereby prepared to withstand nutritional deprivation.

a. (p)ppGpp Mediates the Stringent Response

The stringent response is correlated with a rapid intracellular accumulation of two unusual nucleotides, ***ppGpp*** *and* ***pppGpp*** *[known collectively as* ***(p)ppGpp****], and their prompt decay when amino acids become available.* The observation that mutants, designated *relA*⁻, which do not exhibit the stringent response (they are said to have **relaxed control**) lack (p)ppGpp suggests that these substances mediate the stringent response. This idea was corroborated by *in vitro* studies demonstrating, for example, that (p)ppGpp inhibits the transcription of rRNA genes but stimulates the transcription of the *trp* and *lac* operons as does the stringent response *in vivo.* Apparently, (p)ppGpp acts by somehow altering RNAP's promoter specificity at stringently controlled operons, a hypothesis that is supported by the isolation of RNAP mutants that exhibit reduced responses to (p)ppGpp. In addition, (p)ppGpp causes an increased frequency of pausing in RNAPs engaged in elongation, thereby reducing the rate of transcription.

The protein encoded by the wild-type *relA* gene, named **stringent factor (RelA),** catalyzes the reaction

$$\text{ATP} + \text{GTP} \rightleftharpoons \text{AMP} + \text{pppGpp}$$

and, to a lesser extent,

$$\text{ATP} + \text{GDP} \rightleftharpoons \text{AMP} + \text{ppGpp}$$

However, several ribosomal proteins convert pppGpp to ppGpp so that ppGpp is the stringent response's usual effector. Stringent factor is only active in association with a ribosome that is actively engaged in translation. (p)ppGpp synthesis occurs when a ribosome binds its mRNA-specified but uncharged tRNA (lacking an appended amino acid residue). The binding of a specified and charged tRNA greatly reduces the rate of (p)ppGpp synthesis. *The ribosome apparently signals the shortage of an amino acid by stimulating the synthesis of (p)ppGpp which, acting as an intracellular messenger, influences the rates at which a great variety of operons are transcribed.*

(p)ppGpp degradation is catalyzed by the *spoT* gene product. The *spoT*⁻ mutants show a normal increase in (p)ppGpp level on amino acid starvation but an abnormally slow decay of (p)ppGpp to basal levels when amino acids again become available. The *spoT*⁻ mutants therefore exhibit a sluggish recovery from the stringent response. *The (p)ppGpp level is apparently regulated by the countervailing activities of stringent factor and the spoT gene product.*

4 ■ POSTTRANSCRIPTIONAL PROCESSING

The immediate products of transcription, the **primary transcripts,** are not necessarily functional entities. In order to acquire biological activity, many of them must be specifically altered in several ways: (1) by the exo- and endonucleolytic removal of polynucleotide segments; (2) by appending nucleotide sequences to their 3′ and 5′ ends; and (3) by the modification of specific nucleosides. The three major classes of RNAs, mRNA, rRNA, and tRNA, are altered in different ways in prokaryotes and in eukaryotes. In this section we shall outline these **posttranscriptional modification** processes.

A. *Messenger RNA Processing: Caps, Tails, and Splicing*

In prokaryotes, most primary mRNA transcripts function in translation without further modification. Indeed, as we have seen, ribosomes in prokaryotes usually commence translation on nascent mRNAs. In eukaryotes, however, mRNAs are synthesized in the cell nucleus whereas translation occurs in the cytosol. Eukaryotic mRNA transcripts can therefore undergo extensive posttranscriptional processing while still in the nucleus.

a. Eukaryotic mRNAs Are Capped

Eukaryotic mRNAs have a peculiar enzymatically appended ***cap structure*** *consisting of a 7-methylguanosine residue joined to the transcript's initial (5′) nucleoside via a 5′–5′ triphosphate bridge (Fig. 31-43).* The cap, which is added to the growing transcript before it is ~30 nucleotides long, defines the eukaryotic translational start site (Section 32-3C). A cap may be $O^{2'}$-methylated at the transcript's leading nucleoside (**cap-1,** the predominant cap in multi-

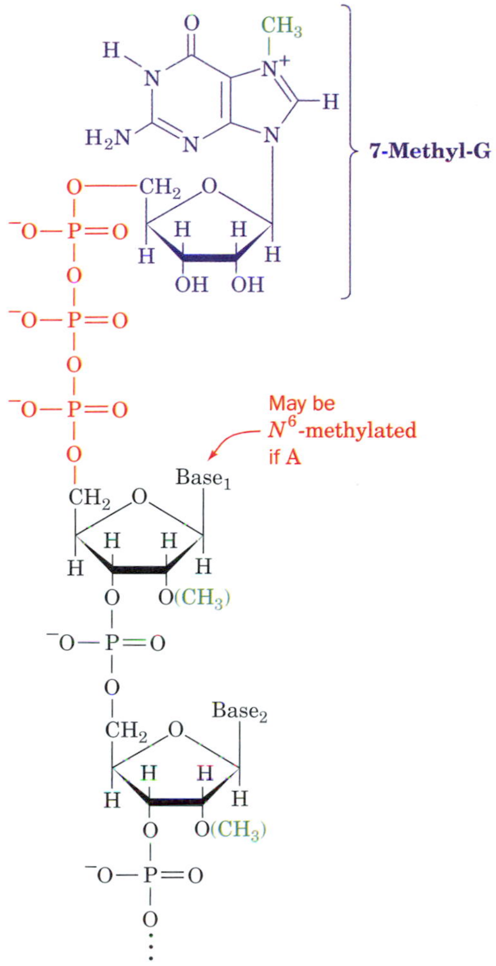

FIGURE 31-43 The structure of the 5′ cap of eukaryotic mRNAs. It is known as cap-0, cap-1, or cap-2, respectively, if it has no further modifications, if the leading nucleoside of the transcript is $O^{2'}$-methylated, or if its first two nucleosides are $O^{2'}$-methylated.

cellular organisms), at its first two nucleosides **(cap-2),** or at neither of these positions (**cap-0,** the predominant cap in unicellular eukaryotes). If the leading nucleoside is adenosine (it is usually a purine), it may also be N^6-methylated.

Capping involves several enzymatic reactions: (1) the removal of the leading phosphate group from the mRNA's 5′ terminal triphosphate group by an **RNA triphosphatase;** (2) the guanylation of the mRNA by **capping enzyme,** which requires GTP and yields the 5′–5′ triphosphate bridge and PP_i; (3) the methylation of guanine by **guanine-7-methyltransferase** in which the methyl group is supplied by *S*-adenosylmethionine (SAM); and possibly (4) the O2′ methylation of the mRNA's first and perhaps its second nucleotide by a SAM-requiring **2′-*O*-methyltransferase.** Both the capping enzyme and the guanine-7-methyltransferase bind to RNAP II's phosphorylated CTD (Section 31-2E). Hence it is likely that capping marks the completion of RNAP II's switch from transcription initiation to elongation.

b. Eukaryotic mRNAs Have Poly(A) Tails

Eukaryotic mRNAs, in contrast to those of prokaryotes, are invariably monocistronic. Yet, the sequences signaling transcriptional termination in eukaryotes have not been identified. This is largely because the termination process is imprecise; that is, the primary transcripts of a given structural gene have heterogeneous 3′ sequences. Nevertheless, mature eukaryotic mRNAs have well-defined 3′ ends; *almost all of them in mammals have 3′-poly(A) tails of ~250 nucleotides (~80 in yeast).* The poly(A) tails are enzymatically appended to the primary transcripts in two reactions that are mediated by a 500- to 1000-kD complex that consists of at least six proteins:

1. A transcript is cleaved to yield a free 3′-OH group at a specific site that is 15 to 25 nucleotides past an AAUAAA sequence and within 50 nucleotides before a U-rich or G + U-rich sequence. The AAUAAA sequence is highly conserved in higher eukaryotes (but not yeast) in which its mutation abolishes cleavage and polyadenylation. The precision of the cleavage reaction has apparently eliminated the need for accurate transcriptional termination. Nevertheless, the identity of the endonuclease that cleaves the RNA is uncertain although **cleavage factors I** and **II (CFI** and **CFII)** are required for this process.

2. The poly(A) tail is subsequently generated from ATP through the stepwise action of **poly(A) polymerase (PAP).** This enzyme, which by itself only weakly binds RNA, is recruited by **cleavage and polyadenylation specificity factor (CPSF)** on this heterotetramer's recognition of the AAUAAA sequence, which it does with almost no tolerance for sequence variation. The downstream G + U-rich element is recognized by the heterotrimer **cleavage stimulation factor (CstF),** which increases the affinity with which CPSF binds the AAUAAA sequence. However, once the poly(A) tail has grown to ~10 residues, the AAUAAA sequence is no longer required for further chain elongation. This suggests that CPSF becomes disengaged from its recognition site in a manner reminiscent of the way σ factor is released from the transcriptional initiation site once the elongation of prokaryotic mRNA is under way (Section 31-2C). The final length of the poly(A) tail is controlled by **poly(A)-binding protein II (PAB II),** multiple copies of which bind to successive segments of poly(A). PAB II also increases the processivity of PAP.

Both CPSF and CstF bind to the phosphorylated RNAP II CTD (Section 31-2E); deleting the CTD inhibits polyadenylation. Evidently, the CTD couples polyadenylation to transcription.

PAP is a template-independent RNA polymerase that elongates an mRNA primer with a free 3′-OH group. The X-ray structures of yeast and bovine PAPs in complex with 3′-dATP (cordycepin), respectively determined by Andrew Bohm and Sylvie Doublié, reveal that these closely similar monomeric proteins each consist of three domains that enclose an ~20 × 25 Å U-shaped cleft (Fig.

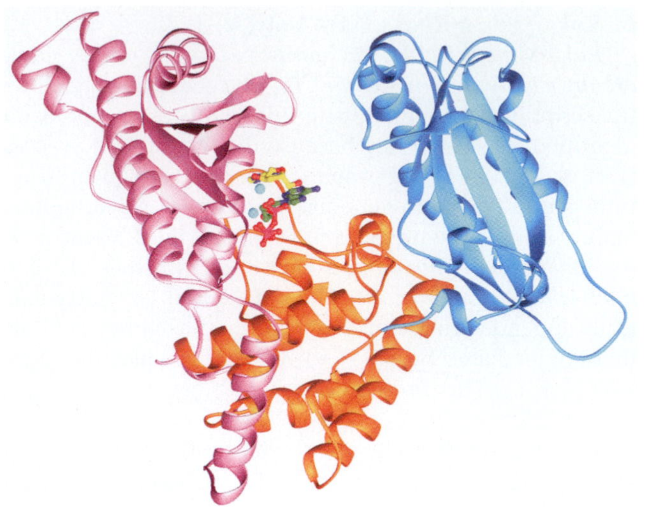

FIGURE 31-44 X-Ray structure of the 568-residue yeast poly(A) polymerase (PAP) in complex with two molecules of 3′-dATP. The N-terminal (palm) domain is lavender, the central (fingers) domain is orange, and the C-terminal domain is light blue. The 3′-dATP molecules are drawn in ball-and-stick form and colored according to atom type (incoming nucleotide C green, mRNA 3′ end C yellow, N blue, O red, and P magenta). The two Mn^{2+} ions that are bound at the active site are represented by cyan balls. [Based on an X-ray structure by Andrew Bohm, Tufts University School of Medicine. PDBid 1FA0.]

31-44). Hence they superficially appear to have the hand-like domain arrangement of the DNA polymerases (Section 30-2A). Indeed, PAP's N-terminal domain, which contains the enzyme's active site, is structurally homologous to the palm domain of DNA polymerase β, although it forms one side of the cleft rather than its base. PAP's central domain, which forms the base of the cleft, is functionally analogous to the polymerase fingers domain in that it interacts with the β and γ phosphates of the incoming nucleotide. However, the C-terminal domain shows no resemblance to a thumb domain. Rather, it is topologically similar to the **RNA-recognition motif (RRM)** that occurs in >200 different RNA-binding proteins (see below). Yeast PAP binds two molecules of 3′-dATP: One occupies the position of the incoming nucleotide and the other, whose triphosphate group is not observed, is presumed to mimic the 3′ end of the mRNA primer on the basis of its similarity to the X-ray structure of DNA polymerase β. The incoming base interacts with the protein so as to differentiate adenine from other bases, whereas in templated polymerases, the incoming base only contacts the template (Section 30-2A).

In vitro studies indicate that a poly(A) tail is not required for mRNA translation. Rather, the observations that an mRNA's poly(A) tail shortens as it ages in the cytosol and that unadenylated mRNAs have abbreviated cytosolic lifetimes suggest that poly(A) tails have a protective role. In fact, the only mature mRNAs that lack poly(A) tails, those of histones (which, with few exceptions, lack the AAUAAA cleavage–polyadenylation signal), have lifetimes of <30 min in the cytosol, whereas most other mRNAs last hours or days. The poly(A) tails are specifically complexed in the cytosol by **poly(A) binding protein (PABP;** not related to PAB II), which organizes poly(A)-bearing mRNAs into ribonucleoprotein particles. PABP is thought to protect mRNA from degradation as is suggested, for example, by the observation that the addition of PABP to a cell-free system containing mRNA and mRNA-degrading nucleases greatly reduces the rate at which the mRNAs are degraded and the rate at which their poly(A) tails are shortened.

All known PABPs contain four tandem and highly conserved RNA-recognition motifs (RRMs) followed by a less conserved Pro-rich C-terminal segment of variable length. A variety of evidence suggests that PABP's first two RRMs support most of the biochemical functions of full length PABP. The X-ray structure of the first two RRMs of human PABP (RRM1/2; the N-terminal 190 residues of this 636-residue protein) in complex with A_{11}, determined by Stephen Burley, reveals that RRM1/2 forms a continuous trough-shaped surface in which the poly(A) binds in an extended conformation via interactions with conserved residues (Fig. 31-45). Each RRM, as also seen in the structures of a variety of other RNA-binding proteins, consists of a compact globule made of a 4-stranded antiparallel sheet that forms the RNA-binding surface backed by two helices.

c. Eukaryotic Genes Consist of Alternating Expressed and Unexpressed Sequences

The most striking difference between eukaryotic and prokaryotic structural genes is that the coding sequences of most eukaryotic genes are interspersed with unexpressed regions. Early investigations of eukaryotic structural gene transcription found, quite surprisingly, that primary transcripts are highly heterogeneous in length (from ~2000 to well over 20,000 nucleotides) and are much larger than is expected from the known sizes of eukaryotic proteins. Rapid labeling experiments demonstrated that little of this so-called **heterogeneous nuclear RNA (hnRNA)** is ever transported to the cytosol; most of it is quickly turned over (degraded) in the nucleus. Yet, the hnRNA's 5′ caps and 3′ tails eventually appear in cytosolic mRNAs. *The straightforward explanation of these observations, that* ***pre-mRNAs*** *are processed by the excision of internal sequences, seemed so bizarre that it came as a great surprise in 1977 when Phillip Sharp and Richard Roberts independently demonstrated that this is actually the case.* In fact, pre-

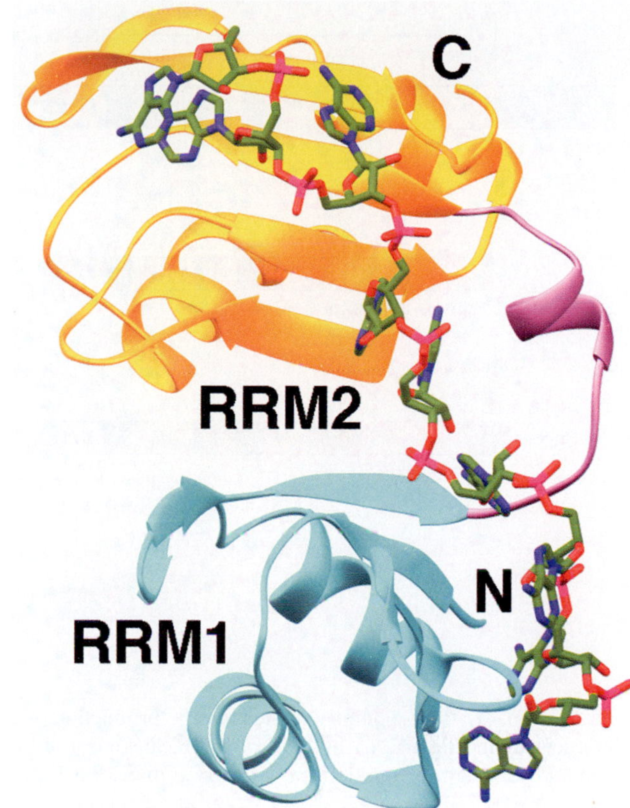

FIGURE 31-45 X-Ray structure of the N-terminal two RNA-recognition motifs (RRMs) of human PABP in complex with A_{11}. RRM1 is cyan, RRM2 is gold, and their linking segment is lavender. The poly(A), only nine of whose nucleotides are observed, is drawn in stick form with C green, N blue, O red, and P magenta. [Based on an X-ray structure by Stephen Burley, The Rockefeller University. PDBid 1CVJ.]

mRNAs typically contain eight noncoding **intervening sequences (introns)** whose aggregate length averages four to ten times that of its flanking **expressed sequences (exons).** This situation is graphically illustrated in Fig. 31-46, which is an electron micrograph of chicken **ovalbumin** mRNA hybridized to the antisense strand of the ovalbumin gene (ovalbumin is the major protein component of egg white).

Exons have lengths that range up to 17,106 nt (in the gene encoding the 29,926-residue muscle protein **titin,** the largest known single-chain protein; Section 35-3A) but with most <300 nt (and averaging 150 nt in humans). Introns, in contrast, are usually much longer, with lengths averaging ~3500 nt and as high 2.4 million nt (in the gene encoding the muscle protein dystrophin; Section 35-3A) with no obvious periodicity. Moreover, the corresponding introns from genes in two vertebrate species can vary extensively in both length and sequence so as to bear little resemblance to one another. The number of introns in a gene averages 7.8 in the human genome and varies from none to 234 (with the latter number also occurring in the gene encoding titin).

The formation of eukaryotic mRNA begins with the transcription of an entire structural gene, including its in-

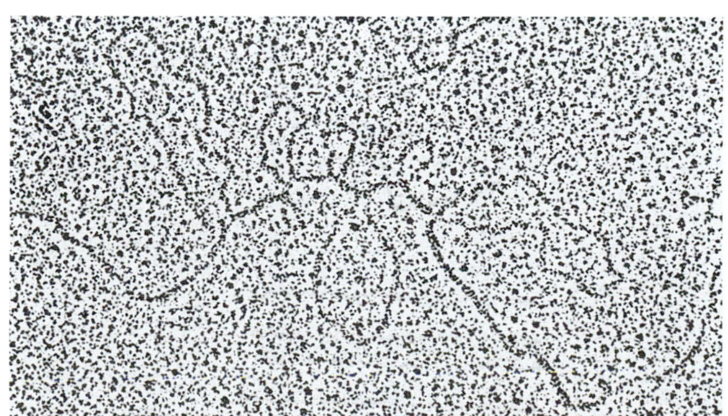

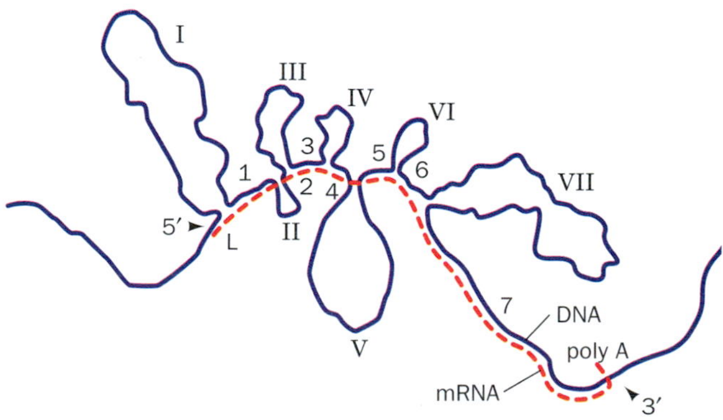

FIGURE 31-46 An electron micrograph and its interpretive drawing of a hybrid between the antisense strand of the chicken ovalbumin gene and its corresponding mRNA. The complementary segments of the DNA (*purple line in drawing*) and mRNA (*red dashed line*) have annealed to reveal the exon positions (*L*, 1–7). The looped-out segments (I–VII), which have no complementary sequences in the mRNA, are the introns. [From Chambon, P., *Sci. Am.* **244**(5), 61 (1981).]

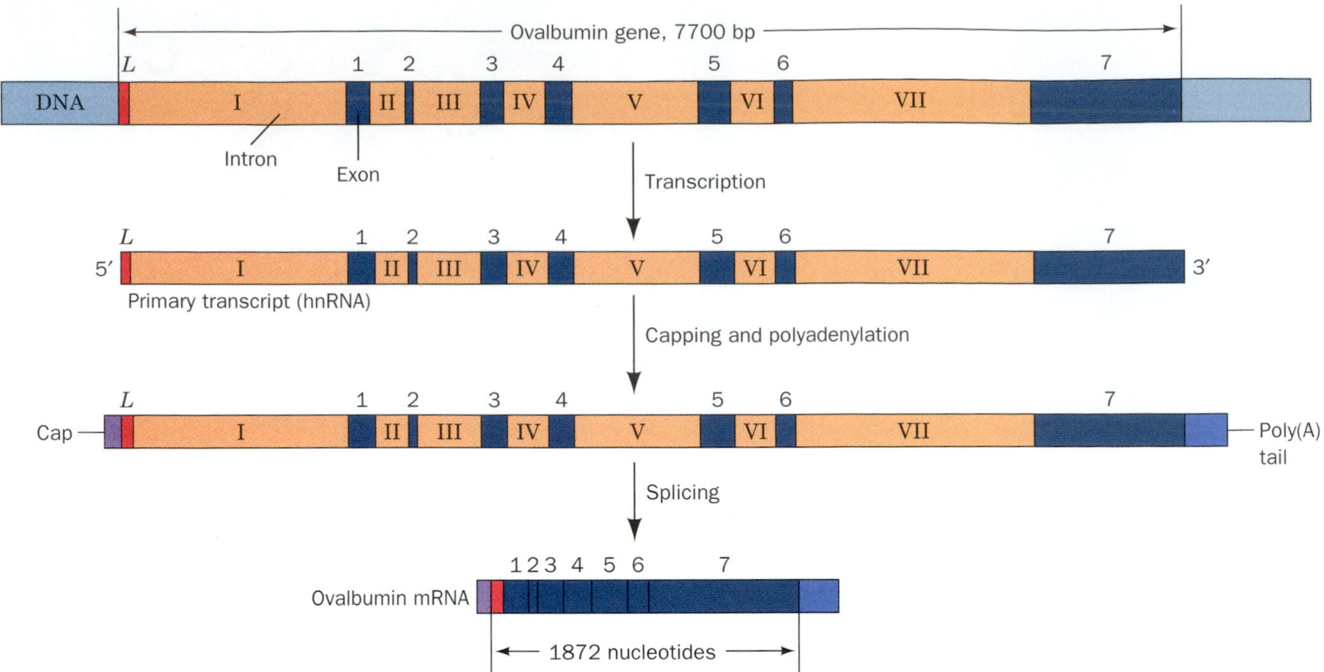

FIGURE 31-47 The sequence of steps in the production of mature eukaryotic mRNA as shown for the chicken ovalbumin gene. Following transcription, the primary transcript is capped and polyadenylated. The introns are then excised and the exons spliced together to form the mature mRNA. However, splicing may also occur cotranscriptionally.

trons, to form pre-mRNA (Fig. 31-47). Then, following capping, the introns are excised and their flanking exons are connected, a process called **gene splicing** or just **splicing,** that often occurs cotranscriptionally. *The most striking aspect of gene splicing is its precision; if one nucleotide too few or too many were excised, the resulting mRNA could not be translated properly (Section 32-1B). Moreover, exons are never shuffled, their order in the mature mRNA is exactly the same as that in the gene from which it is derived.*

d. Exons Are Spliced in a Two-Stage Reaction

Sequence comparisons of exon–intron junctions from a diverse group of eukaryotes indicate that they have a high degree of homology (Fig. 31-48, including, as Richard Breathnach and Pierre Chambon first pointed out, *an invariant GU at the intron's 5′ boundary and an invariant AG at its 3′ boundary. These sequences are necessary and sufficient to define a splice junction:* Mutations that alter the sequences interfere with splicing, whereas mutations that change a nonjunction to a consensus-like sequence can generate a new splice junction.

Investigations of both cell free and *in vivo* splicing systems by Argiris Efstradiadis, Tom Maniatis, Michael Rosbash, and Sharp established that intron excision occurs via two transesterification reactions that are remarkably similar from yeast to humans (Fig. 31-49):

1. The formation of a 2′,5′-phosphodiester bond between an intron adenosine residue and its 5′-terminal phosphate group with the concomitant liberation of the 5′ exon's 3′-OH group. *The intron thereby assumes a novel* ***lariat structure.*** The adenosine residue at the lariat branch has been identified in yeast as the last A in the highly conserved sequence UACUAAC and in vertebrates as the A in the equivalent but more permissive sequence YNCURAY [where R represents purines (A or G), Y represents pyrimidines (C or U), and N represents any nucleotide]. In yeast and vertebrates, the branch point A occurs ~50 and 18 to 40 residues upstream of the associated 3′ splice site, respectively. In yeast, which have relatively few introns, mutations that change this branch point A residue abolish splicing at that site. However, in higher eukaryotes, the mutation or deletion of a branch site often activates a so-called

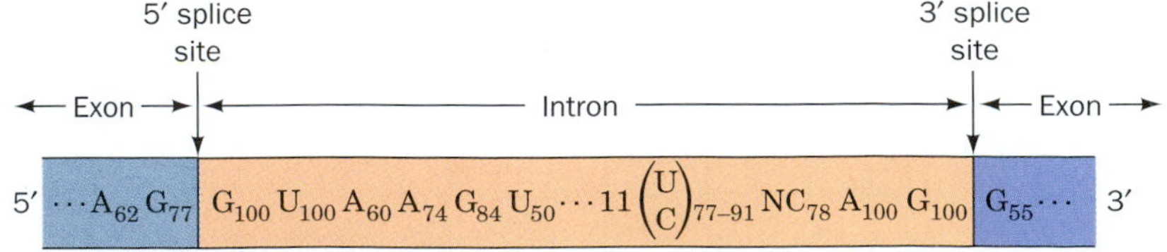

FIGURE 31-48 The consensus sequence at the exon–intron junctions of vertebrate pre-mRNAs. The subscripts indicate the percentage of pre-mRNAs in which the specified base(s) occurs. Note that the 3′ splice site is preceded by a tract of 11 predominantly pyrimidine nucleotides. [Based on data from Padgett, R.A., Grabowski, P.J., Konarska, M.M., Seiler, S.S., and Sharp, P.A., *Annu. Rev. Biochem.* **55,** 1123 (1986).]

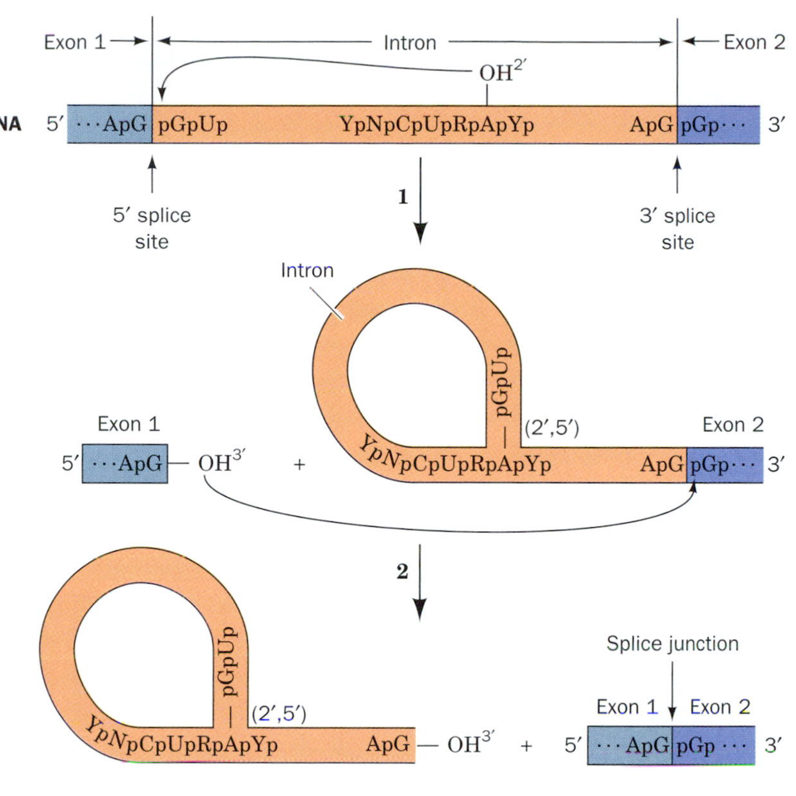

FIGURE 31-49 The sequence of transesterification reactions that splice together the exons of eukaryotic pre-mRNAs. The exons and introns are drawn in blue and orange, and R and Y represent purine and pyrimidine residues. **(1)** The 2′-OH group of a specific intron A residue nucleophilically attacks the 5′-phosphate at the 5′ intron boundary to yield an unusual 2′,5′-phosphodiester bond and thus form a lariat structure. **(2)** The liberated 3′-OH group forms a 3′,5′-phosphodiester bond with the 5′ terminal residue of the 3′ exon, thereby splicing the two exons together and releasing the intron in lariat form with a free 3′-OH.

cryptic branch site that is also near the 3′ splice site. Evidently, the branch site functions to identify the nearest 3′ splice site as a target for linkage to the 5′ splice site.

2. The now free 3′-OH group of the 5′ exon forms a phosphodiester bond with the 5′-terminal phosphate of the 3′ exon yielding the spliced product and releasing the intron lariat with a free 3′-OH group. The intron lariat is then debranched (linearized) and, *in vivo*, is rapidly degraded. Mutations that alter the conserved AG at the 3′ splice site block this second step, although they do not interfere with lariat formation.

Note that the splicing process proceeds without free energy input; its transesterification reactions preserve the free energy of each cleaved phosphodiester bond through the concomitant formation of a new one.

The sequences required for splicing are the short consensus sequences at the 3′ and 5′ splice sites and at the branch site. Nevertheless, these sequences are poorly conserved. However, other short sequence elements within exons that are known as **exonic sequence enhancers (ESEs)** also play important roles in splice site selection although their characteristics are poorly understood (even highly sophisticated computer programs are only ~50% successful in predicting actual splice sites over apparently equally good candidates that are not). In contrast, large portions of most introns can be deleted without impeding splicing.

e. Some Eukaryotic Genes Are Self-Splicing

It is now recognized that there are eight distinct types of introns, seven of which occur in eukaryotes (Table 31-4). **Group I introns** occur in the nuclei, mitochondria, and

TABLE 31-4. Types of Introns

Intron Type	Where Found
GU–AG introns	Eukaryotic nuclear pre-mRNA
AU–AC introns	Eukaryotic nuclear pre-mRNA
Group I	Eukaryotic nuclear pre-mRNA, organelle RNAs, a few bacterial RNAs
Group II	Organelle RNAs, a few prokaryotic RNAs
Group III	Organelle RNAs
Twintrons (composites of two and/or more group II or III introns)	Organelle RNAs
Pre-tRNA introns	Eukaryotic nuclear pre-tRNAs
Archaeal introns	Various RNAs

Source: Brown, T.A., *Genomes* (2nd ed.), Wiley-Liss, *p.* 287 (2002).

chloroplasts of diverse eukaryotes (but not vertebrates), and even in some bacteria. Thomas Cech's study of how group I introns are spliced in the ciliated protozoan *Tetrahymena thermophila* led to an astonishing discovery: *RNA can act as an enzyme. When the isolated pre-rRNA of this organism is incubated with guanosine or a free guanine nucleotide (GMP, GDP, or GTP), but in the absence of protein, its single 421-nucleotide intron excises itself and splices together its flanking exons; that is, this pre-rRNA is self-splicing.* The three-step reaction sequence of this process (Fig. 31-50) resembles that of mRNA splicing:

1. The 3′-OH group of the guanosine forms a phosphodiester bond with the intron's 5′ end, liberating the 5′ exon.

2. The 3′-terminal OH group of the newly liberated 5′ exon forms a phosphodiester bond with the 5′-terminal phosphate of the 3′ exon, thereby splicing together the two exons and releasing the intron.

3. The 3′-terminal OH group of the intron forms a phosphodiester bond with the phosphate of the nucleotide 15 residues from the intron's 5′ end, yielding the 5′-terminal fragment with the remainder of the intron in cyclic form.

This self-splicing process consists of a series of transesterifications and therefore does not require free energy input. Cech further established the enzymatic properties of the *Tetrahymena* intron, which presumably stem from its three-dimensional structure, by demonstrating that it catalyzes the *in vitro* cleavage of poly(C) with an enhancement factor of 10^{10} over the rate of spontaneous hydrolysis. Indeed, this RNA catalyst even exhibits Michaelis–Menten kinetics (K_M = 42 μ*M* and k_{cat} = 0.033 s^{-1} for C_5). Such RNA enzymes have been named **ribozymes.**

Although the idea that an RNA can have enzymatic properties may seem unorthodox, *there is no fundamental reason why an RNA, or any other macromolecule, cannot have catalytic activity* (recall that it was likewise once generally accepted that nucleic acids lack the complexity to carry hereditary information; Section 5-2). Of course, in order to be an efficient catalyst, a macromolecule must be able to assume a stable structure but, as we shall see below and in Sections 32-2B and 32-3A, RNAs, including tRNAs and rRNAs, can do so. [Synthetic ssDNAs are also known to have catalytic properties although such "deoxyribozymes" are unknown in biology.]

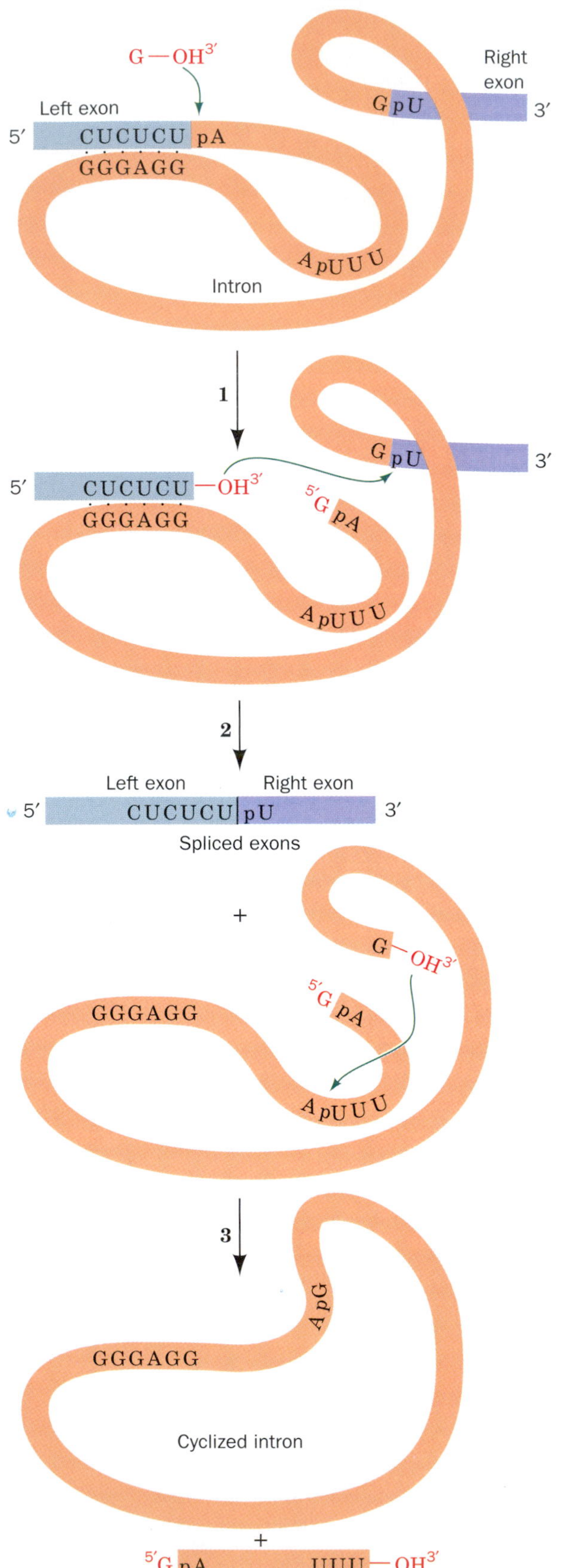

FIGURE 31-50 The sequence of reactions in the self-splicing of *Tetrahymena* group I intron. (1) The 3′-OH group of a guanine nucleotide attacks the intron's 5′-terminal phosphate so as to form a phosphodiester bond and release the 5′ exon. **(2)** The newly generated 3′-OH group of the 5′ exon attacks the 5′-terminal phosphate of the 3′ exon, thereby splicing the two exons and releasing the intron. **(3)** The 3′-OH group of the intron attacks the phosphate of the nucleotide that is 15 residues from the 5′ end so as to cyclize the intron and release its 5′-terminal fragment. Throughout this process, the RNA maintains a folded, internally hydrogen bonded conformation that permits the precise excision of the intron.

The **group II introns,** which occur in the mitochondria of fungi and plants and comprise the majority of the introns in chloroplasts, are also self-splicing. They react via a lariat intermediate and do not utilize an external nucleotide, a process that resembles the splicing of nuclear pre-mRNAs (Fig. 31-49). We shall see below that nuclear pre-mRNA splicing is mediated by complex ribonucleoprotein particles known as **spliceosomes.** The chemical similarities of the pre-mRNA and group II intron splicing reactions therefore suggest that *spliceosomes are ribozymal systems whose RNA components have evolved from primordial self-splicing RNAs and that their protein components serve mainly to fine-tune ribozymal structure and function.* Similarly, the RNA components of ribosomes, which are two-thirds RNA and one-third protein, clearly have a catalytic function in addition to the structural and recognition roles traditionally attributed to them (Section 32-3). Thus, the observations that nucleic acids but not proteins can direct their own synthesis, that cells contain batteries of protein-based enzymes for manipulating DNA but few for processing RNA, and that many coenzymes are ribonucleotides (e.g., ATP, NAD^+, and CoA), led to the hypothesis that *RNAs were the original biological catalysts in precellular times (the so-called* ***RNA world****) and that the chemically more versatile proteins were relative latecomers in macromolecular evolution (Section 1-5C).*

f. The X-Ray Structures of a Group I Ribozyme

The sequence of the *Tetrahymena* group I intron, together with phylogenetic comparisons, indicates that it contains nine double helical segments that are designated P1 through P9 (Fig. 31-51*a*). Such analysis further indicates that the conserved catalytic core of group I introns consists of sets of coaxially stacked helices interspersed with internal loops that are organized into three domains, P1-P2, P4-P6 and P3-P9. Chemical protection experiments suggest that the isolated P4-P6 domain of the *Tetrahymena*

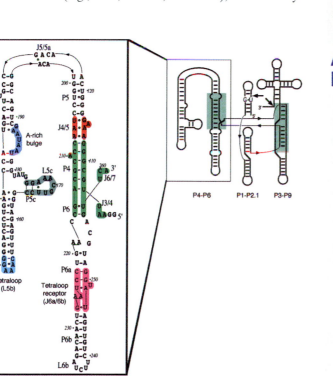

(*a*)

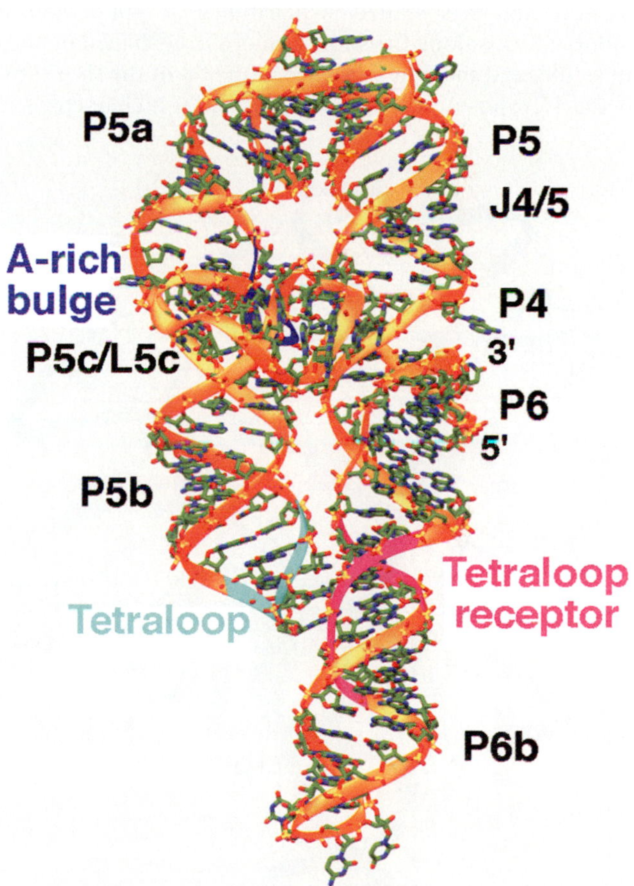

(*b*)

FIGURE 31-51 The self-splicing group I intron from *Tetrahymena thermophila.* (*a*) The secondary structure of the entire 413-nt intron is shown on the right with its phylogenetically conserved catalytic core shaded in blue-gray. Helical regions are numbered sequentially along the intron's sequence with P for base *p*aired segment, J for *j*oining region, and L for *l*oop; the arrows indicate the 5′ and 3′ splice sites. The sequence of the independently folded 160-nt P4-P6 domain is enlarged at left. Segments of functional interest are highlighted as follows: the GAAA tetraloop is cyan; the conserved tetraloop receptor is magenta; the A-rich bulge, which is required for the proper folding of P4-P6, is dark blue; segments of the conserved core are light green and red; and P5c is blue-gray. Watson–Crick and non-Watson–Crick base pairing interactions are represented by short horizontal lines and small circles. (*b*) The X-ray structure of P4-P6 viewed as in Part *a.* The structure is drawn in stick form with C green, N blue, O red, and P yellow. The sugar–phosphate backbone is traced by a ribbon that is colored as in Part *a* for the tetraloop, tetraloop receptor, and A-rich bulge and is gold elsewhere. Note the numerous interactions between the various segments of this RNA molecule. [Part *a* based on a drawing by and Part *b* based on an X-ray structure by Jennifer Doudna, Yale University. PDBid 1GID.] **See the Interactive Exercises**

group I intron folds as an independent unit. Indeed, when the P4-P6 domain is combined with the remaining portion of the *Tetrahymena* intron, it forms a catalytically active complex.

Jennifer Doudna and Cech determined the X-ray structure of the 160-nt P4-P6 domain of the *Tetrahymena* group I intron (Fig. 31-51*b*). When this structure was reported in 1997, it was over twice the size of the largest RNA molecule whose structure had previously been determined, that of a 76-nt tRNA (Section 32-2B). For the most part, the P4-P6 domain consists of two coaxially stacked sets of A-RNA-like helices, one of 29 bp and the other of 23 bp. Its dimensions are around $25 \times 50 \times 110$ Å, the first two dimensions being the widths of one A-RNA helix and two side-by-side A-RNA helices.

P4-P6 was the first RNA of known structure that is large enough to exhibit side-by-side helical packing. Of particular note are its so-called A-rich bulge, a 7-nt sequence about halfway along the short arm of the U-shaped macromolecule, and the 6-nt sequence at the tip of the short arm of the U, whose central GAAA assumes a characteristic conformation known as a **tetraloop.** In both of these substructures, the bases are splayed outward so as to stack on each other and to associate in the minor groove of specific segments of the long arm of the U via hydrogen bonding interactions involving ribose residues as well as bases. In the interaction involving the A-rich bulge, the close packing of phosphates from adjacent helices is mediated by hydrated Mg^{2+} ions. Throughout this structure, the defining characteristic of RNA, its 2′-OH group, is both a donor and an acceptor of hydrogen bonds to phosphates, bases, and other 2′-OH groups.

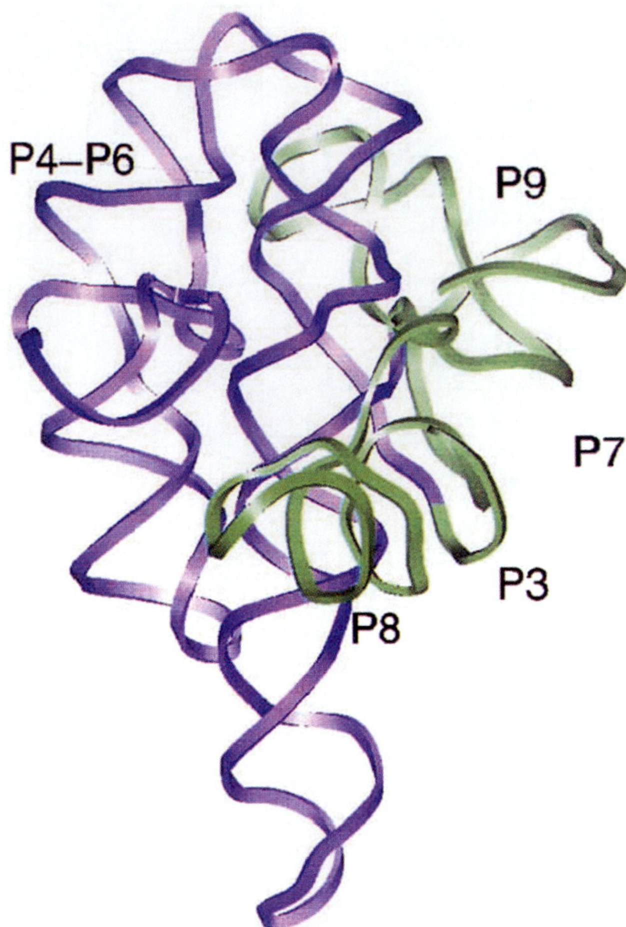

FIGURE 31-52 Low resolution X-ray structure of the *Tetrahymena* group I intron encompassing its P4-P6 (*violet*) and its P3-P9 (*green*) domains. The ribozyme, which is represented by a ribbon drawn through its phosphate groups, is oriented such that its P4-P6 domain is viewed similarly to that in Fig. 31-51*b*. [Courtesy of Thomas Cech, University of Colorado. PDBid 1GRZ.]

Subsequently, Cech designed a 247-nt RNA that encompasses both the P4-P6 and the P3-P9 domains of the *Tetrahymena* group I intron, with the addition of a 3′ G, which functions as an internal guanosine nucleophile. This RNA is catalytically active; it binds the P1-P2 domain via tertiary interactions and, with the assistance of its 3′ G, cleaves P1 in a manner similar to the intact intron.

The X-ray structure of this ribozyme was determined to a resolution of 5 Å (Fig. 31-52). At this low resolution, the sugar–phosphate backbone can be traced and stacked bases often appear as continuous tubes of electron density, but such atomic-level features as hydrogen bonding interactions are not apparent. Within this structure the P4-P6 domain appears to be essentially unchanged from that of the domain alone (Fig. 31-51*b*). Moreover, the structure is compatible with a large body of biochemical data. The close packing of the two domains forms a shallow cleft that appears capable of binding the short helix that contains the 5′ splice site. The intron's guanosine binding site is located on P7, which deviates significantly from A-form geometry in a way that provides a snug binding site for the guanosine substrate. Evidently, this ribozyme is largely preorganized for substrate binding and catalysis, much as are protein enzymes.

g. Hammerhead Ribozymes Catalyze an In-Line Nucleophilic Attack

The simplest and perhaps best characterized ribozymes, which are embedded in the RNAs of certain plant viruses, are named **hammerhead ribozymes** due to the superficial resemblance of their secondary structures, as customarily laid out, to a hammer (Fig. 31-53*a*). Hammerhead ribozymes have three duplex stems and a conserved core of two nonhelical segments.

Hammerhead ribozymes catalyze a transesterification reaction in which the 3′,5′-phosphodiester bond between nucleotides C-17 and A-1.1 is cleaved so as to yield a cyclic 2′,3′-phosphodiester on C-17 with inversion of configuration about the P atom, together with a free 5′-OH on A-1.1, much like the intermediate product in the RNA hydrolysis reaction catalyzed by RNase A (Section 15-1A). This suggests that the reaction proceeds via an "in-line" mechanism such as that diagrammed in Fig. 16-6*b* in which the transition state forms a trigonal bipyramidal intermediate in which the attacking nucleophile, the 2′-OH group (Y in Fig. 16-6*b*), and the leaving group, which forms the free 5′-OH group (X in Fig. 16-6*b*), occupy the axial positions. Under physiological conditions the reaction requires the

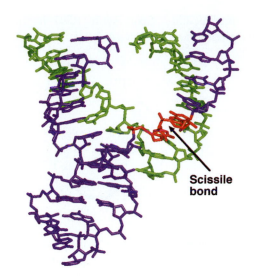

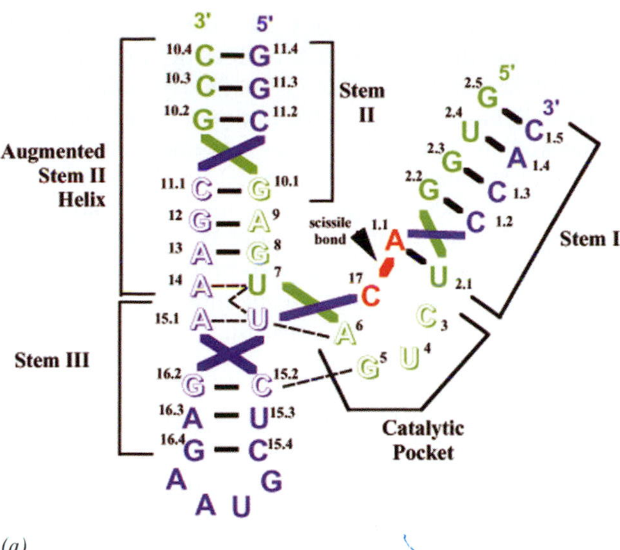

Scissile bond

(b)

FIGURE 31-53 X-Ray structure of a hammerhead ribozyme. (*a*) The sequence and schematic structural representation of the ribozyme drawn with its 16-nt enzyme strand green, with its 25-nt substrate strand blue, but with the nucleotides spanning its cleavage site (C-17 and A-1.1) red. Essential and highly conserved nucleotides are represented by hollow letters, and the universal numbering scheme is provided. Watson–Crick as well as two G · A Hoogsteen base pairing interactions (Section 29-2C) are shown as single black dashes, and single hydrogen bonds between bases or between bases and backbone riboses are indicated by black dashed lines. (*b*) A stick model of the ribozyme in its ground state colored as in Part *a*. [Courtesy of William Scott, University of California at Santa Cruz, PDBid 1MME.] **See the Interactive Exercises**

presence of a divalent cation, preferentially Mg^{2+} or Mn^{2+}, in roughly millimolar concentrations. However, the ribozyme is active in the absence of divalent cations when the concentration of monovalent cations is very high (e.g., 4M Na^+, Li^+, or NH_4^+), which suggests that metal ions serve a structural rather than a catalytic role in the hammerhead ribozyme.

In order to crystallize a hammerhead ribozyme in its native conformation without it self-destructing, William Scott and Aaron Klug synthesized it with C-17 at its cleavage site replaced by 2′-methoxy-C, thereby blocking the formation of the 2′,3′-cyclic reaction product at this position. The X-ray structure of this catalytically inactive ribozyme (Fig. 31-53*b*) reveals that it has the expected secondary structure of three A-form helical segments although its overall shape more closely resembles a wishbone than a hammer. The nucleotides in the helical stems form normal Watson–Crick base pairs, whereas nucleotides U-7 through A-9 form non-Watson–Crick base pairs with nucleotides G-12 through A-14 in which ribose oxygens participate as both hydrogen bond donors and acceptors. This explains the observations that most helical positions can be occupied by any Watson–Crick base pair but that few core bases can be changed without reducing ribozymal activity. The absolutely conserved CUGA tetranucleotide loop forms a catalytic pocket into which the cleavage site base, that of C-17, is inserted.

The X-ray structure of this so-called ground state ribozyme (Fig. 31-53*b*) indicates that its C-17 and A-1.1 have the standard A-RNA conformation, which is incompatible with an in-line attack of the C-17 O2′ atom on the scissile phosphate group to form the cyclic 2′,3′-phosphodiester product (Fig. 31-54*a*). In order to trap the hammerhead ribozyme in a conformation capable of forming this

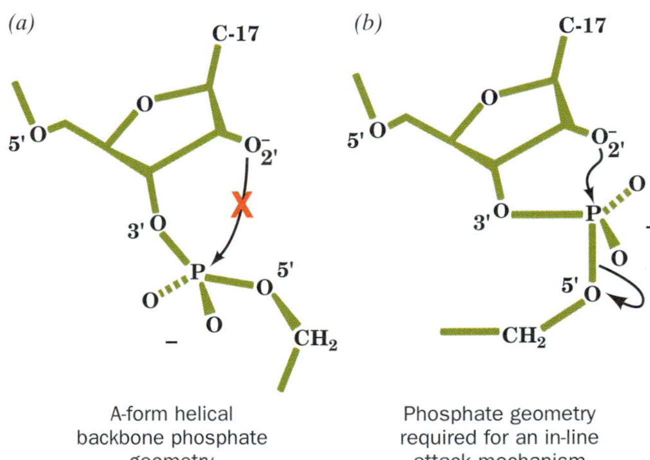

FIGURE 31-54 The conformation required for an in-line nucleophilic attack in the hammerhead ribozyme. (*a*) In the ribozyme's ground state conformation, C-17 has the standard A-RNA conformation in which its O2′ atom is 90° from the proper position for an in-line attack on the scissile phosphate group. (*b*) In the ribozyme's kinetically trapped active conformation, the scissile phosphate group has moved into the proper position for an in-line attack by O2′.

covalent intermediate, Scott synthesized it with a 5′-*C*-methyl-ribose modification on A-1.1:

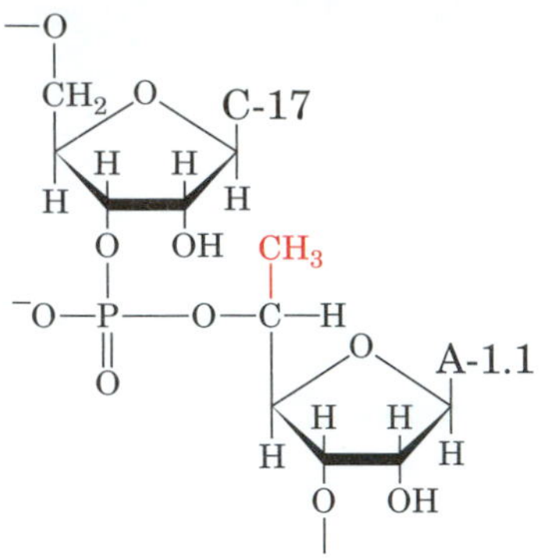

This creates a "kinetic bottleneck" for the reaction by stabilizing the otherwise scissile bond, presumably by altering the electronic properties of the leaving group. A crystal of this modified ribozyme that was soaked in a Co^{2+}-containing buffer at pH 8.5 for 30 min and then flash-frozen to near liquid nitrogen temperatures (−196°C; which arrests all molecular motion) had an X-ray structure that was indistinguishable from that of the ground state ribozyme. However, the X-ray structure of a crystal that was flash-frozen after a 2.5 hour soak revealed extensive changes in the region of the active site (Fig. 31-55). In particular, the base and ribose of C-17 have rotated by ~60° so as to cause the base to stack on that of A-6 (which remains stacked on the base of G-5), a movement of 8.7 Å by the C-17 base. Moreover, the furanose oxygen of A-1.1 has stacked on the C-17 base. This movement has resulted in a conformational change at the scissile phosphate so as to properly position it for an in-line nucleophilic attack by the C-17 O2′ atom (Figs. 31-54*b* and 31-55). In fact, similar conformation changes had been predicted by molecular dynamics simulations of the hammerhead ribozyme, starting from its ground state structure.

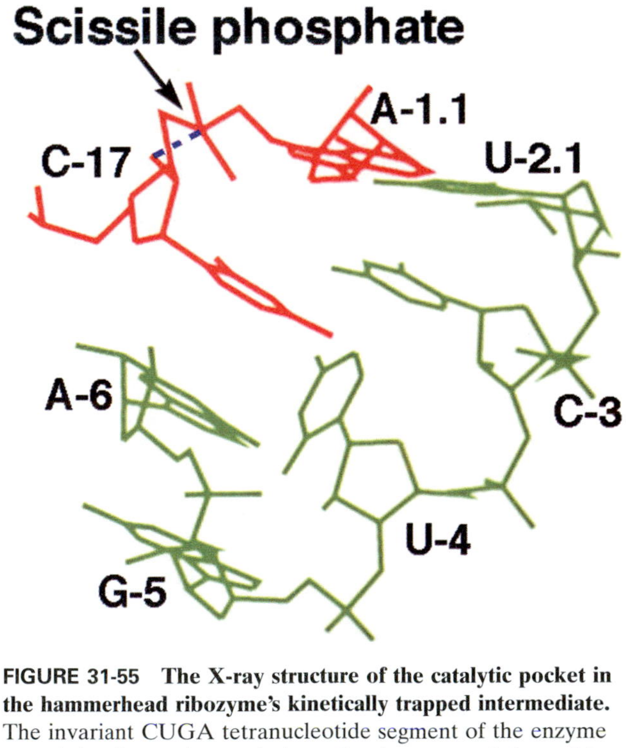

FIGURE 31-55 The X-ray structure of the catalytic pocket in the hammerhead ribozyme's kinetically trapped intermediate. The invariant CUGA tetranucleotide segment of the enzyme strand that forms the catalytic pocket is green and the residues on the substrate strand that span the scissile phosphate group are red. The dashed blue line, which marks the trajectory that atom O2′ of C-17 must take to nucleophilically attack the phosphorus atom of the scissile phosphate group, is in line with the bond joining this phosphorus atom to atom O5′ of A-1.1, the leaving group. [Adapted from Murray, J.B., Terwey, D.P., Maloney, L., Karpeisky, A., Usman, N., Beigleman, L., and Scott, W.G., *Cell* **92,** 665 (1998). PDBid 379D.]

h. Splicing of Pre-mRNAs Is Mediated by snRNPs in the Spliceosome

How are the splice junctions of pre-mRNAs recognized and how are the two exons to be joined brought together in the splicing process? Part of the answer to this question was established by Joan Steitz going on the assumption that one nucleic acid is best recognized by another. The eukaryotic nucleus, as has been known since the 1960s, contains numerous copies of several highly conserved 60- to 300-nucleotide RNAs called **small nuclear RNAs (snRNAs),** which form protein complexes termed **small nuclear ribonucleoproteins (snRNPs;** pronounced "snurps"). Steitz recognized that the 5′ end of one of these snRNAs, **U1-snRNA** (so called because it is a member of a U-rich subfamily of snRNAs), is partially complementary to the consensus sequence of the 5′ splice site. The consequent hypothesis, that *U1-snRNA recognizes the 5′ splice site,* was corroborated by the observations that splicing is inhibited by the selective destruction of the U1-snRNA sequences that are complementary to the 5′ splice site or by the presence of anti-U1-snRNP antibodies (produced by patients suffering from **systemic lupus erythematosus,** an often fatal autoimmune disease). Three other snRNPs are also implicated in splicing: **U2-snRNP, U4–U6-snRNP** (in which the **U4-** and **U6-snRNAs** associate via base pairing), and **U5-snRNP.**

Splicing takes place in an as yet poorly characterized ~45S particle dubbed the ***spliceosome*** *(Fig. 31-56).* The spliceosome brings together a pre-mRNA, the foregoing four snRNPs, and a variety of pre-mRNA binding proteins. Note that the spliceosome, which consists of 5 RNAs and ~65 polypeptides, is comparable in size and complexity to the ribosome (which in *E. coli* consists of 3 RNAs and 52 polypeptides; Section 32-3A). Although it had been generally accepted that the spliceosome's component snRNPs assembled anew on each pre-mRNA substrate, John Abelson has recently demonstrated that, at least with yeast, this is an experimental artifact arising from the use of unphysiologically high salt concentrations. Rather, it appears that the preassembled spliceosome associates as a whole with the pre-mRNA. The spliceosome is, nevertheless, a highly dynamic entity whose machinations in carrying out the splicing process are ATP-driven. For example, to carry out the first transesterification reaction yielding the lariat structure (Fig. 31-49), the spliceosome undergoes a complex series of rearrangements that are schematically diagrammed in Fig. 31-57. Similarly extensive rearrange-

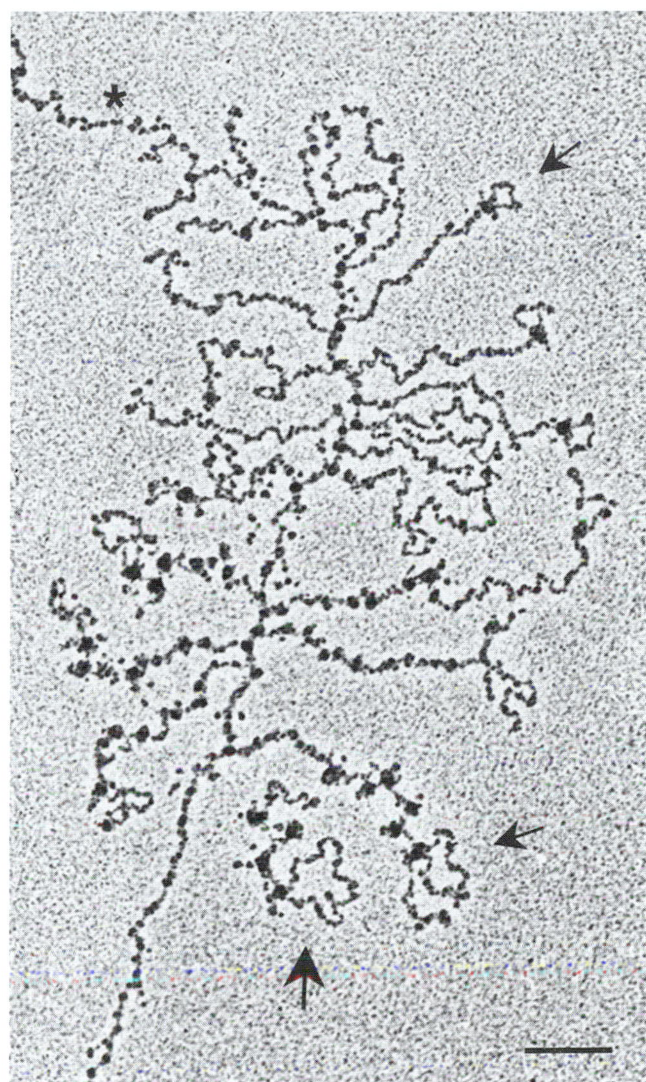

FIGURE 31-56 An electron micrograph of spliceosomes in action. A *Drosophila* gene that is ~6 Kb long enters from the upper left of the micrograph and exits at the lower left. Transcription initiates near the point marked by an asterisk. The growing RNA chains appear as fibrils of increasing lengths that emanate from the DNA. The transcripts are undergoing cotranslational splicing as revealed by the progressive formation and loss of intron loops near the 5′ ends of the RNA transcripts (*arrows*). The beads at the base of each intron loop as well as elsewhere on the transcripts are the spliceosomes. The large arrow points to a transcript near the 3′ end of the gene that is no longer attached to the DNA template and hence appears to have recently been terminated and released. The bar is 200nm long. [Courtesy of Ann Beyer and Yvonne Osheim, University of Virginia.]

ments are required to carry out the second transesterification reaction and to recycle the spliceosome for subsequent splicing reactions.

Although spliceosomal transesterification reactions were initially assumed to be mediated by protein catalysts, their chemical resemblance to the reactions carried out by the self-splicing group II introns suggests, as is noted above, that it is really the snRNAs that catalyze the splicing of pre-mRNAs (pre-mRNA introns have such varied sequences outside of their splice and branch sites that they are unlikely to play an active role in splicing). In fact, James Manley has shown that, in the absence of protein, segments of human U2- and U6-snRNAs catalyze an Mg^{2+}-dependent reaction in an intron branch site sequence-containing RNA that resembles splicing's first transesterification reaction.

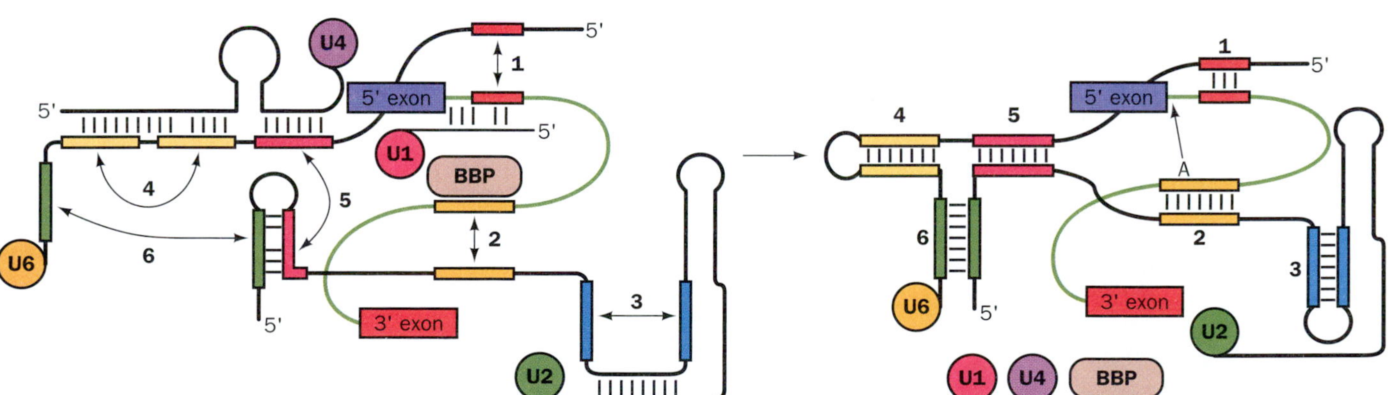

FIGURE 31-57 A schematic diagram of six rearrangements that the spliceosome undergoes in mediating the first transesterification reaction in pre-mRNA splicing. The RNA is color coded to indicate segments that become base paired. The black and green lines represent snRNA and pre-mRNA, and BBP stands for branch point-binding protein. U5, which participates in the second transesterification reaction, has been omitted for clarity. **(1)** Exchange of U1 for U6 in base pairing to the intron's 5′ splice site. **(2)** Exchange of BBP for U2 in binding to the intron's branch site. **(3)** Intramolecular rearrangement in U2. **(4)** Disruption of a base paired stem between U4 and U6 to form a stem–loop in U6. **(5)** Disruption of a second stem between U4 and U6 to form a stem between U2 and U6. **(6)** Disruption of a stem–loop in U2 to form a second stem between U2 and U6. The order of these rearrangements is unclear. The transesterification reaction is represented by the arrow from the A in the yellow segment of the pre-mRNA (*right panel*) to the 3′ end of the 5′ exon. [Adapted from Staley, J.P. and Guthrie, C., *Cell* **92,** 315 (1998).]

i. Splicing Also Requires the Participation of Splicing Factors

A variety of proteins known as **splicing factors** that are extrinsic to spliceosomes also participate in splicing. Among them are **branch point-binding protein [BBP;** also known as **splicing factor 1 (SF1)]** and **U2-snRNP auxiliary factor (U2AF),** which cooperate to select the intron's branch point. U2AF binds to the polypyrimidine tract upstream of the 3′ splice site (Fig. 31-48), whereas BBP recognizes the nearby branch point sequence (Figs. 31-49 and 31-57). The NMR structure of the 131-residue RNA-binding segment of the 638-residue BBP in complex with an 11-nt RNA containing a branch point sequence, determined by Michael Sattler, reveals that the RNA assumes an extended conformation and is largely buried in a groove that is lined with both aliphatic and basic residues (Fig. 31-58). The branch point adenosine, whose mutation abolishes BBP binding, is deeply buried and binds to BBP via hydrogen bonds that mimic Watson–Crick base pairing with uracil.

Other splicing factors include **SR** proteins and several members of the **heterogeneous nuclear ribonucleoprotein (hnRNP)** family. SR proteins each have one or more RRMs near their N-terminus and a distinctive C-terminal domain that is rich in Ser and Arg (SR) and which participates in protein–protein interactions. SR proteins specifically bind to exonic splicing enhancers (ESEs) and thereby recruit the splicing machinery to the flanking 5′ and 3′ splice sites. hnRNP proteins are highly abundant RNA-binding proteins that lack RS domains and whose functions are discussed below.

A simplistic interpretation of Fig. 31-49 suggests that any 5′ splice site could be joined with any following 3′ splice site, thereby eliminating all the intervening exons together with the introns joining them. However, such **exon skipping** does not normally take place (but see below). Rather, all of a pre-mRNA's introns are individually excised in what appears to be a largely fixed order that more or less proceeds in the 5′ → 3′ direction. This occurs, at least in part, because splicing occurs cotranscriptionally. Thus, as a newly synthesized exon emerges from an RNAP II, it is bound by splicing factors that are also bound to the RNAP II's highly phosphorylated C-terminal domain (CTD; Section 31-2E). This tethers the exon and its associated spliceosome to the CTD so as to ensure that splicing occurs when the next exon emerges from the RNAP II.

j. Spliceosomal Structures

All four snRNPs involved in pre-mRNA splicing contain the same so-called **snRNP core protein,** which consists of seven **Sm proteins** (so called because they react with autoantibodies of the Sm serotype from patients with systemic lupus erythematosis), which are named **B/B′, D_1, D_2, D_3, E, F,** and **G proteins** [B and B′ are alternatively spliced products of a single gene (see below) that differ only in their C-terminal 11 residues]. Each of these Sm proteins contains two conserved segments, Sm1 and Sm2, that are separated by a linker of variable length. The seven Sm proteins collectively bind to a conserved RNA sequence, the **Sm RNA motif,** which occurs in U1-, U2-, U4-, and U5-snRNAs and which has the single-stranded sequence AAUUUGUGG. However, in the absence of a U-snRNA,

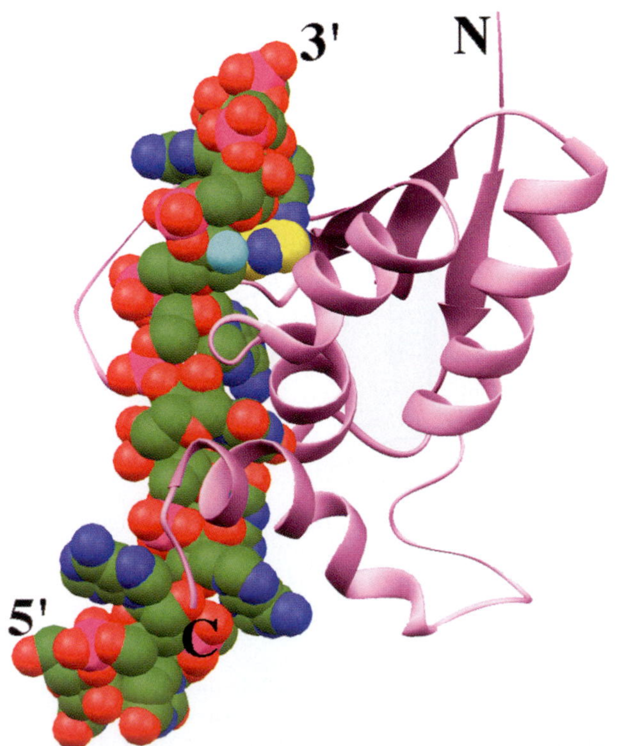

FIGURE 31-58 The NMR structure of the RNA binding portion of human branch point-binding protein (BBP) in complex with its target RNA. The 11-nt RNA has the sequence 5′-UAUACUAACAA-3′ in which the branch site sequence for both yeast and vertebrates is underlined. The RNA is drawn in space-filling form with C green, N blue, O red, and P magenta except for the C's of the branch point adenine, which are yellow, and the branch point O2′, which is cyan. [Based on an NMR structure by Michael Sattler, European Molecular Biology Laboratory, Heidelberg, Germany. PDBid 1K1G.]

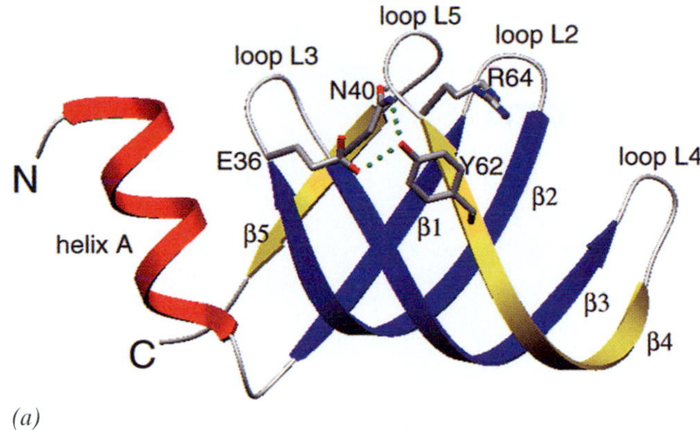

(a)

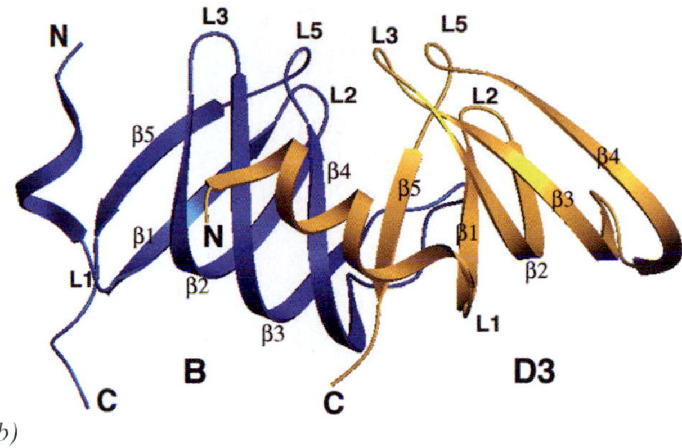

(b)

FIGURE 31-59 X-Ray structures of Sm proteins. (a) The structure of D_3 protein. The N-terminal helix and the β strands of its Sm1 domain are red and blue and the β strands of its Sm2 domain are yellow. The B, D_1, and D_2 Sm proteins have similar structures with their L4 loops and N-terminal segments, including helix A, comprising their most variable portions. Several highly conserved residues are shown in stick form (with C gray, N blue, and O red), and a conserved hydrogen bonding network is represented by green dotted lines. (b) The D_3B dimer with D_3 gold and B blue. The β5 strand of D_3 associates with the β4 strand of B to form a continuous antiparallel β sheet. Note that their corresponding loops extend in similar directions. [Courtesy of Kiyoshi Nagai, MRC Laboratory of Molecular Biology, Cambridge, U.K. PDBid 1D3B.]

the Sm proteins form three stable complexes consisting of D_1 and D_2; D_3 and B/B′; and E, F, and G. None of these complexes alone bind U-snRNA. However, the D_1D_2 and EFG complexes form a stable subcore snRNP with U-snRNA, to which D_3B binds to form the complete **Sm core domain.**

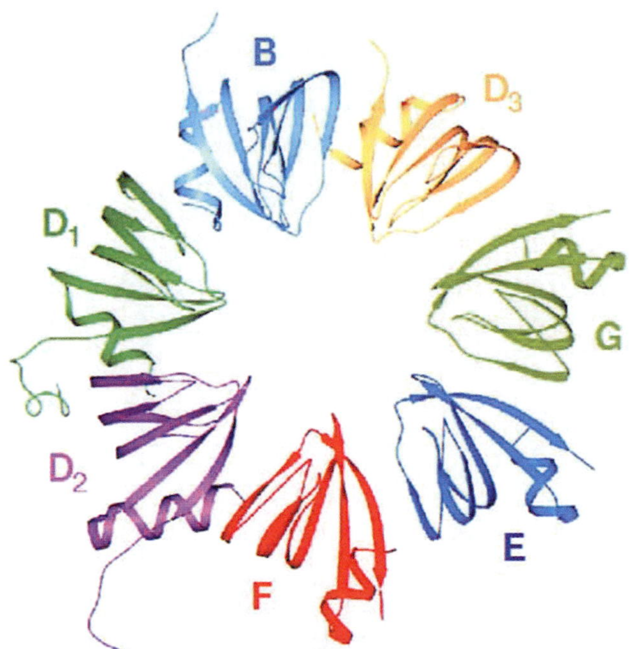

FIGURE 31-60 A model of the snRNP core protein. Its seven Sm proteins, which are each differently colored, are arranged in a 7-membered ring based on the structures of its component D_3B and D_1D_2 complexes with other pairwise interactions deduced from biochemical and mutagenic evidence. This heptameric ring has an outer diameter of 70 Å and a central hole with a diameter of 20 Å when only the main chain atoms are considered. [Courtesy of Kiyoshi Nagai, MRC Laboratory of Molecular Biology, Cambridge, U.K.]

The X-ray structures of the D_3B and D_1D_2 heterodimers, determined by Reinhard Lührmann and Kiyoshi Nagai, reveals that these four proteins share a common fold which consists of an N-terminal helix followed by a 5-stranded antiparallel β sheet that is strongly bent so as to form a hydrophobic core (Fig. 31-59a). The subunits of both dimers associate in a similar manner with the β5 strands of D_3 and D_1 binding to the β4 strands of B and D_2, respectively, so as to join their β sheets (Fig. 31-59b).

The structures of the D_1D_2 and D_3B complexes suggest that all seven Sm proteins could form a heptameric ring. Biochemical and mutagenic evidence indicates that in the EFG complex, E protein is centrally located and binds to the β5 side of F and β4 side of G. In addition, yeast two-hybrid experiments (Section 19-3C) indicate that D_2 and F interact as do D_3 and G. This has led to the model for the snRNP core protein that is drawn in Fig. 31-60. Its funnel-shaped central hole is positively charged and is large enough to permit the passage of single-stranded but not double-stranded RNA. The loops L2, L3, L4, and L5 from all seven proteins protrude into the central hole where they appear poised to bind ligands. This model is corroborated by the X-ray structure of an Sm-like protein from the hyperthermophilic archeon *Pyrobaculum aerophilum,* determined by David Eisenberg, that forms a homoheptameric ring that is structurally similar to the heteroheptameric model. This structure also supports the hypothesis that the seven eukaryotic Sm proteins arose through a series of duplications of an archaeal Sm-like protein gene.

Mammalian U1-snRNP consists of U1-snRNA and ten proteins, the seven Sm proteins that are common to all U-snRNPs, as well as three that are specific to U1-snRNP:

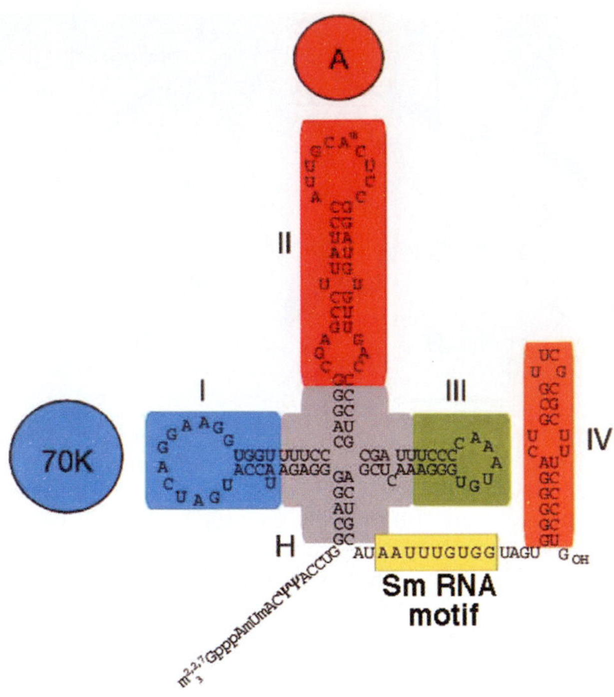

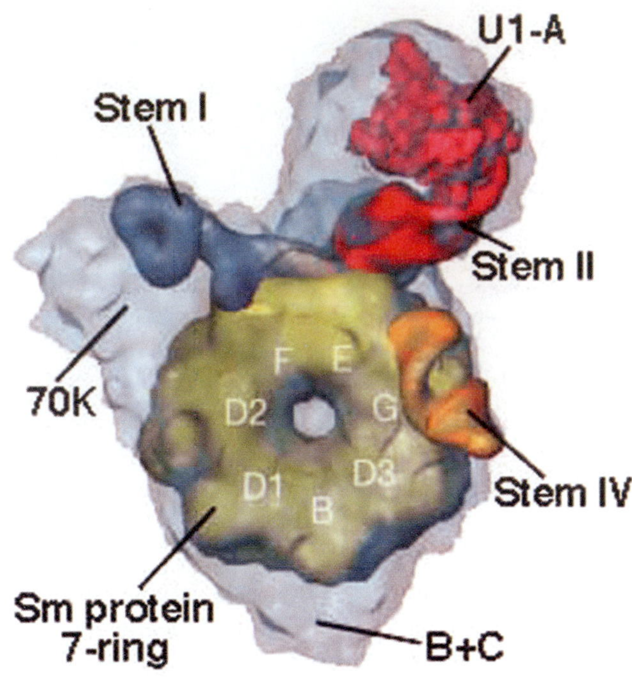

(a)

(b)

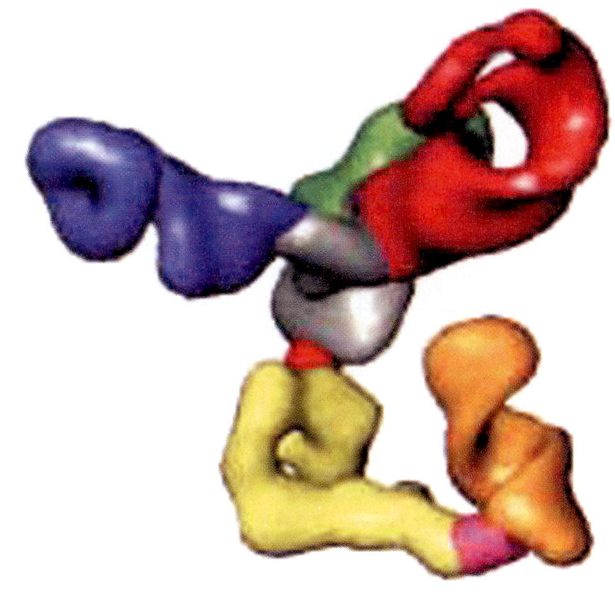

(c)

FIGURE 31-61 The electron microscopy–based structure of U1-snRNP at 10 Å resolution. (*a*) The predicted secondary structure of U1-snRNA with the positions at which the proteins U1-70K and U1-A bind the RNA indicated. (*b*) The molecular outline of U1-snRNP in light blue with its component ring-shaped Sm core protein yellow (and viewed oppositely from Fig. 31-60) and U1-snRNA colored as in Part *a*. (*c*) The U1-snRNA colored as in Part *a*. [Courtesy sof Holgar Stark, Max-Planck-Institut für biophysikalische Chemie, Göttingen, Germany.]

U1-70K, U1-A, and **U1-C.** The predicted secondary structure of the 165-nt U1-snRNA (Fig. 31-61*a*) contains five double helical stems, four of which come together at a 4-way junction. U1-70K and U1-A bind directly to RNA stem–loops I and II, respectively, whereas U1-C is bound by other proteins.

The 10-Å-resolution electron microscopy–based image of U1 snRNP (Fig. 31-61*b*) was elucidated by Holgar Stark and Lührmann. Its most obvious feature is a ring-shaped body that is 70 to 80 Å in diameter with a funnel-shaped central hole that closely matches the model of the Sm core protein in Fig. 31-60. Proteins were assigned to U1-snRNP's various protuberances based on cross-linking and binding studies as well electron micrographs of U1-snRNP lacking U1-A or U1-70K. The positions of the U1-snRNA's various structural elements (Fig. 31-61*c*), which were identified on the basis of known protein–RNA interactions, indicates that the Sm RNA motif, in fact, passes through the central hole in Sm core protein.

k. The Significance of Gene Splicing

The analysis of the large body of known DNA sequences reveals that introns are rare in prokaryotic structural genes, uncommon in lower eukaryotes such as yeast (which has a total of 239 introns in its ~6000 genes and, with two exceptions, only one intron per polypeptide), and abundant in higher eukaryotes (the only known vertebrate structural genes lacking introns are those encoding histones and the antiviral proteins known as interferons). Pre-mRNA introns, as we have seen, can be quite long and many genes contain large numbers of them. Consequently, unexpressed sequences constitute ~80% of a typical vertebrate structural gene and >99% of a few of them.

The argument that introns are only molecular parasites **(junk DNA)** seems untenable since it would then be difficult to rationalize why the evolution of complex splicing machinery offered any selective advantage over the elimination of the split genes. What then is the function of gene splicing? Although, since its discovery, the significance of gene splicing has been often vehemently debated, two important roles for it have emerged: (1) It is an agent for rapid protein evolution; and (2) through **alternative splicing,** it permits a single gene to encode several (sometimes many) proteins that may have significantly different functions. In the following paragraphs, we discuss these aspects of gene splicing.

l. Many Eukaryotic Proteins Consist of Modules That Also Occur in Other Proteins

The 839-residue LDL receptor is a plasma membrane protein that functions to bind low-density lipoprotein (LDL) to coated pits for transport into the cell via endocytosis (Section 12-5B). LDL receptor's 45-kb gene contains 18 exons, most of which encode specific functional domains of the protein. *Moreover, 13 of these exons specify polypeptide segments that are homologous to segments in other proteins:*

1. Five exons encode a 7-fold repeat of a 40-residue sequence that occurs once in **complement C9** (an immune system protein; Section 35-2F).

2. Three exons each encode a 40-residue repeat similar to that occurring four times in **epidermal growth factor (EGF;** Section 19-3C) and once each in three blood clotting system proteins: **factor IX, factor X,** and **protein C** (Section 35-1).

3. Five exons encode a 400-residue sequence that is 33% identical with a polypeptide segment that is shared only with EGF.

Evidently, the LDL receptor gene is modularly constructed from exons that also encode portions of other proteins. Numerous other eukaryotic proteins are similarly constituted including, as we have seen, many of the proteins involved in signal transduction (e.g., those containing SH2 and SH3 domains; Chapter 19). *It therefore appears that the genes encoding these modular proteins arose by the stepwise collection of exons that were assembled by (aberrant) recombination between their neighboring introns.*

m. Alternative Splicing Greatly Increases the Number of Proteins Encoded by Eukaryotic Genomes

The expression of numerous cellular genes is modulated by the selection of alternative splice sites. Thus, certain exons in one type of cell may be introns in another. For example, a single rat gene encodes seven tissue-specific variants of the muscle protein **α-tropomyosin** (Section 35-3B) through the selection of alternative splice sites (Fig. 31-62).

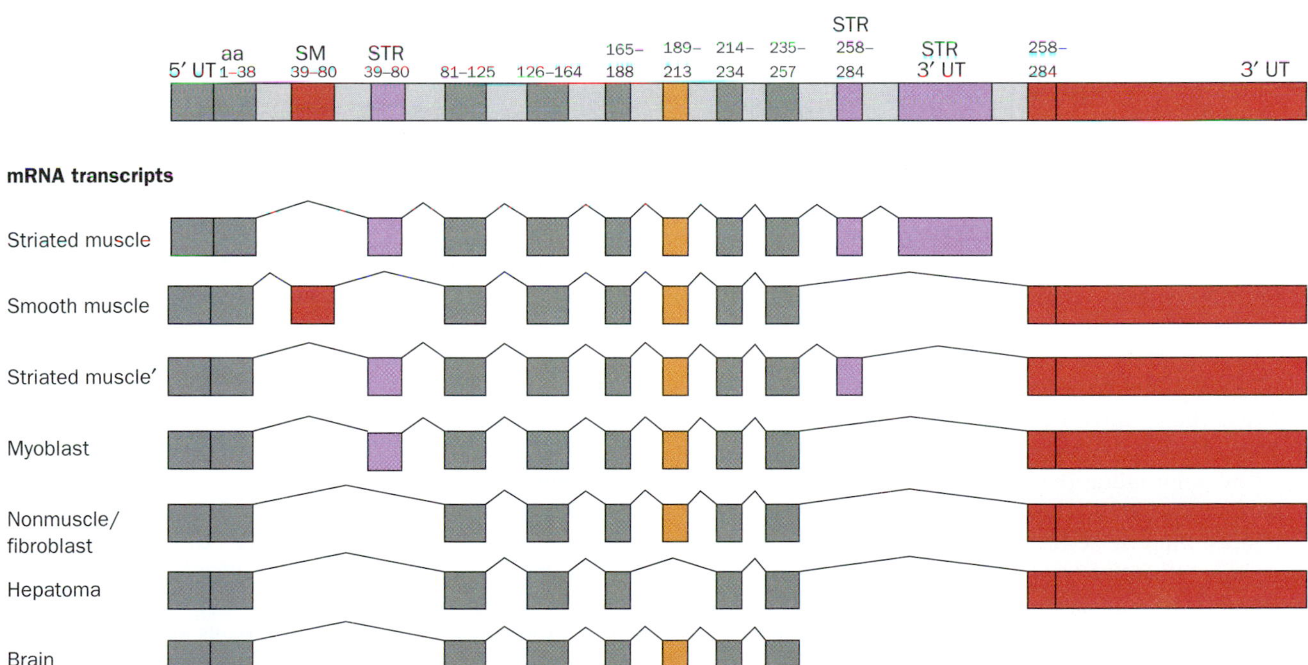

FIGURE 31-62 The organization of the rat α-tropomyosin gene and the seven alternative splicing pathways that give rise to cell-specific α-tropomyosin variants. The thin kinked lines indicate the positions occupied by the introns before they are spliced out to form the mature mRNAs. Tissue-specific exons are indicated together with the amino acid (aa) residues they encode: "constitutive" exons (those expressed in all tissues) are green, those expressed only in smooth muscle (SM) are brown, those expressed only in striated muscle (STR) are purple, and those variably expressed are yellow. Note that the smooth and striated muscle exons encoding amino acid residues 39 to 80 are mutually exclusive; likewise, there are alternative 3′-untranslated (UT) exons. [After Breitbart, R.E., Andreadis, A., and Nadal-Ginard, B., *Annu. Rev. Biochem.* **56,** 481 (1987).]

Alternative splicing occurs in all metazoa and is especially prevalent in vertebrates. In fact, it is estimated that up to 60% of human structural genes are subject to alternative splicing. This perhaps rationalizes the discrepancy between the ~30,000 genes identified in the human genome (Section 7-2B) and earlier estimates that it contains 50,000 to 140,000 structural genes.

The variation in mRNA sequence can take several different forms: Exons can be retained in an mRNA or they can be skipped; introns may be excised or retained; and the positions of 5′ and 3′ splice sites can be shifted to make exons shorter or longer. Alterations in the transcriptional start site and/or the polyadenylation site can further contribute to the diversity of the mRNAs that are transcribed from a single gene. In a particularly striking example, the *Drosophila* **DSCAM** protein, which functions in neuronal development, is encoded by 24 exons of which there are 12 mutually exclusive variants of exon 4, 48 of exon 6, 33 of exon 9, and 2 of exon 17 (which are therefore known as **cassette exons**) for total of 38,016 possible variants of this protein (compared to ~13,000 identified genes in the *Drosophila* genome). Although it is unknown if all possible DSCAM variants are produced, experimental evidence suggests that the *Dscam* gene expresses many thousands of them. Clearly, the number of genes in an organism's genome does not by itself provide an adequate assessment of its protein diversity. Indeed, it has been estimated that, on average, each human structural gene encodes three different proteins.

The types of changes that alternative splicing confers on expressed proteins spans the entire spectrum of protein properties and functions. Entire functional domains or even single amino acid residues may be inserted into or deleted from a protein, and the insertion of a stop codon may truncate the expressed polypeptide. Splice variations may, for example, control whether a protein is soluble or membrane bound, whether it is phosphorylated by a specific kinase, the subcellular location to which it is targeted, whether an enzyme binds a particular allosteric effector, and the affinity with which a receptor binds a ligand. Changes in an mRNA, particularly in its noncoding regions, may also influence the rate at which it is transcribed and its susceptibility to degradation. Since the selection of alternative splice sites is both tissue- and developmental stage-specific, splice site choice must be tightly regulated in both space and time. In fact, ~15% of human genetic diseases are caused by point mutations that result in pre-mRNA splicing defects. Some of these mutations delete functional splice sites, thereby activating nearby pre-existing **cryptic splice sites.** Others generate new splice sites that are used instead of the normal ones. In addition, tumor progression is correlated with changes in levels of proteins implicated in alternative splice site selection.

How are alternative splice sites selected? The best understood examples of such processes occur in the pathway responsible for sex determination in *Drosophila,* two of which we discuss here:

1. Exon 2 of *transformer (tra)* pre-mRNA contains two alternative 3′ splice sites (which suceed the excised intron), with the proximal (close; to exon 1) site used in males and the distal (far) site used in females (Fig. 31-63*a*). The region between these two sites contains a stop codon (UAG). In males, the splicing factor U2AF binds to the proximal 3′ splice site to yield an mRNA containing this premature stop codon, which thereby directs the synthesis of truncated and hence nonfunctional **TRA** protein. In females, however, the proximal 3′ splice site is bound by the female-specific **SXL** protein, the product of the *sex-lethal (sxl)* gene (which is only expressed in females), so as to block the binding of U2AF, which then binds to the distal 3′ splice site, thereby excising the UAG and inducing the expression of functional TRA protein.

2. In *doublesex (dsx)* pre-mRNA, the first three exons are constitutively spliced in both males and females. However, the branch site immediately upstream of exon 4 has a suboptimal pyrimidine tract to which U2AF does not bind (Fig. 31-63*b*). Hence in males, exon 4 is not included in *dsx* mRNA, leading to the synthesis of male-specific **DSX-M** protein that functions as a repressor of female-specific genes. However, in females, TRA protein promotes the cooperative binding of the SR protein **RBP1** and the SR-like protein **TRA2** [the product of the *transformer 2 (tra-2)* gene] to six copies of an exonic splice enhancer (ESE) within exon 4. This heterotrimeric complex recruits the splicing machinery to the upstream 3′ splice site of exon 4, leading to its inclusion in *dsx* mRNA. The resulting female-specific **DSX-F** protein is a repressor of male-specific genes.

Thus, the synthesis of functional TRA protein involves the repression of a splice site, whereas the synthesis of female-specific DSX-F protein involves the activation of a splice site. Similar mechanisms of alternative splice site selection have been identified in vertebrates.

n. AU–AC Introns Are Excised by a Novel Spliceosome

A small fraction of introns (~0.3%) have AU rather than GU at their 5′ ends and AC rather than AG at their 3′ ends, but are nevertheless excised via a lariat structure to an internal intron A. These so-called **AU–AC introns** (alternatively, **AT–AC introns** after their DNA sequences), which occur in organisms as diverse as *Drosophila,* plants, and humans, are excised by a novel so-called **AU–AC spliceosome** (alternatively, an **AT–AC spliceosome)** that has one snRNP, U5, in common with the major (GU–AG) spliceosome, and three others, **U11, U12,** and **U4atac–U6atac,** which are distinct from but structurally and functionally analogous to U1, U2, and U4–U6. Curiously, all genes known to contain AU–AC introns also contain multiple major class introns. Moreover, AU–AC introns are not conserved in either length or position in their host genes. Thus, the functional and evolutionary significance of the AU–AC spliceosome and introns is obscure.

o. Trans-Splicing

The types of splicing we have so far considered occur within single RNA molecules and hence are known as **cis-**

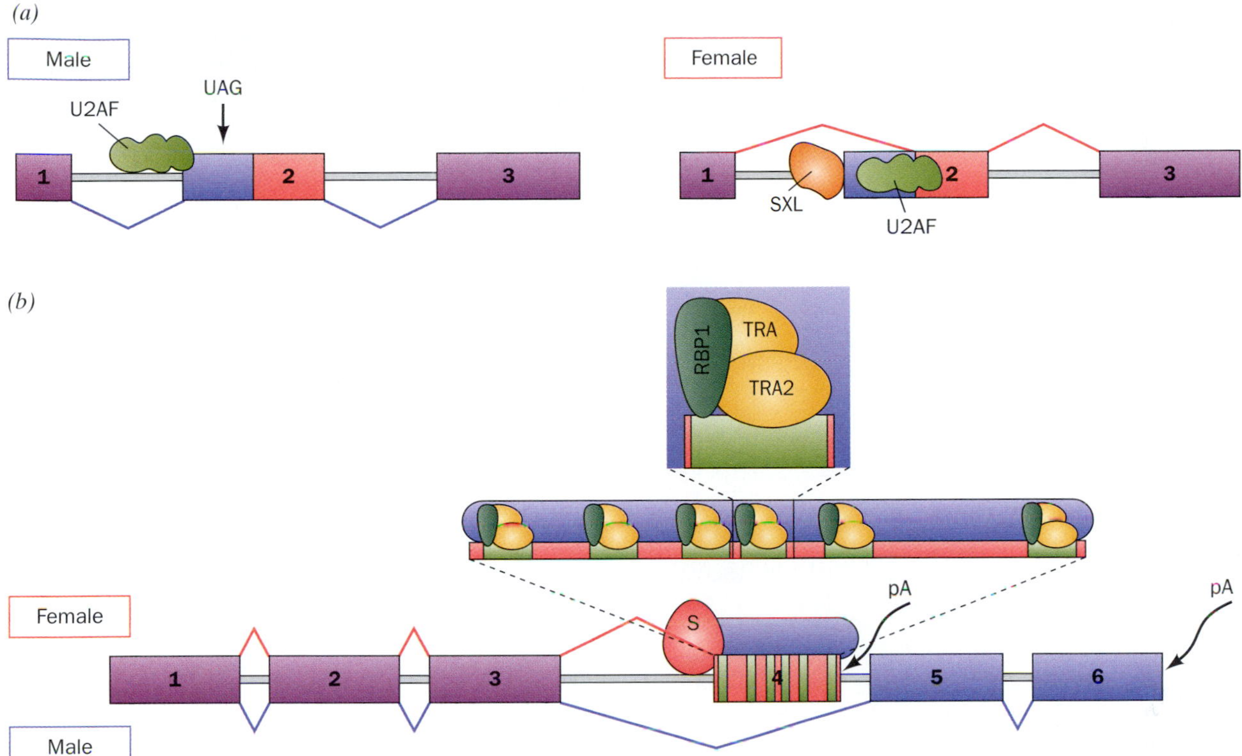

FIGURE 31-63 Mechanisms of alternative splice site selection in the *Drosophila* sex-determination pathway as described in the text. In all panels, exons are represented by colored rectangles and introns are shown as pale gray lines. (*a*) Alternative splicing in *tra* pre-mRNA. UAG is a stop codon. (*b*) Alternative splicing in *dsx* pre-mRNA. The six ESEs (exonic splice enhancers) in exon 4 are indicated by green rectangles and S represents the splicing machinery. In females, polyadenylation (pA) of *dsx* mRNA occurs downstream of exon 4, whereas in males, it occurs downstream of exon 6. [After a drawing by Maniatis, T. and Tasic, B., *Nature* **418,** 236 (2002).]

splicing. The chemistry of the spliceosomal cis-splicing reaction, however, is the same as would occur if the two exons to be joined initially resided on two different RNA molecules, a process called **trans-splicing.** This, in fact, occurs in trypanosomes (kinetoplastid protozoa; the cause of African sleeping sickness). Trypanosomal mRNAs all have the same 35-nt noncoding leader sequence, although this leader sequence is not present in the corresponding genes. Rather, this sequence is part of a so-called **spliced leader (SL) RNA** that is transcribed from an independent gene. The 5′ splice site that succeeds the SL RNA leader sequence, and the branch site and 3′ splice site that precede the exon sequence have the same consensus sequences as occur in the RNAs spliced by the major spliceosome. Consequently, the SL RNA leader and the pre-mRNA are joined in a trans-splicing reaction that resembles the spliceosomal cis-splicing reaction (Fig. 31-49) with the exception that the product of the first transesterification reaction is necessarily Y-shaped rather than lariat-shaped (Fig. 31-64). Trypanosomes, whose pre-mRNAs lack introns, nevertheless have U2- and U4–U6-snRNPs but lack U1- and U5-snRNPs. However, the SL RNA, which is predicted to fold into three stem–loops and a single-stranded Sm RNA-like motif as does U1-snRNA (Fig. 31-61*a*), apparently carries out the functions of U1-snRNA in the trans-splicing reaction.

Trans-splicing has been shown to occur in nematodes (roundworms; e.g., *C. elegans*) and flatworms. These organisms also carry out cis-splicing and, indeed, perform both types of splicing on the same pre-mRNA. There are also several reports that trans-splicing occurs in higher eukaryotes such as *Drosophila* and vertebrates, but if it does occur, it does so in only a few pre-mRNAs and at a very low level.

p. mRNA Is Methylated at Certain Adenylate Residues

During or shortly after the synthesis of vertebrate pre-mRNAs, ~0.1% of their A residues are methylated at their N6 atoms. These m^{6}A's tend to occur in the sequence RRm^{6}ACX, where X is rarely G. Although the functional significance of these methylated A's is unknown, it should be noted that a large fraction of them are components of the corresponding mature mRNAs.

q. hnRNP Proteins Coat mRNAs

Throughout their residency in the nucleus, hnRNAs (pre-mRNAs) are coated with a great variety of proteins, thereby forming hnRNPs. Although the functions of these

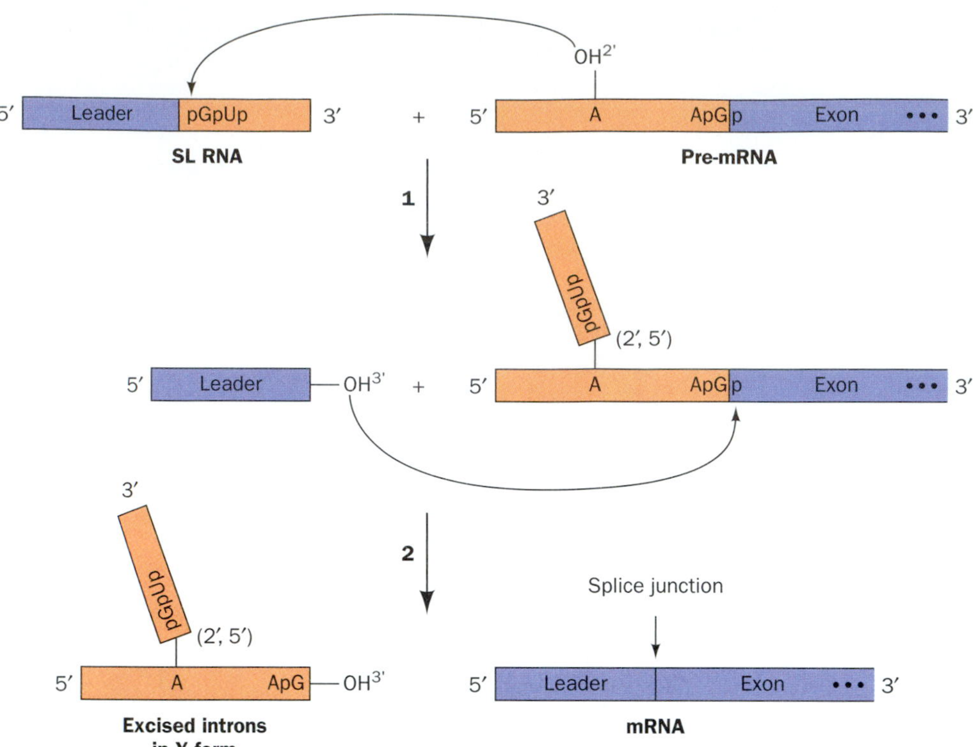

FIGURE 31-64 **The sequence of transesterification reactions that occurs in trans-splicing.** The chemistry is closely similar to that of pre-mRNA cis-splicing (Fig. 31-49).

hnRNP proteins are only beginning to come into focus, it is clear that they facilitate the processing of the mRNAs, including, as described above, constitutive and alternative splicing, their stability, and their transport to different regions of the nucleus and, ultimately, as Gideon Dreyfuss demonstrated, to the cytoplasm, where they regulate the localization, translation, and turnover of their associated mRNAs.

r. RNA Can Be Edited by the Insertion or Deletion of Specific Nucleotides

Certain mRNAs from a variety of eukaryotic organisms have been found to differ from their corresponding genes in several unexpected ways, including C → U and U → C changes, the insertion or deletion of U residues, and the insertion of multiple G or C residues. The most extreme examples of this phenomenon, which occur in the mitochondria of trypanosomes (whose DNA encodes only 20 genes), involve the addition and removal of up to hundreds of U's to and from 12 otherwise untranslatable mRNAs. The process whereby a transcript is altered in this manner is called **RNA editing** because it originally seemed that the required enzymatic reactions occurred without the direction of a nucleic acid template and hence violated the central dogma of molecular biology (Fig. 5-21). Eventually, however, a new class of trypanosomal mitochondrial transcripts called **guide RNAs (gRNAs)** was identified. gRNAs, which consist of 50 to 70 nucleotides, have 3′ oligo(U) tails, an internal segment that is precisely complementary to the edited portion of the pre-edited mRNA (if G · U pairs, which are common in RNAs, are taken to be complementary), and a 10- to 15-nt so-called anchor sequence near the 5′ end that is largely complementary in the Watson–Crick sense to a segment of the mRNA that is not edited.

An unedited transcript presumably associates with the corresponding gRNA via its anchor sequence (Fig. 31-65). Then, in a process mediated by the appropriate enzymatic machinery in an ~20S RNP named the **editosome,** the gRNA's internal segment is used as a template to "correct" the transcript, thereby yielding the edited mRNA. Insertion editing requires at least three enzymatic activities that, somewhat surprisingly, are encoded by nuclear genes (Fig. 31-66*a*): (1) an endonuclease at a mismatch between the gRNA and the pre-edited mRNA to cleave the pre-edited mRNA on the 5′ side of the insertion point; (2) **terminal uridylyltransferase (TUTase)** to insert the new U(s); and (3) an **RNA ligase** to reseal the RNA. Deletion requires similar enzymatic apparatus with the exceptions that the endonuclease cleaves the RNA being edited on the 3′ side of the U(s) to be deleted and TUTase is replaced by **3′-U-exonuclease (3′-U-exo),** which excises the U(s) at the deletion site (Fig. 31-66*b*). A single gRNA mediates the editing of a block of 1 to 10 sites. Thus, the genetic information specifying an edited mRNA is derived from two or more genes. The functional advantage of this complicated process, either presently or more likely in some ancestral organism, is obscure.

s. RNA Can Be Edited by Base Deamination

Humans express two forms of **apolipoprotein B (apoB): apoB-48,** which is made only in the small intestine and

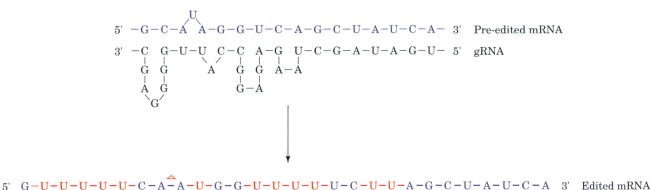

```
              U
             / \
5' -G-C-A   A-G-G-U-C-A-G-C-U-A-U-C-A- 3'  Pre-edited mRNA
3' -C  G-U-U  C-C  A-G  U-C-G-A-U-A-G-U- 5'  gRNA
    |  |     \ /   |  |  |
    G  G      A    G  G  A-A
    |  |           |  |
    A  G           G-A
     \ /
      G
```

↓

```
5'  G-U-U-U-U-U-C-A-A-U-G-G-U-U-U-U-U-C-U-U-A-G-C-U-A-U-C-A  3'  Edited mRNA
3'  C-G-A-G-G-G-G-U-U-A-C-C-G-G-A-G-A-G-A-A-U-C-G-A-A-A-G-U  5'  gRNA
```

FIGURE 31-65 A schematic diagram indicating how gRNAs direct the editing of trypanosomal pre-edited mRNAs. The red U's in the edited mRNA are insertions and the triangle (△) marks a deletion. Several gRNAs may be necessary to direct the editing of consecutive segments of a pre-edited mRNA. [After Bass, B.L., *in* Gesteland, R.F. and Atkins, J.F. (Eds.), *The RNA World, p.* 387, Cold Spring Harbor Laboratory Press (1993).]

functions in chylomicrons to transport triacylglycerols from the intestine to the liver and peripheral tissues; and **apoB-100,** which is made only in the liver and functions in VLDL, IDL, and LDL to transport cholesterol from the liver to the peripheral tissues (Sections 12-5A and 12-5B). ApoB-100 is an enormous 4536-residue protein, whereas apoB-48 consists of apoB-100's N-terminal 2152 residues and therefore lacks the C-terminal domain of apoB-100 that mediates LDL receptor binding.

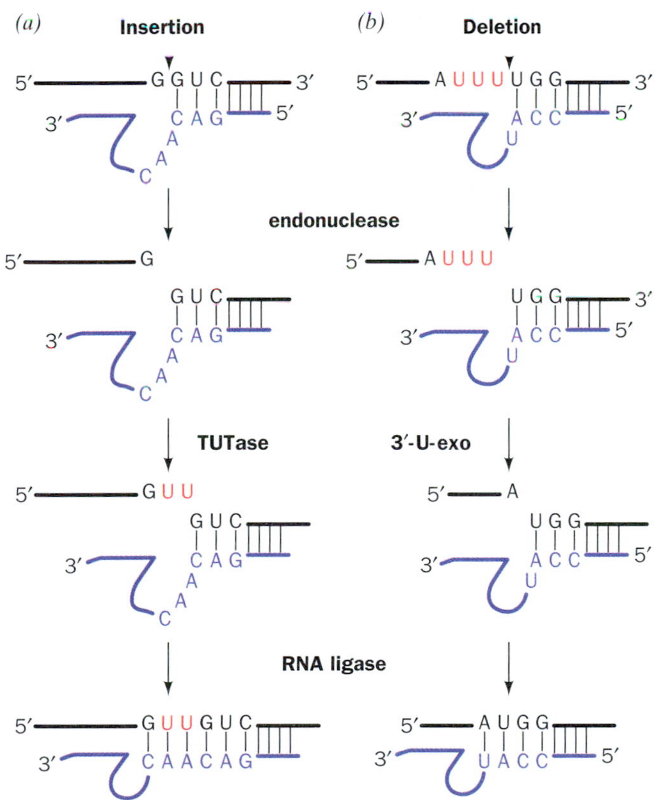

FIGURE 31-66 Trypanosomal RNA editing pathways. The RNAs being edited (*black*) are shown base paired to the gRNAs (*blue*) with the U's that are (*a*) inserted by TUTase or (*b*) deleted by 3′-U-exo drawn in red. The arrowheads indicate the positions that are cleaved by the endonuclease. [After Madison-Antenucci, S., Grams, J., and Hajduk, S.L., *Cell* **108,** 435 (2002).]

Despite their differences, both apoB-48 and apoB-100 are expressed from the same gene. How does this occur? Comparison of the mRNAs encoding the two proteins indicates that they differ by a single C → U change: The codon for Gln 2153 (CAA) in apoB-100 mRNA is, in apoB-48 mRNA, a UAA stop codon. The activity that catalyzes this conversion is a protein: It is destroyed by proteases and protein-specific reagents but not by nucleases. When apoB mRNA is synthesized with $[\alpha\text{-}^{32}P]$CTP, *in vitro* editing yields a $[^{32}P]$UMP residue solely at the editing site. Evidently, the editing activity is a site-specific **cytidine deaminase.** This type of RNA editing differs in character from that in trypanosomal mitochondria, which inserts and deletes multiple U's into mRNAs under the direction of gRNAs. ApoB mRNA editing therefore falls into a different class of RNA editing that is called **substitutional editing.**

The several other known examples of pre-mRNA substitutional editing all occur on pre-mRNAs that encode ion channels and G protein-coupled receptors in nerve tissue. Among them is vertebrate brain **glutamate receptor** pre-mRNA, which undergoes an A → I deamination [where I is inosine (guanosine lacking its 2-amino group), which the translational apparatus reads as G] that transforms a Gln codon (CAG) to that of a functionally important Arg (CIG; normally CGG). The enzymes that catalyze such A → I RNA editing, **ADAR1** (1200 residues) and **ADAR2** (729 residues; ADAR for *a*denosine *d*eaminases *a*cting on *R*NA), have the curious requirement that their target A residues must be members of RNA double helices that are formed between the editing site and a complementary sequence that is usually located in a downstream intron (Fig.

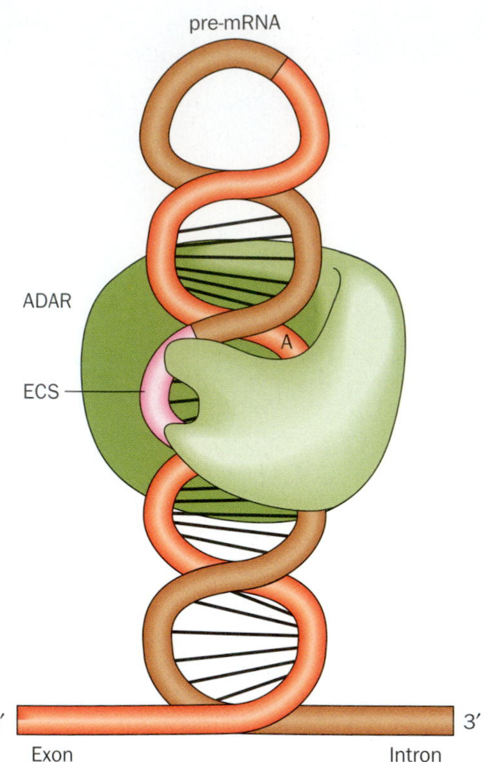

FIGURE 31-67 The recognition of ADAR editing sites. Both ADAR1 and ADAR2 bind to a 9- to 15-bp double-stranded RNA that is formed between the editing site (*orange*) on a pre-mRNA exon and a so-called **editing site complementary sequence** (**ECS;** *magenta*) that is often located in a downstream intron (*brown*). A represents the adenosine that the ADAR (*green*) converts to inosine. [After Keegan, L.P., Gallo, A., and O'Connell, M.A., *Nature Rev. Genet.* **2,** 869 (2001).]

31-67). Hence, ADAR-mediated editing must precede splicing.

Substitutional editing may contribute to protein diversity. For example, *Drosophila cacophony* pre-mRNA that encodes a voltage-gated Ca^{2+} channel subunit contains 10 different substitutional editing sites and hence has the potential of generating 1000 different isoforms in the absence of alternative splicing.

Substitutional editing can also generate alternative splice sites. For example, rat ADAR2 edits its own pre-mRNA by converting an intronic AA dinucleotide to AI, which mimics the AG normally found at 3′ splice sites (Fig. 31-49). The consequent new splice site adds 47 nucleotides near the 5′ end of the *ADAR2* mRNA so as to generate a new translational initiation site. The resulting ADAR2 isozyme is catalytically active but is produced in smaller amounts than that from unedited transcripts, perhaps due to a less efficient translational initiation site. Thus, rat ADAR2 appears to regulate its own rate of expression.

ADAR1 contains an N-terminal Z-DNA-binding domain, Zab, that is composed of two subdomains, Zα and Zβ. We have seen that in the X-ray structure of Zα in complex with Z-DNA (Fig. 29-5), Zα binds Z-DNA via sequence-independent complementary surfaces (Section 29-1B). What is the function of Zab? Alexander Rich has proposed that since the negative supercoiling of the DNA immediately behind actively transcribing RNAP stimulates the transient formation of Z-DNA (recall that Z-DNA has a left-handed helix), Zab targets ADAR1 to genes that are undergoing transcription. This would facilitate rapid A → I editing, which must take place before the next splicing reaction occurs.

t. RNA Interference

In recent years it has become increasingly clear that noncoding RNAs can have important roles in controlling gene expression. One of the first indications of this phenomenon occurred in Richard Jorgensen's attempt to genetically engineer more vividly purple petunias by introducing extra copies of the gene that directs the synthesis of the purple pigment. Surprisingly, the resulting transgenic plants had variegated and often entirely white flowers. Apparently, the purple-making genes somehow switched each other off. Similarly, it is well known that **antisense RNA** (RNA that is complementary to at least a portion of an mRNA) prevents the translation of the corresponding mRNA because the ribosome cannot translate double-stranded RNA (Section 32-4E). Yet, injecting **sense RNA** (RNA with the same sequence as an mRNA) into the nematode *C. elegans* also blocks protein production. Since the added RNA somehow interferes with gene expression, this phenomenon is known as **RNA interference [RNAi; posttranscriptional gene silencing (PTGS)** in plants]. RNAi/PTGS is now known to occur in all eukaryotes except perhaps yeast.

The mechanism of RNAi began to come to light in 1998 when Andrew Fire and Craig Mello showed that double-stranded RNA **(dsRNA)** was substantially more effective in causing RNAi in *C. elegans* than were either of its component strands alone. RNAi is induced by only a few molecules of dsRNA per affected cell, suggesting that RNAi is a catalytic rather than a stoichiometric effect. Further investigations, in large part in *Drosophila,* have led to the elucidation of the following pathway mediating RNAi (Fig. 31-68):

1. The trigger dsRNA, as Phillip Zamore discovered, is chopped up into ~21- to 23-nt-long double-stranded fragments known as **small interfering RNAs (siRNAs),** each of whose strands has a 2-nt overhang at its 3′ end and a 5′ phosphate. This reaction is mediated by an ATP-dependent RNase named **Dicer,** a homodimer of 2249-residue subunits that is a member of the **RNase III** family of double-strand –specific RNA endonucleases.

2. An siRNA is transferred to a 250- to 500-kD multisubunit complex known as **RNA-induced silencing complex (RISC),** which contains an endoribonuclease that is distinct from Dicer. The antisense strand of the siRNA guides the RISC complex to an mRNA with the complementary sequence.

3. RISC cleaves the mRNA, probably opposite the bound siRNA. The cleaved mRNA is then further degraded by cellular nucleases, thereby preventing its translation.

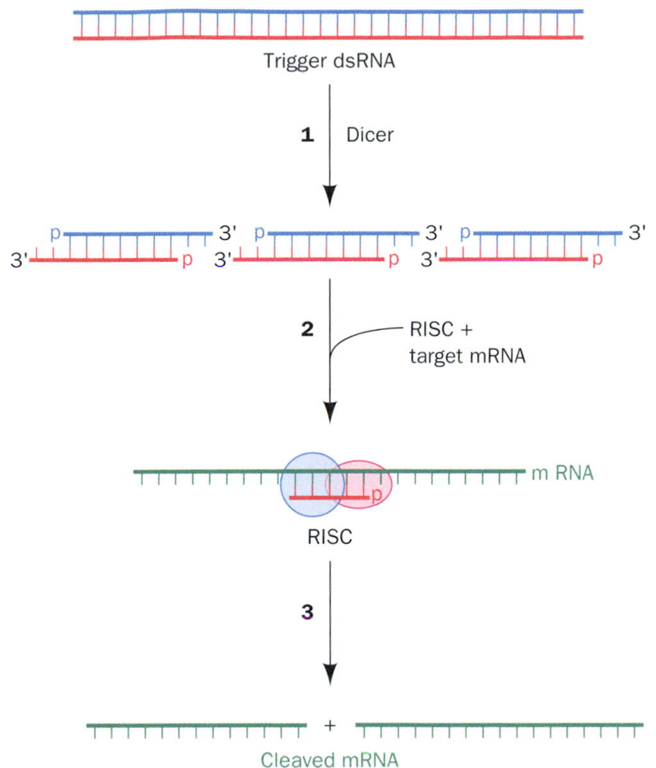

FIGURE 31-68 A model for RNA interference (RNAi). See the text for details.

RNAi requires that the trigger dsRNA be copied so as to permit the siRNAs to reach sufficient concentrations to cleave the target mRNAs. This amplification process is mediated by an **RNA-dependent RNA polymerase (RdRP).** Moreover, an siRNA strand can act as a primer for the RdRP-catalyzed synthesis of secondary trigger dsRNA, which is subsequently "diced" to yield secondary siRNAs (Fig. 31-69). Since the secondary trigger dsRNA may extend beyond the sequence complementary to the original trigger dsRNA, some of the resulting secondary siRNAs could be complementary to segments of other mRNAs that have no complementarity to the original trigger dsRNA. This would cause the silencing of genes with segments resembling portions of the original target mRNA but with no similarity to any portion of the original trigger dsRNA, a phenomenon known as **transitive RNAi.**

The ease with which RNAi/PTGS may be induced has made it the method of choice for generating null mutant (knockout) phenotypes of specific genes in plants and nonvertebrates, although care must be taken that transitive RNAi does not cause spurious results. It also appears that RNAi can be of similar use in mammalian systems, even though mammals lack the mechanisms that amplify silencing in plants and nonvertebrates so that the effects of RNAi in mammals are transient. But what is the physiological function of RNAi/PTGS? Since most eukaryotic viruses store and replicate their genomes as RNA (Chapter 33), it seems likely that RNAi arose as a defense against viral infections. Indeed, many plant viruses contain genes that suppress various steps of PTGS and which are essential for pathogenesis. RNAi/PTGS may also inhibit the movement of retrotransposons (Section 30-6B). Finally, the exquisite specificity of RNAi may make it possible to prevent viral infections and to silence disease-causing mutant genes such as oncogenes.

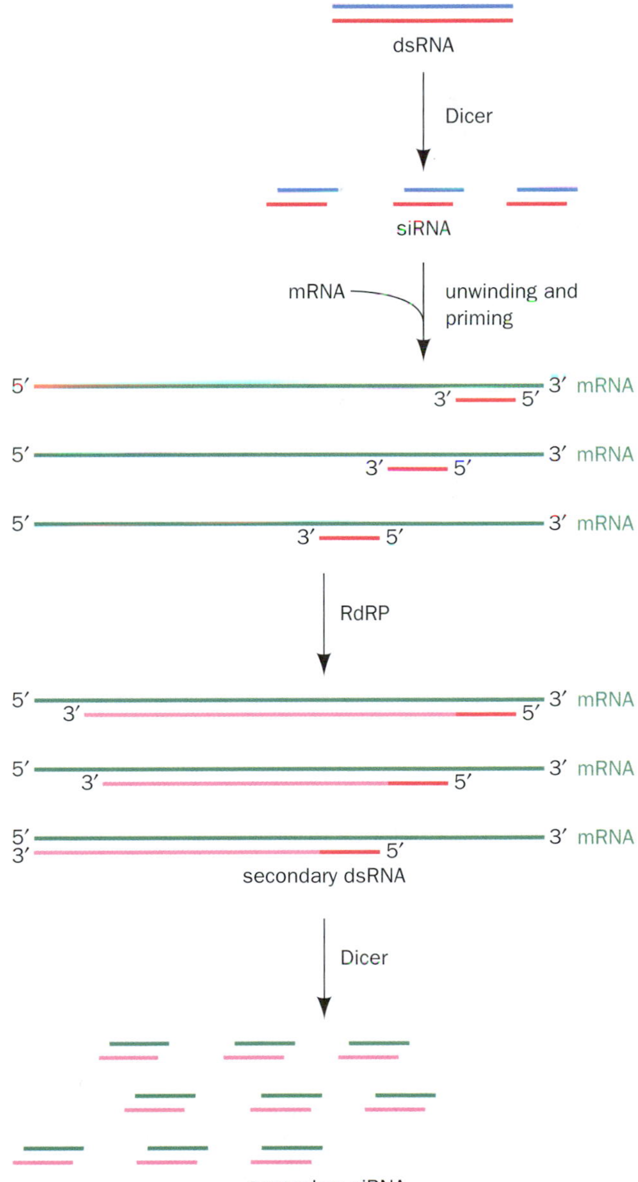

FIGURE 31-69 A model for transitive RNAi. The siRNA resulting from the action of Dicer on dsRNA is unwound and binds its target mRNA. There it acts as a primer for RNA-directed RNA polymerase (RdRP), which extends it in the $5' \rightarrow 3'$ direction (as do all known RNA and DNA polymerases). The resulting secondary dsRNA is subsequently cleaved by Dicer to form secondary siRNA. Note that those secondary siRNAs that contain segments of the mRNA downstream of those complementary to the original dsRNA may silence genes containing sequences similar to those of these downstream mRNA segments but yet have no sequence in common with the original dsRNA.

B. Ribosomal RNA Processing

The seven *E. coli* rRNA operons all contain one (nearly identical) copy of each of the three types of rRNA genes (Section 32-3F). Their polycistronic primary transcripts, which are >5500 nucleotides in length, contain 16S rRNA at their 5′ ends followed by the transcripts for 1 or 2 tRNAs, 23S rRNA, 5S rRNA, and, in some rRNA operons, 1 or 2 more tRNAs at the 3′ end (Fig. 31-70). The steps in processing these primary transcripts to mature rRNAs were elucidated with the aid of mutants defective in one or more of the processing enzymes.

The initial processing, which yields products known as **pre-rRNAs,** commences while the primary transcript is still being synthesized. It consists of specific endonucleolytic cleavages by **RNase III, RNase P, RNase E,** and **RNase F** at the sites indicated in Fig. 31-70. The base sequence of the primary transcript suggests the existence of several base paired stems. The RNase III cleavages occur in a stem consisting of complementary sequences flanking the 5′ and 3′ ends of the 23S segment (Fig. 31-71) as well as that of the 16S segment. Presumably, certain features of these stems constitute the RNase III recognition site.

The 5′ and 3′ ends of the pre-rRNAs are trimmed away in secondary processing steps (Fig. 31-70) through the action of **RNases D, M16, M23,** and **M5** to produce the mature rRNAs. These final cleavages only occur after the pre-rRNAs become associated with ribosomal proteins.

a. Ribosomal RNAs Are Methylated

During ribosomal assembly, the 16S and 23S rRNAs are methylated at a total of 24 specific nucleosides. The methylation reactions, which employ *S*-adenosylmethionine (Section 26-3E) as a methyl donor, yield N^6,N^6-dimethyladenine and $O^{2'}$-methylribose residues. $O^{2'}$-methyl groups may protect adjacent phosphodiester bonds from degradation by intracellular RNases (the mechanism of RNase hydrolysis involves utilization of the free 2′-OH group of ribose to eliminate the substituent on the 3′-phosphoryl group via the formation of a 2′,3′-cyclic phosphate intermediate; Figs. 5-3 and 15-3). However, the function of base methylation is unknown.

b. Eukaryotic rRNA Processing Is Guided by snoRNAs

The eukaryotic genome typically has several hundred tandemly repeated copies of rRNA genes that are contained in small, dark-staining nuclear bodies known as **nucleoli** (the site of rRNA transcription and processing and ribosomal subunit assembly; Fig. 1-5; note that nucleoli are not membrane enveloped). The primary rRNA transcript is an ~7500-nucleotide 45S RNA that contains, starting from its 5′ end, the 18S, 5.8S, and 28S rRNAs separated by spacer sequences (Fig. 31-72). In the first stage of its processing, 45S RNA is specifically methylated at numerous sites (106 in humans) that occur mostly in its rRNA sequences. About 80% of these modifications yield $O^{2'}$-methylribose residues and the remainder form methylated bases such as N^6,N^6-dimethyladenine and 2-methylguanine. In addition, many pre-rRNA U's (95 in humans) are converted to pseudouridines (Ψ's) (Section 30-5B), which may contribute to the rRNA's tertiary stability through hydrogen bonding involving its newly acquired ring NH group. The subsequent cleavage and trimming of the 45S RNA superficially resembles that of prokaryotic rRNAs. In fact, enzymes exhibiting RNase III- and RNase P-like activities occur in eukaryotes. The 5S eukaryotic rRNA is separately processed in a manner resembling that of tRNA (Section 31-4C).

The methylation sites in eukaryotic rRNAs occur exclusively within conserved domains that are therefore likely to participate in fundamental ribosomal processes. Indeed, the methylation sites generally occur in invariant sequences among yeast and vertebrates although the methylations themselves are not always conserved. These

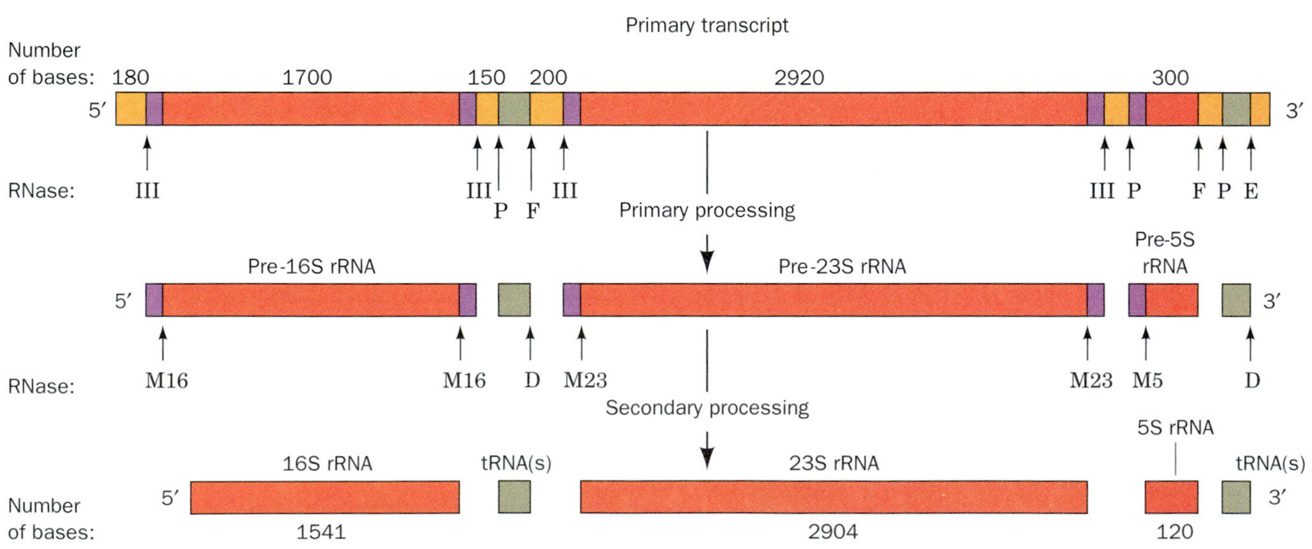

FIGURE 31-70 The posttranscriptional processing of *E. coli* rRNA. The transcriptional map is shown approximately to scale. The labeled arrows indicate the positions of the various nucleolytic cuts and the nucleases that generate them. [After Apiron, D., Ghora, B.K., Plantz, G., Misra, T.K., and Gegenheimer, P., *in* Söll, D., Abelson, J.N., and Schimmel P.R. (Eds.), *Transfer RNA: Biological Aspects,* p. 148, Cold Spring Harbor Laboratory Press (1980).]

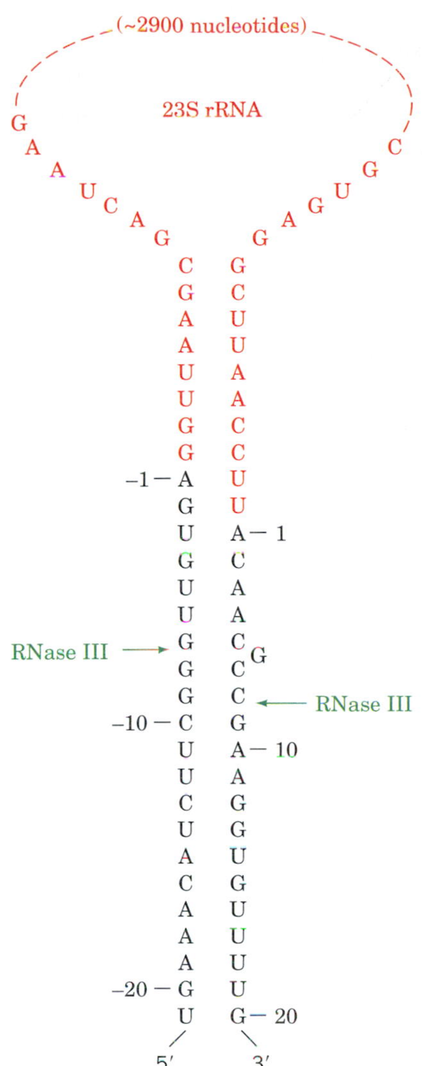

FIGURE 31-71 The proposed stem-and-giant-loop secondary structure in the 23S region of the *E. coli* primary rRNA transcript. The RNase III cleavage sites are indicated. [After Young, R.R., Bram, R.J., and Steitz, J.A., *in* Söll, D., Abelson, J.N., and Schimmel, P.R. (Eds.), *Transfer RNA: Biological Aspects, p.* 102, Cold Spring Harbor Laboratory Press (1980).]

methylation sites do not appear to have a consensus structure that might be recognized by a single methyltransferase. How, then, are these methylation sites targeted?

An important clue as to how the methylation sites on rRNA are selected came from the observation that pre-rRNA interacts with the members of a large family of **small nucleolar RNAs** (**snoRNAs;** ~100 in yeast and ~200 in mammals). These snoRNAs, whose lengths vary from 70 to 100 nt, contain segments of 10 to 21 nt that are precisely complementary to segments of the mature rRNAs that contain the O2′-methylation sites. These snoRNA sequences are located between the conserved sequence motifs known as box C (RUGAUGA) and box D (CUGA), which are respectively located on the 5′ and 3′ sides of the complementary segments. In intron-rich organisms such as vertebrates, most snoRNAs are encoded by the introns of structural genes so that not all excised introns are discarded.

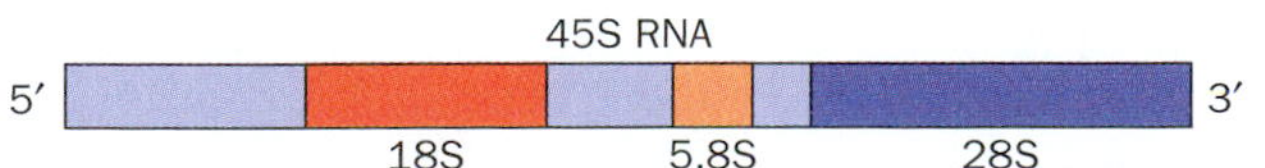

FIGURE 31-72 The organization of the 45S primary transcript of eukaryotic rRNA.

The snoRNA nucleotide that pairs with the nucleotide to be O2′-methylated always precedes box D by exactly 5 nt. Evidently, each of these so-called **box C/D snoRNAs** act to guide the methylation of a single site. In fact, in those cases in which two adjacent ribose residues are methylated, two box C/D snoRNAs with overlapping sequences occur. The methylation is mediated by a complex of at least six nucleolar proteins, including **fibrillarin** (~325 residues), the likely methyltransferase, which together with a box C/D snoRNA form **snoRNPs.** The conversion of specific rRNA U's to Ψ's is similarly mediated by a different subgroup of snoRNAs, the **box H/ACA snoRNAs,** so called because they contain the sequence motifs ACANNN at the snoRNA's 3′ end and box H (ANANNA) at its 5′ end, so as to flank a sequence that partially base pairs to the pre-rRNA segment containing the U to be converted to Ψ. Archaea also modify their rRNAs via RNA-guided methylations and U to Ψ conversions, but interestingly, the analogous reactions in eubacteria are mediated by protein enzymes that lack RNA.

C. *Transfer RNA Processing*

tRNAs, as we discuss in Section 32-2A, consist of ~80 nucleotides that assume a secondary structure with four base paired stems known as the **cloverleaf structure** (Fig. 31-73). All tRNAs have a large fraction of modified bases

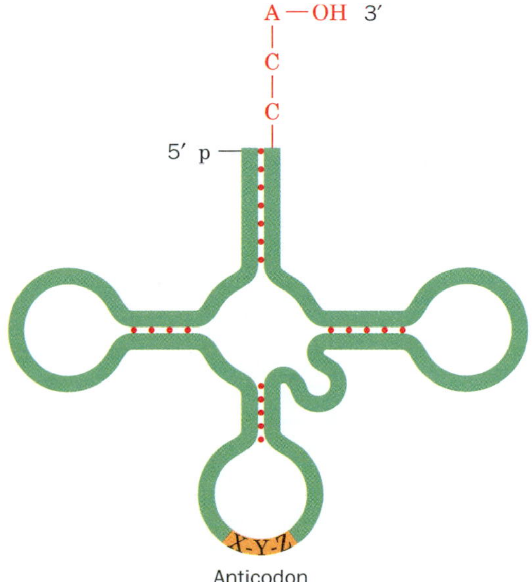

FIGURE 31-73 A schematic diagram of the tRNA cloverleaf secondary structure. Each dot indicates a base pair in the hydrogen bonded stems. The position of the anticodon triplet and the 3′-terminal —CCA are indicated.

(whose structure, function, and synthesis is also discussed in Section 32-2A) and each has the 3′-terminal sequence —CCA to which the corresponding amino acid is appended in the amino acid–charged tRNA. The **anticodon** (which is complementary to the codon specifying the tRNA's corresponding amino acid) occurs in the loop of the cloverleaf structure opposite the stem containing the terminal nucleotides.

The *E. coli* chromosome contains ~60 tRNA genes. Some of them are components of rRNA operons (Section 31-4B); the others are distributed, often in clusters, throughout the chromosome. The primary tRNA transcripts, which contain from one to as many as four or five identical tRNA copies, have extra nucleotides at the 3′ and 5′ ends of each tRNA sequence. The excision and trimming of these tRNA sequences resemble those for *E. coli* rRNAs (Section 31-4B) in that both processes employ some of the same nucleases.

a. RNase P Is a Ribozyme

RNase P, which generates the 5′ ends of tRNAs (Fig. 31-70), is a particularly interesting enzyme because it has, in *E. coli*, a 377-nucleotide RNA component (~125 kD vs 14 kD for its 119-residue protein subunit) that is essential for its enzymatic activity. The enzyme's RNA was, quite understandably, first proposed to function in recognizing the substrate RNA through base pairing and to thereby guide the protein subunit, which was presumed to be the actual nuclease, to the cleavage site. However, Sidney Altman demonstrated that *the RNA component of RNase P is, in fact, the enzyme's catalytic subunit* by showing that protein-free RNase P RNA catalyzes the cleavage of substrate RNA at high salt concentrations. RNase P protein, which is basic, evidently functions at physiological salt concentrations to electrostatically reduce the repulsions between the polyanionic ribozyme and substrate RNAs. The argument that trace quantities of RNase P protein are really responsible for the RNase P reaction was disposed of by showing that catalytic activity is exhibited by RNase P RNA that has been transcribed in a cell-free system. RNase P activity occurs in eukaryotes (nuclei, mitochondria, and chloroplasts) as well as in prokaryotes although eukaryotic nuclear RNase P's have 9 or 10 protein subunits. Indeed, RNase P mediates one of the two ribozymal activities that occur in all cellular life, the other being associated with ribosomes (Section 32-3D).

The 400-residue RNase P from *B. subtilis,* which differs greatly in sequence from that of *E. coli,* is predicted to form two secondary structural domains (Fig. 31-74*a*): the specificity domain, which consists of nucleotides 86 to 239, and the catalytic domain, which comprises the remainder of the molecule. The X-ray structure of the specificity domain, determined by Alfonso Mondragón, is in excellent agreement with this secondary structural prediction (Fig. 31-74*b*). It consists mainly of two sets of stacked helical stems (P7-P10-P12 and P8-P9) coming together at a junction together with an unusual folded module (J11/12-J12/11) that links stems P11 and P12. Sequence analysis reveals that many of this RNA domain's conserved residues participate in interactions that are necessary for its proper

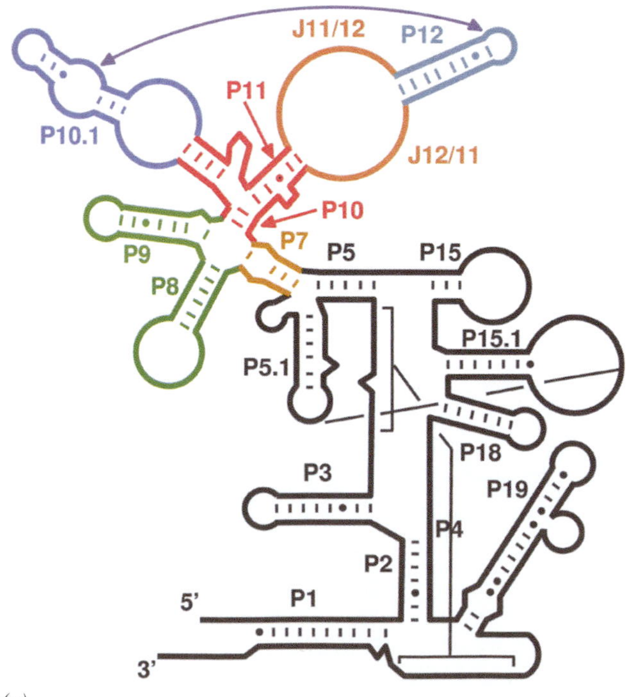

(*a*)

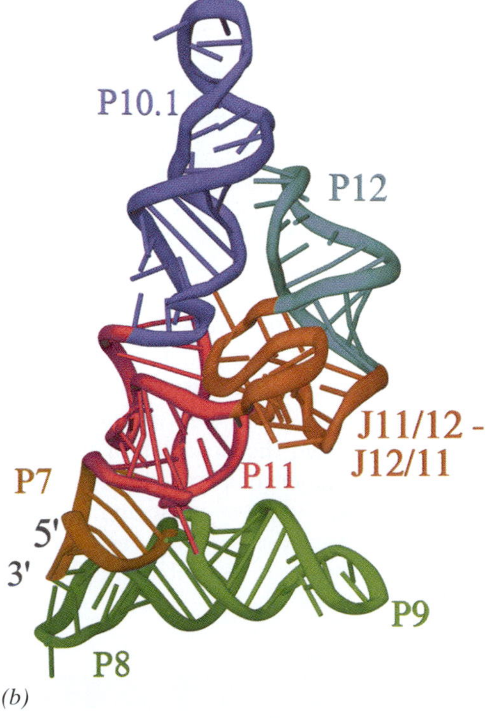

(*b*)

FIGURE 31-74 The structure of the RNA of B. *subtilis* RNase P. (*a*) Its predicted secondary structure in which the specificity domain is drawn in various colors and the catalytic domain is black. (*b*) The X-ray structure of the specificity domain in which its various segments are colored as in Part *a*. [Courtesy of Alfonso Mondragón, Northwestern University, PDBid 1NBS.]

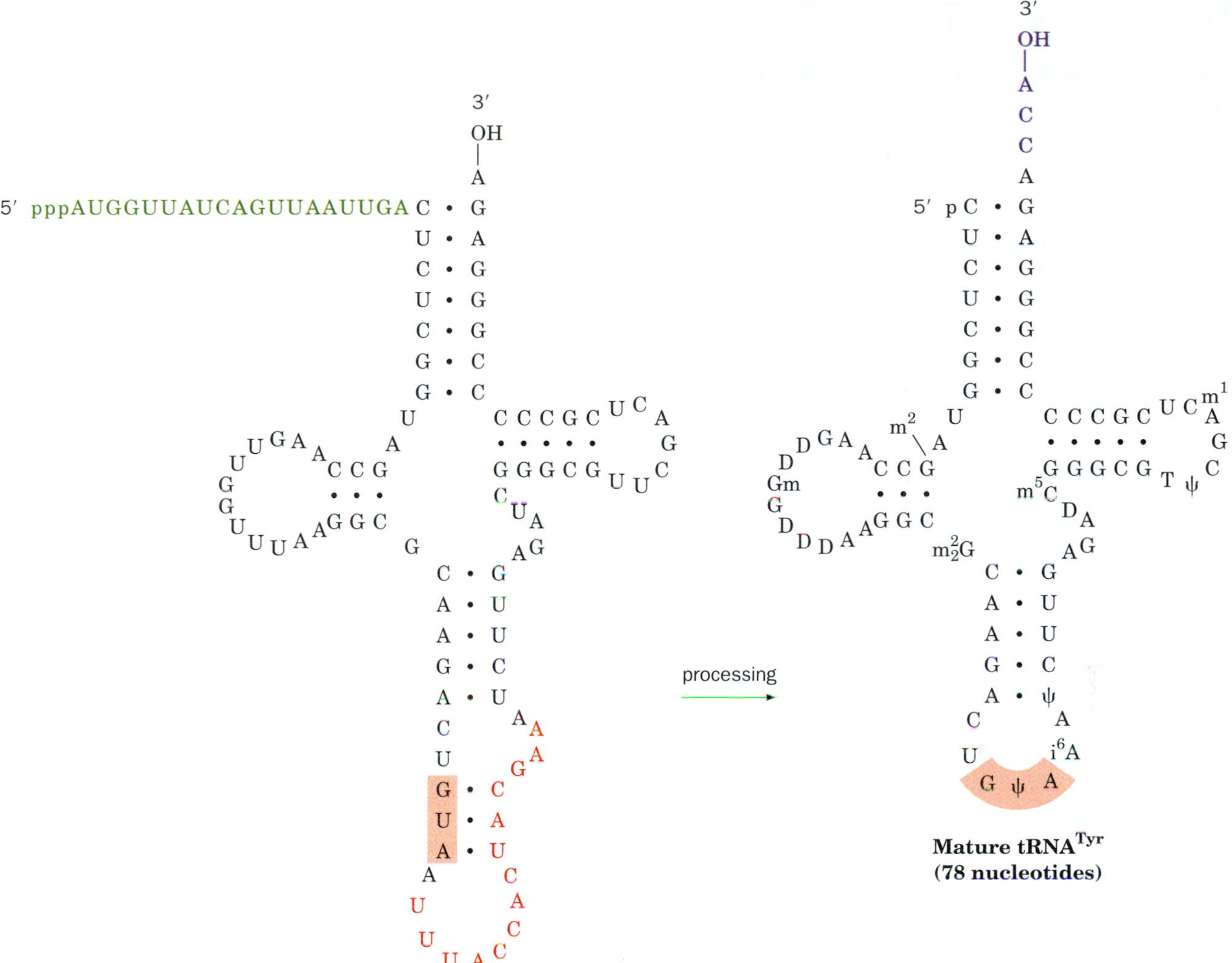

FIGURE 31-75 The posttranscriptional processing of yeast tRNA^{Tyr}. A 14-nucleotide intervening sequence and a 19-nucleotide 5′-terminal sequence are excised from the primary transcript, a —CCA is appended to the 3′ end, and several of the bases are modified (their symbols are defined in Fig. 32-13) to form the mature tRNA. The anticodon is shaded. [After DeRobertis, E.M. and Olsen, M.V., *Nature* **278,** 142 (1989).

folding. Chemical modification studies indicate that the binding of pre-tRNA to RNase P protects the stacked bases of A130 and A230 that respectively bulge out of the P9 and P11 stems, together with the base of G220, which extends from J12/11. This identifies the side of the specificity domain to which its pre-tRNA substrate binds as that facing the viewer in Fig. 31-74*b*.

b. Many Eukaryotic Pre-tRNAs Contain Introns

Eukaryotic genomes contain from several hundred to several thousand tRNA genes. Many eukaryotic primary tRNA transcripts, for example, that of yeast tRNA^{Tyr} (Fig. 31-75), contain a small intron adjacent to their anticodons as well as extra nucleotides at their 5′ and 3′ ends. Note that this intron is unlikely to disrupt the tRNA's cloverleaf structure.

c. The —CCA Ends of Eukaryotic tRNAs Are Posttranscriptionally Appended

Eukaryotic tRNA transcripts lack the obligatory —CCA sequence at their 3′ ends. This is appended to the immature tRNAs by the enzyme **tRNA nucleotidyltransferase,** which sequentially adds two C's and an A to tRNA using CTP and ATP as substrates. This enzyme also occurs in prokaryotes, although, at least in *E. coli,* the tRNA genes all encode a —CCA terminus. The *E. coli* tRNA nucleotidyltransferase is therefore likely to function in the repair of degraded tRNAs.

CHAPTER SUMMARY

1 ■ The Role of RNA in Protein Synthesis The central dogma of molecular biology states that "DNA makes RNA makes protein" (although RNA can also "make" DNA). There is, however, enormous variation among the rates at which the various proteins are made. Certain enzymes, such as those of the *lac* operon, are synthesized only when the substances whose metabolism they catalyze are present. The *lac* operon consists of the control sequences *lacP* and *lacO* followed by the tandemly arranged genes for β-galactosidase *(lacZ),* galactoside permease *(lacY),* and thiogalactoside transacetylase *(lacA).* In the absence of inducer, physiologically allolactose, the *lac* repressor, the product of the *lacI* gene, binds to operator *(lacO)* so as to prevent the transcription of the *lac* operon by RNA polymerase. The binding of inducer causes the repressor to release the operator that allows the *lac* structural genes to be transcribed onto a single polycistronic mRNA. The mRNAs transiently associate with ribosomes so as to direct them to synthesize their encoded polypeptides.

2 ■ RNA Polymerase The holoenzyme of *E. coli* RNA polymerase (RNAP) has the subunit structure $\alpha_2\beta\beta'\omega\sigma$. It initiates transcription on the antisense (template) strand of a gene at a position designated by its promoter. The most conserved region of the promoter is centered at about the -10 position and has the consensus sequence TATAAT. The -35 region is also conserved in efficient promoters. DMS footprinting studies indicate that the RNAP holoenzyme forms an "open" initiation complex with the promoter. *Taq* RNAP holoenzyme has the shape of a crab claw with its σ subunit extended along its "top." In the closed complex, the DNA binds across the holoenzyme's "top" with all sequence-specific contacts made by the σ subunit. This suggests a model of the open complex in which the template strand in the transcription bubble passes through a tunnel in the protein to the active site channel where it pairs with incoming ribonucleotides. After the initiation of RNA synthesis, the σ subunit dissociates from the core enzyme, which then autonomously catalyzes chain elongation in the $5' \rightarrow 3'$ direction. RNA synthesis is terminated by a segment of the transcript that forms a G + C-rich hairpin with an oligo(U) tail that spontaneously dissociates from the DNA. Termination sites that lack these sequences require the assistance of Rho factor for proper chain termination.

In the nuclei of eukaryotic cells, RNAPs I, II, and III, respectively, synthesize rRNA precursors, mRNA precursors, and tRNAs + 5S RNA. The structure of yeast RNAP II resembles that of *Taq* RNAP but is somewhat larger and has more subunits. The structure of its transcribing complex reveals a one-turn segment of RNA · DNA hybrid helix at the active site, which is in contact with the solvent via a pore leading into a funnel through which NTPs presumably pass. The minimal RNA polymerase I promoter extends between nucleotides -31 and $+6$. Many RNA polymerase II promoters contain a conserved TATAAAA sequence, the TATA box, located around position -27. Enhancers are transcriptional activators that can have variable positions and orientations relative to the transcription start site. RNA polymerase III promoters are located within the transcribed regions of their gene between positions $+40$ and $+80$.

3 ■ Control of Transcription in Prokaryotes Prokaryotes can respond rapidly to environmental changes, in part because the translation of mRNAs commences during their transcription and because most mRNAs are degraded within 1 to 3 min of their synthesis. The ordered expression of sets of genes in some bacteriophages and bacteria is controlled by a succession of σ factors. The *lac* repressor is a tetrameric protein of identical subunits that, in the absence of inducer, nonspecifically binds to duplex DNA but binds much more tightly to *lac* promoter. The promoter sequence that *lac* repressor protects from nuclease digestion has nearly palindromic symmetry. Yet, methylation protection and mutational studies indicate that repressor is not symmetrically bound to promoter. *lac* repressor prevents RNA polymerase from properly initiating transcription at the *lac* promoter.

The presence of glucose represses the transcription of operons specifying certain catabolic enzymes through the mediation of cAMP. On binding cAMP, which accumulates only in the absence of glucose, catabolite gene activator protein (CAP) binds at or immediately upstream of the promoters of these operons, including the *lac* operon, thereby activating their transcription through the binding to the C-terminal domain of the associated RNAP's α subunit (αCTD). CAP's two symmetry equivalent DNA-binding domains each bind in the major groove of their target DNA via a helix–turn–helix (HTH) motif that also occurs in numerous prokaryotic repressors. The binding between these repressors and their target DNAs is mediated by mutually favorable associations between these macromolecules rather than any specific interactions between particular base pairs and amino acid side chains analogous to Watson–Crick base pairing. Sequence-specific interactions between the *met* repressor and its target DNA occur through a 2-fold symmetric antiparallel β ribbon that this protein inserts into the DNA's major groove. *araBAD* transcription is controlled by the levels of L-arabinose and CAP–cAMP through a remarkable complex of the control protein AraC to two binding sites, *ara*O_2 and *ara*I_1, that forms an inhibitory DNA loop. On binding L-arabinose and when CAP–cAMP is adjacently bound, AraC releases *ara*O_2 and instead binds *ara*I_2, thereby releasing the loop and activating RNA polymerase to transcribe the *araBAD* operon. The expression of the *lac* operon is also in part controlled by DNA loop formation. The *lac* repressor is a dimer of dimers, one of which binds to the operator *lac*O_1 and the other to *lac*O_2 or *lac*O_3 to form a DNA loop that may interfere with RNAP binding to the *lac* promoter. The binding of an inducer such as IPTG to a *lac* repressor dimer core domain alters the angle between its two attached DNA-binding domains such that they cannot simultaneously bind to the *lac* operator, thereby weakening the repressor's grip on the DNA.

The expression of the *E. coli trp* operon is regulated by both attenuation and repression. On binding tryptophan, its corepressor, *trp* repressor binds to the *trp* operator, thereby blocking *trp* operon transcription. When tryptophan is available, much of the *trp* transcript that has escaped repression is prematurely terminated in the *trpL* sequence because its transcript contains a segment that forms a normal terminator structure. When tryptophanyl–$tRNA^{Trp}$ is scarce, ribosomes stall at the transcript's two tandem Trp codons. This permits the newly synthesized RNA to form a base paired stem and loop that prevents the formation of the terminator structure. Several other operons are similarly regulated by attenuation. The stringent response is another mechanism by which *E. coli* match the rate of transcription to charged tRNA availability. When a specified charged tRNA is scarce, stringent factor on active ribosomes synthesizes ppGpp, which inhibits the transcription of rRNA and some mRNAs while stimulating the transcription of other mRNAs.

4 ■ Posttranscriptional Processing Most prokaryotic mRNA transcripts require no additional processing. However, eukaryotic mRNAs have an enzymatically appended 5′ cap and, in most cases, an enzymatically generated poly(A) tail. Moreover, the introns of eukaryotic mRNA primary transcripts (hnRNAs) are precisely excised via lariat intermediates and their flanking exons are spliced together. Group I and Group II introns are self-splicing, that is, their RNAs function as ribozymes (RNA enzymes). Ribozymes, such as the *Tetrahymena* pre-rRNA and hammerhead ribozymes, have complex structures containing several base paired stems. Pre-mRNAs are spliced by large and complex particles named spliceosomes that consist of four different small nuclear ribonucleoproteins (snRNPs) and which are assisted by the participation of a variety of protein splicing factors. Many eukaryotic proteins consist of modules that also occur in other proteins and hence appear to have evolved via the stepwise collection of exons through recombination events. The alternative splicing of pre-mRNAs greatly increases the variety of proteins expressed by eukaryotic genomes. Certain mRNAs are subject to RNA editing, either by the replacement, insertion, or deletion of specific bases in a process that is directed by guide RNAs (gRNAs), or by substitutional editing mediated by cytidine deaminases or adenosine deaminases. In RNA interference (RNAi), dsRNA is cleaved by the endoribonuclease Dicer to small interfering RNAs (siRNAs) that guide the hydrolytic cleavage of the complementary mRNAs by the RNA-induced silencing complex (RISC), thereby preventing the mRNAs' transcription.

The primary transcripts of *E. coli* rRNAs contains all three rRNAs together with some tRNAs. These are excised and trimmed by specific endonucleases and exonucleases. The eukaryotic 18S, 5.8S, and 28S rRNAs are similarly transcribed as a 45S precursor, which is processed in a manner resembling that of *E. coli* rRNAs. Eukaryotic rRNAs are modified by the methylation of specific nucleosides, as are prokaryotic rRNA, and by the conversion of certain U's to pseudouridines (Ψ's). These processes are guided by small nucleolar RNAs (snoRNAs). Prokaryotic tRNAs are excised from their primary transcripts and trimmed in much the same way as are rRNAs. In RNase P, one of the enzymes mediating this process, the catalytic subunit is an RNA. Eukaryotic tRNA transcripts also require the excision of a short intron and the enzymatic addition of a 3′-terminal —CCA to form the mature tRNA.

REFERENCES

GENERAL

Adams, R.L.P., Knowler, J.T., and Leader, D.P., *The Biochemistry of the Nucleic Acids* (11th ed.), Chapters 9–11, Chapman and Hall (1992).

Brown, T.A., *Genomes* (2nd ed.), Chapter 10, Wiley-Liss (2002).

Gesteland, R.F., Cech, T.R., and Atkins, J.F. (Eds.), *The RNA World* (2nd ed.), Cold Spring Harbor Laboratory (1999). [A series of authoritative articles on the nature of the prebiotic "RNA world" as revealed by the RNA "relics" in modern organisms.]

Hodgson, D.A. and Thomas, C.M. (Eds.), *Signals, Switches, Regulons and Cascades: Control of Bacterial Gene Expression,* Cambridge University Press (2002).

Lewin, B., *Genes VII,* Chapters 5, 9, 10, and 20, Oxford (2000).

THE GENETIC ROLE OF RNA

Brachet, J., Reminiscences about nucleic acid cytochemistry and biochemistry, *Trends Biochem. Sci.* **12,** 244–246 (1987).

Brenner, S., Jacob, F., and Meselson, M., An unstable intermediate carrying information from genes to ribosomes for protein synthesis, *Nature* **190,** 576–581 (1960). [The experimental verification of mRNA's existence.]

Crick, F., Central dogma of molecular biology, *Nature* **227,** 561–563 (1970).

Hall, B.D. and Spiegelman, S., Sequence complementarity of T2-DNA and T2-specific RNA, *Proc. Natl. Acad. Sci.* **47,** 137–146 (1964). [The first use of RNA–DNA hybridization.]

Jacob, F. and Monod, J., Genetic regulatory mechanisms in the synthesis of proteins, *J. Mol. Biol.* **3,** 318–356 (1961). [The classic paper postulating the existence of mRNA and operons and explaining how the transcription of operons is regulated.]

Pardee, A.B., Jacob, F., and Monod, J., The genetic control and cytoplasmic expression of "inducibility" in the synthesis of β-galactosidase by *E. coli, J. Mol. Biol.* **1,** 165–178 (1959). [The PaJaMo experiment.]

Thieffry, D., Forty years under the central dogma, *Trends Biochem. Sci.* **23,** 312–316 (1998).

RNA POLYMERASE AND mRNA

Campbell, E.A., Korzheva, N., Mustaev, A., Murakami, K., Nair, S., Goldfarb, A., and Darst, S.A., Structural mechanism for rifampicin inhibition of bacterial RNA polymerase, *Cell* **104,** 901–912 (2001).

Cramer, P., Multisubunit RNA polymerases, *Curr. Opin. Struct. Biol.* **12,** 89–97 (2002).

Cramer, P., Bushnell, D.A., and Kornberg, R.D., Structural basis of transcription: RNA polymerase at 2.8 Å resolution, *Science* **292,** 1863–1876 (2001); *and* Gnatt, A.L., Cramer, P., Fu, J., Bushnell, D.A., and Kornberg, R.D., Structural basis of transcription: An RNA polymerase II elongation complex at 3.3 Å resolution, *Science* **292,** 1876–1882 (2001).

Dahmus, M.E., Reversible phosphorylation of the C-terminal domain of RNA polymerase II, *J. Biol. Chem.* **271,** 19009–19012 (1996).

Darst, S.A., Bacterial RNA polymerase, *Curr. Opin. Struct. Biol.* **11,** 155–162 (2001).

Das, A., Control of transcription termination by RNA-binding proteins, *Annu. Rev. Biochem.* **62,** 893–930 (1993).

DeHaseth, P.L., Zupancic, M.L., and Record, M.T., Jr., RNA polymerase-promoter interactions: the comings and goings of RNA polymerase, *J. Bacteriol.* **180,** 3019–3025 (1998).

Erie, D.A., Yager, T.D., and von Hippel, P.H., The single nucleotide addition cycle in transcription, *Annu. Rev. Biophys. Biomol. Struct.* **21,** 379–415 (1992).

Estrem, S.T., Gaal, T., Ross, W., and Gourse, R.L., Identification of an UP element consensus sequence for bacterial promoters, *Proc. Natl. Acad. Sci.* **95,** 9761–9766 (1998).

Futcher, B., Supercoiling and transcription, or vice versa? *Trends Genet.* **4,** 271–272 (1988).

Gannan, F., O'Hare, K., Perrin, F., LePennec, J.P., Benoist, C., Cochet, M., Breathnach, R., Royal, A., Garapin, A., Cami, B., and Chambon, P., Organization and sequences of the 5′ end of a cloned complete ovalbumin gene, *Nature* **278,** 428–434 (1979).

Geiduschek, E.P. and Tocchini-Valentini, G.P., Transcription by RNA polymerase III, *Annu. Rev. Biochem.* **57,** 873–914 (1988).

Harada, Y., Ohara, O., Takatsuki, A., Itoh, H., Shimamoto, N., and Kinosita, K., Jr., Direct observation of DNA rotation during transcription by *Escherichia coli* RNA polymerase, *Nature* **409,** 113–115 (2001).

Huffman, J.L. and Brennan, R.G., Prokaryotic transcriptional regulators: more than just the helix-turn-helix motif, *Curr. Opin. Struct. Biol.* **12,** 98–106 (2002).

Kamitori, S. and Takusagawa, F., Crystal structure of the 2:1 complex between d(GAAGCTTC) and the anticancer drug actinomycin D, *J. Mol. Biol.* **225,** 445–456 (1992).

Khoury, G. and Gruss, P., Enhancer elements, *Cell* **33,** 313–314 (1983).

Murikami, K.S., Masuda, S., and Darst, S., Structural basis of transcription initiation: RNA polymerase at 4 Å resolution, *Science* **296,** 1280–1284 (2002); Murikami, K.S., Masuda, S., Campbell, E.A., Muzzin, O., and Darst, S., Structural basis of transcription initiation: An RNA polymerase holoenzyme-DNA complex, *Science* **296,** 1285–1290 (2002); *and* Marakami, K.S. and Darst, S.A., Bacterial RNA polymerases: the whole story, *Curr. Opin. Struct. Biol.* **13,** 31–39 (2003).

Reynolds, R., Bermúdez-Cruz, R.M., and Chamberlin, M.J., Parameters affecting transcription termination by *Escherichia coli* RNA. I. Analysis of 13 rho-independent terminators, *J. Mol. Biol.* **224,** 31–51 (1992); *and* Reynolds, R. and Chamberlin, M.J., Parameters affecting transcription termination by *Escherichia coli* RNA. II. Construction of hybrid terminators, *J. Mol. Biol.* **224,** 53–63 (1992).

Richardson, J.P., Transcription termination, *Crit. Rev. Biochem. Mol. Biol.* **28,** 1–30 (1993).

Richardson, J.P., Structural organization of transcription termination factor rho, *J. Biol. Chem.* **271,** 1251–1254 (1996).

Shilatifard, A., Conway, R.C., and Conway, J.W., The RNA polymerase II elongation complex, *Annu. Rev. Biochem.* **72,** 693–715 (2003).

Skordalakes, E. and Berger, J.M., The structure of Rho transcription terminator: Mechanism of mRNA recognition and helicase loading, *Cell.* **114,** 135–146 (2003).

Uptain, S.M., Kane, C.M., and Chamberlin, M.J., Basic mechanisms of transcription elongation and its regulation, *Annu. Rev. Biochem.* **66,** 117–172 (1997).

Willis, I.M., RNA polymerase III, *Eur. J. Biochem.* **212,** 1–11 (1993).

Zhang, G., Campbell, E.A., Minakhin, L., Richter, C., Severinov, K., and Darst, S.A., Crystal structure of *Thermus aquaticus* core RNA polymerase at 3.3 Å resolution, *Cell* **98,** 811–824 (1999).

CONTROL OF TRANSCRIPTION

Anderson, J.E., Ptashne, M., and Harrison, S.C., The structure of the repressor–operator complex of bacteriophage 434, *Nature* **326,** 846–852 (1987).

Bell, C.E. and Lewis, M., The Lac repressor: a second generation of structural and functional studies, *Curr. Opin. Struct. Biol.* **11,** 19–25 (2001).

Bennoff, B., Yang, H., Lawson, C.L., Parkinson, G., Liu, J., Blatter, E., Ebright, Y.W., Berman, H.M., and Ebright, R.H., Structural basis of transcription activation: The CAP-αCTD-DNA complex, *Science* **297,** 1562–1566 (2002).

Busby, S. and Ebright, R.H., Transcription activation by catabolite activator protein (CAP), *J. Mol. Biol.* **293,** 199–213 (1999).

Gallant, J.A., Stringent control in *E. coli, Annu. Rev. Genet.* **13,** 393–415 (1979).

Gartenberg, M.R. and Crothers, D.M., Synthetic DNA bending sequences increase the rate of *in vitro* transcription initiation at the *Escherichia coli lac* promoter, *J. Mol. Biol.* **219,** 217–230 (1991).

Gilbert, W. and Müller-Hill, B., Isolation of the lac repressor, *Proc. Natl. Acad. Sci.* **56,** 1891–1898 (1966).

Harmor, T., Wu, M., and Schleif, R., The role of rigidity in DNA looping–unlooping by AraC, *Proc. Natl. Acad. Sci.* **98,** 427–431 (2001).

Kolb, A., Busby, S., Buc, H., Garges, S., and Adhya, S., Transcriptional regulation by cAMP and its receptor protein, *Annu. Rev. Biochem.* **62,** 749–795 (1993).

Kolter, R. and Yanofsky, C., Attenuation in amino acid biosynthetic operons, *Annu. Rev. Genet.* **16,** 113–134 (1982).

Lamond, A.I. and Travers, A.A., Stringent control of bacterial transcription, *Cell* **41,** 6–8 (1985).

Lee, J. and Goldfarb, A., *lac* repressor acts by modifying the initial transcribing complex so that it cannot leave the promoter, *Cell* **66,** 793–798 (1991).

Lewis, M., Chang, G., Horton, N.C., Kercher, M.A., Pace, H.C., Schumacher, M.A., Brennan, R.G., and Lu, P., Crystal structure of the lactose operon repressor and its complexes with DNA and inducer, *Science* **271,** 1247–1254 (1996).

Lobel, R.B. and Schleif, R.F., DNA looping and unlooping by AraC protein, *Science* **250,** 528–532 (1990).

Luisi, B.F. and Sigler, P.B., The stereochemistry and biochemistry of the *trp* repressor-operator complex, *Biochim. Biophys. Acta* **1048,** 113–126 (1990).

McKnight, S.L. and Yamamoto, K.R. (Eds.), *Transcriptional Regulation,* Cold Spring Harbor Laboratory Press (1992). [A two-volume compendium that contains authoritative articles on many aspects of prokaryotic transcriptional control.]

Mondragón, A. and Harrison, S.C., The phage 434 Cro/O_R1 complex at 2.5 Å resolution, *J. Mol. Biol.* **219,** 321–334 (1991); *and* Wolberger, C., Dong, Y., Ptashne, M., and Harrison, S.C., Structure of phage 434 Cro/DNA complex, *Nature* **335,** 789–795 (1988).

Oehler, S., Eismann, E.R., Krämer, H., and Müller-Hill, B., The three operators of the *lac* operon cooperate in repression, *EMBO J.* **9,** 973–979 (1990).

Pace, H.C., Kercher, M.A., Lu, P., Markiewicz, P., Miller, J.H., Chang, G., and Lewis, M., *Lac* repressor genetic map in real space, *Trends Biochem. Sci.* **22,** 334–339 (1997).

Reeder, T. and Schleif, R., AraC protein can activate transcription from only one position and when pointed in only one direction, *J. Mol. Biol.* **231,** 205–218 (1993).

Rogers, D.W. and Harrison, S.C., The complex between phage 434 repressor DNA-binding domain and operator site O_R3: structural differences between consensus and non-consensus half-sites, *Structure* **1,** 227–240 (1993).

Schleif, R., DNA looping, *Annu. Rev. Biochem.* **61,** 199–223 (1992).

Schleif, R., Regulation of the L-arabinose operon of *Escherichia coli, Trends Genet.* **16,** 559–565 (2000).

Schultz, S.C., Shields, G.C., and Steitz, T.A., Crystal structure of a CAP-DNA complex: The DNA is bent by 90°, *Science* **253,** 1001–1007 (1991).

Shakked, Z., Guzikevich-Guerstein, G., Frolow, F., Rabinovich, D., Joachimiak, A., and Sigler, P.B., Determinants of repressor/operator recognition from the structure of the *trp* operator binding site, *Nature* **368,** 469–473 (1994).

Soisson, S.M., MacDougall-Shackleton, B., Schleif, R., and Wolberger, C., Structural basis for ligand-regulated oligomerization of AraC, *Science* **276,** 421–425 (1997). [The X-ray structure of AraC alone and in complex with arabinose.]

Somers, W.S. and Phillips, S.E.V., Crystal structure of the *met* repressor-operator complex at 2.8 Å resolution reveals DNA recognition by β-strands, *Nature* **359,** 387–393 (1992).

Spronk, C.A.E.M., Bonvin, A.M.J.J., Radha, P.K., Melacini, G., Boelens, R., and Kaptein, R., The solution structure of *Lac* repressor headpiece 62 complexed to a symmetrical *lac* operator, *Structure* **7,** 1483–1492 (1999).

Steitz, T.A., Structural studies of protein–nucleic acid interaction: the sources of sequence-specific binding, *Quart. Rev. Biophys.* **23,** 205–280 (1990). [Also published as a book of the same title by Cambridge University Press (1993).]

Yanofsky, C., Transcription attenuation, *J. Biol. Chem.* **263,** 609–612 (1988); *and* Attenuation in the control of expression of bacterial operons, *Nature* **289,** 751–758 (1981).

POSTTRANSCRIPTIONAL PROCESSING

Apiron, D. and Miczak, A., RNA processing in prokaryotic cells, *BioEssays* **15,** 113–119 (1993).

Bachellerie, J.-P. and Cavaillé, J., Guiding ribose methylation of rRNA, *Trends Biochem. Sci.* **22,** 257–261 (1997).

Bard, J., Zhelkovsky, A.M., Helmling, S., Earnest, T.N., Moore, C.L. and Bohm, A., Structure of yeast poly(A) polymerase alone and in complex with 3′-dATP, *Science* **289,** 1346–1349 (2000); *and* Martin, G., Keller, W., and Doublié, S., Crystal structure of mammalian poly(A) polymerase in complex with an analog of ATP, *EMBO J.* **19,** 4193–4203 (2000).

Bass, B.L., RNA editing by adenosine deaminases that act on RNA, *Annu. Rev. Biochem.* **71,** 817–846 (2002).

Black, D.L., Mechanism of alternative pre-messenger RNA splicing, *Annu. Rev. Biochem.* **72,** 291–336 (2003).

Brantl, S., Antisense regulation and RNA interference, *Biochim. Biophys. Acta* **1575,** 15–25 (2002).

Cate, J.H., Gooding, A.R., Podell, E., Zhou, K., Golden, B.L., Kundrot, C.E., Cech, T.R., and Doudna, J.A., Crystal structure of a group I ribozyme domain: principles of RNA packing, *Science* **273,** 1678–1690 (1996); Cate, J.H., Gooding, A.R., Podell, E., Zhou, K., Golden, B.L., Szewczak, A.A., Kundrot, C.E., Cech, T.R., and Doudna, J.A., RNA tertiary mediation by adenosine platforms, *Science* **273,** 1696–1699 (1996); *and* Cate, J.H. and Doudna, J.A., Metal-binding sites in the major groove of a large ribozyme domain, *Structure* **4,** 1221–1229 (1996).

Cech, T.R., Self-splicing of group I introns, *Annu. Rev. Biochem.* **59,** 543–568 (1990).

Chambon, P., Split genes, *Sci. Am.* **244**(5), 60–71 (1981).

Davis, R.E., Spliced leader RNA *trans*-splicing in metazoa, *Parasitology Today* **12,** 33–40 (1996).

Decatur, W.A. and Fournier, M.J., RNA-guided nucleotide modification of ribosomal and other RNAs, *J. Biol. Chem.* **278,** 695–698 (2003).

Denli, A.M. and Hannon, G.J., RNAi: An evergrowing puzzle, *Trends Biochem. Sci.* **28,** 196–201 (2003).

Doherty, E.A. and Doudna, J.A., Ribozyme structures and mechanisms, *Annu. Rev. Biophys. Biomol. Struct.* **30,** 457–475 (2001); and *Annu. Rev. Biochem.* **69,** 597–615 (2000).

Dreyfuss, G., Kim, V.N., and Kataoka, N., Messenger-RNA-binding proteins and the messages they carry. *Nature Rev. Cell Biol.* **3,** 195–205 (2002) ; *and* Shyu, A.-B. and Wilkinson, M.F., The double lives of shuttling mRNA binding proteins, *Cell* **102,** 135–138 (2000).

Ehretsmann, C.P., Carpousis, A.J., and Krisch, H.M., mRNA degradation in prokaryotes, *FASEB J.* **6,** 3186–3192 (1992).

Frank, D.N. and Pace, N.R., Ribonuclease P: unity and diversity in a tRNA processing ribozyme, *Annu. Rev. Biochem.* **67,** 153–180 (1998).

Gerber, A.P. and Keller, W., RNA editing by base deamination: more enzymes, more targets, new mysteries, *Trends Biochem. Sci.* **26,** 376–384 (2001).

Golden, B.L., Gooding, A.R., Podell, E.R., and Cech, T.R., A pre-organized active site in the crystal structure of the *Tetrahymena* ribozyme, *Science* **282,** 259–264 (1998).

Gopalan, V., Vioque, A., and Altman, S., RNase P: variations and uses, *J. Biol. Chem.* **277,** 6759–6762 (2002).

Gott, J.M. and Emeson, R.B., Functions and mechanisms of RNA editing, *Annu. Rev. Genet.* **34,** 499–531 (2000).

Hannon, G.J., RNA interference, *Nature* **418,** 244–251 (2002).

Grosjean, H. and Benne, R. (Eds.), *Modification and Editing of RNA,* ASM Press (1998).

Kambach, C., Walke, S., Young, R., Avis, J.M., de la Fortelle, E., Raker, V.A., Lührmann, R., and Nagai, K., Crystal structures of two Sm protein complexes and their implications for the assembly of the spliceosomal snRNPs, *Cell* **96,** 375–387 (1999).

Keegan, L.P., Gallo, A., and O'Connell, M.A., The many roles of an RNA editor, *Nature Rev. Genet.* **2,** 869–878 (2001).

Krämer, A., The structure and function of proteins involved in mammalian pre-mRNA splicing, *Annu. Rev. Biochem.* **65,** 367–409 (1996).

Krasilnikov, A.S., Yang, X., Pan, T., and Mondragón, A., Crystal structure of the specificity domain of ribonuclease P, *Nature* **421,** 760–764 (2003).

Li, Y. and Breaker, R.R., Deoxyribozymes: new players in an ancient game of biocatalysis, *Curr. Opin. Struct. Biol.* **9,** 315–323 (1999).

Liu, Z., Luyten, I., Bottomley, M.J., Messais, A.C., Houngninou-Molango, S., Sprangers, R., Zanier, K., Krämer, A., and Satler, M., Structural basis for recognition of the intron branch site RNA by splicing factor 1, *Science* **294,** 1098–1102 (2001).

Maas, S., Rich, A., and Nishikura, K., A-to-I RNA editing: Recent news and residual mysteries, *J. Biol. Chem.* **278,** 1391–1394 (2003); *and* Blanc, V. and Davidson, N.O., C-to-U RNA editing: Mechanisms leading to genetic diversity, *J. Biol. Chem.* **278,** 1395–1398 (2003).

Madison-Antenucci, S., Grams, J., and Hajduk, S.L., Editing machines: the complexities of trypanosome editing, *Cell* **108,** 435–438 (2002).

Maniatis, T. and Tasic, B., Alternative pre-mRNA splicing and proteome expansion in metazoans, *Nature* **418,** 236–243 (2002).

McManus, M.T. and Sharp, P.A., Gene silencing in mammals by small interfering RNAs, *Nature Rev. Genet.* **3,** 737–747 (2002).

Mura, C., Cascio, D., Sawaya, M.R., and Eisenberg, D.S., The crystal structure of a heptameric archaeal Sm protein: Implication for the eukaryotic snRNP core, *Proc. Natl. Acad. Sci.* **98,** 5532–5537 (2001).

Nishikura, K., A short primer on RNAi: RNA-directed RNA polymerase acts as a key catalyst, *Cell* **107,** 415–418 (2001).

Proudfoot, N., Connecting transcription to messenger RNA processing, *Trends Biochem. Sci.* **25,** 290–293 (2000); *and* Proudfoot, N.J., Furger, A., and Dye, M.J., Integrating mRNA processing with transcription, *Cell* **108,** 501–512 (2002).

Rio, D.C., RNA processing, *Curr. Opin. Cell Biol.* **4,** 444–452 (1992).

Scott, W.G., Biophysical and biochemical investigations of RNA catalysis in the hammerhead ribozyme, *Quart. Rev. Biophys.* **32,** 241–284 (1999); Murray, J.B., Terwey, D.P., Maloney, L., Karpeisky, A., Usman, N., Beigleman, L., and Scott, W.G., The structural basis of hammerhead ribozyme self-cleavage, *Cell* **92,** 665–673 (1998); *and* Scott, W.G., Murray, J.B., Arnold, J.R.P., Stoddard, B.L., and Klug, A., Capturing the structure of a catalytic RNA intermediate: the hammerhead ribozyme, *Science* **274,** 2065–2069 (1996).

Sharp, P.A., Split genes and RNA splicing, *Cell* **77,** 805–815 (1994).

Smith, C.W.J. and Valcárcel, J., Alternative pre-mRNA splicing: the logic of combinatorial control, *Trends Biochem. Sci.* **25,** 381–388 (2000).

Staley, J.P. and Guthrie, C., Mechanical devices of the spliceosome: motors, clocks, springs, and things, *Cell* **92,** 315–326 (1998).

Stark, H., Dube, P., Lührmann, R., and Kastner, B., Arrangement of RNA and proteins in the spliceosomal U1 small nuclear ribonucleoprotein particle, *Nature* **409,** 539–543 (2001).

Stevens, S.W., Ryan, D.E., Ge, H.Y., Moore, R.E., Young, M.K., Lee, T.D., and Abelson, J., Composition and functional characterization of the yeast spliceosomal penta-snRNP, *Molec. Cell* **9,** 31–44 (2002).

Tanaka Hall, T.M., Poly(A) tail synthesis and regulation: recent structural insights, *Curr. Opin. Struct. Biol.* **12,** 82–88 (2002).

Tarn, W.-Y. and Steitz, J.A., Pre-mRNA splicing: the discovery of a new spliceosome doubles the challenge, *Trends Biochem. Sci.* **22,** 132–137 (1997).

Valadkhan, S. and Manley, J.L. Splicing-related catalysis by protein-free snRNAs, *Nature* **413,** 701–707 (2001).

Wahle, E. and Kühn, U., The mechanism of cleavage and polyadenylation of eukaryotic pre-RNA, *Prog. Nucl. Acid Res. Mol. Biol.* **57,** 41–71 (1997); *and* Wahle, E. and Keller, W., The biochemistry of polyadenylation, *Trends Biochem. Sci.* **21,** 247–250 (1996).

Weinstein, L.B. and Steitz, J.A., Guided tours: from precursor to snoRNA to functional snoRNP, *Curr. Opin. Cell Biol.* **11,** 378–384 (1999).

Xiao, S., Scott, F., Fierke, C.A., and Enelke, D.R., Eukaryotic ribonuclease P: A plurality of ribonucleoprotein enzymes, *Annu. Rev. Biochem.* **71,** 165–189 (2002).

Zamore, P.D., Ancient pathways programmed by small RNAs, *Science* **296,** 1265–1269 (2002); *and* Hutvágner and Zamore, P.D., RNAi: Nature abhors a double strand, *Carr. Opin. Genet. Dev.* **12,** 225–232 (2002).

PROBLEMS

1. Indicate the phenotypes of the following *E. coli lac* partial diploids in terms of inducibility and active enzymes synthesized.

a. $I^-P^+O^+Z^+Y^-/I^+P^-O^+Z^+Y^+$
b. $I^-P^+O^cZ^+Y^-/I^+P^+O^+Z^-Y^+$
c. $I^-P^+O^cZ^+Y^+/I^-P^+O^+Z^+Y^+$
d. $I^+P^-O^cZ^+Y^+/I^-P^+O^cZ^-Y^-$

2. Superrepressed mutants, I^S, encode *lac* repressors that bind operator but do not respond to the presence of inducer. Indicate the phenotypes of the following genotypes in terms of inducibility and enzyme production.

a. $I^SO^+Z^+$ b. $I^SO^cZ^+$ c. $I^+O^+Z^+/I^SO^+Z^+$

3. Why do *lacZ*⁻ *E. coli* fail to show galactoside permease activity after the addition of lactose in the absence of glucose? Why do *lac Y*⁻ mutants lack β-galactosidase activity under the same conditions?

4. What is the experimental advantage of using IPTG instead of 1,6-allolactose as an inducer of the *lac* operon?

5. Indicate the −10 region, the −35 region, and the initiating nucleotide on the sense strand of the *E. coli* tRNA^{Tyr} promoter shown below.

5′ CAACGTAACACTTTACAGCGGCGCGTCATTTGATATGATGCGCCCCGCTTCCCGATA 3′
3′ GTTGCATTGTGAAATGTCGCCGCGCAGTAAACTATACTACGCGGGGCGAAGGGCTAT 5′

6. Why are *E. coli* that are diploid for rifamycin resistance and rifamycin sensitivity (rif^R/rif^S) sensitive to rifamycin?

7. What is the probability that the 4026-nucleotide DNA sequence coding for the β subunit of *E. coli* RNA polymerase will be transcribed with the correct base sequence? Perform the calculations for the probabilities of 0.0001, 0.001, and 0.01 that each base is incorrectly transcribed.

8. If an enhancer is placed on one plasmid and its corresponding promoter is placed on a second plasmid that is catenated (linked) with the first, initiation is almost as efficient as when the enhancer and promoter are on the same plasmid. However, initiation does not occur when the two plasmids are unlinked. Explain.

9. What is the probability that the symmetry of the *lac* operator is merely accidental?

10. Why does the inhibition of DNA gyrase in *E. coli* inhibit the expression of catabolite-sensitive operons?

11. Describe the transcription of the *trp* operon in the absence of active ribosomes and tryptophan.

12. Why can't eukaryotic transcription be regulated by attenuation?

13. Charles Yanofsky and his associates have synthesized a 15-nucleotide RNA that is complementary to segment 1 of *trpL* mRNA (but only partially complementary to segment 3). What is its effect on the *in vitro* transcription of the *trp* operon? What is its effect if the *trpL* gene contains a mutation in segment 2 that destabilizes the 2 · 3 stem and loop?

14. Why are *relA*⁻ mutants defective in the *in vivo* transcription of the *his* and *trp* operons?

15. Why aren't primary rRNA transcripts observed in wild-type *E. coli?*

16. Why can't hammerhead ribozymes catalyze the cleavage of ssDNA?

Chapter

32 Translation

In this chapter we consider **translation,** the mRNA-directed biosynthesis of polypeptides. Although peptide bond formation is a relatively simple chemical reaction, the complexity of the translational process, which involves the coordinated participation of over 100 macromolecules, is mandated by the need to link 20 different amino acid residues accurately in the order specified by a particular mRNA. Note that we previewed this process in Section 5-4B.

We begin by discussing the **genetic code,** the correspondence between nucleic acid sequences and polypeptide sequences. Next, we examine the structures and properties of **tRNAs,** the amino acid–bearing entities that mediate the translation process. Following this, we consider the structure and functions of **ribosomes,** the complex molecular machines that catalyze peptide bond formation between the mRNA-specified amino acids. Peptide bond formation, however, does not necessarily yield a functional protein; many polypeptides must first be posttranslationally modified as we discuss in the subsequent section. Finally, we study how cells degrade proteins, a process that must balance protein synthesis.

1 ■ THE GENETIC CODE

How does DNA encode genetic information? According to the one gene–one polypeptide hypothesis, the genetic message dictates the amino acid sequences of proteins. Since the base sequence of DNA is the only variable element in this otherwise monotonously repeating polymer, the amino acid sequence of a protein must somehow be specified by the base sequence of the corresponding segment of DNA.

A DNA base sequence might specify an amino acid sequence in many conceivable ways. With only 4 bases to code for 20 amino acids, a group of several bases, termed a **codon,** is necessary to specify a single amino acid. A triplet code, that is, one with 3 bases per codon, is minimally required since there are $4^3 = 64$ different triplets of bases, whereas there can be only $4^2 = 16$ different doublets, which is insufficient to specify all the amino acids. In a triplet code, as many as 44 codons might not code for amino acids. On the other hand, many amino acids could be specified by more than one codon. Such a code, in a term borrowed from mathematics, is said to be **degenerate.**

Another mystery was, how does the polypeptide synthesizing apparatus group DNA's continuous sequence of bases into codons? For example, the code might be overlapping; that is, in the sequence

ABCDEFGHIJ· · ·

ABC might code for one amino acid, BCD for a second, CDE for a third, and so on. Alternatively, the code might be nonoverlapping, so that ABC specifies one amino acid, DEF a second, GHI a third, and so on. The code might also contain internal "punctuation" such as in the nonoverlapping triplet code

ABC,DEF,GHI,· · ·

in which the commas represent particular bases or base sequences. A related question is, how does the genetic code specify the beginning and the end of a polypeptide chain?

The genetic code is, in fact, a nonoverlapping, comma-free, degenerate, triplet code. How this was determined and how the genetic code dictionary was elucidated are the subjects of this section.

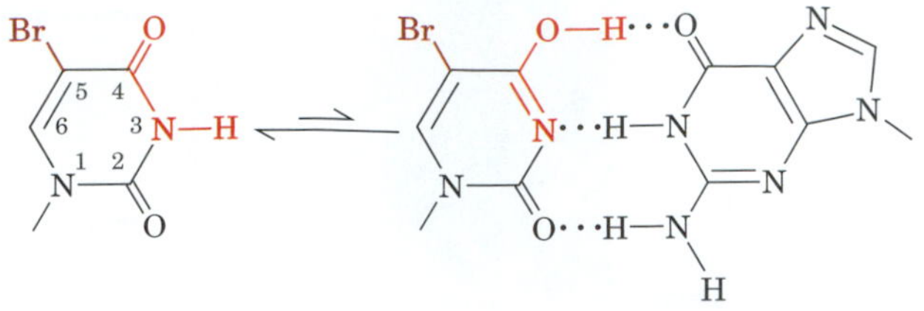

5-Bromouracil (5BU) (keto tautomer) **5BU (enol tautomer)** **Guanine**

FIGURE 32-1 5-Bromouracil. Its keto form (*left*) is its most common tautomer. However, it frequently assumes the enol form (*right*), which base pairs with guanine.

A. *Chemical Mutagenesis*

The triplet character of the genetic code, as we shall see below, was established through the use of **chemical mutagens,** substances that chemically induce mutations. We therefore precede our study of the genetic code with a discussion of these substances. There are two major classes of mutations:

1. Point mutations, in which one base pair replaces another. These are subclassified as

(a) Transitions, in which one purine (or pyrimidine) is replaced by another.

(b) Transversions, in which a purine is replaced by a pyrimidine or vice versa.

2. Insertion/deletion mutations, in which one or more nucleotide pairs are inserted in or deleted from DNA.

A mutation in any of these three categories may be reversed by a subsequent mutation of the same but not another category.

a. Point Mutations Are Generated by Altered Bases

Point mutations can result from the treatment of an organism with base analogs or with substances that chemically alter bases. For example, the base analog **5-bromouracil (5BU)** sterically resembles thymine (5-methyluracil) but, through the influence of its electronegative Br atom, frequently assumes a tautomeric form that base pairs with guanine instead of adenine (Fig. 32-1). Consequently, when 5BU is incorporated into DNA in place of thymine, as it usually is, it occasionally induces an A · T → G · C transition in subsequent rounds of DNA replication. Occasionally, 5BU is also incorporated into DNA in place of cytosine, which instead generates a G · C → A · T transition.

The adenine analog **2-aminopurine (2AP),** normally base pairs with thymine (Fig. 32-2*a*) but occasionally forms an undistorted but singly hydrogen bonded base pair with cytosine (Fig. 32-2*b*). Thus 2AP generates A · T → G · C transitions.

In aqueous solutions, **nitrous acid** (HNO_2) oxidatively deaminates aromatic primary amines so that it converts cytosine to uracil (Fig. 32-3*a*) and adenine to the guanine-like **hypoxanthine** (which forms two of guanine's three hydrogen bonds with cytosine; Fig. 32-3*b*). Hence, treatment of DNA with nitrous acid, or compounds such as **nitrosamines**

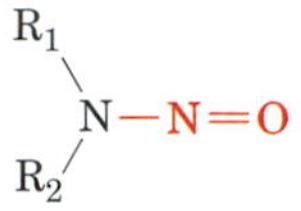

$$R_1R_2N{-}N{=}O$$

Nitrosamines

that react to form nitrous acid, results in both A · T → G · C and G · C → A · T transitions.

Nitrite, the conjugate base of nitrous acid, has long been used as a preservative of prepared meats such as frankfurters. However, the observation that many mutagens are also carcinogens (Section 30-5F) suggests that the consumption of nitrite-containing meat is harmful to humans. Proponents of nitrite preservation nevertheless argue that to stop it would

(*a*) **2-Aminopurine (2AP)** **Thymine**

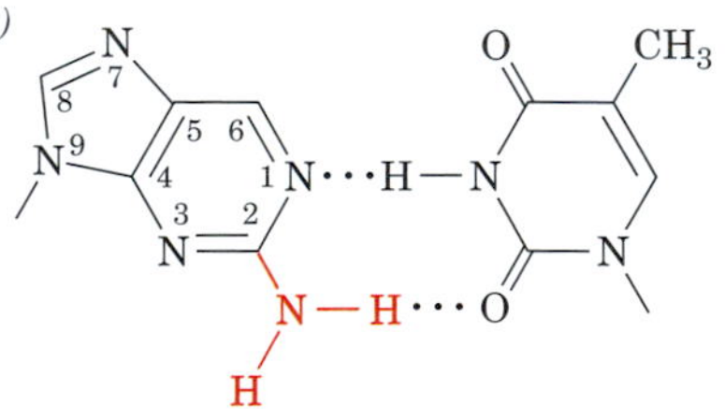

(*b*) **2AP** **Cytosine**

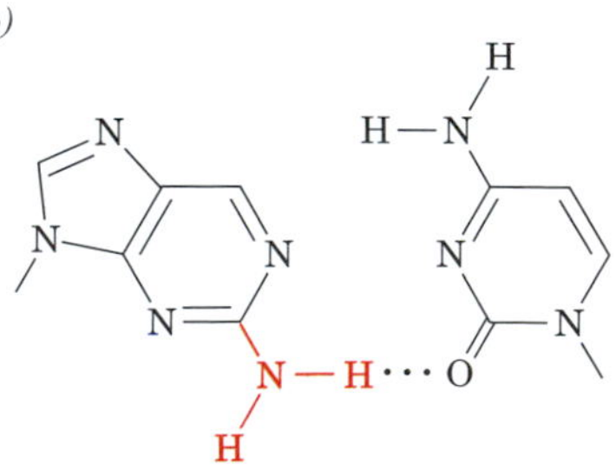

FIGURE 32-2 Base pairing by the adenine analog 2-aminopurine. It normally base pairs with thymine (*a*) but occasionally also does so with cytosine (*b*).

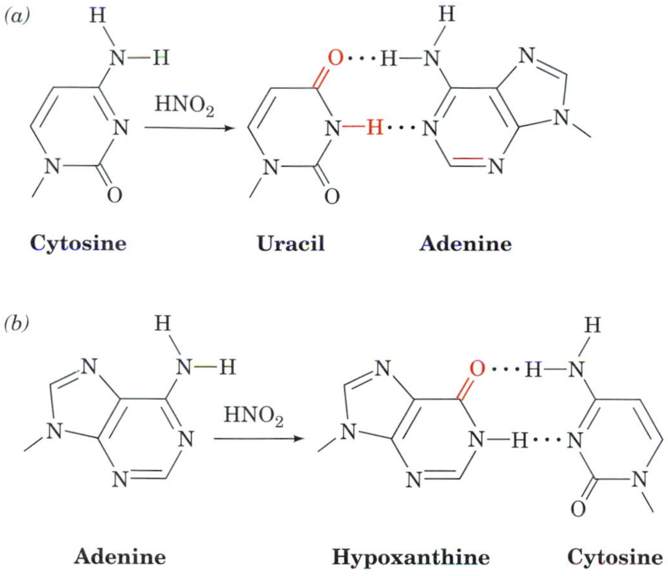

(a) Cytosine → (HNO_2) Uracil · Adenine

(b) Adenine → (HNO_2) Hypoxanthine · Cytosine

FIGURE 32-3 Oxidative deamination by nitrous acid. (*a*) Cytosine is converted to uracil, which base pairs with adenine. (*b*) Adenine is converted to hypoxanthine, a guanine derivative (it lacks guanine's 2-amino group) that base pairs with cytosine.

result in far more fatalities. This is because lack of such treatment would greatly increase the incidence of **botulism,** an often fatal form of food poisoning caused by the ingestion of protein neurotoxins secreted by the anaerobic bacterium *Clostridium botulinum* (Section 12-4D).

Hydroxylamine (NH_2OH) also induces G · C → A · T transitions by specifically reacting with cytosine to convert it to a compound that base pairs with adenine (Fig. 32-4).

The use of alkylating agents such as dimethyl sulfate, **nitrogen mustard,** and **ethylnitrosourea**

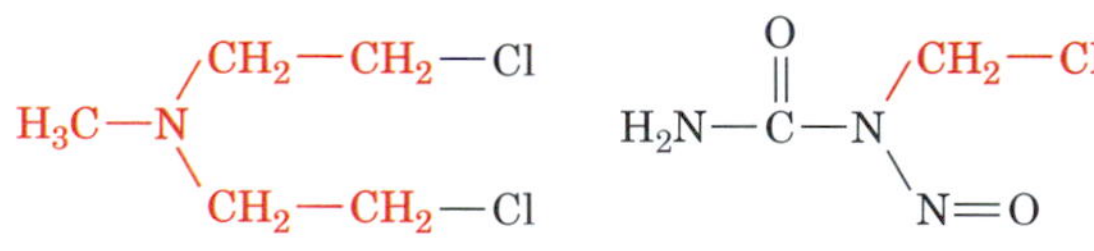

$$H_3C{-}N(CH_2{-}CH_2{-}Cl)_2 \qquad H_2N{-}\overset{O}{\overset{\|}{C}}{-}N(N{=}O){-}CH_2{-}CH_3$$

Nitrogen mustard **Ethylnitrosourea**

often generates transversions. The alkylation of the N7 position of a purine nucleotide causes its subsequent depurination. The resulting gap in the sequence is filled in by an error-prone repair system (Section 30-5D). Transversions arise when the missing purine is replaced by a pyrimidine. The repair of DNA that has been damaged by UV radiation may also generate transversions.

b. Insertion/Deletion Mutations Are Generated by Intercalating Agents

Insertion/deletion mutations may arise from the treatment of DNA with intercalating agents such as acridine orange or proflavin (Section 6-6C). The distance between two consecutive base pairs is doubled by the intercalation of such a molecule between them. The replication of such a

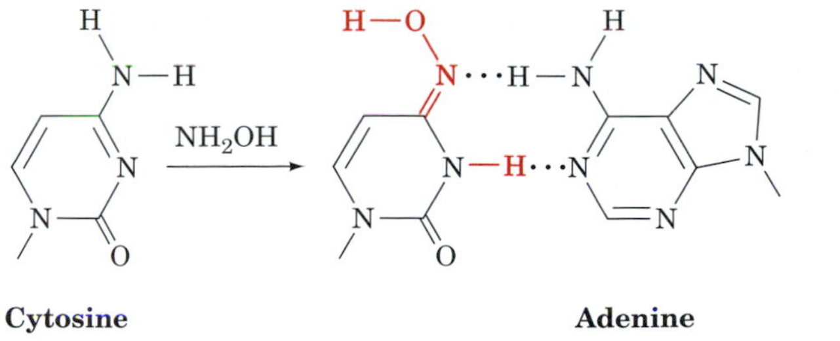

Cytosine → (NH_2OH) · Adenine

FIGURE 32-4 Reaction with hydroxylamine converts cytosine to a derivative that base pairs with adenine.

distorted DNA occasionally results in the insertion or deletion of one or more nucleotides in the newly synthesized polynucleotide. (Insertions and deletions of large DNA segments generally arise from aberrant crossover events; Section 34-2C.)

B. *Codons Are Triplets*

In 1961, Francis Crick and Sydney Brenner, through genetic investigations into the previously unknown character of proflavin-induced mutations, determined the triplet character of the genetic code. In bacteriophage T4, a particular proflavin-induced mutation, designated *FC*0, maps in the *rIIB* cistron (Section 1-4E). The growth of this mutant phage on a permissive host (*E. coli* B) resulted in the occasional spontaneous appearance of phenotypically wild-type phages as was demonstrated by their ability to grow on a restrictive host [*E. coli* K12(λ); recall that *rIIB* mutants form characteristically large plaques on *E. coli* B but cannot lyse *E. coli* K12(λ); Section 1-4E]. Yet, these doubly mutated phages are not genotypically wild type; the simultaneous infection of a permissive host by one of them and true wild-type phage yielded recombinant progeny that have either the *FC*0 mutation or a new mutation designated *FC*1. Thus the phenotypically wild-type phage is a double mutant that actually contains both *FC*0 and *FC*1. *These two genes are therefore* **suppressors** *of one another; that is, they cancel each other's mutant properties. Furthermore, since they map together in the rIIB* cistron, they are mutual **intragenic suppressors** (suppressors in the same gene).

The treatment of *FC*1 in a manner identical to that described for *FC*0 provided similar results: the appearance of a new mutant, *FC*2, that is an intragenic suppressor of *FC*1. By proceeding in this iterative manner, Crick and Brenner collected a series of different *rIIB* mutants, *FC3, FC4, FC5,* etc., in which each mutant *FC*(*n*) is an intragenic suppressor of its predecessor, *FC*(*n* − 1). Recombination studies showed, moreover, that odd-numbered mutations are intragenic suppressors of even-numbered mutations, but neither pairs of different odd-numbered mutations nor pairs of different even-numbered mutations suppress each other. However, recombinants containing three odd-numbered mutations or three even-numbered mutations all are phenotypically wild type.

Crick and Brenner accounted for these observations by the following set of hypotheses:

1. The proflavin-induced mutation *FC*0 is either an insertion or a deletion of one nucleotide pair from the *rIIB* cistron. If it is a deletion then *FC*1 is an insertion, *FC*2 is a deletion, and so on, and vice versa.

2. *The code is read in a sequential manner starting from a fixed point in the gene.* The insertion or deletion of a nucleotide shifts the **frame** (grouping) in which succeeding nucleotides are read as codons (insertions or deletions of nucleotides are therefore also known as **frameshift mutations**). Thus the code has no internal punctuation that indicates the reading frame; that is, *the code is comma free.*

3. *The code is a triplet code.*

4. All or nearly all of the 64 triplet codons code for an amino acid; that is, *the code is degenerate.*

These principles are illustrated by the following analogy. Consider a sentence (gene) in which the words (codons) each consist of three letters (bases).

THE BIG RED FOX ATE THE EGG

(Here the spaces separating the words have no physical significance; they are only present to indicate the reading frame.) The deletion of the fourth letter, which shifts the reading frame, changes the sentence to

THE IGR EDF OXA TET HEE GG

so that all words past the point of deletion are unintelligible (specify the wrong amino acids). An insertion of any letter, however, say an X in the ninth position,

THE IGR EDX FOX ATE THE EGG

restores the original reading frame. Consequently, only the words between the two changes (mutations) are altered. As in this example, such a sentence might still be intelligible (the gene could still specify a functional protein), particularly if the changes are close together. Two deletions or two insertions, no matter how close together, would not suppress each other but just shift the reading frame. However, three insertions, say X, Y, and Z in the fifth, eighth, and twelfth positions, respectively, would change the sentence to

THE BXI GYR EDZ FOX ATE THE EGG

which, after the third insertion, restores the original reading frame. The same would be true of three deletions. As before, if all three changes were close together, the sentence might still retain its meaning.

Crick and Brenner did not unambiguously demonstrate that the genetic code is a triplet code because they had no proof that their insertions and deletions involved only single nucleotides. Strictly speaking, they showed that a codon consists of $3r$ nucleotides where r is the number of nucleotides in an insertion or deletion. Although it was generally assumed at the time that $r = 1$, proof of this assertion had to await the elucidation of the genetic code (Section 32-1C).

C. *Deciphering the Genetic Code*

The genetic code could, in principle, be determined by simply comparing the base sequence of an mRNA with the amino acid sequence of the polypeptide it specifies. In the 1960s, however, techniques for isolating and sequencing mRNAs had not yet been developed. The elucidation of the genetic code dictionary therefore proved to be a difficult task.

a. UUU Specifies Phe

The major breakthrough in deciphering the genetic code came in 1961 when Marshall Nirenberg and Heinrich Matthaei established that UUU is the codon specifying Phe. They did so by demonstrating that the addition of poly(U) to a cell-free protein synthesizing system stimulates only the synthesis of poly(Phe). The cell-free protein synthesizing system was prepared by gently breaking open *E. coli* cells by grinding them with powdered alumina and centrifuging the resulting cell sap to remove the cell walls and membranes. This extract contained DNA, mRNA, ribosomes, enzymes, and other cell constituents necessary for protein synthesis. When fortified with ATP, GTP, and amino acids, the system synthesized small amounts of proteins. This was demonstrated by the incubation of the system with ^{14}C-labeled amino acids followed by the precipitation of its proteins by the addition of trichloroacetic acid. The precipitate proved to be radioactive.

A cell-free protein synthesizing system, of course, produces proteins specified by the cell's DNA. On addition of DNase, however, protein synthesis stops within a few minutes because the system can no longer synthesize mRNA, whereas the mRNA originally present is rapidly degraded. Nirenberg found that crude mRNA-containing fractions from other organisms were highly active in stimulating protein synthesis in a DNase-treated protein synthesizing system. This system is likewise responsive to synthetic mRNAs.

The synthetic mRNAs that Nirenberg used in subsequent experiments were synthesized by the *Azotobacter vinelandii* enzyme **polynucleotide phosphorylase.** This enzyme, which was discovered by Severo Ochoa and Marianne Grunberg-Manago, links together nucleotides in the reaction

$$(\text{RNA})_n + \text{NDP} \rightleftharpoons (\text{RNA})_{n+1} + \text{P}_i$$

In contrast to RNA polymerase, however, polynucleotide phosphorylase does not utilize a template. Rather, it randomly links together the available NDPs so that the base composition of the product RNA reflects that of the reactant NDP mixture.

Nirenberg and Matthaei demonstrated that poly(U) stimulates the synthesis of poly(Phe) by incubating poly(U) and a mixture of 1 radioactive and 19 unlabeled amino acids in a DNase-treated protein synthesizing

TABLE 32-1 Amino Acid Incorporation Stimulated by a Random Copolymer of U and G in Mole Ratio 0.76:0.24

Codon	Probability of Occurrence	Relative Incidence[a]	Amino Acid	Relative Amount of Amino Acid Incorporated
UUU	0.44	100	Phe	100
UUG	0.14	32	Leu	36
UGU	0.14	32	Cys	35
GUU	0.14	32	Val	37
UGG	0.04	9	Trp	14
GUG	0.04	9		
GGU	0.04	9	Gly	12
GGG	0.01	2		

[a]Relative incidence is defined here as 100 × probability of occurrence/0.44.

Source: Matthaei, J.H., Jones, O.W., Martin, R.G., and Nirenberg, M., *Proc. Natl. Acad. Sci.* **48,** 666 (1962).

system. Significant radioactivity appeared in the protein precipitate only when phenylalanine was labeled. *UUU must therefore be the codon specifying Phe.* In similar experiments using poly(A) and poly(C), it was found that poly(Lys) and poly(Pro), respectively, were synthesized. Thus *AAA specifies* Lys *and CCC specifies Pro.* [Poly(G) cannot function as a synthetic mRNA because, even under denaturing conditions, it aggregates to form a four-stranded helix (Section 30-4D). An mRNA must be single stranded to direct its translation; Section 32-2D.]

Nirenberg and Ochoa independently employed ribonucleotide copolymers to further elucidate the genetic code. For example, in a poly(UG) composed of 76% U and 24% G, the probability of a given triplet being UUU is 0.76 × 0.76 × 0.76 = 0.44. Likewise, the probability of a particular triplet consisting of 2U's and 1G, that is, UUG, UGU, or GUU, is 0.76 × 0.76 × 0.24 = 0.14. The use of this poly(UG) as an mRNA therefore indicated the base compositions, but not the sequences, of the codons specifying several amino acids (Table 32-1). Through the use of copolymers containing two, three, and four bases, the base compositions of codons specifying each of the 20 amino acids were inferred. Moreover, *these experiments demonstrated that the genetic code is degenerate since, for example, poly(UA), poly(UC), and poly(UG) all direct the incorporation of Leu into a polypeptide.*

b. The Genetic Code Was Elucidated through Triplet Binding Assays and the Use of Polyribonucleotides with Known Sequences

In the absence of GTP, which is necessary for protein synthesis, trinucleotides but not dinucleotides are almost as effective as mRNAs in promoting the ribosomal binding of specific tRNAs. This phenomenon, which Nirenberg and Philip Leder discovered in 1964, permitted the various codons to be identified by a simple binding assay. Ribosomes, together with their bound tRNAs, are retained by a nitrocellulose filter but free tRNA is not. The bound tRNA was identified by using charged tRNA mixtures in which only one of the pendent amino acid residues was radioactively labeled. For instance, it was found, as expected, that UUU stimulates the ribosomal binding of only Phe tRNA. Likewise, UUG, UGU, and GUU stimulate the binding of Leu, Cys, and Val tRNAs, respectively. Hence UUG, UGU, and GUU must be codons that specify Leu, Cys, and Val, respectively. In this way, the amino acids specified by some 50 codons were identified. For the remaining codons, the binding assay was either negative (no tRNA bound) or ambiguous.

The genetic code dictionary was completed and previous results confirmed through H. Gobind Khorana's synthesis of polyribonucleotides with specified repeating sequences (Section 7-6A). In a cell-free protein synthesizing system, UCUCUCUC· · ·, for example, is read

UCU CUC UCU CUC UCU C···

so that it specifies a polypeptide chain of two alternating amino acid residues. In fact, it was observed that this mRNA stimulated the production of

Ser—Leu—Ser—Leu—Ser—Leu—···

which indicates that either UCU or CUC specifies Ser and the other specifies Leu. This information, together with the tRNA-binding data, permitted the conclusion that UCU codes for Ser and CUC codes for Leu. These data also prove that codons consist of an odd number of nucleotides, thereby relieving any residual suspicions that codons consist of six rather than three nucleotides.

Alternating sequences of three nucleotides, such as poly(UAC), specify three different homopolypeptides because ribosomes may initiate polypeptide synthesis on these synthetic mRNAs in any of the three possible reading frames (Fig. 32-5). Analyses of the polypeptides specified by various alternating sequences of two and three nucleotides confirmed the identity of many codons and filled out missing portions of the genetic code.

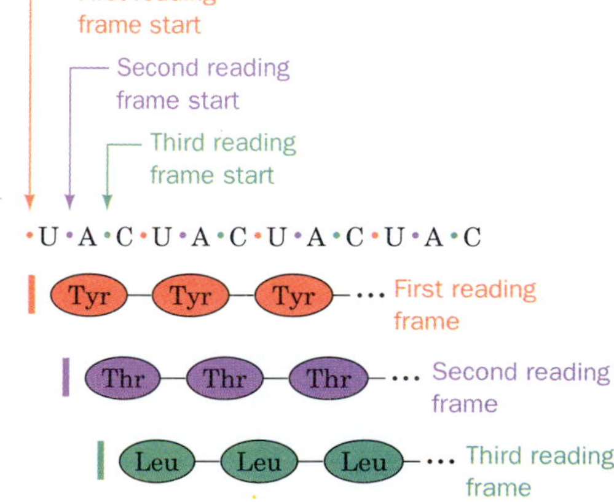

FIGURE 32-5 The three potential reading frames of an mRNA. Each reading frame would yield a different polypeptide.

c. mRNAs Are Read in the 5′ → 3′ Direction

The use of repeating tetranucleotides indicated the reading direction of the code and identified the chain termination codons. Poly(UAUC) specifies, as expected, a polypeptide with a tetrapeptide repeat:

5′ UAU CUA UCU AUC UAU CUA ··· 3′

Tyr — Leu — Ser — Ile — Tyr — Leu — ···

The amino acid sequence of this polypeptide indicates that the mRNA's 5′ end corresponds to the polypeptide's N-terminus; that is, *mRNA is read in the 5′ → 3′ direction.*

d. UAG, UAA, and UGA Are Stop Codons

In contrast to the above results, poly(AUAG) yields only dipeptides and tripeptides. This is because *UAG is a signal to the ribosome to terminate protein synthesis:*

AUA GAU AGA UAG AUA GAU ···

Ile — Asp — Arg Stop Ile — Asp — ···

Likewise, poly(GUAA) yields dipeptides and tripeptides because UAA is also a chain termination signal:

GUA AGU AAG UAA GUA AGU ···

Val — Ser — Lys Stop Val — Ser — ···

UGA is a third stop signal. These Stop codons, whose existence was first inferred from genetic experiments, are known, somewhat inappropriately, as **nonsense codons** because they are the only codons that do not specify amino acids. UAG, UAA, and UGA are sometimes referred to as ***amber, ochre,*** and ***opal*** codons. [They were so named as the result of a laboratory joke: The German word for amber is Bernstein, the name of an individual who helped discover *amber* mutations (mutations that change some other codon to UAG); *ochre* and *opal* are puns on *amber.*]

e. AUG and GUG Are Chain Initiation Codons

The codons AUG, and less frequently GUG, form part of the chain initiation sequence (Section 32-3C). However, they also specify the amino acid residues Met and Val, respectively, at internal positions of polypeptide chains. (Nirenberg and Matthaei's discovery that UUU specifies Phe was only possible because ribosomes indiscriminately initiate polypeptide synthesis on an mRNA when the Mg^{2+} concentration is unphysiologically high as it was, serendipitously, in their experiments.)

TABLE 32-2 The "Standard" Genetic Code[a]

First position (5′ end)	Second position: U	C	A	G	Third position (3′ end)
U	UUU Phe	UCU Ser	UAU Tyr	UGU Cys	U
	UUC Phe	UCC Ser	UAC Tyr	UGC Cys	C
	UUA Leu	UCA Ser	UAA Stop	UGA Stop	A
	UUG Leu	UCG Ser	UAG Stop	UGG Trp	G
C	CUU Leu	CCU Pro	CAU His	CGU Arg	U
	CUC Leu	CCC Pro	CAC His	CGC Arg	C
	CUA Leu	CCA Pro	CAA Gln	CGA Arg	A
	CUG Leu	CCG Pro	CAG Gln	CGG Arg	G
A	AUU Ile	ACU Thr	AAU Asn	AGU Ser	U
	AUC Ile	ACC Thr	AAC Asn	AGC Ser	C
	AUA Ile	ACA Thr	AAA Lys	AGA Arg	A
	AUG Met[b]	ACG Thr	AAG Lys	AGG Arg	G
G	GUU Val	GCU Ala	GAU Asp	GGU Gly	U
	GUC Val	GCC Ala	GAC Asp	GGC Gly	C
	GUA Val	GCA Ala	GAA Glu	GGA Gly	A
	GUG Val	GCG Ala	GAG Glu	GGG Gly	G

[a]Nonpolar amino acid residues are tan, basic residues are blue, acidic residues are red, and nonpolar uncharged residues are purple.
[b]AUG forms part of the initiation signal as well as coding for internal Met residues.

D. *The Nature of the Code*

The genetic code dictionary, as elucidated by the above methods, is presented in Table 32-2 as well as in Table 5-3. Examination of the table indicates that the genetic code has several remarkable features:

1. *The code is highly degenerate.* Three amino acids, Arg, Leu, and Ser, are each specified by six codons, and most of the rest are specified by either four, three, or two codons. Only Met and Trp, two of the least common amino acids in proteins (Table 4-1), are represented by a single codon. Codons that specify the same amino acid are termed **synonyms.**

2. *The arrangement of the code table is nonrandom.* Most synonyms occupy the same box in Table 32-2; that is, they differ only in their third nucleotide. The only exceptions are Arg, Leu, and Ser, which, having six codons each, must occupy more than one box. XYU and XYC always specify the same amino acid; XYA and XYG do so in all but two cases. Moreover, changes in the first codon position tend to specify similar (if not the same) amino acids, whereas codons with second position pyrimidines encode mostly hydrophobic amino acids (tan in Table 32-2), and those with second position purines encode mostly polar amino acids (blue, red, and purple in Table 32-2). Apparently *the code evolved so as to minimize the deleterious effects of mutations.*

Many of the mutations causing amino acid substitutions in a protein can be rationalized, according to the genetic code, as single point mutations. *As a consequence of the genetic code's degeneracy, however, many point mutations at a third codon position are phenotypically silent; that is, the mutated codon specifies the same amino acid as the wild type.* Degeneracy may account for as much as 33% of the

25 to 75% range in the G + C content among the DNAs of different organisms (Section 5-1B). The frequent occurrence of Arg, Ala, Gly, and Pro also tends to give a high G + C content, whereas Asn, Ile, Lys, Met, Phe, and Tyr contribute to a low G + C content.

a. Some Phage DNA Segments Contain Overlapping Genes in Different Reading Frames

Since any nucleotide sequence may have three reading frames, it is possible, at least in principle, for a polynucleotide to encode two or even three different polypeptides. This idea was never seriously entertained, however, because it seemed that the constraints on even two overlapping genes in different reading frames would be too great for them to evolve so that both could specify sensible proteins. It therefore came as a great surprise, in 1976, when Frederick Sanger reported that the 5386-nucleotide DNA of bacteriophage φX174 (which, at the time, was the largest DNA to have been sequenced) contains two genes that are completely contained within larger genes of different reading frames (Fig. 32-6). Moreover, the end of the overlapping D and E genes contains the control sequence for the ribosomal initiation of the J gene so that this short DNA segment performs triple duty. Bacteria also exhibit such coding economy; the ribosomal initiation sequence of one gene in a polycistronic mRNA often overlaps the end of the preceding gene. Nevertheless, completely overlapping genes have only been found in small single-stranded DNA phages, which presumably must make maximal use of the little DNA that they can pack inside their capsids.

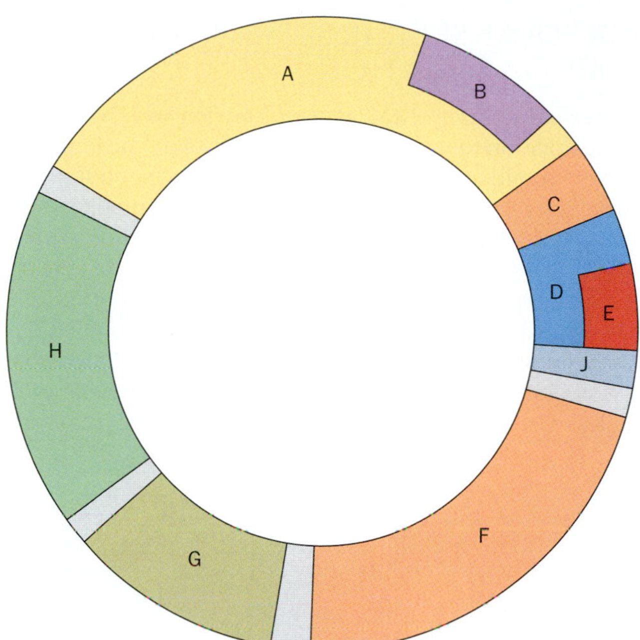

FIGURE 32-6 Genetic map of bacteriophage φX174 as determined by DNA sequence analysis. Genes are labeled A, B, C, etc. Note that gene B is wholly contained within gene A and gene E is wholly contained within gene D. These pairs of genes are read in different reading frames and therefore specify unrelated proteins. The unlabeled regions correspond to untranslated control sequences.

b. The "Standard" Genetic Code Is Widespread but Not Universal

For many years it was thought that the "standard" genetic code (that given in Table 32-2) is universal. This assumption was, in part, based on the observations that one kind of organism (e.g., *E. coli*) can accurately translate the genes from quite different organisms (e.g., humans). This phenomenon is, in fact, the basis of genetic engineering. Once the "standard" genetic code had been established, presumably during the time of prebiotic evolution (Section 1-5B), any mutation that would alter the way the code is translated would result in numerous, often deleterious, protein sequence changes. Undoubtedly there is strong selection against such mutations. DNA sequencing studies in 1981 nevertheless revealed that *the genetic codes of certain mitochondria (mitochondria contain their own genes and protein synthesizing systems but produce only a few mitochondrial proteins; Section 12-4E) are variants of the "standard" genetic code (Table 32-3).* For example, in mammalian mitochondria, AUA, as well as the standard AUG, is a Met/initiation codon; UGA specifies Trp rather than "Stop"; and AGA and AGG are "Stop" rather than Arg. Note that all mitochondrial genetic codes except those of plants simplify the "standard" code by increasing its degeneracy. For example, in the mammalian mitochondrial code, each amino acid is specified by at least two codons that differ only in their third nucleotide. Apparently the constraints preventing alterations of the genetic code are eased by the small sizes of mitochondrial genomes. More recent studies, however, have revealed that in ciliated protozoa, the codons UAA and UAG specify Gln rather than "Stop." Perhaps UAA and UAG were sufficiently rare codons in a primordial ciliate (which molecular phylogenetic studies indicate branched off very early in eukaryotic evolution) to permit the code change without unacceptable deleterious effects. At any rate, *the "standard" genetic code, although very widely utilized, is not universal.*

TABLE 32-3 Mitochondrial Deviations from the "Standard" Genetic Code

Mitochondrion	UGA	AUA	CUN[a]	AG$^{A}_{G}$	CGG
Mammalian	Trp	Met[b]		Stop	
Baker's yeast	Trp	Met[b]	Thr		
Neurospora crassa	Trp				
Drosophila	Trp	Met[b]		Ser[c]	
Protozoan	Trp				
Plant					Trp
"Standard" code	Stop	Ile	Leu	Arg	Arg

[a]N represents any of the four nucleotides.
[b]Also acts as part of an initiation signal.
[c]AGA only; no AGG codons occur in *Drosophila* mitochondrial DNA.

Source: Mainly Breitenberger, C.A. and RajBhandary, U.L., *Trends Biochem. Sci.* **10,** 481 (1985).

2 ■ TRANSFER RNA AND ITS AMINOACYLATION

The establishment of the genetic function of DNA led to the realization that cells somehow "translate" the language of base sequences into the language of polypeptides. Yet, nucleic acids originally appeared unable to bind specific amino acids [more recently RNA aptamers for specific amino acids have been generated; aptamers are nucleic acids that have been selected for their ability to bind specific ligands (Section 7-6C)]. In 1955, Crick, in what became known as the **adaptor hypothesis,** postulated that translation occurs through the mediation of "adaptor" molecules. Each adaptor was postulated to carry a specific enzymatically appended amino acid and to recognize the corresponding codon (Fig. 32-7). Crick suggested that these adaptors contain RNA because codon recognition could then occur by complementary base pairing. At about this time, Paul Zamecnik and Mahlon Hoagland discovered that in the course of protein synthesis, ^{14}C-labeled amino acids became transiently bound to a low molecular mass fraction of RNA. Further investigations indicated that these RNAs, which at first were called "soluble RNA" or "sRNA" but are now known as **transfer RNA (tRNA),** are, in fact, Crick's putative adaptor molecules.

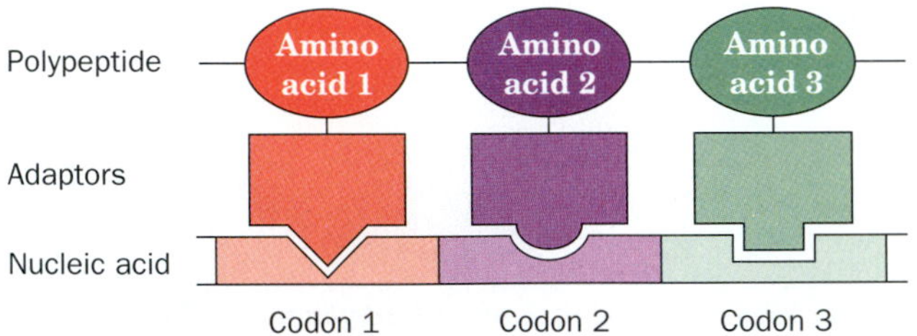

FIGURE 32-7 The adaptor hypothesis. It postulates that the genetic code is read by molecules that recognize a particular codon and carry the corresponding amino acid.

A. *Primary and Secondary Structures of tRNA*

See Guided Exploration 26: The Structure of tRNA. In 1965, after a 7-year effort, Robert Holley reported the first known base sequence of a biologically significant nucleic acid, that of yeast **alanine tRNA ($tRNA^{Ala}$; ** Fig. 32-8). To do so Holley had to overcome several major obstacles:

1. All organisms contain many species of tRNAs (usually at least one for each of the 20 amino acids) which, because of their nearly identical properties (see below), are not easily separated. Preparative techniques had to be developed to provide the gram or so of pure yeast $tRNA^{Ala}$ Holley required for its sequence determination.

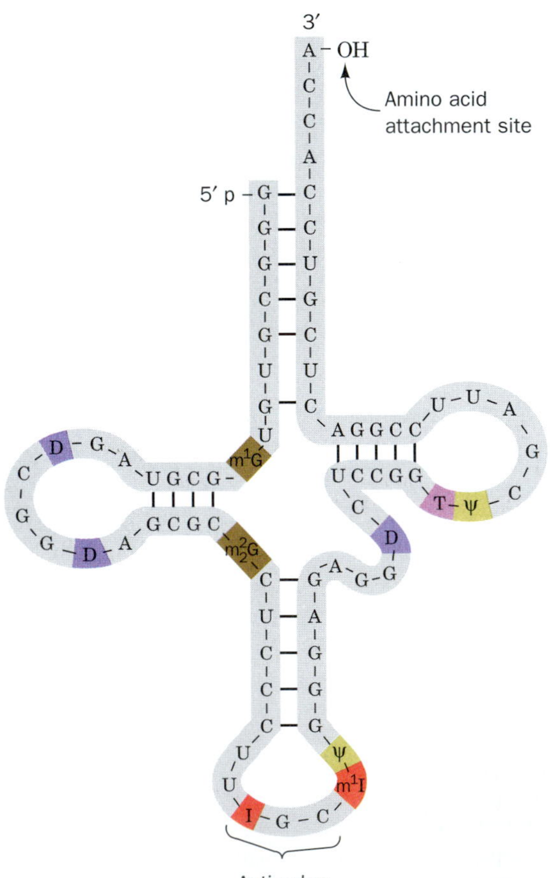

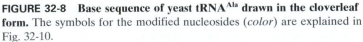

FIGURE 32-8 Base sequence of yeast $tRNA^{Ala}$ drawn in the cloverleaf form. The symbols for the modified nucleosides (*color*) are explained in Fig. 32-10.

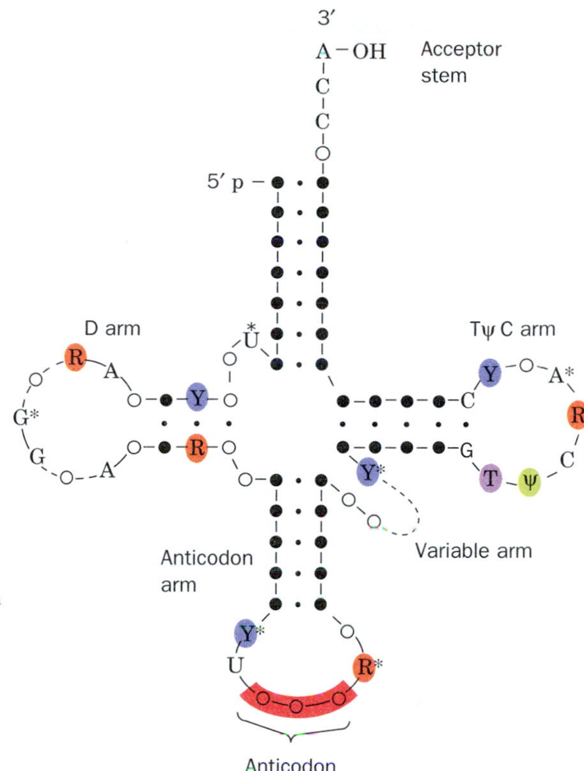

FIGURE 32-9 Cloverleaf secondary structure of tRNA. Filled circles connected by dots represent Watson–Crick base pairs, and open circles in the double-helical regions indicate bases involved in non-Watson–Crick base pairing. Invariant positions are indicated: R and Y represent invariant purines and pyrimidines and ψ signifies pseudouridine. The starred nucleosides are often modified. The dashed regions in the D and variable arms contain different numbers of nucleotides in the various tRNAs.

2. Holley had to invent the methods that were initially used to sequence RNA (Section 7-2).

3. Ten of the 76 bases of yeast $tRNA^{Ala}$ are modified (see below). Their structural formulas had to be elucidated although they were never available in more than milligram quantities.

Since 1965, the techniques for tRNA purification and sequencing have vastly improved. A tRNA may now be sequenced in a few days' time with only ~1 μg of material. Presently, the base sequences of more than 4000 tRNAs from over 200 organisms and organelles are known (most from their corresponding DNA sequences). They vary in length from 54 to 100 nucleotides (18–28 kD) although most have ~76 nucleotides.

Almost all known tRNAs, as Holley first recognized, may be schematically arranged in the so-called cloverleaf secondary structure (Fig. 32-9). Starting from the 5′ end, they have the following common features:

1. A 5′-terminal phosphate group.

2. A 7-bp stem that includes the 5′-terminal nucleotide and that may contain non-Watson–Crick base pairs such as G · U. This assembly is known as the **acceptor** or **amino acid stem** because the amino acid residue carried by the tRNA is appended to its 3′-terminal OH group (Section 32-2C).

3. A 3- or 4-bp stem ending in a loop that frequently contains the modified base **dihydrouridine (D;** see below). This stem and loop are therefore collectively termed the **D arm.**

4. A 5-bp stem ending in a loop that contains the **anticodon,** the triplet of bases that is complementary to the codon specifying the tRNA. These features are known as the **anticodon arm.**

5. A 5-bp stem ending in a loop that usually contains the sequence TψC (where ψ is the symbol for **pseudouridine;** see below). This assembly is called the **TψC or T arm.**

6. All tRNAs terminate in the sequence CCA with a free 3′-OH group. The —CCA may be genetically specified or enzymatically appended to immature tRNA (Section 31-4C).

7. There are 15 invariant positions (always have the same base) and 8 **semi-invariant** positions (only a purine or only a pyrimidine) that occur mostly in the loop regions. These regions also contain **correlated invariants,** that is, pairs of nonstem nucleotides that are base paired in all tRNAs. The purine on the 3′ side of the anticodon is invariably modified. The structural significance of these features is examined in Section 32-2B.

The site of greatest variability among the known tRNAs occurs in the so-called **variable arm.** It has from 3 to 21 nucleotides and may have a stem consisting of up to 7 bp. The D loop also varies in length from 5 to 7 nucleotides.

a. Transfer RNAs Have Numerous Modified Bases

One of the most striking characteristics of tRNAs is their large proportion, up to 25%, of posttranslationally

Uracil derivatives

Pseudouridine (ψ)

Dihydrouridine (D)

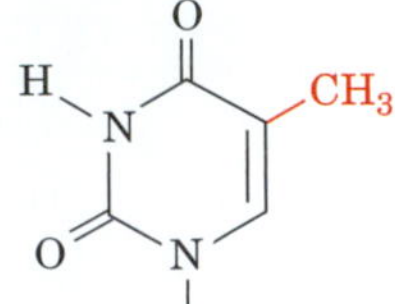

Ribothymidine (T)

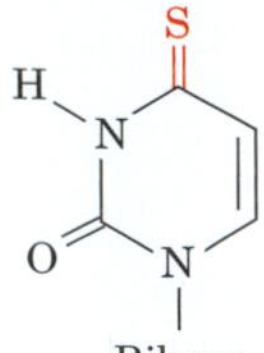

4-Thiouridine (s^4U)

Cytosine derivatives

3-Methylcytidine (m^3C)

N^4-Acetylcytidine (ac^4C)

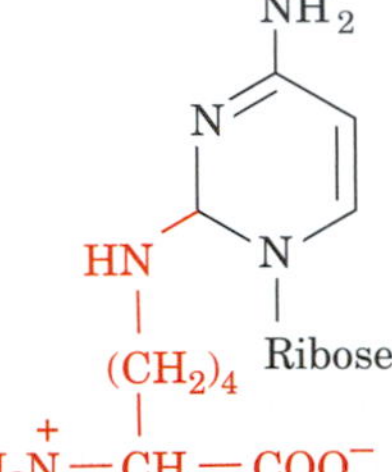

Lysidine (L)

Adenine derivatives

1-Methyladenosine (m^1A)

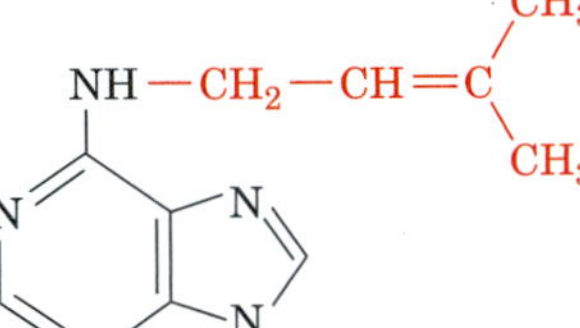

N^6-Isopentenyladenosine (i^6A)

Inosine (I)

Guanine derivatives

N^7-Methylguanosine (m^7G)

N^2,N^2-Dimethylguanosine (m_2^2G)

R = H Wyosine (Wyo)

$R = CH_2CH_2CH(\overset{O}{\overset{\|}{C}}OCH_3)_2$ Y

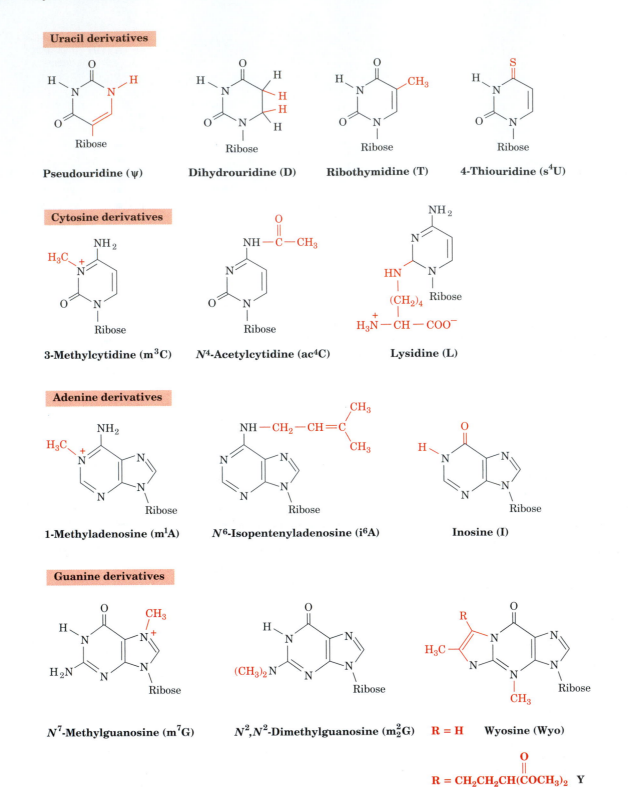

FIGURE 32-10 A selection of the modified nucleosides that occur in tRNAs together with their standard abbreviations. Note that although inosine chemically resembles guanosine, it is biochemically derived from adenosine. Nucleosides may also be methylated at their ribose 2′ positions to form residues symbolized, for instance, by Cm, Gm, and Um.

modified or hypermodified bases. Nearly 80 such bases, found at >60 different tRNA positions, have been characterized. A few of them, together with their standard abbreviations, are indicated in Fig. 32-10. Hypermodified nucleosides, such as i^6A, are usually adjacent to the anticodon's 3′ nucleotide when it is A or U. Their low polarities probably strengthen the otherwise relatively weak pairing associations of these bases with the codon, thereby increasing translational fidelity. Conversely, certain methylations block base pairing and hence prevent inappropriate structures from forming. Some of these modifications form important recognition elements for the enzyme that attaches the correct amino acid to a tRNA (Section 32-2C). However, none of them are essential for maintaining a tRNA's structural integrity (see below) or for its proper binding to the ribosome. Nevertheless, mutant bacteria unable to form certain modified bases compete poorly against the corresponding normal bacteria.

B. *Tertiary Structure of tRNA*

See Guided Exploration 26: The Structure of tRNA. The earliest physicochemical investigations of tRNA indicated that it has a well-defined conformation. Yet, despite numerous hydrodynamic, spectroscopic, and chemical cross-linking studies, its three-dimensional structure remained an enigma until 1974. In that year, the 2.5-Å resolution X-ray crystal structure of yeast **tRNA^{Phe}** was separately elucidated by Alexander Rich in collaboration with Sung Hou Kim and, in a different crystal form, by Aaron Klug. *The molecule assumes an L-shaped conformation in which one leg of the L is formed by the acceptor and T stems folded into a continuous A-RNA-like double helix (Section 29-1B) and the other leg is similarly composed of the D and anticodon stems* (Fig. 32-11). Each leg of the L is ~60 Å long and the anticodon and amino acid acceptor sites are at opposite ends of the molecule, some 76 Å apart. The narrow 20- to 25-Å width of native tRNA is essential to its biological function: During protein synthesis, two RNA molecules must simultaneously bind in close proximity at adjacent codons on mRNA (Section 32-3D).

a. tRNA's Complex Tertiary Structure Is Maintained by Hydrogen Bonding and Stacking Interactions

The structural complexity of yeast tRNA^{Phe} is reminiscent of that of a protein. Although only 42 of its 76 bases occur in double helical stems, *71 of them participate in*

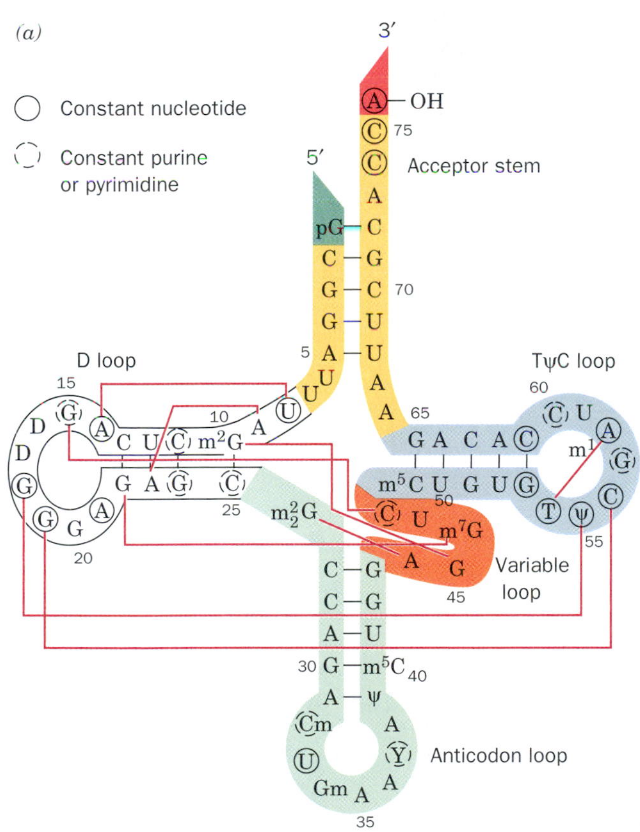

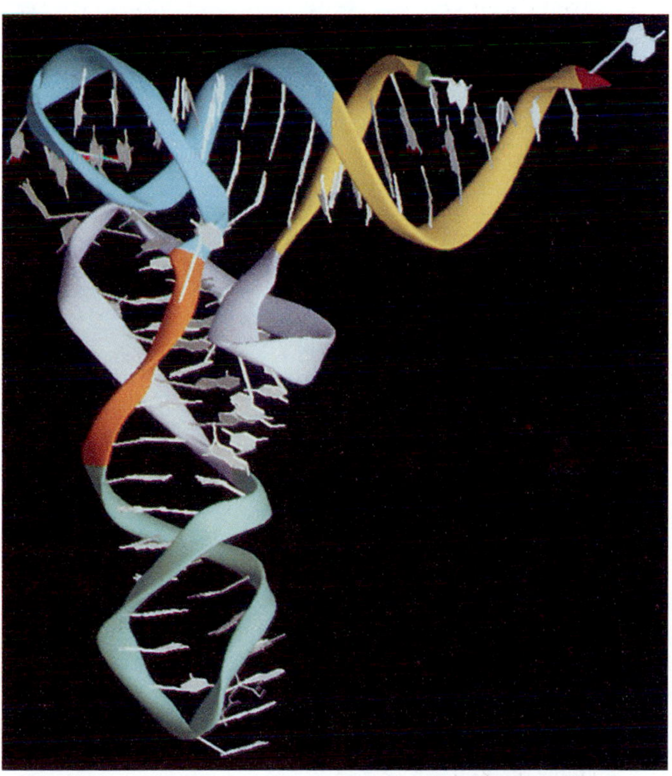

FIGURE 32-11 **Structure of yeast tRNA^{Phe}.** (*a*) The base sequence drawn in cloverleaf form. Tertiary base pairing interactions are represented by thin red lines connecting the participating bases. Bases that are conserved or semiconserved in all tRNAs are circled by solid and dashed lines, respectively. The 5′ terminus is colored bright green, the acceptor stem is yellow, the D arm is white, the anticodon arm is light green, the variable arm is orange, the TψC arm is cyan, and the 3′ terminus is red. (*b*) The X-ray structure drawn to show how its base paired stems are arranged form the L-shaped molecule. The sugar–phosphate backbone is represented by a ribbon with the same color scheme as that in Part *a*. [Courtesy of Mike Carson, University of Alabama at Birmingham. PDBid 6TNA.] **See the Kinemage Exercise 20-1**

stacking associations (Fig. 32-12). The structure also contains 9 base pairing interactions that cross link its tertiary structure (Figs. 32-11*a* and 32-12). Remarkably, all but one of these tertiary interactions, which appear to be the mainstays of the molecular structure, are non-Watson–Crick associations. Moreover, most of the bases involved in these interactions are either invariant or semi-invariant, which strongly suggests that all tRNAs have similar conformations (see below). The structure is also stabilized by several unusual hydrogen bonds between bases and either phosphate groups or the 2′-OH groups of ribose residues.

The compact structure of yeast $tRNA^{Phe}$ results from its large number of intramolecular associations, which renders most of its bases inaccessible to solvent. The most notable exceptions to this are the anticodon bases and those of the amino acid–bearing —CCA terminus. Both of these groupings must be accessible in order to carry out their biological functions.

The observation that the molecular structures of yeast $tRNA^{Phe}$ in two different crystal forms are essentially identical lends much credence to the supposition that its crystal structure closely resembles its solution structure. Transfer RNAs other than yeast $tRNA^{Phe}$ have, unfortunately, been notoriously difficult to crystallize. As yet, the X-ray structures of only three other uncomplexed tRNAs have been reported (although the X-ray structures of numerous tRNAs in complex with the enzymes that append their corresponding amino acids and with ribosomes have been elucidated; Sections 32-2C and 32-3D). The major structural differences among them result from an apparent flexibility in the anticodon loop and the —CCA terminus as well as from a hingelike mobility between the two legs of the L that gives, for instance, yeast $\mathbf{tRNA^{Asp}}$ a boomerang-like shape. Such observations are in accord with the expectation that all tRNAs fit into the same ribosomal cavities.

C. *Aminoacyl–tRNA Synthetases*

See Guided Exploration 27: The Structures of aminoacyl-tRNA synthetases and their interaction with tRNAs. *Accurate translation requires two equally important recognition steps: (1) the choice of the correct amino acid for covalent attachment to a*

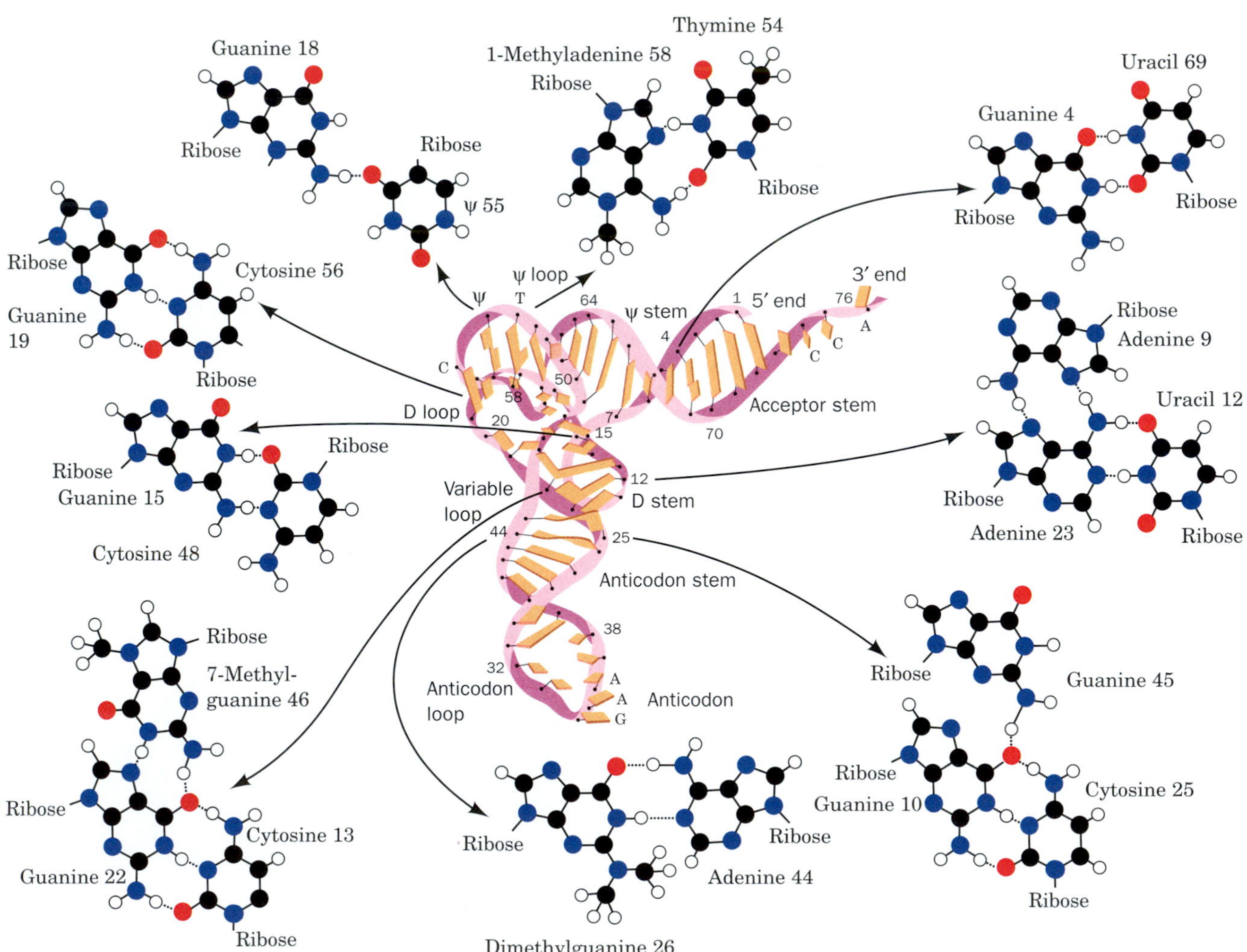

FIGURE 32-12 Tertiary base pairing interactions in yeast $tRNA^{Phe}$. Note that all but one of these nine interactions involve non-Watson–Crick pairs and that they are all located near the corner of the L. [After Kim, S.H., *in* Schimmel, P.R., Söll, D., and Abelson, J.N. (Eds.), *Transfer RNA: Structure, Properties and Recognition,* p. 87, Cold Spring Harbor Laboratory Press (1979). Drawing of tRNA copyrighted © by Irving Geis.] **See the Kinemage Exercise 20-3**

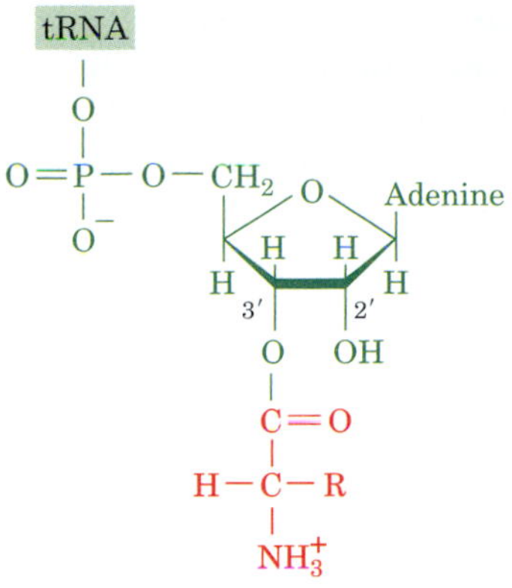

tRNA–O–PO_2^-–O–CH_2–(ribose, Adenine; 3′-O–C(=O)–CH(R)–NH_3^+; 2′-OH)

Aminoacyl–tRNA

FIGURE 32-13 An Aminoacyl–tRNA. The amino acid residue is esterified to the tRNA's 3′-terminal nucleoside at either its 3′-OH group, as shown here, or its 2′-OH group.

tRNA; and (2) the selection of the amino acid-charged tRNA specified by mRNA. The first of these steps, which is catalyzed by amino acid–specific enzymes known as **aminoacyl–tRNA synthetases (aaRSs),** appends an amino acid to the 3′-terminal ribose residue of its cognate tRNA to form an **aminoacyl–tRNA** (Fig. 32-13). This otherwise unfavorable process is driven by the hydrolysis of ATP in two sequential reactions that are catalyzed by a single enzyme.

1. The amino acid is first "activated" by its reaction with ATP to form an **aminoacyl–adenylate:**

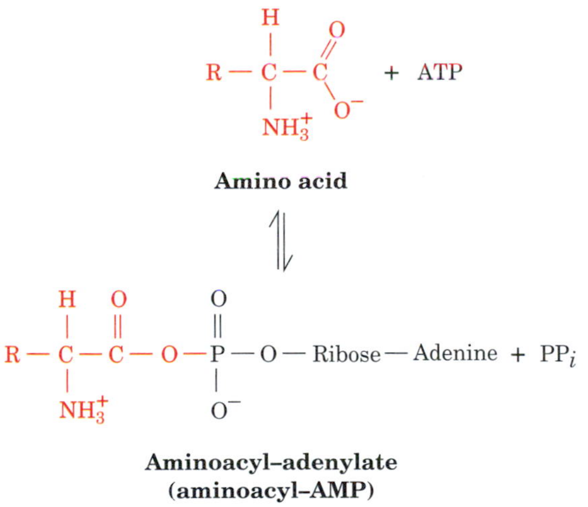

$$\text{R}-\text{CH}(\text{NH}_3^+)-\text{COO}^- + \text{ATP}$$

Amino acid

$$\rightleftharpoons$$

$$\text{R}-\text{CH}(\text{NH}_3^+)-\text{C}(=\text{O})-\text{O}-\text{P}(=\text{O})(\text{O}^-)-\text{O}-\text{Ribose}-\text{Adenine} + \text{PP}_i$$

Aminoacyl–adenylate (aminoacyl–AMP)

which, with all but three aaRSs, can occur in the absence of tRNA. Indeed, this intermediate may be isolated although it normally remains tightly bound to the enzyme.

2. This mixed anhydride then reacts with tRNA to form the aminoacyl–tRNA:

$$\text{Aminoacyl–AMP} + \text{tRNA} \rightleftharpoons \text{aminoacyl–tRNA} + \text{AMP}$$

Some aaRSs exclusively append an amino acid to the terminal 2′-OH group of their cognate tRNAs, and others do so at the 3′-OH group. This selectivity was established with the use of chemically modified tRNAs that lack either the 2′- or 3′-OH group of their 3′-terminal ribose residue. The use of these derivatives was necessary because, in solution, the aminoacyl group rapidly equilibrates between the 2′ and 3′ positions.

The overall aminoacylation reaction is

$$\text{Amino acid} + \text{tRNA} + \text{ATP} \rightleftharpoons \text{aminoacyl–tRNA} + \text{AMP} + \text{PP}_i$$

These reaction steps are readily reversible because the free energies of hydrolysis of the bonds formed in both the aminoacyl–adenylate and the aminoacyl–tRNA are comparable to that of ATP hydrolysis. The overall reaction is driven to completion by the inorganic pyrophosphatase-catalyzed hydrolysis of the PP_i generated in the first reaction step. Amino acid activation therefore chemically resembles fatty acid activation (Section 25-2A); the major difference between these two processes, which were both elucidated by Paul Berg, is that tRNA is the acyl acceptor in amino acid activation, whereas CoA performs this function in fatty acid activation.

a. There Are Two Classes of Aminoacyl–tRNA Synthetases

Most cells have one aaRS for each of the 20 amino acids. The similarity of the reactions catalyzed by these enzymes and the structural resemblance of all tRNAs suggests that all aaRSs evolved from a common ancestor and should therefore be structurally related. This is not the case. In fact, *the aaRSs form a diverse group of enzymes.* The over 1000 such enzymes that have been characterized each have one of four different types of subunit structures, α, α_2 (the predominant forms), α_4, and $\alpha_2\beta_2$, with known subunit sizes ranging from ~300 to ~1200 residues. Moreover, there is little sequence similarity among synthetases specific for different amino acids. Quite possibly, aminoacyl–tRNA synthetases arose very early in evolution, before the development of the modern protein synthesis apparatus other than tRNAs.

Detailed sequence and structural comparisons of aminoacyl–tRNA synthetases by Dino Moras indicate that these enzymes form two unrelated families, termed **Class I** and **Class II aaRSs,** that each have the same 10 members in nearly all organisms (Table 32-4). The Class I enzymes, although of largely dissimilar sequences, share two homologous polypeptide segments, not present in other proteins, that have the consensus sequences His-Ile-Gly-His (HIGH) and Lys-Met-Ser-Lys-Ser (KMSKS). The X-ray structures of Class I enzymes indicate that both of these segments are components of a dinucleotide-binding fold (Rossmann fold, which is also possessed by many NAD^+- and ATP-binding proteins; Section 8-3B) in which they participate in ATP binding and are implicated in catalysis. The Class II synthetases lack the foregoing sequences but have three other sequences in common. Their X-ray structures reveal that these sequences occur in a so-called signature motif, a fold found only in Class II enzymes that consists of a 7-stranded antiparallel β sheet with three flanking helices, which forms the core of their catalytic domains.

Many Class I aaRSs require anticodon recognition to aminoacylate their cognate tRNAs. In contrast, several

TABLE 32-4 Characteristics of Bacterial Aminoacyl–tRNA Synthetases

Amino Acid	Quaternary Structure	Number of Residues
Class I		
Arg	α	577
Cys	α	461
Gln	α	553
Glu	α	471
Ile	α	939
Leu	α	860
Met	α, α_2	676
Trp	α_2	325
Tyr	α_2	424
Val	α	951
Class II		
Ala	α, α_4	875
Asn	α_2	467
Asp	α_2	590
Gly	$\alpha_2\beta_2$	303/689
His	α_2	424
Lys	α_2	505
Pro	α_2	572
Phe	$\alpha_2\beta_2$, α	327/795
Ser	α_2	430
Thr	α_2	642

Source: Mainly Carter, C.W., Jr., *Annu. Rev. Biochem.* **62,** 715 (1993).

Class II enzymes, including **AlaRS** and **SerRS,** do not interact with their bound tRNA's anticodon. Indeed, several class II aaRSs accurately aminoacylate "microhelices" derived from only the acceptor stems of their cognate tRNAs. Another difference between Class I and Class II synthetases is that all Class I enzymes aminoacylate their bound tRNA's 3′-terminal 2′-OH group, whereas Class II enzymes, with the exception of **PheRS,** all charge the 3′-OH group. The amino acids for which the Class I synthetases are specific tend to be larger and more hydrophobic than those used by Class II synthetases. Finally, as Table 32-4 indicates, Class I aaRSs are mainly monomers, whereas most Class II aaRSs are homodimers.

LysRS has been classified as a Class II aaRS. However, a search of the genome sequences of *Methanococcus jannaschii* and *Methanobacterium thermoautotrophicum* failed to reveal the presence of such a LysRS. This led to the discovery that the LysRSs expressed by these archaebacteria are Class I rather than Class II enzymes. This raises the interesting question of how Class I LysRS evolved.

Prokaryotic aaRSs occur as individual protein molecules. However, in many higher eukaryotes (e.g., *Drosophila* and mammals), 9 aaRSs, some of each class, associate to form a multienzyme particle in which the glutamyl and prolyl synthetase functions are fused into a single polypeptide named **GluProRS.** The advantages of these systems are unknown.

b. The Structural Features Recognized by Aminoacyl–tRNA Synthetases May Be Quite Simple

As we shall see in Section 32-2D, ribosomes select aminoacyl–tRNAs only via codon–anticodon interactions, not according to the identities of their aminoacyl groups. *Accurate translation therefore requires not only that each tRNA be aminoacylated by its cognate aaRS but that it not be aminoacylated by any of its 19 noncognate aaRSs.* Considerable effort has therefore been expended, notably by LaDonne Schulman, Paul Schimmel, Olke Uhlenbeck, and John Abelson, in elucidating how aaRSs manage this feat, despite the close structural similarities of nearly all tRNAs. The experimental methods employed involved the use of specific tRNA fragments, mutationally altered tRNAs, chemical cross-linking agents, computerized sequence comparisons, and X-ray crystallography. The most common synthetase contact sites on tRNA occur on the inner (concave) face of the L. Other than that, there appears to be little regularity in how the various tRNAs are recognized by their cognate synthetases. Indeed, as we shall see, some aaRSs recognize only their cognate tRNA's acceptor stem, whereas others also interact with its anticodon region. Additional tRNA regions may also be recognized.

Genetic manipulations by Schimmel revealed that the tRNA features recognized by at least one type of aaRS are surprisingly simple. Numerous sequence alterations of *E. coli* tRNA^{Ala} do not appreciably affect its capacity to be aminoacylated with alanine. Yet, most base substitutions in the G3 · U70 base pair located in the tRNA's acceptor stem (Fig. 32-14*a*) greatly diminish this reaction. Moreover, the introduction of a G · U base pair into the analogous position of **tRNA^{Cys}** and tRNA^{Phe} causes them to be aminoacylated with alanine even though there are few other sequence identities between these mutant tRNAs and tRNA^{Ala} (e.g., Fig. 32-15). In fact, *E. coli* AlaRS even efficiently aminoacylates a 24-nt "microhelix" derived from only the G3 · U70-containing acceptor stem of *E. coli* tRNA^{Ala}. Since the only known *E. coli* tRNAs that normally have a G3 · U70 base pair are the tRNA^{Ala}, and this base pair is also present in the tRNA^{Ala} from many organisms including yeast (Fig. 32-8), the foregoing observations strongly suggest that *the G3 · U70 base pair is a major feature recognized by AlaRSs.* These enzymes presumably recognize the distorted shape of the G · U base pair (Fig. 32-12), a hypothesis corroborated by the observation that base changes at G3 · U70 which least affect the acceptor identity of tRNA^{Ala} yield base pairs that structurally resemble G · U.

The elements of three other tRNAs, which are recognized by their cognate tRNA synthetases, are indicated in Fig. 32-14. As with tRNA^{Ala}, these identity elements appear to comprise only a few bases. Note that the anticodon forms an identity element in two of these tRNAs. In another example of an anticodon identifier, the *E. coli* **tRNA^{Ile}** specific for the codon AUA has the anticodon LAU, where L is **lysidine,** a modified cytosine whose 2-keto group is replaced by the amino acid lysine (Fig. 32-10). The L in this context pairs with A rather than G, a rare instance of base modification altering base pairing specificity. The replacement of this L with unmodified C,

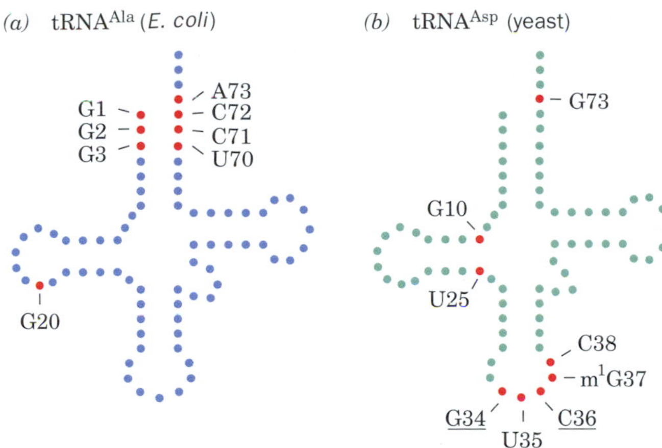

(c) tRNA^{Gln} (E. coli) — G73, G2, C71, G3, C70, G10, C25, Um32, ψ38, U33, m^{2}A37, C34, U35, G36

(d) tRNA^{Ser} — G73, G1, C72, G2, C71, AU3, UA70, C11, G24

FIGURE 32-14 Major identity elements in four tRNAs. Each base in the tRNA is represented by a filled circle. Red circles indicate positions that have been shown to be identity elements for the recognition of the tRNA by its cognate aminoacyl–tRNA synthetase. The anticodon bases that are identity elements are underlined. In each case, additional identity elements may yet be discovered. The base at position 73, which is an identity element in all four tRNAs shown here, is known as the **discriminator base.**

as expected, yields a tRNA that recognizes the Met codon AUG (codons bind anticodons in an antiparallel fashion). Surprisingly, however, this altered tRNA^{Ile} is also a much better substrate for **MetRS** than it is for **IleRS.** Thus, both the codon and the amino acid specificity of this tRNA are changed by a single posttranscriptional modification. The N^1-methylation of G37 in yeast tRNA^{Asp} (Fig. 32-14*b*) provides another example of a base modification forming an identity element. In the absence of this N^1-methyl group, tRNA^{Asp} is recognized by **ArgRS,** largely via its C36 and G37, whereas ArgRS normally recognizes only **tRNA^{Arg},** mainly via its C35 and U36.

The available experimental evidence has largely located the various tRNA identifiers in the acceptor stem and the anticodon loop (Fig. 32-16). The X-ray structures of sev-

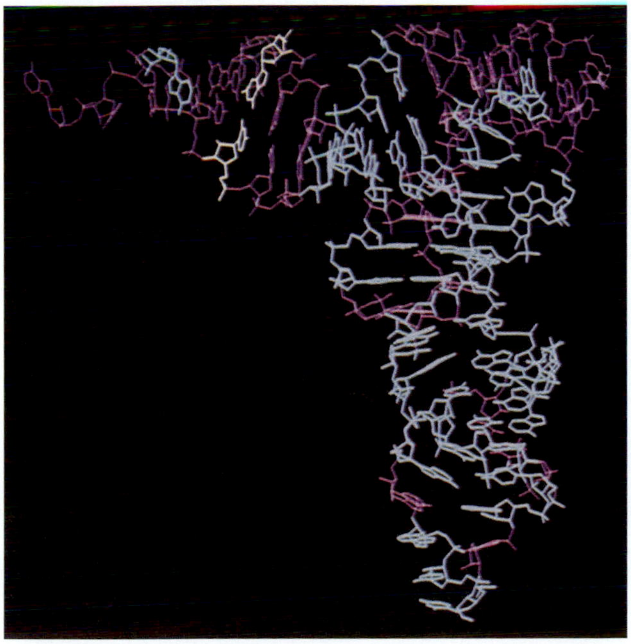

FIGURE 32-15 Three-dimensional model of *E. coli* tRNA^{Ala}. This model is based on the X-ray structure of yeast tRNA^{Phe} (Fig. 32-11*b*) in which the nucleotides that are different in *E. coli* tRNA^{Cys} are highlighted in cyan and the G3 · U70 base pair is highlighted in ivory. [Courtesy of Ya-Ming Hou, MIT.]

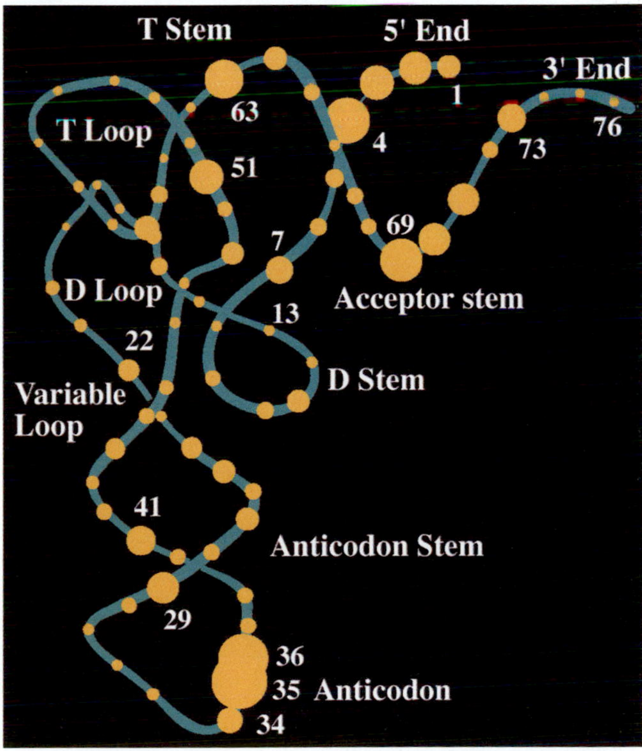

FIGURE 32-16 Experimentally observed identity elements of tRNAs. The tRNA backbone is cyan and each of its nucleotides is represented by a yellow circle whose diameter is proportional to the fraction of the 20 tRNA acceptor types for which the nucleotide is an observed determinant. [Courtesy of William McClain, University of Wisconsin.]

eral aaRS · tRNA complexes, which we consider next, have structurally rationalized some of these observations.

c. The X-Ray Structure of GlnRS · tRNA^{Gln}, a Class I Complex

The X-ray structures of all but two (Ala and Leu) of the 20 different amino acid–specific aaRSs have been determined, many of which are in complex with ATP, their cognate amino acids, or their analogs. These structures reveal that the active sites of these enzymes bind the ATP and target amino acid in optimal positions for in-line nucleophilic displacement (Section 16-2B) during amino acid activation and that the specificity of an aaRS for its target amino acid is determined by idiosyncratic contacts with the side chain of the amino acid.

The X-ray structures of 12 different aaRSs in their complexes with their cognate tRNAs have so far been reported. The first of them to be elucidated, that of *E. coli* **GlnRS,** a Class I synthetase, in its complex with **tRNA^{Gln}** and ATP (Fig. 32-17), was determined by Thomas Steitz. The tRNA^{Gln} assumes an L-shaped conformation that resembles those of tRNAs of known structures (e.g., Fig. 32-11*b*). GlnRS, a 553-residue monomeric protein that consists of four domains arranged to form an elongated molecule, interacts with the tRNA along the entire inside face of the L such that the anticodon is bound near one end of the protein and the acceptor stem is bound near its other end.

Genetic and biochemical data indicate that the identity elements of tRNA^{Gln} are largely clustered in its anticodon loop and acceptor stem (Fig. 32-14*c*). The anticodon loop of tRNA^{Gln} is extended by two novel non-Watson–Crick base pairs (2′-*O*-methyl-U32 · ψ38 and U33 · m^{2}A37), thereby causing the bases of the anticodon to unstack and splay outward in different directions so as to bind in separate recognition pockets of GlnRS. These structural features suggest that GlnRS uses all seven bases of the anticodon loop to discriminate among tRNAs. Indeed, changes to any one of the bases of residues C34 through ψ38 yield tRNAs with decreases in k_{cat}/K_M for aminoacylation by GlnRS by factors ranging from 70 to 28,000.

The GCCA at the 3′ end of the tRNA^{Gln} makes a hairpin turn toward the inside of the L rather than continuing

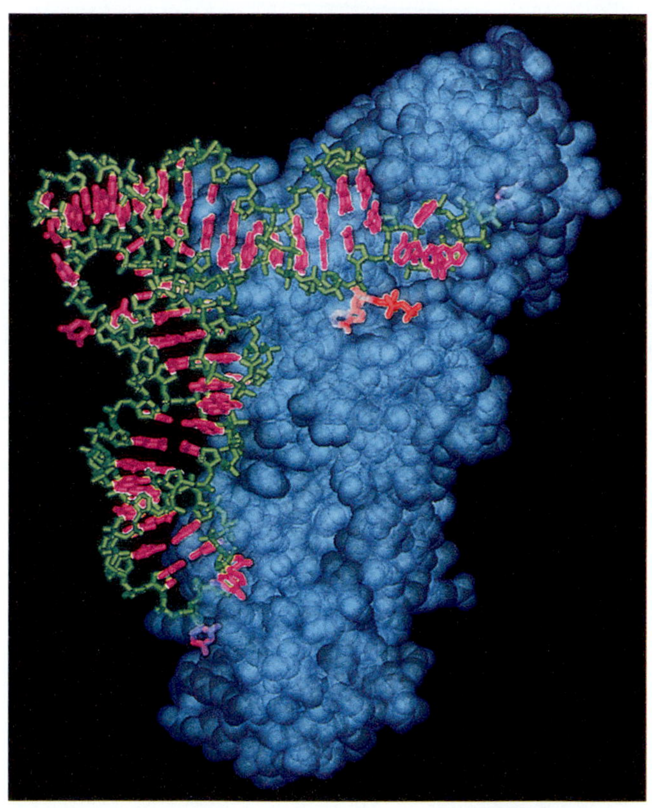

(*a*)

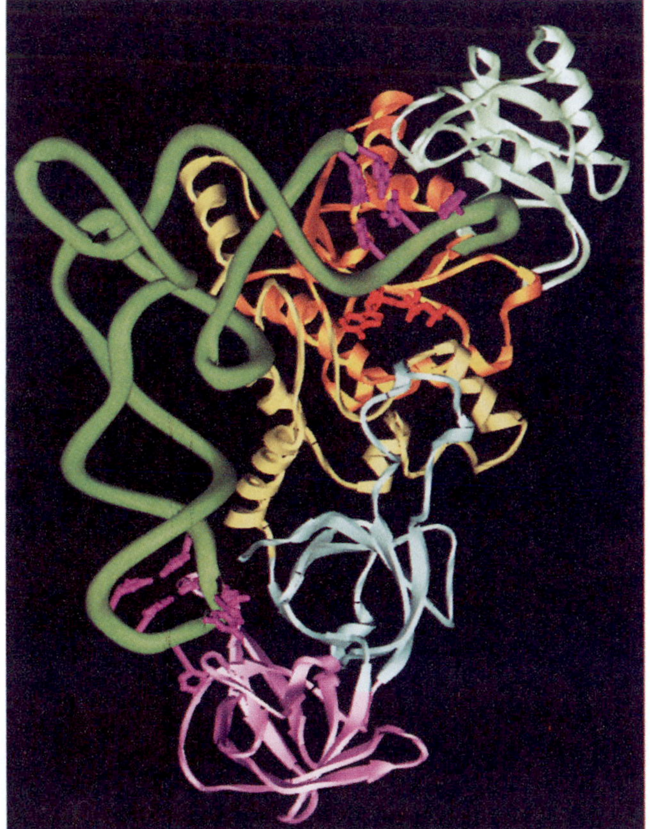

(*b*)

FIGURE 32-17 X-Ray structure of *E. coli* GlnRS · tRNA^{Gln} · ATP. (*a*) The tRNA and ATP are shown in skeletal form with the tRNA sugar–phosphate backbone green, its bases magenta, and the ATP red. The protein is represented by a translucent cyan space-filling model that reveals the buried portions of the tRNA and ATP. Note that both the 3′ end of the tRNA (*top right*) and its anticodon bases (*bottom*) are inserted into deep pockets in the protein. (*b*) A ribbon drawing of the complex viewed as in Part *a*. The tRNA's sugar–phosphate backbone is represented by a green worm and the bases forming its identity elements (Fig. 32-14*c*) are magenta. The protein's four domains are differently colored with the dinucleotide binding fold gold and the remainder of the catalytic domain that contains it yellow. The ATP is shown in skeletal form (*red*). [Based on an X-ray structure by Thomas Steitz, Yale University. PDBid 1GSG.] **See the Kinemage Exercise 21**

helically onward (as does the ACCA at the 3′ end in the X-ray structure of tRNA^{Phe}; Fig. 32-11*b*). This conformation change is facilitated by the insinuation of a Leu side chain between the 5′ and 3′ ends of the tRNA so as to disrupt the first base pair of the acceptor stem (U1 · A72). The GlnRS reaction is therefore relatively insensitive to base changes in these latter two positions except when base pairing is strengthened by their conversion to G1 · C72. The GCCA end of the tRNA^{Gln} plunges deeply into a protein pocket that also binds the enzyme's ATP and glutamine substrates. Three protein "fingers" are inserted into the minor groove of the acceptor stem to make sequence-specific interactions with base pairs G2 · C71 and G3 · C70 [recall that double helical RNA has an A-DNA-like structure (Section 29-1B) whose wide minor groove readily admits protein but whose major groove is normally too narrow to do so].

The GlnRS domain that binds glutamine, ATP, and the GCCA end of tRNA^{Gln}, the so-called catalytic domain, contains, as we previously discussed, a dinucleotide-binding fold. Much of this domain is nearly superimposable with and thus evolutionarily related to the corresponding domains of other Class I aaRSs.

d. The X-Ray Structure of AspRS · tRNA^{Asp}, a Class II Complex

Yeast **AspRS,** a Class II synthetase, is an α_2 dimer of 557-residue subunits. Its X-ray structure in complex with tRNA^{Asp}, determined by Moras, reveals that the protein symmetrically binds two tRNA molecules (Fig. 32-18).

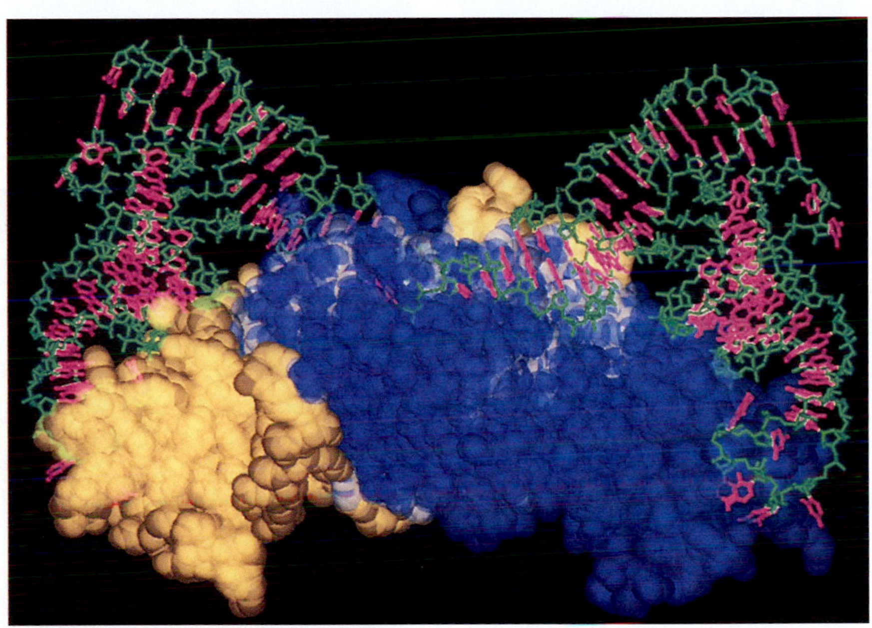

(*a*)

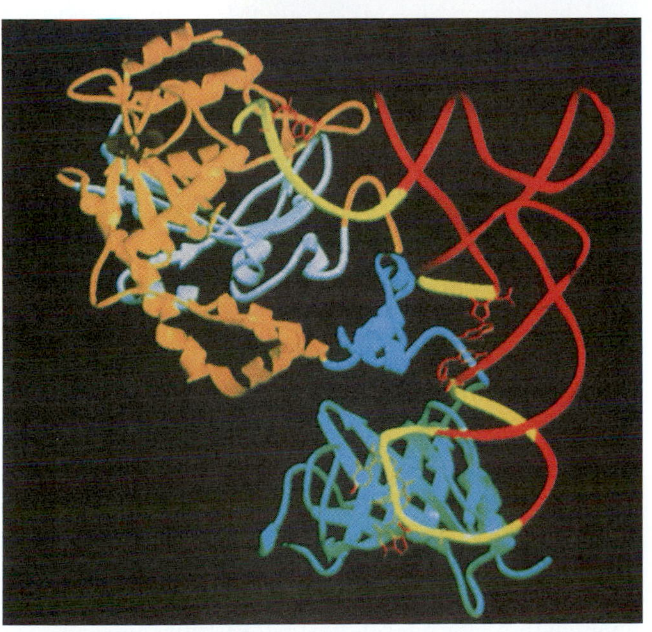

(*b*)

FIGURE 32-18 X-Ray structure of yeast AspRS · tRNA^{Asp} · ATP. (*a*) The homodimeric enzyme with its two symmetrically bound tRNAs viewed with its 2-fold axis approximately vertical. The tRNAs are shown in skeletal form with their sugar–phosphate backbones green and their bases magenta. The two protein subunits are represented by translucent yellow and blue space-filling models that reveal buried portions of the tRNAs. (*b*) A ribbon diagram of the AspRS · tRNA^{Asp} · ATP monomer. The tRNA's contact regions with the protein are yellow and its identity elements are shown in stick form in red as is the ATP. The protein's N-terminal domain is blue-green, the central domain is cyan, and the C-terminal catalytic domain is orange with its component signature motif (the 7-stranded antiparallel β sheet with 3 flanking helices characteristic of Type II aaRSs) in white. [Part *a* based on an X-ray structure by and Part *b* courtesy of Dino Moras, CNRS/INSERM/ULP, Illkirch Cédex, France. PDBid 1ASY.]

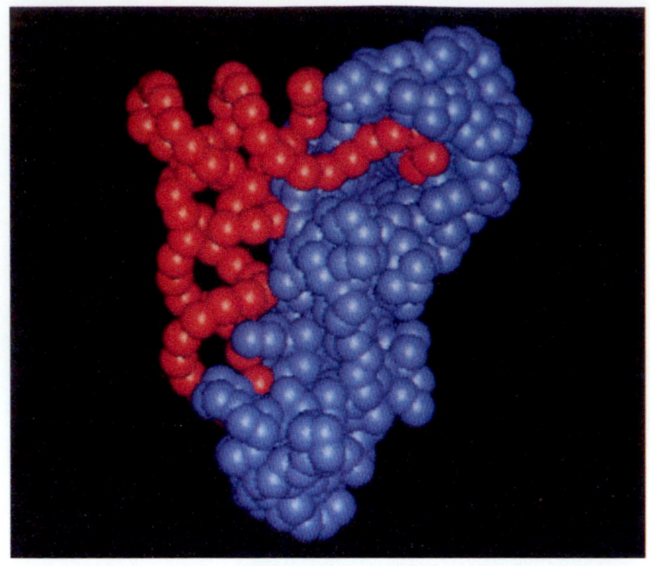

(a)

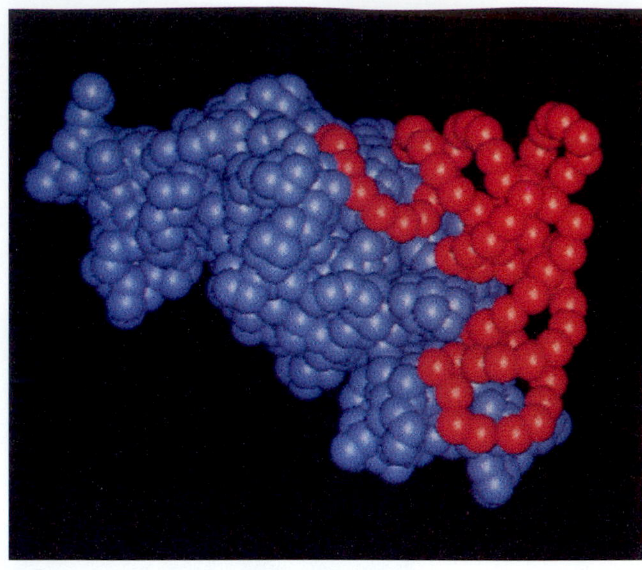

(b)

FIGURE 32-19 Comparison of the modes by which GlnRS and AspRS bind their cognate tRNAs. The proteins and tRNAs are represented by blue and red spheres centered on their C_α and P atom positions. Note how GlnRS (*a*), a Class I synthetase, binds $tRNA^{Gln}$ from the minor groove side of its acceptor stem so as to bend its 3′ end into a hairpin conformation. In contrast, AspRS (*b*), a Class II synthetase, binds $tRNA^{Asp}$ from the major groove side of its acceptor stem so that its 3′ end continues its helical path on entering the active site. [Courtesy of Dino Moras, CNRS/INSERM/ULP, Illkirch Cédex, France.]

Like GlnRS, AspRS principally contacts its bound tRNA both at the end of its acceptor stem and in its anticodon region. The contacts in these two enzymes are, nevertheless, quite different in character (Fig. 32-19): Although both tRNAs approach their cognate synthetases along the inside of their L shapes, $tRNA^{Gln}$ does so toward the direction of the minor groove of its acceptor stem, whereas $tRNA^{Asp}$ does so toward the direction of its major groove. The GCCA at the 3′ end of $tRNA^{Asp}$ thereby continues its helical track as it plunges into AspRS's catalytic site, whereas, as we saw, the GCCA end of $tRNA^{Gln}$ bends backward into a hairpin turn that opens up the first base pair (U1 · A72) of its acceptor stem. Although the deep major groove of an A-RNA helix is normally too narrow to admit groups larger than water molecules (Section 29-1B), the major groove at the end of the acceptor stem in AspRS · $tRNA^{Asp}$ is sufficiently widened for its base pairs to interact with a protein loop.

The anticodon arm of $tRNA^{Asp}$ is bent by as much as 20 Å toward the inside of the L relative to that in the X-ray structure of uncomplexed $tRNA^{Asp}$ and its anticodon bases are unstacked. The hinge point for this bend is a G30 · U40 base pair in the anticodon stem which, in nearly all other species of tRNA, is a Watson–Crick base pair. The anticodon bases of $tRNA^{Gln}$ are also unstacked in contacting GlnRS but with a backbone conformation that differs from that in $tRNA^{Asp}$. Evidently, the conformation of a tRNA in complex with its cognate synthetase appears to be dictated more by its interactions with the protein (induced fit) than by its sequence.

Structural analyses of complexes of AspRS · $tRNA^{Asp}$ with ATP and aspartic acid, and of GlnRS · $tRNA^{Gln}$ with ATP, have permitted models of the aminoacyl–AMP complexes of these enzymes to be independently formulated. Comparison of these models reveals that the 3′-terminal A residues of $tRNA^{Gln}$ and $tRNA^{Asp}$ (to which the aminoacyl groups are appended; Fig. 32-13) are positioned on opposite sides of the enzyme-bound aminoacyl–AMP intermediate (Fig. 32-20). The 3′-terminal ribose residues are puckered C2′-*endo* for $tRNA^{Asp}$ and C3′-*endo* for $tRNA^{Gln}$; see Fig. 29-10) such that the 2′-hydroxyl group of $tRNA^{Gln}$ (Class I) is stereochemically positioned to attack the aminoacyl–AMP's carboxyl group, whereas for $tRNA^{Asp}$ (Class II), only the 3′ hydroxyl group is situated to do so. This clearly explains the different aminoacylation specificities of the Class I and Class II aaRSs.

e. Proofreading Enhances the Fidelity of Amino Acid Attachment to tRNA

The charging of a tRNA with its cognate amino acid is a remarkably accurate process: aaRSs display an overall error rate of about 1 in 10,000. We have seen that aaRSs bind only their cognate tRNAs through an intricate series of specific contacts. But how do they discriminate among the various amino acids, some of which are quite similar?

Experimental measurements indicate, for example, that IleRS transfers as many as 40,000 isoleucines to $\mathbf{tRNA^{Ile}}$ for every valine it so transfers. Yet, as Linus Pauling first pointed out, *there are insufficient structural differences between Val and Ile to permit such a high degree of discrimination in the direct generation of aminoacyl–tRNAs.* The X-ray structure of *Thermus thermophilus* IleRS, a monomeric Class I aaRS, in complex with isoleucine, determined by Shigeyuki Yokoyama and Schimmel, indicates that isoleucine fits snugly into its binding site in the enzyme's Rossmann fold domain and hence that this binding site would sterically exclude leucine as well as larger amino acids. However, valine, which differs from isoleucine by only the lack of a single methylene group, fits into this isoleucine-binding site. The binding free energy of a methylene group is estimated to be ~12 kJ · mol^{-1}. Equation [3.16] indicates that the ratio f of the equilibrium constants, K_1 and K_2, with which two substances bind to a given binding site is given by

$$f = \frac{K_1}{K_2} = \frac{e^{-\Delta G_1^{\circ\prime}/RT}}{e^{-\Delta G_2^{\circ\prime}/RT}} = e^{-\Delta\Delta G^{\circ\prime}/RT} \qquad [32.1]$$

where $\Delta\Delta G^{\circ\prime} = \Delta G_1^{\circ\prime} - \Delta G_2^{\circ\prime}$ is the difference between the free energies of binding of the two substances. It is therefore estimated that isoleucyl–tRNA synthetase could discriminate between isoleucine and valine by no more than a factor of ~100.

Berg resolved this apparent paradox by demonstrating that, in the presence of tRNA^{Ile}, IleRS catalyzes the nearly quantitative hydrolysis of valyl-aminoacyl–adenylate to valine + AMP rather than forming Val-tRNA^{Ile}. Moreover, the few Val-tRNA^{Ile} molecules that do form are hydrolyzed to valine + tRNA^{Ile}. Thus, *IleRS subjects both aminoacyl–adenylate and aminoacyl–tRNA^{Ile} to a* ***proofreading*** *or* ***editing*** *step that occurs at a separate catalytic site.* This site binds Val residues but excludes the larger Ile residues. *The enzymes's overall selectivity is therefore the product of the selectivities of its synthesis and proofreading steps, thereby accounting for the high fidelity of aminoacylation. Note that in this so-called* ***double-sieve*** *mechanism, editing occurs at the expense of ATP hydrolysis, the thermodynamic price of high fidelity (increased order).*

The X-ray structure of *Staphylococcus aureus* IleRS in complex with tRNA^{Ile} and the clinically useful antibiotic **mupirocin**

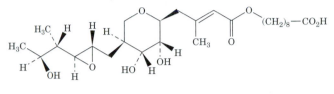

Mupirocin

(a product of *Pseudomonas fluorescens* that acts by specifically binding to bacterial IleRS so as to inhibit bacterial protein synthesis), determined by Steitz, suggests how IleRS carries out its editing process. The X-ray structure (Fig. 32-21) reveals that this complex resembles the GlnRS · tRNA^{Gln} · ATP complex (Fig. 32-17) but with IleRS having an additional editing domain (also called CP1 for *c*onnective *p*eptide 1) inserted in its Rossmann fold domain. The two 3′ terminal residues of the tRNA^{Ile}, C75 and A76, are disordered but, when modeled so as to continue

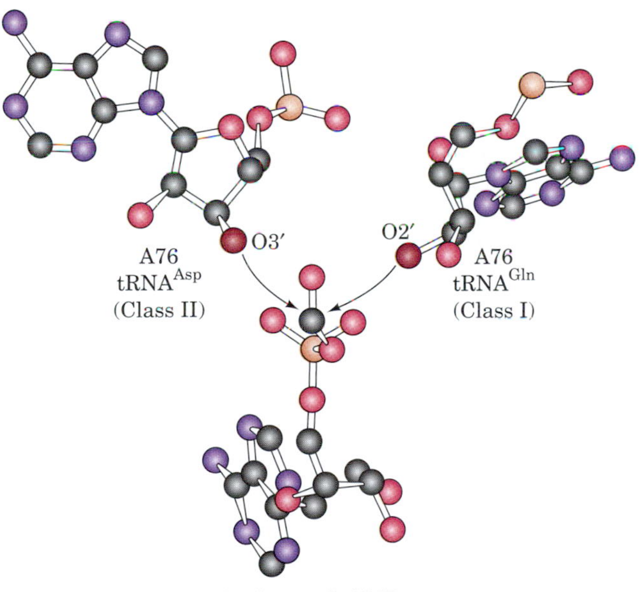

FIGURE 32-20 Comparison of the stereochemistries of aminoacylation by Class I and Class II aaRSs. The positions of the 3′ terminal adenosine residues (A76) of AspRS (Class II, *left*) and GlnRS (Class I, *right*) are drawn relative to that of the enzyme-bound aminoacyl–AMP (*below*; only the carbonyl group of its aminoacyl residue is shown). Note how only O3′ of tRNA^{Gln} and O2′ of tRNA^{Asp} are suitably positioned to attack the aminoacyl residue's carbonyl group and thereby transfer the aminoacyl residue to the tRNA. [After Cavarelli, J., Eriani, G., Rees, B., Ruff, M., Boeglin, M., Mitschler, A., Martin, F., Gangloff, J., Thierry, J.-C., and Moras, D., *EMBO J.* **13,** 335 (1994).]

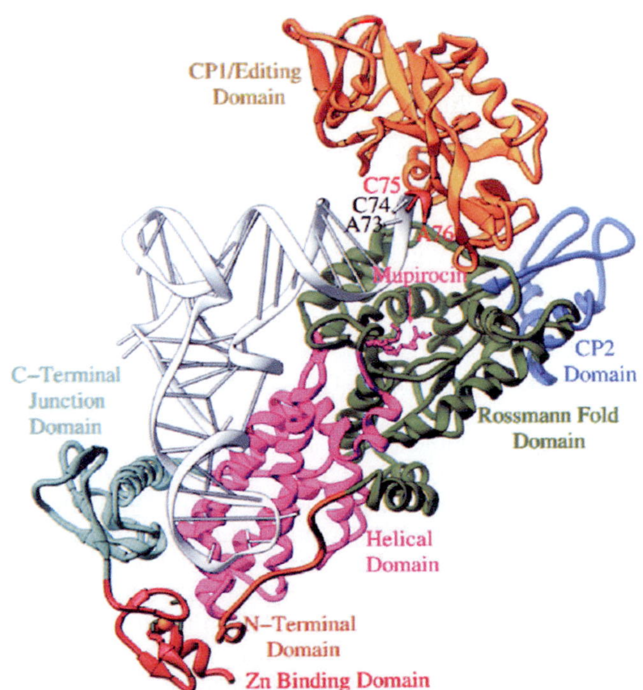

FIGURE 32-21 X-Ray structure of *T. thermophilus* isoleucyl–tRNA synthetase in complex with tRNA^{Ile} and mupirocin. The tRNA is white, the protein is colored by domain, and the mupirocin is shown in stick form in pink. [Courtesy of Thomas Steitz, Yale University. PDBid 1QU2.]

(*a*)

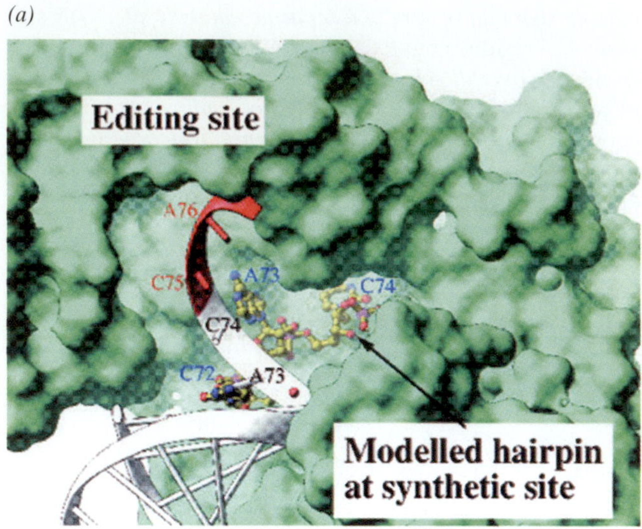

FIGURE 32-22 Comparison of the putative aminoacylation and editing modes of IleRS · tRNA^{Ile}. (*a*) The superposition of tRNA^{Ile} in these two binding modes on the solvent-accessible surface of IleRS (*cyan*). The acceptor strand of tRNA^{Ile} in the editing mode observed in the X-ray structure of IleRS · tRNA^{Ile} · mupirocin (Fig. 32–21) is drawn in ribbon form in white with the modeled positions of C75 and A76 in red. This places the tRNA's 3′ end in the editing site. In contrast, the three 3′ terminal residues of tRNA^{Ile}, as positioned through homology modeling based on the X-ray structure of GlnRS · tRNA^{Gln} · ATP (Fig. 32-17) and drawn in ball-and-stick form with C yellow, N blue, O red, and P magenta, places the tRNA's 3′ end in the synthetic (aminoacylation) site, 34 Å distant from its position in the editing site. Note that there is a cleft running between the editing and synthetic sites and that the 3′ end of the tRNA continues its A-form helical path in the editing mode but assumes a hairpin conformation in the synthetic mode. (*b*) A cartoon comparing the positions of the 3′ end of tRNA^{Ile} in its complex with IleRS in its synthetic mode (*left*) and in its editing mode (*right*). [Part *a* courtesy of and Part *b* based on a drawing by Thomas Steitz, Yale University.]

(*b*)

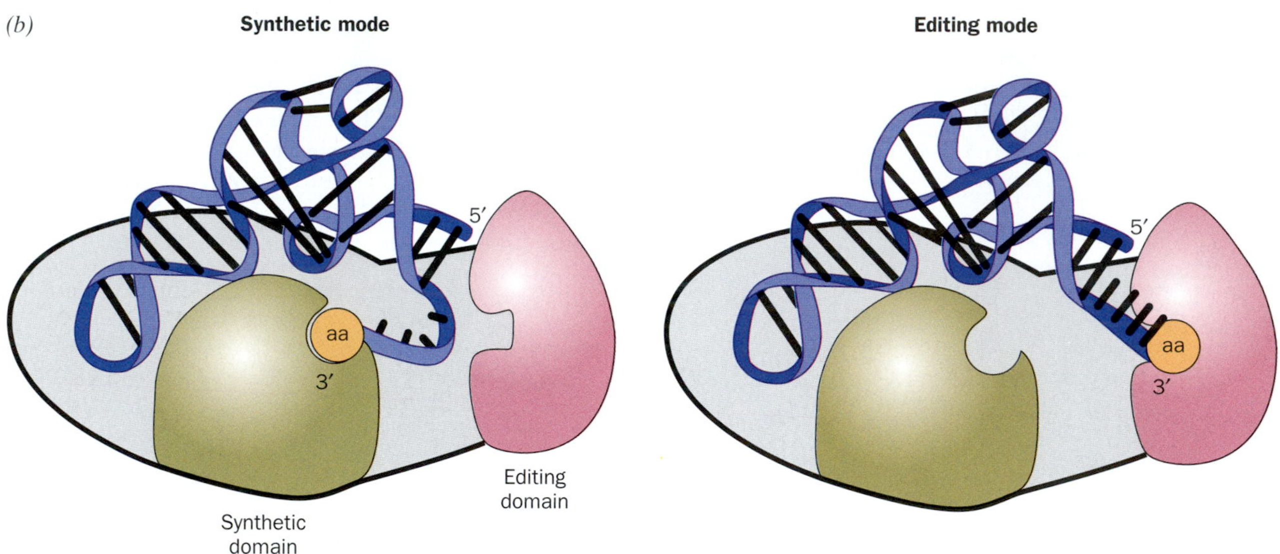

the acceptor stem's stacked A-form helix, extend into a cleft in the editing domain that has been implicated as its hydrolytic site (Fig. 32-22*a, left*). Thus, this IleRS complex appears to resemble an "editing complex" instead of a "transfer complex" as seen in the GlnRS structure. However, a transfer complex would form if the 3′ ending segment of the tRNA^{Ile} assumes a hairpin conformation (Fig. 32-22*a, right*) similar to that in the GlnRS structure (Fig. 32-17*b;* recall that IleRS and GlnRS are both Class I aaRSs). Steitz has therefore postulated that the aminoacyl group is shuttled between the IleRS's aminoacylation site and its editing site by such a conformational change (Fig. 32-22*b*). This process functionally resembles the way in which DNA polymerase I edits its newly synthesized strand (Section 30-2A), which Steitz also elucidated.

ValRS is a monomeric Class I aaRS that resembles IleRS. The X-ray structure of the complex of *T. thermophilus* ValRS, **tRNA^{Val}**, and the nonhydrolyzable **valyl-aminoacyl–adenylate** analog **5′-*O*′-[*N*-(L-valyl)sulfamoyl]adenosine (Val-AMS),**

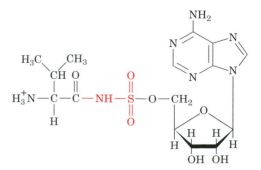

H$_3$C CH$_3$ / CH / $H_3\overset{+}{N}$—C(H)—C(=O)—NH—S(=O)(=O)—O—CH$_2$—ribose(OH, OH)—adenine (NH$_2$)

5′-*O*-[*N*-(L-valyl)sulfamoyl]adenosine (Val-AMS)

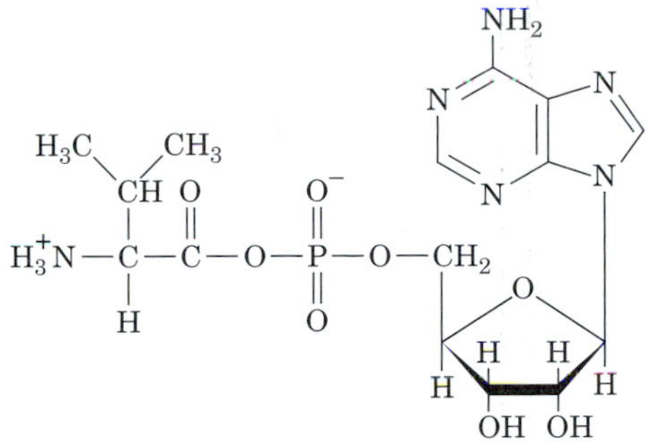

Valyl-aminoacyl–adenylate

determined by Yokoyama, reveals that the Val-AMS is bound in the aminoacylation pocket in the Rossmann fold domain, which accommodates the isosteric Val and Thr moieties but sterically excludes Ile. Modeling studies based on the IleRS · $tRNA^{Ile}$ · mupirocin structure indicate that the Thr side chain would fit into the ValRS editing pocket with its side chain hydroxyl group hydrogen bonded to the side chain of Asp 279 of ValRS, which protrudes into the pocket in contrast to the corresponding Asp 328 of IleRS, which does not. Consequently, a Val side chain would be excluded from the ValRS editing pocket because it cannot form such a hydrogen bond, thereby explaining why this editing pocket hydrolyzes **threonyl-aminoacyl–adenylate** and Thr-$tRNA^{Val}$ but not the corresponding Val derivatives. The ValRS · $tRNA^{Val}$ structure also indicates that ValRS and $tRNA^{Val}$ together form a tunnel connecting the ValRS's aminoacylation pocket with its editing pocket. Improperly formed threonyl-aminoacyl–adenylate is proposed to be channeled through this tunnel for hydrolysis in the editing pocket, thereby explaining why $tRNA^{Val}$ must be bound to ValRS for this pretransfer editing reaction to occur. Valyl-aminoacyl–adenylate is presumably channeled through the similar IleRS · $tRNA^{Ile}$ complex for its hydrolysis.

ThrRS, a Class II homodimer, has the opposite problem of ValRS: It must synthesize **Thr–$tRNA^{Thr}$** but not Val–$tRNA^{Thr}$. The X-ray structure of *E. coli* ThrRS that lacks its N-terminal domain but remains catalytically active in a complex with either threonine or the threonyl-aminoacyl–adenylate analog **Thr-AMS,** determined by Moras, reveals that ThrRS's aminoacylation pocket contains a Zn^{2+} ion that is coordinated by the side chain hydroxyl and amino groups of the threonyl group as well as by three protein side chains. The isosteric valine could not coordinate the Zn^{2+} ion in this way and hence does not undergo adenylylation by ThrRS. However, what prevents ThrRS from synthesizing Ser–$tRNA^{Thr}$? In fact, the truncated ThrRS synthesizes Ser–$tRNA^{Thr}$ at more than half the rate it synthesizes Thr–$tRNA^{Thr}$, thereby indicating that the N-terminal domain of wild-type ThrRS contains the enzyme's editing site. Mutational analysis of ThrRS has localized this editing site to a cleft in the N-terminal domain of wild-type ThrRS, whose X-ray structure in complex with $tRNA^{Thr}$ was also determined by Moras. In this latter structure, the tRNA's 3′ end follows a regular helical path similar to that seen in the X-ray structure of AspRS · $tRNA^{Asp}$ · ATP (Fig. 32-18) so as to enter the

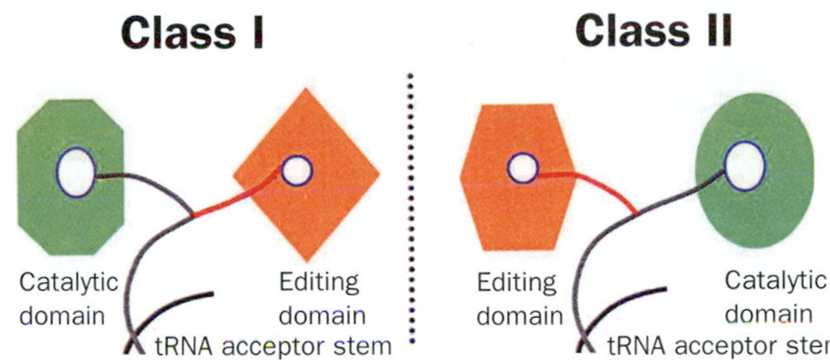

FIGURE 32-23 Schematic diagram of the aminoacylation and editing mechanisms of Class I and Class II aaRSs emphasizing the "mirror symmetry" of their overall mechanisms. With Class I aaRSs (*left;* e.g., IleRS), the 3′ end of the bound tRNA's acceptor stem assumes a hairpin conformation in the synthetic mode and a helical conformation in the editing mode, whereas the converse occurs with Class II aaRSs (*right;* e.g., ThrRS). [Courtesy of Dino Moras, CNRS/INSERM/ULP, Illkirch Cedex, France.]

aminoacylation site. However, if the 3′ end of the bound $tRNA^{Thr}$ assumed a hairpin conformation similar to that seen in X-ray structure of $tRNA^{Gln}$ in complex with the Class I enzyme GlnRS and ATP (Fig. 32-17), its covalently linked aminoacyl group would enter the editing site. This indicates an intriguing "mirror symmetry" (Fig. 32-23): In Class I aaRSs that mediate a double-sieve editing mechanism, the 3′ end of the bound cognate tRNA assumes a hairpin conformation when it enters the aminoacylation site and a helical conformation when it enters the editing site, whereas the converse holds for Class II aaRSs. Finally, ThrRS does not appear to mediate pretransfer editing (does not hydrolyze **seryl-aminoacyl–adenylate**), and, in fact, the ThrRS · $tRNA^{Thr}$ complex lacks a channel connecting its aminoacylation and editing sites such as is seen in the ValRS · $tRNA^{Val}$ complex.

Synthetases that have adequate selectivity for their corresponding amino acid lack editing functions. Thus, for example, the TyrRS aminoadenylylation site discriminates between tyrosine and phenylalanine through hydrogen bonding with the tyrosine —OH group. The cell's other amino acids, standard as well as nonstandard, have even less resemblance to tyrosine which rationalizes why TyrRS lacks an editing site.

f. Gln–$tRNA^{Gln}$ May Be Formed via an Alternative Pathway

Although it was long believed that each of the 20 standard amino acids is covalently linked to a tRNA by its corresponding aaRS, it is now clear that gram-positive bacteria, archaebacteria, cyanobacteria, mitochondria, and chloroplasts all lack GlnRS. Rather glutamate is linked to $tRNA^{Gln}$ by the same GluRS that synthesizes **Glu–$tRNA^{Glu}$.** The resulting **Glu–$tRNA^{Gln}$** is then transamidated to Gln–$tRNA^{Gln}$ by the enzyme **Glu–$tRNA^{Gln}$ amidotransferase (Glu-AdT)** in an ATP-requiring reaction in which glutamine is the amide donor. Some microorganisms use a similar transamidation pathway for the synthesis of Asn–$tRNA^{Asn}$ from **Asp–$tRNA^{Asn}$.**

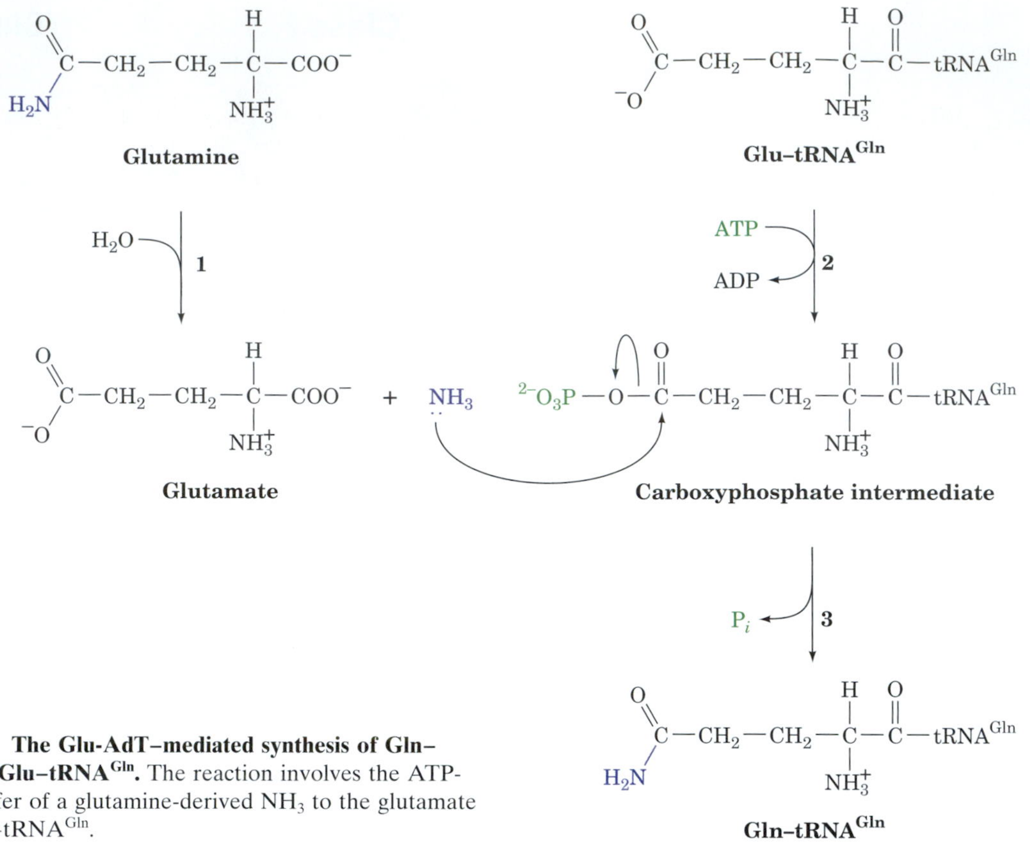

FIGURE 32-24 The Glu-AdT–mediated synthesis of Gln–tRNA^{Gln} from Glu–tRNA^{Gln}. The reaction involves the ATP-activated transfer of a glutamine-derived NH_3 to the glutamate moiety of Glu–tRNA^{Gln}.

The overall reaction catalyzed by Glu-AdT occurs in three stages (Fig. 32-24): (1) Glutamine is hydrolyzed to glutamate and the resulting NH_3 sequestered; (2) ATP reacts with the Glu side chain of Glu–tRNA^{Gln} to yield an activated acylphosphate intermediate and ADP; and (3) the acylphosphate intermediate reacts with the NH_3 to yield Gln–tRNA^{Gln} + P_i. Glu-AdT from *Bacillus subtilis,* which was characterized by Söll, is a heterotrimeric protein, none of whose subunits exhibit significant sequence similarity to GlnRS. The genes encoding these subunits, *gatA, gatB,* and *gatC,* form a single operon whose disruption is lethal, thereby demonstrating that *B. subtilis* has no alternative pathway for Gln–tRNA^{Gln} production. The **GatA** subunit of Glu-AdT appears to catalyze the activation of the side chain carboxyl of glutamic acid via a reaction resembling that catalyzed by carbamoyl phosphate synthetase (Section 26-2A). Nevertheless, GatA exhibits no sequence similarity with other known glutamine amidotransferases (members of the triad or Ntn families; Section 26-5A). The **GatB** subunit may be used to select the correct tRNA substrate. The role of the **GatC** subunit is unclear, although the observation that its presence is necessary for the expression of GatA in *E. coli* suggests that it participates in the modification, folding, and/or stabilization of GatA.

Since Glu is not misincorporated into *B. subtilis* proteins in place of Gln, the Glu–tRNA^{Gln} product of the above aminoacylation reaction must not be transported to the ribosome. It is likely that this does not occur because, as has been shown in chloroplasts, **EF-Tu,** the elongation factor that binds and transports most aminoacyl–tRNAs to the ribosome in a GTP-dependent process (Section 32-3D), does not bind Glu–tRNA^{Gln}. It is unclear why two independent routes have evolved for the synthesis of Gln–tRNA^{Gln}.

g. Some Archaebacteria Lack a Separate CysRS

The genomes of certain archaebacteria such as *M. jannaschii* lack an identifiable gene for CysRS. This is because the enzyme responsible for synthesizing **Pro–tRNA^{Pro}** in these organisms also synthesizes **Cys–tRNA^{Cys}.** Interestingly, this enzyme, which is named **ProCysRS,** does not synthesize Pro–tRNA^{Cys} or Cys–tRNA^{Pro}. Although ProCysRS synthesizes **cysteinyl-aminoacyl–adenylate** only in the presence of tRNA^{Cys}, it synthesizes **prolyl-aminoacyl–adenylate** in the absence of tRNA^{Pro}. The binding of tRNA^{Cys} to ProCysRS blocks the activation of proline so that only cysteine can be activated. Conversely, the activation of proline facilitates the binding of tRNA^{Pro} while preventing the binding of tRNA^{Cys}. However, the mechanism through which ProCysRS carries out these mutually exclusive syntheses is unknown. In any case, it appears that some organisms can get by with as few as 17 different aaRSs; they may lack GlnRS, AspRS, and a separate CysRS.

D. *Codon–Anticodon Interactions*

In protein synthesis, the proper tRNA is selected only through codon–anticodon interactions; the aminoacyl group does not participate in this process. This phenomenon was demonstrated as follows. Cys–tRNA^{Cys}, in which the Cys residue was ^{14}C labeled, was reductively desulfurized with Raney nickel so as to convert the Cys residue to Ala:

$$\text{HS}-\text{CH}_2-\underset{\text{NH}_3^+}{\overset{\text{H}}{\text{C}}}-\overset{\text{O}}{\overset{\|}{\text{C}}}-\text{O}-\text{tRNA}^{\text{Cys}} + \text{Ni(H)}_x$$

Cys–tRNA^{Cys} **Raney nickel**

$$\rightleftharpoons$$

$$\text{H}-\text{CH}_2-\underset{\text{NH}_3^+}{\overset{\text{H}}{\text{C}}}-\overset{\text{O}}{\overset{\|}{\text{C}}}-\text{O}-\text{tRNA}^{\text{Cys}} + \text{H}_2\text{S} + \text{Ni}$$

Ala–tRNA^{Cys}

The resulting ^{14}C-labeled hybrid, Ala–tRNA^{Cys}, was added to a cell-free protein synthesizing system extracted from rabbit reticulocytes. The product hemoglobin α chain's only radioactive tryptic peptide was the one that normally contains the subunit's only Cys. No radioactivity was found in the peptides that normally contain Ala but no Cys. Evidently, *only the anticodons of aminoacyl–tRNAs participate in codon recognition.*

a. Genetic Code Degeneracy Is Largely Due to Variable Third Position Codon–Anticodon Interactions

One might naively guess that each of the 61 codons specifying an amino acid would be read by a different tRNA. Yet, even though most cells contain several groups of **isoaccepting tRNAs** (different tRNAs that are specific for the same amino acid), *many tRNAs bind to two or three of the codons specifying their cognate amino acids.* For example, yeast tRNA^{Phe}, which has the anticodon GmAA, recognizes the codons UUC and UUU (remember that the anticodon pairs with the codon in an antiparallel fashion),

```
            3'          5'     3'          5'
Anticodon:  —A—A—Gm—          —A—A—Gm—
             :  :  :           :  :  :
            5'          3'     5'          3'
Codon:      —U—U—C—           —U—U—U—
```

and yeast tRNA^{Ala}, which has the anticodon IGC, recognizes the codons GCU, GCC, and GCA.

```
            3'          5'     3'          5'
Anticodon:  —C—G—I—           —C—G—I—
             :  :  :           :  :  :
            5'          3'     5'          3'
Codon:      —G—C—U—           —G—C—C—

            3'          5'
Anticodon:  —C—G—I—
             :  :  :
            5'          3'
Codon:      —G—C—A—
```

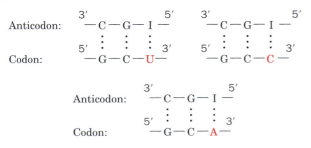

It therefore seems that non-Watson–Crick base pairing can occur at the third codon–anticodon position (the anticodon's first position is defined as its 3′ nucleotide), the site of most codon degeneracy (Table 32-2). Note also that the third (5′) anticodon position commonly contains a modified base such as Gm or I.

b. The Wobble Hypothesis Structurally Accounts for Codon Degeneracy

By combining structural insight with logical deduction, Crick proposed, in what he named the **wobble hypothesis,** how a tRNA can recognize several degenerate codons. He assumed that the first two codon–anticodon pairings have normal Watson–Crick geometry. The structural constraints that this places on the third codon–anticodon pairing ensure that its conformation does not drastically differ from that of a Watson–Crick pair. Crick then proposed that there could be a small amount of play or "wobble" in the third codon position which allows limited conformational adjustments in its pairing geometry. This permits the formation of several non-Watson–Crick pairs such as U · G and I · A (Fig. 32-25*a*). The allowed "wobble" pairings are indicated in Fig. 32-25*b*. Then, by analyzing the known pat-

(*a*)

U · G

I · A

(*b*)

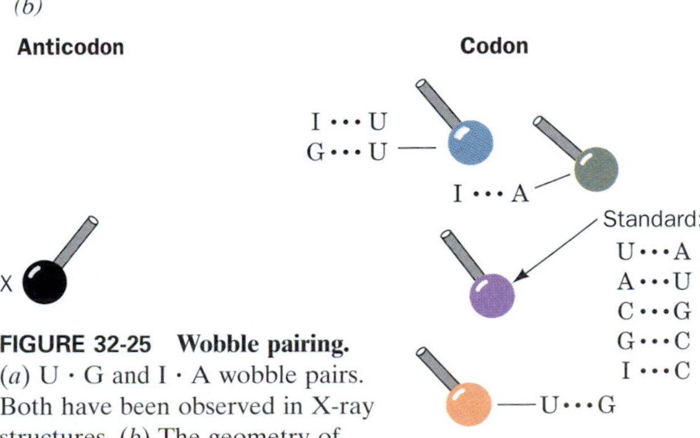

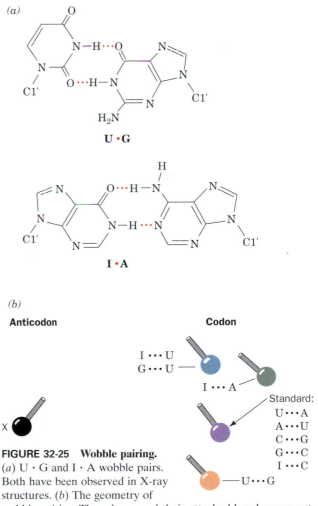

FIGURE 32-25 Wobble pairing. (*a*) U · G and I · A wobble pairs. Both have been observed in X-ray structures. (*b*) The geometry of wobble pairing. The spheres and their attached bonds represent the positions of ribose C1′ atoms with their accompanying glycosidic bonds. X (*left*) designates the nucleoside at the 5′ end of the anticodon (tRNA). The positions on the right are those of the 3′ nucleoside of the codon (mRNA) in the indicated wobble pairings. [After Crick, F.H.C., *J. Mol. Biol.* **19,** 552 (1966).]

TABLE 32-5 Allowed Wobble Pairing Combinations in the Third Codon–Anticodon Position

5′-Anticodon Base	3′-Codon Base
C	G
A	U
U	A or G
G	U or C
I	U, C, or A

tern of codon–anticodon pairing, Crick deduced the most plausible sets of pairing combinations in the third codon–anticodon position (Table 32-5). Thus, an anticodon with C or A in its third position can only pair with its Watson–Crick complementary codon. If U, G, or I occupies the third anticodon position, two, or three codons are recognized, respectively.

No prokaryotic or eukaryotic cytoplasmic tRNA is known to participate in a nonwobble pairing combination. There is, however, no known instance of such a tRNA with an A in its third anticodon position, which suggests that the consequent A · U pair is not permitted. The structural basis of wobble pairing is poorly understood, although it is clear that it is influenced by base modifications.

A consideration of the various wobble pairings indicates that at least 31 tRNAs are required to translate all 61 coding triplets of the genetic code (there are 32 tRNAs in the minimal set because translational initiation requires a separate tRNA; Section 32-3C). Most cells have >32 tRNAs, some of which have identical anticodons. In fact, mammalian cells have >150 tRNAs. Nevertheless, *all isoaccepting tRNAs in a cell are recognized by a single aminoacyl–tRNA synthetase.*

c. Some Mitochondrial tRNAs Have More Permissive Wobble Pairings than Other tRNAs

The codon recognition properties of mitochondrial tRNAs must reflect the fact that mitochondrial genetic codes are variants of the "standard" genetic code (Table 32-3). For instance, the human mitochondrial genome, which consists of only 16,569 bp, encodes 22 tRNAs (together with 2 ribosomal RNAs and 13 proteins). Fourteen of these tRNAs each read one of the synonymous pairs of codons indicated in Tables 32-2 and 32-3 (MNX, where X is either C or U or else A or G) according to normal G · U wobble rules: The tRNAs have either a G or a modified U in their third anticodon position that, respectively, permits them to pair with codons having X = C or U or else X = A or G. The remaining 8 tRNAs, which, contrary to wobble rules, each recognize one of the groups of four synonymous codons (MNY, where Y = A, C, G, or U), all have anticodons with a U in their third position. Either this U can somehow pair with any of the four bases or these tRNAs read only the first two codon positions and ignore the third. Thus, not surprisingly, many mitochondrial tRNAs have unusual structures in which, for example, the GTΨCRA sequence (Fig. 32-9) is missing, or, in the most bizarre case, a $tRNA^{Ser}$ lacks the entire D arm.

d. Frequently Used Codons Are Complementary to the Most Abundant tRNA Species

The analysis of the base sequences of several highly expressed structural genes of *S. cerevisiae* has revealed a remarkable bias in their codon usage. Only 25 of the 61 coding triplets are commonly used. *The preferred codons are those that are most nearly complementary, in the Watson–Crick sense, to the anticodons in the most abundant species in each set of isoaccepting tRNAs.* Furthermore, codons that bind anticodons with two consecutive G · C pairs or three A · U pairs are avoided so that the preferred codon–anticodon complexes all have approximately the same binding free energies. A similar phenomenon occurs in *E. coli,* although several of its 22 preferred codons differ from those in yeast. The degree with which the preferred codons occur in a given gene is strongly correlated, in both organisms, with the gene's level of expression (the measured rates of aminoacyl–tRNA selection in *E. coli* span a 25-fold range). This, it has been proposed, permits the mRNAs of proteins that are required in high abundance to be rapidly and smoothly translated.

e. Selenocysteine Is Carried by a Specific tRNA

Although it is widely stated, even in this text, that proteins are synthesized from the 20 "standard" amino acids, that is, those specified by the "standard" genetic code, some organisms, as Theresa Stadtman discovered, use a twenty-first amino acid, **selenocysteine (Sec;** alternatively **SeCys),** in synthesizing a few of their proteins:

```
 |
 NH
 |
 CH—CH2—Se—H
 |
 C=O
 |
```

The selenocysteine (Sec) residue

Selenium, a biologically essential trace element, is a component of several enzymes in both prokaryotes and eukaryotes. These include thioredoxin reductase (Section 28-3A) and the **thyroid hormone deiodinases** (which participate in thyroid hormone synthesis; Section 19-1D) in mammals and three forms of **formate dehydrogenases** in *E. coli,* all of which contain Sec residues. The Sec residues are ribosomally incorporated into these proteins by a unique tRNA, **$tRNA^{Sec}$,** bearing a UCA anticodon that is specified by a particular (in the mRNA) UGA codon (normally the *opal* Stop codon). The Sec–$tRNA^{Sec}$ is synthesized by the aminoacylation of $tRNA^{Sec}$ with L-serine by the same SerRS that charges $tRNA^{Ser}$, followed by the enzymatic selenylation of the resulting Ser residue.

How does the ribosomal system differentiate a Sec-specifying UGA codon from a normal opal Stop codon? As we saw to be the case with Glu–$tRNA^{Gln}$ (Section 32-2C), EF-Tu, the elongation factor that conducts most aminoacyl–tRNAs to the ribosome in a GTP-dependent process, does not bind Sec–$tRNA^{Sec}$. Instead it is bound by a specific elongation factor named **SELB,** which, in the presence of GTP, recognizes a ribosomally bound mRNA hairpin structure on the 3′ side of the UGA codon specifying Sec.

E. *Nonsense Suppression*

Nonsense mutations are usually lethal when they prematurely terminate the synthesis of an essential protein. An organism with such a mutation may nevertheless be "rescued" by a second mutation on another part of the genome. For many years after their discovery, the existence of such **intergenic suppressors** was quite puzzling. It is now known, however, that they usually arise from mutations in a tRNA gene that cause the tRNA to recognize a nonsense codon. Such a **nonsense suppressor** tRNA appends its amino acid (which is the same as that carried by the corresponding wild-type tRNA) to a growing polypeptide in response to the recognized Stop codon, thereby preventing chain termination. For example, the *E. coli amber* suppressor known as *su*3 is a $tRNA^{Tyr}$ whose anticodon has mutated from the wild-type GUA (which reads the Tyr codons UAU and UAC) to CUA (which recognizes the *amber* Stop codon UAG). An $su3^+$ *E. coli* with an otherwise lethal *amber* mutation in a gene coding for an essential protein would be viable if the replacement of the wild-type amino acid residue by Tyr does not inactivate the protein.

There are several well-characterized examples of *amber* (UAG), *ochre* (UAA), and *opal* (UGA) suppressors in *E. coli* (Table 32-6). Most of them, as expected, have mutated anticodons. UGA-1 tRNA, however, differs from the wildtype only by a G → A mutation in its D stem, which changes a G · U pair to a stronger A · U pair. This mutation apparently alters the conformation of the tRNA's CCA anticodon so that it can form an unusual wobble pairing with UGA as well as with its normal codon, UGG. Nonsense suppressors also occur in yeast.

a. Suppressor tRNAs Are Mutants of Minor tRNAs

How do cells tolerate a mutation that both eliminates a normal tRNA and prevents the termination of polypeptide synthesis? They survive because the mutated tRNA is usually a minor member of a set of isoaccepting tRNAs and because nonsense suppressor tRNAs must compete for Stop codons with the protein factors that mediate the termination of polypeptide synthesis (Section 32-3E). Consequently, the rate of suppressor-mediated synthesis of active proteins with either UAG or UGA nonsense mutations rarely exceeds 50% of the wild-type rate, whereas mutants with UAA, the most common termination codon, have suppression efficiencies of <5%. Many mRNAs, moreover, have two tandem Stop codons so that even if their first Stop codon were suppressed, termination could occur at the second. Nevertheless, many suppressor-rescued mutants grow relatively slowly because they cannot make an otherwise prematurely terminated protein as efficiently as do wild-type cells.

Other types of suppressor tRNAs are also known. **Missense suppressors** act similarly to nonsense suppressors but substitute one amino acid in place of another.

TABLE 32-6 Some *E. coli* Nonsense Suppressors

Name	Codon Suppressed	Amino Acid Inserted
*su*1	UAG	Ser
*su*2	UAG	Gln
*su*3	UAG	Tyr
*su*4	UAA, UAG	Tyr
*su*5	UAA, UAG	Lys
*su*6	UAA	Leu
*su*7	UAA	Gln
UGA-1	UGA	Trp
UGA-2	UGA	Trp

Source: Körner, A.M., Feinstein, S.I., and Altman, S., *in* Altman, S. (Ed.), *Transfer RNA*, p. 109, MIT Press (1978).

Frameshift suppressors have eight nucleotides in their anticodon loops rather than the normal seven. They read a four base codon beyond a base insertion thereby restoring the wild-type reading frame.

3 ■ RIBOSOMES AND POLYPEPTIDE SYNTHESIS

Ribosomes were first seen in cellular homogenates by dark-field microscopy in the late 1930s by Albert Claude who referred to them as "microsomes." It was not until the mid-1950s, however, that George Palade observed them in cells by electron microscopy, thereby disposing of the contention that they were merely artifacts of cell disruption. The name ribosome derives from the fact that these particles in *E. coli* consist of approximately two-thirds RNA and one-third protein. (**Microsomes** are now defined as the artifactual vesicles formed by the endoplasmic reticulum on cell disruption. They are easily isolated by differential centrifugation and are rich in ribosomes.) The correlation between the amount of RNA in a cell and the rate at which it synthesizes protein led to the suspicion that ribosomes are the site of protein synthesis. This hypothesis was confirmed in 1955 by Paul Zamecnik, who demonstrated that ^{14}C-labeled amino acids are transiently associated with ribosomes before they appear in free proteins. Further research showed that ribosomal polypeptide synthesis has three distinct phases: (1) chain initiation, (2) chain elongation, and (3) chain termination.

In this section we examine the structure of the ribosome and then outline the ribosomal mechanism of polypeptide synthesis. In doing so we shall compare the properties of ribosomes from prokaryotes with those of eukaryotes.

A. *Ribosome Structure*

The *E. coli* ribosome, which has a particle mass of ~2.5×10^6 D and a sedimentation coefficient of 70S, is a spheroidal particle that is ~250 Å across in its largest dimension. It may be dissociated, as James Watson discovered,

TABLE 32-7 Components of *E. coli* Ribosomes

	Ribosome	Small Subunit	Large Subunit
Sedimentation coefficient	70S	30S	50S
Mass (kD)	2520	930	1590
RNA			
Major		16S, 1542 nucleotides	23S, 2904 nucleotides
Minor			5S, 120 nucleotides
RNA mass (kD)	1664	560	1104
Proportion of mass	66%	60%	70%
Proteins		21 polypeptides	31 polypeptides
Protein mass (kD)	857	370	487
Proportion of mass	34%	40%	30%

into two unequal subunits (Table 32-7). The small (30S) subunit consists of a 16S rRNA molecule and 21 different polypeptides, whereas the large (50S) subunit contains a 5S and a 23S rRNA together with 31 different polypeptides. The up to 20,000 ribosomes in an *E. coli* cell account for ~80% of its RNA content and ~10% of its protein.

Structural studies of the ribosome began soon after its discovery through electron microscopy. Three-dimensional (3D) structures of the ribosome and its subunits at low (~50 Å) resolution first became available in the 1970s through image reconstruction techniques, pioneered by Klug, in which electron micrographs of a single particle or ordered sheets of particles taken from several directions are combined to yield its 3D image. The small subunit is a roughly mitten-shaped particle, whereas the large subunit is spheroidal with three protuberances on one side (Fig. 32-26).

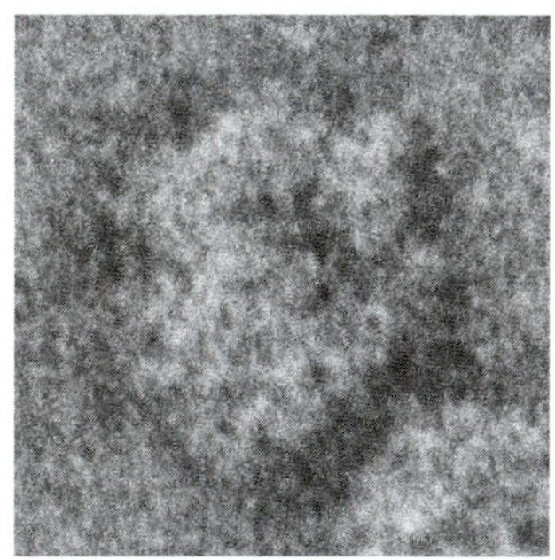

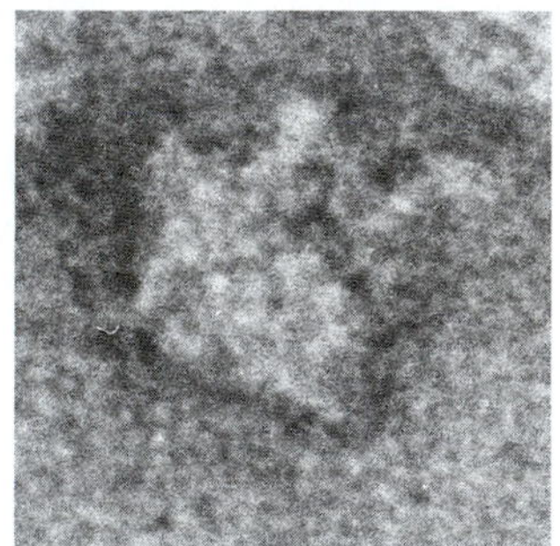

(*a*)

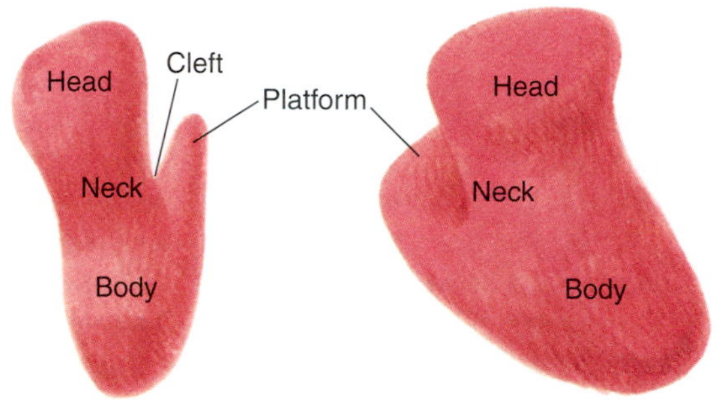

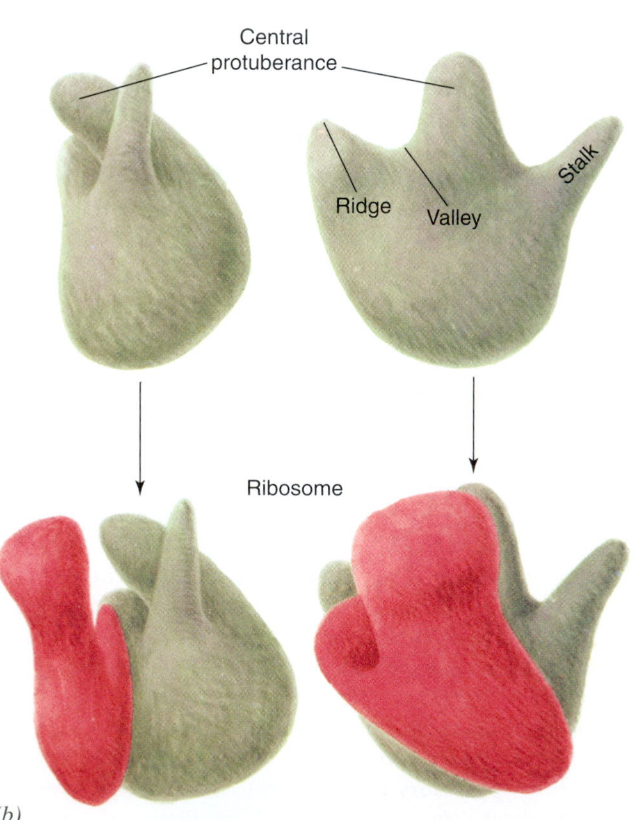

(*b*)

FIGURE 32-26 *E. coli* ribosome at low resolution. (*a*) Transmission electron micrographs of negatively stained ribosomes. In **negative staining,** the particle being imaged is embedded in electron-absorbing heavy metal salts, thereby providing contrast between the relatively electron-transparent particle and the background. [Courtesy of James Lake, UCLA.] (*b*) A three-dimensional model of the *E. coli* ribosome that was mathematically deduced from series of two-dimensional electron micrographs such as those in Part *a* taken from different directions, a process known as **image reconstruction.** The small subunit (*top*) combines with the large subunit (*middle*) to form the complete ribosome (*bottom*). The two views of the complete ribosome match those seen in Part *a*.

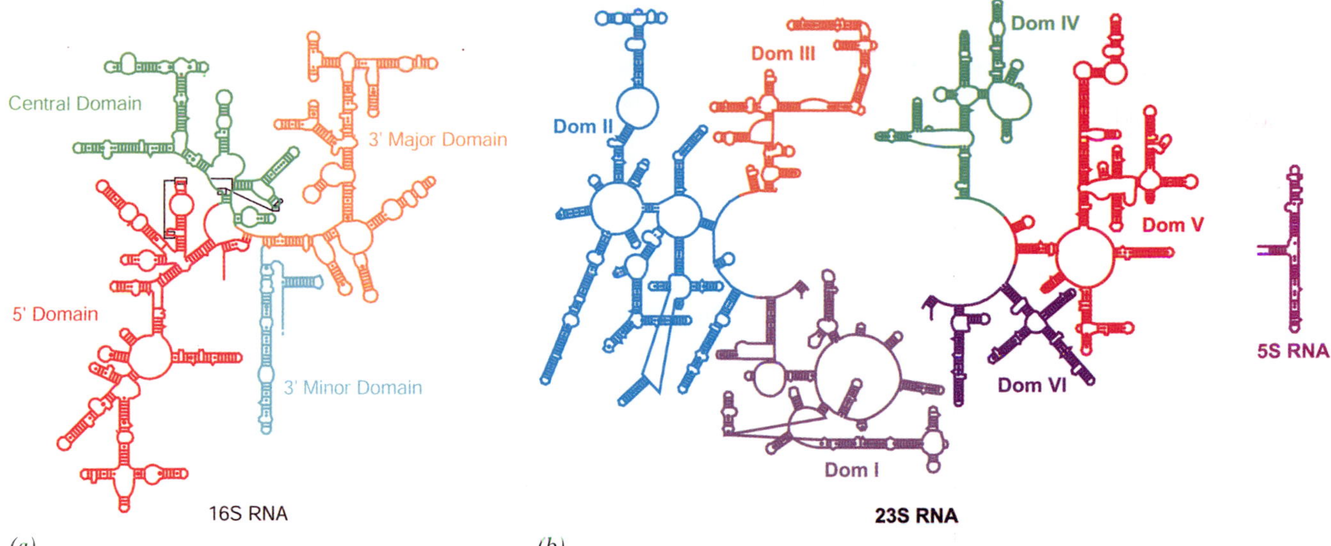

FIGURE 32-27 Secondary structures of the *E. coli* ribosomal RNAs. (*a*) 16S RNA and (*b*) 23S and 5S RNAs. The rRNAs are colored by domain with short lines spanning a stem representing Watson–Crick base pairs, small dots representing G · U base pairs, and large dots representing other non-Watson–Crick base pairs. Note the flowerlike series of stems and loops forming each domain. [Courtesy of V. Ramakrishnan, MRC Laboratory of Molecular Biology, Cambridge, U.K., and Peter Moore, Yale University. Adapted from diagrams in http://www.rna.icmb.utexas.edu.]

a. Ribosomal RNAs Have Complicated Secondary Structures

The *E. coli* 16S rRNA, which was sequenced by Harry Noller, consists of 1542 nucleotides. A computerized search of this sequence for stable double helical segments yielded many plausible but often mutually exclusive secondary structures. However, the comparison of the sequences of 16S rRNAs from several prokaryotes, under the assumption that their structures have been evolutionarily conserved, led to the flowerlike secondary structure for 16S rRNA seen in Fig. 32-27*a*. This four-domain structure, which is 54% base paired, is reasonably consistent with the results of nuclease digestion and chemical modification studies. Its double helical stems tend to be short (<8 bp) and many of them are imperfect. Intriguingly, electron micrographs of the 16S rRNA resemble those of the complete 30S subunit, thereby suggesting that the 30S subunit's overall shape is largely determined by the 16S rRNA. The large ribosomal subunit's 5S and 23S rRNAs, which consist of 120 and 2904 nucleotides, respectively, have also been sequenced. As with the 16S rRNA, they have extensive secondary structures (Fig. 32-27*b*).

b. Ribosomal Proteins Have Been Partially Characterized

Ribosomal proteins are difficult to separate because most of them are insoluble in ordinary buffers. By convention, ribosomal proteins from the small and large subunits are designated with the prefixes S and L, respectively, followed by a number indicating their position, from upper left to lower right, on a two-dimensional gel electrophoretogram (roughly in order of decreasing molecular mass; Fig. 32-28). Only protein S20/L26 appears to be common to both subunits. One of the large subunit proteins is partially acetylated at its N-terminus so that it gives rise to two electrophoretic spots (L7/L12). Four copies of this protein, a dimer of dimers, are present in the large subunit. Moreover, these four copies of L7/L12 aggregate with L10 to form a stable complex that was initially thought to be a unique protein, "L8." All the other ribosomal proteins occur in only one copy per subunit.

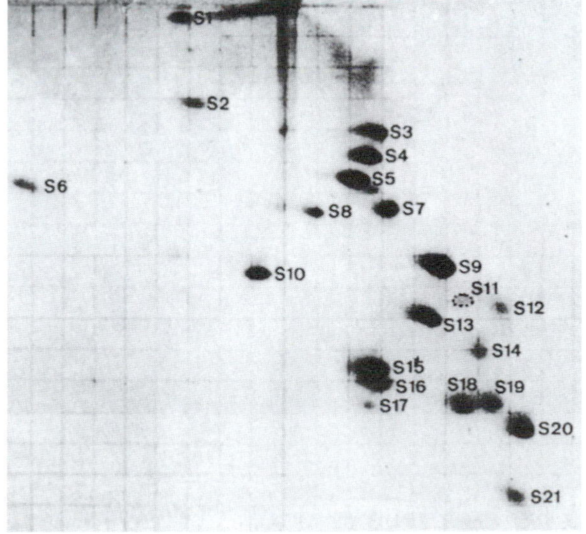

FIGURE 32-28 Two-dimensional gel electrophoretogram of *E. coli* small ribosomal subunit proteins. First dimension (*vertical*): 8% acrylamide, pH 8.6; second dimension (*horizontal*): 18% acrylamide, pH 4.6. [From Kaltschmidt, E. and Wittmann, H.G., *Proc. Natl. Acad. Sci.* **67,** 1277 (1970).]

The amino acid sequences of all 52 *E. coli* ribosomal proteins were elucidated, mainly by Heinz-Günter Wittmann and Brigitte Wittmann-Liebold. They range in size from 46 residues for L34 to 557 residues for S1. Most of these proteins, which exhibit little sequence similarity with one another, are rich in the basic amino acids Lys and Arg and contain few aromatic residues as is expected for proteins that are closely associated with polyanionic RNA molecules.

The X-ray and NMR structures of around half of the ribosomal proteins or their fragments have been independently determined. These proteins form a wide variety of structural motifs although most of their folds occur in other proteins of known structure. Around one-third of these ribosomal proteins contain the **RNA-recognition motif** (**RRM;** Fig. 32-29), which occurs in >200 RNA-binding proteins including rho protein (the transcriptional termination factor, which contains four such motifs; Section 31-2D), poly(A) polymerase, poly(A)-binding protein (PABP), several proteins involved in gene splicing (Section 31-4A), and the translational initiation factor **eIF4B** (Section 32-3C). All of these proteins presumably evolved from an ancient RNA-binding protein.

c. Ribosomal Subunits Are Self-Assembling

Ribosomal subunits form, under proper conditions, from mixtures of their numerous macromolecular components. *Ribosomal subunits are therefore self-assembling entities.* Masayasu Nomura determined how this occurs through partial reconstitution experiments. If one macromolecular component is left out of an otherwise self-assembling mixture of proteins and RNA, the other components that fail to bind to the resulting partially assembled subunit must somehow interact with the omitted component. Through the analysis of a series of such partial reconstitution experiments, Nomura constructed an assembly map of the small (30S) subunit (Fig. 32-30). This map indicates that the initial steps in small subunit assembly are the independent binding to naked 16S rRNA of six so-called primary binding proteins (S4, S7, S8, S15, S17, and S20). The resulting assembly intermediates provide the molecular scaffolding for binding secondary binding proteins, which in turn generate the attachment sites for tertiary binding proteins. At one stage in the assembly process, an intermediate particle must undergo a marked conformational change before assembly can continue. The large subunit self-assembles in a similar manner. The observation that similar assembly intermediates occur *in vivo* and *in vitro* suggests that *in vivo* and *in vitro* assembly processes are much alike.

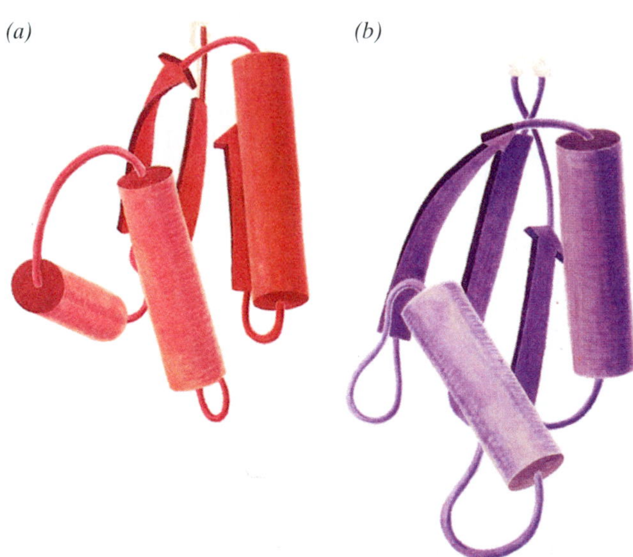

FIGURE 32-29 X-Ray structures of two ribosomal proteins. (*a*) The 74-residue C-terminal fragment of *E. coli* L7/L12. (*b*) *Bacillus stearothermophilus* L30 (61 residues). The two protein molecules are oriented so as to show their closely similar RNA-recognition motifs (RRMs, *darker shading*). [After Leijonmarck, M., Appelt, K., Badger, J., Liljas, A., Wilson, K.S., and White, S.W., *Proteins* **3,** 244 (1988).]

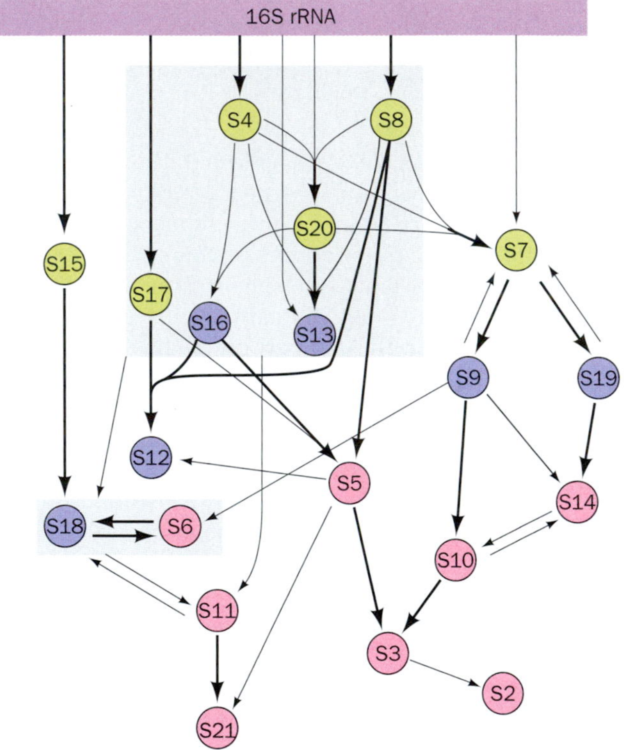

FIGURE 32-30 Assembly map of the *E. coli* small subunit. Primary, secondary, and tertiary binding proteins are represented by green, blue, and red circles, respectively, and thick and thin arrows between components indicate strong and weak facilitation of binding. For example, the thick arrow from the 16S rRNA to S15 indicates that S15 binds directly to the 16S rRNA in the absence of other proteins and is therefore a primary binding protein, the thick arrow from S15 to S18 indicates that S18 is a secondary binding protein, and the arrows between S18 and S6 and S11 indicate that S6 and S11 are tertiary binding proteins. The arrows from the shaded boxes to S11 indicate that the proteins in the boxes collectively bind S11. [After Held, W.A., Ballou, B., Mizushima, S., and Nomura, M., *J. Biol. Chem.* **249,** 3109 (1974).]

d. The Atomic Structure of the Prokaryotic Ribosome Has Been Long in Coming

The elucidation of the ribosome's atomic structure has been a tortuous affair extending over four decades in which slow incremental improvements were occasionally punctuated by significant technical gains. The process began in the 1960s with shadowy transmission electron micrographs such as Fig. 32-26*a* that provided only rough 2D shapes. This was followed in the 1970s by image reconstruction techniques that generated 3D models although still at low resolution (Fig. 32-26*b*). Later in the 1970s, the sites of many of the ribosome's proteins were determined by James Lake and Georg Stöffler through **immune electron microscopy,** a technique in which antibodies raised against a particular ribosomal protein are used to mark its position in electron micrographs of the antibody complexed to a ribosomal subunit. These results were improved and extended in the 1980s by neutron scattering experiments conducted by Donald Engleman and Peter Moore on the 30S subunit, which indicated the distances between the centers of mass of its component proteins and hence their three-dimensional distribution. These structural studies were supplemented by a variety of chemical cross-linking and fluorescence transfer studies that demonstrated the proximity of various ribosomal components.

The molecular structure of the prokaryotic ribosome began to come into focus in the mid-1990s through the development of **cryoelectron microscopy (cryo-EM).** In this technique, the sample is cooled to near liquid N_2 temperatures (−196°C) so rapidly (in a few milliseconds) that the water in the sample does not have time to crystallize but, rather, assumes a vitreous (glasslike) state. Consequently, the sample remains hydrated and hence retains its native shape to a greater extent than in conventional electron microscopy (in which the sample is vacuum dried). Studies, carried out in large part by Joachim Frank, revealed the positions where tRNAs and mRNA as well as various soluble protein factors bind to the ribosome (Fig. 32-31). The highest resolution achieved by cryo-EM of ribosomes has gradually improved over the years to ~10 Å.

Ribosomal subunits were first crystallized by Ada Yonath in 1980 although they diffracted X-rays poorly. Over the course of several years, however, the quality of these crystals were incrementally improved until, in 1991, Yonath reported crystals of the 50S subunit that diffracted X-rays to 3-Å resolution. It was not until later in the 1990s, however, that technology was up to the task of determining the X-ray structures of these gargantuan molecular complexes. In 2000, the *annus mirabilis* (miracle year) of ribosomology, Moore and Steitz reported the X-ray structure of the 50S ribosomal subunit of the halophilic (salt-loving) bacterium *Haloarcula marismortui* at atomic (2.4-Å) resolution and V. Ramakrishnan and Yonath independently reported the X-ray structure of the 30S subunit of *T. thermophilus* at ~3-Å resolution. In 2001, Noller reported the 5.5-Å resolution structure of the entire *T. thermophilus* ribosome. In the following paragraphs we discuss the properties of these ground-breaking structures. We consider their functional implications starting in Section 32-3C.

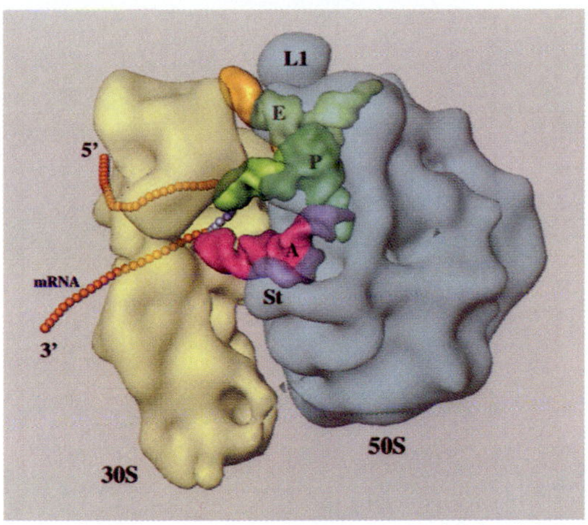

FIGURE 32-31 Cryoelectron microscopy–based image of the *E. coli* ribosome at ~25 Å resolution. In this semitransparent 3D model, the 30S subunit (*yellow*) is on the left and the 50S subunit (*blue*) is on the right. The tRNAs that occupy the A, P, and E sites (Section 32-3B) are colored magenta, green, and gold. The inferred path of the mRNA is represented by a chain of orange beads with the six nucleotides contacting the A and P sites colored blue and purple, respectively. A portion of the tunnel in the 50S subunit through which the growing polypeptide chain is extruded is visible (*center right*). The symbols L1 and St indicate the positions of the L1 protein and the stalk in the 50S subunit (Fig. 32-26*b*). [Courtesy of Joachim Frank, State University of New York at Albany.]

e. Ribosomal Architecture

Several generalizations can be made about ribosomal architecture based on the structures of the 30S and 50S subunits:

1. Both the 16S and 23S rRNAs are assemblies of helical elements connected by loops, most of which are irregular extensions of helices (Fig. 32-32). These structures, which are in close accord with previous secondary structure predictions (Fig. 32-27), are stabilized by interactions between helices such as minor groove to minor groove packing, which has also been seen in the structure of the group I intron (Section 31-4A; recall that A-form RNA has a very shallow minor groove); the insertion of a phosphate ridge into a minor groove; and adenines that are distant in sequence but often highly conserved that are inserted into minor grooves. Although the determination of the structures of the 30S and 50S ribosomal subunits increased the amount of RNA structure that is known at atomic resolution by ~10-fold, nearly all of the secondary structural motifs seen in the ribosome also occur in these smaller RNA structures. This suggests that the repertoire of RNA secondary structural motifs is limited.

2. Each of the 16S RNA's four domains, which extend out from a central junction (Fig. 32-27*a*), forms a morphologically distinct portion of the 30S subunit (Fig. 32-32*a*): The 5′ domain forms most of the body (Fig. 32-26*b*), the central domain forms the platform, the 3′ major domain forms the entire head, and the 3′ minor domain, which consists of just two helices, is located at the

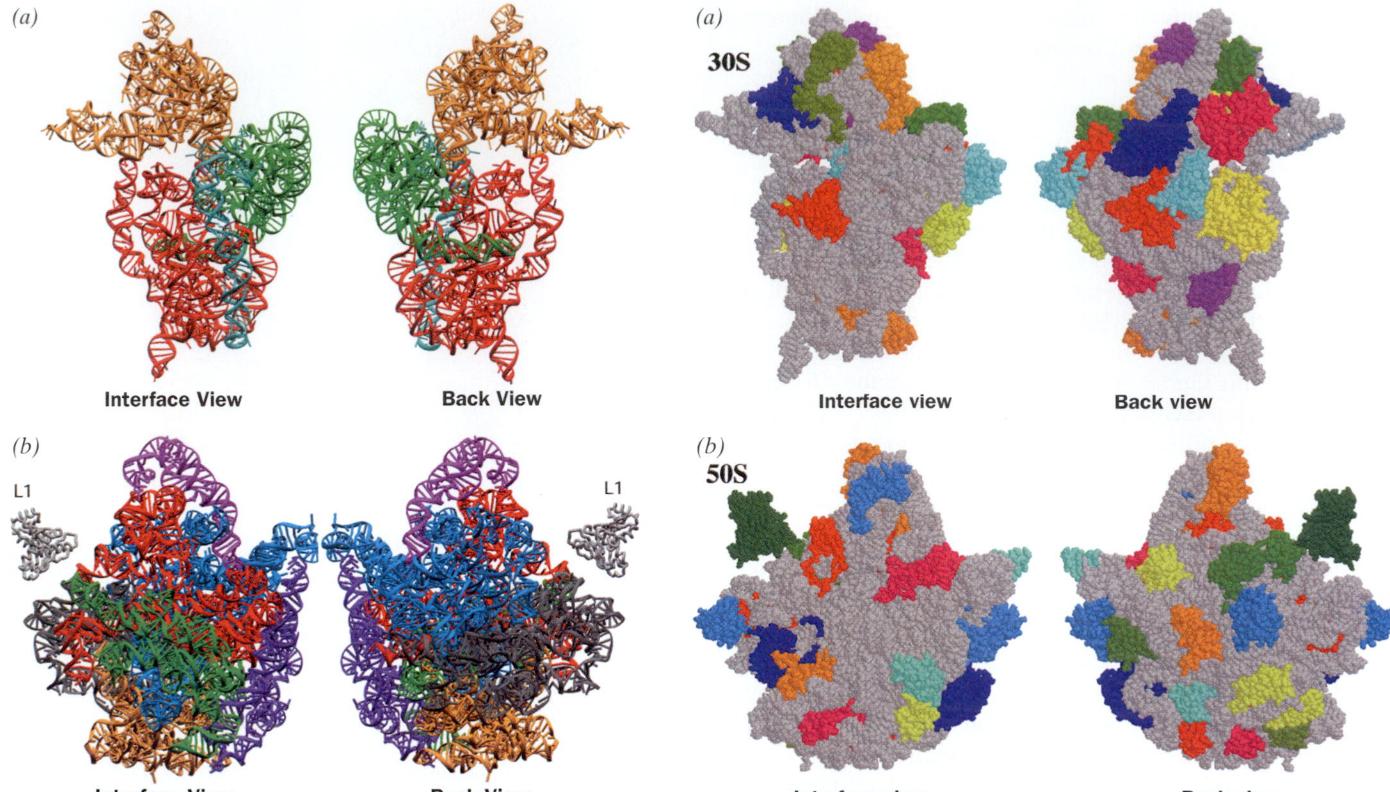

FIGURE 32-32 Tertiary structures of the ribosomal RNAs. (*a*) The 16S rRNA of *T. thermophilus.* (*b*) The 23S rRNA of *H. marismortui.* The rRNAs are colored according to domain as in Fig. 32-27. The interface view of a ribosomal subunit (*left*) is toward its surface that associates with the other subunit in the whole ribosome and the back view (*right*) is from the opposite (solvent-exposed) side. Note that the secondary structure domains of the 16S rRNA fold as separate tertiary structure domains, whereas in the 23S rRNA the secondary structure domains are convoluted together. [Courtesy of V. Ramakrishnan, MRC Laboratory of Molecular Biology, Cambridge, U.K., and Peter Moore, Yale University. PDBids

FIGURE 32-33 Distribution of protein and RNA in the ribosomal subunits. (*a*) The 30S subunit of *T. thermophilus.* (*b*) The 50S subunit of *H. marismortui.* The subunits are drawn in space-filling form with their RNAs gray and their proteins in various colors. Note that the interface side of each subunit is largely free of protein, particularly in its regions that interact with mRNA and tRNAs. [Part *a* based on an X-ray structure by V. Ramakrishnan, MRC Laboratory of Molecular Biology, Cambridge, U.K. Part *b* based on an X-ray structure by Peter Moore and Thomas Steitz, Yale University. PDBids 1J5E and 1JJ2.]

interface between the 30S and 50S subunits. In contrast, the 23S RNA's six domains (Fig. 32-27*b*) are intricately intertwined in the 50S subunit (Fig. 32-32*b*). Since the ribosomal proteins are embedded in the RNA (see below), this suggests that the domains of the 30S subunit can move relative to one another during protein synthesis, whereas the 50S subunit appears to be rigid.

3. The distribution of the proteins in the two ribosomal subunits is not uniform (Fig. 32-33). The vast majority of the ribosomal proteins are located on the back and sides of their subunits. In contrast, the face of each subunit that forms the interface between the two subunits, particularly those regions that bind the tRNAs and mRNA (see below), is largely devoid of proteins.

4. Most ribosomal proteins consist of a globular domain, which is, for the most part, located on a subunit surface (Fig. 32-33), and a long segment that is largely devoid of secondary structure and unusually rich in basic residues that infiltrates between the RNA helices into the subunit interior (Fig. 32-34). Indeed, a few ribosomal proteins lack a globular domain altogether (e.g., L39e in Fig. 32-34*b*). Ribosomal proteins make far fewer base-specific interactions than do other known RNA-binding proteins. They tend to interact with the RNA through salt bridges between their positively charged side chains and the RNAs' negatively charged phosphate oxygen atoms, thereby neutralizing the repulsive charge–charge interactions between nearby RNA segments. This is consistent with the hypothesis that the primordial ribosome consisted entirely of RNA (the RNA world) and that the proteins that were eventually acquired stabilized its structure and fine-tuned its function.

The X-ray structure of the entire *T. thermophilus* ribosome in complex with three tRNAs and a 36-nt mRNA fragment was determined by Noller at 5.5 Å resolution. At this low resolution, the RNA backbones can be confidently traced and proteins of known structure can be properly positioned. Stereo diagrams of this enormous molecular machine are presented in Fig. 32-35. The structures of the

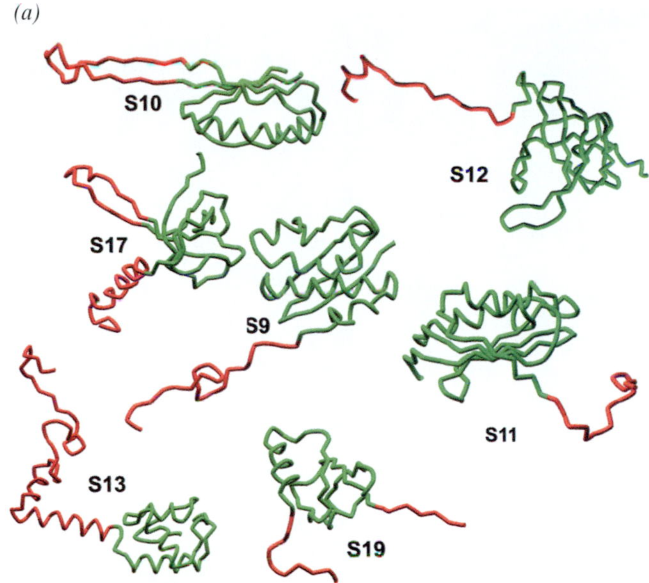

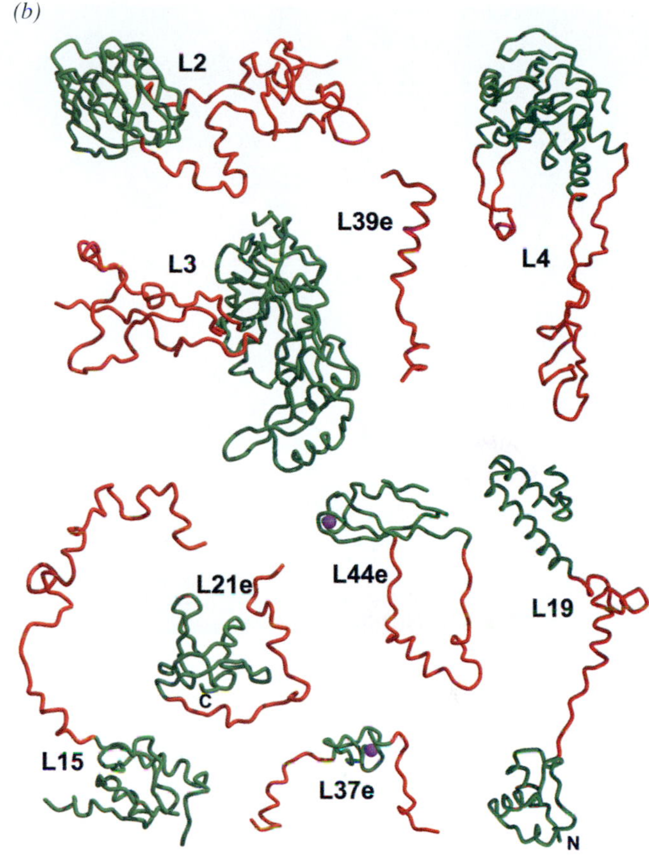

FIGURE 32-34 Gallery of ribosomal protein structures. Proteins from (*a*) the 30S subunit and (*b*) the 50S subunit. The proteins are represented by their backbones with their globular portions green and their highly extended segments red. The globular portions are exposed on the surface of their associated subunit (Fig. 32-33), whereas the extended segments are largely buried in the RNA. The Zn^{2+} ions bound by L37e and L44e are represented by magenta spheres. [Courtesy of V. Ramakrishnan, MRC Laboratory of Molecular Biology, Cambridge, U.K., and Peter Moore, Yale University.]

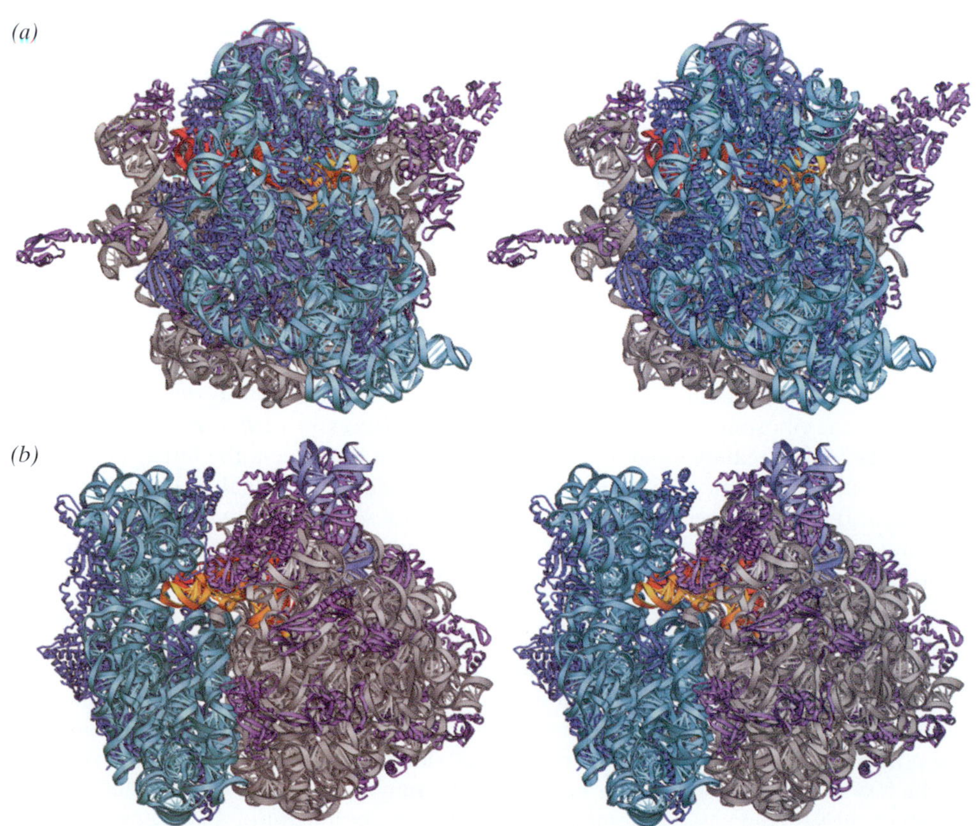

FIGURE 32-35 X-Ray structure of the *T. thermophilus* 70S ribosome in complex with three tRNAs and an mRNA fragment. In these stereo diagrams (whose viewing is described in the Appendix to Chapter 8), the 16S RNA is cyan, the 23S RNA is gray, the 5S RNA is light blue, the small subunit proteins are dark blue, the large subunit proteins are violet, and the tRNAs bound to the A, P, and E sites (which are largely occluded) are gold, orange, and red, respectively. (*a*) View similar to that on the lower right of Fig. 32-26*b* in which the small subunit is in front of the large subunit. (*b*) A view rotated 90° around the vertical axis relative to Part *a* which resembles that in Fig. 32-31. Here the A-site tRNA is more clearly visible at the bottom of a funnel in which elongation factors bind (Section 32-3D). [Courtesy of Harry Noller, University of California at Santa Cruz. PDBids 1GIX and 1GIY.]

See the Interactive Exercises

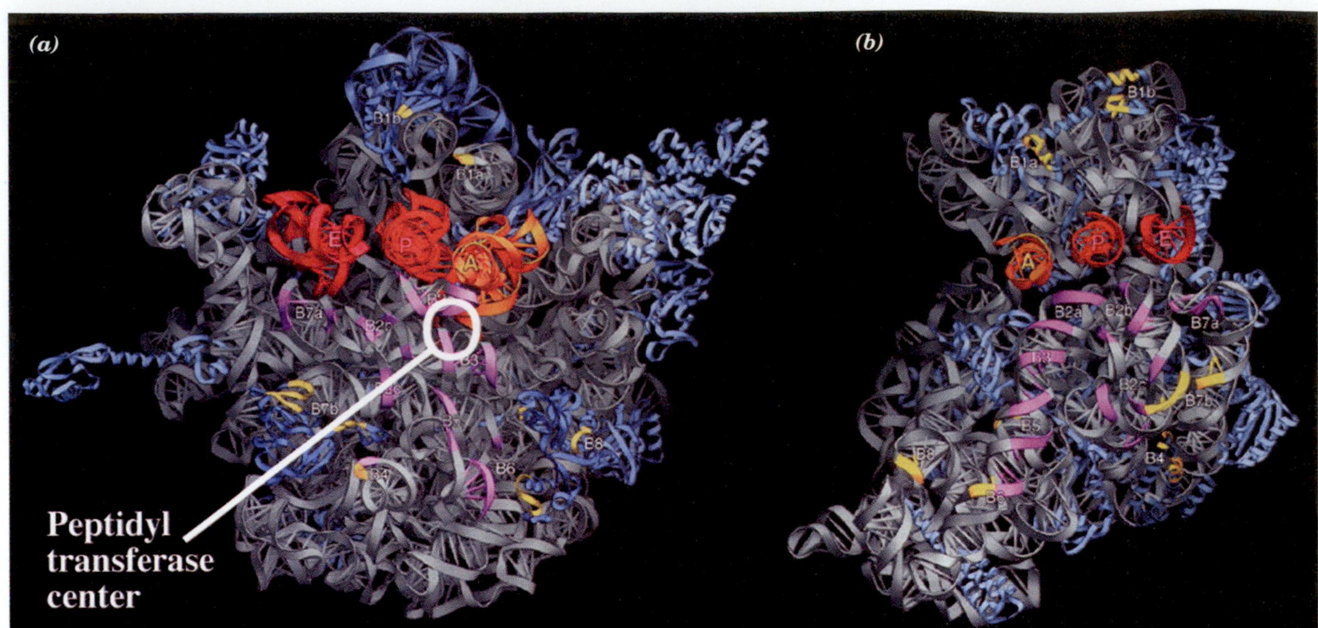

FIGURE 32-36 Ribosomal subunits in the X-ray structure of the *T. thermophilus* 70S ribosome in complex with three tRNAs and an mRNA. (*a*) Interface view of the large subunit (similar to Fig. 32-33*b, left*). (*b*) Interface view of the small subunit (similar to Fig. 32-33*a, left*). Here the RNA is gray with its segments that participate in intersubunit contacts magenta and the protein is blue with its segments that participate in intersubunit contacts yellow. The tRNAs bound in the A, P, and E sites are gold, orange, and red, respectively. [Courtesy of Harry Noller, University of California at Santa Cruz. PDBids 1GIX and 1GIY.]

associated 30S and 50S subunits closely resemble those of the isolated subunits although there are several regions at the subunit interface that exhibit significant conformational shifts (between 3.5 and 10 Å), which suggests that these changes occur as a consequence of subunit association. In addition, several disordered portions of the isolated *H. marismortui* 50S subunit are ordered in the intact *T. thermophilus* ribosome, although this may be a consequence of the latter's greater thermal stability.

The two subunits in the intact ribosome contact each other at 12 positions via RNA–RNA, protein–protein, and RNA–protein bridges (Fig. 32-36). These intersubunit bridges have a distinct distribution: The RNA–RNA bridges are centrally located adjacent to the three bound tRNAs, whereas the protein–protein and RNA–protein bridges are peripherally located away from the ribosome's functional sites. The RNA–RNA contacts consist mainly of minor groove–minor groove interactions although major groove, loop, and backbone contacts also occur. In the RNA–protein bridges, the proteins contact nearly all types of RNA features including major groove, minor groove, backbone, and loop elements.

Ribosomes, as we shall see in Section 32-3B, have three functionally distinct tRNA-binding sites known as the A, P, and E sites. The ribosome binds all three tRNAs in a similar manner with their anticodon stem–loops bound to the 30S subunit and their remaining portions, the D-stem, elbow, and acceptor stem, bound to the 50S subunit. These interactions, which mainly consist of RNA–RNA contacts, are made to the tRNAs' universally conserved segments, thereby permitting the ribosome to bind different species of tRNAs in the same way. Nevertheless, the three tRNAs have somewhat different conformations: The A-site tRNA closely resembles crystallized tRNA^{Phe} (Fig. 32-11*b*), the P-site tRNA is slightly kinked around the junction of the D and anticodon stems so as to angle the anticodon loop toward the A site, and the E-site tRNA is further distorted with its elbow angle more open and its anticodon loop making an unusually sharp turn.

We discuss the path of the mRNA and how it interacts with the tRNAs in Section 32-3D. There we shall see that *the large subunit is mainly involved in mediating biochemical tasks such as catalyzing the reactions of polypeptide elongation, whereas the small subunit is the major actor in ribosomal recognition processes such as mRNA and tRNA binding* (although, as we have seen, the large subunit also participates in tRNA binding). We shall also see that *rRNA has the major functional role in ribosomal processes* (recall that RNA has demonstrated catalytic properties; Sections 31-4A and 31-4C).

f. Eukaryotic Ribosomes Are Larger and More Complex than Prokaryotic Ribosomes

Although eukaryotic and prokaryotic ribosomes resemble one another in both structure and function, they differ in nearly all details. Eukaryotic ribosomes have particle masses in the range 3.9 to 4.5 $\times$ 10^6 D and have a nominal sedimentation coefficient of 80S. They dissociate

TABLE 32-8 Components of Rat Liver Cytoplasmic Ribosomes

	Ribosome	Small Subunit	Large Subunit
Sedimentation coefficient	80S	40S	60S
Mass (kD)	4220	1400	2820
RNA			
Major		18S, 1874 nucleotides	28S, 4718 nucleotides
Minor			5.8S, 160 nucleotides 5S, 120 nucleotides
RNA mass (kD)	2520	700	1820
Proportion of mass	60%	50%	65%
Proteins		33 polypeptides	49 polypeptides
Protein mass (kD)	1700	700	1000
Proportion of mass	40%	50%	35%

into two unequal subunits that have compositions that are distinctly different from those of prokaryotes (Table 32-8; compare with Table 32-7). The small **(40S)** subunit of the rat liver cytoplasmic ribosome, the most well-characterized eukaryotic ribosome, consists of 33 unique polypeptides and an **18S rRNA.** Its large **(60S)** subunit contains 49 different polypeptides and three rRNAs of **28S, 5.8S,** and **5S.** The additional complexity of the eukaryotic ribosome relative to its prokaryotic counterpart is presumably due to the eukaryotic ribosome's additional functions: Its mechanism of translational initiation is more complex (Section 32-3C); it must be transported from the nucleus, where it is formed, to the cytoplasm, where translation occurs; and the machinery with which it participates in the secretory pathway is more complicated (Section 12-4B).

Sequence comparisons of the corresponding rRNAs from various species indicates that evolution has conserved their secondary structures rather than their base sequences (Figs. 32-27*a* and 32-37). For example, a G · C in a base paired stem of *E. coli* 16S rRNA has been replaced by an A · U in the analogous stem of yeast 18S rRNA. The **5.8S rRNA,** which occurs in the large eukaryotic subunit in base paired complex with the **28S rRNA,** is homologous in sequence to the 5′ end of prokaryotic 23S rRNA. Apparently 5.8S RNA arose through mutations that altered rRNA's posttranscriptional processing producing a fourth rRNA.

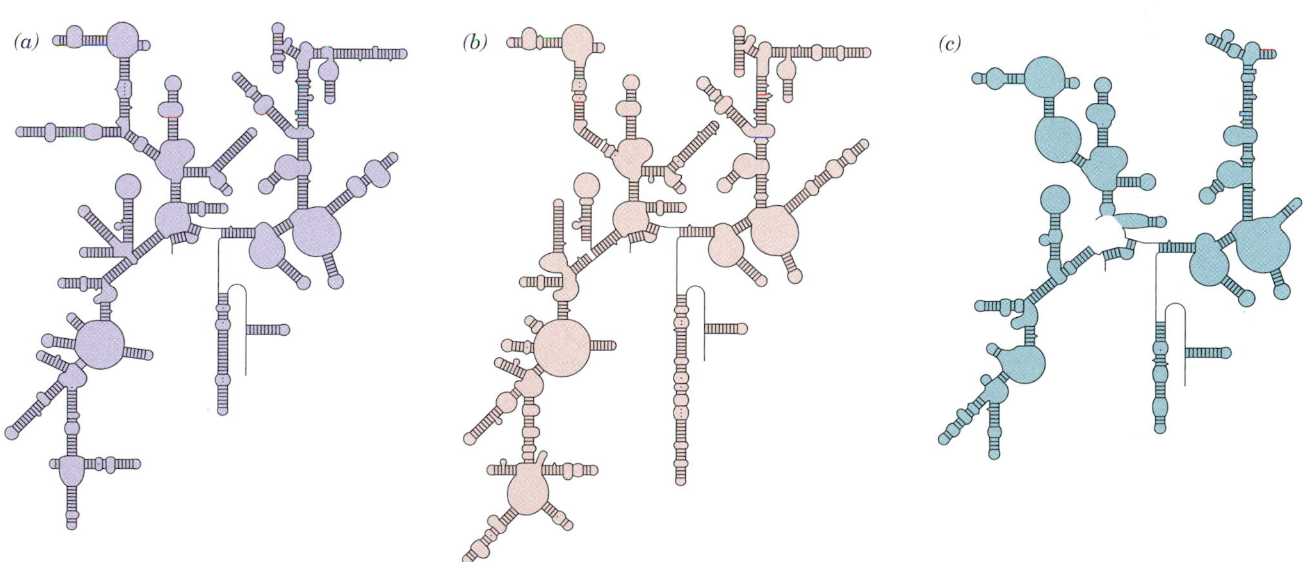

FIGURE 32-37 Predicted secondary structures of evolutionarily distant 16S-like rRNAs. (*a*) Archaebacteria (*Halobacterium volcanii*), (*b*) eukaryotes (*S. cerevisiae*), and (*c*) mammalian mitochondria (bovine). Compare them with Fig. 32-27*a*, the secondary structure of 16S RNA from eubacteria (*E. coli*). Note the close similarities of these assemblies; they differ mostly by insertions and deletions of stem-and-loop structures. The 23S-like rRNAs from a variety of species likewise have similar secondary structures. [After Gutell, R.R., Weiser, B., Woese, C.R., and Noller, H.F., *Prog. Nucleic Acid Res. Mol. Biol.* **32,** 183 (1985).]

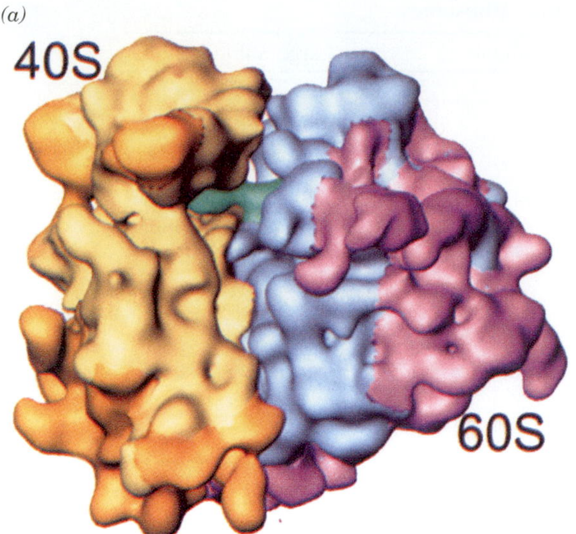

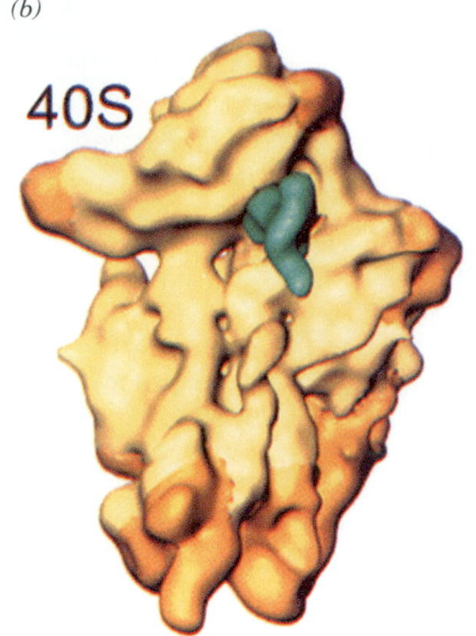

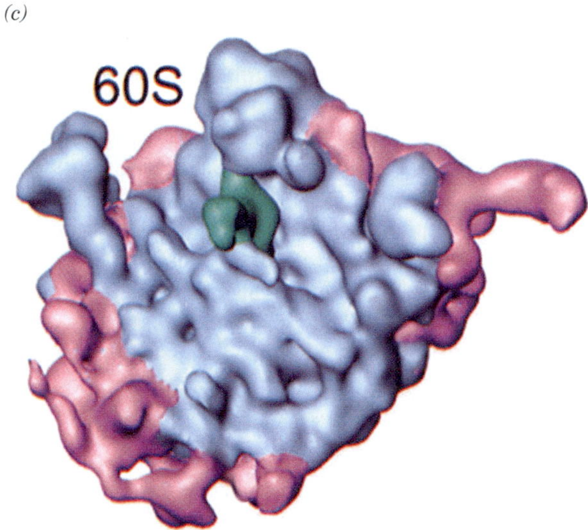

FIGURE 32-38 Cryo-EM–based image of the yeast 80S ribosome at 15 Å resolution. (*a*) The ribosome shown in side view analogous to Fig. 32-31 of the *E. coli* ribosome. The small (40S) subunit is yellow, the large (60S) subunit is blue, and the tRNA that is bound in the ribosomal P site is green. Portions of this ribosome that are not homologous to the RNA or proteins of the *E. coli* ribosome are shown in gold for the small subunit and magenta for the large subunit. (*b*) The computationally isolated small subunit shown in interface view analogous to the left panel of Fig. 32-33*a*. (*c*) The computationally isolated large subunit shown in interface view analogous to the left panel of Fig. 32-33*b*. [Courtesy of Joachim Frank, State University of New York at Albany.]

The cryo-EM–based image of the yeast 80S ribosome (Fig. 32-38), determined at 15 Å resolution by Andrej Sali, Günter Blobel, and Frank, reveals that there is a high degree of structural conservation between eukaryotic and prokaryotic ribosomes. Although the yeast 40S subunit (which consists of a 1798-nt 18S rRNA and 32 proteins) contains an additional 256 nt of RNA and 11 proteins relative to the *E. coli* 30S subunit (Table 32-8; 15 of the *E. coli* proteins are homologous to those of yeast), both exhibit a similar division into head, neck, body, and platform (Fig. 32-38*b* vs Figs. 32-26*b* and 32-33*a*). Many of the differences between these two small ribosomal subunits are accounted for by the 40S subunit's additional RNA and proteins, although their homologous portions exhibit several distinct conformational differences. Similarly, the yeast 60S subunit (Fig. 32-38*c;* which consists of an aggregate of 3671 nt and 45 proteins) structurally resembles the considerably smaller (Table 32-7) prokaryotic 50S subunit (Figs. 32-26*b* and 32-33*b*). The yeast ribosome exhibits 16 intersubunit bridges, 12 of which match the 12 that were observed in the X-ray structure of the *T. thermophilus* ribosome (Fig. 32-36), a remarkable evolutionary conservation that indicates the importance of these bridges. Moreover, the tRNA that occupies the P site of the yeast ribosome has a conformation that more closely resembles that of the P-site tRNA in the *T. thermophilus* ribosome than that of crystallized $tRNA^{Phe}$.

B. *Polypeptide Synthesis: An Overview*

Before we commence our detailed discussion of polypeptide synthesis, it will be helpful to outline some of its major features.

a. Polypeptide Synthesis Proceeds from N-Terminus to C-Terminus

The direction of ribosomal polypeptide synthesis was established, in 1961 by Howard Dintzis, through radioactive labeling experiments. He exposed reticulocytes that were actively synthesizing hemoglobin to ^{3}H-labeled leucine for times less than that required to make an entire polypeptide. The extent to which the tryptic peptides from the soluble (completed) hemoglobin molecules were labeled increased with their proximity to the C-terminus (Fig. 32-39). Incoming amino acids must therefore be appended to a growing polypeptide's C-terminus; that is, *polypeptide synthesis proceeds from N-terminus to C-terminus.*

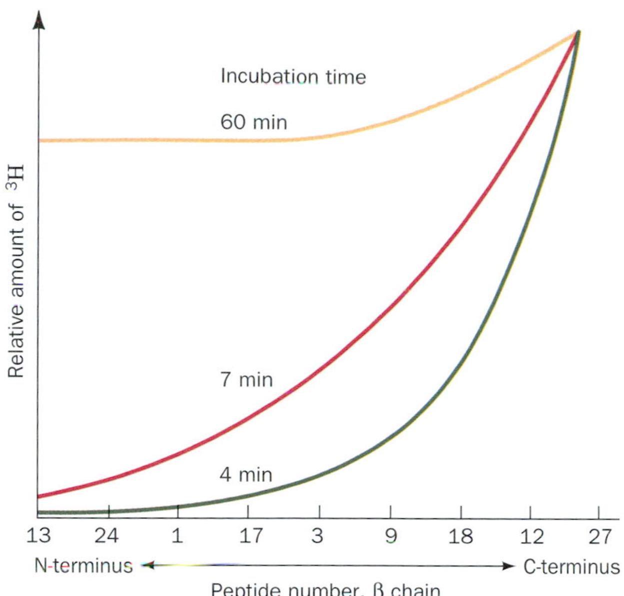

FIGURE 32-39 Demonstration that polypeptide synthesis proceeds from the N-terminus to the C-terminus. Rabbit reticulocytes were incubated with [^{3}H]leucine. The curves show the distribution of [^{3}H]Leu among the tryptic peptides from the β subunit of soluble rabbit hemoglobin after the indicated incubation times. The numbers on the horizontal axis are peptide identifiers arranged from N-terminus to C-terminus. [After Dintzis, H.M., *Proc. Natl. Acad. Sci.* **47,** 255 (1961).]

b. Ribosomes Read mRNA in the 5′ → 3′ direction

The direction in which the ribosome reads mRNAs was determined through the use of a cell-free protein synthesizing system in which the mRNA was poly(A) with a 3′-terminal C.

$$5'\ \mathrm{A—A—A—\cdots—A—A—A—C}\ \ 3'$$

Such a system synthesizes a poly(Lys) that has a C-terminal Asn.

$$\mathrm{H_3\overset{+}{N}—Lys—Lys—Lys—\cdots—Lys—Lys—Asn—COO^-}$$

This, together with the knowledge that AAA and AAC code for Lys and Asn and the polarity of polypeptide synthesis, indicates that *the ribosome reads mRNA in the 5′ → 3′ direction.* Since mRNA is synthesized in the 5′ → 3′ direction, this accounts for the observation that, in prokaryotes, ribosomes initiate translation on nascent mRNAs (Section 31-3).

c. Active Translation Occurs on Polyribosomes

Electron micrographs, as Rich discovered, reveal that ribosomes engaged in protein synthesis are tandemly arranged on mRNAs like beads on a string (Figs. 32-40 and 31-24). The individual ribosomes in these **polyribosomes (polysomes)** are separated by gaps of 50 to 150 Å so they have a maximum density on mRNA of ~1 ribosome per 80 nucleotides. Polysomes arise because once an active ribosome has cleared its initiation site, a second ribosome can initiate at that site.

d. Chain Elongation Occurs by the Linkage of the Growing Polypeptide to the Incoming tRNA's Amino Acid Residue

During polypeptide synthesis, amino acid residues are sequentially added to the C-terminus of the nascent, ribosomally bound polypeptide chain. If the growing polypeptide is released from the ribosome by treatment with high

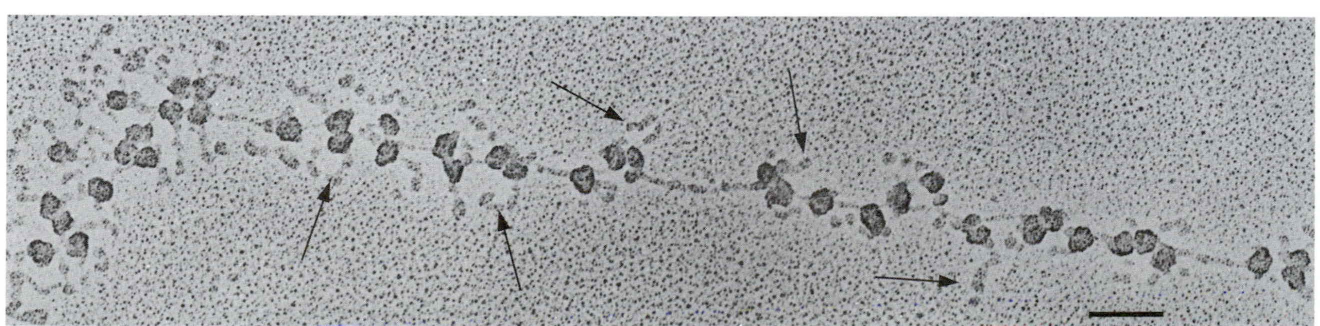

FIGURE 32-40 Electron micrographs of polysomes from silk gland cells of the silkworm *Bombyx mori.* The 3′ end of the mRNA is on the left. Arrows point to the silk fibroin polypeptides. The bar represents 0.1 μm. [Courtesy of Oscar L. Miller, Jr. and Steven L. McKnight, University of Virginia.]

salt concentrations, its C-terminal residue is invariably esterified to a tRNA molecule as a **peptidyl–tRNA:**

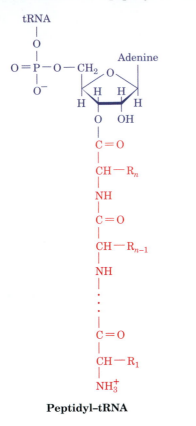

tRNA–O–P(=O)(–O^-)–O–CH_2–(ribose: H, H, H, H; Adenine)–O–C(=O)–CH–R_n–NH–C(=O)–CH–R_{n-1}–NH– ⋮ –C(=O)–CH–R_1–NH_3^+

Peptidyl–tRNA

The nascent polypeptide must therefore grow by being transferred from the peptidyl–tRNA to the incoming aminoacyl–tRNA to form a peptidyl–tRNA with one more residue (Fig. 32-41). Apparently, the ribosome has at least two tRNA-binding sites: the so-called **peptidyl** or **P site,** which binds the peptidyl–tRNA, and the **aminoacyl** or **A site,** which binds the incoming aminoacyl–tRNA (Fig. 32-41). Consequently, after the formation of a peptide bond, the newly deacylated P-site tRNA must be released and replaced by the newly formed peptidyl–tRNA from the A site, thereby permitting a new round of peptide bond formation. The finding by Knud Nierhaus that each ribosome can bind up to three deacylated tRNAs but only two aminoacyl–tRNAs indicates, however, that the ribosome has a third tRNA-binding site, the **exit** or **E site,** which transiently binds the outgoing deacylated tRNA. All three sites, as we shall see, extend over both ribosomal subunits.

The details of the chain elongation process are discussed in Section 32-3D. Chain initiation and chain termination, which are special processes, are examined in Sections 32-3C and 32-3E, respectively. In all of these sections we shall first consider the process of interest in *E. coli* and then compare it with the analogous eukaryotic activity.

C. *Chain Initiation*

a. fMet Is the N-Terminal Residue of Prokaryotic Polypeptides

The first indication that the initiation of translation requires a special codon, since identified as AUG (and, in prokaryotes, occasionally GUG), was the observation that almost half of the *E. coli* proteins begin with the otherwise uncommon amino acid Met. This was followed by the discovery of a peculiar form of Met–$tRNA^{Met}$ in which the Met residue is *N*-formylated:

HC(=O)–NH–CH(CH_2–CH_2–S–CH_3)–C(=O)–O–$tRNA_f^{Met}$

***N*-Formylmethionine–$tRNA_f^{Met}$ (fMet–$tRNA_f^{Met}$)**

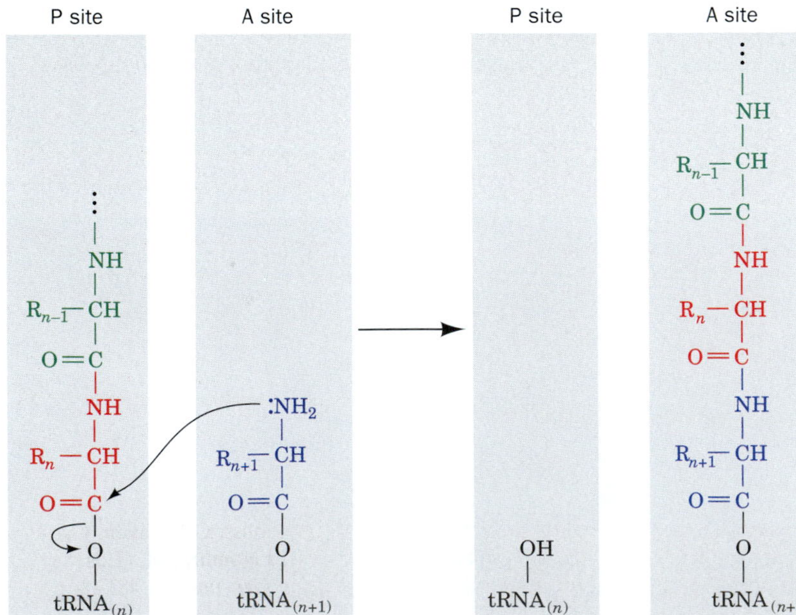

FIGURE 32-41 Ribosomal peptidyl transferase reaction forming a peptide bond. The ribosome catalyzes the nucleophilic attack of the amino group of the aminoacyl–tRNA in the A site on the peptidyl–tRNA ester in the P site, thereby forming a new peptide bond and transferring the nascent polypeptide to the A-site tRNA, while displacing the P-site tRNA.

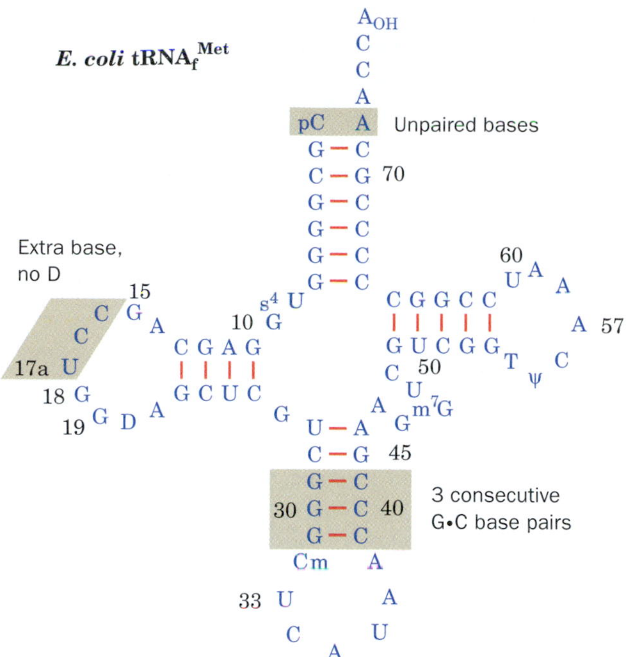

FIGURE 32-42 Nucleotide sequence of *E. coli* tRNA$_f^{Met}$ shown in cloverleaf form. The shaded boxes indicate the significant differences between this initiator tRNA and noninitiator tRNAs such as yeast tRNA^{Ala} (Fig. 32-8). [After Woo, N.M., Roe, B.A., and Rich, A., *Nature* **286,** 346 (1980).]

The ***N*-formylmethionine** residue **(fMet)** already has an amide bond and can therefore only be the N-terminal residue of a polypeptide. In fact, polypeptides synthesized in an *E. coli*–derived cell-free protein synthesizing system always have a leading fMet residue. *fMet must therefore be E. coli's initiating residue.*

The tRNA that recognizes the initiation codon, **tRNA$_f^{Met}$** (Fig. 32-42), differs from the tRNA that carries internal Met residues, **tRNA$_m^{Met}$**, although they both recognize the same codon. In *E. coli,* uncharged (deacylated) tRNA$_f^{Met}$ is first aminoacylated with methionine by the same MetRS that charges tRNA$_m^{Met}$. The resulting **Met–tRNA$_f^{Met}$** is specifically *N*-formylated to yield **fMet–tRNA$_f^{Met}$** in an enzymatic reaction that employs N^{10}-formyltetrahydrofolate (Section 26-4D) as its formyl donor. The formylation enzyme does not recognize Met–tRNA$_m^{Met}$. The X-ray structures of *E. coli* tRNA$_f^{Met}$ and yeast tRNA^{Phe} (Fig. 32-11*b*) are largely similar but differ conformationally in their acceptor stems and anticodon loops. Perhaps these structural differences permit tRNA$_f^{Met}$ to be distinguished from tRNA$_m^{Met}$ in the reactions of chain initiation and elongation (see Section 32-3D).

E. coli proteins are posttranslationally modified by deformylation of their fMet residue and, in many proteins, by the subsequent removal of the resulting N-terminal Met. This processing usually occurs on the nascent polypeptide, which accounts for the observation that mature *E. coli* proteins all lack fMet.

b. Base Pairing between mRNA and the 16S rRNA Helps Select the Translational Initiation Site

AUG codes for internal Met residues as well as the initiating Met residue of a polypeptide. Moreover, mRNAs usually contain many AUGs (and GUGs) in different reading frames. Clearly, *a translational initiation site must be specified by more than just an initiation codon.* This occurs in two ways: (1) the masking of AUGs that are not initiation codons by mRNA secondary structure; and (2) interactions between the mRNA and the 16S rRNA that select the initiating AUG as we now discuss.

The 16S rRNA contains a pyrimidine-rich sequence at its 3′ end. This sequence, as John Shine and Lynn Dalgarno pointed out in 1974, is partially complementary to a purine-rich tract of 3 to 10 nucleotides, the **Shine–Dalgarno sequence,** that is centered ~10 nucleotides upstream from the start codon of nearly all known prokaryotic mRNAs (Fig. 32-43). *Base pairing interactions between an mRNA's Shine–Dalgarno sequence and the 16S rRNA apparently*

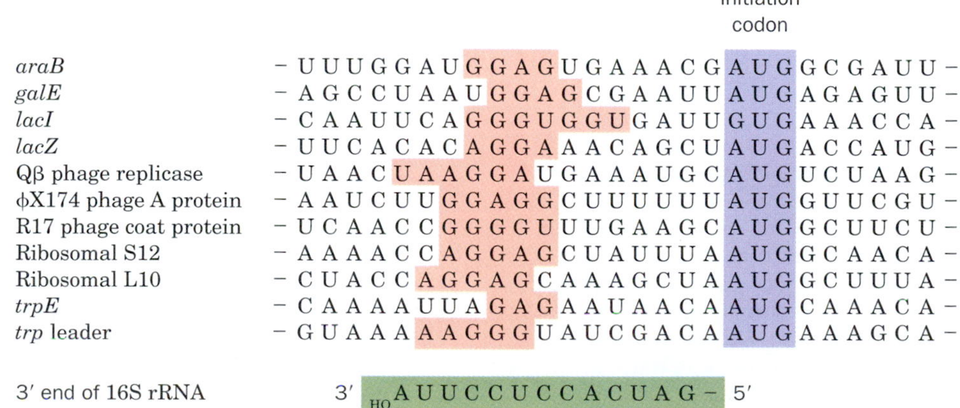

FIGURE 32-43 Some translational initiation sequences recognized by *E. coli* ribosomes. The mRNAs are aligned according to their initiation codons (*blue shading*). Their Shine–Dalgarno sequences (*red shading*) are complementary, counting G · U pairs, to a portion of the 16S rRNA's 3′ end (*below*). [After Steitz, J.A., *in* Chambliss, G., Craven, G.R., Davies, J., Davis, K., Kahan, L., and Nomura, M. (Eds.), *Ribosomes. Structure, Function and Genetics,* pp. 481–482, University Park Press (1979).]

permit the ribosome to select the proper initiation codon. Thus ribosomes with mutationally altered anti-Shine–Dalgarno sequences often have greatly reduced ability to recognize natural mRNAs, although they efficiently translate mRNAs whose Shine–Dalgarno sequences have been made complementary to the altered anti-Shine–Dalgarno sequences. Moreover, treatment of ribosomes with the bactericidal protein **colicin E3** (produced by *E. coli* strains carrying the E3 plasmid), which specifically cleaves a 49-nucleotide fragment from the 3′ terminus of 16S rRNA, yields ribosomes that cannot initiate new polypeptide synthesis but can complete the synthesis of a previously initiated chain.

The X-ray structure of the 70S ribosome reveals, in agreement with Fig. 32-31, that an ~30-nt segment of the mRNA is wrapped in a groove that encircles the neck of the 30S subunit (Fig. 32-44). The mRNA codons in the A and P sites are exposed on the interface side of the 30S subunit, whereas its 5′ and 3′ ends are bound in tunnels composed of RNA and protein. The mRNA's Shine–Dalgarno sequence, which is located near its 5′ end, is base paired, as expected, with the 16S RNA's anti-Shine–Dalgarno sequence, which is situated close to the E site. The resulting double helical segment is accommodated in a cleft formed by both RNA and protein elements of the 16S subunit's head, neck, and platform.

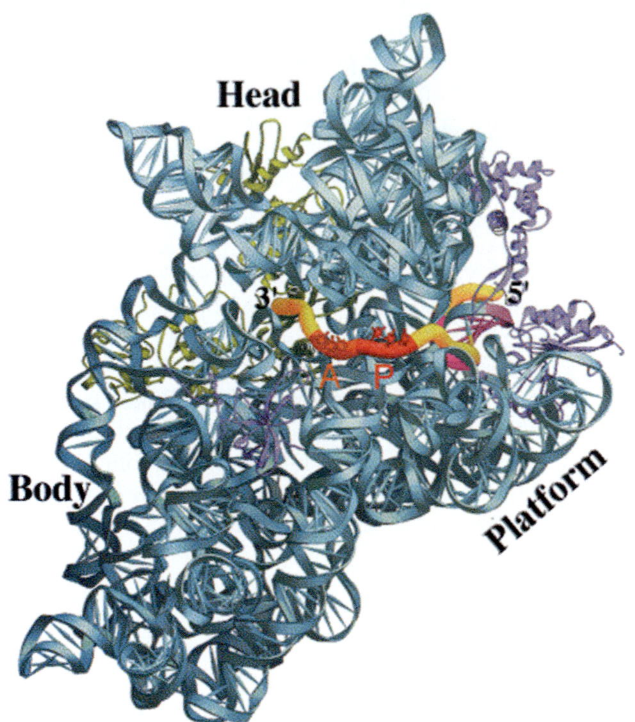

FIGURE 32-44 Path of mRNA through the ribosomal 30S subunit as viewed from its interface side. The 16S RNA is cyan, the mRNA is represented in worm form with its A- and P-site codons orange and red, the Shine–Dalgarno helix (which includes a segment of 16S RNA) magenta, and its remaining segments yellow. The S3, S4, and S5 proteins are green, the S7, S11, and S12 proteins are purple, and the remaining ribosomal proteins have been omitted for clarity. The S3, S4, and S5 proteins, which in part form the tunnel through which the mRNA enters the ribosome, may function as a helicase to remove secondary structure from the mRNA that would otherwise interfere with tRNA binding. [Courtesy of Gloria Culver, Iowa State University. Based on an X-ray structure by Harry Noller, University of California at Santa Cruz. PDBid 1JGO.]

c. Prokaryotic Initiation Is a Three-Stage Process That Requires the Participation of Soluble Protein Initiation Factors

See Guided Exploration 28: Translational Initiation. Intact ribosomes do not directly bind mRNA so as to initiate polypeptide synthesis. Rather, *initiation is a complex process in which the two ribosomal subunits and fMet–tRNA$_f^{Met}$ assemble on a properly aligned mRNA to form a complex that is competent to commence chain elongation. This assembly process also requires the participation of protein **initiation factors** that are not permanently associated with the ribosome.* Initiation in *E. coli* involves three initiation factors designated **IF-1, IF-2,** and **IF-3** (Table 32-9). Their existence was discovered when it was found that

TABLE 32-9 The Soluble Protein Factors of *E. coli* Protein Synthesis

Factor	Number of Residues[a]	Function
Initiation Factors		
IF-1	71	Assists IF-3 binding
IF-2	890	Binds initiator tRNA and GTP
IF-3	180	Releases mRNA and tRNA from recycled 30S subunit and aids new mRNA binding
Elongation Factors		
EF-Tu	393	Binds aminoacyl–tRNA and GTP
EF-Ts	282	Displaces GDP from EF-Tu
EF-G	703	Promotes translocation through GTP binding and hydrolysis
Release Factors		
RF-1	360	Recognizes UAA and UAG Stop codons
RF-2	365	Recognizes UAA and UGA Stop codons
RF-3	528	Stimulates RF-1/RF-2 release via GTP hydrolysis
RRF	185	Together with EF-G, induces ribosomal dissociation to small and large subunits

[a]All *E. coli* translational factors are monomeric proteins.

washing small ribosomal subunits with 1*M* ammonium chloride solution, which removes the initiation factors but not the "permanent" ribosomal proteins, prevents initiation.

The initiation sequence in *E. coli* ribosomes has three stages (Fig. 32-45):

1. On completing a cycle of polypeptide synthesis, the 30S and 50S subunits remain associated as an inactive 70S ribosome (Section 32-3E). IF-3 binds to the 30S subunit of this complex so as to promote its dissociation. The X-ray structure of the 30S subunit in complex with the C-terminal domain of IF-3 (which by itself prevents the association of the 30S and 50S subunits), determined by Yonath and François Franceschi, indicates that IF-3 binds to the upper end of the platform (Fig. 32-26) on its solvent (back) side. Hence IF-3 does not function by physically blocking the binding of the 50S subunit.

2. mRNA and IF-2 in a ternary complex with GTP and fMet–tRNA$_f^{Met}$ that is accompanied by IF-1 subsequently bind to the 30S subunit in either order. Hence, fMet–tRNA$_f^{Met}$ recognition must not be mediated by a codon–anticodon interaction; it is the only tRNA–ribosome association that does not require one. This interaction, nevertheless, helps bind fMet–tRNA$_f^{Met}$ to the ribosome. IF-1 binds in the A site where it may function to prevent the inappropriate or premature binding of a tRNA. IF-3 also functions in this stage of the initiation process: it destabilizes the binding of tRNAs that lack the three G · C pairs in the anticodon stem of RNA$_f^{Met}$ (Fig. 32-42) and helps discriminate between matched and mismatched codon–anticodon interactions.

3. Last, in a process that is preceded by IF-1 and IF-3 release, the 50S subunit joins the 30S initiation complex in a manner that stimulates IF-2 to hydrolyze its bound GTP to GDP + P_i. This irreversible reaction conformationally rearranges the 30S subunit and releases IF-2 for participation in further initiation reactions.

IF-2 is a member of the superfamily of regulatory GTPases such as Ras and hence is a **G-protein** (Section 19-2A). The 30S initiation complex therefore functions as its **GAP** (GTPase-activating protein; Section 19-3C).

Initiation results in the formation of an fMet–tRNA$_f^{Met}$ · mRNA · ribosome complex in which the fMet–tRNA$_f^{Met}$ occupies the ribosome's P site while its A site is poised to accept an incoming aminoacyl–tRNA (an arrangement analogous to that at the conclusion of a round of elongation; Section 32-3D). In fact, tRNA$_f^{Met}$ is the only tRNA that directly enters the P site. All other tRNAs must do so via the A site during chain elongation (Section 32-3D). This arrangement was established through the use of the antibiotic **puromycin** as is discussed in Section 32-3D.

d. Eukaryotic Initiation Is Far More Complicated than That of Prokaryotes

Although translational initiation in eukaryotes superficially resembles that in prokaryotes, it is, in fact, a far more complicated process. Whereas prokaryotic initiation only requires the assistance of three monomeric initiation factors, that in eukaryotes involves the participation of at least 11 initiation factors (designated eIF*n*; "e" for eukaryotic)

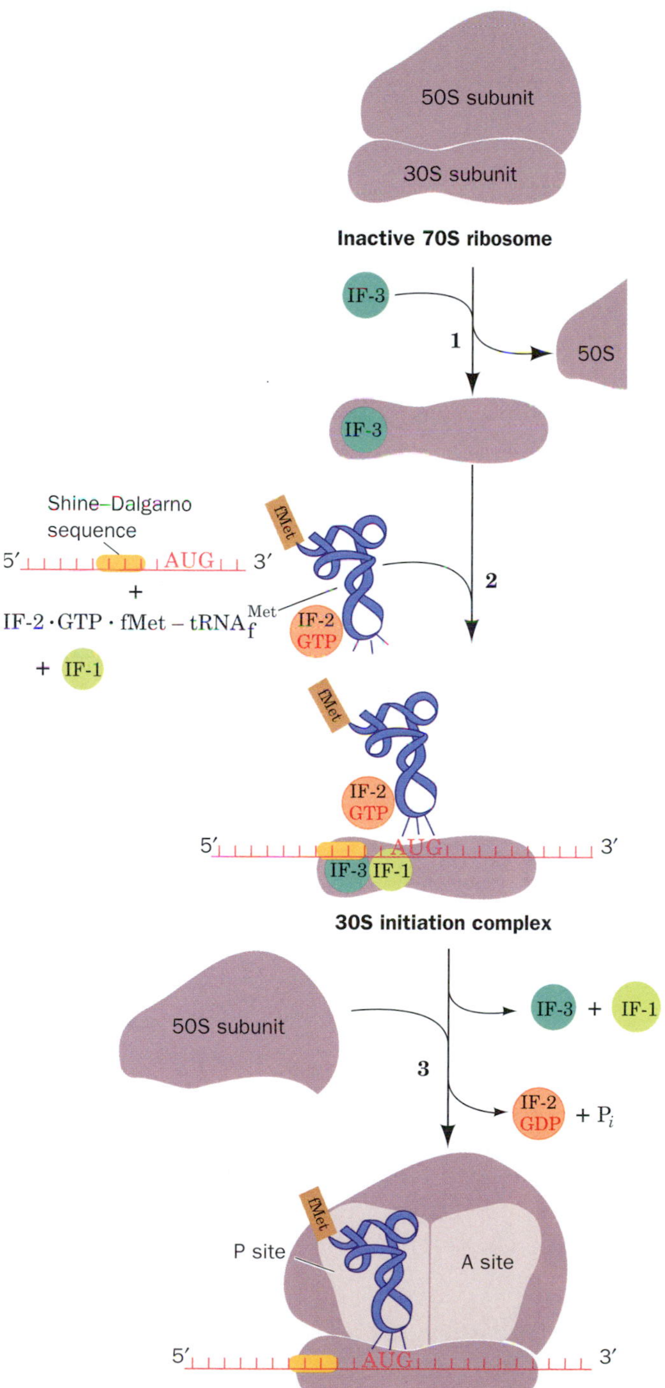

FIGURE 32-45 **Translational initiation pathway in *E. coli*.**

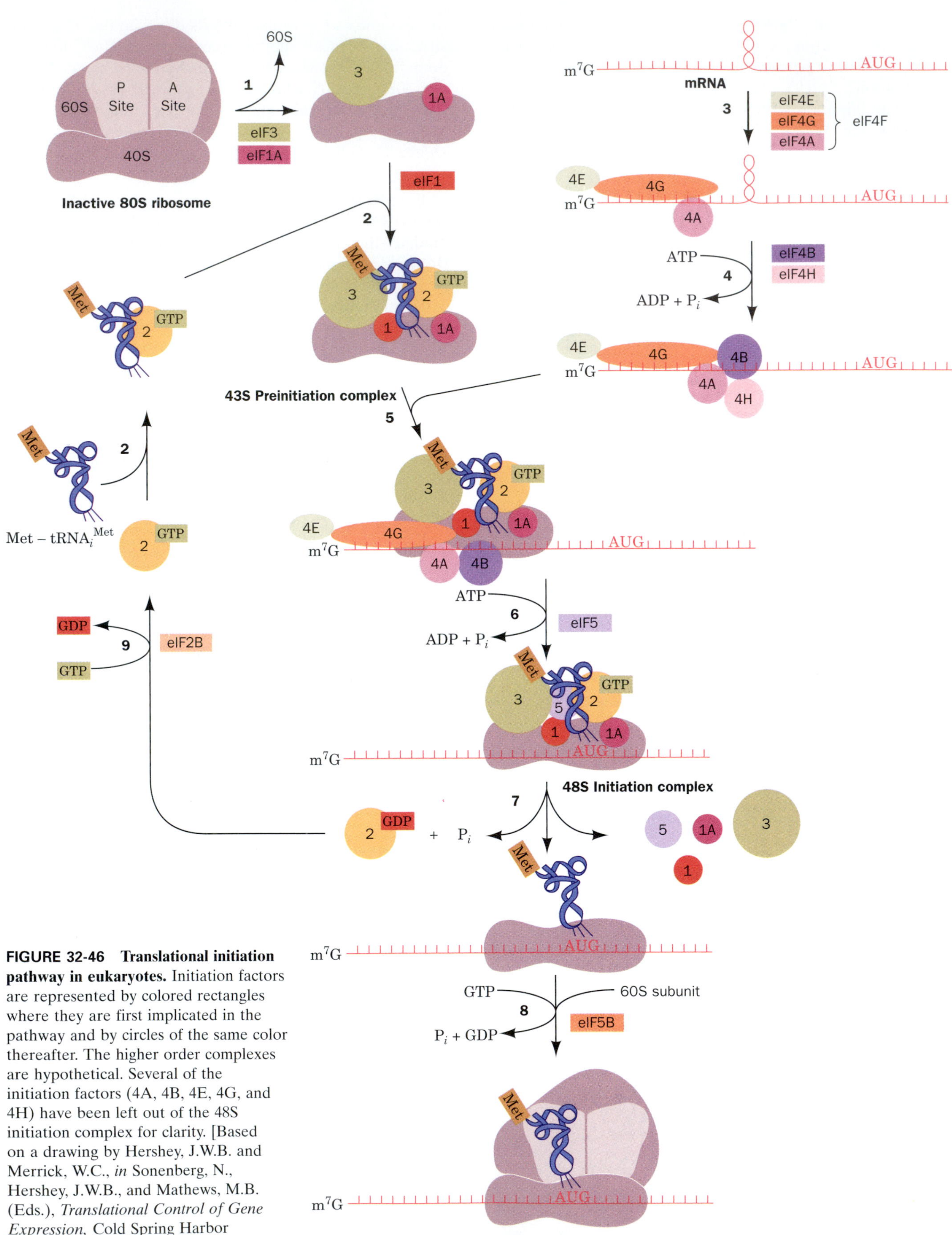

FIGURE 32-46 Translational initiation pathway in eukaryotes. Initiation factors are represented by colored rectangles where they are first implicated in the pathway and by circles of the same color thereafter. The higher order complexes are hypothetical. Several of the initiation factors (4A, 4B, 4E, 4G, and 4H) have been left out of the 48S initiation complex for clarity. [Based on a drawing by Hershey, J.W.B. and Merrick, W.C., *in* Sonenberg, N., Hershey, J.W.B., and Mathews, M.B. (Eds.), *Translational Control of Gene Expression,* Cold Spring Harbor Laboratory Press (2000).]

that consist of at least 26 polypeptide chains. Eukaryotic initiation occurs as follows (Fig. 32-46):

1. The process begins with the binding of **eIF3** (which consists of 11 different subunits) and **eIF1A** (a monomer and homolog of bacterial IF-1) to the 40S subunit in the inactive 80S ribosome (which had terminated elongation in its previous elongation cycle) so that it releases the 50S subunit.

2. The ternary complex of **eIF2** (a heterotrimer), GTP, and **Met–tRNA$_i^{Met}$** binds to the 40S ribosomal subunit accompanied by **eIF1** (a monomer) to form the so-called **43S preinitiation complex.** Here the subscript "i" on tRNA$_i^{Met}$ distinguishes this eukaryotic initiator tRNA, whose appended Met residue is never *N*-formylated, from that of prokaryotes; both species are, nevertheless, readily interchangeable *in vitro*.

3. Eukaryotic mRNAs lack the complementary sequences to bind to the 18S rRNA in the Shine–Dalgarno manner. Rather, they have an entirely different mechanism for recognizing the mRNA's initiating AUG codon. *Eukaryotic mRNAs, nearly all of which have an m^{7}G cap and a poly(A) tail (Section 31-4A), are invariably monocistronic and almost always initiate translation at their leading AUG.* This AUG, which occurs at the end of a 5′-untranslated region of 50 to 70 nt, is embedded in the consensus sequence GCCRCC**AUG**G, with changes in the purine (R) 3 nt before the AUG and the G immediately following it reducing translational efficiency by ~10-fold each and with other changes having much smaller effects. In addition, secondary structure (stem–loops) in the mRNA upstream of the initiation site may affect initiation efficiency. The recognition of the initiation site begins by the binding of **eIF4F** to the m^{7}G cap. eIF4F is a heterotrimeric complex of **eIF4E, eIF4G,** and **eIF4A** (all monomers), in which eIF4E **(cap-binding protein)** recognizes the mRNA's m^{7}G cap and eIF4G serves as a scaffold to join eIF4E with eIF4A. Both the X-ray and NMR structures of eIF4E in complex with **m^{7}GDP,** determined by Nahun Sonenberg and Stephen Burley and by Sonenberg and Gerhard Wagner, reveal that the protein binds the m^{7}G base by intercalating it between two highly conserved Trp residues (Fig. 32-47*a*) in a region that is adjacent to a positively charged cleft that forms the putative mRNA-binding site (Fig. 32-47*b*). The m^{7}G base is specifically recognized by hydrogen bonding to protein side chains in a manner reminiscent of G · C base pairing. eIF4G also binds poly(A)-binding protein (PABP; Section 31-4A) bound to the mRNA's poly(A) tail, thereby circularizing the mRNA (not shown in Fig. 32-46). Although this explains the synergism between an mRNA's m^{7}A cap and its poly(A) tail in stimulating translational initiation, the func-

(a)

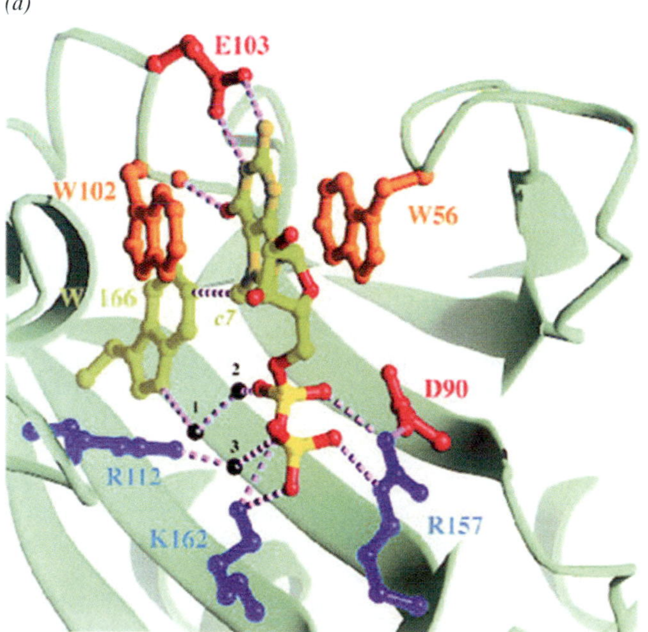

(b)

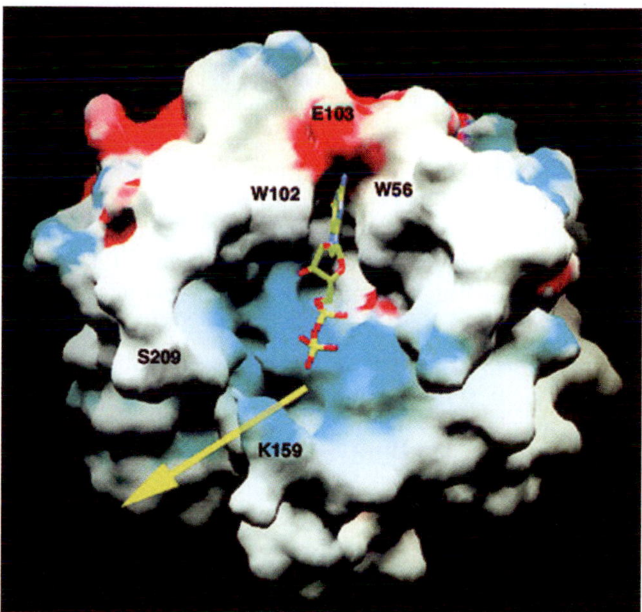

FIGURE 32-47 X-Ray structure of murine eIF4E in complex with the m^{7}G cap analog m^{7}GDP. (*a*) The m^{7}GDP-binding site with the m^{7}GDP and the side chains that bind it drawn in ball-and-stick form with the atoms of the m^{7}GDP colored according to type (C green, N dark yellow, O red, and P bright yellow) and the protein side chains with which the m^{7}GDP interacts drawn in various colors. Hydrogen bonds, salt bridges, and van der Waals interactions are represented by dashed lines and bridging water molecules are drawn as black spheres. The m^{7}G base is intercalated between the indole rings of Trp 56 and Trp 102, where it specifically interacts with protein side chains through hydrogen bonds and van der Waals interactions. The GDP's phosphate groups interact directly and indirectly with three basic side chains. (*b*) The solvent-accessible surface of eIF4E colored according to its electrostatic potential (red negative, blue positive, and white neutral) and viewed approximately as in Part *a*. The m^{7}GDP is drawn in ball-and-stick form colored as in Part *a*. The mRNA presumably binds in the positively charged cleft (*yellow arrow*) that is adjacent to the m^{7}G binding site and which passes between Lys 159 and Ser 209. [Courtesy of Nahum Sonenberg, McGill University, Montréal, Québec, Canada. PDBid 1EJ1.]

tion of this circle is unclear. However, an attractive hypothesis is that it enables a ribosome that has finished translating the mRNA to reinitiate translation without having to disassemble and then reassemble. Another possibility is that it prevents the translation of incomplete (broken) mRNAs.

4. eIF4B (an RRM-containing homodimer) and **eIF4H** (a monomer) join the eIF4F–mRNA complex where they stimulate the RNA helicase activity of eIF4A to unwind the mRNA's helical segments in an ATP-dependent process. This presumably also strips away the proteins that are bound to the mRNA (Section 32-4A). eIF4A is the prototype of the so-called **DEAD-box family** of proteins, which is named after one of the sequence motifs shared by the diverse members of this family, all of which have NTPase activity.

5. The eIF4F–mRNA–eIF4B–eIF4H complex joins the 43S preinitiation complex through a protein–protein interaction between eIF4G and the 40S subunit-bound eIF3. This differs substantially from the corresponding prokaryotic process (Fig. 32-45) in which the mRNA is bound to the 30S ribosomal subunit via associations between RNA molecules (that involving the Shine–Dalgarno sequence and the codon–anticodon interaction).

6. eIF5 (a monomer) joins the growing assembly. The 43S preinitiation complex then translocates along the mRNA, an ATP-dependent process called **scanning,** until it encounters the mRNA's AUG initiation codon, thereby yielding the **48S initiation complex.** The recognition of the AUG occurs mainly through base pairing with the CUA anticodon on the bound Met–$tRNA_i^{Met}$, as was demonstrated by the observation that mutating this anticodon results in the recognition of the new cognate codon instead of AUG.

7. The formation of the 48S initiation complex induces eIF2 to hydrolyze its bound GTP to GDP + P_i, which results in the release of all the initiation factors, thereby leaving the Met–$tRNA_i^{Met}$ in the small subunit's P site. The hydrolysis reaction is stimulated by eIF5, acting as a GAP (Section 19-3C).

8. The 60S subunit then joins the mRNA-bound Met–$tRNA_i^{Met}$–40S subunit complex in a GTPase reaction mediated by **eIF5B** (a monomer and homolog of bacterial IF-2), thereby yielding the 80S ribosomal initiation complex. Thus eukaryotic translation initiation consumes two GTPs versus one for prokaryotic initiation (Fig. 32-45).

9. What remains is to recycle the eIF2 · GDP complex by exchanging its GDP for GTP. This reaction is mediated by **eIF2B** (a heteropentamer), which therefore functions as eIF2's **GEF** (guanine nucleotide exchange factor; Section 19-3C).

Many eukaryotic initiation factors are subject to phosphorylation/dephosphorylation and are therefore likely to participate in the control of eukaryotic translation, a subject we discuss in Section 32-4.

Although the initiation sites on most eukaryotic mRNAs are identified by the above-described scanning mechanism, a few mRNAs have an **internal ribosome entry site (IRES)** to which the 40S subunit can directly bind in a process reminiscent of prokaryotic initiation. However, little is yet known about the mechanism of IRES-based initiation. Indeed, IRESs lack clearly identifiable consensus sequences.

D. *Chain Elongation*

See Guided Exploration 29: Translational Elongation. *Ribosomes elongate polypeptide chains in a three-stage reaction cycle that adds amino acid residues to a growing polypeptide's C-terminus (Fig. 32-48):*

1. Decoding, in which the ribosome selects and binds an aminoacyl–tRNA, whose anticodon is complementary to the mRNA codon in the A site.

2. Transpeptidation, in which the peptidyl group on the P-site tRNA is transferred to the aminoacyl group in the A site through the formation of a peptide bond (Fig. 32-41).

3. Translocation, in which A-site and P-site tRNAs are respectively transferred to the P site and E site accompanied by their bound mRNA; that is, the mRNA, together with its base paired tRNAs, is ratcheted through the ribosome by one codon.

This process, which occurs at a rate of 10 to 20 residues/s, involves the participation of several nonribosomal proteins known as **elongation factors** (Table 32-9). We describe these processes in the following paragraphs.

a. Decoding

In the decoding stage of the *E. coli* elongation cycle, a binary complex of GTP with the elongation factor **EF-Tu** (also called **EF1A**) combines with an aminoacyl–tRNA. The resulting ternary complex binds to the ribosome, and, in a reaction that hydrolyzes the GTP to GDP + P_i, the aminoacyl–tRNA is bound in a codon–anticodon complex to the ribosomal A site and EF-Tu · GDP + P_i is released. In the remainder of this stage, the bound GDP is replaced by GTP in a reaction mediated the elongation factor **EF-Ts** (also called **EF1B**). EF-Tu, as are several other GTP-binding ribosomal factors, is a G-protein, and hence the ribosome functions as its GAP and EF-Ts is its GEF.

Aminoacyl–tRNAs can bind to the ribosomal A site without the mediation of EF-Tu but at a rate too slow to support cell growth. The importance of EF-Tu is indicated by the fact that it is the most abundant *E. coli* protein; it is present in ~100,000 copies per cell (>5% of the cell's protein), which is approximately the number of tRNA molecules in the cell. Consequently, *the cell's entire complement of aminoacyl–tRNAs is essentially sequestered by EF-Tu.*

b. EF-Tu Is Sterically Prevented from Binding Initiator tRNA

The X-ray structure of the Phe–$tRNA^{Phe}$ · EF-Tu · GDPNP ternary complex (GDPNP is a nonhydrolyzable GTP analog; Section 19-3C), determined by Brian Clark and Jens Nyborg, reveals that these two macromolecules associate to form a corkscrew-shaped complex in which the

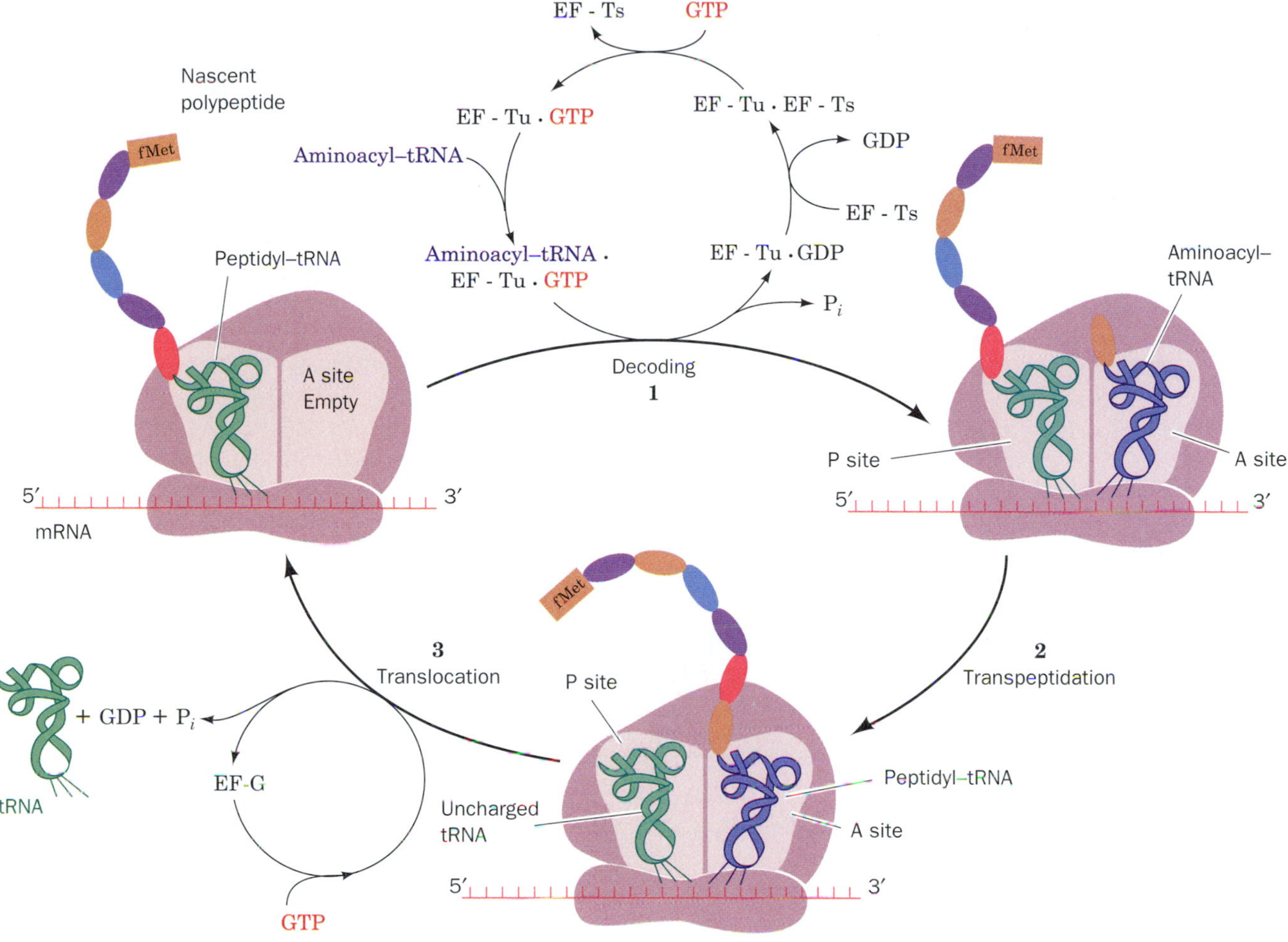

FIGURE 32-48 Elongation cycle in *E. coli* ribosomes. The E site, to which discharged tRNAs are transferred before being released to solution, is not shown. Eukaryotic elongation follows a similar cycle but EF-Tu and EF-Ts are replaced by a single multisubunit protein, eEF-1, and EF-G is replaced by eEF2.

EF-Tu and the tRNA's acceptor stem form a knoblike handle and the tRNA's anticodon helix forms the screw (Fig. 32-49). The conformation of the $tRNA^{Phe}$ closely resembles that of the uncomplexed molecule (Fig. 32-11*b*). The EF-Tu folds into three distinct domains that are connected by flexible peptides, rather like beads on a string. The N-terminal domain 1, which binds guanine nucleotides and catalyzes GTP hydrolysis, structurally resembles other known G-proteins.

The two macromolecules associate rather tenuously via three major regions: (1) the CCA—Phe segment at the 3′ end of the Phe–$tRNA^{Phe}$ binds in a cleft between domains 1 and 2 of the EF-Tu (red and green in Fig. 32-49) that ends in a pocket large enough to accommodate all amino acid residues; (2) the 5′-phosphate of the tRNA binds in a depression at the junction of EF-Tu's three domains; and (3) one side of the tRNA's TψC stem contacts exposed main chain and side chains of EF-Tu domain 3 (blue in Fig. 32-49). The tight association of the aminoacyl group with EF-Tu appears to greatly increase the affinity of EF-Tu for the otherwise loosely bound tRNA, which explains why EF-Tu does not bind uncharged elongator tRNAs.

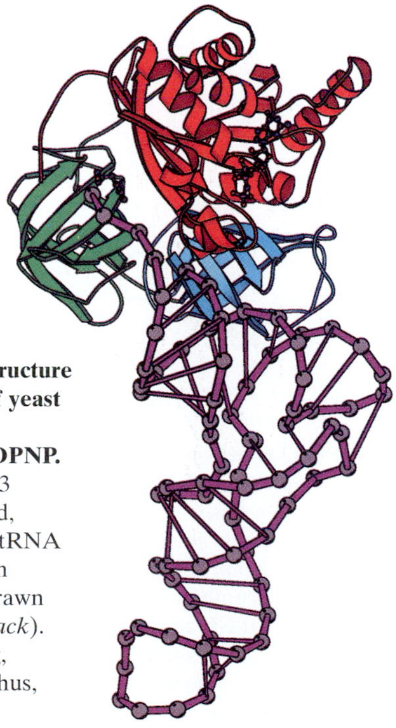

FIGURE 32-49 X-Ray structure of the ternary complex of yeast Phe–$tRNA^{Phe}$, *Thermus aquaticus* EF-Tu, and GDPNP. EF-Tu domains 1, 2, and 3 (N- to C-terminal) are red, green, and cyan, and the tRNA is shown in ladder form in purple. The GDPNP is drawn in ball-and-stick form (*black*). [Courtesy of Jens Nyborg, University of Aarhus, Århus, Denmark. PDBid 1TTT.]

EF-Tu binds neither formylated aminoacyl–tRNAs nor unformylated Met–tRNA$_f^{Met}$, which is why the initiator tRNA never reads internal AUG or GUG codons. The first base pair of tRNA$_f^{Met}$ is mismatched (Fig. 32-42) and hence this initiator tRNA has a 3′ overhang of 5 nt vs 4 nt for an elongator tRNA. It seems likely that this mismatch, together with the formyl group, prevents fMet–tRNA$_f^{Met}$ from binding to EF-Tu. Indeed, EF-Tu binds to *E. coli* tRNA$_f^{Met}$ whose 5′-terminal C residue has been deaminated by bisulfite treatment (Section 30-7), which reestablishes the "missing" base pair as U · A. Similarly, Sec–tRNA^{Sec}, which is also not bound by EF-Tu (but rather by SELB; Section 32-2D), has 8 bp in its acceptor stem vs 7 bp in those of other elongator tRNAs. However, initiator tRNAs from several sources have fully base paired acceptor stems, and the U1 · A72 base pair of tRNA^{Gln} is opened up on binding to GlnRS (Section 32-2C).

c. EF-Tu Undergoes a Major Conformational Change on Hydrolyzing GTP

Morten Kjeldgaard and Nyborg determined the X-ray structures of *T. aquaticus* EF-Tu (405 residues) in complex with GDPNP and the 70% identical *E. coli* EF-Tu (393 residues) in complex with GDP (Fig. 32-50). The conformation of EF-Tu in its complex with only GDPNP closely resembles that in its ternary complex with Phe–tRNA^{Phe} and GDPNP (Fig. 32-49). However, comparison of the GDPNP and GDP complexes indicates that, on hydrolyzing its bound GTP, EF-Tu undergoes a major structural reorganization. Its greatest local conformation changes occur in the Switch I and Switch II regions of domain 1, which in all G-proteins signal the state of the bound nucleotide to interacting partners (Section 19-2B; here domains 2 and 3): Switch I converts from a β hairpin to a short α helix and the α helix of Switch II shifts toward the C-terminus by 4 residues. As a consequence, this latter helix reorients by 42°, which results in domain 1 rigidly changing its orientation with respect to domains 2 and 3 by a dramatic 91° rotation. The tRNA binding site is thereby eliminated.

d. EF-Ts Disrupts the Binding of GDP to EF-Tu

EF-Tu has a 100-fold higher affinity for GDP than GTP. Hence, replacement of the EF-Tu–bound GDP by GTP must be facilitated by the interaction of EF-Tu with EF-Ts (Fig. 32-48, *top*). The X-ray structure of the EF-Tu · EF-Ts complex, determined by Stephen Cusack and Reuben Leberman, reveals that the EF-Tu has a conformation resembling that of its GDP complex (Fig. 32-51) but with its domains 2 and 3 swung away from domain 1 by ~18°. EF-Ts is an elongated molecule that binds along the right side of EF-Tu as shown in Fig. 32-51, where it contacts EF-Tu's domains 1 and 3. The intrusive interactions of EF-Ts side chains with the GDP binding pocket on EF-Tu disrupts the Mg^{2+} ion binding site. This reduces the affinity of EF-Tu for GDP, thereby facilitating its exchange for GTP (after EF-Ts has dissociated), which has a 10-fold higher concentration in the cell than does GDP (the GEF-containing segment of Sos similarly interferes with Mg^{2+} binding and hence guanine nucleotide binding by Ras; Section 19-3C). EF-Tu's subsequent binding of a charged elongator tRNA increases its affinity for GTP.

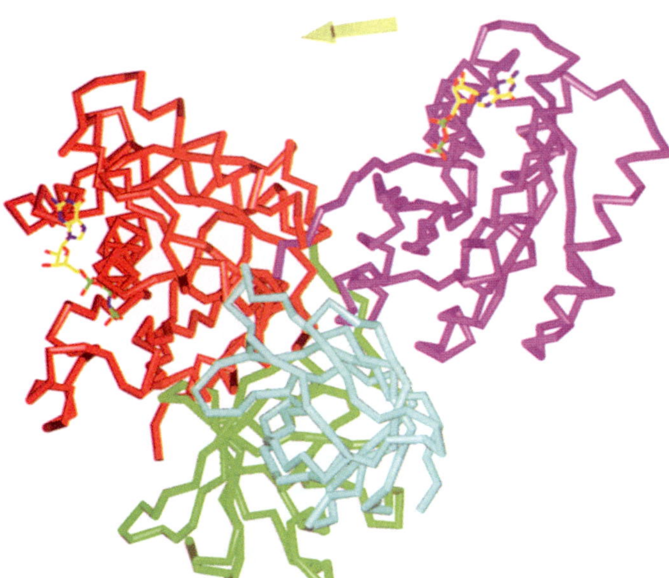

FIGURE 32-50 Comparison of the X-ray structures of EF-Tu in its complexes with GDP and GDPNP. The protein is represented by its C_α backbone with domain 1, its GTP-binding domain, purple in the GDP complex and red in the GDPNP complex. Domain 2 and domain 3, which have the same orientation in both complexes, are green and cyan. The bound GDP and GDPNP are shown in stick form with C yellow, N blue, O red, and P green. [Courtesy of Morten Kjeldgaard and Jens Nyborg, University of Aarhus, Århus, Denmark. PDBid 1EFT.] **See the Interactive Exercises**

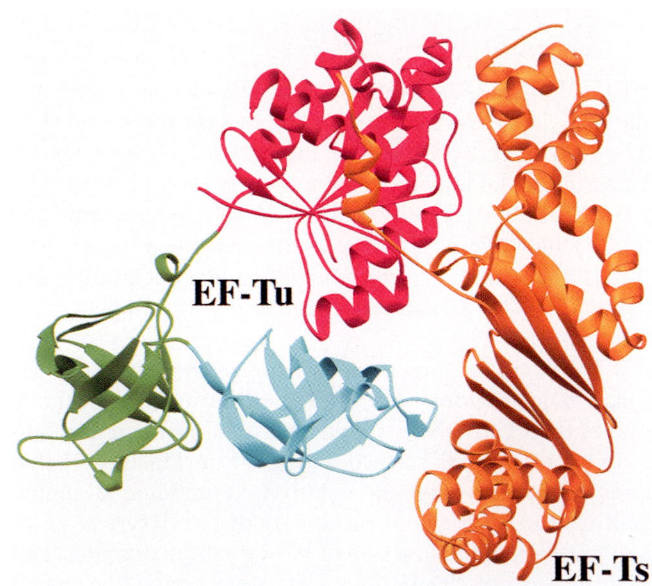

FIGURE 32-51 X-Ray structure of the *E. coli* EF-Tu · EF-Ts complex. Domains 1, 2, and 3 of EF-Tu are magenta, green, and cyan, respectively, and EF-Ts is orange. [Based on an X-ray structure by Stephen Cusack and Reuben Leberman, EMBL, Grenoble Cedex, France. PDBid 1EFU.]

e. Transpeptidation

In the transpeptidation stage of the elongation cycle (Fig. 32-48), the peptide bond is formed through the nucleophilic displacement of the P-site tRNA by the amino group of the 3′-linked aminoacyl–tRNA in the A site (Fig 32-41). The nascent polypeptide chain is thereby lengthened at its C terminus by one residue and transferred to the A-site tRNA. The reaction occurs without the need of activating cofactors such as ATP because the ester linkage between the nascent polypeptide and the P-site tRNA is a "high-energy" bond. The **peptidyl transferase** center that catalyzes peptide bond formation is located entirely on the large subunit as is demonstrated by the observation that in high concentrations of organic solvents such as ethanol, the large subunit alone catalyzes peptide bond formation. The organic solvent apparently distorts the large subunit in a way that mimics the effect of small subunit binding.

f. Puromycin Is an Aminoacyl–tRNA Analog

The ribosomal elongation cycle was originally characterized through the use of the antibiotic **puromycin** (Fig. 32-52). This product of *Streptomyces alboniger,* which resembles the 3′ end of Tyr–tRNA, causes the premature termination of polypeptide chain synthesis. Puromycin, in competition with the mRNA-specified aminoacyl–tRNA but without the need of elongation factors, binds to the ribosomal A site which, in turn, catalyzes a normal transpeptidation reaction to form peptidyl–puromycin. Yet, the ribosome cannot catalyze the transpeptidation reaction in the next elongation cycle because puromycin's "amino acid residue" is linked to its "tRNA" via an amide rather than an ester bond. Polypeptide synthesis is therefore aborted and the peptidyl–puromycin is released.

In the absence of the elongation factor EF-G (see below), an active ribosome cannot bind puromycin because its A site is at least partially occupied by a peptidyl–tRNA. A newly initiated ribosome, however, violates this rule; it catalyzes fMet–puromycin formation. *These observations demonstrated the functional existence of the ribosomal P and A sites and established that fMet–tRNA$_f^{Met}$ binds directly to the P site, whereas other aminoacyl–tRNAs must first enter the A site.*

g. The Ribosome Is a Ribozyme

What is the nature of the peptidyl transferase center, that is, does it consist of RNA, protein, or both? Since all proteins, including those associated with ribosomes, are ribosomally synthesized, the primordial ribosome must have preceded the primordial proteins and hence consisted entirely of RNA. Despite this (in hindsight) obvious evolutionary argument, the idea that rRNA functions catalytically was not seriously entertained until after it had been discovered that RNA can, in fact, act as a catalyst (Section 31-4A). Several other observations further indicate that the ribosome is a ribozyme:

1. The absence of any of all but three ribosomal proteins (L2, L3, and L4) from the 50S subunit does not abolish peptidyl transferase function.

2. rRNAs are more highly conserved throughout evolution than are ribosomal proteins.

3. Most mutations that confer resistance to antibiotics that inhibit protein synthesis occur in genes encoding rRNAs rather than ribosomal proteins.

Nevertheless, the unambiguous demonstration that rRNA functions catalytically in polypeptide synthesis proved to be surprisingly elusive. Noller succeeded in showing that the *T. thermophilus* large ribosomal subunit from which ~95% of the protein had been removed by treatment with SDS and **proteinase K** followed by phenol extraction (which denatures proteins; Section 6-6A) maintained >80% of its peptidyl transferase activity in a model reaction. Moreover, this activity was abolished by RNase treatment. However, since the remaining protein was due to several intact ribosomal proteins (which are presumably

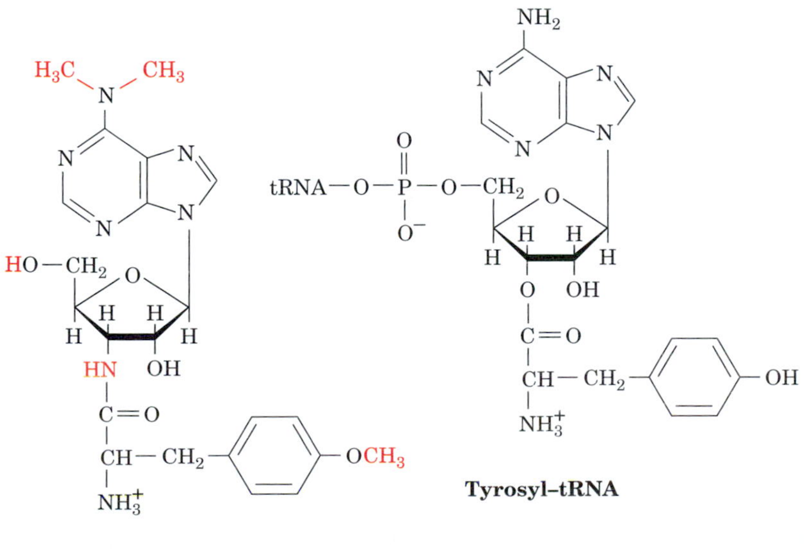

Puromycin **Tyrosyl–tRNA**

FIGURE 32-52 Puromycin. This antibiotic (*left*) resembles the 3′-terminus of tyrosyl–tRNA (*right*).

sequestered within the 23S RNA), it could be argued that these proteins are essential for ribosomal catalytic function, a reasonable expectation in light of the >3.5 billion years that ribosomal proteins and RNAs have coevolved.

The nature of the peptidyl transferase center was unequivocally revealed through its identification in the X-ray structure of the 50S subunit. Peptide bond formation presumably resembles the reverse of peptide bond hydrolysis such as that catalyzed by serine proteases (Section 15-3C). The ribosomal reaction's tetrahedral intermediate (Fig. 32-53*a*) is mimicked by a compound synthesized by Michael Yarus that consists of the trinucleotide CCdA linked to puromycin via a phosphoramidite group (Fig. 32-53*b*). This compound, which is named **CCdA-p-Puro,** binds tightly to the ribosome so as to inhibit its peptidyl transferse activity. The X-ray structure of the 50S subunit in complex with CCdA-p-Puro reveals that the inhibitor binds at the bottom of a deep cleft (Fig. 32-36*a*) at the entrance to the 100-Å-long polypeptide exit tunnel that runs through to the back of the subunit (Fig. 12-50*b*). There, *the inhibitor is completely enveloped in RNA with no protein side chain approaching closer than ~18 Å to the inhibitor's phosphoramidite group.* Moreover, all the nucleotides that contact the CCdA-p-Puro are >95% conserved among all three kingdoms of life. Clearly, *the ribosomal transpeptidase reaction is catalyzed by RNA.*

How does the ribosome catalyze the transpeptidase reaction? Certainly the ribosome's greatest catalytic influence, as is true of all enzymes, is that it correctly positions its substrates for reaction. In addition, inspection of the peptidyl transferase active site reveals that the only acid–base group within 5 Å of the bound CCdA-p-Puro's phosphoramidite group is atom N3 of the invariant rRNA base A2486 (A2451 in *E. coli*). It is ~3 Å from and hence hydrogen bonded to the phosphoramidite oxygen that corresponds to the oxyanion of the tetrahedral intermediate. Moore and Steitz therefore postulated that A2486-N3 acts as a general base in the peptidyl transferase reaction (Fig. 32-54). This would require A2486-N3 to have a p*K* of at least 7 (recall that proton transfers between hydrogen bonded groups only occur at physiologically significant rates when the p*K* of the proton acceptor is not less than 2 or 3 pH units below that of the proton donor; Section 15-3D). Yet, the p*K* of N3 in AMP is <3.5. However, since the pH of the 50S subunit crystals is 5.8, A2486-N3 could only be hydrogen bonded to the phosphoramidite oxygen if its p*K* is >6. Moreover, kinetic investigations of 70S ribosomes have identified a titratable group with a p*K* of 7.5 that affects catalysis and which disappears when A2486 is mutated to U, a mutation that reduces the rate of peptide bond formation by 130-fold. Apparently, the p*K* of A2486-N3 is greatly increased by its ribosomal environment. It is postulated that this occurs via a charge relay system in which the phosphate group of A2485, one of the most solvent-inaccessible phosphate groups in the 50S subunit, electrostatically increases the p*K* of A2486-N3 through a hydrogen bonded network involving the highly conserved G2482 (Fig. 32-55).

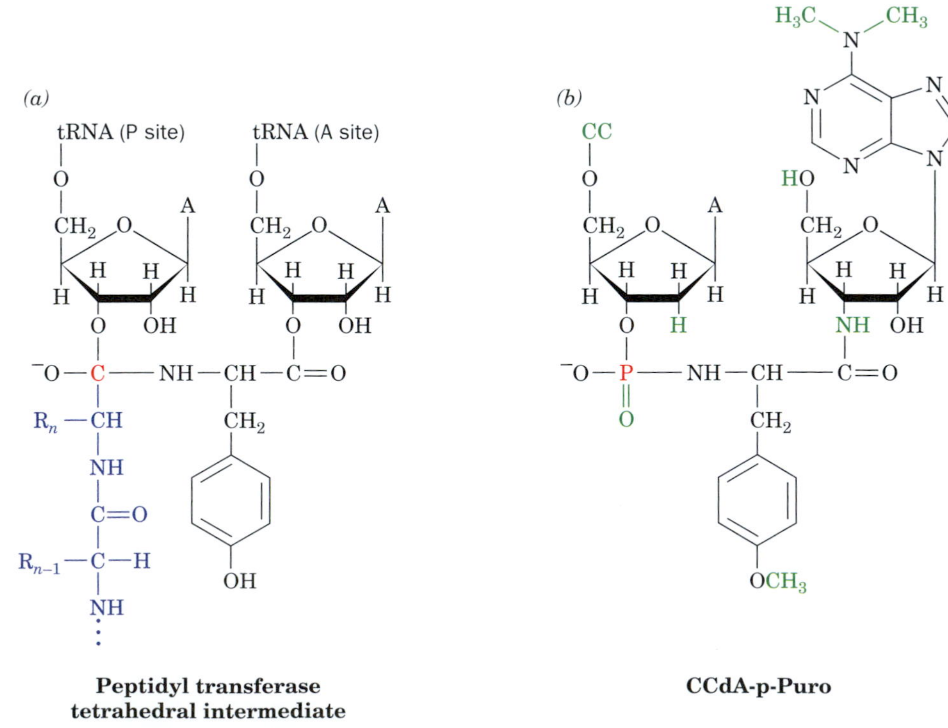

(*a*)

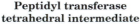

Peptidyl transferase tetrahedral intermediate

(*b*)

CCdA-p-Puro

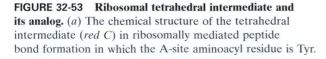

FIGURE 32-53 Ribosomal tetrahedral intermediate and its analog. (*a*) The chemical structure of the tetrahedral intermediate (*red C*) in ribosomally mediated peptide bond formation in which the A-site aminoacyl residue is Tyr. (*b*) CCdA-p-Puro, the transition state analog of the tetrahedral intermediate in Part *a* produced by linking the 3′-OH group of CCdA to the amino group of puromycin's *O*-methyltyrosine residue via a phosphoryl group.

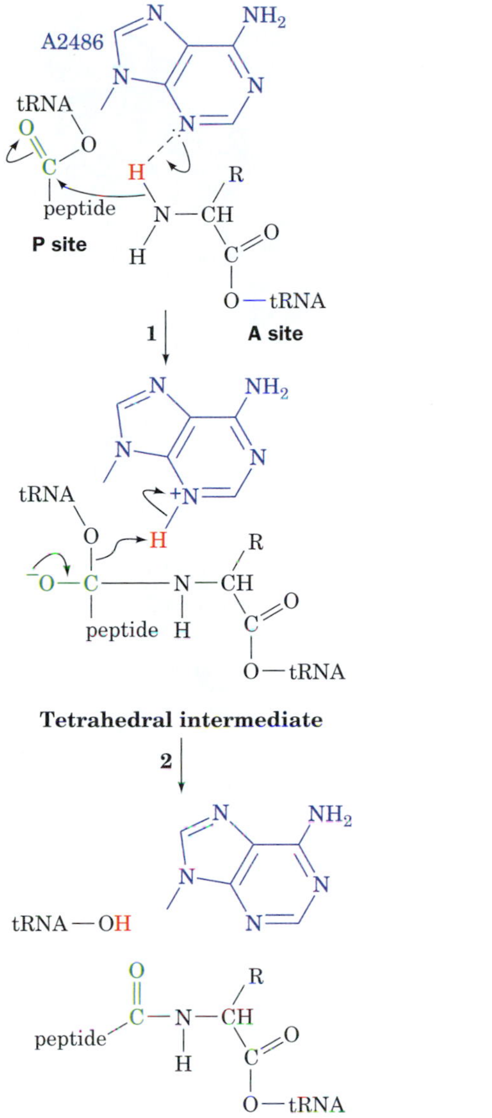

FIGURE 32-54 Proposed mechanism of ribosomal peptide synthesis. The reaction is catalyzed, in part, through the abstraction of a proton from the attacking amino group by A2486-N3 (general base catalysis) in Step 1 and its donation of this proton to the tetrahedral intermediate (general acid catalysis) in Step 2.

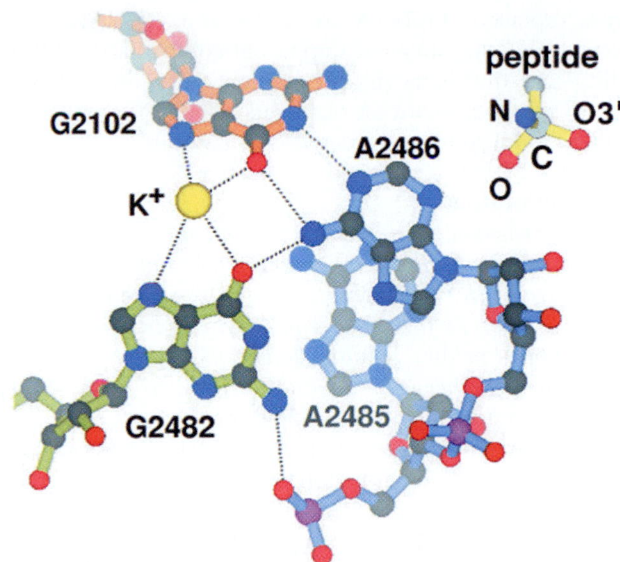

FIGURE 32-55 Catalytic apparatus of the peptidyl transferase active site. Atoms are colored according to type with C gray, N blue, O red, and P magenta. A K^+ ion is drawn as a yellow sphere and hydrogen bonds are represented by dotted lines. Note that the buried phosphate group of A2485 is hydrogen bonded to the base of G2482, which in turn is hydrogen bonded to the base of A2486. It is these interactions that are proposed to relay the electronic charge from the anionic phosphate group to A2486-N3, thereby greatly increasing the latter atom's pK and enabling it to function as a general base. [Courtesy of Peter Moore and Thomas Steitz, Yale University.]

The X-ray structures of the 50S subunit in its complexes with several substrate, reaction intermediate, and product analogs have led to a reaction model that begins with the reactant complex shown in Fig. 32-56. Here, the 2′ O atom of the P-site tRNA's A76 acts to properly orient the attacking amino group via a hydrogen bond. Indeed, a P-site substrate with a 2′-deoxyribose residue at A76 is unreactive (which is why the CCdA-p-Puro inhibitor was synthesized with dA rather than A). However, this model also indicates that the reaction intermediate's oxyanion points away from and hence is not stabilized by A2486-N3 (Fig. 32-54; the hydrogen bond between A2486-N3 and the phosphoramidite oxyanion of CCdA-p-Puro occurs because this inhibitor assumes a conformation in the 50S subunit that would be sterically prevented if its dA residue

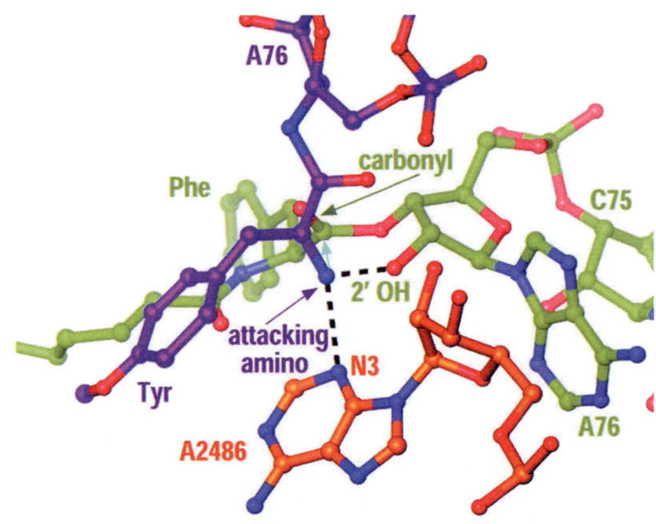

FIGURE 32-56 Model of the substrate complex of the 50S ribosomal subunit. Atoms are colored according to type with the A-site substrate C and P purple, P-site substrate C and P green, 23S rRNA C and P orange, N blue, and O red. The attacking amino group of the A-site aminoacyl residue is held in position for nucleophilic attack (*cyan arrow*) on the carbonyl C of the P-site aminoacyl ester through hydrogen bonds (*dashed black lines*) to A2486-N3 and the 2′-O of the P-site A76. [Courtesy of Peter Moore and Thomas Steitz, Yale University.]

were replaced by ribo-A, as occurs in all tRNAs). Thus, considering the ribosome's apparent conformational flexibility and the many pitfalls in identifying an enzymatic group responsible for an observed pK, the foregoing catalytic model should be treated with caution.

h. Translocation

In the translocation stage of the elongation cycle, the now uncharged P-site tRNA (at first $tRNA_f^{Met}$ *but subsequently an elongator tRNA) is transferred to the E site (not shown in Fig. 32-48), its former occupant having been previously expelled (see below). Simultaneously, the peptidyl–tRNA in the A site, together with its bound mRNA, is moved to the P site.* This prepares the ribosome for the next elongation cycle. The maintenance of the peptidyl–tRNA's codon–anticodon association is no longer necessary for amino acid specification. Rather, it probably acts as a place-keeper that permits the ribosome to precisely step off the three nucleotides along the mRNA required to preserve the reading frame. Indeed, the observation that frameshift suppressor tRNAs induce a four-nucleotide translocation (Section 32-2E) indicates that mRNA movement is directly coupled to tRNA movement.

i. EF-G Structurally Mimics the EF-Ts · tRNA Complex

The translocation process requires the participation of elongation factor **EF-G** (also called **EF2**), which binds to the ribosome together with GTP and is only released on hydrolysis of the GTP to GDP + P_i (Fig. 32-48). EF-G release is a prerequisite for beginning the next elongation cycle because EF-G and EF-Tu bind to the same site of the ribosome and hence their binding is mutually exclusive.

The X-ray structure of *Thermus thermophilus* EF-G · GDP, determined by Steitz and Moore, reveals a tadpole-shaped monomeric protein that consists of five domains (Fig. 32-57). The first two domains closely resemble those in EF-Tu · GDPNP rather than those in EF-Tu · GDP (Fig. 32-50). This, it is argued, is because the two elongation factors have reciprocal functions with EF-Tu · GTP facilitating the conversion of the ribosome from its post- to its pretranslocational state and EF-G · GTP promoting the reverse transition. This idea is supported by the intriguing observation that the Phe–$tRNA^{Phe}$ · EF-Tu · GDPNP and EF-G · GDP complexes are almost identical in appearance: EF-G's three C-terminal domains (magenta in Fig. 32-57), which have no counterparts in EF-Tu, closely resemble the EF-Tu–bound tRNA in shape, a remarkable case of **macromolecular mimicry.**

EF-G is unusual among G-proteins in that it has no corresponding GEF. However, its N-terminal guanine nucleotide-binding domain contains a unique α helical insert (dark blue in Fig. 32-57) that contacts the domain's conserved core at sites analogous to those in EF-Tu that interact with EF-Ts. This suggests that this subdomain acts as an internal GEF.

j. Translocation Occurs via Intermediate States

Chemical footprinting studies (Section 31-2A) by Noller revealed that certain bases in the 16S rRNA are protected by tRNAs bound in the ribosomal A and P-sites and that certain bases in the 23S rRNA are protected by tRNAs in the A, P, and E sites. Almost all of these protected bases are absolutely conserved in evolution and many of them have been implicated in ribosomal function through biochemical or genetic studies.

FIGURE 32-57 X-Ray structure of EF-G in complex with GDP. Domain 1 is red with its α helical insert dark blue, domain 2 is green, and domains 3, 4, and 5 are magenta. The GDP is drawn in ball-and-stick form (*black*). A 25-residue segment of EF-G's domain 1 is not visible in the X-ray structure and domain 3 is poorly defined, thereby accounting for the several peptide segments seen in this model. Note the remarkable resemblance in shape between this structure and that of Phe–$tRNA^{Phe}$ · EF-Tu · GDPNP (Fig. 32-49). [Courtesy of Jens Nyborg, University of Aarhus, Århus, Denmark. Based on an X-ray structure by Thomas Steitz and Peter Moore, Yale University. PDBid 2EFG.]

Variations in chemical footprinting patterns during the elongation cycle together with the more recently determined X-ray and cryo-EM studies indicate that the translocation of tRNA occurs in several discrete steps (Fig. 32-58):

1. Let us start with the ribosome in its **posttranslocational state:** a deacylated tRNA bound to the E subsites of both the 30S and 50S subunits (the E/E binding state), a peptidyl–tRNA bound in the P subsites of both subunits (the P/P state), and the A site empty. An aminoacyl–tRNA (aa–tRNA) in ternary complex with EF-Tu and GTP binds to the A site accompanied by the release of the E-site tRNA (but see below). This yields a complex in which the incoming aa–tRNA is bound in the 30S subunit's A subsite via a codon–anticodon interaction (recall that the mRNA is bound to the 30S subunit) but with the EF-Tu

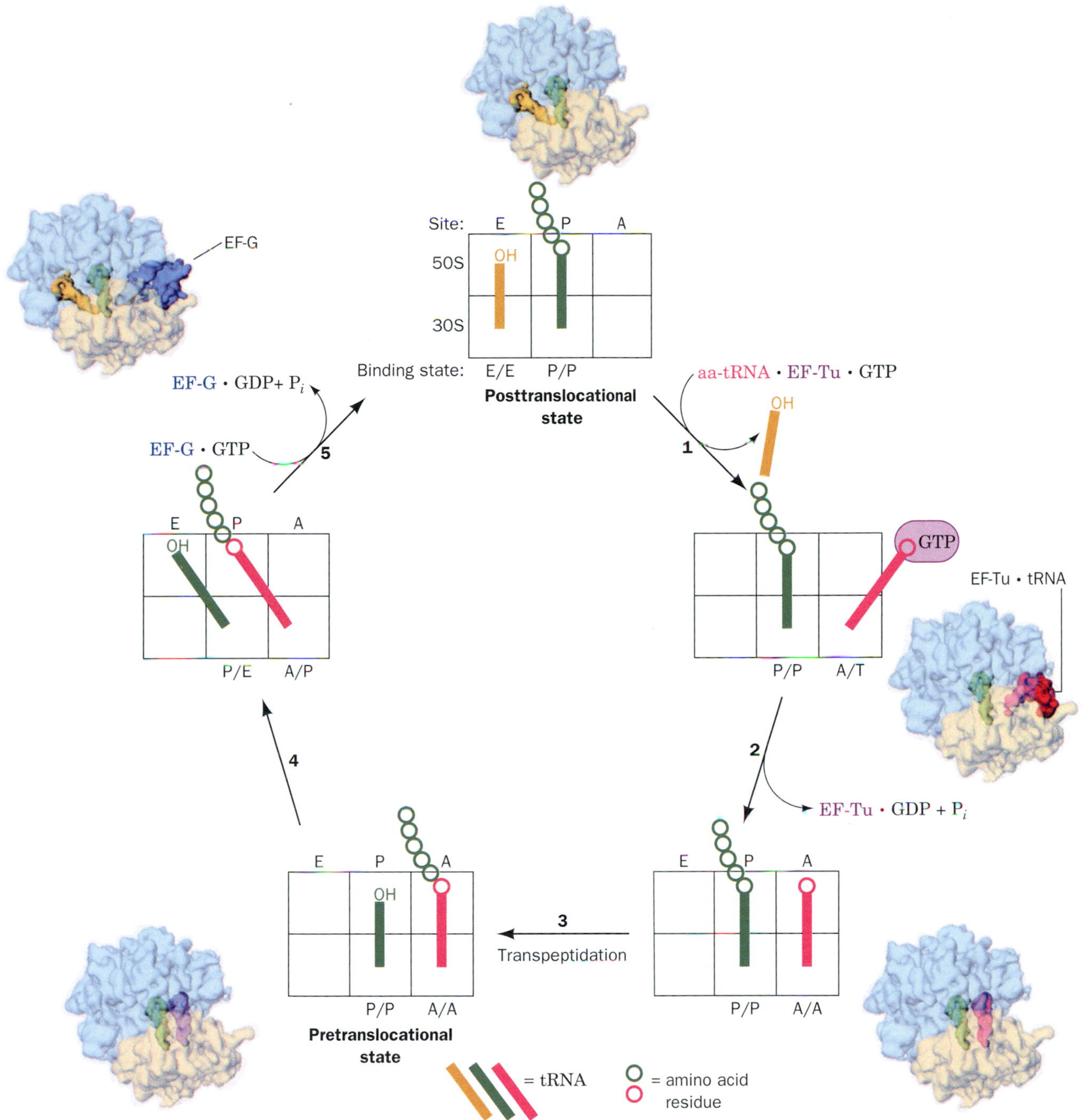

FIGURE 32-58 Ribosomal binding states in the elongation cycle. Note how this scheme elaborates the classical elongation cycle diagrammed in Fig. 32-48. The drawings are accompanied by 17-Å-resolution cryo-EM–based images of the *E. coli* 70S ribosome in the corresponding binding states in which the 30S subunit is transparent yellow, the 50S subunit is transparent blue, and the tRNAs and elongation factors are colored as in the drawing they accompany. [Cryo-EM images courtesy of Knud Nierhaus, Max-Planck-Institut für Molekulare Genetik, Berlin, Germany, and Joachim Frank, Wadsworth Center, State University of New York at Albany.]

preventing the entry of the tRNA's aminoacyl end into the 50S subunit's A subsite, an arrangement termed the A/T state (T for EF-*T*u).

2. EF-Tu hydrolyzes its bound GTP to GDP + P_i and is released from the ribosome, permitting the aa–tRNA to fully bind to the A site (the A/A state), a process called **accommodation.**

3. The transpeptidation reaction occurs, yielding the **pretranslocational state.**

4. The acceptor end of the new peptidyl–tRNA shifts from the A subsite of the 50S subunit to its P-subsite, while the tRNA's anticodon end remains associated with the A subsite of the 30S subunit (yielding the A/P hybrid binding state). The acceptor end of the newly deacylated tRNA simultaneously moves from the P subsite to the E subsite of the 50S subunit while its anticodon end remains associated with the P subsite of the 30S subunit (the P/E state).

5. The ribosomal binding of the EF-G · GTP complex and the subsequent GTP hydrolysis impel the anticodon ends of these tRNAs, together with their bound mRNA, to move relative to the small ribosomal subunit such that the peptidyl–tRNA assumes the P/P state and the deacylated tRNA assumes the E/E state (the posttranslocational state), thereby completing the elongation cycle

The binding of tRNA to the A and E sites, as Nierhaus has shown, exhibits negative allosteric cooperativity. In the pretranslocational state, the E site binds the newly deacylated tRNA with high affinity, whereas the empty A site has low affinity for aminoacyl–tRNA. However, in the posttranslocational state, the ribosome has undergone a conformational change that converts the A site to a high-affinity state and the E site to a low-affinity state, which consequently releases the deacylated tRNA when aa–tRNA · EF-Tu · GTP binds to the A site. Thus, the E site is not simply a passive holding site for spent tRNAs but performs an essential function in the translation process. The GTP hydrolysis by the elongation factors EF-Tu and EF-G as well as the transpeptidation reaction apparently function to reduce the activation barriers between these conformational states. The unidirectional A → P → E flow of tRNAs through the ribosome is thereby facilitated.

Certain aspects of the foregoing mechanism are not fully resolved. Thus, Noller has proposed that the E-site tRNA is not released from the ribosome until Step 4 of Fig. 32-58, when it is displaced by the preceding tRNA. This latter model is largely based on the X-ray structure of the 70S ribosome in complex with three tRNAs (Figs. 32-35 and 32-36). However, Nierhaus and Frank argue that this complex was crystallized in the presence of an unphysiologically high tRNA concentration and, since the tRNA sites are only partially occupied, the X-ray structure is likely to be a superposition of different tRNA binding states (e.g., E/E + P/P and P/P + A/A). Whatever the case, it is clear that the changes in binding states result in large scale tRNA movements, in some instances >50 Å. Moreover, cryo-EM studies indicate that on binding EF-G · GDP(CH_2)P (like GDPNP but with a CH_2 group rather than an NH group bridging its β and γ phosphates), the 30S subunit rotates with respect to the 50S subunit by 6° clockwise when viewed from the 30S subunit's solvent side, which results in a maximum displacement of ~19 Å at the periphery of the ribosome. This rotation is accompanied by many smaller conformational changes in both subunits, particularly in the regions about the entrance and exit to the mRNA channel. Clearly, we are only beginning to understand how the ribosome works at the molecular level.

k. The Eukaryotic Elongation Cycle Resembles That of Prokaryotes

The eukaryotic elongation cycle closely resembles that of prokaryotes. In eukaryotes, the functions of EF-Tu and EF-Ts are respectively assumed by the eukaryotic elongation factors **eEF1A** and **eEF1B,** with yeast eEF1B consisting of two subunits: **eEF1Bα,** which catalyzes nucleotide exchange, and **eEF1Bγ,** which has unknown function (in higher eukaryotes, eEF1B contains a third subunit, **eEF1Bβ,** that possesses a nucleotide exchange activity similar to that of eEF1Bα). Likewise, **eEF2** functions in a manner analogous to EF-G. However, the corresponding eukaryotic and prokaryotic elongation factors are not interchangeable.

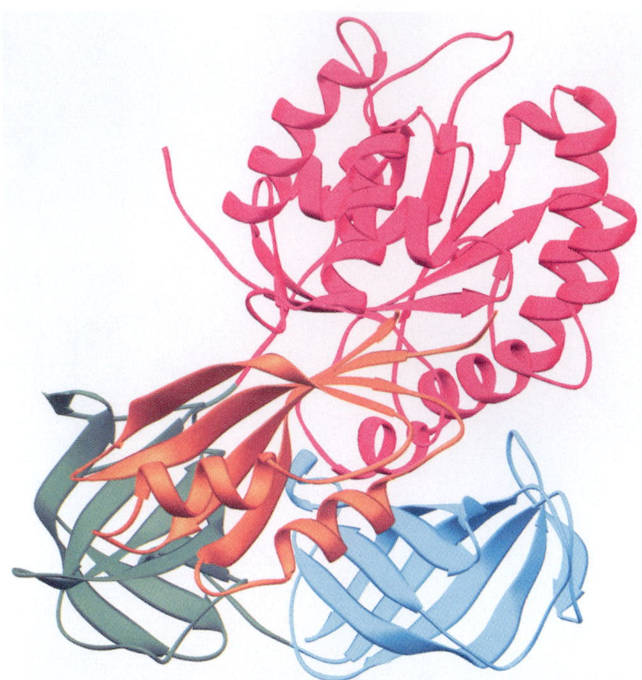

FIGURE 32-59 X-Ray structure of yeast eEF1A · eEF1Bα Domains 1, 2, and 3 of eEF1A are magenta, green, and cyan, respectively, and eEF1Bα is orange. The complex is oriented so as to emphasize the structural resemblance between eEF1A and the similarly colored EF-Tu in its complex with EF-Ts (Fig. 32-51). Note the lack of resemblance between eEF1Bα and EF-Ts. [Based on an X-ray structure by Morten Kjeldgaard and Jens Nyborg, University of Aarhus, Århus, Denmark. PDBid 1F60.]

The X-ray structure of yeast eEF1A · eEF1Bα (Fig. 32-59), determined by Kjeldgaard and Nyborg, reveals that eEF1A structurally resembles the homologous EF-Tu (Fig. 32-51), whereas eEF1Bα exhibits no resemblance to EF-Ts, either in sequence or in structure. Nevertheless, eEF1Bα functionally interacts with eEF1A much as EF-Ts interacts with EF-Tu: Both GEFs associate with their corresponding G-protein so as to disrupt the Mg^{2+} binding site associated with its bound guanine nucleotide.

E. *Chain Termination*

Polypeptide synthesis under the direction of synthetic mRNAs such as poly(U) terminates with a peptidyl–tRNA in association with the ribosome. However, *the translation*

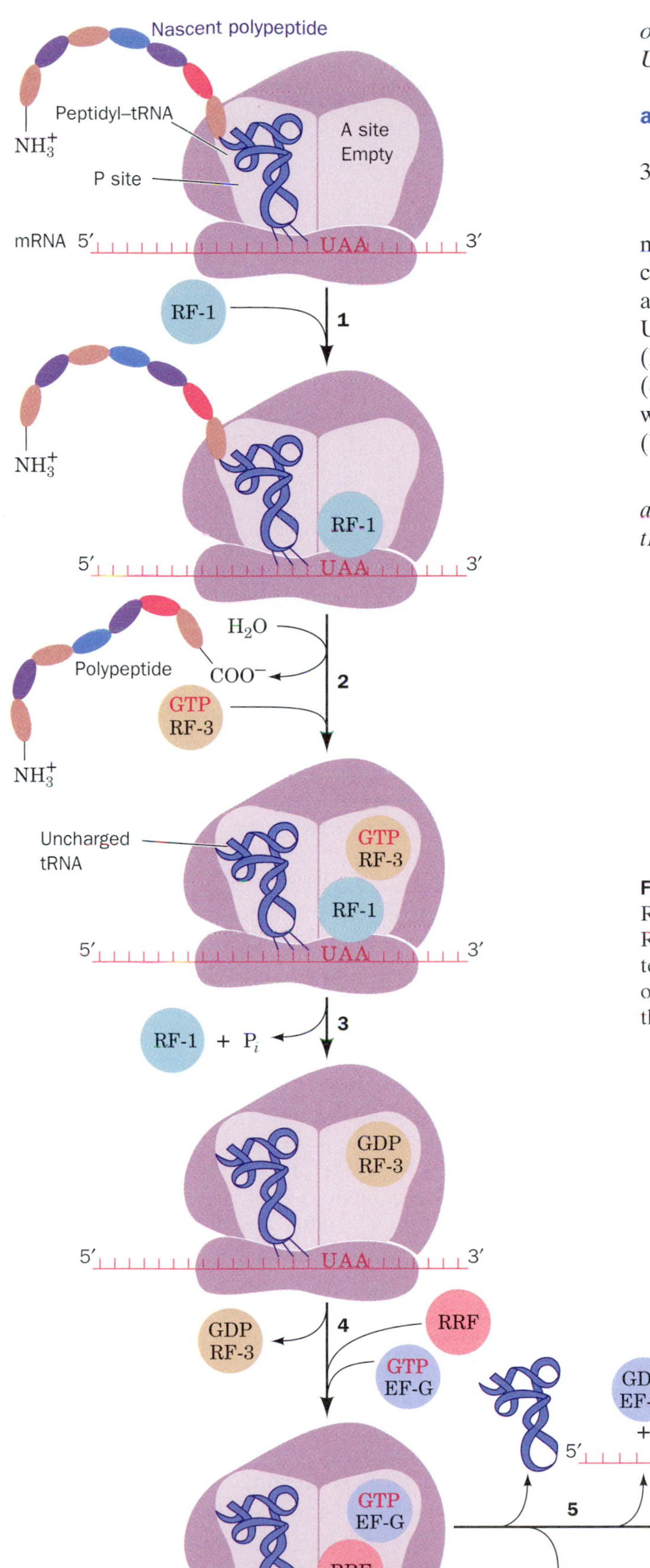

FIGURE 32-60 Termination pathway in *E. coli* ribosomes. RF-1 recognizes the Stop codons UAA and UAG, whereas RF-2 (not shown) recognizes UAA and UGA. Eukaryotic termination follows an analogous pathway but requires only a single class I release factor, eRF1, that recognizes all three Stop codons.

of natural mRNAs, which contain the Stop codons UAA, UGA, or UAG, results in the release of free polypeptides.

a. Prokaryotic Termination

In *E. coli,* chain termination has several stages (Fig. 32-60):

1. The termination codons, the only codons that normally have no corresponding tRNAs, are recognized by class I **release factors** (Table 32-9): **RF-1** recognizes UAA and UAG, whereas the 39% identical **RF-2** recognizes UAA and UGA. Swapping a conserved Pro-Ala-Thr (PAT) tripeptide in RF-1 with a conserved Ser-Pro-Phe (SPF) in RF-2 interchanges their Stop codon specificities, which suggests that these tripeptides mimic anticodons (but see below).

2. *On binding to their corresponding Stop codon, RF-1 and RF-2 induce the transfer of the peptidyl group from tRNA to water rather than to an aminoacyl–tRNA, thereby*

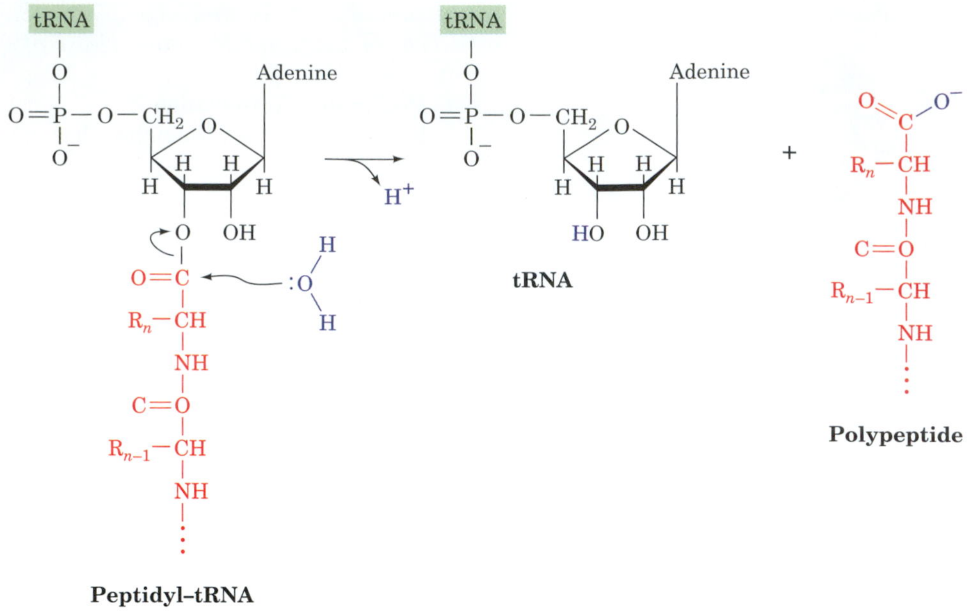

FIGURE 32-61 Ribosome-catalyzed hydrolysis of peptidyl–tRNA to form a polypeptide and free tRNA.

releasing the completed polypeptide (Fig 32-61). The class I release factors act at the ribosomal A site as is indicated by the observations that they compete with suppressor tRNAs for termination codons and that they cannot bind to the ribosome simultaneously with EF-G.

3. Once the newly synthesized polypeptide has been released from the ribosome, the class II release factor, **RF-3,** a G-protein, binds to the ribosome in complex with GTP and, on hydrolyzing its bound GTP to GDP + P_i, induces the ribosome to release its bound class I release factor. RF-3 is not required for cell viability although it is necessary for maximum growth rate.

4. Ribosomal recycling factor (RRF) binds in the ribosomal A site followed by EF-G · GTP. RRF, which was characterized largely by Akira Kaji, is essential for cell viability.

5. EF-G hydrolyzes its bound GTP, which causes RRF to be translocated to the P site and the tRNAs previously in the P and E sites (the latter not shown in Fig. 32-60) to be released. Finally, the RRF and then the EF-G · GDP and mRNA are released, yielding an inactive 70S ribosome ready for reinitiation (Fig. 32-45).

b. Eukaryotic Termination

Chain termination in eukaryotes resembles that in prokaryotes, but has only one class I release factor, **eRF1,** that recognizes all three Stop codons. It is unrelated in sequence to RF-1 and RF-2. However, the eukaryotic class II release factor, **eRF3,** resembles RF-3 in both sequence and function. Nevertheless, eRF3 is essential for eukaryotic cell viability.

c. RF-2, eRF1, and RRF Are Structural but Not Functional Mimics of tRNA

The class I release factors' functional mimicry of tRNAs suggests that they structurally mimic tRNAs. Indeed the X-ray structure of human eRF1 (Fig. 32-62*a*), determined by David Barford, and *E. coli* RF-2 (Fig. 32-62*b*), determined by Richard Buckingham, Nyborg, and Kjeldgaard, indicate that these two unrelated proteins both structurally resemble tRNA. However, the SPF tripeptide of RF-2 that was postulated to be an anticodon mimic (see above) is located in a different domain of RF-2 than the putative anticodon loop mimic (Fig. 32-62*b*). Moreover, cryo-EM studies of the *E. coli* ribosome in complex with RF2, independently carried out by Frank and Marin van Heel, indicate that RF2 undergoes a large conformational change on binding to the ribosome so that it no longer mimics tRNA. The X-ray structure of *Thermatoga maritima* RRF (Fig. 32-62*c*), determined by Kaji and Anders Liljas, reveals that it too resembles tRNA in appearance. Nevertheless, footprinting studies of RRF bound to the 70S ribosome indicate that RRF binds to the A site in an orientation that differs markedly from any previously observed for a tRNA. Evidently, things are not necessarily what they appear to be.

d. GTP Hydrolysis Speeds Up Ribosomal Processes

What is the role of the GTP hydrolysis reactions mediated by the various GTP-binding factors (IF-2, EF-Tu, EF-G, and RF-3 in *E. coli*)? Translation occurs in the absence of GTP, albeit slowly, so that the free energy of the transpeptidation reaction is sufficient to drive the entire translational process. Moreover, none of the GTP hydrolysis reactions yields a "high-energy" covalent intermediate as does, say, ATP hydrolysis in numerous biosynthetic reactions. It therefore appears that the ribosomal binding of

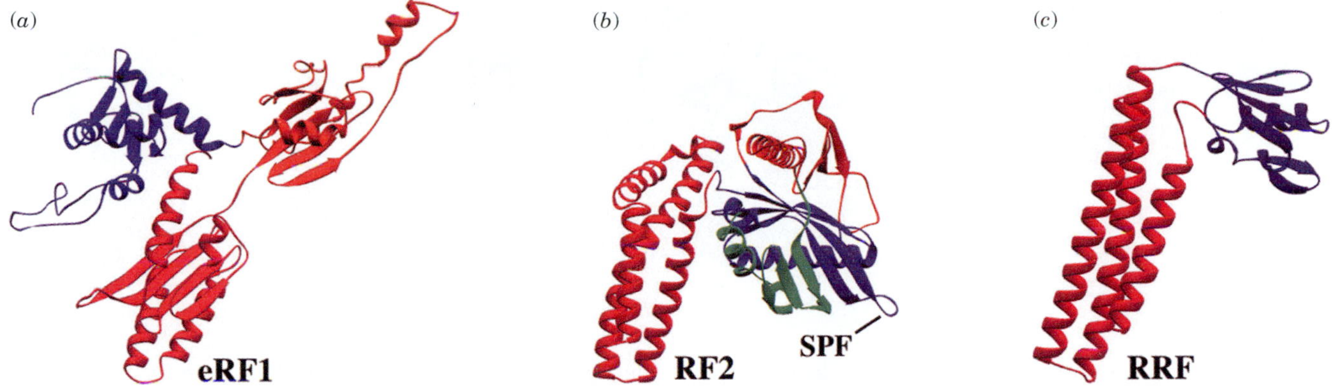

FIGURE 32-62 X-Ray structures of putative tRNA mimics that participate in translational termination. (*a*) Human eRF1, (*b*) *E. coli* RF-2, and (*c*) *T. maritima* RRF. The various domains in these proteins are differently colored with the domains that appear to mimic the tRNA anticodon stem drawn in red. The position of the RF-2 SPF tripeptide is indicated. Compare these structures to Figs. 32-49 and 32-57. [Courtesy of V. Ramakrishnan, MRC Laboratory of Molecular Biology, Cambridge, U.K. Part *a* is based on an X-ray structure by David Barford, Institute of Cancer Research, London, U.K.; Part *b* is based on an X-ray structure by Richard Buckingham, CNRS, Paris, France, and Jens Nyborg and Morten Kjeldgaard, University of Aarhus, Århus, Denmark; and Part *c* is based on an X-ray structure by Akira Kaji, University of Pennsylvania, and Anders Liljas, Lund University, Lund, Sweden. PDBids (*a*) 1DT9, (*b*) 1GQE, and (*c*) 1DD5.]

a GTP-binding factor (G-protein) in complex with GTP allosterically facilitates a change in ribosomal conformation so as to permit a particular process such as tRNA binding to occur (e.g., Fig. 32-50). This conformation change also catalyzes GTP hydrolysis, which, in turn, permits the ribosome to relax to its initial conformation with the concomitant release of products, including GDP + P_i (with the aid of EF-Ts for EF-Tu). *The high rate and irreversibility of the GTP hydrolysis reaction therefore ensures that the various complex ribosomal processes to which it is coupled, initiation, elongation, and termination, will themselves be fast and irreversible.* In essence, G-protein · GTP complexes act as Maxwell's demons to trap the ribosome in functionally productive conformations. Hence, the ribosome utilizes the free energy of GTP hydrolysis to gain a more ordered (lower entropy) state rather than a higher energy state as often occurs in ATP-dependent processes. In the same way, GTP hydrolysis facilitates translational accuracy (see below).

F. *Translational Accuracy*

The genetic code is normally expressed with remarkable fidelity. We have already seen that transcription and tRNA aminoacylation both proceed with high accuracy (Sections 31-2C and 32-2C). The accuracy of ribosomal mRNA decoding was estimated from the rate of misincorporation of [^{35}S]Cys into highly purified **flagellin,** an *E. coli* protein (Section 35-3G) that normally lacks Cys. These measurements indicated that the mistranslation rate is ~10^{-4} errors per codon. This rate is greatly increased in the presence of **streptomycin,** an antibiotic that increases the rate of ribosomal misreading (Section 32-3G). From the types of reading errors that streptomycin is known to induce, it was concluded that the mistranslation arose almost entirely from the confusion of the Arg codons CGU and CGC for the Cys codons UGU and UGC. The above error rate is therefore largely caused by mistakes in ribosomal decoding.

An aminoacyl–tRNA is selected by the ribosome only according to its anticodon. Yet the binding energy loss arising from a single base mismatch in a codon–anticodon interaction is estimated to be ~12 kJ · mol^{-1}, which, according to Eq. [32.1], cannot account for a ribosomal decoding accuracy of less than ~10^{-2} errors per codon. Moreover, the base pairing interaction between the UUU codon for Phe and the GAA anticodon of tRNA^{Phe} would be naively expected to be less stable than the incorrect pairing between the UGC codon for Ser and the GCG anticodon of tRNA^{Arg}. This is because both interactions have one G · U base pair and the former correct interaction's remaining two A · U base pairs are weaker than the latter incorrect interaction's remaining two G · C base pairs. Evidently, the ribosome has some sort of proofreading mechanism that increases its overall decoding accuracy.

a. The Ribosome Monitors the Formation of a Correct Codon–Anticodon Complex

As we have seen (Fig. 32-58), the aminoacyl–tRNA · EF-Tu · GTP ternary complex initially binds to the ribosome with the tRNA in the A/T binding state. The tRNA only assumes the fully bound A/A state (accommodation) after the GTP has been hydrolyzed and the EF-G · GDP complex has been released from the ribosome. These two states presumably permit the ribosome to double-check (proofread) the codon–anticodon complex that the mRNA makes with the incoming tRNA.

The X-ray structure of the *T. thermophilus* 30S subunit in complex with a U_6 hexanucleotide mRNA and a 17-nt RNA consisting of the tRNA^{Phe} anticodon stem–loop (Fig. 32-11, although its nucleotides are unmodified), deter-

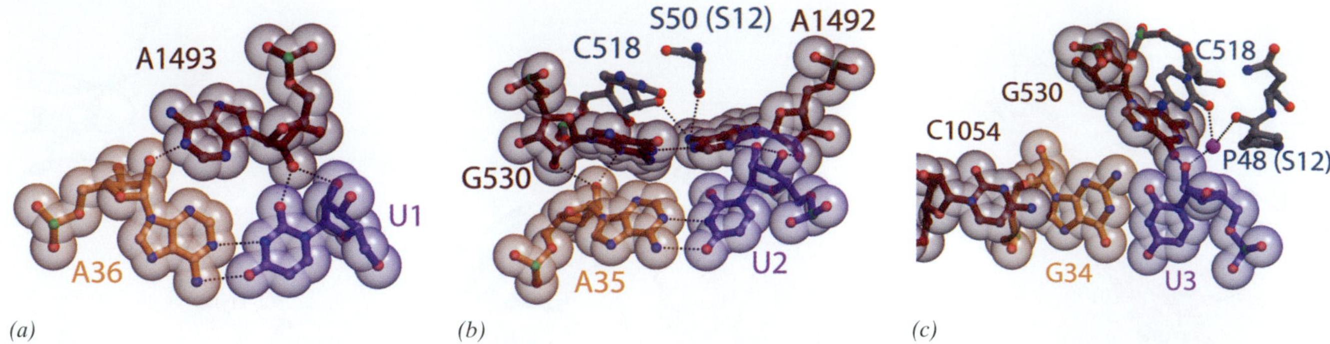

(a) (b) (c)

FIGURE 32-63 Codon–anticodon interactions in the ribosome. The (*a*) first, (*b*) second, and (*c*) third codon–anticodon base pairs as seen in the X-ray structure of the *T. thermophilus* 30S subunit in complex with U_6 (a model mRNA) and the 17-nt anticodon stem–loop of $tRNA^{Phe}$ (whose anticodon is GAA). The structures are drawn in ball-and-stick form embedded in their semitransparent van der Waals surfaces. Codons are purple, anticodons are yellow, and rRNA is brown or gray with non-C atoms colored according to type (N blue, O red, and P green). Protein segments are gray and Mg^{2+} ions are represented by magenta spheres. [Courtesy of V. Ramakrishnan, MRC Laboratory of Molecular Biology, Cambridge, U.K. PDBid 1IBM.]

mined by Ramakrishnan, reveals how an mRNA-specified tRNA initially binds to the ribosome. The codon–anticodon association is stabilized by its interactions with three universally conserved ribosomal bases, A1492, A1493, and G530 (Fig. 32-63):

1. The first codon–anticodon base pair, that between mRNA U1 and tRNA A36, is stabilized by the binding of the rRNA A1493 base in the base pair's minor groove (Fig. 32-63*a*).

2. The second codon–anticodon base pair, that between U2 and A35, is bolstered by A1492 and G530, which both bind in this base pair's minor groove (Fig. 32-63*b*).

3. The third codon–anticodon base pair (the wobble pair; Section 32-2D), that between U3 and G34, is reinforced through minor groove binding by G530 (Fig. 32-63*c*). This latter interaction appears to be less stringent than those in the first and second codon–anticodon positions, which is consistent with the need for the third codon–anticodon pairing to tolerate non-Watson–Crick base pairs (Section 32-2D).

Comparison of this structure with that of the 30S subunit alone reveals that the foregoing rRNA nucleotides undergo conformational changes on the formation of a codon–anticodon complex (Fig. 32-64). In the absence of tRNA, the bases of A1492 and A1493 stack in the interior of an RNA loop but flip out of this loop to form the codon–anticodon complex, whereas the G530 base switches from the syn to the anti conformation (Section 29-2A). These interactions enable the ribosome to monitor whether an incoming tRNA is cognate to the codon in the A site; a non-Watson–Crick base pair could not bind these ribosomal bases in the same way. Indeed, any mutation of A1492 or A1493 is lethal because pyrimidines in these positions could not reach far enough to interact with the codon–anticodon complex or G530 and because a G in either position would be unable to form the required hydrogen bonds and its N2 would be subjected to steric collisions. An incorrect codon–anticodon provides insufficient free energy to bind the tRNA to the ribosome and it therefore dissociates from it, still in its ternary complex with EF-Tu and GTP.

b. GTP Hydrolysis by EF-Tu Is a Thermodynamic Prerequisite to Ribosomal Proofreading

A proofreading step must be entirely independent of the initial selection step. Only then can the overall probability of error be equal to the product of the probabilities of error of the individual selection steps. We have seen that DNA polymerases and aminoacyl–tRNA synthetases maintain the independence of their two selection steps by carrying them out at separate active sites (Sections 30-2A and 32-2C). Yet the ribosome only recognizes the incoming aminoacyl–tRNA according to its anticodon's complementarity to the codon in the A site. Consequently, the ribosome must somehow examine this codon–anticodon interaction in two separate ways.

The formation of a correct codon–anticodon complex triggers EF-Tu to hydrolyze its bound GTP, although how this occurs is unclear (note that EF-Tu's GTPase domain is bound in the 50S subunit which, together with the observation that GTP hydrolysis requires an intact tRNA, suggests that the hydrolysis signal is at least in part transmitted through the tRNA). The resulting conformational change in EF-Tu (Fig. 32-50) presumably swings its bound tRNA into the A/A state, a process in which the codon–anticodon interaction is subjected to a second screening that only permits a cognate aminoacyl–tRNA to enter the peptidyl transferase center. The irreversible GTPase reaction must precede this proofreading step because otherwise the dissociation of a noncognate tRNA (the release of its anticodon from the codon) would simply be the reverse of the initial binding step, that is, it would be part of the initial selection step rather than proofreading. *GTP hydrolysis therefore provides the second context necessary for proofreading; it is the entropic price the system must pay for*

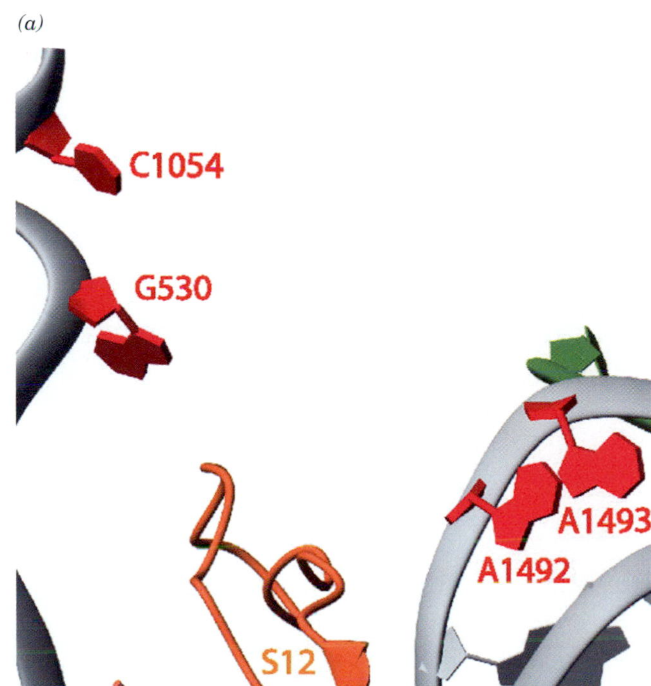

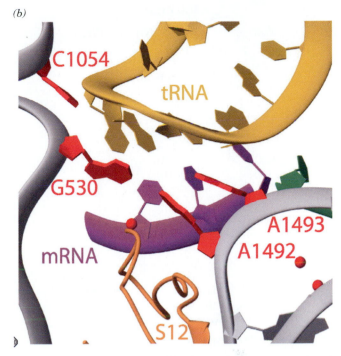

FIGURE 32-64 Ribosomal decoding site. The X-ray structures of *T. thermophilus* 30S subunit (*a*) alone and (*b*) in its complex with U_6 and the 17-nt anticodon stem–loop of $tRNA^{Phe}$. The RNAs are drawn as ribbons with their nucleotides in paddle form with tRNA gold, A-site mRNA purple, P-site mRNA green, rRNA gray, and nucleotides that undergo conformational changes red. Protein S12 is tan and Mg^{2+} ions are represented by red spheres. Compare Part *b* with Figure 32-63. [Courtesy of V. Ramakrishnan, MRC Laboratory of Molecular Biology, Cambridge, U.K. PDBids (*a*) 1FJF and (*b*) 1IBM.]

accurate tRNA selection. Further high resolution structural studies will be necessary to elucidate how this occurs.

G. *Protein Synthesis Inhibitors: Antibiotics*

Antibiotics are bacterially or fungally produced substances that inhibit the growth of other organisms. Antibiotics are known to inhibit a variety of essential biological processes, including DNA replication (e.g., ciprofloxacin, Section 29-3C), transcription (e.g., rifamycin B; Section 31-2B), and bacterial cell wall synthesis (e.g., penicillin; Section 11-3B). However, *the majority of known antibiotics, including a great variety of medically useful substances, block translation.* This situation is presumably a consequence of the translational machinery's enormous complexity, which makes it vulnerable to disruption in many ways. Antibiotics have also been useful in analyzing ribosomal mechanisms because, as we have seen for puromycin (Section 32-3D), the blockade of a specific function often permits its biochemical dissection into its component steps. Table 32-10

TABLE 32-10 Some Ribosomal Inhibitors

Inhibitor	Action
Chloramphenicol	Inhibits peptidyl transferase on the prokaryotic large subunit
Cycloheximide	Inhibits peptidyl transferase on the eukaryotic large subunit
Erythromycin	Inhibits translocation by the prokaryotic large subunit
Fusidic acid	Inhibits elongation in prokaryotes by binding to EF-G · GDP in a way that prevents its dissociation from the large subunit
Paromomycin	Increases the ribosomal error rate
Puromycin	An aminoacyl–tRNA analog that causes premature chain termination in prokaryotes and eukaryotes
Streptomycin	Causes mRNA misreading and inhibits chain initiation in prokaryotes
Tetracycline	Inhibits the binding of aminoacyl–tRNAs to the prokaryotic small subunit
Diphtheria toxin	Catalytically inactivates eEF2 by ADP-ribosylation
Ricin/abrin/α-sarcin	**Ricin** and **abrin** are poisonous plant glycosidases that catalytically inactivate the eukaryotic large subunit by hydrolytically depurinating a specific highly conserved A residue of the 28S RNA, which is located on the so-called **sarcin–ricin loop** that forms a critical part of the ribosomal factor-binding center; **α-sarcin** is a fungal protein that cleaves a specific phosphodiester bond in the sarcin–ricin loop

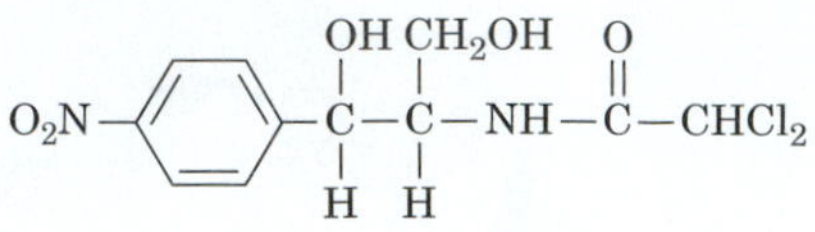

Chloramphenicol

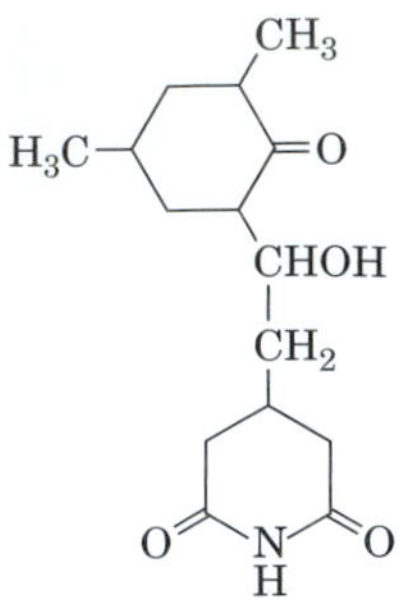

Cycloheximide

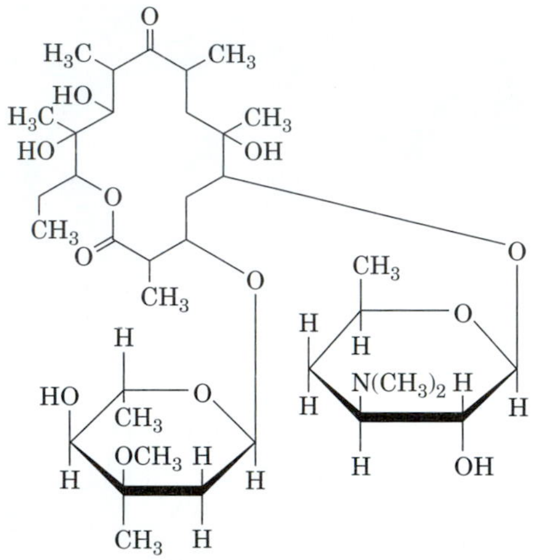

Erythromycin

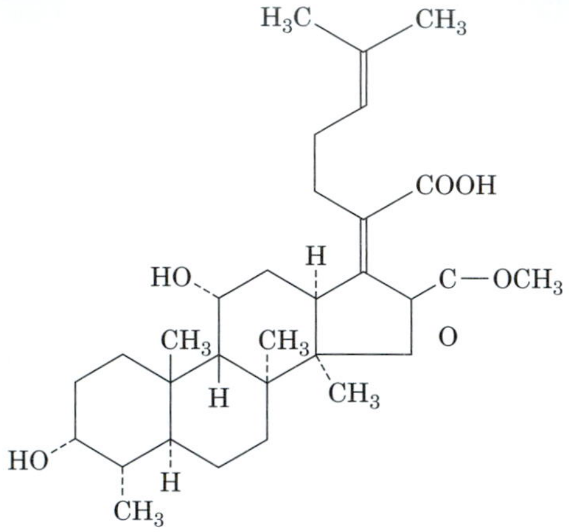

Fusidic acid

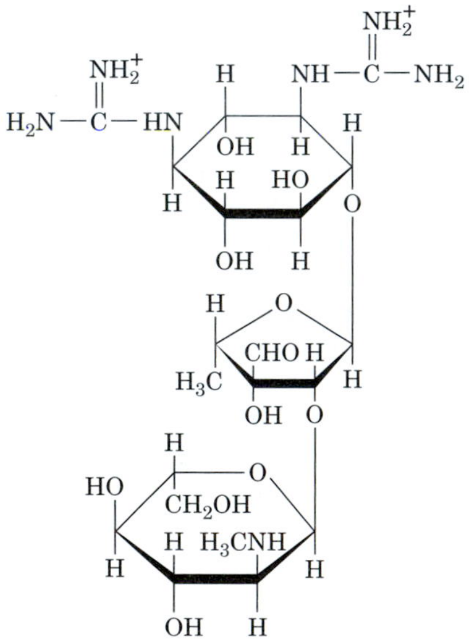

Streptomycin

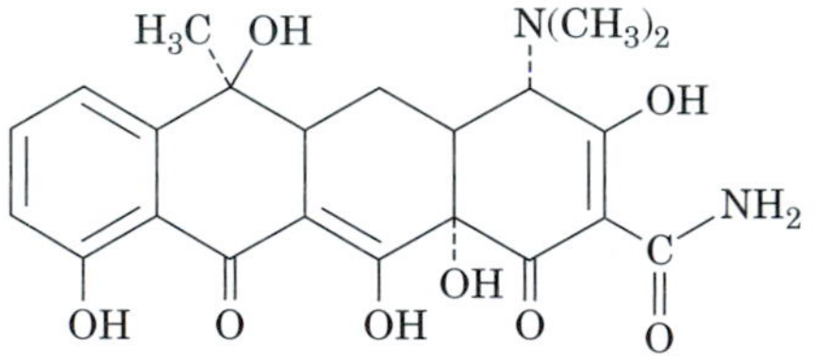

Tetracycline

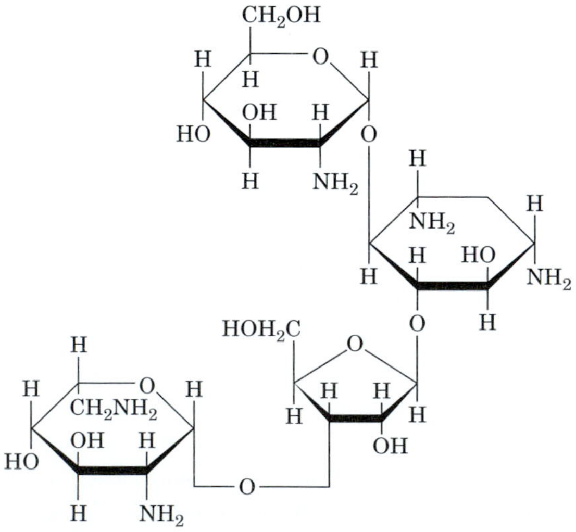

Paromomycin

FIGURE 32-65 **Selection of antibiotics that act as translational inhibitors.**

and Fig. 32-65 present several medically significant and/or biochemically useful translational inhibitors. We study the mechanisms of a few of the best characterized of them below.

a. Streptomycin

Streptomycin, which was discovered in 1944 by Selman Waksman, is a medically important member of a family of antibiotics known as **aminoglycosides** that inhibit prokaryotic ribosomes in a variety of ways. At low concentrations, streptomycin induces the ribosome to characteristically misread mRNA: One pyrimidine may be mistaken for the other in the first and second codon positions and either pyrimidine may be mistaken for adenine in the first position. This inhibits the growth of susceptible cells but does not kill them. At higher concentrations, however, streptomycin prevents proper chain initiation and thereby causes cell death.

Certain streptomycin-resistant mutants (str^R) have ribosomes with an altered protein S12 compared with streptomycin-sensitive bacteria (str^S). Intriguingly, a change in base C912 of 16S rRNA (which lies in its central loop; Fig. 32-27*a*) also confers streptomycin resistance. (Some mutant bacteria are not only resistant to streptomycin but dependent on it; they require it for growth.) In partial diploid bacteria that are heterozygous for streptomycin resistance (str^R/str^S), streptomycin sensitivity is dominant. This puzzling observation is explained by the finding that, in the presence of streptomycin, str^S ribosomes remain bound to initiation sites, thereby excluding str^R ribosomes from these sites. Moreover, the mRNAs in these blocked complexes are degraded after a few minutes, which allows the str^S ribosomes to bind to newly synthesized mRNAs as well.

b. Chloramphenicol

Chloramphenicol, the first of the "broad-spectrum" antibiotics, inhibits the peptidyl transferase activity on the large subunit of prokaryotic ribosomes. However, its clinical uses are limited to only severe infections because of its toxic side effects, which are caused, at least in part, by the chloramphenicol sensitivity of mitochondrial ribosomes. The 23S RNA is implicated in chloramphenicol binding by the observation that some of its mutants are chloramphenicol resistant. Indeed, X-ray studies indicate that chloramphenicol binds in the large subunit's polypeptide exit tunnel in the vicinity of the A site. This explains why chloramphenicol competes for binding with the 3′ end of aminoacyl–tRNAs as well as with puromycin (whose ribosomal binding site overlaps that of chloramphenicol) but not with peptidyl–tRNAs. These observations suggest that chloramphenicol inhibits peptidyl transfer by interfering with the interactions of ribosomes with A site–bound aminoacyl–tRNAs.

c. Paromomycin

Paromomycin, a clinically useful aminoglycoside antibiotic, increases the ribosomal error rate. The X-ray structure of the 30S subunit in complex with paromomycin

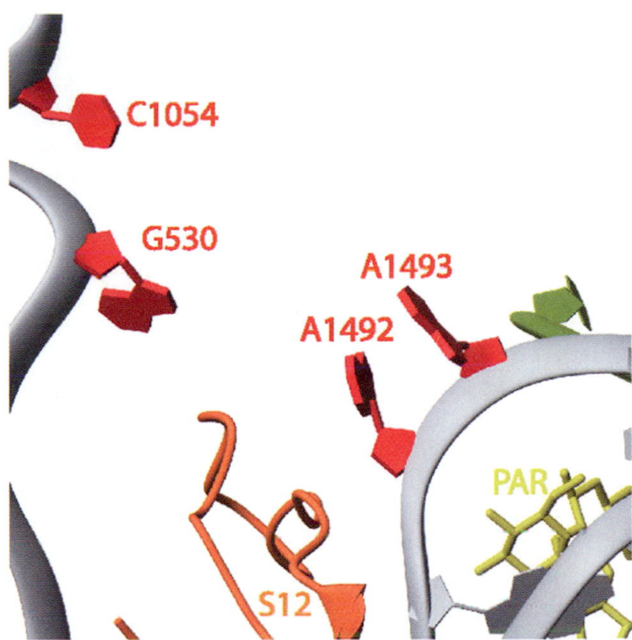

FIGURE 32-66 X-Ray structure of the 30S ribosome in complex with the antibiotic paromomycin. The view and coloring are the same as those in Fig. 32-64 with the paromomycin (PAR) drawn in stick form in yellow-green. [Courtesy of V. Ramakrishnan, MRC Laboratory of Molecular Biology, Cambridge, U.K. PDBid 1IBK.]

(Fig. 32-66) reveals that it binds to the interior of the RNA loop in which the bases of A1492 and A1493 are normally stacked (Fig. 32-64*a*). This causes these bases to flip out of the loop and assume a conformation resembling that in the codon–anticodon–30S subunit complex (Fig. 32-64*b*). Indeed, this codon–anticodon–30S subunit complex is not significantly disturbed by the binding of paromomycin. As we have seen in Section 32-3F, the 30S subunit employs A1492 and A1493 to ascertain whether the first two codon–anticodon base pairs are Watson–Crick base pairs, that is, whether the incoming tRNA is cognate to the codon in the A site. Noncognate tRNAs normally have insufficient codon–anticodon binding energy to flip A1492 and A1493 out of the loop and consequently are rejected by the ribosome. However, the binding of paromomycin to the 30S subunit pays the energetic price of these base flips. This facilitates the ribosomal acceptance (stabilizes the binding) of near-cognate aminoacyl–tRNAs and hence the erroneous incorporation of their amino acid residues into the polypeptide being synthesized.

d. Tetracycline

Tetracycline and its derivatives are broad-spectrum antibiotics that bind to the small subunit of prokaryotic ribosomes, where they inhibit aminoacyl–tRNA binding. An X-ray structure of tetracycline in complex with the 30S subunit reveals that tetracycline mainly binds in a crevice comprised of only the 3′ major domain of 16S RNA (Fig. 32-27*a*) and which is located in the neck of the 30S subunit just above its A site. This permits the initial screening of the aminoacyl–tRNA to proceed but physically blocks its accommodation into the peptidyl transferase (A/A) site after EF-Tu–catalyzed GTP hydrolysis has occurred, resulting in the release of the tRNA. Hence, in addition to preventing protein synthesis, tetracycline binding causes the unproductive hydrolysis of GTP which, since this occurs every time a cognate aminoacyl–tRNA binds to the ribosome, poses an enormous energetic drain on the cell. The nucleotides forming the tetracycline binding site are poorly conserved in eukaryotic ribosomes, thereby accounting for tetracycline's bacterial specificity.

Tetracycline also blocks the stringent response (Section 31-3H) by inhibiting (p)ppGpp synthesis. This indicates that deacylated tRNA must bind to the A site in order to activate stringent factor.

Tetracycline-resistant bacterial strains have become quite common, thereby precipitating a serious clinical problem. Resistance is often conferred by a decrease in bacterial cell membrane permeability to tetracycline rather than any alteration of ribosomal components.

e. Diphtheria Toxin

Diphtheria is a disease resulting from bacterial infection by *Corynebacterium diphtheriae* that harbor the bacteriophage **corynephage β.** Diphtheria was a leading cause of childhood death until the late 1920s when immunization became prevalent. Although the bacterial infection is usually confined to the upper respiratory tract, the bacteria secrete a phage-encoded protein, known as **diphtheria toxin (DT),** which is responsible for the disease's lethal effects. *Diphtheria toxin specifically inactivates the eukaryotic elongation factor eEF2, thereby inhibiting eukaryotic protein synthesis.*

The pathogenic effects of diphtheria are prevented, as was discovered in the 1880s, by immunization with **toxoid** (formaldehyde-inactivated toxin). Individuals who have contracted diphtheria are treated with antitoxin from horse serum, which binds to and thereby inactivates DT, as well as with antibiotics to combat the bacterial infection.

DT is a member of the family of bacterial toxins that includes cholera toxin (CT) and pertussis toxin (PT; Section 19-2C). It is a monomeric 535-residue protein that is readily cleaved past its Arg residues 190, 192, and 193 by trypsin and trypsinlike enzymes. This hydrolysis occurs around the time diphtheria toxin encounters its target cell, yielding two fragments, A and B, which, nevertheless, remain linked by a disulfide bond. The B fragment's C-terminal domain binds to a specific receptor on the plasma membrane of susceptible cells, thereby inducing DT's uptake into the endosome (Fig. 12-79) via receptor-mediated endocytosis (Section 12-5B; free fragment A is devoid of toxic activity). The endosome's low pH of 5 triggers a conformational change in the B fragment's N-terminal domain, which then inserts into the endosomal membrane so as to facilitate the entry of the A fragment into the cytoplasm. The disulfide bond linking the A and B subunits is then cleaved by the cytoplasm's reducing environment.

Within the cytosol, the A fragment catalyzes the **ADP-ribosylation** of eEF2 by NAD^+,

$$\underset{(active)}{\text{eEF2}} + \text{NAD}^+ \xrightarrow{\text{diphtheria toxin}} \underset{(inactive)}{\text{ADP-ribosyl-eEF2}} + \text{Nicotinamide} + H^+$$

thereby inactivating this elongation factor. Since the A fragment acts catalytically, *one molecule is sufficient to ADP-ribosylate all of a cell's eEF2s, which halts protein synthesis and kills the cell.* Only a few micrograms of diphtheria toxin are therefore sufficient to kill an unimmunized individual.

Diphtheria toxin specifically ADP-ribosylates a modified His residue on eEF2 known as **diphthamide:**

His residue

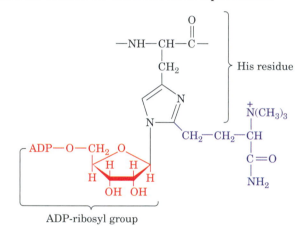

ADP-ribosyl group

ADP-Ribosylated diphthamide

Diphthamide occurs only in eEF2 (not even in its bacterial counterpart, EF-G), which accounts for the specificity of diphtheria toxin in exclusively modifying eEF2 (recall that CT ADP-ribosylates a specific Arg residue on $G_{s\alpha}$ and PT ADP-ribosylates a specific Cys residue on $G_{i\alpha}$; Section 19-2C). Since diphthamide occurs in all eukaryotic eEF2s, it probably is essential to eEF2 activity. Yet, certain mutant cultured animal cells, which have unimpaired capacity to synthesize proteins, lack the enzymes that post-translationally modify His to diphthamide (although mutating the diphthamide His to Asp, Lys, or Arg inactivates translation). Perhaps the diphthamide residue has a control function.

4 ■ CONTROL OF EUKARYOTIC TRANSLATION

The rates of ribosomal initiation on prokaryotic mRNAs vary by factors of up to 100. For example, the proteins specified by the *E. coli lac* operon, β-galactosidase, galactose permease, and thiogalactoside transacetylase, are synthesized in molar ratios of 10:5:2. This variation is probably a consequence of their different Shine–Dalgarno sequences. Alternatively, ribosomes may attach to *lac* mRNA only at its β-galactosidase gene and occasionally detach in response to a chain termination signal, thereby accounting for the decreasing translational rates along the operon. In any case, there is no evidence that prokaryotic translation rates are responsive to environmental changes. *Genetic expression in prokaryotes is therefore almost entirely transcriptionally controlled (Section 31-3).* Of course, since their mRNAs have lifetimes of only a few minutes, it would seem that prokaryotes have little need of translational controls.

It is clear, however, that eukaryotic cells can respond to at least some of their needs through translational control. This is feasible because the lifetimes of eukaryotic mRNAs are generally hours or days. In this section, we examine how translation is regulated via the phosphorylation/dephosphorylation of eIF2 and eIF4E. We then consider translational control by mRNA masking and cytoplasmic polyadenylation and end by discussing the uses of antisense oligonucleotides.

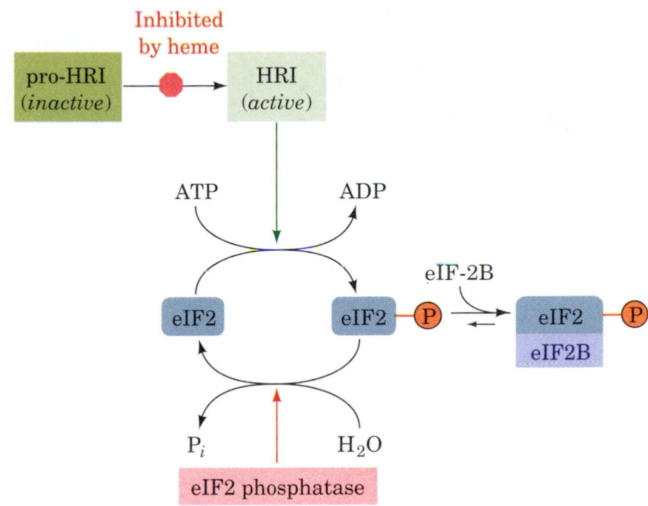

FIGURE 32-67 Model for heme-controlled protein synthesis in reticulocytes.

A. *Regulation of eIF2*

Four important pathways for the regulation of translation in eukaryotes involve the phosphorylation of the conserved Ser 51 on the α subunit of eIF2 (**eIF2α;** recall that eIF2 is an αβγ trimer that conducts Met–$tRNA_i^{Met}$ to the 40S ribosomal subunit, and the resulting complex scans the bound mRNA for the initiating AUG codon to form the 48S initiation complex; Section 32-3C). The so-called **eIF2α kinases** that do so share a conserved kinase domain but have unique regulatory domains

a. Heme Availability Controls Globin Translation

Reticulocytes synthesize protein, almost exclusively hemoglobin, at an exceedingly high rate and are therefore a favorite subject for the study of eukaryotic translation. Hemoglobin synthesis in fresh reticulocyte lysates proceeds normally for several minutes but then abruptly stops because of the inhibition of translational initiation and the consequent polysome disaggregation. This process is prevented by the addition of heme [a mitochondrial product (Section 26-4A) that this *in vitro* system cannot synthesize], thereby indicating that *globin synthesis is regulated by heme availability.* The inhibition of globin translational initiation is also reversed by the addition of the eukaryotic initiation factor eIF2 and by high levels of GTP.

In the absence of heme, reticulocyte lysates accumulate an eIF2α kinase named **heme-regulated inhibitor [HRI;** also called heme-controlled repressor **(HCR)]**. HRI is a homodimer whose 629-residue subunits each contain two heme-binding sites. When heme is plentiful, both of these sites are occupied and the protein, which is autophosphorylated at several Ser and Thr residues, is inactive. However, when heme is scarce, one of these sites loses its bound heme, thereby activating HRI to autophosphorylate itself at several additional sites and to phosphorylate Ser 51 of eIF2α.

Phosphorylated eIF2 can participate in the ribosomal initiation process in much the same way as unphosphorylated eIF2. This puzzling observation was clarified by the discovery that GDP does not dissociate from phosphorylated eIF2 at the completion of the initiation process as it normally does through a process facilitated by eIF2B acting as a GEF (Fig. 32-46). This is because phosphorylated eIF2 forms a much tighter complex with eIF2B than does unphosphorylated eIF2. This sequesters eIF2B (Fig. 32-67), which is present in lesser amounts than eIF2, thereby preventing the regeneration of the eIF2 · GTP required for translational initiation. The presence of heme reverses this process by inhibiting HRI, whereupon the phosphorylated eIF2 molecules are reactivated through the action of **eIF2 phosphatase,** which is unaffected by heme. The reticulocyte thereby coordinates its synthesis of globin and heme.

b. Interferons Protect against Viral Infection

***Interferons** are cytokines that are secreted by virus-infected vertebrate cells. On binding to surface receptors of other cells, interferons convert them to an antiviral state, which inhibits the replication of a wide variety of RNA and DNA viruses.* Indeed, the discovery of interferons in the 1950s arose from the observation that virus-infected individuals are resistant to infection by a second type of virus.

There are three families of interferons: **type α** or **leukocyte interferon** (165 residues; leukocytes are white blood cells), the related **type β** or **fibroblast interferon** (166 residues; fibroblasts are connective tissue cells), and **type γ** or **lymphocyte interferon** (146 residues; lymphocytes are immune system cells). *Interferon synthesis is induced by the double-stranded RNA (dsRNA) that is generated during infection by both DNA and RNA viruses, as well as by the synthetic dsRNA poly(I) · poly(C).* Interferons are effective antiviral agents in concentrations as low as $3 \times 10^{-14} M$, which makes them among the most potent biological substances known. Moreover, they have far wider specificities than antibodies raised against a particular virus. They have therefore elicited great medical interest, particularly since some cancers are virally induced (Section 19-3B). Indeed, they are in clinical use against certain tumors and viral infections. These treatments are made possible by the production of large quantities of these otherwise quite scarce proteins through recombinant DNA techniques (Section 5-5D).

Interferons prevent viral proliferation largely by inhibiting protein synthesis in infected cells (lymphocyte interferon also modulates the immune response). They do so in two independent ways (Fig. 32-68):

1. Interferons induce the production of an eIF2α kinase, **double-stranded RNA-activated protein kinase [PKR;** also known as **double-stranded RNA-activated inhibitor (DAI);** 551 residues], which on binding dsRNA, dimerizes and autophosphorylates itself. This activates PKR to phosphorylate eIF2α at its Ser 51, thereby inhibiting ribosomal initiation and hence the proliferation of viruses in virus-infected cells. The importance of PKR to cellular antiviral defense is indicated by the observation that many viruses express inhibitors of PKR.

2. Interferons also induce the synthesis of **(2′,5′)-oligoadenylate synthetase (2,5A synthetase).** In the presence of dsRNA, this enzyme catalyzes the synthesis from ATP of the unusual oligonucleotide **pppA(2′p5′A)$_n$** where n = 1 to 10. *This compound,* ***2,5-A,*** *activates a preexisting endonuclease,* ***RNase L,*** *to degrade mRNA, thereby inhibiting protein synthesis.* 2,5-A is itself rapidly degraded by an enzyme named **(2′,5′)-phosphodiesterase** so that it must be continually synthesized to maintain its effect.

The independence of the 2,5-A and PKR systems is demonstrated by the observation that the effect of 2,5-A on protein synthesis is reversed by added mRNA but not by added eIF2. [Recall that RNA interference (RNAi; Section 31-4A) constitutes an alternative dsRNA-based antiviral defense.]

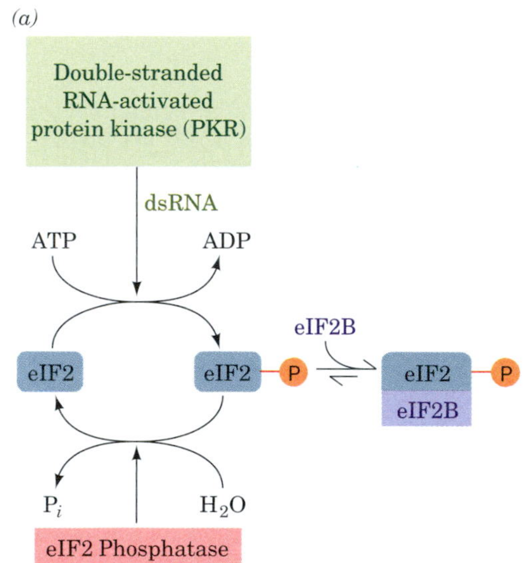

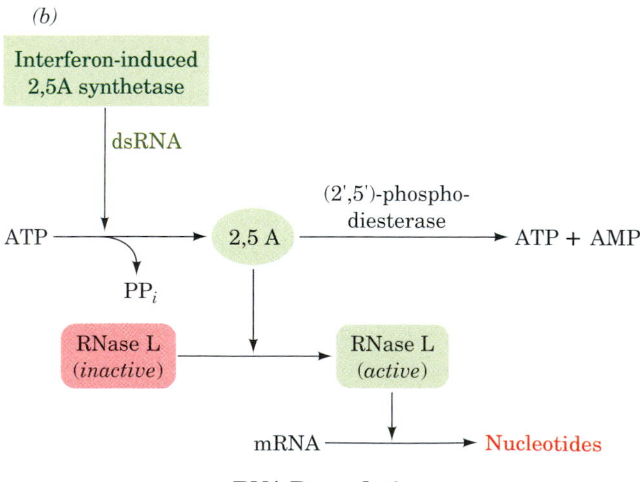

FIGURE 32-68 The action of interferon. In interferon-treated cells, the presence of dsRNA, which normally results from a viral infection, causes (*a*) the inhibition of translational initiation and (*b*) the degradation of mRNA, thereby blocking translation and preventing virus replication.

c. PERK Prevents the Buildup of Unfolded Proteins in the ER

PKR-like endoplasmic reticulum kinase (PERK), a 1087-residue transmembrane protein, resides in the endoplasmic reticulum (ER) membrane of all multicellular eukaryotes. It is repressed by its binding to the ER-resident chaperone BiP (Section 12-4B). When the ER contains an excessive amount of unfolded proteins (caused by various forms of stress such as high temperatures), BiP dissociates from PERK, thereby activating PERK to phosphorylate eIFα at its Ser 51 and hence inhibit translation. Thus PERK functions to protect the cell from the irreversible damage caused by the accumulation of unfolded proteins in the ER.

Wolcott–Rallison syndrome is a genetic disease characterized mainly by insulin-dependent (type I) diabetes that develops in early infancy (type I diabetes usually first appears in childhood; Section 27-3B). It is caused by mutations in the catalytic domain of PERK. This results in the death of pancreatic β cells, in which PERK is particularly abundant. Multiple systemic disorders subsequently occur including **osteoporosis** (reduction in the quantity of bone) and growth retardation.

d. GCN2 Regulates Amino Acid Biosynthesis

GCN2 (1590 residues), the sole eIF2α kinase in yeast, is a transcriptional activator of the gene encoding **GCN4,** a transcriptional activator of numerous yeast genes, many of which encode enzymes that participate in amino acid biosynthetic pathways. The C-terminal domain of GCN2, which resembles histidyl–tRNA synthetase (HisRS), preferentially binds uncharged tRNAs (whose presence is indicative of an insufficient supply of amino acids). The binding of an uncharged tRNA to this HisRS-like domain activates the adjacent eIF2α kinase domain and thereby inhibits translational initiation, although at only a modest level.

Despite this inhibition of yeast protein synthesis, activated GCN2 induces the expression of GCN4. This seemingly paradoxical property of GCN2, as Alan Hinnebusch explained, arises from the fact that GCN4 mRNA contains four short so-called **upstream open reading frames (uORFs),** uORF1 to uORF4, in its 5′ leader that precedes the sequence encoding GCN4. Under the normal nutrient conditions in which GCN2 is inactive, the ribosome binds to the mRNA near its 5′ cap and scans for the nearest AUG initiation codon (which is in uORF1), where it forms the 48S initiation complex (Fig. 32-46) and commences the translation of uORF1 (Section 32-3C). On terminating translation at uORF1's Stop codon, the presence of the surrounding A + U-rich sequences causes the ribosome to resume scanning for the next AUG codon, where it initiates the translation of uORF2. This process repeats until the ribosome terminates at the end of uORF4, where its Stop codon's surrounding G + C-rich sequences induce the ribosome to disengage from the mRNA. Hence GNC4 is only expressed at a low basal level. However, under the low nutrient conditions in which GCN2 phosphorylates eIF2α at its Ser 51, the resulting reduced level of the eIF2 · Met–$tRNA_i^{Met}$ · GTP ternary complex causes the 40S subunit to scan longer distances before it can form the 48S initiation complex. Consequently, ~50% of the ribosomes scan past uORF2, uORF3, and uORF4 and only initiate translation at the *GCN4* AUG codon, which is therefore translated at a high level (uORF2 and uORF3 can be mutationally eliminated without significantly affecting translational control).

Mammalian homologs of GCN2 are activated under conditions of amino acid starvation. This suggests that the foregoing process has been conserved throughout eukaryotic evolution.

B. *Regulation of eIF4E*

eIF4E (cap-binding protein) binds to the m^7G cap of eukaryotic mRNAs and thereby participates in translational initiation by helping to identify the initiating AUG codon (Section 32-3C). When mammalian cells are treated with hormones, cytokines, **mitogens** (substances that induce mitosis), and/or growth factors, Ser 209 of human eIF4E is phosphorylated via a Ras-activated MAP kinase cascade (Sections 19-3C and 19-3D), thereby increasing eIF4E's affinity for capped mRNA and hence stimulating translational initiation. Ser 209 occupies a surface position on eIF4E adjacent to the binding site for the β phosphate group of the m^7GDP and flanking the putative binding cleft for mRNA (Fig. 32-47*b*). The structure of eIF4E suggests that the phosphoryl group of phosphorylated Ser 209 forms a salt bridge with Lys 159, which occupies the other side of the putative mRNA-binding cleft, so as to form a clamp that would help stabilize the bound mRNA. The importance of regulating eIF4E activity is indicated by the observations that the overexpression of eIF4E causes the malignant transformation of rodent cell lines and that eIF4E expression is elevated in several human cancers.

The homologous ~120-residue proteins known as **4E-BP1, 4E-BP2,** and **4E-BP3** (BP for *b*inding *p*rotein; the first two are also known as **PHAS-I** and **PHAS-II**) inhibit cap-dependent translation. They do so by binding on the opposite side of eIF4E from its mRNA-binding site, presumably to a patch of seven highly conserved surface residues, and hence do not prevent eIF4E from binding the m^7G cap. Rather, they block eIF4E from binding to eIF4G and thereby interfere with the formation of the eIF4F complex that positions the 40S ribosomal subunit-bound Met–$tRNA_i^{Met}$ on the mRNA's initiating AUG codon (Section 32-3C). In fact, the 4E-BPs and eIF4G all possess the sequence motif YXXXXLϕ (where ϕ is an aliphatic residue, most often L but also M or F) through which they bind to eIF4E.

The treatment of responsive cells with insulin or any of several protein growth factors causes the 4E-BPs to dissociate from eIF4E. This is because the presence of these hormones induces the phosphorylation of the 4E-BPs at six Ser/Thr residues via the signal transduction pathway involving PI3K, PKB, and mTOR (Fig. 19-64). Evidently, the phosphorylation of eIF4E and the 4E-BPs have similar if not synergistic effects in the hormonal regulation of translation in eukaryotes.

C. *mRNA Masking and Cytoplasmic Polyadenylation*

It has been known since the nineteenth century that early embryonic development in animals such as sea urchins, insects, and frogs is governed almost entirely by information present in the oocyte (egg) before fertilization. Indeed, sea urchin embryos exposed to sufficient actinomycin D (Section 31-2C) to inhibit RNA synthesis without blocking DNA replication develop normally through their early stages without a change in their protein synthesis program. This is in part because an unfertilized egg contains large quantities of mRNA that is "masked" by associated proteins to form ribonucleoprotein particles, thereby preventing the mRNAs' association with the ribosomes that are also present. On fertilization, this mRNA is "unmasked" in a controlled fashion, quite possibly by the dephosphorylation of the associated proteins, and commences directing protein synthesis. Development of the embryo can therefore start immediately on fertilization rather than wait for the synthesis of paternally specified mRNAs. Thus, gene expression in the early stages of development is entirely translationally controlled; transcriptional control only becomes important when transcription is initiated.

a. Cytoplasmic Polyadenylation

Another mechanism of translational control in oocytes and early embryos involves the polyadenylation of mRNAs in the cytoplasm (polyadenylation usually occurs in the nucleus, following which the mRNA is exported to the cytoplasm; Section 31-4A). A substantial number of maternally supplied mRNAs in oocytes have relatively short poly(A) tails (20–40 nt versus a usual length of ~250 nt). The 3′ untranslated region of these mRNAs contains both the AAUAAA polyadenylation signal (which is required for polyadenylation in the nucleus; Section 31-4A) together with a so-called **cytoplasmic polyadenylation element (CPE),** which has the consensus sequence UUUUUAU. The CPE is recognized by **CPE-binding protein (CPEB),** which contains two RNA recognition motifs (RRMs) as well as a **zinc finger** (Fig. 9-27*a*) that contribute to its binding to the mRNA. Joel Richter discovered that CPEB recruits a 931-residue protein named **maskin** which, in turn, binds the eIF4E (cap-binding protein) that is bound to the mRNA's 5′ cap (Fig. 32-69*a*). Maskin contains the same YXXXXLϕ motif through which the 4E-BPs and eIF4G bind to eIF4E (Section 32-4B), thereby blocking the binding of eIF4G to eIF4E and hence preventing the formation of the 48S initiation complex (Fig. 32-46).

In the maturation of *Xenopus laevis* oocytes, a process that precedes fertilization and is stimulated by the steroid hormone progesterone (Section 19-1I), a variety of mRNAs, including those encoding several cyclins (which participate in cell cycle control; Section 30-4A) are translationally activated. Soon after exposure to progesterone, a protein kinase named **aurora** phosphorylates the mRNA-bound CPEB at its Ser 174. This increases CPEB's affinity for cleavage and polyadenylation specificity factor (CPSF; Section 31-4A), which then binds to the mRNA's AAUAAA sequence, where it recruits poly(A) polymerase (PAP) to lengthen the mRNA's poly(A) tail (Fig. 32-69*b*).

Translational initiation and cytoplasmic polyadenylation occur simultaneously, which suggests that these processes are linked. Indeed, Richter has shown that this occurs through the binding to poly(A) of poly(A)-binding protein (PABP; Section 31-4A), which as we saw (Section 32-3C), also binds to eIF4G to circularize the mRNA. The eIF4G in this complex displaces maskin from eIF4E, thereby permitting the formation of the 48S initiation complex and hence the mRNA's translation (Fig. 32-69*b*).

Mammalian cells also exhibit cell cycle–dependent cytoplasmic polyadenylation of mRNAs. This suggests that translational control by polyadenylation is a general feature in animal cells.

D. *Antisense Oligonucleotides*

Since ribosomes cannot translate double-stranded RNA or DNA–RNA hybrid helices, the translation of a given mRNA can be inhibited by a segment of its complementary RNA or DNA, that is, an **antisense RNA** or an **antisense oligodeoxynucleotide,** which are collectively known as **antisense oligonucleotides.** Moreover, endogenous RNase H's (enzymes that cleave the RNA strand of an RNA–DNA duplex; Section 31-4C) cleave an mRNA–oligodeoxynucleotide duplex on its mRNA strand, leaving the antisense oligodeoxynucleotide intact for binding to another mRNA. Alternatively, an antisense RNA could be incorporated into a ribozyme that would destroy the target mRNA (Section 31-4A).

Since the human genome consists of ~3.2 billion bp, an ~15-nt oligonucleotide (which is easily synthesized; Section 7-6A) should ideally be able to target any segment of the human genome. This exquisite specificity provides the delivery of an antisense oligonucleotide to, or its expression in a selected tissue or organism with enormous biomedical and biotechnological potential. However, care must be taken that an antisense oligonucleotide does not also eliminate nontarget mRNAs.

Methods for the delivery of a therapeutically useful antisense oligonucleotide to a target tissue are as yet in their infancy. This is in large part because oligonucleotides are readily degraded by the many nucleases present in an organism and because they do not readily pass through cell membranes. Moreover, a target mRNA is likely to be associated with cellular proteins and hence not available for binding to other molecules. The nuclease resistance of oligonucleotides can be increased by derivitizing them, for example, by replacing a nonbridging oxygen at each phosphate group with a methyl group or an S atom so as to

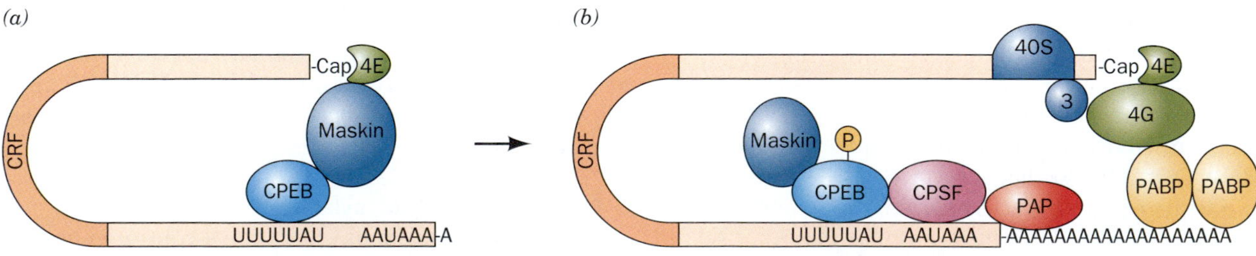

FIGURE 32-69 CPEB-mediated translational control. (*a*) In immature *Xenopus* oocytes, an mRNA containing the CPE (UUUUUAU) is bound by CPEB, which binds maskin, which then binds eIF4E so as to prevent it from binding eIF4G, thereby maintaining the mRNA in a translationally dormant (masked) state. (*b*) In the maturation process, CPEB is phosphorylated by an aurora protein kinase. The phosphorylated CPEB binds CPSF, which recruits PAP to extend the mRNA's heretofore short poly(A) tail. PABP binds to the newly lengthened poly(A) tail and simultaneously binds to eIF4G so as to displace maskin. This permits the 48S initiation complex to assemble and hence the translation of the mRNA to proceed. [Based on a drawing by Mendez, R. and Richer, J.D., *Nature Rev. Mol. Cell Biol.* **2,** 521 (2001).]

yield **methylphosphonate** or **phosphorothioate oligonucleotides,** although this reduces their antisense activity. The expression of antisense oligonucleotides in the specified tissues would, of course, circumvent the delivery problem but has all the difficulties associated with gene therapy (Section 5-5H).

Despite the foregoing, antisense technology is beginning to show success. **Fomivirsen** (trade name **Vitravene**), a 21-nt phosphorothioate oligonucleotide that is complementary to an mRNA expressed by **cytomegalovirus (CMV),** is effective in the treatment of retinitis (inflammation of the retina) caused by CMV infection in individuals with AIDS (CMV is an opportunistic pathogen that rarely infects individuals with normal immune systems). It was approved for human use in 1998 by the FDA, the first antisense drug so approved. In addition, a number of antisense oligonucleotides that are mainly targeted against genes that are overexpressed in specific cancers and autoimmune diseases as well as other viral infections are in Phase II or Phase III clinical trials (Section 15-4B).

Antisense technology has also had some success in the arena of biotechnology. For example, in tomatoes and other fruits, the enzyme **polygalacturonase (PG),** which is expressed during ripening, depolymerizes the pectin (mainly polygalacturonic acid) in the cell wall. This results in a softening of tomatoes to the point that vine-ripened (and hence better tasting) tomatoes are unable to withstand the rigors of shipping and hence must be picked before they are ripe. The introduction into a tomato, via genetic engineering techniques, of a gene expressing antisense PG RNA yielded the so-called Flavr Savr tomato that had substantially reduced PG expression and hence remained firm after vine ripening.

5 ■ POSTTRANSLATIONAL MODIFICATION

To become mature proteins, polypeptides must fold to their native conformations, their disulfide bonds, if any, must form, and, in the case of multisubunit proteins, the subunits must properly combine. Moreover, as we have seen throughout this text, many proteins are modified in enzymatic reactions that proteolytically cleave certain peptide bonds and/or derivatize specific residues. In this section we shall review some of these **posttranslational modifications.**

A. *Proteolytic Cleavage*

Proteolytic cleavage is the most common type of posttranslational modification. Probably all mature proteins have been so modified, if by nothing else than the proteolytic removal of their leading Met (or fMet) residue shortly after it emerges from the ribosome. Many proteins, which are involved in a wide variety of biological processes, are synthesized as inactive precursors that are activated under proper conditions by limited proteolysis. Some examples of this phenomenon that we have encountered are the conversion of trypsinogen and chymotrypsinogen to their active forms by tryptic cleavages of specific peptide bonds (Section 15-3E), and the formation of active insulin from the 84-residue proinsulin by the excision of its internal 33-residue C chain (Section 9-1A). Inactive proteins that are activated by removal of polypeptides are called **proproteins,** whereas the excised polypeptides are termed **propeptides.**

a. Propeptides Direct Collagen Assembly

Collagen biosynthesis is illustrative of many facets of posttranslational modification. Recall that collagen, a major extracellular component of connective tissue, is a fibrous triple-helical protein whose polypeptides each contain the amino acid sequence $(\text{Gly-X-Y})_n$ where X is often Pro, Y is often 4-hydroxyproline (Hyp), and $n \approx 340$ (Section 8-2B). The polypeptides of **procollagen** (Fig. 32-70) differ from those of the mature protein by the presence of both N-terminal and C-terminal propeptides of ~100 residues whose sequences, for the most part, are unlike those of mature collagen. The procollagen polypeptides rapidly assemble, *in vitro* as well as *in vivo,* to form a collagen triple helix. In contrast, polypeptides extracted from mature collagen will reassemble only over a period of days, if at all. *The collagen propeptides are apparently necessary for proper procollagen folding.*

The N- and C-terminal propeptides of procollagen are respectively removed by **amino-** and **carboxyprocollagen**

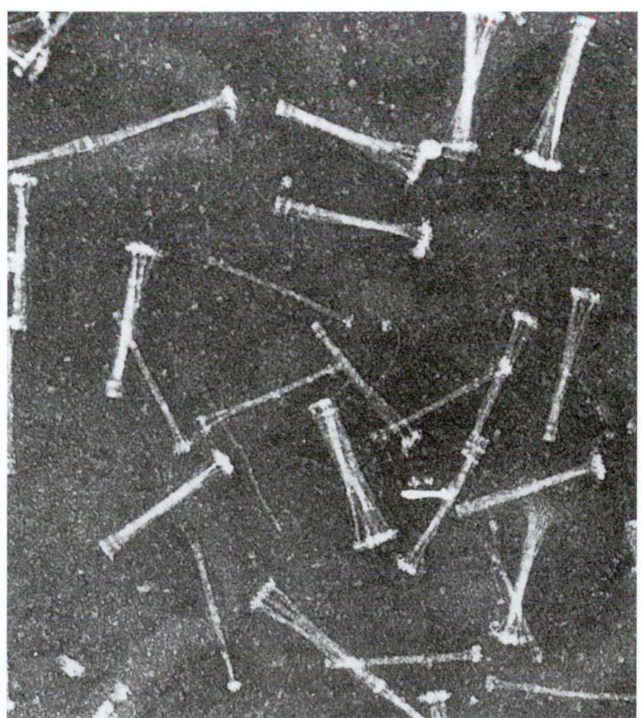

FIGURE 32-70 Electron micrograph of procollagen aggregates that have been secreted by fibroblasts into the extracellular medium. [Courtesy of Jerome Gross, Massachusetts General Hospital, Harvard Medical School.]

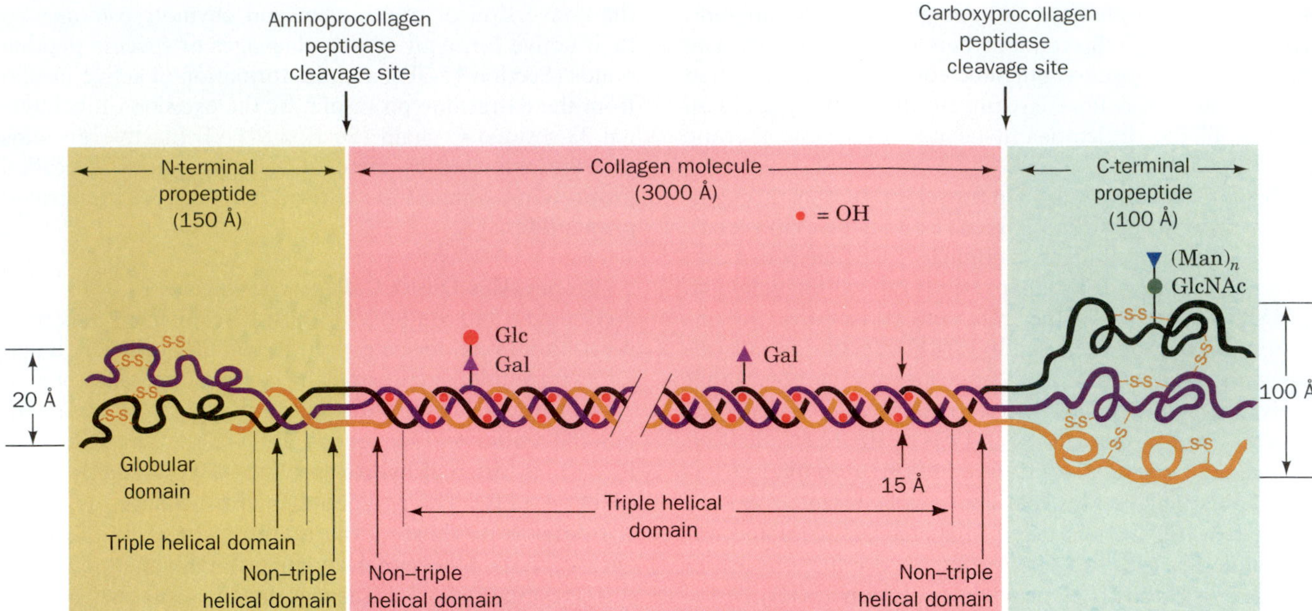

FIGURE 32-71 Schematic representation of the procollagen molecule. Gal, Glc, GlcNAc, and Man, respectively, denote galactose, glucose, *N*-acetylglucosamine, and mannose residues. Note that the N-terminal propeptide has intrachain disulfide bonds while the C-terminal propeptide has both intrachain and interchain disulfide bonds. [After Prockop, D.J., Kivirikko, K.I., Tuderman, L., and Guzman, N.A., *New Engl. J. Med.* **301,** 16 (1979).]

peptidases (Fig. 32-71), which may also be specific for the different collagen types. An inherited defect of aminoprocollagen peptidase in cattle and sheep results in a bizarre condition, **dermatosparaxis,** that is characterized by extremely fragile skin. An analogous disease in humans, **Ehlers–Danlos syndrome VII,** is caused by a mutation in one of the procollagen polypeptides that inhibits the enzymatic removal of its aminopropeptide. Collagen molecules normally spontaneously aggregate to form collagen fibrils (Figs. 8-31 and 8-32). However, electron micrographs of dermatosparaxic skin show sparse and disorganized collagen fibrils. *The retention of collagen's aminopropeptides apparently interferes with proper fibril formation.* (The dermatosparaxis gene was bred into some cattle herds because heterozygotes produce tender meat.)

b. Signal Peptides Are Removed from Nascent Proteins by a Signal Peptidase

Many transmembrane proteins or proteins that are destined to be secreted are synthesized with an N-terminal **signal peptide** of 13 to 36 predominantly hydrophobic residues. As described by the **signal hypothesis** (Section 12-4B), a signal peptide is recognized by a **signal recognition particle (SRP).** The SRP binds a ribosome synthesizing a signal peptide to a protein pore known as the **translocon** that is embedded in the membrane [the rough endoplasmic reticulum (RER) in eukaryotes and the plasma membrane in bacteria] and conducts the signal peptide and its following nascent polypeptide through the translocon.

Proteins bearing a signal peptide are known as **preproteins** or, if they also contain propeptides, as **preproproteins.** Once the signal peptide has passed through the membrane, it is specifically cleaved from the nascent polypeptide by a membrane-bound **signal peptidase.** Both insulin and collagen are secreted proteins and are therefore synthesized with leading signal peptides in the form of **preproinsulin** and **preprocollagen.** These and many other proteins are therefore subject to three sets of sequential proteolytic cleavages: (1) the deletion of their initiating Met residue, (2) the removal of their signal peptides, and (3) the excision of their propeptides.

c. Polyproteins

Some proteins are synthesized as segments of **polyproteins,** polypeptides that contain the sequences of two or more proteins. Examples include many polypeptide hormones (Section 34-3C); the proteins synthesized by many viruses, including those causing polio (Section 33-2C) and AIDS (Section 15-4C); and **ubiquitin,** a highly conserved eukaryotic protein involved in protein degradation (Section 32-6B). Specific proteases posttranslationally cleave polyproteins to their component proteins, presumably through the recognition of the cleavage site sequences. Some of these proteases are conserved over remarkable evolutionary distances. For instance, ubiquitin is synthesized as several tandem repeats **(polyubiquitin)** that *E. coli* properly cleave even though prokaryotes lack ubiquitin. Other proteases have more idiosyncratic cleavage sequences. This has allowed medicinal chemists to design inhibitors of **HIV protease** (which catalyzes an essential step in the viral life cycle) that have been highly effective in attenuating if not preventing the progression of AIDS (Section 32-6B).

B. *Covalent Modification*

Proteins are subject to specific chemical derivatizations, both at the functional groups of their side chains and at their terminal amino and carboxyl groups. Over 150 different types of side chain modifications, involving all side chains but those of Ala, Gly, Ile, Leu, Met, and Val, are known (Section 4-3A). These include acetylations, glycosylations, hydroxylations, methylations, nucleotidylations, phosphorylations, and ADP-ribosylations as well as numerous "miscellaneous" modifications.

Some protein modifications, such as the phosphorylation of glycogen phosphorylase (Section 18-1A) and the ADP-ribosylation of eEF-2 (Section 32-3G), modulate protein activity. Several side chain modifications covalently bond cofactors to enzymes, presumably to increase their catalytic efficiency. Examples of linked cofactors that we have encountered are N^{ε}-lipoyllysine in dihydrolipoyl transacetylase (Section 21-2A) and 8α-histidylflavin in succinate dehydrogenase (Section 21-3F). The attachment of complex carbohydrates, which occur in almost infinite variety, alter the structural properties of proteins and form recognition markers in various types of targeting and cell–cell interactions (Sections 11-3C, 12-3E, and 23-3B). Modifications that cross-link proteins, such as occur in collagen (Section 8-2B), stabilize supramolecular aggregates. The functions of most side chain modifications, however, remain enigmatic.

a. Collagen Assembly Requires Chemical Modification

Collagen biosynthesis (Fig. 32-72) is illustrative of protein maturation through chemical modification. As the nascent procollagen polypeptides pass into the RER of the fibroblasts that synthesized them, the Pro and Lys residues are hydroxylated to Hyp, 3-hydroxy-Pro, and 5-hydroxy-Lys (Hyl). The enzymes that do so are sequence specific: **Prolyl 4-hydroxylase** and **lysyl hydroxylase** act only on the Y residues of the Gly-X-Y sequences, whereas **prolyl 3-hydroxylase** acts on the X residues but only if Y is Hyp. Glycosylation, which also occurs in the RER, subsequently attaches sugar residues to Hyl residues (Section 8-2B). The folding of three polypeptides into the collagen triple helix must follow hydroxylation and glycosylation because the hydroxylases and glycosyl transferases do not act on helical substrates. Moreover, the collagen triple helix denatures below physiological temperatures unless stabilized by hydrogen bonding interactions involving Hyp residues (Section 8-2B). Folding is also preceded by the formation of specific interchain disulfide bonds between the carboxyl-propeptides. This observation bolsters the previously discussed conclusion that collagen propeptides help select and align the three collagen polypeptides for proper folding.

The procollagen molecules pass into the Golgi apparatus where they are packaged into **secretory vesicles** (Sections 12-4C, 12-4D, and 23-3B) and secreted into the extracellular spaces of connective tissue. The amino-propeptides are excised just after procollagen leaves the cell and the carboxypropeptides are removed sometime later. The collagen molecules then spontaneously assemble into fibrils, which suggests that an important propeptide function is to prevent intracellular fibril formation. Finally, after the action of the extracellular enzyme lysyl oxidase, the collagen molecules in the fibrils spontaneously cross-link (Fig. 8-33).

C. *Protein Splicing: Inteins and Exteins*

Protein splicing is a posttranslational modification process in which an *in*ternal pro*tein* segment (an **intein**) excises itself from a surrounding *ex*ternal pro*tein*, which it ligates to form the mature **extein.** The portions of the unspliced extein on the N- and C-terminal sides of the intein are called the **N-extein** and the **C-extein.** Over 130 putative inteins, ranging in length from 134 to 600 residues, have so far been identified in archaebacteria, eubacteria, and single-celled

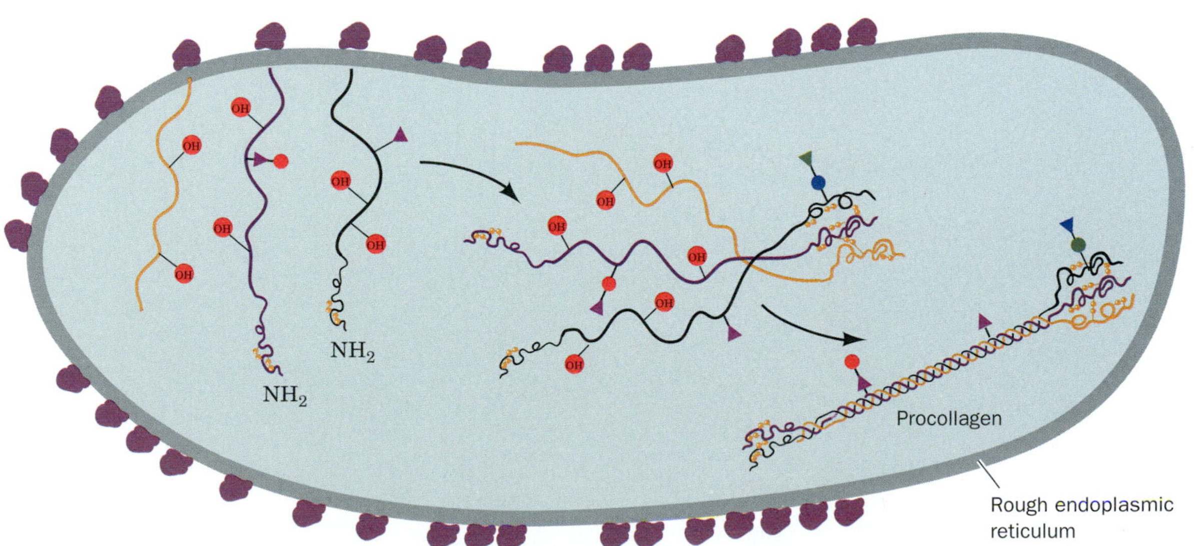

FIGURE 32-72 Schematic representation of procollagen biosynthesis. The diagram does not indicate the removal of signal peptides. [After Prockop, D.J., Kivirikko, K.I., Tuderman, L., and Guzman, N.A., *New Engl. J. Med.* **301,** 18 (1979).]

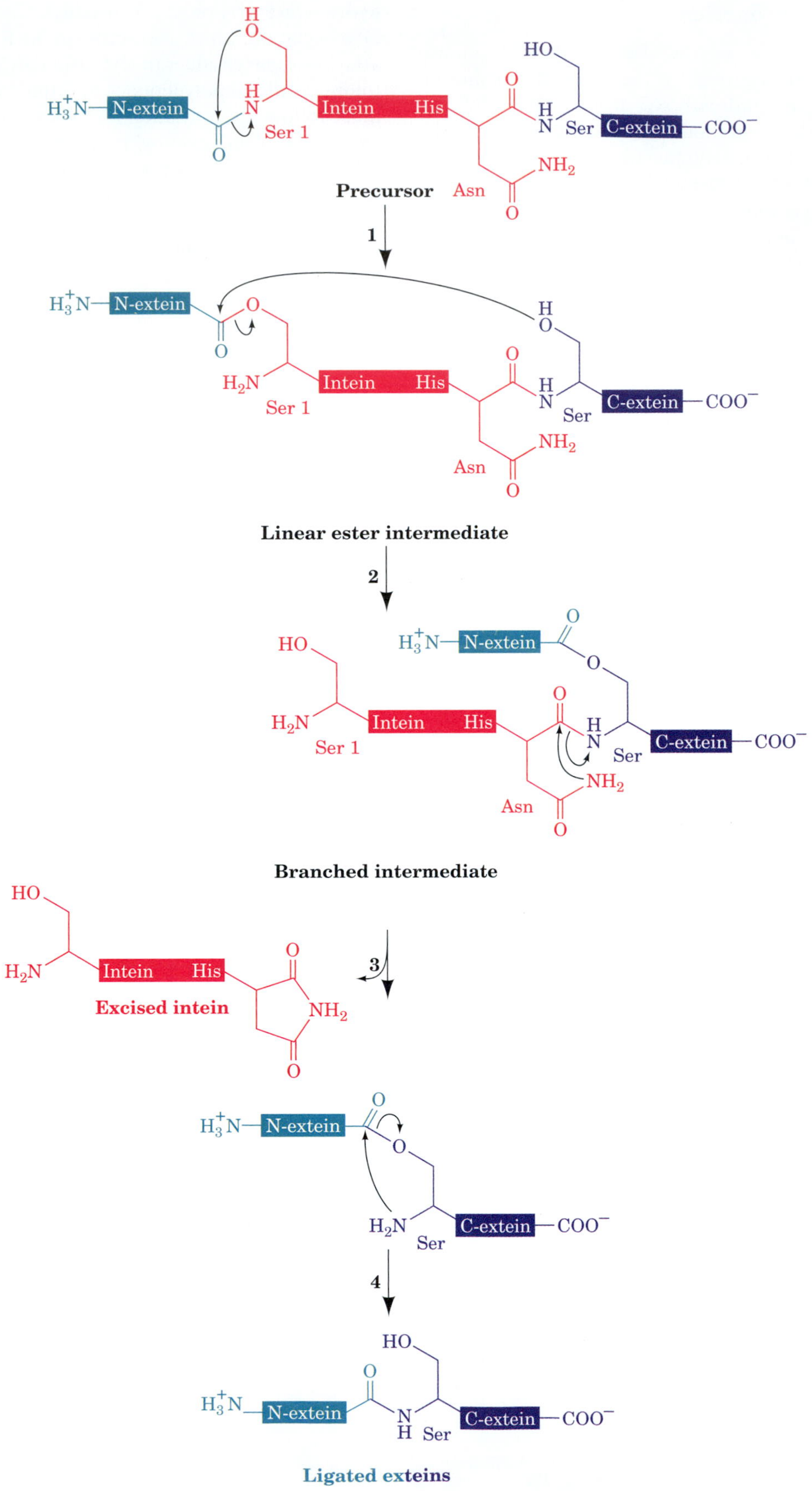

Precursor

1

Linear ester intermediate

2

Branched intermediate

Excised intein

3

4

Ligated exteins

FIGURE 32-73 Series of reactions catalyzed by inteins to splice themselves out of a polypeptide chain.

eukaryotes (and are registered in the Intein Database at http://www.neb.com/neb/inteins.html/). The various exteins in which these inteins are embedded exhibit no significant sequence similarity and, in fact, can be replaced by other polypeptides, thereby indicating that exteins do not contain the catalytic elements that mediate protein splicing. In contrast, the ~150-residue splicing elements of inteins exhibit significant sequence similarity. All of them have four conserved splice-junction residues: (1) a Ser/Thr/Cys at the intein's N-terminus; and (2 and 3) a His–Asn/Gln dipeptide at the intein's C-terminus; which is immediately followed by (4) a Ser/Thr/Cys at the N-terminus of the C-extein.

Protein splicing occurs via a reaction sequence that involves four successive nucleophilic displacements, the first three of which are catalyzed by the intein (Fig. 32-73):

1. Attack by the N-terminal intein residue (Ser, Thr, or Cys; shown in Fig. 32-73 as Ser) on its preceding carbonyl group, yielding a linear (thio)ester intermediate.

2. A transesterification reaction in which the -OH or -SH group on the C-extein's N-terminal residue (shown in Fig. 32-73 as Ser) attacks the above (thio)ester linkage, thereby yielding a branched intermediate in which the N-extein has been transferred to the C-extein.

3. Cleavage of the amide linkage connecting the intein to the C-extein by cyclization of the intein's C-terminal Asn or Gln (shown in Fig. 32-73 as Asn).

4. Spontaneous rearrangement of the (thio)ester linkage between the ligated exteins to yield the more stable peptide bond.

The X-ray structure of the 198-residue **GyrA intein** from *Mycobacterium xenopi,* determined by James Sacchetini, indicates how this intein catalyzes the foregoing splicing reactions. This intein's N-terminal residue, Cys 1, was replaced by an Ala–Ser dipeptide with the expectation that the mutant protein would resemble the intein's presplicing state (the new N-terminal residue, Ala 0, presumably represents the C-terminal residue of the N-extein). The X-ray structure reveals that this monomeric protein consists primarily of β strands, two of which curve about the periphery of the entire protein to give it the shape of a flattened horseshoe (Fig. 32-74). The intein's catalytic site is located at the bottom of a broad and shallow cleft near the center of this so-called **β-horseshoe,** where the intein's N-terminal and C-terminal residues are in close proximity. The Ala 0—Ser 1 peptide bond, the bond cleaved in Reaction 1 of the protein splicing process (Fig. 32-73) assumes the cis conformation (Fig. 8-2), a rare high-energy conformation (except when the peptide bond is followed by Pro) that destabilizes this bond. Its amide nitrogen atom is hydrogen bonded to the side chain of the highly conserved His 75. Hence His 75 is well positioned to donate a proton that would promote the breakdown of the tetrahedral intermediate in Reaction 1. The side chains of Thr 72 and Asn 74 appear well positioned to stabilize this tetrahedral intermediate in a manner resembling that of the

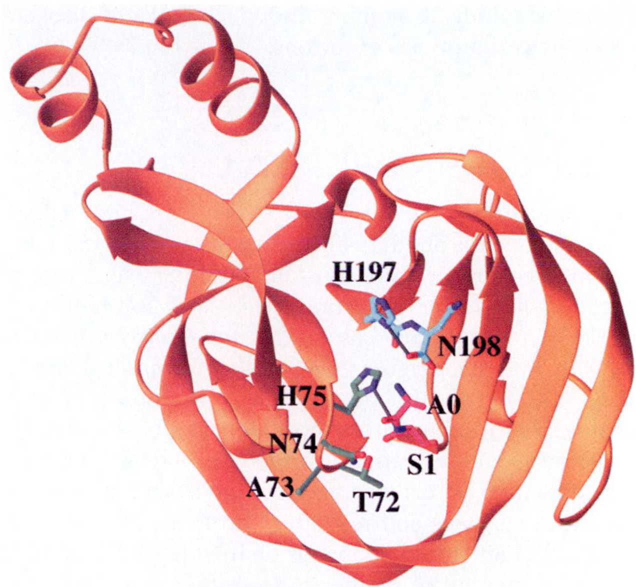

FIGURE 32-74 X-Ray structure of the *M. xenopi* Gyr A intein in which Cys 1 was replaced by an Ala 0–Ser 1 dipeptide. The protein is drawn in ribbon form with its N-terminal Ala 0–Ser 1 dipeptide and its C-terminal His 197–Asn 198 dipeptide as well as the side chains of residues 72 through 75 drawn in stick form colored according to atom type (C of residues 0–1 magenta, C of residues 72–75 green, C of residues 197–198 cyan, N blue, and O red). Hydrogen bonds are represented by thin gray bonds. [Based on an X-Ray structure by James Sacchetini, Texas A&M University. PDBid 1AM2.]

oxyanion hole in serine proteases (Section 15-3D). The position of Ser 1 and a modeled Thr at the intein's C-terminus is consistent with Reaction 2 of the splicing process. The side chain of the invariant His 197 is hydrogen bonded to the carboxylate of the C-terminal Asn 198 and hence is positioned to protonate the peptide bond cleaved in Reaction 3.

a. Most Inteins Encode a Homing Endonuclease

What is the biological function of inteins? Nearly all inteins contain polypeptide inserts forming so-called **homing endonucleases.** These are site-specific endonucleases that make a double-strand break in genes that are homologous to their corresponding extein but which lack inteins. The break initiates the double-strand break repair of the DNA via recombination (Section 30-6A). Since the intein-containing gene is likely to be the only other gene in the cell containing extein-like sequences, the intein gene is copied into the break. Thus, most inteins mediate a highly specific transposition or "homing" of the genes that insert them in similar sites. The intein's protease and endonuclease activities appear to have a symbiotic relationship: The protease activity excises the intein from the host protein, thereby preventing deleterious effects on the host, whereas the endonuclease activity assures the mobility of the intein gene. Thus intein genes appear to be molecular parasites (junk DNA) that only function to propagate themselves. Indeed, homing endonucleases are also encoded by certain types of introns.

6 ■ PROTEIN DEGRADATION

The pioneering work of Henry Borsook and Rudolf Schoenheimer around 1940 demonstrated that the components of living cells are constantly turning over. Proteins have lifetimes that range from as short as a few minutes to weeks or more. In any case, *cells continuously synthesize proteins from and degrade them to their component amino acids.* The function of this seemingly wasteful process is twofold: (1) to eliminate abnormal proteins whose accumulation would be harmful to the cell, and (2) to permit the regulation of cellular metabolism by eliminating superfluous enzymes and regulatory proteins. Indeed, since the level of an enzyme depends on its rate of degradation as well as its rate of synthesis, *controlling a protein's rate of degradation is as important to the cellular economy as is controlling its rate of synthesis.* In this section we consider the processes of intracellular protein degradation and their consequences.

TABLE 32-11 Half-lives of Some Rat Liver Enzymes

Enzyme	Half-life (h)
Short-Lived Enzymes	
Ornithine decarboxylase	0.2
RNA polymerase I	1.3
Tyrosine aminotransferase	2.0
Serine dehydratase	4.0
PEP carboxylase	5.0
Long-Lived Enzymes	
Aldolase	118
GAPDH	130
Cytochrome *b*	130
LDH	130
Cytochrome *c*	150

Source: Dice, J.F. and Goldberg, A.L., *Arch. Biochem. Biophys.* **170,** 214 (1975).

A. *Degradation Specificity*

Cells selectively degrade abnormal proteins. For example, hemoglobin that has been synthesized with the valine analog **α-amino-β-chlorobutyrate**

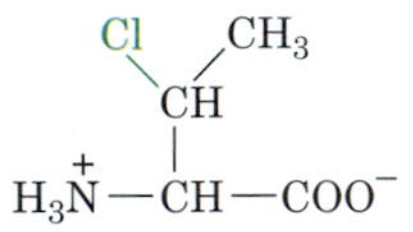

$$\mathrm{H_3\overset{+}{N}{-}CH(CH(Cl)CH_3){-}COO^-}$$

α-Amino-β-Chlorobutyrate

$$\mathrm{H_3\overset{+}{N}{-}CH(CH(CH_3)CH_3){-}COO^-}$$

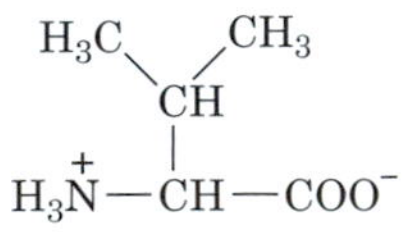

Valine

has a half-life in reticulocytes of ~10 min, whereas normal hemoglobin lasts the 120-day lifetime of the red cell (which makes it perhaps the longest lived cytoplasmic protein). Likewise, unstable mutant hemoglobins are degraded soon after their synthesis, which, for reasons explained in Section 10-3A, results in the hemolytic anemia characteristic of these molecular disease agents. Bacteria also selectively degrade abnormal proteins. For instance, *amber* and *ochre* mutants of β-galactosidase have half-lives in *E. coli* of only a few minutes, whereas the wild-type enzyme is almost indefinitely stable. Most abnormal proteins, however, probably arise from the chemical modification and/or spontaneous denaturation of these fragile molecules in the cell's reactive environment rather than by mutations or the rare errors in transcription or translation. *The ability to eliminate damaged proteins selectively is therefore an essential recycling mechanism that prevents the buildup of substances that would otherwise interfere with cellular processes.*

Normal intracellular proteins are eliminated at rates that depend on their identities. A given protein is eliminated with first-order kinetics, indicating that the molecules being degraded are chosen at random rather than according to their age. The half-lives of different enzymes in a given tissue vary substantially as is indicated for rat liver in Table 32-11. Remarkably, *the most rapidly degraded enzymes all occupy important metabolic control points, whereas the relatively stable enzymes have nearly constant catalytic activities under all physiological conditions. The susceptibilities of enzymes to degradation have evidently evolved together with their catalytic and allosteric properties so that cells can efficiently respond to environmental changes and metabolic requirements.* The criteria through which native proteins are selected for degradation are considered in Section 32-6B.

The rate of protein degradation in a cell also varies with its nutritional and hormonal state. Under conditions of nutritional deprivation, cells increase their rate of protein degradation so as to provide the necessary nutrients for indispensable metabolic processes. The mechanism that increases degradative rates in *E. coli* is the stringent response (Section 31-3H). A similar mechanism may be operative in eukaryotes since, as happens in *E. coli,* increased rates of degradation are prevented by antibiotics that block protein synthesis.

B. *Degradation Mechanisms*

Eukaryotic cells have dual systems for protein degradation: lysosomal mechanisms and ATP-dependent cytosolically based mechanisms. We consider both mechanisms below.

a. Lysosomes Mostly Degrade Proteins Nonselectively

Lysosomes are membrane-encapsulated organelles (Section 1-2A) that contain ~50 hydrolytic enzymes, including a variety of proteases known as **cathepsins.** The lysosome maintains an internal pH of ~5 and its enzymes have acidic pH optima. This situation presumably protects the cell against accidental lysosomal leakage since lysosomal enzymes are largely inactive at cytosolic pH's.

Lysosomes recycle intracellular constituents by fusing with membrane-enclosed bits of cytoplasm known as **autophagic vacuoles** and subsequently breaking down their contents. They similarly degrade substances that the cell takes up via endocytosis (Section 12-5B). The existence of these processes has been demonstrated through the use of lysosomal inhibitors. For example, the antimalarial drug **chloroquine**

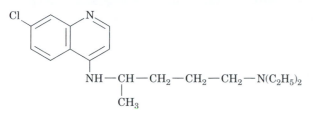

$$\text{Cl-quinoline-4-}NH{-}CH(CH_3){-}CH_2{-}CH_2{-}CH_2{-}N(C_2H_5)_2$$

Chloroquine

is a weak base that, in uncharged form, freely penetrates the lysosome where it accumulates in charged form, thereby increasing the intralysosomal pH and inhibiting lysosomal function. The treatment of cells with chloroquine reduces their rate of protein degradation. Similar effects arise from treatment of cells with cathepsin inhibitors such as the polypeptide antibiotic **antipain.**

$$^{-}OOCCHNH{-}\overset{O}{\overset{\|}{C}}{-}NHCH\overset{O}{\overset{\|}{C}}{-}NHCH\overset{O}{\overset{\|}{C}}{-}NHCH\overset{O}{\overset{\|}{C}}{-}H$$

(side chains: Phe, Arg, Val, Arg)

Antipain

Lysosomal protein degradation in well nourished cells appears to be nonselective. Lysosomal inhibitors do not affect the rapid degradation of abnormal proteins or short-lived enzymes. Rather, they prevent the acceleration of nonselective protein breakdown on starvation. However, the continued nonselective degradation of proteins in starving cells would rapidly lead to an intolerable depletion of essential enzymes and regulatory proteins. Lysosomes therefore also have a selective pathway, which is activated only after a prolonged fast, that takes up and degrades proteins containing the pentapeptide Lys-Phe-Glu-Arg-Gln (KFERQ) or a closely related sequence. Such KFERQ proteins are selectively lost in fasting animals from tissues that atrophy in response to fasting (e.g., liver and kidney) but not from tissues that do not do so (e.g., brain and testes). KFERQ proteins are specifically bound in the cytosol and delivered to the lysosome by a 73-kD **peptide recognition protein (prp73),** a member of the 70-kD heat shock protein (Hsp70) family (Section 9-2C).

Both normal and pathological processes are associated with increased lysosomal activity. **Diabetes mellitus** (Section 27-3B) stimulates the lysosomal breakdown of proteins. Similarly, muscle wastage caused by disuse, denervation, or traumatic injury arises from increased lysosomal activity. The regression of the uterus after childbirth, in which this muscular organ reduces its mass from 2 kg to 50 g in 9 days, is a striking example of this process. Many chronic inflammatory diseases, such as **rheumatoid arthritis,** involve the extracellular release of lysosomal enzymes that break down the surrounding tissues.

b. Ubiquitin Marks Proteins Selected for Degradation

It was initially assumed that protein degradation in eukaryotic cells is primarily a lysosomal process. Yet, reticulocytes, which lack lysosomes, selectively degrade abnormal proteins. The observation that protein breakdown is inhibited under anaerobic conditions led to the discovery of a cytosolically based ATP-dependent proteolytic system that is independent of the lysosomal system. This phenomenon was thermodynamically unexpected since peptide hydrolysis is an exergonic process.

The analysis of a cell-free rabbit reticulocyte system demonstrated that **ubiquitin** (Fig. 32-75) is required for ATP-dependent protein degradation. *This 76-residue monomeric protein, so named because it is ubiquitous as well as abundant in eukaryotes, is the most highly conserved protein known:* It is identical in such diverse organisms as humans, toad, trout, and *Drosophila* and differs in only three residues between humans and yeast. Evidently, ubiquitin is all but uniquely suited to an essential cellular function.

Proteins that are selected for degradation are so marked by covalently linking them to ubiquitin. This process, which is reminiscent of amino acid activation (Section 32-2C), oc-

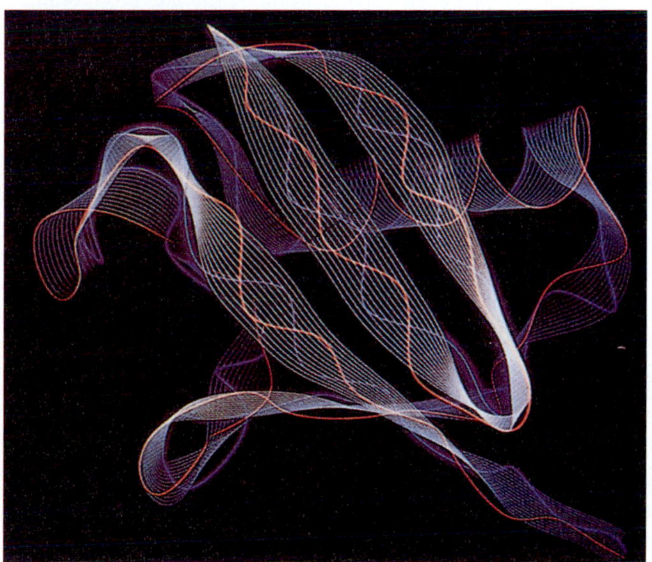

FIGURE 32-75 X-Ray structure of ubiquitin. The multistranded white ribbon represents the polypeptide backbone, and the red and blue curves, respectively, indicate the directions of the carbonyl and amide groups. [Courtesy of Mike Carson, University of Alabama at Birmingham. X-Ray structure determined by Charles Bugg, University of Alabama at Birmingham. PDBid 1UBQ.] **See the Interactive Exercises**

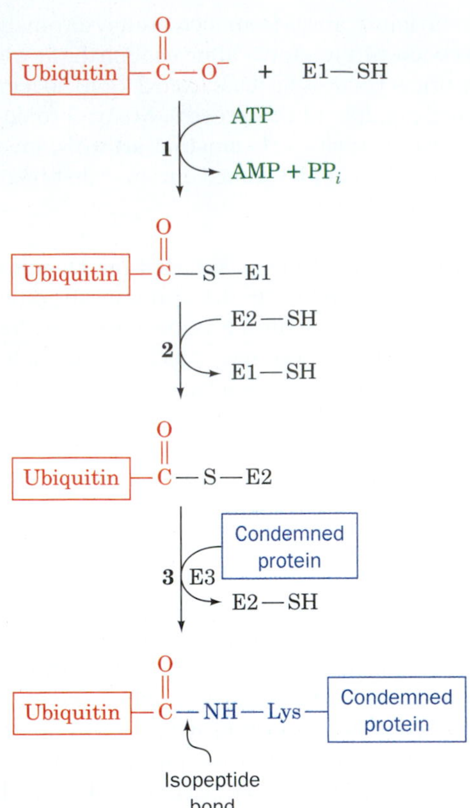

FIGURE 32-76 Reactions involved in the attachment of ubiquitin to a protein. In the first part of the process, ubiquitin's terminal carboxyl group is joined, via a thioester linkage, to E1 in a reaction driven by ATP hydrolysis. The activated ubiquitin is subsequently transferred to a sulfhydryl group of an E2 and then, in a reaction catalyzed by an E3, to a Lys ϵ-amino group on a condemned protein, thereby flagging the protein for proteolytic degradation by the 26S proteasome.

curs in a three-step pathway that was elucidated notably by Avram Hershko and Aaron Ciechanover (Fig. 32-76):

1. In an ATP-requiring reaction, ubiquitin's terminal carboxyl group is conjugated, via a thioester bond, to **ubiquitin-activating enzyme (E1),** a homodimer of ~1050-residue subunits. Most organisms, including yeast and humans, have only one type of E1.

2. The ubiquitin is then transferred to a specific Cys sulfhydryl group on one of numerous proteins named **ubiquitin-conjugating enzymes** (**E2s;** 11 in yeast and over 20 in mammals). The various E2's are characterized by ~150-residue catalytic cores containing the active site Cys that exhibit at least 25% sequence identities and which mainly vary by the presence or absence of N- and/or C-terminal extensions that exhibit little sequence identity to each other. The X-ray and NMR structures of several species of E2 reveal that their catalytic cores all assume closely similar α/β structures (e.g., Fig. 32-77) in which most of the identical residues are clustered on one surface near the ubiquitin-accepting Cys residue, where they presumably interact with ubiquitin and E1.

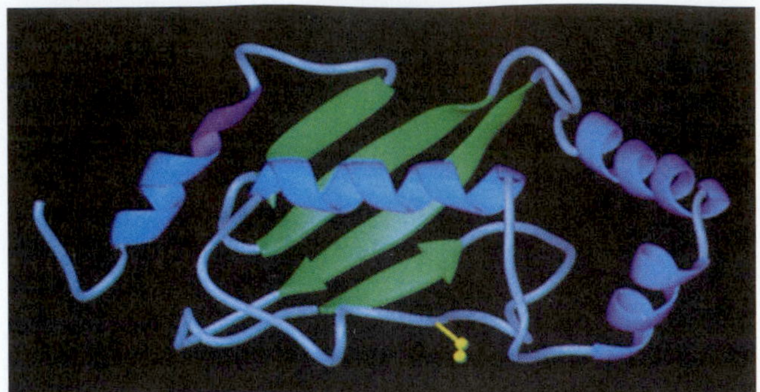

FIGURE 32-77 X-Ray structure of an E2 protein from *Arabidopsis thaliana*. α Helices are blue, the 3_{10} helical segment is purple, β strands are green, and the remainder of the molecule is light blue. The side chain of Cys 88, to which ubiquitin is covalently linked, is shown in ball-and-stick form in yellow. [Courtesy of William Cook, University of Alabama at Birmingham. PDBid 2AAK.]

3. ***Ubiquitin–protein ligase (E3)*** *transfers the activated ubiquitin from E2 to a Lys ε-amino group of its target protein, thereby forming an* ***isopeptide bond.*** Cells contain many species of E3s, each of which mediates the ubiquitination (alternatively, ubiquitylation) of a specific set of proteins and thereby marks them for degradation. Each E3 is served by one or a few specific E2s. The known E3s are members of two unrelated families, those containing a **HECT** domain (HECT for *h*omologous to *E*6AP *C-t*erminus) and those containing a so-called **RING finger** (RING for *r*eally *i*nteresting *n*ew *g*ene), although some E2s react well with members of both families. HECT domain E3s are modularly constructed with a unique N-terminal domain that interacts with its target proteins via their so-called **ubiquitination signals** (usually short polypeptide segments; see below) and an ~350-residue HECT domain that mediates E2 binding and catalyzes the ubiquitination reaction. The RING finger, which is implicated in recognizing the substrate protein's ubiquitination signal, is a 40- to 60-residue motif that binds two structurally but not catalytically implicated Zn^{2+} ions via a total of 8 Cys and His residues in a characteristic consensus sequence (much like the zinc finger motifs in certain DNA-binding proteins, Section 34-3B). RING finger–containing E3s may consist of a single subunit or may be multisubunit proteins in which the RING finger is contained in one subunit. RING finger E3s mediate the direct transfer of ubiquitin from E2 to a substrate protein's Lys residue, whereas HECT E3s do so via the intermediate transfer of ubiquitin from E2 to a catalytically essential Cys residue that is located ~35 residues from the N-terminus of the HECT domain.

In order for a target protein to be efficiently degraded, it must be linked to a chain of at least four tandemly linked ubiquitin molecules in which Lys 48 of each ubiquitin forms an isopeptide bond with the C-terminal carboxyl group of the succeeding ubiquitin (Fig. 32-78). These **polyubiquitin (polyUb)** chains, which can reach lengths of 50 or more ubiquitin molecules, are generated by the E3s, although how they

(a)

(b)

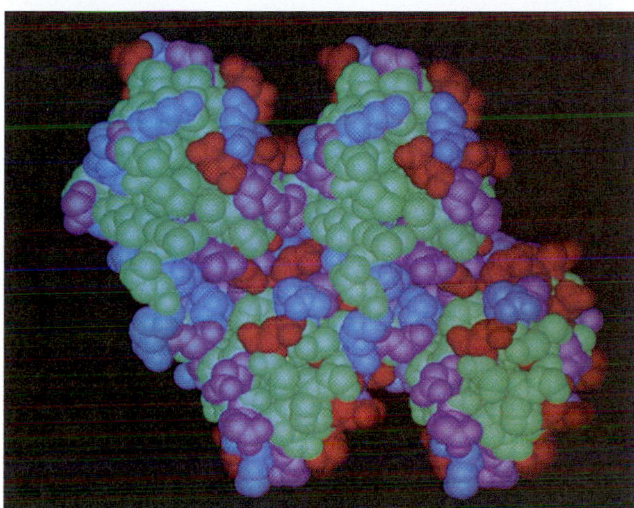

FIGURE 32-78 X-Ray structure of tetraubiquitin. (*a*) A ribbon drawing in which the isopeptide bonds connecting successive ubiquitin molecules, together with the Lys side chains making them, are orange. However, since the isopeptide bond connecting ubiquitins 2 and 3 is not visible in the X-ray structure, it is represented by a stick bond (this isopeptide bond nevertheless exists, as was demonstrated by SDS–PAGE of dissolved crystals). It seems likely that the monomer units in a multiubiquitin chain of any length would be arranged with the repeating symmetry of the tetraubiquitin structure. (*b*) A space-filling model, viewed as in Part *a*, in which basic residues (Arg, Lys, His) are blue, acidic residues (Asp, Glu) are red, uncharged polar residues (Gly, Ser, Thr, Asn, Gln) are purple, and hydrophobic residues (Ile, Leu, Val, Ala, Met, Phe, Tyr, Pro) are green. Note the unusually large solvent-exposed surface occupied by hydrophobic residues. [Courtesy of William Cook, University of Alabama at Birmingham. PDBid 1TBE.]

switch from transferring a ubiquitin to the target protein to processively synthesizing a polyubiquitin chain is unknown.

c. Ubiquitinated Proteins Are Hydrolyzed in the Proteasome

A ubiquitinated protein is proteolytically degraded to short peptides in an ATP-dependent process mediated by a large (2000 kD, 26S) multisubunit protein complex named the ***26S proteasome*** *(sometimes spelled "proteosome") that electron micrographic studies reveal has the shape of a bicapped hollow barrel (Fig. 32-79).* Proteolysis occurs inside the barrel, which permits this process to be extensive and processive, while preventing nonspecific proteolytic damage to other cellular components. PolyUb chains are the signals that target a protein to the proteasome; the identity of the target protein has little effect on the efficiency with which it is degraded by the proteasome. Nevertheless, the proteasome does not degrade ubiquitin molecules; they are returned to the cell. The size and functional complexity of this entire proteolytic system, which occurs in the nucleus as well as the cytosol, rivals that of the ribosome (Section 32-3) and the spliceosome (Section 31-4A) and hence is indicative of the importance of properly managing protein degradation. Indeed, ~5% of the proteins expressed by yeast participate in protein degradation. We discuss the structure and function of the 26S proteasome below.

d. Many E3s Have Elaborate Modular Structures

The proto-oncogene product **c-Cbl** (906 residues) is a single-subunit, RING finger–containing E3 that functions to ubiquitinate certain activated receptor tyrosine kinases (RTKs; Section 19-3A), thereby terminating their signaling. Nikola Pavletich determined the X-ray structure of the N-terminal half of c-Cbl (residues 47–447) in its ternary complex with the E2 protein **UbcH7** (which consists of little more than the ~150-residue E2 catalytic core) and an 11-residue peptide containing the ubiquitination signal from a nonreceptor tyrosine kinase (NRTK) named

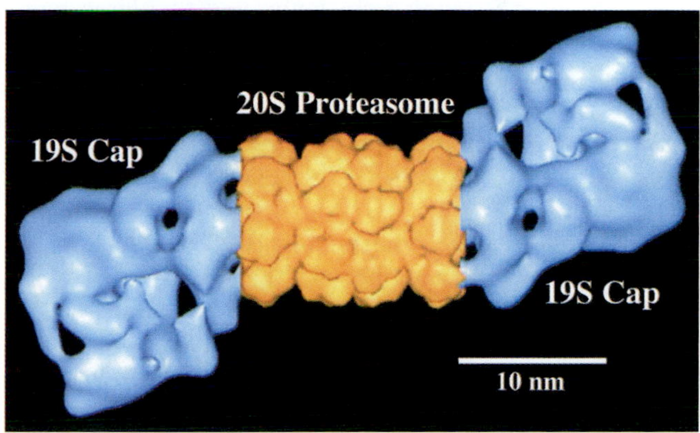

FIGURE 32-79 Electron microscopy–based image of the *Drosophila melanogaster* 26S proteasome. The complex is around 450 × 190 Å. The central portion of this 2-fold symmetric multiprotein complex (*yellow*), the **20S proteasome,** consists of four stacked 7-membered rings of subunits that form a hollow barrel in which the proteolysis of ubiquitin-linked proteins occurs. The **19S caps** (*blue*), which may attach to one or both ends of the 20S proteasome, control the access of condemned proteins to the 20S proteasome (see below). [Courtesy of Wolfgang Baumeister, Max-Planck-Institut für Biochemie, Martinsried, Germany.]

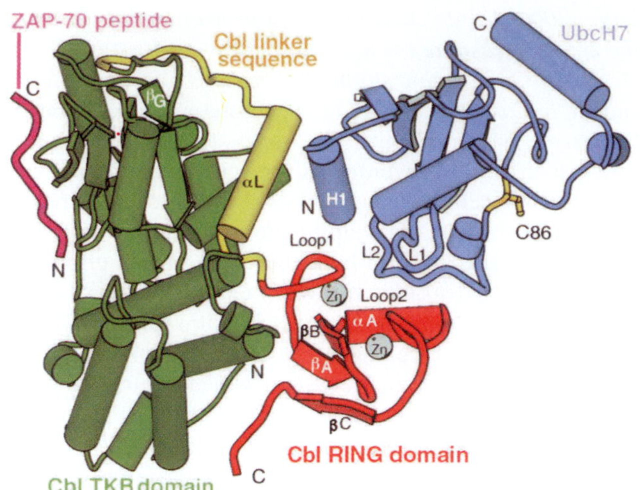

FIGURE 32-80 X-Ray structure of the human c-Cbl–UbcH7–ZAP-70 peptide ternary complex. UbcH7, an E2 that consists almost entirely of the E2 catalytic core, is colored blue with its active site Cys 86 shown in stick form (*yellow*). The 11-residue ubiquitination site of the RTK ZAP-70 is magenta. c-Cbl (residues 47–447 of the 903-residue protein), a RING finger, single-subunit E3, is colored according to domain with its RING finger domain red, its TKB domain green, and the linker joining them yellow. The RING finger's two bound Zn^{2+} ions are represented by gray spheres. [Courtesy of Nikola Pavletich, Memorial Sloan-Kettering Cancer Center, New York, New York. PDBid 1FBV.]

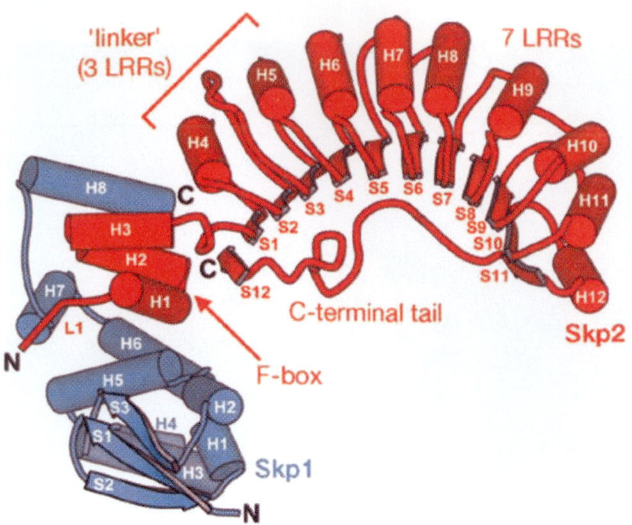

FIGURE 32-81 X-Ray structure of the human Skp1–Skp2 complex. Skp1 is blue and Skp2 is red. Skp2 consists of an N-terminal F-box that forms three helices, followed by 3 noncanonical so-called linker leucine-rich repeats (LRRs) that are contiguous with 7 LRRs that were predicted from their amino acid sequences for a total of 10 LRRs. After the tenth LRR, Skp2's ~30-residue C-terminal tail extends back past the first LRR by packing under the concave surface of the LRR domain. [Courtesy of Nikola Pavletich, Memorial Sloan-Kettering Cancer Center, New York, New York. PDBid 1FQB.]

ZAP-70 (Fig. 19-42). The structure (Fig. 32-80) reveals that UbcH7, and c-Cbl's RING finger and SH2-containing tyrosine kinase-binding (TKB) domains interact with one another across multiple interfaces to form a compact and apparently rigid structure. The RING finger domain consists of a 3-stranded β sheet, an α helix, and two large loops that are held together by the two tetrahedrally coordinated Zn^{2+} ions. UbcH7 adopts the characteristic α/β fold of other E2s of known structure (Fig. 32-77). The ZAP-70 peptide is bound on the opposite side of the TKB domain from the UbcH7 active site Cys residue (Cys 86) and is ~60 Å distant from it.

SCF complexes are multisubunit RING finger E3s that consist of **Cul1** (a member of the **cullin** family; 776 residues), **Rbx1** (which contains the complex's RING finger domain; 108 residues), **Skp1** (163 residues), and a member of the **F-box protein** family (~430 to >1000 residues; SCF for *S*kp1–*c*ullin–*F*-box protein). Rbx1 and Cul1 form the complex's catalytic core that binds E2; F-box proteins consist of an ~40-residue **F-box** that binds Skp1 followed by protein–protein interaction modules such as **leucine-rich repeats (LRRs)** or WD40 repeats (Section 19-2C) that bind substrate protein; and Skp1 functions as an adapter that links the F-box to Cul1. Cells contain numerous different F-box proteins (at least 38 in humans) that presumably permit the specific ubiquitination of a diverse variety of protein substrates (see below).

Pavletich has also determined the X-ray structures of two segments of the **SCF^{Skp2}** complex (where the superscript identifies the complex's F-box protein, here **Skp2;** 436 residues). The structure of the Skp1–Skp2 complex (Fig. 32-81) reveals that it has the shape of a sickle with the Skp1 and the 3-helix F-box of Skp2 forming the handle and its 10 LRRs (~26 residues each) forming the curved blade. The structure of the Cul1–Rbx1–Skp1–F-box^{Skp2} quaternary complex (Fig. 32-82) shows that Cul1 is an elongated protein that consists of a long stalk formed by

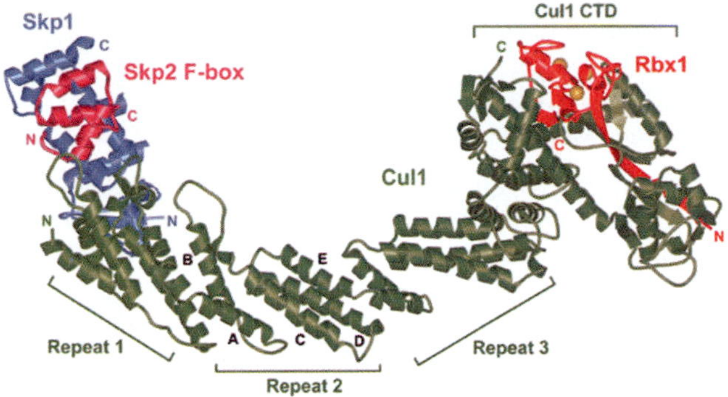

FIGURE 32-82 X-Ray structure of the human Cul1–Rbx1–Skp1–F-box^{Skp2} quaternary complex. Cul1, Rbx1, Skp1, and the Skp2 F-box are respectively colored green, red, blue, and magenta. The three cullin repeats of Cul1 are indicated with the five helices of the second repeat labeled A through E. The Zn^{2+} ions bound to Rbx1 are represented by yellow spheres. [Courtesy of Nikola Pavletich, Memorial Sloan-Kettering Cancer Center, New York, New York. PDBid 1LDK.]

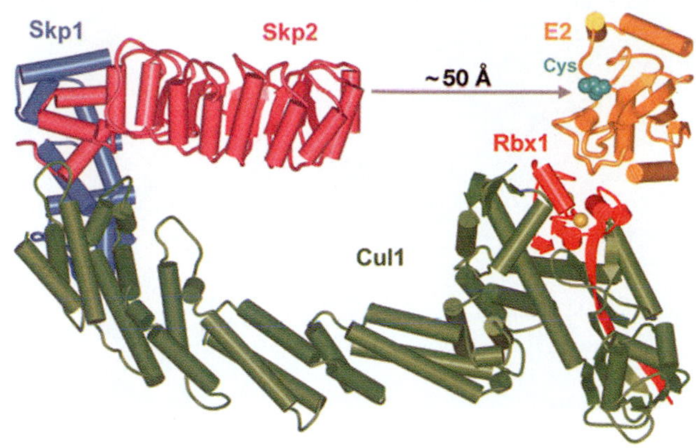

FIGURE 32-83 Model of the SCF^{Skp2}–E2 complex. The model, which is based on the X-ray structures in Figs. 32-80, 32-81, and 32-82, is colored and viewed as in Fig. 32-82. E2 is yellow with its active site Cys residue, to which ubiquitin would be covalently linked, drawn in space-filling form in cyan. The Zn^{2+} ions associated with the Rbx1 RING finger are represented by yellow spheres. The gray arrow indicates the 50-Å gap between the tip of the Skp2 LRR domain and the E2 active site. [Courtesy of Nikola Pavletich, Memorial Sloan-Kettering Cancer Center, New York, New York.]

three repeats of a novel five-helix motif known as a cullin repeat followed by a globular domain that binds Rbx1. Apparently Cul1 acts like a rigid scaffold that organizes the Skp1–F-box^{Skp2} complex and Rbx1 so as to hold them over 100 Å apart. The Rbx1 RING finger contains a 20-residue insert that forms the binding site for a third tetrahedrally liganded Zn^{2+} ion.

The apparent rigidity of the foregoing three structures has enabled Pavletich to construct a model of the intact SCF^{Skp2}–E2 complex by superimposing Skp1–Skp2 on Cul1–Rbx1–Skp1–F-box^{Skp2} and docking the E2 UbcH7 onto the Rbx1 RING finger based on the c-Cbl–UbcH7 structure (Fig. 32-83). The model indicates that E2 and the LRR-containing domain of Skp2 are on the same side of the SCF complex but separated by a distance of ~50 Å. This suggests that Cul1's long stalk functions to separate the complex's substrate-binding and catalytic sites so that substrates with different sizes and various distances between their ubiquitinated Lys residues and their ubiquitination signals can be accommodated.

e. The Ubiquitin System Has Both Housekeeping and Regulatory Functions

Until the mid-1990s, it appeared that the ubiquitin system functioned mainly in a "housekeeping" capacity to maintain the proper balance among metabolic proteins and to eliminate damaged proteins. Indeed, as Alexander Varshavsky discovered, *the half-lives of many cytoplasmic proteins vary with the identities of their N-terminal residues (Table 32-12)*. Thus, in a selection of 208 cytoplasmic proteins known to be long lived, all have a "stabilizing" residue, Met, Ser, Ala, Thr, Val, or Gly, at their N-termini. This so-called **N-end rule** is true for both eukaryotes and prokaryotes, which suggests the system that selects proteins for degradation is conserved in eukaryotes and prokaryotes, even though prokaryotes lack ubiquitin.

TABLE 32-12 Half-lives of Cytoplasmic Enzymes as a Function of Their N-Terminal Residues

N-Terminal Residue	Half-life
Stabilizing	
Met	>20 h
Ser	
Ala	
Thr	
Val	
Gly	
Destabilizing	
Ile	~30 min
Glu	
Tyr	~10 min
Gln	
Highly destabilizing	
Phe	~3 min
Leu	
Asp	
Lys	
Arg	~2 min

Source: Bachmair, A., Finley, D., and Varshavsky, A., *Science* **234,** 180 (1986).

The N-end rule results from the actions of the single-subunit, RING finger E3 named **E3α** (~1950 residues; also known as **Ubr1**) whose ubiquitination signals are the destabilizing N-terminal residues in Table 32-12. However, it is now clear that the ubiquitin system is far more sophisticated than a simple garbage disposal system. Thus, the growing list of known E3s have a variety of ubiquitination signals that often occur on a quite limited range of target proteins, many of which have regulatory functions. For example, the transcription factor **NF-κB** (NF for *n*uclear *f*actor), which plays a central role in immune and inflammatory responses, is maintained in an inactive state in the cytoplasm through its binding to the inhibitor **IκBα** (Fig. 12-38) in a way that occludes the short internal basic sequence that directs NF-κB's import into the nucleus (its **nuclear localization signal; NLS**). However, the stimulation of cell-surface receptors by certain cytokines initiates a signal transduction phosphorylation cascade (Section 19-3D) that phosphorylates IκBα bound to NF-κB at both Ser residues in the sequence DSGLDS. This phosphorylated sequence is the ubiquitination signal for the SCF complex containing the F-box protein **β-TrCP** (605 residues), which mediates the ubiquitination of the phosphorylated IκBα. The consequent destruction of IκBα exposes the NLS of NF-κB, which is then translocated to the nucleus where it activates the transcription of its target genes (Section 34-3B).

It has long been known that proteins with segments rich in Pro (P), Glu (E), Ser (S), and Thr (T), the so-called

PEST proteins, are rapidly degraded. It is now realized that this is because these PEST elements often contain phosphorylation sites that target their proteins for ubiquitination.

The ubiquitination system also has an essential function in cell cycle progression. The cell cycle, as we have seen in Section 30-4A and will further discuss in Section 34-4D, is regulated by a series of proteins known as cyclins. A given cyclin, which is expressed immediately preceding and/or during a specific phase of the cell cycle, binds to a corresponding cyclin-dependent protein kinase (Cdk), which then phosphorylates its target proteins so as to activate them to carry out the processes of that phase of the cell cycle. In addition, many cyclins also inhibit the transition to the subsequent phase of the cell cycle (e.g., DNA replication or mitosis). Thus, in order for a cell to progress from one phase of the cell cycle to the next, the cyclin(s) governing that phase must be eliminated. This occurs via the specific ubiquitination of the cyclin, thereby condemning it to be destroyed by the proteasome. The E3s responsible for this process are the SCF complexes containing F-box proteins targeted to a corresponding cyclin and a multisubunit complex known as the **anaphase-promoting complex (APC;** alternatively the **cyclosome)**. APC, an ~1500-kD RING finger–containing particle that in yeast consists of 11 subunits, specifically ubiquitinates proteins that contain the 9-residue consensus sequence RTALGDIGN, the so-called **destruction box,** near their N-termini.

Some viruses usurp the ubiquitination system. Oncogenic forms of **human papillomavirus (HPV),** the cause of nearly all cervical cancers (a leading cause of death of women in developing countries), encode the ~150-residue **E6 protein,** which combines with the 875-residue cellular protein named **E6-associated protein (E6AP;** the first E3 known to contain a HECT domain) to ubiquitinate **p53,** thereby marking it for destruction. This latter protein is a transcription factor that monitors genome integrity and hence is important in preventing malignant transformation and the proliferation of cancer cells (Section 34-4C), that is, it is a **tumor suppressor** (a protein whose loss of function is a cause of cancer). Consequently, HPV provokes the uncontrolled growth of the cells it infects and hence its own proliferation. E6AP normally functions to ubiquitinate certain members of the Src family of protein tyrosine kinases (Section 19-3B), including Src itself. The deletion of the segment of chromosome 15 that contains the E6AP gene causes Angelman syndrome, which as we have seen (Section 30-7) is characterized by severe mental retardation and is exclusively maternally inherited due to genomic imprinting.

BRCA1 is a tumor suppressor, many of whose mutations greatly increase the predisposition for a number of cancers, notably breast and ovarian cancers (Section 30-6A; BRCA for *br*east *ca*ncer). Around 20% of BRCA1's clinically relevant mutations occur in this 1863-residue protein's N-terminal 100 residues, which contains a RING finger domain. BRCA1 forms a heterodimer with the RING finger domain-containing protein **BARD1** (for *B*RCA1-*a*ssociated *R*ing *d*omain protein 1; 777 residues) that collectively functions as a ubiquitin-protein ligase (E3), whose

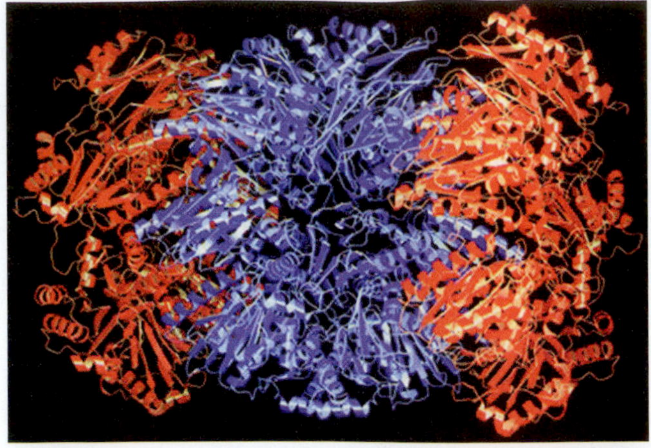

(a)

(b)

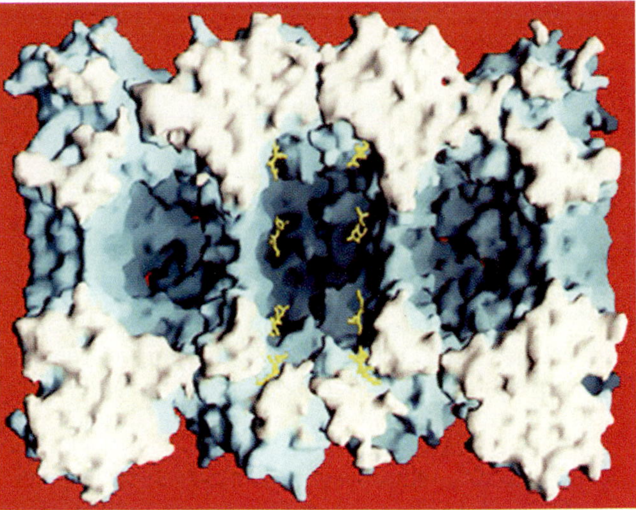

(c)

FIGURE 32-84 X-Ray structure of the *T. acidophilum* 20S proteasome. (*a*) A ribbon diagram, in which α subunits are red and β subunits are blue, viewed with a 2-fold axis tilted to the left and its 7-fold axis horizontal. Only subunits with a forward exposure are shown for clarity. (*b*) A diagram in which subunits are represented by equal-sized spheres that is viewed and colored as in Part *a*. (*c*) A cutaway diagram of the solvent-accessible surface showing its three internal cavities. The active sites on the β subunits are marked by the bound inhibitor, LLnL, which is drawn in stick form in yellow. [Courtesy of Robert Huber, Max-Planck-Institut für Biochemie, Martinsried, Germany. PDBid 1PMA.]

cognate ubiquitin-conjugating enzyme (E2), **UbcH5c,** binds only to the BRCA1 RING finger domain. All known cancer-associated mutations of the BRCA1 RING finger domain abrogate the heterodimer's E3 activity *in vitro.*

f. The 20S Proteasome Catalyzes Proteolysis Inside a Hollow Barrel

The 26S proteasome (Fig. 32-79) is an ~2100-kD multisubunit protein that catalyzes the ATP-dependent hydrolysis of ubiquitin-linked proteins. This yields oligopeptides averaging from 7 to 9 residues in length that are subsequently degraded to their component amino acids by cytosolic exopeptidases. The 26S proteasome consists of a **20S proteasome** (~670 kD), the barrel-shaped catalytic core of the 26S proteasome, and its **19S caps** (~700 kD; also known as **PA700** and the **19S regulator**), which associate with the ends of the 20S proteasome and stimulate its activity (PA for *p*roteasome *a*ctivator). The 20S proteasome only hydrolyzes unfolded proteins in an ATP-independent manner; the 19S caps function to identify and unfold the ubiquitinated protein substrates.

The 20S proteasome occurs in the nuclei and cytosol of all eukaryotic cells and in all Archaebacteria yet examined. However, the only eubacteria in which it occurs are those of the class Actinomyces (Fig. 1-4), which suggests that they obtained it via horizontal gene transfer from some other organism.

The 20S proteasome of *Thermoplasma acidophilum* (an archaebacterium) consists of 14 copies each of α and β subunits (233 and 203 residues) that electron microscopy studies reveal form a 150-Å long and 110-Å in-diameter barrel in which the subunits are arranged in four stacked rings (as is evident in the central portion of the 26S proteasome seen in Fig. 32-79). The α and β subunits are 26% identical in sequence except for an ~35-residue N-terminal segment of the α subunit, which the β subunit lacks. Eukaryotic 20S proteasomes are more complex in that they consist of 7 different α-like and 7 different β-like subunits versus only one of each type for the *T. acidophilum* 20S proteasome.

The X-ray structure of the *T. acidophilum* 20S proteasome, determined by Wolfgang Baumeister and Robert Huber, reveals that its two inner rings each consist of 7 β subunits and its two outer rings each consist of 7 α subunits arranged with D_7 symmetry (Fig. 32-84). Thus the overall structure of the 20S proteasome superficially resembles that of the unrelated molecular chaperone GroEL (Section 9-2C). The structures of the α and β subunits are remarkably similar (Fig. 32-85) except, of course, for the α subunit's N-terminal segment, which contacts the N-terminal segment in an adjacent α subunit. This accounts for the observation that α subunits alone spontaneously assemble into 7-membered rings (a capacity that is abolished by the deletion of their N-terminal 35 residues), whereas β subunits alone remain monomeric.

The central channel of the *T. acidophilum* 20S proteasome, which has a maximum diameter of 53 Å, consists of three large chambers (Fig. 32-84*c*: Two are located at the interfaces between adjoining rings of α and β subunits, with

(*a*)

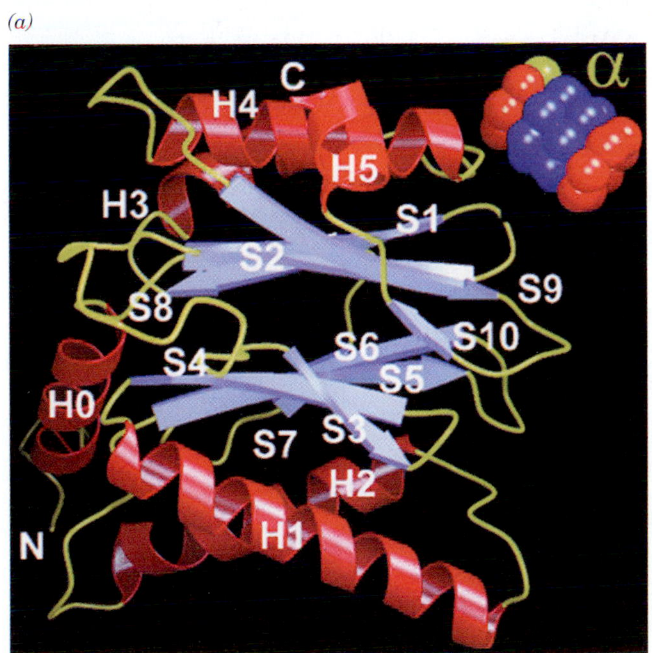

(*b*)

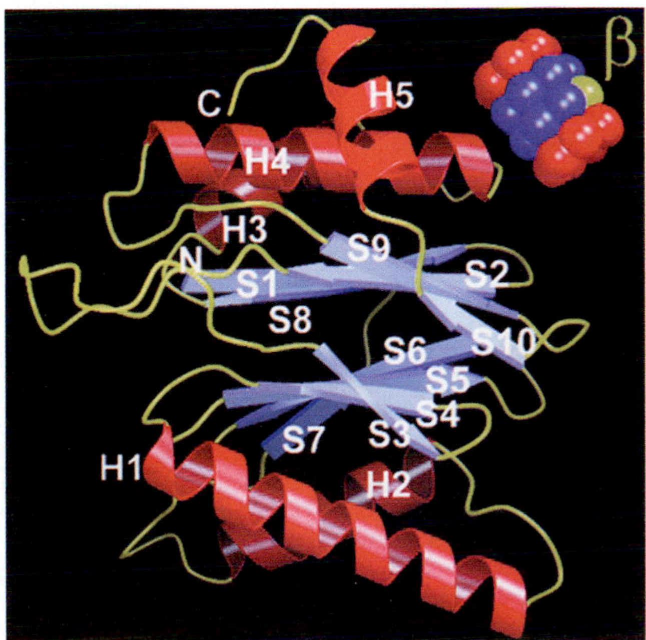

FIGURE 32-85 X-Ray structures of the subunits of the *T. acidophilum* 20S proteasome. (*a*) The α subunit which is colored according to secondary structure with helices red, β strands blue, and other segments yellow. The N-terminal helix, H0, occupies a position at the end of the proteasome. The identically oriented subunit is represented by the yellow sphere in the inset diagram of the entire 20S subunit (*upper right*). (*b*) The similarly colored and oriented β subunit. [Courtesy of Robert Huber, Max-Planck-Institut für Biochemie, Martinsried, Germany. PDBid 1PMA.]

the third, larger chamber centrally located between the two rings of β subunits. Unfolded polypeptide substrates appear to enter the central chamber of the barrel (where the proteasome's active sites are located; see below) through ~13-Å in diameter axially located apertures in the α rings that are lined with hydrophobic residues. This allows only unfolded proteins to enter the central chamber, thereby protecting properly folded proteins from indiscriminant degradation by this omnivorous protein-dismantling machine.

The X-ray structure of the yeast 20S proteasome, also determined by Huber, demonstrates that its outer and inner rings respectively consist of seven different α-type subunits and seven different β-type subunits, all of which are uniquely arranged (Fig. 32-86). The α-like subunits have folds that are similar to one another as well as to that of the *T. acidophilum* 20S proteasome and likewise for the β-like subunits. Consequently, this 28-subunit, 6182-residue protein complex has exact 2-fold rotational symmetry relating the two pairs of rings but only pseudo-7-fold rotational symmetry relating the subunits within each ring. The narrow axial apertures in the α rings through which unfolded polypeptides almost certainly enter the hydrolytic chamber in the *T. acidophilum* 20S proteasome (Fig. 32-84*c*) are closed in the yeast 20S proteasome (Fig. 32-86*c*) by a plug formed by the interdigitation of its α subunits' N-terminal tails. This suggests that the 19S caps of the 26S proteasome, which have been shown to activate the 20S proteasome, control the access to it by inducing conformational changes in its α rings. The X-ray structure of the bovine 20S proteasome, determined by Tomitake Tsukihara, reveals that its arrangement of seven α-type and seven β-type subunits is similar to that in yeast.

g. The Proteasome Catalyzes Peptide Hydrolysis via a Novel Mechanism

The X-ray structure of the *T. acidophilum* 20S proteasome in complex with the aldehyde inhibitor **acetyl-Leu-Leu-norleucinal (LLnL)**

$$CH_3-\overset{\overset{\displaystyle O}{\|}}{C}-\text{Leu}-\text{Leu}-\text{NH}-\underset{}{\overset{\overset{\displaystyle CH_3}{|}}{\overset{(CH_2)_3}{\overset{|}{CH}}}}-\overset{\overset{\displaystyle O}{\|}}{CH}$$

Acetyl-Leu-Leu-norleucinal (LLnL)

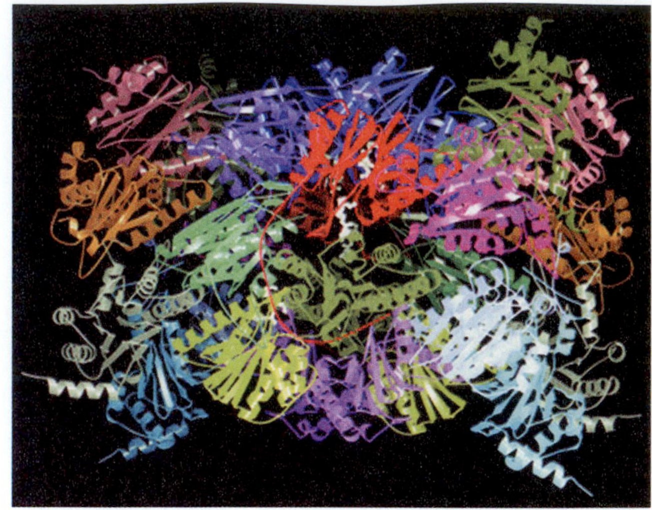

(*a*)

(*b*)

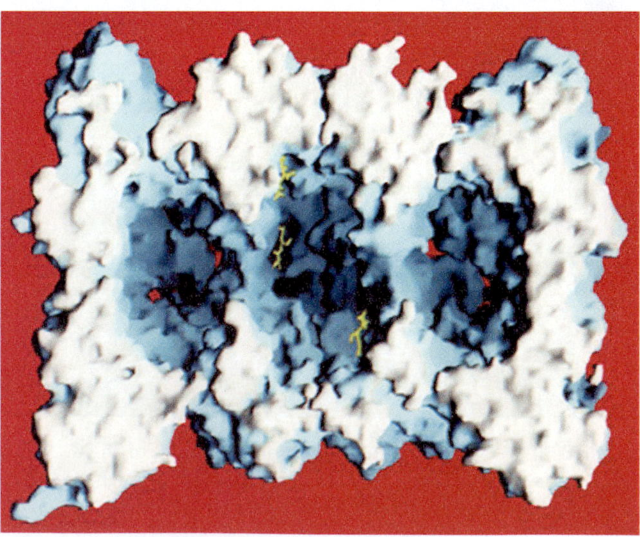

(*c*)

FIGURE 32-86 X-Ray structure of the yeast 20S proteasome. (*a*) A ribbon diagram in which the various subunits are differently colored and viewed with a 2-fold axis tilted to the left and its pseudo-7-fold axis horizontal. Only subunits with a forward exposure are shown for clarity. (*b*) A diagram in which subunits are represented by equal-sized spheres that are viewed and colored as in Part *a*. (*c*) A cutaway diagram of the solvent-accessible surface showing its three internal cavities. The active sites on the β1, β2, and β5 subunits are marked by the bound inhibitor, LLnL, which is drawn in stick form in yellow. Compare these diagrams with those in Fig. 32-84. [Courtesy of Robert Huber, Max-Planck-Institut für Biochemie, Martinsried, Germany. PDBid 1RYP.]

reveals that its active sites are on the inner surfaces of its rings of β subunits, with the aldehyde function of the LLnL close to the side chain of the highly conserved Thr 1β. Deletion of this Thr or its mutation to Ala yields properly assembled 20S proteasomes that are completely inactive. Evidently, 20S proteasomes catalyze peptide hydrolysis by a novel mechanism in which the hydroxyl group of its Thr 1β is the attacking nucleophile. This as yet poorly understood mechanism in which the amino group at the N-terminus presumably acts to increase the nucleophilicity of the hydroxyl sidechain, is now known to be employed by other hydrolases (e.g., glutamate synthase; Section 26-5A), which are collectively known as the **N-terminal nucleophile (Ntn)** family of hydrolases. The *T. acidophilum* β subunits preferably cleave polypeptides after hydrophobic residues. However, in the yeast and the bovine 20S proteasomes, only subunits β1, β2, and β5 are catalytically active. Their respective preferences for cleavage after acidic, basic (trypsin-like), and hydrophobic (chymotrypsin-like) residues are explained by the basic, acidic, and nonpolar characters of their pockets that bind the side chain of the residue preceding the scissile peptide bond. The functions of the four different catalytically inactive β subunits is unknown, although mutagenically modifying an inactive β subunit can abolish the catalytic activity of an active β subunit.

h. The 19S Caps Control the Access of Ubiquitinated Proteins to the 20S Proteasome

The 20S proteasome probably does not exist alone *in vivo;* it is most often in complex with two 19S caps that function to recognize ubiquitinated proteins, unfold them, and feed them into the 20S proteasome in an ATP-dependent manner (it may also associate with other regulatory complexes; see below). The 19S cap, which consists of ~18 different subunits, is poorly characterized due in large part to its low intrinsic stability. Its so-called base complex consists of 9 different subunits, 6 of which are ATPases that form a ring that abuts the α ring of the 20S proteasome (Fig. 32-79). Each of these ATPases contains an ~230-residue **AAA** module (AAA for *A*TPases *a*ssociated with a variety of *a*ctivities, a highly diverse family of proteins; also called **AAA+**). Cecile Pickart has demonstrated via cross-linking experiments that one of these ATPases, named **S6′** (alternatively **Rpt5**), contacts the polyUb signal that targets a condemned protein to the 26S proteasome. This suggests that the recognition of the polyUb chain as well as substrate protein unfolding are ATP-driven processes. Moreover, the ring of ATPases must function to open (gate) the otherwise closed axial aperture of the 20S proteasome so as to permit the entry of the unfolded substrate protein.

Eight additional subunits form the so-called lid complex, the portion of the 19S cap that is more distal to (distant from) the 20S proteasome. The functions of the lid subunits are largely unknown, although a truncated 26S proteasome that lacks the lid subunits is unable to degrade polyubiquitinated substrates. Several other subunits may be transiently associated with the 19S cap and/or with the 20S proteasome.

i. Deubiquitinating Enzymes Have Several Functions

The enzymes that hydrolytically cleave the isopeptide bonds linking successive ubiquitin units in polyUb are known as **deubiquitinating enzymes (DUBs).** Cells contain a surprisingly large number of DUBs (at least 17 in yeast and nearly 100 in humans). Nearly all known DUBs are **cysteine proteases,** enzymes whose catalytic mechanism resembles that of serine proteases (Section 15-3C) but whose attacking nucleophile is Cys—S$^-$ rather than Ser—OH.

DUBs may release entire polyUb chains from a condemned protein or sequentially release ubiquitin units from the chain terminus. It has been proposed that this latter process functions as a clock to time the protein degradation process. If a polyUb chain is trimmed to less than four ubiquitin units before degradation begins, then its attached protein is likely to escape destruction. This would spare proteins that had been inappropriately tagged with only short polyUb chains.

The mammalian 19S lid subunit known as **POH1** (**Rpn11** for the 65% identical yeast subunit) appears to be responsible for the deubiquitination of target proteins prior to their degradation; its inactivation prevents target protein degradation. Curiously, this DUB is a Zn^{2+}-dependent protease (as is carboxypeptidase A; Fig. 15-42) rather than a cysteine protease.

Certain DUBs function to dismember polyUb chains that have been released from substrate proteins by sequentially removing ubiquitin units from the end of the chain that is nearest to the substrate protein (that with a free C-terminus). Consequently, these DUBs cannot remove ubiquitin units from polyUb chains that are still attached to substrate proteins, thereby preventing their premature removal.

Cells express ubiquitin as polyproteins containing several ubiquitin units or with ubiquitin fused to certain ribosomal subunits (there is no gene that encodes a single ubiquitin unit). These polyproteins are rapidly processed by as yet unidentified DUBs to yield free ubiquitin.

j. The 11S Regulator Forms a Heptameric Barrel That Opens the 20S Proteasome

Higher eukaryotes contain an **11S regulator** that functions to open the channel into the 20S proteasome in an ATP-independent manner so as to permit the entrance of polypeptides (but not folded proteins). The mammalian 11S regulator, which functions in the generation of peptides for presentation to the immune system (Section 35-2E), is named **REG** (alternatively **PA28**). It is a heteroheptameric complex of two ~245-residue subunits, **REGα** and **REGβ,** that exhibit ~50% sequence identity except for a highly variable internal 18-residue segment that is thought to confer subunit-specific properties. Indeed REGα alone forms a heptamer whose biochemical properties are similar to that of REG (although both subunits must be present *in vivo*).

The trypanosome *Trypanosoma brucei,* which lacks 19S caps, expresses a homoheptameric 11S regulator named **PA26** that is only 14% identical to human REGα. Nevertheless, the various 11S regulators activate 20S proteasomes from widely divergent species. Thus, rat 20S proteasome is activated by PA26 and the yeast 20S proteasome is activated by human REGα despite the fact that yeast lacks 11S regulators.

The X-ray structure of PA26 in complex with the yeast 20S proteasome, determined by Christopher Hill, reveals that each PA26 monomer consists of an up–down–up–down 4-helix bundle. These monomers form a 7-fold symmetric heptameric barrel that is 90 Å in diameter, 70 Å long, and has a 33-Å in diameter central pore (Fig. 32-87*a*) and which closely resembles the previously determined X-ray structure of human REGα. Two PA26 barrels associate coaxially with the 20S proteasome, one at each end (Fig. 32-87*b*). The conformation of the 20S proteasome in this complex, for the most part, is closely similar to that of the 20S proteasome alone (Fig. 32-86). However, the C-terminal tails of the PA26 subunits insert into pockets on the 20S proteasome's α subunits in a way that induces conformational changes in its N-terminal tails that clear the 20S proteasome's otherwise blocked central aperture, thus permitting unfolded polypeptides to enter the proteasome's central chamber.

k. Bacteria Contain a Variety of Self-Compartmentalized Proteases

Nearly all eubacteria lack 20S proteasomes. Nevertheless they have ATP-dependent proteolytic assemblies that share the same barrel-shaped architecture and carry out similar functions. For example, in *E. coli,* two proteins known as **Lon** and **Clp** mediate up to 80% of the bacterium's protein degradation, with additional contributions from at least three other proteins including **heat shock locus UV (HslUV).** Thus, *all cells appear to contain proteases whose active sites are only available from the inner cavity of a hollow particle to which access is controlled.* These so-called **self-compartmentalized proteases** appear to have arisen early in the history of cellular life, before the ad-

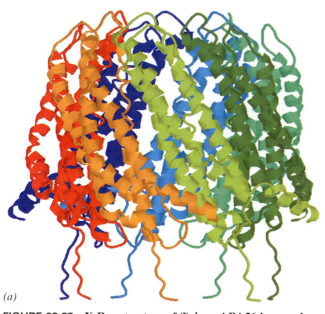

(*a*)

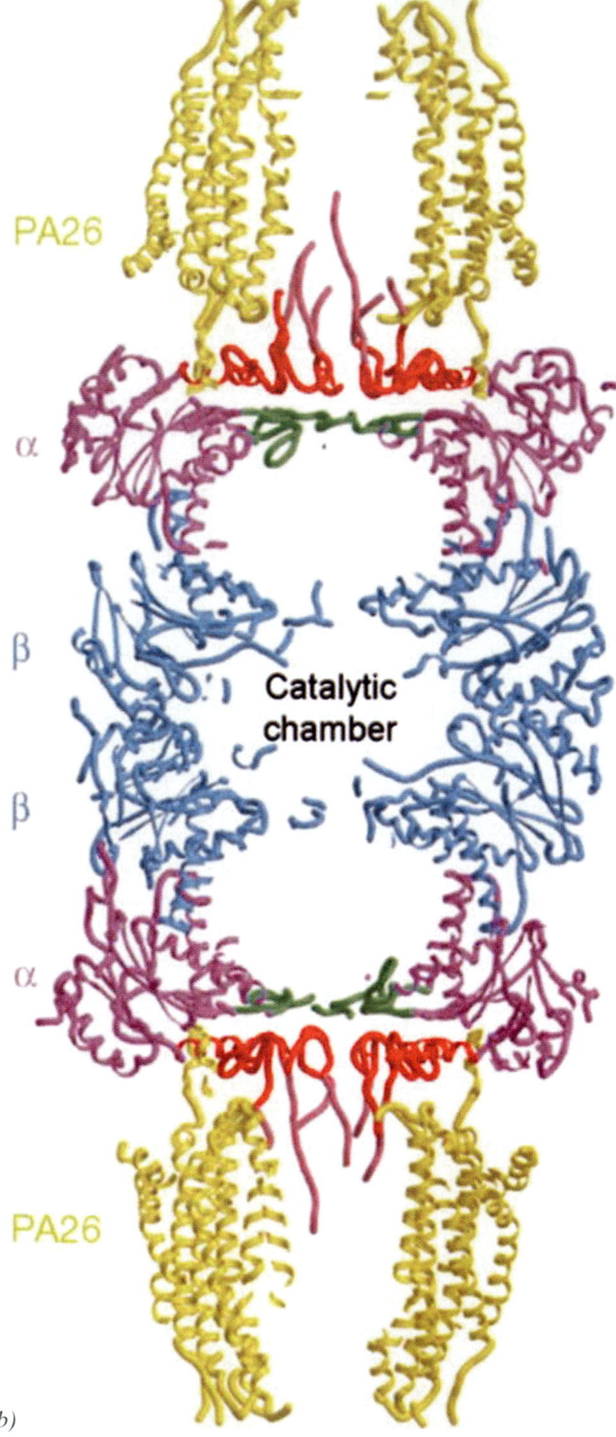

FIGURE 32-87 X-Ray structure of *T. brucei* PA26 in complex with the yeast 20S proteasome. (*a*) The PA26 heptamer viewed with its 7-fold axis vertical. Each of its subunits are differently colored. (*b*) Cutaway diagram of the entire complex viewed with its 7-fold axis vertical. The PA26 is yellow, the α and β subunits of the 20S proteasome are magenta and blue, its α-annulus is green, and its N-terminal segments that are ordered and partially disordered are red and pink. [Part *a* based on an X-ray structure by and Part *b* courtesy of Christopher Hill, University of Utah. PDBid 1FNT.]

vent of eukaryotic membrane-bound organelles such as the lysosome, which similarly carry out degradative processes in a way that protects the cell contents from indiscriminant destruction.

Clp protease consists of two components, the proteolytically active **ClpP** and one of several ATPases, which in *E. coli* are **ClpA** and **ClpX.** The X-ray structure of ClpP, determined by John Flanagan, reveals that it oligomerizes to form a ~90-Å long and wide hollow barrel that consists of two back-to-back 7-fold symmetric rings of 193-residue subunits and thereby has the same D_7 symmetry as does the 20S proteasome (Fig. 32-88). Nevertheless, the ClpP subunit has a novel fold that is entirely different from that of the 20S proteasome's homologous α and β subunits. The ClpP active site, which is only exposed on the inside of the barrel, contains a catalytic triad composed of Ser 97, His 122, and Asp 171, and hence is a serine protease (Section 15-3B).

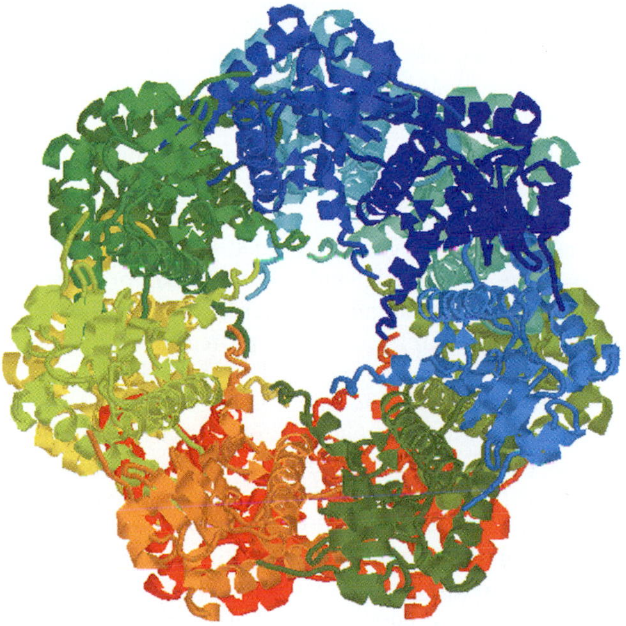

(*a*)

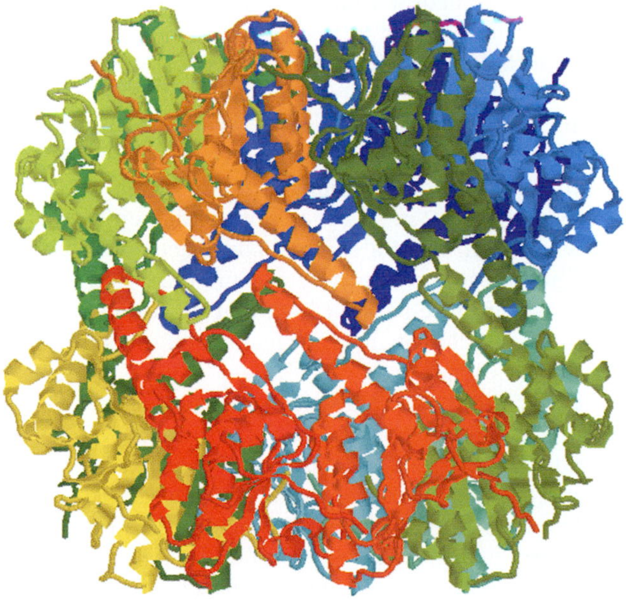

(*b*)

FIGURE 32-88 X-Ray structure of *E. coli* ClpP. (*a*) View of the heptameric complex along its 7-fold axis with each of its subunits differently colored. (*b*) View along the complex's 2-fold axis (rotated 90° about a horizontal axis with respect to Part *a*). [Based on an X-ray structure by John Flannagan, Brookhaven National Laboratory, Upton, New York. PDBid 1TYF.]

HslUV protease appears to be a hybrid of Clp and the 26S proteasome. Its **HslV** subunits in *E. coli* (145 residues) are 18% identical to the β subunits of the *T. acidophilum* 20S proteasome, whereas its regulatory **HslU** caps (443 residues) have ATPase activity and are homologous to *E. coli* ClpX. The X-ray structure of HslUV, determined by Huber, indicates that HslV forms a dimer of hexameric rather than heptameric rings (Fig. 32-89). A hexameric ring

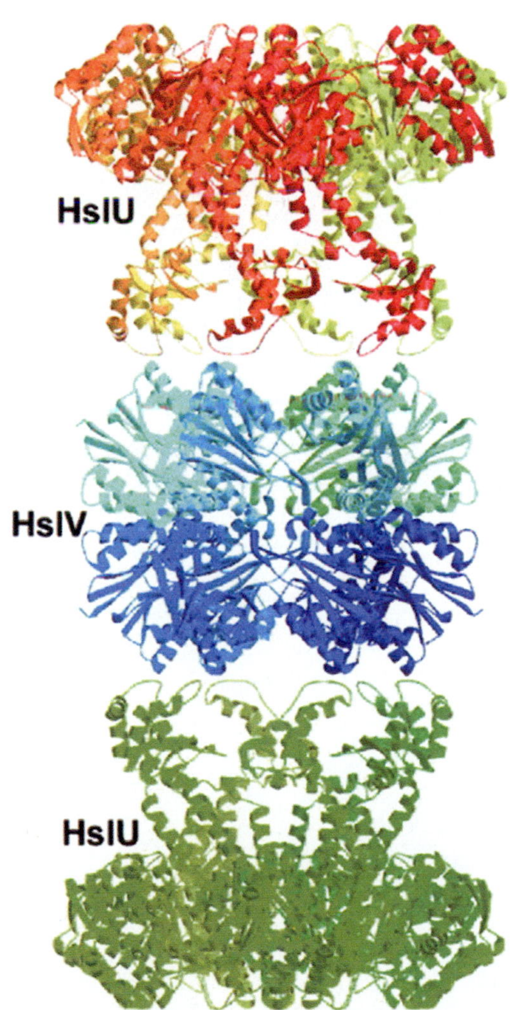

FIGURE 32-89 X-Ray structure of *E. coli* HslVU. The 820-kD complex is viewed along a 2-fold axis with its 6-fold axis vertical. The D_6 symmetric dodecamer of HslV subunits is coaxially bound at both ends by C_6 symmetric HslU hexamers to yield a complex with overall D_6 symmetry. [Courtesy of Robert Huber, Max-Planck-Institut für Biochemie, Martinsried, Germany. PDBid 1E94.]

(a)

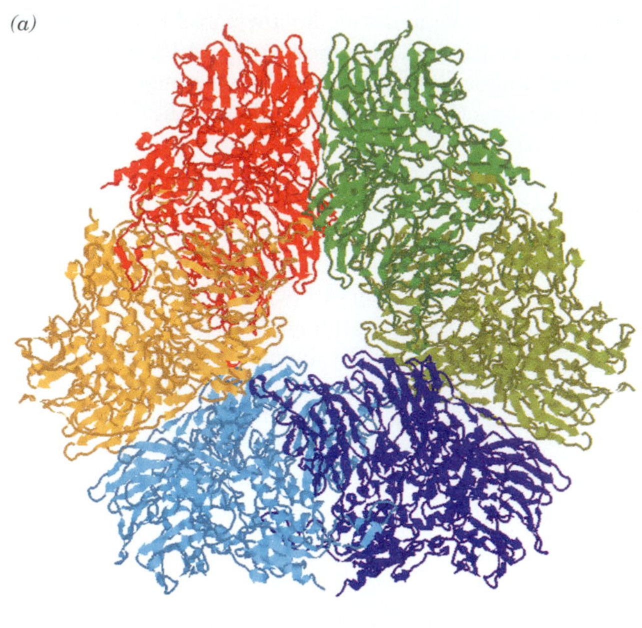

(b)

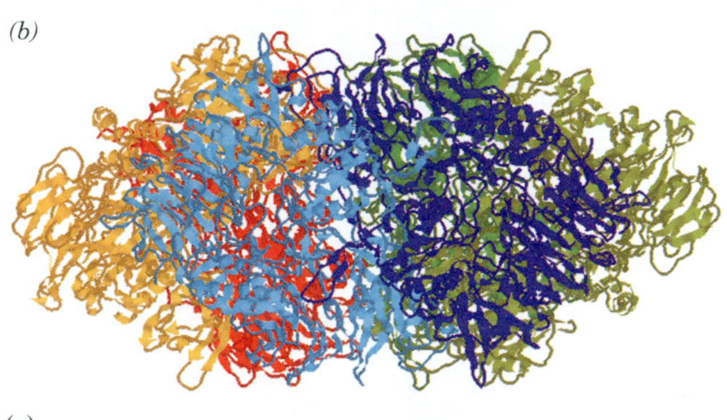

(c)

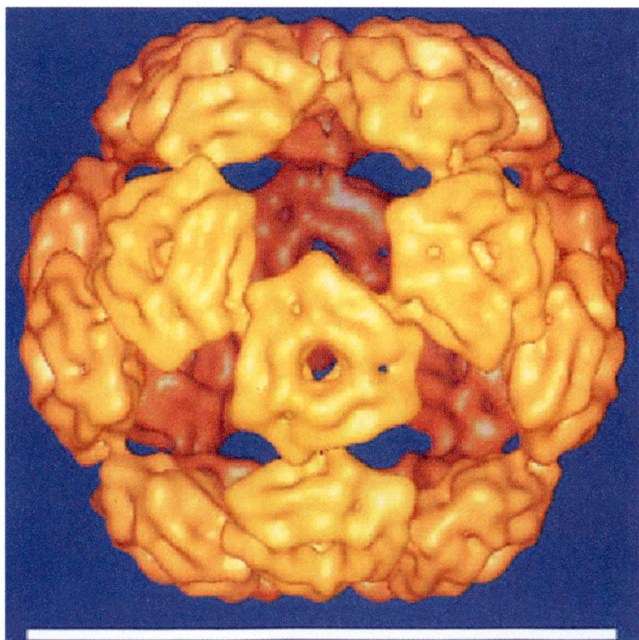

FIGURE 32-90 Structure of the tricorn protease. (*a*) The X-ray structure of the hexameric complex as viewed along its 3-fold axis with its subunits differently colored. (*b*) View along the hexameric complex's 2-fold axis (rotated 90° about a horizontal axis with respect to Part *a*). (*c*) A cryo-EM–based image of the icosahedral complex. Each of its component "plates" represents a hexameric complex such as those drawn in Parts *a* and *b*. Note how neighboring plates are related by both 3-fold axes (which are coincident with those of the hexameric complexes) and 5-fold axes. The white scale bar is 50 nm long. [Parts *a* and *b* are based on an X-ray structure by Robert Huber, Max-Planck-Institut für Biochemie, Martinsried, Germany. PDBid 1K32. Part *c* is courtesy of Wolfgang Baumeister, Max-Planck-Institut für Biochemie, Martinsried, Germany.]

of HslU subunits binds to both ends of the HslV dodecamer to form a 24-subunit assembly with D_6 symmetry, rather than the D_7 symmetry of the 20S proteasome. Nevertheless, both the fold and the intersubunit contacts of the HslV subunits are closely similar to those of the 20S proteasome β subunits. In addition, both have N-terminal Thr residues. Thus, HslV can be regarded as the eubacterial homolog of archaebacterial and eukaryotic 20S proteasomes.

T. acidophilum contains another large proteolytic complex, which is unrelated to the proteasome. The X-ray structure of this protease (Fig. 32-90*a,b*), determined by Huber, indicates that it forms a 730-kD toroidal hexameric ring with D_3 symmetry that has a peculiar triangular shape reminiscent of a tricorn (a hat whose brim is turned up on three sides) and hence was named **tricorn protease.** Cryo-EM studies indicate that 20 of these tricorn hexamers associate to form a 14,600-kD hollow icosahedron (Fig. 32-90*c*; an icosahedron is shown in Fig. 8-64*c*), making it by far the largest homooligomeric enzyme complex known (it is even larger than some virus particles, many of which also have icosahedral symmetry; Section 33-2A).

I. Ubiquitinlike Modifiers Participate in a Variety of Regulatory Processes

Eukaryotic cells express several proteins that are related in sequence to ubiquitin and are similarly conjugated to other proteins. These **ubiquitinlike modifiers (Ubls),** which participate in a variety of fundamental cellular processes, each have a corresponding activating enzyme (E1), at least one conjugating enzyme (E2), and in many cases, one or more ligases (E3s), that function to link the Ub1 to its target protein(s) in a manner closely resembling that of ubiquitin.

Two of the most extensively studied Ubls are **SUMO** (*s*mall *u*biquitin-related *mo*difier; 18% identical to ubiquitin) and **RUB1** (*r*elated-to-*ub*iquitin 1; called **NEDD8** in vertebrates; 50% identical to ubiquitin), proteins that are highly conserved from yeast to humans. One of SUMO's target proteins is IκBα, which binds to the transcription factor NF-κB so as to occlude its nuclear localization signal (see above). IκBα is sumoylated at the same residue (Lys 21) at which it is ubiquitinated, thereby blocking its ubiquitination and subsequent degradation and hence preventing the translocation of NF-κB to the nucleus. Evidently, there is a complex regulatory interplay between the ubiquitination and sumoylation of IκBα. SUMO also modifies two mammalian glucose transporters, GLUT1 and

GLUT4 (Section 20-2E), and in doing so, increases the availability of GLUT4 but decreases that of GLUT1.

All known RUB1 targets are cullins, all of which are subunits of SCF complexes, the multisubunit RING finger E3s (see above). In fact, β-TrCP, the E3 that directs the ubiquitination of IκBα, must be conjugated to RUB1 before it can do so, thereby adding further complexity to the control of NF-κB.

CHAPTER SUMMARY

1 ■ The Genetic Code Point mutations are caused by either base analogs that mispair during DNA replication or by substances that react with bases to form products that mispair. Insertion/deletion (frameshift) mutations arise from the association of DNA with intercalating agents that distort the DNA structure. The analysis of a series of frameshift mutations that suppress one another has established that the genetic code is an unpunctuated triplet code. In a cell-free protein synthesizing system, poly(U) directs the synthesis of poly(Phe), thereby demonstrating that UUU is the codon specifying Phe. The genetic code was elucidated through the use of polynucleotides of known composition but random sequence, by the ability of defined triplets to promote the ribosomal binding of tRNAs bearing specific amino acids, and through the use of synthetic mRNAs of known alternating sequences. The latter investigations have also demonstrated that the 5′ end of mRNA corresponds to the N-terminus of the polypeptide it specifies and have established the sequences of the Stop codons. Degenerate codons differ mostly in the identities of their third base. Small single-stranded DNA phages such as ϕX174 contain overlapping genes in different reading frames. The genetic code used by mitochondria differs in several codons from the "standard" genetic code.

2 ■ Transfer RNA and Its Aminoacylation Transfer RNAs consist of 54 to 100 nucleotides that can be arranged in the cloverleaf secondary structure. As many as 10% of a tRNA's bases may be modified. Yeast $tRNA^{Phe}$ forms a narrow, L-shaped, three-dimensional structure that resembles that of other tRNAs. Most of the bases are involved in stacking and base pairing associations including nine tertiary interactions that appear to be essential for maintaining the molecule's native conformation. Amino acids are appended to their cognate tRNAs in a two-stage reaction catalyzed by the corresponding aminoacyl–tRNA synthetase (aaRS). There are two classes of aaRSs, each containing 10 members. Class I aaRSs have two conserved sequence motifs that occur in the Rossmann fold common to the catalytic domain of these enzymes. Class II aaRSs have three conserved sequence motifs that occur in the 7-stranded antiparallel β sheet-containing fold that forms the core of their catalytic domains. In binding only their cognate tRNAs, aaRSs recognize only an idiosyncratic but limited number of bases (identity elements) that are, most often, located at the anticodon and in the acceptor stem. The great accuracy of tRNA charging arises from the proofreading of the bound amino acid by certain aminoacyl–tRNA synthetases via a double-sieve mechanism and at the expense of ATP hydrolysis.

Many organisms and organelles lack a GlnRS and instead synthesize Gln–$tRNA^{Gln}$ by the GluRS-catalyzed charging of $tRNA^{Gln}$ with glutamate followed by its transamidation using glutamine as the amido group source in a reaction mediated by Glu–$tRNA^{Gln}$ amidotransferase (Glu-AdT). Ribosomes select tRNAs solely on the basis of their anticodons. Sets of degenerate codons are read by a single tRNA through wobble pairing. The UGA codon, which is normally the *opal* Stop codon may, depending on its context in mRNA, specify a selenoCys (Sec) residue, which is carried by a specific tRNA ($tRNA^{Sec}$), thereby forming a selenoprotein. Nonsense mutations may be suppressed by tRNAs whose anticodons have mutated to recognize a Stop codon.

3 ■ Ribosomes and Polypeptide Synthesis The ribosome consists of a small and a large subunit whose complex shapes have been revealed by cryoelectron microscopy and X-ray crystallography. The three RNAs and 52 proteins comprising the *E. coli* ribosome self-assemble under proper conditions. Both ribosomal subunits consist of an RNA core in which the proteins are embedded, mainly as globular domains on the back and sides of the particle, with long basic polypeptide segments that infiltrate between the RNA helices so as to neutralize their anionic charges. Eukaryotic ribosomes are larger and more complex than those of prokaryotes.

Ribosomal polypeptide synthesis proceeds by the addition of amino acid residues to the C-terminal end of the nascent polypeptide. The mRNAs are read in the 5′ → 3′ direction. mRNAs are usually simultaneously translated by several ribosomes in the form of polysomes. The ribosome has three tRNA-binding sites: the A site, which binds the incoming aminoacyl–tRNA; the P-site, which binds the peptidyl–tRNA; and the E site, which transiently binds the outgoing deacylated tRNA. During polypeptide synthesis, the nascent polypeptide is transferred to the aminoacyl–tRNA, thereby lengthening the nascent polypeptide by one residue. The newly deacylated tRNA is translocated to the E site and the new peptidyl–tRNA, with its associated codon, is translocated to the P site. In prokaryotes, the initiation sites on mRNA are recognized through their Shine–Dalgarno sequences and by their initiating codon. Prokaryotic initiating codons specify fMet–$tRNA_f^{Met}$. Initiation involves the participation of three initiation factors that induce the assembly of the ribosomal subunits with fMet–$tRNA_f^{Met}$ in the P site and mRNA. Eukaryotic initiation is a far more complicated process that requires the participation of at least 11 initiation factors. The system binds the mRNA's 5′ cap and scans along the mRNA until it finds its AUG initiation codon, usually the mRNA's first AUG, through codon–anticodon interactions with the initiating tRNA, Met–$tRNA_i^{Met}$.

Polypeptides are elongated in a three-part cycle, consisting of aminoacyl–tRNA decoding, transpeptidation, and translocation, that requires the participation of elongation factors and is vectorially driven by GTP hydrolysis. EF-Tu, which functions to escort aminoacyl–tRNA into the ribosomal A site, undergoes a major conformational change on hydrolyzing its bound GTP. The X-ray structure of the 50S subunit clearly shows that the ribosomal peptidyl transferase center is distant from any protein and hence that the ribosome is a ribozyme. A2486 appears to be properly positioned and activated to function as a general base in the peptidyl transferase reaction. Translocation is motivated through the EF-G–catalyzed hydrolysis of GTP. EF-G · GDP, which binds to the same ribosomal site as aminoacyl–tRNA · EF-Tu · GTP, is a macromolecular mimic of this complex. Translocation occurs via intermediate states, the A/P and P/E states, in which the newly formed peptidyl–tRNA and the newly deacylated tRNA are respectively bound to the A and P subsites of the 30S subunit and to the P and E subsites of the 50S subunit, following which EF-G hydrolyzes its bound GTP and shifts these tRNAs to the P/P and E/E states. Termination codons bind release factors that induce the peptidyl transferase to hydrolyze the peptidyl–tRNA bond. Eukaryotic elongation and termination resemble that of prokaryotes.

The ribosome initially selects an aminoacyl–tRNA whose anticodon is cognate to its A-site–bound codon through interactions involving three universally conserved 30S subunit bases while the tRNA is in the A/T binding state. The codon–anticodon interaction is then proofread in an independent process that follows the hydrolysis of the EF-Tu–bound GTP and which occurs when the tRNA has shifted to the A/A binding state. Ribosomal inhibitors, many of which are antibiotics, are medically important and biochemically useful in elucidating ribosomal function. Streptomycin causes mRNA misreading and inhibits prokaryotic chain initiation, chloramphenicol inhibits prokaryotic peptidyl transferase, paromomycin causes codon misreading, tetracycline inhibits aminoacyl–tRNA binding to the prokaryotic 30S subunit, and diphtheria toxin ADP-ribosylates eEF2.

4 ■ Control of Eukaryotic Translation Several mechanisms of translational control have been elucidated in eukaryotes. eIF2α kinases catalyze the phosphorylation of eIF2α, which then tightly binds eIF2B so as to prevent it from exchanging eIF2-bound GDP for GTP and hence inhibits translational initiation. These eIF2α kinases include heme-regulated inhibitor (HRI), which functions to coordinate globin synthesis with heme availability; double-stranded RNA-activated protein kinase (PKR), an interferon-induced protein that functions to inhibit viral proliferation; and PKR-like endoplasmic reticulum kinase (PERK), which functions to protect the cell from the irreversible damage caused by the accumulation of unfolded proteins in the ER. GCN2, in contrast, is an eIF2α kinase that, when amino acids are scarce, stimulates the translation of the transcriptional activator GCN4 by causing the 40S ribosomal subunit to scan across four upstream open reading frames (uORFs) in the *GCN4* mRNA, thereby permitting the ribosome to initiate translation at the GCN4 coding sequence. The phosphorylation of eIF4E (cap-binding protein) by a MAP kinase cascade increases eIF4E's affinity for capped mRNA and thereby stimulates translational initiation. The binding of 4E-BPs to eIF4E blocks its binding of eIF4G and hence prevents initiation. However, the insulin-induced phosphorylation of the 4E-BPs causes them to dissociate from eIF4E. mRNAs in certain animal oocytes are masked by the binding of proteins, which prevents their translation.

Many oocyte mRNAs have short poly(A) tails that are preceded by a cytoplasmic polyadenylation element (CPE) that is bound by CPE-binding protein (CPEB). CPEB binds maskin which binds eIF4E, thereby inhibiting translational initiation. However, when CPEB is phosphorylated, it recruits poly(A) polymerase (PAP), which extends the mRNA's poly(A) tail such that it is bound by poly(A)-binding protein (PABP). PABP then binds to eIF4G, which in turn displaces maskin from eIF4E, thereby permitting the translation of the mRNA. Antisense oligonucleotides can be used to inhibit the translation of their complementary mRNAs. Although the delivery of antisense oligonucleotides to their sites of action has proved to be a difficult problem, their use is starting to show some medical and biotechnological successes.

5 ■ Posttranslational Modification Proteins may be posttranslationally modified in a variety of ways. Proteolytic cleavages, usually by specific peptidases, activate proproteins. The signal peptides of preproteins are removed by signal peptidases. Covalent modifications alter many types of side chains in a variety of ways that modulate the catalytic activities of enzymes, provide recognition markers, and stabilize protein structures. Protein splicing occurs via the intein-catalyzed self-excision between an N-extein and a C-extein accompanied by the ligation of the N- and C-exteins via a peptide bond. Most inteins contain a homing endonuclease that makes a double-strand nick in a gene similar to that encoding the corresponding extein, thereby triggering a recombinational double-strand DNA repair process that copies the gene encoding the intein into the break. Inteins therefore appear to be molecular parasites.

6 ■ Protein Degradation Proteins in living cells are continually turning over. This controls the level of regulatory enzymes and disposes of abnormal proteins that would otherwise interfere with cellular processes. Proteins are degraded by lysosomes via a nonspecific process as well via a process specific for KFERQ proteins that is stimulated during starvation. A cytosolically based ATP-dependent system degrades normal as well as abnormal proteins in a process that flags these proteins by the covalent attachment of polyubiquitin chains to their Lys residues. This process is mediated by three consecutively acting enzymes: ubiquitin-activating enzyme (E1), ubiquitin-conjugating enzyme (E2), and ubiquitin–protein ligase (E3). Most cells have one species of E1, several species of E2, and many species of E3, each of which is served by one or a few E2s. The polyubiquitinated protein is proteolytically degraded in the 26S proteasome.

E3s can have complicated modular structures, each having different specificities for target proteins. SCF complexes, one of whose several subunits contains a RING finger, are particularly elaborate. The RING finger E3 known as E3α functions to ubiquitinate proteins that satisfy the N-end rule. The transcription factor NF-κB is activated through the ubiquitination and subsequent destruction of its otherwise bound inhibitor IκBα by the SCF β-TrCP, which is activated through phosphorylation via a signal transduction cascade. Cyclins, which mediate the cell cycle, are destroyed in a programmed manner through ubiquitination by their cognate E3s, one of which is anaphase-promoting complex (APC).

The 26S proteasome consists of a hollow protein barrel formed by two rings of seven α subunits flanking two rings of seven β subunits known as the 20S proteasome, which is bound at each end by 19S caps that each consist of ~18 subunits. The active sites of the β subunits, which are members of the N-terminal nucleophile (Ntn) family of hydrolases, are inside the barrel. Ubiquitinated proteins are selected by the 19S caps, which unfold them in an ATP-dependent matter and then feed them into the 20S proteasome via its axial channel.

The polyubiquitin (polyUb) chains are excised from the condemned protein by proteasome-associated deubiquitinating enzymes (DUBs), while other DUBs dismember the polyUb chains to their component ubiquitin units, thereby recycling them. The 11S regulator is a heptameric complex that, on binding to one end of a 20S proteasome, opens its axial channel in an ATP-independent manner, thereby permiting polypeptides, but not folded proteins, to enter the 20S proteasome. Eubacteria, nearly all of which lack proteasomes, nevertheless express a variety of self-compartmentalized proteases, including ClpP, heat shock locus UV (HslUV), and tricorn protease, that function to proteolytically dispose of their cellular proteins. Ubiquitinlike modifiers (Ubls), such as SUMO and RUB1, participate in numerous regulatory processes.

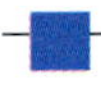

REFERENCES

GENERAL

Adams, R.L.P., Knowler, J.T., and Leader, D.P., *The Biochemistry of the Nucleic Acids* (11th ed.), Chapter 12, Chapman & Hall (1992).

Lewin, B., *Genes VII,* Chapters 5–7, Wiley (2000).

THE GENETIC CODE

Attardi, G., Animal mitochondrial DNA: an extreme example of genetic economy, *Int. Rev. Cytol.* **93,** 93–145 (1985).

Benzer, S., The fine structure of the gene, *Sci. Am.* **206**(1), 70–84 (1962). *The Genetic Code, Cold Spring Harbor Symp. Quant. Biol.* **31** (1966). [A collection of papers describing the establishment of the genetic code. See especially the articles by Crick, Nirenberg, and Khorana.]

Crick, F.H.C., The genetic code, *Sci. Am.* **207**(4), 66–74 (1962) [The structure of the code as determined by phage genetics]; *and* The genetic code: III, *Sci. Am.* **215**(4), 55–62 (1966). [A description of the nature of the code after its elucidation was almost complete.]

Crick, F.H.C., Burnett, L., Brenner, S., and Watts-Tobin, R.J., General nature of the genetic code for proteins, *Nature* **192,** 1227–1232 (1961).

Fox, T.D., Natural variation in the genetic code, *Annu. Rev. Genet.* **21,** 67–91 (1987).

Judson, J.F., *The Eighth Day of Creation,* Expanded Edition, Part II, Cold Spring Harbor Laboratory Press (1996). [A fascinating historical narrative on the elucidation of the genetic code.]

Khorana, H.G., Nucleic acid synthesis in the study of the genetic code, *Nobel Lectures in Molecular Biology, 1933–1975,* pp. 303–331, Elsevier (1977).

Knight, R.D., Freeland, S.J., and Landweber, L.F., Selection, history and chemistry: the three faces of the genetic code, *Trends Biochem. Sci.* **24,** 241–247 (1999).

Nirenberg, M., The genetic code, *Nobel Lectures in Molecular Biology, 1933–1975,* pp. 335–360, Elsevier (1977).

Nirenberg, M.W., The genetic code: II, *Sci. Am.* **208,** 80–94 (1963). [Discusses the use of synthetic mRNAs to analyze the genetic code.]

Nirenberg, M. and Leder, P., RNA code words and protein synthesis, *Science* **145,** 1399–1407 (1964). [The determination of the genetic code by the ribosomal binding of tRNAs using specific trinucleotides.]

Nirenberg, M.W. and Matthaei, J.H., The dependence of cell-free protein synthesis in *E. coli* upon naturally occurring or synthetic polyribonucleotides, *Proc. Natl. Acad. Sci.* **47,** 1588–1602 (1961). [The landmark paper reporting the finding that poly(U) stimulates the synthesis of poly(Phe).]

Singer, B. and Kusmierek, J.T., Chemical mutagenesis, *Annu. Rev. Biochem.* **51,** 655–693 (1982).

TRANSFER RNA AND ITS AMINOACYLATION

Alexander, R.W. and Schimmel, P., Domain–domain communication in aminoacyl–tRNA synthetases, *Prog. Nucleic Acid Res. Mol. Biol.* **69,** 317–349 (2001).

Björk, G.R., Ericson, J.U., Gustafsson, C.E.D., Hagervall, T.G., Jösson, Y.H., and Wikström, P.M., Transfer RNA modification, *Annu. Rev. Biochem.* **56,** 263–287 (1987).

Böck, A., Forschhammer, K., Heider, J., and Baron, C., Selenoprotein synthesis: an expansion of the genetic code, *Trends Biochem. Sci.* **16,** 463–467 (1991).

Carter, C.W., Jr., Cognition, mechanism, and evolutionary relationships in aminoacyl–tRNA synthetases, *Annu. Rev. Biochem.* **62,** 715–748 (1993).

Crick, F.H.C., Codon–anticodon pairing: the wobble hypothesis, *J. Mol. Biol.* **19,** 548–555 (1966).

Cusack, S., Aminoacyl–tRNA synthetases, *Curr. Opin. Struct. Biol.* **7,** 881–889 (1997).

Fukai, S., Nureki, O., Sekine, S., Shimada, A., Tao, J., Vassylyev, D.G., and Yokoyama, S., Structural basis for double-sieve discrimination of L-valine from L-isoleucine and L-threonine by the complex of $tRNA^{Val}$ and valyl–tRNA synthetase, *Cell* **103,** 793–803 (2000).

Geigé, R., Puglisi, J.D., and Florentz, C., tRNA structure and aminoacylation efficiency, *Prog. Nucleic Acid Res. Mol. Biol.* **45,** 129–206 (1993). [A detailed review.]

Gesteland, R.F., Weiss, R.B., and Atkins, J.F., Recoding: Reprogrammed genetic coding, *Science* **257,** 1640–1641 (1992). [Discusses contextual signals in mRNA that alter the way the ribosome reads certain codons.]

Hatfield, D.L., Lee, B.J., and Pirtle, R.M. (Eds.), *Transfer RNA in Protein Synthesis,* CRC Press (1992). [Contains articles on such subjects as the role of modified nucleosides in tRNAs, variations in reading the genetic code, patterns of codon usage, and tRNA identity elements.]

Hou, Y.-M., Discriminating among the discriminator bases of tRNAs, *Chem. Biol.* **4,** 93–96 (1997).

Ibba, M. and Söll, D., Aminoacyl–tRNA synthesis, *Annu. Rev. Biochem.* **69,** 617–650 (2000).

Ibba, M., Becker, H.D., Stathopoulos, C., Tumbula, D.L., and Söll, D., The adaptor hypothesis revisited, *Trends Biochem. Sci.* **25,** 311–316 (2000).

Ibba, M., Morgan, S., Curnow, A.W., Pridmore, D.R., Vothknecht, U.C., Gardner, W., Lin, W., Woese, C.R., and Söll, D., A euryarchaeal lysyl–tRNA synthetase: Resemblance to class I synthetases, *Science* **278,** 1119–1122 (1997).

Jacquin-Becker, C., Ahel, I., Ambrogelly, A., Ruan, B., Söll, D., and Stathopoulos, C., Cysteinyl–tRNA formation and prolyl-tRNA synthetase, *FEBS Lett.* **514,** 34–36 (2002).

Kim, S.H., Suddath, F.L., Quigley, G.J., McPherson, A., Sussman, J.L., Wang, A.M.J., Seeman, N.C., and Rich, A., Three-dimensional tertiary structure of yeast phenylalanine transfer RNA, *Science* **185,** 435–440 (1974); *and* Robertus, J.D., Ladner, J.E., Finch, J.T., Rhodes, D., Brown, R.S., Clark, B.F.C., and Klug, A., Structure of yeast phenylalanine tRNA at 3 Å resolution, *Nature* **250,** 546–551 (1974). [The landmark papers describing the high-resolution structure of a tRNA.]

McClain, W.H., Rules that govern tRNA identity in protein synthesis, *J. Mol. Biol.* **234,** 257–280 (1993).

Moras, D., Structural and functional relationships between aminoacyl–tRNA synthetases, *Trends Biochem. Sci.* **17,** 159–169 (1992). [Discusses Class I and Class II enzymes.]

Nureki, O., Vassylyev, D.G., Tateno, M., Shimada, A., Nakama, T., Fukai, S., Konno, M., Hendrickson, T.L., Schimmel, P., and Yokoyama, S., Enzyme structure with two catalytic sites for double-sieve selection of substrate, *Science* **280,** 578–582 (1998). [The X-ray structures of IleRS in complexes with isoleucine and valine.]

Rould, M.A., Perona, J.J., and Steitz, T.A., Structural basis of anticodon loop recognition by glutaminyl–tRNA synthetase, *Nature* **352,** 213–218 (1991).

Ruff, M., Krishnaswamy, S., Boeglin, M., Poterszman, A., Mitschler, A., Podjarny, A., Rees, B., Thierry, J.C., and Moras, D., Class II aminoacyl transfer RNA synthetases: Crystal structure of yeast aspartyl–tRNA synthetase complexed with $tRNA^{Asp}$, *Science* **252,** 1682–1689 (1991).

Saks, M.E., Sampson, J.R., and Abelson, J.N., The transfer identity problem: A search for rules, *Science* **263,** 191–197 (1994).

Sankaranarayanan, R., Dock-Bregeon, A.-C., Rees, B., Bovee, M., Caillet, J., Romby, P., Francklyn, C.S., and Moras, D., Zinc ion-mediated amino acid discrimination by threonyl–tRNA synthetase, *Nature Struct. Biol.* **7,** 461–465 (2000); *and* Dock-Bregeon, A.-C., Sankaranarayanan, R., Romby, P., Caillet, J., Springer, P., Rees, B., Francklyn, C.S., Ehresmann, C., and Moras, D., Transfer RNA-mediated editing in threonyl–tRNA synthetase: the Class II solution to the double discrimination problem, *Cell* **103,** 877–884 (2000).

Schimmel, P., Giegé, R., Moras, D., and Yokoyama, S., An operational RNA code for amino acids and possible relationship to genetic code, *Proc. Natl. Acad. Sci.* **90,** 8763–8768 (1993).

Schulman, L.H., Recognition of tRNAs by aminoacyl–tRNA synthetases, *Prog. Nucleic Acid Res. Mol. Biol.* **41,** 23–87 (1991).

Silvian, L.F., Wang, J., and Steitz, T.A., Insights into editing from an Ile-tRNA synthetase structure with $tRNA^{Ile}$ and mupirocin, *Science* **285,** 1074–1077 (1999).

Söll, D. and RajBhandary, U.L. (Eds.), *tRNA: Structure, Biosynthesis, and Function,* ASM Press (1995).

Stadtman, T.C., Selenocysteine, *Annu. Rev. Biochem.* **65,** 83–100 (1996).

Steege, D.A. and Söll, D.G., Suppression, *in* Goldberger, R.F. (Ed.), *Biological Regulation and Development,* Vol. 1, pp. 433–485, Plenum Press (1979).

RIBOSOMES AND POLYPEPTIDE SYNTHESIS

Agrawal, R.K., Spahn, C.MT., Penczek, P., Grassuci, R.A., Nierhaus, K.H., and Frank, J., Visualization of tRNA movements in the *Escherichia coli* 70S ribosome during the elongation cycle, *J. Cell Biol.* **150,** 447–459 (2000).

Ban, N., Nissen, P., Hansen, J., Moore, P.B., and Steitz, T., The complete atomic structure of the large ribosomal subunit at 2.4 Å resolution, *Science* **289,** 905–920 (2000).

Bell, C.E. and Eisenberg, D.E., Crystal structure of diphtheria toxin bound to nicotinamide adenine dinucleotide, *Biochemistry* **35,** 1137–1149 (1996).

Brodersen, D.E., Clemons, W.M., Jr., Carter, A.P., Morgan-Warren, R.J., Wimberly, B.T., and Ramakrishnan, V., The structural basis for the action of the antibiotics tetracycline, pactamycin, and hygromycin B on the 30S ribosomal subunit, *Cell* **103,** 1143–1154 (2000).

Czworkowski, J., Wang, J., Steitz, J.A., and Moore, P.B., The crystal structures of elongation factor G complexed with GDP, at 2.7 Å resolution; *and* Ævarsson, A., Brazhnikov, E., Garber, M., Zheltonosova, J., Chirgadze, Yu., Al-Karadaghi, S., Svensson, L.A., and Liljas, A., Three-dimensional structure of the ribosomal translocase: elongation factor G from *Thermus thermophilus, EMBO J.* **13,** 3661–3668 *and* 3669–3677 (1994).

Dintzis, H.M., Assembly of the peptide chains of hemoglobin, *Proc. Natl. Acad Sci.* **47,** 247–261 (1961). [The determination of the direction of polypeptide biosynthesis.]

Fersht, A., *Structure and Mechanism in Protein Science,* Chapter 13, Freeman (1999). [A discussion of enzymatic specificity and editing mechanisms.]

Frank, J., Single-particle imaging of macromolecules by cryo-electron microscopy, *Annu. Rev. Biophys. Biomol. Struct.* **31,** 303–319 (2002).

Frank, J. and Agrawal, R.K., A ratchet-like inter-subunit reorganization of the ribosome during translocation, *Nature* **406,** 318–322 (2000).

Gingras, A.-C., Raught, B., and Sonnberg, N., eIF4 initiation factors: effectors of mRNA recruitment to ribosomes and regulators of translation, *Annu. Rev. Biochem.* **68,** 913–963 (1999).

Green, R. and Lorsch, J.R., The path to perdition is paved with protons, *Cell* **110,** 665–668 (2002). [Discusses the difficulties in characterizing the catalytic mechanism of the ribosomal peptidyl transferase.]

Hansen, J., Schmeing, T.M., Moore, P.B., and Steitz, T., Structural insights into peptide bond formation, *Proc. Natl. Acad. Sci.* **99,** 11670–11675 (2002); *and* Nissen, P., Hansen, J., Ban, N., Moore, P.B., and Steitz, T., The structural basis of ribosome activity in peptide bond synthesis, *Science* **289,** 920–930 (2000).

Held, W.A., Ballou, B., Mizushima, S., and Nomura, M., Assembly mapping of 30S ribosomal proteins from *Escherichia coli, J. Biol. Chem.* **249,** 3103–3111 (1974).

Jenni, S. and Ban, N., The chemistry of protein synthesis and voyage through the ribosomal tunnel, *Curr. Opin. Struct. Biol.* **13,** 212–219 (2003).

Kawashima, T., Berthet-Colominas, C., Wulff, M., Cusack, S., and Leberman, R., The structure of the *Escherichia coli* EF-Tu · EF-Ts complex at 2.5 Å resolution, *Nature* **379,** 511–518 (1996).

Kisselev, L.L. and Buckingham, R.H., Translation termination comes of age, *Trends Biochem. Sci.* **25,** 561–566 (2000).

Kjeldgaard, M. and Nyborg, J., Refined structure of elongation factor EF-Tu from *Escherichia coli, J. Mol. Biol.* **223,** 721–742 (1992); *and* Kjeldgaard, M., Nissen, P., Thirup, S., and Nyborg, J., The crystal structure of elongation factor EF-Tu from *Thermus aquaticus* in the GTP conformation, *Structure* **1,** 35–50 (1993).

Lake, J.A., Evolving ribosome structure: domains in archaebacteria, eubacteria, eocytes and eukaryotes, *Annu. Rev. Biochem.* **54,** 507–530 (1985).

Lancaster, L., Kiel, M.C., Kaji, A., and Noller, H.F., Orientation of ribosome recycling factor in the ribosome from directed hydroxyl radical probing, *Cell* **111,** 129–140 (2002).

Marcotrigiano, J., Gingras, A.-C., Sonenberg, N., and Burley, S.K., Cocrystal structure of the messenger RNA 5′ cap-binding protein (eIF4E) bound to 7-methyl-GDP, *Cell* **89,** 951–961 (1997).

Moazed, D. and Noller, H.F., Intermediate states in the movement of transfer RNA in the ribosome, *Nature* **342,** 142–148 (1989).

Moore, P.B. and Steitz, T.A., The involvement of RNA in ribosome function, *Nature* **418,** 229–235 (2002); After the ribosome: How does peptidyl transferase work, *RNA* **9,** 155–159 (2003); *and* The structural basis of large ribosomal subunit function, *Annu. Rev. Biochem.* **72,** 813–850 (2003).

Nakamura, Y., Ito, K., and Ehrenberg, M., Mimicry grasps reality in translation terminator, *Cell* **101,** 349–352 (2000).

Nissen, P., Kjeldgaard, M., Thirup, S., Polekhina, G., Reshetnikova, L., Clark, B.F.C., and Nyborg, J., Crystal structure of the ternary complex of Phe–$tRNA^{Phe}$, EF-Tu, and a GTP analog, *Science* **270,** 1464–1472 (1995).

Noller, H.F., Hoffarth, V., and Zimniak, L., Unusual resistance of peptidyl transferase to protein extraction procedures, *Science* **256,** 1416–1419 (1992); *and* Noller, H.F., Peptidyl transferase: protein, ribonucleoprotein, or RNA? *J. Bacteriol.* **175,** 5297–5300 (1993).

Nollar, H.F., Yusupov, M.M., Yusupova, G.Z., Baucom, A., and Cate, J.H.D., Translocation of tRNA during protein synthesis, *FEBS Lett.* **514,** 11–16 (2002).

Ogle, J.M., Brodersen, D.E., Clemons, W.M., Jr., Tarry, M.J., Carter, A.P., and Ramakrishnan, V., Recognition of cognate transfer RNA by the 30S ribosomal subunit, *Science* **292,** 897–902 (2001); *and* Ogle, J.M., Carter, A.P., and Ramakrishnan, V., Insights into the decoding mechanism from recent ribosome structures, *Trends Biochem. Sci.* **28,** 259–266 (2003).

Pioletti, M., et al., Crystal structure of complexes of the small ribosomal subunit with tetracycline, edeine and IF3, *EMBO J.* **20,** 1829–1839 (2001).

Poole, E. and Tate, W., Release factors and their role as decoding proteins: specificity and fidelity for termination of protein synthesis, *Biochim. Biophys. Acta* **1493,** 1–11 (2000).

Ramakrishnan, V., Ribosome structure and the mechanism of translocation, *Cell* **108,** 557–572 (2002). [A detailed and incisive review.]

Ramakrishnan, V. and Moore, P.B., Atomic structure at last: the ribosome in 2000, *Curr. Opin. Struct. Biol.* **11,** 144–154 (2001).

Rané, H.A., Klootwijk, J., and Musters, W., Evolutionary conservation of structure and function of high molecular weight ribosomal RNA, *Prog. Biophys. Mol. Biol.* **51,** 77–129 (1988).

Rawat, U.B.S., Zavialov, A.V., Sengupta, J., Valle, M., Grassuci, R.A., Linde, J., Vestergaard, B., Ehrenberg, M., and Frank, J., A cryo-electro microscopic study of ribosome-bound termination factor RF2; *and* Klaholz, B.P., Pape, T., Zavialov, A.V., Myasnikov, A.G., Orlova, E.V., Vestergaard, B., Ehrenberg, M., and van Heel, M., Structure of Eschericia coli ribosomal termination complex with release factor 2, *Nature* **421,** 87–90 and 90–94 (2003).

Rodnina, M.V. and Wintermeyer, W., Ribosome fidelity: tRNA discrimination, proofreading and induced fit, *Trends Biochem. Sci.* **26,** 124–130 (2001); Fidelity of aminoacyl–tRNA selection on the ribosome: kinetic and structural mechanisms, *Annu. Rev. Biochem.* **70,** 415–435 (2001); *and* Peptide bond formation on the ribosome: Structure and mechanism, *Curr. Opin. Struct. Biol.* **13,** 334–340 (2003).

Schluenzen, F., Tocilj, A., Zarivach, R., Harms, J., Gluehmann, M., Janell, D., Bashan, A., Bartels, H., Agmon, I., Franceschi, F., and Yonath, A., Structure of functionally activated small ribosomal subunit at 3.3 Å resolution, *Cell* **102,** 615–623 (2000).

Selmer, M., Al-Karadaghi, S., Hirokawa, G., Kaji, A., and Liljas, A., Crystal structure of *Thermatoga maritima* ribosome recycling factor: a tRNA mimic, *Science* **286,** 2349–2352 (1999).

Sonenberg, N. and Dever, T.E., Eukaryotic translation initiation factors and regulators, *Curr. Opin. Struct. Biol.* **13,** 56–63 (2003).

Song, H., Mugnier, P., Das, A.K., Webb, H.M., Evans, D.R., Tuite, M.F., Hemmings, B.A., and Barford, D., The crystal structure of human eukaryotic release factor eRF1—Mechanism of stop codon recognition and peptidyl–tRNA hydrolysis, *Cell* **100,** 311–321 (2000).

Spahn, C.M.T., Beckmann, R., Eswar, N., Penczek, P.A., Sali, A., Blobel, G., and Frank, J., Structure of the 80S ribosome from *Saccharomyces cerevisiae*—tRNA-ribosome and subunit-subunit interactions, *Cell* **107,** 373–386 (2001).

Spirin, A.S., Ribosome as a molecular machine, *FEBS Lett.* **514,** 2–10 (2002). [Discusses the role in GTP hydrolysis in ribosomal processes.]

Steitz, J.A. and Jakes, K., How ribosomes select initiator regions in mRNA: base pair formation between the 3′ terminus of 16S RNA and the mRNA during initiation of protein synthesis in *Escherichia coli, Proc. Natl. Acad. Sci.* **72,** 4734–4738 (1975).

The Ribosome, Cold Spring Harbor Symposium on Quantitative Biology, Volume LXVI, Cold Spring Harbor Laboratory Press (2001). [The latest "bible" of ribosomology.]

Vestergaard, B., Van, L.B., Andersen, G.R., Nyborg, J., Buckingham, R.H., and Kjeldgaard, M., Bacterial polypeptide release factor RF2 is structurally distinct from eukaryotic eRF1, *Mol. Cell* **8,** 1375–1382 (2001).

Wimberly, B.T., Broderson, D.E., Clemons, W.M., Jr., Morgan-Warren, R., von Rhein, C., Hartsch, T., and Ramakrishnan, V., Structure of the 30S ribosomal subunit, *Nature* **407,** 327–339 (2000); *and* Broderson, D.E., Clemons, W.M., Jr., Carter, A.P., Wimberly, B.T., and Ramakrishnan, V., Crystal structure of the 30S ribosomal subunit from *Thermus thermophilus:* Structure of the proteins and their interactions with 16S RNA, *J. Mol. Biol.* **316,** 725–768 (2002).

Yonath, A., The search and its outcome: High resolution structures of ribosomal particles from mesophilic, thermophilic, and halophilic bacteria at various functional states, *Annu. Rev. Biophys. Biomol. Struct.* **31,** 257–273 (2002).

Yusupova, G.Z., Yusupov, M.M., Cate, J.D.H., and Noller, H.F., The path of messenger RNA through the ribosome, *Cell* **106,** 233–241 (2001).

CONTROL OF TRANSLATION

Branch, A.D., A good antisense molecule is hard to find, *Trends Biochem. Sci.* **23,** 45–50 (1998).

Calkhoven, C.F., Müller, C., and Leutz, A., Translational control of gene expression and disease, *Trends Molec. Med.* **8,** 577–583 (2002).

Chen, J.-J., and London, I.M., Regulation of protein synthesis by heme-regulated eIF-2α kinase, *Trends Biochem. Sci.* **20,** 105–108 (1995).

Clemens, M.J., PKR—A protein kinase regulated by double-stranded RNA, *Int. J. Biochem. Cell Biol.* **29,** 945–949 (1997).

Dever, T.E., Gene-specific regulation by general translation factors, *Cell* **108,** 545–556 (2002).

Gray, N.K. and Wickens, M., Control of translation initiation in animals, *Annu. Rev. Cell Dev. Biol.* **14,** 399–458 (1998).

Hershey, J.W.B., Translational control in mammals, *Annu. Rev. Biochem.* **60,** 717–755 (1991).

Kozak, M., Regulation of translation in eukaryotic systems, *Annu. Rev. Cell Biol.* **8,** 197–225 (1992).

Lawrence, J.C., Jr. and Abraham, R.T., PHAS/4E-BPs as regulators of mRNA translation and cell proliferation, *Trends Biochem. Sci.* **22,** 345–349 (1997).

Lebedeva, I. and Stein, C.A., Antisense oligonucleotides: promise and reality, *Annu. Rev. Pharmacol. Toxicol.* **41,** 403–419 (2001).

Mendez, R. and Richter, J.D., Translational control by CPEB: a means to the end, *Nature Rev. Mol. Cell Biol.* **2,** 521–529 (2001).

Phillips, M.I. (Ed.), *Antisense technology: Part A. General Methods, Methods of Delivery, and RNA Studies;* and *Part B. Applications, Meth. Enzymol.* **313** *and* **314** (2000).

Sen, G.C. and Lengyel, P., The interferon system, *J. Biol. Chem.* **267,** 5017–5020 (1992).

Sheehy, R.E., Kramer, M., and Hiatt, W.R., Reduction of polygalacturonase activity in tomato fruit by antisense RNA, *Proc. Natl. Acad. Sci.* **85,** 8805–8809 (1988).

Sonenberg, N., Hershey, J.W.B., and Mathews, M.B., *Translational Control of Gene Expression,* Cold Spring Harbor Laboratory Press (2000). [A compendium of authoritative articles.]

Tafuri, S.R. and Wolffe, A.P., Dual roles for transcription and translation factors in the RNA storage particles of *Xenopus* oocytes, *Trends Cell. Biol.* **3,** 94–98 (1993).

Tamm, I., Dörken, B., and Hartmann, G., Antisense therapy in oncology: new hope for an old idea? *Lancet* **358,** 489–497 (2001).

Weiss, B., Davidkova, G., and Zhou, L.-W., Antisense RNA therapy for studying and modulating biological processes, *Cell. Mol. Life Sci.* **55,** 334–358 (1999).

POSTTRANSLATIONAL MODIFICATION

Fessler, J.H. and Fessler, L.I., Biosynthesis of procollagen, *Annu. Rev. Biochem.* **47,** 129–162 (1978).

Harding, J.J., and Crabbe, M.J.C. (Eds.), *Post-Translational Modifications of Proteins,* CRC Press (1992).

Klabunde, T., Sharma, S., Telenti, A., Jacobs, W.R., Jr., and Sacchetini, J.C., Crystal structure of Gyr A protein from *Mycobacterium xenopi* reveals structural basis of splicing, *Nature Struct. Biol.* **5,** 31–36 (1998).

Liu, X.-Q., Protein-splicing intein: genetic mobility, origin, and evolution, *Annu. Rev. Genet.* **34,** 61–76 (2000).

Noren, C.J., Wang, J., and Perler, F.B., Dissecting the chemistry of protein splicing and its applications, *Angew. Chem. Int. Ed.* **39,** 450–466 (2000).

Wold, F., In vivo chemical modification of proteins, *Annu. Rev. Biochem.* **50,** 783–814 (1981).

Wold, F. and Moldave, K. (Eds.), Posttranslational Modifications, Parts A and B, *Methods Enzymol.* **106** and **107** (1984). [Contains extensive descriptions of the amino acid "zoo."]

PROTEIN DEGRADATION

Bochtler, M., Ditzel, L., Groll, M., Hartmann, C., and Huber, R., The proteasome, *Annu. Rev. Biophys. Biomol. Struct.* **28,** 295–317 (1999).

Brandstetter, H., Kim, J.-S., Groll, M., and Huber, R., Crystal structure of the tricorn protease reveals a protein disassembly line, *Nature* **414,** 466–470 (2001); Walz, J., Tamura, T., Tamura, N., Grimm, R., Baumeister, W., and Koster, A.J., Tricorn protease exists as an icosahedral supermolecule *in vivo, Mol. Cell* **1,** 59–65 (1997); *and* Walz, J., Koster, A.J., Tamura, T., and Baumeister, W., Capsids of tricorn protease studied by cryomicroscopy, *J. Struct. Biol.* **128,** 65–68 (1999).

Cook, W.J., Jeffrey, L.C., Kasperek, E., and Pickart, C.M., Structure of tetraubiquitin shows how multiubiquitin chains can be formed, *J. Mol. Biol.* **236,** 601–609 (1994).

Cook, W.J., Jeffrey, L.C., Sullivan, M.L., and Vierstra, R.D., Three-dimensional structure of a ubiquitin-conjugating enzyme (E2), *J. Biol. Chem.* **267,** 15116–15121.

Deshaies, R.J., SCF and cullin/RING H2-based ubiquitin ligases, *Annu. Rev. Cell Dev. Biol.* **15,** 435–467 (1999).

Dice, F., Peptide sequences that target cytosolic proteins for lysosomal proteolysis, *Trends Biochem. Sci.* **15,** 305–309 (1990).

Ferrell, K., Wilkinson, C.R.M., Dubiel, W., and Gordon, C., Regulatory subunit interactions of the 26S proteasome, a complex problem, *Trends Biochem. Sci.* **25,** 83–88 (2000).

Glickman, M.H. and Ciechanover, A., The ubiquitin-proteasome proteolytic pathway: destruction for the sake of construction, *Physiol. Rev.* **82,** 373–428 (2002).

Goldberg, A.L., and Rock, K.L., Proteolysis, proteasomes and antigen presentation, *Nature* **357,** 375–379 (1992).

Hartmann-Petersen, R., Seeger, M., and Gordon, C., Transferring substrates to the 26S proteasome, *Trends Biochem. Sci.* **28,** 26–31 (2003).

Hershko, A. and Ciechanover, A., The ubiquitin system, *Annu. Rev. Biochem.* **67,** 425–479 (1998).

Jentsch, S. and Pyrowalakis, G. Ubiquitin and its kin: how close are the family ties, *Trends Cell Biol.* **10,** 335–342 (2003). [Discusses Ubls.]

Lam, Y.A., Lawson, T.G., Velayutham, M., Zweier, J.L., and Pickart, C.M., A proteasomal ATPase subunit recognizes the polyubiquitin degradation signal, *Nature* **416,** 763–767 (2002).

Laney, J.D. and Hochstrasser, M., Substrate targeting in the ubiquitin system, *Cell* **97,** 427–430 (1999).

Löwe, J., Stock, D., Jap, B., Zwicki, P., Baumeister, W., and Huber, R., Crystal structure of the 20S proteasome from the archeon *T. acidophilum* at 3.4 Å resolution, *Science* **268,** 533–539 (1995); *and* Groll, M., Ditzel, L., Löwe, J., Stock, D., Bochtler, M., Bartunik, H.D., and Huber, R., Structure of 20S proteasome from yeast at 2.4 Å resolution, *Nature* **386,** 463–471 (1997).

Page, A.M. and Hieter, P., The anaphase-promoting complex: new subunits and regulators, *Annu. Rev. Biochem.* **68,** 583–609 (1999).

Pickart, C.M., Mechanisms underlying ubiquitination, *Annu. Rev. Biochem.* **70,** 503–533 (2001); *and* Ubiquitin in chains, *Trends Biochem. Sci.* **25,** 544 (2000).

Schwartz, A.L. and Ciechanover, A., The ubiquitin-proteasome pathway and the pathogenesis of human disease, *Annu. Rev. Med.* **50,** 57–74 (1999).

Senahdi, V.-J., Bugg, C.E., Wilkinson, K.D., and Cook, W.J., Three-dimensional structure of ubiquitin at 2.8 Å resolution, *Proc. Natl. Acad. Sci.* **82,** 3582–3585 (1985).

Song, H.K., Hartmann, C., Ramachandran, R., Bochtler, M., Behrendt, R., Moroder, L., and Huber, R., Mutational studies on HslU and its docking mode with HslV, *Proc. Natl. Acad. Sci.* **97,** 14103–14108 (2000).

Unno, M., Mizushima, T., Morimoto, Y., Tomisugi, Y., Tanaka, K., Yasuoka, N., and Tsukihara, T., The structure of the mammalian proteasome at 2.75 Å resolution, *Structure* **10,** 609–618 (2002).

VanDemark, A.P. and Hill, C.P., Structural basis of ubiquitylation, *Curr. Opin. Struct. Biol.* **12,** 822–830 (2002).

Varshavsky, A., Turner, G., Du, F., and Xie, Y., The ubiquitin system and the N-end rule, *Biol. Chem.* **381,** 779–789 (2000); *and* Varshavsky, A., The N-end rule, *Cell* **69,** 725–735 (1992).

Voges, D., Zwickl, P., and Baumeister, W., The 26S proteasome: a molecular machine designed for controlled proteolysis, *Annu. Rev. Biochem.* **68,** 1015–1068 (1999).

Wang, J., Hartling, J.A., and Flanagan, J.M., The structure of ClpP at 2.3 Å resolution suggests a model for ATP-dependent proteolysis, *Cell* **91,** 447–456 (1997).

Whitby, F.G., Masters, E.I, Kramer, L., Knowlton, J.R., Yao, Y., Wang, C.C., and Hill, C.P., Structural basis for the activation of 20S proteasomes by 11S regulators, *Nature* **408,** 115–120 (2000).

Yao, T. and Cohen, R.E., A cryptic protease couples deubiqitination and degradation by the proteasome, *Nature* **419,** 403–407 (2002); and Verma, R., Aravind, L., Oania, R., McDonald, W.H., Yates, J.R., III, Koonin, E.V., and Deshaies, R.J., Role of Rpn11 metalloprotease in deubiquitination and degradation by the 26S proteasome, *Science* **298,** 611–615 (2002).

Zheng, N., et al., Structure of the Cul1–Rbx1–Skp1–F-box^{Skp2} SCF ubiquitin ligase complex, *Nature* **41,** 703–709 (2002);

Schulman, B.A., et al., Insights into SCF ubiquitin ligases from the structure of the Skp1–Skp2 complex, *Nature* **408,** 381–386 (2000); *and* Zheng, N., Wang, P., Jeffrey, P.D., and Pavletich, N.P., Structure of a c-Cbl–UbcH7 complex: RING domain function in ubiquitin-protein ligases, *Cell* **102,** 533–539 (2000).

Zwickl, P., Seemüller, E., Kapelari, B., and Baumeister, W., The proteasome: A supramolecular assembly designed for controlled proteolysis, *Adv. Prot. Chem.* **59,** 187–222 (2002).

PROBLEMS

1. What is the product of the reaction of guanine with nitrous acid? Is the reaction mutagenic? Explain.

2. What is the polypeptide specified by the following DNA antisense strand? Assume translation starts after the first initiation codon.

5′-TCTGACTATTGAGCTCTCTGGCACATAGCA-3′

***3.** The fingerprint of a protein from a phenotypically revertant mutant of bacteriophage T4 indicates the presence of an altered tryptic peptide with respect to the wild type. The wild-type and mutant peptides have the following sequences:

Wild type	Cys-Glu-Asp-His-Val-Pro-Gln-Tyr-Arg
Mutant	Cys-Glu-Thr-Met-Ser-His-Ser-Tyr-Arg

Indicate how the mutant could have arisen and give the base sequences, as far as possible, of the mRNAs specifying the two peptides. Comment on the function of the peptide in the protein.

4. Explain why the various classes of mutations can reverse a mutation of the same class but not a different class.

5. Which amino acids are specified by codons that can be changed to an *amber* codon by a single point mutation?

6. The mRNA specifying the α chain of human hemoglobin contains the base sequence

· · ·UCCAAAUACCGUUAAGCUGGA· · ·

The C-terminal tetrapeptide of the normal α chain, which is specified by part of this sequence, is

-Ser-Lys-Tyr-Arg

In hemoglobin Constant Spring, the corresponding region of the α chain has the sequence

-Ser-Lys-Tyr-Arg-Gln-Ala-Gly-· · ·

Specify the mutation that causes hemoglobin Constant Spring.

7. Explain why a minimum of 32 tRNAs are required to translate the "standard" genetic code.

8. Draw the wobble pairings not in Fig. 32-25*a*.

9. A colleague of yours claims that by exposing *E. coli* to HNO_2 she has mutated a $tRNA^{Gly}$ to an *amber* suppressor. Do you believe this claim? Explain.

***10.** Deduce the anticodon sequences of all suppressors listed in Table 32-6 except UGA-1 and indicate the mutations that caused them.

11. How many different types of macromolecules must be minimally contained in a cell-free protein synthesizing system from *E. coli?* Count each type of ribosomal component as a different macromolecule.

12. Why do oligonucleotides containing Shine–Dalgarno sequences inhibit translation in prokaryotes? Why don't they do so in eukaryotes?

13. Why does m^7GTP inhibit translation in eukaryotes? Why doesn't it do so in prokaryotes?

14. What would be the distribution of radioactivity in the completed hemoglobin chains on exposing reticulocytes to ^3H-labeled leucine for a short time followed by a chase with unlabeled leucine?

15. Design an mRNA with the necessary prokaryotic control sites that codes for the octapeptide Lys-Pro-Ala-Gly-Thr-Glu-Asn-Ser.

16. Indicate the translational control sites in and the amino acid sequence specified by the following prokaryotic mRNA.

5′-CUGAUAAGGAUUUAAAUUAUGUGUCAAUCACGA-
AUGCUAAUCGAGGCUCCAUAAUAACACUU CGAC-3′

17. What is the energetic cost, in ATP equivalents, for the *E. coli* synthesis of a polypeptide chain of 100 residues starting from amino acids and mRNA? Assume that no losses are incurred as a result of proofreading.

***18.** It has been suggested that Gly-tRNA synthetase does not require an editing mechanism. Why?

19. An antibiotic named fixmycin, which you have isolated from a fungus growing on ripe passion fruit, is effective in curing many types of venereal disease. In characterizing fixmycin's mode of action, you have found that it is a bacterial translational inhibitor that binds exclusively to the large subunit of *E. coli* ribosomes. The initiation of protein synthesis in the presence of fixmycin results in the generation of dipeptides that remain associated with the ribosome. Suggest a mechanism of fixmycin action.

20. Heme inhibits protein degradation in reticulocytes by allosterically regulating ubiquitin-activating enzyme (E1). What physiological function might this serve?

21. Genbux Inc., a biotechnology firm, has cloned the gene encoding an industrially valuable enzyme into *E. coli* such that the enzyme is produced in large quantities. However, since the firm wishes to produce the enzyme in ton quantities, the expense of isolating it would be greatly reduced if the bacterium could be made to secrete it. As a high-priced consultant, what general advice would you offer to solve this problem?

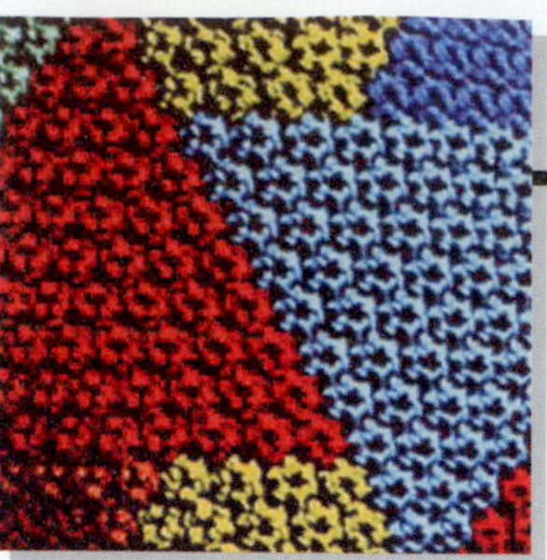

Chapter **33**

Viruses: Paradigms For Cellular Functions

***Viruses** are parasitic entities, consisting of nucleic acid molecules with protective coats, that are replicated by the enzymatic machinery of suitable host cells.* Since they lack metabolic apparatus, viruses are not considered to be alive (although this is a semantic rather than a scientific distinction). They range in complexity from **satellite tobacco necrosis virus** (**STNV;** Section 33-2B), whose genome encodes only one protein, to **Paramecium bursaria Chlorella virus** (**PBCV;** Section 33-2G), which encodes 377 proteins [to put this latter number into perspective, the genome of the smallest free-living organism, *Mycoplasma genitalium* (Table 7-3), encodes ~470 proteins].

Viruses were originally characterized at the end of the nineteenth century as infectious agents that could pass through filters that held back bacteria. Yet viral diseases, varying in severity from smallpox and rabies to the common cold, have no doubt plagued mankind since before the dawn of history. It is now known that viruses can infect plants and bacteria as well as animals. Each viral species has a very limited **host range;** that is, it can reproduce in only a small group of closely related species.

An intact virus particle, which is referred to as a **virion,** consists of a nucleic acid molecule encased by a protein **capsid.** In some of the more complex virions, the capsid is surrounded by a lipid bilayer and glycoprotein-containing **envelope,** which is derived from a host cell membrane. Since the small size of a viral nucleic acid severely limits the number of proteins that can be encoded by its genome, its capsid, as Francis Crick and James Watson pointed out in 1957, must be built up of one or a few kinds of protein subunits that are arranged in a symmetrical or nearly symmetrical fashion. There are two ways that this can occur:

1. In the **helical viruses** (Section 33-1), the coat protein subunits associate to form helical tubes.

2. In the **icosahedral viruses** (also known as **spherical viruses;** Section 33-2), coat proteins aggregate as closed polyhedral shells.

In both cases, the viral nucleic acid occupies the capsid's central region. In many viruses, the coat protein subunits may be "decorated" by other proteins so that the capsid exhibits spikes and, in larger bacteriophages, a complex tail. These assemblies are involved in recognizing the host cell and delivering the viral nucleic acid into its interior. Figure 33-1 is a "rogues' gallery" of viruses of varying sizes and morphologies.

The great simplicity of viruses in comparison to cells makes them invaluable tools in the elucidation of gene structure and function, as well as our best characterized models for the assembly of biological structures. Although all viruses use ribosomes and other host factors for the RNA-instructed synthesis of proteins, their modes of genome replication are far more varied than that of cellular life. In contrast to cells, in which the hereditary molecules are invariably double-stranded DNA, viruses contain either single- or double-stranded DNA or RNA. In RNA viruses, the viral RNA may be directly replicated or act as a template in the synthesis of DNA. The RNA of single-stranded RNA viruses may be the positive strand (the mRNA) or the negative strand (complementary to the mRNA). Viral

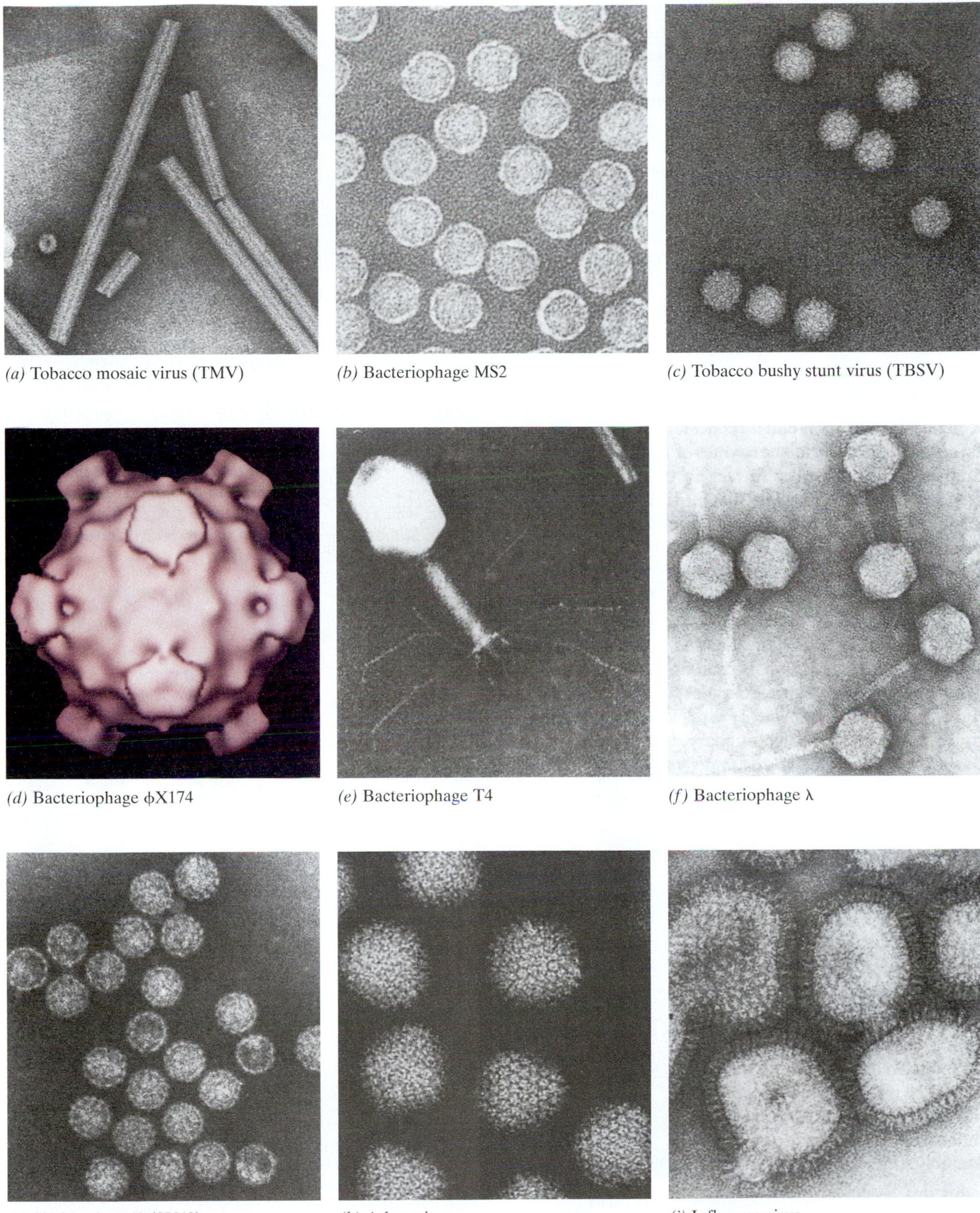

(a) Tobacco mosaic virus (TMV)
(b) Bacteriophage MS2
(c) Tobacco bushy stunt virus (TBSV)
(d) Bacteriophage φX174
(e) Bacteriophage T4
(f) Bacteriophage λ
(g) Simian virus 40 (SV40)
(h) Adenovirus
(i) Influenza virus

FIGURE 33-1 Electron micrographs of a selection of viruses. TMV, MS2, TBSV, and influenza virus are single-stranded RNA viruses; φX174 is a single-stranded DNA virus; and λ, T4, SV40, and adenovirus are double-stranded DNA viruses. Bacteriophage M13, a filamentous, single-stranded DNA coliphage, is shown in Fig. 5-45. [Parts *a–c* and *f–i* courtesy of Robley Williams, University of California at Berkeley, and Harold Fisher, University of Rhode Island; Part *d* courtesy of Michael Rossmann, Purdue University; and Part *e* courtesy of John Finch, Cambridge University.]

DNA may replicate autonomously or be inserted in the host chromosome for replication with the host DNA. The DNA of eukaryotic viruses is either replicated and transcribed in the cell nucleus by cellular enzymes or in the cytoplasm by virally specified enzymes. In fact, in the case of negative strand RNA viruses, enzymes that mediate viral RNA transcription must be carried by the virion because most cells lack the ability to transcribe RNA.

This chapter is a discussion of the structures and biology of a variety of viruses. In it, we examine mainly **tobacco mosaic virus (TMV),** a helical RNA virus; several **icosahedral viruses; bacteriophage λ,** a tailed DNA bacteriophage; and **influenza virus,** an enveloped RNA virus. These examples have been chosen to illustrate important aspects of viral structure, assembly, molecular genetics, and evolutionary strategy. *Much of this information is relevant to the understanding of the corresponding cellular phenomena.*

1 ■ TOBACCO MOSAIC VIRUS

Tobacco mosaic virus causes leaf mottling and discoloration in tobacco and many other plants. It was the first virus to be discovered (by Dmitri Iwanowsky in 1892), the first virus to be isolated (by Wendell Stanley in 1935), and even now is among the most extensively investigated and well-understood viruses from the standpoint of structure and assembly. In this section, we discuss these aspects of TMV.

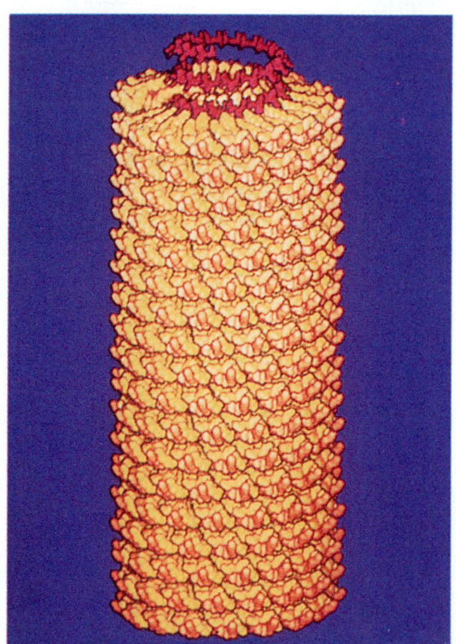

FIGURE 33-2 Model of tobacco mosaic virus (TMV) illustrating the helical arrangement of its coat protein subunits and RNA molecule. The RNA is represented by the red chain exposed at the top of the viral helix. Only 18 turns (415 Å) of the TMV helix are shown, which represent ~14% of the TMV rod. [Courtesy of Gerald Stubbs and Keiichi Namba, Vanderbilt University; and Donald Caspar, Brandeis University.]

A. *Structure*

TMV is a rod-shaped particle (Fig. 33-1*a*) that is ~3000 Å long, 180 Å in diameter, and has a particle mass of 40 million D. Its ~2130 identical copies of coat protein subunits (158 amino acid residues; 17.5 kD) are arranged in a hollow right-handed helix that has 161/3 subunits/turn, a pitch (rise per turn) of 23 Å, and a 40-Å-diameter central cavity (Fig. 33-2). TMV's single RNA strand (~6400 nt; 2 million D) is coaxially wound within the turns of the coat protein helix such that 3 nt are bound to each protein subunit (Fig. 33-2).

a. TMV Coat Protein Aggregates to Form Viruslike Helical Rods

The aggregation state of TMV coat protein is both pH and ionic strength dependent (Fig. 33-3). At slightly alkaline pH's and low ionic strengths, the coat protein forms complexes of only a few subunits. At higher ionic strengths, however, the subunits associate to form a double-layered disk of 17 subunits/layer, a number that is nearly equal to the number of subunits per turn in the intact virion. At neutral pH and low ionic strengths, the subunits form short helices of slightly more than two turns (39 ± 2 subunits) termed "protohelices" (also known as "lockwashers"). If

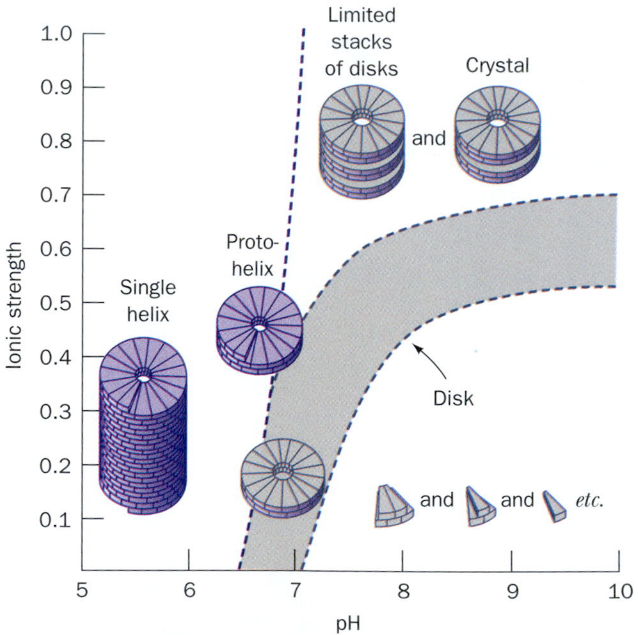

FIGURE 33-3 Aggregation state of TMV coat protein as a function of pH and ionic strength. Under basic conditions, the subunits aggregate into small clusters. Around neutrality and at high ionic strengths, the protein forms a 34-subunit double-layered disk. Under acidic conditions and at low ionic strengths, the subunits form protohelices that stack to form long helices. At neutral pH and low ionic strength, which resembles physiological conditions, the protein forms helices only in the presence of TMV RNA. [After Durham, A.C.H., Finch, J.T., and Klug, A., *Nature New Biol.* **229,** 38 (1971).]

the pH of these protohelices is shifted to ~5, they stack in imperfect register and eventually anneal to form indefinitely long helical rods that, although they lack RNA, resemble intact virions (Fig. 33-4). These observations, as we shall see below, lead to the explanation of how TMV assembles.

b. TMV Coat Protein Interacts Flexibly with Viral RNA

X-Ray studies of TMV have been pursued on two fronts. The virus itself does not crystallize but forms a highly oriented gel of parallel viral rods. The X-ray analysis of this gel by Kenneth Holmes and Gerald Stubbs yielded a structure of sufficient resolution (2.9 Å) to re-

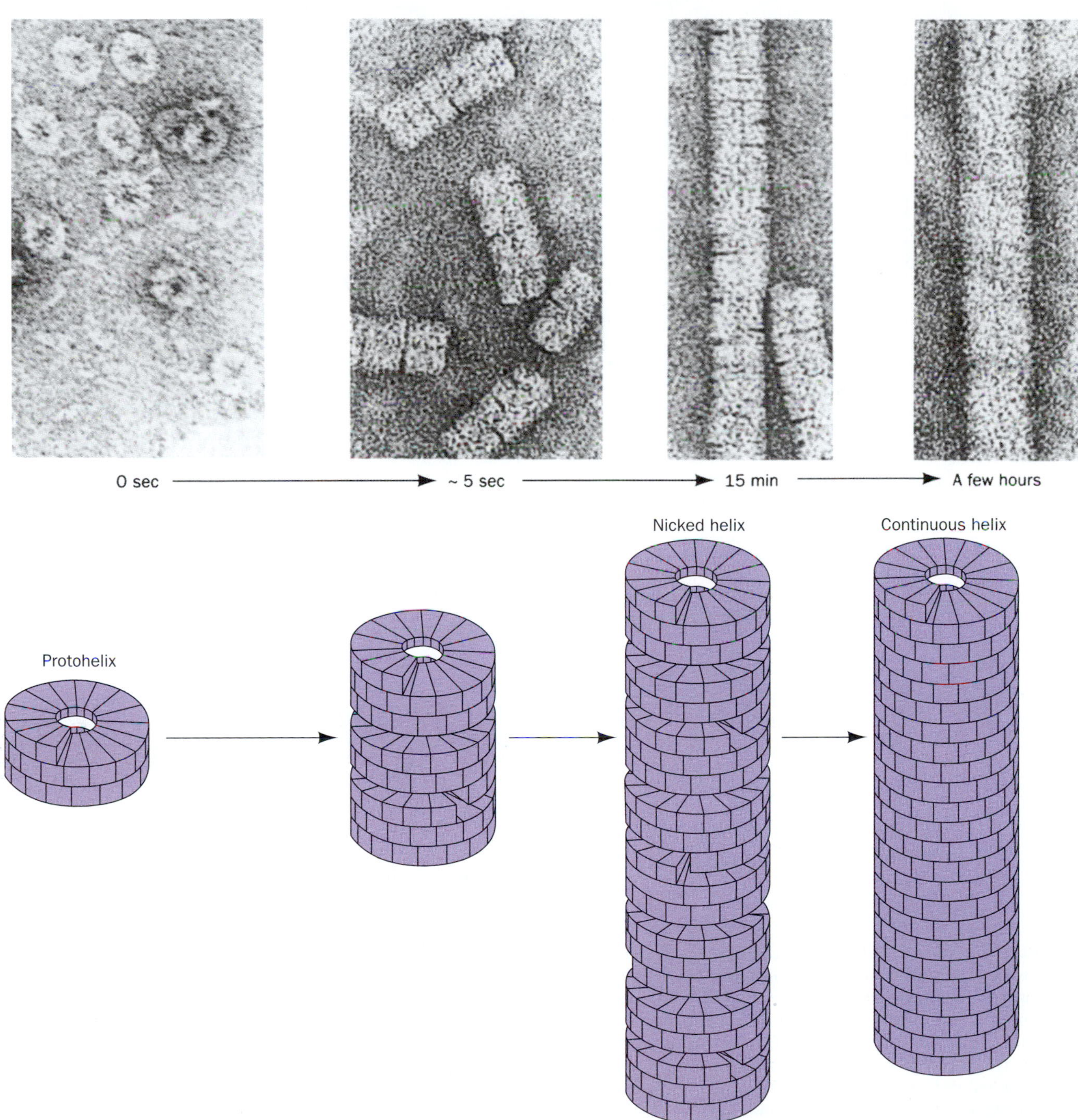

FIGURE 33-4 Growth of TMV coat protein rods. Electron micrographs (*above*) and their interpretive diagrams (*below*) show TMV coat protein aggregates following a rapid change in pH from 7 to 5 at low ionic strength. This pH shift causes the protohelices to form "nicked" (imperfectly stacked) helices that, within a few hours, anneal to yield continuous helical protein rods. [Courtesy of Aaron Klug, MRC Laboratory of Molecular Biology, Cambridge, U.K.]

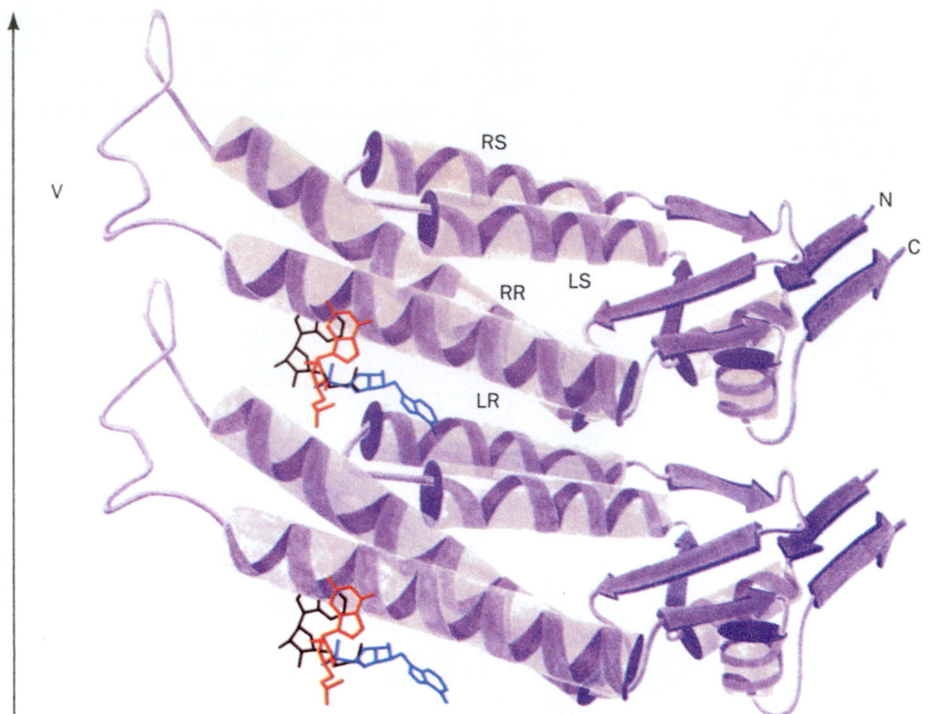

FIGURE 33-5 X-Ray structure of two vertically stacked TMV subunits. The structure is viewed perpendicular to the virus helix axis (*vertical arrow on the left*). Each subunit has four approximately radially extending helices (LR, RR, LS, and RS), as well as a short vertical segment (V), which comprises part of the flexible loop in the disk structure (dashed lines in Fig. 33-7). Portions of two successive turns of RNA are shown passing through their binding sites. Each subunit binds three nucleotides, here represented by GAA with each of its nucleotides differently colored, such that their three bases lie flat against the LR helix so as to grasp it in a clawlike manner. [After Namba, K., Pattanayek, R., and Stubbs, G., *J. Mol. Biol.* **208,** 314 (1989). PDBid 2TMV.]

veal the folding of the protein and the RNA (Figs. 33-5 and 33-6). This study is complemented by Aaron Klug's X-ray crystal structure determination, at 2.8-Å resolution, of the 34-subunit coat protein disk (Fig. 33-7).

A major portion of each subunit consists of a bundle of four alternately parallel and antiparallel α helices that project more or less radially from the virus axis (Figs. 33-5 to 33-7). In the disk, one of the inner connections between these α helices, a 24-residue loop (residues 90–113; dashed line in Fig. 33-7), is not visible, apparently because it is highly mobile. This disordered loop is also present in the protohelix as shown by NMR studies. In the virus, however, the loop adopts a definite conformation containing

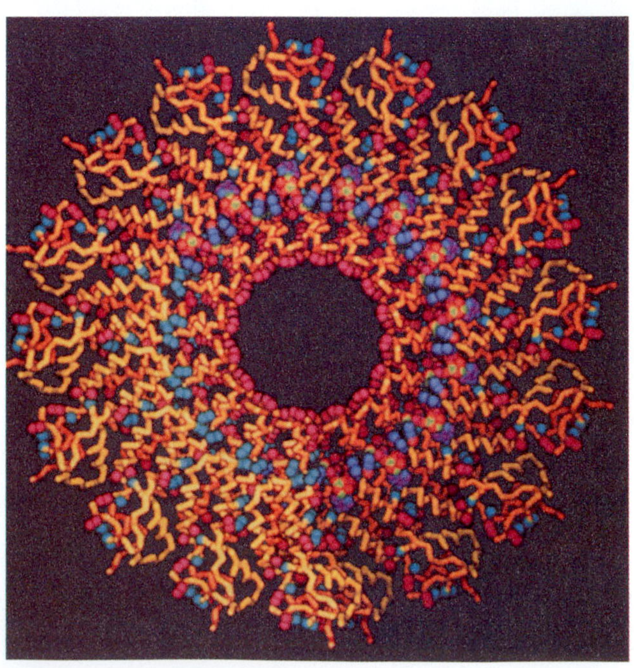

FIGURE 33-6 Top view of 17 TMV coat protein subunits comprising slightly more than one helical turn in complex with a 33-nucleotide RNA segment. The protein is represented by its C_α atoms, shown as connected yellow rods, together with its acidic side chains (Asp and Glu) as red balls and its basic side chains (Arg and Lys) as blue balls. The RNA's phosphate atoms are green and its bases are magenta. Note that the acidic side chains form a 25-Å-radius helix that lines the virion's inner cavity and the basic side chains form a 40-Å-radius helix that interacts with the RNA's anionic sugar–phosphate chain. [Courtesy of Gerald Stubbs and Keiichi Namba, Vanderbilt University; and Donald Caspar, Brandeis University. PDBid 2TMV.]

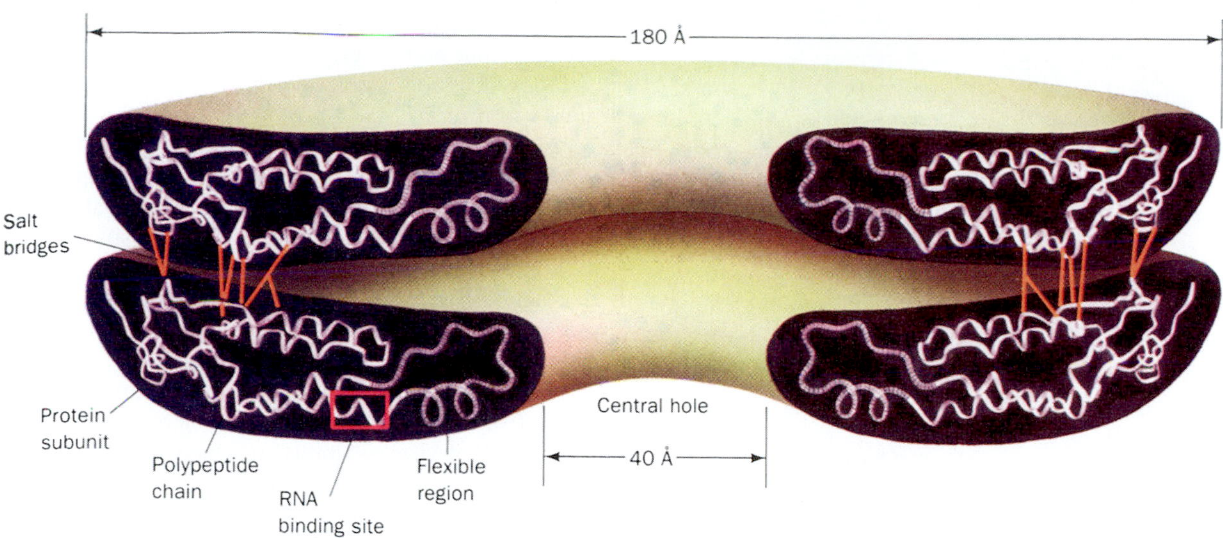

FIGURE 33-7 Structure of the TMV protein disk in cross section showing its polypeptide chains as ribbon diagrams. The dashed lines represent disordered loops of polypeptide chain that are therefore not visible in the disk X-ray structure. The stacked protein rings interact along their outer rims through a system of salt bridges (*red lines*). [After Butler, P.J.G. and Klug, A., *Sci. Am.* **239**(5): 67 (1978). Copyright © 1978 by Scientific American, Inc.]

a series of reverse turns arranged such that the overall direction of this polypeptide segment is approximately parallel to the virus axis (V in Fig. 33-5). This conformational change, as we shall see, is an important aspect of virus assembly.

In the virus, the RNA is helically wrapped between the coat protein subunits at a radius of ~40 Å. The triplet of bases binding to each subunit forms a clawlike structure around one of the radial helices (LR in Fig. 33-5) with each base occupying a hydrophobic pocket in which it lies flat against LR. Arg residues 90 and 92, which are invariant in the several known TMV strains and which are part of the disk and protohelix's disordered loop, as well as Arg 41, form salt bridges to the RNA phosphate groups.

B. *Assembly*

How is the TMV virion assembled from its component RNA and coat protein subunits? *The assembly of any large molecular aggregate, such as a crystal or a virus, generally occurs in two stages: (1)* ***nucleation,*** *the largely random aggregation of subunits to form a quasi-stable nucleation complex, which is almost always the rate-determining step of the assembly process; followed by (2)* ***growth,*** *the cooperative addition of subunits to the nucleation complex in an orderly arrangement that usually proceeds relatively rapidly.* For TMV, it might reasonably be expected that the nucleation complex minimally consists of the viral RNA in association with the 17 or 18 subunits necessary to form a stable helical turn, which could then grow by the accumulation of subunits at one or both ends of the helix. The low probability for the formation of such a complicated nucleation complex from disaggregated subunits accounts for the observed 6-h time necessary to complete this *in vitro* assembly process. Yet, the *in vivo* assembly of TMV probably occurs much faster. A clue as to the nature of this *in vivo* process was provided by the observation that if protohelices rather than disaggregated subunits are mixed with TMV RNA, complete virus particles are formed in 10 min. Other RNAs do not have this effect. Evidently, *the in vivo nucleation complex in TMV assembly is the association of a protohelix with a specific segment of TMV RNA.* (Although it was originally assumed that the double-layered disk rather than the protohelix formed the nucleating complex, experimental evidence indicates that the disk does not form under physiological conditions and that its rate of conversion to the protohelix under these conditions is too slow to account for the rate of TMV assembly. Other experiments, however, suggest that it is the disk that predominates at pH 7.0, the pH at which TMV most rapidly assembles from its component protein and RNA. Thus, keep in mind that the question as to whether TMV assembles from protohelices, as we state here, or from double-layered disks has not been fully resolved.)

a. TMV Assembly Proceeds by the Sequential Addition of Protohelices

The specific region of the TMV RNA responsible for initiating the virus particle's growth was isolated using the now classic nuclease protection technique. The RNA is mixed with a small amount of coat protein so as to form a nucleation complex that cannot grow because of the lack of coat protein. The RNA that is not protected by coat protein is then digested away by RNase, leaving intact only the initiation sequence. This RNA fragment forms a hairpin loop whose 18-nucleotide apical sequence, AGAAGAAGUUGUUGAUGA, has a G at every third residue

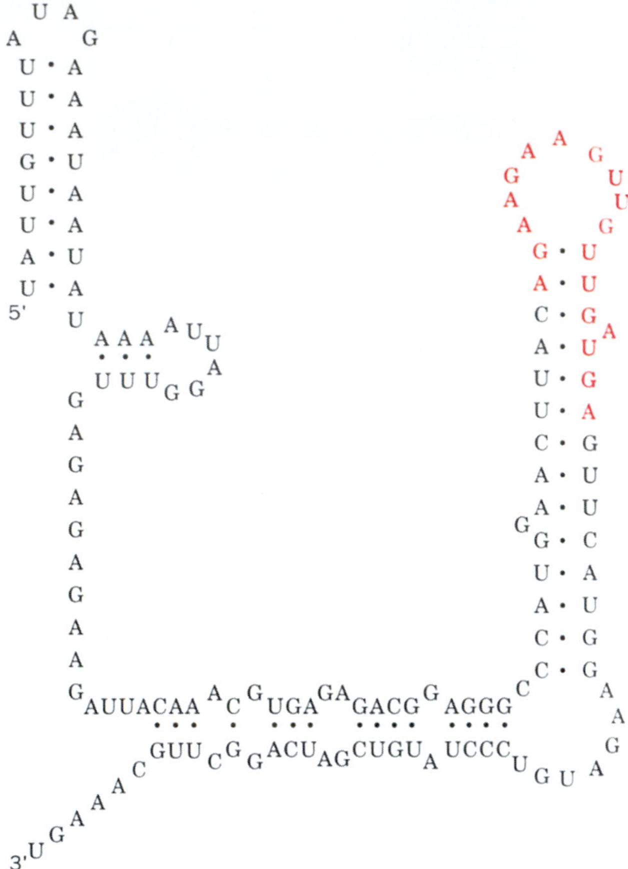

FIGURE 33-8 Initiation segment of TMV RNA. It probably forms a weakly base paired hairpin, as drawn, that is thought to begin TMV assembly by specifically binding to a coat protein protohelix. Note that this RNA's loop region has an 18-nt segment (*red*) with a G every third residue (each coat protein subunit binds three nucleotides) but no C's.

(recall that each coat protein subunit binds three nucleotides) but no C's (Fig. 33-8). Site-directed mutagenesis studies have confirmed that this initiation sequence is sufficient to direct TMV assembly and that the regularly spaced G's and lack of C's are important for its function. TMV's high binding affinity for this initiation sequence is explained, in part, by the observations that coat protein subunits bind every third nucleotide in the unusual syn conformation and that G assumes this conformation more easily than any other nucleotide (Section 29-2A). The lack of C's perhaps prevents the involvement of these G's in base pairing associations.

The above initiation complex is located some 1000 nucleotides from the 3′ end of the TMV RNA. Hence, the simple model of viral assembly in which the RNA is sequentially coated by protein from one end to the other cannot be correct. Rather, the RNA initiation hairpin must insert itself between the protohelix's protein layers from its central cavity (Fig. 33-9*a*). The RNA binding, for reasons explained below, induces the ordering of the disordered loop, thereby trapping the RNA (Fig. 33-9*b*). Growth then proceeds by a repetition of this process at the "top" of the complex, thereby incrementally pulling the RNA's 5′ end up through the central cavity of the growing viral helix (Fig. 33-9*c*).

The above assembly model has been corroborated by several experimental observations:

1. Electron micrographs reveal that partially assembled rods (Fig. 33-10) have two RNA "tails" projecting from one end.

2. The length of the longer tail, presumably the 5′ end, decreases linearly with the length of the rod, whereas the shorter tail maintains a more or less constant length.

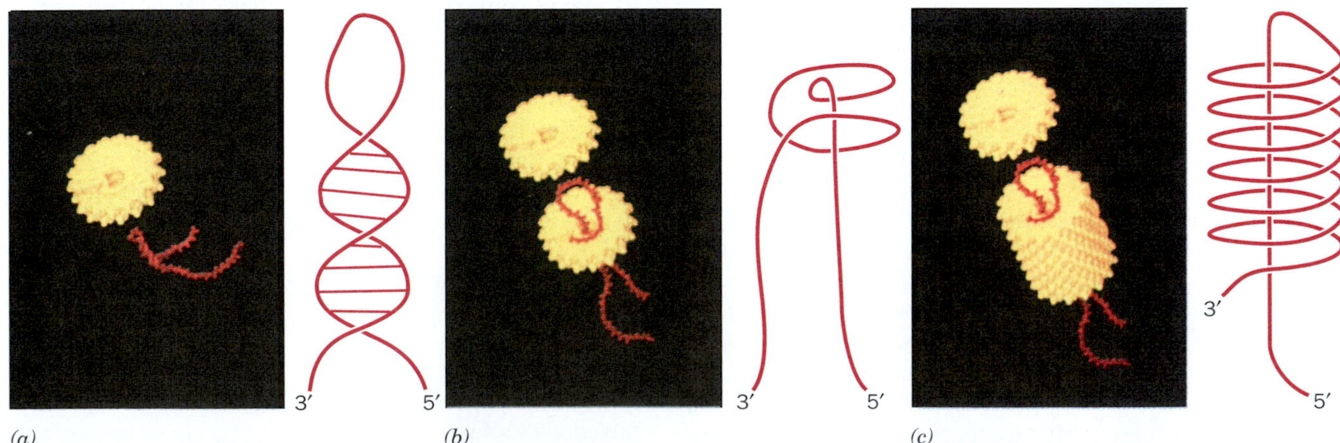

FIGURE 33-9 Assembly of TMV. (*a*) The process begins by the insertion of the hairpin loop formed by the initiation sequence of the viral RNA into the protohelix's central cavity. (*b*) The RNA then intercalates between the layers of the protohelix, thereby ordering the disordered loop and trapping the RNA. (*c*) Elongation proceeds by the stepwise addition of protohelices to the "top" of the viral rod. The consequent binding of the RNA to each protohelix, which converts it to the helical form, pulls the RNA's 5′ end up through the virus' 40-Å-diameter central cavity to form a traveling loop at the viral rod's growing end. [Viral images courtesy of Hong Wang and Gerald Stubbs, Vanderbilt University.]

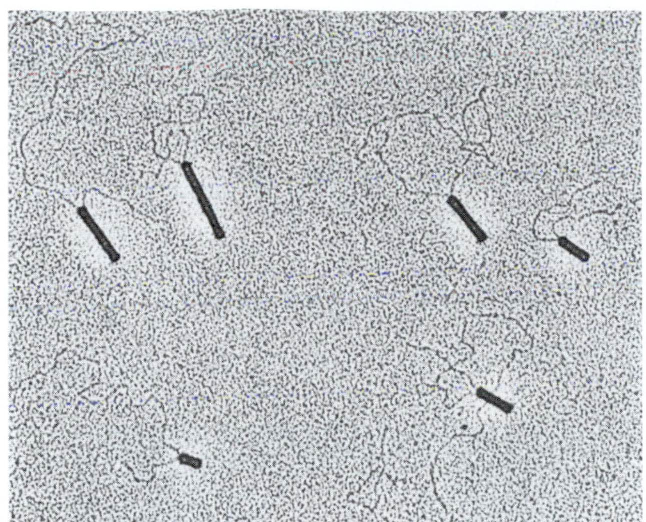

FIGURE 33-10 Electron micrograph of partially reconstituted TMV particles showing that their two RNA tails emerge from the same end of the growing viral rod. An analysis of these particles indicates that the length of one of the tails, probably the 3′ end, is constant (720 ± 80 nucleotides), whereas that of the other tail is inversely proportional to the length of its incomplete rod. [Courtesy of K.E. Richards, CNRS, France.]

3. Nuclease digestion experiments on partially assembled rods indicate that the RNA is protected in increments of ~100 nucleotides, as is expected for elongation steps consisting of the addition of a protohelix to the growing rod.

The coating of the 3′ end of the RNA is a much slower process than the coating of its 5′ end and hence probably occurs by the successive addition of single subunits. The RNA, which acts as the viral mRNA, carries the gene specifying the coat protein near its 3′ end. Perhaps this assembly mechanism allows coat protein synthesis during all but the final stages of assembly, thereby permitting the completion of this process.

b. Electrostatic Repulsions and Steric Interactions Prevent Helix Formation in the Absence of RNA

What is the mechanism that prevents the formation of TMV coat protein helices in the absence of viral RNA but triggers virus assembly in its presence (and, conversely, how does intact TMV disassemble to initiate an infection)? Structural considerations suggest that the coat protein subunit's disordered loop sterically prevents the protohelix from growing longer. Moreover, as we have seen (Fig. 33-3), the state of coat protein aggregation varies with pH. Titration studies show that each subunit has two ionizations with p*K*'s near 7, which must each be attributed to anomalously basic carboxyl groups because coat protein has no His residues. The most plausible candidates for these anomalously basic carboxyls are two intersubunit pairs of carboxyl groups: Glu 95–Glu 106, disordered loop members which interact across a side-to-side subunit interface; and Glu 50–Asp 77, which interact across a top-to-bottom subunit interface. Moreover, Asp 116 is close to an RNA phosphate group. The electrostatic repulsions between these closely spaced negative charges promotes the formation of the disordered loop and therefore favors the protohelix conformation. The binding of the RNA initiation sequence to the protein apparently provides sufficient free energy to overcome these repulsions, thereby triggering helix formation (a process that partially protonates the anomalously basic carboxyl groups; recall the similar conformationally induced p*K* changes in the Bohr effect of hemoglobin; Section 10-2E). Indeed, site-directed mutagenesis of Glu 50 → Gln or Asp 77 → Asn both increases virion stability and decreases its infectivity (presumably by inhibiting viral disassembly). Further growth of the viral rod can occur on RNA segments that lack this sequence as a consequence of the additional binding interactions between adjacent protohelices. *The carboxyl groups evidently act as a negative switch to prevent the formation of a protein helix in the absence of RNA under physiological conditions.*

2 ■ ICOSAHEDRAL VIRUSES

The simpler **icosahedral viruses,** being uniform molecular assemblies, crystallize in much the same way as proteins. The techniques of X-ray crystallography can therefore be brought to bear on determining virus structures. In this section we consider the results of such studies.

A. *Virus Architecture*

The very limited genomic resources of the simpler viruses in many cases limit them to having but one type of protein in their capsid. Since these coat protein subunits are chemically identical, they must all assume the same or nearly the same conformations and have similar interactions with their neighbors. What geometrical constraints does this limitation impose on viral architecture?

We have already seen that TMV solves this problem by assuming a helical geometry (Fig. 33-2). The coat protein subunits in such a long but finite helix, although geometrically distinguishable, have, with the exception of the subunits at the helix ends, virtually identical environments. Such subunits are said to be **quasi-equivalent** to indicate that they are not completely indistinguishable as they would be in an object whose elements are all related by exact symmetry.

a. Icosahedral Viruses Have Icosahedral Capsids

A second arrangement of equivalent subunits that can encapsulate a nucleic acid is that of a polyhedral shell. There are only three polyhedral symmetries in which all the elements are indistinguishable: those of a tetrahedron, a cube, and an icosahedron (Fig. 8-64*c*). Capsids with these symmetries would have 12, 24, or 60 subunits identically arranged on the surface of a sphere. For example, an

(a)

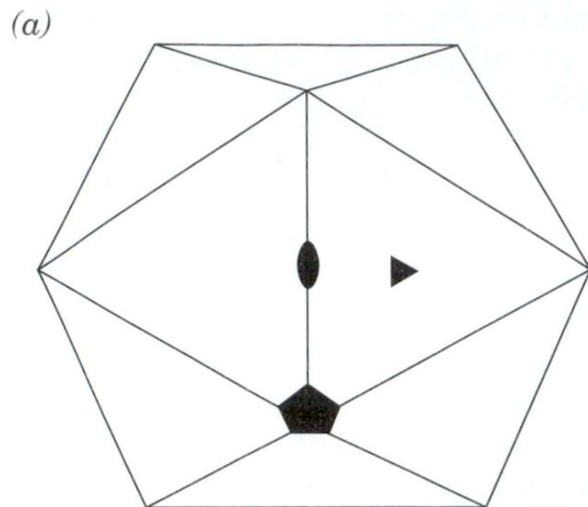

(b)

FIGURE 33-11 Icosahedron. (*a*) This regular polyhedron has 12 vertices, 20 equilateral triangular faces of identical size, and 30 edges. It has a 5-fold axis of symmetry through each vertex, a 3-fold axis through the center of each face, and a 2-fold axis through the center of each edge (also see Fig. 8-64*c*). (*b*) A drawing of 60 identical subunits (*lobes*) arranged with icosahedral symmetry. [Illustration, Irving Geis/Geis Archives Trust, Copyright Howard Hughes Medical Institute. Reproduced with permission.]

(a)

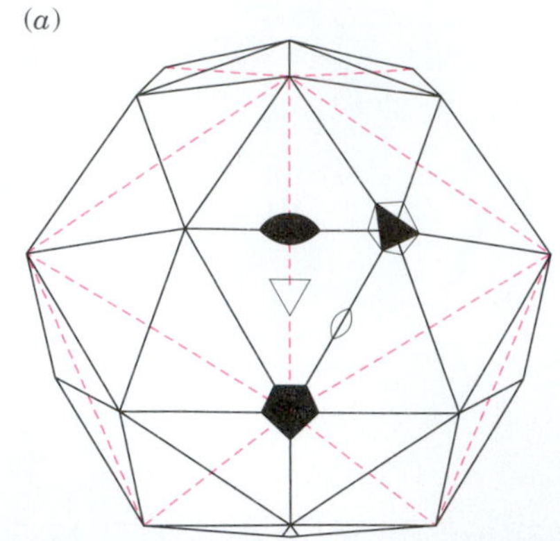

(b)

FIGURE 33-12 $T = 3$ icosadeltahedron. (*a*) This polyhedron has the exact rotational symmetry of an icosahedron (*solid symbols*) together with local 6-fold, 3-fold, and 2-fold rotational axes (hollow symbols). Note that the edges of the underlying icosahedron (*dashed red lines*), are not edges of this polyhedron and that its local 6-fold axes are coincident with its exact 3-fold axes. (*b*) A drawing of a $T = 3$ icosadeltahedron showing its arrangement of 3 quasi-equivalent sets of 60 icosahedrally related subunits (*lobes*). The A lobes (*orange*) pack about the icosadeltahedron's exact 5-fold axes, whereas the B and C lobes (*blue and green*) alternate about its local 6-fold axes. TBSV's chemically identical coat protein subunits are arranged in this manner. [Illustration, Irving Geis/Geis Archives Trust, Copyright Howard Hughes Medical Institute. Reproduced with permission.]

icosahedron (Fig. 33-11*a*) has 20 triangular faces, each with 3-fold symmetry, for a total of $20 \times 3 = 60$ equivalent positions (each represented by a lobe in Fig. 33-11*b*). Of these polyhedra, the icosahedron encloses the greatest volume per subunit. Indeed, electron microscopy of the so-called icosahedral viruses (such as Fig. 33-1*b–h*) first demonstrated that *they have icosahedral symmetry.*

b. Viral Capsids Resemble Geodesic Domes

A viral nucleic acid, if it is to be protected effectively against a hostile environment, must be completely covered by coat protein. Yet, many viral nucleic acids occupy so large a volume that their coat protein subunits would have to be prohibitively large if their capsids were limited to the 60 subunits required by exact icosahedral symmetry. In fact, nearly all viral capsids have considerably more than 60 chemically identical subunits. How is this possible?

Donald Caspar and Klug pointed out the solution to this dilemma. *The triangular faces of an icosahedron can be subdivided into integral numbers of equal sized equilateral triangles (e.g, Fig. 33-12a).* The resulting polyhedron, an **icosadeltahedron,** has "local" symmetry elements relating its subunits (lobes in Fig. 33-12*b*) in addition to its exact icosahedral symmetry. By local symmetry, we mean that the symmetry is only approximate so that, in contrast to the case for exact symmetry, it breaks down over larger distances. For instance, the subunits (lobes) in Fig. 33-12*b* that are distributed about each exact triangular vertex form clusters whose members are related by a local 6-fold axis of symmetry. *Adjacent subunits in these clusters are not exactly equivalent; they are quasi-equivalent.* In contrast, the subunits clustered about the twelve 5-fold axes of icosahedral symmetry are exactly equivalent. The interac-

FIGURE 33-13 Geodesic dome built on the plan of a $T = 36$ icosadeltahedron. Two of its pentagonal vertices are visible in this photograph. [Stanley Schoenberger/Grant Heilman.]

tions between the subunits clustered about the local 6-fold axes are therefore essentially distorted versions of those about the exact 5-fold axes. Consequently, *the coat protein subunits of any viral capsid with icosadeltahedral symmetry must make alternative sets of intersubunit associations and/or have sufficient conformational flexibility to accommodate these distortions.*

Icosadeltahedra are familiar figures. The faceted surface of a soccer ball is an icosadeltahedron. Likewise, **geodesic domes** (Fig. 33-13), which were originally designed by Buckminster Fuller, are portions of icosadeltahedra. It was, in fact, Fuller's designs that inspired Caspar and Klug. *Geodesic domes are inherently rigid shell-like structures that are constructed from a few standard parts, make particularly efficient use of structural materials, and can be rapidly and easily assembled. Presumably the evolution of icosahedral virus capsids was guided by these very principles.*

The number of subunits in an icosadeltahedron is $60T$, where T is called the **triangulation number.** The permissible values of T are given by $T = h^2 + hk + k^2$, where h and k are positive integers. An icosahedron, the simplest icosadeltahedron, has $T = 1$ ($h = 1$, $k = 0$) and therefore 60 subunits. The icosadeltahedron with the next level of complexity has a triangulation number of $T = 3$ ($h = 1$, $k = 1$) and hence 180 subunits (Fig. 33-12). A capsid with this geometry has three different sets of icosahedrally related subunits that are quasi-equivalent to each other (lobes A, B, and C in Fig. 33-12*b*). The X-ray structures of viruses with capsids consisting of $T = 1$, 3, 4, 7, and 13 icosadeltahedra have been determined. Some of the larger icosahedral viruses form icosadeltahedra with even greater triangulation numbers (see below). However, some of them are based on somewhat different assembly principles (Section 33-2D). The T value for any particular capsid, presumably, depends on its subunit's innate curvature.

B. *Tomato Bushy Stunt Virus*

Tomato Bushy Stunt Virus (TBSV); Fig. 33-1*c* is a $T = 3$ icosahedral virus that is ~175 Å in radius. It consists of 180 identical coat protein subunits, each of 386 residues (43 kD), encapsulating a single-stranded RNA molecule of ~4800 nt (1500 kD; the positive or message strand) and a single copy of an ~85-kD protein. The X-ray crystal structure of TBSV, the first of a virus to be determined at high resolution, was reported in 1978 by Stephen Harrison. TBSV's coat protein subunits have three domains (Fig. 33-14): P, the C-terminal domain, which projects outward from the virus; S, which forms the viral shell; and R, the protein's inwardly extending N-terminal domain, which is attached to the S domain via a connecting arm. The S domain is almost entirely composed of an 8-stranded antiparallel β barrel, which we shall see occurs in the coat proteins of the majority of icosahedral viruses with known structures.

a. TBSV's Identical Subunits Associate through Nonidentical Contacts

The chemically identical TBSV coat protein subunits occupy three symmetrically distinct environments denoted

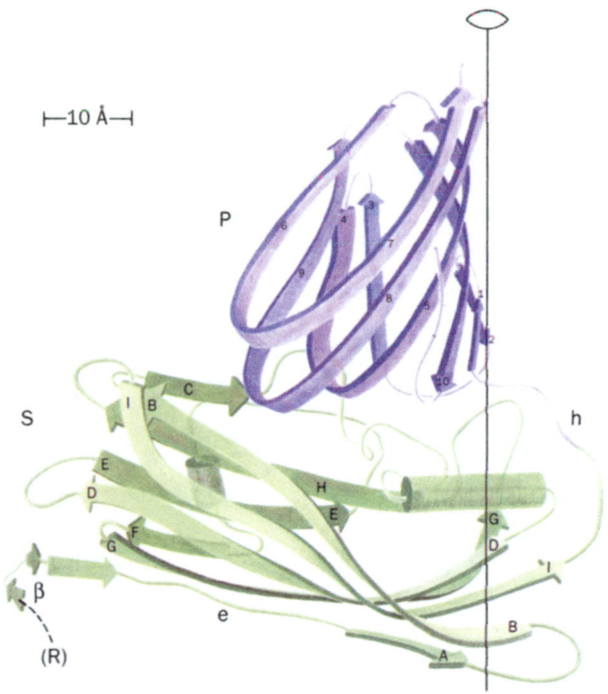

FIGURE 33-14 X-Ray structure of the tomato bushy stunt virus (TBSV) coat protein subunit. It consists of three domains: P, which projects from the virion's surface (*purple*); S, which forms the capsid (*green*); and R, which extends below the capsid surface where it participates in binding the viral RNA. The S domain is largely comprised of an 8-stranded antiparallel β barrel that has the jelly roll or Swiss roll topology (Section 8-3B). The P domain is also composed largely of an antiparallel β sheet, whereas the R domain is not visible in the viral X-ray structure so its tertiary structure is unknown. [After Olsen, A.J., Bricogne, G., and Harrison, S.C., *J. Mol. Biol.* **171,** 78 (1983). PDBid 2TBV.]

A, B, and C (Fig. 33-15). How does the protein accommodate the different contacts required by its several sets of analogous but nonidentical associations? TBSV's structure reveals that *analogous intersubunit contacts vary both through alternative sets of interactions and by conformational distortions of the same interactions.* Perhaps the most remarkable alternative interaction is the interdigitation of the arms connecting the R and S domains of the C subunits. These arms extend toward each icosahedral 3-fold axis (quasi-6-fold axis) in the clefts between the adjacent C and B subunits and then spiral downward about this 3-fold axis to form a β-sheetlike arrangement that resembles the overlapping flaps of a cardboard carton: chain 1 over chain 2 over chain 3 over chain 1 (Fig. 33-16*a*). This interaction, together with a strong association between neighboring C subunits across the icosahedral 2-fold axis (Fig. 33-15), organizes the 60 C subunits into a coherent network (Fig. 33-16*b*) that determines the triangulation number of the TBSV capsid: *The C subunits can be thought of as forming a $T = 1$ icosahedral shell whose gaps are filled in by the A and B subunits.* In response, the three sets of quasi-equivalent subunits assume somewhat different conformations: The three- or four-residue "hinge" connecting the S and P domains (h in Fig. 33-14) has an ~30° greater dihedral angle in the A and B subunits than in the C subunits (Fig. 33-15, *right*). This, in turn, permits the interactions between P domains to be identical in the AB and CC dimers (projecting dimeric knobs in Fig. 33-15). Evidently, interdomain associations between subunits are stronger in TBSV than those within subunits.

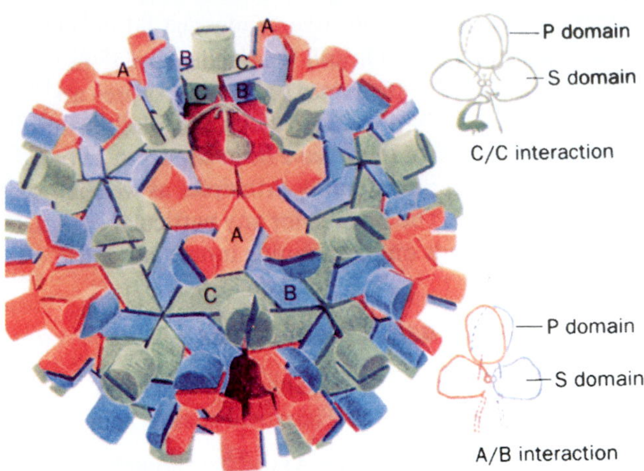

FIGURE 33-15 $T = 3$ icosadeltahedral arrangement of TBSV's coat protein subunits. The subunits occur in three quasi-equivalent packing environments, A, B, and C. The A subunits (*orange*) pack around exact 5-fold axes, whereas the B subunits (*blue*) alternate with the C subunits (*green*) about the exact 3-fold axes (local 6-fold axes). The C subunits are also disposed about the strict 2-fold axes, whereas the A and B subunits are related by local 2-fold axes. The subunits respond to the different conformational requirements of their three quasi-equivalent positions through flexion at the hinge region between their S and P domains (*right and in cutouts*). Compare this drawing to Fig. 33-12. [After Harrison, S.C., *Trends Biochem. Sci.* **9,** 348, 349 (1984).]

b. TBSV's RNA-Containing Core Is Disordered

The entire connecting arm between the R and S domains in the A and B subunits, as well as their first few residues in the C subunits, are not visible in TBSV's X-ray structure, thereby indicating that these polypeptide segments have no fixed conformations. The R domains are therefore flexibly tethered to the S domains so that they are also absent from the X-ray structure, even though these domains probably have a fixed conformation. Neutron scattering studies, nevertheless, suggest that protein, con-

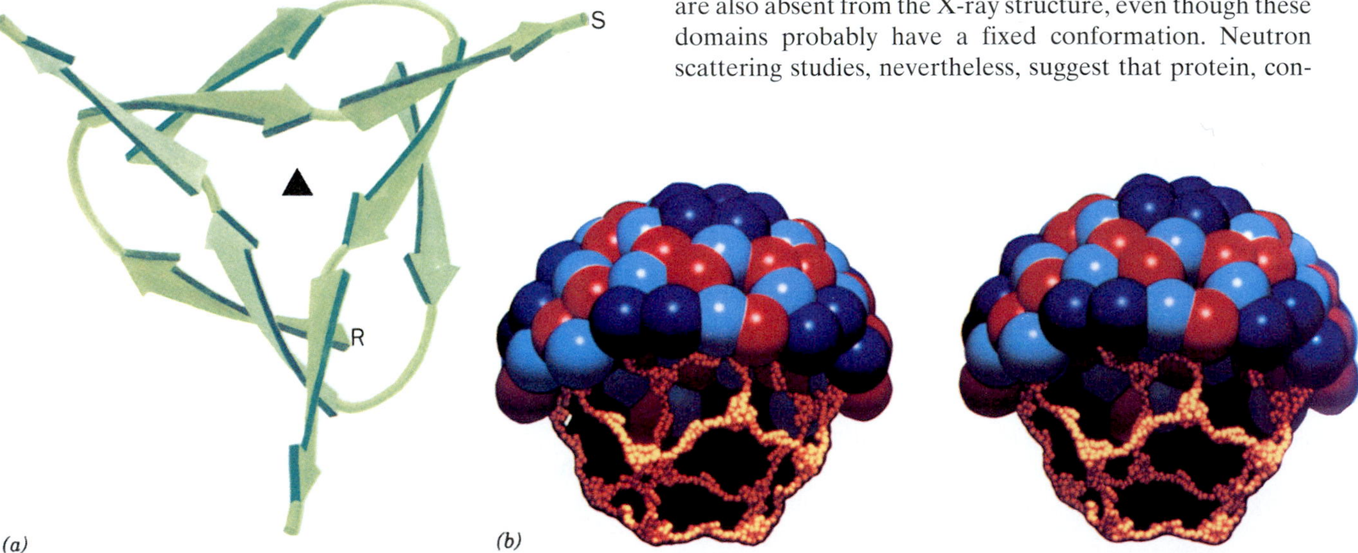

FIGURE 33-16 Architecture of the TBSV capsid. (*a*) The C subunit arms of TBSV protein pack about the capsid's exact 3-fold axes (*triangle*) and associate as β sheets. The view is from outside the capsid. (*b*) A stereo cutaway drawing showing the capsid's internal scaffolding of C subunit arms. The chemically identical A (*blue*), B (*cyan*), and C subunits (*red*) are represented by large spheres, whereas the residues comprising the C subunit arms are represented by small yellow spheres. The C subunit arms associate to form an icosahedral ($T = 1$) framework that apparently plays a major role in holding together the viral capsid. Directions for viewing stereo drawings are given in the Appendix to Chapter 8. [Part *a* after a drawing by Jane Richardson, Duke University, and Part *b* courtesy of Arthur Olson, The Scripps Research Institute, La Jolla, California. Based on an X-ray structure by Stephen Harrison, Harvard University. PDBid 2TBV.]

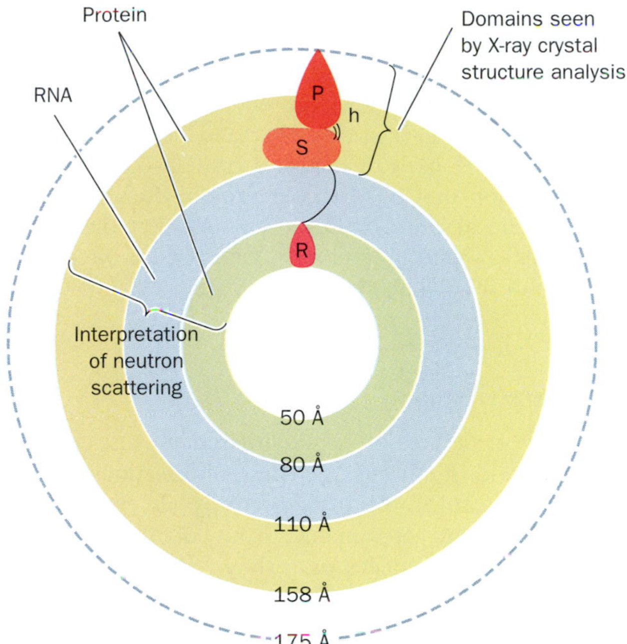

FIGURE 33-17 Radial organization of TBSV indicating the distribution of its protein and RNA components. The R domain positions are inferred from their known chain length. Only about half of the R domains are contained in the inner protein shell. [After Harrison, S.C., *Biophys. J.* **32,** 140 (1980).]

stituting perhaps half of the R domains, forms a 50- to 80-Å radius inner shell. The remaining R domains are thought to project into the space between the inner and outer shells.

The viral RNA is absent from the X-ray structure, which indicates that it too is disordered. The above neutron scattering studies reveal that this RNA is sandwiched between the virus' inner and outer protein shells (Fig. 33-17). The volume constraints imposed by this arrangement require that the RNA be tightly packed. This packing is made possible because most of the negative charges of the RNA phosphate groups are neutralized by the numerous positively charged Arg and Lys residues of the R domains, the inner faces of the S domains, and their connecting arms.

c. Many Other RNA Viruses Are Remarkably Similar to TBSV

The structures of numerous other RNA plant viruses have been elucidated, including those of **southern bean mosaic virus (SBMV)** by Michael Rossmann and **satellite tobacco mosaic virus (STMV)** by Alexander McPherson. SBMV is a $T = 3$ virus that closely resembles TBSV in its quaternary structure. Moreover, SBMV's 260-residue coat protein subunit, although it entirely lacks a P domain, has an S domain whose polypeptide backbone is nearly superimposable on that of TBSV (Fig. 33-18*a*). The RNA in SBMV, as is that in TBSV, is disordered.

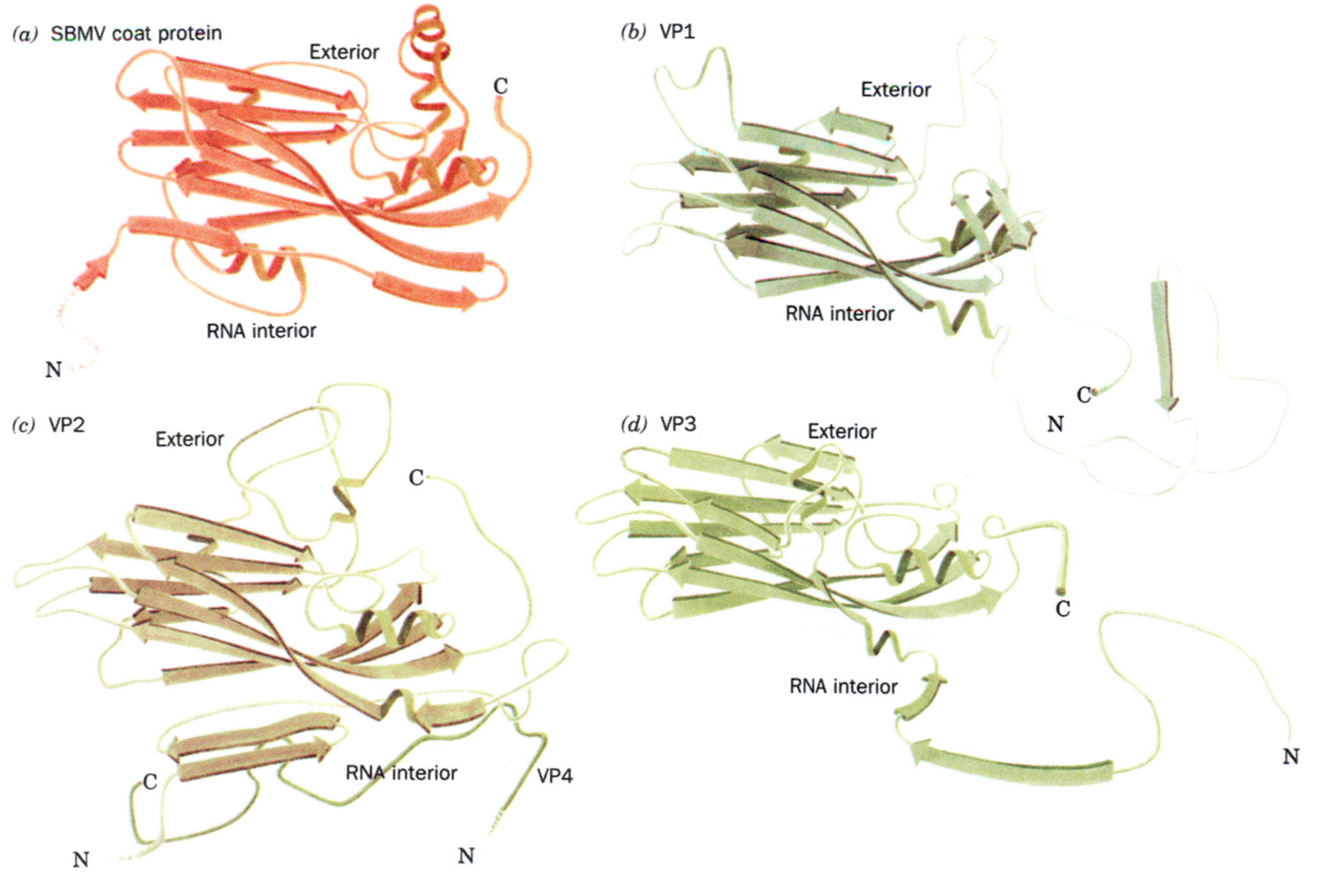

FIGURE 33-18 Comparison of the X-ray structures of southern bean mosaic virus (SBMV) and human rhinovirus coat proteins. (*a*) SBMV coat protein, and (*b*) VP1, (*c*) VP2 (together with VP4), and (*d*) VP3 proteins of human rhinovirus. Note the close structural similarities of their 8-stranded β-barrel cores and that of TBSV's S domain (Fig. 33-14). The VP1, VP2, and VP3 proteins of poliovirus also have this fold. [After Rossmann, M.G., et al., *Nature* **317,** 148 (1985). PDBids 4BSV and 4RHV.]

STMV's quaternary structure differs from those of TBSV or SBMV: It is a $T = 1$ RNA virus. This 172-Å-diameter particle, which is among the smallest of known virions, encloses a 1058-nt RNA that encodes only one protein, its 196-residue viral coat protein (STMV can only multiply in cells that are coinfected with the more complex TMV, the only known example of such a parasitic relationship between an icosahedral virus and a rod-shaped virus). Nevertheless, STMV's coat protein, which also lacks a P domain, has an S domain that structurally resembles those of SBMV and TBSV. Evidently, these biochemically dissimilar viruses arose from a common ancestor.

d. Most of STMV's RNA Is Visible

The most striking aspect of the STMV structure is that nearly 80% of its RNA is visible (Fig. 33-19). The RNA takes the form of 30 largely double helical segments that lie on the icosahedron's 2-fold axes and which are linked by mainly disordered single-stranded regions. Computerized searches indicate up to 68% of STMV RNA could simultaneously form base pairs. However, it seems unlikely that a unique structure with extensive and nonrepetitive pairing between bases that are distant in sequence could be made to fold in a manner consistent with STMV's icosahedral symmetry. Rather, it appears that the double helical segments seen in Fig. 33-19 represent a series of somewhat different local stem–loop structures. Thus, inside the viral capsid, STMV RNA assumes a structure that is probably not its lowest free energy state. More likely, it assumes one of numerous relatively low free energy states that transiently form during viral assembly and become trapped through interactions with the viral protein coat. Indeed, the observation that STMV coat protein does not form capsids in the absence of RNA suggests that the RNA, which lacks icosahedral symmetry, nevertheless directs the formation of the icosahedral viral particle.

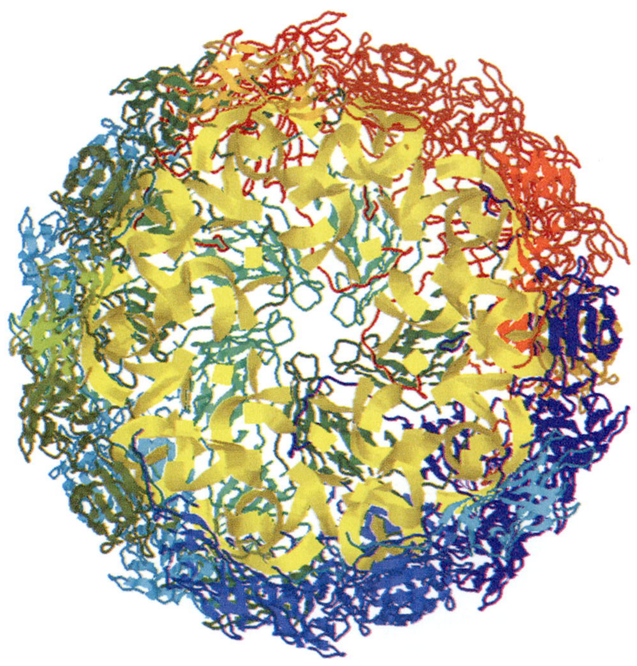

FIGURE 33-19 X-Ray structure of satellite tobacco mosaic virus (STMV). The virion is shown in cutaway view along its icosahedral ($T = 1$) 5-fold axis. Its coat protein subunits, which form a shell between the radii of 57 to 86 Å, are drawn in different colors, whereas its 30 mainly double helical RNA segments that are located on the icosahedral 2-fold axes are yellow. [Based on an X-ray structure by Alexander McPherson, University of California at Irvine. PDBid 1A34.]

C. *Picornaviruses*

The X-ray structures of **poliovirus,** the cause of **poliomyelitis,** and **rhinovirus,** the cause of **infectious rhinitis** (the common cold), were respectively determined by James Hogle and Rossmann. Both of these human pathogens are **picornaviruses,** a large family of animal viruses that also includes the agents causing human **hepatitis A** and **foot-and-mouth disease.** Picornaviruses (*pico,* small + *rna*) are among the smallest RNA-containing animal viruses: They have a particle mass of $\sim 8.5 \times 10^6$ D of which ~30% is a single-stranded RNA of ~7500 nucleotides. Their icosahedral protein shell, which is ~300 Å in diameter, contains 60 protomers, each consisting of four structural proteins, **VP1, VP2, VP3,** and **VP4.** These four proteins are synthesized by an infected cell as a single polyprotein, which is cleaved to the individual subunits during virion assembly. Picornaviruses can be highly specific as to the cells they infect; for example, poliovirus binds to receptors that occur only on certain types of primate cells.

The structures of poliovirus, rhinovirus, and **foot-and-mouth disease virus** (**FMDV;** determined by David Stuart) are remarkably alike, both to each other and to TBSV and SBMV. Although VP1, VP2, and VP3 of picornaviruses have no apparent sequence similarities with each other or with the coat proteins of TBSV and SBMV, these proteins all exhibit striking structural similarities (Figs. 33-14 and 33-18; VP4, which is much smaller than the other subunits, forms, in effect, an N-terminal extension of VP2). Indeed, the picornaviruses' chemically distinct VP1, VP2, and VP3 subunits are pseudosymmetrically related by pseudo-3-fold axes passing through the center of each triangular face of the icosahedral ($T = 1$) virion, which therefore has pseudo-$T = 3$ symmetry (Fig. 33-20). The chemically identical but conformationally distinct A, B, and C subunits of the $T = 3$ plant viruses are likewise quasi-symmetrically related by analogously located local 3-fold axes (Fig. 33-15). These structural similarities strongly suggest that the picornaviruses and the icosahedral plant viruses all diverged from a common ancestor.

The protein capsids of poliovirus, rhinovirus, and FMDV form a hollow shell enclosing a disordered core composed of the viral RNA and some protein, much as in the icosahedral plant viruses. This arrangement is vividly illustrated in Fig. 33-21, which shows both the inner and outer views of the poliovirus capsid. Note that VP4 largely

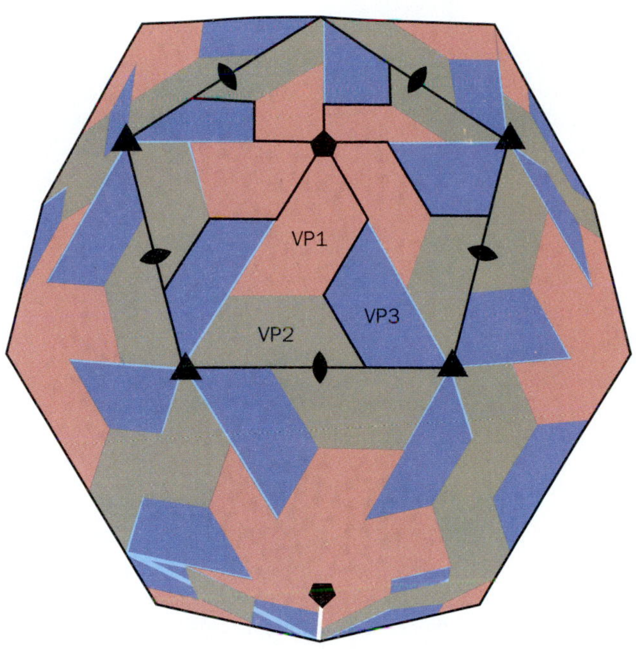

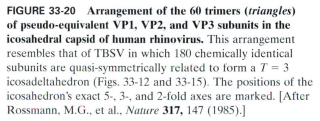

FIGURE 33-20 Arrangement of the 60 trimers (*triangles*) of pseudo-equivalent VP1, VP2, and VP3 subunits in the icosahedral capsid of human rhinovirus. This arrangement resembles that of TBSV in which 180 chemically identical subunits are quasi-symmetrically related to form a $T = 3$ icosadeltahedron (Figs. 33-12 and 33-15). The positions of the icosahedron's exact 5-, 3-, and 2-fold axes are marked. [After Rossmann, M.G., et al., *Nature* **317,** 147 (1985).]

lies inside of the capsid. Also note the rugged topography of the capsid's outer surface. Some of its crevices form the receptor-binding site through which the virus is targeted to specific cells.

D. *Simian Virus 40 (SV40)*

Simian virus 40 (SV40) is a **polyomavirus,** the simplest class of viruses containing double-stranded DNA. This ~500-Å external diameter icosahedral virus (Fig. 33-1*g*) functions to transfer a 5243-bp circular "minichromosome" (DNA in complex with histone-containing particles known as **nucleosomes;** Section 34-1B) from the nucleus of one cell to that of another. The viral capsid consists of 360 copies of a 361-residue protein, **VP1,** that are arranged with icosahedral symmetry. However, this number of particles cannot be arranged with the icosadeltahedral symmetry characteristic of TBSV, for example, because $T = 360/60 = 6$ is a forbidden value for icosadeltahedra (for which $T = h^2 + hk + k^2$). Rather, as Caspar demonstrated through low-resolution X-ray studies of polyomaviruses, VP1 exclu-

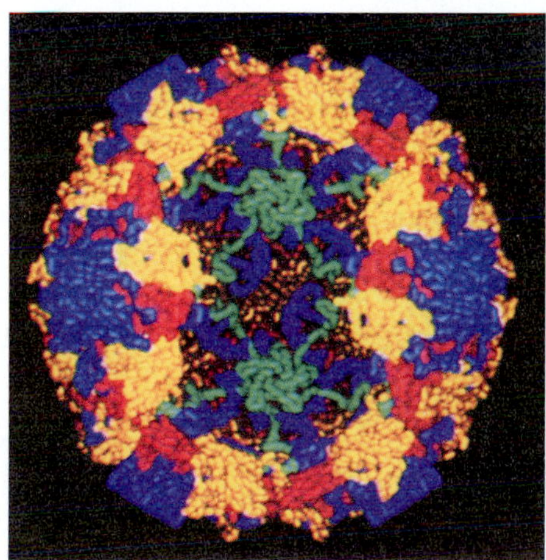
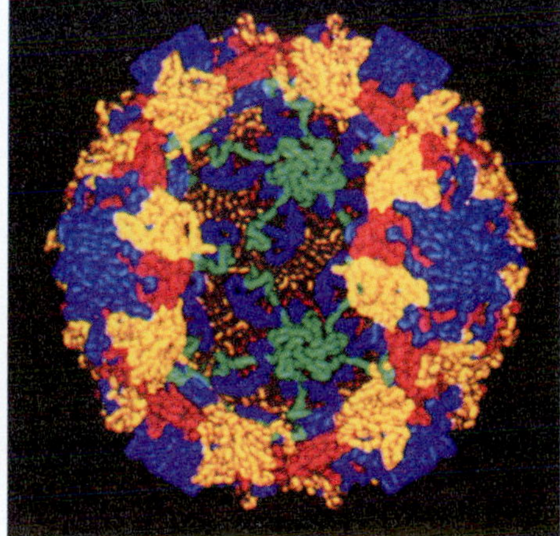

FIGURE 33-21 Stereo diagram of the poliovirus capsid in which the inner surface is revealed by the removal of two pentagonal faces. Here, the polypeptide chain is represented by a folded tube that approximates the volume of the protein and which is blue in VP1, yellow in VP2, red in VP3, and green in VP4. The VP4 subunits, which line the capsid's inner surface, associate about its 5-fold axes of symmetry to form a framework similar to although geometrically distinct from that formed by the C subunit arms in TBSV (Fig. 33-16). [Courtesy of Arthur Olson, The Scripps Research Institute, La Jolla, California. Based on an X-ray structure by James Hogle, Harvard Medical School. PDBid 2PLV.]

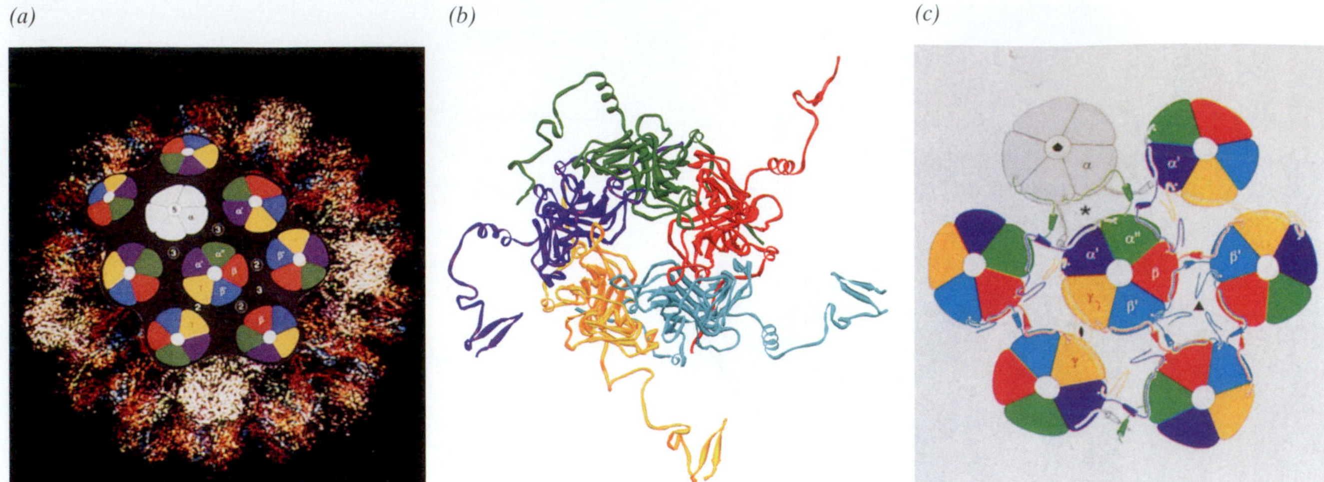

FIGURE 33-22 X-Ray structure of simian virus 40 (SV40). (*a*) The SV40 virion consists of 360 copies of VP1 that are organized into 72 pentamers of which 12 (*white*) are 5-coordinated and 60 (*colored*) are 6-coordinated. Three types of interpentamer clustering are indicated on the schematic part of the drawing: The white (α), purple (α'), and green (α'') subunits form a 3-fold interaction ③; the red (β) and cyan (β') subunits form one type of 2-fold interaction ②; and the yellow subunits (γ) form a second type of 2-fold interaction (2). The icosahedral axes of symmetry are indicated by the numerals 5, 3, and 2. (*b*) A 6-coordinated pentamer as viewed from outside the virion. The VP1 subunits are colored as in Part *a*. Note the C-terminal arms extending out from each subunit. (*c*) Schematic diagram showing how the C-terminal arms tie the pentamers together. The C-terminal arms are represented by lines and small cylinders (helices). The icosahedral particle's exact 5-, 3-, and 2-fold axes are represented by the conventional symbols, whereas the asterisk indicates a local 2-fold axis relating 5- and 6-coordinated pentamers. [Parts *a* and *c* courtesy of and Part *b* based on an X-ray structure by Stephen Harrison, Harvard University. PDBid 1SVA.]

sively forms pentamers that take up two nonequivalent positions (Fig. 33-22*a*). In fact, the SV40 capsid consists of 72 VP1 pentamers that are centered on the vertices of a $T = 7$ icosadeltahedron. Twelve of these pentamers lie on the icosahedron's twelve 5-fold rotation axes, each surrounded by 5 pentamers of a different geometric class. This latter class of 60 pentamers (which in a true $T = 7$ icosadeltahedron would have to be hexamers) are each surrounded by 6 pentamers, 5 of its own class and one of the former class. As a consequence, each capsid contains six symmetry-inequivalent classes of the chemically identical VP1 subunits. What conformational adjustments must the subunits make to form such a structure and, in particular, how does a pentameric structure coordinate with 6 other such pentamers?

The X-ray structure of SV40, determined by Harrison, indicates that VP1 consists of three modules: (1) an N-terminal arm that extends across the inside of the pentamer beneath the clockwise neighboring subunit (looking from the outside in) and whose first 15 residues are not visible in the structure (they probably extend inward to interact with the minichromosome which is likewise not visible); (2) an antiparallel β barrel with the same topology as that in RNA plant viruses and picornaviruses (Figs. 33-14 and 33-18), although oriented more or less radially with respect to the capsid rather than tangentially; and (3) a 45- to 50-residue C-terminal arm, the site of the only major conformational variation among the six symmetry-inequivalent sets of VP1 subunits. The C-terminal arms form the principal interpentamer contacts by extending from their pentamer of origin so as to invade a neighboring pentagon (Fig. 33-22*b,c*). Each pentamer thereby receives five invading arms from adjacent pentamers as well as donating five such arms. *It is the differing patterns of C-terminal arm exchange among the various pentamers that determines how they associate in forming the capsid.* Since these C-terminal arms are probably flexible and unstructured on a free pentamer, the capsid's pentameric building blocks probably behave, so to speak, as if they are tied together with ropes rather than being cemented together across extended complementary surfaces. Indeed, deletion of the C-terminal arms from recombinant VP1 subunits does not prevent their associating into pentamers but precludes these pentamers from assembling into the viruslike shells that they would otherwise form.

There are many other viruses of known structure whose capsid proteins contain the foregoing 8-stranded antiparallel β barrel. These include the $T = 1$ bacteriophage ϕX174 (a single-stranded DNA virus; Section 30-3B; Fig. 33-1*d*), the $T = 3$ **Norwalk virus** (which is responsible for >96% of nonbacterial gastroenteritis in the United States) and **black beetle virus** (both single-stranded RNA viruses), the $T = 4$ **Nudaurelia ω Capensis virus** (a single-stranded RNA insect virus), and the $T = 13$ **bluetongue virus** (Section 33-2F). Information concerning all icosahedral viruses of known X-ray structure is available over the Web from the Virus Particle ExploreR (VIPER) at http://chagall.scripps.edu/viper/.

E. *Bacteriophage MS2*

The RNA bacteriophage MS2 infects only F^+ (male) *E. coli* (Section 31-1A) because infection is initiated by viral attachment to bacterial F pili. The 275-Å-diameter MS2 virion (Fig. 33-1*b*) consists of 180 identical 129-residue coat protein subunits arranged with $T = 3$ icosadeltahedral symmetry encapsidating a 3569-nt single-stranded RNA molecule. The virion also contains a single copy of the 44-kD A-protein, which is thought to be responsible for viral attachment to the F pili and must therefore be exposed on the phage surface.

The X-ray structure of MS2, determined by Karin Valegård and Lars Liljas, reveals that it has a $T = 3$ protein shell formed by 60 icosahedrally related triangular protomers, each of which consists of three chemically identical subunits with slightly different conformations, much as in TBSV (Fig. 33-15). However, *the MS2 coat protein does not contain the 8-stranded antiparallel β barrel present in all the previously discussed icosahedral viruses.* Rather, each subunit consists of a 5-stranded antiparallel β sheet facing the interior of the particle overlaid with a short β hairpin and two α helices facing the viral exterior (Fig. 33-23). This protein fold resembles those of several other bacteriophage coat proteins.

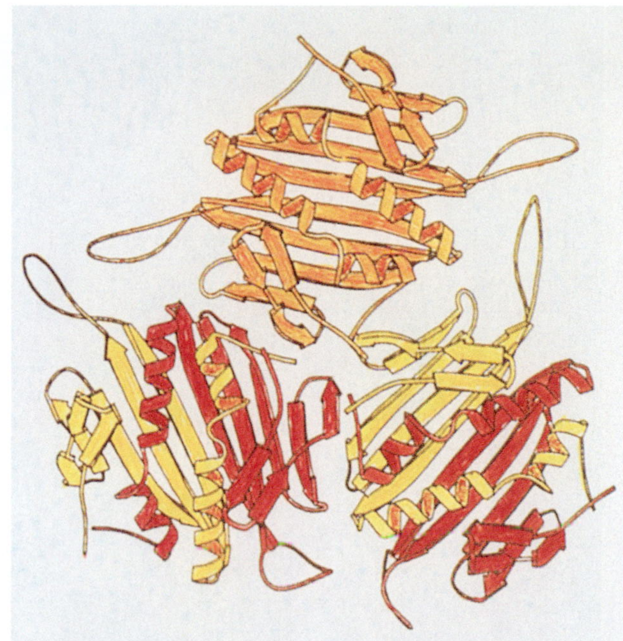

FIGURE 33-23 X-Ray structure of bacteriophage MS2 showing three dimers related by a quasi-3-fold axis of the $T = 3$ icosadeltahedral particle. The A, B, and C subunits, as defined in Fig. 33-12*b*, are, respectively, yellow, red, and orange (those in Fig. 33-12*b* are differently colored). The two C monomers shown are related by the particle's exact 2-fold axis, whereas closely associated A and B monomers are related by quasi-2-fold axes. In all cases, each monomer's five-stranded antiparallel β sheet is extended across the 2-fold axis and its helices interlock with those of its dimeric mate. Note the lack of structural resemblance between the MS2 subunits and the eight-stranded antiparallel β barrels that form the coat proteins of nearly all other icosahedral viruses with known structures (e.g., Figs. 33-14 and 33-18). [Courtesy of Karin Valegård, Uppsala University, Sweden. PDBid 2MS2.]

F. *Bluetongue Virus*

Bluetongue virus is a member of the **orbivirus** genus within the ***Reoviridae*** family, one of the largest families of viruses. Members of the *Reoviridae* are responsible for significant levels of child mortality in developing countries (by the **rotaviruses,** which cause diarrhea) as well as a variety of economically important diseases of both plants and animals. The orbiviruses are icosahedral viruses whose capsids are made of two shells: an outer shell consisting of the viral proteins **VP2** and **VP5,** which is lost on cell entry; and a transcriptionally active core that is released into the cytoplasm. The core consists of two layers, an inner $T = 2$ shell constructed from 120 copies of the 100-kD protein **VP3(T2),** and an outer $T = 13$ shell consisting of 780 copies of the 38-kD protein **VP7(T13).** The capsid contains an aggregate of ~20 kb of double-stranded RNA (dsRNA) in usually 10 different segments, nearly all of which encode only one protein each. During infection, the dsRNA is maintained within the core because its release into the cell would trigger the interferon-mediated shutdown of translation (Section 32-4A), which would prevent viral proliferation. Consequently, the core also contains multiple copies of virally encoded **dsRNA-dependent RNA polymerase [VP1(Pol)],** helicase **[VP6(Hel)],** and capping enzyme **[VP4(Cap)],** which are associated with each dsRNA segment so as to form active transcription complexes. The resulting mRNAs are extruded into the host cell's cytoplasm, where they direct the ribosomal synthesis of viral proteins. In addition, the mRNAs are encapsidated within growing cores, where they act as templates for the synthesis of negative strand RNA segments, thus forming progeny dsRNAs. Nevertheless, how each core is packaged with precisely one copy of each dsRNA segment is unknown.

Bluetongue virus **(BTV)** infects ungulates (hoofed mammals; e.g., sheep) and is transmitted by certain blood-feeding insects. It is named for the cyanotic tongues (due to swelling) that many BTV-infected animals have. The X-ray structure of the BTV core, determined by Stuart, reveals both layers of the 700-Å-diameter, ~55,000-kD particle and hence BTV is the largest particle of known X-ray structure (although the ribosome constitutes the largest asymmetric assembly of known X-ray structure; Section 32-3A). The asymmetric portion of the outer shell consists of 13 independent copies of VP7(T13) arranged as five geo-

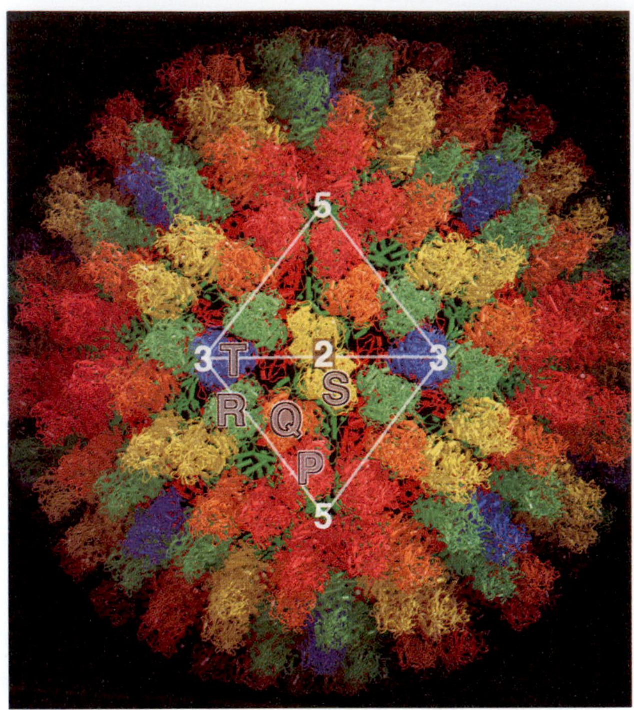

(a)

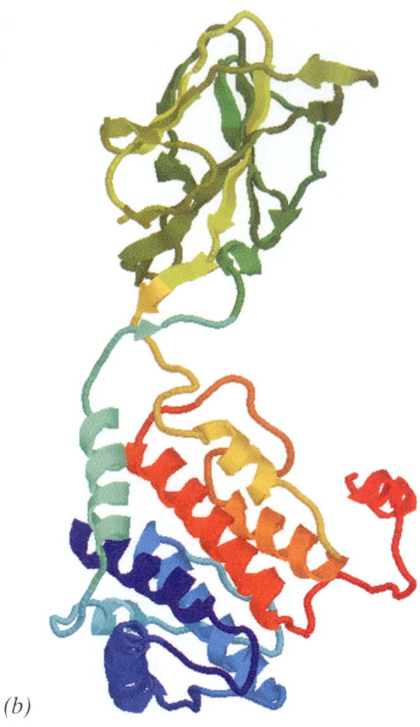

(b)

FIGURE 33-24 X-Ray structure of bluetongue virus core. (*a*) Its $T = 13$ outer shell. The triangular icosahedral asymmetric unit, whose edges (*white lines*) link the icosahedron's symmetry axes, contains 13 copies of VP7 arranged as five trimers, P, Q, R, S, and T, which are colored red, orange, green, yellow, and blue, respectively. Trimer T sits on an icosahedral 3-fold axis and hence contributes a monomer to the asymmetric unit. (*b*) The structure of VP7, which is color-ramped in rainbow order from its N-terminus (*blue*) to its C-terminus (*red*). The 8-stranded antiparallel β barrel domain (*above*) forms the viral core's outer projections and the helical domain (*below*) forms its outer shell. [Part *a* courtesy of and Part *b* based on an X-ray structure by David Stuart, Oxford University, U.K. PDBid 2BTV.]

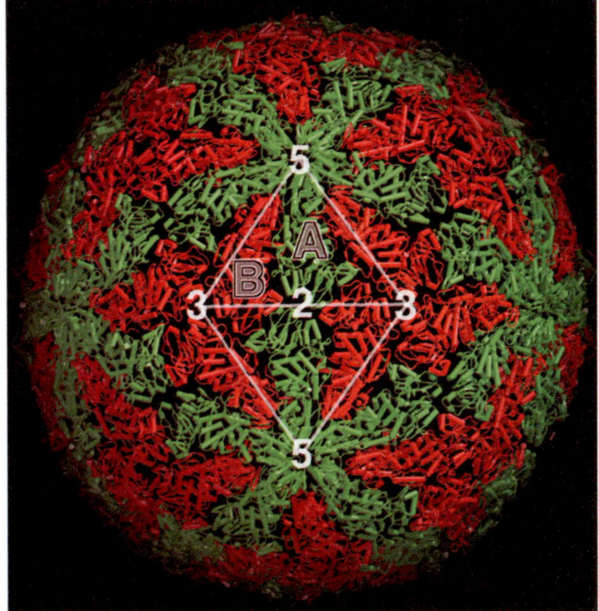

(a)

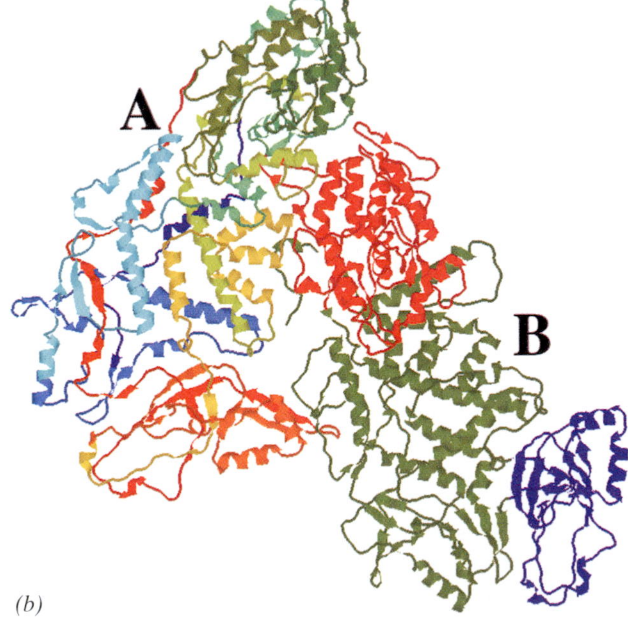

(b)

FIGURE 33-25 X-Ray structure of bluetongue virus core. (*a*) Its $T = 2$ inner shell. It is constructed from homodimers of VP3 subunits arranged with icosahedral ($T = 1$) symmetry so that the two subunits forming the homodimer, A (*green*) and B (*red*), are symmetically inequivalent. The triangular icosahedral asymmetric unit is indicated (*white lines*). Compare this structure with that of the $T = 13$ outer shell (Fig. 33-24*a*). (*b*) The structure of the VP3 asymmetric dimer. Its A subunit is color-ramped in rainbow order from its N-terminus (*blue*) to its C-terminus (*red*) and its B subunit is colored according to domain with its apical domain red, its carapace domain green, and its dimerization (across the 2-fold axis) domain blue. Note the somewhat different conformations and the very different structural environments of these two chemically identical subunits. [Part *a* courtesy of and Part *b* based on an X-ray structure by David Stuart, Oxford University, U.K. PDBid 2BTV.]

metrically distinct trimers, P, Q, R, S, and T (Fig. 33-24*a*). The 349-residue VP7(T13) consists of two domains (Fig. 33-24*b*): an 8-stranded antiparallel β barrel common to many icosahedral viruses that forms the core's outer projections (and is responsible for its hedgehoglike appearance) and presumably contacts the virion's outer layer; and a helical domain that forms the core's outer shell. Note that in most icosahedral viruses that we have encountered, the β barrel domain forms the viral shell (Section 33-2B). The various geometrically distinct copies of VP7(T13) have nearly identical conformations, with maximum deviations between pairs of equivalent C_α positions of only 0.3 Å, even though there are significant differences in the way neighboring subunits contact one another.

BTV's inner shell consists of 60 asymmetric homodimers of VP3(T2) subunits, A and B, arranged with icosahedral ($T = 1$) symmetry (Fig. 33-25*a*). The 901-residue VP3(T2) consists of three domains (Fig. 33-25*b*): an apical domain, which contains 11 helices and 10 β strands in A but 10 helices and 11 β strands in B; a carapace domain, which contains 20 helices in A and 21 in B; and a dimerization domain, which contains 5 helices and 13 β strands in A and 4 helices and 14 β strands in B. The inner shell is relatively smooth and has few charged residues. It has 9-Å-wide pores at its icosahedral 5-fold vertices. These pores, which are lined with conserved Arg residues, are too narrow to permit the exit of mRNA, but in the presence of Mg^{2+} they open sufficiently to do so (see below).

The BTV core's outer and inner shells interact through relatively flat, predominantly hydrophobic surfaces. The symmetry mismatch between these two shells requires that they have 13 different sets of contacts, which makes the closely similar conformations of the 13 geometrically distinct VP7(T13) subunits all the more remarkable. VP3(T2) self-assembles to form a subcore but VP7(T13), although it forms trimers in solution, does not self-assemble to form an icosahedral shell. It is therefore likely that the inner shell forms a permanent scaffolding on which 260 VP7(T13) trimers crystallize in two dimensions to form the outer shell.

The X-ray structure of BTV also reveals the paths of ~80% of its 19,219-bp dsRNA (Fig. 33-26*a*). The dsRNA appears to be partially ordered about the icosahedral 5-fold axes through its interactions with the inside of the VP3(T2) shell (although note that the dsRNA, whose 10 segments range in size from 822 to 3954 bp, cannot be

(*a*)

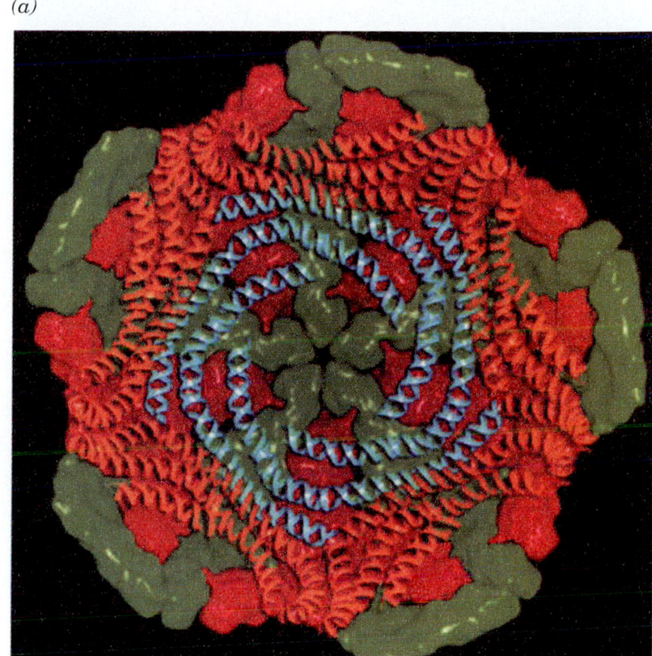

(*b*)

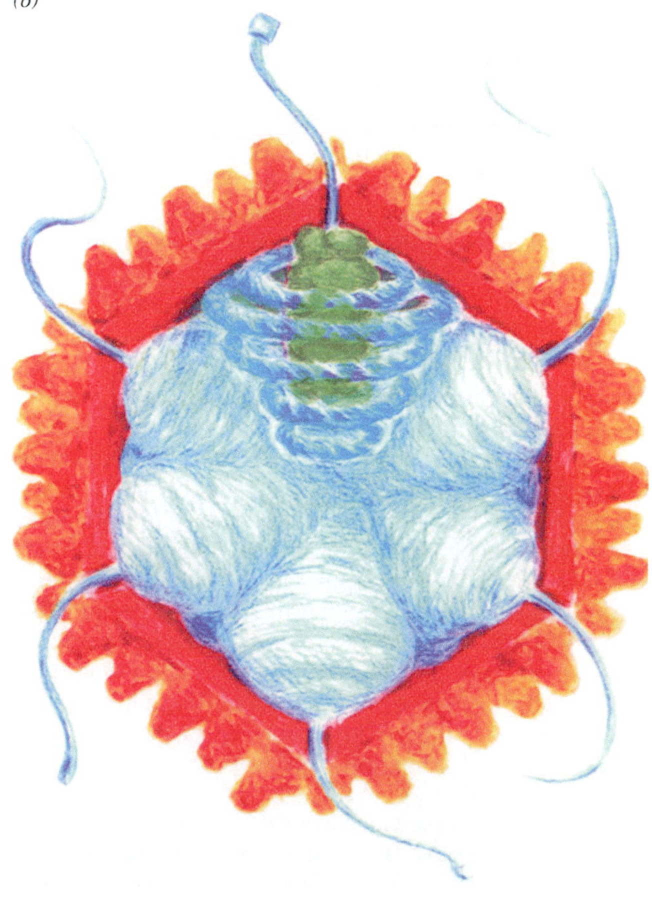

FIGURE 33-26 Arrangement of RNA in BTV. (*a*) The X-ray structure–based packing of dsRNA in the BTV core's inner shell as viewed along an icosahedral 5-fold axis and showing only the core's lower hemisphere. The relatively poorly resolved electron density has been modeled as A-form RNA with the RNA that is packed about the centrally located 5-fold axis blue and that packed about other 5-fold axes orange. The A and B subunits of the VP3(T2) forming the core's inner shell are green and red. (*b*) A cartoon model for the arrangement of RNA (*blue*) in the BTV core. The dsRNA is drawn as a coil that is wrapped about its associated transcription complex (*green*). Newly synthesized mRNA tails are shown exiting the pores at the core's 5-fold vertices. Note that each transcription complex is associated with one 5-fold vertex. [Courtesy of David Stuart, Oxford University, U.K.]

(a)

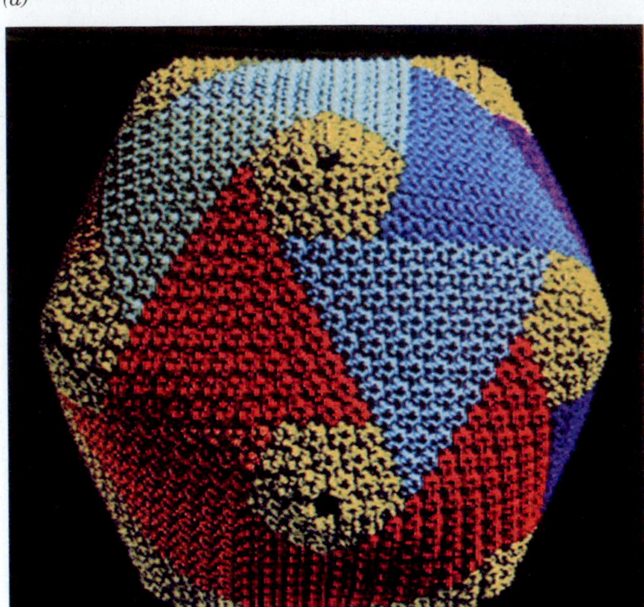

(b)

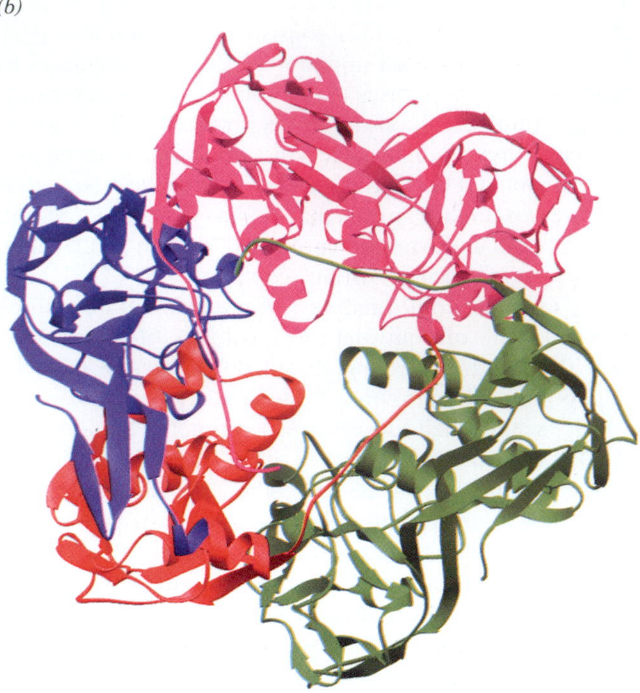

FIGURE 33-27 Structure of the PBCV-1 capsid. (*a*) A quasi-atomic model based on fitting the X-ray structure of Vp54 to the cryo-EM–based image of the capsid. The pentasymmetrons are yellow and the trisymmetrons are variously colored. (*b*) The X-ray structure of the Vp54 homotrimer as viewed along its 3-fold axis. The N-terminal β barrel domain of the leftmost subunit is red, its C-terminal β barrel is blue, and the remaining monomers are green and magenta. [Part *a* courtesy of and Part *b* based on an X-ray structure by Michael Rossmann, Purdue University. PDBid 1M4X.]

arranged with true icosahedral symmetry). This suggests that each transcription complex is organized about a 5-fold axis near the inner surface of the core with its associated dsRNA segment spiraling toward the center of the core. Indeed, X-ray structures of BTV crystals that had been soaked in solutions containing 20-nt oligonucleotides reveal electron density emanating from the viral core along its 5-fold axes that presumably mimics newly synthesized mRNAs (Fig. 33-26*b*).

G. *Paramecium bursaria Chlorella Virus*

Viruses of the **chlorovirus** genus are among the largest and most complex known icosahedral viruses. These viruses have a layered structure comprising dsDNA surrounded by a protein core, a lipid membrane, and finally an icosahedral protein shell. **Paramecium bursaria Chlorella virus type 1 (PBCV-1)** infects certain *Chlorella*-like algae. It attaches to its host cell and, through the mediation of viral enzymes to digest the host cell wall around the point of attachment, injects its dsDNA into the cell, leaving the empty capsid on the cell surface. PBCV-1, which has a molecular mass of $\sim 10^9$ D, has a 331-kb genome that encodes 377 proteins and 10 tRNAs. Its major capsid protein, **Vp54** (a 437-residue glycoprotein), accounts for ~40% of the virion's protein mass.

The cryo-EM–based image of PBCV-1 (Fig. 33-27*a*), determined by Timothy Baker and Rossmann, reveals that its outer shell forms a 1900-Å-diameter, $T = 169$ icosadeltahedron ($h = 7$ and $k = 8$). This enormous capsid is constructed from 20 triangular units named **trisymmetrons** and 12 pentagonal caps named **pentasymmetrons,** which respectively consist of pseudohexagonal arrays of 66 and 30 trimers of Vp54 for a total of 1680 trimers and hence 5040 monomers of Vp54 in the capsid (see below). The trisymmetrons do not correspond to the capsid's icosahedral faces. Instead, they bend around the edges of the icosahedron, leaving openings at its 5-fold vertices that are filled by the pentasymmetrons. Each pentasymmetron also contains a pentamer of a different protein at its 5-fold vertex.

The X-ray structure of Vp54 (Fig. 33-27*b*) reveals that it forms cyclic trimers in which each monomer consists of two consecutive antiparallel β barrels similar to those in other icosahedral viruses. The two β barrels in the Vp54 monomer are related by a 53° rotation about the trimer's 3-fold axis and hence the trimer has pseudohexagonal symmetry. These trimers have been fitted to the cryo-EM–based image of the PBCV-1 capsid, thereby yielding its quasi-atomic model (Fig. 33-27*a*).

3 ■ BACTERIOPHAGE λ

Bacteriophage λ (Figs. 33-1*f* and 33-28), a midsize (58 million D) **coliphage** (bacteriophage that infects *E. coli*), has a 55-nm-diameter icosahedral head and a flexible 15- to 135-nm-long tail that bears a single thin fiber at its end. The virion contains a 48,502-bp linear double-stranded

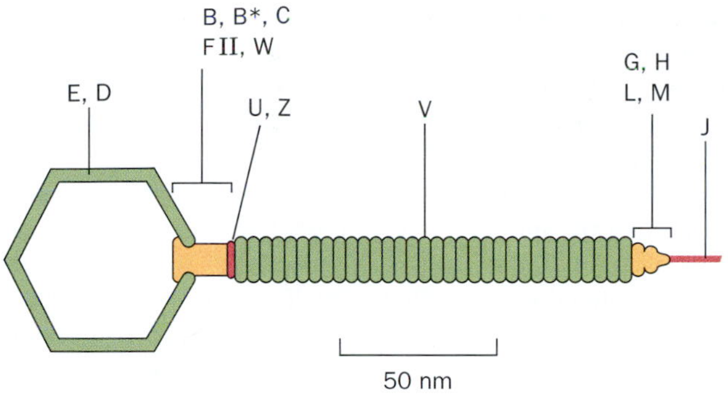

FIGURE 33-28 A sketch of bacteriophage λ indicating the locations of its protein components. The letters refer to specific proteins (gene products; see text). The bar represents 50 nm. [After Eiserling, F.A., *in* Fraenkel-Conrat, H. and Wagner, R.R. (Eds.), *Comparative Virology*, Vol. 13, *p.* 550, Plenum (1979).]

B-DNA molecule of known sequence. Phage λ is among the most extensively characterized complex viruses with respect to its molecular biology. Indeed, as we shall see in this section, *its genetic regulatory mechanisms form one of our best paradigms for the control of development in higher organisms and its assembly is among our most well-characterized examples of the morphogenesis of biological structures.*

Bacteriophage λ adsorbs to *E. coli* through a specific interaction between the viral tail fiber and **maltoporin** (Section 20-2D; the product of the *E. coli lamB* gene), which is a component of the bacterium's outer membrane. This interaction initiates a complex and poorly understood process in which the phage DNA is injected through the viral tail into the host cell. Soon after entering the host, the λ DNA, which has complementary single-stranded ends of 12 nucleotides (cohesive ends), circularizes and is covalently closed and supertwisted by the host DNA ligase and DNA gyrase (Fig. 33-29, Stages 1 and 2).

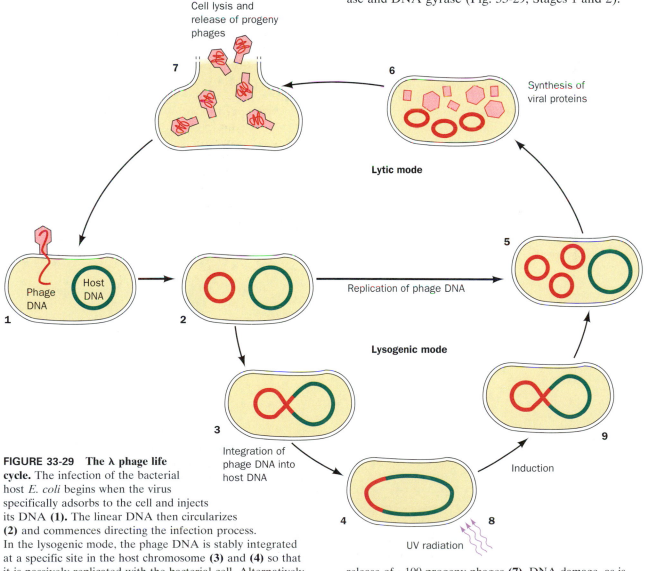

FIGURE 33-29 The λ phage life cycle. The infection of the bacterial host *E. coli* begins when the virus specifically adsorbs to the cell and injects its DNA **(1).** The linear DNA then circularizes **(2)** and commences directing the infection process. In the lysogenic mode, the phage DNA is stably integrated at a specific site in the host chromosome **(3)** and **(4)** so that it is passively replicated with the bacterial cell. Alternatively, the phage may take up the lytic mode in which the DNA directs its own replication **(5),** as well as the synthesis of viral proteins **(6)** so as to result in the lysis of the host cell with the release of ~100 progeny phages **(7).** DNA damage, as is caused, for example, by UV radiation **(8),** induces the excision of the prophage DNA from the lysogenic bacterial chromosome **(9)** and causes the phage to take up the lytic mode.

At this stage the virus has a "choice" of two alternative life styles (Fig. 33-29):

1. It can follow the familiar **lytic** mode in which the phage is replicated by the host such that, after 45 min at 37°C, the host lyses to release ~100 progeny phages.

2. *The phage may take up the so-called* ***lysogenic*** *life cycle, in which its DNA is inserted at a specific site in the host chromosome such that the phage DNA passively replicates with the host DNA. Nevertheless, even after many bacterial generations, if conditions warrant, the phage DNA will be excised from the host DNA to initiate a lytic cycle in a process known as* ***induction.***

How the phage chooses between the lytic and lysogenic modes is the subject of Section 33-3D.

Phage DNA that is following a lysogenic life cycle is described as a **prophage,** whereas its host is called a **lysogen.** An intriguing property of lysogens is that they cannot be reinfected by phages of the type with which they are lysogenized: *They are* ***immune*** *to superinfection.* A bacteriophage that can follow either a lytic or a lysogenic life style is known as a **temperate phage,** whereas those that have only a lytic mode are said to be **virulent.** Bacteriophages that are reproducing lytically are said to be engaged in **vegetative growth.**

Over 90% of the thousands of known types of phages are temperate and, conversely, most bacteria in nature are lysogens. Yet, the presence of prophages has frequently gone unnoticed because they have little apparent affect on their hosts. For example, the K12 strain of *E. coli* had been the subject of intensive investigations for >20 years before 1951 when Ester Lederberg found it to be lysogenic for bacteriophage λ (which marks the discovery of this phage as well as the phenomenon of lysogeny).

The advantage of lysogeny is clear. A parasite that can form a stable association with its host has a better chance of long-term survival than one that invariably destroys its host. A virulent phage, on encountering a colony of its host bacteria, will multiply prodigiously. After the colony has been wiped out, however, it may be some time, if at all, before any of the progeny encounter another suitable host

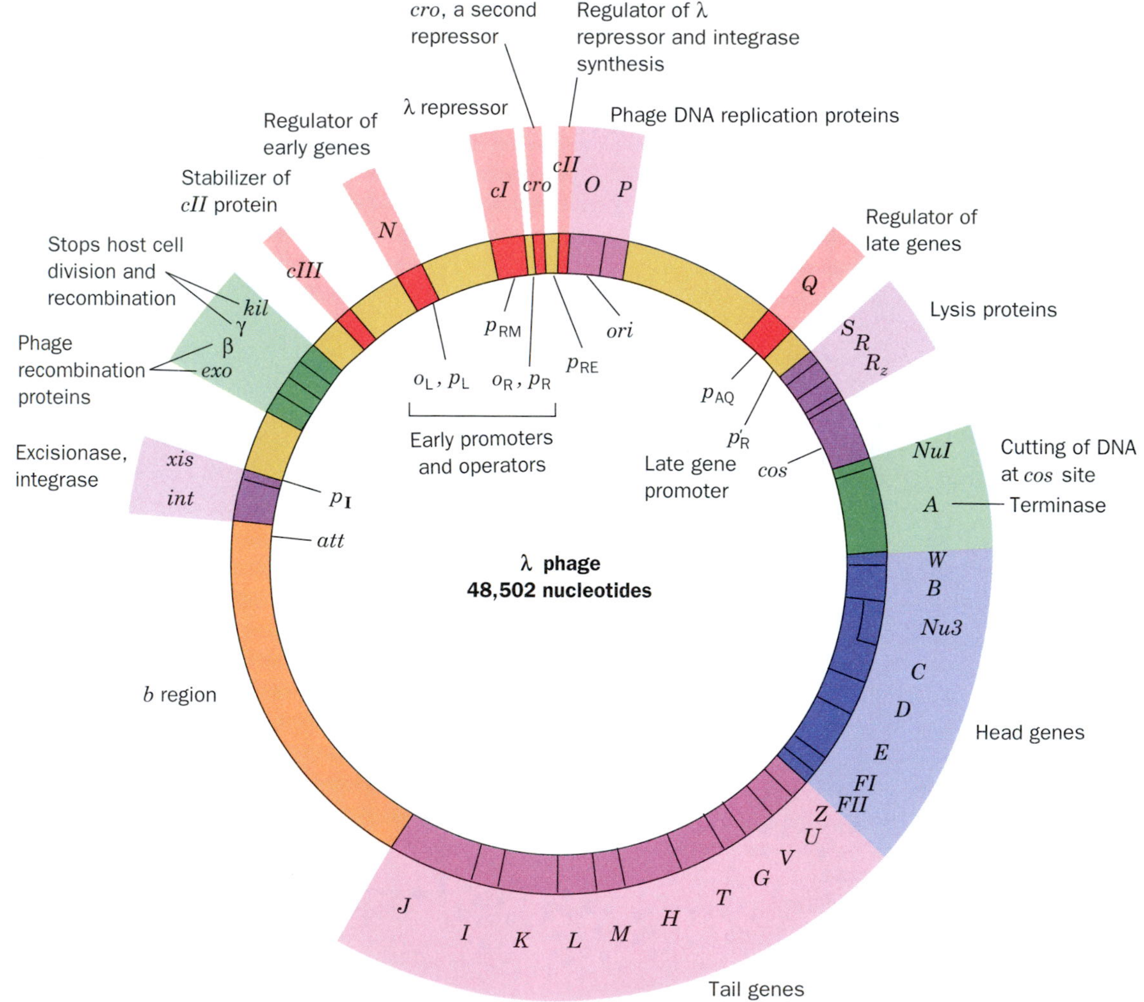

FIGURE 33-30 A genetic map of bacteriophage λ. Most of its structural genes (indicated outside the circle) and control sites (indicated inside the circle) are shown, with the genes encoding regulatory proteins shaded in red. On packaging into the virion, the circular chromosome is cut at the *cos* site yielding a linear DNA.

in a generally hostile world. In contrast, a prophage will multiply with its host indefinitely so long as the host remains viable.

But what if the host is fatally injured? Does the parasite die with the host? In the case of bacteriophage λ, it is precisely such traumatic conditions, exposure to agents that damage the host DNA or disrupt its replication, that induce the lytic phase. This has been described as the "lifeboat" response: The prophage escapes a doomed host through the formation of infectious viral particles that have at least some chance of further replication. Conversely, lysogeny is triggered by poor nutritional conditions for the host (phages can lytically replicate only in an actively growing host) or a large number of phages infecting each host cell (which signals that the phages are on the verge of eliminating the host).

This section describes the genetic system that controls the orderly formation of phage particles in the lytic mode, the mechanism through which these phage particles are assembled, and the regulatory mechanism through which bacteriophage λ selects and maintains its life cycle. *Analogous systems are believed to underlie many cellular processes.*

A. *The Lytic Pathway*

The bacteriophage λ genome, as its genetic map indicates (Fig. 33-30), encodes ~50 gene products and contains numerous control sites. Note the λ chromosome's organization. Its genes are clustered according to function. For example, the genes concerned with the synthesis of phage tail proteins are tandemly arranged on the bottom of Fig. 33-30. This organization, as we shall see, enables these genes to be transcribed together, that is, as an operon. The functions of many of the λ genes and control sites, together with those of the host that are important in phage function, are tabulated in Table 33-1.

In the lytic replication of phage λ, as in love and war, proper timing is essential. This is because the DNA must be replicated in sufficient quantity before it is made unavailable by packaging into phage particles and because packaging must be completed before the host cell is enzymatically lysed. The transcription of the λ genome, which is carried out by host RNA polymerase, is controlled in both the lytic and the lysogenic programs by the regulatory genes that are shaded in red in Fig. 33-30.

a. The Lytic Mode Has Early, Delayed Early, and Late Phases

The lytic transcriptional program has three phases (Fig. 33-31):

1. Early transcription *Soon after phage infection or induction, E. coli RNA polymerase commences "leftward" transcription of the phage DNA starting at the promoter p_L and "rightward" transcription (and thus from the opposite DNA strand) from the promoters p_R and p'_R (Fig. 33-31a):*

(i) The "leftward" transcript, L1, which terminates at termination site t_{L1}, encodes the ***N*** gene.

TABLE 33-1 Important Genes and Genetic Sites for Bacteriophage λ

Gene or Site	Function
Phage genes	
cI	λ repressor; establishment and maintenance of lysogeny
cII, cIII	Establishment of lysogeny
cro	Repressor of *cI* and early genes
N, Q	Antiterminators for early and delayed early genes
O, P	Origin recognition in DNA replication
γ	Inhibits host RecBCD
int	Prophage integration and excision
xis	Prophage excision
B, C, D, E, W, Nu3, FI, FII	Head assembly
G, H, I, J, K, L, M, U, V, Z	Tail assembly
A, Nu1	DNA packaging
R, R_z, S	Host lysis
b	Accessory gene region
Phage sites	
*att*P	Attachment site for prophage integration
*att*L, *att*R	Prophage excision sites
cos	Cohesive end sites in linear duplex DNA
o_L, o_R	Operators
p_I, p_L, p_R, p_{RM}, p_{RE}, p'_R	Promoters
t_{L1}, t_{R1}, t_{R2}, t_{R3}, t'_R	Transcriptional termination sites
*nut*L, *nut*R	*N* utilization sites
qut	*Q* utilization site
ori	DNA replication origin
Host genes[a]	
lamB	Host recognition protein
dnaA, dnaB	DNA replication initiation
lig	DNA ligase
gyrA, gyrB	DNA gyrase
rpoA, rpoB, rpoC	RNA polymerase core enzyme
rho	Transcription termination factor
nusA, nusB, nusE	Necessary for gp*N* function
groEL, groES	Head assembly
himA, himD	Integration host factor
hflA, hflB	Degrades gp*cII*
cap, cya	Catabolite repressor system
*att*B	Prophage integration site
*rec*A	Induction of lytic growth

[a]The genes encoding DNA polymerase I and the subunits of DNA polymerase III (Table 30-2) and the primosome (Table 30-4) are also required.

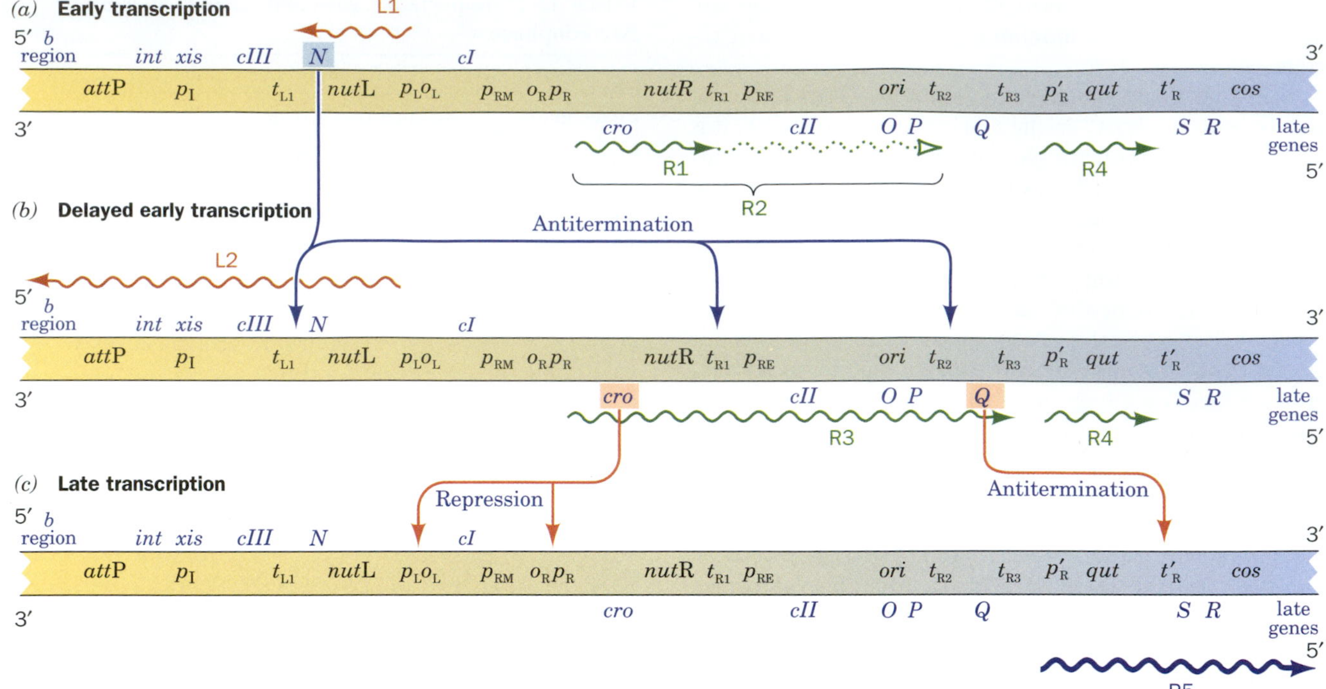

FIGURE 33-31 Gene expression in the lytic pathway of phage λ. Genes specifying proteins that are transcribed to the "left" and "right" are shown above and below the phage chromosome. Control sites are indicated between the DNA strands. The genetic map is not drawn to scale and not all of the genes or control sites are indicated. Transcripts are represented by wiggly arrows pointing in the direction of mRNA elongation; the actions of regulatory proteins are denoted by arrows pointing from each regulatory protein to the site(s) it controls. The lytic pathway has three transcriptional phases: (*a*) early transcription, (*b*) delayed early transcription, and (*c*) late transcription. Gene expression in each of the latter two phases is regulated by proteins synthesized in the preceding phase as is explained in the text. [After Arber, W., *in* Hendrix, R.W., Roberts, J.W., Stahl, F.W., and Weisberg, R.A. (Eds.), *Lambda II, p.* 389, Cold Spring Harbor Laboratory (1983).]

(ii) "Rightward" transcription from p_R terminates with ~50% efficiency at t_{R1}, to yield transcript R1, and otherwise at t_{R2} to yield transcript R2. R1 contains only the ***cro*** gene transcript, whereas R2 also contains the ***cII, O,*** and ***P*** gene transcripts.

(iii) "Rightward" transcription from p'_R terminating at t'_R yields a short transcript, R4, that specifies no protein.

L1, R1, and R2 are translated by host ribosomes to yield proteins whose functions are described below.

2. Delayed early transcription The second transcriptional phase commences as soon as a significant quantity of the protein **gp*N*** (gp for gene product) accumulates. *This protein, through a mechanism considered below, acts as a* ***transcriptional antiterminator*** *at termination sites* t_{L1}, t_{R1}, *and* t_{R2} *(Fig. 33-31b):*

(i) Leftward transcript L1 is extended to form L2, which additionally contains the transcripts of the ***cIII, xis,*** and ***int*** genes (which encode proteins involved in switching between the lytic and lysogenic modes; Sections 33-3C and 33-3D) together with the ***b* region** gene transcripts (which specify the so-called **accessory proteins** that, although not essential for lytic growth, increase its efficiency).

(ii) Transcript R2 is extended to form transcript R3, which additionally encodes a second antiterminator, **gp*Q*,** whose function is discussed below. The continuing translation of R2 and later R3 to yield **gp*O*** and **gp*P*,** proteins that are both required for λ DNA replication, stimulates viral DNA production. Similarly, the translation of R1 and later R3 yields **Cro protein (gp*cro*),** a repressor of both the "rightward" and "leftward" genes (see below; *cro* stands for *c*ontrol of *r*epressor and *o*ther things).

At this stage, ~15 min postinfection, Cro protein has accumulated in sufficient quantity to bind to operators o_L *and* o_R, *thereby shutting off transcription from* p_L *and* p_R. This is more than just efficient use of resources; the overexpression of the early genes, as occurs in λ*cro*⁻ phage, poisons the lytic cycle's late phase.

3. Late transcription In the final transcriptional phase (Fig. 33-31*c*), *the antiterminator gpQ acts to extend transcript R4 through* t'_R *to form transcript R5.* The "gene dosage" effect of the ~30 copies of phage DNA that have accumulated by the beginning of this stage results in the rapid synthesis of the capsid-forming proteins (which are all encoded by late genes; their assembly to form mature phage particles is described in Section 33-3B), as well as

gp*R*, gp*R*$_z$, and **gp*S*,** which catalyze host cell lysis [gp*R* is a transglycosidase that cleaves the bond between NAG and NAM in the host cell wall peptidoglycan (Section 11-3B); gp*R*$_z$ is an endopeptidase that hydrolyzes a peptidoglycan peptide bond; and gp*S* forms pores in the cell membrane, thereby providing gp*R* and gp*R*$_z$ with access to their peptidoglycan substrate]. The first phage particle is completed ~22 min postinfection.

b. Antitermination Requires the Action of Several Proteins

Transcriptional control in the λ lytic phase is exerted by gpN- and gpQ-mediated antitermination rather than by repressor binding at an operator site through which, for example, *lac* operon expression (Section 31-1B) is regulated. gp*N* (107 residues) acts at both rho-dependent and rho-independent termination sites [t_{L1} and t_{R1} are rho dependent (and are, in fact, the terminators with which rho was originally identified), whereas t_{R2} is rho independent; transcriptional termination is discussed in Section 31-2D]. Yet, gp*N* does not act at just any transcriptional termination site. Rather, genetic analysis of mutant phage defective for antitermination has established the existence of two so-called *nut* (for *N utilization*) sites that are required for antitermination: ***nut*L,** which is located between p_L and *N*, and ***nut*R,** which occurs between *cro* and t_{R1} (Fig. 33-31). These sites have closely similar sequences consisting of two elements, *box*B, whose transcripts can form hydrogen-bonded hairpin loops, and *box*A (Fig. 33-32*a*).

What is the mechanism of gp*N*-mediated antitermination? The observation that some *E. coli* defective in antitermination have mutations that map in the *rpoB* gene (which encodes the RNA polymerase β subunit) suggests that gp*N* acts at *nut* sites to render core RNA polymerase (lacking a σ subunit) resistant to termination. Indeed, gp*N*-modulated RNA polymerase will pass over many different terminators that it encounters either naturally or by experimental design. A variety of evidence, including the observation that covering *nut* RNA with ribosomes prevents antitermination, indicates that gp*N* recognizes this site on RNA, not DNA.

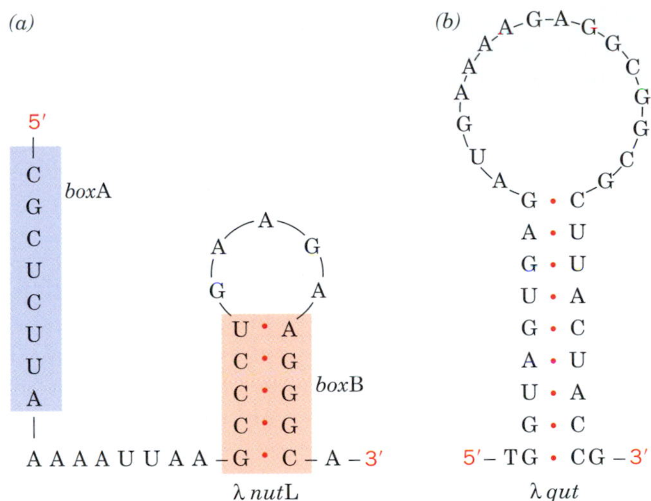

FIGURE 33-32 The RNA sequences of the phage λ control sites: (*a*) *nut*L, which closely resembles *nut*R, and (*b*) *qut*. Each of these control sites is thought to form a base-paired hairpin.

Genetic analyses have revealed that antitermination requires several other host factors termed **Nus** (for *N utilization substance*) **proteins** (Fig. 33-33): **NusA,** which specifically binds to both gp*N* and RNA polymerase; **NusE**

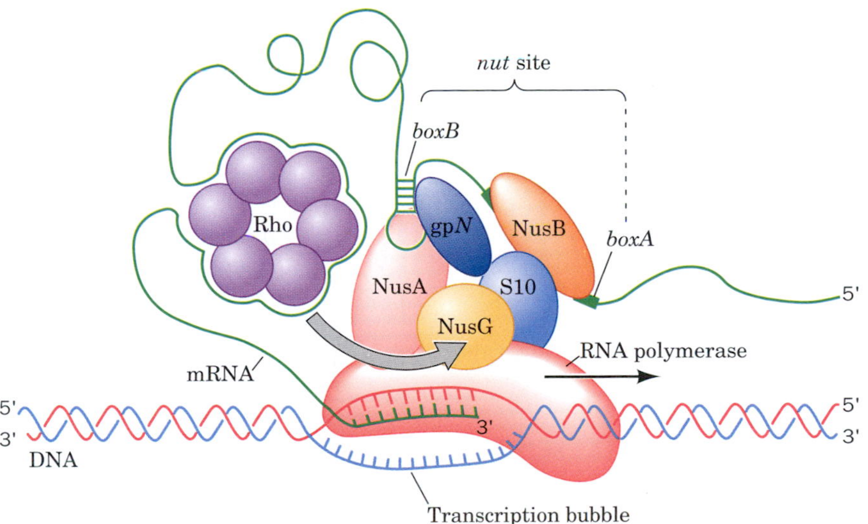

FIGURE 33-33 Schematic model of the antitermination complex between transcribing RNA polymerase, gp*N*, and Nus proteins. gp*N* and the Nus proteins form a complex on a *nut* site of the nascent RNA that binds to the transcribing RNA polymerase further along the looped-out RNA. This complex inhibits RNA polymerase from pausing at a transcriptional termination site, which may prevent rho factor from overtaking the RNA polymerase so as to release the transcript. Another possibility is that transcript release may be inhibited by a gp*N*-modulated direct interaction between NusG and rho factor (*curved arrow*). [After Greenblatt, J., Nodwell, J.R., and Mason, S.W., *Nature* **364,** 402 (1993).]

(which, interestingly, is ribosomal protein S10) and **NusG,** which both bind to RNA polymerase; and **NusB,** which binds to S10. On encountering a *nut* site, gp*N* forms a complex with the Nus proteins and RNA polymerase that travels with this enzyme during elongation and inhibits it from pausing at termination sites. At rho-independent terminators, this deters the release of the transcript at the terminator's weakly bound poly(U) segment, whereas at rho-dependent terminators, it may prevent rho factor from overtaking RNA polymerase, thereby stopping it from unwinding and thus releasing the transcript at the transcription bubble. Alternatively, since it has been shown that NusG binds directly to rho, this interaction, as modulated by gp*N*, may inhibit rho from releasing the nascent transcript.

The *boxB* RNA is recognized by gp*N* via the latter's ~18-residue, Arg-rich, N-terminal segment. The NMR structure of the 15-nt *boxB* hairpin from the lambdoid (λ-like) **bacteriophage P22** (which grows on *Salmonella typhimurium*), in complex with the 20-residue N-terminal segment of its gp*N*, was determined by Dinshaw Patel. It reveals that the peptide forms a helix that binds against the major groove face of the *boxB* hairpin via electrostatic and hydrophobic interactions (Fig. 33-34). This presumably orients the opposite face of the RNA for interactions with host factors.

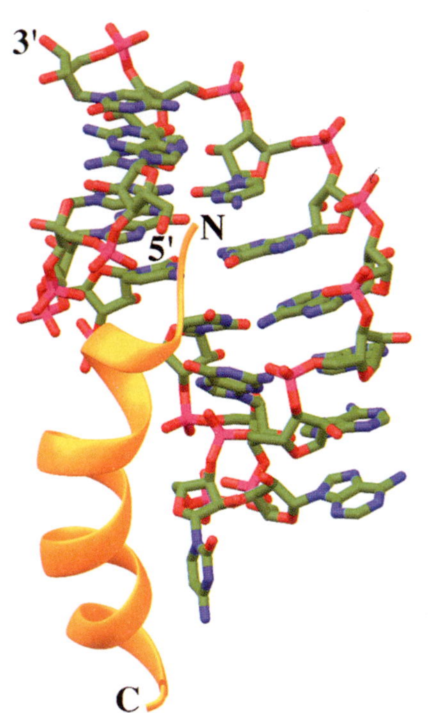

FIGURE 33-34 NMR structure of the bacteriophage P22 *boxB* RNA in complex with the 20-residue, Arg rich, N-terminal segment of its gp*N*. The peptide is represented by a gold ribbon and the RNA is drawn in stick form colored according to atom type (C green, N blue, O red, and P magenta). [Based on an NMR structure by Dinshaw Patel, Memorial Sloan-Kettering Cancer Center, New York, New York. PDBid 1A4T.]

Transcriptional antitermination is not limited only to certain bacteriophage. Indeed, the 7 ribosomal RNA (*rrn*) operons of λ's host organism, *E. coli* (which encode its 5S, 16S, and 23S RNAs; Section 31-4B), each contain a *box*A-like element which, together with the Nus proteins, mediates antitermination at *rrn* (which probably explains the function of S10 as a Nus protein). This suggests that λ *box*A is a defective form of *rrn box*A that requires the presence of gp*N* bound to *box*B in addition to the Nus proteins to inhibit termination.

gp*Q* (207 residues), which overrides t_R' to permit late transcription, acts at a ***qut*** site (analogous to the *nut* sites) that is located some 20 bp downstream from p_R' and that can form an RNA hairpin similar to those of the *nut* sites (Fig. 33-32*b*). Curiously, however, gp*Q*-mediated antitermination occurs via a mechanism that is quite different from that mediated by gp*N*. In fact, gp*Q* binds specifically to *qut* DNA, not to RNA, where together with NusA it binds to RNA polymerase holoenzyme that is paused at p_R' during the initiation phase, thereby accelerating it out of this promoter site and somehow inducing it not to terminate transcription at t_R'.

c. gp*O* and gp*P* Participate in λ DNA Replication

The course of DNA replication in phage λ is diagrammed in Fig. 33-35. Electron microscopy indicates that in the early stages of lytic infection, λ DNA replication occurs via the bidirectional θ mode (Section 30-1A) from a single replication origin **(*ori*).** However, by the late stage of the lytic program, when ~50 λ DNA circles have been synthesized, θ mode DNA replication ceases, probably due to exhaustion of one or more of the required host proteins. At this point, around 3 of the ~50 DNA circles commence replication via the rolling circle (σ) mode (Section 30-3B), with the accompanying synthesis of the complementary strand, although the mechanism of the switchover between the two modes of DNA replication is unclear. The host RecBCD protein (Section 30-6A), a nuclease that would rapidly fragment the resulting concatemeric (consisting of tandemly linked identical units) linear duplex DNA, is inactivated by the phage **γ protein.**

In the process of phage assembly (Section 33-3B), the concatemeric DNA is specifically cleaved in its ***cos*** (for *co*hesive-end *s*ite) site to yield the linear duplex DNA with complementary 12-nt single-stranded ends that are contained in mature phage particles. The staggered double-stranded scission is made by the so-called **terminase,** which is a complex of the phage proteins **gp*A*** (641 residues) and **gp*Nu1*** (181 residues).

Phage λ DNA is replicated by the host DNA replication machinery (Sections 30-1, to 30-3) with the participation of only two phage proteins, gp*O* (333 residues) and gp*P* (233 residues). gp*O*, presumably as dimers, specifically binds to four repeated 18-bp palindromic segments within the phage DNA *ori* region, whereas gp*P* interacts with both gp*O* and the DnaB protein of the host primosome. gp*O* and gp*P*, it is thought, act analogously to host DnaA and DnaC proteins, which as we saw, are required for the initiation of replication of *E. coli* DNA (Section 30-3C).

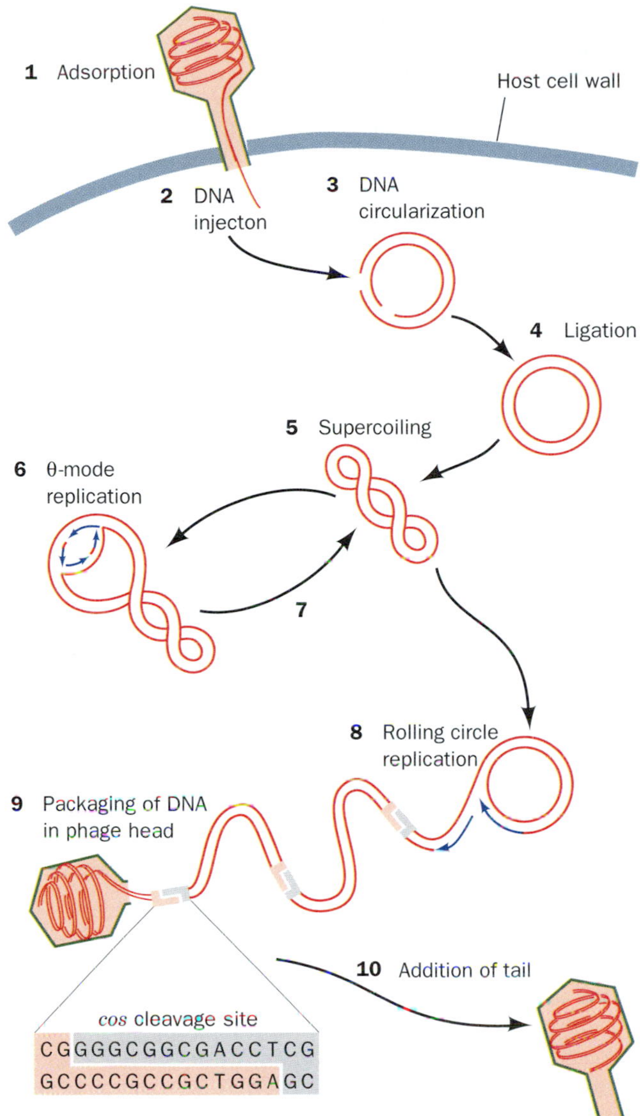

FIGURE 33-35 DNA replication in the lytic mode of bacteriophage λ. The phage particle adsorbs to the host cell **(1)** and injects its linear duplex DNA chromosome **(2).** The DNA circularizes by base pairing at its complementary single-stranded ends **(3),** and the resulting nicked circle is covalently closed **(4)** and supercoiled **(5)** by the sequential actions of host DNA ligase and host DNA gyrase. DNA replication commences according to both the bidirectional θ mode (**6** and **7**) and the rolling circle mode **(8)** but in the later stages of infection occurs exclusively by the rolling circle mode. Here curved blue arrows indicate the most recently synthesized DNA at the replication forks and the arrowheads represent the 3′ ends of the growing DNA chains. The concatemeric DNA produced by the rolling circle mode is specifically cleaved at its *cos* sites (*shaded boxes*) and is packaged into phage heads **(9).** The addition of tails **(10)** completes the assembly of the mature phage particles, which are each capable of initiating a new round of infection. [After Furth, M.E. and Wickner, S.H., *in* Hendrix, R.W., Roberts, J.W., Stahl, F.W., and Weisberg, R.A. (Eds.), *Lambda II, p.* 146, Cold Spring Harbor Laboratory (1983).]

Nevertheless, DnaA is required for λ DNA replication. The gp*O* and gp*P* proteins apparently function to recognize the λ *ori* site, which, curiously, lies within the *O* gene.

B. *Virus Assembly*

The mature λ phage head contains two major proteins: **gp*E*** (341 residues), which forms its polyhedral shell, and **gp*D*** (110 residues), which "decorates" its surface. Electron microscopy indicates that these proteins, which are present in equal numbers, are arranged on the surface of a $T = 7$ icosadeltahedron. However, the λ head also contains four major proteins, **gp*B*, gp*C*, gp*FII*,** and **gp*W*,** which form a cylindrical structure that attaches the tail to the head. This **head–tail connector** occurs at one of the head's 5-fold vertices and thereby breaks its icosahedral symmetry. Hence, gp*E* and gp*D* are present in somewhat fewer than the 420 copies/phage in a perfect $T = 7$ icosadeltahedron.

The tail is a tubular entity that consists of 32 stacked hexagonal rings of **gp*V*** (246 residues) for a total of 192 subunits. The tail begins with a complex adsorption organelle composed of five different proteins, **gp*G*, gp*H*, gp*L*, gp*M*,** and **gp*J*,** and ends with an assembly of **gp*U*** and **gp*Z*** (Fig. 33-28).

The study of complex virus assembly has been motivated by the conviction that it will provide a foundation for understanding the assembly of cellular organelles. Phage assembly is studied through a procedure developed by Robert Edgar and William Wood that combines genetics, biochemistry, and electron microscopy. Conditionally lethal mutations (either temperature-sensitive mutants, which appear normal at low temperatures but exhibit a mutant phenotype at higher temperatures; or suppressor-sensitive *amber* mutants, Section 32-2E) are generated that, under nonpermissive conditions, block phage assembly at various stages. This process results in the accumulation of intermediate assemblies or side products that can be isolated and structurally characterized through electron microscopy. The mutant protein can be identified, through a process known as ***in vitro* complementation** (in analogy with *in vivo* genetic complementation; Section 1-4C), by mixing cell-free extracts containing these structural intermediates with the corresponding normal protein to yield infectious phage particles.

The assembly of bacteriophage λ occurs through a branched pathway in which the phage heads and tails are formed separately and then joined to yield mature virions.

a. Phage Head Assembly

λ Phage head assembly occurs in five stages (Fig. 33-36, *right*):

1. Two phage proteins, gp*B* (533 residues) and **gp*Nu3*,** together with two host-supplied chaperonin proteins, GroEL and GroES, interact to form an "initiator" that consists of 12 copies of gp*B* arranged in a ring with a central orifice. This precursor of the mature phage head–tail connector (Fig. 33-28) apparently organizes the phage

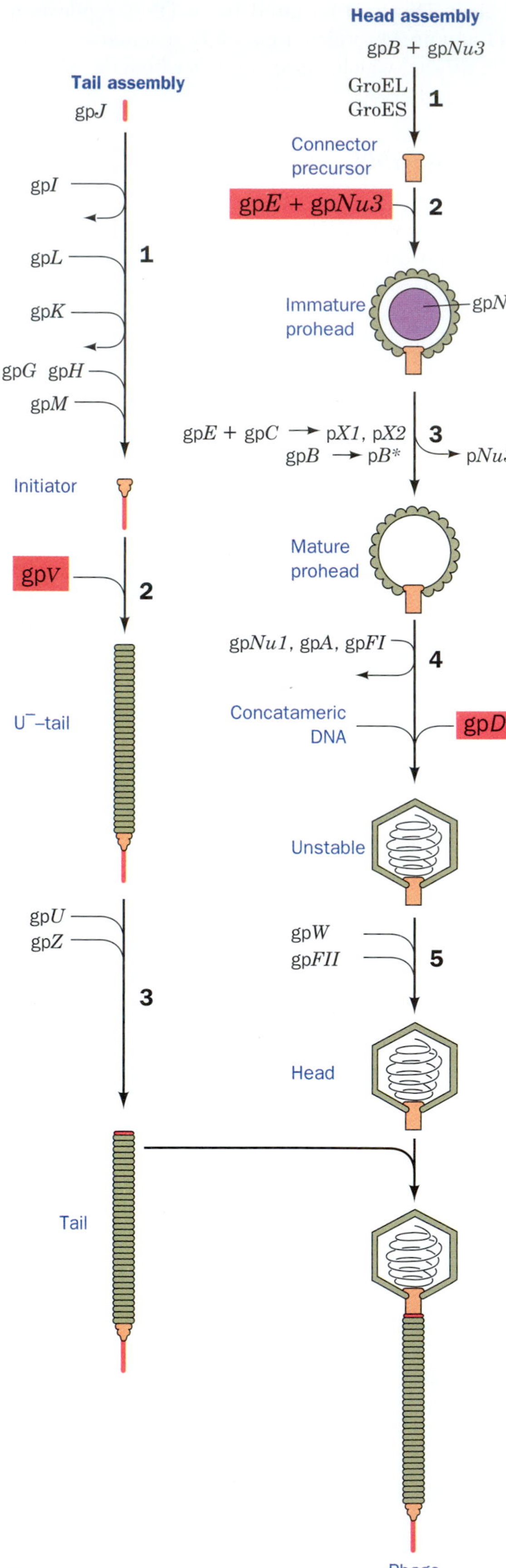

FIGURE 33-36 The assembly of bacteriophage λ. The heads and tails are assembled in separate pathways before joining to form the mature phage particle. Within each pathway the order of the various reactions is obligatory for proper assembly to occur. gp*E*, gp*Nu3*, gp*D*, and gp*V* are highlighted in red boxes to indicate that relatively large numbers of these proteins are required for phage assembly. The numbered steps are described in the text.

head's subsequent formation. GroEL and GroES, it will be recalled, provide a protected environment that facilitates the proper folding and assembly of proteins and protein complexes such as the connector precursor (Section 9-2C). In fact, these chaperonins were discovered through their role in λ assembly. gp*Nu3*, as we shall see, also functions as a molecular chaperone in that it has but a transient role in phage head assembly.

2. gp*E* and additional gp*Nu3* associate to form a structure called an immature **prohead.** If gp*B*, GroEL, or GroES is defective or absent, some gp*E* assembles into spiral or tubular structures, which indicates that the missing proteins guide the formation of a proper shell. The absence of gp*Nu3* results in the formation of but a few shells that contain only gp*E*. gp*Nu3* evidently facilitates proper shell construction and promotes the association of gp*E* with gp*B*.

3. In the formation of the mature prohead, the N-terminal 22-residue segment of ~75% of the gp*B* is excised to form **gp*B****; the gp*Nu3* is degraded and lost from the structure; and 10 copies of gp*C* (439 residues) participate in a fusion–cleavage reaction with 10 additional copies of gp*E* to yield the hybrid proteins **p*X1*** and **p*X2*** (p for protein), which form the collar that apparently holds the connector in place. This maturation process, which involves only phage gene products that are part of the immature prohead, requires that all of the prohead components be present and functional, that is, that the immature prohead be correctly assembled to start with. The enzyme(s) that catalyzes this process has, nevertheless, not been identified.

4. The concatemeric viral DNA is packaged in the phage head and cleaved by mechanisms discussed below. During this process, the capsid proteins undergo a conformational change that results in an expansion of the phage head to twice its original volume (a process that occurs in 4*M* urea in the absence of DNA). gp*D* then attaches to newly exposed binding sites on gp*E*, thereby partially stabilizing the capsid's expanded structure.

5. In the final stage of phage head assembly, gp*W* (68 residues) and gp*FII* (117 residues) add in that order to stabilize the head and form the tail-binding site.

These stages of phage head assembly, as well as some of their component reactions, must proceed in an obligatory order for proper assembly to occur. Of particular interest is that *the components of the mature phage head are not entirely self-assembling as are, for example, TMV (Section 33-1B) and ribosomes (Section 32-3A)*. Rather, the

E. coli proteins GroEL and GroES, as we saw, facilitate head–tail connector assembly. Moreover, gp*Nu3*, which occurs in ~200 copies inside the immature prohead but is absent from the mature prohead, evidently acts as a "scaffolding" protein that organizes gp*E* to form a properly assembled phage head. Finally, *since phage assembly involves several proteolytic reactions, it must also be considered to occur via enzyme-directed processes.*

b. DNA Is Tightly Packed in the Phage Head

An intriguing question of tailed phage assembly is, how does a relatively small phage head (55 nm in diameter in λ) package a far longer (16,500 nm in λ), stiff dsDNA molecule? Cryo-EM studies of DNA-filled phage heads from **bacteriophage T7** (a tailed coliphage) appear to answer this question. The images of these phage heads seen in axial view (along the line through the capsid vertex containing the head–tail connector to the center of the particle) reveal a striking pattern of at least 10 concentric rings, with only the outer ring, which is slightly thicker than the others, representing the protein shell (Fig. 33-37*a*). In contrast, the side views of these phage heads show only punctate (marked with dots or spots) patterns that in some places form linear features. Computer modeling indicates that these patterns can be accounted for by the spooling of the DNA in concentric shells around the axis through the connector (Fig. 33-37*b*). Six such shells are required to accommodate the entire 40-kb T7 DNA into its 55-nm-diameter T7 phage head. This does not contradict the observation of at least nine concentric rings of DNA because, in a given shell, the DNA is coiled more tightly toward the poles of the phage head and therefore appears in projection as multiple rings of lower radii. Since the DNA linearly enters the phage prohead through the head–tail connector (see below), it has been proposed that its stiffness causes it to first coil against the inner wall of the rigid protein shell and then to wind concentrically inward, much like a spool of twine. Nevertheless, the DNA's detailed winding path varies randomly from particle to particle as is indicated by the observation that packaged DNA can be cross-linked to the capsid along its entire length.

c. DNA Is "Pumped" into the Phage Head by an ATP-Driven Process

The packaging of λ DNA begins when terminase (gp*A* + gp*Nu1*) binds to its recognition sequences on a randomly selected ~200-bp *cos* site. The resulting complex then binds to the prohead so as to introduce the DNA into it through the orifice in its head–tail connector. The "left" end of the DNA chromosome enters the prohead first as is indicated by the observation that only this end of the chromosome is packaged by an *in vitro* system when λ DNA restriction fragments are used. Whether the cutting of the initial *cos* site precedes or follows the initiation of packaging is unknown. However, at least *in vitro,* this process requires the binding to *cos* of the *E. coli* histone-like protein known as **integration host factor (IHF).** IHF binds specific sequences of duplex DNA, which it wraps around its surface, thereby inducing a sharp bend in the DNA (see below).

(*a*)

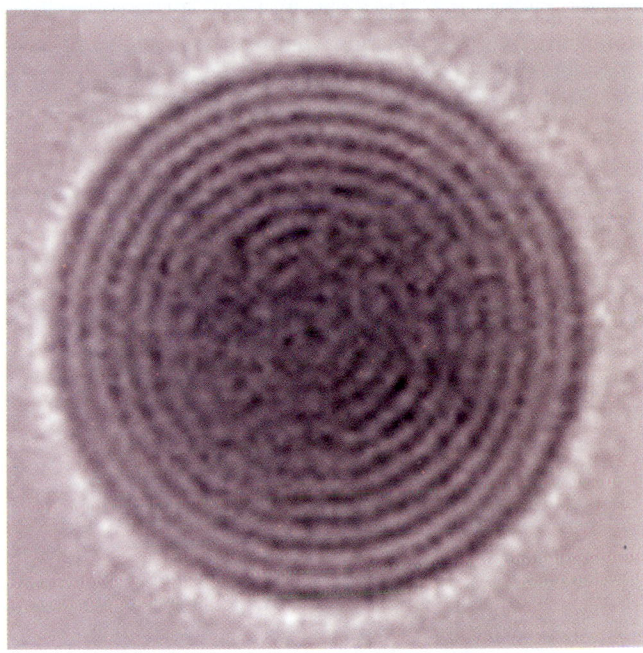

(*b*)

FIGURE 33-37 The packing of dsDNA inside a T7 phage head. (*a*) A cryo-EM–based image of a dsDNA-filled T7 phage head as viewed along a line from the head–tail connector to the center of the particle. The outer somewhat thicker ring represents the phage's protein capsid and the nine inner rings, whose spacing is 2.5 nm, represent coiled dsDNA. [Courtesy of Alasdair Steven, NIH, Bethesda, Maryland.] (*b*) A drawing of the concentric shell model in which the DNA is wound inward like a spool of twine about the phage's long axis. [After Harrison, S.C., *J. Mol. Biol.* **171,** 579 (1983).]

The packing of double-stranded DNA (dsDNA) inside a phage head must be an enthalpically as well as entropically unfavorable process because of dsDNA's stiffness and its intramolecular charge repulsions. The observation that DNA packaging requires the presence of ATP therefore strongly suggests that dsDNA is actively "pumped" into the phage head by an ATP-driven process. The injection of λ DNA into a host bacterium by a mature phage is presumably a spontaneous process that, once it has been triggered, is driven by the free energy stored in the compacted DNA.

The structure of the head–tail connector of bacteriophage λ is unknown. However, Baker and Rossmann have determined the X-ray structure of the head–tail connector of **bacteriophage φ29** (a tailed dsDNA-containing phage that infects *Bacillus subtilis*), which as in bacteriophage λ is the portal through which dsDNA enters and exits the phage head. The structure reveals that the φ29 head–tail connector consists of a funnel-shaped cyclic dodecamer of identical 309-residue subunits that is 75 Å high, has a maximum width of 69 Å, and encloses a central channel whose diameter is 36 Å at its narrow end and 60 Å at its wide end (Fig. 33-38). Two Arg-rich peptide segments of each subunit that project into the central channel are disordered, and are therefore presumed to flexibly interact with the anionic dsDNA as it is translocated through the channel.

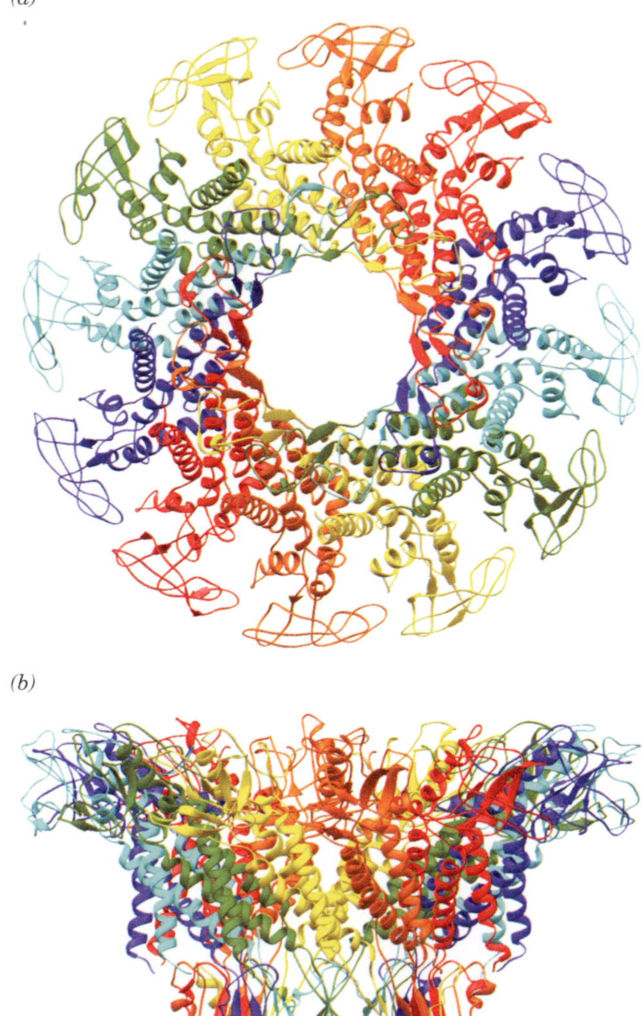

FIGURE 33-38 X-Ray structure of the bacteriophage φ29 head–tail connector. The dodecameric protein is shown in ribbon form with its identical subunits in different colors. (*a*) View along the 12-fold axis. (*b*) View perpendicular to Part *a*. [Based on an X-ray structure by Timothy Baker and Michael Rossmann, Purdue University. PDBid 1FOU.]

Cryo-EM studies of φ29 indicate that the connector is mounted at a pentagonal vertex of the phage prohead with its narrow end protruding into the exterior. The prohead, in addition, contains a virally encoded 174-nt RNA known as **pRNA** (p for *p*rohead), whose presence is essential for DNA packaging but which is absent in the mature virion. Cryo-EM studies indicate that the pRNA forms a cyclic homopentamer that surrounds the narrow end of the dodecameric head–tail connector (Fig. 33-39) and, moreover, suggest that the viral ATPase that drives DNA packaging is associated with the prohead's pentagonal vertex. This led Baker and Rossmann to propose that this entire assembly together with the dsDNA being pumped into the prohead acts as a rotary engine with the prohead–pRNA–ATPase complex as its stator (the stationary part of a rotary machine), the head–tail connector as its ball-race (a groove along which a ball bearing slides), and the dsDNA as its spindle. In this model, the five ATPases around the stator successively fire so as to motivate a traveling wave of conformational changes about the connector that, in essence, causes it to walk along the dsDNA's helical groove(s) so as to translocate it through the connector into the prohead (a mechanism that is reminiscent of that postulated for hexagonal helicases such as T7 gene 4 helicase/primase; Section 30-2C). This would translocate the DNA by one-fifth of its helical pitch for every ATP hydrolyzed, which for canonical B-DNA (which has 10 bp/turn) is the height of two base pairs, a quantity that is consistent with the observed ATP consumption of φ29 during packaging. The

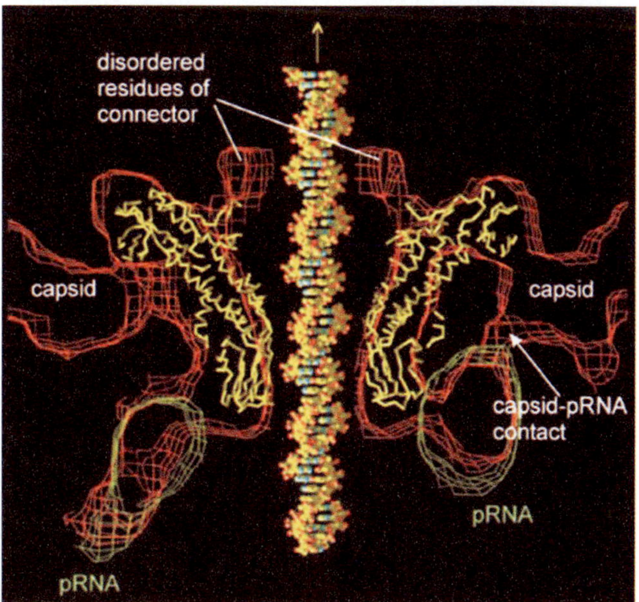

FIGURE 33-39 Cross section of the cryo-EM–based electron density of the bacteriophage φ29 prohead. The capsid (*red meshwork*) is fitted with the C_α backbone of the head–tail connector (*yellow*) and the pRNA (*green meshwork*). Canonical B-DNA is placed in the connector's central channel. The positions of the partially disordered and highly basic peptide segments (residues 229–246 and 287–309) that presumably contact the DNA are indicated. [Courtesy of Timothy Baker and Michael Rossmann, Purdue University.]

DNA packaging engine of bacteriophage λ most likely has a similar mechanism, although φ29 and its close relatives are the only phages known to have a pRNA.

The final step in the bacteriophage λ DNA packaging process is the recognition and cleavage of the next *cos* site (Fig. 33-35) on the concatemeric DNA by terminase, possibly with the participation of **gp*FI***. Phage λ therefore contains a unique segment of DNA (in contrast to some phages in which the amount of DNA packaged is limited by a "headful" mechanism that results in their containing somewhat more DNA than an entire chromosome). Indeed, the λ packaging system will efficiently package a DNA that is 75 to 105% the length of the wild-type λ DNA as long as it is flanked by *cos* sites (the central third of the phage DNA, which encodes the dispensable accessory genes, can be replaced by other sequences, thereby making phage λ a useful cloning vector; Section 5-5B).

d. Tail Assembly

Tail assembly, which occurs independently of head assembly, proceeds, as a comparison of Figs. 33-28 and 33-36 indicates, from the 200-Å-long tail fiber toward the head-binding end. This strictly ordered series of reactions can be considered to have three stages (Fig. 33-36, *left*):

1. The formation of the "initiator," which ultimately becomes the adsorption organelle, requires the sequential actions on gp*J* (the tail fiber protein) of the products of phage genes ***I, L, K, G, H,*** and ***M***. Of these, only **gp*I*** and **gp*K*** are not components of the mature tail.

2. The initiator forms the nucleus for the polymerization of gp*V*, the major tail protein, to form a stack of 32 hexameric rings. The length of this stack is thought to be regulated by gp*H* (853 residues), which, the available evidence suggests, becomes extended along the length of the growing tail and somehow limits its growth. λ tail length is apparently specified in much the same way that the helical length of TMV is governed (Section 33-1B), although in TMV the regulating template is an RNA molecule rather than a protein.

3. In the termination and maturation stage of tail assembly, **gp*U*** attaches to the growing tail, thereby preventing its further elongation. The resultant immature tail has the same shape as the mature tail and can attach to the head. In order to form an infectious phage particle, however, the immature tail must be activated by the action of **gp*Z*** before joining the head.

The completed tail then spontaneously attaches to a mature phage head to form an infectious λ phage particle (Fig. 33-36, *bottom*).

e. The Assembly of Other Double-Stranded DNA Phages Resembles That of λ

The assembly of several other double-stranded DNA bacteriophages has been studied in detail, notably that of coliphages **T4** (Fig. 33-1*e*) and T7 and phage P22. All of them are formed in assembly processes that closely resemble that of phage λ. For example, their head assembly processes proceed in obligatory reaction sequences through an initiation stage; the scaffolded assembly of a prohead; an ATP-driven DNA packaging process, in which the DNA assumes a tightly packed conformation and the prohead undergoes an expansion; and a final stabilization. The mature phages then form by the attachment of separately assembled tails to the completed and DNA-filled heads.

C. The Lysogenic Mode

Lysogeny is established by the integration of viral DNA into the host chromosome accompanied by the shutdown of all lytic gene expression. With phage λ, integration takes place through a site-specific recombination process that differs from homologous recombination (Section 30-6A) in that it occurs only between the chromosomal sites designated *att*P on the phage and *att*B on the bacterial host (Fig. 33-40). These two *att*achment sites have a 15-bp identity

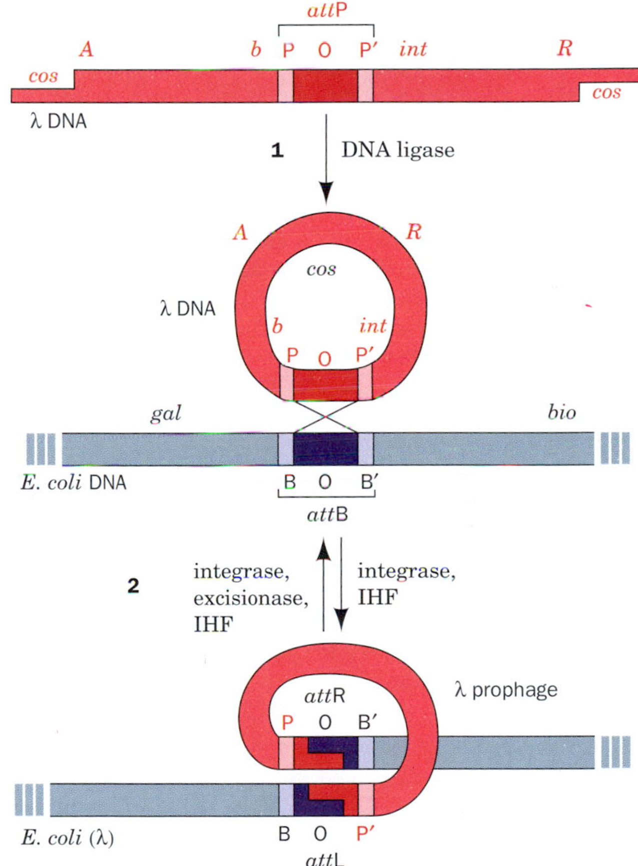

FIGURE 33-40 Site-specific recombination in bacteriophage λ. This schematic diagram shows **(1)** the circularization of the linear phage λ DNA through base pairing between its complementary ends to form the *cos* site; and **(2)** the integration/excision of this DNA into/from the *E. coli* chromosome through site-specific recombination between the phage *att*P and host *att*B sites. The darker colored regions in the *att* sites represent the identical 15-bp crossover sequences (O), whereas the lighter colored regions symbolize the unique sequences of bacterial (B and B′) and phage (P and P′) origin. [After Landy, A. and Weisberg, R.A., *in* Hendrix, R.W., Roberts, J.W., Stahl, F.W., and Weisberg, R.A. (Eds.), *Lambda II*, p. 212, Cold Spring Harbor Laboratory (1983).]

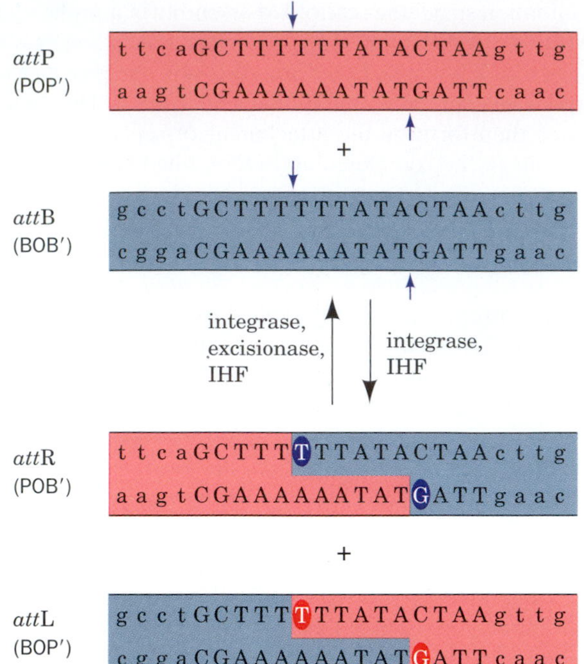

FIGURE 33-41 The site-specific recombination process that inserts/excises phage λ DNA into/from the chromosome of its *E. coli* host. Exchange occurs between the phage *att*P site (*red*) and the bacterial *att*B site (*blue*), and the prophage *att*L and *att*R sites. The strand breaks occur at the approximate positions indicated by the short blue arrows. The sources of the more darkly shaded bases in *att*R and *att*L are uncertain. The uppercase letters represent bases in the O region common to the phage and bacterial DNAs, whereas lowercase letters symbolize bases in the flanking B, B′, P, and P′ sites.

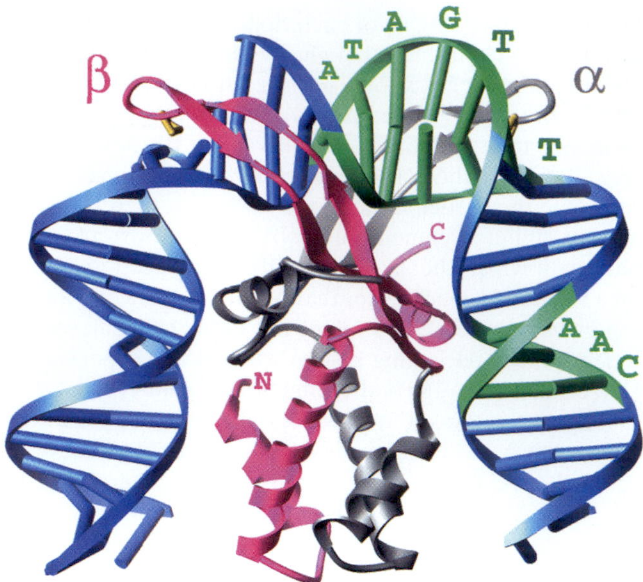

FIGURE 33-42 X-Ray structure of integration host factor (IHF) in complex with a 35-bp target DNA. The structure is viewed with its pseudo-2-fold axis vertical. The α subunit of IHF is gray and its β subunit is magenta. The "top" strand of the 35-bp *att* site DNA, which is shown in ladder form, was synthesized in two segments. The consensus sequence to which IHF binds is highlighted in green and interacts mainly with the β ribbon arm of the α subunit and the body of the β subunit. The Pro side chains near the tip of each arm that intercalate between base pairs so as to kink the DNA are drawn in yellow. [Courtesy of Phoebe Rice, University of Chicago. PDBid 1IHF.]

(Fig. 33-41), so they can be represented as having the sequences POP′ for *att*P and BOB′ for *att*B, where O denotes their common sequence. Phage integration occurs through a process that yields the inserted phage chromosome flanked by the sequence BOP′ on the "left" (the *att*L site) and POB′ on the "right" (the *att*R site; Fig. 33-40). The nature of the crossover site was determined through the use of ^{32}P-labeled bacterial DNA and unlabeled phage DNA. The crossover site occurs at a unique position on each strand that is displaced with respect to its complementary strand so as to form a staggered recombination joint (Fig. 33-41).

a. Integrase Mediates λ DNA Integration, whereas Excisionase Is Additionally Required for λ DNA Excision

Phage integration is mediated by **λ integrase** (356 residues; the ***λ int*** gene product) acting in concert with IHF. The λ integrase is homologous to and mediates a similar site-specific recombination reaction as does the Cre recombinase of bacteriophage P1, whose structure and mechanism are discussed in Section 30-6B. This, together with a variety of biochemical and genetic evidence, indicates that λ integrase functions via a similar mechanism.

IHF, a heterodimer of 30% identical 99- and 94-residue subunits, has no demonstrable endonuclease or topoisomerase activity but specifically binds to a dsDNA bearing an *att* sequence. The X-ray structure of IHF in complex with a 35-bp target DNA, determined by Phoebe Rice and Howard Nash, reveals that the pseudo-2-fold symmetric protein wraps the DNA around it in a >160° bend (Fig. 33-42). Most of this bend arises from two large kinks, separated by 9 bp, that are each formed by the intercalation of a highly conserved Pro side chain between two consecutive base pairs. IHF presumably facilitates the action of λ integrase by bending the DNA in a U-turn so as to bring the two DNA segments of an *att* site which λ integrase binds into close proximity. [The bacterial protein **HU** (Section 30-3C) is closely related to IHF, both in sequence and structure, but functionally differs from it in that HU binds to dsDNA nonspecifically.]

Since viral integration is not an energy-consuming process, why is phage integration not readily reversible? The answer is that the prophage excision requires the participation of **excisionase** (72 residues; the ***λ xis*** gene product) in concert with integrase, IHF, and **Fis** (a DNA-binding host protein that also stimulates Hin-mediated gene inversion; Section 30-6B). Apparently the λ recombination system has an inherent asymmetry that ensures the kinetic stability of the lysogenic integration product.

The mechanism by which excisionase reverses the integration process in unknown, although it has been shown that this protein specifically binds to POB′, where it induces a sharp bend in this DNA.

b. The Relative Levels of Cro Protein and cI Repressor Determine the λ Phage Life Cycle

The establishment of lysogeny in phage λ is triggered by high concentrations of **gpcII** *(see below).* This early gene product stimulates "leftward" transcription from two promoters, p_I (I for integrase) and p_{RE} (RE for *r*epressor *e*stablishment; Fig. 33-43*a*):

1. Transcription initiated from p_I, which is located within the *xis* gene, results in the production of integrase but not excisionase. λ DNA is consequently integrated into the host chromosome to form the prophage.

2. The transcript initiated from p_{RE} encodes the ***cI*** gene whose product is called the **λ** or **cI repressor.** The λ repressor, as does Cro protein (Section 33-3A), binds to the o_L and o_R operators, thereby blocking transcription from p_L and p_R, respectively (Fig. 33-43; note that these operators are upstream from their corresponding promoters rather than downstream as in the *lac* operon; Fig. 31-2). *Both repressors therefore act to shut down the synthesis of early gene products, including Cro protein and gpcII.*

gp*cII* is metabolically unstable with a half-life of ~1 min (see below) so that *cI* transcription from p_{RE} soon ceases. λ repressor bound at o_R, but not Cro protein, however, stimulates "leftward" transcription of *cI* from p_{RM} (RM for *r*epressor *m*aintenance; Fig. 33-43*b*). In other words, *Cro protein represses all mRNA synthesis, whereas λ repressor stimulates transcription of its own gene while repressing all other mRNA synthesis. This conceptually simple difference between the actions of λ repressor and Cro protein forms the basis of a genetic switch that stably maintains phage λ in either the lytic or the lysogenic state.* The molecular mechanism of this switch is described in Section 33-3D. In the following paragraphs we discuss how this switch is "thrown" from one state to another. You should recognize, however, that, *once the switch is thrown in favor of the lytic cycle, that is, when Cro protein occupies o_L and o_R, the phage is irrevocably committed to at least one generation of lytic growth.*

c. gp*cII* Is Activated when Phage Multiplicity Is High or Nutritional Conditions Are Poor

The reason why a high gp*cII* concentration is required to establish lysogeny is that this early gene product can stimulate transcription from p_I and p_{RE} only when it is in oligomeric form. This phenomenon accounts for the observation that lysogeny is induced when the **multiplicity of infection** (ratio of infecting phages to bacteria) is large (≥10) since this gene dosage effect results in gp*cII* being synthesized at a high rate.

gp*cII* is metabolically unstable because it is preferentially proteolyzed by host proteins, notably **gp*hflA*** and **gp*hflB*.** However, gp*cIII* somehow protects gp*cII* from the action of gp*hflA*, which is why its presence enhances lysogenation (Fig. 33-43*a*). The activity of gp*hflA* is dependent on the host cAMP-activated catabolite repression system (Section 31-3C) as is indicated by the observation that *E. coli* mutants defective in this system lysogenize with less than normal frequency. Yet, if these mutant strains are also *hflA*$^-$, they lysogenize with greater than normal frequency. Apparently the *E. coli* catabolite repression system, which is known to regulate the transcription of many bacterial genes, controls *hflA* activity, perhaps by directly repressing this protein's synthesis at high cAMP concentrations. *This explains why poor host nutrition, which results in elevated cAMP concentrations, stimulates lysogenation.*

Once a prophage has been integrated in the host chromosome, lysogeny is stably maintained from generation to generation by λ repressor. This is because λ repressor stimulates its own synthesis at a rate sufficient to maintain

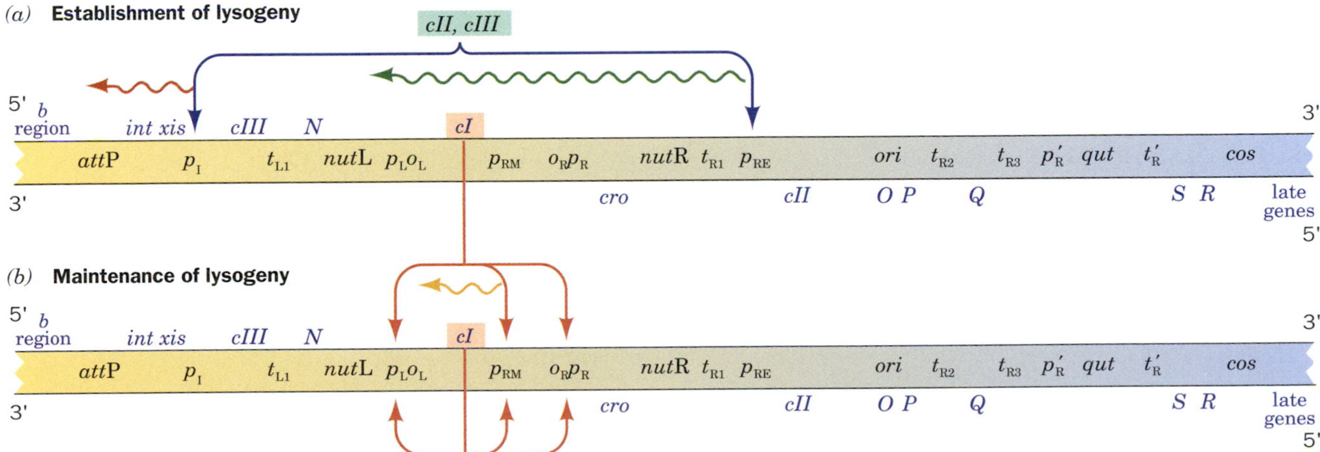

FIGURE 33-43 Control of gene expression in bacteriophage λ. (*a*) The establishment lysogeny. (*b*) The maintenance of lysogeny. The symbols used are described in the legend of Fig. 33-31. [After Arber, W., *in* Hendrix, R.W., Roberts, J.W., Stahl, F.W., and Weisberg, R.A. (Eds.), *Lambda II, p.* 389, Cold Spring Harbor Laboratory (1983).]

lysogeny in the bacterial progeny while repressing the transcription of all other phage genes. In fact, *λ repressor is synthesized in sufficient excess to also repress transcription from superinfecting λ phage, thereby accounting for the phenomenon of immunity.* We shall see below how induction occurs.

D. *Mechanism of the λ Switch*

The lysogenic cycle is a highly stable mode of phage λ replication; under normal conditions lysogens spontaneously induce only about once per 10^5 cell divisions. Yet, transient exposure to inducing conditions triggers lytic growth in almost every cell of a lysogenic bacterial culture. In this section, we consider how this genetic switch, whose mechanism was largely elucidated by Mark Ptashne, can so tightly repress lytic growth and yet remain poised to turn it on efficiently.

a. o_R Consists of Three Homologous Palindromic Subsites

Both of the operators to which λ repressor and Cro protein bind, o_L and o_R, consist of three subsites (Fig. 33-44). These are designated o_{L1}, o_{L2}, and o_{L3} for o_L, and o_{R1}, o_{R2}, and o_{R3} for o_R. *Each of these subsites consists of a similar 17-bp segment that has approximate palindromic symmetry.* However, as we shall see, o_L plays only a minor role in the λ switch relative to that of o_R.

b. λ Repressor and Cro Protein Structurally Resemble Other Repressors

λ repressor binds to DNA as a dimer so that its 2-fold symmetry matches those of the operator subsites to which it binds. The monomer's 236-residue polypeptide chain is folded into two roughly equal sized domains connected by an ~30-residue segment that is readily cleaved by proteolytic enzymes. The isolated N-terminal domains retain their ability to bind specifically to operators (although with only half of the binding energy of the intact repressor) but do not dimerize in solution. The C-terminal domains can still dimerize but lack the capacity to bind DNA. Evidently, *repressor's N-terminal domain binds operator, whereas its C-terminal domain provides the contacts for dimer formation.*

Although the λ repressor has not been crystallized, its N-terminal domain comprising residues 1 to 92, as excised by treatment with the papaya protease **papain,** does crystallize. The X-ray structure of this protein, both alone and in complex with a 20-bp DNA containing the o_{L1} sequence, was determined by Carl Pabo. The N-terminal domain crystallizes as a symmetric dimer with each subunit containing an N-terminal arm and five α helices (Fig. 33-45*a*). Two of these helices, α2 and α3, form a helix–turn–helix (HTH) motif, much like those in other prokaryotic repressors of known structure (Section 31-3D). The α3 helix, the recognition helix, protrudes from the protein surface such that the two α3 helices of the dimeric protein fit into successive major grooves of the operator DNA. Similar associations are observed in the X-ray structures of the closely related **bacteriophage 434 repressor** N-terminal fragment in complex with a 20-bp DNA containing its operator sequence (Fig. 31-29). The X-ray structure of the C-terminal domain (residues 132–236) of the λ repressor, determined by Mitchell Lewis, reveals how this domain dimerizes (Fig. 33-45*b*). The intact λ repressor presumably dimerizes on its target DNA as is drawn in Fig. 33-45*c*.

Cro protein also forms dimers. In contrast to λ or 434 repressor, however, this 66-residue polypeptide forms but one domain that contains both its operator recognition site and its dimerization contacts. The X-ray structure of Cro in complex with a 17-bp tight-binding operator DNA, determined by Brian Matthews, reveals that this dimer likewise contains a pair of HTH units (Fig. 33-46), but which bind to the DNA such that they induce it to bend about the protein by 40°. The sequence-specific binding predicted by this structure is supported by Robert Sauer's genetic studies indicating that mutant varieties of Cro, in which the proposed DNA-contacting residues have been changed, are defective in operator binding. Moreover, this structure closely resembles that of the related **phage 434 Cro protein** in complex with a 20-bp DNA containing its operator sequence (Fig. 31-30).

FIGURE 33-44 Base sequences of the operator regions of the phage λ chromosome. (*a*) o_L and (*b*) o_R. Each of these operators consists of three homologous 17-bp subsites separated by short AT-rich spacers. Each subsite has approximate palindromic (2-fold) symmetry as is demonstrated by the comparison of the two sets of red letters in each subsite. The wiggly arrows mark the transcriptional start sites and directions at the indicated promoters.

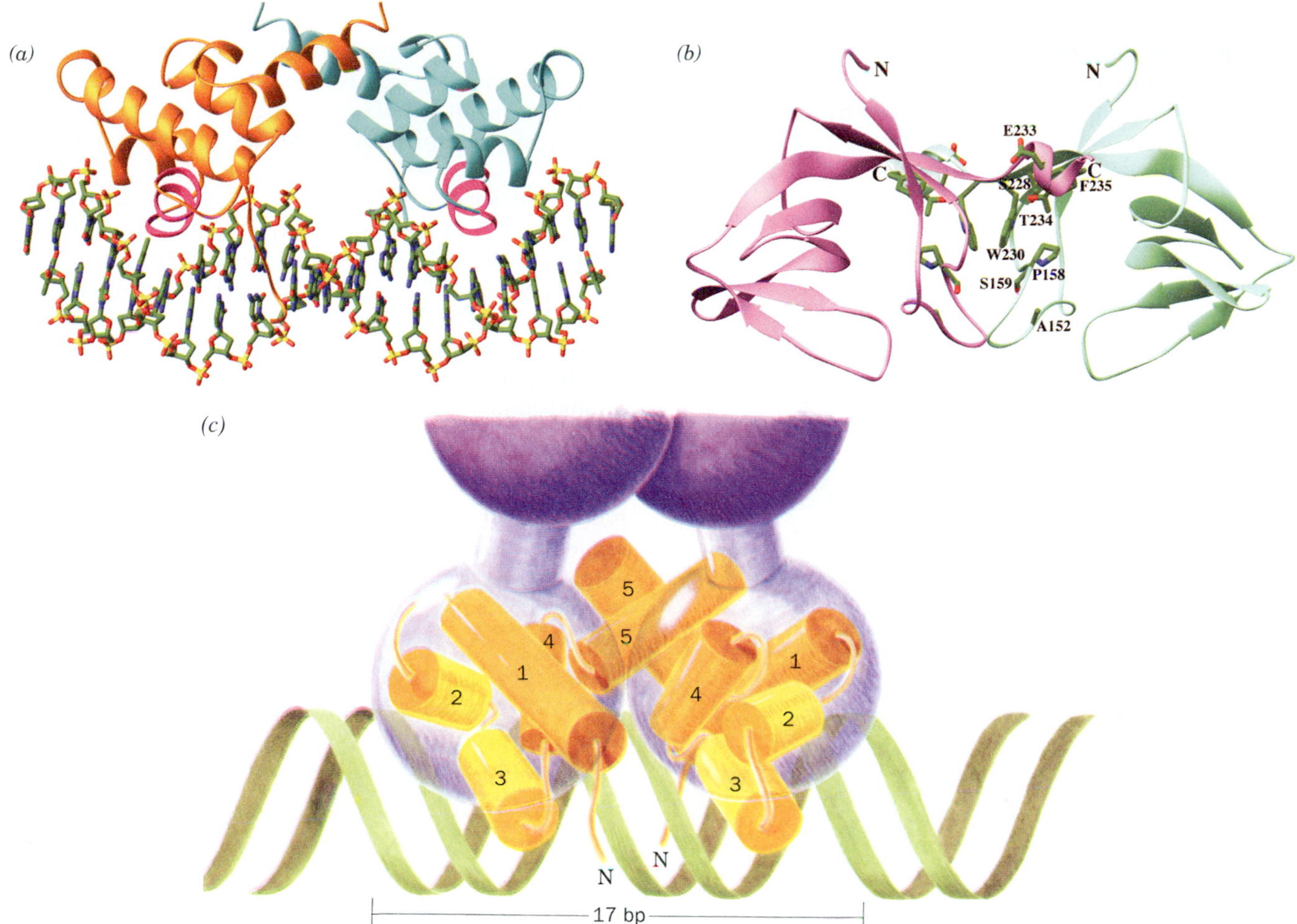

FIGURE 33-45 **Structure of the λ repressor.** (*a*) The X-ray structure of the N-terminal domain homodimer in complex with B-DNA. The DNA is drawn in stick form colored according to atom type (C green, N blue, O red, and P yellow). The two protein subunits are drawn as orange and cyan ribbons with their recognition helices magenta. Note that the protein's N-terminal arms wrap around the DNA. This accounts for the observation that the G residues in the major groove on the repressor–operator complex's "back side" are protected from methylation only when these N-terminal arms are intact. [Based on an X-Ray structure by Carl Pabo, The Johns Hopkins University. PDBid 1LMB.] (*b*) The X-ray structure of the C-terminal domain dimer. Mutations of the residues that are drawn in ball-and-stick form (and which are labeled for the green subunit) interfere with dimerization. [Based on an X-ray structure by Mitchell Lewis, University of Pennsylvania. PDBid 1F39.] (*c*) An interpretive drawing indicating how contacts between the repressor's C-terminal domains (*upper lobes*) maintain the intact protein's dimeric character. The λ repressor binds to the 17-bp operator subsites of o_L and o_R as symmetric dimers with the N-terminal domain of each subunit specifically binding to a half-subsite. Note how the α3 recognition helices of the symmetry related α2–α3 HTH units (*light yellow*) fit into successive turns of the DNA's major groove. [After Ptashne, M., *A Genetic Switch* (2nd ed.), *p.* 38, Cell Press & Blackwell Scientific Publications (1992).] **See the Interactive Exercises**

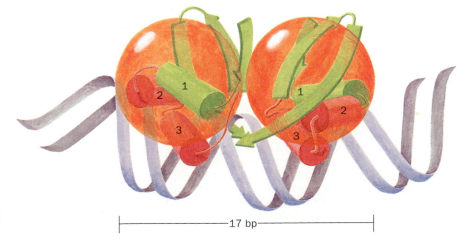

FIGURE 33-46 **X-Ray structure of the Cro protein dimer in its complex with B-DNA.** Note that the λ repressor (Fig. 33-45), although otherwise dissimilar, contains HTH units that also bind in successive turns of the DNA's major groove. [After Ptashne, M., *A Genetic Switch* (2nd ed.), *p.* 40, Cell Press & Blackwell Scientific Publications (1992).] **See the Interactive Exercises**

c. Repressor Stimulates Its Own Synthesis While Repressing All Other λ Genes

Chemical and nuclease protection experiments have indicated that λ repressor has the following order of intrinsic affinities for the subsites of o_R (Fig. 33-47):

$$o_{R1} > o_{R2} > o_{R3}$$

Despite this order, o_{R1} and o_{R2} are filled nearly together. This is because *λ repressor bound at o_{R1} cooperatively binds repressor at o_{R2} through associations between their C-terminal domains (Fig. 33-47c).* o_{R1} and o_{R2} are therefore both occupied at low λ repressor concentrations, whereas o_{R3} becomes occupied only at higher repressor concentrations.

The binding of λ repressor to o_R, as we previously mentioned, abolishes transcription from p_R and stimulates it from p_{RM} (Fig. 33-47c). At high concentrations of λ repressor, however, transcription from p_{RM} is also repressed (Fig. 33-47*d*). These phenomena have been clearly demonstrated through the construction of a series of hybrid operons that permit the effect of λ repressor on a promoter to be studied in a controlled manner. The system has two elements (Fig. 33-48):

1. A plasmid bearing the *lacI* gene (which encodes *lac* repressor; Section 31-1A) and the *lac* operator–promoter sequence fused to the *cI* gene. This construct permits the amount of λ repressor produced to be directly controlled by varying the concentration of the *lac* inducer IPTG (Section 31-1A).

2. A prophage containing o_R and either p_{RM}, as Fig. 33-48 indicates, or p_R fused to the *lacZ* gene. The amount of the *lacZ* gene product, β-galactosidase, produced, which can be readily assayed, reflects the activity of p_{RM} (or p_R).

The manipulation of these systems has demonstrated that at intermediate λ repressor concentrations (when o_{R1} and

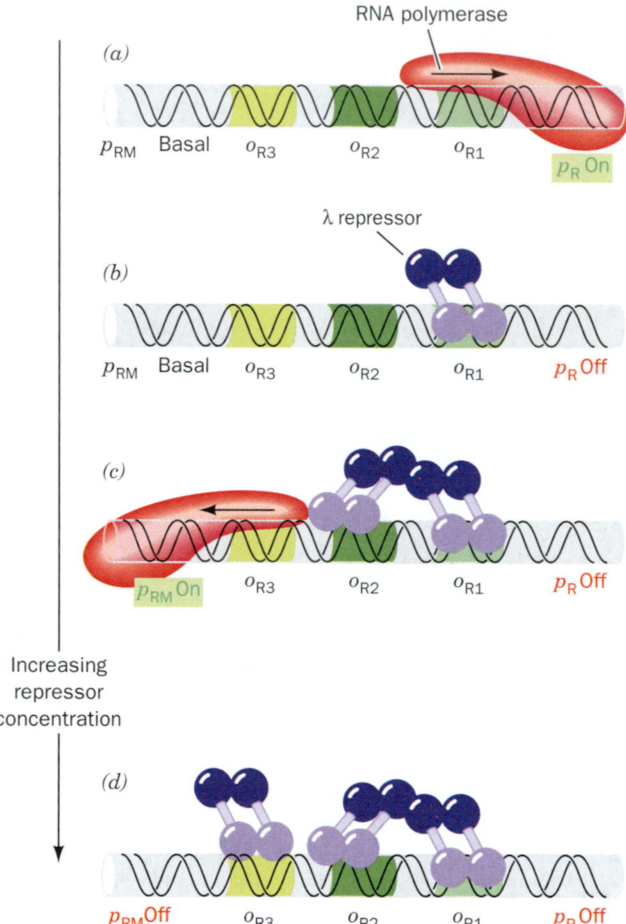

FIGURE 33-47 The binding of λ repressor to the three subsites of o_R. (*a*) In the absence of repressor, RNA polymerase initiates transcription at a high level from p_R (*right*) and at a basal level from p_{RM}. (*b*) Repressor has ~10 times higher affinity for o_{R1} than it does for o_{R2} or o_{R3}. Repressor dimer therefore first binds to o_{R1} so as to block transcription from p_R. (*c*) A second repressor dimer binds to o_{R2} at only slightly higher repressor concentrations due to specific binding between the C-terminal domains of neighboring repressors. In doing so, it stimulates RNA polymerase to initiate transcription from p_{RM} at a high level (*left*). (*d*) At high repressor concentrations, repressor binds to o_{R3} so as to block transcription from p_{RM}. Note that although Parts *c* and *d* are drawn with interdimer contacts between the C-terminal domains of only two repressor monomers, this interaction may involve contacts between the C-terminal domains of all four repressor monomers. [After Ptashne, M., *A Genetic Switch* (2nd ed.), *p.* 23, Cell Press & Blackwell Scientific Publications (1992).]

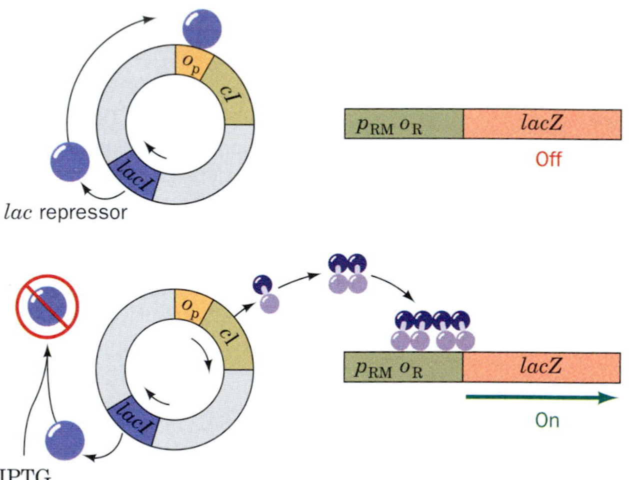

FIGURE 33-48 The genetic system used to study the effect of λ repressor on p_{RM}. The bacterium contains two hybrid operons. The first (*left*) is a plasmid bearing the *lac* operator–promoter (*Op*) fused to the λ *cI* gene so as to provide a source of repressor. The *lacI* gene, which encodes *lac* repressor, is also incorporated in the plasmid so that the level of λ repressor in the bacterium may be controlled by the concentration of the *lac* inducer IPTG. The second operon (*right*) is carried on a prophage that contains the promoter p_{RM} fused to the *lacZ* gene. The level of β-galactosidase (gp*lacZ*) in these cells therefore reflects the activity of p_{RM}. In similar experiments, the *cro* gene was substituted for λ *cI* and/or p_{RM} was replaced by p_R. [After Ptashne, M., *A Genetic Switch* (2nd ed.), *p.* 89, Cell Press & Blackwell Scientific Publications (1992).]

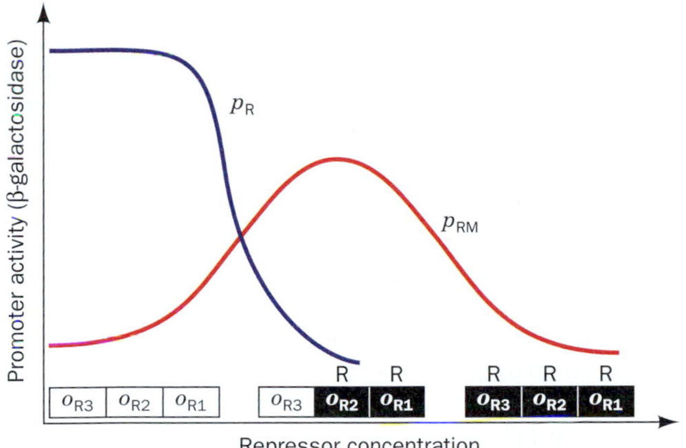

FIGURE 33-49 The response of p_{RM} and p_R to the λ repressor level. The p_{RM} curve was derived using the system diagrammed in Fig. 33-48, whereas the p_R curve was obtained using a similar system but with p_R rather than p_{RM} fused to *lacZ*. The amount of λ repressor that maximally stimulates p_{RM} is approximately that which occurs in a λ lysogen. At least 5-fold more repressor is required to half-maximally repress p_{RM}. The boxes indicate the states of each o_R subsite at the various repressor concentrations; black represents repressor occupancy. [After Ptashne, M., *A Genetic Switch* (2nd ed.), *p.* 90, Cell Press & Blackwell Scientific Publications (1992).]

o_{R2} are occupied), transcription from p_R is indeed repressed, whereas that from p_{RM} is stimulated (Fig. 33-49). Transcription from p_{RM} only becomes repressed at high levels of λ repressor (when o_{R3} is also occupied). The stimulation of transcription from p_{RM} is abolished by mutations in o_{R2} that prevent repressor binding, whereas its repression at high repressor concentrations is relieved by mutations in o_{R3}. Thus, *occupancy of o_{R2} by λ repressor stimulates transcription from p_{RM}, whereas occupancy of o_{R3} prevents it by excluding RNA polymerase from p_{RM} (Fig. 33-47c,d). By the same token, occupancy of o_{R1} and/or o_{R2} prevents transcription from p_R.* In this way, λ repressor prevents the synthesis of all phage gene products but itself. Yet, at high repressor concentrations, its synthesis is also repressed, thereby maintaining the repressor concentration within reasonable limits.

What is the basis of λ repressor's remarkable property of inhibiting transcription from one promoter while stimulating it from another? Knowledge of the sizes and shapes of repressor and RNA polymerase, as well as their positions on the DNA as demonstrated by chemical protection experiments, indicate that repressor at o_{R2} and RNA polymerase at p_{RM} are in contact (Fig. 33-50). Evidently, *repressor stimulates RNA polymerase activity through their cooperative binding to DNA.* This model was corroborated by the analysis of repressor mutants that bind normally (or nearly so) to operators but fail to stimulate the binding of RNA polymerase: All of the mutated residues occur either in helix α2 or in the link connecting it to helix α3 and lie on the surface of the protein that is thought to face the RNA polymerase-binding site (Fig. 33-50).

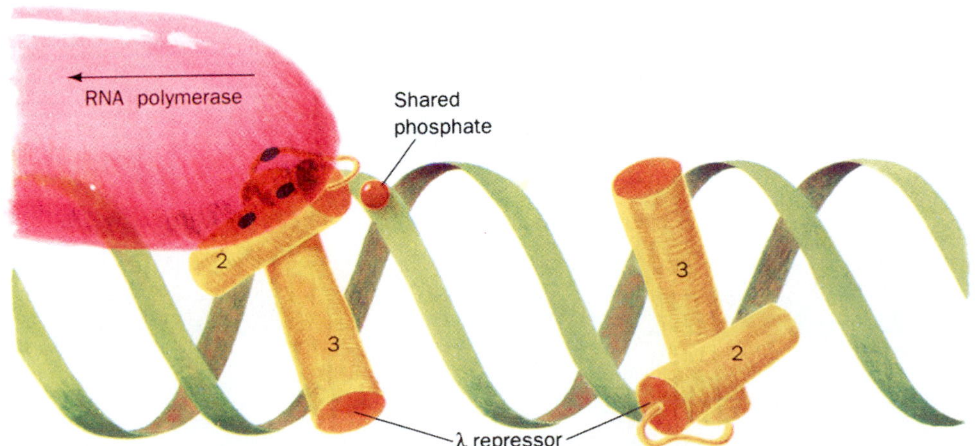

FIGURE 33-50 Interactions between the λ repressor and RNA polymerase. Repressor bound at o_{R2} is proposed to stimulate transcription at p_{RM} through a specific association with RNA polymerase that helps the polymerase bind to the promoter. This model is supported by the locations of the altered residues (*blue dots*) in three mutant repressors that bind normally to o_{R2} but fail to stimulate transcription at p_{RM}. The relative positions of repressor and RNA polymerase are established by the location of a phosphate group (*orange sphere*) whose ethylation interferes with the binding of both proteins to the DNA. For the sake of clarity, only the α_2–α_3 helix–turn–helix units of the repressor dimer are shown.

d. Cro Protein Binding to o_R Represses All λ Genes

Cro protein binds to the subsites of o_R in an order opposite to that of λ repressor (Fig. 33-51):

$$o_{R3} > o_{R2} \approx o_{R1}$$

This binding is noncooperative. Through experiments similar to that diagrammed in Fig. 33-48, but with *cro* in place of *cI,* the binding of Cro protein to o_{R3} was shown to abolish transcription from p_{RM}. Additional Cro binding to o_{R2} and/or o_{R1} turns off transcription from p_R.

e. The SOS Response Induces the RecA-Mediated Cleavage of λ Repressor

A final piece of information allows us to understand the workings of the λ switch. *The lytic phase is induced by agents that damage host DNA or inhibit its replication.* These are just the conditions that induce *E. coli*'s SOS response: The resulting fragments of single-stranded DNA activate RecA protein to stimulate the self-cleavage of LexA protein, the SOS gene repressor, at an Ala—Gly bond (Section 30-5D). *Activated RecA protein likewise stimulates the autocatalytic cleavage of λ repressor monomer's Ala 111—Gly 112 bond, which occurs in the polypeptide segment linking the λ repressor's two domains.* The ability of λ repressor to cooperatively bind to o_{R2} is thereby abolished (Fig. 33-52*a,b*)*;* the C-terminal domains can still dimerize but they no longer link the DNA-binding N-terminal domains. The consequent reduction in concentration of intact free monomers shifts the monomer–dimer equilibrium such that the operator-bound dimers dissociate to form monomers, which are then cleaved through the influence of activated RecA before they can rebind to their target DNA.

In the absence of repressor at o_R, the λ early genes, including *cro*, are transcribed (Fig. 33-52*c*). As Cro accumulates, it first binds to o_{R3} so as to block even basal levels of λ repressor synthesis (Fig. 33-52*d*). Thus, *there being no mechanism for selectively inactivating Cro, the phage irreversibly enters the lytic mode:* The λ switch, once thrown, cannot be reset. The prophage is subsequently excised from the host chromosome by the integrase and excisionase that are produced in the delayed early phase (Fig. 33-31).

f. The λ Switch's Responsiveness to Conditions Arises from Cooperative Interactions among Its Components

The complexity of the above switch mechanism endows it with a sensitivity that is not possible in simpler systems. The degree of repression at p_R is a steep function of repressor concentration (Fig. 33-53, *right*): The repression of p_R in a lysogen is normally 99.7% complete but drops to half this level on inactivation of 90% of the repressor. This steep sigmoid binding curve arises from the much greater operator affinity of repressor dimers compared to monomers. This situation, in turn, results from the cooperative linking of the monomer–dimer equilibrium, the binding of dimer to operator, and the association of dimers bound at o_{R1} and o_{R2} to form a tetramer. In fact, this cooperative effect is further enhanced by the similar binding of repressor tetramer to o_{L1} and o_{L2}, which through DNA looping, forms an octamer with the repressor tetramer bound at o_{R1} and o_{R2} (a phenomenon that likewise increases the repression of p_L). In contrast, a 99.7% repressed promoter controlled by a stably oligomeric repressor binding to a single operator site, such as occurs in the *lac* system, requires 99% repressor inactivation for 50% expression (Fig. 33-53, *left*). *The cooperativity of λ repressor oligomerization and multiple operator site binding are therefore responsible for the remarkable responsiveness of the λ switch to the health of its host.*

4 ■ INFLUENZA VIRUS

Influenza is one of the few common infectious diseases that is poorly controlled by modern medicine. Its annual epidemics, one of which was recorded by Hippocrates in

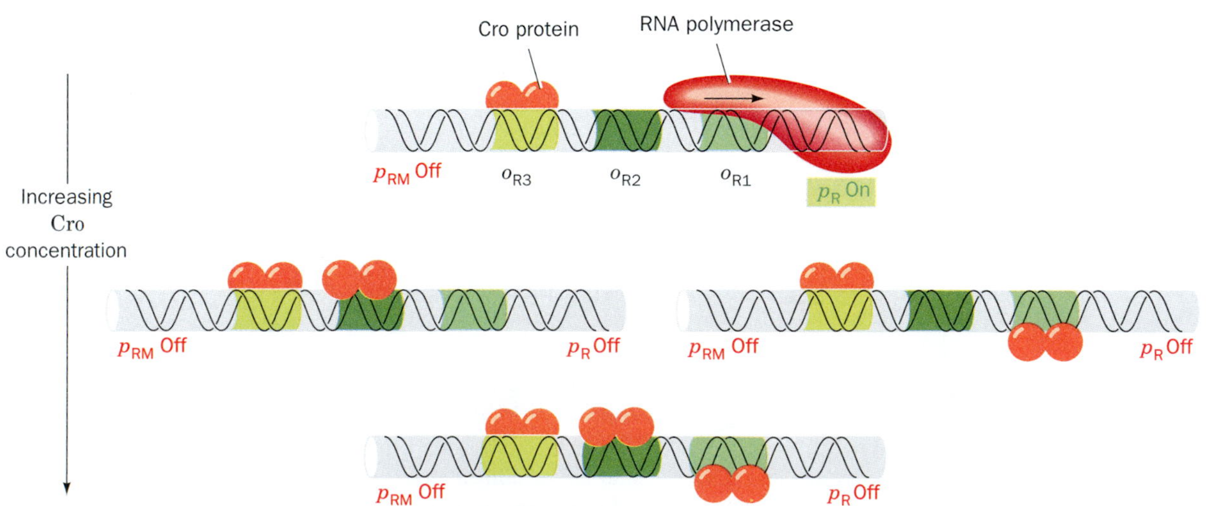

FIGURE 33-51 The binding of Cro protein to the three o_R subsites. o_{R3} binds Cro ~10 times more tightly than does o_{R1} or o_{R2}. Cro dimer therefore first binds to o_{R3}. A second dimer then binds to either o_{R1} or o_{R2} and in each case blocks transcription from p_R. At high Cro concentrations, all three operator subsites are occupied. Compare this binding sequence with that of λ repressor (Fig. 33-47). [After Ptashne, M., *A Genetic Switch* (2nd ed.), *p.* 27, Cell Press & Blackwell Scientific Publications (1992).]

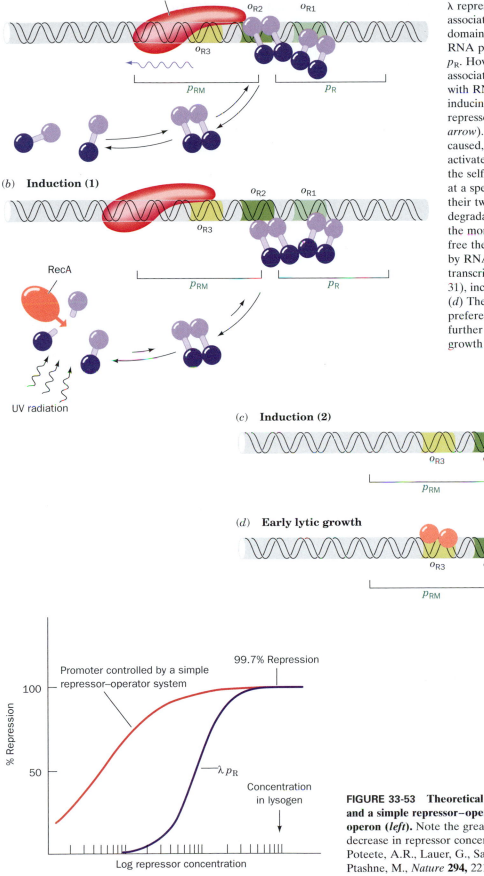

FIGURE 33-52 The λ switch. (*a*) In the lysogenic mode, two dimeric molecules of λ repressor cooperatively bind, through associations between their C-terminal domains, to o_{R1} and o_{R2}. This blocks host RNA polymerase from gaining access to p_R. However, the repressor bound to o_{R2} associates, through its N-terminal domain, with RNA polymerase at p_{RM}, thereby inducing the transcription of *cI*, the λ repressor gene, from this promoter (*wiggly arrow*). (*b*) Damage to the host DNA, as caused, for example, by UV radiation, activates host RecA protein to stimulate the self-cleavage of λ repressor monomers at a specific Ala—Gly bond between their two domains. (*c*) The consequent degradation of repressor monomers shifts the monomer–dimer equilibrium so as to free the o_{R1} and o_{R2} subsites for binding by RNA polymerase. This results in the transcription of the early genes (Fig. 33-31), including *cro*, from p_R (*wiggly arrow*). (*d*) The Cro protein thus synthesized preferentially binds at o_{R3} so as to block further transcription of *cI* from p_{RM}. Lytic growth is thereby irreversibly induced.

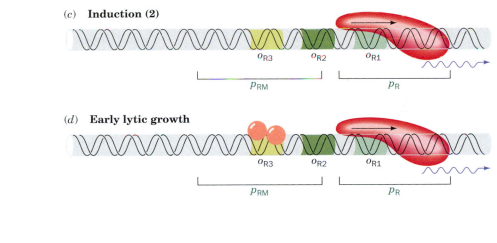

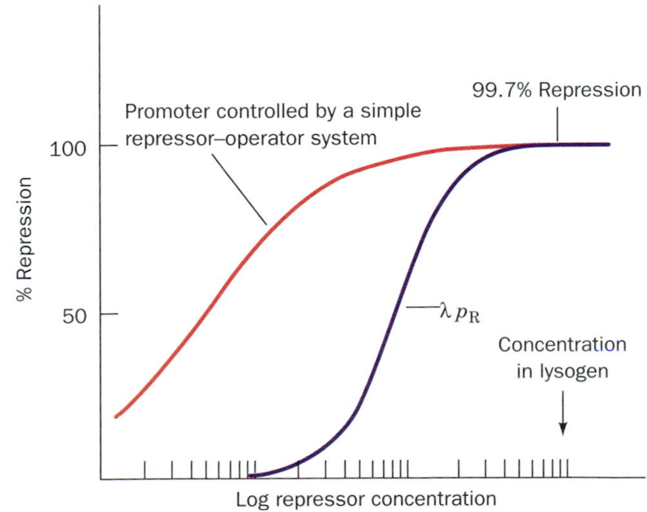

FIGURE 33-53 Theoretical repression curves for λ p_R (*right*) and a simple repressor–operator system such as that of the *lac* operon (*left*). Note the greater sensitivity of the λ system to a decrease in repressor concentration. [After Johnson, A.D., Poteete, A.R., Lauer, G., Sauer, R.T., Ackers, G.K., and Ptashne, M., *Nature* **294,** 221 (1981).]

412 B.C., are occasionally punctuated by devastating pandemics that infect 20 to 40% of the world's population. For example, the influenza pandemic of 1918, the so-called Spanish flu, which killed 40 to 50 million people worldwide (often previously healthy young adults; around 2% of the world's population at the time) was among the most lethal plagues ever recorded (it lowered the average life expectancy in the United States by 10 years). Since that time there have been three other pandemics of lesser severity, the so-called Asian flu of 1957, the Hong Kong flu of 1968, and the Russian flu of 1977 (the historical record suggests that there have been 12 pandemics in the past 400 years). All of these pandemics were characterized by the appearance of a new strain of influenza virus to which the human population had little resistance and against which previously existing influenza virus vaccines were ineffective. Moreover, between pandemics, influenza virus undergoes a gradual antigenic variation that degrades the level of immunological resistance against renewed infection. Even in nonpandemic years, influenza is responsible for the deaths of one-half to one million mainly elderly people; it is among the ten leading causes of death in the United States. What characteristics of the influenza virus permit it to evade human immunological defenses? In this section we shall discuss this question and, in doing so, examine the structure and life cycle of the influenza virus.

A. *Virus Structure and Life Cycle*

Electron micrographs of influenza virus (Fig. 33-1*i*) reveal a collection of nonuniform spheroidal particles that are ~100 nm in diameter and whose surfaces are densely studded with radially projecting "spikes." The influenza virion, which grows by budding from the plasma membrane of an infected cell (Fig. 33-54), is an example of an **enveloped virus.** *Its outer envelope consists of a lipid bilayer of cellular origin that is pierced by virally specified integral membrane glycoproteins, the "spikes."* There are two types of these surface spikes (Fig. 33-55):

1. A rod-shaped spike composed of **hemagglutinin (HA),** so named because it causes erythrocytes to agglutinate (clump together). HA mediates influenza target cell recognition by specifically binding to cell-surface receptors (glycophorin A molecules in erythrocytes; Section 12-3A) bearing terminal *N*-acetylneuraminic acid (sialic acid; Fig. 11-11) residues. Each virion contains ~500 copies of HA.

2. A mushroom-shaped spike known as **neuraminidase (NA),** which catalyzes the hydrolysis of the linkage joining a terminal sialic acid residue to a D-galactose or a D-galactosamine residue. NA probably facilitates the transport of the virus to and from the infection site by permitting its passage through mucin (mucus) and preventing viral self-aggregation. Each virion incorporates ~100 copies of NA.

In addition, the membranous outer envelope contains small amounts of **matrix protein 2 (M2).**

Just beneath the viral membrane is a 6-nm-thick protein shell composed of ~3000 copies of **matrix protein 1 (M1),** the virion's most abundant protein. M1 interacts with **nuclear export protein [NEP;** formerly known as **nonstructural protein 2 (NS2)].**

The influenza virus genome is unusual in that it consists of eight different sized segments of single-stranded RNA. These RNA molecules are negative strands; that is, they are complementary to the viral mRNAs. In the viral core,

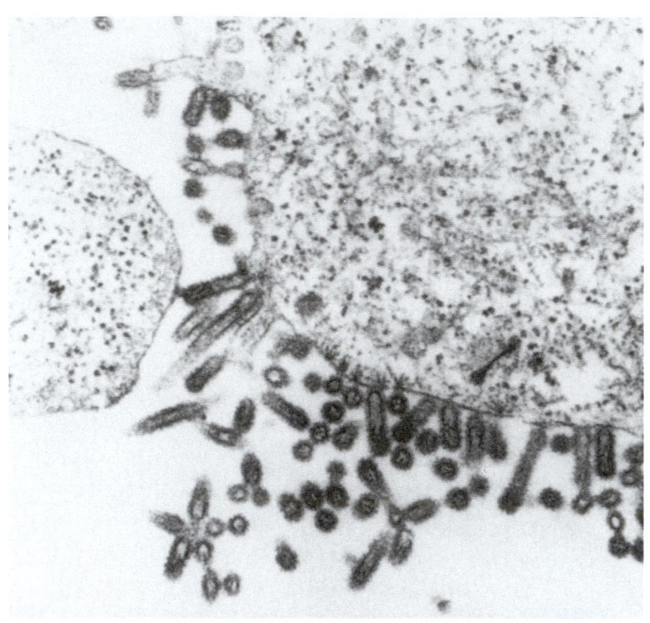

FIGURE 33-54 Electron micrograph of influenza viruses budding from infected chick embryo cells. [From Sanders, F.K., *The Growth of Viruses,* p. 15, Oxford University Press (1975).]

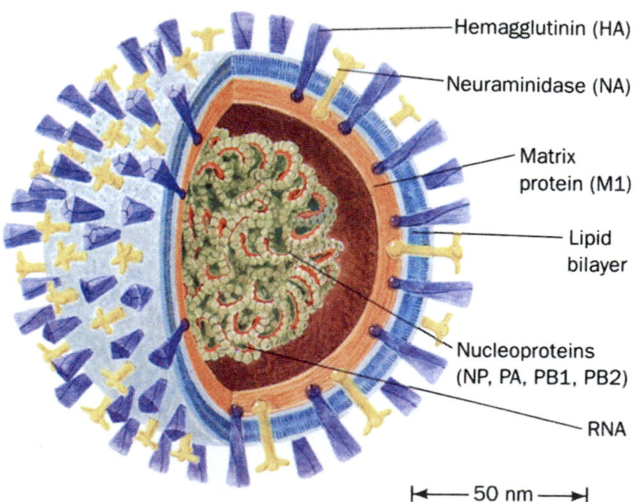

FIGURE 33-55 Cutaway diagram of the influenza virion. The HA and NA spikes are embedded in a lipid bilayer that forms the virion's outer envelope. M1 coats the underside of this membrane. The virion core contains the eight single-stranded RNA segments that comprise its genome in complex with the proteins NP, PA, PB1, and PB2 to form helical structures named nucleocapsids. [After Kaplan, M.M. and Webster, R.G., *Sci. Am.* **237**(6): 91 (1977). Copyright © 1977 by Scientific American, Inc.]

these RNAs occur in complex with four different proteins: **nucleocapsid protein (NP),** which occurs in ~1000 copies, and **polymerase acidic protein (PA), polymerase basic protein 1 (PB1),** and **polymerase basic protein 2 (PB2),** which are present in 30 to 60 copies each. The resulting **nucleocapsids** have the appearance of flexible rods.

The eight viral RNAs, which vary in length from 890 to 2341 nucleotides, have all been sequenced. They encode the virus' nine structural proteins (HA, NA, M1, M2, NEP, NP, PA, PB1, and PB2) and **nonstructural protein 1 (NS1),** which occurs only in infected cells. The sizes of the RNAs and the proteins they encode are listed in Table 33-2. About 10% of the viral mRNAs encoding M1 and NS1 are processed by the host cell splicing machinery to yield smaller mRNAs that respectively encode M2 and NEP but in mainly different reading frames from M1 and NS1.

TABLE 33-2 The Influenza Virus Genome

RNA Segment	Length (nt)	Polypeptide(s) Encoded
1	2341	PB2
2	2341	PB1
3	2233	PA
4	1778	HA
5	1565	NP
6	1413	NA
7	1027	M1, M2
8	890	NS1, NEP

Source: Lamb, R.A. and Choppin, P.W., *Annu. Rev. Biochem.* **52,** 473 (1983).

a. Virus Life Cycle

The influenza infection of a susceptible cell begins with the HA-mediated adsorption of the virus to specific cell-surface receptors. The virus is then taken into the cell via endocytosis (Section 12-5B), whereupon the endocytotic vesicle fuses with the endosome (Fig. 12-79). In the acidic (pH ~5) medium of the endosome, the viral M2 protein, a proton channel, admits protons into the virion, which induces the separation of the nucleocapsids from M1. The viral and endosome membranes then fuse through a mechanism discussed in Section 33-4C, thereby introducing the nucleocapsids into the cytosol. By ~20 min postinfection, in a process mediated by NP, the still intact nucleocapsids have been transported to the cell nucleus, where they commence transcription of the viral RNAs **(vRNAs).** Cellular enzyme systems are incapable of mediating such RNA-directed RNA synthesis. Rather, it is carried out by a viral RNA transcriptase system that consists of the nucleocapsid proteins.

The transcription of the influenza virus genome is terminated if infected cells are treated with inhibitors of RNA polymerase II (which synthesizes cellular mRNA precursors; Section 31-2E) such as actinomycin D or α-amanitin. Yet, none of these agents affects the viral transcriptase's *in vitro* activity. The resolution of this seeming paradox is that *in vivo* viral mRNA synthesis is primed by newly synthesized cellular mRNA fragments consisting of a 7-methyl-G cap (Section 31-4A) followed by a 9- to 17-nt chain ending in A or G (Fig. 33-56, *top*). Viral mRNAs, as do most mature cellular mRNAs, have poly(A) tails appended to their 3′ ends by the cellular polyadenylation machinery (Section 31-4A).

The synthesis of the viral mRNAs is terminated 16- to 17-nt from the 5′ ends of their vRNA templates by the presence of a sequence of 5 to 7 U's on the vRNAs. Consequently, the viral mRNAs cannot act as templates in vRNA replication. Rather, in an alternative transcription process that begins some 30 min postinfection, complete vRNA complements are synthesized. These so-called **cRNAs,** whose synthesis does not require a primer, begin with pppA at their 5′ ends and lack poly(A) tails (Fig. 33-56, *bottom*). Hence cRNAs, unlike viral mRNAs, do not associate with ribosomes in infected cells. The synthesis of cRNAs, in contrast to that of viral mRNAs, requires the

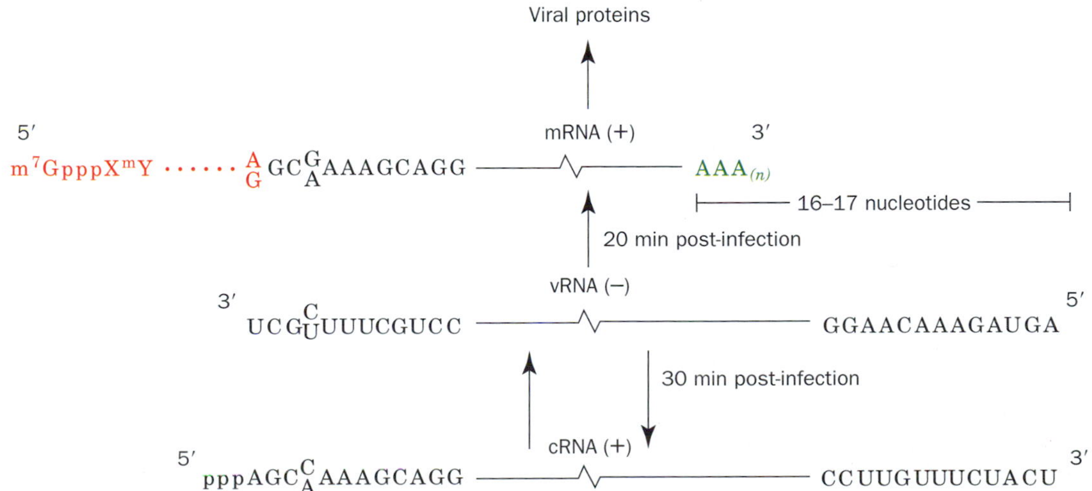

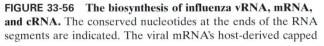

FIGURE 33-56 The biosynthesis of influenza vRNA, mRNA, and cRNA. The conserved nucleotides at the ends of the RNA segments are indicated. The viral mRNA's host-derived capped 5′ head and 3′ poly(A) tail are shown in color. [After Lamb, R.A. and Choppin, P.W., *Annu. Rev. Biochem.* **52,** 490 (1983).]

presence of NS1, which inhibits the processing of cellular pre-mRNAs and interferes with the synthesis of poly(A) tails. The cRNAs are the templates for vRNA synthesis. The resulting dsRNA would normally induce an interferon-mediated antiviral state in the infected cell (Section 32-4A). However, NS1 also functions as an interferon antagonist, thereby permitting viral proliferation.

The influenza virus transcription complex is a trimer consisting of PB1, PB2, and PA, which, together with NP, bind the RNA template. PB1 is the polymerase that catalyzes both the initiation and the elongation of the RNA transcript. PB2 binds to the 5′ caps of cellular pre-mRNAs, although the endonuclease function that cleaves the capped primers from them appears to reside on PB1. Mutational experiments indicate that PA is required for vRNA but not mRNA synthesis, although its role in vRNA synthesis is poorly understood. The abundance of NP suggests that it has a structural role in the nucleocapsid, although it has also been implicated in the antitermination required to synthesize cRNAs rather than vRNAs.

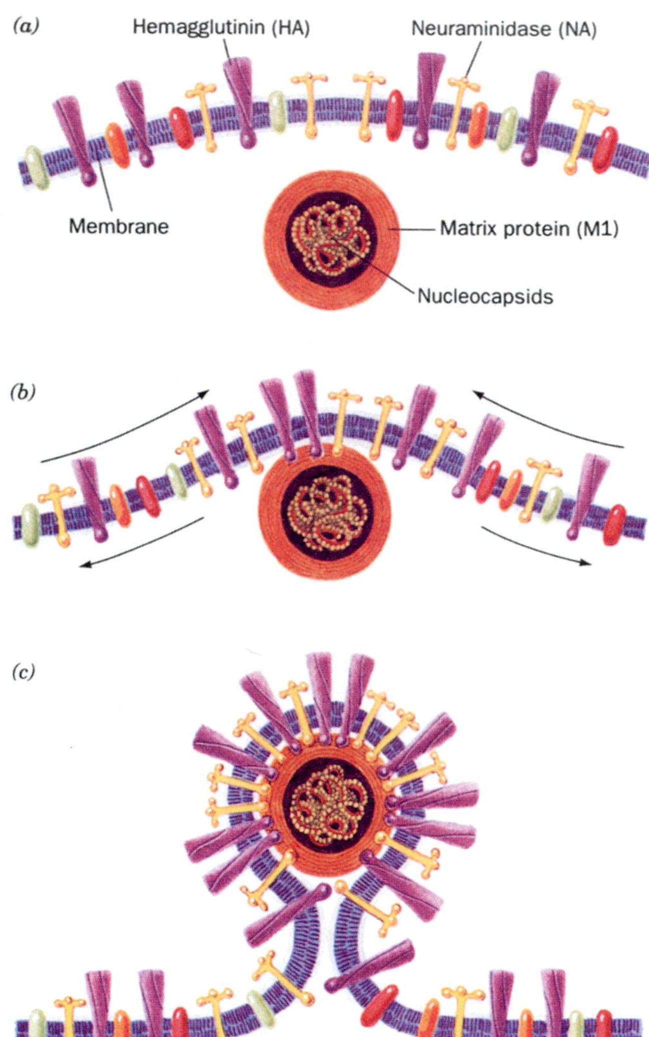

FIGURE 33-57 The budding of influenza virus from the host cell membrane. (*a*) The viral glycoproteins, HA and NA, are inserted into the plasma membrane of the host cell and the matrix protein, M1, forms the nucleocapsid-containing shell. (*b*) The binding of the matrix protein to the cytoplasmic domains of HA and NA results in the aggregation of these glycoproteins so as to exclude host cell membrane proteins (*arrows*). (*c*) This binding process induces the membrane to envelop the matrix protein shell such that the mature virion buds from the host cell surface. [After Wiley, D.C., Wilson, I.A., and Skehel, J.J., *in* Jurnak, F.A. and McPherson, A. (Eds.), *Biological Macromolecules and Assemblies,* Vol. 1: *Virus Structures,* Wiley (1984).]

The mechanism of influenza virus assembly is not well characterized. The viral spike glycoproteins, HA and NA, are ribosomally synthesized on the rough endoplasmic reticulum, further processed in the Golgi apparatus (Section 12-4B), and then transported, presumably in clathrin-coated vesicles, to areas of the plasma membrane containing lipid rafts (Section 12-4D). There, they aggregate in sufficient numbers to exclude host proteins (Fig. 33-57*a,b*). In the nucleus, the vRNAs combine with PA, PB1, and PB2 to form the nucleocapsids, which then interact with M1 protein. Nuclear export protein (NEP), as its name implies, mediates the export of nucleocapsids from the nucleus, which it does in partnership with both M1 and a cellular export factor. M1 then forms a nucleocapsid-enclosing shell that binds to HA and NA on the inside of the plasma membrane (Fig. 33-57*b*). This binding process causes the entire assembly to bud from the cell surface, thereby forming the mature virion (Fig. 33-57*c*). The complete infection cycle occupies ~8 to 12 h.

One of the mysteries of influenza virus assembly is how each virion acquires a complete set of the eight vRNAs. There is no evidence that the newly formed nucleocapsids are physically linked. On the contrary, in mixed infections with various influenza strains, the reassortment of their genomic segments occurs with high frequency. It has therefore been suggested that the nucleocapsids are randomly selected but that each virion contains sufficient numbers of vRNAs to ensure a reasonable probability that a given particle be infectious. This proposal is in agreement with the observation that aggregates of influenza virus have enhanced infectivity, a process that presumably occurs through the complementation of their vRNAs. Alternatively, the eight vRNAs may be selected by an ordered process, a hypothesis that is supported by the observation that mature viruses, but not infected cells, contain roughly equimolar amounts of the vRNAs.

B. *Mechanism of Antigenic Variation*

Influenza viruses are classified into three immunological types, A, B, and C, depending on the antigenic properties of their differing nucleoproteins and matrix proteins. The A virus has caused all of the major pandemics in humans and has therefore been more extensively investigated than the B and C viruses. The B and C viruses infect mainly humans. However, the A virus infects a wide variety of mammalian and avian species in addition to humans. Indeed, it is thought that migratory birds (and, more recently, jet planes) are the major vectors that transport influenza A viruses around the world. The species specificity of a particular viral strain presumably arises from the binding specificity of its HA for cell-surface glycolipids.

a. HA Residue Changes Are Responsible for Most of the Antigenic Variation in Influenza Viruses

HA, being the influenza virus' major surface protein, is largely responsible for stimulating the production of the antibodies that neutralize the virus. Consequently, the different influenza virus subtypes arise mainly through the variation of HA. Antigenic variation in NA, the virus' other major surface protein, also occurs but this has lesser immunological consequences.

Two distinct mechanisms of antigenic variation have been observed in influenza A viruses:

1. Antigenic shift, in which the gene encoding one HA species is replaced by an entirely new one. This change may or may not be accompanied by a replacement of NA. It is thought that these new viral strains arise from the reassortment of genes among animal and human flu viruses. *Antigenic shift is responsible for influenza pandemics because the human population's immunity against previously existing viral strains is ineffective against the newly generated strain.* Evidently, these viruses had retained the (largely unknown) genetic traits responsible for their virulence in humans.

2. Antigenic drift, which occurs through a succession of point mutations in the HA gene, resulting in an accumulation of amino acid residue changes that attenuate the host's immunity. This process occurs in response to the selective pressure brought about by the buildup in the human population of immunity to the extant viral strains. HA varies in this manner by an average of 3.5 accepted amino acid changes per year.

Influenza A viruses are classified into subtypes according to the similarities of their HA and NA. There are 15 known subtypes of HA (H1 through H15) and 9 of NA (N1 through N9) that occur in mammals and birds. Avian virus subtypes occur in nearly all combinations, whereas only a few combinations have been found in humans. For example, human influenza A viruses circulating before 1957 were designated H1N1, those of the 1957 pandemic were H2N2, those of the 1968 pandemic were H3N2, and those of the 1977 pandemic were again H1N1 (and hence affected mainly young people who not been exposed to pre-1918 viruses). Since 1977, H3N2 and H1N1 viruses have been cocirculating.

Humans are rarely infected by avian flu viruses and such viruses do not appear to be transmitted between humans. However, phylogenetic studies indicate that pigs (swine) and birds can exchange influenza viruses as can pigs and humans. This suggests that pigs serve as "mixing vessels" for the creation of new pandemic flu viruses, which in turn, explains why Southeast Asia, where humans, pigs, and birds (ducks and chickens) often live in close proximity, is where most flu pandemics appear to have originated.

In early 1976, at Fort Dix, New Jersey, there was an outbreak of an H1N1 influenza strain whose HA subtype occurs in swine flu virus. This viral strain is thought to have caused the great pandemic of 1918 (although influenza virus was not isolated until 1933, individuals who had contracted influenza during the 1918 pandemic have antibodies against swine flu virus in their serum; the sequencing, in 1999, by Ann Reid and Jeffery Taubenberger, of the HA gene obtained from preserved tissues of individuals who had died of the 1918 flu confirmed that its HA resembles that of swine flu). If this new strain had been virulent, no one under the age of 50 at the time would have been immune to it. There was, consequently, grave concern that a deadly influenza pandemic would ensue. This situation led to a crash program in which well over one million people deemed to be at high risk (such as pregnant women and the elderly) were vaccinated against swine flu. Fortunately, the 1976 swine flu was not virulent; it did not spread beyond Fort Dix.

b. HA Is an Elongated Trimeric Transmembrane Glycoprotein

HA plays a central role in both the viral infection process and in the immunological measures and countermeasures taken in the continuing biological contest between host and parasite. This has motivated considerable efforts to elucidate the structural basis of its properties. HA is a homotrimer of 550-residue subunits that is 19% carbohydrate by weight. The protein has three domains (Fig. 33-58):

1. A large hydrophilic, carbohydrate-bearing N-terminal domain that occupies the viral membrane's external surface and that contains its sialic acid–binding site.

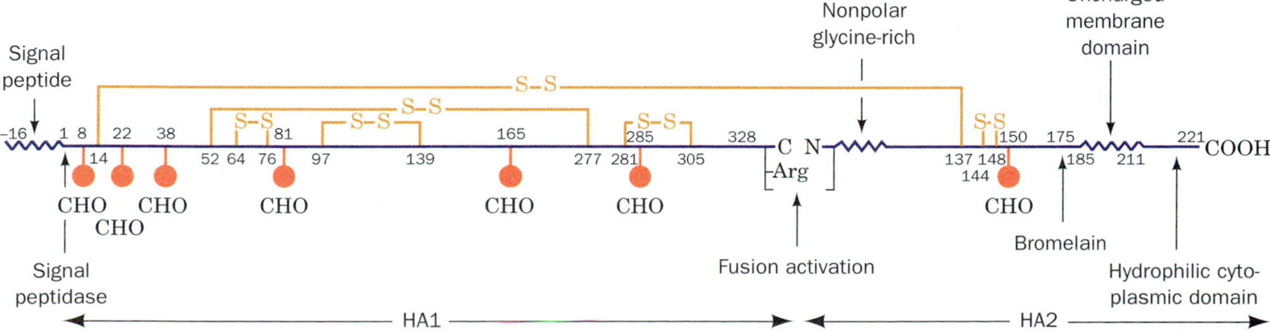

FIGURE 33-58 The primary structure of the 1968 Hong Kong influenza virus hemagglutinin. Its external domain (all of HA1 and HA2 through 185), its membrane anchoring domain (185–211 of HA2), and its cytoplasmic domain (212–221 of HA2) are indicated as are the positions of the signal peptide directing the protein's insertion into the membrane, the S—S bridges, the carbohydrate (CHO) attachment sites, the fusion activation site, and the bromelain cleavage site. [After Wilson, I.A., Skehel, J.J., and Wiley, D.C., *Nature* **289,** 367 (1981).]

2. A hydrophobic 24- to 28-residue membrane-spanning domain that is located near the polypeptide's C-terminus.

3. A hydrophilic C-terminal domain that occurs on the membrane's inner side and that consists of the protein's 10 C-terminal residues.

HA, which is synthesized as a single polypeptide designated HA0, is posttranslationally cleaved by host-secreted proteases by the excision of Arg 329, thereby yielding two chains, HA1 and HA2, that are linked by a disulfide bond. This cleavage, which does not affect HA's receptor-binding affinity, is required for the fusion of the virus with the host cell and therefore activates viral infectivity (see below). Indeed, the cleavability of HA is one of the major factors determining the virulence of influenza viruses.

HA can be removed from the virion by treatment with detergent but the resulting solubilized protein has not been made to crystallize. However, treatment of HA from a Hong Kong–type (H3) virus with the pineapple protease **bromelain,** which cleaves the polypeptide 9 residues before the membrane-spanning segment, yields a water-soluble protein named **BHA** that has been crystallized. X-Ray analysis of these crystals by John Skehel and Don Wiley revealed an unusual structure (Fig. 33-59). The monomer consists of a long fibrous stalk extending from the membrane surface on which is perched a globular region. The fibrous stalk consists of segments from HA1 and HA2 and includes a remarkable 76-Å-long (53 residues in 14 turns) α helix. The globular region, which is comprised of only HA1 residues, contains an 8-stranded antiparallel β-sheet

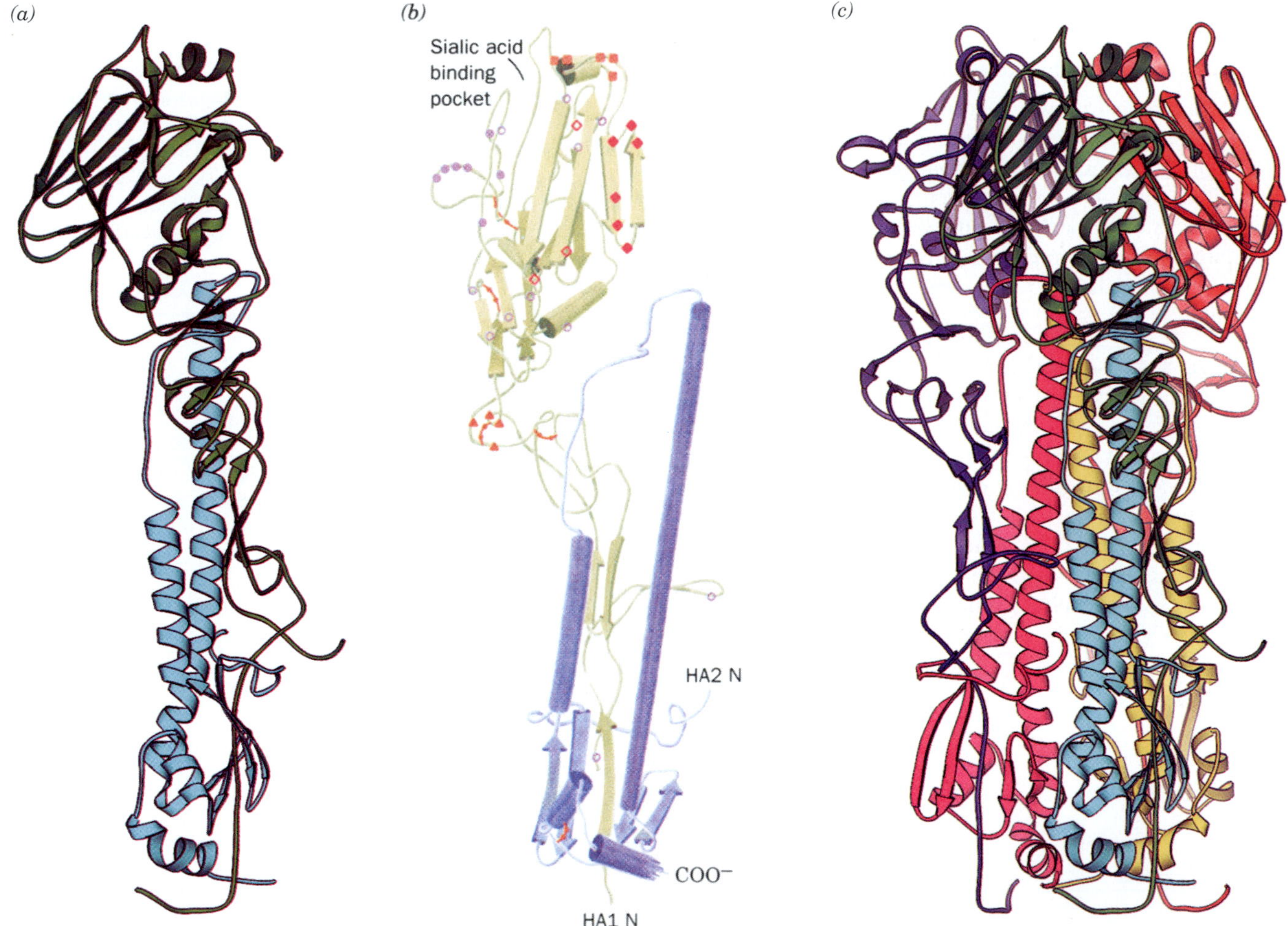

FIGURE 33-59 X-Ray structure of influenza hemagglutinin. (*a*) The polypeptide backbone of the monomer drawn as a ribbon. HA1 is green and HA2 is cyan. (*b*) A cartoon diagram of the monomer from a somewhat different point of view than Part *a* but similarly colored. The pairs of linked, small, filled circles represent disulfide groups. The positions of the mutant residues at the four antigenic sites are indicated by filled circles, squares, triangles, and diamonds. Open symbols represent antigenically neutral residues. Note the position of the sialic acid–binding pocket. (*c*) A ribbon diagram of the HA trimer. Each HA1 and HA2 chain is drawn in a different color. The orientation of the green HA1 and cyan HA2 are the same as in Part *a*. [Based on an X-ray structure by John Skehel, National Institute for Medical Research, London, U.K., and Don Wiley, Harvard University. Parts *a* and *c* courtesy of Michael Carson, University of Alabama at Birmingham; Part *b* after a drawing by Hidde Ploegh, *in* Wilson, I.A., Skehel, J.J., and Wiley, D.C., *Nature* **289,** 366 (1981). PDBid 4HMG.]

structure (a distorted jelly roll barrel; Section 8-3B) that forms the sialic acid–binding pocket.

The dominant interaction stabilizing BHA's trimeric structure is a triple-stranded coiled coil consisting of the 76-Å α helices from each of its subunits (Fig. 33-59*c*). The BHA trimer is therefore an elongated molecule, some 135 Å in length, with a triangular cross-section that varies in radius from 15 to 40 Å. The carbohydrate chains, which are attached to the protein via *N*-glycosidic linkages at each of its subunit's seven Asn-X-Thr/Ser sequences (Section 11-3C), are located almost entirely along the trimer's lateral surfaces. The role of the carbohydrates is unclear despite the fact that they cover some 20% of the protein's surface. However, the observation that the mutational generation of a new oligosaccharide attachment site blocks antibody binding to HA suggests that carbohydrates modulate HA's antigenicity.

c. Antigenic Variation Results from Surface Residue Changes

HA's antigenic sites have been identified by mapping HA sequence changes on the protein's three-dimensional structure. The HA residues that mutated in an antigenically significant manner in Hong Kong–type viruses during the period 1968 to 1977 are indicated in Fig. 33-59*b*. *These residues all occur on the protein's surface, often in polypeptide loops, where their mutational variation affects the protein's surface character but apparently not its overall structure or stability. The variable residues are clustered in four sites surrounding HA's receptor-binding pocket, which is formed from amino acid residues that are largely conserved in numerous influenza virus strains.* The strains responsible for the major flu epidemics between 1968 and 1975 had at least one mutation in each of these four antigenic sites. This degree of antigenic variation appears necessary to reinfect individuals previously infected with the same viral type. Evidently, *antibodies directed against even conserved residues in HA's receptor-binding pocket, which would otherwise prevent HA from binding to its receptor, are dislodged by the antigenic variation that so readily occurs about the rim of this binding pocket (we study antibody–antigen interactions in Section 35-2B).*

d. NA Is a Tetrameric Transmembrane Glycoprotein

Influenza virus neuraminidase (NA) is a homotetrameric glycoprotein of 469-residue subunits. It has a box-shaped globular head attached to a slender stalk that is anchored in the viral membrane. Pronase digestion cleaves NA before residues 74 or 77, after the membrane attachment site, to yield an enzymatically active and crystallizable protein. The X-ray structure of this protein (Fig. 33-60), determined by Peter Colman and Graeme Laver, shows it to have 4-fold symmetry. Each subunit is composed of six topologically identical 4-stranded antiparallel β sheets arranged like the blades of a propeller. This so-called β propeller structurally resembles the 7-bladed β propellers of the clathrin heavy chain (Fig. 12-56*b*) and the heterotrimeric G protein G_β subunit (Fig. 19-18*b*). Sugar residues are linked to NA at four of its five potential Asn-X-Ser/Thr *N*-glycosylation sites.

NA's sialic acid–binding site is located in a large pocket on the top of each monomer (star on the upper right subunit in Fig. 33-60). Its bound sialic acid residue interacts, through an extensive hydrogen bonding network, with 16 polar residues that are conserved in all known NA sequences (HA, in contrast, has only 2 polar residues in its sialic acid–binding site). Sequence changes in antigenic variants of NA occur in 7 chain segments that form a nearly continuous surface that encircles the catalytic site (squares on the lower right subunit in Fig. 33-60) in a manner similar to that of HA's receptor-binding site. Between 1968 and 1975, NA exhibited the same number of residue changes in its putative antigenic determinants as did HA. Antibodies against NA, nevertheless, do not neutralize infectivity. Rather, they restrict multiple cycles of viral replication and thus probably attenuate illness.

The realization that the structure of the NA catalytic site is strain-invariant together with the X-ray structure of NA in complex with sialic acid has led to the design of clinically effective inhibitors of NA (drug design is discussed in Section 15-4). For example, **zanamivir** (trade name, **Relenza**),

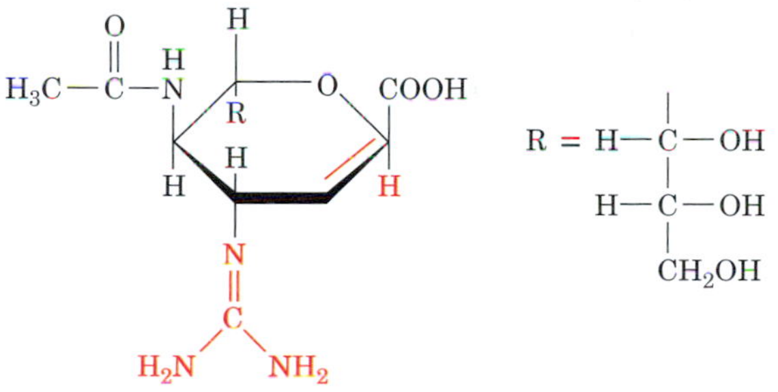

R = H—C—OH, H—C—OH, CH_2OH

Zanamivir (Relenza)

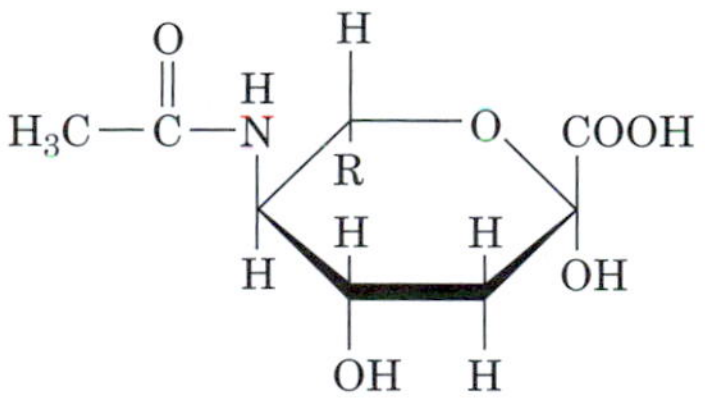

***N*-Acetylneuraminic acid (sialic acid)**

a sialic acid mimic and a potent inhibitor of NA (K_I = 0.1 nM), is an effective antiviral agent, both in tissue culture and when delivered as an orally inhaled powder. Although the use of zanamivir does not prevent influenza infection, if it is administered within 2 days of the onset of flulike symptoms, it significantly diminishes the length and severity of these symptoms and reduces the incidence of secondary bacterial infections.

C. *Mechanism of Membrane Fusion*

An influenza virus infection begins with the binding of HA to its cell-surface receptor. Then, as we discussed in Section 12-4B, the bound virus is taken into the cell via receptor-

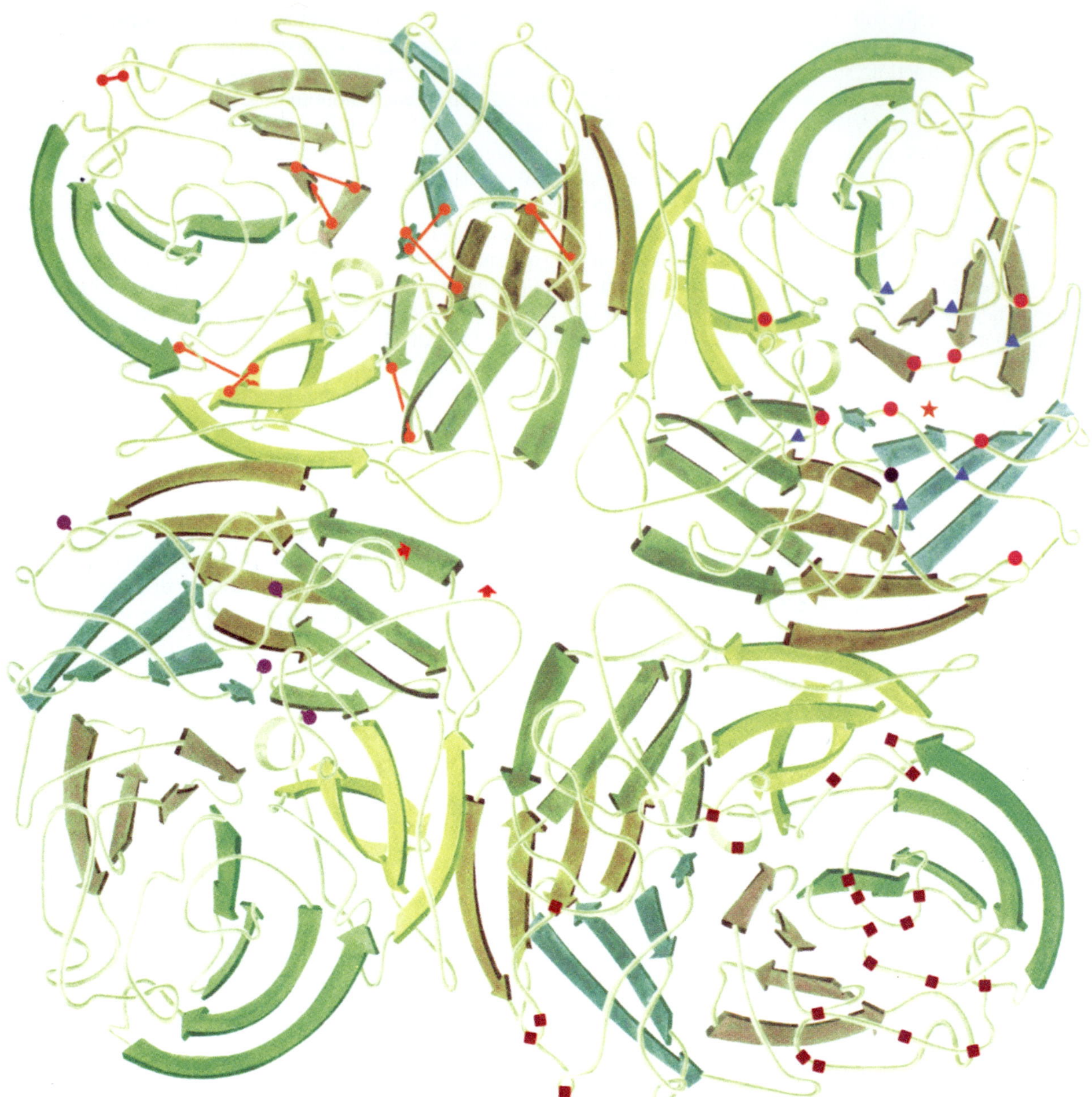

FIGURE 33-60 X-Ray structure of the influenza neuraminidase tetramer. The view is along the 4-fold axis of the homotetramaric protein, looking toward the viral membrane. In each monomer unit, each of the six topologically equivalent 4-stranded antiparallel β sheets are differently shaded. The positions of the disulfide bonds are indicated in the upper left subunit. In the lower left subunit, the four carbohydrate attachment sites are indicated by filled purple circles and the Asp residues that ligand Ca^{2+} ions are represented by red arrows. In the upper right monomer, the filled red circles and blue triangles, respectively, represent the conserved acidic and basic residues surrounding the enzyme's sialic acid–binding site, which is represented by a red star. In the lower right monomer, the positions of the mutated residues in NA's antigenic variants are flagged by filled brown squares. [After Varghese, J.N., Laver, W.G., and Colman, P.M., *Nature* **303,** 35 (1983). PDBid 1NN2.]

mediated endocytosis, a process that is accompanied by the fusion of the virus' enveloping membrane with that of the cell, thereby injecting the viral nucleocapsids into the cytosol. What is the mechanism of this membrane fusion?

Membrane fusion is mediated by HA but only after it has been exposed to the ~5.0 pH of the endosomal vesicle (Fig. 12-79). A variety of studies have implicated an ~25-residue conserved hydrophobic segment at the N-terminus of HA2, the so-called fusion peptide, with mediating membrane fusion by inserting into the cellular membrane. Yet, BHA's X-ray structure indicates that the fusion peptide is buried in the protein's hydrophobic interior, ~100 Å from the receptor binding site at the "top" of protein, the region in closest proximity to the cellular membrane. Thus, HA must undergo an extensive conformational change before it can initiate membrane fusion.

At pH 5.0, BHA indeed undergoes a conformational change but one that causes it to aggregate in a manner un-

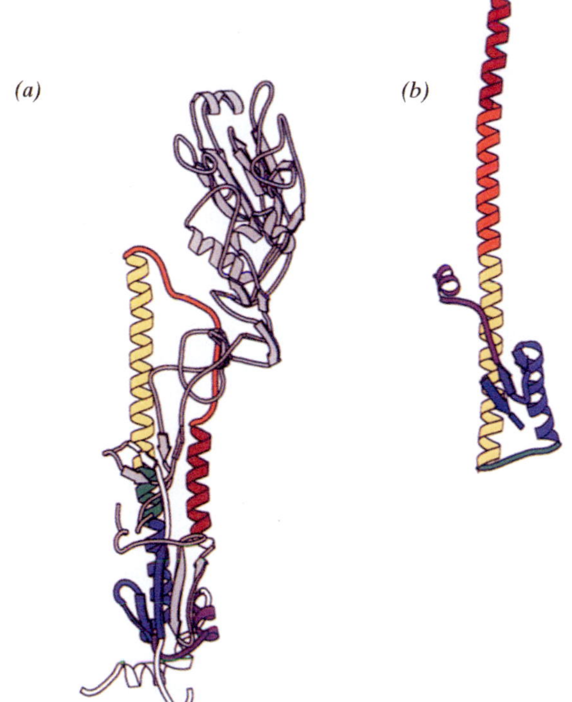

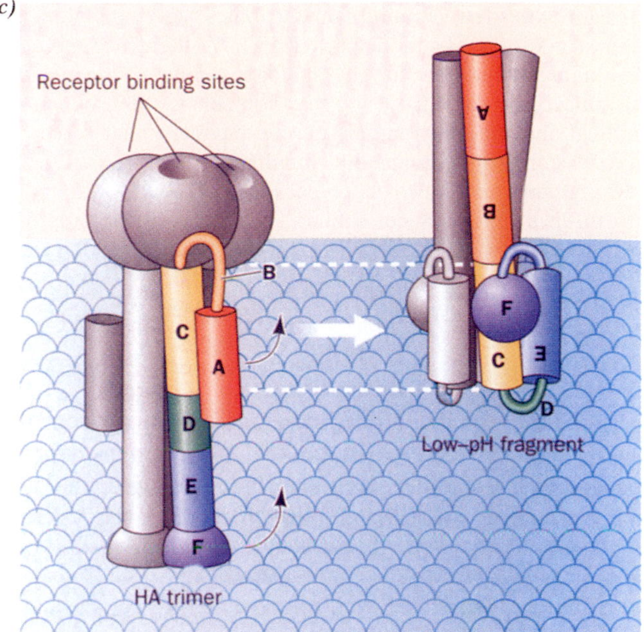

FIGURE 33-61 Comparison of the X-ray structures of BHA and $TBHA_2$. (*a*) Ribbon diagram of BHA in which the structural elements of the HA2 chain in $TBHA_2$ are colored in rainbow order (*red to violet*) from N- to C-terminus and the HA1 segment of $TBHA_2$, which is disulfide-linked to HA2, is blue. Regions of BHA that are proteolytically excised to form $TBHA_2$ are gray and those that are apparently disordered in $TBHA_2$ are white. (*b*) Ribbon diagram of $TBHA_2$ colored as in Part *a*. The heights of the various structural element relative to the yellow helix segment, which is common to both BHA and $TBHA_2$, are indicated. (*c*) Schematic diagram showing the positions and heights above the viral membrane surface of $TBHA_2$'s various structural elements in the HA trimer and in the low-pH fragment. The structural elements are colored as in Parts *a* and *b*. In the low-pH fragment, the fusion peptide (*not shown*) would protrude well above the receptor-binding heads where it would presumably insert itself into the cellular membrane. [Parts *a* and *b* courtesy of Don Wiley, Harvard University. PDBids 4HMG and 1HTM.

suitable for crystallographic studies. However, the successive proteolytic digestion of BHA at pH 5.0 by trypsin and thermolysin (Table 7-2) yields a crystallizable protein fragment named **$TBHA_2$,** which consists of residues 1 to 27 of HA1 and residues 38 to 175 of HA2 that are linked by a disulfide bond.

The X-ray structure of $TBHA_2$, determined by Skehel and Wiley, reveals that this protein has dramatically refolded relative to BHA in a way that involves extensive changes in both its secondary and tertiary structural elements (Fig. 33-61). Thus, segments A and B at the N-terminus of $TBHA_2$ (Fig. 33-61*c;* the red and orange segments in Fig. 33-61) undergo a jackknife-like movement of ~100 Å in a way that extends the top of the long helix by ~10 helical turns toward the cell membrane [although the long helix is shortened from the bottom by a similar but not so extensive shift of segment D (green in Fig. 33-61) so as to partially replace the flipped out helix A]. This conformational change is irreversible (HA does not revert to its original form when the pH is raised) and hence has been described as occurring via a "spring-loaded" mechanism. The rearrangement of segments A and B had been predicted by Peter Kim who noted that segments A, B, and C have the heptad repeat characteristic of coiled coils (Section 8-2A). Such conformational shifts in intact HA translocate the fusion peptide (which would extend beyond the N-terminus of $TBHA_2$ at the top of segment A) by at least 100 Å from its position in BHA. There it could bridge the viral and cellular membranes so as to facilitate their fusion in a manner similar to that postulated for SNARE complexes (Section 12-4D). The conformational change is probably triggered, at least in part, by the protonation at pH 5 of the six Asp side chains in the 19-residue segment B, thereby reducing the charge–charge repulsions that apparently prevent this segment from forming a coiled coil at higher pH's.

CHAPTER SUMMARY

1 ■ Tobacco Mosaic Virus Viruses are complex molecular aggregates that exhibit many attributes of living systems. Their structural and genetic properties have therefore served as valuable paradigms for the analogous cellular functions. The tobacco mosaic virus (TMV) virion consists of a helix of identical and therefore largely quasi-equivalent coat protein subunits enclosing a coaxially wound single strand of RNA. X-Ray studies of TMV gels reveal that this RNA is bound, with three nucleotides per subunit, between the subunits of the protein helix. In the absence of TMV RNA, the subunits aggregate at high ionic strengths to form double-layered disks, and at low ionic strengths to form protohelices, which stack to form helical rods under acidic conditions. The virus' innermost polypeptide loop is disordered in both the disk and the protohelix. Virus assembly is initiated when a protohelix (or possibly a double-layered disk) binds to the initiation sequence of TMV RNA, which is located ~1000 nucleotides from the RNA's 3′ end. Interactions between the RNA and protohelix trigger the ordering of the disordered loop, thereby converting the protohelix to the helical form. Elongation of the virus particle then proceeds by the sequential addition of protohelices (disks) to the "top" of the assembly so as to pull the 5′ end of the RNA up through the center of the growing viral helix.

2 ■ Icosahedral Viruses Viral capsids are formed from one or a few types of coat protein subunits. These must be either helically arranged, as in TMV, or quasi-equivalently arranged in a polyhedral shell so as to enclose the viral nucleic acid. The coat proteins of many icosahedral viruses are arranged in icosadeltahedra consisting of $60T$ subunits, where T is the triangulation number. The coat protein of tomato bushy stunt virus (TBSV) is arranged in a $T = 3$ icosadeltahedron so that TBSV subunits occupy three symmetrically distinct positions. The subunits must therefore associate through several sets of nonidentical intersubunit contacts. Some of the R domains form a structurally disordered inner protein shell. The viral RNA together with the remaining R domains are tightly packed in the space between the inner and outer protein shells. Other spherical plant viruses, southern bean mosaic virus (SBMV) and satellite tobacco mosaic virus (STMV), have tertiary and quaternary structures that are clearly related to those of TBSV. STMV's single-stranded RNA appears to form a series of stem and loop structures with the viral capsid. The structurally similar VP1, VP2, and VP3 coat proteins of poliovirus, rhinovirus, and foot-and-mouth disease virus (FMDV) are likewise icosahedrally arranged. However, the simian virus 40 (SV40) capsid consists of 72 pentagons of identical subunits in two different environments that are linked together by differering arrangements of their C-terminal arms in a nonicosadeltahedral arrangement. Although the coat proteins of most spherical viruses consist mainly of structurally similar 8-stranded antiparallel β barrels, that of bacteriophage MS2, a $T = 3$ virion, has an unrelated fold. Among the large icosahedral viruses of known structure are the bluetongue virus (BTV) core, which has a $T = 13$ outer shell and a $T = 2$ inner shell that envelops its largely visible dsRNA genome; and Paramecium bursaria Chlorella virus type 1 (PBCV-1), which has a $T = 169$ outer shell consisting of 1680 trimers of Vp54.

3 ■ Bacteriophage λ Lytic growth of bacteriophage λ in *E. coli* is controlled by the sequential syntheses of antiterminators, which inhibit both rho-independent and rho-dependent transcriptional terminators. Thus gp*N*, which is synthesized in the early stage of growth, permits the synthesis of gp*Q* in the delayed early stage which, in turn, permits the synthesis of the capsid proteins in the late stage. Early gene transcription is repressed in the delayed early stage by Cro protein. DNA replication, which commences in the early stage, is mediated by the host DNA replication machinery with the aid of the phage proteins gp*O* and gp*P*. DNA synthesis initially occurs by both the θ and rolling circle (σ) modes but eventually switches entirely to the rolling circle mode.

The λ virion heads and tails are separately assembled. Head assembly is a complex process involving the participation of many phage gene products, not all of which are part of the mature virion. Phage heads are not self-assembling in that their formation is guided by host chaperonins and a viral scaffolding protein and requires several enzymatically catalyzed protein modification reactions. The mature phage head is a $T = 7$ icosadeltahedron of gp*E*, which is decorated by an equal number of gp*D* subunits. Just before the final stage of its assembly, the phage head is filled with a linear double strand of DNA in a process that is driven by ATP hydrolysis. The packaged DNA appears to be wound in a spool that winds from outside to inside. Tail assembly occurs in a stepwise process from the tail fiber to the head-binding end. The body of the tail consists of a stack of hexameric rings of gp*V*. The completed heads and tails spontaneously join to form the mature virion.

Lysogeny is established by site-specific recombination between the phage *att*P site and the bacterial *att*B sites in a process mediated by phage integrase (gp*int*) and integration host factor (IHF). Induction, in which this process is reversed, requires the additional action of phage excisionase (gp*xis*). Lysogeny is established by a high level of gp*cII*, which stimulates the transcription of *int* and the λ repressor gene, *cI*. Repressor, as does Cro, binds to the o_L and o_R operators to shut down early gene transcription, including that of *cro* and *cII*. Each of these dimeric proteins, like other repressors of known structure, contains two symmetrically related helix–turn–helix (HTH) units that bind in successive turns of B-DNA's major groove. However, repressor, but not Cro, induces its own synthesis from the promoter p_{RM} by binding to o_{R2} so as to interact with RNA polymerase. The induction of repressor synthesis therefore throws the genetic switch that stably maintains the phage in the lysogenic state from generation to generation. Damage to host DNA, nevertheless, stimulates host RecA protein to mediate λ repressor cleavage so as to release repressor from o_L and o_R. This initiates the synthesis of early gene products, including gp*int* and gp*xis*, from p_L and p_R and thus triggers induction. If sufficient Cro protein is then synthesized to repress the synthesis of repressor, the phage becomes irrevocably committed to at least one generation of lytic growth. The tripartite character of o_R, the site of the λ switch, together with the cooperative nature of repressor binding to o_R, confers the λ switch with a remarkable sensitivity to the health of its host.

4 ■ Influenza Virus The influenza virion's enveloping membrane is studded with protein spikes consisting of hemagglutinin (HA), which mediates host recognition, and neuraminidase (NA), which facilitates the passage of the virus to and from the infection site. Inside the membrane is a shell of matrix protein that contains the virus' genome of eight single-stranded RNAs, each in a separate protein complex known as a nucleocapsid. These vRNAs are templates for the transcription of mRNAs as catalyzed by the nucleocapsid proteins. This process is primed by host-derived 7-methyl-G-capped mRNA fragments. The viral mRNAs, which have poly(A) tails, lack the sequences complementary to the vRNA's 5′ ends. The vRNAs, however, also act as templates for the transcription of the corresponding cRNAs which, in turn, are the templates for vRNA synthesis. The virus is assembled in and near the plasma membrane and forms by budding from the cell surface.

Influenza viruses infect a variety of mammals besides humans as well as many birds. Variation in the antigenic character of HA has been mainly responsible for the different influenza subtypes. Antigenic variation in HA occurs by either antigenic shift, in which the HA gene from an animal virus replaces that from a human virus, or antigenic drift, which occurs by a succession of point mutations in the HA gene. NA may vary in a similar fashion. HA is an elongated trimeric glycoprotein. Its surface has four antigenic sites that surround its sialic acid–binding pocket and that, in the viruses which caused the major epidemics between 1968 and 1975, all exhibit at least one mutational change. NA is a mushroom-shaped tetrameric glycoprotein. Its antigenic variations occur on a surface that also encircles its active site. HA mediates the fusion of the viral and host endosome membranes through a dramatic conformational change that translocates its fusion peptides to the vicinity of the endosome membrane into which they then insert.

REFERENCES

GENERAL

Cann, A.J., *Principles of Modern Virology,* Academic Press (1993).

Chiu, W., Burnett, R.M., and Garcea, R.L. (Eds.), *Structural Biology of Viruses,* Oxford University Press (1997).

Dimmock, N.J., Easton, A.J., and Leppard, K.N., *Introduction to Modern Virology* (5th ed.), Blackwell Science (2001).

Levine, A.J., *Viruses,* Scientific American Library (1992).

Radetsky, P., *The Invisible Invaders. The Story of the Emerging Age of Viruses,* Little, Brown and Co. (1991).

Voyles, B.A., *The Biology of Viruses,* Mosby (1993).

TOBACCO MOSAIC VIRUS

Bloomer, A.C., Champness, J.N., Bricogne, G., Staden, R., and Klug, A., Protein disk of tobacco mosaic virus at 2.8 Å showing the interactions within and between subunits, *Nature* **276,** 362–368 (1978).

Butler, P.J.G., Self-assembly of tobacco mosaic virus: The role of an intermediate aggregate in generating both specificity and speed, *Phil. Trans. R. Soc. Lond.* **B354,** 537–550 (1999).

Butler, P.J.G., Bloomer, A.C., and Finch, J.T., Direct visualization of the structure of the "20 S" aggregate of coat protein of tobacco mosaic virus, *J. Mol. Biol.* **224,** 381–394 (1992). [Evidence indicating that the TMV coat protein double-layered disk predominates over the protohelix at pH 7.0.]

Butler, P.J. and Klug, A., The assembly of a virus, *Sci. Am.* **239**(5): 62–69 (1978).

Klug, A., The tobacco mosaic virus particle: Structure and assembly, *Phil. Trans. R. Soc. Lond.* **B354,** 531–535 (1999).

Lomonosoff, G.P. and Wilson, T.M.A., Structure and in vitro assembly of tobacco mosaic virus, *in* Davis, J.W. (Ed.), *Molecular Plant Virology,* Vol. I, *pp.* 43–83, CRC Press (1985).

Namba, K., Pattanayek, R., and Stubbs, G., Visualization of protein–nucleic acid interactions in a virus. Refined structure of intact tobacco mosic virus at 2.9 Å by X-ray fiber diffraction, *J. Mol. Biol.* **208,** 307–325 (1989).

Raghavendra, K., Kelly, J.A., Khairallah, L., and Schuster, T.M., Structure and function of disk aggregates of the coat protein of tobacco mosaic virus, *Biochemistry* **27,** 7583–7588 (1988). [Evidence indicating that the TMV coat protein disks do not convert to the protohelices that nucleate TMV assembly.]

Stubbs, G., Molecular structures of viruses from the tobacco mosaic group, *Sem. Virol.* **1,** 405–412 (1990).

Stubbs, G., Tobacco mosaic virus particle structure and the initiation of disassembly, *Phil. Trans. R. Soc. Lond.* **B354,** 551–557 (1999).

ICOSAHEDRAL VIRUSES

Abad-Zapetero, C., Abdel-Meguid, S.S., Johnson, J.E., Leslie, A.G.W., Rayment, I., Rossmann, M.G., Suck, D., and Tsukihara, T., Structure of southern bean mosaic virus at 2.8 Å resolution, *Nature* **286,** 33–39 (1980).

Acharya, R., Fry, E., Stuart, D., Fox, G., and Brown, F., The three-dimensional structure of foot-and-mouth disease virus at 2.9 Å resolution, *Nature* **337,** 709–716 (1989).

Arnold, E. and Rossmann, M.G., Analysis of the structure of a common cold virus, human rhinovirus 14, refined at a resolution of 3.0 Å, *J. Mol. Biol.* **211,** 763–801 (1990); *and* Rossmann, M.G., et al., Structure of a human common cold virus and relationship to other picornaviruses, *Nature* **317,** 145–153 (1985).

Caspar, D.L.D. and Klug, A., Physical principles in the construction of regular viruses, *Cold Spring Harbor Symp. Quant. Biol.* **27,** 1–24 (1962). [The classic paper formulating the geometric principles governing the construction of icosahedral viruses.]

Dokland, T., Freedom and restraint: Themes in virus capsid assembly, *Structure* **8,** R157–R162 (2000).

Grimes, J.M., Burroughs, J.N., Gouet, P., Diprose, J.M., Malby, R., Ziéntara, S., Mertens, P.P.C., and Stuart, D.I., The atomic structure of the bluetongue virus core, *Nature* **395,** 470–478 (1998); Gouet, P., Diprose, J.M., Grimes, J.M., Malby, R., Burroughs, J.N., Ziéntara, S., Stuart, D.I., and Mertens, P.P.C., The highly ordered double-stranded RNA genome of bluetongue virus revealed by crystallography, *Cell* **97,** 481–490 (1999); *and* Diprose, J.M., et al., Translocation portals for the substrates and products of a viral transcription complex: the bluetongue virus core, *EMBO J.* **20,** 7229–7239 (2001).

Harrison, S.C., Common features in the structures of some icosahedral viruses: a partly historical view, *Sem. Virol.* **1,** 387–403 (1990).

Harrison, S.C., The familiar and unexpected in structures of icosahedral viruses, *Curr. Opin. Struct. Biol.* **11,** 195–199 (2001).

Harrison, S.C., Olson, A.J., Schutt, C.E., Winkler, F.K., and

Bricogne, G., Tomato bushy stunt virus at 2.9 Å resolution, *Nature* **276,** 368–373 (1978). [The first report of a high-resolution virus structure.]

Hogle, J.M., Chow, M., and Filman, D.J., The structure of poliovirus, *Sci. Am.* **256**(3): 42–49 (1987); *and* Three-dimensional structure of poliovirus at 2.9 Å resolution, *Science* **229,** 1358–1365 (1985).

Hurst, C.J., Benton, W.H., and Enneking, J.M., Three dimensional model of human rhinovirus type 14, *Trends Biochem. Sci.* **12,** 460 (1987). [A "paper doll"-type cutout with accompanying assembly directions for constructing an icosahedral model of human rhinovirus. This useful learning device may also be taken as a $T = 3$ icosadeltahedron.]

Larson, S.B. and McPherson, A., Satellite tobacco mosaic virus RNA: structure and implications for assembly, *Curr. Opin. Struct. Biol.* **11,** 59–65 (2001); *and* Larson, S.B., Day, J., Greenwood, A., and McPherson, A., Refined structure of satellite tobacco mosaic virus at 1.8 Å resolution, *J. Mol. Biol.* **277,** 37–59 (1998).

Munshi, S., Liljas, L., Cavarelli, J., Bomu, W., McKinney, B., Reddy, V, and Johnson, J.E., The 2.8 Å structure of a $T = 4$ animal virus and its implications for membrane translocation of RNA, *J. Mol. Biol.* **261,** 1–10 (1996). [The X-ray structure of Nudaurelia ω Capensis virus.]

Nandhagopal, N., Simpson, A.A., Gurnon, J.R., Yan, X., Baker, T.S., Graves, M.V., Van Etten, J.L., and Rossmann, M.G., The structure and evolution of the major capsid protein of a large, lipid-containing DNA virus, *Proc. Natl. Acad. Sci.* **99,** 14758–14763 (2002); *and* Yan, X., Olson, N.H., Van Etten, J.L, Bergoin, M., Rossmann, M.G., and Baker, T.S., Structure and assembly of large lipid-containing dsDNA viruses, *Nature Struct. Biol.* **7,** 101–103 (2000). [The structure of PBCV-1.]

Rossmann, M.G. and Johnson, J.E., Icosahedral RNA virus structure, *Annu. Rev. Biochem.* **58,** 533–573 (1989).

Stehle, T., Gamblin, S.J., Yan, Y., and Harrison, S.C., The structure of simian virus 40 refined at 3.1 Å, *Structure* **4,** 165–182 (1996); *and* Liddington, R.C., Yan, Y., Moulai, J., Sahli, R., Benjamin, T.L., and Harrison, S.C., Structure of simian virus 40 at 3.8-Å resolution, *Nature* **354,** 278–284 (1991).

Valegård, K., Liljas, L., Fridborg, K., and Unge, T., The three-dimensional structure of the bacterial virus MS2, *Nature* **345,** 36–41 (1990); *and* Golmohammadi, R., Valegård, K., Fridborg, K., and Liljas, L., The refined structure of bacteriophage MS2 at 2.8 Å resolution, *J. Mol. Biol.* **234,** 620–639 (1993).

BACTERIOPHAGE λ

Albright, R.A. and Matthews, B.W., Crystal structure of λ-Cro bound to a consensus operator at 3.0 Å resolution, *J. Mol. Biol.* **280,** 137–151 (1998); *and* Brennan, R.G., Roderick, S.L., Takeda, Y., and Matthews, B.W., Protein–DNA conformational changes in the crystal structure of λ Cro–operator complex, *Proc. Natl. Acad. Sci.* **87,** 8165–8169 (1990).

Azaro, M.A. and Landy, A., λ Integrase and λ Int family, *in* Craig, N.L., Craigie, R., Gellert, M. and Lambowitz, A.M. (Eds.), *Mobile DNA II,* 118–148, ASM Press (2002).

Beamer, L.J. and Pabo, C.O., Refined 1.8 Å crystal structure of the λ repressor–operator complex, *J. Mol. Biol.* **227,** 177–196 (1992); *and* Jordan, S.R. and Pabo, C.O., Structure of the lambda complex at 2.5 Å resolution: details of the repressor–operator interactions, *Science* **242,** 893–899 (1988).

Bell, C.E., Frescura, P., Hochschild, A., and Lewis, M., Crystal structure of the λ repressor C-terminal domain provides a model for cooperative operator binding, *Cell* **101,** 801–811 (2000).

Brüssow, H. and Hendrix, R.W., Phage genomics: Small is beautiful, *Cell* **108,** 13–16 (2002).

Cai, Z., Gorin, A., Frederick, R., Ye, X., Hu, W., Majumdar, A., Kettani, A., and Patel, D.J., Solution structure of P22 transcriptional antitermination N peptide–box B RNA complex, *Nature Struct. Biol.* **5,** 203–212 (1998); *and* Legault, P., Li, J., Mogridge, J., Kay, L.E., and Greenblatt, J., NMR structure of the bacteriophage λ N peptide/*boxB* RNA complex: Recognition of a GNRA fold by an arginine-rich motif, *Cell* **93,** 289–299 (1998).

Cerritelli, M.E., Cheng, N., Rosenberg, A.H., McPherson, C.E., Booy, F.P., and Steven, A.C., Encapsidated conformation of bacteriophage T7 DNA, *Cell* **91,** 271–280 (1997).

Echols, H., Bacteriophage λ development: temporal switches and the choice of lysis or lysogeny, *Trends Genet.* **2,** 26–30 (1986).

Greenblatt, J., Nodwell, J.R., and Mason, S.W., Transcriptional antitermination, *Nature* **364,** 401–406 (1993).

Hendrix, R.W., and Garcea, R.L., Capsid assembly of dsDNA viruses, *Sem. Virol.* **5,** 15–26 (1994).

Hendrix, R.W., Roberts, J.W., Stahl, F.W., and Weisberg, R.A. (Eds.), *Lambda II,* Cold Spring Harbor Laboratory (1982). [A compendium of review articles on many aspects of bacteriophage λ.]

Murialdo, H., Bacteriophage lambda DNA maturation and packaging, *Annu. Rev. Biochem.* **60,** 125–153 (1991).

Oppenheim, A.B., Kornitzer, D., and Altuvia, S., Posttranscriptional control of the lysogenic pathway in bacteriophage lambda, *Prog. Nucleic Acid Res. Mol. Biol.* **46,** 37–49 (1993).

Pabo, C.O. and Lewis, M., The operator-binding domain of λ repressor: structure and DNA recognition, *Nature* **298,** 443–447 (1982).

Ptashne, M., *A Genetic Switch* (2nd ed.), Cell Press & Blackwell Scientific Publications (1992). [An authoritative review of the λ switch.]

Rice, P.A., Yang, S., Mizuuchi, K., and Nash, H., Crystal structure of an IHF-DNA complex: A protein-induced DNA turn, *Cell* **87,** 1295–1306 (1996); *and* Rice, P.A., Making DNA do a U-turn: IHF and releated proteins, *Curr. Opin. Struct. Biol.* **7,** 86–93 (1997).

Roberts, J.W., RNA and protein elements of *E. coli* and λ transcription antitermination complexes, *Cell* **72,** 653–655 (1993).

Simpson, A.A., et al., Structure of the bacteriophage ϕ29 DNA packaging motor, *Nature* **409,** 745–750 (2000).

Taylor, K. and Wegrzyn, G., Replication of coliphage lambda DNA, *FEMS Microbiol. Rev.* **17,** 109–119 (1995).

INFLUENZA VIRUS

Air, G.M. and Laver, W.G., The molecular basis of antigenic variation in influenza virus, *Adv. Virus Res.* **31,** 53–102 (1986).

Bullough, P.A., Hughson, F.M., Skehel, J.J., and Wiley, D.C., Structure of influenza haemagglutinin at the pH of membrane fusion, *Nature* **371,** 37–43 (1994).

Carr, C.M., and Kim, P.S., A spring-loaded mechanism for the conformational change of influenza hemagglutinin, *Cell* **73,** 823–832 (1994).

Colman, P., Influenza virus neuraminidase: Structure, antibodies, and inhibitors, *Protein Sci.* **3,** 1687–1696 (1994).

Colman, P.M., Neuraminidase inhibitors as antivirals, *Vaccine* **20,** S55–S58 (2002).

Cox, N.J. and Subbaro, K., Global epidemiology of influenza: Past and present, *Annu. Rev. Med.* **51,** 407–421 (2000).

Eckert, D.M. and Kim, P.S., Mechanisms of viral membrane fusion and its inhibition, *Annu. Rev. Biochem.* **70,** 777–810 (2001).

Kolata, G.B., *Flu: The Story of the Great Influenza Pandemic of 1918 and the Search for the Virus that Caused It,* Farrar, Straus and Giroux (1999).

Nicholson, K.G., Webster, R.G., and Hay, A.J., *Textbook of Influenza,* Blackwell Science (1998).

Potter, C.W. (Ed.), *Influenza,* Elsevier (2002).

Skehel, J.J. and Wiley, D.C., Receptor binding and membrane fusion in virus entry: The influenza hemagglutinin, *Annu. Rev. Biochem.* **69,** 531–569 (2000).

Skehel, J.J., Stevens, D.J., Daniels, R.S., Douglas, A.R., Knossow, M., Wilson, I.A., and Wiley, D.C., A carbohydrate side chain on hemagglutinins of Hong Kong influenza viruses inhibits recognition by a monoclonal antibody, *Proc. Natl. Acad Sci.* **81,** 1779–1783 (1984).

Varghese, J.N., Laver, W.G., and Colman, P.M., Structure of the influenza virus glycoprotein antigen neuramimidase at 2.9 Å resolution, *Nature* **303,** 35–40 (1983); *and* Colman, P.M., Varghese, J.N., and Laver, W.G., Structure of the catalytic and antigenic sites in influenza virus neuraminidase, *Nature* **303,** 41–44 (1983).

Varghese, J.N., McKimm-Breschkin, J.L., Caldwell, J.B., Kortt, A.A., and Colman, P.M., The structure of the complex between influenza virus neuraminidase and sialic acid, the viral receptor, *Proteins* **14,** 327–332 (1992).

Webster, R.G., Laver, W.G., Air, G.M., and Schild, G.C., Molecular mechanisms of variation in influenza viruses, *Nature* **296,** 115–121 (1982).

Weis, W., Brown, J.H., Cusack, S., Paulson, J.C., Skehel, J.J., and Wiley, D.C., Structure of the influenza virus haemagglutinin complexed with its receptor, sialic acid, *Nature* **333,** 426–431 (1988).

Wilson, I.A., Skehel, J.J., and Wiley, D.C., Structure of the haemagglutinin membrane glycoprotein of influenza virus at 3 Å resolution, *Nature* **289,** 366–373 (1981); *and* Wiley, D.C., Wilson, I.A., and Skehel, J.J., Structural identification of the antibody-binding sites of Hong Kong influenza haemaglutinin and involvement in antigenic variation, *Nature* **289,** 373–378 (1981).

PROBLEMS

1. Why does a pH shift from 7 to 5 at low ionic strengths cause TMV double-layered disks to aggregate as helical rods?

2. Can a nucleic acid encode a monomeric protein large enough to enclose it? Explain.

***3.** Explain why the number of vertices in an icosadeltahedron always ends in the numeral "2" (e.g., 12 for $T = 1$).

4. Sketch a $T = 9$ icosadeltahedron.

5. The coat protein pentagons of SV40 are arranged at the vertices of a $T = 7$ icosadeltahedron. Yet the SV40 virion cannot have icosadeltahedral symmetry. Explain.

6. Why is it necessary to use conditionally lethal mutations in studying phage assembly rather than just lethal mutations?

7. Compare the volume contained by a λ phage head to that of λ DNA.

8. Virulent phages form clear plaques on a bacterial lawn, whereas bacteriophage λ forms turbid (cloudy) plaques. Explain.

9. What is the mutual consensus sequence of the o_L and o_R half-subsites?

10. λ repressor binds cooperatively to o_{R1} and o_{R2} but independently to o_{R3}. However, if o_{R1} is mutationally altered so that it does not bind repressor, then repressor binds cooperatively to o_{R2} and o_{R3}. Explain.

11. Bacteriophage 434 is a lambdoid phage that has both a repressor and a Cro protein. You have constructed a hybrid repressor that consists of the 434 repressor with its α3 helix replaced by that from 434 Cro protein. Compare the pattern of contacts this hybrid protein makes with its operator, as indicated by chemical protection experiments, with those of the native 434 repressor and Cro proteins.

12. What is the probability that an influenza virion will have its proper complement of eight different RNAs if it has room for only eight nucleocapsids and binds them at random?

Chapter 34

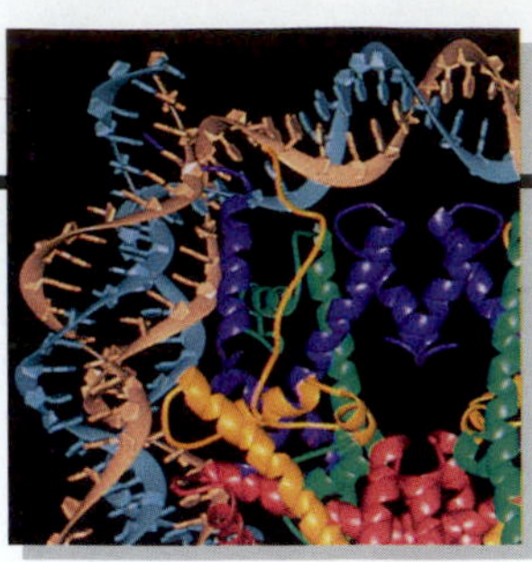

Eukaryotic Gene Expression

How does a fertilized ovum give rise to a highly differentiated multicellular organism? This question, of course, is just a sophisticated version of one that every child has asked: Where did I come from? Biologists began rational attempts to answer this question in the late nineteenth century and since that time have assembled an impressive body of knowledge concerning the general patterns of cellular differentiation and organismal development. Yet we have had the technical ability to study embryogenesis on the molecular level only in the last 30 years or so.

In order to understand cellular differentiation we must first understand the workings of the eukaryotic cell. Eukaryotic cells are, for the most part, much larger and far more complex than prokaryotic cells (Section 1-2). However, *the basic difference between these two types of cells is that eukaryotes have a nuclear membrane that separates their chromosomes from their cytoplasm, thereby physically divorcing the eukaryotic transcriptional process from that of translation.* In contrast, the prokaryotic chromosome is embedded in the cytosol so that the initiation of protein synthesis often occurs on mRNAs that are still being transcribed. The transcriptional and translational control processes in eukaryotes are consequently fundamentally different from those of prokaryotes. This situation is reflected in both the packaging and the genetic organization of eukaryotic DNA in comparison with that of prokaryotes. We therefore begin this chapter with a physical description of the eukaryotic chromosome. We then consider how the eukaryotic genome is organized and how it is expressed. Finally, we discuss cell differentiation, its aberration, cancer, how the cell cycle is controlled, and programmed cell death. In all these subjects, as we shall see, our knowledge is quite fragmentary. Eukaryotic molecular biology is under such intense scrutiny, however, that significant advances in its understanding are made almost daily. Thus, perhaps more so than for other subject matter considered in this text, it is important that the reader supplement the material in this chapter with that in the recent biochemical literature.

1 ■ CHROMOSOME STRUCTURE

Eukaryotic chromosomes, which consist of a complex of DNA, RNA, and protein called **chromatin,** are dynamic entities whose appearance varies dramatically with the stage of the cell cycle. The individual chromosomes assume their familiar condensed forms (Figs. 1-18 and 34-1) only during cell division (M phase of the cell cycle; Section 30-4A). During interphase, the remainder of the cell cycle, when the chromosomal DNA is transcribed and replicated, the chromosomes of most cells become so highly dispersed that they cannot be individually distinguished (Fig. 34-2). Cytologists have long recognized that there are two types of this dispersed chromatin: a less densely packed variety named **euchromatin** and a more densely packed variety termed **hetero-**

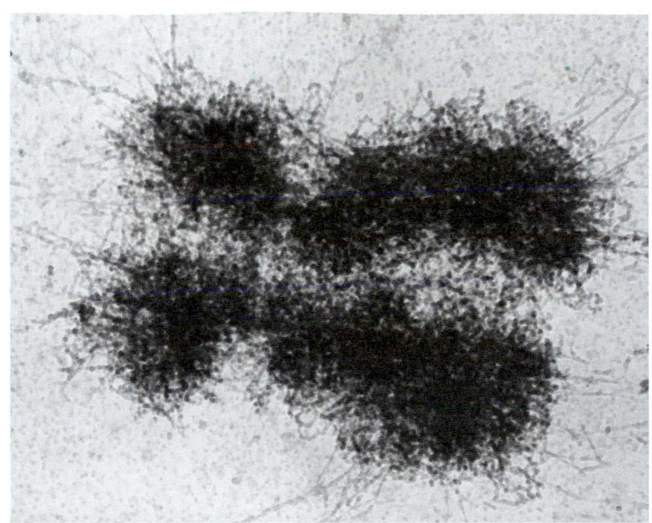

FIGURE 34-1 Electron micrograph of a human metaphase chromosome. It consists of two sister (identical) chromatids joined at their centromeres (the constricted portion near the left end of the chromosome). [Courtesy of Gunther Bahr, Armed Forces Institute of Pathology.]

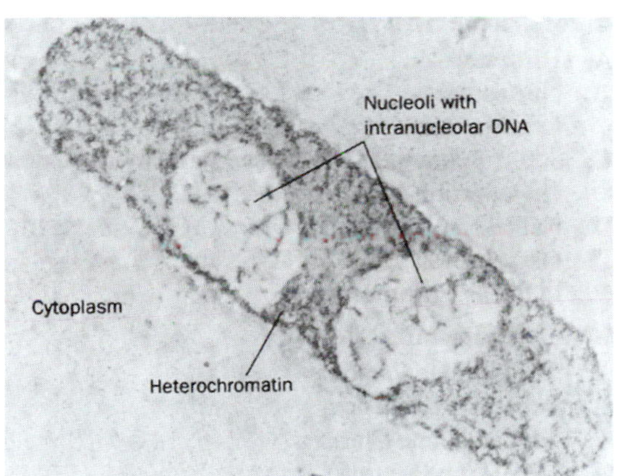

FIGURE 34-2 Thin section through a cell nucleus treated with Feulgen reagent (which reacts with DNA to form an intense red stain). Heterochromatin appears as dark-staining regions near the nucleolus and the nuclear membrane. The less darkly staining material is euchromatin. [Courtesy Edmund Puvion, CNRS, France.]

chromatin (Fig. 34-2). These two types of chromatin differ, as we shall see, in that euchromatin is genetically expressed, whereas heterochromatin is not expressed.

The 46 chromosomes in a human cell each contain between 44 and 246 million bp, so their DNAs, which are continuous (Section 5-3D), have contour lengths between 1.5 and 8.4 cm (3.4 Å/bp). Yet in metaphase, their most condensed state (Fig. 34-1), these chromosomes range in length from 1.3 to 10 μm. *Chromosomal DNA therefore has a* ***packing ratio*** *(ratio of its contour length to the length of its container) of >8000.* How does the DNA in chromatin attain such a high degree of condensation? Structural studies have revealed that this results from three levels of folding. We discuss these levels below, starting with the lowest level. We begin, however, by studying the proteins responsible for much of this folding.

A. Histones

The protein components of chromatin, which comprise somewhat more than half its mass, consist mostly of ***histones,*** which were discovered in 1884 by Albrecht Kossel and for many years were believed to be the genetic material itself. There are five major classes of these proteins, **histones H1, H2A, H2B, H3,** and **H4,** all of which have a large proportion of positively charged residues (Arg and Lys; Table 34-1). These proteins therefore ionically bind DNA's negatively charged phosphate groups. Indeed, histones may be extracted from chromatin by 0.5*M* NaCl, a salt solution of sufficient concentration to interfere with these electrostatic interactions.

a. Histones Are Evolutionarily Conserved

The amino acid sequences of histones H2A, H2B, H3, and H4 have remarkably high evolutionary stability (Table 34-1). For example, histones H4 from cows and peas, species that diverged 1.2 billion years ago, differ by only two conservative residue changes (Fig. 34-3), which makes histone H4, the most invariant histone, among the most evolutionarily conserved proteins known (Section 7-3B). *Such rigid evolutionary stability implies that the above four histones have critical functions to which their structures are so well tuned that they are all but intolerant to change.* The fifth histone, histone H1, is more variable than the other histones; we shall see below that its role differs from that of the other histones.

TABLE 34-1 Calf Thymus Histones

Histone	Number of Residues	Mass (kD)	% Arg	% Lys	UEP[a] ($\times 10^{-6}$ year)
H1	215	23.0	1	29	8
H2A	129	14.0	9	11	60
H2B	125	13.8	6	16	60
H3	135	15.3	13	10	330
H4	102	11.3	14	11	600

[a]*U*nit *e*volutionary *p*eriod: The time for a protein's amino acid sequence to change by 1% after two species have diverged (Section 7-3B).

Ac—Ser—Gly—Arg—Gly—Lys—Gly—Gly—Lys—Gly—Leu—10
Gly—Lys—Gly—Gly—Ala—Lys—Arg—His—Arg—Lys—20
Val—Leu—Arg—Asp—Asn—Ile—Gln—Gly—Ile—Thr—30
Lys—Pro—Ala—Ile—Arg—Arg—Leu—Ala—Arg—Arg—40
Gly—Gly—Val—Lys—Arg—Ile—Ser—Gly—Leu—Ile—50
Tyr—Glu—Glu—Thr—Arg—Gly—Val—Leu—Lys—Val—60
Phe—Leu—Glu—Asn—Val—Ile—Arg—Asp—Ala—Val—70
Thr—Tyr—Thr—Glu—His—Ala—Lys—Arg—Lys—Thr—80
Val—Thr—Ala—Met—Asp—Val—Val—Tyr—Ala—Leu—90
Lys—Arg—Gln—Gly—Arg—Thr—Leu—Tyr—Gly—Phe—100
Gly—Gly 102

FIGURE 34-3 The amino acid sequence of calf thymus histone H4. This 102-residue protein's 25 Arg and Lys residues are indicated in red. Pea seedling H4 differs from that of calf thymus by conservative changes at the two shaded residues: Val 60 → Ile and Lys 77 → Arg. The underlined residues are subject to posttranslational modification: Ser 1 is invariably *N*-acetylated and may also be *O*-phosphorylated; Lys residues 5, 8, 12, and 16 may be *N*-acetylated; and Lys 20 may be mono- or di-*N*-methylated. [After DeLange, R.J., Fambrough, D.M., Smith, E.L., and Bonner, J., *J. Biol. Chem.* **244,** 5678 (1969).]

b. Histones May Be Modified

Histones are subject to posttranslational modifications that include methylations, acetylations, and phosphorylations of specific Arg, His, Lys, and Ser side chains. These modifications, many of which are reversible, all decrease the histones' positive charges, thereby significantly altering histone–DNA interactions. Yet, despite the histones' great evolutionary stability, their degree of modification varies enormously with the species, tissue, and the stage of the cell cycle. A particularly intriguing modification is that 10% of the H2As have an isopeptide bond between the ε-amino group of their Lys 119 and the terminal carboxyl group of the protein ubiquitin. Although ubiquitination, really polyubiquitination, marks cytosolic proteins for degradation by cellular proteases (Section 32-6B), this is not the case for H2A. Rather, as we shall see in Section 34-3B, monoubiquitination as well as the other histone modifications serve to modulate eukaryotic gene expression.

Many, if not all, eukaryotes have genetically distinct subtypes of histones H1, H2A, H2B, and H3, many of whose syntheses are switched on or off during specific stages of embryogenesis and in the development of certain cell types. The sequence variations of these subtypes are limited to only a few residues in H2A, H2B, and H3 but are much more extensive in H1. Indeed, the erythroid cells of chick embryos contain an H1 variant that differs so greatly from other H1s that it is named **histone H5** (avian erythrocytes, unlike those of mammals, have nuclei). Histone switching seems to be related to cell differentiation, but the nature of this relationship is unknown.

B. *Nucleosomes: The First Level of Chromatin Organization*

The first level of chromatin organization was pointed out by Roger Kornberg in 1974 through the synthesis of several lines of evidence:

1. Chromatin contains roughly equal numbers of molecules of histones H2A, H2B, H3, and H4, and no more than half that number of histone H1 molecules.

2. X-Ray diffraction studies indicate that chromatin fibers have a regular structure that repeats about every 10 nm along the fiber direction. This same X-ray pattern is observed when purified DNA is mixed with equimolar amounts of all the histones except histone H1.

3. Electron micrographs of chromatin (Fig. 34-4) reveal that it consists of ~10-nm-diameter particles connected by thin strands of apparently naked DNA, rather like beads on a string. These particles are presumably responsible for the foregoing X-ray pattern.

4. Brief digestion of chromatin by **micrococcal nuclease** (which cleaves double-stranded DNA) cleaves the DNA between some of the above particles (Fig. 34-5*a*); apparently the particles protect the DNA closely associated with them from nuclease digestion. Gel electrophoresis indicates that each particle *n*-mer contains ~200*n* bp of DNA (Fig. 34-5*b*).

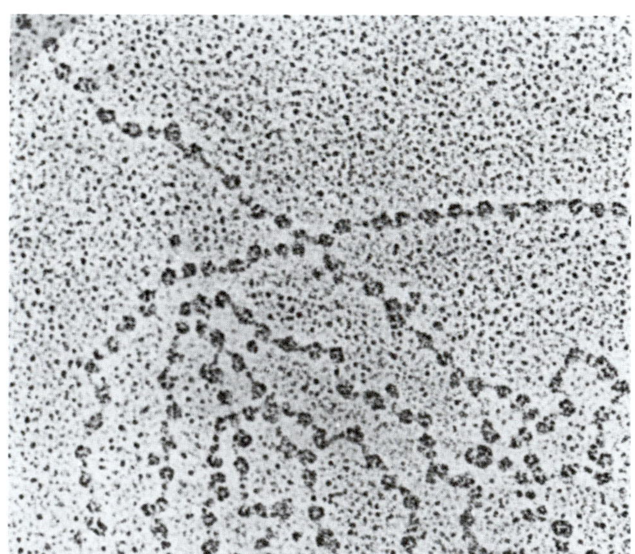

FIGURE 34-4 Electron micrograph of *D. melanogaster* chromatin showing that its 10-nm fibers are strings of closely spaced nucleosomes. [Courtesy of Oscar L. Miller, Jr., University of Virginia.]

(a)

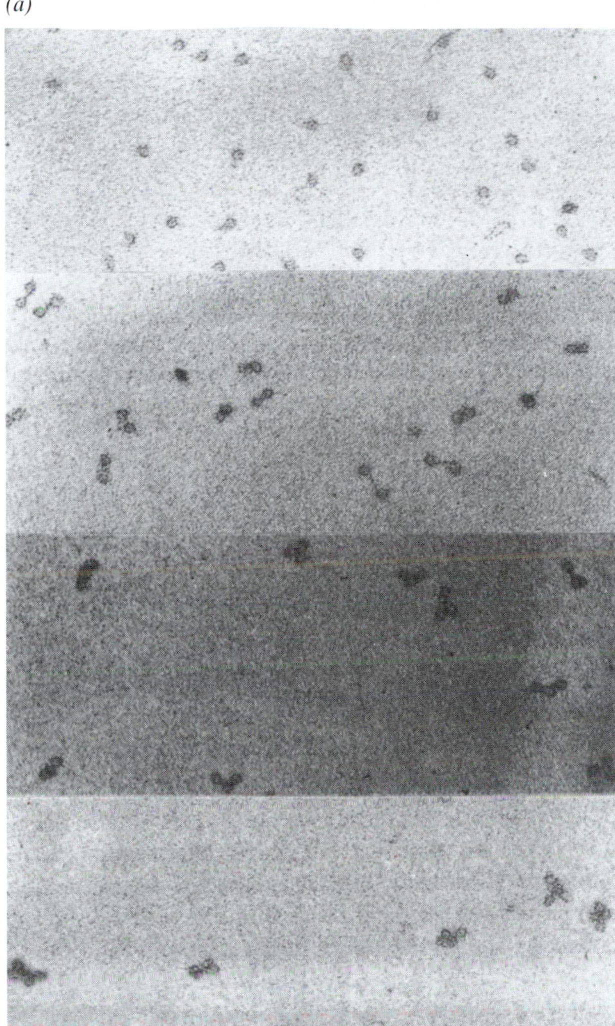

(b)

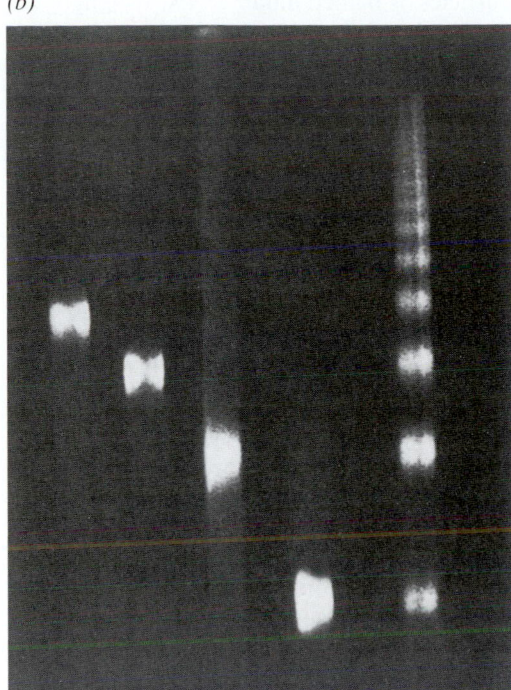

FIGURE 34-5 Defined lengths of calf thymus chromatin obtained by sucrose density gradient ultracentrifugation of chromatin that had been partially digested by micrococcal nuclease. (*a*) Electron micrographs of sucrose density gradient fractions containing, from top to bottom, nucleosome monomers, dimers, trimers, and tetramers. (*b*) Gel electrophoresis of DNA extracted from the nucleosome multimers indicates that they are the corresponding multiples of ~200 bp. The rightmost lane contains DNA from the unfractionated nuclease digest. [Courtesy of Roger Kornberg, Stanford University School of Medicine.]

5. Chemical cross-linking experiments, such as are described in Section 8-5C, indicate that histones H3 and H4 associate to form the tetramer $(H3)_2(H4)_2$ (Fig. 34-6).

These observations led Kornberg to propose that *the chromatin particles, which are called* ***nucleosomes,*** *consist of the octamer* $(H2A)_2(H2B)_2(H3)_2(H4)_2$ *in association with ~200 bp of DNA.* The fifth histone, H1, was postulated to be associated in some manner with the outside of the nucleosome (see below).

a. DNA Coils around a Histone Octamer to Form the Nucleosome Core Particle

Micrococcal nuclease, as described above, initially degrades chromatin to particles known as **chromatosomes** that each consist of 166 bp of DNA in complex with a histone octamer and one molecule of histone H1. On further digestion, some of the chromatosome's DNA is trimmed away in a process that releases histone H1. This yields the 205-kD **nucleosome core particle,** which consists of a 145- to 147-bp strand of DNA in association with the above histone octamer. The DNA cumulatively removed by this digestion, which had previously joined neighboring nucle-

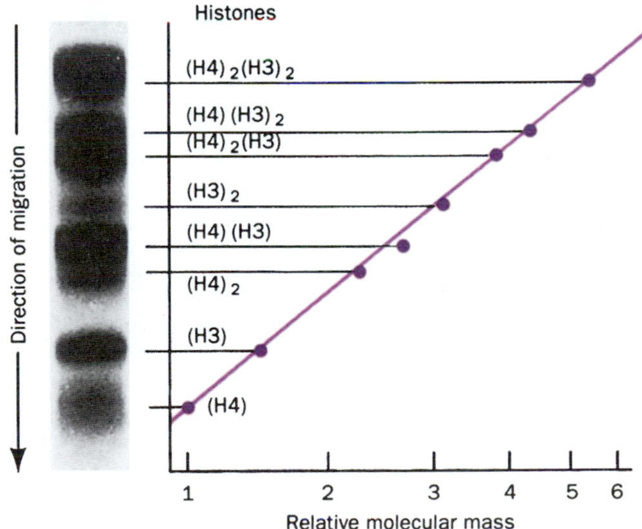

FIGURE 34-6 SDS–gel electrophoresis of a mixture of calf thymus histones H3 and H4 that had been cross-linked by dimethylsuberimidate. The electrophoretogram contains all the bands expected from an $(H3)_2(H4)_2$ tetramer. [Courtesy of Roger Kornberg, Stanford University School of Medicine.]

osome core particles, is named **linker DNA.** Its length has been found to vary between 8 and 114 bp from organism to organism and tissue to tissue although it is usually ~55 bp.

The X-ray structures of nucleosome core particles containing 146- or 147-bp palindromic DNAs of defined sequence and histones from *X. laevis,* chicken, and *S. cerevisiae* were respectively determined by Timothy Richmond, Gerard Bunick, and Karolin Luger. These structures reveal the nucleosome core particle to be a nearly 2-fold symmetric wedge-shaped disk that has a diameter of ~110 Å and a maximum thickness of ~60 Å. The DNA, which assumes the B form, is wrapped around the outside of the histone octamer in 1.65 turns of a left-handed superhelix (Fig. 34-7). This is the origin of supercoiling in eukaryotic DNA.

Despite having only weak sequence similarity, all four types of histones in the histone octamer share a similar ~70-residue **histone fold** near their C-termini in which a long central helix is flanked on each side by a loop and a shorter helix (Fig. 34-8). Pairs of histone folds interdigitate in head-to-tail arrangements to form the crescent-shaped heterodimers H2A–H2B and H3–H4, each of which binds 2.5 turns (27–28 bp) of duplex DNA that curves around it in a 140° arc. Successive arcs are joined by 3- or 4-bp segments. The H3–H4 pairs interact, via a bundle of four helices from the two H3 histones, to form an $(H3–H4)_2$ tetramer with which each H2A–H2B pair interacts, via a similar four-helix bundle between H2B and H4, to form the histone octamer (Fig. 34-7*b*).

The histones bind exclusively to the inner face of the DNA, primarily via its sugar–phosphate backbones, through hydrogen bonds, salt bridges, and helix dipoles (their positive N-terminal ends), all interacting with phosphate oxygens, as well as through hydrophobic interactions with the deoxyribose rings. There are few contacts between the histones and the bases, in accord with the nucleosome's lack of sequence specificity. However, an Arg side chain is inserted into the DNA's minor groove at each of the 14 positions it faces the histone octamer. The DNA superhelix has an average radius of 42 Å and a pitch (rise per turn) of 26 Å. However, the DNA does not follow a uniform superhelical path but, rather, is bent fairly sharply at several locations due to outward bulges of the histone core. Moreover, the DNA double helix exhibits considerable conformational variation along its length such that its twist, for example, varies from 7.5 to 15.2 bp/turn with an average value of 10.4 bp/turn (vs 10.4 bp/turn for B-DNA in solution). ~75% of the DNA surface is accessible to solvent and hence appears to be available for interactions with DNA-binding proteins.

The histones of the nucleosome core contain N-terminal tails that emanate from their central histone folds (Fig. 34-8) and vary in length from 23 to 43 residues (they comprise ~25% of the mass of these histones). These highly positively charged polypeptide segments, in agree-

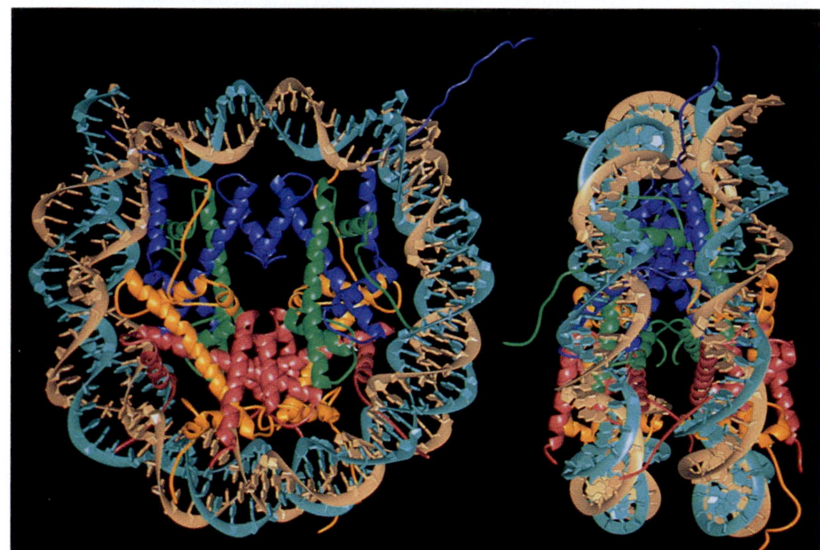

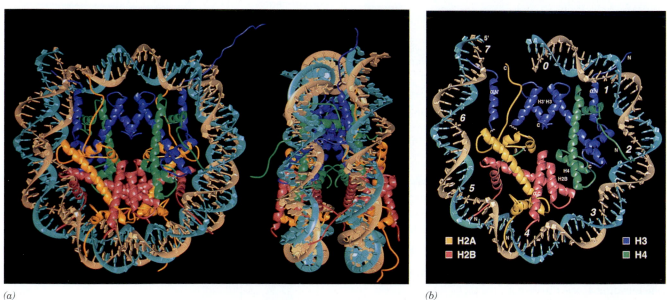

(*a*) (*b*)

FIGURE 34-7 X-Ray structure of the nucleosome core particle. (*a*) The entire core particle as viewed (*left*) along its superhelical axis and (*right*) rotated 90° about the vertical axis. The proteins of the histone octamer are drawn in ribbon form with H2A yellow, H2B red, H3 blue, and H4 green. The sugar–phosphate backbones of the 146-bp DNA are drawn as tan and cyan ribbons whose attached bases are represented by polygons of the same color. In both views, the pseudo-twofold axis is vertical and passes through the DNA center at the top. (*b*) The top half of the nucleosome core particle as viewed in Part *a, left,* and identically colored. The numbers 0 through 7 arranged about the inside of the 73-bp DNA superhelix mark the positions of sequential double helical turns. Those histones that are drawn in their entirety are primarily associated with this DNA segment, whereas only fragments of H3 and H2B from the other half of the particle are shown. The two four-helix bundles shown are labeled H3′ H3 and H2B H4. [Courtesy of Timothy Richmond, Eidgenössische Technische Hochschule, Zürich, Switzerland. PDBid 1AOI.]

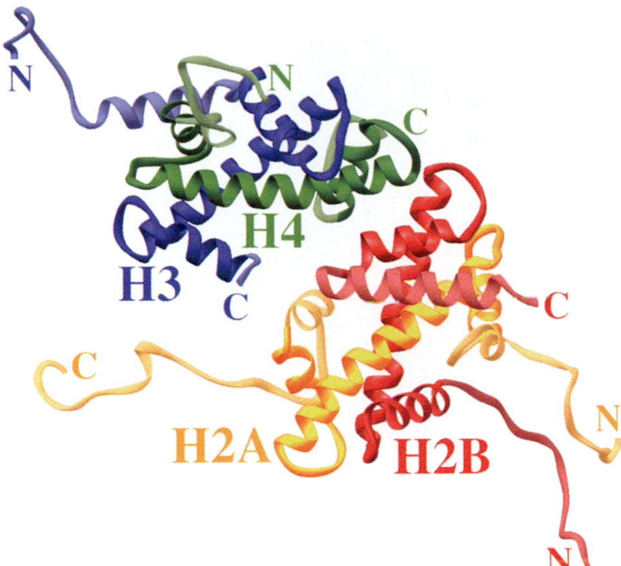

FIGURE 34-8 X-Ray structure of a histone octamer within the nucleosome core particle. Those portions of H2A, H2B, H3, and H4 that form the histone folds are yellow, red, blue, and green, respectively, with their N- and C-terminal tails colored in lighter shades. [Based on an X-ray structure by Gerard Bunick, University of Tennessee and Oak Ridge National Laboratory, Oak Ridge, Tennessee. PDBid 1EQZ.]

ment with previous biochemical studies, extend beyond the DNA; they exit the nucleosome between the gyres of the DNA superhelix, with those from H2B and H3 doing so in channels formed by two vertically aligned minor grooves. Those portions of the N-terminal tails that extend past the DNA are largely unstructured, that is, they are devoid of secondary structure, and substantial portions of their N-terminal segments are disordered. Nevertheless, one of the H4 N-terminal tails makes multiple hydrogen bonds and salt bridges with a highly negatively charged region on an H2A–H2B dimer of an adjacent nucleosome in the crystal structure. In addition, an H2A N-terminal tail of one nucleosome interacts with both the DNA and the H2A N-terminal tail of a neighboring nucleosome. Solution studies indicate, moreover, that the N-terminal tails interact with linker DNA. We shall see in Section 34-3B that the modulation of these interactions by the extensive and varied posttranslational modifications of the N-terminal tails listed above are implicated in facilitating chromatin unfolding to make its component DNA available to participate in such essential processes as transcription, DNA replication, recombination, and DNA repair. Among the foregoing histones, only H2A has an extensive C-terminal tail (39 residues), although it is entirely contained within the body of the nucleosome core and hence is unlikely to participate in internucleosomal interactions.

The archaeon *Methanothermus fervidus* expresses two closely related proteins that form a spheroidal complex with DNA, which presumably functions to prevent the thermal denaturation of this hyperthermophile's DNA. These proteins are ~30% identical in sequence to the histone fold domains of the histone octamer but lack their N- and C-terminal tails. Evidently, these tails have been added to the histone fold during the course of evolution.

b. Histone H1 "Seals Off" the Nucleosome

In the micrococcal nuclease digestion of nucleosomes, the ~200-bp DNA is first degraded to 166 bp. Then there is a pause before histone H1 is released from the chromatosomes and the DNA is further shortened to 146 bp. The 2-fold symmetry of the core particle suggests that the reduction in length of the 166-bp DNA comes about by the removal of 10 bp from each of its two ends. Since the 146-bp DNA of the core particle makes 1.8 superhelical turns, the 166-bp intermediate should be able to make two full superhelical turns, which would bring its two ends as close together as possible. Aaron Klug therefore proposed that histone H1 binds to nucleosomal DNA in a cavity formed by the central segment of its DNA and the segments that enter and leave the core particle (Fig. 34-9). This model is supported by the observation that in chromatin filaments containing H1, the DNA enters and leaves the nucleosome on the same side (Fig. 34-10*a*), whereas in H1-depleted chromatin, the entry and exit points are more randomly distributed and tend to occur on opposite sides of the nucleosome (Fig. 34-10*b*). The model also suggests that the length of the linker DNA is controlled by the subspecies of histone H1, which are collectively known as **linker histones,** bound to it.

Histone H5 is a variant of histone H1 that has several Lys → Arg substitutions and binds chromatin more tightly. The observations that the expression of histone H5 in rat sarcoma cells inhibits DNA replication, thereby arresting cells in the G1 phase of the cell cycle, and that histone H5

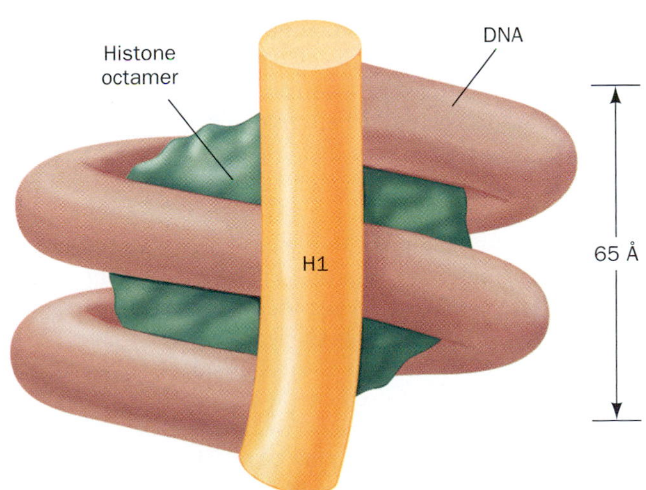

FIGURE 34-9 Model of the interaction of histone H1 with the DNA of the 166-bp chromatosome. The DNA's two complete superhelical turns enable H1 to bind to the DNA's two ends and its middle. Here the histone octamer is represented by the central spheroid (*green*) and the H1 molecule is represented by the cylinder (*yellow*).

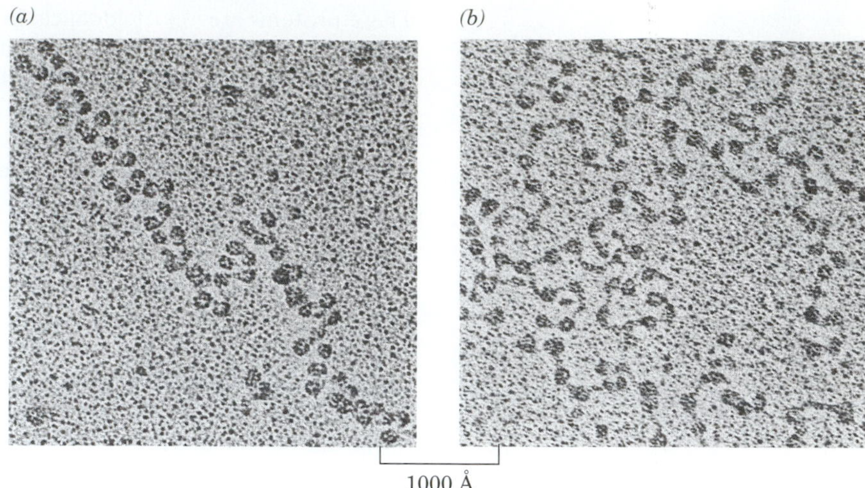

FIGURE 34-10 Electron micrographs of chromatin. (*a*) H1-containing chromatin and (*b*) H1-depleted chromatin, both in 5 to 15 m*M* salt. [Courtesy of Fritz Thoma, Eidgenössische Technische Hochschule, Zürich, Switzerland.]

more closely resembles **histone H1°** (a histone H1 variant that occurs in terminally differentiated cells) than does histone H1 itself, suggest that histone H5 is associated with replicationally and transcriptionally inactive chromatin.

Linker histones consist of a highly conserved globular, trypsin-resistant domain that is flanked by extended N- and C-terminal arms that are rich in basic residues. These basic arms, which comprise more than half of the intact protein, are therefore thought to interact with the linker DNA connecting adjacent nucleosomes even though it is the globular domain that is required for the binding of histone H1 to the nucleosome.

c. The Globular Domain of Histone H5 Structurally Resembles CAP Protein

V. Ramakrishnan has determined the X-ray structure of **GH5,** an 89-residue polypeptide that contains the 81-residue globular domain of histone H5 (although its five N-terminal and eleven C-terminal residues are disordered). The polypeptide chain folds into a 3-helix bundle with a 2-stranded β sheet at its C-terminus (Fig. 34-11). This structure and, in particular, its 3-helix bundle, is strikingly similar in conformation to that of the helix–turn–helix (HTH) motif-containing DNA-binding domain of *E. coli* catabolite activator protein (CAP; Fig. 31-28). Thus, even though there is little sequence identity between GH5 and CAP, their similar structures suggest that GH5 binds DNA in a manner analogous to CAP. Indeed, a model of the GH5–DNA complex based on the known X-ray structure of the CAP–DNA complex (Section 31-3C) positions GH5's highly conserved Lys 69, Arg 73, and Lys 85 side chains to interact with the DNA (Fig. 34-11). These residues, which all have counterparts in CAP, are protected against chemical modification in chromatin. Moreover, GH5 contains a cluster of four conserved basic residues on the opposite face of the protein from its "recognition helix," which could interact with a second segment of duplex

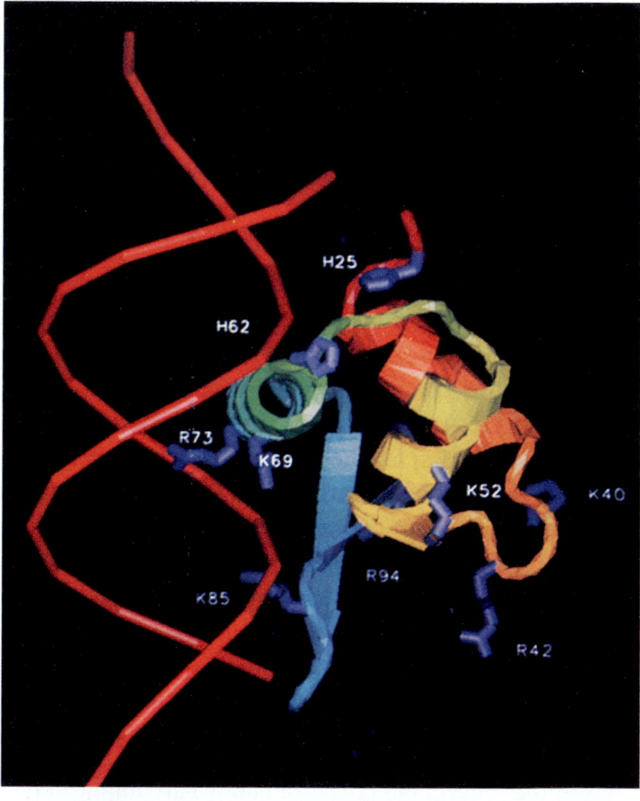

FIGURE 34-11 X-Ray structure of GH5 shown in hypothetical complex with DNA. This model was constructed by superimposing the structure of GH5 on that of CAP in the CAP–DNA structure (Fig. 31-28*a*). However, to avoid any presumptions about the nature of the DNA, that in the CAP structure, which is bent, was replaced by ideal B-DNA, which is represented here by its phosphate backbone (*red*). GH5 is shown in ribbon form and is color-ramped from red to blue going from its N- to its C-terminus. Conserved basic residues, as well as two His residues that have been cross-linked to DNA, are shown in stick form (*blue*). [Courtesy V. Ramakrishnan, MRC Laboratory of Molecular Biology, Cambridge, U.K. PDBid 1HST.]

DNA in agreement with the experimental evidence that GH5 simultaneously binds two DNA duplexes.

d. Parental Nucleosomes Are Transferred to Daughter Duplexes on DNA Replication

The *in vivo* replication of eukaryotic DNA is accompanied by its packaging into chromatin; that is, it is the chromatin that actually is replicated. What, then, is the fate of the histone octamers originally associated with the parental DNA? There are several possibilities: The "parental" octamers may remain associated with either the leading strand or the lagging strand, or they may be partitioned between the two daughter DNA duplexes, either at random or in some systematic way. Attempts to resolve this issue have yielded contradictory results. However, the weight of the evidence now indicates that parental octamers are distributed at random between the daughter duplexes. Moreover, the parental octamers remain associated with DNA during the replication process instead of dissociating from the parental DNA and later rebinding the daughter duplexes. Thus, nucleosomes either open up to permit the passage of a replication fork or parental histone octamers immediately in front of an advancing replication fork are somehow transferred to the daughter duplexes immediately behind the replication fork.

e. Nucleosome Assembly Is Facilitated by Molecular Chaperones

How are nucleosomes formed *in vivo*? *In vitro,* at high salt concentrations, nucleosomes self-assemble from the proper mixture of DNA and histones. In fact, when only H3, H4, and DNA are present, the mixture forms nucleosome-like particles by the deposition of $(H3)_2(H4)_2$ tetramers onto the DNA. Nucleosome cores are then formed by the recruitment of H2A–H2B dimers to these particles.

At physiological salt concentrations, *in vitro* nucleosome assembly occurs much more slowly than at high salt concentrations and, unless the histone concentrations are carefully controlled, is accompanied by considerable histone precipitation. However, in the presence of **nucleoplasmin,** an acidic protein that has been isolated from *X. laevis* oocyte nuclei, and DNA topoisomerase I (Section 29-3C), nucleosome assembly proceeds rapidly without histone precipitation. Nucleoplasmin binds to histones but neither to DNA nor to nucleosomes. Evidently, *nucleoplasmin functions as a molecular chaperone (Section 9-2C) to bring histones and DNA together in a controlled fashion, thereby preventing their nonspecific aggregation through their otherwise strong electrostatic interactions.* The topoisomerase I, no doubt, acts to provide the nucleosome with its preferred level of supercoiling.

C. *30-nm Filaments: The Second Level of Chromatin Organization*

The 166-bp nucleosomal DNA has a packing ratio of ~7 (its 560-Å contour length is wound into an ~80-Å-high supercoil). Clearly, the filament of nucleosomes, which only occurs at low ionic strengths (and hence is unlikely to have an independent existence *in vivo*), represents only the first level of chromosomal DNA compaction. Only at physiological ionic strengths does the next level of chromosomal organization become apparent.

As the salt concentration is raised, the H1-containing nucleosome filament initially folds to a zigzag conformation (Fig. 34-10*a*), whose appearance suggests that nucleosomes interact through contacts between their H1 molecules. Then, as the salt concentration approaches the physiological range, chromatin forms a 30-nm-thick filament in which the nucleosomes are visible (Fig. 34-12). Klug proposed that the 30-nm filament is constructed by winding the 10-nm nucleosome filament into a solenoid with ~6 nucleosomes per turn and a pitch of 110 Å (the diameter of a nucleosome; Fig. 34-13). The solenoid is stabilized by H1 molecules whose relatively variable, extended N-terminal and C-terminal arms (which are absent in GH5; Fig. 34-11) are thought to contact adjacent nucleosomes, at least in part by interacting with neighboring H1s in a head-to-tail fashion. This model, which is consistent with the X-ray diffraction pattern of the 30-nm filaments, has a packing ratio of ~40 (6 nucleosomes, each with ~200 bp DNA, rising a total of 110 Å). Note, however, that several other plausible models for the 30-nm chromatin filament have also been formulated. Indeed, Kensal van Holde has argued that the 30-nm filament does not have a regular structure but rather, because of the varying lengths of the presumably straight and stiff linker DNAs connecting the nucleosome cores, has an irregular helix-like structure that simulations indicate forms a filament with an average diameter of 30 nm. This would account for

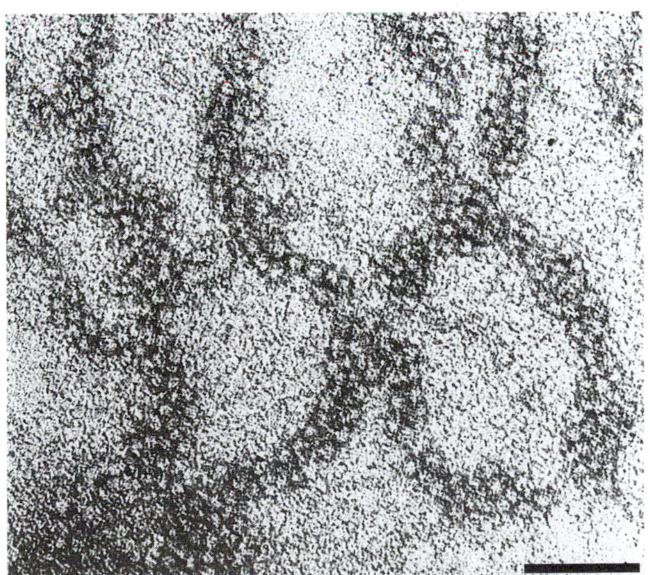

FIGURE 34-12 Electron micrograph of the 30-nm chromatin filaments. Note that the filaments are two to three nucleosomes across. The bar represents 1000 Å. [Courtesy of Jerome B. Rattner, University of Calgary, Canada.]

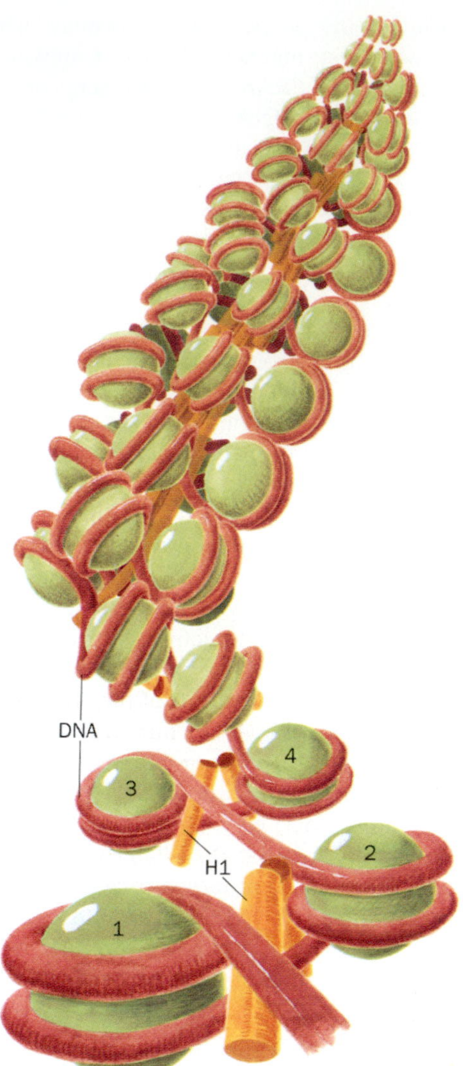

FIGURE 34-13 Model of the 30-nm chromatin filament. The filament is represented (*bottom to top*) as it might form with increasing salt concentration. The zigzag pattern of nucleosomes (*1, 2, 3, 4*) closes up to form a solenoid with ~6 nucleosomes per turn. The H1 molecules (*yellow cylinders*), which stabilize the structure, are thought to form a helical polymer running along the center of the solenoid.

the difficulty in experimentally determining the structure of the 30-nm filament despite numerous attempts to do so over nearly three decades.

D. *Radial Loops: The Third Level of Chromatin Organization*

Histone-depleted metaphase chromosomes exhibit a central fibrous protein matrix or scaffold surrounded by an extensive halo of DNA (Fig. 34-14*a*). The strands of DNA that can be followed are observed to form loops that enter and exit the scaffold at nearly the same point (Fig. 34-14*b*). Most of these loops have lengths in the range 15 to 30 μm (which corresponds to 45–90 kb), so that when condensed as 30-nm filaments they would be ~0.6 μm long. Electron

(*a*)

(*b*)

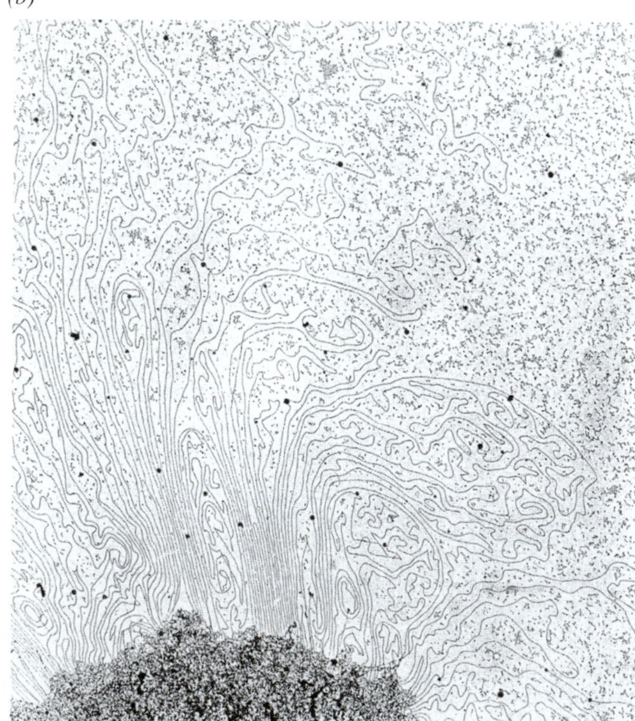

FIGURE 34-14 Electron micrographs of a histone-depleted metaphase human chromosome. (*a*) The central protein matrix (scaffold) serves to anchor the surrounding DNA. (*b*) At higher magnification it can be seen that the DNA is attached to the scaffold in loops. [Courtesy of Ulrich Laemmli, University of Geneva, Switzerland.]

micrographs of chromosomes in cross section, such as Fig. 34-15*a*, strongly suggest that the chromatin fibers of metaphase chromosomes are radially arranged. If the observed loops correspond to these radial fibers, they would each contribute 0.3 μm to the diameter of the chromosome (a fiber must double back on itself to form a loop). Taking into account the 0.4-μm width of the scaffold, this model predicts the diameter of the metaphase chromosome to be 1.0 μm, in agreement with observation (Fig. 34-15*b*). A typical human chromosome, which contains ~140 million bp, would therefore have ~2000 of these ~70-kb radial loops. The 0.4-μm-diameter scaffold of such a chromosome has sufficient surface area along its 6-μm length to bind this number of radial loops. The radial loop model therefore accounts for DNA's observed packing ratio in metaphase chromosomes.

The radial DNA loops are attached to the matrix via AT-rich **matrix-associated regions [MARs;** alternatively, **scaffold attachment regions (SARs)].** The radial loops are therefore also known as **structural domains.** Nevertheless, little is known about how the matrix is composed, how the radial loops are organized, and how metaphase chromosomes and the far more dispersed interphase chromosomes interconvert. Certainly, **nonhistone proteins,** whose thousands of varieties constitute ~10% of chromosomal proteins, must participate in these processes. Moreover, since the protein machinery controlling gene expression in a given structural domain is unlikely to directly affect expression in a neighboring structural domain, the structural domains probably comprise the chromosomal transcriptional units.

(a)

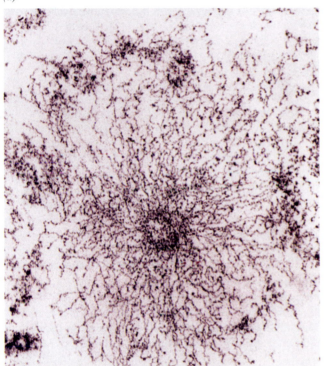

(b)

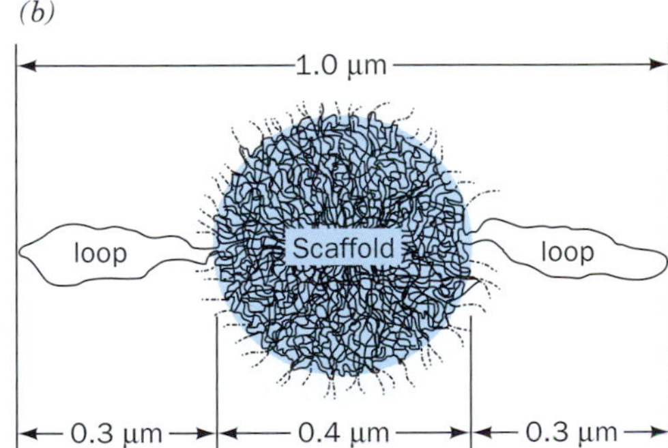

FIGURE 34-15 Organization of DNA in a metaphase chromosome. (*a*) Electron micrograph of a human metaphase chromosome in cross section. Note the mass of chromatin fibers radially projecting from the central scaffold. [Courtesy of Ulrich Laemmli, University of Geneva, Switzerland.] (*b*) Interpretive diagram indicating how the 0.3-μm-long radial loops are thought to combine with the 0.4-μm-wide scaffold to form the 1.0-μm-diameter metaphase chromosome.

E. *Polytene Chromosomes*

The diffuse structure of most interphase chromosomes (Fig. 34-2) makes it all but impossible to characterize them at the level of individual genes. Nature, however, has greatly ameliorated this predicament through the production of "giant" banded chromosomes in certain terminally differentiated (nondividing) secretory cells of dipteran (two-winged) flies (Fig. 34-16). These chromosomes, of which those from the salivary glands of *D. melanogaster* larvae are the most extensively studied, are produced by multiple replications of a synapsed (joined in parallel) diploid pair in which the replicas remain attached to one another and in register (see below). Each diploid pair may replicate in this manner as many as nine times so that the final **polytene chromosome** contains up to $2 \times 2^9 = 1024$ DNA strands. The function(s) of polytene chromosomes is unknown although perhaps this permits a greatly increased rate of transcription of certain genes.

D. melanogaster's four giant chromosomes have an aggregate length of ~2 mm so that its haploid genome of 1.37×10^8 bp has an average packing ratio in these chromosomes of almost 25. About 95% of this DNA is concentrated in chromosomal bands (Fig. 34-17). These bands (more properly, **chromomeres**), as microscopically visualized through staining, form a pattern that is characteristic of each *D. melanogaster* strain. Indeed, chromosomal rearrangements such as duplications, deletions, and inversions result in a corresponding change in the banding pattern. *A polytene chromosome's banding pattern therefore forms a cytological map that parallels its genetic map.*

The characteristic banding pattern of each polytene chromosome suggests that its component DNA molecules are precisely aligned. This hypothesis was corroborated by the application of ***in situ*** (on site) **hybridization.** In this technique, developed by Mary Lou Pardue and Joseph Gall, an immobilized chromosome preparation is treated with NaOH to denature its DNA; it is then hybridized with a purified species of radioactively labeled mRNA (or its

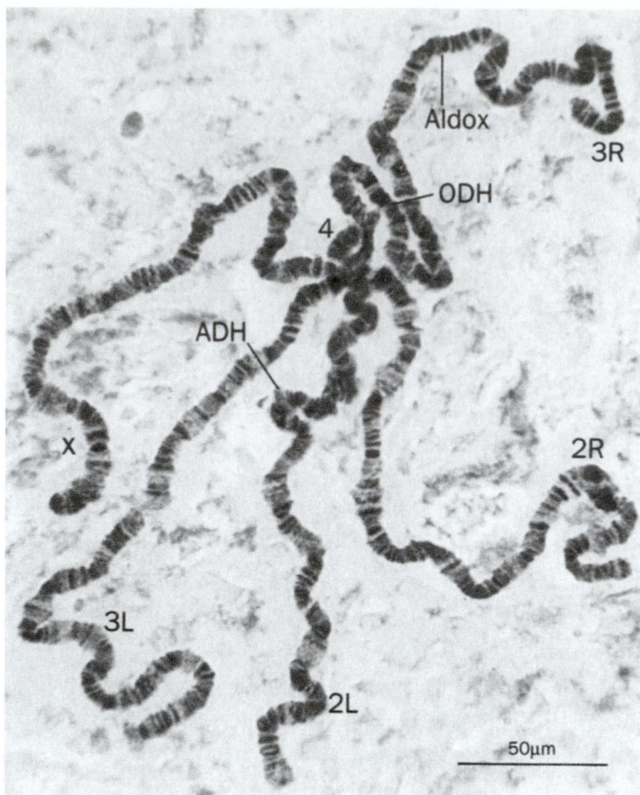

FIGURE 34-16 Photomicrograph of the stained polytene chromosomes from the *D. melanogaster* salivary gland. Such chromosomes consist of darkly staining bands interspersed with light-staining interband regions. All four chromosomes in a single cell are held together by their centromeres. The chromosomal positions for the genes specifying alcohol dehydrogenase (ADH), aldehyde oxidase (Aldox), and octanol dehydrogenase (ODH) are indicated. [Courtesy of B.P. Kaufmann, University of Michigan.]

corresponding cDNA), and the chromosomal binding site of the radioactive probe is determined by autoradiography. A given mRNA hybridizes with one, or no more than a few, chromosomal bands (Fig. 34-18).

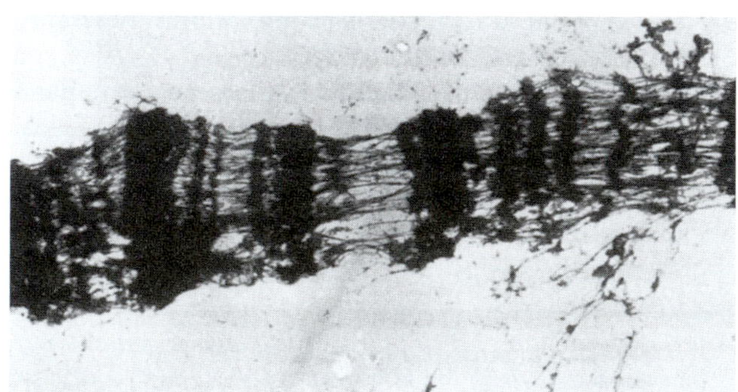

FIGURE 34-17 Electron micrograph of a segment of polytene chromosome from *D. melanogaster*. Note that its interband regions consist of chromatin fibers that are more or less parallel to the long axis of the chromosome, whereas its bands, which contain ~95% of the chromosome's DNA, are much more highly condensed. [Courtesy of Gary Burkholder, University of Saskatechewan, Canada.]

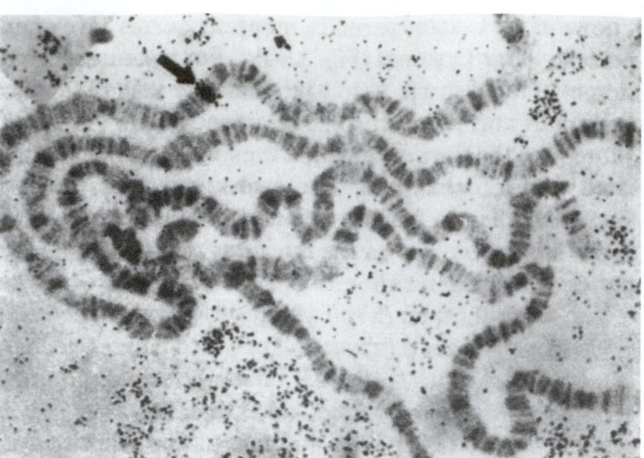

FIGURE 34-18 Autoradiograph of a *D. melanogaster* polytene chromosome that has been *in situ* hybridized with yolk protein cDNA. The dark grains (*arrow*) identify the chromosomal location of the yolk protein gene. [From Barnett, T., Pachl, C., Gergen, J.P., and Wensink, P.C., *Cell* **21,** 735 (1980). Copyright © 1980 by Cell Press.]

D. melanogaster's four polytene chromosomes exhibit an aggregate of ~5000 bands. It originally appeared that the number of *D. melanogaster* genes was roughly equal to this number of bands and hence it was thought that each band corresponds to a single gene. However, the recently determined genome sequence of *D. melanogaster* indicates that it has ~13,000 genes, nearly three times its number of bands. In fact, genes have been shown to be located in both band and interband regions, with some bands containing several genes and others containing none. Thus, it is likely that the banding pattern of polytene chromosomes is a consequence of different levels of gene expression due to variations in chromatin structure (see Section 34-3B), with the genes in the relatively open interband regions presumably more highly expressed than those in the more condensed and hence less accessible bands.

2 ■ GENOMIC ORGANIZATION

Higher organisms contain a great variety of cells that differ not only in their appearances (e.g., Fig. 1-10) but in the proteins they synthesize. Pancreatic acinar cells, for example, synthesize copious quantities of digestive enzymes, including trypsin and chymotrypsin, but no insulin, whereas the neighboring pancreatic β cells produce large amounts of insulin but no digestive enzymes. Clearly, each of these different types of cells expresses different genes. Yet most of a multicellular organism's somatic cells contain the same genetic information as the fertilized ovum from which they are descended (a phenomenon described as **totipotency**) as is demonstrated, for example, by the ability to raise a mammal such as a sheep, cow, or mouse from an enucleated oocyte into which the nucleus from an adult cell had been inserted. Similarly, a single cell from a plant can give rise to the normal plant. Evidently, cells have enormous expressional flexibility. Nevertheless, only a small fraction of the DNA in higher eukaryotic genomes is expressed. What is the nature of the remaining unexpressed sequences and do they have any function? In this section we describe the genetic organization of the eukaryotic chromosome. How eukaryotic gene expression is controlled is the subject of Section 34-3.

A. *The C-Value Paradox*

One might reasonably expect the morphological complexity of an organism to be roughly correlated with the amount of DNA in its haploid genome, its **C value.** After all, the morphological complexity of an organism must reflect an underlying genetic complexity. Nevertheless, in what is known as the **C-value paradox,** many organisms have unexpectedly large C values (Fig. 34-19). For instance, the genomes of lungfish are 10 to 15 times larger than of those of mammals and those of some salamanders are yet larger. Moreover, the C-value paradox even applies to closely related species; for example, the C values for several species of *Drosophila* have a 2.5-fold spread. Does the "extra" DNA in the larger genomes have a function, and if not, why is it preserved from generation to generation?

The 4.6 million-bp *E. coli* genome encodes ~4300 proteins. In contrast, the 3.2 billion-bp haploid human genome, which is ~700 times larger than that of *E. coli,* is estimated to encode ~30,000 proteins; that is, humans have only ~7 times as many structural genes as do *E. coli.* Certainly the control of genetic expression in eukaryotes is a far more elaborate process than it is in prokaryotes. Yet does much of the unexpressed DNA in the human genome, at least 98% of the total, function in the control of genetic expression? The recent and ongoing eukaryotic genome sequence determinations are beginning to provide answers to the foregoing questions.

a. C_0t Curve Analysis Indicates DNA Complexity

The rate at which DNA renatures is indicative of the lengths of its unique sequences. If DNA is sheared into uniform fragments of 300 to 10,000 bp (Section 5-3D), denatured, and kept at a low concentration so that the effects of mechanical entanglement are small, the rate-determining step in renaturation is the collision of complementary sequences. Once the complementary sequences have found each other through random diffusion, they rapidly zip up to form duplex molecules. The rate of re-

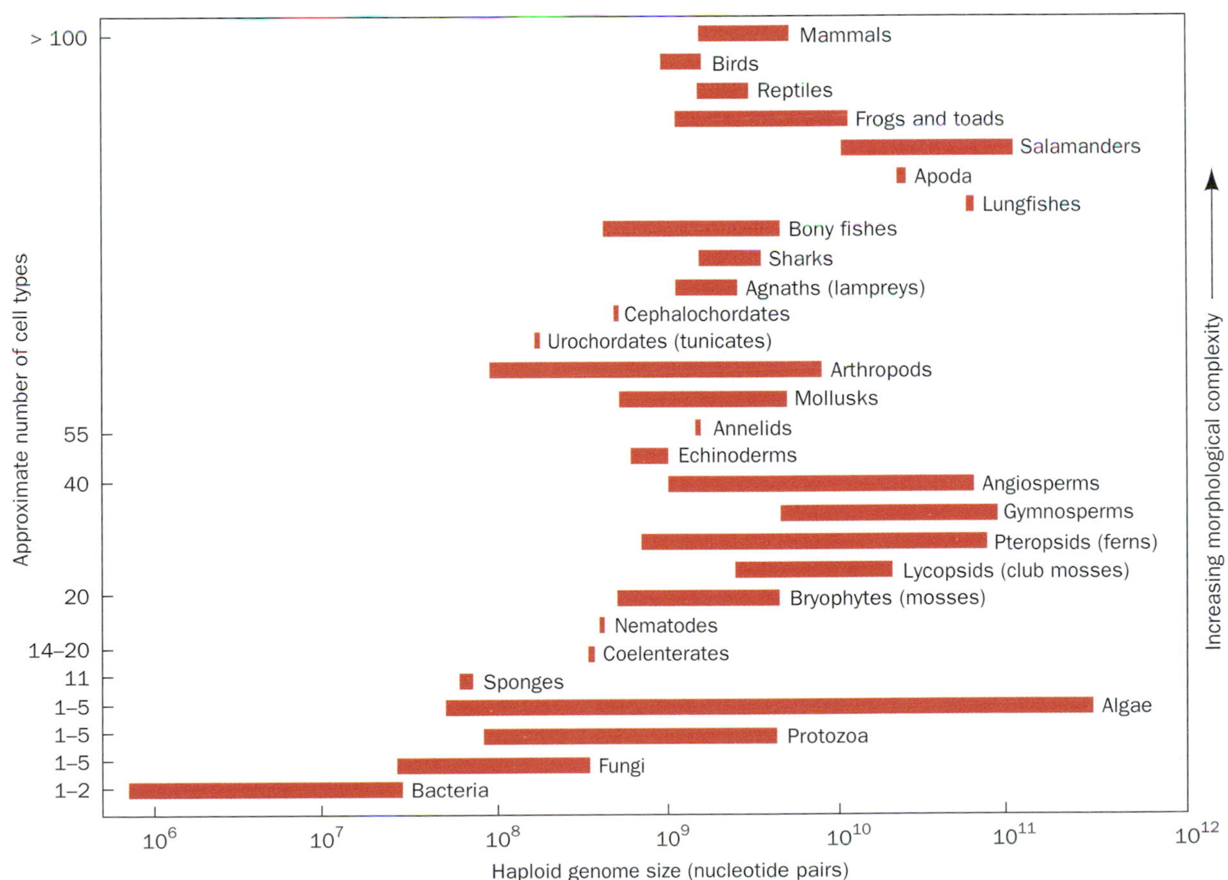

FIGURE 34-19 The range of haploid genome DNA contents in various categories of organisms indicating the C-value paradox. The morphological complexity of the organisms, as estimated according to their number of cell types, increases from bottom to top. [After Raff, R.A. and Kaufman, T.C., *Embryos, Genes, and Evolution, p.* 314, Macmillan (1983).]

naturation of denatured DNA is therefore expressed

$$\frac{d[\mathrm{A}]}{dt} = -k[\mathrm{A}][\mathrm{B}] \quad [34.1]$$

where A and B represent complementary single-stranded sequences and k is a second-order rate constant (Section 14-1B). Since [A] = [B] for duplex DNA, Eq. [34.1] integrates to

$$\frac{1}{[\mathrm{A}]} = \frac{1}{[\mathrm{A}]_0} + kt \quad [34.2]$$

where $[\mathrm{A}]_0$ is the initial concentration of A.

It is convenient to measure the fraction f of unpaired strands:

$$f = \frac{[\mathrm{A}]}{[\mathrm{A}]_0} \quad [34.3]$$

Combining Eqs. [34.2] and [34.3] yields

$$f = \frac{1}{1 + [\mathrm{A}]_0 kt} \quad [34.4]$$

The concentration terms in these equations refer to unique sequences since the collision of noncomplementary sequences does not lead to renaturation. Hence, if C_0 is the initial concentration of base pairs in solution, then

$$[\mathrm{A}]_0 = \frac{C_0}{x} \quad [34.5]$$

where x is the number of base pairs in each unique sequence and is known as the DNA's **complexity.** For example, the repeating sequence $(\mathrm{AGCT})_n$ has a complexity of 4, whereas an *E. coli* chromosome, which consists of ~4.6 million bp of unrepeated sequence, has a complexity of ~4.6 million. Combining Eqs. [34.4] and [34.5] yields

$$f = \frac{1}{1 + C_0 kt/x} \quad [34.6]$$

When half of the molecules in the sample have renatured, $f = 0.5$, so that

$$C_0 t_{1/2} = \frac{x}{k} \quad [34.7]$$

where $t_{1/2}$ is the time for this to occur. The rate constant k is characteristic of the rate at which single strands collide in solution under the conditions employed, so it is independent of the complexity of the DNA and, for reasonably short DNA fragments, the length of a strand. Consequently, *for a given set of conditions, the value of* $C_0 t_{1/2}$ *depends only on the complexity x of the DNA.* This situation is indicated in Fig. 34-20, which is a series of plots of f versus $C_0 t$ for various DNAs. Such plots are referred to as $C_0 t$ (pronounced "cot") curves. The complexities of the DNAs in Fig. 34-20 vary from 1 for the synthetic duplex poly(A) · poly(U) to ~3 × 10⁹ for some fractions of mammalian DNAs. Their corresponding values of $C_0 t_{1/2}$ vary accordingly.

The speed and sensitivity of $C_0 t$ curve analysis is greatly enhanced through the hydroxyapatite fractionation of the renaturing DNA. Hydroxyapatite, it will be recalled (Section 6-6B), binds double-stranded DNA at a higher phosphate concentration than it binds single-stranded DNA. The single- and double-stranded DNAs in a solution of renaturing DNA may therefore be separated by hydroxyapatite chromatography and the amounts of each measured. The single-stranded DNA can then be further renatured and the process repeated. If the renaturing DNA is radioactively labeled, much smaller quantities of it can be detected than is possible by spectroscopic means. Thus,

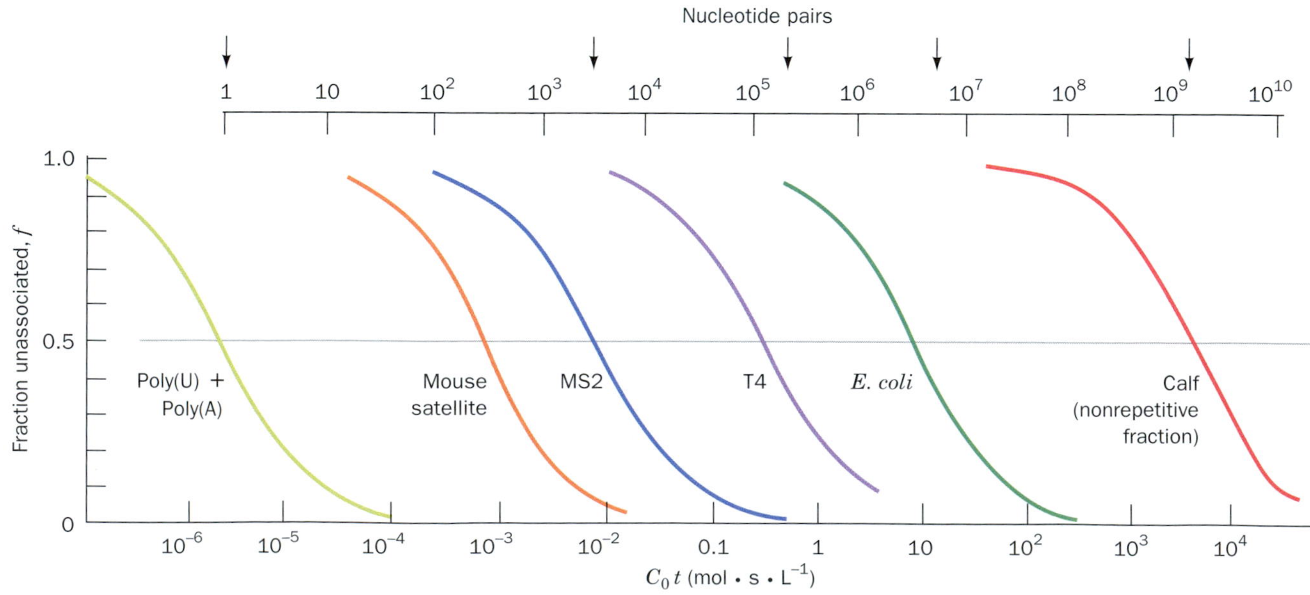

FIGURE 34-20 The reassociation ($C_0 t$) curves of duplex DNAs from the indicated sources. The DNA was dissolved in a solution containing 0.18*M* Na^+ and sheared to an average length of 400 bp. The upper scale indicates the genome sizes of some of the DNAs (**MS2** and **T4** are bacteriophages). [After Britten, R.J. and Kohne, D.E., *Science* **161,** 530 (1968).]

through the hydroxyapatite chromatography of radioactively labeled DNA, the C_0t curve analysis of a DNA of such a high complexity that its $t_{1/2}$ is days or weeks can be conveniently measured in a small fraction of that time.

B. *Repetitive Sequences*

Consider a sample of DNA that consists of sequences with varying degrees of complexity. Its C_0t curve, Fig. 34-21 for example, is the sum of the individual C_0t curves for each complexity class of DNA. *C_0t curve analysis (and more recently, genome sequencing) has demonstrated that viral and prokaryotic DNAs have few, if any, repeated sequences (e.g., Fig. 34-20 for MS2, T4, and E. coli).* In contrast, *eukaryotic DNAs exhibit complicated C_0t curves (e.g., Fig. 34-22)* that must arise from the presence of DNA segments of several different complexities.

Eukaryotic C_0t curves may be attributed to the presence of five somewhat arbitrarily defined classes of DNAs: (1) **inverted repeats,** (2) **highly repetitive sequences** ($>10^6$ copies per haploid genome), (3) **moderately repetitive sequences** ($<10^6$ copies per haploid genome), (4) **segmental duplications** (blocks of 1–200 kb that have been copied to one or more regions of the genome that may be within the same chromosome or on different chromosomes; they constitute ~5% of the human genome), and (5) **unique sequences** (~1 copy per haploid genome). The sequences and chromosomal distributions of these DNA segments vary with the species, so a unifying description of their arrangements cannot be made. Nevertheless, several broad generalizations are possible, as we shall see below.

a. Inverted Repeats Form Foldback Structures

The most rapidly reassociating eukaryotic DNA, which represents as much as 10% of some genomes, renatures with first-order kinetics. This DNA contains inverted (self-complementary) sequences in close proximity, which can

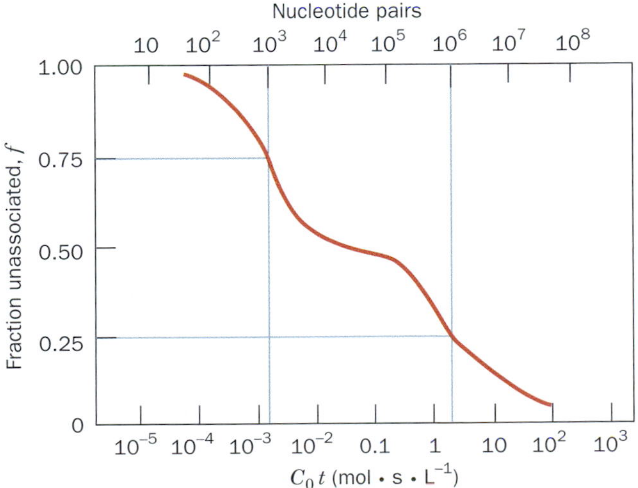

FIGURE 34-21 C_0t curve of a hypothetical DNA molecule. Before fragmentation, this DNA was 2 million bp in length and consisted of a unique sequence of 1 million bp and 1000 copies of a 1000-bp sequence. Note the curve's biphasic nature.

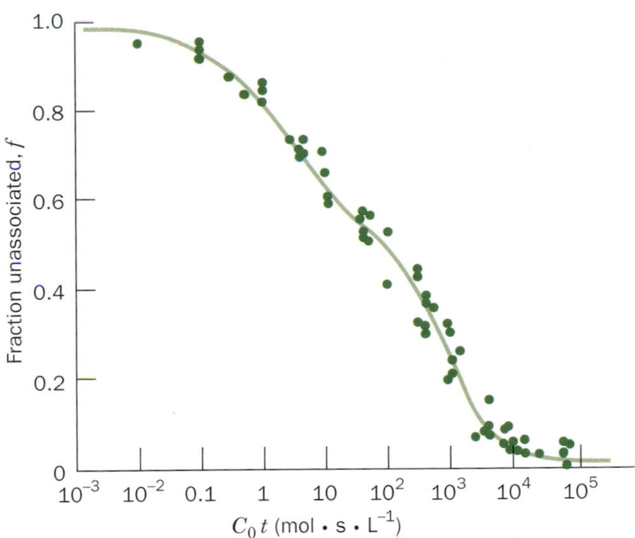

FIGURE 34-22 The C_0t curve of *Strongylocentrotus purpuratus* (a sea urchin) DNA. [After Galau, G.A., Britten, R.J., and Davidson, E.H., *Cell* **2,** 11 (1974).]

fold back on themselves to form hairpinlike **foldback structures** (Fig. 34-23*a*). Inverted sequences may be isolated by adsorbing the duplex DNA formed at very low C_0t values to hydroxyapatite and subsequently degrading its single-stranded loop and tails with **S1 nuclease** (an endonuclease from *Aspergillus oryzae* that preferentially cleaves single strands). The resulting inverted repeats range in length from 100 to 1000 bp, sizes much too large to have evolved at random. *In situ* hybridization studies on metaphase chromosomes using these inverted repeats as probes indicate that they are distributed at many chromosomal sites.

The function of inverted repeats, some 2 million of which occur in the human genome, is unknown. However, since the cruciform structures formed by paired foldback structures (Fig. 34-23*b*) are only slightly less stable than the corresponding normal duplex DNA, it has been suggested that the inverted repeats function in chromatin as some sort of molecular switch.

b. Highly Repetitive DNA Is Clustered at Telomeres and Centromeres

Highly repetitive DNA consists of short sequences that are tandemly repeated, either perfectly or slightly imperfectly, often thousands of times. Such **simple sequence repeats [(SSRs);** alternatively, **short tandem repeats (STRs)]** can often be separated from the bulk of the chromosomal DNA by shear degradation followed by density gradient ultracentrifugation in CsCl since their distinctive base compositions cause them to form "satellites" to the main DNA band (Fig. 34-24; recall that the buoyant density of DNA in CsCl increases with its G + C content; Section 6-6D). The sequences of these SSRs, which are also known as **satellite DNAs,** are species specific (SSRs with a short repeat unit of n = 1–13 nt are often called **microsatellites,** whereas those with n = 14–500 are often called **minisatellites**). For example, the crab *Cancer borealis* has an SSR comprising 30% of its genome in which the repeating unit is the dinucleotide

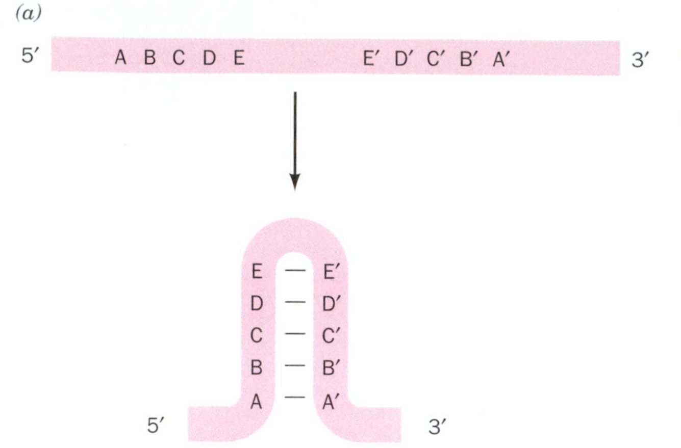

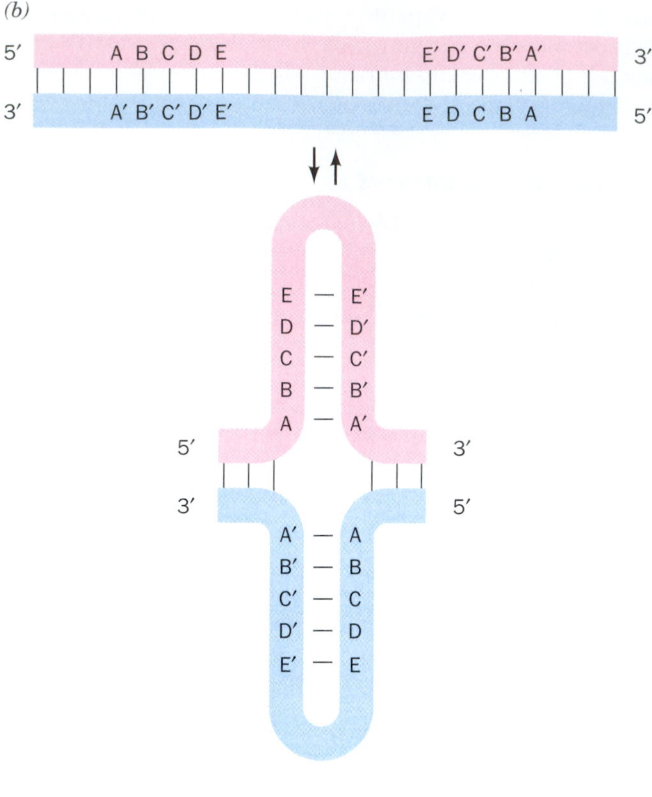

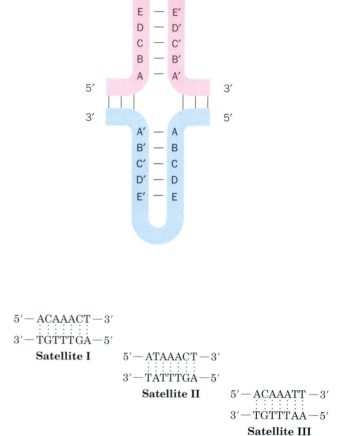

FIGURE 34-23 Foldback structures in DNA. (*a*) Single-stranded DNA containing an inverted repeat will, under renaturing conditions, form a base paired loop known as a foldback structure. Here A is complementary to A′, B is complementary to B′, etc. (*b*) An inverted repeat in duplex DNA could assume a cruciform conformation consisting of two opposing foldback structures. The stability of this structure would be less than that of the corresponding duplex but only by the loss of the base pairing energy in the unpaired loops.

AT. The DNA of *Drosophila virilis* exhibits three satellite bands (Fig. 34-24), which each consist of a different although closely related repeating heptanucleotide sequence:

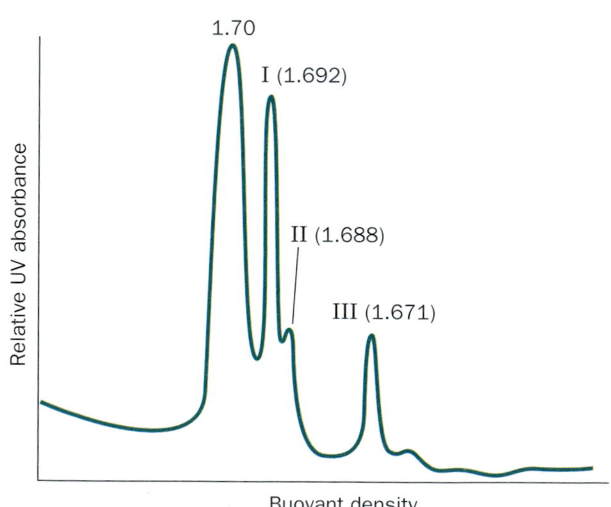

FIGURE 34-24 The buoyant density pattern of *Drosophila virilis* DNA centrifuged to equilibrium in neutral CsCl. Three prominent bands of satellite DNA (ρ = 1.692, 1.688, and 1.671 g/cm^3) are present, in addition to the main DNA band (ρ = 1.70 g/cm^3). [After Gall, J.G., Cohen, E.H., and Atherton, D.D., *Cold Spring Harbor Symp. Quant. Biol.* **38,** 417 (1973).]

5′—ACAAACT—3′
3′—TGTTTGA—5′
Satellite I

5′—ATAAACT—3′
3′—TATTTGA—5′
Satellite II

5′—ACAAATT—3′
3′—TGTTTAA—5′
Satellite III

These comprise 25, 8, and 8% of the 3.1 × 10^8-bp *D. virilis* genome, so that these sequences are repeated 11, 3.6, and 3.6 million times, respectively.

Telomeres, as we have seen (Section 30-4D), consist of G-rich SSRs. In addition, the *in situ* hybridization of mouse chromosomes with ^{3}H-labeled RNA synthesized on mouse simple sequence DNA templates established that SSR DNA is concentrated in the heterochromatic region associated with the chromosomal centromere [Fig. 34-25; the centromere is the constricted segment of the chromosome at which sister chromatids are joined (Fig. 34-1) and at which the chromosome attaches to the mitotic spindle (Fig. 1-19)]. This observation suggests that centromeric SSR DNA, which is not transcribed *in vivo,* functions to align homologous chromosomes during meiosis (Fig. 1-20) and/or to facilitate their recombination. This hypothesis is supported by the observation that satellite DNAs are largely or entirely eliminated from the somatic cells of a variety of eukaryotes (which are consequently no longer totipotent) but not from their germ cells.

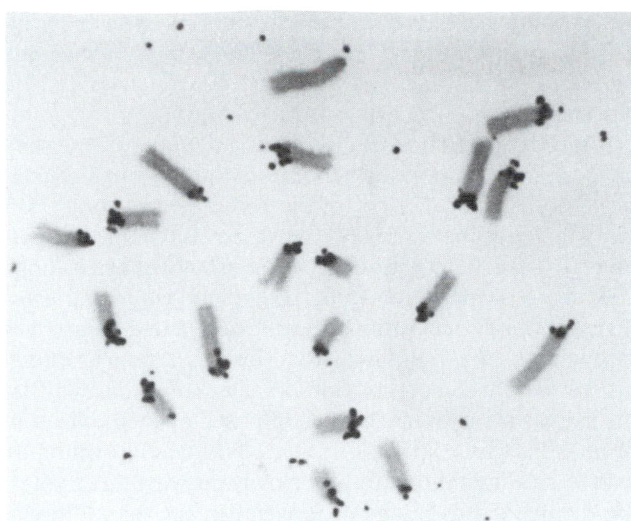

FIGURE 34-25 Autoradiograph of mouse chromosomes showing the centromeric location of their SSR DNA through *in situ* hybridization. Note that the centromeres in mouse chromosomes are all located at one end of the chromosome (no genes lie beyond the mouse centromeres). In human and yeast chromosomes, however, centromeres occupy more internal positions (e.g., Fig. 34-1), whereas *D. melanogaster* has chromosomes of both types. [Courtesy of Joseph Gall, Carnegie Institution of Washington.]

SSRs comprise ~3% of the human genome, with the greatest contribution (0.5%) provided by dinucleotide repeats, most frequently $(CA)_n$ and $(TA)_n$. SSRs appear to have arisen by template slippage during DNA replication. This occurs more frequently with short repeats, which therefore have a high degree of length polymorphism in the human population. Consequently, genetic markers based on the lengths of SSRs, particularly $(CA)_n$ repeats, have been a mainstay of human genetic studies (Section 5-5F).

c. Moderately Repetitive DNAs Are Arranged in Dispersed Repeats

Moderately repetitive DNAs occur in segments of 100 to several thousand base pairs that are interspersed with larger blocks of unique DNA. Some of this repetitive DNA consists of tandemly repeated groups of genes that specify products that cells require in large quantities, such as rRNAs, tRNAs, and histones. The organization of these repeated genes is discussed in Section 34-2D.

Around 42% of the human genome consists of retrotransposons (transposable elements that propagate through the intermediate synthesis of RNA; Section 30-6B). Three major types of retrotransposons inhabit the human genome (Table 34-2):

1. Long interspersed nuclear elements (LINEs), which comprise 20.4% of the human genome, are 6- to 8-kb-long segments that encode the proteins that mediate their transposition (Section 30-6B), although the vast majority (>99%) of LINEs have accumulated mutations that render them transpositionally inactive. LINEs, which are derived from RNA polymerase II–generated transcripts, are dispersed throughout the genomes of all mammals, which suggests that the ancestral LINE became associated with the mammalian genome very early in its evolution. The

TABLE 34-2 Moderately Repetitive Sequences in the Human Genome[a]

Type of Repeat	Number of Copies (× 1000)	Total Number of Nucleotides (Mb)	Fraction of the Genome Sequence (%)
LINEs	868	559	20.4
L1	516	462	16.9
L2	315	88	3.2
L3	37	8	0.3
SINEs	1558	360	13.1
Alu	1090	290	10.6
MIR	393	60	2.2
MIR3	75	9	0.3
LTR Retrotransposons	443	227	8.3
ERV-class I	112	79	2.9
ERV(K)-class II	8	8	0.3
ERV(L)-class III	83	40	1.4
MaLR	240	100	3.6
DNA Transposons	294	78	2.8
HAT group	195	42	1.6
Tc-1 group	75	32	1.2
Unclassified	22	3.2	0.1
Total		1227	44.8

[a]These numbers are approximate and are likely to be underestimates.

Source: International Human Genome Sequencing Consortium, *Nature* **409,** 880 (2001).

most common LINE in the human genome, **LINE-1 (L1),** consists of ~6.1 kb containing a 5′ untranslated region (UTR), two open reading frames (ORFs), the second of which contains a reverse transcriptase gene, and a 3′ UTR ending in a poly(A) tail. However, the average L1 has a length of ~900 bp because most L1s consist of truncated fragments of the full length retrotransposon. Two other LINEs occur in the human genome, **L2** and **L3,** which are distantly related to L1. However, L1 is the only LINE in the human genome that is still transpositionally active.

2. Short interspersed nuclear elements (SINEs), which comprise 13.1% of the human genome, consist of 100- to 400-bp elements that are derived from RNA polymerase III–generated transcripts. SINEs each contain an RNA polymerase III promoter but, in contrast to LINEs, do not encode proteins; they are apparently propagated by LINE-encoded enzymes. The most common SINEs in the human genome are members of the ***Alu* family,** which are so named because most of their ~300-bp segments contain a cleavage site for the restriction endonuclease *Alu*I (AGCT; Table 5-4). *Alu* elements consist of two imperfect tandem repeats that are ~90% identical in sequence to portions of the 7S RNA of the signal recognition particle (SRP; Section 12-4B) but that both end in poly(A) segments that are not present in SRP 7S RNA. *Alu* elements occur only in primates, which indicates that they are of relatively recent origin. However, *Alu*-like elements occur in such distantly related organisms as slime molds, echinoderms, amphibians, and birds. All other types of SINEs are derived from tRNA sequences.

3. LTR retrotransposons, which contain long terminal repeats (LTRs) flanking *gag* and *pol* genes, are propagated via cytoplasmic retrovirus-like particles (Section 30-6B). They comprise 8.3% of the human genome. Only the vertebrate-specific **endogenous retroviruses (ERVs)** appear to have been active in the mammalian genome.

In addition, the human genome contains **DNA transposons** (Table 34-2) that resemble bacterial transposons (Section 30-6B). They comprise 2.8% of the human genome. Hence, *a total of ~45% of the human genome consists of widely dispersed and almost entirely inactive transposable elements.*

d. Moderately Repetitive DNAs Are Probably Selfish DNA

It would seem likely, considering their ranges of segment lengths and copy numbers, that nonexpressed, moderately repetitive DNAs have several different functions. There is, however, little experimental evidence in support of any of the various proposals that have been put forward in this regard. The proposal that is usually given the most credence is that moderately repetitive DNAs function as control sequences that participate in coordinately activating nearby genes. Another possibility, which is based on the observation that *Alu* elements contain a segment that is homologous to the **papovavirus** replication origin, is that certain families of moderately repetitive DNAs act as DNA replication origins. A third class of proposed functions for moderately repetitive DNAs is that they increase the evolutionary versatility of eukaryotic genomes by facilitating chromosomal rearrangements and/or forming reservoirs from which new functional sequences can be recruited.

Considering both the enormous amount of repetitive DNA in most eukaryotic genomes and the dearth of confirmatory evidence for any of the above proposals, a possibility that must be seriously entertained is that most repetitive DNA serves no useful purpose whatever for its host. Rather, it is **selfish** or **junk DNA,** molecular parasites that, over many generations, have disseminated themselves throughout the genome via various transpositional processes. The theory of natural selection indicates that the increased metabolic burden imposed by the replication of an otherwise harmless selfish DNA would eventually lead to its elimination. Yet, for slowly growing eukaryotes, the relative disadvantage of replicating, say, an additional 1000 bp of selfish DNA in an ~1 billion-bp genome would be so slight that its rate of elimination would be balanced by its rate of propagation. The C-value paradox may therefore simply indicate that a significant fraction, if not the great majority, of each eukaryotic genome is selfish DNA.

C. *Distribution of Genes*

The major goal of the human genome project is to provide a catalog of all human genes and their encoded proteins. Even with the finished sequence now in hand, this is by no means a simple task. In organisms with small genomes, such as bacteria and yeast, gene identification is quite straightforward because these genomes contain relatively little unexpressed DNA. However, *only 1.1 to 1.4% of the human genome consists of expressed sequences,* with ~24% of the genome consisting of introns and ~75% consisting of **intragenic sequences** (untranscribed sequences between genes). Consequently, our incomplete knowledge of the features through which cells recognize genes combined with the fact that human genes consist of relatively short exons (averaging ~150 nt) interspersed by much longer introns (averaging ~3500 nt and often much longer) greatly increases the difficulty (decreases the signal-to-noise ratio) of identifying genes. Hence, computer programs for sequence-based gene identification have had but limited success. Gene prediction algorithms therefore rely on sequence alignments with **expressed sequence tags (ESTs;** cDNAs that have been reverse transcribed from mRNAs; Section 7-2B) together with alignments with known genes from other organisms (which is often successful for highly conserved genes but is less so for genes that are rapidly evolving).

An important clue as to the occurrence of a gene is provided by the presence of a **CpG island.** 5-Methylcytosine (m^5C), as we have seen, occurs largely in the CG dinucleotides of various eukaryotic palindromic sequences, where it is implicated in switching off gene expression (Section 30-7). Since the spontaneous deamination of m^5C yields a normal T and thereby often results in a CG → TA mutation, CG dinucleotides occur in the human genome at about one-fifth of their randomly expected frequency

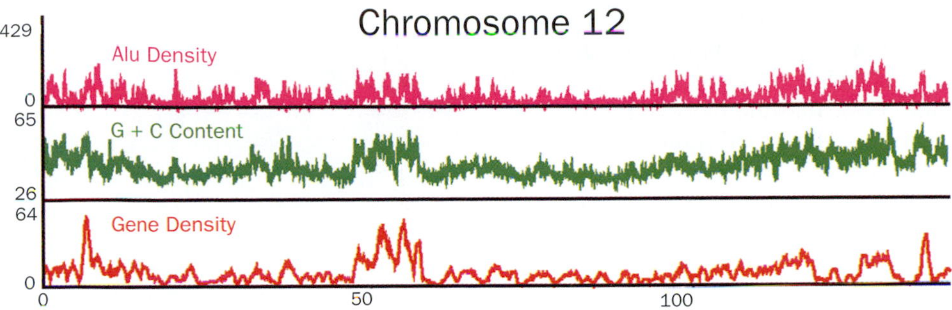

FIGURE 34-26 Density of structural features along the length of human chromosome 12. Gene density (*orange*) in this 133-megabase pair **(Mb)** chromosome is calculated per 1-Mb window, the percent G + C content (*green*) is calculated per 100-kb window, and the density of *Alu* elements (*magenta*) is calculated per 100-kb window. [Courtesy of Craig Venter, Celera Genomics, Rockville, Maryland.]

(which, since human DNA is 42% G + C, is $0.21 \times 0.21 \times 100 = 4\%$ of dinucleotides). Nevertheless, the human genome contains ~29,000 ~1-kb regions known as CpG islands in which unmethylated CG dinucleotides occur at close to their expected frequency. About 56% of human genes are associated with CpG islands, which overlap the promoter regions of these genes and extend up to ~1 kb into their coding regions. Hence the presence of a CpG island in a vertebrate chromosome is strongly indicative of the presence of an associated gene.

Around 30,000 putative genes have been identified in the human genome. The discrepancy between this number and previous estimates of 50,000 to 140,000 genes is largely attributed to a much greater prevalence of alternative splicing than had previously been surmised (Section 31-4A). The gene density along the lengths of the various chromosomes is highly variable. Thus, although the average gene frequency in the human genome is ~1 gene per 100 kb of DNA, this value varies from 0 to 64 genes per 100 kb (e.g., Fig. 34-26).

Genes may be classified as those that encode proteins (structural genes) and those that are transcribed to RNAs that are not translated. These latter so-called **noncoding RNAs (ncRNAs)** consist of tRNAs, rRNAs, small nuclear RNAs (snRNAs, which are components of spliceosomes; Section 31-4A), small nucleolar RNAs (snoRNAs, which participate in nucleolar RNA processing and base modification; Section 31-4B), as well as a variety of miscellaneous RNAs including the RNA components of the signal recognition particle (Section 12-4B), RNase P (Section 31-4B), and telomerase (Section 30-4D). The distribution of the rRNA and tRNA genes is discussed in Section 34-2D.

A total of 26,383 predicted structural genes in the human genome have been classified according to molecular function through sequence comparisons at both the level of protein families and of domains (Fig. 34-27). Note that

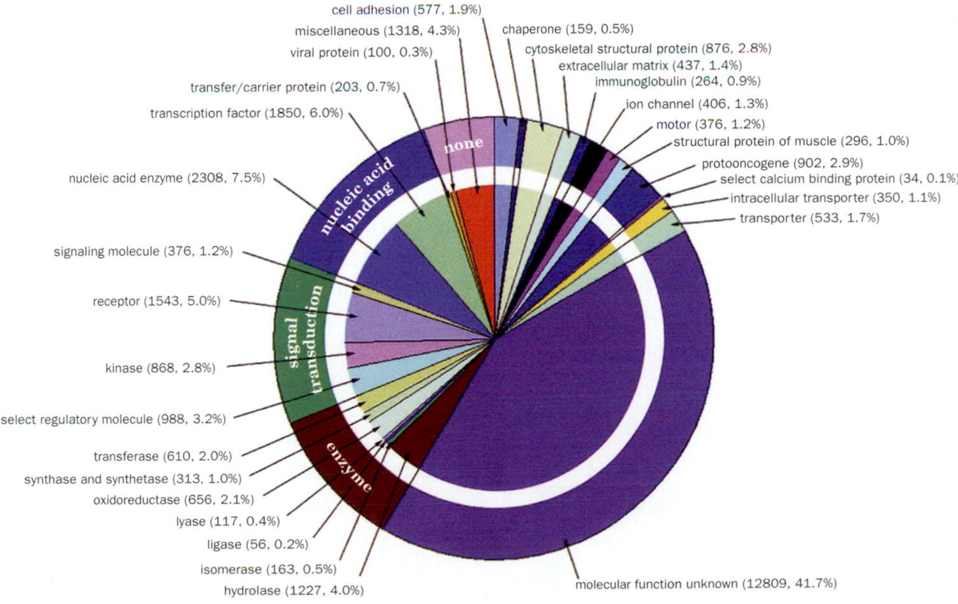

FIGURE 34-27 Distribution of molecular functions of 26,383 putative structural genes in the human genome. Each wedge of this pie chart lists, in parentheses, the number and percentage of the genes assigned to the indicated category of molecular function. The outer circle indicates the general functional categories whereas the inner circle provides a more detailed breakdown of these categories. [Courtesy Craig Venter, Celera Genomics, Rockville, Maryland.]

nearly 42% of them are classified as having unknown functions, as is likewise the case with most other genomes of known sequence, including those of prokaryotes. It can be seen from Fig. 34-27 that the most common molecular functions are those of transcriptions factors, proteins that mediate nucleic acid metabolism (nucleic acid enzymes), and receptors. Other common functions are those of kinases, hydrolases (most of which are proteases), proto-oncogenes (Section 19-3B), and select regulatory proteins (proteins that participate in signal transduction). Comparison of the structural genes in the human genome with those in the genomes of *D. melanogaster* and the nemotode worm *Caenorhabditis elegans* reveals that the greatest expansions of gene families occurred in those encoding proteins involved in developmental regulation (Section 34-4B), neuronal structure and function, hemostasis (blood clotting and related processes; Section 35-1), the acquired immune response (Section 35-2), and cytoskeletal complexity.

D. *Tandem Gene Clusters*

Most genes occur but once in an organism's haploid genome. This is sufficient, even for genes specifying proteins required in large amounts, through the accumulation of their corresponding mRNAs. However, the great cellular demand for rRNAs (which comprise ~80% of a cell's RNA) and tRNAs, which are all ncRNAs, can only be satisfied through the expression of multiple copies of the genes specifying them. In this subsection we discuss the organization of the genes coding for rRNAs and tRNAs. We shall also consider the organization of histone genes, the only protein-encoding genes that occur in multiple identical copies.

a. rRNA Genes Are Organized into Repeating Sets

We have seen in Sections 31-4B and 31-4C that even the *E. coli* genome, which otherwise consists of unique sequences, contains multiple copies of rRNA and tRNA genes. In eukaryotes, the genes specifying the 18S, 5.8S, and 28S rRNAs are invariably arranged in this order, reading 5′ → 3′ on the RNA strand, and separated by short transcribed spacers to form a single transcription unit of ~7500 bp (Fig. 34-28). (Recall that the primary transcript of this gene cluster is a 45S RNA from which the mature rRNAs are derived by posttranscriptional cleavage; Section 31-4B.) *Indeed, this rRNA gene arrangement is universal since the 5′ end of prokaryotic 23S rRNA is homologous to eukaryotic 5.8S rRNA (Section 32-3A).*

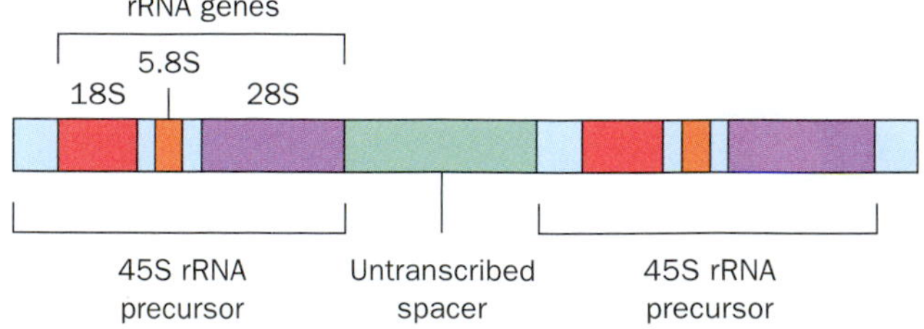

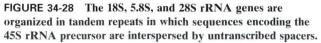

FIGURE 34-28 The 18S, 5.8S, and 28S rRNA genes are organized in tandem repeats in which sequences encoding the 45S rRNA precursor are interspersed by untranscribed spacers.

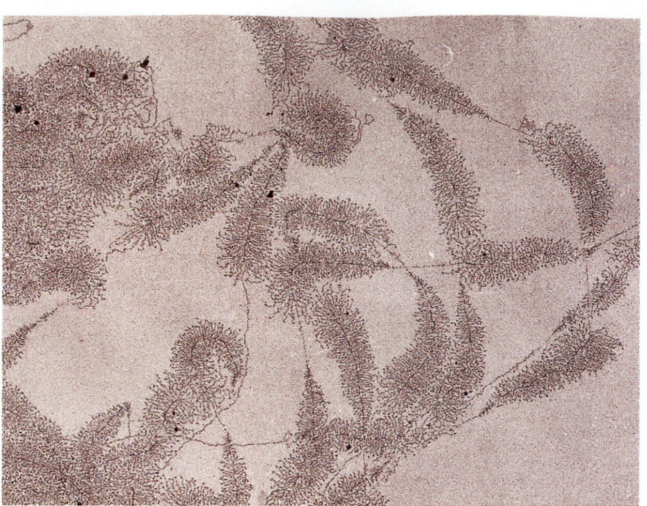

FIGURE 34-29 Electron micrograph of tandem arrays of actively transcribing 18S, 5.8S, and 28S rRNA genes from the nucleoli of the newt *Notophthalmus viridescens.* The axial fibers are DNA. The fibrillar "Christmas tree" matrices, which consist of newly synthesized RNA strands in complex with proteins, outline each transcriptional unit. Note that the longest ribonucleoprotein branches of each "Christmas tree" are only ~10% the length of their corresponding DNA stem. Apparently, the RNA strands are compacted through secondary structure interactions and/or protein associations. The matrix-free segments of DNA are the untranscribed spacers. [Courtesy of Oscar L. Miller, Jr., and Barbara R. Beatty, University of Virginia.]

Electron micrographs, such as Fig. 34-29, indicate that *the blocks of transcribed eukaryotic rRNA genes are arranged in tandem repeats that are separated by untranscribed spacers (Fig. 34-28).* These tandem repeats are typically ~12,000 bp in length, although the untranscribed spacer varies in length between species and, to a lesser extent, from gene to gene. Quantitative measurements of the amounts of radioactively labeled rRNAs that can hybridize with the corresponding nuclear DNA **(rDNA)** and more recently genomic sequencing indicate that these rRNA genes, which may be distributed among several chromosomes, vary in haploid number from less than 50 to over 10,000, depending on the species. Humans, for example, have 150 to 200 blocks of rDNA spread over 5 chromosomes.

b. The Nucleolus Is the Site of rRNA Synthesis and Ribosome Assembly

In a typical interphase cell nucleus, the rDNA condenses to form a single nucleolus (Fig. 1-5). There, as Fig. 34-29 suggests, these genes are rapidly and continuously transcribed by RNA polymerase I (Section 31-2E). The nucleolus, as demonstrated by radioactive labeling experiments, is also the site where these rRNAs are posttranscriptionally processed and assembled with cytoplasmically synthesized ribosomal proteins into immature ribosomal subunits. Final assembly of the ribosomal subunits only occurs as they are being transferred to the cytoplasm, which presumably prevents the premature translation of partially processed mRNAs (hnRNAs) in the nucleus.

c. 5S rRNA and tRNA Genes Occur in Multiple Clusters

The genes encoding the 120-nucleotide 5S rRNAs, much like the other rRNA genes, are arranged in clusters that contain a total of several hundred to several hundred thousand tandem repeats distributed among one or more chromosomes. In *X. laevis,* for example, the repeating unit consists of the 5S rRNA gene, a nearby **pseudogene** (a 101-bp segment of the 5S rRNA gene that, curiously, is not transcribed), and an untranscribed spacer of variable length but averaging ~400 bp (Fig. 34-30). The 5S rRNA genes are transcribed outside of the nucleolus by RNA polymerase III (Section 31-2E). 5S rRNA must therefore be transported into the nucleolus for incorporation into the large ribosomal subunit.

The 497 tRNA genes that have been identified in the human genome are likewise transcribed by RNA polymerase III. They are also multiply reiterated and clustered, with >25% of them occurring in a 4-Mb region on chromosome 6 and most of the remainder clustered on numerous but not all chromosomes.

d. Histone Genes Are Reiterated

Histone mRNAs have relatively short cytoplasmic lifetimes because of their lack of the poly(A) tails that are appended to other eukaryotic mRNAs (Section 31-4A). Yet histones must be synthesized in large amounts during S phase of the cell cycle (when DNA is synthesized). *This process is made possible through the multiple reiteration of histone genes, which in most organisms are the only identically repeated genes that code for proteins.* This organization, it is thought, permits the sensitive control of histone synthesis through the coordinate transcription of sets of histone genes. Histone genes also differ from nearly all other eukaryotic genes in that almost all histone sequences lack introns. The significance of this observation is unknown.

There is little relationship between a genome's size and its total number of histone genes. For example, birds and mammals have 10 to 20 copies of each of the five histone genes, *D. melanogaster* has ~100, and sea urchins have several hundred. This suggests that the efficiency of histone gene expression varies with species. In many organisms the histone genes are organized into tandemly repeated quintets consisting of a gene coding for each of the five different histones interspersed by untranscribed spacers (Fig. 34-31). The gene order and the direction of transcription in these quintets are preserved over large evolutionary distances. Corresponding spacer sequences vary widely among species and, to a limited extent, among the repeating quintets within a genome. In birds and mammals, this repetitious organization has broken down; their histone genes occur in clusters but in no particular order.

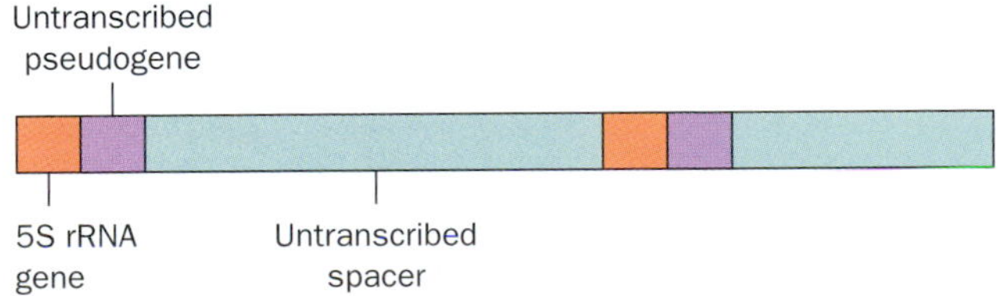

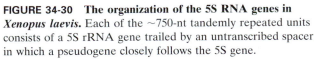

FIGURE 34-30 The organization of the 5S RNA genes in *Xenopus laevis.* Each of the ~750-nt tandemly repeated units consists of a 5S rRNA gene trailed by an untranscribed spacer in which a pseudogene closely follows the 5S gene.

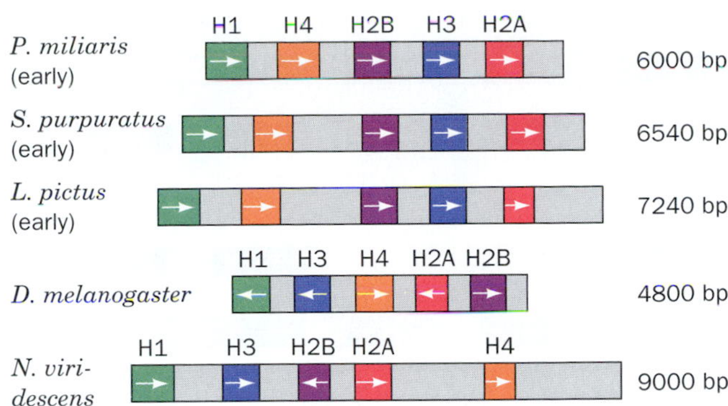

FIGURE 34-31 The organization and lengths of the histone gene cluster repeating units in a variety of organisms. Coding regions are indicated in color and spacers are gray. The arrows denote the directions of transcription (the top three organisms are distantly related sea urchins).

e. Reiterated Sequences May Be Generated and Maintained by Unequal Crossovers and/or Gene Conversion

How do reiterated genes maintain their identity? The usual mechanism of Darwinian selection would seem ineffective in accomplishing this since deleterious mutations in a few members of a multiply repeated set of identical genes would have little phenotypic effect. Indeed, many mutations do not affect the function of a gene product and are therefore selectively neutral. Reiterated gene sets must therefore maintain their homogeneity through some additional mechanism. Two such mechanisms seem plausible:

1. In the **unequal crossover** mechanism (Fig. 34-32*a*), recombination occurs between homologous segments of misaligned chromosomes, thereby excising a segment from one of the chromosomes and adding it to the other. Computer simulations indicate that such repeated expansions and contractions of a chromosome will, by random processes, generate a cluster of reiterated sequences that have been derived from a much smaller ancestral cluster. Unequal crossing-over is also thought to be the mechanism that generated segmental duplications (Section 34-2B).

2. In the **gene conversion** mechanism (Fig. 34-32*b*), one member of a reiterated gene set "corrects" a nearby variant through a process resembling recombination repair (Section 30-6A).

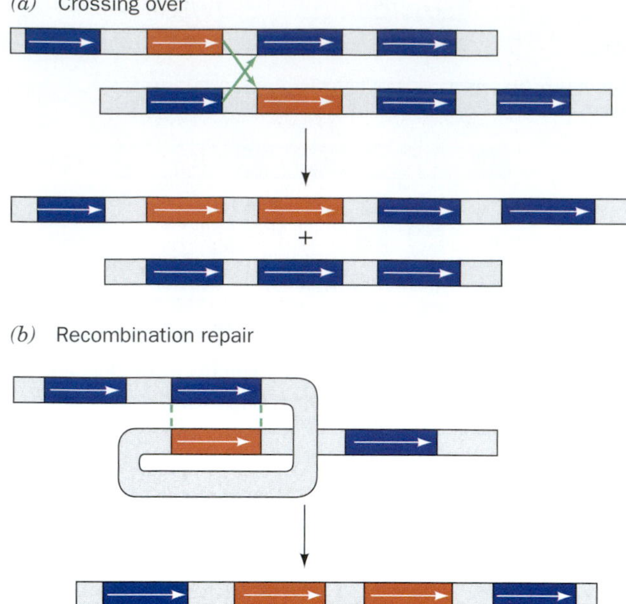

FIGURE 34-32 Two possible mechanisms for maintaining the homogeneity of a tandem multigene family. (*a*) Unequal crossing-over between mispaired but similar genes results in an unpaired DNA segment being deleted from one chromosome and added to the other. (*b*) Gene conversion "corrects" one member of a tandem array with respect to the other via a recombination repair mechanism. Repeated cycles of either process may either eliminate a variant gene or spread it throughout the entire tandem array.

Since point mutations are rare events compared to crossovers, either mechanism would eventually result in a newly arisen variant copy of a repeated sequence either being eliminated or taking over the entire cluster. If a mutation that has been so concentrated is deleterious, it will be eliminated by Darwinian selection. In contrast, variant spacers, which are not as subject to selective pressure, would be eliminated at a slower rate. The existence of reiterated sets of identical genes separated by somewhat heterogeneous spacers may therefore be reasonably attributed to either homogenization model.

E. *Gene Amplification*

The selective replication of a particular set of genes, a process known as **gene amplification,** normally occurs only at specific stages of the life cycle of certain organisms. In the following subsections, we outline what is known about this phenomenon.

a. rRNA Genes Are Amplified during Oogenesis

The rate of protein synthesis during the early stages of embryonic growth is so great that in some species the normal genomic complement of rRNA genes cannot satisfy the demand for rRNA. In these species, notably certain insects, fish, and amphibians, the rDNA is differentially replicated in developing oocytes (immature egg cells). In one of the most spectacular examples of this process, the rDNA in *X. laevis* oocytes is amplified by ~1500 times its amount in somatic cells to yield some 2 million sets of rRNA genes comprising nearly 75% of the total cellular DNA. The amplified rDNA occurs as extrachromosomal circles, each containing one or two transcription units, that are organized into hundreds of nucleoli (Fig. 34-33). Mature *Xenopus* oocytes therefore contain ~10^{12} ribosomes, 200,000 times the number in most larval cells. This is so many that mutant zygotes (fertilized ova) that lack nucleoli (and thus cannot synthesize new ribosomes; the oocyte's extra nucleoli are destroyed during its first meiotic division) survive to the swimming tadpole stage with only their maternally supplied ribosomes.

What is the mechanism of rDNA amplification? An important clue is that the untranscribed spacers from a given extrachromosomal nucleolus all have the same length, whereas we have seen that the corresponding chromosomal spacers exhibit marked length heterogeneities. This observation suggests that the rDNA circles in a single nucleolus are all descended from a single chromosomal gene. Gene amplification has been shown to occur in two stages: A low level of amplification in the first stage followed by massive amplification in the second stage. It therefore seems likely that, in the first stage, no more than a few chromosomal rRNA genes are replicated by an unknown mechanism and the daughter strands are released as extrachromosomal circles. Then, in the second stage, these circles are multiply replicated by the rolling circle mechanism (Section 30-3B). In support of this hypothesis are electron micrographs of amplified genes showing the

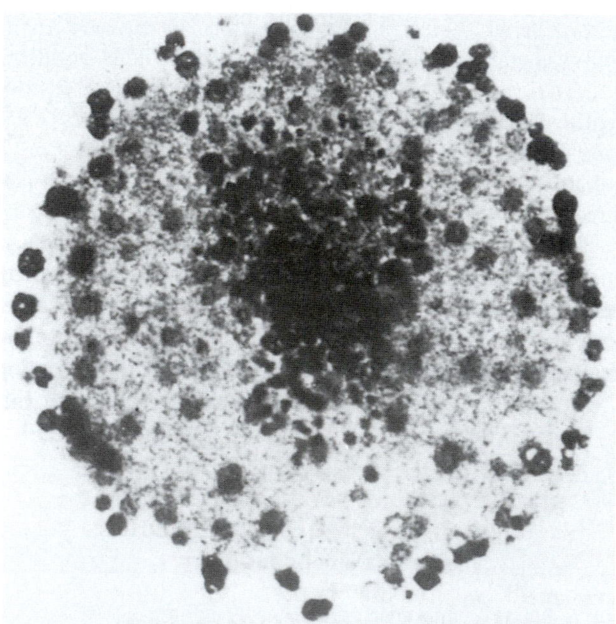

FIGURE 34-33 Photomicrograph of an isolated oocyte nucleus from *X. laevis*. Its several hundred nucleoli, which contain amplified rRNA genes, appear as darkly staining spots. [Courtesy of Donald Brown, Carnegie Institution of Washington.]

"lariat" structures postulated to be rolling circle intermediates (Fig. 30-25).

b. Chorion Genes Are Amplified

The only other known example of programmed gene amplification is that of the *D. melanogaster* ovarian follicle cell genes that code for **chorion** (egg shell) **proteins** (ovarian follicle cells surround and nourish the maturing egg). Prior to chorion synthesis, the entire haploid genome of each ovarian follicle cell is replicated 16-fold. This process is followed by an ~10-fold selective replication of only the chorion genes to form a multiply branched (partially polytene) structure in which the amplified chorion genes remain part of the chromosome (Fig. 34-34). Interestingly, chorion gene amplification does not occur in silk moth oocytes. Rather, this organism's genome has multiple copies of chorion genes.

c. Drug Resistance Can Result from Gene Amplification

In cancer chemotherapy, a common observation is that the continued administration of a cytotoxic drug causes an initially sensitive tumor to become increasingly drug resistant to the point that the drug loses its therapeutic efficacy. One mechanism by which a cell line can acquire such drug resistance is through the overproduction of the drug's target enzyme. Such a process can be observed, for example, by exposing cultured animal cells to the dihydrofolate analog methotrexate. This substance, it will be recalled, all but irreversibly binds to dihydrofolate reductase (DHFR), thereby inhibiting DNA synthesis (Section 28-3B). Slowly increasing the methotrexate dose yields surviving cells that ultimately contain up to 1000 copies of the DHFR gene and are thereby capable of tremendous overproduction of this enzyme, a clear laboratory demonstration of Darwinian selection. Members of some of these cell lines contain extrachromosomal elements known as **double minute chromosomes** that each bear one or more copies of the DHFR gene, whereas in other cell lines the additional DHFR genes are chromosomally integrated. The mechanism of gene amplification in either cell type is not well understood, although it is worth noting that this phenomenon is only known to occur in cancer cells. Both types of amplified genes are genetically unstable; further cell growth in the absence of methotrexate results in the gradual loss of the extra DHFR genes.

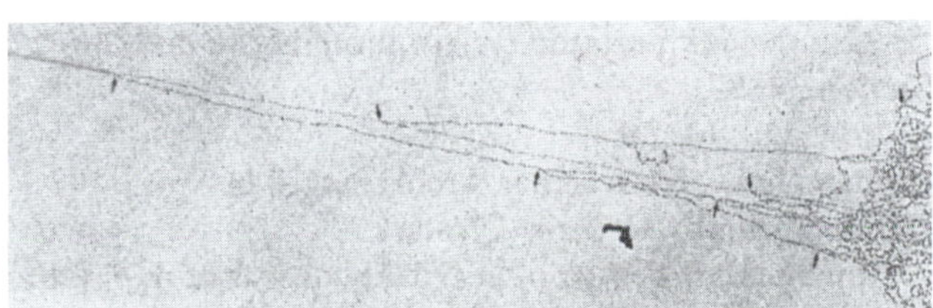

FIGURE 34-34 An electron micrograph of a chorion gene-containing chromatin strand from an oocyte follicle cell of *D. melanogaster*. The strand has undergone several rounds of partial replication (*arrows at replication forks*) to yield a multiforked structure containing several parallel copies of chorion genes. [Courtesy of Oscar L. Miller, Jr. and Yvonne Osheim, University of Virginia.]

F. *Clustered Gene Families: Hemoglobin Gene Organization*

Few proteins in a given organism are really unique. Rather, like the digestive enzymes trypsin, chymotrypsin, and elastase (Section 15-3), or the various collagens (Section 8-2B), they are usually members of families of structurally and functionally related proteins. In many cases, the family of genes specifying such proteins are clustered together in a single chromosomal region. In the following subsections, we consider the organization of two of the best characterized clustered gene families, those encoding the two types of human hemoglobin subunits. The clustered gene families that encode immune system proteins are discussed in Section 35-2C.

a. Human Hemoglobin Genes Are Arranged in Two Developmentally Ordered Clusters

Human adult hemoglobin (HbA) consists of $\alpha_2\beta_2$ tetramers in which the α and β subunits are structurally related. The first hemoglobin made by the human embryo, however, is a $\zeta_2\varepsilon_2$ tetramer **(Hb Gower 1)** in which ζ and ε are α- and β-like subunits, respectively (Fig. 34-35). By around 8 weeks postconception, the embryonic subunits have been supplanted (in newly formed erythrocytes) by the α subunit and the β-like γ subunit to form fetal hemoglobin (HbF), $\alpha_2\gamma_2$ (the hemoglobins present during the changeover period, $\alpha_2\varepsilon_2$ and $\zeta_2\gamma_2$, are named **Hb Gower 2** and **Hb Portland,** respectively). The γ subunit is gradually superseded by β starting a few weeks before birth. Adult blood normally contains ~97% HbA, 2% **HbA$_2$** ($\alpha_2\delta_2$ in which δ is a β variant), and 1% HbF.

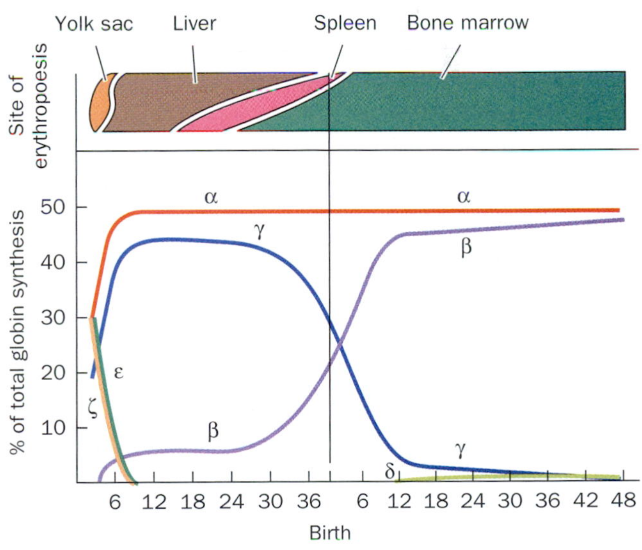

FIGURE 34-35 The progression of human globin chain synthesis with embryonic and fetal development. Note that any red blood cell contains only one type each of α- and β-like subunits. The progression in the sites of **erythropoiesis** (red cell formation), which is indicated in the upper panel, corresponds roughly to the major switches in hemoglobin types. [After Weatherall, D.J. and Clegg, J.B., *The Thalassaemia Syndromes* (3rd ed.), *p*. 64, Blackwell Scientific Publications (1981).]

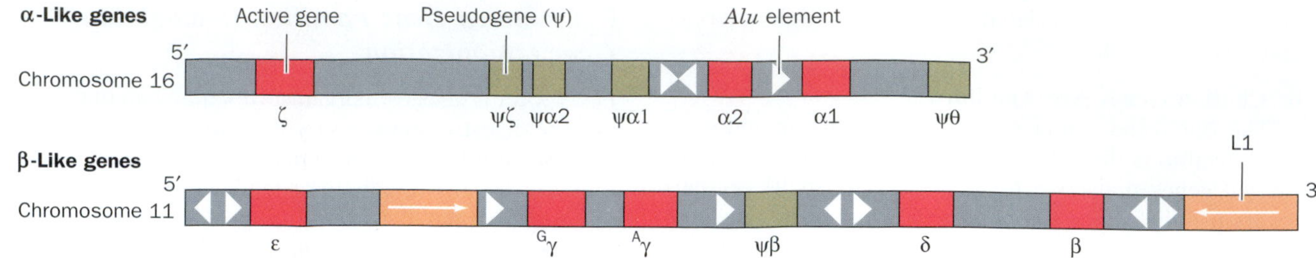

FIGURE 34-36 The organization of human globin genes on their respective sense strands. Red boxes represent active genes; green boxes represent pseudogenes; yellow boxes represent L1 sequences, with the arrows indicating their relative orientations; and triangles represent *Alu* elements in their relative orientations. [After Karlsson, S. and Nienhuis, A.W., *Annu. Rev. Biochem.* **54,** 1074 (1985).]

In mammals, the genes specifying the α- and β-like hemoglobin subunits form two different gene clusters that occur on separate chromosomes. In humans and many other mammals, the genes in each globin cluster are arranged, $5' \rightarrow 3'$ on the coding strands, in the order of their developmental expression (Fig. 34-36). This ordering is common in mammals but not universal; in the mouse β gene cluster, for instance, the adult genes precede the embryonic genes.

The β-globin gene cluster (Fig. 34-36), which spans ~100 kb, contains five functional genes: the embryonic ε gene, two fetal genes, $^{G}\gamma$ and $^{A}\gamma$ (duplicated genes that encode polypeptides differing only by having either Gly or Ala at their positions 136), and the two adult genes, δ and β. The β-globin cluster also contains one **pseudogene,** ψβ (an untranscribed relic of an ancient gene duplication that is ~75% identical to the β gene), eight *Alu* elements, and two L1 elements (Section 34-2B).

The α-globin gene cluster (Fig. 34-36), which spans ~28 kb, contains three functional genes: the embryonic ζ gene and two slightly different α genes, α1 and α2, which encode identical polypeptides. The α cluster also contains four pseudogenes, ψζ, ψα2, ψα1, and ψθ, and three *Alu* elements.

b. Hemoglobin Genes All Have the Same Exon–Intron Structure

Protein-coding sequences represent <5% of either globin gene cluster. This situation is largely a consequence of the heterogeneous collection of untranscribed spacers separating the genes in each cluster. In addition, *all known vertebrate globin genes, including that of myoglobin and most hemoglobin pseudogenes, consist of three nearly identically placed exons separated by two somewhat variable introns (Fig. 34-37).* This gene structure apparently arose quite early in vertebrate history, well over 500 million years ago. Indeed, much of this structure even predates the divergence of plants and animals. The structure of the gene encoding leghemoglobin (a plant globin that functions in legumes to protect nitrogenase from O_2 poisoning; Section 26-6) differs from that of vertebrates only in that the central exon of vertebrate globins is split by a third intron in the leghemoglobin gene. Quite possibly the central exon in vertebrate globins arose through the fusion of the two interior exons in a leghemoglobin-like ancestral gene.

c. DNA Polymorphisms Can Establish Genealogies

Unexpressed sequences, which are subject to little selective pressure, evolve so much faster than expressed

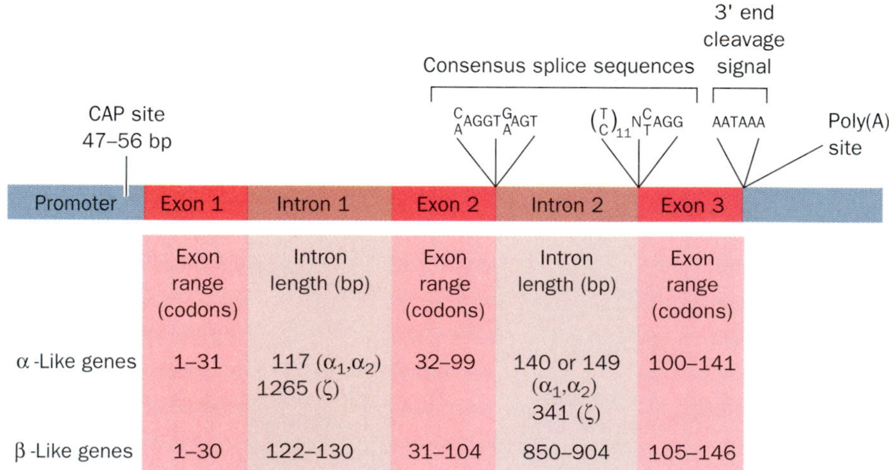

	Exon range (codons)	Intron length (bp)	Exon range (codons)	Intron length (bp)	Exon range (codons)
α-Like genes	1–31	117 (α_1,α_2) 1265 (ζ)	32–99	140 or 149 (α_1,α_2) 341 (ζ)	100–141
β-Like genes	1–30	122–130	31–104	850–904	105–146

FIGURE 34-37 Structure of the prototypical hemoglobin gene. The conserved sequences at the exon–intron boundaries (splice sequences) and at the 3′ end of the gene (polyadenylation site) are indicated. The length range of each exon (in codons) and each intron (in base pairs) is given. [After Karlsson, S. and Nienhuis, A.W., *Annu. Rev. Biochem.* **54,** 1079 (1985).]

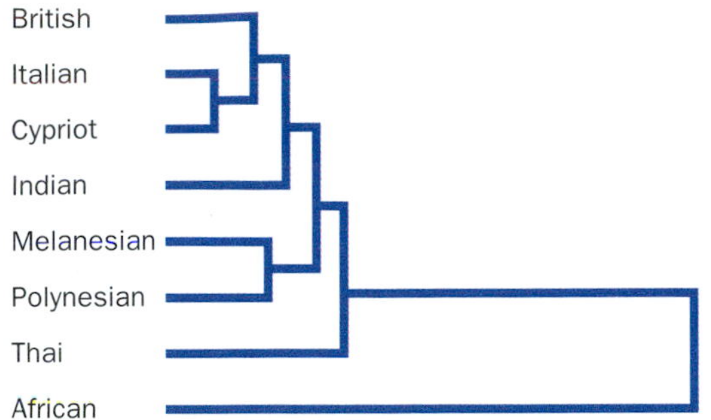

FIGURE 34-38 A family tree showing the lines of descent among eight human population groups. These were determined from the distribution of five restriction-fragment length polymorphisms (RFLPs) in their β-globin gene clusters. The horizontal axis is indicative of the genetic distances between related populations and therefore of the times between their divergence. [After Wainscoat, J.S., Hill, A.V.S., Boyce, A.L., Flint, J., Hernandez, M., Thein, S.L., Old, J.M., Lynch, J.R., Falusi, A.G., Weatherall, D.J., and Clegg, J.B., *Nature* **319,** 493 (1986).]

sequences that they even accumulate significant numbers of sequence **polymorphisms** (variations) within a single species. Consequently, the evolutionary relationships among populations within a species can be established by determining how a series of polymorphic DNA sequences are distributed among them. For example, the genealogy of several diverse human populations has been inferred from the presence or absence of certain restriction sites [restriction-fragment length polymorphisms (RFLPs); Section 5-5A] in five segments of their β-globin gene clusters. This study has led to the construction of a "family tree" (Fig. 34-38), which indicates that non-African (Eurasian) populations are much more closely related to each other than they are to African populations. Fossil evidence indicates that anatomically modern man arose in Africa about 100,000 years ago and rapidly spread throughout that continent. This family tree therefore suggests that all Eurasian populations are descended from a surprisingly small "founder population" (perhaps only a few hundred individuals) that left Africa ~50,000 years ago. A similar analysis indicates that the sickle-cell variant of the β gene arose on at least three separate occasions in geographically distinct regions of Africa.

G. *The Thalassemias: Genetic Disorders of Hemoglobin Synthesis*

The study of mutant hemoglobins (Section 10-3) has provided invaluable insights into structure–function relationships in proteins. Likewise, the study of defects in hemoglobin expression has greatly facilitated our understanding of eukaryotic gene expression.

The most common class of inherited human disease results from the impaired synthesis of hemoglobin subunits. These anemias are named **thalassemias** (Greek: *thalassa,* sea) because they commonly occur in the region surrounding the Mediterranean Sea (although they are also prevalent in Central Africa, India, and the Far East). The observation that malaria is or was endemic in these same areas (Fig. 7-20) led to the realization that heterozygotes for thalassemic genes (who appear normal or are only mildly anemic; a condition known as **thalassemia minor**) are resistant to malaria. Thus, as we have seen in our study of sickle-cell anemia (Section 10-3B), mutations that are seriously debilitating or even lethal in homozygotes (who are said to suffer from **thalassemia major**) may offer sufficient selective advantage to heterozygotes to ensure the propagation of the mutant gene.

Thalassemia can arise from many different mutations, each of which causes a disease state of characteristic severity. In α^0- and β^0-thalassemias, the indicated globin chain is absent, whereas in α^+- and β^+-thalassemias, the normal globin subunit is synthesized in reduced amounts. In what follows, we shall consider thalassemias that are illustrative of several different types of genetic lesions.

a. α-Thalassemias

Most α-thalassemias are caused by the deletion of one or both of the α-globin genes in an α gene cluster (Fig. 34-36). A variety of such mutations have been cataloged. In the absence of equivalent numbers of α chains, the fetal γ chains and the adult β chains form homotetramers: **Hb Bart's** (γ_4) and **HbH** (β_4). Neither of these tetramers exhibits any cooperativity or Bohr effect (Sections 10-1C and 10-1D), which makes their oxygen affinities so high that they cannot release oxygen under physiological conditions. Consequently, α^0-thalassemia occurs with four degrees of severity depending on whether an individual has 1, 2, 3, or 4 missing α-globin genes:

1. **Silent-carrier state:** The loss of one α gene is an asymptomatic condition. The rate of expression of the remaining α genes largely compensates for the less than normal α gene dosage so that, at birth, the blood contains only ~1 to 2% Hb Bart's.

2. **α-Thalassemia trait:** With two missing α genes (either one each deleted from both α gene clusters or both deleted from one cluster), only minor anemic symptoms occur. The blood contains ~5% Hb Bart's at birth.

3. **Hemoglobin H disease:** Three missing α genes results in a mild to moderate anemia. Affected individuals can usually lead normal or nearly normal lives.

4. **Hydrops fetalis:** The lack of all four α genes is lethal. Unfortunately, the synthesis of the embryonic ζ chain continues well past the 8 weeks postconception when it normally ceases (Fig. 34-35), so the fetus usually survives until around birth.

α-Thalassemias caused by nondeletion mutations are relatively uncommon. One of the best characterized such lesions changes the UAA stop codon of the α2-globin gene to CAA (a Gln codon), so that protein synthesis continues for the 31 codons beyond this site to the next UAA. The resultant **Hb Constant Spring** is produced in only small amounts because, for unknown reasons, its mRNA is rapidly degraded in the cytosol. Another point mutation in the α2 gene changes Leu H8(125)α to Pro, which no doubt disrupts the H helix. The consequent α^+-thalassemia results from the rapid degradation of this abnormal **Hb Quong Sze.**

b. β-Thalassemias

Heterozygotes of β-thalassemias are usually asymptomatic. Homozygotes become so severely anemic, however, that once their HbF production has diminished, many require frequent blood transfusions to sustain life and all require them to prevent the severe skeletal deformities caused by bone marrow expansion. The anemia results not only from the lack of β chains but also from the surplus of α chains. The latter form insoluble membrane-damaging precipitates that cause premature red cell destruction (Section 10-3A). The coinheritance of α-thalassemia therefore tends to lessen the severity of β-thalassemia major.

In β-thalassemia, there may be an increased production of the δ and γ chains, so that the consequent extra HbA_2 and HbF can compensate for some of the missing HbA. In **δβ-thalassemia,** the neighboring δ and β genes have both been deleted, so that only increased production of the γ chain is possible. Yet many adult δβ-thalassemics, for reasons that are not understood, produce so much HbF that they are asymptomatic. Such individuals are said to have **hereditary persistence of fetal hemoglobin (HPFH).** This condition is therefore of medical interest because it could also alleviate the symptoms of β-thalassemia and sickle-cell anemia.

The so-called Greek form of HPFH is associated with a G → A mutation at position −117 of the γ-globin gene (its promoter region). In an effort to establish whether this mutation does, in fact, cause HPFH, the mutated γ-globin gene was introduced into mice. The resulting fetal and adult transgenic animals synthesized γ-globin at a high level, with a concomitant decrease in the synthesis of the β-globin gene. These changes in gene expression correlate with the loss of binding of the transcription factor **GATA-1** to the γ-globin promoter, thereby suggesting that this protein is a negative regulator of the γ-globin gene expression in normal human adults (transcription factors are discussed in Section 34-3B).

β^0-Thalassemias caused by deletions are rare compared to those causing α^0-thalassemias. This is probably because the long repeated sequences in which the α-globin genes are embedded make them more prone to unequal crossing-over than the β-globin gene. Nevertheless, a β-thalassemic lesion causing the production of **Hb Lepore** is a particularly clear instance of this deletion mechanism. This lesion, the consequence of a deletion extending from within the δ gene to the corresponding position of its neighboring β gene, yields a δ/β hybrid subunit. Such deletions almost certainly arose through unequal crossovers between the β gene on one chromosome and the δ gene on another (Fig. 34-39; the two genes are 93% identical in sequence). The second product of such crossovers, a chromosome containing a β/δ hybrid flanked by normal δ and β genes (Fig. 34-39), is known as **Hb anti-Lepore.** Homozygotes for Hb Lepore have symptoms similar to those of β-thalassemia major, whereas homozygotes for Hb anti-Lepore, which have the full complement of normal globin genes, are symptom free and have only been detected through blood tests.

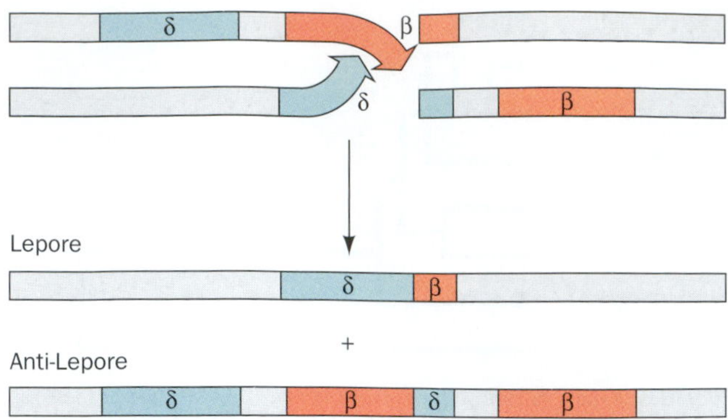

FIGURE 34-39 The formation of Hb Lepore and Hb anti-Lepore. This occurs by unequal crossing-over between the β-globin gene on one chromosome and the δ-globin gene on its homolog.

Most β-thalassemias are caused by a wide variety of point mutations that affect the production of β chains. These include:

1. Nonsense mutations that convert normal codons to the stop codon UAG.

2. Frameshift mutations that insert/delete one or more base pairs into/from an exon.

3. Point mutations in the β gene's promoter region, either in its TATA box or in its CACCC box (Section 31-2E). These attenuate transcriptional initiation.

4. Point mutations that alter the sequence at an exon–intron junction (Section 31-4A). These diminish/abolish splicing and/or activate a **cryptic splice site** (an exon–intron junction-like sequence that normally is not spliced) to pair with the altered intron's unaltered end.

5. A point mutation that alters an intron's lariat-branch site (Section 31-4A). This activates a cryptic 3′ splice upstream of the original site, leading to the excision of a shorter than normal intron.

6. Point mutations that create new splice sites. These either compete with the neighboring normal splice site or pair with a nearby cryptic splice site.

7. A point mutation that alters the AAUAAA cleavage signal at the mRNA's 3′ end (Section 31-4A).

Consideration of the effects of these mutations, particularly those involving gene splicing, has confirmed and extended our understanding of how eukaryotic genes are constructed and expressed.

3 ■ CONTROL OF EXPRESSION

The elucidation of the mechanisms controlling gene expression in eukaryotes had lagged at least 20 years behind that of prokaryotes. In addition to the far greater complexity of eukaryotic systems, this is largely because the types of genetic analyses that have been so useful in characterizing prokaryotic systems (which require the detection of very rare events) are precluded in metazoa (multicellular animals) by their much slower reproductive rates.

Compounding this problem are the difficulties in selecting for mutations in essential genes; the missing product of a defective enzyme in a metazoan usually cannot be replaced by simply adding that product to the diet as is often possible with, say, *E. coli.* This latter difficulty can be partially overcome by the rather laborious task of growing cells from metazoa in tissue culture. Since somatic cells do not normally undergo genetic recombination, however, genetic manipulations cannot be carried out in tissue culture the way they can in a bacterial culture.

What has made genetic manipulations of metazoa feasible is the development, in the 1970s, of molecular cloning techniques (Section 5-5). The gene encoding a particular eukaryotic protein can be identified in genomic or cDNA libraries through Southern blotting (Section 5-5D) or PCR (Section 5-5F) using an oligonucleotide probe or primer whose sequence encodes a segment of the protein (a process termed reverse genetics; Section 7-2C). Alternatively, if the organism's genome has been sequenced, the gene may be identified *in silico* (computationally). The gene may then be modified, for example, through site-directed mutagenesis (Section 5-5G), and the effects of the modification analyzed in an expression vector such as *E. coli* or yeast, or alternatively, *in vitro.*

The expression of foreign genes in metazoans (a gain of function) has been made possible through the development of a process in which DNA is microinjected into the nucleus of a fertilized ovum (Fig. 5-59). Such DNA often integrates into the chromosome of the resulting zygote, that then undergoes normal development to form a **transgenic** individual whose cells each contain the foreign genes (in *Xenopus,* this merely involves allowing the transfected egg to hatch, whereas in mice the fertilized ovum must be implanted in the uterus of a properly prepared foster mother; see Fig. 5-5 for a striking example of a transgenic mouse). Alternatively, a normal gene may be selectively inactivated (knocked out; a loss of function) through the use of DNA specifying the defective gene, which then recombines with the normal gene. The genome of an already multicellular organism may be altered, in a technique that holds great promise for gene therapy, through the use of defective (unable to reproduce) retroviruses that contain the genes to be transferred (Section 5-5H). Thus, the genomes of metazoans can now be manipulated, albeit with considerable clumsiness. We are, however, rapidly becoming more adept at these procedures as we gain further understanding of how eukaryotic chromosomes are organized and expressed.

Single-celled eukaryotes, particularly yeasts, are exceptions to the foregoing discussion because they can be grown and manipulated in much the same way as bacteria. Indeed, much of our knowledge of eukaryotic molecular biology has been obtained through molecular genetic analyses of budding (baker's) yeast (*Saccharomyces cerevisiae*). In this chapter, unless otherwise indicated, the term "yeast" refers to *S. cerevisiae.*

In this section we consider the molecular basis of the enormous expressional variation that eukaryotic cells exhibit. In doing so, we shall first study the nature of transcriptionally active chromatin, then discuss how genetic expression in eukaryotes is mainly regulated through the control of transcriptional initiation, and finally consider the other means by which eukaryotes control genetic expression. Eukaryotic gene regulation, as we shall see, is an astoundingly complex process that requires the participation of well over 100 polypeptides that form assemblies with molecular masses of several million daltons. In the following section, we take up the molecular basis of normal cell differentiation, its aberration, cancer, and programmed cell death.

A. *Chromosomal Activation and Deactivation*

Interphase chromatin, as is mentioned in Section 34-1, may be classified in two categories: the highly condensed and transcriptionally inactive heterochromatin, and the diffuse and transcriptionally active or activatable euchromatin (Fig. 34-2). Two types of heterochromatin have been distinguished:

1. Constitutive heterochromatin, which is permanently condensed in all cells and consists mostly of the highly repetitive sequences clustered near the chromosomal centromeres (Section 34-2B) and telomeres (Section 30-4D). Constitutive heterochromatin is transcriptionally inert.

2. Facultative heterochromatin, which varies in a tissue-specific manner. Presumably the condensation of facultative heterochromatin functions to transcriptionally inactivate large chromosomal blocks.

a. Most Mammalian Cells Have Only One Active X Chromosome

Female mammalian cells contain two X chromosomes, whereas male cells have one X and one Y chromosome. *Female somatic cells, however, maintain only one of their X chromosomes in a transcriptionally active state.* Consequently, males and females make approximately equal amounts of X chromosome–encoded gene products, a phenomenon known as **dosage compensation.** The inactive X chromosome is visible during interphase as a heterochromatin structure known as a **Barr body** (Fig. 34-40). In marsupials (pouched mammals), the Barr body is always the paternally inherited X chromosome, an epigenetic phenomenon (Section 30-7). In placental mammals, however, one randomly selected X chromosome in every somatic cell is inactivated when the embryo consists of only a few cells. The progeny of each of these cells epigenetically maintain the same inactive X chromosomes. *Female placental mammals are therefore mosaics composed of clonal groups of cells in which the active X chromosome is either paternally or maternally inherited.* This situation is particularly evident in human females who are heterozygotes for the X-linked congenital sweat gland deficiency **anhidrotic ectodermal dysplasia.** The skin of these women consists of patches lacking sweat glands, in which only the X chromosome containing the mutant gene is active, alternating with normal patches in which only the other X chromosome is active. Similarly, calico cats, whose coats consist of patches of black fur and yellow fur, are almost always females whose two X chromosomes are allelic for black and yellow furs.

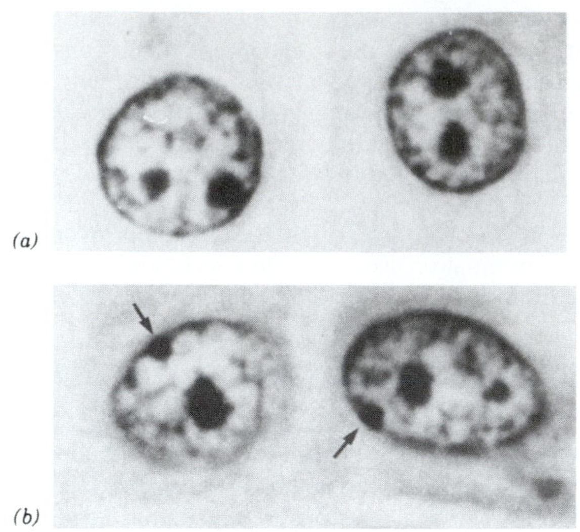

FIGURE 34-40 Photomicrographs of stained nuclei from human oral epithelial cells. (*a*) From a normal XY male showing no Barr body. (*b*) From a normal XX female showing a single Barr body (*arrow*). The presence of Barr bodies permits the rapid determination of an individual's chromosomal sex. [From Moore, K.L. and Barr, M.L., *Lancet* **2,** 57 (1955).]

The mechanism of X chromosome inactivation is only beginning to come to light. Inactivation appears to be triggered by the transcription of the ***Xist*** gene in the inactive chromosome only. The consequent *Xist* RNA "paints" the inactive X chromosome over its entire length but does not bind to the active X chromosome. This localized *Xist* RNA, through a cascade of poorly understood changes, recruits DNA-binding proteins that repress transcription (Section 34-3B) as well as variant histones, particularly the H2A variant **macroH2A1,** which has a large C-terminal globular domain that H2A lacks. X inactivation also results in modification of its bound core histones, particularly the methylation of H3 Lys 4, the demethylation of H3 Lys 9, and the hypoacetylation of histones H3 and H4 (histone methylation and acetylation are discussed in Section 34-3B). In addition, the DNA of the *Xist* gene on the active X chromosome becomes methylated, whereas the active *Xist* gene on the otherwise inactive X chromosome remains unmethylated (DNA methylation is discussed in Section 30-7). Conversely, the CpG islands within many promoters on the inactive X chromosome become methylated, whereas those on the active X chromosome remain unmethylated. Some or all of these changes are essential for X chromosome inactivation and, moreover, almost certainly provide the epigenetic imprint that is responsible for maintaining the inactive X chromosome's state of inactivity in subsequent cell generations.

b. Chromosome Puffs and Lampbrush Chromosomes Are Transcriptionally Active

The condensed state of facultative heterochromatin presumably renders it transcriptionally inactive by making its DNA inaccessible to the proteins mediating transcription. Conversely, *transcriptionally active chromatin must have a relatively open structure.* Such decondensed chromatin occurs in the **chromosome puffs** that emanate from single bands of giant polytene chromosomes (Fig. 34-41). These puffs reproducibly form and regress as part of the normal larval development program and in response to such physiological stimuli as hormones and heat. Autoradiographic studies with ^{3}H-labeled uridine and immunofluorescence studies using antibodies against RNA polymerase II clearly demonstrate that *puffs are the major sites of RNA synthesis in polytene chromosomes.*

In amphibian oocytes, the analogous decondensation of nonpolytene chromosomes occurs most conspicuously in the so-called **lampbrush chromosomes** (Fig. 34-42). During their prolonged meiotic prophase I (Fig. 1-20), these previously condensed chromosomes loop out segments of transcriptionally active DNA that electron micrographs such as Fig. 34-43 indicate are often single transcription units.

B. *Regulation of Transcriptional Initiation*

The foregoing observations suggest that selective transcription is mainly responsible for the differential protein synthesis among the various types of cells in the same organism. It was not until 1981, however, that James Darnell actually demonstrated this to be the case, as follows. Experimentally useful amounts of mouse liver genes were obtained by inserting the cDNAs of mouse liver mRNAs (some 95% of which are cytosolic) into plasmids and replicating them in *E. coli* (Section 5-5B). By hybridizing the resulting cloned cDNAs with radioactively labeled mRNAs from various mouse cell types, the *E. coli* colonies

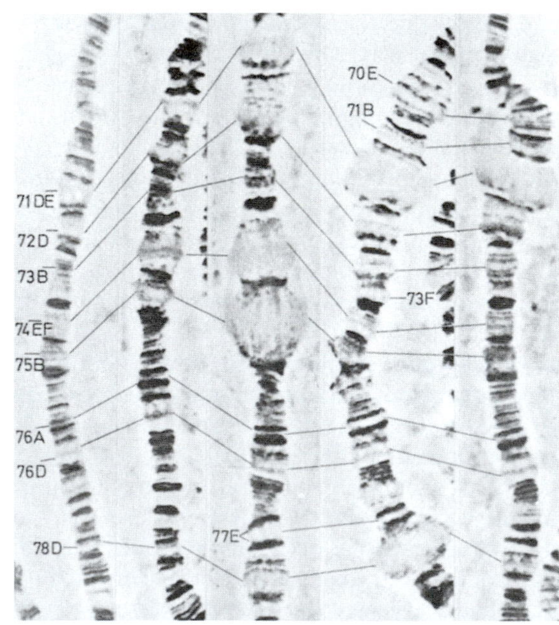

FIGURE 34-41 Formation and regression of chromosome puffs (*lines*) in a *D. melanogaster* polytene chromosome over a 22-h period of larval development. The very large puffs in this series of photomicrographs are also known as **Balbiani rings.** [Courtesy of Michael Ashburner, Cambridge University.]

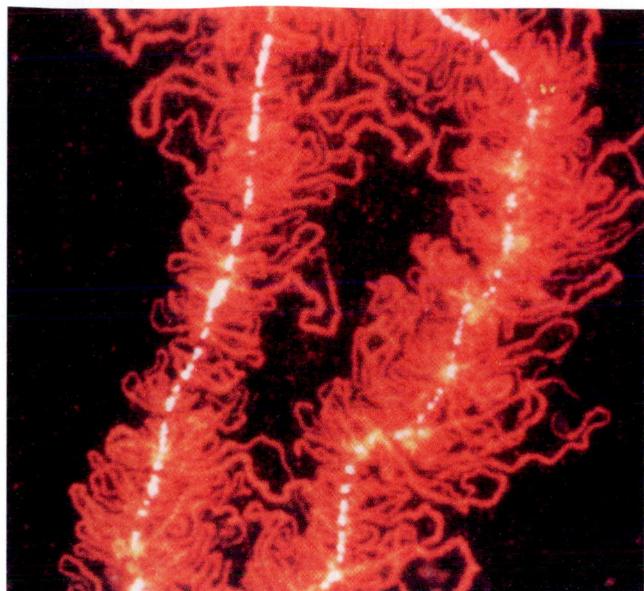

FIGURE 34-42 Immunofluorescence micrograph of a lampbrush chromosome from an oocyte nucleus of the newt *Notophthalmus viridescens*. The chromosome's numerous transcriptionally active loops give rise to the name "lampbrush" (an obsolete implement for cleaning kerosene lamps). [From Roth, M.B. and Gall, J.G., *J. Cell Biol.* **105,** 1049 (1987). Copyright © 1987 by Rockefeller University Press.]

containing liver-specific genes were distinguished from colonies containing genes common to most mouse cells. In this way, 12 liver-specific cDNA clones and three common cDNA clones were obtained. The question was then asked, does a eukaryotic cell transcribe only the genes encoding the proteins it synthesizes, or does it transcribe all of its genes but only process properly the transcripts it translates? This question was answered by hybridizing the cloned mouse genes with freshly synthesized and therefore unprocessed RNAs (hnRNAs) obtained from the nuclei of mouse liver, kidney, and brain cells (Fig. 34-44). Only the

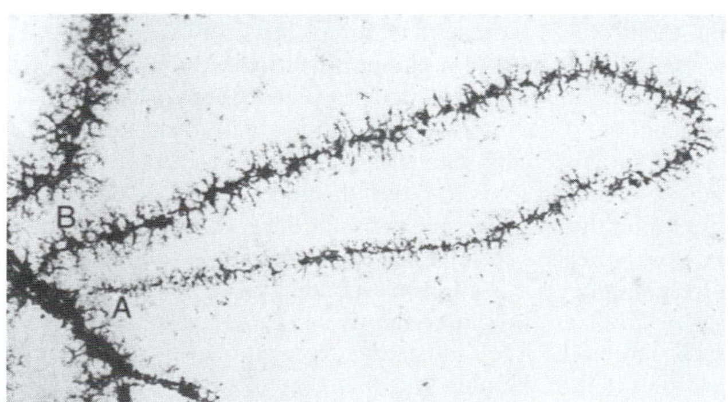

FIGURE 34-43 Electron micrograph of a single loop of a lampbrush chromosome. The ribonucleoprotein matrix coating the loop increases in thickness from one end of the loop (A) to the other (B), which indicates that the loop comprises a single transcriptional unit. [Courtesy of Oscar L. Miller, Jr., University of Virginia.]

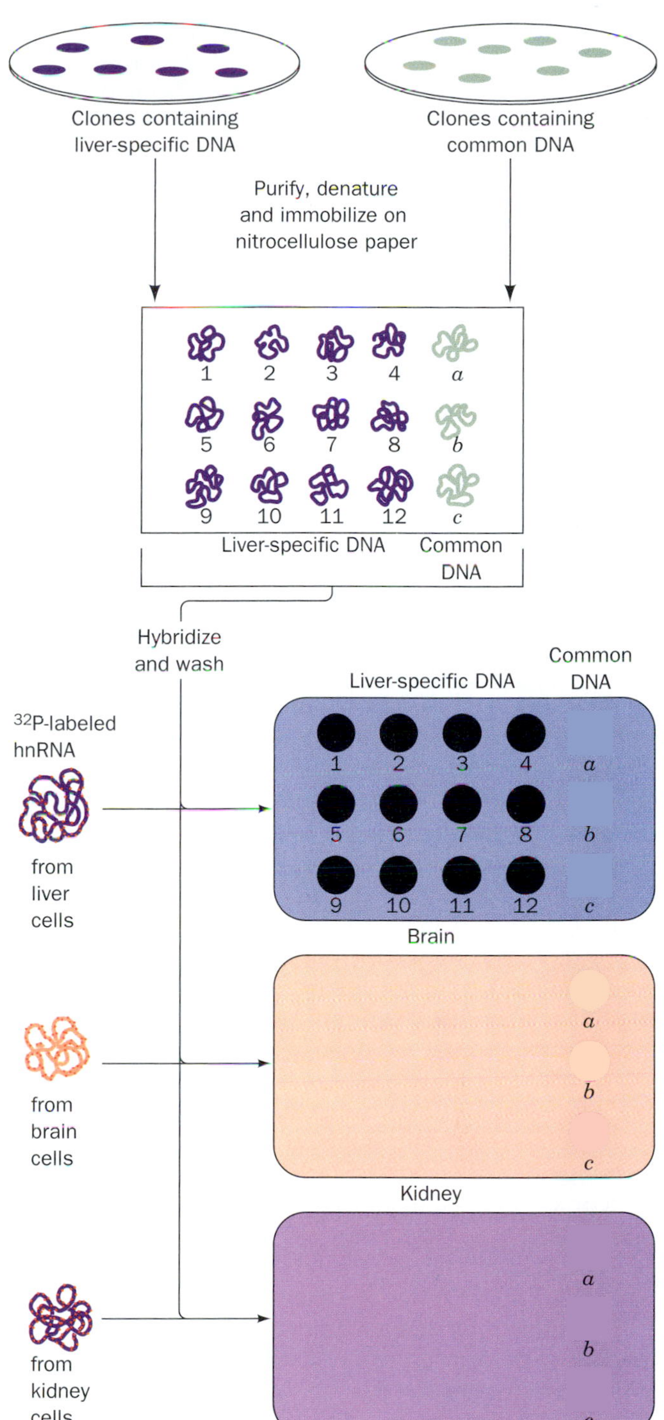

FIGURE 34-44 Determination of the primary role of selective transcription in the control of eukaryotic gene expression. This was established as follows. Cloned cDNAs encoding 12 different mouse liver-specific proteins (1–12) and 3 different proteins common to most mouse cells (*a–c*) were purified, denatured, and spotted onto filter paper (*top*). The DNAs were hybridized with newly formed and therefore unprocessed radioactively labeled RNAs produced by either mouse liver, kidney, or brain nuclei (*lower left*). Autoradiography showed that the liver RNAs hybridized with all 12 liver-specific cDNAs and all 3 common cDNAs but that the kidney and brain RNAs only hybridized with the common cDNAs (*right*).

RNAs extracted from liver nuclei hybridized with the 12 liver-specific cDNAs that were probed. The RNAs from all three cell types, however, hybridized with the DNA from the three clones containing the common mouse genes. Evidently, *liver-specific genes are not transcribed by brain or kidney cells. This strongly suggests that the control of genetic expression in eukaryotes is primarily exerted at the level of transcription.*

In more recent times, the use of DNA microarray technology (DNA chips; Section 7-6B) has enormously increased the number of genes whose levels of transcription may be simultaneously monitored as well as greatly reduced the effort required to do so. For example, **hepatocellular carcinoma (HCC),** the most common liver malignancy, is among the five leading causes of cancer deaths in the world and is closely associated with chronic infections by hepatitis B or C viruses although the nature of this association is unclear. David Botstein and Patrick Brown characterized the gene expression patterns in HCC tumors and normal liver tissues by isolating their mRNAs and reverse transcribing them to cDNAs, which were then coupled to a fluorescent dye. These labeled cDNAs were then hybridized to DNA microarrays containing ~17,400 human genes. The level of transcription of each of these genes in a given tissue sample was determined from the fluorescence intensity at the corresponding position on the microarray relative to that of a reference cDNA that had a different fluorescent label. The results of this exhaustive analysis are presented in Fig. 34-45, which clearly indicates that HCC tumors have transcriptional patterns that differ from those of nontumor liver tissues. Surprisingly, however, different HCC nodules from the same patient exhibited gene expression patterns whose similarities were no greater than those of tumors from different patients. Nevertheless, certain genes are consistently expressed at high levels in HCC tumors, and hence those whose products are secreted or membrane associated may serve as serological markers for the early detection of liver cancers and/or as potential therapeutic targets.

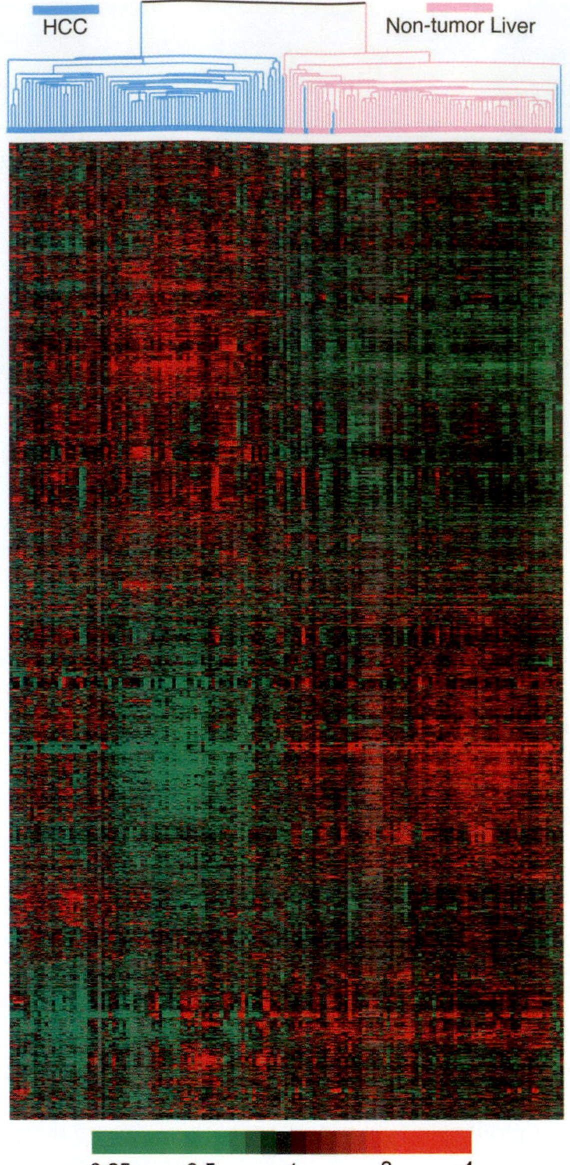

FIGURE 34-45 Relative transcriptional activities of the genes in hepatocellular carcinoma (HCC) tumors as determined using DNA microarrays. The data are presented in matrix form, with each column representing one of 156 tissue samples (82 HCC tumors and 74 nontumor liver tissues) and each row representing one of 3180 genes (those of the ~17,400 genes on the DNA microarray with the greatest variation in transcriptional activity among the various tissue samples). The data are arranged so as to group the genes as well as the tissue samples on the basis of similarities of their expression patterns. The color of each cell indicates the expression level of the corresponding gene in the corresponding tissue relative to its mean expression level in all the tissue samples with bright red, black, and bright green indicating expression levels of 4, 1, and 1/4 times that of the mean for that gene (as indicated on the scale below). The dendrogram at the top of the matrix indicates the similarities in expression patterns among the various tissue samples. [Courtesy of David Botstein and Patrick Brown, Stanford University School of Medicine.]

a. The Transcriptional Initiation of Structural Genes Involves Three Classes of Transcription Factors

Transcriptional initiation in eukaryotes has been most widely studied in protein-encoding genes, that is, genes that are transcribed by RNA polymerase II **(RNAP II).** In the following paragraphs, we concentrate on the major findings of these studies.

Differentiated eukaryotic cells possess a remarkable capacity for the selective expression of specific genes. The synthesis rates of a particular protein in two cells of the same organism may differ by as much as a factor of 10^9; that is, unexpressed eukaryotic genes are completely turned off. In contrast, simply repressible prokaryotic systems such as the *E. coli lac* operon (Section 31-3B) exhibit no more than a thousand-fold range in their transcriptional rates; they have significant basal levels of expression. Nevertheless, as we shall see below, *the basic mechanism of expressional control in eukaryotes resembles that in prokaryotes: the selective binding of proteins to specific genetic control sequences so as to modulate the rate of transcriptional initiation.*

RNAP II, unlike prokaryotic RNA polymerase holoenzyme (Section 31-2), has little if any inherent ability to bind to its promoters. Rather, three different classes of so-called **transcription factors** have been implicated in regulating transcriptional initiation by RNAP II:

1. General transcription factors (GTFs), which are required for the synthesis of all mRNAs, select the transcriptional initiation site and deliver RNAP II to it, thereby forming a complex that initiates transcription at a basal rate.

2. Upstream transcription factors are proteins that bind to specific DNA sequences upstream of the initiation site so as to stimulate or repress transcriptional initiation by GTF-complexed RNAP II. The binding of upstream factors to DNA is unregulated; that is, they bind to any available DNA containing their target sequence. Those that are present in a cell vary with its developmental state and its needs; their synthesis is also regulated.

3. Inducible transcription factors function similarly to upstream transcription factors but must be activated (or inhibited), either by phosphorylation or by specific ligands, in order bind to their target DNA sites and influence transcriptional initiation. They are synthesized and/or activated in specific tissues at particular times and therefore mediate gene expression in a positionally and temporally specific manner.

We discuss these transcription factors below.

b. The Preinitiation Complex Is a Large and Complex Assembly

Extensive research in numerous laboratories has revealed that *the accurate transcriptional initiation of most structural genes requires the presence of six GTFs, most of which are multiprotein complexes, named **TFIIA, TFIIB, TFIID, TFIIE, TFIIF,** and **TFIIH*** (Table 34-3; TF for transcription *f*actor and II for **class II genes,** those that are transcribed by RNAP II). These GTFs combine, in an ordered pathway, with RNAP II and promoter-containing DNA near the transcriptional start site to form a so-called **preinitiation complex (PIC)** that supports a basal level of transcription. The so-called **core promoters** in structural genes are largely upstream of the transcriptional start site and often contain a **TATA box,** a segment of the sense strand (the DNA strand with the same sequence as its corresponding mRNA; Section 31-2A) that is centered at around position −27 and whose consensus sequence is TATA$^{A}_{T}$ A$^{A}_{T}$ (Fig. 31-23). The sequence motifs in a typical core promoter are indicated in Fig. 34-46. The remaining portions of the promoter, to which various transcription factors are idiosyncratically targeted, are known as **upstream activation sequences (UASs).** An entire eukaryotic promoter typically extends over ~100 bp.

*The assembly of the preinitiation complex, which is diagrammed in Fig. 34-47, begins with the binding of **TATA box–binding protein (TBP)** to the TATA box, thereby identifying the transcriptional start site (recall that eliminating the TATA box does not necessarily eliminate transcription but does result in heterogeneities in the transcriptional start site; Section 31-2E). TBP is then joined by a series of ~10 **TBP-associated factors** (**TAFs;** previously called **TAF$_{II}$s** to indicate they are associated with class II genes) to form the ~700-kD multisubunit complex TFIID.* TFIIA then binds to the TFIID–DNA complex so as to stabilize it, followed by TFIIB. At this point, TFIIF recruits RNAP II to the promoter in a manner reminiscent of the way that σ factors interact with core RNA polymerase in bacteria (Sections 31-2A and 31-2B). Indeed, the smaller of human TFIIF's two

TABLE 34-3 Properties of the General Transcription Factors

Factor	Number of Unique Subunits in Yeast	Mass in Yeast (kD)	Number of Unique Subunits in Humans	Mass in Humans (kD)	Functions
TFIIA	2	46	3	69	Stabilizes TBP and TAF binding
TFIIB	1	38	1	35	Stabilizes TBP binding; recruits RNAP II; influences start site selection
TFIID					Recognizes TATA box; recruits TFIIA and TFIIB; has positive and negative regulatory functions
TBP	1	27	1	38	
TAFs	14	~1050	≥12[a]	≥960	
TFIIE	2	184	2	165	An $\alpha_2\beta_2$ heterotetramer; recruits TFIIH and stimulates its helicase activity; enhances promoter melting
TFIIF	3	156	2	87	Facilitates promoter targeting; stimulates elongation
TFIIH	9	518	9	470	Contains an ATP-dependent helicase that functions in promoter melting and clearance

[a]Although only 12 human TAFs have been identified versus 14 yeast TAFs, the close correspondence between each known human TAF and a yeast TAF suggests that two more human TAFs are yet to be identified.

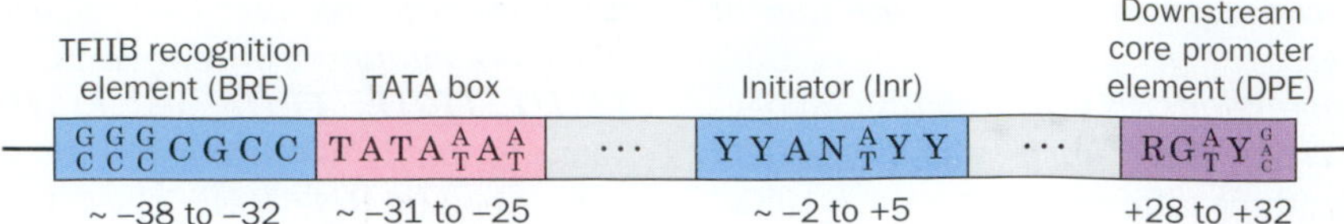

FIGURE 34-46 The sequence elements in a typical class II core promoter. The approximate positions of its various elements relative to the transcriptional initiation site (+1) are indicated below. Note that any or all of these elements may be absent in a given class II promoter.

subunits exhibits substantial sequence homology with σ^{70} (the predominant bacterial σ factor) and, moreover, can specifically interact with bacterial RNAPs (although it does not participate in promoter recognition). Finally, TFIIE and TFIIH join, in that order, to form the PIC. Once this complex has been assembled, an ATP-dependent activation step, probably mediated by TFIIH's helicase function to melt the promoter, is required to initiate transcription at a basal rate. You should note that the human PIC, exclusive of the ~12-subunit, ~600-kD RNAP II, contains at least 25 subunits with an aggregate mass of ~1600 kD. Indeed, many of the proteins in the PIC are the targets of transcriptional regulators.

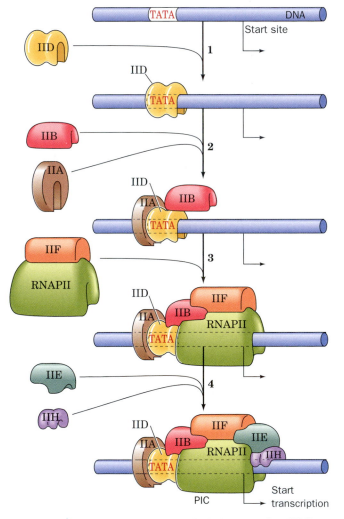

FIGURE 34-47 Assembly of the preinitiation complex (PIC) on a TATA box–containing promoter. (1) TFIID assembles on the TATA box beginning with the binding of TATA box–binding protein (TBP) to the TATA box. **(2)** TFIIA and TFIIB then bind to the growing complex. **(3)** TFIIF then binds to RNAP II and escorts it to the complex. **(4)** Finally, TFIIE and TFIIH are sequentially recruited to the complex, thereby completing the PIC. [After Zawel, L. and Reinberg, D., *Curr. Opin. Cell Biol.* **4,** 490 (1992).]

c. TBP Greatly Distorts Its Bound TATA Box DNA

TBP has a highly conserved (81% identical between yeast and humans) C-terminal domain of 180 residues that contains two ~40% identical direct repeats of 66 or 67 residues separated by a highly basic segment. In contrast, TBP's N-terminal domain is widely divergent, both in length and sequence, and, in fact, is unnecessary for TBP function *in vitro*. Curiously, the human N-terminal domain contains an uninterrupted run of 38 Gln residues, whereas that of *D. melanogaster* contains two blocks of 6 and 8 Gln residues separated by 32 residues, and that of yeast entirely lacks such sequences. Perhaps the N-terminal domains of TBPs have evolved to satisfy species-specific functions.

The X-ray structures of TBP from yeast (only its C-terminal domain) and from the flowering plant *Arabidopsis thaliana* (whose N-terminal domain consists of only 18 residues), by Kornberg and by Stephen Burley, reveal a saddle-shaped molecule (Fig. 34-48*a*) that consists of two structurally similar and topologically identical domains, each composed of one of the direct repeats. These are arranged with pseudo-twofold symmetry such that the protein consists of a 10-stranded antiparallel β sheet, 5 strands from each domain, flanked at each end by two α helices and a loop that is reminiscent of a stirrup hanging from the protein saddle. The curvature of the β pleated sheet saddle is such that it appears that TBP, in agreement with biochemical and genetic evidence, could fit snugly astride the DNA. However, the X-ray structures of the DNA complexes tell quite a different story.

Two closely similar X-ray structures of TBP–DNA complexes have been determined: one by Paul Sigler of yeast TBP in complex with a 27-nt DNA that forms an 11-bp TATA box–containing stem whose ends are joined by a 5-nt loop; and one by Burley of *Arabidopsis* TBP in complex with a 14-bp TATA box–containing duplex DNA. The DNA indeed binds to the concave surface of TBP but with its duplex axis nearly perpendicular rather than parallel to the saddle's "cylindrical" axis (Fig. 34-48*b*). The DNA is kinked by ~45° between the first two and the last two base pairs of its 8-bp TATA element. Between these kinks, the DNA is severely, although smoothly, bent with a radius of

(a)

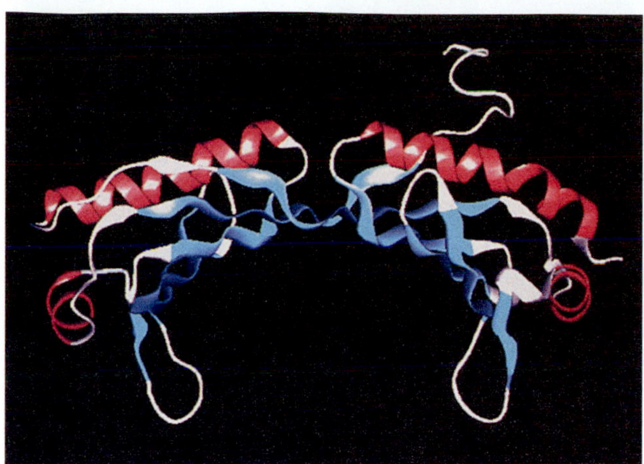

(b)

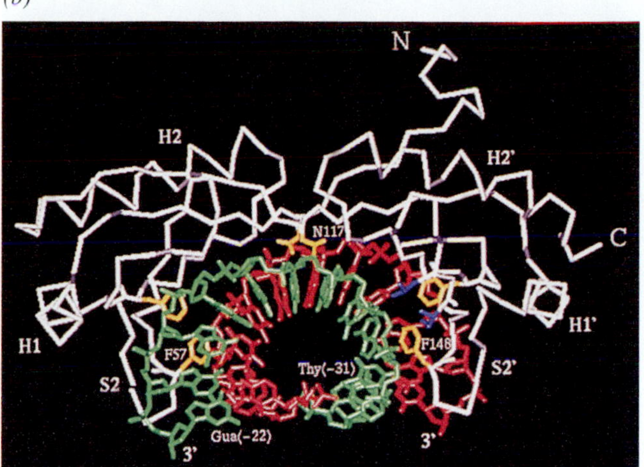

FIGURE 34-48 X-Ray structure of *Arabidopsis thaliana* TATA box–binding protein (TBP). *(a)* A ribbon diagram of the protein in the absence of DNA in which α helices are red, β strands are blue, and the remainder of the polypeptide backbone is white. The protein's pseudo-twofold axis of symmetry is vertical. Note that the protein seems to be precisely the proper size and shape to sit astride a 20-Å-diameter cylinder of B-DNA. This, however, is not what happens. *(b)* TBP in complex with a 14-bp TATA box–containing segment of the adenovirus major late promoter (single-strand sequence GC**TATAAAAG**GGCA, with its TATA box in bold) viewed as in Part *a*. The protein is represented as its C_α backbone (*white*); together with the side chains of Phe residues 57, 74, 148, and 165 (*yellow*), which induce sharp kinks in the DNA; Asn residues 27 and 117 (*also yellow*), which make hydrogen bonds in the minor groove; and Ile 152 and Leu 163 (*blue*), which are implicated in specific DNA recognition. The DNA is drawn in stick form with the sense and antisense strands in green and red, respectively. B-form DNA enters its binding site with the 5′ end of the sense strand below the saddle on the right and exits on the left with its helix axis nearly perpendicular to the page (the last two base pairs have been removed for clarity). Between the kinks, which are located at each end of the TATA box, the DNA is partially unwound with the protein's central 8 strands of its 10-stranded antiparallel β sheet inserted into the DNA's greatly widened minor groove. [Courtesy of Stephen Burley, Structural GenomiX, Inc., San Diego, California.]

curvature of ~25 Å and unwound by ~1/3 of a turn. This permits the protein's antiparallel β sheet to bind in the DNA's greatly widened and more shallow minor groove through hydrogen bonding and van der Waals interactions (the protein does not contact the DNA's major groove). A noteworthy aspect of this remarkable structure is that each kink in the DNA is stabilized by a wedge of two Phe side chains extending from the adjacent stirrup that pries apart the base pairs flanking the kink from their minor groove side and severely buckles the interior base pair. As a result of these unprecedented distortions to the DNA (the protein undergoes only slight conformational adjustments on binding DNA), there is an ~100° angle and a lateral 18-Å displacement between the helix axes of the B-form DNA entering and leaving TBP's binding site, thereby giving the DNA a cranklike shape. The DNA, nevertheless, maintains normal Watson–Crick pairing throughout the distorted region.

d. TFIIA and TFIIB Both Bind to DNA and TBP

The X-ray structures of ternary complexes of yeast TFIIA, TBP, and a TATA box–containing promoter DNA were independently determined by Richmond and Sigler and those of ternary complexes of human TFIIB, human or *Arabadopsis* TBP, and a TATA box–containing promoter DNA were independently determined by Sigler and by Robert Roeder and Burley. The TBP–DNA complexes in all the foregoing binary and ternary complexes are closely similar. A plausible model of the TFIIA–TFIIB–TBP–DNA quaternary complex was therefore constructed by superimposing the TBP–DNA complexes in TFIIA- and TFIIB-containing ternary complexes (Fig. 34-49). TFIIA, a heterodimer in yeast, consists of a 6-stranded β barrel and a 4-helix bundle that, together, have a bootlike shape. The β barrel domain of TFIIA binds to TBP's N-terminal stirrup so as to extend TBP's β sheet to form a continuous 16-stranded β sheet. In addition, the TFIIA β barrel binds to the DNA over its major groove through salt bridges between four of its Lys and Arg side chains and the DNA's phosphate groups. TFIIB, a monomer, consists of two similar α helical domains that are rotated by 90° with respect to one another so as to form a cleft that clamps the TBP's C-terminal stirrup. TFIIB binds to the DNA via both its domains through several salt bridges with the DNA's phosphate groups as well as base-specific contacts in both the major and minor grooves to the consensus sequence $^{\rm GGG}_{\rm CCC}$CGCC, which occurs just upstream of the TATA box in many core promoters (Fig. 34-46). The formation of these interactions requires the distortions that TBP binding imposes on the DNA structure and hence TFIIB binding is synergistic with TBP binding. Since the pseudosymmetric TBP has been shown to bind to the TATA box in either orientation, it appears that the base-specific interactions between TFIIB and the promoter at its so-called **TFIIB recognition element (BRE;** Fig. 34-46) function to position TFIIB to properly orient

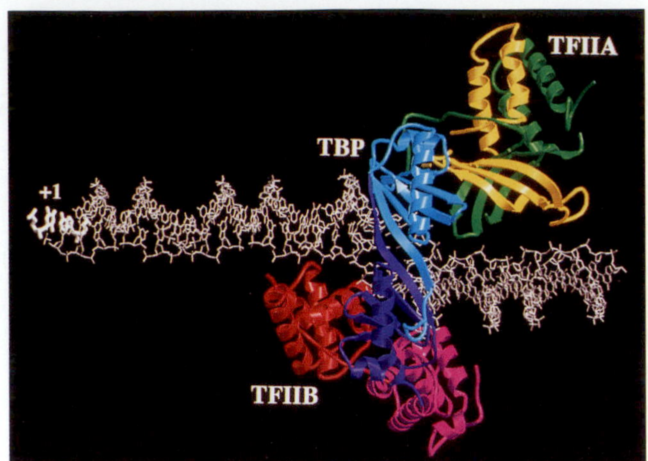

FIGURE 34-49 Model of the TFIIA–TFIIB–TBP–TATA box–containing DNA quaternary complex. The arrangement of the proteins (*ribbons*) and DNA (*white stick model*) is based on the independently determined X-ray structures of the TFIIA–TBP–DNA and TFIIB–TBP–DNA ternary complexes. In the model, the DNA has been extended in both directions beyond the TATA box, with its transcription start site (+1) on the left. The TBP's pseudosymmetrically related N- and C-terminal domains are cyan and purple, the TFIIA's two subunits are yellow and green, and the TFIIB's similar N- and C-terminal domains are red and magenta. The TBP binds to both TFIIA and TFIIB and all three proteins bind to the DNA at independent sites. [Courtesy of Stephen Burley, Structural GenomiX, Inc., San Diego, California. Based on PDBids 1YTF and 1VOL.]

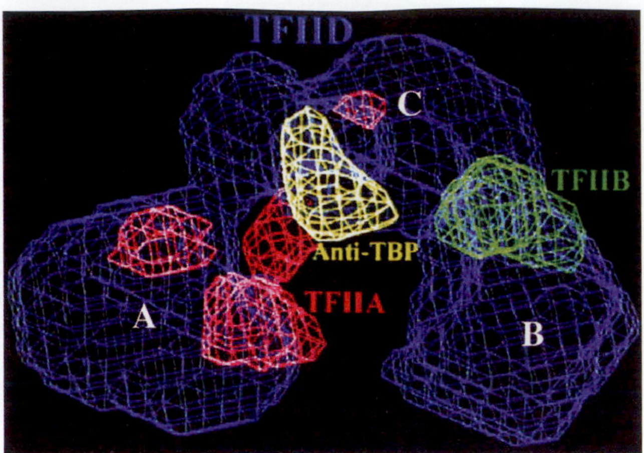

FIGURE 34-50 EM-based image of the human TFIID–TFIIA–TFIIB complex at 35-Å resolution. The blue mesh outlines the entire ternary complex, which consists of three domains, A, B, and C, arranged in a horseshoe shape and roughly 200 Å wide, 135 Å high, and 110 Å thick. The red and green meshes indicate the positions of TFIIA and TFIIB as determined by comparison with the EM-based images of the TFIID–TFIIA complex and TFIID alone. The yellow mesh indicates the binding position of an anti-TBP antibody. The different shapes of TFIIA here and in Fig. 34-49 are probably due to the fact that the TFIIA in Fig. 34-49 consists only of residues 56 to 209 of the 286-residue protein. [Courtesy of Eva Nogales, University of California at Berkeley.]

the TBP on the promoter. The model of the quaternary complex (Fig. 34-49) indicates that its three proteins all bind to the DNA upstream of the transcriptional start site, leaving ample room for the additional protein–DNA and protein–protein interactions that regulate the frequency with which RNAP II is recruited to the promoter.

e. TFIID Is a Horseshoe-Shaped Complex That Probably Contains a Histonelike Octamer

The electron microscopy–based structure of the human TFIID–TFIIA–TFIIB complex was determined at 35-Å resolution by Robert Tjian and Eva Nogales. TFIID is a horseshoe-shaped trilobal complex to which TFIIA and TFIIB are bound on opposite lobes that flank the central cavity (Fig. 34-50). This, together with the foregoing model of the TFIIA–TFIIB–TBP–DNA quaternary complex (Fig. 34-49), strongly suggests that TBP is located at the top of the cavity where it can contact both TFIIA and TFIIB and that the core promoter DNA passes through the cavity, where it is bound by TBP, TFIIA, and TFIIB. Indeed, the EM-based image of an anti-TBP antibody in complex with TFIID reveals that the antibody binds to TFIID in the expected position (Fig. 34-50).

The various TAFs are highly conserved from yeast to humans. Moreover, portions of 9 of the 14 known species of TAFs are homologous to nonlinker histones. For example, segments consisting of residues 17 to 86 of the 268-residue **dTAF9** (d for *Drosophila;* previously called **dTAF$_{II}$42,** where the number indicates its nominal molecular mass in kD) and residues 1 to 70 of the 592-residue **dTAF6** (previously called **dTAF$_{II}$60**) are, respectively, homologous to histones H3 and H4. The X-ray structure of the dTAF9(17-86)–dTAF6(1-70) complex, determined by Roeder and Burley, reveals that both of these polypeptide segments assume the histone fold (Fig. 34-51*a*): two short helices flanking a long central helix (Fig. 34-8). In fact, TAF9(17-86) and TAF6(1-70) associate quite similarly to H3 and H4 in the nucleosome (Fig. 34-7) to form an $\alpha_2\beta_2$ heterotetramer (Fig. 34-51*a*). In addition, hTAF12(57-128) (h for human; **hTAF12,** which has 161 residues, was previously called **hTAF$_{II}$20**), which is homologous to histone H2B, forms a complex with hTAF4(870-943) **(hTAF4,** which has 1083 residues, was previously called **hTAF$_{II}$135),** which is homologous to H2A. The X-ray structure of this complex, determined by Dino Moras, reveals that it forms a histonelike heterodimer (Fig. 34-51*b*) but not a histonelike tetramer. **TAF11** and **TAF13** also form a histonelike heterodimer.

Gel filtration chromatography and sedimentation measurements (Sections 6-3B and 6-5A) by Stephen Buratowski and Song Tan indicate that the heterotetramer of **yTAF6** (y for yeast) and **yTAF9** associates with two heterodimers of **yTAF12** and **yTAF4** to form a heterooctamer. The mutation to Ala or Tyr of the highly conserved Leu 464 of yTAF12 (the homolog of H2B residue Leu 77, which is located near the C-terminus of this histone's long

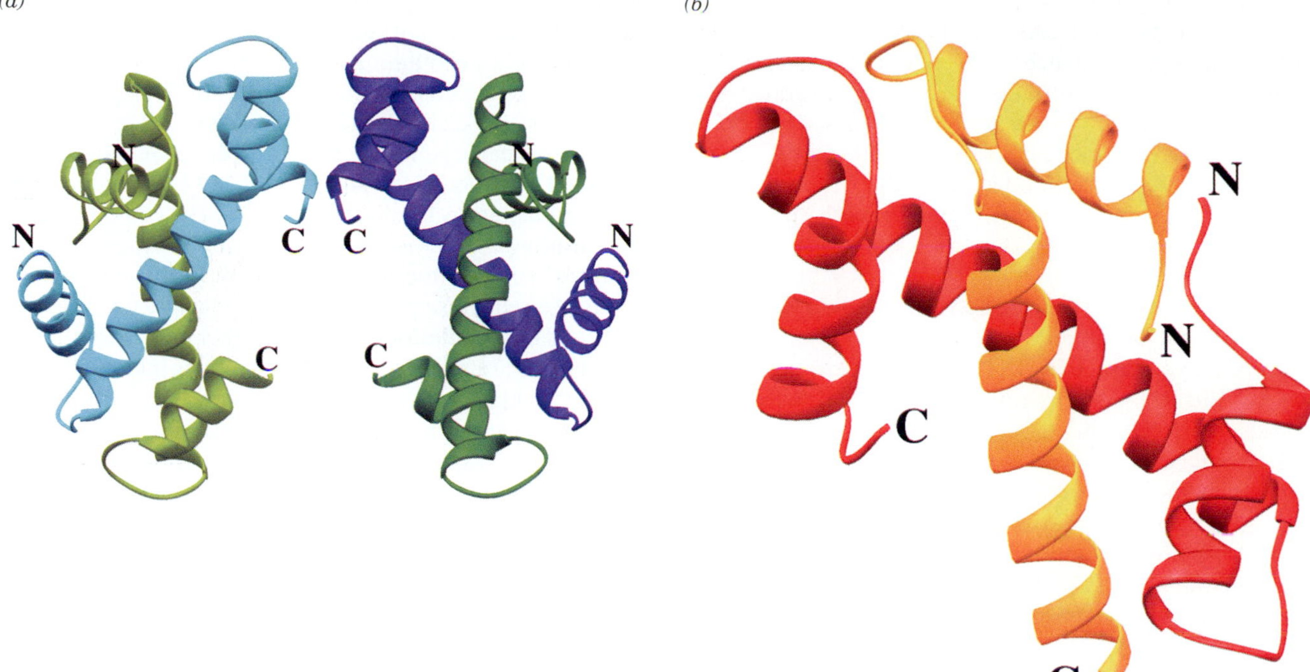

FIGURE 34-51 X-Ray structures of TAFs that form histonelike complexes. (*a*) The dTAF9(17-86)–dTAF6(1-70) $\alpha_2\beta_2$ heterotetramer as viewed with its twofold axis vertical. Note how the H3-like TAF9 segments (*blue and cyan*) and the H4-like TAF6 segments (*green and olive*) all assume the histone fold, how TAF9–TAF6 pairs interdigitate in head-to-tail arrangements to form heterodimers, and how the two TAF9 segments interact via a four-helix bundle to form the heterotetramer, much as do histones H3 and H4 in nucleosome cores (Figs. 34-7 and 34-8). [Based on an X-ray structure by Robert Roeder and Stephen Burley, The Rockefeller University. PDBid 1TAF.] (*b*) The hTAF12(57-128)–hTAF4(870-943) heterodimer. Note how the H2B-like TAF12 segment (*red*) forms a regular histone fold but that the H2A-like TAF4 segment (*gold*) lacks the histone fold's C-terminal loop and helix. This is because TAF4 residues 918 to 943 are disordered. Nevertheless, the two subunits interdigitate to form a heterodimer, much as do histones H2B and H2A in nucleosome cores (Fig. 34-8). [After an X-ray structure by Dino Moras, CNRS/INSERM/ULP, Illkirch Cédex, France. PDBid 1H3O.]

central helix and hence occupies the hydrophobic core of the H4–H2B four-helix bundle; Fig. 34-8) prevents the formation of this octamer. This suggests that the octamer is held together by 4-helix bundles between yTAF6 and yTAF12 similar to those between H4 and H2B in nucleosomes (Figs. 34-7*b* and 34-8). Indeed, a model of this interface constructed from the above two X-ray structures suggests that its putative 4-helix bundle is remarkably similar to that of the H4–H2B interface. Nevertheless, it seems unlikely that this putative TAF octamer is wrapped with DNA in the PIC as is the histone octamer in the nucleosome. This is because most of the histone residues that make critical contacts with DNA in the nucleosome have not been conserved in the foregoing TAFs and, in fact, many of them have been replaced in these TAFs by highly conserved (in the TAFs) acidic residues, which would repel the anionic DNA.

f. Many Class II Core Promoters Lack a TATA Box

The core promoters of 65% of class II genes lack TATA boxes. They are mostly "housekeeping" genes; that is, genes that are constitutively expressed in all cells at relatively low rates. How can RNAP II properly initiate transcription at these TATA-less promoters? Investigations have shown that TATA-less promoters often contain a so-called **initiator (Inr)** element that extends from positions −6 to +11 and that contains the loose consensus sequence YYAN$^{\text{A}}_{\text{T}}$YY, where Y is a pyrimidine (C or T), N is any nucleotide, and A is the initiating (+1) nucleotide (Fig. 34-46). The presence of the Inr element is sufficient to direct RNAP II to the correct start site. These systems require the participation of many of the same GTFs that initiate transcription from TATA box–containing promoters. Surprisingly, they also require TBP. This suggests that with TATA-less promoters, Inr recruits TFIID such that its component TBP binds to the −30 region in a sequence-nonspecific manner. Indeed, in Inr-containing promoters that also contain a TATA box, the two elements act synergistically to promote transcriptional initiation. Nevertheless, a mutant TBP that is defective in TATA box binding will support efficient transcription from some TATA-less promoters although not from others. This suggests that the former promoters do not require a stable interaction with TBP. Some TATA-less promoters have a so-called **downstream core promoter element (DPE),** which has the consensus sequence RG$^{\text{A}}_{\text{T}}$Y$^{\text{G}}_{\text{A,C}}$, where R is a purine (A or G) and is located precisely from +28 to +32 (Fig. 34-46).

The foregoing suggests that there are variants of at least some of the GTFs and TAFs. In fact, the human genome contains multiple sequences related to TFIIA and TFIID subunits as well as alternative genes for several TAFs. Some of these variant genes are only expressed in certain cell types and/or at specific developmental stages. The resulting variant transcription factors probably recognize alternative core promoter elements and/or mediate selective interactions with upstream transcription factors.

g. Class I and Class III Genes Also Require TBP for Transcriptional Initiation

RNA polymerase I (**RNAP I,** which synthesizes most rRNAs) and RNA polymerase III (**RNAP III,** which synthesizes 5S rRNA and tRNAs) require different sets of GTFs from each other and from RNAP II to initiate transcription at their respective promoters. This is not unexpected considering the very different organizations of these three classes of promoters (Section 31-2E). Indeed, the promoters recognized by RNAP I (class I promoters) and nearly all those recognized by RNAP III (class III promoters) lack TATA boxes. Thus, it came as a surprise when it was demonstrated that *TBP is required for initiation by both RNAP I and RNAP III.* It participates by combining with different sets of TAFs to form the GTFs **SLI** (with class I promoters) and **TFIIIB** (with class III promoters). As with certain class II TATA-less promoters, a TBP mutant that is defective for TATA-box binding can still support *in vitro* transcriptional initiation by both RNAP I and RNAP III. Clearly, TBP, the only known universal transcription factor, is an unusually versatile protein.

h. Transcriptional Initiation of Class II Genes Is Mediated by Cell-Specific Upstream Transcription Factors Bound to Promoter and Enhancer Elements

The use of molecular cloning procedures has permitted the demonstration that *eukaryotic promoter and enhancer elements mediate the expression of cell-specific genes* (recall that an enhancer is a gene sequence that is required for the full activity of its associated promoter but that may have a variable position and orientation with respect to that promoter; Section 31-2E). For example, William Rutter linked the 5′-flanking sequences of either the insulin or the chymotrypsin gene to the sequence encoding **chloramphenicol acetyltransferase (CAT),** an easily assayed enzyme not normally present in eukaryotic cells. A plasmid containing the insulin gene recombinant elicits expression of the CAT gene only when introduced into cultured cells that normally produce insulin. Likewise, the chymotrypsin recombinant is only active in chymotrypsin-producing cells. Dissection of the insulin control sequence indicates that the segment between its positions −103 and −333 contains an enhancer: In insulin-producing cells only, it stimulates the transcription of the CAT gene with little regard to the enhancer's position and orientation relative to its promoter.

The foregoing indicates that cells contain specific transcription factors, the upstream transcription factors, that recognize the promoters and enhancers in the genes they transcribe. For instance, Tjian isolated a protein, **Sp1** (for *s*pecificity *p*rotein-1), from cultured human cells that stimulates, by factors of 10 to 50, the transcription of cellular and viral genes containing at least one properly positioned GC box [GGGCGG (Section 31-2E); Fig. 34-52]. This protein binds, for example, to the 5′-flanking region of the SV40 virus early genes so as to protect its GC boxes from DNase I digestion (Fig. 34-53*a*; **DNase I footprinting**) and from methylation by dimethyl sulfate (Fig. 34-53*b*; **DMS footprinting**). Likewise, Sp1 specifically interacts with the four GC boxes in the upstream region of the mouse dihydrofolate reductase gene and with the single GC boxes in the human **metallothionein I$_A$** and **II$_A$** promoters (metallothioneins are metal ion–binding proteins that participate in heavy metal ion detoxification processes and whose synthesis is triggered by heavy metal ions).

Upstream transcription factors are essential participants in controlling the differential expression of the various globin genes in the human embryo, fetus, and adult (Section 34-2F). A typical β-globin gene promoter, in addition to its TATA box, has two positive-acting promoter elements: a CCAAT box near the −70 to −90 region and a CACCC motif at variable sites but often near positions −95 to −120 (Section 31-2E). Their importance is demonstrated by the observations that individuals with point mutations in their TATA or CACCC elements have reduced β-globin levels. These promoter elements are specifically bound by upstream transcription factors. Thus, the CCAAT box is bound by the ubiquitous transcription factor **CP1** and the CACCC element is bound by Sp1, which also binds to other globin promoter sequences that resemble Sp1's consensus binding sequence. Four erythroid-specific upstream transcription factors have also been implicated in globin gene expression: **GATA-1** (so named because it binds to sequences that contain the conserved core GATA), **NF-E2** (NF-E for *n*uclear *f*actor-*e*rythroid), **NF-E3,** and **NF-E4** (GATA-1 was previously named NF-E1).

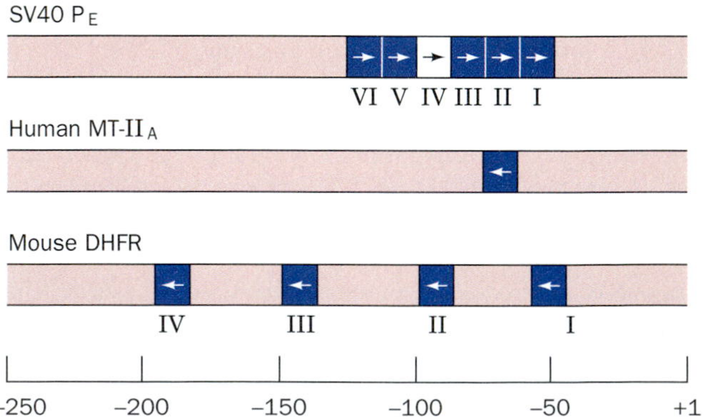

FIGURE 34-52 Arrangement and relative orientations of the GC boxes in the indicated promoters. Each arrow indicates the relative orientation of a GC box, which has the sequence NGGGCGGNNN. The blue boxes represent Sp1-binding sites, whereas SV40 GC box IV is shown as a white box because Sp1 bound at GC box V prevents this transcription factor from efficiently binding to GC box IV. The transcription start site is designated by +1. DHFR = dihydrofolate reductase; MT = metallothionein. [After Kadonaga, J.T., Jones, K.A., and Tjian, R., *Trends Biochem. Sci.* **11,** 21 (1986).]

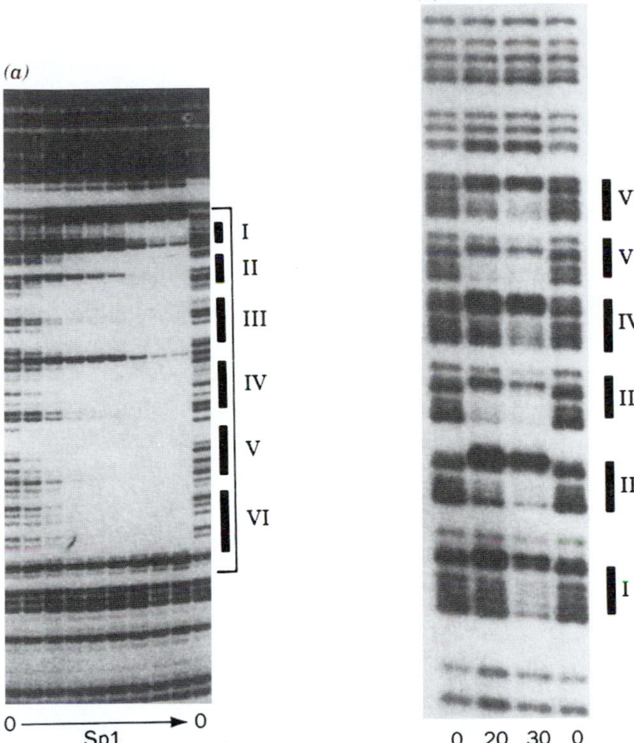

FIGURE 34-53 Identification of the Sp1-binding sites on the SV40 early promoter (Fig. 34-52, *top*). (*a*) Pancreatic DNase I is a relatively nonspecific endonuclease. In a DNase I footprinting assay, a DNA segment that is ^{32}P end-labeled on one strand is incubated with a binding protein and then lightly digested with DNase I such that, on average, each labeled DNA strand is cleaved only once. The DNA is then denatured, the resulting labeled fragments separated according to size by electrophoresis on a sequencing gel (Section 7-2A), and detected by autoradiography. Unprotected DNA is cleaved more or less at random and therefore appears as a "ladder" of bands, each representing an additional nucleotide (as in a sequencing ladder; Figs. 7-14 and 7-15). In contrast, the DNA sequences that the protein protects from DNase I cleavage have no corresponding bands. In the above footprint, the lanes labeled "0" are the DNase I digestion pattern in the absence of Sp1 and in the other lanes the amount of Sp1 increases from left to right. The footprint boundary is delineated by the bracket and the positions of SV40 GC boxes I to VI are indicated. [From Kadonaga, J.T., Jones, K.A., and Tjian, R., *Trends Biochem. Sci.* **11,** 21 (1986). Copyright © 1986 by Elsevier Biomedical Press.] (*b*) Dimethyl sulfate (DMS) methylates DNA's G residues at their N7 positions, which on treatment with weak base, excises the methylated G nucleosides from the DNA, thereby cleaving its sugar–phosphate backbone. In DMS footprinting, a protein-complexed ^{32}P end-labeled DNA segment is lightly treated with DMS such that each labeled DNA strand is, on average, cleaved only once. The resulting fragments are electrophoretically separated on a sequencing gel and detected by autoradiography. The DNA regions that the protein protects from methylation are not cleaved by this procedure and therefore are not represented in the resulting G residue "ladder." In the above autoradiogram, the number below each lane indicates the amount, in μL, of an Sp1 fraction added to a fixed quantity of SV40 early promoter DNA. The positions of its GC boxes are indicated. [From Gidoni, D., Katonaga, J.T., Barrera-Saldana, H., Takahashi, K., Chambon, P., and Tjian, R., *Science* **230,** 516 (1985). Copyright © 1985 by the American Society for the Advancement of Science.]

Analysis of hereditary persistence of fetal hemoglobin (HPFH), a syndrome characterized by the inappropriate expression of γ genes in human adults (Section 34-2G), has provided valuable insights into the basis of stage-specific globin expression. There are several HPFH variants that differ from normal only by a point mutation in the γ gene promoter. Such mutations might result in either tighter binding of a positive transcription factor or looser binding of a negative regulator. Thus, an HPFH mutation at position −117, which is located in the more upstream of the γ gene's two CCAAT boxes, increases the resemblance of this site to CP1's consensus binding sequence and results in a twofold tighter binding of CP1 to the mutant site. Similarly, HPFH mutations in a GC-rich region close to position −200 result in tighter Sp1 binding.

i. Upstream Transcription Factors Interact Cooperatively with Each Other and the PIC

How do upstream transcription factors stimulate (or inhibit) transcription? *Evidently, when these proteins bind to their target DNA sites in the vicinity of a PIC (in some cases, many thousands of base pairs distant), they somehow activate (or repress) its component RNAP II to initiate transcription.* Transcription factors may bind cooperatively to each other and/or the PIC in a manner resembling the binding of two λ repressor dimers and RNA polymerase to the o_R operator of bacteriophage λ (Section 33-3D), thereby synergistically stimulating (or repressing) transcriptional initiation. Indeed, molecular cloning experiments indicate that many enhancers and **silencers** (the analogs of enhancers that function in the transcriptional repression of their associated gene) consist of segments (modules) whose individual deletion reduces but does not eliminate enhancer/silencer activity. *Such complex arrangements presumably permit transcriptional control systems to respond to a variety of stimuli in a graded manner.* In some cases, however, several transcription factors together with so-called **architectural proteins** cooperatively assemble on an ~100-bp enhancer to form a multisubunit complex, known as an **enhanceosome,** in which the absence of a single subunit all but eliminates its ability to stimulate transcriptional initiation at the associated promoter. Thus, enhancesomes function more like on/off switches rather than providing a graded response. Architectural proteins function to bend and/or otherwise deform enhancers so as to promote the assembly of the other enhanceosome proteins. Enhanceosomes may also contain **coactivators** and/or **corepressors,** proteins that do not bind to DNA but, rather, interact with proteins that do so to activate or repress transcription.

The functional properties of many upstream transcription factors are surprisingly simple. They appear to have (at least) two domains:

1. A DNA-binding domain that binds to the protein's target DNA sequence (and whose structural properties are discussed below).

2. A domain containing the transcription factor's activation function. Sequence analysis indicates that many of these **activation domains** (also called **transactivation domains** because they act in trans with the genes they control) have con-

spicuously acidic surface regions whose negative charges, if mutationally increased or decreased, respectively raise or lower the transcription factor's activity. This suggests that the associations between these transcription factors and a PIC are mediated by relatively nonspecific electrostatic interactions rather than by conformationally more demanding hydrogen bonds. Other types of activation domains have also been characterized, including those with Gln-rich regions, such as Sp1, and those with Pro-rich regions.

The DNA-binding and activation functions of eukaryotic transcription factors can be physically separated (which is why they are thought to occur on different domains). Thus, a genetically engineered hybrid protein, containing the DNA-binding domain of one transcription factor and the activation domain of a second, activates the same genes as the first transcription factor. Indeed, it makes little functional difference as to whether the activation domain is placed on the N-terminal side of the DNA-binding domain or on its C-terminal side. This geometric permissiveness in the binding between the activation domain and its target protein is also indicated by the observation that transcription factors are largely insensitive to the orientations and positions of their corresponding enhancers relative to the transcriptional start site [Section 31-2E; it is also the basis of the two-hybrid system for identifying proteins that interact *in vivo* (Section 19-3C)]. Of course, *the DNA between an enhancer and its distant transcriptional start site must be looped around for an enhancer-bound transcription factor to interact with the promoter-bound PIC (Section 31-2E).*

The synergy (cooperativity) of multiple transcription factors in initiating transcription may be understood in terms of a simple recruitment model. Suppose an enhancer-bound transcription factor increases the affinity with which a PIC binds to the enhancer's associated promoter so as to increase the rate at which the PIC initiates transcription there by a factor of 10. Then, if another transcription factor binding to a different enhancer subsite likewise increases the initiation rate by a factor of 20, both transcription factors acting together will increase the initiation rate by a factor of 200. *In this way, a limited number of transcription factors can support a much larger number of transcription patterns.* Transcriptional activation, according to this model, is essentially a mass action effect: The binding of a transcription factor to an enhancer increases the transcription factor's effective concentration at the associated promoter (the DNA holds the transcription factor in the vicinity of the promoter), which consequently increases the rate at which the PIC binds to the promoter. This explains why a transcription factor that is not bound to DNA (or even lacks a DNA-binding domain) inhibits transcriptional initiation. Such unbound transcription factors compete with DNA-bound transcription factors for their target sites and thereby reduce the rate at which the PIC is recruited to the associated promoter. This phenomenon, which is known as **squelching,** is apparently why transcription factors in the nucleus are almost always bound to inhibitors unless they are actively engaged in transcriptional initiation.

j. Steroid Receptors Are Examples of Inducible Transcription Factors

Eukaryotic cells express many cell-specific proteins in response to the presence of various hormones. Many of these hormones are **steroids** (Section 25-6C), cholesterol derivatives that mediate a wide variety of physiological and developmental responses (Section 19-1G). For example, the administration of **estrogens** (female sex hormones) such as **β-estradiol**

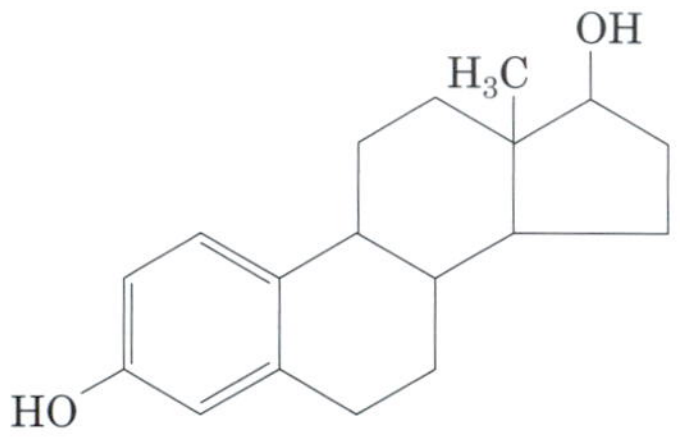

causes chicken oviducts to increase their ovalbumin mRNA level from ~10 to ~50,000 molecules per cell, and the amount of ovalbumin they produce rises from unde-

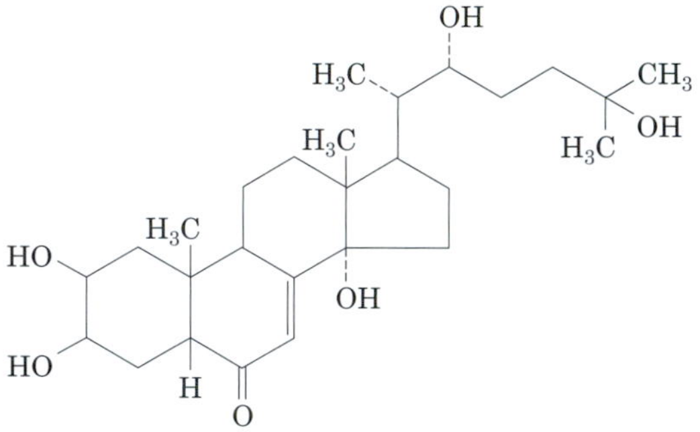

β-Estradiol

Ecdysone

tectable levels to a majority of their newly synthesized protein. Similarly, the insect steroid hormone **ecdysone** mediates several aspects of larval development (the temporal sequence of chromosome puffing shown in Fig. 34-41 can be induced by ecdysone administration).

Steroids, which are nonpolar molecules, spontaneously pass through the membranes of their target cells to bind to their corresponding steroid receptors. In the absence of their cognate steroid, these receptors are bound in large multiprotein complexes that contain chaperone proteins such as Hsp90 and Hsp70 as well as immunophilins (Sections 9-2B and 9-2C), which presumably function to maintain the receptor in its native conformation, ready to bind its cognate steroid. Depending on the identity of the receptor, these complexes mainly inhabit the nucleus or the cytosol. Steroid binding releases the receptors from these complexes, whereupon they dimerize. In the case of cytosolically located receptors, steroid binding is thought to also unmask their previously sequestered nuclear localization signals (NLS; Section 32-6B), thereby causing the steroid–receptor complexes to be transported to the nucleus. [Most NLSs consist of a 48-residue segment of mainly basic residues or two such segments separated by an 8- to 12-residue linker that is mutation-resistant; the precise location of an NLS within a polypeptide is unimportant, unlike the case for other types of signal peptides (Section 12-4).]

In the nucleus, steroid–receptor complexes bind to specific segments of chromosomal enhancers known as ***hormone response elements (HREs)*** *so as to induce, or in some cases repress, the transcription of their associated genes.* For example, receptors for **glucocorticoids** (a class of steroids that affect carbohydrate metabolism; Section 19-1G) bind to specific 15-bp **glucocorticoid response elements (GREs)** in the upstream regions of many genes, including those of metallothioneins. Thus, eukaryotic steroid receptors are inducible transcription factors: Their actions resemble those of prokaryotic transcriptional regulators such as the *E. coli* CAP–cAMP complex (Section 31-3C). However, eukaryotic systems are much more complex. For instance, different cell types may have the same receptor for a given steroid hormone and yet synthesize different proteins in response to the hormone. Apparently, only some of the genes inducible by a given steroid are made available for activation in each type of cell responsive to that steroid. Consequently, a given eukaryotic sequence-specific regulator may function as an activator or a repressor depending on the identities of the proteins with which it is interacting. The structures of steroid receptors are discussed below.

k. Eukaryotic Transcription Factors Have a Great Variety of DNA-Binding Motifs

How do DNA-binding transcription factors recognize their target DNA sequences? In prokaryotes, as we have seen (Section 31-3D), most repressors and activators do so via helix–turn–helix (HTH) motifs and, in a few cases, via β ribbon motifs. Eukaryotes, as we shall see, employ a far greater variety of DNA-binding motifs in their transcription factors. In the following paragraphs we discuss the structures of several of the more common of these motifs and how they bind their target DNAs.

l. Zinc Finger DNA-Binding Motifs

The first of the predominantly eukaryotic DNA-binding motifs was discovered by Aaron Klug in *Xenopus* **transcription factor IIIA (TFIIIA),** a protein that binds to the internal control sequence of the 5S rRNA gene (Section 31-2E). This complex then sequentially binds TFIIIB (which contains TBP), **TFIIIC,** and RNA polymerase III, which, in turn, initiates transcription of the 5S rRNA gene. The 344-residue TFIIIA contains nine similar, tandemly repeated, ~30-residue modules, each of which contains two invariant Cys residues, two invariant His residues, and several conserved hydrophobic residues (Fig. 34-54*a*). Each of these units binds a Zn^{2+} ion, which X-ray absorption measurements indicate is tetrahedrally liganded by the invariant Cys and His residues. Sequence analyses have since revealed that these so-called **zinc fingers** occur from 2 to over 60 times in a variety of eukaryotic transcription factors, including Sp1, several *D. melanogaster* developmental regulators (Section 34-4B), and certain proto-oncogene proteins (proteins whose mutant forms promote cancerous growth; Section 19-3B), as well as the *E. coli* UvrA protein (Section 30-5B). In fact, it is estimated that ~1% of mammalian proteins contain zinc fingers. In some zinc fingers, the two Zn^{2+}-liganding His residues are replaced by two additional Cys residues, whereas others have six Cys residues liganding two Zn^{2+} ions. Indeed, as we shall see, structural diversity is a hallmark of zinc finger proteins. In all cases, however, the Zn^{2+} ions appear to knit together relatively small globular domains, thereby elim-

(a)

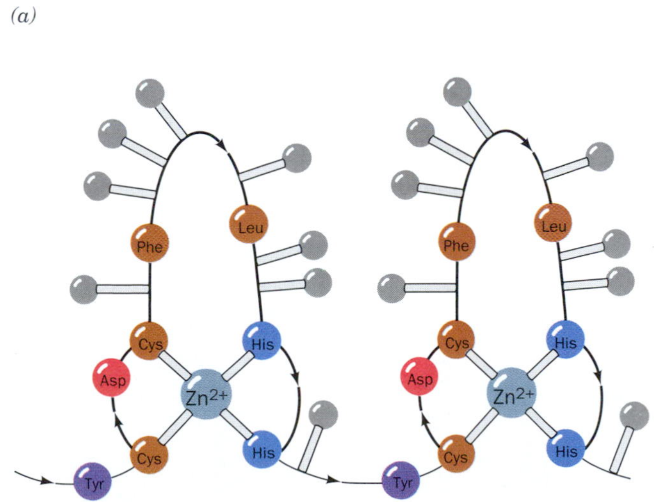

(b)

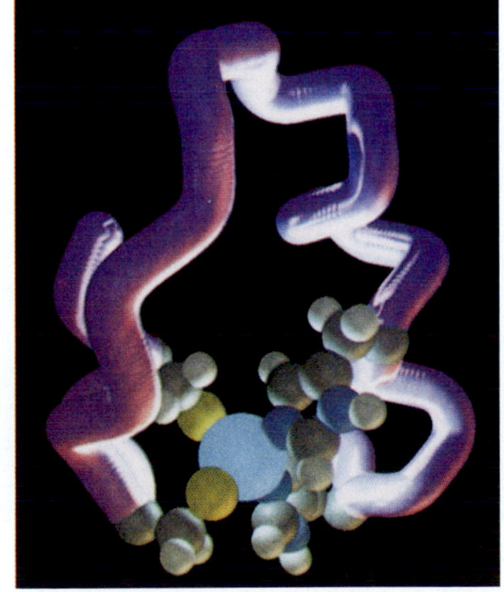

FIGURE 34-54 Zinc fingers. (*a*) A schematic diagram of tandemly repeated zinc finger motifs indicating their tetrahedrally liganded Zn^{2+} ions. Conserved amino acid residues are labeled. Gray balls represent the most probable DNA-binding side chains. [After Klug, A. and Rhodes, D., *Trends Biochem. Sci.* **12,** 465 (1988).] (*b*) The NMR structure of a single zinc finger from the *Xenopus* protein Xfin. The Zn^{2+} ion together with the atoms of its His and Cys ligands are represented as spheres with Zn^{2+} cyan, C gray, N blue, S yellow, and H white. [Courtesy of Michael Pique, The Scripps Research Institute, La Jolla, California. Based on an NMR structure by Peter E. Wright, The Scripps Research Institute. PDBid 1ZNF.]

inating the need for much larger hydrophobic cores (although see Section 9-3C).

m. Cys$_2$–His$_2$ Zinc Fingers: Xfin and Zif268 Proteins

The first reported zinc finger structure, an NMR structure by Peter Wright of the 31st of the 37 tandemly repeated zinc fingers in the *Xenopus* **Xfin protein,** revealed that this 25-residue peptide forms a compact globule containing a 2-stranded antiparallel β sheet and one α helix (a ββα unit) that are held together by the tetrahedrally liganded Zn^{2+} ion (Fig. 34-54*b*). This was followed by Carl Pabo's X-ray structure of a 72-residue segment of the mouse protein **Zif268** that incorporates the protein's three zinc fingers in complex with a DNA segment containing the protein's 9-bp consensus binding sequence. The structures of the three Zif268 zinc finger motifs (Fig. 34-55*a*) are closely superimposable and are nearly identical to that of the Xfin zinc finger (Fig. 34-54*b*). The three Zif268 zinc fingers are arranged as separate domains in a C-shaped structure that fits snugly into the DNA's major groove (Fig. 34-55*b*). Each zinc finger interacts in a conformationally identical manner with successive 3-bp segments of the DNA, predominantly through hydrogen bonding interactions between the zinc finger's α helix and one strand of the DNA (here, a G-rich strand). Each zinc finger makes specific hydrogen bonding contacts with two bases in the major groove. Interestingly, five of these six associations involve interactions between Arg and G residues. In addition to these sequence-specific interactions, each zinc finger hydrogen bonds with the DNA's phosphate groups via conserved Arg and His residues.

The Cys$_2$–His$_2$ zinc finger broadly resembles the prokaryotic HTH motif as well as most other DNA-binding motifs we shall encounter (including other types of zinc finger modules) in that *all of these DNA-binding motifs provide a platform for inserting an α helix into the major groove of B-DNA.* However, Cys$_2$–His$_2$ zinc finger proteins, unlike those containing other DNA-binding motifs, possess repeated protein modules that each contact successive DNA segments. Such a modular system can recognize extended asymmetric base sequences.

n. Cys$_2$–Cys$_2$ Zinc Fingers: The Glucocorticoid Receptor and Estrogen Receptor DNA-Binding Domains

The **nuclear receptor superfamily,** which occurs in animals ranging from worms to humans, is composed of >150 proteins that bind a variety of hormones such as steroids (glucocorticoids, mineralocorticoids, progesterone, estrogens, and androgens; Section 19-1G), thyroid hormones (Section 19-1D), vitamin D (Section 19-1E), and **retinoids** (Section 34-4B). However, the ligands, if any, that many superfamily members bind are, as yet, unknown and hence these proteins are known as **orphan receptors.** The nuclear receptors, many of which activate distinct but overlapping sets of genes, share a conserved modular organization that includes, from N- to C-terminus, a poorly conserved transactivation domain, a highly conserved DNA-binding domain, a connecting hinge region, and a ligand-binding domain. The DNA-binding domains contain 8 Cys residues that, in groups of four, tetrahedrally coordinate two Zn^{2+} ions. Many members of the nuclear receptor superfamily recognize hormone re-

(*a*)

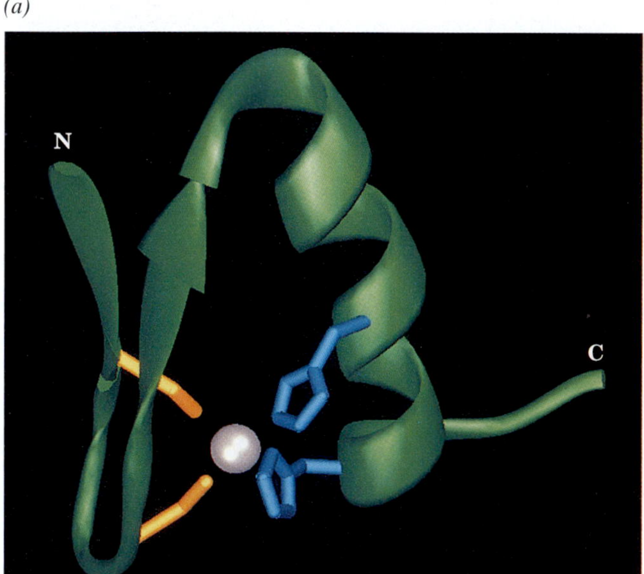

(*b*)

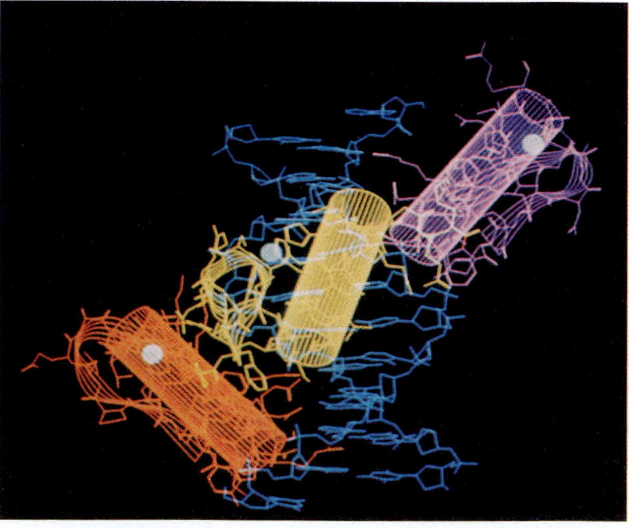

FIGURE 34-55 X-Ray structure of a three-zinc finger segment of Zif268 in complex with a 10-bp DNA. (*a*) A ribbon diagram of a single zinc finger motif (finger 1) with its Zn^{2+} ion's tetrahedrally liganding His (*cyan*) and Cys (*yellow*) side chains shown in stick form and its Zn^{2+} ion represented by a silver sphere. (*b*) The complex of the entire protein segment with DNA. The protein and DNA are shown in stick form, with superimposed cylinders and ribbons marking the protein's α helices and β sheets. Finger 1 is orange, finger 2 is yellow, finger 3 is pink, the DNA is blue, and the Zn^{2+} ions are represented by white spheres. Note how the N-terminal end of each zinc finger's helix extends into the DNA's major groove to contact three base pairs. [Part *a* based on an X-ray structure by and Part *b* courtesy of Carl Pabo, MIT. PDBid 1ZAA.]

(a)

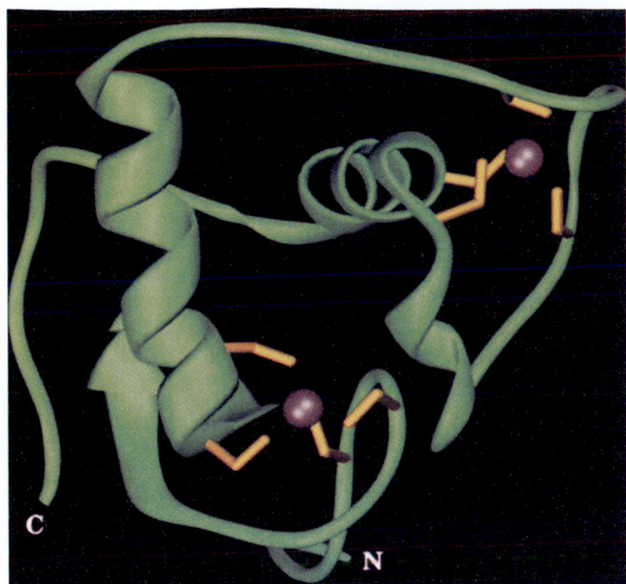

(b)

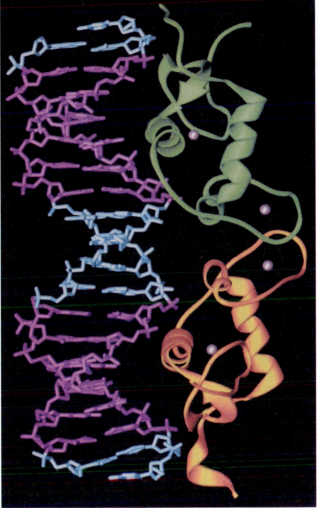

FIGURE 34-56 X-Ray structure of the dimeric glucocorticoid receptor (GR) DNA-binding domain in complex with an 18-bp DNA. The DNA contains two inverted repeats of the 6-bp glucocorticoid response element (GRE) half-sites (5′-AGAACA-3′) separated by a 4-bp spacer (GRE_{4S}). (*a*) A ribbon diagram of a single subunit of the GR with its two Zn^{2+} ions represented by silver spheres and their tetrahedrally liganding Cys side chains shown in stick form (*yellow*). Compare this structure with Fig. 34-55*a*. (*b*) The complex of the dimeric protein with GRE_{4S} DNA as viewed with its approximate 2-fold molecular axis horizontal. The protein is shown in ribbon form with its two subunits differently colored and its bound Zn^{2+} ions represented by silver spheres. The DNA is drawn in stick form with its two 6-bp GRE half-sites magenta and the remainder cyan. Note how the GR's two N-terminal helices are inserted into adjacent major grooves of the DNA. However, only the upper (*green*) subunit binds to the DNA in a sequence-specific manner; the lower (*gold*) subunit binds to the palindromic DNA one base pair closer to the center of the DNA molecule than does the upper subunit and hence does not make sequence-specific contacts with the DNA. [Based on an X-ray structure by Paul Sigler, Yale University. PDBid 1GLU.]

sponse elements that have the half-site consensus sequences 5′-AGAACA-3′ for steroid receptors and 5′-AGGTCA-3′ for other nuclear receptors. These sequences are arranged in direct repeats ($\rightarrow n \rightarrow$), inverted repeats ($\rightarrow n \leftarrow$), and everted repeats ($\leftarrow n \rightarrow$), where n represents a 0- to 8-bp spacer (usually 1–5 bp) to whose length a specific receptor is targeted. Steroid receptors bind to their hormone response elements as homodimers, whereas other nuclear receptors do so as homodimers, heterodimers, and in a few cases, as monomers.

The X-ray structures of two related DNA segments complexed with the 86-residue DNA-binding domain of rat **glucocorticoid receptor (GR)** were determined by Sigler and Keith Yamamoto. One segment, designated GRE_{4S}, contains two ideal 6-bp glucocorticoid response element (GRE) half-sites arranged in inverted repeats about a 4-bp (nonnative) spacer ($n = 4$), whereas the other DNA, GRE_{3S}, differs from GRE_{4S} in that its spacer has the naturally occurring length of $n = 3$ bp. In both complexes, the protein forms a symmetric dimer involving protein–protein contacts even though it exhibits no tendency to dimerize in the absence of DNA (NMR measurements indicate that the contact region is flexible in solution).

The X-ray structure of the DNA-binding domain of the GR subunit complexed to DNA resembles that of its NMR structure in the absence of DNA: It consists of two structurally distinct modules, each nucleated by a Zn^{2+} coordination center, that closely associate to form a compact globular fold (Fig. 34-56*a*). The C-terminal module provides the entire dimerization interface as well as making several contacts with the phosphate groups of the DNA backbone. The N-terminal module, which is also anchored to the phosphate backbone, makes all of the GR's sequence-specific interactions with the GRE via three side chains extending from the N-terminal α helix, its recognition helix, which is inserted into the GRE's major groove.

In the GRE_{3S} complex, a subunit of the GR DNA-binding domain binds to each GRE half-site in a structurally identical manner, making sequence-specific contacts even though the odd number of base pairs in its spacer, which the protein does not contact, renders the DNA sequence nonpalindromic. However, in the GRE_{4S} complex (Fig. 34-56*b*), the protein dimer maintains a structure that is essentially identical to that in the GRE_{3S} complex so that only one of its subunits can bind to a GRE half-site in a manner resembling that in the GRE_{3S} complex. The other subunit is shifted out of register with the GRE sequence by 1 bp and hence only makes nonspecific contacts with the DNA. The dimer interactions are apparently stronger than the protein–DNA interactions, a surprising finding in light of the protein's failure to dimerize in the absence of DNA. Thus, the two subunits and the DNA associate in a cooperative fashion that favors the binding of the glucocorticoid receptor to targets with properly spaced half-sites.

The **estrogen response element (ERE),** the DNA segment to which the **estrogen receptor (ER)** specifically binds, differs from the GRE only by changes in the central two base pairs in their otherwise identical 6-bp half-sites. The X-ray structure of the ER DNA-binding domain in complex with an ERE-containing DNA segment, determined by Daniela Rhodes, closely resembles that of the GR–GRE complex. However, the side chains that make base-specific contacts with each ERE half-site are quite differently arranged from those contacting the GRE_{4S} half-sites. Evidently, the discrimination of a half-site sequence is not simply a matter of substituting one or more different amino acid residues into a common framework but, rather, involves considerable side chain rearrangement.

Members of the nuclear receptor superfamily often recognize hormone response elements with similar or even identical half-site sequences as well as different spacings. The foregoing observations provide a structural basis for the graded affinities of these receptors toward their various target genes.

o. Binuclear Cys_6 Zinc Fingers: The GAL4 DNA-Binding Domain

The yeast protein **GAL4** is a transcriptional activator of several genes that encode galactose-metabolizing proteins. This 881-residue protein binds to a 17-bp DNA segment as a homodimer. Residues 1 to 65, which contain six Cys residues that collectively ligand two Zn^{2+} ions (Fig. 34-57), have been implicated in DNA binding; residues 65 to 94 participate in dimerization (although, as we shall see, residues 50–64 also have a weak dimerization function); and residues 94 to 106, 148 to 196, and 768 to 881 function as acidic transcriptional activating regions. The X-ray crystal structure of the 65-residue N-terminal fragment of GAL4 in complex with a symmetrical 19-bp DNA containing GAL4's palindromic 17-bp consensus sequence has been determined by Mark Ptashne, Ronen Mamorstein, and Stephen Harrison.

The protein binds to the DNA as a symmetric dimer (Fig. 34-57*a*), although in the absence of DNA it is only monomeric. Each subunit folds into three distinct modules: a compact Zn^{2+}-liganding domain that binds specific sequences of DNA (residues 8–40), an extended linker (residues 41–49), and a short a helical dimerization element (residues 50–64). In the Zn^{2+}-liganding module (Fig. 34-57*b* and top and bottom of Fig. 34-57*a*), the two Zn^{2+} ions are each tetrahedrally coordinated by four of the six Cys residues, with two of these residues ligating both metal ions so as to form a binuclear cluster. This module's polypeptide chain forms two short α helices connected by a loop such that the module, together with its bound Zn^{2+} ions, has pseudo-2-fold symmetry. The N-terminal helix is inserted into the DNA's major groove, thereby making sequence-specific contacts with a highly conserved CCG sequence at each end of the consensus sequence. The DNA's conformation deviates little from that of ideal B-DNA.

The dimerization helices (center of Fig. 34-57*a*) associate to form a short segment of parallel coiled coil in which the contact region between the coiled coil's component helices is hydrophobically stabilized by three pairs of Leu residues and a pair of Val residues (an arrangement similar to that in the so-called **leucine zipper** described below). The coiled coil is positioned over the minor groove of the DNA such that its superhelix axis coincides with the DNA's 2-fold axis. The linkers connecting the coiled coil to the DNA-binding modules wrap around the DNA, largely following its minor groove while making several nonspecific contacts with DNA phosphate groups until, on reaching the DNA-binding module, they shift over into the DNA's major groove. The two symmetrically related DNA-binding modules thereby approach the major groove from opposite sides of the DNA, ~1.5 helical turns apart, rather than from the same side of the DNA, ~1 helical turn apart, as do, for example, HTH motifs and the glucocorticoid receptor. The resulting relatively open structure could permit other proteins to bind simultaneously to the DNA.

p. Leucine Zippers Mediate Transcription Factor Dimerization

Transcriptional activation requires, as we have seen, the cooperative association of several proteins that bind to specific sequences on DNA. Steven McKnight discovered one way in which such associations occur. We have seen (Section 8-2A) that α helices with the 7-residue pseudorepeating sequence $(a\text{-}b\text{-}c\text{-}d\text{-}e\text{-}f\text{-}g)_n$, in which the *a* and *d* residues are hydrophobic, have a hydrophobic strip along one side, which induces them to dimerize so as to form a coiled coil. McKnight noticed that the rat liver transcription factor named **C/EBP** (for *C*CAAT/*e*nhancer *b*inding *p*rotein), which specifically binds to the CCAAT box (Section 31-2E), has a Leu at every seventh position of a 28-residue segment in its DNA-binding domain. Similar heptad repeats occur in a number of known dimeric DNA-binding proteins, including the yeast transcriptional activator **GCN4** and several DNA-binding proteins encoded by proto-oncogenes (Section 34-4C). McKnight suggested that these proteins form coiled coils in which the Leu side chains are interdigitated, much like the teeth of a zipper. He therefore named this motif the **leucine zipper.** The leucine zipper, as we shall see, mediates both the homodimerization and the heterodimerization of DNA-binding proteins (but note that it is not, in itself, a DNA-binding motif).

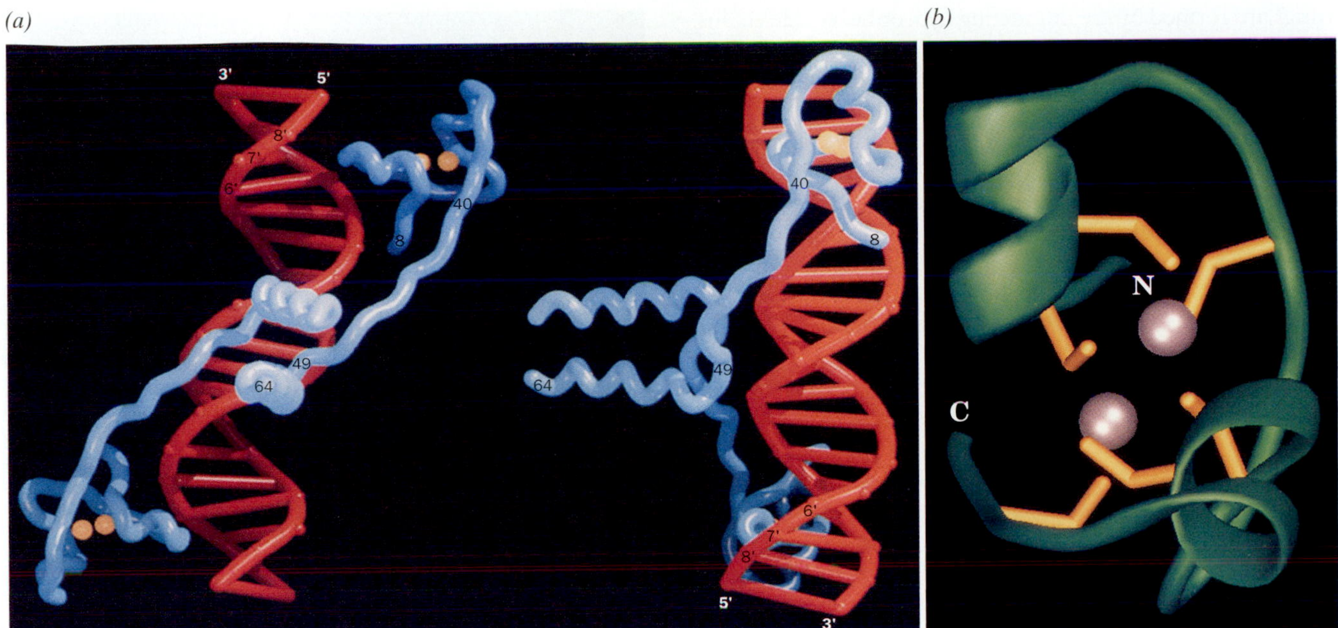

FIGURE 34-57 X-Ray structure of the yeast GAL4 DNA-binding domain in complex with a palindromic 19-bp DNA (except for the central base pair) containing the protein's consensus binding sequence. (*a*) The complex of the dimeric protein with the DNA as shown in tube form and with the DNA red, the protein backbone cyan, and the Zn^{2+} represented by yellow spheres. The views are along the complex's 2-fold axis (*left*) and turned 90° with the 2-fold axis horizontal (*right*). Note how the C-terminal end of each subunit's N-terminal helix extends into the DNA's major groove. (*b*) A ribbon diagram of the protein's zinc finger domain (residues 8–40) with the Cys side chains of its $Zn_2^{2+}Cys_6$ complex shown in stick form (*yellow*) and its Zn^{2+} ions shown as silver spheres. Compare this structure with Figs. 34-55*a* and 34-56*a*. [Part *a* courtesy of and Part *b* based on an X-ray structure by Stephen Harrison and Ronen Mamorstein, Harvard University. PDBid 1D66.]

The X-ray structure of the 33-residue polypeptide corresponding to the leucine zipper of the 281-residue GCN4 was determined by Peter Kim and Thomas Alber. Its first 30 residues, which contain ~3.6 heptad repeats (Fig. 34-58*a*), coil into an ~8-turn α helix that dimerizes as McKnight predicted to form ~1/4 turn of a parallel left-handed coiled coil (Fig. 34-58*b*). The dimer can be envisioned as a twisted ladder whose sides consist of the helix backbones and whose

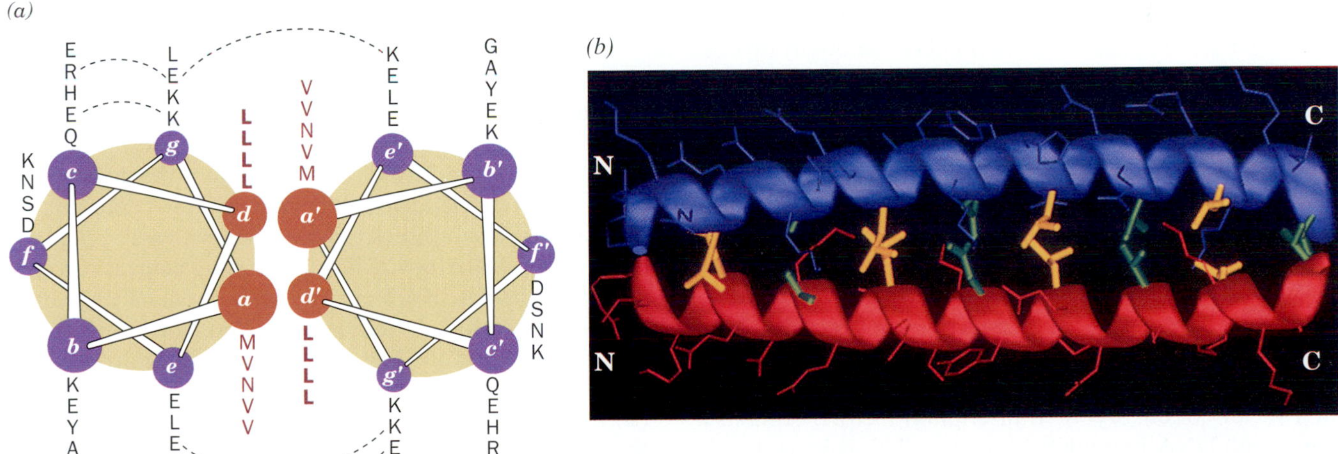

FIGURE 34-58 The GCN4 leucine zipper motif. (*a*) A helical wheel representation of the motif's two helices as viewed from their N-termini. The sequences of residues at each position are indicated by the adjacent column of one-letter codes. Residues that form ion pairs in the crystal structure are connected by dashed lines. Note that all residues at positions *d* and *d'* are Leu (L), those at positions *a* and *a'* are mostly Val (V), and those at other positions are mostly polar. [After O'Shea, E.K., Klemm, J.D., Kim, P.S., and Alber, T., *Science* **254,** 540 (1991).] (*b*) The X-ray structure, in side view, in which the helices are shown in ribbon form. Side chains are shown in stick form with the contacting Leu residues at positions *d* and *d'* yellow and residues at positions *a* and *a'* green. [Based on an X-ray structure by Peter Kim, MIT, and Tom Alber, University of Utah School of Medicine. PDBid 2ZTA.]

rungs are formed by the interacting hydrophobic side chains. The conserved Leu residues at heptad position *d*, which comprise every second rung, are not interdigitated as McKnight originally suggested but, instead, make side-to-side contacts. The alternate rungs are likewise formed by the *a* residues of the heptad repeat (which are mostly Val) in side-to-side contact. Each Leu side chain at position *d*, in addition to packing against the symmetry-related Leu side chain, d', from the other polypeptide, packs against the side chain of the succeeding residue, e'. Similarly, each side chain at position *a* packs between its symmetry mate, a', and the preceding residue, g'. These two sets of alternating layers thereby form an extensive hydrophobic interface between the coiled coil's component helices.

q. bZIP Motifs: The GCN4 DNA-Binding Domain

In many but not all leucine zipper proteins, a DNA-binding region, which is rich in basic residues, is immediately N-terminal to the leucine zipper. Sequence comparisons among 11 of these so-called **basic region leucine zipper (bZIP) proteins** revealed that the 16-residue basic sequence invariably ends 7 residues before the leucine zipper's N-terminal Leu residue. Moreover, all of these basic regions, as well as the 6-residue segment linking them to the leucine zipper, are devoid of the two strongest helix-destabilizing residues, Pro and Gly (Section 9-3A), thereby suggesting that each bZIP polypeptide is entirely α helical.

The C-terminal 56 residues of GCN4 constitute its bZIP element. Harrison and Kevin Struhl determined the X-ray structure of this polypeptide segment in complex with a 19-bp-containing duplex DNA whose central 9 bp consist of GCN4's symmetrized target sequence (Fig. 34-59). The bZIP element forms a symmetric dimer in which each subunit consists, almost entirely, of a continuous α helix. The C-terminal 25 residues of two such helices associate via a leucine zipper whose geometry closely resembles that of the 33-residue GCN4 leucine zipper element alone (Fig. 34-58*b*). Past this point, the two α helices smoothly diverge to bind in the DNA's major groove on opposite sides of the helix, thereby clasping the DNA in a sort of scissors grip. The DNA, whose helix axis is nearly perpendicular to that of the coiled coil, maintains what is essentially a straight and undistorted B-form conformation. The basic region residues that are conserved in bZIP proteins thereby make numerous contacts with both the bases and with phosphate oxygens of the DNA target sequence.

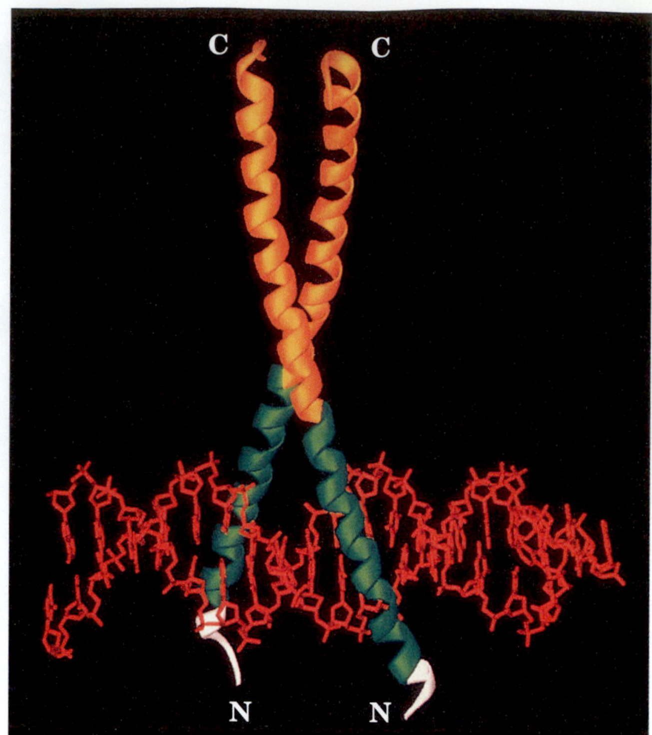

FIGURE 34-59 X-Ray structure of the GCN4 bZIP region in complex with its target DNA. The DNA (*red*) is represented in stick form and is viewed with its molecular 2-fold axis vertical. It consists of a 19-bp segment with a single nucleotide overhang at each end and contains the protein's palindromic (except for the central base pair) 7-bp target sequence. The two identical subunits, shown in ribbon form, each contain a continuous 52-residue α helix. At their C-terminal ends (*yellow*), the two subunits associate in a parallel coiled coil (a leucine zipper), and at their basic regions (*green*), they smoothly diverge to each engage the DNA in its major groove at the target sequence. The N-terminal ends are white. [Based on an X-ray structure by Stephen Harrison, Harvard University. PDBid 1YSA.]

r. bHLH Motifs: The Max DNA-Binding Domain

The **basic helix–loop–helix (bHLH) motif,** which occurs in a variety of eukaryotic transcription factors, contains a conserved DNA-binding basic region. This is immediately followed by two amphipathic helices connected by a loop that mediates the protein's dimerization. The bHLH motif in many proteins is followed by a conserved leucine zipper (Z) motif that presumably augments protein dimerization. The transcription factor **Max** is such a **bHLH/Z** protein, which, *in vivo,* forms a heterodimer with the protooncogene protein **Myc** and is required for both its normal and cancer-inducing activities. Max, by itself, readily homodimerizes and binds DNA with high affinity but Myc does not do so.

The X-ray structure of a truncated version of the 160-residue Max, Max(22-113), which contains the parent protein's bHLH and leucine zipper elements, was determined, by Edward Ziff and Burley, in complex with a 22-bp quasi-palindromic DNA containing Max's 6-bp central recognition element. Each subunit of this homodimeric protein consists of two long α helices connected by a loop to form a novel protein fold (Fig. 34-60). The N-terminal α helix (b/H1) contains residues from the protein's basic region (b) followed, without interruption, by those of the HLH motif's leading helix (H1). The C-terminal α helix (H2/Z), which is composed of the second HLH helix (H2) and the leucine zipper (Z), mediates the protein's homodimerization through the formation of a parallel left-handed coiled coil similar to that in GCN4 (Fig. 34-59). Each of the dimer's two b/H1 helices projects from the resulting parallel 4-helix bundle to engage the DNA in a manner reminiscent of a pair of forceps by binding in its major groove on opposite sides of the helix (much like the way GCN4

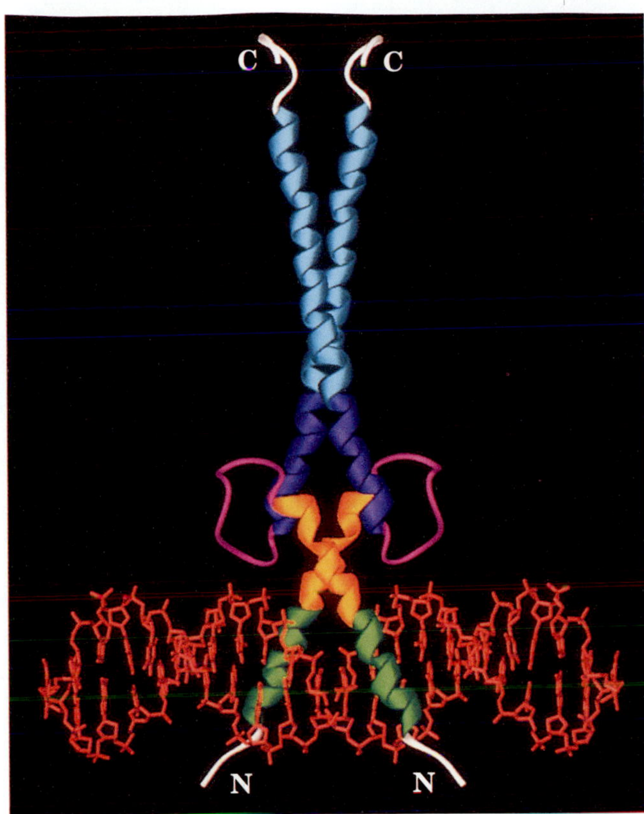

FIGURE 34-60 X-Ray structure of the Max(22-113) dimer in complex with a 22-bp DNA containing the protein's palindromic 6-bp target sequence. The DNA (*red*) is shown in stick form and the homodimeric protein is shown in ribbon form. The protein's N-terminal basic region (*green*) forms an α helix that engages its target sequence in the DNA's major groove and then merges smoothly with the H1 helix (*yellow*) of the helix–loop–helix (HLH) motif. Following the loop (*magenta*), the protein's two H2 helices (*purple*) of the HLH motif form a parallel left-handed four-helix bundle with the two H1 helices. Each H2 helix then merges smoothly with the leucine zipper (Z) motif (*cyan*) to form a parallel coiled coil. The protein's N- and C-terminal ends are white. [Based on an X-ray structure by Edward Ziff and Stephen Burley, The Rockefeller University. PDBid 1AN2.]

grips its target DNA, although GCN4's bZIP element consists of only two α helices rather than the four of Max). The DNA helix is essentially straight with only small deviations from the ideal B-DNA structure. Each basic region makes several sequence-specific interactions with the bases of the DNA's 6-bp recognition element as well as numerous contacts with its phosphate groups. Side chains of both the loop and the N-terminal end of the H2 helix also contact DNA phosphate groups.

s. NF-κB Binds DNA Differently from Other Transcription Factors

Nuclear factor κB (NF-κB), a transcription factor that was originally identified as an inducible nuclear activity that binds to the κB sequence in the immunoglobulin κ light chain gene enhancer (immunoglobulin genes are discussed in Section 35-2), is present in nearly all animal cells, although its role is particularly prominent in the immune system. It is present *in vivo* mainly as a DNA-binding heterodimer of the **p50** and **p65** (alternatively **RelA**) proteins (p for *p*rotein with the number indicating its nominal molecular mass in kD), both of which contain an ~300-residue segment known as the **Rel homology region (RHR)** because it also occurs in the product of the ***rel* oncogene.** However, p50 and p65 can also form DNA-binding homodimers. RHRs, which mediate protein dimerization and DNA binding, and which contain nuclear localization signals (NLSs), are present in a variety of proteins that serve as regulators of cellular defense mechanisms against stress, injury, and external pathogens, as well as of differentiation. Moreover, certain viruses, including HIV, have subverted RHRs to activate the expression of their genes. There are two classes of **Rel proteins:** those such as p65, **c-Rel,** and the *Drosophila* morphogen proteins (proteins that mediate development; Section 34-4B) **Dorsal** and **Dif,** whose N-terminal domains contain an RHR and whose highly variable C-terminal domains are strong transcriptional activators; and those such as p50 and the closely related **p52,** which are generated by the proteolytic processing of larger precursors and lack transactivation domains so that their homodimers function primarily as repressors.

The activity of NF-κB is largely regulated by proteins known as **inhibitor-κBs (IκBs),** which by binding to an NF-κB mask its NLS so that IκB–NF-κB complexes reside in the cytoplasm. The IκBs contain multiple ankyrin repeats (Section 12-3D) through which they bind the NF-κBs. The extracellular presence of a remarkable variety of external stimuli, including certain bacterial and viral products, several cytokines (Section 19-1L), phorbol esters (Section 19-4C), and oxidative and physical stress (e.g, free radicals and UV radiation), results, via signaling cascades, in IκBs being phosphorylated by **IκB kinase (IKK).** This, in turn, induces ubiquitination of the IκBs and their subsequent degradation by the proteosome (Section 32-6B). The liberated NF-κB is thereupon translocated to the nucleus, where it mediates transcriptional initiation by binding to 10-bp κB DNA segments that have the consensus sequence GGGRNNYYCC. Additional specificity may be achieved through the synergistic interaction of the NF-κB with other DNA-bound transcription factors such as Sp1. This activation process is self-limiting: The transcription of the gene encoding the most common IκB protein, **IκBα** (whose X-ray structure is shown in Fig. 12-38), is induced by the binding of NF-κB to the κB sites in this gene's promoter. The resulting newly synthesized IκBα enters the nucleus, where it releases the NF-κB from its complex with DNA and directs its export to the cytoplasm.

In a related mode of NF-κB activation, p50 is synthesized as the N-terminal domain of **p105,** a protein whose C-terminal domain is an IκB. The IκB domain of p105 prevents both the nuclear localization and the DNA binding of p105 as well as other RHR-containing proteins. The above external stimuli also accelerate the proteolytic processing of p105 to yield a free NF-κB and the IκB-containing C-terminal domain of p105, which as discussed above, is phosphorylated and proteolytically degraded.

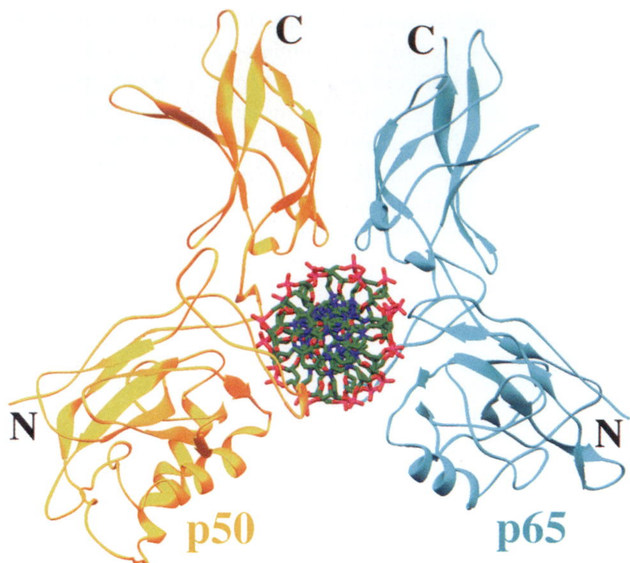

FIGURE 34-61 X-Ray structure of the mouse NF-κB p50–p65 heterodimer bound to κB DNA from the interferon β enhancer. The structure is viewed along the helix axis of the DNA, whose two strands have the sequences 5′-TGGGAAATTCCT-3′ and 5′-AAGGAATTTCCC-3′ (the duplex DNA consists of 11 bp with a 1-nt overhang at each end) and are drawn in stick form colored according to atom type (C green, N blue, O red, and P magenta). The protein is represented by ribbons with p50 (residues 39–364 of 435 residues) gold and p65 (residues 19–291 of 549 residues) cyan. [Based on an X-ray structure by Gourisankar Ghosh, University of California at San Diego. PDBid 1LE5.]

The X-ray structure of the heterodimer of mouse p50 and p65 in complex with the κB segment of the β-interferon enhancer, determined by Gourisankar Ghosh, bears a striking resemblance to a butterfly with its homologous protein subunits forming its outspread wings and the DNA its torso (Fig. 34-61). The two protein subunits have similar structures that each consists of two domains, with the C-terminal domains forming the dimerization interface and both domains interacting with the DNA. Both their N- and C-terminal domains have immunoglobulin-like folds (a sandwich of a 3- and a 4-stranded antiparallel β sheet; Section 35-2B) and interact with the DNA exclusively through 10 loops, 5 from each subunit, that link their β strands and fill the DNA's major groove. The p50 and p65 bind to 5-bp and 4-bp subsites at the 5′ and 3′ ends of the consensus sequence, respectively, with the two subsites separated by a single base pair. The protein's DNA-binding surface is much more extensive than those of other transcription factors, which accounts for the unusually high affinity of NF-κBs for their target sequences. This also explains the inability of deletion mutagenesis to localize NF-κB's DNA-binding region, since changes anywhere in its structure are likely to affect the disposition of its DNA-binding loops.

Comparisons of the X-ray structures of the p50–p65 heterodimer bound to several κB DNA segments with different sequences, all determined by Ghosh, reveal small but significant structural differences among these various complexes. These arise mainly from the different degrees of bending of the various κB DNAs as well as the different interactions of the proteins with the different sequences of bases, all of which result in small conformational differences among the chemically identical proteins in these complexes. This is probably why the substitution in an enhancer of one κB segment for another does not produce the same level of transcription, even though the NF-κB binds to the isolated κB segments with equal affinity. Evidently, the way in which NF-κB interacts with other proteins that are bound to the enhancer (such as the glucocorticoid receptor, which interacts with p65) affects its activational potency and hence fine-tunes the expressional levels of its target genes.

t. Mediator Provides the Interface between Transcriptional Activators and RNAP II

Eukaryotic genomes encode as many as several thousand transcriptional regulators for class II genes (e.g., Fig. 34-27). How does the binding of these various regulators to their cognate enhancers/silencers influence the rate at which RNAP II initiates transcription? Genetic studies have implicated TFIIB, TFIID, and TFIIH in this process *in vivo.* Nevertheless, activators fail to stimulate transcription by a reconstituted PIC *in vitro.* Evidently, an additional factor is required to do so. *Indeed, genetic studies in yeast by Kornberg led him to discover an ~20-subunit, ~1000-kD complex named* ***Mediator,*** *whose presence is required for transcription from nearly all class II gene promoters in yeast.* Mediator, which is therefore considered to be a coactivator, binds to the C-terminal domain (CTD) of RNAP II's β′ subunit (Section 31-2E) to form the so-called **RNAP II holoenzyme.** Further investigations revealed that metazoans contain several multisubunit complexes that function similarly to yeast Mediator. These include complexes known as **CRSP, NAT, ARC/DRIP, TRAP/SMCC, mMED,** and **PC2,** which share many of their numerous subunits. Moreover, many of their subunits are related, albeit distantly, to those of yeast Mediator. *Mediators apparently function as adaptors that bridge DNA-bound transcriptional regulators and RNAP II so as to influence (induce or inhibit) the formation of a stable PIC at the associated promoter. They thereby function to integrate the various signals implied by the binding of these transcriptional regulators to their target DNAs.* The different metazoan mediators presumably relay signals from different sets of transcriptional regulators.

Kornberg and Francisco Asturias have determined the EM-based low resolution (30–35 Å) structure of yeast Mediator and human TRAP (Fig. 34-62). The two particles are similarly shaped with nearly perpendicular "head" and "middle-tail" domains. The EM-based image of yeast RNAP II holoenzyme (Fig. 34-63) reveals that Mediator has assumed a more extended conformation in which the "middle" and "tail" domains are clearly separated. The head domain interacts closely with the RNAP II, although >75% of RNAP II's surface remains accessible for interaction with other components of the PIC. However, the tail domain appears not to contact RNAP II at all.

Altogether, then, the transcriptional machinery for class II genes comprises nearly 60 polypeptides with an aggregate

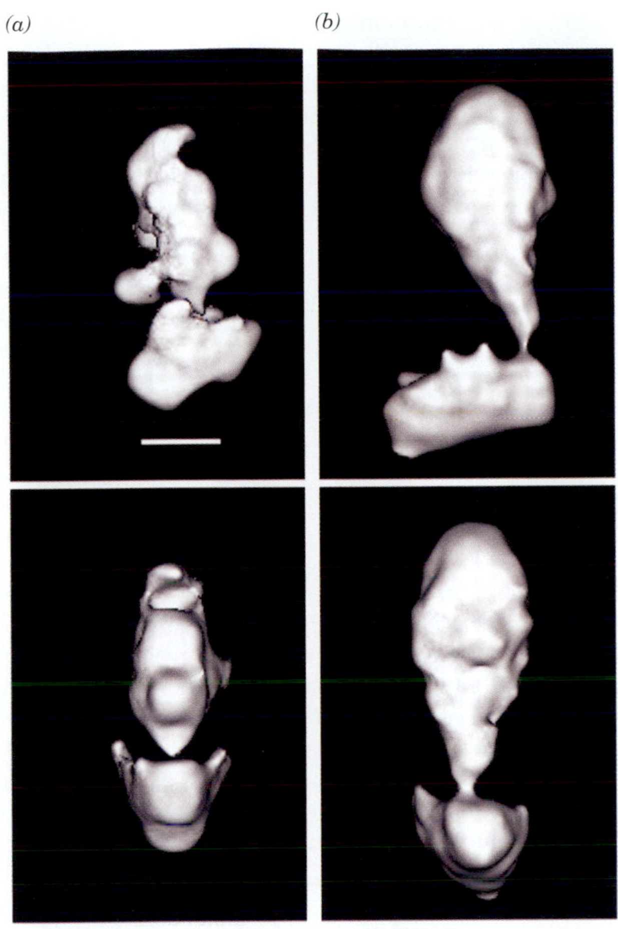

FIGURE 34-62 EM-based structures of (*a*) yeast Mediator and (*b*) human TRAP complex. The orientations of the complexes in the upper and lower rows differ by 90° rotations about the vertical direction. The bottom part of each image forms the "head" domain of the complex, which is nearly perpendicular to its top portion, the "middle-tail" domain. The bar is 100 Å in length. [Courtesy of Francisco Asturias, The Scripps Research Institute, La Jolla, California.]

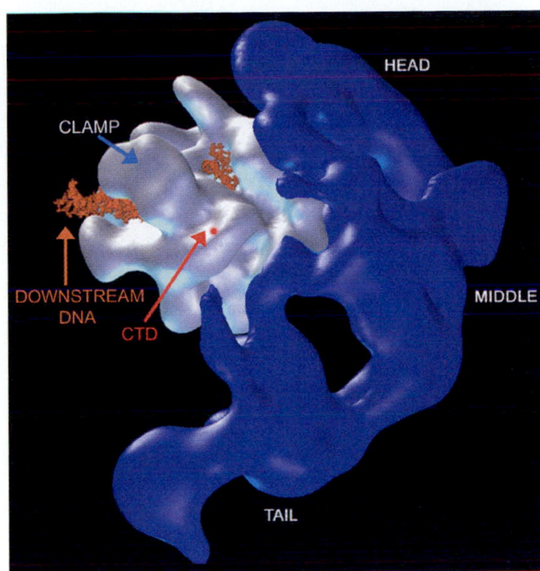

FIGURE 34-63 EM-based projection of the yeast RNAP II holoenzyme at 35-Å resolution. Mediator (*blue*) assumes a more extended conformation than that in Fig. 34-62*a,* such that its head, middle, and tail domains are clearly distinguishable. The independently determined EM-based image of yeast RNAP II (*white*) is oriented to best match the RNAP II density in the lower resolution holoenzyme image. Promoter DNA (*orange*) was modeled in, based on the structure of the yeast RNAP II elongation complex (Fig. 31-21*b*). Note that the RNAP II's DNA binding cleft remains fully accessible in the holoenzyme complex. [Courtesy of Francisco Asturias, The Scripps Research Institute, La Jolla, California, and Roger Kornberg, Stanford University School of Medicine.]

molecular mass of ~3 million D. Nevertheless, as we see below, this ribosome-sized assembly (the eukaryotic ribosome has a molecular mass of ~4.2 million; Table 32-8) requires considerable assistance from yet other large macromolecular assemblies to gain access to the DNA in chromatin.

u. Transcriptionally Active Chromatin Is Sensitive to Nuclease Digestion

Early research on the mechanism of eukaryotic transcription largely ignored the influence of chromatin. Yet, as we have seen (Section 34-2A), euchromatin but not heterochromatin is transcriptionally active. *Indeed, investigations over the past decade have revealed that eukaryotic cells contain elaborate systems that participate in controlling transcriptional initiation by altering chromatin structure.* In the remainder of this subsection we discuss the nature of these systems.

The open structure of transcriptionally active chromatin presumably gives the transcriptional machinery access to the active genes. This hypothesis was corroborated by Harold Weintraub's demonstration that *transcriptionally active chromatin is about an order of magnitude more susceptible to cleavage by DNase I than is transcriptionally inactive chromatin.* For example, globin genes from chicken erythrocytes (avian red cells have nuclei) are more sensitive to DNase I digestion than are those from chicken oviduct (where eggs are made), as was indicated by the loss of the abilities of these genes to hybridize with a complementary DNA probe after DNase I treatment. Conversely, the gene encoding **ovalbumin** (the major egg white protein) from oviduct is more sensitive to DNase I than is that from erythrocytes. Thus, nuclease sensitivity appears to delineate chromatin's **functional domains,** although their relationship to chromatin's structural domains (Section 34-3A) is unclear. Nevertheless, nuclease sensitivity reflects a gene's potential for transcription rather than transcription itself: The DNase I sensitivity of the oviduct ovalbumin gene is independent of whether or not the oviduct has been hormonally stimulated to produce ovalbumin.

The variation of a given gene's transcriptional activity with the cell in which it is located indicates that chromosomal proteins participate in the gene activation process. Yet histones' chromosomal abundance and lack of variety make it highly unlikely that they have the specificity required for this role. Among the most common nonhistone proteins are the members of the **high mobility group (HMG),** so named because of their high electrophoretic mobilities in polyacrylamide gels (and possibly because they were discovered by H.M. Goodwin). These highly conserved, low molecular mass (<30 kD) proteins, which have the unusual amino acid composition of ~25% basic side chains and 30% acidic side chains, are relatively abundant, with ~1 HMG molecule per 10 to 15 nucleosomes. The HMG proteins can be eluted from chick erythrocyte chromatin by 0.35*M* NaCl without gross structural changes to the nucleosomes. This treatment eliminates the preferential nuclease sensitivity of the erythrocyte globin genes.

v. HMG Proteins Are Architectural Proteins That Participate in Regulating Gene Expression

The HMG proteins consist of three superfamilies, **HMGB, HMGA,** and **HMGN,** which have the following properties:

1. The mammalian HMGB proteins, **HMGB1** and **HMGB2** (~210 residues; previously known as **HMG1** and **HMG2**), which bind DNA without regard to sequence, each consist of two tandem ~80-residue **HMG boxes,** A and B, followed by an acidic tail consisting of ~30 (HMGB1) or ~20 (HMGB2) consecutive Asp or Glu residues. However, *Drosophila* **HMG-D** and yeast **NHP6A** proteins each contain only one HMG box, which is closely similar to the B domain of HMGB1. The NMR structure of NHP6A in complex with a 15-bp DNA, determined by Juli Feigon, reveals, in agreement with the structures of several other HMG box–containing proteins, that the HMG box consists of three helices arranged in an L-shape with the inside of the L inserted into the minor groove of the DNA (Fig. 34-64). This induces the DNA to bend by as much as 130° toward its major groove. Apparently, nuclear HMGB proteins function as architectural proteins that induce the binding of other proteins, including various steroid receptors, to DNA and hence facilitate the assembly of nucleoprotein complexes. Indeed, NHP6A and HMG-D can functionally replace the bacterial DNA-bending protein HU even though HMGB and HU proteins have no structural or sequence similarity (HU is discussed in Section 33-3C). HMG boxes also occur in several sequence-specific transcription factors, including the mammalian male sex determining factor SRY (Section 19-1G).

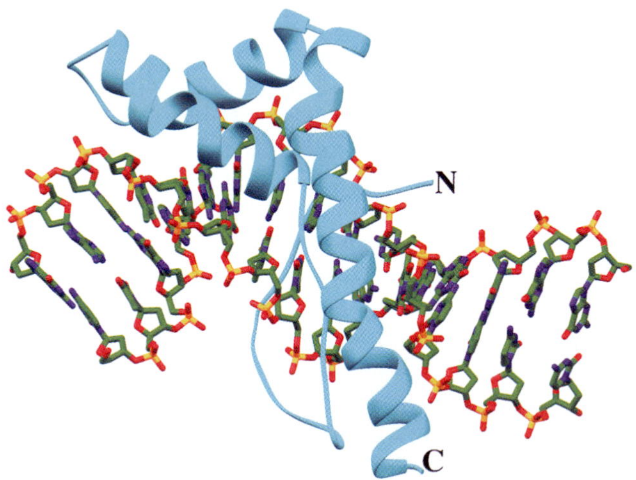

FIGURE 34-64 NMR structure of yeast NHP6A protein in complex with a 15-bp DNA. The protein is drawn in ribbon form (*cyan*) and the DNA is drawn in stick form colored according to atom type (C green, N blue, O red, and P yellow). The NHP6A's L-shaped HMG box binds in the DNA's minor groove so as to bend the DNA by ~70° toward its major groove. [Based on an NMR structure by Juli Feigon, University of California at Los Angeles. PDBid 1J5N.]

2. The HMGA superfamily consists of four proteins: the 107-, 96-, and 179-residue splice variants **HMGA1a, HMGA1b,** and **HMGA1c** (previously named **HMG-I, HMG-Y,** and **HMG-I/R**) and the homologous 109-residue **HMGA2** (previously named **HMG-C**). Each of these proteins contains three similar so-called **AT hooks** that have the invariant core sequence Arg-Gly-Arg-Pro flanked by positively charged residues and that bind to AT-rich DNA sequences. The NMR spectrum of a truncated form of HMGA1a that contains only its second and third AT hooks (residues 51–90) is indicative of a random coil. However, the NMR structure of this truncated HMGA1a in complex with a 12-bp DNA containing an AT-rich segment of the β-interferon enhancer, determined by Angela Gronenborn and Marius Clore, reveals that each of its AT hooks binds in an extended conformation in the minor groove of a separate DNA molecule (Fig. 34-65). Despite the relatively undistorted DNA in this structure, it has been shown that full-length HMGA proteins can bend, straighten, unwind, and induce loop formation in dsDNA. HMGA proteins have been implicated in regulating the transcription of numerous genes. For example, HMGA1 proteins recruit the transcription factors NF-κB and **c-Jun** (Section 34-4C) to the enhanceosome at the β-interferon enhancer by a combination of DNA bending and protein–protein interactions.

3. The HMGN proteins, **HMGN1** and **HMGN2** (98 and 89 residues; previously known as **HMG14** and **HMG17**), occur in mammals but not in *Drosophila* or yeast. They are 60% identical in sequence and consist of three functional motifs: a bipartite nuclear localization signal (NLS; see above), a conserved ~30-residue, positively charged **nucleosome-binding domain (NBD),** and a **chromatin-unfolding domain (CHUD).** The ~30-residue, positively charged NBD, as its name implies, targets HMGN proteins to bind to nucleosome core particles as homodimers of HMGN1 or HMGN2 (but not as heterodimers) without preference for the underlying DNA sequence. This stabilizes the nucleosome core particle by bridging its two adjacent dsDNA strands. Nevertheless, HMGN proteins increase the rate of transcription and DNA replication, presumably because they loosen the structure of chromatin fibers. This apparently occurs because the CHUD domain interacts with the N-terminal tail of histone H3 (see below) and because nucleosome-bound HMGN proteins compete with histone H1 for its nucleosomal binding site. HMGN-containing nucleosomes occur as clusters averaging six adjacent nucleosomes, thereby confirming that they alter internucleosomal structure. The presumably decondensed chromatin in these clusters could provide gateways through which regulatory proteins gain access to their target DNAs.

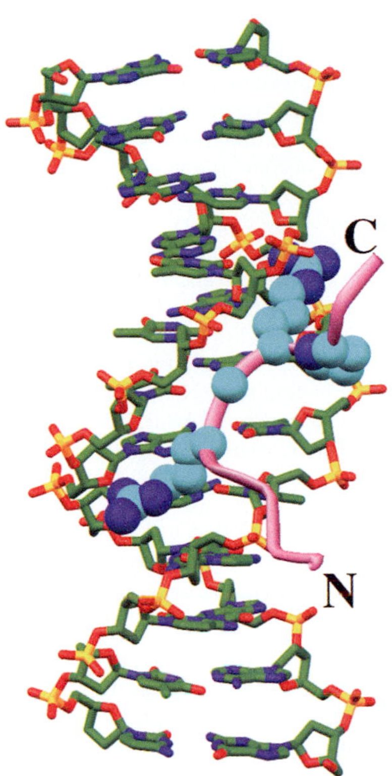

FIGURE 34-65 NMR structure of a truncated HMGA1a consisting of only its second and third AT hooks in complex with a 12-bp AT-rich DNA. The protein is drawn in ribbon form (*lavender*) with the side chains and C_α atoms of the invariant core sequence of its AT hook, Arg-Gly-Arg-Pro, drawn in space-filling form with C cyan and N blue. The DNA is drawn in stick form colored according to atom type (C green, N blue, O red, and P yellow). The protein's two 10-residue AT hooks bind to separate DNA dodecamers. Nevertheless, only one set of DNA resonances was observed, which indicates that the two AT hook–DNA structures are closely similar. The peptide segment that links the two AT hooks is not observed and hence must be highly mobile. Consequently, only the structure shown was observed. Note that the AT hook binds in the DNA's minor groove but does not cause it to bend. [Based on an NMR structure by Angela Gronenborn and Marius Clore, NIH, Bethesda, Maryland. PDBid 2EZF.]

w. Nucleosome Cores Are Transferred Out of the Path of an Advancing RNA Polymerase

Since nucleosomes bind their component DNA tightly and quite stably, how does an actively transcribing RNA polymerase, which is roughly the size of a nucleosome and must separate the strands of duplex DNA to transcribe it, get access to the DNA? Two classes of models have been proposed: The advancing RNA polymerase either (1) induces a conformational change in the nucleosome that permits its DNA to be transcribed while still associated with the nucleosome or (2) displaces the nucleosome from the DNA. These models were differentiated by Gary Felsenfeld as follows: A single nucleosome core was assembled onto a short DNA segment of defined sequence. Then, under conditions in which nucleosome cores are stable (don't decompose or move) in the absence of transcription, the resulting assembly was ligated into a plasmid between a promoter and terminators for the RNA polymerase from **bacteriophage SP6** and the DNA between these two sites was transcribed by this enzyme. This treatment caused the nucleosome to move to a different site on the same plasmid, with a small preference for the untranscribed region preceding the promoter. However, the use of a very short (227-bp) DNA template containing the SP6 promoter and a bound nucleosome revealed that nucleosome transfer occurred only to the same template molecule, 40 to 95 bp upstream of its original site, even in the presence of a large excess of competitor DNA. Evidently, the histone octamer somehow steps around a transcribing RNA polymerase so as to transfer to a nearby segment of the same DNA. Felsenfeld has proposed that this occurs via a DNA looping mechanism in which the histone octamer incrementally spools onto its new position behind the advancing RNA polymerase as the polymerase peels the octamer away from its original position (Fig. 34-66).

How does RNA polymerase displace nucleosomes from DNA? SP6 RNA polymerase, being a phage enzyme, cannot have evolved to interact with histones but, nevertheless, appears to do so. Other prokaryotic RNA polymerases can likewise transcribe through nucleosomes. A plausible mechanism for this phenomenon is that it is promoted by the transcriptionally induced supercoiling of DNA. A moving transcription bubble, it will be recalled (Section 31-2C), generates positive supercoils in the DNA ahead of it and negative supercoils behind it. However, nucleosomal DNA is wound around its histone core in a left-handed toroidal coil and is therefore negatively supercoiled (Section 29-3A). Consequently, an advancing RNA polymerase molecule should destabilize the nucleosomes ahead of it while facilitating nucleosome assembly in its wake, precisely what is observed (although, as we shall see below, the cell employs several methods for loosening the grip of nucleosomes on DNA). Subsequent investigations revealed that yeast RNA polymerase III, which is considerably larger than SP6 RNA polymerase, interacts with nucleosomes in a similar manner.

x. Locus Control Regions Are Nuclease Hypersensitive

The very light digestion of transcriptionally active chromatin with DNase I and other nucleases has revealed the presence of **DNase I hypersensitive sites** that are about an order of magnitude more susceptible to cleavage by DNase I than are DNase I sensitive sites. These specific 100- to 200-bp DNA segments are mostly located in the 5′-flanking regions of transcriptionally active or activatable genes as well as in sequences involved in replication and recombination. Nuclease hypersensitive sites, as we shall see, are apparently the "open windows" that allow the transcriptional machinery access to DNA control sequences. This is because *DNase I hypersensitive gene segments are free of nucleosomes.* For example, in SV40-infected cells, none of the ~24 nucleosomes that are complexed to the virus' 5.2-kb circular DNA (Fig. 34-67) incorporate the ~250-bp viral transcription initiation site, thereby rendering that site nuclease hypersensitive.

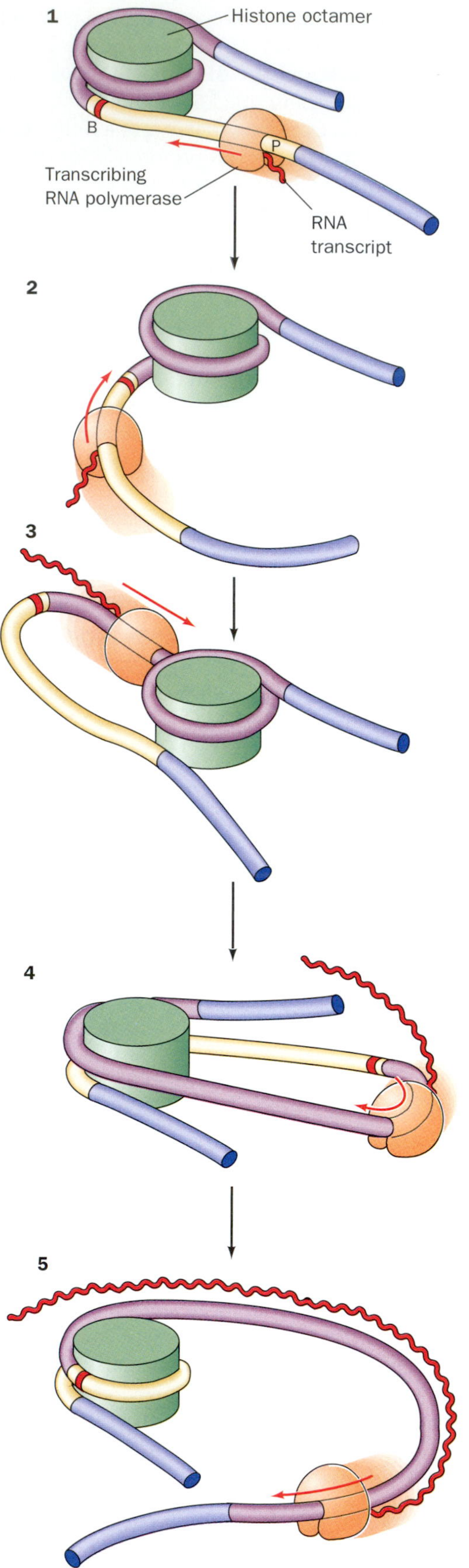

FIGURE 34-66 Spooling model for transcription through a nucleosome. **(1)** RNA polymerase commences transcription at a promoter, P; the border of the nucleosome is indicated by B. **(2)** As the RNA polymerase approaches the nucleosome, it induces the dissociation of the proximal (nearest) DNA, thereby exposing part of the histone octamer surface. **(3)** The exposed histone surface binds to the DNA behind the RNA polymerase, thus forming a loop. Note that this loop is topologically isolated from the rest of the DNA and, consequently, is subject to the superhelical stress that the advancing RNA polymerase generates (see the text). **(4)** As the RNA polymerase continues to advance, the DNA ahead of it peels off the histone octamer while the trailing DNA spools onto it. **(5)** The nucleosome is thereby re-formed behind the RNA polymerase, thus permitting the transcript to be completed. [After Studitsky, V.M., Clark, D.J., and Felsenfeld, G., *Cell* **76,** 379 (1994).]

However, since naked DNA is not DNase I hypersensitive, the special properties of nuclease hypersensitive chromatin must arise from the sequence-specific binding of proteins so as to exclude nucleosomes.

The human β-globin cluster (Section 34-2F) has five nuclease hypersensitive sites in a region 6 to 22 kb on the 5′ side of the ε gene as well as one hypersensitive site 20 kb on the 3′ side of the β gene (Fig. 34-68). These hypersensitive sites appear to demarcate the boundaries of a large segment of transcriptionally active chromatin. Individuals with an extensive upstream deletion that eliminates the 5′ hypersensitive sites, the so-called Hispanic deletion, but with normal β-like genes, have **$(\gamma\delta\beta)^0$-thalassemia** (severely reduced synthesis of γ, δ, and β globins). Similarly, mice that are transgenic for the human β-globin gene together with its local regulatory sites either fail to express or express very low levels of human β-globin. This is because a DNA segment that is randomly inserted into a genome will most often occupy a position in transcriptionally inactive heterochromatin, a phenomenon known as a **position effect.** However, mice transgenic for the entire region of the human β-globin cluster between its hypersensitive sites express high levels of human β-globin in

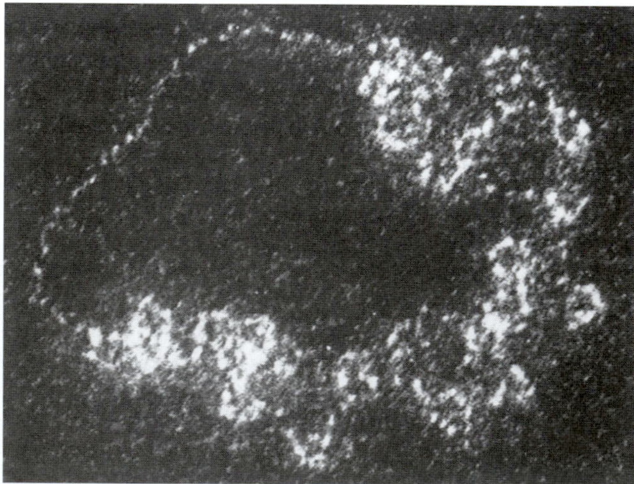

FIGURE 34-67 Electron micrograph of an SV40 minichromosome that has a nucleosome-free DNA segment. [Courtesy of Moshe Yaniv, Institut Pasteur, Paris, France.]

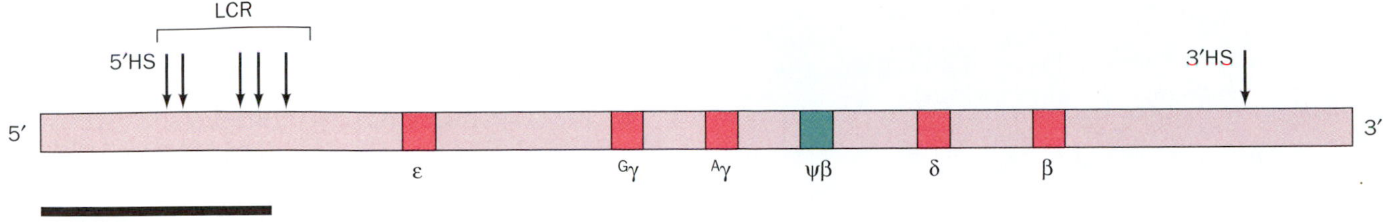

FIGURE 34-68 The β-globin cluster showing the positions (*arrows*) of its genes and DNase I hypersensitive sites (HSs). The hypersensitive sites on the 5′ side of the ε-globin gene (5′HSs) form the locus control region (LCR), whose presence is required for the expression of the β-like genes. The deletion of the LCR, as occurs in the Hispanic deletion, all but eliminates the expression of the β-like genes. The products of the β-globin cluster are discussed in Section 34-2F.

erythroid tissues. Thus the β-globin cluster's 5′ nuclease hypersensitive sites, which are collectively known as a **locus control region (LCR),** function to suppress position effects over large distances (e.g., the ~100 kb length of the β-globin cluster). LCRs have enhancerlike properties but, unlike enhancers, are orientation- and position-specific.

LCRs are apparently activated by proteins that are expressed only in specific cell lineages (e.g., only in erythroid cells for genes controlled by the β-globin LCR) so as to render the gene(s) under an LCR's control susceptible to activation by transcription factors. In support of this contention, it has been shown that nonglobin genes that have been put under the control of the β-globin cluster LCR are expressed in erythroid cells but not in nonerythroid cells. LCRs occur in a growing list of mammalian genes. However, the way that they permit the expression of the genes under their control remains largely conjectural.

y. Insulators Isolate Genes from Distant Regulatory Elements

We have seen that enhancers function independently of their position and oriention. But then, what prevents an enhancer from affecting the transcription of all the genes in its chromosome? Conversely, the formation of heterochromatin appears to be self-nucleating. What prevents heterochromatin from spreading into neighboring segments of euchromatin so as to prevent the transcription of their component genes? In many cases, this appears to be the job of short (<2 kb) DNA sequences known as **insulators** that thereby define the boundaries of functional domains.

Among the best characterized insulators are the *Drosophila* sequences **scs** and **scs′** (for *s*pecialized *c*hromatin *s*tructure), which normally flank two consecutive *hsp70* heat shock genes. The transformation into *Drosophila* of the *white* gene (which confers white eye color; Section 1-4C) together with a minimal promoter yielded lines of flies that varied in eye color, a manifestation of the position effect. However, when the construct was flanked by scs and scs′, it yielded only flies with white eyes. Evidently, these insulators overcome the position effect. In addition, if scs or scs′ is inserted between a gene and its upstream regulatory sequences, then the expression of this gene is no longer influenced by these sequences. Several other insulators have been characterized, both in *Drosophila* and in vertebrates.

The foregoing indicates that insulators resemble LCRs in that they suppress position effects. However, unlike LCRs, insulators have no enhancerlike properties; that is, they do not have positive or negative effects on the expression of the genes they control. Rather, *insulators only function to prevent regulatory elements outside the region they control from influencing the expression of the genes inside the region.* LCRs lack this property; they do not protect their associated genes from the influence of control sequences that are upstream of the LCR. In fact, the chicken β-globin cluster has an upstream insulator named **HS4** (for *h*ypersensitive *s*ite *4*) that prevents regulatory elements that are further upstream from influencing the expression of its genes.

The way that insulators work remains enigmatic. Presumably, it is not the insulators themselves but the proteins that bind to them that form the active insulator elements. For example, in *Drosophila,* the insertion of the transposable element *gypsy* between the promoter of the gene *yellow* (which gives flies a pale yellow body rather than the wild-type yellow-brown) and its upstream enhancers prevents these enhancers from activating *yellow* but does not affect downstream enhancers. The 12-zinc finger protein named **Su(Hw)** (for *su*ppressor of *h*airy *w*ing) specifically binds to the *gypsy* insulator and is required for its enhancer-blocking properties. Su(Hw) also binds to the protein **Mod(mdg4)** (for *mod*ifier of *mdg4*), and together they bind to the **nuclear matrix** (the nuclear equivalent of the cytoskeleton). In fact, immunostaining studies by Victor Corces indicate that Su(Hw) and Mod(mdg4) colocalize to several hundred sites in *Drosophila* polytene chromosomes (Fig. 34-69) and are distributed in a punctuate pattern around the nuclear matrix. Similar distributions were seen with the protein **BEAF-32** (for *b*oundary *e*lement-*a*ssociated *f*actor of *32* kD), which specifically binds to scs′ but not to scs. These observations suggest that these proteins each bind to numerous insulator sites on *Drosophila* chromosomes, which, in turn, suggests that insulators function as matrix-associated regions (MARs; Section 34-1D) to form structural domains. These proteins may prevent enhancers that are outside such a domain from influencing the expression of the genes that are inside the domain and, furthermore, may inhibit heterochromatin from encroaching on and thereby transcriptionally inactivating the domain. Similarly, enhancer blocking in the chicken β-globin cluster is associated with the binding of the 11-zinc finger

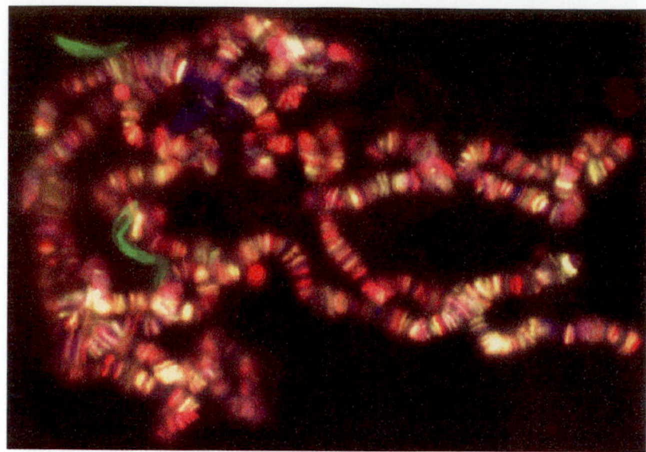

FIGURE 34-69 Colocalization of the Su(Hw) and Mod(mdg4) proteins on the polytene chromosomes of *Drosophila* larvae. The DNA has been stained blue, whereas the proteins have been immunostained such that sites containing Su(Hw) are green, those containing Mod(mdg4) are red, and sites where both proteins colocalize are yellow. [Courtesy of Victor Corces, The Johns Hopkins University.]

protein **CTCF** to HS4 and, moreover, CTCF-binding sites occur throughout the genome.

z. Chromatin Immunoprecipitation Reveals the DNA Binding Sites of Proteins

Throughout this section, we discuss the DNA sequences in chromatin to which specific proteins bind. In many cases, these DNA sequences have been identified using a procedure known as **chromatin immunoprecipitation (ChIP).** ChIP involves the following steps:

1. Living cells are treated with formaldehyde, which rapidly cross-links the amino and imino groups of Arg, His, and Lys residues to nearby (within ~2 Å) amino groups of bases, primarily those of adenine and cytosine, while preserving the chromatin structure:

$$R_1{-}NH_2 \;+\; \underset{\textbf{Formaldehyde}}{H{-}\overset{\overset{\Large O}{\|}}{C}{-}H} \;+\; H_2N{-}R_2 \;\rightleftharpoons\; R_1{-}NH{-}CH_2{-}HN{-}R_2$$

2. The cells are lysed and the cross-linked chromatin is sheared into manageable (~500-nt) fragments by sonication.

3. The chromatin fragments are treated with antibodies (Section 35-2B) raised against the protein of interest (which may be a histone with a specific modification such as an acetylation or a methylation at a particular site; see below). The mixture is then absorbed to agarose gel beads to which *Staphylococcus aureus* **Protein A** has been cross-linked (Section 6-3C). Protein A binds antibodies, but only when they are bound to their target antigens, thereby permitting the isolation of only those chromatin fragments in which the DNA is cross-linked to the antibody-bound protein.

4. The DNA is released from the protein to which it is cross-linked by acidification, which reverses the formaldehyde cross-linking reaction, and the DNA is isolated.

5. The DNA is identified by PCR using specific primers (Section 5-5F), by Southern blotting (Section 5-5D), or by hybridization to a DNA microarray (Section 7-6B), thereby revealing the DNA segments to which the protein of interest binds.

aa. Histone Modification and Remodeling Play an Essential Role in Transcriptional Activation

The DNA packaged by chromatin must be accessible by the transcriptional machinery in order for it to be expressed. This, as we shall see, occurs via two types of processes acting in synergy: (1) the posttranslational modifications of core histones, mainly their N-terminal tails, and (2) the remodeling of chromatin through the ATP-driven alteration of the position and/or properties of its nucleosomes.

The posttranslational modifications to which core histones are subject include the acetylation/deacetylation of specific Lys side chains, the methylation of specific Lys and Arg side chains, the phosphorylation/dephosphorylation of specific Ser and possibly Thr side chains, and the ubiquitination of specific Lys side chains (Fig. 34-70). Moreover, Lys side chains can be mono-, di-, and trimethylated and Arg

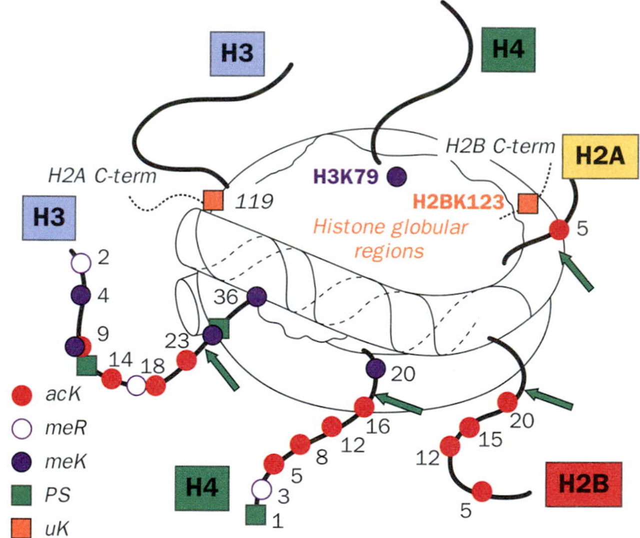

FIGURE 34-70 Histone modifications on the nucleosome core particle. Posttranslational modification sites are indicated by the residue numbers and the colored symbols, which are defined in the key at the lower left (*acK* = acetyl-Lys, *meR* = methyl-Arg, *meK* = methyl-Lys, *PS* = phospho-Ser, and *uK* = ubiquitinated Lys). Note that H3 Lys 9 can be either methylated or acetylated. The N-terminal tail modifications are shown on only one of the two copies of H3 and H4 and only one molecule each of H2A and H2B are shown. The C-terminal tails of one H2A and one H2B are represented by dashed lines. The green arrows indicate the sites in intact nucleosomes that are susceptible to trypsin cleavage. This cartoon summarizes data from several organisms, some of which may lack particular modifications. [Courtesy of Bryan Turner, University of Birmingham School of Medicine, U.K.]

side chains can be mono- and both symmetrically and asymmetrically dimethylated. The core histones' N-terminal tails, as we have seen (Section 34-1B), are implicated in stabilizing the structures of both core nucleosomes and higher order chromatin. All of these modifications but methylations reduce (make more negative) the electronic charge of the side chains to which they are appended and hence are likely to weaken histone–DNA interactions so as to promote chromatin decondensation, although as we shall see, this is not always the case. Methyl groups, in contrast, increase the basicity and hydrophobicity of the side chains to which they are linked and hence tend to stabilize chromatin structure. Modified histone tails also interact with specific chromatin-associated nonhistone proteins in a way that changes the transcriptional accessibility of their associated genes.

The characterization of a variety of histone tail modifications led David Allis to hypothesize that *there is a "histone code" in which specific modifications evoke certain chromatin-based functions and that these modifications act sequentially or in combination to generate unique biological outcomes.* For example, uncondensed and hence transcriptionally active chromatin is associated with the acetylation of H3 Lys 9 and 14 and H4 Lys 5 and the methylation of H3 Lys 4 and H4 Arg 3; condensed and hence transcriptionally inactive chromatin is associated with the acetylation of H4 Lys 12 and the methylation of H3 Lys 9; and nucleosome deposition is associated with the phosphorylation of H3 Ser 10 and 28. It can be seen from Fig. 34-70 that there are a vast number of possible combinations of histone modifications.

The growth of a multicellular organism requires the proliferation of the cells in its various tissues without changing their identities (the process whereby cells progressively and irreversibly change their identities, which is known as differentiation, is discussed in Section 34-4). A particular cell type is largely defined by its characteristic pattern of gene expression. Since most cells in a multicellular organism have the same complement of DNA, how do cells maintain their identities (patterns of gene expression) from one cell generation to the next? Evidently, histone modifications are largely preserved between cell generations, that is, they are epigenetic markings in much the same way as are the methylation patterns of DNA (Section 30-7). The way in which a cell confers its histone epigenetic markings on its progeny is unknown, although it almost certainly involves the recruitment of histone-modifying enzymes to newly assembled nucleosomes on recently replicated DNA.

bb. Histone Acetyltransferases (HATs) Are Components of Multisubunit Transcriptional Coactivators

Although Vincent Allfrey discovered, in the late 1960s, that histone acetylation and deacetylation are respectively correlated with transcriptional activation and repression, it was not until the mid-1990s that the proteins that mediate histone acetylation and deacetylation were identified and characterized. Histone Lys side chains are acetylated in a sequence-specific manner by enzymes known as **histone acetyltransferases (HATs),** all of which employ acetyl-CoA (Fig. 21-2) as their acetyl group donors:

$$\underset{\textbf{Acetyl-CoA}}{\text{CoA}-\text{S}-\overset{\overset{\text{O}}{\|}}{\text{C}}-\text{CH}_3} + \underset{\textbf{Histone Lys}}{\text{Lys}-(\text{CH}_2)_4-\overset{+}{\text{N}}\text{H}_3} \xrightarrow{\text{HAT}} \text{CoA}-\text{SH} + \underset{\textbf{Histone Acetyl-Lys}}{\text{Lys}-(\text{CH}_2)_4-\text{NH}-\overset{\overset{\text{O}}{\|}}{\text{C}}-\text{CH}_3}$$

The large number of known HATs are members of five families: (1) the **GNAT family** (for *G*cn5-relatated *N*-*a*cetyl*t*ransferase; **Gcn5,** first found in yeast, is one of the best characterized HATs), whose members include Gcn5, its homologs **Gcn5L** (for *Gcn5-l*ike protein) and **PCAF** [for *p*300/*C*BP-*a*ssociated *f*actor; **p300** and **CBP** (for *c*AMP response *b*inding element *p*rotein) are homologous transcriptional coactivators], and **Hat1** (which acetylates histones in the cytoplasm before they are imported to the nucleus); (2) the **MYST family** (named for its founding members, **MOZ, Ybf2/Sas3, Sas2,** and **Tip60**); (3) the **p300/CBP family;** (4) the **TAF1 family** (TAF1, the largest subunit of TFIID, was formerly named $\textbf{TAF}_{\textbf{II}}\textbf{250}$); and (5) the **SRC family** (for *s*teroid *r*eceptor *c*oactivator). Most HATs besides Hat1 function as transcriptional coactivators or silencers but some are implicated in regulating cell cycle progression, DNA regulation, and transcriptional elongation.

Most if not all HATs function *in vivo* as members of often large (10–20 subunits) multisubunit complexes, many of which were initially characterized as transcriptional regulators. These include **SAGA** (for *S*pt/*A*da/*G*cn5/*a*cetyltransferase), the closely similar **PCAF complex** (which contains PCAF), **STAGA** (*S*pt3/*TAF*/*G*cn5L *a*cetyltransferase; its HAT is Gcn5L), **ADA** (transcriptional *ada*ptor), TFIID (which contains TAF1), **TFTC** (*T*BP-*f*ree *TA*F-*c*ontaining complex), **NuA3,** and **NuA4** (*nu*cleosomal *a*cetyltransferases of H*3* and H*4*).

Many **HAT complexes** share subunits. For example, three of ADA's four subunits, Gcn5, **Ada2,** and **Ada3,** are common to the 14-subunit SAGA. Likewise, SAGA and NuA4 both contain **Tra1,** a homolog of the phosphoinositide 3-kinases (Section 19-4D) that interacts with specific transcriptional activators, including **Myc** (Section 34-4C). Intriguingly, several HAT complexes besides TFIID and TFTC contain TAFs. For example, SAGA contains **TAF5,** TAF6, TAF9, **TAF10,** and TAF12, as does the PCAF complex with the exception that TAF5 and TAF6 are replaced in the PCAF complex by their close homologs **PAF65β** and **PAF65α** (PAF for *P*CAF *a*ssociated *f*actor). TAF6, TAF9, and TAF12, as we discussed above, are structural homologs of histones H3, H4, and H2B, respectively. Consequently, these TAFs probably associate to form an architectural element that is common to TFIID, SAGA, and the PCAF complex and hence these complexes are likely to interact with TBP in a similar manner.

The various HAT complexes presumably target their component HATs to the promoters of active genes. Moreover, they alter the specificities of these HATs. For example, a general property of HATs is that although they can acetylate at least one type of free histone, they can only acetylate histones in nucleosomes as members of HAT complexes. Thus, free Gcn5 acetylates H3 Lys 14 and, to a lesser extent, H4 Lys 8 and 16. However, Gcn5 in SAGA expands its H3 sites to Lys 9, 14, and 18 and also acetylates H2B, whereas Gcn5 in ADA acetylates H3 at Lys 14 and 18, as well as H2B. Neither complex acetylates H4.

The X-ray structures of several HATs have been determined. That of the HAT domain of *Tetrahymena thermophila* GCN5 (residues 48–210 of the 418-residue protein) in complex with a bisubstrate inhibitor was determined by Ronen Marmorstein. The bisubstrate inhibitor (Fig. 34-71*a*) consists of CoA covalently linked from its S atom via an isopropionyl group (which mimics an acetyl group) to the side chain of Lys 14 of the 20-residue N-terminal segment of histone H3. The structure (Fig. 34-71*b*) reveals the enzyme to be deeply clefted and to contain a core region common to all HATs of known structure (magenta in Fig. 34-71*b*) that consists of a 3-stranded antiparallel β sheet connected via an α helix to a fourth β strand that forms a parallel interaction with the β sheet. Only 6 residues of the histone tail, Gly 12 through Arg 17, are visible in the X-ray structure. The CoA moiety binds in the enzyme's cleft such that it is mainly contacted by core residues. The comparison of this structure with other Gcn5-containing structures indicates that the cleft has closed down about the CoA moiety.

cc. Bromodomains Recruit Coactivators to Acetylated Lys Residues in Histone Tails

The different patterns of histone acetylation required for different functions (the histone code) suggest that the function of histone acetylation is more complex than merely attenuating the charge–charge interactions between the cationic histone N-terminal tails and anionic DNA. In fact, there is growing evidence that specific acetylation patterns are recognized by protein modules of transcriptional coactivators in much the same way that specific phosphorylated sequences are recognized by protein modules such as the SH2 and PTB domains that mediate signal transduction via protein kinase cascades (Section 19-3). Thus, nearly all HAT-associated transcriptional coactivators contain ~110-residue modules known as **bromodomains** that specifically bind acetylated Lys residues on histones. For example, Gcn5 essentially consists of its HAT domain followed by a bromodomain, whereas TAF1 consists mainly of an N-terminal kinase domain followed by a HAT domain and two tandem bromodomains.

The X-ray structure of human TAF1's double bromodomain (residues 1359–1638 of the 1872-residue protein), determined by Tjian, reveals that it consists of two nearly

(*a*)

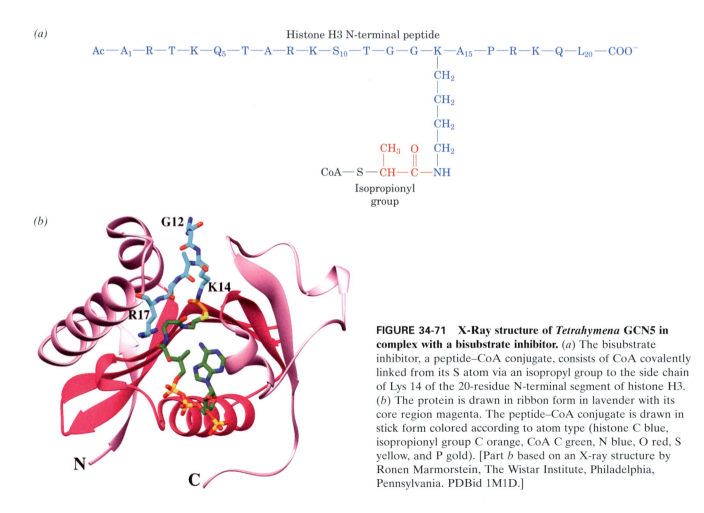

Histone H3 N-terminal peptide

$$\text{Ac—A}_1\text{—R—T—K—Q}_5\text{—T—A—R—K—S}_{10}\text{—T—G—G—K—A}_{15}\text{—P—R—K—Q—L}_{20}\text{—COO}^-$$

Lys 14 side chain: $\text{K—CH}_2\text{—CH}_2\text{—CH}_2\text{—CH}_2\text{—NH—C(=O)—CH(CH}_3\text{)—S—CoA}$

$$\text{CoA—S—CH(CH}_3\text{)—C(=O)—NH—}$$

Isopropionyl group

(*b*)

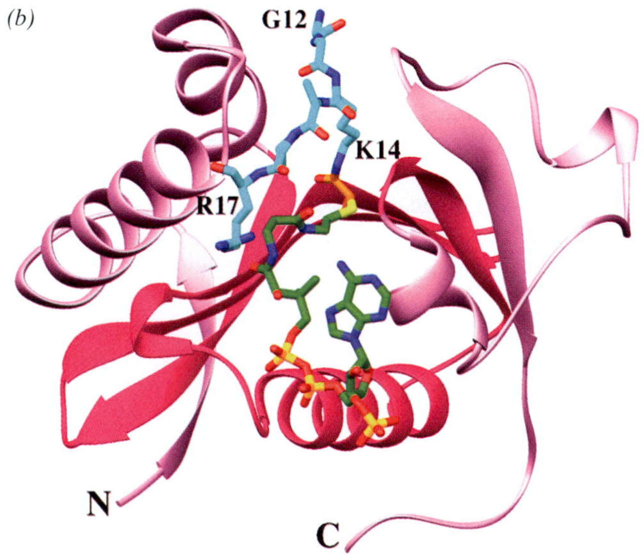

FIGURE 34-71 X-Ray structure of *Tetrahymena* GCN5 in complex with a bisubstrate inhibitor. (*a*) The bisubstrate inhibitor, a peptide–CoA conjugate, consists of CoA covalently linked from its S atom via an isopropyl group to the side chain of Lys 14 of the 20-residue N-terminal segment of histone H3. (*b*) The protein is drawn in ribbon form in lavender with its core region magenta. The peptide–CoA conjugate is drawn in stick form colored according to atom type (histone C blue, isopropionyl group C orange, CoA C green, N blue, O red, S yellow, and P gold). [Part *b* based on an X-ray structure by Ronen Marmorstein, The Wistar Institute, Philadelphia, Pennsylvania. PDBid 1M1D.]

identical antiparallel 4-helix bundles (Fig. 34-72). A variety of evidence, including NMR structures of single bromodomains in complex with their target acetyl-Lys–containing peptides, indicates that the acetyl-Lys binding site of each bromodomain occurs in a deep hydrophobic pocket that is located at the end of its 4-helix bundle opposite its N- and C-termini. The double bromodomain's two binding pockets are separated by ~25 Å, which makes them ideally positioned to bind two acetyl-Lys residues that are separated by 7 or 8 residues. In fact, the N-terminal tail of histone H4 contains Lys residues at its positions 5, 8, 12, and 16 (Fig. 34-70), whose acetylation is correlated with increased transcriptional activity. Moreover, the 36-residue N-terminal peptide of histone H4, when fully acetylated, binds to the TAF1 double bromodomain in 1:1 ratio with 70-fold higher affinity than do single bromodomains but fails to bind when it is unacetylated.

The foregoing structure suggests that the TAF1 bromodomains serve to target TFIID to promoters that are within or near nucleosomes (in contrast to the widely held notion that TFIID targets PICs to nucleosome-free regions). Tjian has therefore postulated that the transcriptional initiation process may begin with the recruitment of a HAT-containing coactivator complex by an upstream DNA-binding protein (Fig. 34-73). The HAT could then acetylate the N-terminal histone tails of nearby nucleosomes, which would recruit TFIID to an appropriately located promoter via the binding of its TAF1 bromodomains to the acetyl-Lys residues. Moreover, the TAF1 HAT activity could acetylate other nearby nucleosomes, thereby initiating a cascade of acetylation events that would render the DNA template competent for transcriptional initiation.

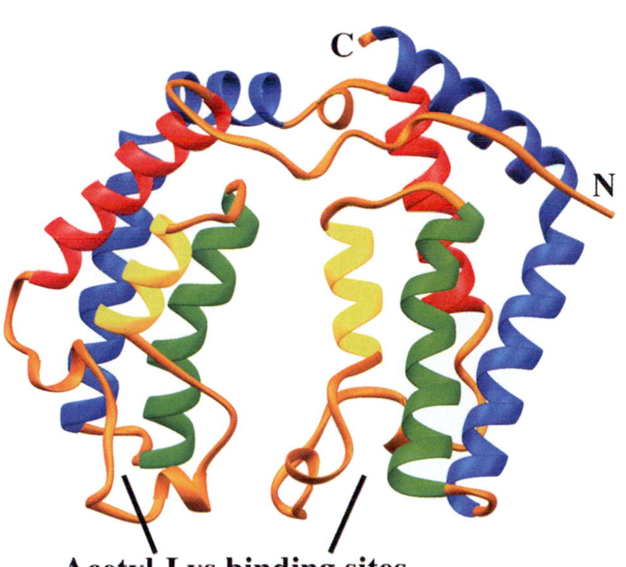

FIGURE 34-72 X-Ray structure of the human TAF1 double bromodomain. Each bromodomain consists of an antiparallel four-helix bundle whose helices are colored, from N- to C-termini, red, yellow, green, and blue, with the remaining portions of the protein orange. The two four-helix bundles are related by an ~108° rotation about an axis that is approximately parallel to the principal axes of the four-helix bundles (the vertical direction in this drawing). The acetyl-Lys binding sites occupy deep hydrophobic pockets at the end of each four-helix bundle opposite its N- and C-termini. [Based on an X-ray structure by Robert Tjian, University of California at Berkeley. PDBid 1EQF.]

dd. Histone Deacetylases (HDACs)

Histone acetylation is a reversible process. The enzymes that remove the acetyl groups from histones, the **histone deacetylases (HDACs),** promote transcriptional repression and gene silencing. Eukaryotic cells from yeast to

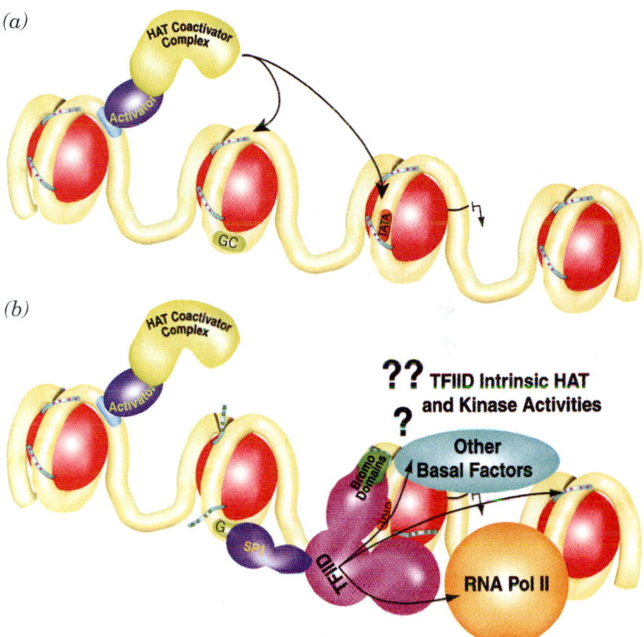

FIGURE 34-73 Simplified model for the assembly of a transcriptional initiation complex on chromatin-bound templates. Here the DNA is represented by a yellow worm, the histone octamers around which the DNA wraps to form nucleosomes are shown as red spheres, and their N-terminal histone tails are drawn as short cyan rods with the red and green dots representing unacetylated and acetylated Lys residues. The transcription initiation site is represented by the black ring about the DNA from which the squared-off arrow points downstream. (*a*) The process begins by the recruitment of a HAT-containing transcriptional coactivator complex (*yellow*) through its interactions with a DNA-binding activator protein (*purple*) that is bound to an upstream enhancer (*light blue*). The coactivator HAT is thereby positioned to acetylate the N-terminal tail on nearby nucleosomes (*curved arrows*). (*b*) The binding of TAF1's bromodomains to the acetylated histone tails could then help recruit TFIID (*magenta*) to a nearby TATA box (*orange patch*). Further acetylation of nearby histone tails by TAF1's HAT domain could help recruit other basal factors (*cyan*) and RNAP II (*orange*) to the promoter, thus stimulating PIC formation. Note that this model does not preclude other activation pathways such as the binding of enhancer-bound SP1 (*purple*) to TFIID. [Courtesy of Robert Tjian, University of California at Berkeley.]

humans typically contain numerous different HDACs; 10 HDACs have been identified in yeast and 17 in humans. The HDACs consist of three protein families: Class I, which in humans contains **HDAC1, 2, 3,** and **8;** Class II, which in humans contains **HDAC4–7, 9,** and **10;** and Class III, which in humans contains the so-called **sirtuins, SIRT1–7** (SIR for *s*ilent *i*nformation *r*egulator). Most, if not all, of the Class I HDACs are members of several multisubunit complexes. Thus, HDAC1 and HDAC2 form the catalytic cores of three complexes, **Sin3, NuRD** (*nu*cleosome *r*emodeling histone *d*eacetylase), and **CoREST** (*co*repressor to *RE*1 *s*ilencing *t*ranscription factor), whereas HDAC3 is the catalytic core of **N-CoR** (*n*uclear hormone receptor *co*repressor) and **SMRT** (*s*ilencing *m*ediator of *r*etinoid and *t*hyroid hormone receptor). These complexes serve as **transcriptional corepressors** for numerous transcriptional repressors as well as cooperating with each other. For example, the repressor **REST** (neuron-*rest*rictive repressor), on binding to its target DNA site, recruits CoREST and Sin3, which together repress transcription from nearby nucleosomes. Many if not all of the Class II HDACs function as transcriptional corepressors although few of them appear to be members of multisubunit complexes.

The Class III HDACs, the sirtuins, are unusual in that they contain an essential NAD^+ cofactor. Rather than simply hydrolyzing the amide bond linking the acetyl group to their target Lys side chain, they transfer it to the ADP–ribosyl group of NAD^+, thereby yielding ***O*-acetyl-ADP-ribose,** nicotinamide, and the deacetylated Lys residue:

NAD$^+$ + **Acetyl-Lys residue** ($-$Lys$-$ with side chain $(CH_2)_4$–NH–C(=O)–CH_3) $\xrightarrow{H_2O}$ ***O*-Acetyl-ADP-ribose** + $-$Lys$-$ (side chain $(CH_2)_4$–NH_3^+) + **Nicotinamide**

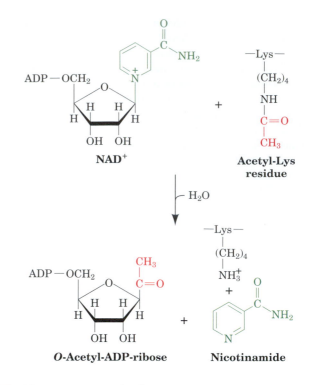

The X-ray structure of only one eukaryotic HDAC has yet been elucidated, that of human SIRT2, a component of transcriptionally silent chromatin that is required for gene silencing. This structure, determined by Nikola Pavletich, reveals that the 323-residue catalytic core of this 389-residue monomer consists of an *N*-terminal, NAD^+-binding, Rossmann fold–like domain and a smaller

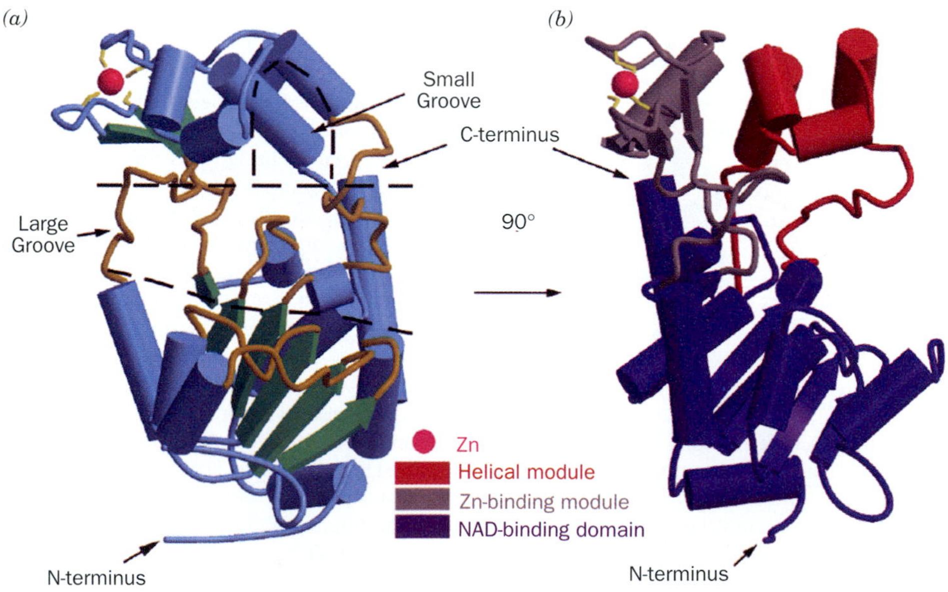

FIGURE 34-74 X-Ray structure of the human sirtuin SIRT2. (*a*) View of the protein in which the β-strands are green, the helices and most loops are cyan, the loops forming the major structural elements of the large groove are tan, the bound Zn^{2+} ion is represented by a magenta sphere, and its four liganding Cys side chains are drawn in stick form in yellow. The position of the large groove, which forms the enzyme's catalytic site, is indicated as is that of a smaller groove. (*b*) A view of the protein rotated by 90° about the vertical axis relative to that of Part *a,* in which the NAD^+-binding domain is blue, the helical module is red, and the Zn^{2+}-binding module is gray. [Courtesy of Nikola Pavletich, Memorial Sloan-Kettering Cancer Center, New York, New York. PDBid 1J8F.]

C-terminal domain that consists of a helical module and a Zn^{2+}-binding module (Fig. 34-74). Mutagenic studies indicate that a large groove between the two domains, which is lined with conserved hydrophobic residues, is the enzyme's likely catalytic site.

ee. Histone Methylation

Histone methylation at both the Lys and Arg side chains of histone H3 and H4 N-terminal tails (Fig. 34-70) tends to silence the associated genes by inducing the formation of heterochromatin. The enzymes mediating these methylations, the **histone methyltransferases (HMTs),** all utilize *S*-adenosylmethionine (SAM; Section 26-3E) as their methyl donor. Thus, the lysine HMTs, the most extensively characterized HMTs, catalyze the reaction

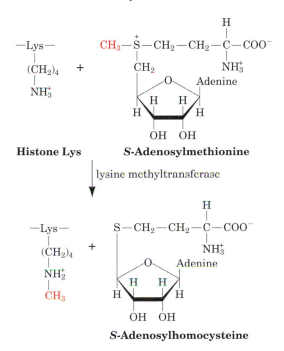

$$\text{—Lys—}(CH_2)_4\text{—}NH_3^+ \;+\; CH_3\text{—}\overset{+}{S}(\text{—}CH_2\text{—Adenosine})\text{—}CH_2\text{—}CH_2\text{—}CH(NH_3^+)\text{—}COO^-$$

Histone Lys **S-Adenosylmethionine**

↓ lysine methyltransferase

$$\text{—Lys—}(CH_2)_4\text{—}NH_2^+\text{—}CH_3 \;+\; S(\text{—}CH_2\text{—Adenosine})\text{—}CH_2\text{—}CH_2\text{—}CH(NH_3^+)\text{—}COO^-$$

S-Adenosylhomocysteine

These enzymes all have a so-called **SET domain** [*S*u(var)3-9, *E*(Z), *T*rithorax], which contains their catalytic sites.

The human lysine HMT named **SET7/9** monomethylates Lys 4 of histone H3. The X-ray structure of the SET domain of SET7/9 (residues 108–366 of the 366-residue protein) in complex with SAM and the N-terminal decapeptide of histone H3 in which Lys 4 is monomethylated, was determined by Steven Gamblin. Interestingly, SAM and the peptide substrate bind to opposite sides of the protein (Fig. 34-75). However, there is a narrow tunnel through the protein into which the Lys 4 side chain is inserted such that its amine group is properly positioned for methylation by SAM. The arrangement of the hydrogen bonding acceptors for the Lys amine group stabilizes the methyl-Lys side chain in its observed orientation about the C_ε—N_ζ bond, thus sterically precluding the methyl-Lys group from assuming a conformation in which it could be further methylated by SAM.

No enzymes that demethylate histones have been identified despite considerable effort to do so. This suggests that histone methylation is irreversible. However, there are several known instances in which demethylation appears to occur, albeit at a low level. For example, in yeast, H3 Lys 4 is trimethylated at active promoters but dimethylated at repressed promoters, which is indicative that histones can reversibly change their methylation states. This demethylation may be mediated by an as yet uncharacterized demethylase, by the replacement of a methylated histone by one that is unmodified, or possibly by the proteolytic excision of a methylated histone tail, although this may be the first step in histone replacement.

Methylated histones are recognized by so-called **chromodomains.** For example, methylated H3 Lys 9 is bound by the chromodomain-containing **heterochromatin protein 1 (HP1),** which thereupon recruits proteins that control chromatin structure and gene expression. The NMR structure of the mouse HP1 chromodomain (residues 8–80 of the 185-residue protein) in complex with the N-terminal 18-residue tail of H3 in which Lys 4 and 9 are dimethylated (Fig. 34-76), determined by Natalia Murzina and Earnest Laue, reveals that HP1 binds the H3 tail in an extended β-strandlike conformation in a groove on its surface (Fig. 34-76). The chromodomain buries the side chain of H3 Lys 9 (but not that of H3 Lys 4) such that its two methyl groups are contained in a hydrophobic box formed by three conserved aromatic residues. In contrast, the unmethylated H3 tail does not bind to HP1.

As previously mentioned, heterochromatin has a tendency to spread, thus silencing the newly heterochromatized genes. One way in which this appears to occur is via the binding of HP1 to nucleosomes whose H3 Lys 9 residues have been methylated (which is associated with transcriptionally inactive chromatin). The bound HP1 recruits the HMT **Suv39h,** which methylates nearby nucleosomes at their H3 Lys 9 residues, which thereupon recruit additional HP1, etc. Such heterochromatin spreading, as we discussed above, is prevented by the presence of an insulator. The HS4 insulator in the chicken β-globin cluster recruits HATs that acetylate H3 Lys 9 on nearby nucleosomes (which is associated with transcriptional activity), thereby blocking their methylation. Note that this activity is distinct from the enhancer-blocking function of HS4, a process that is mediated by the binding of CTCF to a different subsite of HS4 than that to which HATs bind.

ff. Histone Ubiquitination Functions to Regulate Transcription

Although ubiquitination mainly functions to mark proteins for destruction by the proteasome (Section 32-6B), it is also implicated in the control of transcription. In yeast, for example, the monoubiquitination of H2B Lys 123 (in contrast to polyubiquitination, which marks proteins for destruction), which is mediated by the ubiquitin-conjugating enzyme (E2) **Rad6** and the RING-finger–containing ubiquitin-protein ligase (E3) **Bre1,** is a prerequisite for the methylation of H3 Lys 4 and Lys 79. Together, these modifications are implicated in the silencing of genes located near telomeres. It has therefore been suggested that H2B ubiquitination functions as a master switch that controls

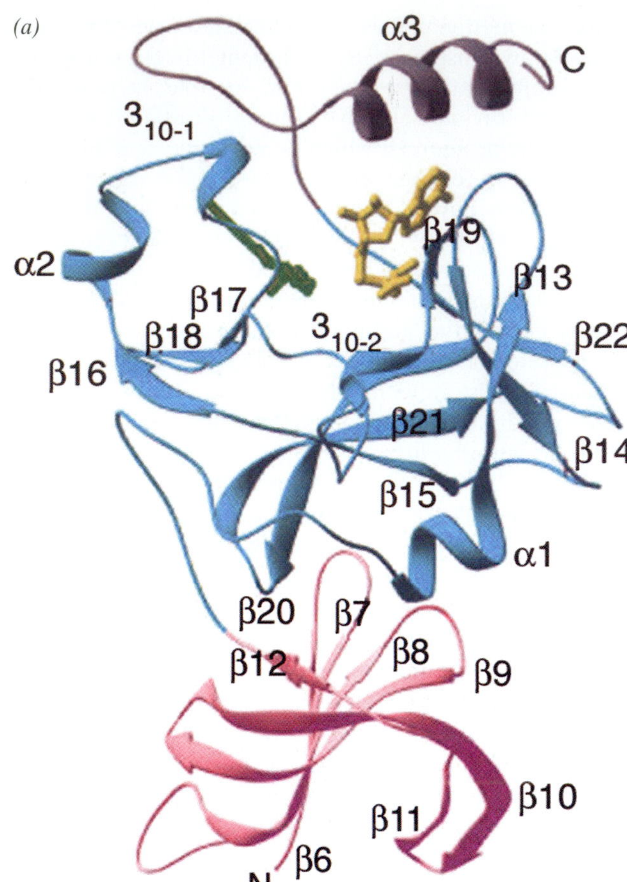

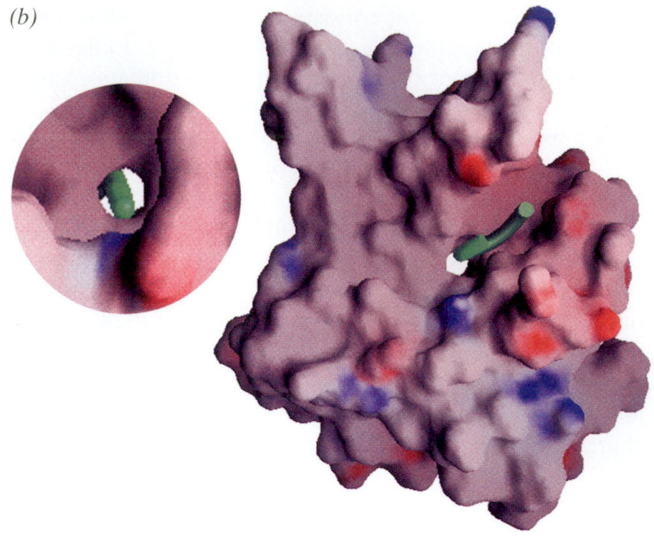

FIGURE 34-75 X-Ray structure of the human histone methyltransferase SET7/9 in complex with SAM and the histone H3 N-terminal decapeptide with its Lys 4 monomethylated. (*a*) Ribbon diagram in which the N-terminal domain is pink, the SET domain is blue, and the C-terminal segment is gray. The H3 N-terminal decapeptide with its methylated Lys 4 is green. The SAM cofactor is drawn in stick form in yellow. (*b*) Surface representation of the protein as seen from the back side of Part *a*. The protein surface is colored according to charge (blue most positive, red most negative, and gray neutral) and the H3 decapeptide is represented by a green ribbon. Note the narrow tunnel through the protein in which the methyl-Lys side chain is inserted. The inset at the left shows a close-up of this Lys access channel containing the methyl-Lys side chain (*green*) as viewed from the SAM-binding site. Note that the dimethylated side chain of Lys 9 is bound to the protein whereas that of Lys 4 is not. [Courtesy of Steven Gamblin, National Institute for Medical Research, London, U.K.]

the site-selective histone methylation patterns responsible for telomeric gene silencing. Similarly, the TAF1 subunit of TFIID, functioning as both a ubiquitin-activating enzyme (E1) and an E2, monoubiquitinates H1, a post-translational modification that is required for the expression of genes in the correct order during *Drosophila* development. Conversely, a histone deubiquitinating enzyme (DUB) is associated with the SAGA chromatin modifying complex. Thus, although the role of histone ubquitination is only beginning to come to light, it is clear that it is an essential transcriptional regulator.

gg. Chromatin-Remodeling Complexes

Sequence-specific DNA-binding proteins must gain access to their target DNAs before they can bind to them. Yet nearly all DNA in eukaryotes is sequestered by nucleosomes if not by higher order chromatin. How then do the proteins that bind to DNA segments gain access to their target DNAs? The answer, which has only become apparent since the mid-1990s, is that *chromatin contains ATP-driven complexes that remodel nucleosomes,* that is, they somehow disrupt the interactions between histones and DNA in nucleosomes to make the DNA more accessible. This may cause the histone octamer to slide along the DNA strand to a new location (a cis transfer) or relocate to a different DNA (a trans transfer). Thus, *these* ***chromatin-remodeling complexes*** *impose a "fluid" state on chromatin that maintains its DNA's overall packaging but transiently exposes individual sequences to interacting factors.*

Chromatin-remodeling complexes consist of multiple subunits. The first of them to be characterized was the yeast **SWI/SNF** complex, so-called because it was discovered through genetic screens as being essential for the expression of the *HO* gene, which is required for mating type switching (SWI for *swi*tching defective), and for the expression of the *SUC2* gene, which is required for growth on sucrose (SNF for *s*ucrose *n*on*f*ermenter). SWI/SNF, an 1150-kD complex of 11 different types of subunits, is only essential for the expression of ~3% of yeast genes and is not required for cell viability. However, a related complex named **RSC** (for *r*emodels the *s*tructure of *c*hromatin) is ~100 times more abundant in yeast and is required for cell viability. RSC shares two subunits with SWI/SNF and many of their remaining subunits are homologs, including their ATPase subunits, which are named **Swi2/Snf2** in SWI/SNF and **Sth1** in RSC. The Swi2/Snf2 ATPase, as well as two of RSC's subunits, contain bromodomains that are likely to facilitate the binding of their complexes to acetylated histones.

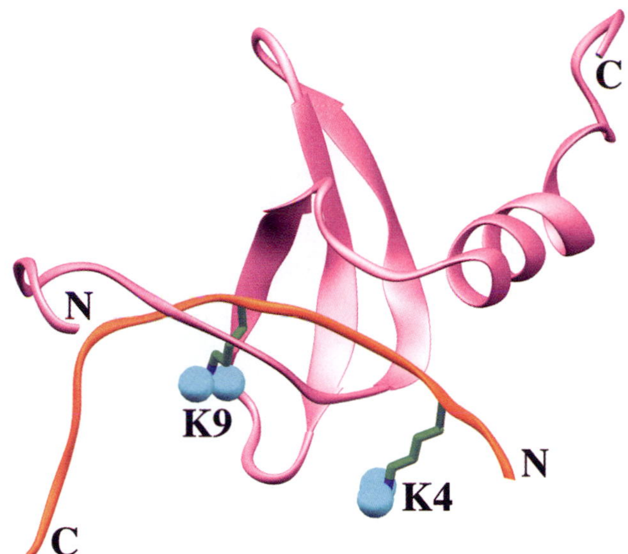

FIGURE 34-76 X-Ray structure of the mouse HP1 chromodomain in complex with the 18-residue N-terminal tail of histone H3 in which Lys 4 and Lys 9 are dimethylated. The 80-residue chromodomain is lavender, the H3 N-terminal tail is orange, and its two dimethyl-Lys side chains are drawn in stick form with C green, N blue, and with their methyl groups represented by cyan spheres. The side chain of H3 Lys 9, but not that of H3 Lys 4, is buried by the chromodomain. [Based on an X-ray structure by Natalia Murzina and Earnest Laue, University of Cambridge, U.K. PDBid 1GUW.]

All eukaryotes contain multiple chromatin-remodeling complexes. They have been classified into three main groups on the basis of the similarities of their component ATPase subunits: (1) the SWI/SNF complexes, whose ATPases are homologous to yeast Swi2/Snf2 and include yeast RSC, *Drosophila* **Brahma,** and human **BRM** (*Brahma* protein homolog) and **BRG1** (*B*rahma-*r*elated *g*ene *1*); (2) the **ISWI** (for *i*mitation *swi*tch) complexes, whose ATPases are homologs of yeast **ISW1** and include the yeast **ISW1** and **ISW2** complexes, *Drosophila* **ACF** (*A*TP-utilizing *c*hromatin assembly and remodeling *f*actor), **CHRAC** (*chr*omatin *ac*cessibility complex), **NURF** (*nu*cleosome-*r*emodeling *f*actor), and human **RSF** (*r*emodeling and *s*pacing *f*actor); and (3) the **Mi-2** complexes, whose ATPases are homologs of the *Xenopus* **Mi-2** complex and include human NuRD (which, as is discussed above, also contains HDAC1 and HDAC2). Many of these complexes contain bromodomains, chromodomains, and/or AT hooks that presumably recruit the complexes to their target genes. Moreover, some complexes are bound by specific transcriptional activators.

An electron microscopy–based image of yeast RSC, determined by Asturias and Kornberg at ~28 Å resolution, reveals that it consists of four modules surrounding a central cavity (Fig. 34-77*a*). Biochemical studies indicate that RSC binds tightly to nucleosomes in a 1:1 complex. Indeed, the size and shape of RSC's central cavity appear to be appropriate for binding a single nucleosome core particle, as Fig. 34-77*b* indicates. This would explain how, in the presence of ATP, RSC could loosen the DNA in a nucleosome without the loss of its associated histones.

The simultaneous release of all of the many interactions holding DNA to a histone octamer would require an enormous free energy input and hence is unlikely to occur. Then, how do chromatin-remodeling complexes function? Their various ATPase subunits share a region of homology with helicases (Section 30-2C), although they lack helicase activity. Nevertheless, it seems plausible that, like helicases, chromatin-remodeling complexes "walk" up DNA strands as driven by ATP hydrolysis. If such a complex were directly or indirectly tethered to a histone, this would put torsional strain on the DNA in the nucleosome, thereby decreasing its local twist (DNA supercoiling is discussed in Section 29-3A). The region of decreased twist could diffuse along the DNA wrapped around the nucleosome, thereby transiently loosening the histone octamer's grip on a segment of DNA. The torsional strain might also

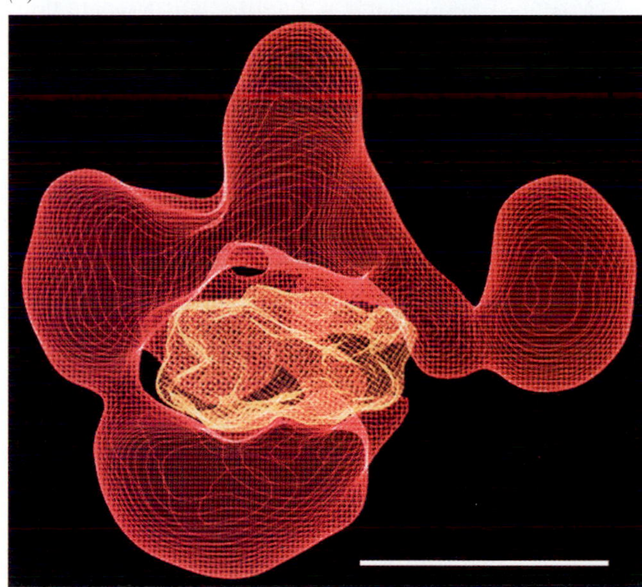

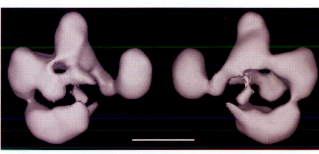

FIGURE 34-77 EM-based image of yeast RSC. (*a*) Two views of the structure (*front and back*) at ~28 Å resolution revealing that it consists of four modules surrounding a central cavity. (*b*) A model made by manually fitting the X-ray structure of the core nucleosome (Fig. 34-7) reduced to 25 Å resolution into the central cavity of the RSC structure in Part *a*. The complex is shown in mesh outline with RSC red and the nucleosome, which is viewed edgewise, yellow. The nucleosome fits snugly into the cavity with no steric clash. The scale bar in both parts represents 100 Å. [Courtesy of Francisco Asturias, The Scripps Research Institute, La Jolla, California.]

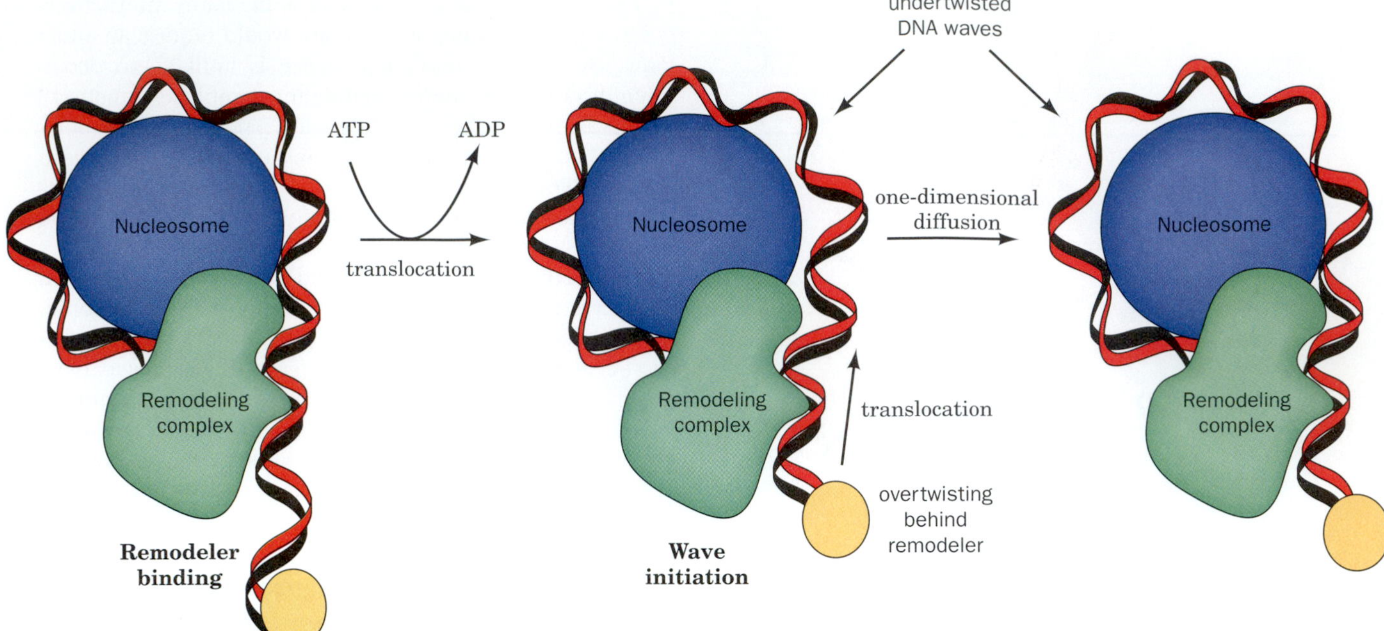

FIGURE 34-78 Model for nucleosome remodeling by chromatin-remodeling complexes. The chromatin-remodeling complex (*green*) couples the free energy of ATP hydrolysis to the translocation and concomitant twisting of the DNA in the nucleosome (*blue*, only half of which is shown for clarity) as depicted by the movement of a fixed point on the DNA (*yellow ellipsoid*). This locally breaks the contacts between the histones and the DNA. The position of the undertwisted and/or bulged DNA propagates around the nucleosome in a one-dimensional wave that transiently releases the DNA from the histone as it passes, thereby providing DNA-binding factors access to the DNA. [After a drawing by Saha, A., Wittmeyer, J., and Cairns, B.R., *Genes Dev*. **16,** 2120 (2002).]

be partially accommodated as a writhe, which would lift a segment of DNA off the nucleosome's surface. In either case, the resulting DNA distortion could diffuse around the surface of the nucleosome in a wave that would locally and transiently release the DNA from the histone octamer as it passed (Fig. 34-78) and hence permit the DNA to bind to its cognate DNA-binding factors. This latter mechanism resembles that proposed for the passage of RNAPs through nucleosomes (Fig. 34-66). Note that multiple cycles of ATP hydrolysis would send multiple DNA-loosening waves around the nucleosome, thereby sliding the nucleosome along the DNA.

hh. Afterword

As we have seen, eukaryotic transcriptional initiation is an astoundingly complex process that involves the synergistic participation of numerous multisubunit complexes comprising several hundred often loosely or sequentially interacting polypeptides (i.e., histones of various types and subtypes; the PIC; Mediator-like complexes; a variety of transcription factors, architectural factors, coactivators, and corepressors that in some cases form enhanceosomes; several types of histone modification complexes; and chromatin-remodeling complexes), as well as large segments of DNA. Intensive investigations in many laboratories over the past two decades have, as we have discussed, identified many of these complexes, characterized their component polypepides, and in many cases, elucidated their general functions. However, we are far from having more than a rudimentary understanding of how these various components interact *in vivo* to transcribe only those genes required by their cell under its particular circumstances in the appropriate amounts and with the proper timing. It is likely to require several additional decades of research to gain a detailed understanding of how this remarkable molecular machinery functions.

C. *Other Expressional Control Mechanisms*

The expression of many eukaryotic genes is regulated only by the control of transcriptional initiation. However, many cellular and viral genes additionally respond to other types of control processes. The various mechanisms employed by these secondary systems are outlined below.

1. Selection of Alternative Initiation Sites: *The expression of several eukaryotic genes is controlled, in part, through the selection of alternative transcriptional initiation sites.* For example, identical molecules of α-amylase are produced by mouse liver and salivary gland but the corresponding mRNAs synthesized by these two organs differ at their 5′ ends. Comparison of the sequences of these mRNAs with that of their corresponding genomic DNA indicates that the different mRNAs arise from separate initiation sites that are ~2.8 kb apart (Fig. 34-79). Thus, after being spliced, the liver and salivary gland α-amylase mRNAs have different untranslated 5′ leaders but the same coding sequences. The two initiation sites, it is

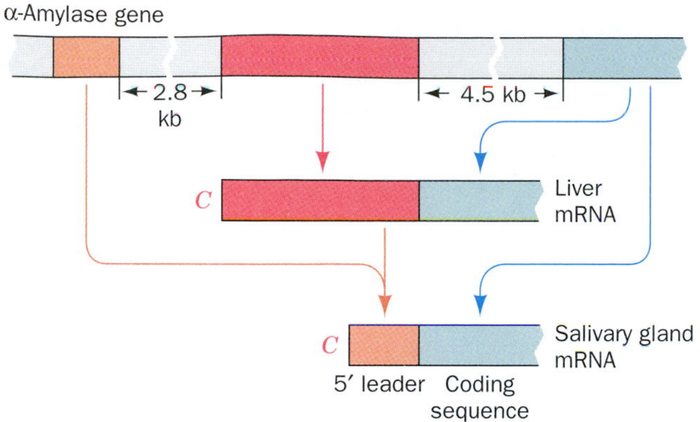

FIGURE 34-79 The transcription start site of the mouse α-amylase gene is subject to tissue-specific selection so as to yield mRNAs with different cap (*C*) and leader segments but the same coding sequences. [After Young, R.A., Hagenbüchle, O., and Schibler, U., *Cell* **23,** 454 (1981).]

thought, support different rates of initiation. This hypothesis accounts for the observation that α-amylase mRNA comprises 2% of the polyadenylated mRNA in salivary gland but only 0.02% of that in liver.

2. Selection of Alternative Splice Sites: Numerous cellular genes, as we have seen (Section 31-4A), are subject to alternative splicing. Thus, certain exons in one type of cell may be introns in another (e.g., Fig. 31-62).

3. Translocational Control: The observation that only ~5% of nuclear RNA ever makes its way to the cytosol, probably less than can be accounted for by gene splicing, suggests that differential mRNA translocation to the cytosol may be an important expressional control mechanism in eukaryotes. Evidence is accumulating that this is, in fact, the case. Cellular RNA is never "naked" but rather is always in complex with a variety of conserved proteins (Section 31-4A). Intriguingly, nuclear and cytosolic mRNAs are associated with different sets of proteins, indicating that there is protein exchange on translocating mRNA out of the nucleus.

4. Control of mRNA Degradation: The rates at which eukaryotic mRNAs are degraded in the cytosol vary widely. Whereas most have half-lives of hours or days, some are degraded within 30 min of entering the cytosol. A given mRNA may also be subject to differential degradation. For example, the major egg yolk protein **vitellogenin** is synthesized in chicken liver in response to estrogens (in roosters as well as in hens) and transported via the bloodstream to the oviduct. Radioactive-labeling experiments established that estrogen stimulation increases the rate of vitellogenin mRNA transcription by several hundredfold and that this mRNA has a cytosolic half-life of 480 h. When estrogen is withdrawn, the synthesis of vitellogenin mRNA returns to its basal rate and its cytosolic half-life falls to 16 h.

The poly(A) tails appended to nearly all eukaryotic mRNAs apparently help protect them from degradation (Section 31-4A). For example, histone mRNAs, which lack poly(A) tails, have much shorter half-lives than most other mRNAs. Histones, in contrast to most other cellular proteins, are largely synthesized during the relatively short S phase of the cell cycle, when they are required in massive amounts for chromatin replication (the small amounts of histones synthesized during the rest of the cell cycle are thought to be used for repair purposes). The short half-lives of histone mRNAs ensure that the rate of histone synthesis closely parallels the rate of histone gene transcription.

A structural feature that increases the rate at which mRNAs are degraded is the presence of certain AU-rich sequences in the untranslated 3′ segments. These sequences, when grafted to mRNAs that lack them, decrease the mRNAs' cytosolic lifetimes. By and large, however, the nature of the signals through which mRNAs are selected for degradation are poorly understood, in part, no doubt, because the nucleases that do so have not been identified.

5. Control of Translational Initiation Rates: The rates of translational initiation of eukaryotic mRNAs, as we have seen (Section 32-4), are responsive to the presence of certain substances, including heme (in reticulocytes) and interferon, as well as to mRNA masking.

6. Selection of Alternative Posttranslational Processing Pathways: Polypeptides synthesized in both prokaryotes and eukaryotes are subject to proteolytic cleavage and covalent modification (Section 32-5). These posttranslational processing steps are important regulators of enzyme activity (e.g., see Section 15-3E) and, in the case of glycosylations, are major determinants of a protein's final cellular destination (Sections 12-4C and 23-3B). The selective degradation of proteins (Section 30-6) is also a significant factor in eukaryotic gene expression.

In addition to the foregoing, most eukaryotic polypeptide hormones (whose functions are discussed in Section 19-1) are synthesized as segments of large precursor polypeptides known as **polyproteins.** These are posttranslationally cleaved to yield several, not necessarily different, polypeptide hormones. *The cleavage pattern of a particular polyprotein may vary among different tissues so that the same gene product can yield different sets of polypeptide hormones.* For example, the polyprotein **pro-opiomelanocortin (POMC),** which, in the rat, is synthesized in both the anterior and intermediate lobes of the pituitary, contains seven different polypeptide hormones (Fig. 34-80). In both of these lobes, which are functionally separate glands, posttranslational processing of POMC yields an N-terminal fragment, **ACTH** and **β-LPH.** Processing in the anterior lobe ceases at this point. In the intermediate lobe, however, the N-terminal fragment is further cleaved to yield **γ-MSH,** ACTH is converted to **α-MSH** and **CLIP,** and β-LPH is split to **γ-LPH** and **β-END** (Fig. 34-80). These various hormones have different activities, so that the products of the anterior and intermediate lobes of the pituitary are physiologically distinct.

Most of the cleavage sites in POMC and other polyproteins consist of pairs of basic amino acid residues, Lys–Arg,

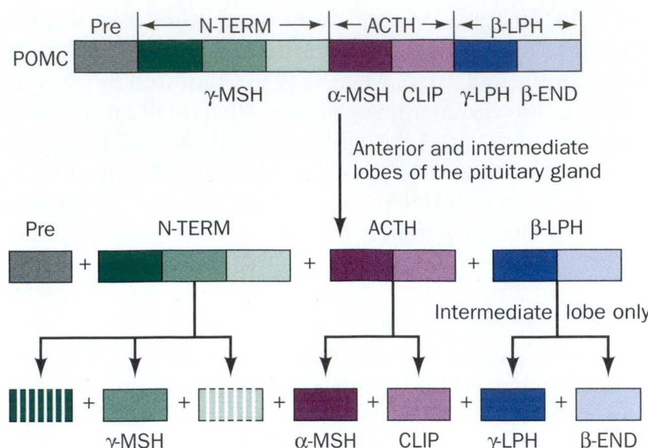

FIGURE 34-80 The tissue-specific posttranscriptional processing of POMC yields two different sets of polypeptide hormones. In both the anterior and intermediate lobes of the pituitary gland, POMC is proteolytically cleaved to yield its N-terminal fragment (N-TERM), **adrenocorticotropic hormone (ACTH;** Section 19-H) and **β-lipotropin (β-LPH).** In the intermediate lobe only, these polypeptide hormones are further cleaved to yield **γ-melanocyte stimulating hormone (γ-MSH), α-MSH, corticotropin-like intermediate lobe peptide (CLIP), γ-LPH,** and **β-endorphin (β-END;** Section 19-K). [After Douglass, J., Civelli, O., and Herbert, E., *Annu. Rev. Biochem.* **53,** 698 (1984).]

for example, which suggests that cleavage is mediated by enzymes with trypsin-like activity. Indeed, the enzymes that process POMC also activate other prohormones such as proinsulin. Moreover, the observation that a yeast protease that normally functions to activate a yeast prohormone also properly processes POMC suggests that prohormone processing enzymes are evolutionarily conserved.

4 ■ CELL DIFFERENTIATION AND GROWTH

Perhaps the most awe inspiring event in biology is the growth and development of a fertilized ovum to form an extensively differentiated multicellular organism. No outside instruction is required to do so; *fertilized ova contain all the information necessary to form complex multicellular organisms such as human beings.* Since, contrary to the beliefs of the earliest microscopists, zygotes do not contain miniature adult structures, these structures must somehow be generated through genetic specification. We begin this section by outlining how embryos develop, followed by a discussion of the best understood example of embryological development, that of the basic body plan in *Drosophila.* We then consider the genetic basis of cancer, a group of diseases caused by the proliferation of cells that have lost some of their developmental constraints. We end by discussing how the cell cycle is controlled and how unneeded or irreparably damaged cells commit suicide through programmed cell death.

A. *Embryological Development*

The formation of multicellular animals can be considered as occurring in four somewhat overlapping stages (Fig. 34-81):

1. **Cleavage,** in which the zygote undergoes a series of rapid mitotic divisions to yield many smaller cells arranged in a hollow ball known as a **blastula.**

2. **Gastrulation,** whereby the blastula, through a structural reorganization that includes the blastula's invagination, forms a triple-layered bilaterally symmetric structure called a **gastrula.** Cleavage and gastrulation together take from a few hours to several days depending on the organism.

3. **Organogenesis,** in which the body structures are formed in a process requiring various groups of proliferating cells to migrate from one part of the embryo to another in a complicated but reproducible choreography. Organogenesis occupies hours to weeks.

4. Maturation and growth, whereby the embryonic structures achieve their final sizes and functional capacities. This stage stretches into and sometimes throughout adulthood.

a. Cell Differentiation Is Mediated by Developmental Signals

As an embryo develops, its cells become progressively and irreversibly committed to specific lines of development. What this means is that these cells undergo sequences of self-perpetuating internal changes that distinguish them and their progeny from other cells. A cell and its descendents therefore "remember" their developmental changes even when placed in a new environment. For example, the dorsal (upper) ectoderm (outer layer) of an amphibian embryo (Fig. 34-82) is normally fated to give rise to brain tissue, whereas its ventral (lower) ectoderm becomes epidermis. If a block of an early gastrula's dorsal ectoderm is cut out and exchanged with a block of its ventral ectoderm, both blocks develop according to their new locations to yield a normal adult. If, however, this experiment is performed on the late gastrula, the transplanted tissues will differentiate as they had originally been fated, that is, as misplaced brain and epidermal tissues. Evidently, the dorsal and ventral ectoderms become committed to form brain and epidermal tissues sometime between the early and late gastrula stages.

How are developmental changes triggered; that is, what are the signals that induce two cells with identical genomes to follow different developmental pathways? To begin with, the zygote is not spherically symmetric. Rather, its yolk, as well as other substances, is concentrated toward one end. Consequently, the various cells in the early cleavage stages inherit different cytoplasmic determinants that apparently govern their further development. Even as early as an embryo's eight-cell stage, some of its cells are demonstrably different in their developmental potential from others. However, as the above transplantation ex-

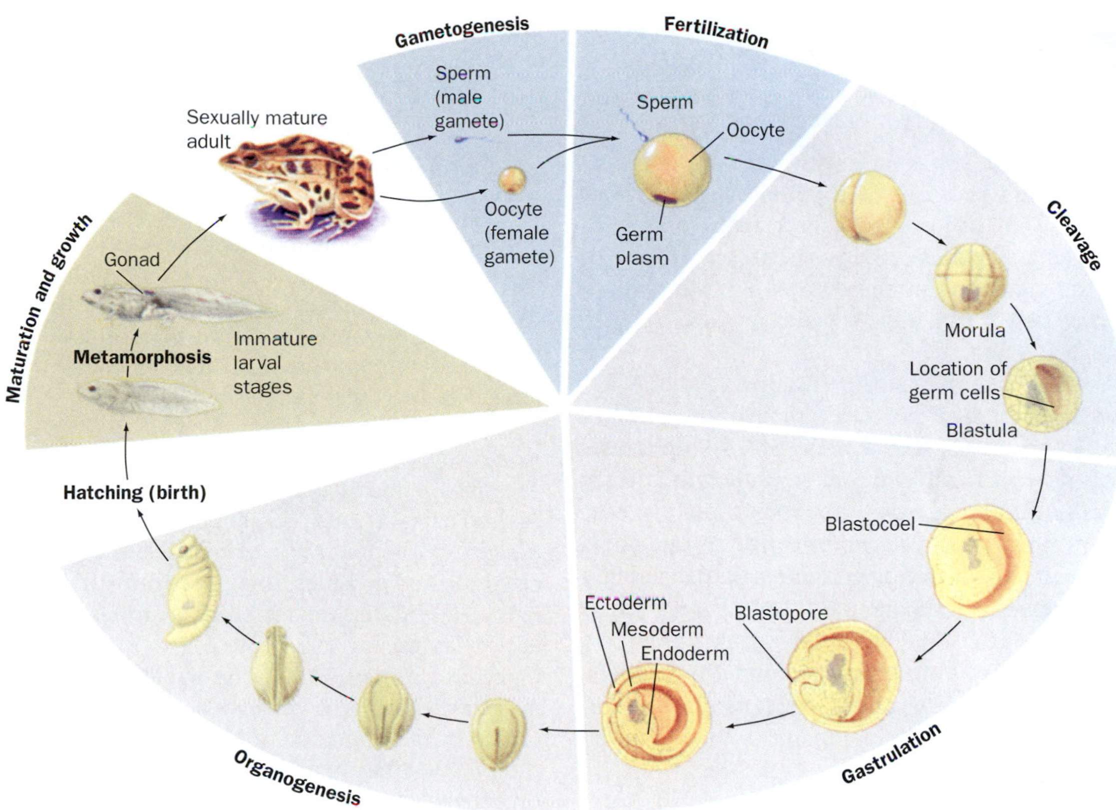

FIGURE 34-81 **Embryogenesis in a representative animal, the frog.**

periments indicate, cells in later stages of development also obtain developmental cues from their embryonic positions.

Cells may obtain spatial information in two ways:

1. Through direct intercellular interactions.

2. From the gradients of diffusible substances called **morphogens** released by other cells.

For most developmental programs, the interacting tissues must be in direct contact, but this is not always the case. For example, mouse ectoderm fated to become eye lens will only do so in the presence of mesenchyme (embryonic tissue that gives rise to the muscle, skeleton, and connective tissue) but this process still occurs if the interacting tissues are separated by a porous filter. Lens development must therefore be mediated by diffusible substances.

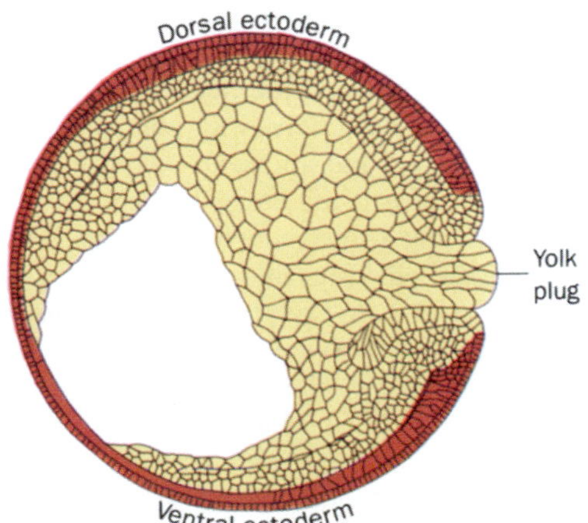

FIGURE 34-82 **The dorsal and ventral ectoderm of an amphibian embryo.**

Developmental signals may be recognized over great evolutionary distances. For instance, the epidermis from the back of a chick embryo, through interactions with the underlying dermis, forms feather buds that are arrayed in a characteristic hexagonal pattern. If embryonic chick epidermis is instead combined with dermis from the whiskered region of mouse embryo snout, the chick epidermis still forms feather buds but arranged in the pattern of mouse whiskers.

Even though mammals and birds diverged ~300 million years ago, mouse inducers can still activate the appropriate chicken genes, although, of course, they cannot alter the products these genes specify. In an intriguing example of this phenomenon, combining epithelium from the jaw-forming region of a chick embryo with molar mesenchyme from mouse embryo induces the chick tissue to grow teeth that are unlike those of mammals (Fig. 34-83). Apparently chickens, whose ancestors have been toothless for ~60 million years (a primordial bird, *Archaeopterix,* had teeth), retain the genetic potential to grow teeth even though they lack the developmental capacity to activate these genes. This observation corroborates the hypothesis that organismal evolution proceeds largely via mutations that alter developmental programs rather than the structural genes whose expression they control (Section 7-3B).

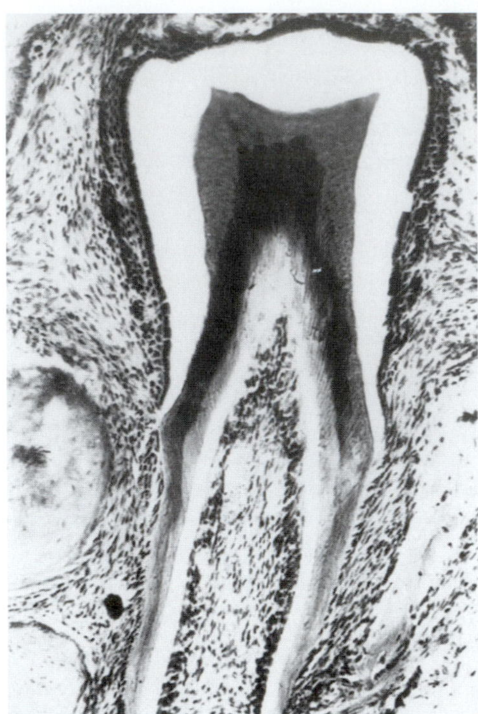

FIGURE 34-83 The proverbial "hen's tooth" forms in chick embryo jaw-forming epithelium under the influence of mouse embryo molar mesenchyme tissue. [Courtesy of Edward Kollar, University of Connecticut Health Center.]

b. Developmental Signals Act in Combination

An additional developmental stimulus to a previously determined cell will modulate, but not reverse, its developmental state. Consider, for example, what happens in a chicken embryo if undifferentiated tissue from the base of a leg bud, which normally gives rise to part of the thigh, is transplanted beneath the end of a wing bud, which normally develops into the handlike wing tip. The transplant does not become a wing tip or even misplaced thigh tissue; instead it forms a foot (Fig. 34-84). Apparently the same stimulus that causes the end of a wing bud to form a wing tip causes tissue that is already committed to be part of a leg to form a leg's morphological equivalent to a wing tip, a foot. Evidently, the many different tissues of a higher organism do not each form in response to a tissue-specific developmental stimulus. Rather, *a given tissue results from the effects of a particular combination of relatively nonspecific developmental stimuli.* This situation, of course, greatly reduces the number of different developmental stimuli necessary to form a complex organism and therefore simplifies the regulation of the developmental process.

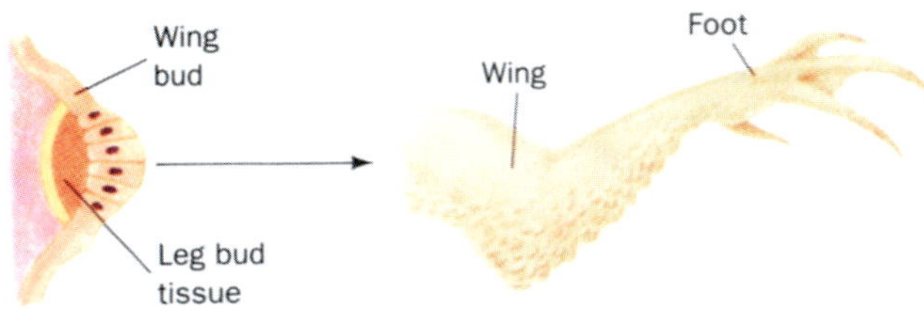

FIGURE 34-84 Presumptive thigh tissue from a chicken leg bud develops into a misplaced foot when implanted beneath the tip of a chicken wing bud.

B. *The Molecular Basis of Development*

The study of the molecular basis of cell differentiation has only become possible in recent decades with the advent of modern methods of molecular genetics. Much of what we know about this subject is based on studies of the fruit fly *D. melanogaster*. We therefore begin this section with a synopsis of embryogenesis in this genetically best characterized multicellular organism.

a. *Drosophila* Development

Almost immediately after the *Drosophila* egg (Fig. 34-85*a*) is laid (which, rather than the earlier fertilization, triggers development), it commences a series of synchronized nuclear divisions, one every 6 to 10 min. The DNA must therefore be replicated at a furious rate, among the fastest known for eukaryotes. Most probably each of its replicons (Section 30–4B) are simultaneously active. The nuclear division process is unusual in that it is not accompanied by the formation of new cell membranes; the nuclei continue sharing their common cytoplasm to form a so-called **syncytium** (Fig. 34-85*b*), which facilitates the rapid pace of nuclear division because there is no need to increase cell mass. After the 8th round of nuclear division, the ~256 nuclei begin to migrate toward the cortex (outer layer) of the egg where, by around the 11th nuclear division, they have formed a single layer surrounding a yolk-rich core known as a **syncytial blastoderm** (Fig. 34-85*c*). At this stage, the mitotic cycle time begins to lengthen while the nuclear genes, which have heretofore been fully engaged in DNA replication, become transcriptionally active (a freshly laid egg contains an enormous store of mRNA that has been contributed, via cytoplasmic bridges, by the developing oocyte's 15 surrounding "nurse" cells). In the 14th nuclear division cycle, which lasts ~60 min, the egg's plasma membrane invaginates around each of the ~6000 nuclei to yield a cellular monolayer surrounding a yolk-rich core called a **cellular blastoderm** (Fig. 34-85*d*). At this point, after ~2.5 h of development, genomic transcriptional activity reaches its maximum in the embryo, mitotic synchrony is lost, the cells become motile, and gastrulation begins.

Until the cellular blastoderm is formed, most of the embryo's nuclei maintain the ability to colonize any portion of the cortical cytoplasm and hence to form any part of the larva or adult except its germ cells [the germ cell progenitors, the five **pole cells** (Fig. 34-85*c*), are set aside after the 9th nuclear division]. *It is therefore a nucleus' location within the syncytium that determines the types of cells its descendents will be become. Once the cellular blastoderm has formed, however, its cells become progressively committed to ever narrower lines of development.* This has been demonstrated, for example, by tracing the developmental fates of small clumps of cells by excising them or ablating (destroying) them with a laser microbeam and characterizing the resultant deformity.

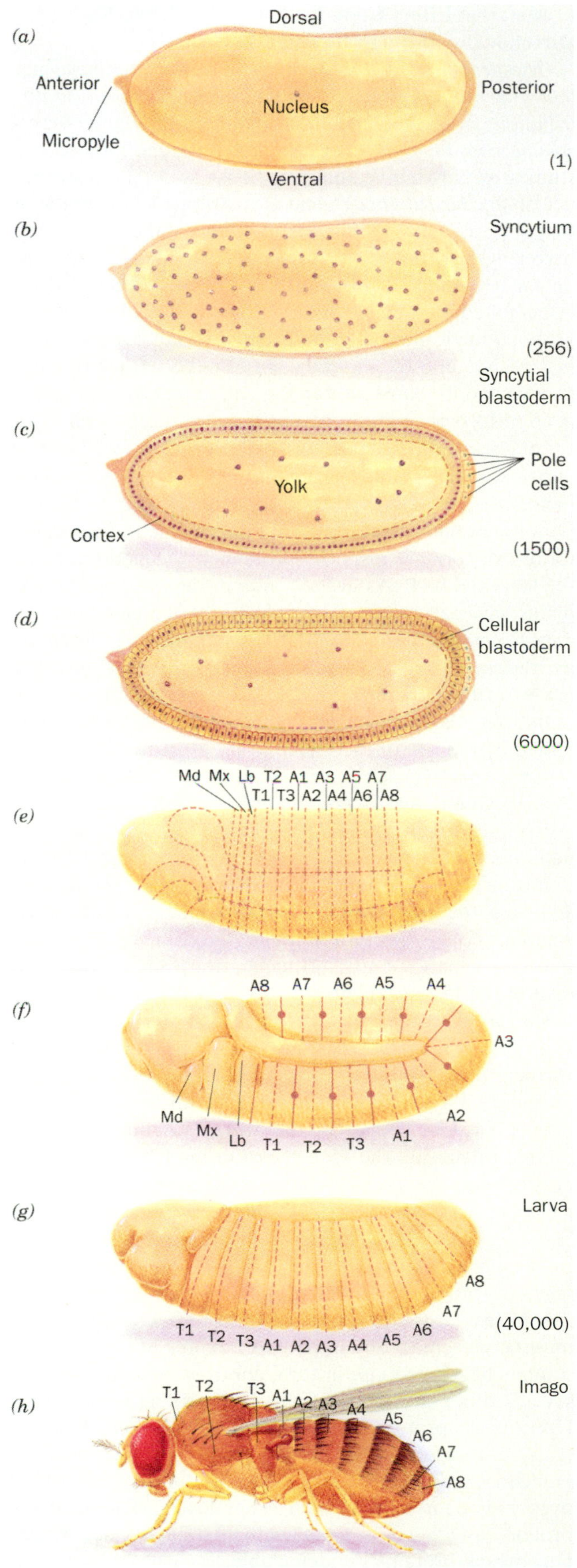

FIGURE 34-85 Development in *Drosophila*. The various stages are explained in the text. Note that the embryos and newly hatched larva are all the same size, ~0.5 mm long. The adult is, of course, much larger. The approximate number of cells in the early stages of development are given in parentheses.

During the embryo's next few hours, it undergoes gastrulation and organogenesis. A striking aspect of this remarkable process, in *Drosophila* as well as in higher animals, is the division of the embryo into a series of segments corresponding to the adult organism's organization (Fig. 34-85*e*). The *Drosophila* embryo has three segments that eventually merge to form its head (Md, Mx, and Lb for mandibulary, maxillary, and labial), three thoracic segments (T1–T3), and eight abdominal segments (A1–A8). As development continues, the embryo elongates and several of its abdominal segments fold over its thoracic segments (which permits it to fit inside the eggshell; Fig. 35-85*f*). At this stage, the segments become subdivided into anterior (forward) and posterior (rear) compartments. The embryo then shortens and straightens to form a larva that consists of ~40,000 cells that hatches ~20 hours after beginning development (Fig. 34-85*g*). Over the next 5 days, the larva feeds, grows, molts twice, pupates, and commences metamorphosis to form an adult (**imago;** Fig. 34-85*h*). In this latter process, the larval epidermis is almost entirely replaced by the outgrowth of apparently undifferentiated patches of larval epithelium known as **imaginal disks** that are committed to their developmental fates as early as the cellular blastoderm stage. These structures, which maintain the larva's segmental boundaries, form the adult's legs, wings, antennae, eyes, etc. (Fig. 34-86). About

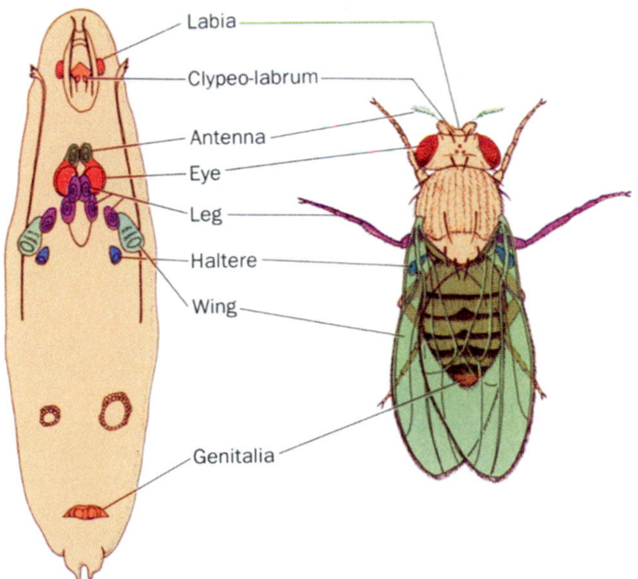

FIGURE 34-86 Locations and developmental fates, in *Drosophila*, of the imaginal disks (*left*), pouches of larval tissue that form the adult's outer structures. [After Fristrom, J.W., Raikow, R., Petri,W., and Stewart, D., *in* Hanly, E.W. (Ed.), *Problems in Biology: RNA in Development, p.* 382, University of Utah Press (1970).]

10 days after commencing development, the adult emerges and, within a few hours, initiates a new reproductive cycle.

b. Developmental Patterns Are Genetically Mediated

What is the mechanism of embryonic pattern formation? In what follows, we discuss only the anteroposterior (head to tail) differentiation system. Keep in mind, however, that *Drosophila* also have a system that imposes dorsoventral (back to belly) differentiation.

Much of what we know about anteroposterior pattern formation stems from genetic analyses of a series of bizarre mutations in three classes of *Drosophila* genes that normally specify progressively finer regions of cellular specialization in the developing embryo:

1. ***Maternal-effect genes,*** *which define the embryo's polarity,* that is, its anteroposterior axis. Mutations of these genes globally alter the embryonic body pattern regardless of the paternal genotype. For instance, females homozygous for the ***dicephalic*** (two-headed) **mutation** lay eggs that develop into nonviable two-headed monsters. These are embryos with two anterior ends pointing in opposite directions and completely lacking posterior structures. Similarly, the ***bicaudal*** (two-tailed) and ***snake*** **mutations** give rise to mirror-symmetric embryos with two abdomens (Fig. 34-87*a*).

2. ***Segmentation genes,*** *which specify the correct number and polarity of embryonic body segments.* Investigations by Christiane Nüsslein-Volhard and Eric Wieschaus led to their subclassification as follows:

- **a. Gap genes,** the first of a developing embryo's to be transcribed, divide the embryo into several broad regions. Gap genes are so named because their mutations result in gaps in the embryo's segmentation pattern. Embryos with defective ***hunchback (hb)*** genes, for example, lack mouthparts and thorax structures.
- **b. Pair-rule genes** specify the division of the embryo's broad gap regions into segments. These genes are so named because their mutations usually delete portions of every second segment. This occurs, for example, in embryos that are homozygous for mutations in the ***fushi tarazu*** (***ftz;*** Japanese for segment deficient) gene (Fig. 34-87*b*).
- **c. Segment polarity genes** specify the polarities of the developing segments. Thus, homozygous ***engrailed*** (***en;*** indented with curved notches) mutants lack the posterior compartment of each segment.

3. ***Homeotic selector genes,*** *which specify segmental identity.* Homeotic mutations transform one body part into another. For instance, ***Antennapedia*** (***antp,*** antenna-foot) mutants have legs in place of antennae (Fig. 34-87*c,d*), whereas the mutations ***bithorax (bx), anteriorbithorax (abx),*** and ***postbithorax (pbx)*** each transform sections of halteres (vestigial wings that function as balancers; Fig. 34-86), which normally occur only on segment T3, to the corresponding sections of wings, which normally occur only on segment T2 (Fig. 34-87*e*).

c. Maternal-Effect Gene Products Specify the Egg's Directionality through Gradient Formation

The properties of maternal-effect gene mutants suggest that maternal-effect genes specify morphogens whose distributions in the egg cytoplasm define the future embryo's spatial coordinate system. Indeed, immunofluorescence studies by Nüsslein-Volhard demonstrated that the product of the ***bicoid (bcd)*** gene is distributed in a gradient that decreases toward the posterior end of the normal embryo (Figs. 34-88 and 34-89*a*), whereas embryos with *bcd*-deficient mothers lack this gradient. The gradient, which is facilitated by the syncytium's lack of cellular boundaries, arises through the secretion, by the ovarian nurse cells, of *bcd* mRNA into the anterior end of the oocyte during oogenesis and its translation in the early embryo. The ***nanos*** gene mRNA is similarly desposited in the egg but it is localized near the egg's posterior pole (Fig. 34-89*a*). The *bcd* and *nanos* gene products, as we shall see, are transcription factors that regulate the expression of specific gap genes. Other maternal-effect genes that participate in anteroposterior axis formation specify proteins that function to trap the localized mRNAs in their area of deposition. This explains why early embryos produced by females homozygous for maternal-effect mutations can often be "rescued" by the injection of cytoplasm, or sometimes just the mRNA, from early wild-type embryos. With some of these mutations, the polarity of the rescued embryo is determined by the site of the injection.

d. Gap Genes Are Expressed in Specific Regions

The mRNA of the gap gene *hunchback* (*hb*) is maternally deposited in a uniform distribution throughout the unfertilized egg (Fig. 34-89*a*). However, **Bicoid protein** activates the transcription of the embryonic *hb* gene, whereas **Nanos protein** inhibits the translation of *hb* mRNA. Consequently, **Hunchback protein** becomes distributed in a gradient that decreases from anterior to posterior (Fig. 34-89*b*).

DNase I footprinting studies have demonstrated that Bicoid protein binds to five homologous sites (consensus sequence TCTAATCCC) in the *hb* gene's upstream promoter region. Nüsslein-Volhard demonstrated the ability of Bicoid protein to activate the *hb* gene by fusing the *hb* promoter upstream of the CAT reporter gene (Section 34-3B) and injecting the resulting construct into early *Drosophila* embryos. CAT was produced in wild-type but not in *bcd*-deficient embryos. Moreover, the use of progressively shorter segments of the *hb*-derived promoter region demonstrated that at least three of the five Bicoid protein-binding sites must be present to obtain full CAT expression.

Hunchback protein, in turn, controls the expression of several other gap genes (Fig. 34-89*c,d*): High levels of Hunchback protein induce ***giant*** expression, ***Krüppel*** (German: cripple) is expressed where the level of Hunchback protein begins to decline, ***knirps*** (German: pygmy) is expressed at even lower levels of Hunchback protein, and *giant* is again activated in regions where Hunchback protein is undetectable.

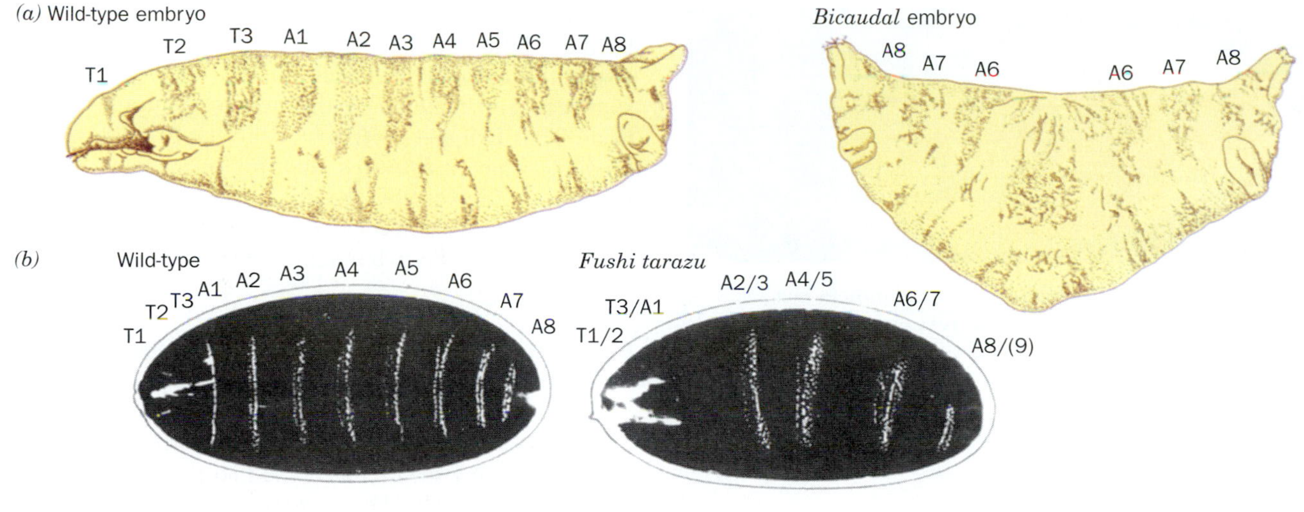

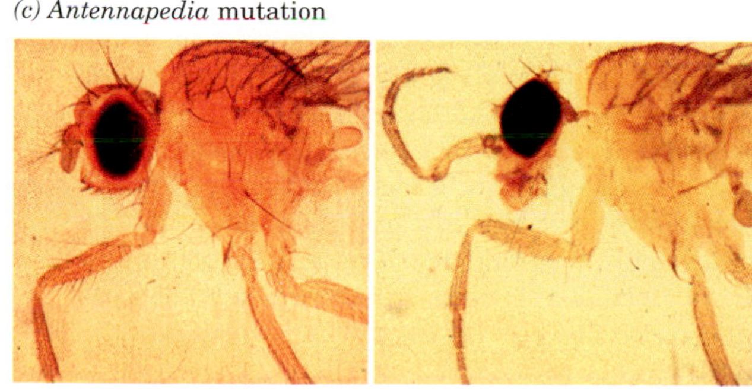

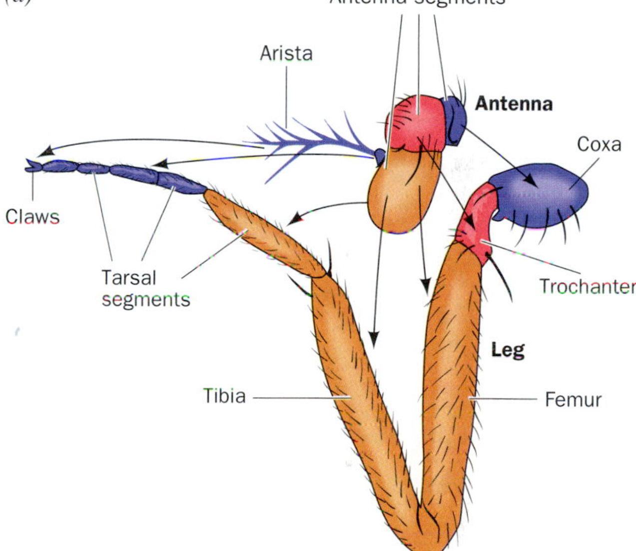

FIGURE 34-87 Developmental mutants of *Drosophila*. (*a*) The cuticle patterns of wild-type embryos (*left*) exhibit 11 body segments, T1 to T3 and A1 to A8 (the head segments have retracted into the body and hence are not visible here). In contrast, the nonviable "monsters" produced by homozygous *bicaudal* mutant females (*right*) develop only abdominal segments arranged with mirror symmetry. [After Gergen, P.J., Coulter, D., and Weischaus, E., *in* Gall, J.G., *Gametogenesis and the Early Embryo, p.* 200, Liss (1986).] (*b*) In the wild-type embryo (*left*), the anterior edge of each of the 11 abdominal and thoracic segments has a belt of tiny projections known as denticles (which help larvae crawl) that appear in these photomicrographs as white stripes. *Fushi tarazu* mutants (*right*) lack portions of alternate segments and the remaining segments are fused together (e.g., A2/3), yielding a nonviable embryo with only half of the normal number of denticle belts. [Courtesy of Walter Gehring, University of Basel, Switzerland.] (*c*) Head and thorax of a wild-type adult fly (*left*) and one that is heterozygous for a mutant form of the homeotic *Antennapedia* (*antp*) gene (*right*). The mutant gene is inappropriately expressed in the imaginal disks that normally form antennae (where the wild-type *antp* gene is not expressed) so that they develop as the legs that normally occur only on segment T2. [Courtesy of Ginés Morata, Universidad Autónoma de Madrid, Spain.] (*d*) The correspondence (*arrows*) between antennae and the legs to which the *Antp* mutation transforms them. [After Postlethwait, J.H. and Schneiderman, H.A., *Devel. Biol.* **25,** 622 (1971).] (*e*) A four-winged *Drosophila* (it normally has two wings; Fig. 34-86) that results from the presence of three mutations in the bithorax complex, *abx, bx*, and *pbx*. These mutations cause the normally haltere-bearing segment T3 to develop as if it were the wing-bearing segment T2. This striking architectural change may reflect evolutionary history: *Drosophila* evolved from more primitive insects that had four wings. [Courtesy of Edward B. Lewis, Caltech.]

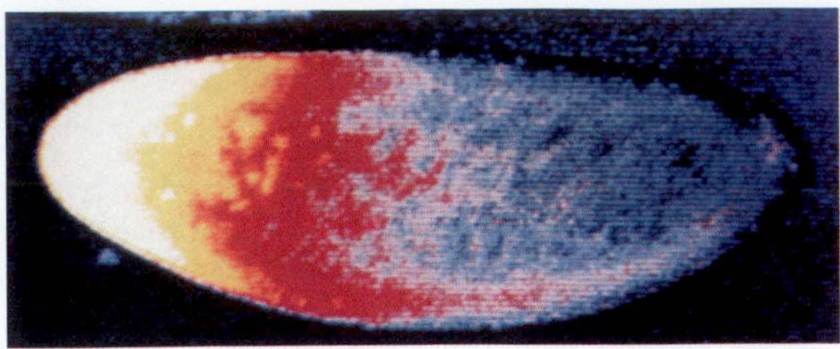

FIGURE 34-88 The distribution of Bicoid protein in a *Drosophila* syncytial blastoderm as revealed by immunofluorescence. High concentrations of the protein are yellow, lower concentrations are red, and its absence is black. [Courtesy of Christiane Nüsslein-Volhard, Max-Planck-Institut für Entwicklungsbiologie, Germany.]

Although the original positions of the proteins encoded by these latter gap genes are elicited by the appropriate concentrations of Hunchback protein, these positions are stabilized and maintained through their mutual interactions. Thus **Krüppel protein** binds to the promoters of the *hb* gene, which it activates, and the *knirps* gene, which it represses. Conversely, **Knirps protein** represses the *Krüppel* gene. This mutual repression is thought to be responsible for the sharp boundaries between the various gap domains.

e. Pair-Rule Genes Are Expressed in "Zebra Stripes"

Pair-rule genes are expressed in sets of 7 stripes, each just a few nuclei wide, along the embryo's anterior–posterior axis. The embryo (which, at this stage, is just beginning to cellularize) is thereby divided into 15 domains (Fig. 34-90). These "zebra stripe" expression patterns for the various pair-rule genes are offset relative to one another.

The gap gene products are transcription factors for three **primary pair-rule genes: *hairy, even-skipped (eve),*** and ***runt.*** The striped pattern of expression arises because the control regions of most primary pair-rule genes comprise a series of enhancer modules, each of which induce their gene's expression in a particular stripe (Fig. 34-91*a*). For example, the transformation of an embryo by the *lacZ* gene preceded by a specific enhancer module in the *eve* gene resulted in *lacZ* transcription in only stripe 1 (Fig. 34-91*b*), whereas a different module did so in only stripe 5 (Fig. 34-91*c*), and both of these modules together induced the production of the *lacZ* transcript in both stripes 1 and 5 (Fig. 34-91*d*). Each of these modules contains a particular arrangement of activating and inhibitory binding sites for the various gap gene proteins so as to enable the expression of the associated pair-rule gene under the particular combination of gap gene proteins present in

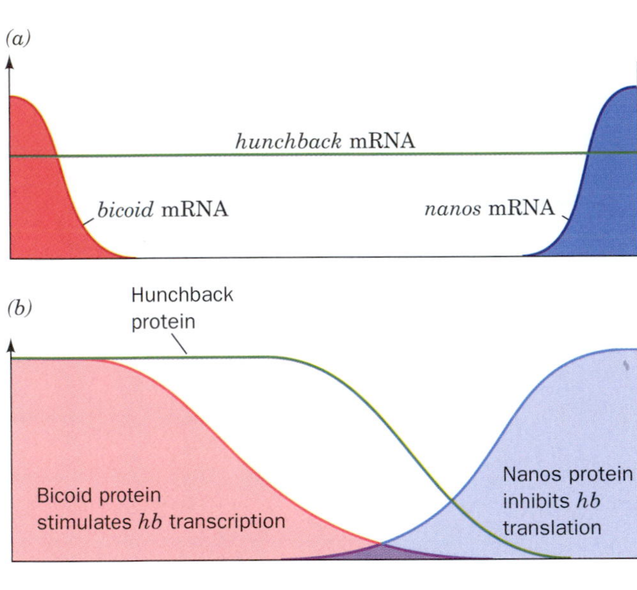

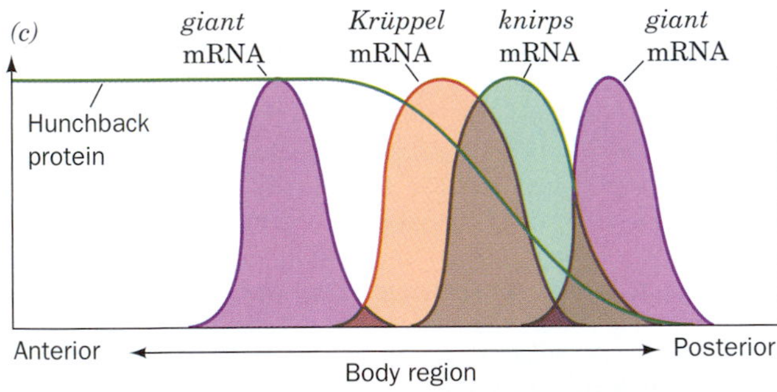

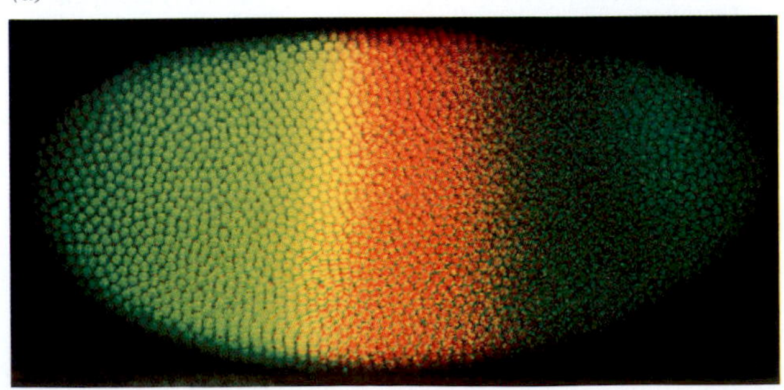

FIGURE 34-89 Formation and effects of the Hunchback protein gradient in *Drosophila* embryos. (*a*) The unfertilized egg contains maternally supplied *bicoid* and *nanos* mRNAs placed at its anterior and posterior poles, together with a uniform distribution of *hunchback* mRNA. (*b*) On fertilization, the three mRNAs are translated. Bicoid and Nanos proteins are not bound to the cytoskeleton as are their mRNAs and hence their gradients are broader than those of the mRNAs. Bicoid protein stimulates the transcription of the *hunchback* gene, whereas Nanos protein inhibits its translation, resulting in a gradient of Hunchback protein that decreases nonlinearly from anterior to posterior. (*c*) Specific concentrations of Hunchback protein induce the transcription of the *giant, Krüppel*, and *knirps* genes. The gradient of Hunchback protein thereby specifies the positions at which these latter mRNAs are synthesized. (*d*) A photomicrograph of a *Drosophila* embryo (*anterior end left*) that has been immunofluorescently stained for both Hunchback (*green*) and Krüppel proteins (*red*). The region where these proteins overlap is yellow. [Parts *a, b*, and *c* after Gilbert, S.F., *Developmental Biology* (4th ed.), p. 543, Sinauer Associates (1994); Part *d* courtesy of Jim Langeland, Steve Paddock, and Sean Carroll, Howard Hughes Medical Institute, University of Wisconsin–Madison.]

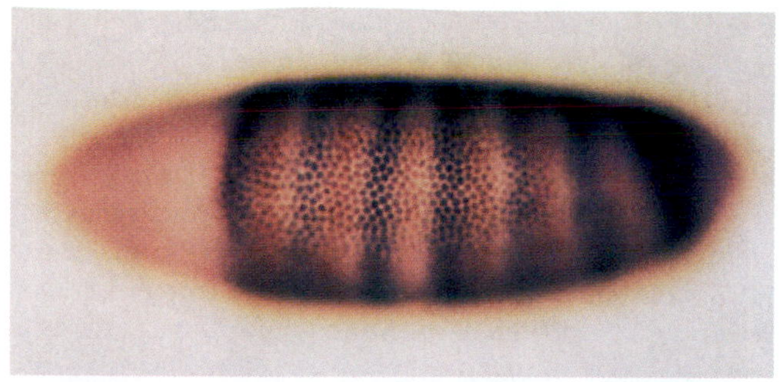

FIGURE 34-90 ***Drosophila*** **embryos stained for Ftz (*brown*) and Eve (*gray*) proteins.** These proteins are each expressed in seven stripes which, at first, are relatively blurred (*left*) but within a short time become sharply defined (*right*). [Courtesy of Peter Lawrence, MRC Laboratory of Molecular Biology, U.K.]

the corresponding stripe. Thus, in *giant*-deficient embryos, the posterior border of stripe 5 is missing (Fig. 34-91*e*). As with the gap genes, the patterns of expression of the primary pair-rule genes become stabilized through interactions among themselves.

The primary pair-rule gene products similarly induce or inhibit the expression of five **secondary pair-rule genes** including *ftz*. Thus, as Walter Gehring demonstrated, *ftz* transcripts first appear in the nuclei lining the cortical cytoplasm during the embryo's 10th nuclear division cycle.

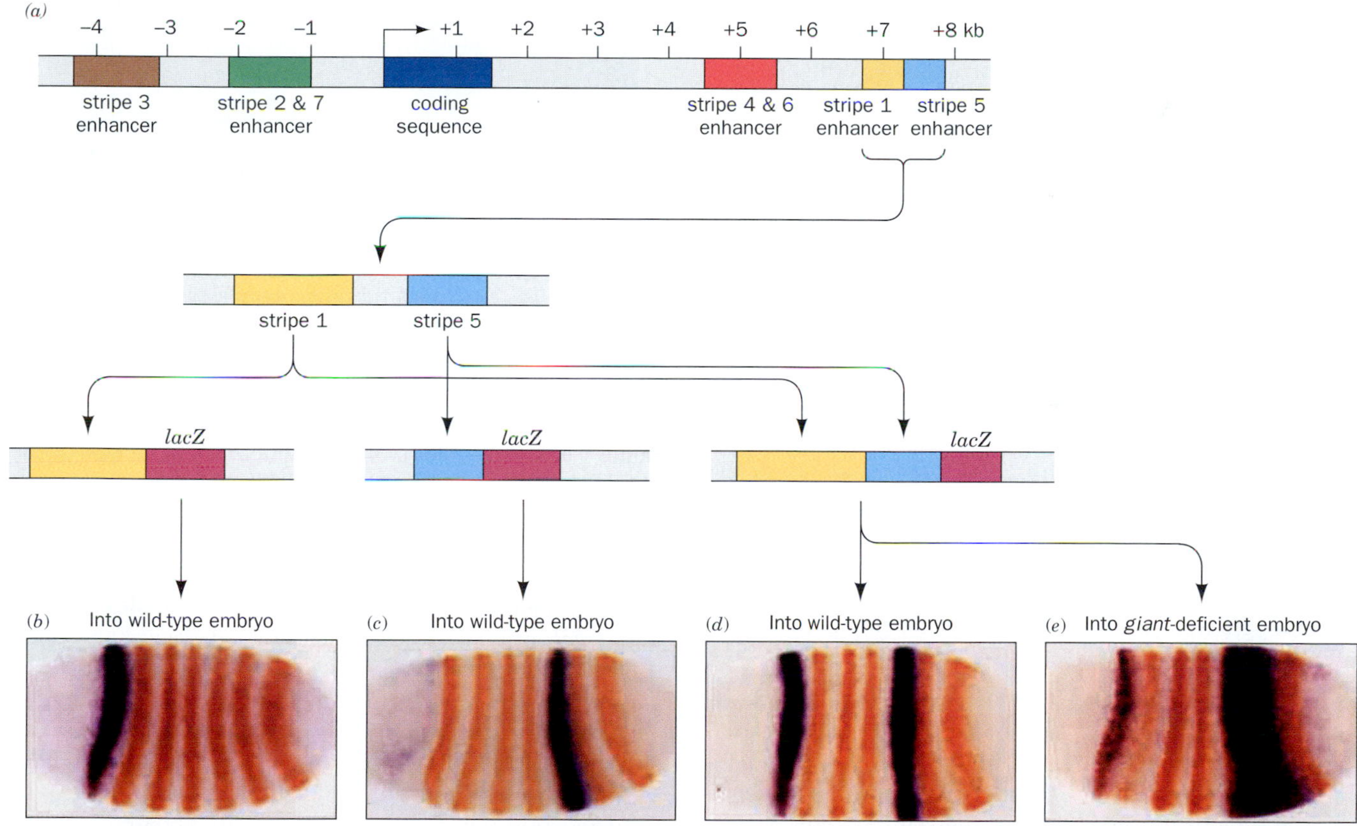

FIGURE 34-91 Expression of the *even-skipped* (*eve*) gene in a pattern of seven stripes in the *Drosophila* embryo. (*a*) Diagram of the *eve* gene, which contains a series of enhancer modules, some upstream of the coding region (*blue*) and others downstream, that on binding the particular combination of gap genes present in their corresponding stripe, induce the expression of *eve* in that stripe. The positions of various elements in the gene, in kb, relative to the transcriptional start site (*squared off arrow*), are indicated. The *lacZ* reporter gene (*magenta*) under control of (*b*) the stripe 1 enhancer (*yellow*), (*c*) the stripe 5 enhancer (*cyan*), or (*d*) both the stripe 1 and 5 enhancers was injected into wild-type *Drosophila* oocytes. The resulting embryos were hybridized *in situ* with dye-labeled *lacZ* antisense RNA to yield a blue band where *lacZ* had been transcribed and then stained with anti-Eve antibodies (*orange*), thereby demonstrating that *lacZ* is expressed only in the corresponding stripe(s). (*e*) When an oocyte deficient in the gap gene *giant* was injected with *lacZ* under the control of the stripe 1 and 5 enhancers, stripe 1 was normal but stripe 5 lacked its posterior border, which indicates that Giant protein normally functions to inhibit *eve* expression past the end of stripe 5. [Part *a* based on a drawing by Scott Gilbert, Swarthmore College. Parts *b, c, d,* and *e* courtesy of James Jaynes, Thomas Jefferson University.]

The rate of *ftz* expression then increases as the embryo develops until the 14th division cycle, when the cellular blastoderm forms. At this stage, as immunochemical staining dramatically shows, *ftz* is expressed in a pattern of 7 belts around the cellular blastoderm, each 3 or 4 cells wide (Fig. 34-90), which correspond precisely to the missing regions in homozygous ftz^- embryos. Then, as the embryonic segments form, *ftz* expression subsides to undetectable levels (although it is later reactivated during the differentiation of specific nerve cells in which it is required to specify their correct "wiring" pattern). Evidently, the *ftz* gene must be expressed in alternate sections of the embryo for normal segmentation to occur.

f. Segment Polarity Genes Define Parasegment Boundaries

The expression of eight known segment polarity genes is initiated by pair-rule gene products. For example, by the 13th nuclear division cycle, as Thomas Kornberg demonstrated, *engrailed (en)* transcripts become detectable but are more or less evenly distributed throughout the embryonic cortex. However, since *en* is expressed in nuclei containing high concentrations of either **Eve** or **Ftz** proteins (Fig. 34-90), by the 14th cycle they form a striking pattern of 14 stripes around the cellular blastoderm (half the spacing of *ftz* expression). Continuing development reveals that these stripes are localized in the primordial posterior compartment of every segment (Fig. 34-92), just those compartments that are missing in homozygous en^- embryos. Thus, much like we saw for *ftz*, the *en* gene product induces the posterior half of each segment to develop in a different fashion from its anterior half.

Another segment polarity gene, ***wingless (wg),*** is expressed simultaneously with *en* but in narrow bands on the anterior side of most *en* bands (Fig. 34-93). Cells expressing *en* and *wg* genes thereby define the boundaries of the so-called **parasegments,** embryonic regions that consist of the posterior portion of one segment and the anterior portion of the segment behind it. Parasegments do not become morphological units in the larva or adult but, nevertheless, are thought to be the embryo's actual developmental units.

g. Homeotic Selector Genes Direct the Development of the Individual Body Segments

The structural components of developmentally analogous body parts, say *Drosophila* antennae and legs, are nearly identical; only their organizations differ (Fig. 34-87*d*). *Consequently, developmental genes must function to control the pattern of structural gene expression rather than simply turning these genes on or off.* Thus, as we saw for the segmentation genes, the expression of the structural genes characteristic of any given tissue must be controlled by a complex network of regulatory genes. The homeotic selector genes, as we shall see, are the "master" genes in the control networks governing segmental differentiation.

Most homeotic mutations in *Drosophila* (which were first described in 1894 by William Bateson, who coined the name "homeosis" to indicate something that has been changed into the likeness of something else) map into eight related genes that are distributed in two clusters that function as a single cluster: the **bithorax complex (*BX-C*),** which controls the development of the thoracic and abdominal segments, and the **antennapedia complex (*ANT-C*),** which primarily affects head and thoracic segments. *Recessive mutations in BX-C, when homozygous, cause one or more segments to develop as if they were more anterior segments.* Thus, the combined **bx, abx,** and **pbx** mutations cause segment T3 to develop as if it were segment T2 (Fig. 34-87*e*). Similarly, the entire deletion of *BX-C* causes all segments posterior to T2 to resemble T2; apparently T2 is the developmental "ground state" of these 10 segments. The evolution of such gene families, it is thought, permitted arthropods (the phylum containing insects) to arise from the more primitive annelids (segmented worms) in which all segments are nearly alike.

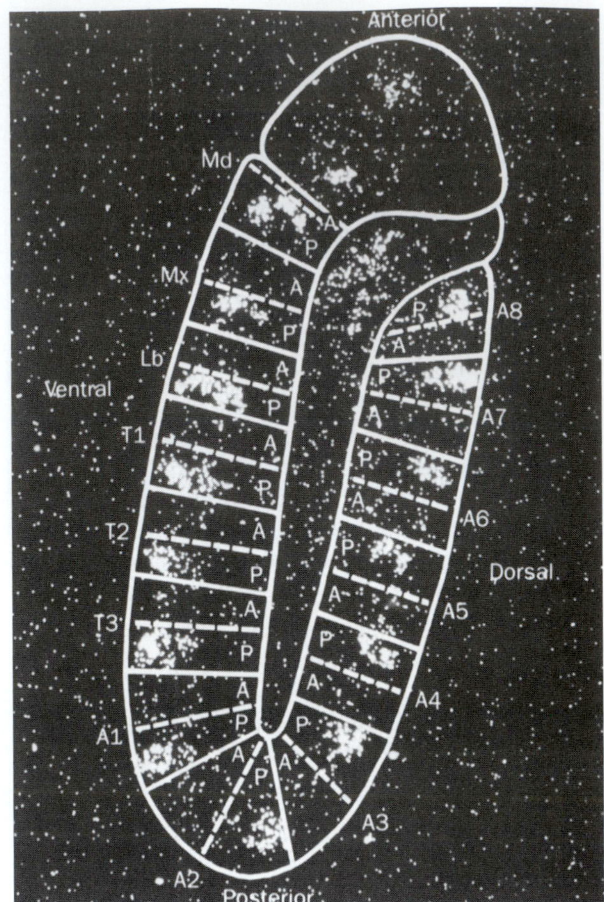

FIGURE 34-92 ***In situ* hybridization demonstrates that the *Drosophila engrailed* gene is expressed in the posterior compartment of every embryonic segment.** [Courtesy of Walter Gehring, University of Basel, Switzerland.]

Detailed genetic analysis of *BX-C* led Edward B. Lewis to formulate a model for segmental differentiation (Fig. 34-94). *BX-C,* Lewis proposed, contains at least one gene for each segment from T3 to A8, which for simplicity are numbered 0 to 8 in Fig. 34-94. These genes, for reasons that are not understood, are arranged in the same order,

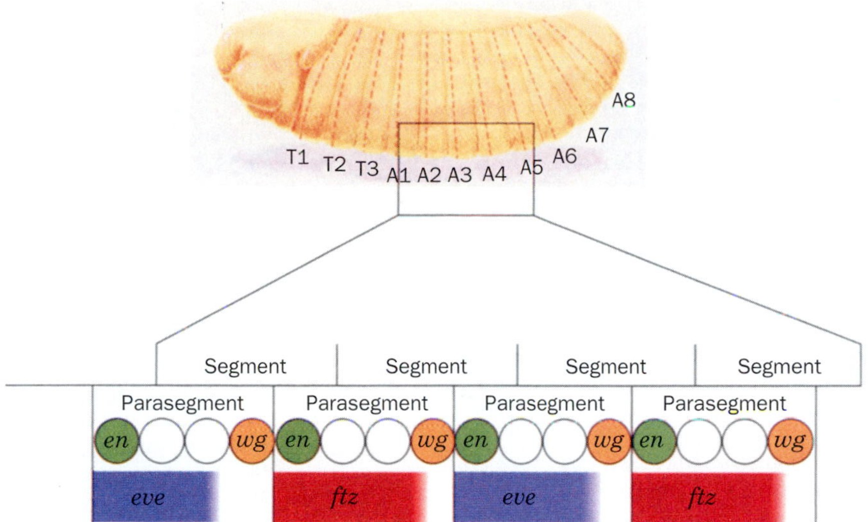

FIGURE 34-93 The pair-rule proteins Eve and Ftz regulate the expression of the segment polarity genes *engrailed* (*en*) and *wingless* (*wg*). When either Eve or Ftz is present, *en* is expressed, whereas when both proteins are absent, *wg* is expressed. The parasegment boundaries are thereby defined. Other pair-rule proteins are thought to inhibit *en* and *wg* expression in nuclei not at the parasegment boundaries.

from "left" to "right," as the segments whose development they influence. Starting with segment T3, progressively more posterior segments express successively more *BX-C* genes until, in segment A8, all of these genes are expressed. The developmental fate of a segment is thereby determined by its position in the embryo.

Sequence analysis of the *BX-C* region led to a difficulty with Lewis' model: The *BX-C* contains only three protein-encoding genes, ***Ultrabithorax* (*Ubx*)**, ***Abdominal-A* (*Abd-A*)**, and ***Abdominal-B* (*Abd-B*)**. However, further analysis indicated, for example, that mutations such as *bx, abx,* and *pdx,* which were previously assumed to occur on separate genes, are actually mutations of enhancer elements that enable the position-specific expression of the *Ubx* gene. Thus, the nine "genes" in Lewis' model have turned out to be enhancer elements on the three *BX-C* genes.

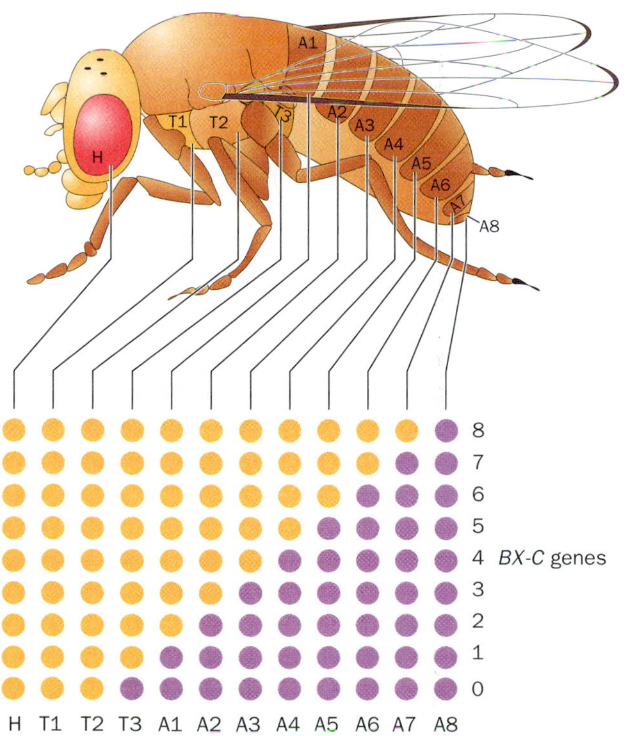

FIGURE 34-94 Model for the differentiation of embryonic segments in *Drosophila* as directed by the genes of the bithorax complex (*BX-C*). Segments T2, T3, and A1–8 in the embryo, as the lower drawing indicates, are each characterized by a unique combination of active (*purple circles*) and inactive (*yellow circles*) *BX-C* "genes." These "genes" (which sequencing studies later demonstrated are really enhancer elements), here numbered 0 to 8, are thought to be sequentially activated from anterior to posterior in the embryo so that segment T2, the developmentally most primitive segment, has no active *BX-C* genes, while in segment A8, all of them are active. Such a pattern of gene expression may result from a gradient in the concentration of a *BX-C* repressor that decreases from the anterior to the posterior of the embryo. [After Ingham, P., *Trends Genet.* **1,** 113 (1985).]

h. Developmental Genes Have Common Sequences

In characterizing the ***Antennapedia* (*Antp*)** gene, Gehring and Matthew Scott independently discovered that *Antp* cDNA hybridizes to both the *Antp* and the *ftz* gene and that, therefore, *these genes share a common base sequence*. This startling observation rapidly led to the discovery that the *Drosophila genome contains numerous such sequences, many of which occur in the homeotic gene complexes ANT-C and BX-C*. DNA sequencing studies of

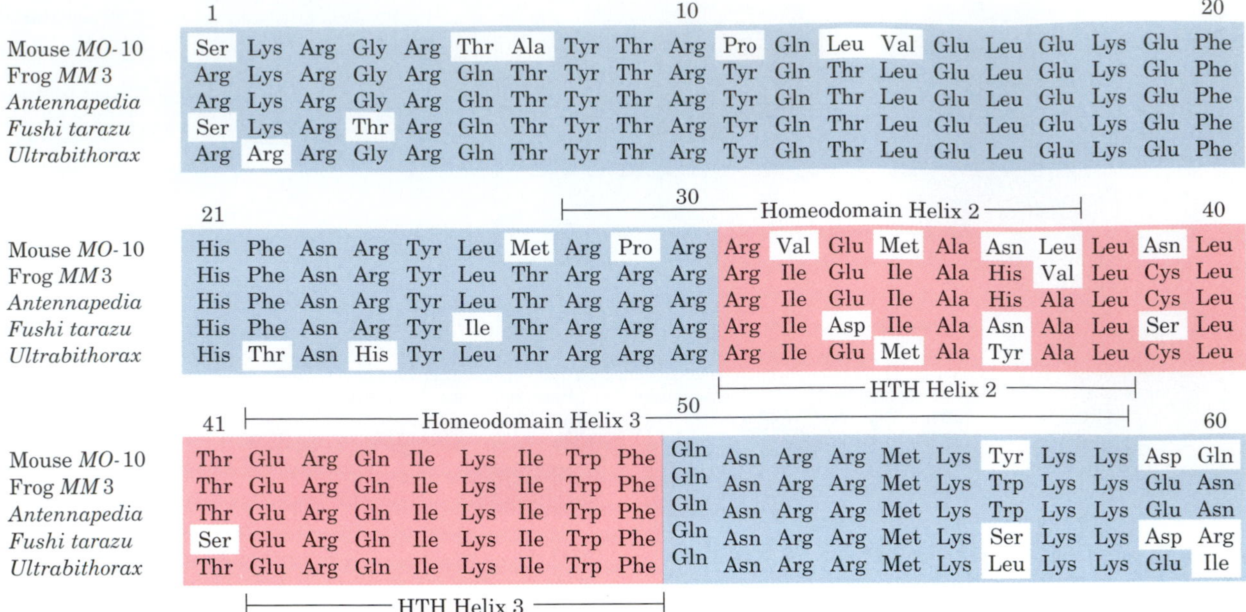

1 10 20
Mouse MO-10 Ser Lys Arg Gly Arg Thr Ala Tyr Thr Arg Pro Gln Leu Val Glu Leu Glu Lys Glu Phe
Frog MM3 Arg Lys Arg Gly Arg Gln Thr Tyr Thr Arg Tyr Gln Thr Leu Glu Leu Glu Lys Glu Phe
Antennapedia Arg Lys Arg Gly Arg Gln Thr Tyr Thr Arg Tyr Gln Thr Leu Glu Leu Glu Lys Glu Phe
Fushi tarazu Ser Lys Arg Thr Arg Gln Thr Tyr Thr Arg Tyr Gln Thr Leu Glu Leu Glu Lys Glu Phe
Ultrabithorax Arg Arg Arg Gly Arg Gln Thr Tyr Thr Arg Tyr Gln Thr Leu Glu Leu Glu Lys Glu Phe
21 30 Homeodomain Helix 2 40
Mouse MO-10 His Phe Asn Arg Tyr Leu Met Arg Pro Arg Arg Val Glu Met Ala Asn Leu Leu Asn Leu
Frog MM3 His Phe Asn Arg Tyr Leu Thr Arg Arg Arg Arg Ile Glu Ile Ala His Val Leu Cys Leu
Antennapedia His Phe Asn Arg Tyr Leu Thr Arg Arg Arg Arg Ile Glu Ile Ala His Ala Leu Cys Leu
Fushi tarazu His Phe Asn Arg Tyr Ile Thr Arg Arg Arg Arg Ile Asp Ile Ala Asn Ala Leu Ser Leu
Ultrabithorax His Thr Asn His Tyr Leu Thr Arg Arg Arg Arg Ile Glu Met Ala Tyr Ala Leu Cys Leu
HTH Helix 2
41 Homeodomain Helix 3 50 60
Mouse MO-10 Thr Glu Arg Gln Ile Lys Ile Trp Phe Gln Asn Arg Arg Met Lys Tyr Lys Lys Asp Gln
Frog MM3 Thr Glu Arg Gln Ile Lys Ile Trp Phe Gln Asn Arg Arg Met Lys Trp Lys Lys Glu Asn
Antennapedia Thr Glu Arg Gln Ile Lys Ile Trp Phe Gln Asn Arg Arg Met Lys Trp Lys Lys Glu Asn
Fushi tarazu Ser Glu Arg Gln Ile Lys Ile Trp Phe Gln Asn Arg Arg Met Lys Ser Lys Lys Asp Arg
Ultrabithorax Thr Glu Arg Gln Ile Lys Ile Trp Phe Gln Asn Arg Arg Met Lys Leu Lys Lys Glu Ile
HTH Helix 3

FIGURE 34-95 Amino acid sequences of the polypeptides encoded by the homeodomains of five genes from mouse, *Xenopus*, and *Drosophila* (*Ultrabithorax* is a *BX-C* gene). Discrepancies between the polypeptide specified by the *Antp* homeobox and those of the other genes lack shading. Each polypeptide has a 19-residue segment (*red shading*), which is homologous to the DNA-binding HTH fold of prokaryotic repressors. The positions of the helices observed in the X-ray and NMR structures of homeodomains, together with the corresponding positions of HTH helices in prokaryotic proteins, are indicated.

these genes revealed that each contains a 180-bp sequence, the so-called **homeodomain** or **homeobox,** which are 70 to 90% identical to one another and which encode even more identical 60-residue polypeptide segments (Fig. 34-95).

Further hybridization studies using homeodomain probes led to the truly astonishing finding that *multiple copies of the homeodomain are also present in the genomes of segmented animals ranging from annelids to vertebrates such as Xenopus, mice, and humans.* In some of these sequences the degree of homology is remarkably high; for example, the homeodomains of the *Drosophila Antp* gene and the *Xenopus* ***MM3*** **gene** encode polypeptides that have 59 of their 60 amino acids in common (Fig. 34-95). The individuality of these homeodomain-containing proteins is presumably imparted by their other segments.

i. The Homeodomain's DNA-Binding Motif Resembles a Helix–Turn–Helix Motif

Since vertebrates and invertebrates diverged over 600 million years ago, this strongly suggest that the gene product of the homeodomain has an essential function. What might this function be? The ~30% Arg + Lys content of homeodomain polypeptides suggests that they bind DNA. Sequence comparisons and NMR studies further suggest that these polypeptide segments form helix–turn–helix (HTH) motifs resembling those of prokaryotic gene regulators such as the *E. coli trp* repressor (Section 31-3D) and the λ Cro protein (Section 33-3D). Indeed, the polypeptide encoded by the homeodomain of the *Drosophila engrailed* gene specifically binds to the DNA sequences just upstream from the transcription start sites of both the *en* and the *ftz* genes. Moreover, fusing the *ftz* gene's upstream sequence to other genes imposes *ftz*'s pattern of stripes (Fig. 34-90) on the expression of these genes in *Drosophila* embryos. *These observations suggest, in agreement with the idea that the products of developmental genes act to regulate the expression of other genes, that homeodomain-containing genes encode transcription factors.* In fact, not all homeodomain-encoded proteins are involved in regulating development. The homeodomain is apparently a widespread genetic motif that specifies the DNA-binding segments of a variety of proteins.

Thomas Kornberg and Pabo have determined the X-ray structure of the 61-residue homeodomain from the *Drosophila* Engrailed protein in complex with a 21-bp DNA (Fig. 34-96). Two copies of the protein bind to the DNA, one near the center of the DNA and the other near one end, where it also contacts a second DNA molecule that, in the crystal, forms a pseudocontinuous helix with the first. The conformations of the two protein molecules, and the contacts they make with the DNA, are nearly identical. The two homeodomains are not in contact so, in contrast to other DNA-binding motifs of known structure, *they bind to their target DNAs as monomers.* The X-ray structure is largely consistent with the NMR structure of the *Antennapedia* homeodomain in complex with a 14-bp DNA determined by Gehring and Kurt Wüthrich.

The homeodomain consists largely of three α helices, the last two of which, as sequence comparisons had previously suggested, form an HTH motif that is closely super-

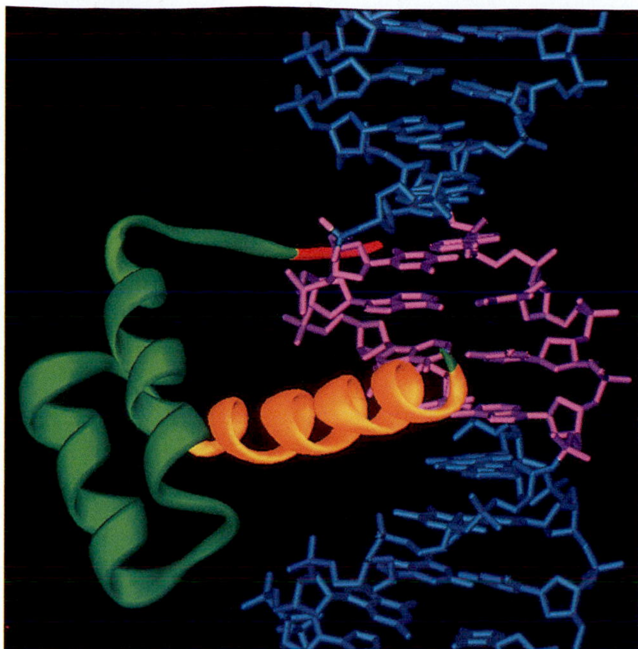

FIGURE 34-96 X-Ray structure of the Engrailed protein homeodomain in complex with a 21-bp DNA containing its target sequence. The 60-residue protein is shown in ribbon form (*green*) with its recognition helix (helix 3, residues 42–58), which is bound in the DNA's major groove, highlighted in gold. The DNA is shown in stick form (*light blue*) with the base pairs comprising its TAAT subsite highlighted in magenta. A second homeodomain that binds to the lower end of the DNA in a nearly identical manner but does not contact the homeodomain shown has been omitted for clarity. Note how the N-terminal segment (*red,* residues 3–5; residues 1 and 2 are disordered) binds in the minor groove of the DNA. [Based on an X-ray structure by Carl Pabo, MIT. PDBid 1HDD.]

imposable with the HTH motifs of prokaryotic repressors such as that of the λ repressor (Fig. 33-45*a*). However, although helix 3, the HTH motif's recognition helix, fits into the major groove of its corresponding DNA, it does so quite differently in the two complexes. In the λ repressor complex, for example, the N-terminal end of the recognition helix is inserted into the DNA's major groove, whereas in the homeodomain complex the DNA is shifted toward the C-terminal end of the helix, which is longer than that of the λ repressor (it extends from residues 42 to 58 in Fig. 34-95). As a consequence, the way in which the first helix of the HTH motif (helix 2; residues 28 to 37 in Fig. 34-95) contacts the DNA also differs between the two complexes.

Most homeodomain binding sites have the subsequence TAAT. The recognition helix in the X-ray structure makes base-specific hydrogen bonding contacts with this subsequence in the major groove through residues that are highly conserved in higher eukaryotic homeodomains. It therefore appears that these interactions function to align the homeodomain with the other bases that it contacts. In addition, two conserved Arg residues located in the N-terminal tail of the homeodomain make base-specific hydrogen bonding contacts with the TAAT subsequence in the minor groove of the DNA. The protein thereby grips the TAAT subsequence from two sides. Note that few other sequence-specific DNA-binding proteins contact bases in the minor groove. Finally, the homeodomain makes extensive contacts with the DNA backbone that, it is presumed, also play an important part in binding and recognition.

j. Homeodomain Genes Function Analogously in Vertebrates and *Drosophila*

Homeodomain-encoding genes have collectively become known as ***Hox* genes.** In vertebrates, they are organized in four clusters of 9 to 11 genes, each located on a separate chromosome and spanning more than 100 kb. In contrast, *Drosophila,* as we saw, has a split *Hox* cluster, whereas in nematodes (roundworms), which are evolutionarily more primitive than insects, the single *Hox* cluster remains unsplit. The genes in the primordial *Hox* cluster presumably arose in some more primitive ancestral organism through a series of gene duplications, as did the four vertebrate *Hox* clusters. The genes in each vertebrate *Hox* cluster, as in *Drosophila,* are activated in the same order, left to right, as they are expressed from the anterior end of the embryo to its posterior end. Perhaps this arrangement is necessary for the homeodomain genes to be activated in the proper order, although, at least in *Drosophila,* gap and pair-rule proteins can still act on *Hox* control regions that have been transplanted to other parts of the genome. Whatever the case, the various *Hox* clusters, as well as their component genes, almost certainly arose through a series of gene duplications and diversifications starting with a single *Hox* gene in a primitive ancestral organism.

Vertebrate *Hox* genes, like those of *Drosophila,* are expressed in specific patterns and at particular stages during embryogenesis. Most *Hox* genes are expressed at a gestational time when organogenesis prevails. That the *Hox* genes directly specify the identities and fates of embryonic cells, that is, are homeotic in character, was shown, for example by the following experiment. Mouse embryos were made transgenic for the *Hox-1.1* gene that had been placed under the control of a promoter that is active throughout the body even though *Hox-1.1* is normally expressed only below the neck. The resulting mice had severe craniofacial abnormalities such as a cleft palate and an extra vertebra and an intervertebral disk at the base of the skull. Some also had an extra pair of ribs in the neck region. Thus, this *Hox* gene's "gain of function" resulted in a homeotic mutation, that is, a change in the development pattern, analogous to those observed in *Drosophila.*

Homozygotic mice resulting from the replacement of their *Hox-3.1* gene coding sequence in embryonic stem cells with that of *lacZ* are born alive but usually die within a few days. They exhibit skeletal deformities in their trunk regions in which several skeletal segments are transformed into the likenesses of more anterior segments. The pattern of β-galactosidase activity (Fig. 34-97), as colorimetrically detected through the use of X-Gal (Section 5-5C), in both homozygotes and heterozygotes, indicates that *Hox-3.1* deletion modifies the properties but not the positions of the embryonic cells that normally express *Hox-3.1.*

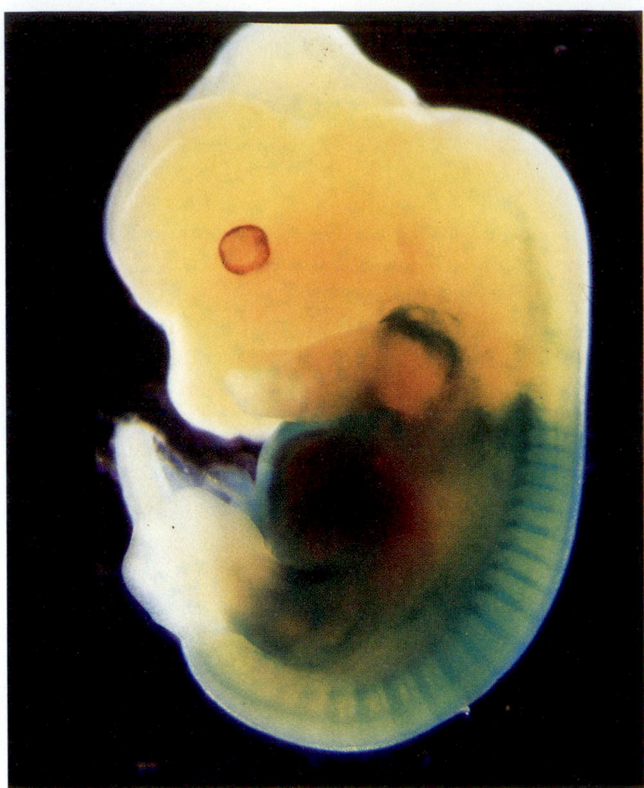

FIGURE 34-97 Pattern of expression of the *Hox-3.1* gene in a 12.5-day postconception mouse embryo. The protein-encoding portion of the embryo's *Hox-3.1* gene was replaced by the *lacZ* gene. The regions of this transgenic embryo in which *Hox-3.1* is expressed are revealed by the blue color that develops on soaking the embryo in X-Gal–containing buffer. [Courtesy of Phillipe Brûlet, Collège de France and the Pasteur Institute, Paris, France.]

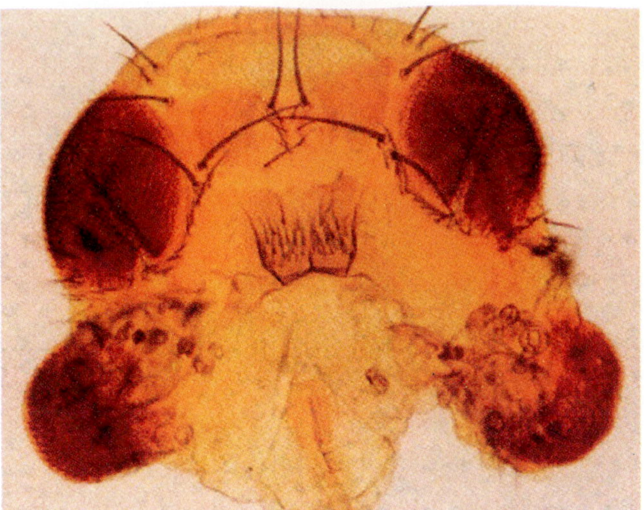

FIGURE 34-98 Ectopic eyes result from the targeted expression of the *Drosophila ey* gene in its imaginal disk primordia. Shown here is the cuticle of an adult *Drosophila* head in which both antennae have formed eye structures that exhibit the morphology and red pigmentation of normal eyes. Such eye structures have been similarly expressed on wings and legs. [Courtesy of Walter Gehring, University of Basel, Switzerland.]

k. Expression of the *Drosophila eyeless* Gene Induces the Ectopic Formation of Eyes

Mutations in the *Drosophila eyeless* (*ey*) gene, first described in 1915, result in flies whose compound eyes are reduced in size or completely absent. The expression of *ey,* which contains a homeodomain, is first detected in the embryonic nervous system and later in the embryonic primordia of the eye. In subsequent larval stages, it is expressed in the developing eye imaginal disks. Mutant forms of four other *Drosophila* genes that have similar phenotypes do not affect the expression of the *ey* gene, which indicates that *ey* acts before these other genes. These observations led to the suggestion that the *ey* gene is the master control gene for eye development.

Genetic engineering studies by Gehring have confirmed this hypothesis. Through the targeted expression of *ey* cDNA in various imaginal disk primordia of *Drosophila,* ectopic (inappropriately positioned) compound eyes were induced to form on the wings, legs, and antennae (Fig. 34-98) of various flies. Moreover, in many cases, these eyes appeared morphologically normal in that they consisted of fully differentiated ommatidia (the simple eye elements that form a compound eye) with a complete set of photoreceptor cells that appear to be electrically active when illuminated (although it is unknown if the flies could see with these ectopic eyes, that is, whether these eyes made appropriate neural connections to the brain).

The mouse *Small eye* (*Sey* or *Pax*-6) gene and the human *Aniridia* gene are closely similar in sequence to the *Drosophila ey* gene and are similarly expressed during morphogenesis. Mice with mutations in one of their two *Sey* genes have underdeveloped eyes, whereas those with mutations in both *Sey* genes are eyeless. Similarly, humans that are heterozygotes for a defective *Aniridia* gene have defects in their iris, lens, cornea, and retina. Evidently, the *ey, Sey,* and *Aniridia* genes all function as master control genes for eye formation in their respective organisms, a surprising result considering the enormous morphological differences between insect and mammalian eyes. Thus, despite the 500 million years since the divergence of insects and mammals, their developmental control mechanisms appear to be closely related.

l. Retinoic Acid Is a Vertebrate Morphogen

Retinoic acid (RA), a derivative of **vitamin A (retinol),**

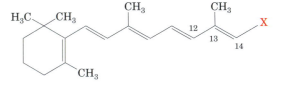

H_3C CH_3 CH_3 CH_3 12 13 14 X CH_3

X = COOH: **Retinoic acid (RA)**

X = CH_2OH: **Retinol (vitamin A)**

has been found to have a graded distribution in developing chick limbs and is therefore thought to be a morphogen. The systematic administration of RA during mouse embryogenesis results in severe malformations, notably skeletal deformities that appear to arise from anterior or posterior shifts of their normal characteristics. A variety of evidence suggests that the expression of *Hox* genes mediates the positional information that RA disrupts. The *Hox* genes are differentially activated by RA according to their positions in their Hox clusters: Those toward the 3′ end of a cluster are maximally induced by as little as $10^{-8}M$ RA, those toward the 5′ end of the cluster require $10^{-5}M$ RA to do so, and those at the 5′ ends are insensitive to RA. Moreover, $10^{-5}M$ RA sequentially activates the *Hox* genes from the 3′ to the 5′ end of a cluster, the same order as their expression patterns in developing axial systems such as the skeleton and the central nervous system.

The foregoing explains why the RA analog **13-*cis*-retinoic acid,** which, taken orally, has been invaluable in the treatment of severe **cystic acne,** induces birth defects if used by pregnant women. The characteristic pattern of cranial deformities in the resulting infants, whose analog is induced in mouse embryos that had been exposed to low concentrations ($2 \times 10^{-6}M$) of this drug, indicates that its presence alters the expression of *Hox* genes early in gestation (~1 month postfertilization in humans, ~9 days in mice).

C. *The Molecular Basis of Cancer*

Cancer, being one of the major human health problems, has received enormous biomedical attention over the past several decades. Around 100 different types of human cancers are recognized, methods of cancer detection and treatment are highly developed, and cancer epidemiology has been extensively characterized. Nevertheless, we are far from fully understanding the biochemical basis of this collection of diseases. In Section 19-3B we discussed the general nature of cancer, its causes, and how tumor viruses cause cancer. In this section we outline how genetic alterations cause cancer.

a. Malignancies May Result from Specific Genetic Alterations

Although much of what we know concerning oncogenes stems from the study of retroviral oncogenes (Section 19-3B), few human cancers are caused by retroviruses. Nevertheless, *it seems likely that all cancers are caused by genetic alterations.* Robert Weinberg demonstrated this to be the case for mouse fibroblasts that had been transformed by a known carcinogen: Normal mouse fibroblasts in culture are transformed on transfection with DNA from the transformed cells. Moreover, these newly transformed cells, when inoculated into mice, form tumors. Similar investigations indicate that DNAs from a wide variety of malignant tumors likewise have transforming activity.

What sorts of genetic changes can give rise to cancer? Several types of changes have been observed:

1. Altered Proteins: *An oncogene, as we have seen, may give rise to a protein product with an anomalous activity relative to that of the corresponding proto-oncogene.* This may even result from a simple point mutation. For example, Weinberg, Michael Wigler, and Mariano Barbacid showed that the *ras* oncogene isolated from a human bladder **carcinoma** (a malignant tumor arising from epithelial tissue) differs from its corresponding proto-oncogene by the mutation of the Gly 12 codon (GGC) to a Val codon (GTC). The resulting amino acid change attenuates the GTPase activity of Ras protein (Section 19-3C), evidently without affecting its ability to stimulate protein phosphorylation, thereby prolonging the time this G-protein remains in the "on" state. Indeed, comparison of the X-ray structures of normal human Ras and its oncogenic counterpart (Gly 12→Val), both in complex with GDP, indicate that the mutation mainly alters the normal protein structure in the vicinity of its GTPase function. Most other *ras* oncogene–activating mutations also change residues close to this site. Ras, which plays a central role in MAP kinase cascades (Fig. 19-38), as might be expected, is one of the most commonly implicated proto-oncogenes in human cancers.

2. Altered Regulatory Sequences: *Malignant transformation can result from the inappropriately high expression of a normal cellular protein.* For example, the proto-oncogene **c-*fos*,** which encodes the transcription factor **Fos** (which is activated by MAP kinase cascades; Fig. 19-38), differs from the retroviral oncogene **v-*fos*** mainly in regulatory sequences: v-*fos* has an efficient enhancer, whereas c-*fos* has a 67-nucleotide AT-rich segment in its unexpressed 3′-terminal end that, when transcribed, promotes rapid mRNA degradation (Section 34-3C). Thus, c-*fos* can be converted to an oncogene by deleting its 3′ end and adding the v-*fos* enhancer.

3. Loss of Degradation Signals: *An oncogene protein that is degraded more slowly than the corresponding normal cellular protein may cause malignant transformation through its consequent inappropriately high concentration in the cell.* For example, the transcription factor **c-Jun** (which is also activated by MAP kinase cascades; Fig. 19-38), but not **v-Jun,** is efficiently multiubiquinated and hence proteolytically degraded by the cell (Section 32-6B). This is because v-Jun lacks a 27-residue segment present in c-Jun that mutagenesis experiments indicate is essential for the efficient ubiquination of c-Jun even though this segment does not contain the protein's principal ubiquitin attachment sites.

4. Chromosomal Rearrangements: *An oncogene may be inappropriately transcribed when brought under the control of a foreign regulatory sequence through chromosomal rearrangement* (a position effect). For example, Carlo Croce found that the human cancer **Burkitt's lymphoma** (a lymphoma is an immune system cell malignancy) is characterized by an exchange of chromosomal segments in which the proto-oncogene **c-*myc*** is translocated from its normal position at one end of chromosome 8 to the end of chromosome 14 adjacent to certain immunoglobulin genes. The misplaced c-*myc* gene is thereby brought under the transcriptional control of the highly active (in immune system cells) immunoglobulin regulatory sequences. The consequent overproduction of the normal c-*myc* gene product **Myc** (a transcription factor that is also activated by MAP kinase cascades and whose transient increase is normally correlated with the onset of cell division), or alternatively, its production at the wrong time in the cell cycle, is apparently a major factor in cell transformation.

5. Gene Amplification: *Oncogene overexpression can also occur when the oncogene is replicated multiple times, either as sequentially repeated chromosomal copies or as extrachromosomal particles.* The amplification of the c-*myc* gene, for example, has been observed in several types of human cancers. Gene amplification is usually an unstable genetic condition that can only be maintained under strong selective pressure such as that conferred by cytotoxic drugs (Section 34-2D). It is not known how oncogene amplification is stably maintained.

6. Viral Insertion into a Chromosome: *Inappropriate oncogene expression may result from the insertion of a viral genome into a cellular chromosome such that the proto-oncogene is brought under the transcriptional control of a viral regulatory sequence.* For instance, **avian leukosis virus,** a retrovirus that lacks an oncogene but that nevertheless induces lymphomas in chickens, has a chromosomal insertion site near c-*myc*. Some DNA tumor viruses also transform cells in this manner.

7. Inappropriate Inactivation or Activation of Chromatin Modification Enzymes: Heterozygotes for a defective *CPB* gene, whose gene product activates the PCAF HAT complex (Section 34-3B), have **Rubinstein–Tabi syndrome,** a condition that predisposes to cancer. In a related example, the **retinoic acid receptor (RAR),** which is important for myeloid (blood-forming) tissue differentiation, helps recruit HDAC complexes such as NCoR and SMRT (Section 34-3B) to **retinoic acid response elements (RAREs),** but on binding ligand, releases them. However, in **promyelocytic leukemia,** a chromosomal translocation yields a defective RAR that binds to RAREs and recruits HDACs but is unresponsive to the presence of retinoids.

8. Loss or Inactivation of Tumor Suppressor Genes: The high incidence of particular cancers in certain families suggests that there are genetic predispositions toward these diseases. A particularly clear-cut example of this phenomenon occurs in **retinoblastoma,** a cancer of the developing retina that therefore afflicts only infants and young children. The offspring of surviving retinoblastoma victims also have a high incidence of this disease, as well as several other types of malignancies. In fact, retinoblastoma is associated with the inheritance of a copy of chromosome 13 from which a particular segment has been deleted. Retinoblastoma develops, as Alfred Knudson first explained, through a somatic mutation in a **retinoblast** (a retinal precursor cell) that alters the same segment of the second, heretofore normal copy of chromosome 13. This is because *the affected chromosomal segment contains a gene, the* ***Rb gene,*** *which specifies a factor that restrains uninhibited cell proliferation; that is, the Rb gene product,* ***pRb,*** *is a* ***tumor suppressor*** *(alternatively, an* ***anti-oncogene*** *protein)*. Indeed, the *Rb* gene is frequently mutated in diverse types of human cancers. The structure and function of pRb is further discussed below.

Several other tumor suppressors have been characterized including **p53,** which is encoded by the most commonly altered gene in human cancers (~50% of cancers contain a mutation in p53 and many other oncogenic mutations occur in genes encoding proteins that directly or indirectly interact with p53; the structure and function of p53 is further discussed in Section 34-4D); **neurofibromatosis type 1 (NF1) protein,** whose defect causes benign tumors of the peripheral nerves, such as those of the famous "Elephant Man" of Victorian England, that occasionally become malignant; **BRCA1,** which forms a portion of a ubiquitin-protein ligase (E3) and whose defect predisposes to breast and ovarian cancers (Section 32-6B); **BRCA2,** a DNA-binding protein that participates in the repair of double-strand breaks and whose defect also predisposes to breast and ovarian cancers; and **PTEN,** an inositol polyphosphate 3-phosphatase, whose structure and function are discussed in Section 19-4E.

Mutations altering normal gene products, causing chromosomal rearrangements and deletions, and perhaps gene amplification can all result from the actions of carcinogens on cellular DNA. Thus, normal cells bear the seeds of their own cancers. To date, over 100 viral and cellular oncogenes and tumor suppressors have been identified.

b. pRb Functions by Binding to Certain Transcription Factors

pRb is a 928-residue DNA-binding protein that is localized in the nucleus of normal retinal cells but is absent in retinoblastoma cells. It is a phosphoprotein that is phosphorylated in a cell cycle–dependent manner, as is discussed in Section 34-4D. Hypophosphorylated forms of pRb form complexes with certain transcription factors, including **E2F,** which regulates the expression of several cellular and viral genes. E2F was first identified as a cellular factor involved in the regulation of the **adenovirus** early ***E2*** gene by the adenovirus oncogene product **E1A,** although further investigation revealed that the adenovirus ***E4*** gene product also participates in this process. E1A protein, which does not bind DNA, promotes the dissociation of pRb from E2F by complexing pRb. It thereby frees E2F protein to combine with **E4 protein** on the adenovirus *E2* promoter so as to stimulate the transcription of the *E2* gene. These observations suggest that *the interaction of pRb with E2F and other transcription factors to which it binds plays an important role in the suppression of cellular proliferation* and that the dissociation of this complex is, at least in part, the means by which E1A inactivates pRb function. Thus, *an additional way that oncogenes can cause cancer is by inactivating the products of normal cellular tumor suppressor genes.* We continue our discussion of pRb below.

D. *The Regulation of the Cell Cycle*

The cell cycle, as we discussed in Section 30-4A, is the sequence of major events that occur during the life of a eukaryotic cell. It is divided into four phases (Fig. 30-38): M phase, during which mitosis and cell division occur; G_1 phase, the main period of cell growth; S phase, during which DNA is synthesized; and G_2 phase, the interval in which the cell prepares for the next M phase. The progression through the cell cycle is regulated by external as well as internal signals. Thus, yeast have a regulatory point known as **START** that occurs late in G_1 beyond which they are committed to enter S phase, that is, replicate their DNA. However, if there are insufficient nutrients available or if the cell has not reached some minimum size, the cell cycle is arrested at START and the cell assumes a resting state until these criteria have been met. Animal cells have a similar decision point in G_1 named the **restriction point,** but it responds mainly to the extracellular presence of the appropriate **mitogens,** protein growth factors that signal the cell to proliferate. The cell cycle also has a series of **checkpoints** that monitor its progress and/or the health of the cell and arrest the cell cycle if certain conditions have not been satisfied. Thus, G_2 has a checkpoint that prevents the initiation of M until all of the cell's DNA has been replicated, thereby ensuring that both daughter cells will receive a full complement of DNA. Similarly, a checkpoint in M prevents mitosis until all chromosomes have properly attached to the mitotic spindle. Checkpoints in G_1 and S, as well as that in G_2, also arrest the cell cycle in response to damaged DNA so as to give the cell time to repair the damage (Section 30-5). In the cells of multicellular organisms, if after a time the checkpoint conditions have not been satisfied, the cell may be directed to commit suicide, a process named **apoptosis** (Section 34-4E), thereby preventing the proliferation of a genetically irreparably damaged and hence dangerous (e.g., cancerous) cell. However, single-celled eukaryotes such as yeast lack such a mechanism, presumably because, in their case, the survival of a genetically damaged cell is preferable in a Darwinian sense to its death.

a. The Activation of Cdk1 Triggers Mitosis

What are the molecular events that drive and coordinate the cell cycle? The first clues to this process came from studies of sea urchin embryos by Tim Hunt, which revealed that a class of proteins named **cyclins** accumulate steadily throughout the cell cycle and then abruptly disappear just before the anaphase portion of mitosis (Fig. 1-19). Homologs of these proteins have since been discovered in all eukaryotic cells examined. Indeed, mammals encode at least 20 different cyclins, many but not all of which participate in cell cycle control or appear and disappear cyclically with the cell cycle. The cyclins form a diverse protein family that are 30 to 50% similar over an ~100-residue segment known as the **cyclin box.**

Further indications as to the way the cell cycle is controlled came from experiments in which human cells at different stages of the cell cycle were fused to yield a single cell with two nuclei. When a cell in G_1 phase was fused with one in S phase, the G_1 nucleus immediately entered S phase, whereas the S nucleus continued replicating its DNA. However, when S-phase and G_2-phase cells were fused, the S nucleus continued replicating its DNA and the G_2 nucleus remained in G_2. Similarly, when G_1 and G_2 cells were fused, the G_1 nucleus entered S phase according to its own schedule and the G_2 nucleus remained in G_2. Evidently, S-phase cells contain a diffusible activator of DNA replication, only G_1 cells can initiate DNA replication, and cells that have transited S phase are unable to re-replicate their DNA until they have passed through M phase.

Many of the proteins that participate in cell cycle regulation were identified in the 1970s by Lee Hartwell through his study of temperature-sensitive mutants in *S. cerevisiae* (budding yeast) that were defective in cell cycle progression and by similar studies in *Schizosaccharomyces pombe* (fission yeast) by Paul Nurse. In what is perhaps the best characterized portion of the cell cycle, that inducing M phase, **cyclin B** combines with **Cdc2** (Cdc for *c*ell *d*ivision *c*ycle), whose sequence clearly indicates that it is a member of the Ser/Thr protein kinase family (Section 19-3C) and which is highly conserved from yeasts to humans. *It is Cdc2 that is the central cell cycle regulator in species ranging from yeasts to humans. It does so by phosphorylating a variety of nuclear proteins, among them histone H1, several oncogene proteins (see below), and proteins involved in nuclear disassembly, cytoskeletal rearrangement, spindle assembly, chromosome condensation, and Golgi fragmentation. This initiates a cascade of cellular events that culminates in mitosis.*

The binding of cyclin B to Cdc2 forms an activated complex that is alternatively called **cyclin-dependent protein kinase 1 (Cdk1)** and **maturation promoting factor (MPF).** This, however, is by no means the entire activation story. Cdc2 is a phosphoprotein that can be phosphorylated on Tyr 15, Thr 161, and in higher eukaryotes, Thr 14. Cdk1 is active only when both Thr 14 and Tyr 15, which occupy the region of its ATP-binding site, are dephosphorylated and when Thr 161 is phosphorylated. Moreover, the phosphorylation of Tyr 15 requires that cyclin B be present. Thus, mitosis is triggered through the following series of events (Fig. 34-99):

1. The cell enters G_1 with cyclin B absent and with Cdc2, which is present at a constant level throughout the cell cycle, dephosphorylated. **Cdk-activating kinase (CAK)** then phosphorylates Cdc2's Thr 161. Curiously, CAK, which consists of the heterodimer **Cdk7–cyclin H,** is also a component of the general transcription factor TFIIH (Section 34-3B), where its function is unknown.

2. In S phase, newly synthesized cyclin B binds to Cdc2, whereupon its Tyr 15 is phosphorylated by **Wee1** (so named because, in fission yeast, its inactivation causes cells to enter mitosis prematurely and hence at an unusually small, that is wee, size) and Thr 14 is phosphorylated by the Wee1 homolog **Myt1,** which can also phosphorylate Tyr 15. The resulting triply phosphorylated cyclin B–Cdc2 complex is enzymatically inactive because Thr 14 and Tyr 15 prevent Cdc2 from binding ATP. Thus, the entire system appears designed to maintain Cdc2 in an inactive state while cyclin B gradually accumulates during S phase.

3. At the cell cycle's G_2/M boundary, Cdc2's Thr 14 and Tyr 15 are rapidly and specifically dephosphorylated by **Cdc25C,** a dual-specificity protein tyrosine phosphatase (PTP; Section 19-3F) that thereby activates Cdk1, which in turn triggers mitosis (M phase). This process is initiated by the activating phosphorylation of Cdc25C by **plk1** (for

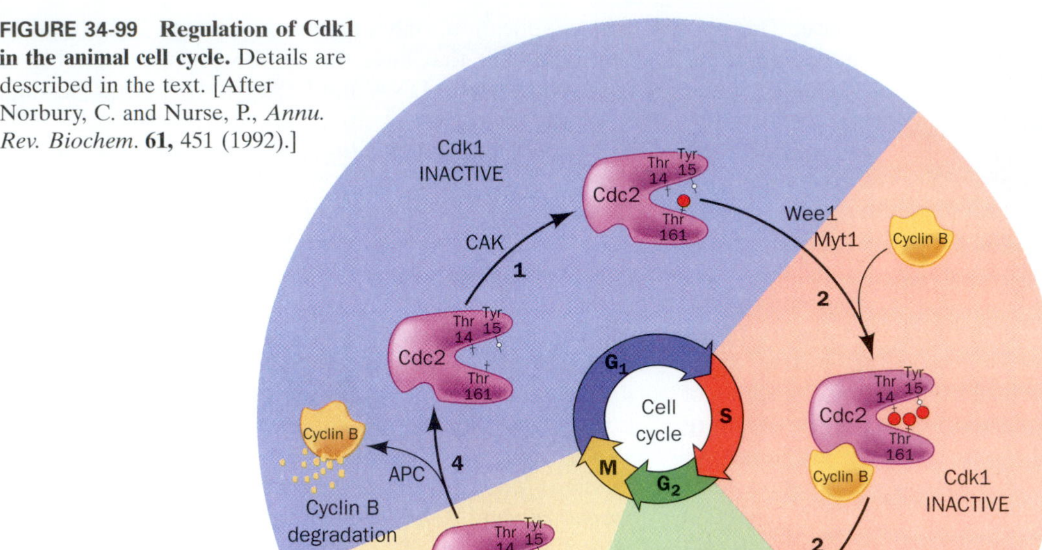

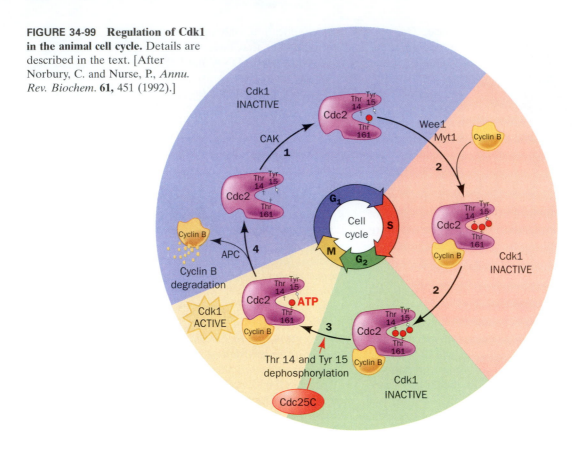

FIGURE 34-99 Regulation of Cdk1 in the animal cell cycle. Details are described in the text. [After Norbury, C. and Nurse, P., *Annu. Rev. Biochem.* **61,** 451 (1992).]

*p*olo-*l*ike *k*inase *1;* so named because it is a homolog of *Drosophila* **polo kinase**). The resulting activated Cdk1 also activates Cdc25C so that, through the intermediacy of Cdc25C, Cdk1 activates itself in a rapid burst. Moreover, Cdk1 inhibits its own inactivation by phosphorylating and thereby inactivating Wee1 and Myt1.

4. Cyclin B is quickly proteolyzed by the proteosome in a ubiquitin-mediated pathway whose E3 is a multisubunit complex known as the **anaphase-promoting complex (APC;** Section 32-6B), followed by the rapid dephosphorylation of Cdc2 Thr 161. This inactivates Cdk1, thereby returning the now divided cell to G_1. APC, which is inactive during S and G_2 phases, is activated, at least in part, by Cdk1, which thereby brings about the destruction of its own cyclin B component.

Other combinations of Cdks and cyclins that similarly mediate specific portions of the cell cycle include the following (Fig. 34-100): **Cdk4** and its close isoform **Cdk6,** which in complex with D-type cyclins **(cyclins D1, D2,** and **D3)** drive events in G_1; **Cdk2** and **cyclin E,** which are required for progression through the G_1/S boundary and for the initiation of DNA synthesis; and Cdk2 and **cyclin A,** which control passage through S phase. Thus, *Cdk–cyclin complexes form the engines that drive the various processes of the cell cycle as well as the clocks that time them.*

b. The X-Ray Structure of Cdk2 Resembles That of PKA

Cdk2, which is closely similar to Cdc2, is activated by the binding of a cyclin A, which is closely similar to cyclin B, followed by the CAK-catalyzed phosphorylation of Thr 160. Cdk2 is also negatively regulated by the phosphorylation of Tyr 15 and, to a lesser extent, the adjacent Thr 14.

The X-ray structure of Cdk2 in complex with ATP (Fig. 34-101*a*), determined by Sung-Hou Kim, indicates that this

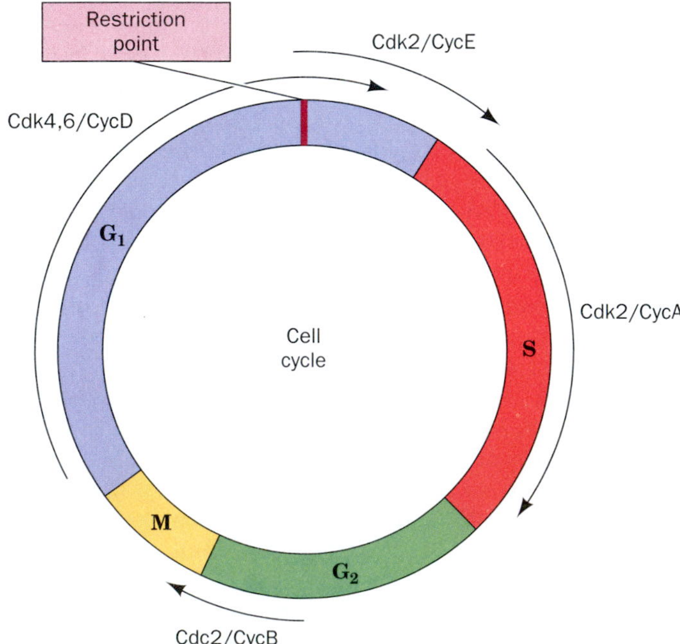

FIGURE 34-100 Complexes of cyclin-dependent kinases (Cdks) and cyclins (Cycs) that mediate passage through specific segments of the cell cycle.

(a)

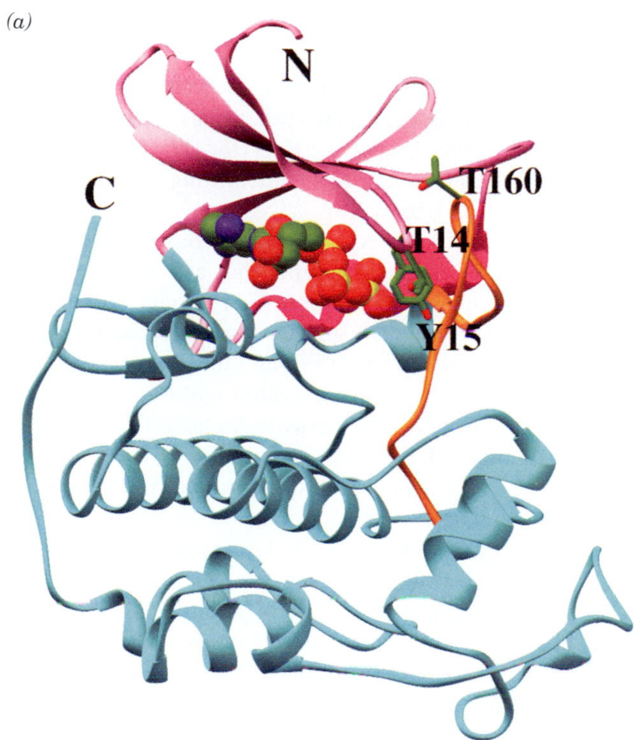

(b)

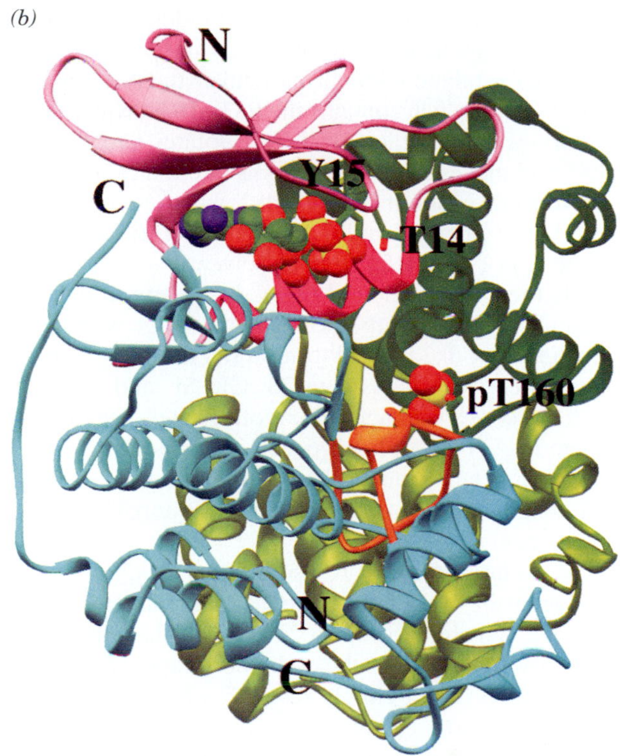

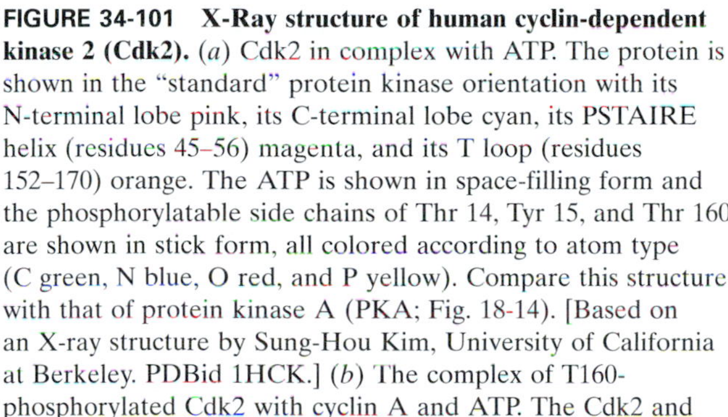

FIGURE 34-101 X-Ray structure of human cyclin-dependent kinase 2 (Cdk2). (*a*) Cdk2 in complex with ATP. The protein is shown in the "standard" protein kinase orientation with its N-terminal lobe pink, its C-terminal lobe cyan, its PSTAIRE helix (residues 45–56) magenta, and its T loop (residues 152–170) orange. The ATP is shown in space-filling form and the phosphorylatable side chains of Thr 14, Tyr 15, and Thr 160 are shown in stick form, all colored according to atom type (C green, N blue, O red, and P yellow). Compare this structure with that of protein kinase A (PKA; Fig. 18-14). [Based on an X-ray structure by Sung-Hou Kim, University of California at Berkeley. PDBid 1HCK.] (*b*) The complex of T160-phosphorylated Cdk2 with cyclin A and ATP. The Cdk2 and ATP are represented as in Part *a* and viewed similarly. The cyclin A is colored yellow-green with its cyclin box (residues 206–306) dark green. The Cdk2 phosphoThr 160 phosphoryl group is drawn in space-filling form. Note how the binding of cyclin A together with the phosphorylation of Cdc2 Thr 160 has caused a major structural reorganization of the T loop together with significant conformational adjustments of the Cdk2 N-terminal lobe, including its PSTAIRE helix. Also note the different conformations of the ATP triphosphate group in the two structures. [Based on an X-ray structure by Nikola Pavletich, Memorial Sloan-Kettering Cancer Center, New York, New York. PDBid 1JST.]

298-residue monomeric Ser/Thr protein kinase closely resembles the catalytic subunit of protein kinase A (PKA; Section 18-3C), a protein whose sequence is 24% identical to that of Cdk2. However, there are functionally significant structural differences between these two kinases:

1. The relative arrangement of ATP's β- and γ-phosphate groups in Cdk2 is likely to greatly reduce the reactivity of the γ-phosphate relative to that in the PKA–ATP complex (stereoelectronic control), thereby rationalizing, in part, why Cdk2 alone is catalytically inactive, whereas the catalytic subunit of PKA alone is catalytically active.

2. Access to the γ-phosphate of Cdk2's bound ATP by its protein substrates appears to be blocked by a 19-residue protein loop (residues 152–170) that has been named the "T loop" because it contains Thr 160.

The X-ray structure also explains why the phosphorylation of Thr 14 inactivates Cdk2: The hydroxyl group of this side chain is close to the ATP's γ-phosphate so that phosphorylation of Thr 14 is likely to disrupt the conformation of the ATP's phosphate groups. It is unclear, however, how the phosphorylation of Tyr 15 affects Cdk2 activity.

c. Cyclin A Binding and Thr 160 Phosphorylation Conformationally Reorganize Cdk2

The X-ray structure of human Cdk2 that is phosphorylated at its Thr 160 in complex with ATP and the C-terminal portion of human cyclin A (its residues 173–432), determined by Pavletich, indicates that cyclin A binds to Cdk2's "back" side (Fig. 34-101*b*). There it interacts with both lobes of the Cdk2 to form an extensive and continuous protein–protein interface. Cyclin A consists mainly of a bundle of 12 α helices (and no β sheets) in which its cyclin box forms helices 2 to 6. Interestingly, helices 7 to 11 form a bundle that is nearly superimposable on that of the cyclin box even though these two motifs exhibit little sequence similarity. Comparison of the X-ray structures of free cyclin A with that in its complex with Cdk2 indicates that cyclin A does not undergo significant conformational change on binding Cdk2. In contrast, cyclin A binding causes Cdk2 to undergo significant conformational shifts in the region around its catalytic cleft. In particular, the N-terminal α helix of Cdk2, which contains the PSTAIRE sequence motif characteristic of the Cdk family, rotates about its axis by 90° and moves several angstroms into the catalytic cleft relative to its position in free Cdk2, where it contacts the cyclin box segment of cyclin A. This movement brings Glu 51 (the E in PSTAIRE) from its solvent-exposed position outside the catalytic cleft of free Cdk2 to a position inside the catalytic cleft, where it forms a salt bridge with Lys 33, which in free Cdk2 instead forms a salt bridge to Asp 145. These three side chains (Lys 33, Glu 51, and Asp 145), which are conserved in all eukaryotic protein kinases, participate in ATP phosphate coordination and Mg^{2+} ion coordination. Their conformational reorientation on cyclin A binding apparently places them in a catalytically active arrangement.

The binding of cyclin A also induces Cdk2's T loop to undergo extensive conformational reorganization involving positional shifts of up to 21 Å, such that the T loop, which now also contacts the cyclin box, adopts a backbone conformation which closely resembles that of the analogous region of the catalytically active PKA. These movements greatly increase the access of a protein substrate to the ATP bound in the catalytic cleft, which has assumed a more reactive conformation than that in free Cdk2. The phosphate group on Thr 160 fits snugly into a positively charged pocket composed of three Arg residues that forms, in part, on cyclin A binding. Indeed, comparison of this structure with that in which Thr 160 is unphosphorylated indicates that Thr 160 phosphorylation induces activating conformational changes in the catalytic cleft of the cyclin A–Cdk2 complex as well as contributing to the reorganization of the T loop.

d. Cdk Inhibitors Function to Arrest the Cell Cycle

In addition to their control by phosphorylation/dephosphorylation and by the binding of the appropriate cyclin, *Cdk activities are regulated by* ***cyclin-dependent kinase inhibitors (CKIs),*** *which induce cell cycle arrest in response to such antiproliferative signals as contact with other cells, DNA damage, terminal differentiation, and senescence (in which cell cycle arrest is permanent).* The known CKIs have been grouped, according to their sequence and functional similarities, into two families: (1) the **Kip/Cip family** (*k*inase *i*nteracting *p*rotein/*c*ytokine-*i*nducible *p*rotein), whose members inhibit most Cdk–cyclin complexes (but not Cdk4/6–cyclin D) and can bind to isolated Cdks or cyclins, although with lower affinity than for Cdk–cyclin complexes; and (2) the **INK4 family** (*in*hibitor of Cd*k4* and Cdk6), whose members specifically inhibit Cdk4 and Cdk6 (which, together with cyclin D, mediate the cell's progression through G_1; Fig. 34-100) and can bind to either the isolated Cdk or its complex with cyclin D. The importance of CKIs is indicated by their frequent alterations in cancer. For example, **p16^{INK4a}** is mutated in about one-third of all human cancers, **p21^{Cip1}** arrests the cell cycle on behalf of p53 (see below), and **p27^{Kip1}** (also known as **p27^{Cip2}**) may be degraded in several types of cancers and its low levels are correlated with a poor clinical prognosis. Moreover, certain herpes viruses, including **Kaposi's sarcoma-associated herpes virus,** express an oncogenic D-type cyclin known as **K-cyclin** that binds to and thereby activates Cdk4/6, thus contributing to the deregulation of the cell cycle.

The members of the Kip/Cip family have homologous N-terminal ~65-residue segments that are necessary and sufficient to bind to and inhibit Cdk–cyclin complexes, whereas their C-terminal segments are divergent in length, sequence, and function. p27^{Kip1} is an inhibitor of Cdk2 that occurs in cells that have been treated with the antimitotic protein named **transforming growth factor β (TGFβ).** The X-ray structure of the N-terminal inhibitory domain (residues 25–93) of the 198-residue human p27^{Kip1} bound to the human cyclin A–Cdk2 complex in which Cdk2 is phosphorylated at Thr 160 was determined by Pavletich (Fig. 34-102). The p27 inhibitory domain is draped across both Cdk2 and cyclin A, where it assumes an extended conformation that does not have a hydrophobic core of its own and whose secondary structural elements do not interact with each other. p27's N-terminal end interacts with cyclin A, whose conformation is essentially unaffected by this interaction. In contrast, the binding of the p27 inhibitory domain's C-terminal segment to Cdk2 causes extensive conformational changes in Cdk2 that appear likely to destabilize its binding of ATP. More importantly, this C-terminal segment extends into Cdk2's active site cleft, where its conserved Tyr 88 side chain mimics the binding

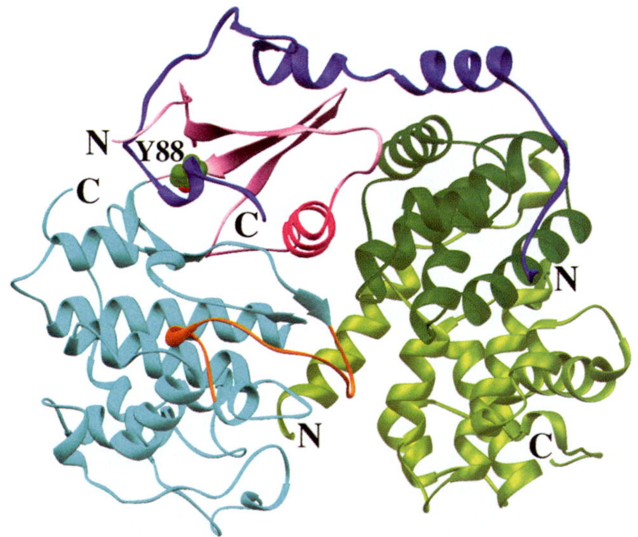

FIGURE 34-102 X-Ray structure of the Cdk2–Cyclin A–p27^{Kip1} ternary complex. The Cdk2 and cyclin A are colored as in Fig. 34-101 and p27^{Kip1} is blue. The complex is viewed along the Cdk2 PSTAIRE helix (approximately from the right of Fig. 34-101). The p27^{Kip1} Tyr 88 side chain, which is drawn in space-filling form with C green and O red, occupies the binding site for the adenine moiety of ATP. [Based on an X-ray structure by Nikola Pavletich, Memorial Sloan-Kettering Cancer Center, New York, New York. PDBid 1JSU.]

of ATP's adenine moiety in both its position and the contacts it makes to the active site groups, thereby eliminating any possibility of ATP binding. Last, p27's N-terminus occupies the peptide-binding groove on cyclin A's conserved cyclin box that is probably a docking site for a number of the cyclin A–Cdk2 complex's tight-binding substrates, thereby reducing the ability of substrate binding to reverse the effects of p27-induced conformational changes.

INK4 proteins can bind to monomeric Cdk4/6 so as to prevent its association with cyclin D or bind to a Cdk4/6–cyclin D binary complex to form a catalytically inactive ternary complex. *In vivo,* INK4–Cdk4/6 binary complexes are more abundant than INK4–Cdk4/6–cyclin D ternary complexes, which suggests that INK4 binding increases the rate of cyclin dissociation from the ternary complex. The X-ray structure of the ternary complex of human **p18^{INK4c},** human Cdk6, and K-cyclin (Fig. 34-103), determined by Pavletich, reveals that p18^{INK4c} binds to the Cdk6–K-cyclin complex in an entirely different manner from the way p27^{Kip1} binds to the Cdk2–cyclin A complex (Fig. 34-102). The 160-residue p18^{INK4c}, which consists of five ankyrin repeats (which participate in protein–protein interactions; Section 12-3E), binds to the 301-residue Cdk6 in the region of its ATP-binding site, where it interacts with Cdk6's N- and C-terminal lobes via the second and third ankyrin repeats of p18^{INK4c}. This rotates the N-terminal lobe by 13° with respect to the C-terminal lobe relative to their orientations in the Cdk2–cyclin A complex (Fig. 34-101*b*) and thereby distorts the Cdk6 ATP-binding site and misaligns its catalytic residues. Moreover, p18^{INK4c} binding distorts Cdk6's cyclin binding site such that their interface is reduced in area by ~30%: The cyclin box of the 253-residue K-cyclin binds to the N-terminal lobe of Cdk6 centered on its PSTAIRE sequence motif (which has the sequence PLSTIRE) but, unlike in the structure of the Cdk2–cyclin A complex (Figs. 34-101*b* and 34-102), there are no significant contacts between the Cdk6 C-terminal lobe and K-cyclin. Apparently, INK4 binding reduces the stability of the Cdk–cyclin interface.

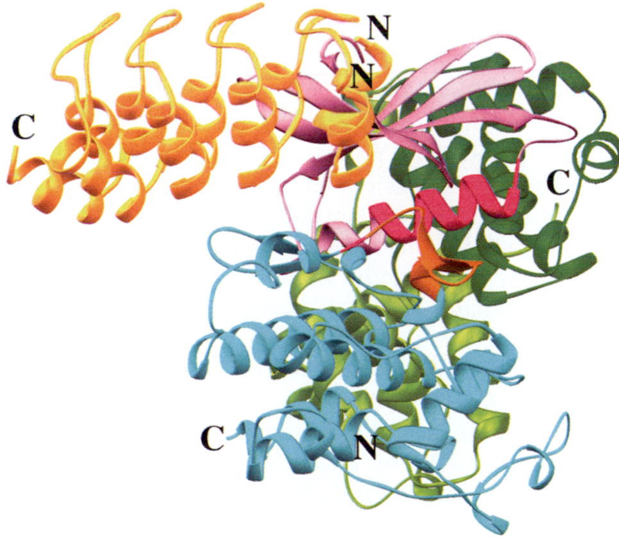

FIGURE 34-103 X-Ray structure of the Cdk6–K-cyclin–p18^{INK4c} ternary complex. The Cdk6 and K-cyclin are colored as in the homologous Cdk2 and cyclin A in Figs. 34-101 and 34-102 and the p18^{INK4c} is gold. The structure is viewed with the Cdk6 in the "standard" protein kinase orientation as in Fig. 34-101. [Based on an X-ray structure by Nikola Pavletich, Memorial Sloan-Kettering Cancer Center, New York, New York. PDBid 1G3N.]

e. Cell Cycle Arrest at the G_2 Checkpoint Is Mediated by a Phosphorylation/Dephosphorylation Cascade

How does failure to satisfy a checkpoint cause cell cycle arrest? For the G_2 checkpoint, the process is initiated, as diagrammed in Fig. 34-104, by at least six poorly characterized **sensor proteins** that have been identified through mutations in the genes ***rad1, rad3, rad9, rad17, rad26,*** and ***hus1*** that are defective in repair and replication checkpoints. The sensor proteins bind to damaged or unreplicated DNA, which causes them to activate two related large (~3000 residue) protein kinases known as **ATM** and **ATR** [ATM for *a*taxia *t*elangiectasia *m*utated (**ataxia telangiectasia** is a rare genetic disease characterized by a progressive loss of motor control, growth retardation, immune system deficiencies, premature aging, and a greatly increased risk of cancer); ATR for *AT*M and *R*ad3-related (**Rad3** is the ATR homolog in *S. pombe*)]. Activated ATM and ATR respectively phosphorylate and thereby activate **Chk2** and **Chk1** (Chk for *ch*eckpoint *k*inase). These latter protein kinases phosphorylate Cdc25C which, it will be recalled, functions to activate Cdk1 by dephosphorylating Cdc2's Thr 14 and Tyr 15. This phosphorylation, which occurs on Ser 216 of Cdc25C, does not inactivate this protein phosphatase. Rather, it provides a binding site for members of the **14-3-3 family** of adaptor proteins (which bind certain phosphorylated motifs in a wide variety of phosphorylated proteins; the name "14-3-3" is based on these proteins' column fractionation and electrophoretic mobility properties), and the resulting

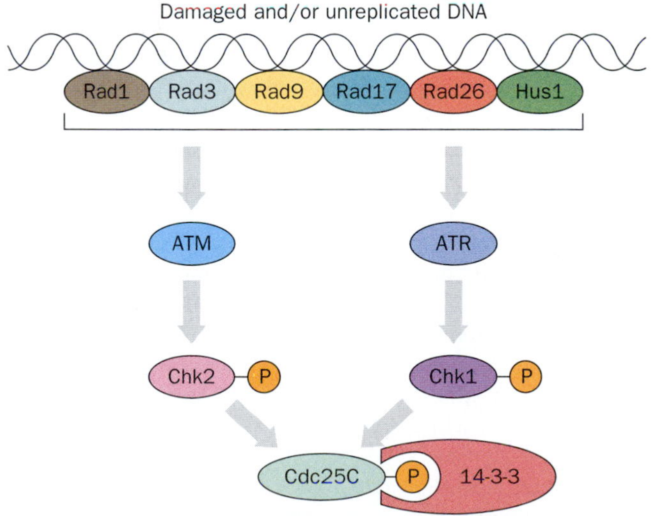

FIGURE 34-104 The G_2 checkpoint phosphorylation/dephosphorylation cascade that results in cell cycle arrest. Details are given in the text.

complex is sequestered in the cytoplasm out of contact with the Cdk1 in the nucleus. Since Cdk1 is the protein kinase that activates mitosis, the cell remains in G_2 until its DNA is repaired and/or fully replicated.

f. p53 Is a Transcriptional Activator That Arrests the Cell Cycle in G_2

The idea that p53 is a tumor suppressor first arose from the discovery that germ line mutations in the *p53* gene often occur in individuals with the rare inherited condition known as **Li–Fraumeni syndrome** that renders them highly susceptible to a variety of malignant tumors, particularly breast cancer, which they often develop before their 30th birthdays. That p53 is indeed a tumor suppressor has been clearly demonstrated in mice in which the *p53* gene has been inactivated. These knockout mice (Section 5-5H) appear to be developmentally normal but spontaneously develop a variety of cancers by the age of 6 months. Indeed, p53 functions as a "molecular policeman" in monitoring genome integrity: On the detection of DNA damage, p53 arrests the cell cycle until the damage is repaired, or failing that, induces apoptosis.

Despite the central role of p53 in preventing tumor formation, the way it does so has only gradually come to light. This tumor suppressor is specifically bound in humans by the homolog of the mouse **Mdm2 protein.** The *mdm2* gene is the dominant transforming oncogene present on *m*ouse *d*ouble *m*inute chromosomes (amplified extrachromosomal segments of DNA; Section 34-2E). Mdm2 protein is a ubiquitin-protein ligase (E3) that specifically ubiquitinates p53, thereby marking it for proteolytic degradation by the proteasome. Consequently, the amplification of the *mdm2* locus, which occurs in >35% of human **sarcomas** (none of which have a mutated *p53* gene; sarcomas are malignancies of connective tissues such as muscle, tendon, and bone), results in an increased rate of degradation of p53, thereby predisposing cells to malignant transformation. Similarly, as we have seen (Section 32-6B), E6 protein from human papilloma virus, which causes the great majority of cervical cancers, functions to ubiquitinate p53. Certain DNA tumor virus oncoproteins, such as SV40 **large T antigen** and adenovirus **E1B protein,** inactivate p53 by specifically binding to it. Thus, *an additional way that oncogenes can cause cancer is by inactivating normal tumor suppressors.*

p53 protein is an efficient transcriptional activator. Indeed, all point mutated forms of p53 that are implicated in cancer have lost their sequence-specific DNA-binding properties. But then, how does p53 function as a tumor suppressor? A clue to this riddle came from the observation that the treatment of cells with DNA-damaging ionizing radiation induces the accumulation of normal p53. This led to the discovery that both ATM and Chk2 phosphorylate p53, which prevents its binding by Mdm2 and hence increases the otherwise low level of this protein in the nucleus. *Although p53 does not initiate cell cycle arrest in G_2, its presence is required to prolong this process. It does so by activating the transcription of the gene encoding the CKI $p21^{Cip1}$, which binds to several Cdk–cyclin complexes so as to inhibit both the G_1/S and G_2/M transitions.* $p21^{Cip1}$ also binds to PCNA, the homotrimeric sliding clamp in DNA replication (Section 30-4B), so as to prevent its participation in DNA replication but not in DNA repair. Thus, $p21^{Cip1}$ has a dual role in cell cycle arrest in that it both blocks cell cycle progression and inhibits DNA replication in S-phase cells.

p53 also induces the transcription of the gene encoding the 14-3-3 family member **14-3-3σ,** which binds to Cdk1, thereby confining it to the cytoplasm. Moreover, the 14-3-3σ–Cdk1 complex binds the protein kinase Wee1 (which, as discussed above, inactivates Cdk1 by phosphorylating its Cdc2 component at its Tyr 15), thereby ensuring that Cdk1 remains in its inactive state. Thus, the disruption of the gene encoding 14-3-3σ is fatal for cells that sustain DNA damage. Chk2 also phosphorylates Wee1, which inhibits its proteasomal degradation. Consequently, germ line mutations in the *chk2* gene are also associated with Li–Fraumeni syndrome. Excessive levels of p53 are toxic, which explains why loss of the *mdm2* gene in mice is lethal unless the *p53* gene is also knocked out. In the absence of p53 activation, cells control the level of p53 through a feedback loop in which p53 stimulates the transcription of the *mdm2* gene.

Cells that are irreparably damaged are induced by p53 to commit suicide via apoptosis (Section 34-4E), thereby preventing the proliferation of potentially cancerous cells. p53 does so by transactivating the expression of several of the proteins that participate in apoptosis (Section 34-4E) and repressing the expression of others that inhibit this process.

g. The X-Ray Structure of p53 Explains Its Oncogenic Mutations

p53 is a tetramer of identical 393-residue subunits. Each subunit consists of four domains: an N-terminal transactivation domain (residues 1–99), a sequence-specific, DNA-binding core domain (residues 100–300, which binds two half-site decamers, each with the consensus palindromic sequence RRRC$^{A}_{T}$$^{T}_{A}$GYYY, that are separated by 0–13 nt and with a p53 dimer binding to each such decamer), a tetramerization domain (residues 301–356), and a nonspecific DNA-binding domain (residues 357–393, which binds a wide variety of DNAs including short single strands, irradiated DNA, Holliday junctions, and insertions/deletions). Although the entire protein has so far resisted crystallization, Pavletich has determined the X-ray structure of the DNA-binding core (residues 102–313) in complex with a 21-bp DNA segment containing its 5-bp target sequence (AGACT). *The vast majority of the >1000 p53 mutations that have been found in human tumors occur in this core.*

The structure of the p53 DNA-binding core domain (Fig. 34-105) contains a sandwich of two antiparallel β pleated sheets, one with four strands and the other with five, and a loop–sheet–helix motif that packs against one edge of the β sandwich. This edge of the β sandwich also contains two large loops running between the two β sheets that are held together through their tetrahedral coordination of a Zn^{2+} ion via one His and three Cys side chains.

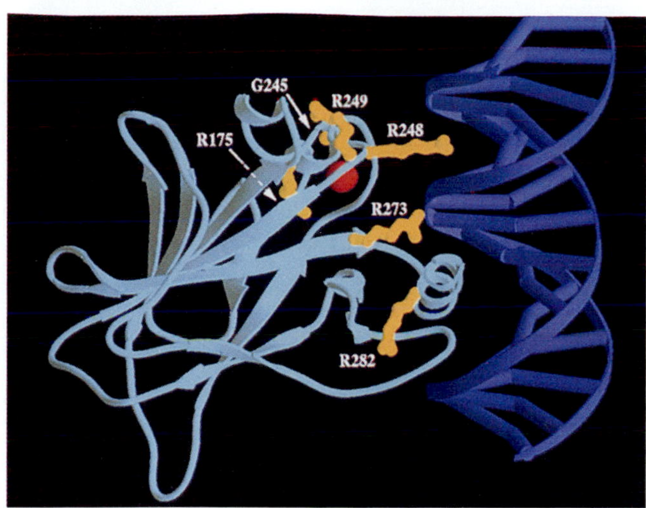

FIGURE 34-105 X-Ray structure of the DNA-binding domain of human p53 in complex with its target DNA. The protein is shown in ribbon form (*cyan*), the DNA in ladder form with its bases represented by cylinders (*blue*), the tetrahedrally liganded Zn^{2+} ion is shown as a red sphere, and the side chains of the six most frequently mutated residues in human tumors are shown in stick form (*yellow*) and identified with their one-letter codes. [Courtesy of Nikola Pavletich, Memorial Sloan-Kettering Cancer Center, New York. PDBid 1TSR.]

The p53 DNA binding motif does not resemble any other that has previously been characterized. The helix and loop from the loop–sheet–helix motif are inserted in the DNA's major groove, where they make sequence-specific contacts with the bases (lower right of Fig. 34-105). One of the large loops provides a side chain (Arg 248) that fits in the minor groove (upper right of Fig. 34-105). The protein also contacts the DNA backbone between the major and minor grooves in this region (notably with Arg 273).

The structure's most striking feature is that *its DNA-binding motif consists of conserved regions comprising the most frequently mutated residues in the p53 variants found in tumors.* Among them are one Gly and five Arg residues (highlighted in yellow in Fig. 34-105) whose mutations collectively account for over 40% of the *p53* variants in tumors. The two most frequently mutated residues, Arg 248 and Arg 273, as we saw, directly contact the DNA. The other four "mutational hotspot" residues appear to play a critical role in structurally stabilizing p53's DNA-binding surface. The relatively sparse secondary structure in the polypeptide segments forming this surface (one helix and three loops) accounts for this high mutational sensitivity: Its structural integrity mostly relies on specific side chain–side chain and side chain–backbone interactions.

h. p53 Is a Sensor That Integrates Information from Several Pathways

p53 may be activated by several other pathways. For example, aberrant growth signals, such as those generated by oncogenic variants of MAP kinase cascade components such as Ras, stimulate the expression of a variety of transcription factors (Fig. 19-38), many of which are proto-oncogene products. One of them, **Myc,** activates the transcription of the gene encoding **p19^{ARF}** (in mice; **p14^{ARF}** in humans), which also encodes **p16^{INK4a}.** This is because these two proteins, which have no sequence similarity, are expressed through alternative splicing of their first exons and share second and third exons that are translated in different reading frames for the two proteins (ARF for *a*lternative *r*eading *f*rame protein), an unprecedented economy of genomic resources in higher eukaryotes, although a common phenomenon in bacteriophage (Section 32-1D). p19^{ARF} binds to Mdm2 and inhibits its activity, thereby preventing the degradation of p53 and hence triggering the p53-dependent transcriptional programs leading to cell cycle arrest as well as apoptosis (Section 34-4E). Evidently, p19^{ARF} acts as part of a p53-dependent fail-safe system to counteract hyperproliferative signals.

A third activation pathway for p53 is induced by a wide variety of DNA-damaging chemotherapeutic agents, protein kinase inhibitors, and UV radiation. These activate ATR to phosphorylate p53 so as to reduce its affinity for Mdm2 in much the same way as do ATM and Chk2.

p53 is also subject to a rich variety of reversible post-translational modifications that markedly influence the expression of its target genes. These include acetylation at several Lys residues, glycosylation, ribosylation, and sumoylation (Section 32-6B), as well as phosphorylation at multiple Ser/Thr residues and ubiquitination. p53 does not bind to short DNA fragments that contain its target sites unless its C-terminal domain has either been deleted or modified by phosphorylation and/or acetylation. Yet NMR evidence indicates that the C-terminal domain does not interact with other p53 domains. This suggests that p53's C-terminal and core domains compete for DNA binding unless the C-terminal domain has been modified.

p53, as we have only glimpsed, is the recipient of a vast number of intracellular signals and, in turn, controls the activities of a large number of downstream regulators. One way to understand the operation of this highly complex and interconnected network is in analogy with the Internet. In the Internet (cell), a small number of highly connected servers or hubs ("master" proteins) transmit information to/from a large number of computers or nodes (other proteins) that directly interact with only a few other nodes (proteins). In such a network, overall performance is largely unperturbed by the inactivation of one of the nodes (other proteins). However, the inactivation of a hub ("master" protein) will greatly impact system performance. p53 is a "master" protein, that is, it is analogous to a hub. Inactivation of one of the many proteins that influences its performance or one of the many proteins whose activity it influences usually has little effect on cellular events due to the system's redundant and highly interconnected components. However, the inactivation of p53 or several of its most closely associated proteins (e.g., Mdm2) disrupts the cell's responses to DNA damage and tumor-predisposing stresses. Nevertheless, a quantitative understanding of the functions and malfunctions of the p53-based network will require a complete description of all the proteins with which p53 directly or indirectly interacts and how they do so under the conditions present in the cell, something that we are far from having. Thus, for the foreseeable future, we will be limited to qualitative descriptions of the functioning of the p53 network. Our understanding of other cellular signal transduction systems is similarly vague (Section 19-4F).

i. pRb Regulates the Cell Cycle's G_1/S Transition

The tumor suppressor pRb **(retinoblastoma-associated protein),** a 928-residue monomer that is localized in the nucleus of normal animal cells but is defective or absent in retinoblastoma cells (Section 34-4C), is an important regulator of the cell cycle's G_1/S transition, that is, the cell's passage through the restriction point. The effects of pRb are largely manifested through its interactions with the members of the **E2F** family of transcription factors, which, in the absence of pRb, activate their target promoters in complex with a member of the **DP** (for E2F *d*imerization *p*artner) family. The mammalian E2F family consists of six ~440-residue members, of which **E2F-1** through **E2F-4** interact with pRb via a conserved 18-residue polypeptide segment contained in their ~70-residue, C-terminal, transactivation domains. The mammalian DP family consists of two ~430-residue members, **DP-1** and **DP-2.** The E2F–DP heterodimers induce the transcription of a variety of genes that encode proteins required for S-phase entry (e.g., Cdc2, Cdk2, and cyclins A and E) and for DNA synthesis [e.g., DNA polymerase α (pol α; Section 30-4B), Orc1 and several Mcm proteins (which participate in initiating DNA replication; Section 30-4B), ribonucleotide reductase (Section 28-3A), thymidylate synthase, and dihydrofolate reductase (Section 28-3B)].

How does pRb mediate cell cycle progression? pRb, which is synthesized throughout the cell cycle, is a phosphoprotein that is phosphorylated at as many as 16 of its Ser/Thr residues by Cdk4/6–cyclin D in mid to late G_1, by Cdk2–cyclin E in late G_1, by Cdk2–cyclin A in S, and by Cdk1 (Cdc2–cyclin B) in M, with different Cdk–cyclin complexes phosphorylating different sets of sites on pRb. *Hypophosphorylated but not hyperphosphorylated pRb binds to the transactivation domain of E2F so as to prevent it from activating transcription at the promoter to which it is bound.* In nonproliferating cells (those in early G_1), pRb remains hypophosphorylated because, unless such cells receive mitogenic signals, the highly unstable D-type cyclins (they have half-lives of ~10 min) do not accumulate to levels sufficient to generate significant amounts of Cdk4/6–cyclin D (mitogens trigger MAP kinase cascades that stimulate the expression of D-type cyclins). Moreover, since hypophosphorylated pRb prevents E2F–DP from activating the expression of Cdk2 and cyclins E and A, Cdk2–cyclin E and Cdk2–cyclin A do not accumulate to sufficient levels to hyperphosphorylate pRb. In fact, the small amounts of Cdk2–cyclin A and Cdk2–cyclin E that are present are inhibited by $p27^{Kip1}$, which occurs in high levels in pre-restriction point G_1 cells.

Mitogens also activate the expression of $p27^{Kip1}$ and $p21^{Cip1}$ which, contrary to what might be expected, do not inhibit Cdk4/6–cyclin D complexes but instead stimulate their activities by enhancing their assembly and promoting their nuclear import. Thus, mitogenic signals break the pRb-imposed blockade to cell cycle progression by inducing the formation of Cdk4/6–cyclin D–$p27^{Kip1}$/$p21^{Cip1}$ complexes that begin the phosphorylation of pRb. This releases a small amount of E2F, which thereupon induces the expression of Cdk2 and cyclins E and A. The Cdk4/6–cyclin D complexes also sequester $p27^{Kip1}$ and $p21^{Cip1}$, which permits the resulting Cdk2–cyclin E complex to catalyze a second wave of pRb phosphorylation and import [although when large amounts of $p21^{Cip1}$ are produced through the influence of activated p53 (see above), it inhibits Cdk2–cyclin E so as to arrest the G_1/S transition]. This frees large amounts of E2F, resulting in a surge in the transcription of the genes that promote cell cycle progression. As the cell cycle continues, pRb is increasingly phosphorylated, first by Cdk2–cyclin A and then by Cdk1, until the exit from M phase, whereupon pRb is abruptly dephosphorylated, probably by the protein Ser/Thr phosphatase PP1 (Section 19-3F), permitting pRb to again arrest cell cycle progression by inhibiting E2F.

A variety of proteins that contain the LXCXE sequence motif bind to pRb. These comprise several cellular proteins, including the D-type cyclins, which may thereby be directed to pRb, and certain viral oncoproteins, whose binding to pRb prevents it from binding E2F (which lacks an LXCXE motif but, as we saw above, binds pRb via an 18-residue sequence). Indeed, E2F was first identified (and named) as a cellular factor involved in the regulation of the adenovirus early *E2* gene by the adenovirus oncogene product **E1A,** although further investigations revealed that the adenovirus *E4* gene product also participates in this process. E1A, which does not bind DNA, binds, via its LXCXE motif, to pRb. This causes the pRb to release its bound E2F, which permits the E2F, in combination with **E4,** to activate the transcription of the *E2* gene from the adenovirus *E2* promoter. The freed E2F also drives the infected cell into S phase, which facilitates adenovirus DNA replication. The SV40 large T antigen and the human papilloma virus **E7** protein, which also contain LXCXE motifs, similarly activate E2F. Over 100 proteins have been reported to bind to pRb, although in most cases by other means than via LXCXE motifs. The functions of these interactions are largely unknown.

j. E2F and the LXCXE Motif Bind to Separate Sites on the pRb Pocket Domain

pRb's so-called **pocket domain** forms the binding site for both E2F and the LXCXE motif and is the major site of genetic alterations in tumors. The pocket domain consists of its conserved A- and B-boxes (residues 379–572 and 646–772) linked by a poorly conserved spacer. However, when the spacer is excised, the A- and B-boxes nevertheless associate noncovalently.

The X-ray structure of the pRb pocket domain lacking its spacer in complex with the 18-residue pRb-binding peptide of E2F was independently determined by Marmorstein and Gamblin and by Yunje Cho. It reveals that the 18-residue E2F peptide binds in a boomerang-shaped con-

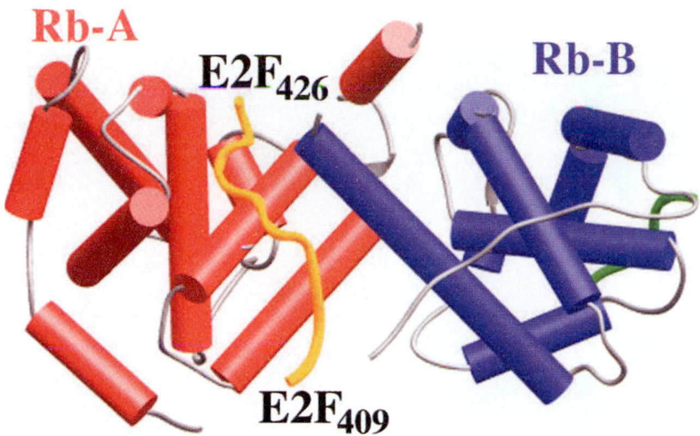

FIGURE 34-106 X-Ray structure of the pRb pocket domain in complex with the 18-residue pRb-binding peptide of E2F. The helices of the A- and B-boxes are respectively drawn as red and blue cylinders and the main chain of the E2F peptide is shown as a gold worm. The superimposed structure of an LXCXE-containing nonapeptide segment of the human papilloma E7 protein in its complex with the pRb pocket domain is represented by a green worm. [Courtesy of Steven Gamblin, National Institute for Medical Research, London, U.K. PDBids 1O9K and 1GUX.]

formation at the highly conserved interface between the A- and B-boxes, both of which contain the five-helix cyclin fold (Fig. 34-106). However, the X-ray structure of the pRb pocket domain in complex with the 9-residue LXCXE-containing peptide from human papilloma virus E7 protein, determined by Pavletich, indicates that the E7 LXCXE peptide binds, in an extended conformation, in a shallow groove on the B-box that is ~30 Å distant from the E2F-binding site (Fig. 34-106). This latter binding site, which is formed by highly conserved residues, closely resembles the primary Cdk2 binding site of cyclin A (Fig. 34-101*b*) and the TBP binding site of the 20% identical TFIIB (Fig. 34-49). The corresponding portion of the A-box participates in forming the A–B interface.

k. pRb Also Represses Transcription by Recruiting HDACs and SWI/SNF Homologs

Binding experiments reveal that pRb associates with the histone deacetylases (HDACs; Section 34-3B) HDAC1 and HDAC2, both of which contain an LXCXE sequence motif (actually IXCXE). Consequently, the presence of human papilloma virus E7 protein or mutations that disrupt the pRb pocket domain abolish the binding of these HDACs to pRb. These observations suggest that pRb also functions to recruit HDAC1 and HDAC2 to DNA-bound E2F so as to facilitate the histone deacetylation and hence the transcriptional inactivation of the chromatin containing E2F's target genes. This explains the observation that pRb can repress transcription from promoters to which it is artificially linked by a DNA-binding domain that differs from that of E2F. HDAC3 also associates with pRb, although it lacks an LXCXE motif.

The human SWI/SNF homologs BRM and BRG1, which both have LXCXE motifs, also bind to pRb. BRM and BRG1, it will be recalled (Section 34-3B), are DNA-dependent ATPases that are components of chromatin-remodeling complexes. Thus, the observation that pRb can simultaneously bind BRG1 and an HDAC (despite both having LXCXE motifs) suggests that pRb recruits chromatin remodeling complexes to facilitate the action of HDACs at E2F promoters.

E. *Apoptosis: Programmed Cell Death*

The maxim that death is part of life is even more appropriate on the cellular level than it is on the organismal level. **Programmed cell death** or **apoptosis** (Greek: falling off, as leaves from a tree), which was first described by John Kerr in the late 1960s, is a normal part of development as well as the maintenance and defense of the adult animal body. For example, in the nematode worm *Caenorhabditis elegans,* a transparent organism whose cell lineages have been microscopically elucidated, precisely 131 of its 1090 somatic cells undergo apoptosis in forming the normal adult body. In many vertebrates, the digits of the developing hands and feet are initially connected by webbing that is eliminated by programmed cell death (Fig. 34-107), as are the tails of tadpoles and the larval tissues of insects during their metamorphoses into adults (Figs. 34-81 and 34-86). Apoptosis is particularly prevalent in the developing mammalian nervous system in which an approximately threefold excess of neurons is produced. However, only those neurons that make adequate synaptic connections are retained; the remainder are eliminated via apoptosis (Fig. 34-108).

In the adult human body, which consists of nearly 10^{14} cells, an estimated 10^{11} cells are eliminated each day through programmed cell death (which closely matches the number of new cells produced by mitosis). Indeed, the mass of the cells that we annually lose in this manner approaches that of our entire body. A particularly obvious manifestation of this phenomenon is the monthly sloughing off of the uterine lining in menstruation (Section 19-1I). Similarly, the immune system cells known as *T* lymphocytes (*T* cells) undergo apoptosis in the thymus if the *T* cell receptors they produce recognize antigens that are normally present in the body or are improperly formed (Sections 35-2A and 35-2D); ~95% of immature *T* cells are eliminated in this way. Autoimmune diseases such as rheumatoid arthritis and insulin-dependent diabetes (Section 27-3B) arise when this process goes awry. Apoptosis is also an essential part of the body's defense systems. The immune system eliminates virus-infected cells, in part, by inducing them to undergo apoptosis, thereby preventing viral replication. Cells with irreparably damaged DNA and hence at risk for malignant transformation undergo apoptosis, thereby protecting the entire organism from cancer. Indeed, *one of the defining characteristics of malignant cells is their ability to evade apoptosis.* Cells that become detached from their normal positions in the body likewise commit suicide. Indeed, as Martin Raff pointed out, *apoptosis appears to be the default option for metazoan cells: Unless they continually receive external*

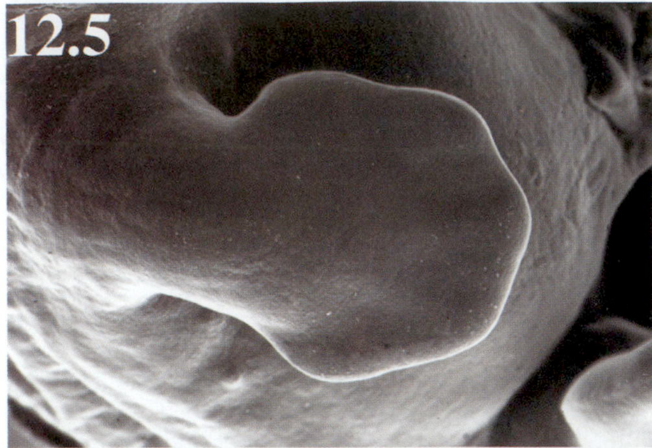

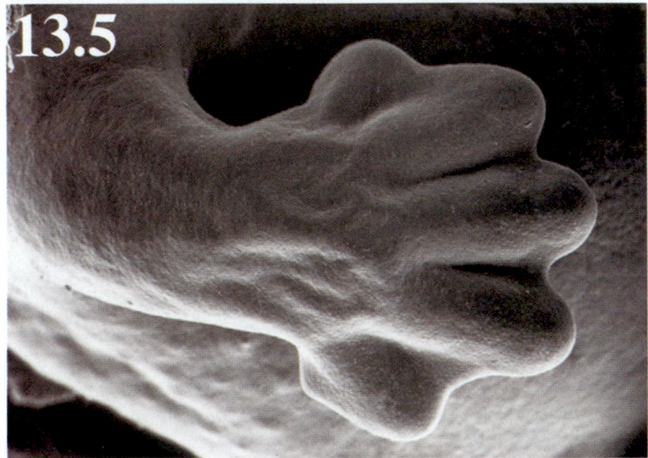

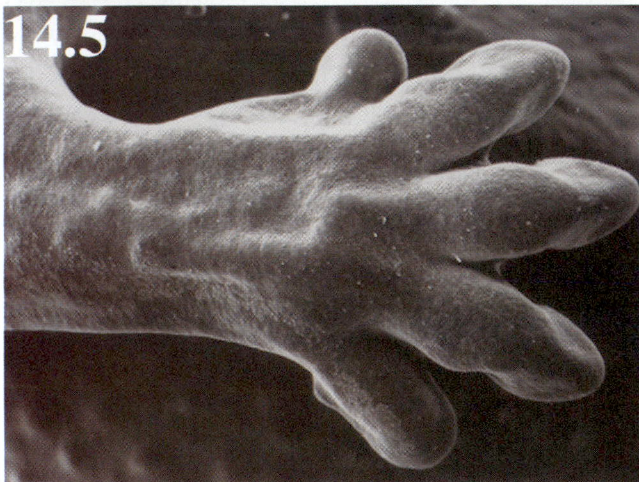

FIGURE 34-107 Programmed cell death in the embryonic mouse paw. At day 12.5 of development, its digits are fully connected by webbing. At day 13.5, the webbing has begun to die. By day 14.5, this apoptotic process is complete. [Courtesy of Paul Martin, University College of London, U.K.]

hormonal and/or neuronal signals not to commit suicide, they will do so. Thus, adult organs maintain their constant size by balancing cell proliferation with apoptosis. Not surprisingly, therefore, inappropriate apoptosis has been implicated in several neurodegenerative diseases including

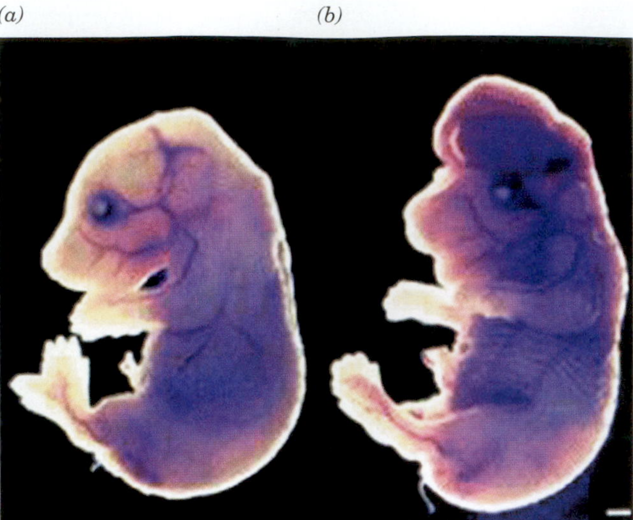

FIGURE 34-108 Brain development in 16.5-day-old mouse embryos. (*a*) A wild-type embryo. (*b*) An embryo in which **caspase-9,** an enzyme that mediates apoptosis (see below) has been knocked out. Note the protruding and morphologically abnormal brain in the knockout embryo due to the overproliferation of brain neurons. [Courtesy of Richard Flavell, Yale University Medical School.]

Alzheimer's disease (Section 9-5B), Parkinson's disease (Section 26-4B), and Huntington's disease (Section 30-7), as well as much of the damage caused by stroke and heart attacks. Consequently, the signaling systems that mediate apoptosis have become targets for therapeutic intervention. In fact, many of the chemotherapeutic agents in present use do not kill their target cancer cells outright but, rather, damage them so as to induce their apoptosis.

Apoptosis is qualitatively different from **necrosis,** the type of cell death caused by trauma (e.g., lack of oxygen, extremes of temperature, and mechanical injury). Cells undergoing necrosis essentially explode: They and their membrane-enclosed organelles swell as water rushes in through their compromised membranes, releasing lytic enzymes that digest the cell contents until the cell lyses, spilling its contents into the surrounding region (Section 22-4B). The cytokines that the cell releases often induce an inflammatory response (which can damage surrounding cells) that attracts **phagocytotic cells** (cells, such as the white blood cells known as **macrophages,** that ingest foreign particles and waste matter) to "mop up" the resulting cell debris. In contrast, apoptosis begins with the loss of intercellular contacts by an apparently healthy cell followed by its shrinkage, the condensation of its chromatin at the nuclear periphery, the collapse of its cytoskeleton, the dissolution of its nuclear envelope, the fragmentation of its DNA, and violent blebbing (blistering) of its plasma membrane. Eventually, the cell disintegrates into numerous membrane-enclosed **apoptotic bodies** that are phagocytosed by neighboring cells as well as by roving macrophages with-

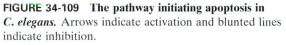

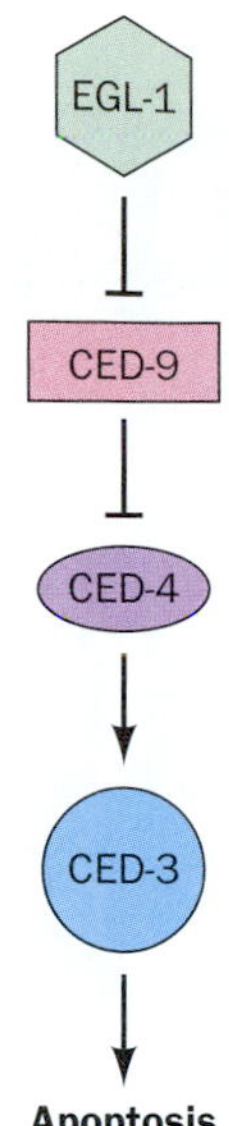

FIGURE 34-109 **The pathway initiating apoptosis in *C. elegans.*** Arrows indicate activation and blunted lines indicate inhibition.

out spilling the cell contents and hence not inducing an inflammatory response.

a. Apoptosis Is Induced by Signaling Cascades

The pathway for apoptosis, which was first elucidated in *C. elegans* through genetic studies by John Sulston and Robert Horvitz, involves three so-called *ced* (for *c*ell *d*eath abnormal) gene products (Fig. 34-109): **CED-4** protein, a protease, activates the protease **CED-3,** which then initiates the destruction of the cell; **CED-9** functions to inactivate CED-4. In fact, mutations that inactivate CED-9 result in numerous embryonic cells that would normally survive in the adult organism to inappropriately activate its CED-4 and CED-3 and hence die, thereby killing the embryo. Conversely, if CED-9 is expressed at abnormally high levels or CED-3 or CED-4 is inactivated, cells that normally die will survive (which, curiously, has little apparent effect on the health of the adult organism). Later investigations revealed that a fourth protein, **EGL-1,** functions to inhibit CED-9 and hence its overexpression induces apoptosis.

Apoptotic pathways in mammals are considerably more complex than that in *C. elegans*. Nevertheless, the above CED proteins and EGL-1 all have counterparts in mammalian pathways:

1. CED-3 is the prototype of a family of proteases known as **caspases** (for *c*ysteinyl *asp*artate-specific prote*ases*) because they are **cysteine proteases** [whose mechanism resembles that of serine proteases (Section 15-3C) but with Cys replacing the active site Ser] that cleave after an Asp residue. Their target cleavage sites are specified mainly by this Asp and its three preceding residues.

2. CED-4 is a scaffolding protein that plays an essential role in caspase activation. Its mammalian counterpart is called **Apaf-1** (for *a*poptotic *p*rotease-*a*ctivating *f*actor-*1*).

3. CED-9 is a member of the **Bcl-2** family (so named because its founding member, Bcl-2, was initially characterized as a gene involved in *B* cell *l*ymphoma). Some of the numerous members of this family, including CED-9, protect cells from death and hence are said to be **anti-apoptotic.** Others promote cell death and are therefore said to be **pro-apoptotic.**

4. EGL-1 is a pro-apoptotic member of the Bcl-2 family.

b. Caspases Have Closely Similar Structures

Caspases are $\alpha_2\beta_2$ heterotetramers that consist of two large α subunits (~300 residues) and two small β subunits (~100 residues). They are expressed as zymogens **(procaspases)** that have three domains (Fig. 34-110): an N-terminal prodomain that is proteolytically excised on activation, followed by sequences comprising the active enzyme's α and β subunits that are proteolytically separated on activation. The activating cleavage sites all follow Asp residues and are, in fact, targets for caspases (the only other eukaryotic protease known to cleave after an Asp

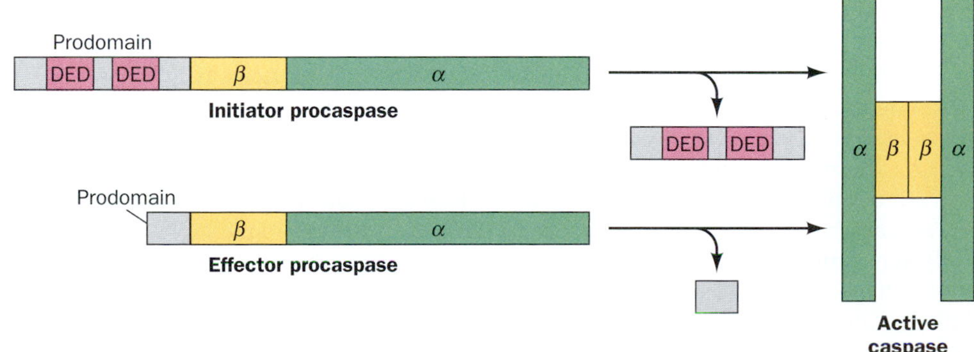

FIGURE 34-110 **Caspase domain structure and activation.** The zymogens of initiator caspases have long N-terminal prodomains, which in several cases contain two death effector domains (DEDs), whereas the zymogens of effector caspases have only short prodomains. Procaspases are activated by proteolytic cleavages that excise their prodomains and separate their α and β subunits to yield the active $\alpha_2\beta_2$ caspases.

residue is **granzyme B,** a chymotrypsin-like serine protease expressed by cytotoxic *T* cells that functions to induce apoptosis in tumor and virally infected cells; Section 35-2A). Thus, caspase activation, as we shall see, may either be autocatalytic or be catalyzed by another caspase.

Humans express eleven caspases, six of which participate exclusively in apoptosis [with the remainder being involved mainly in the cytokine activation and hence the control of inflammation; the founding member, **caspase-1,** is also known as **interleukin-1β-converting enzyme (ICE)** because it proteolytically activates the cytokine **interleukin-1β** (Section 35-2A)]. There are two classes of apoptotic caspases (Fig. 34-110):

1. **Initiator caspases (caspases-8, -9,** and **-10)** are characterized by long prodomains (129–219 residues) that target their zymogens to scaffolding proteins that promote their autoactivation. The prodomains of caspases-8 and -10 each contain two ~80-residue **death effector domains (DEDs),** through which they bind to DEDs on their target adaptor proteins (see below). The prodomain of caspase-9 instead contains the structurally similar ~90-residue **caspase recruitment domain (CARD)** that promotes the interaction of this caspase with certain scaffolding and regulatory proteins.

2. **Effector caspases (caspases -3, -6,** and **-7)** have short (~25 residue) prodomains and are activated by initiator caspases. The activated effector caspases, which have been described as the cell's executioners, cleave a wide variety of cellular proteins (see below), thereby bringing about apoptosis.

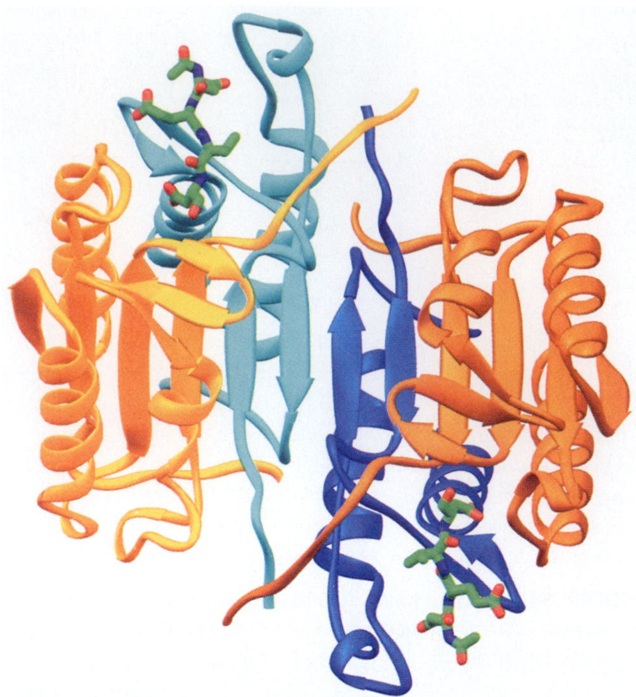

FIGURE 34-111 X-Ray structure of caspase-7 in complex with the tetrapeptide aldehyde inhibitor acetyl-Asp-Glu-Val-Asp-CHO (Ac-DEVD-CHO). The $\alpha_2\beta_2$ heterotetrameric enzyme is viewed along its 2-fold axis with its large (α) subunits orange and gold and its small (β) subunits cyan and light blue. The Ac-DEVD-CHO inhibitor is drawn in stick form with C green, N blue, and O red. [Based on an X-ray structure by Keith Wilson and Paul Charifson, Vertex Pharmaceuticals, Cambridge, Massachusetts. PDBid 1F1J.]

The X-ray structure of caspase-7 (Fig. 34-111), determined by Keith Wilson and Paul Charifson, closely resembles those of the several other caspases of known X-ray structures. Each αβ heterodimer of this twofold symmetric $\alpha_2\beta_2$ heterotetramer contains a six-stranded β sheet, five of whose strands are parallel, and which is flanked by five α helices, two on one side and three on the other, that are approximately parallel to the β strands. The β sheet is continued across the protein's 2-fold axis to form a twisted 12-stranded β sheet. Each αβ heterodimer contains an active site that is located at the C-terminal ends of its parallel β strands and which recognizes a tetrapeptide on the N-terminal side of its Asp–X cleavage site. The structures of the various caspases differ mainly in the conformations of the four loops forming their active sites. Comparison of the X-ray structure of caspase-7 with that of **procaspase-7** (with its active site Cys 186 mutated to Ser to prevent its autoactivation), independently determined by Weigon Shi and Wolfram Bode, reveals that, although the two proteins are otherwise closely superimposable, the four active site loops in procaspase-7 have undergone large conformational changes relative to those in caspase-7 that essentially obliterates the active site. In particular, the loop containing both the catalytically essential Cys residue and the activating cleavage site between the α and β subunits changes its orientation by 90° after this cleavage so as to expose and properly position the previously buried catalytic Cys residue.

c. Caspases Cleave a Wide Variety of Proteins and Activate the Degradation of Chromosomal DNA

Over 60 cellular proteins have been identified as caspase substrates. These include cytoskeletal proteins [e.g., actins (Section 35-3E) and **lamins** (intermediate filaments that form the meshwork lining the inner nuclear envelope)], proteins involved in cell cycle regulation (e.g., cyclin A, Wee1, $p21^{Cip1}$, ATM, and pRb; Section 34-4D), proteins that participate in DNA replication [e.g., topoisomerase I (Section 29-3C) and Mcm3 (Section 30-4B)], transcription factors (e.g., Sp-1 and NF-κB; Section 34-3B), and proteins that participate in signal transduction [e.g., RasGAP (Section 19-3C) and protein kinase C (Section 19-4C)]. Nevertheless, how the cleavage of these numerous proteins causes the morphological changes that cells undergo during apoptosis is unclear.

The induction of apoptosis also causes the rapid degradation of chromosomal DNA. Chromosomal DNA is attached to the chromosomal protein matrix at intervals of ~70 kb via AT-rich matrix-associated regions (MARs; Section 34-1D). During apoptosis, **caspase-activated**

DNase (CAD) cleaves the chromosomal DNA at these sites, which is often followed by its cleavage between nucleosomes to yield a series of DNA fragments that differ in their lengths by increments of ~200 bp. CAD is ubiquitously expressed in all tissues in complex with its inhibitor **ICAD** (*i*nhibitor of *CAD*), which on induction of apoptosis is cleaved by caspases-3 and -7, thereby releasing active CAD. ICAD also functions as a chaperone that must be present when CAD is being ribosomally synthesized in order for CAD to fold to its native conformation. This ensures that native CAD can only form in complex with ICAD and hence prevents inappropriate DNA cleavage. Although the cleavage of a cell's chromosomal DNA would certainly cause its death, cells containing mutant ICAD undergo apoptosis even though their chromosomal DNA remains intact. This suggests that DNA cleavage during apoptosis functions to prevent the cells that have phagocytosed apoptotic bodies from being transformed by the intact viral or damaged chromosomal DNA that the apoptotic bodies might otherwise contain.

d. The Death-Inducing Signal Complex Activates Apoptosis

Apoptosis in a given cell may be induced either by externally supplied signals in the so-called ***extrinsic pathway*** *(death by commission) or by the absence of external signals that inhibit apoptosis in the so-called* ***intrinsic pathway*** *(death by omission).* The extrinsic pathway is initiated by the association of a cell destined to undergo apoptosis with a cell that has selected it to so. In what is perhaps the best characterized such pathway (Fig. 34-112), a 281-residue, single-pass, transmembrane protein named **Fas ligand (FasL)** that projects from the plasma membrane of the inducing cell, a so-called **death ligand,** binds to a 335-residue, single-pass, transmembrane protein known as **Fas** (alternatively, **CD95** and **Apo1**) that projects from the plasma membrane of the apoptotic cell, a so-called **death receptor.** FasL is a cytokine that is predominantly expressed by certain immune system cells, including activated *T* cells (although the association between the apoptotic cell and the immune system cell is mainly mediated by antigen-containing complexes; Section 35-2E); it is a member of the **tumor necrosis factor (TNF)** family (so named because its founding member, **TNFα,** was originally characterized as a cytokine that kills tumor cells—but by inducing their apoptosis, not their necrosis).

FasL is a homotrimeric protein, whose extracellular C-terminal domains associate with the extracellular N-terminal domains of three Fas molecules to form a 3-fold symmetric complex, thereby causing the Fas cytoplasmic domains to trimerize. This is the triggering event of the extrinsic pathway; it can also be induced by cross-linking Fas molecules using antibodies. Fas, which is abundantly expressed in a variety of tissues, is a member of the **TNF receptor (TNFR)** family. Consequently, the arrangement of the FasL–Fas complex almost certainly resembles that observed in the X-ray structure of the **TNFβ** trimer in its complex with the extracellular domains from three **TNF**

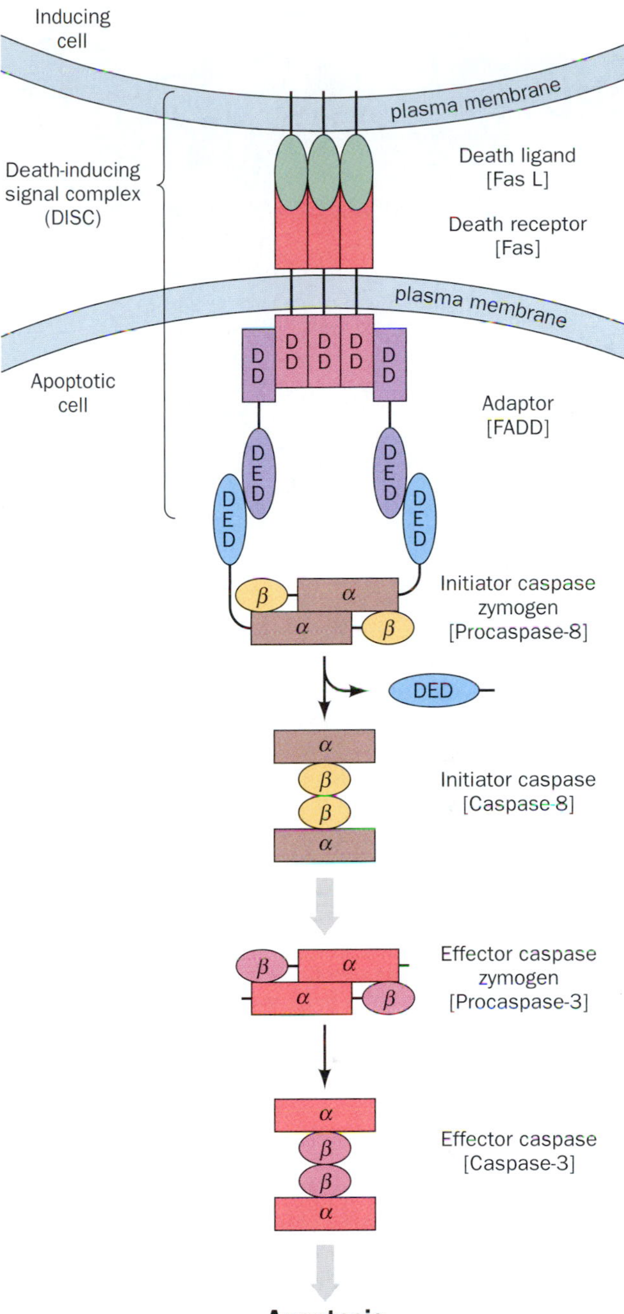

FIGURE 34-112 The extrinsic pathway of apoptosis. Flat arrows indicate activation. The binding of a trimeric death ligand (e.g., FasL) on the inducing cell to the death receptor (e.g., Fas) on the apoptotic cell causes the death receptor's cytoplasmic death domains (DDs) to trimerize. This recruits adaptors (e.g., FADD), which bind via their DDs to the DDs of the death receptor. The adaptors, in turn, recruit initiator procaspases (e.g., procaspase-8) via the interactions between the death effector domains (DEDs) on the adaptors and the initiator procaspases, which induces the autoactivation of the initiator procaspases to form the corresponding heterotetrameric initiator caspases (e.g., caspase-8). The initiator caspases then proteolytically activate effector procaspases (e.g., procaspase-3) to yield the heterotetrameric effector caspases (e.g., caspase-8), which catalyze the proteolytic cleavages, resulting in apoptosis.

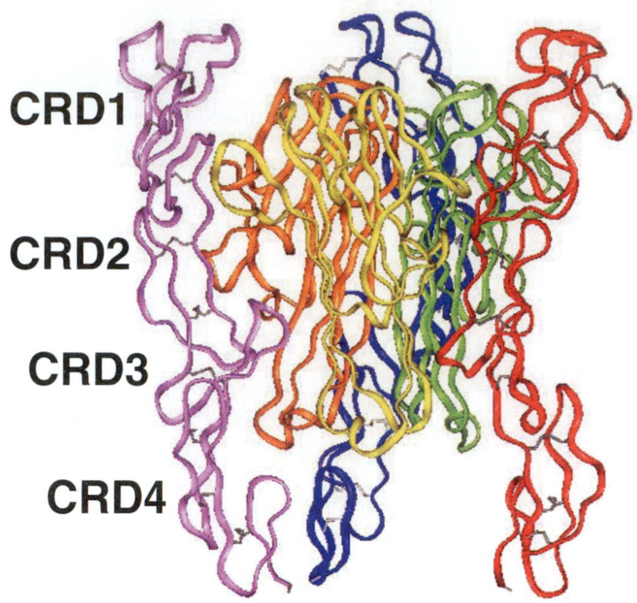

FIGURE 34-113 X-Ray structure of the TNFβ homotrimer in complex with the extracellular domains of three TNFR1 molecules. The centrally located TNFβ subunits are orange, yellow, and green, and the peripherally located TNFR1 domains are red, blue, and magenta. The TNFR1 domains each consist of four ~40-residue pseudorepeats known as **cysteine-rich domains (CRDs)** that each contain three disulfide bonds formed by six Cys residues that are drawn here in stick form (*gray*). The elongated TNFR1 domains each bind at an interface between two TNFβ subunits. This complex, whose formation also induces apoptosis, presumably resembles that between the extracellular domains of the homologous proteins FasL and Fas (whose extracellular domain contains only three CRDs). [Courtesy of Stephen Fesik, Abbott Laboratories, Abbott Park, Illinois. Based on an X-ray structure by David Banner, F. Hoffmann-La Roche Ltd., Basel, Switzerland. PDBid 1TNR.]

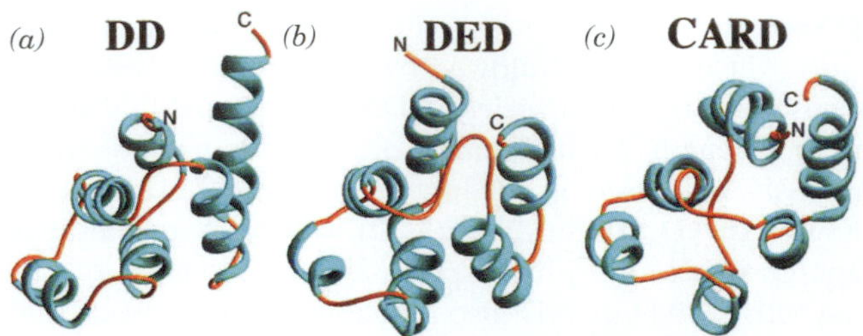

FIGURE 34-114 NMR structures of modules that transduce the death signal. (*a*) The death domain (DD) from Fas. (*b*) The death effector domain (DED) from FADD. (*c*) The caspase recruitment domain (CARD) from **RAIDD,** an adapter protein that is similar to FADD. Each of these domains consists of a bundle of six antiparallel α helices that associates with a domain of the same type but not with one of a different type. Nevertheless, the similarities of their structures and functions suggests that these domains are distantly related. [Courtesy of Stephen Fesik, Abbott Laboratories, Abbott Park, Illinois. Part *c* based on an NMR structure by Gerhard Wagner, Harvard University. PDBids 1DDF, 1A1Z, and 3CRD.]

receptor 1 (TNFR1) molecules that was determined by David Banner (Fig. 34-113). The cytoplasmic C-terminal domain of Fas consists mostly of an ~80-residue **death domain (DD)** that occurs in all of the six known mammalian death receptors (one of which is TNFR1), each of which are TNFR family members. The DD consists of six antiparallel, amphipathic α helices that have an unusual arrangement and whose structure resembles those of the death effector domain (DED) and the caspase recruitment domain (CARD), as Fig. 34-114 indicates.

Trimerized Fas recruits three molecules of the 208-residue adaptor protein known as **FADD** (for *F*as-*a*ssociating *d*eath *d*omain-containing protein; alternatively **MORT1** for *m*ediator *o*f *r*eceptor-induced *t*oxicity *1*) via interactions between FADD's C-terminal DD and that on Fas. The remaining portion of FADD consists almost entirely of a DED that, in turn, recruits procaspases-8 and -10 via the DEDs in their prodomains (Fig. 34-110) to form the **death-inducing signal complex (DISC).** The consequent clustering of the procaspases-8 and -10 molecules results in their proteolytic autoactivation, yielding caspases-8 and -10. These initiator caspases, in turn, activate the effector (executioner) caspase, caspase-3, whose actions cause the cell to undergo apoptosis.

Cells also express a protein named **c-FLIP** [for *c*ellular *FL*ICE *i*nhibitory *p*rotein; **FLICE** (for *F*ADD-*l*ike *ICE*) is an alternative name for procaspase-8] that resembles caspase-8 but is catalytically inactive. It associates with FADD via its two DEDs and thereby inhibits the autoactivation of caspases-8 and -10. FLIP apparently functions to dampen the cell's response to Fas so as to prevent inappropriate apoptosis. Certain herpes viruses and poxviruses encode **v-FLIPs** that function similarly to c-FLIPs to prevent apoptosis, thereby permitting the virus to propagate in the infected cell.

e. The Intrinsic Pathway Is Controlled by Bcl-2 Family Proteins

Most metazoan cells are continuously bathed in an extracellular soup, generated in part by neighboring cells, that contains a wide variety of cytokines that regulate the cell's growth, differentiation, activity, and survival. The withdrawal of this chemical support for its survival or the loss of direct cell–cell interactions induces a cell to undergo apoptosis via the intrinsic pathway. The initial step of this pathway (Fig. 34-115) appears to be the activation of one or more of the cell's several pro-apoptotic Bcl-2 family members.

The 15 known members of the ~180-residue mammalian Bcl-2 family have been classified into three groups (Fig. 34-116):

1. Group I members, which include Bcl-2 and **Bcl-x$_L$**, all have four short regions of homology, **BH1** to **BH4** (BH for *B*cl-2 *h*omology region), and a C-terminal hydrophobic segment that inserts into the outer mitochondrial membrane, or less frequently the endoplasmic reticulum, such that the bulk of these proteins face the cytosol. All Group I Bcl-2 family members are anti-apoptotic.

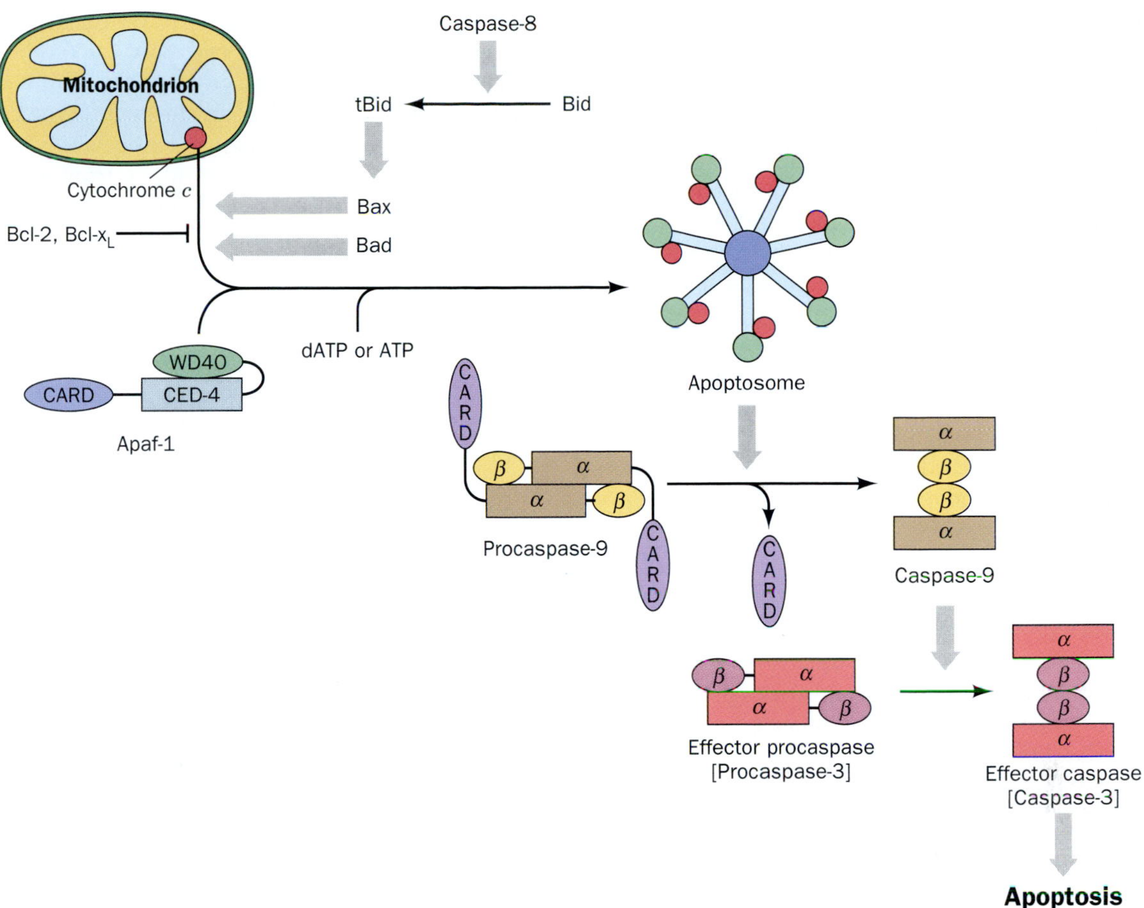

Figure 34-115 The intrinsic pathway of apoptosis. Flat arrows indicate activation and a blunted line indicates inhibition. A variety of stimuli or lack of them causes the mitochondrion to release cytochrome *c* from its intermembrane space. This process is induced by activated pro-apoptotic Bcl-2 family members such as Bid after it has been proteolytically cleaved by caspase-8 to yield tBid or dephosphorylated Bad; the process is inhibited by anti-apoptotic Bcl-2 family members such as Bcl-2 and Bcl-x_L. The liberated cytochrome c binds to Apaf-1, which on additionally binding dATP or ATP, forms the wheel-shaped heptameric apoptosome. The apoptosome binds procaspase-9, which it activates to cleave initiator procaspases (e.g., procaspase-3) to yield the corresponding effector caspases (e.g., caspase-3), which catalyze the proteolytic cleavages, resulting in apoptosis.

2. Group II members, which include **Bax** and **Bak,** resemble Group I proteins but lack a BH4 region. All Group II members are pro-apoptotic.

3. Group III members, which include **Bad, Bid, Bik, Bim,** and **Blk** (and the *C. elegans* protein EGL-1), all possess only one BH region, the ~15-residue BH3, and have

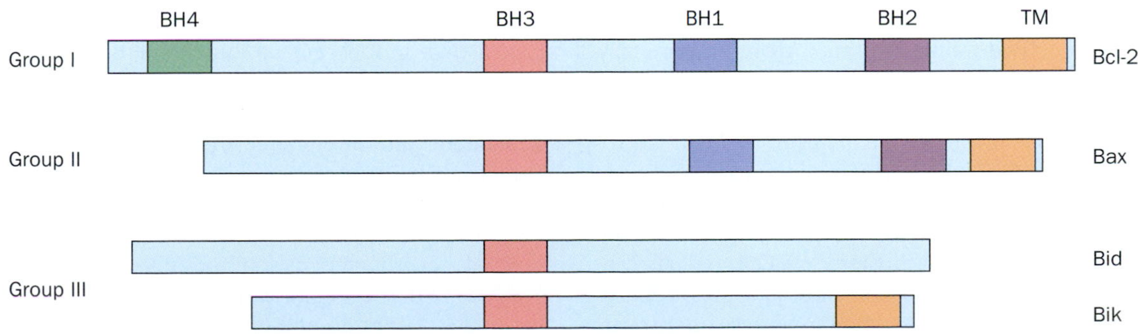

FIGURE 34-116 Sequence comparisons of members of the Bcl-2 family of proteins. The BH1 through BH4 homology regions are blue, purple, red, and green, respectively, and the hydrophobic transmembrane (TM) region is yellow. Group I proteins are anti-apoptotic, whereas Group II and Group III proteins are pro-apoptotic. [After a drawing by Michael Hengartner, *Nature* **407,** 770 (2000).]

no other sequence resemblance to Bcl-2. These so-called **BH3-only proteins** are all pro-apoptotic.

The activities of the various BH3-only proteins are controlled by specific posttranslational modifications. For example, Bad is phosphorylated at two Ser residues by protein kinase A (PKA; Section 18-3C), mitogen-activated protein kinase (MAPK; Section 19-3D), and Akt (Section 19-4D) (which themselves are activated by complex signal transduction pathways), thereby generating a binding site for 14-3-3 proteins that then sequester Bad in the cytosol. On appropriate stimulation, calcineurin and PP1 (Section 19-3F) dephosphorylate Bad, which permits it to interact with the mitochondrion, where it initiates apoptosis (see below). In contrast, Bid is activated by caspase-8–catalyzed proteolytic cleavage to **tBid** (*t*runcated *Bid*), thereby providing a link between the extrinsic and intrinsic pathways of apoptosis.

f. Cytochrome *c* Is an Essential Participant in the Intrinsic Pathway

The association of members of the pro-apoptotic Bcl-2 family with the mitochondrion causes it to release cytochrome c from its intermembrane space into the cytosol. There, as Xiaodong Wang unexpectedly discovered, this well-characterized component of the mitochondrial electron transport chain (Section 22-2C) functions to induce apoptosis. It does so by combining with Apaf-1 and dATP or ATP to form an ~1100-kD complex named the **apoptosome** (Fig. 34-115). The apoptosome binds several molecules of procaspase-9 in a manner that induces their autoactivation to yield caspase-9, which remains bound to the apoptosome. This caspase-9 then activates procaspase-3 to instigate cell death.

g. The Apoptasome Has a Wheel-like Structure

Apaf-1, the apoptosome's major component, is a 1248-residue scaffolding protein that consists of an N-terminal caspase recruitment domain (CARD), a central nucleotide-binding domain that is homolgous to CED-4, and a C-terminal domain that consists of seven WD40 repeats (Section 25-6B), a short linker, and six additional WD40 repeats. Procaspase-9 binds to Apaf-1's CARD in the apoptosome via its CARD, thereby placing several procaspase-9 molecules in close proximity such that they proteolytically activate one another. More importantly, however, caspase-9's association with the apoptosome increases its catalytic activity by three orders of magnitude, presumably via an allosteric mechanism. Indeed, a procaspase-9 mutant (D315A) that is noncleavable between its α and β domains, in complex with the apoptosome, nevertheless efficiently activates procaspase-3. Apaf-1's WD40 repeats function to bind cytochrome *c* (WD40 repeats usually participate in protein–protein interactions); their excision from Apaf-1 permits it to bind and activate procaspase-9 in the absence of cytochrome *c*. This suggests that Apaf-1's WD40 repeats bind to its CARD so as to prevent it from binding procaspase-9 but preferentially bind cytochrome *c* and thereby release the CARD. Evolution presumably selected cytochrome *c* for this function because it is normally absent from the cytosol.

The 27-Å resolution structure of the apoptosome, determined by Chistopher Akey through cryoelectron microscopy, reveals a wheel-like assembly with seven spokes ending in two lobes each that radiate from a central hub (Fig. 34-117*a*). Modeling studies suggest that the larger and smaller of these lobes respectively consist of a 7- and a 6-bladed β propeller (WD40 repeats form β propellers of various sizes; e.g., Fig. 19-18*b*) that are bridged by a cytochrome *c* molecule and that the Apaf-1 CARD occupies the apoptosome's hub region so that the nucleotide-binding domain must form at least the arm of the spoke. The cryoelectron microscopy–based image of the apoptosome in complex with the noncleavable D315A mutant of procaspase-9 revealed a previously unobserved domelike feature above the central hub that presumably represents the bound procaspase-9 (Fig. 34-117*b*). However, this dome is too small to accommodate seven procaspase-9 monomers, which suggests that the procaspase-9 in this complex is partially disordered, perhaps due to a flexible link between its CARD and its α domain.

h. Several Mechanisms Have Been Proposed for Mitochondrial Cytochrome *c* Release

The way in which pro-apoptotic Bcl-2 family members cause the mitochondrion to release its cytochrome *c* is unclear. However, based on largely circumstantial evidence, three models for this process, which are not mutually exclusive, have been proposed. The first model is based on the structural resemblance of the anti-apoptotic Bcl-2 family member Bcl-x_L to membrane-inserting bacterial toxins such as diphtheria toxin (Fig. 34-118; the mechanism of diphtheria toxin is discussed in Section 32-3G). This suggests that one or more molecules of Bcl-x_L and/or its homologs can insert into the outer mitochondrial membrane to form a pore. Indeed, Bcl-x_L, Bcl-2, and Bax have been shown to form pores in synthetic lipid bilayers. Moreover, tBid is targeted to mitochondria, where it triggers Bax oligomerization and its insertion into the outer mitochondrial membrane. However, it is unclear if the resulting pores are large enough to permit the passage of cytochrome *c*; in addition, it is not known how pores formed by anti-apoptotic proteins such as Bcl-x_L would inhibit apoptosis.

In the second model, Bcl-2 family members induce preexisting mitochondrial outer membrane proteins to form channels through which cytochrome *c* is released. An attractive candidate for such a protein is the **voltage-dependent anion channel (VDAC;** alternatively **mitochondrial porin)** because several Bcl-2 family members can bind to it and alter its channel activity. However, the known size of the VDAC channel is too small to allow the passage of cytochrome *c*, so that this model requires that VDAC undergo significant conformational change on binding Bcl-2 family members.

In the third model, Bcl-2 family members perturb or stabilize the preexisting pores through which ATP and ADP are exchanged between the mitochondrial matrix and the cytosol. This exchange process is mediated by the ATP–

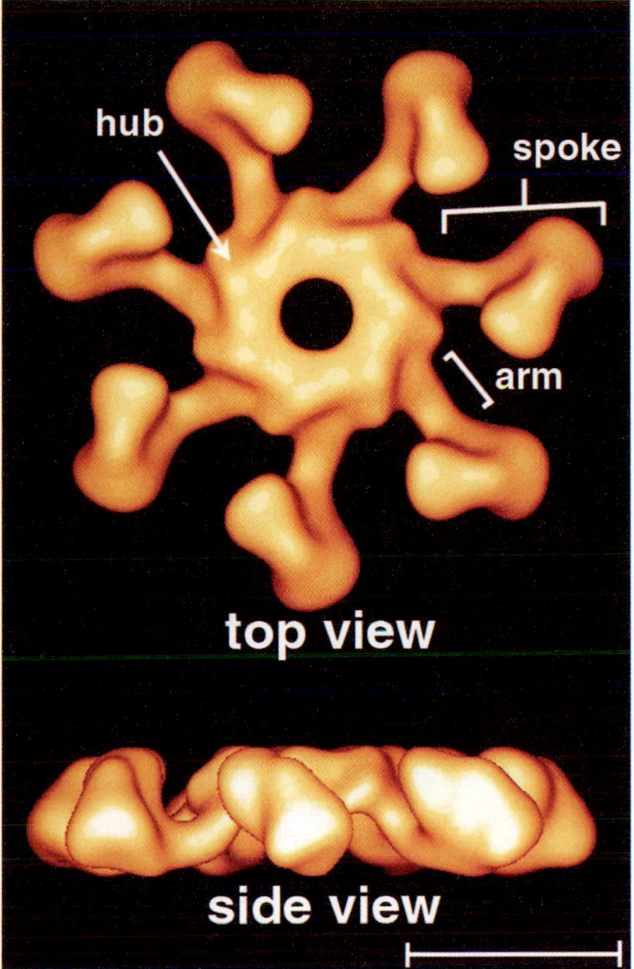

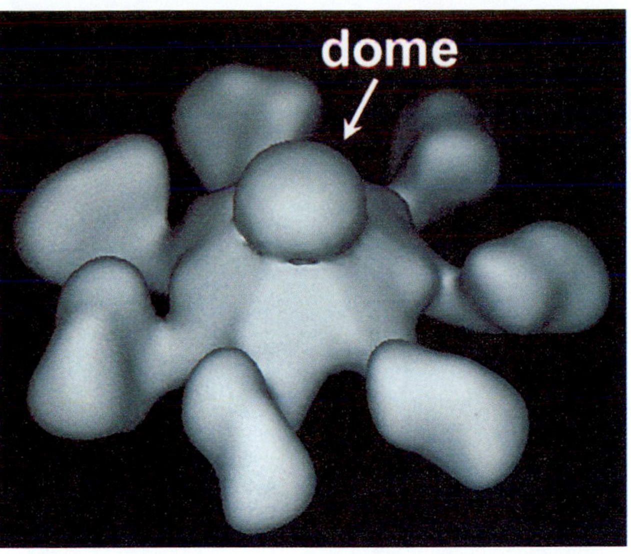

FIGURE 34-117 Cryoelectron microscopy–based images of the apoptosome at 27-Å resolution. (*a*) The free apoptosome. In its top view, the particle is viewed along its 7-fold axis of symmetry. The side view reveals the flattened nature of this wheel-like particle. The scale bar represents 100 Å. (*b*) The apoptosome in complex with a noncleavable mutant of procaspase-9 in oblique top view. Note the domelike feature above the central hub, which presumably represents the bound procaspase-9. [Courtesy of Christopher Akey, Boston University School of Medicine.]

ADP translocator in the inner mitochondrial membrane (Section 20-4C) and by VDAC in the outer mitochondrial membrane. The opening of these pores, it is postulated, would result in chemical equilibration between the cytosol and the matrix, which would cause the highly concentrated matrix to osmotically swell until it ruptured the outer mitochondrial membrane (which has considerably less surface area than the inner mitochondrial membrane due to the latter's cristae; Section 22-1A), thereby releasing cytochrome *c* into the cytosol. Consistent with this model, atractyloside and bongkregic acid (Section 20-4C), inhibitors of the ATP–ADP translocator that respectively open and close its pore, respectively induce and inhibit apoptosis. However, the rupture of the outer mitochondrial membrane has rarely been observed in cells undergoing apoptosis.

The way in which anti-apoptotic Bcl-2 family members antagonize the functions of pro-apoptotic family members is better understood. The members of these opposing factions readily form homodimers in which the BH3 region of the pro-apoptotic protein, which forms an amphipathic α helix, binds in a hydrophobic groove on the anti-apoptotic protein. Such an arrangement occurs in the NMR structure, determined by Fesik, of Bcl-x_L in its complex with the 16-residue segment from the BH3 region of the pro-apoptotic Bcl-2 protein Bak (Fig. 34-119). Since the BH3 regions of pro-apoptotic Bcl-2 proteins are necessary and probably even sufficient for their killing activity, the sequestering of

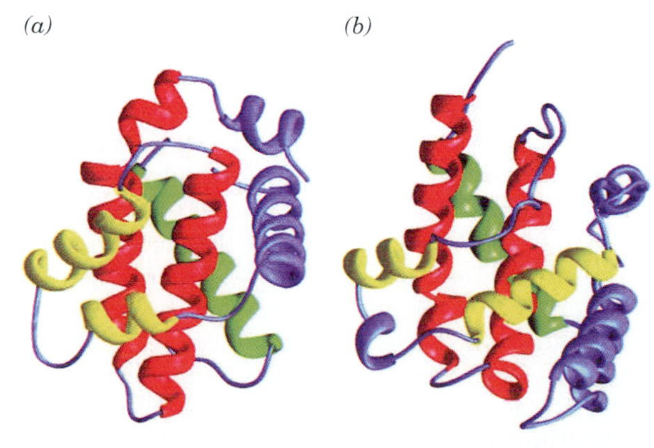

FIGURE 34-118 Comparison of the structures of (*a*) Bcl-x_L and (*b*) the pore-forming domain of diphtheria toxin. In both proteins, the two central hydrophobic helices (*red*) are surrounded by amphipathic α helices. [Courtesy of Stephen Fesik, Abbott Laboratories, Abbott Park, Illinois. Part *b* based on an X-ray structure by David Eisenberg, UCLA. PDBids 1LXL and 1DDT.]

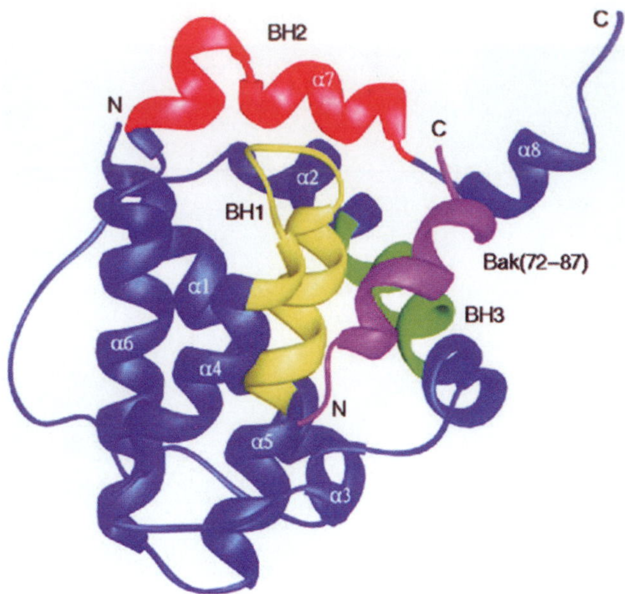

FIGURE 34-119 NMR structure of Bcl-x_L in complex with the 16-residue BH3 region of Bak. The BH1, BH2, and BH3 regions of Bcl-x_L are yellow, red, and green, respectively, and the Bak peptide is violet. [Courtesy of Stephen Fesik, Abbott Laboratories, Abbott Park, Illinois. PDBid 1BXL.]

these BH3 regions by anti-apoptotic Bcl-2 proteins at least partially explains their anti-apoptotic properties.

i. IAPs Regulate Apoptosis by Inhibiting Caspases

As might be expected, cells have elaborate systems that prevent their inadvertent apoptosis. We have previously discussed how the anti-apoptotic Bcl-2 family proteins keep their pro-apoptotic cousins in check, and how c-FLIP inhibits the death-inducing signal complex (DISC)-mediated activation of caspases-8 and -10. In addition, the members of the **IAP** (for *i*nhibitors of *ap*optosis) family of proteins, which are conserved from *Drosophila* to humans, regulate apoptosis by directly inhibiting caspases. Humans express eight IAPs, which vary in length from 236 to 4829 residues.

All IAPs have one to three ~70-residue **BIR domains** (for *b*aculovirus *I*AP *r*epeat; so named because they were discovered in the baculovirus protein **p35,** which functions to inhibit apoptosis in host cells during viral infection). BIRs contain a characteristic signature sequence, $CX_2CX_{16}HX_6C$, which forms a novel Zn^{2+}-binding motif. In addition, many IAPs (five in humans) have a C-terminal RING finger domain [a ubiquitin-protein ligase (E3); Section 32-6B]. **BIR2** [the second BIR domain of the human protein **XIAP** (for *X*-linked *IAP*)] and its surrounding regions specifically bind and inhibit the effector caspases-3 and -7, whereas **BIR3** and its surrounding regions do so for the initiator caspase-9; the function of **BIR1** is unknown. An IAP that has a RING finger domain may also ubiquitinate its bound caspase, thereby condemning it to destruction in the proteasome.

The X-ray structure of caspase-3 in complex with the XIAP BIR2 domain and its 38-residue N-terminal extension, determined by Fesik, Robert Liddington, and Guy Salvesen, reveals, unexpectedly, that the BIR2 domain makes only limited contacts with caspase-3 (Fig. 34-120). Rather, most of the contacts to caspase-3 are made by the N-terminal extension, which spans the enzyme's active site so as to sterically block substrate binding. Curiously, the N-terminal extension extends across that active site in the reverse direction relative to that taken by polypeptide inhibitors of caspases such as Ac-DEVD-CHO (Fig. 34-111). The structures of both caspase-3 and the BIR2 domain in the complex are largely unperturbed relative to their uncomplexed structures.

The induction of apoptosis requires that the inhibitory effects of IAPs on caspases be relieved. This is the task of a homodimeric protein that is alternatively named **Smac** (for *s*econd *m*itochondria-derived *a*ctivator of *c*aspases) and **DIABLO** (for *d*irect *IA*P-*b*inding protein with *lo*w pI). Smac/DIABLO, which binds to the BIR domains of IAPs so as to prevent them from binding to caspases, is released from the mitochondrion together with cytochrome *c*, thereby ensuring that the intrinsic pathway of apoptosis will generate active caspases.

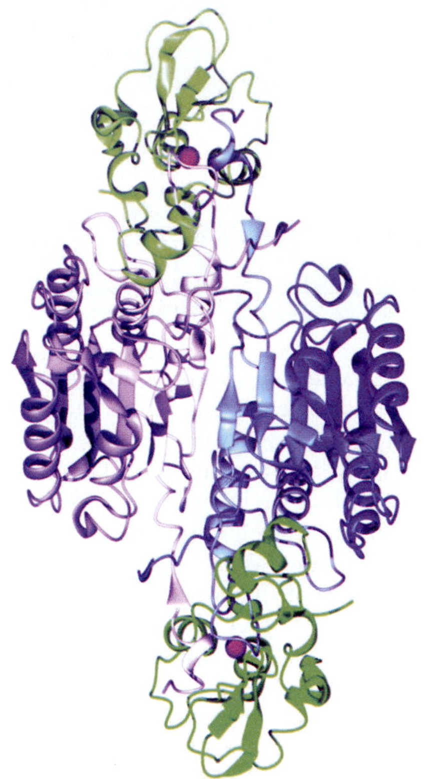

FIGURE 34-120 X-Ray structure of caspase-3 in complex with the XIAP BIR2 domain and its N-terminal extension. The complex is viewed along its 2-fold axis (as is caspase-7 in Fig. 34-111). The two αβ protomers of the heterotetrameric caspase-3 are blue and purple, with its β subunits shaded more lightly than its α subunits. The globular BIR2 domains are green and their bound Zn^{2+} ions represented by magenta spheres. Each of their N-terminal extensions, which begin with an α helix, are bound across the active site face of a caspase protomer, thereby blocking the binding of substrate proteins. [Courtesy of Guy Salvesen, The Burnham Institute, La Jolla, California. PDBid 1I3O.]

CHAPTER SUMMARY

1 ■ **Chromosome Structure** There are two types of eukaryotic chromatin: euchromatin, which can be transcriptionally active, and the more densely packed heterochromatin, which is transcriptionally inactive. Chromatin consists of DNA and proteins, the majority of which are the highly conserved histones. Chromatin is structurally organized in a hierarchical manner. In the first level of chromatin organization, ~200 bp of DNA are doubly wrapped around a histone octamer, $(H2A)_2(H2B)_2(H3)_2(H4)_2$, to form a nucleosome. Each nucleosome is associated with one molecule of histone H1. The passage of transcribing RNA polymerase causes nucleosomes to dissociate from the DNA and then to rebind it in a process that appears to be driven by supercoiling. DNA replication causes the parental nucleosomes to be randomly distributed between the daughter duplexes. The assembly of nucleosomes from their components is mediated by the molecular chaperone nucleoplasmin. In the second level of chromatin organization, the nucleosome filaments coil into 30-nm-thick filaments that probably contain six nucleosomes per turn. Then, in the third and final level of chromatin organization, the 30-nm-thick filaments form 15- to 30-μm-long radial loops that project from the axis of the metaphase chromosome. This accounts for DNA's packing ratio of >8000 in the metaphase chromosome. The larvae of certain dipteran flies, including *Drosophila,* contain banded polytene chromosomes, which consist of up to 1024 identical DNA strands in parallel register.

2 ■ **Genomic Organization** The complexity of a DNA sample can be determined from its renaturation rate through C_0t curve analysis. Eukaryotic DNAs have complex C_0t curves that arise from the presence of unique, moderately repetitive and highly repetitive sequences, as well as from inverted repeats. The function of inverted repeats, which form foldback structures, is unknown. Highly repetitive sequences, which occur in the heterochromatic regions near the chromosomal centromeres, probably function to align homologous chromosomes during meiosis and/or to facilitate their recombination. Moderately repetitive DNAs, which consist largely of inactive retrotransposons, mainly LINEs and SINEs, make up ~42% of the human genome. For the most part, they have unknown functions; they may simply be selfish DNA. Expressed sequences comprise only 1.1 to 1.4% of the human genome, which has made the identification of genes an uncertain process. However, CpG islands are often associated with the 5′ ends of genes.

Around 30,000 putative genes have been identified in the human genome. Some of them are transcribed to noncoding RNAs. The structural genes in the human genome have been classified according to function through sequence comparisons. Around 42% of them have unknown functions. The genes specifying rRNAs and tRNAs are organized into tandemly repeated clusters. The rDNA condenses to form nucleoli, the sites of rRNA transcription by RNA polymerase I and of partial ribosomal assembly. The 5S RNA and tRNAs are transcribed outside the nucleoli by RNA polymerase III. The genes specifying histones, which are required in large quantities only during S phase of the cell cycle, are the only repeated structural genes. The identity of a series of repeated genes is probably maintained through unequal crossing-over and/or gene conversion. Certain genes are amplified such as the *Xenopus* genes for rRNAs during oogenesis, chorion genes in *Drosophila,* and genes targeted by cancer chemotherapy.

Many families of genes specifying related proteins are clustered into gene families. In mammals, the gene clusters encoding the α- and β-like hemoglobin subunits occur on separate chromosomes. Nevertheless, all vertebrate globin genes have the same exon–intron structure: three exons separated by two introns. The thalassemias are inherited diseases caused by the genetic impairment of hemoglobin synthesis. Many α-thalassemias are caused by the deletion of one or more of the α-globin genes, whereas many β-thalassemias arise from point mutations that affect the transcription or the posttranscriptional processing of the β-globin mRNAs.

3 ■ **Control of Expression** Heterochromatin may be subclassified as constitutive heterochromatin, which is never transcriptionally active, and facultative heterochromatin, whose activity varies in a tissue-specific manner. The Barr bodies in the cells of female mammals constitute a common form of facultative heterochromatin: One of each cell's two X chromosomes is permanently condensed via the binding of *Xist* RNA and epigenetically confers its state of inactivity on its progeny through histone modification and DNA methylation. Active chromatin has a relatively open structure that makes it available to the transcriptional machinery. Two well-characterized examples of transcriptionally active chromatin are the chromosome puffs that emanate from single bands in polytene chromosomes and the lampbrush chromosomes of amphibian oocytes.

The differential protein synthesis characteristic of the cells in a multicellular organism largely stems from the selective transcription of the expressed genes. The first step in the transcriptional initiation of RNAP II-transcribed genes is often the binding of TATA-box binding protein (TBP) to the promoter's TATA box, which is located around position −27. This is followed by the addition of TBP-associated factors (TAFs) to form transcription factor IID (TFIID), together with general transcription factors (GTFs) and RNAP II to form the preinitiation complex (PIC), which is capable of a basal rate of transcription. Several TAFs assume the histone fold and associate with one another in TFIID, as do hisones in the histone octamer. TBP, together with other GTFs, is also required for the transcriptional initiation of class I and III genes. Promoters of class II genes that lack a TATA box often contain an initiator (Inr) sequence that spans the transcription start site and may also have a downstream promoter element (DPE).

The cell-specific expression of a gene is mediated by the gene's promoter and enhancer elements. Consequently, cells contain specific upstream transcription factors that recognize these genetic elements. For example, Sp1 binds to the GC box that precedes many genes. Likewise, steroid hormones bind to their cognate receptors, which in turn bind to specific enhancers so as to modulate the transcriptional activitiy of the associated gene. The cooperative binding of several transcriptional factors to their target promoter and enhancer sites stimulates the associated PIC to increase the rate at which it initiates the transcription of the associated gene. The binding of transcription factors to a silencer represses the transcription of the associated gene. Several transcription factors may bind an enhancer and associate with architectural factors and coactivators to form an enhanceosome. Many transcription factors have two domains, a DNA-binding domain targeted to a specific sequence, and an activation domain, which interacts with the PIC in a largely nonspecific manner, often via a negatively charged surface region. Eukaryotic transcription factors have a great variety of DNA-binding motifs, including several types of zinc fingers, the bZIP motif, and the bHLH/Z motif. Many transcription factors, including those with the latter two types of motif, dimerize through the formation of a leucine zipper. Nuclear factor κB (NF-κB) is activated in the cytoplasm by the destruction of its bound inhibitor IκB, whereupon NF-κB is translocated to the nucleus, where it binds to a κB DNA segment so as to activate the associated gene. Mediator is an ~20-subunit yeast complex that binds to the RNAP II β′ subunit's C-terminal domain (CTD), where it influences RNAP II's activity through its binding of DNA-bound transcriptional regulators. Metazoa contain several mediator-like complexes that presumably relay signals from different sets of transcriptional activators.

Transcriptionally poised or active genes contain nuclease-hypersensitive sites that occur in nucleosome-free regions of DNA. Nuclease hypersensitivity is conferred on DNA by the binding of specific proteins that presumably make the genes accessible to the proteins mediating transcriptional initiation. This is largely due to the presence of the nonhistone, DNA-binding, high mobility group (HMG) proteins, which are architectural proteins that function to activate gene expression by decondensing chromatin and recruiting transcription factors. RNA polymerase transcribes through nucleosomes by inducing the histone octamers it encounters to step around it. Locus control regions (LCRs), which are DNase I hypersensitive sites that function to suppress position effects, that is, the encroachment of heterochromatin, are activated by proteins that are expressed only in specific cell lineages such as erythroid cells. Insulators are DNA segments that, through the binding of specific proteins, inhibit heterochromatin from spreading into neighboring segments of euchromatin and prevent regulatory elements outside the region controlled by the insulator from influencing the expression of the genes inside the region.

The transcriptional machinery gains access to the DNA packaged by chromatin through the posttranscriptional modification of the core histones' N-terminal tails and the ATP-driven remodeling of chromatin. The modifications to which histone N-terminal tails are subject include the acetylation/deacetylation of specific Lys side chains, the methylation of specific Lys and Arg side chains, the phosphorylation/dephosphorylation of specific Ser side chains, and the ubiquitination of specific Lys side chains. There appears to be a histone code in which specific modifications, acting sequentially or in combination, evoke certain chromatin-based functions that result in unique biological outcomes such as transcriptional activation or silencing. In addition, some of these histone modifications may act as epigenetic markers through which cells confer their identities on their progeny. Histone acetylation is catalyzed by histone acetyltransferases (HATs) that are components of multisubunit transcriptional activators such as SAGA, PCAF, and TFIID, which all contain histonelike TAFs. Nearly all HAT-associated coactivators contain bromodomains that specifically bind acetylated histone Lys residues and hence are likely to recruit HATs to acetylate the N-terminal tails of nearby nucleosomes. Histone deacetylases (HDACs), many of which are also members of multisubunit complexes, serve as transcriptional corepressors. Histone methylation, which is largely if not entirely irreversible, is catalyzed by histone methyltransferases (HMTs), which often function as transcriptional corepressors. Methylated histones are recognized by chromodomains such as that in heterochromatin protein 1 (HP1). The spreading of heterochromatin appears to be mediated, at least in part, by HP1's recruitment of the HMT Suv39, which methylates nearby nucleosomes such that additional HP1 can bind to them, etc. Histone monoubiquitination functions as an essential transcriptional regulator. Chromatin-remodeling complexes, such as yeast SWI/SNF and RSC, contain helicase-like ATPases that, it appears, "walk" up the DNA in a nucleosome so as to decrease its helical twist. The resulting DNA distortion, it is postulated, diffuses around the nucleosome in a wave that locally and transiently releases the DNA from the histone octamer, thereby permitting the nucleosome to slide along the DNA and providing transcriptional activators access to their target sequences that would otherwise be sequestered by the nucleosome.

Other forms of selective gene expression in eukaryotes include the use of alternative initiation sites in a single gene, the selection of alternative splice sites, the possible regulation of mRNA translocation across the nuclear membrane, the control of mRNA degradation, the control of translational initiation rates, and the selection of alternative posttranslational processing pathways.

4 ■ Cell Differentiation and Growth Embryogenesis occurs in four stages: cleavage, gastrulation, organogenesis, and maturation and growth. One of the most striking characteristics of embryological development is that cells become progressively and irreversibly committed to specific lines of development. The signals that trigger developmental changes, which are recognized over great evolutionary distances, may be transmitted through direct intercellular contacts or from the gradients of substances, known as morphogens, released by other embryonic cells. Developmental signals act combinatorially; that is, the developmental fate of a specific tissue is determined by several not necessarily unique developmental stimuli. In *Drosophila,* early embryonic development is governed by maternal-effect genes whose distribution imposes the embryo's spatial coordinate system. These encode transcription factors that regulate the expression of gap genes, which in turn regulate the expression of pair-rule genes, which in turn regulate the expression of segment polarity genes. Sequentially finer domains of the embryonic body are thereby defined in a way that specifies the number and polarity of the larval and adult body segments. Homeotic selector or *Hox* genes, whose mutations transform one body part into another, then regulate the differentiation of the individual segments. These regulatory genes, which occur in two gene clusters, are, as the preceding genes, selectively expressed in the embryonic tissues whose development they control. They have closely related base sequences that encode ~60-residue polypeptide segments known as homeodomains, which bind their target DNA sequences in a manner similar to but distinct from that of the homologous HTH module. In vertebrates, *Hox* genes occur in four clusters and likewise control development.

Cancers result from specific genetic alterations to cells. The types of genetic changes that give rise to malignancies include the generation of altered proteins such as a Ras variant that lacks GTPase activity; altered regulatory sequences that, for example, result in the overexpression of key transcription factors; the loss of degradation signals that cause an oncogene protein such as Jun to be degraded abnormally slowly; chromosomal rearrangements that place proto-oncogenes such as c-*myc* under the control of inappropriately active regulatory sequences; gene amplification that results in the overexpression of a proto-oncogene; the insertion of a virus into a chromosome such that a proto-oncogene is brought under the control of viral regulatory sequences; the inappropriate activation or inactivation of chromatin modification enzymes such as HAT and HDAC; and the loss or inactivation of tumor suppressor genes such as those encoding p53 and pRb. The mutations causing these gene alterations often arise from the actions of carcinogens on cellular DNA.

The progression of a cell through the cell cycle is regulated mainly by the presence of the appropriate mitogens together with a series of checkpoints that monitor the cell's health as well as its progress through the cell cycle. Checkpoints arrest the cell cycle until the proper conditions for its progression are met, for example, that the replication of DNA has been successfully completed and it is undamaged. The cell cycle is characterized by the accumulation of cyclins, which abruptly disappear at the end of mitosis. For example, M phase is induced when cyclin B combines with Cdc2 to form cyclin-dependent kinase 1 (Cdk1), which is preceded by Cdc2's phosphorylation at Thr 161 by Cdk-activating kinase (CAK) and succeeded by Cdc2's inactivating phosphorylation at Thr 14 and Tyr 15 by Wee1 and Myt1. At the G_2/M boundary, Thr 14 and Tyr 15 are rapidly dephosphorylated by Cdc25C, yielding active Cdk1, which phosphorylates a variety of nuclear proteins. The structures of Cdks resemble those of other protein kinases. However, cyclin binding to a Cdk and its phosphorylation at Thr 160 conformationally reorganize its active site. Members of the Kip/Cip family, such as $p21^{Cip1}$, inhibit most Cdk–cyclin complexes except Cdk4/6–cyclin D, whereas members of the INK4 family, such as $p16^{INK4a}$, inhibit Cdk4/6 –cyclin D. Cell cycle arrest at the G_2 checkpoint is initiated by sensor proteins that bind to damaged and unreplicated DNA. These activate ATM and ATR to respectively phosphorylate Chk2 and Chk1, which phosphorylate Cdc25C, thereby providing a binding site for 14-3-3 proteins such that Cdc25C is sequestered in the cytoplasm, where it cannot dephosphorylate and hence activate Cdc2.

p53, a tumor suppressor that is implicated in 50% of human cancers, is bound by Mdm2, which ubiquitinates it so as to mark it for destruction in the proteasome. Hence the cell normally has a low level of p53. However, when p53 is phosphorylated by ATM or Chk2, it no longer binds to Mdm2 and thereupon transactivates the expression of $p21^{Cip1}$, which binds to several Cdk–cyclin complexes and to PCNA, thereby inhibiting both the G_1/S and G_2/M transitions and DNA replication. If the cell is irreparably damaged, p53 induces it to undergo apoptosis, thereby preventing the proliferation of potentially cancerous cells. The X-ray structure of p53's DNA-binding core in complex with its target DNA reveals that many of its residues that participate in DNA binding are frequently mutated in tumors. p53 is also activated by a variety of pathways such as MAP kinase cascades, the activation of ATR by DNA-damaging agents, and a variety of posttranslational modifications. Thus, p53 is the recipient of numerous intracellular signals and activates a variety of downstream regulators.

The tumor suppressor pRb, a regulator of the cell cycle's G_1/S transition, functions by inhibiting E2F, a transcription factor for many proteins required for S-phase entry. pRb, a phosphoprotein that is phosphorylated at numerous Ser/Thr sites by various Cdk–cyclin complexes, must be in its hypophosphorylated form to bind E2F. A variety of proteins that have an LXCXE sequence motif bind to pRb at a separate site from E2F on pRb's pocket domain, a major site of genetic alteration in tumors. Viral proteins with the LXCXE motif, including adenovirus E1A and papillomavirus E7, cause pRb to release its bound E2F, thereby driving the infected cell into S phase, which facilitates viral DNA replication. The histone deacetylases HDAC1 and HDAC2 each contain an LXCXE motif, which suggests that pRb functions to recruit these proteins to E2F's target promoters, thereby deactivating them. The SWI/SNF homologs BRM and BRG1, which both have LXCXE motifs, can bind to pRb simultaneously with HDACs, which suggests that these chromatin-remodeling complexes are recruited to E2F promoters, where they facilitate the action of HDACs.

Apoptosis (programmed cell death) occurs normally during embryogenesis and in many adult processes. In fact, it is the default option for metazoan cells. Insufficient apoptosis can cause autoimmune diseases and cancer, whereas inappropriate apoptosis is responsible for several neurodegenerative diseases and much of the damage caused by stroke and heart attacks. In apoptosis, the cell dismantles itself in an orderly program to yield membrane-enclosed apoptotic bodies that are phagocytosed by surrounding cells without inducing an inflammatory response. The executioners in apoptosis are cysteine proteases known as caspases that specifically cleave polypeptides after Asp residues. Caspases are synthesized as zymogens called procaspases that are proteolytically activated via apoptotic pathways, ending in the activation of initiator caspases that activate effector caspases, which cleave a wide variety of cellular proteins. Among the latter is ICAD, which is an inhibitor of caspase-activated DNase (CAD) that, in the absence of ICAD, functions to fragment the cell's DNA.

In the extrinsic pathway of apoptosis (death by commission), a trimeric transmembrane cytokine of the tumor necrosis factor (TNF) family, such as Fas ligand (FasL), which is on the inducing cell, binds to a transmembrane so-called death receptor of the TNFR family, such as Fas, which is on the apoptotic cell. The binding of trimeric ligand to a death receptor causes its cytoplasmic death domain (DD) to form a trimer to which three molecules of the adaptor protein FADD then bind via their DDs. FADD, in turn, recruits procaspases-8 and -10 via interactions between the two proteins' death effector domains (DEDs) to form the death-inducing signal complex (DISC). This results in the proteolytic autoactivation of the bound procaspases-8 and -10, which then activate procaspase-3, an effector caspase. In the intrinsic pathway (death by omission), pro-apoptotic members of the Bcl-2 family are activated in various ways, including the withdrawal of cytokines and contact with other cells, to induce the release of cytochrome *c* from the mitochondrion. The cytochrome c binds to the scaffolding protein Apaf-1 to form a wheel-shaped heptameric complex called the apoptosome. The apoptosome binds several molecules of procaspase-9 through interactions between the two proteins' CARD domains, which activates procaspase-9 to activate procaspase-3. Pro-apoptotic Bcl-2 family members are kept in check through their heterodimerization with anti-apoptotic Bcl-2 family members. In addition, the members of the IAP family inhibit apoptosis by directly binding to caspases so as to block their active sites and, in some cases, also ubiquitinating them so as to mark them for destruction in the proteasome. Smac/DIABLO, which is released from the mitochondrion together with cytochrome *c*, reverses this inhibition by binding to IAPs, thereby permitting apoptosis to commence.

REFERENCES

GENERAL

Alberts, B., Johnson, A., Lewis, J., Raff, M., Roberts, K., and Walter, P., *Molecular Biology of the Cell* (4th ed.), Chapters 4, 7, 17, 21, and 23, Garland Publishing (2002).

Brown, T.A., *Genomes 2,* Chapters 1, 2, 8, 9, and 12, Wiley-Liss (2002).

Elgin, S.C.R. and Workman, J.L. (Eds.), *Chromatin Structure and Gene Expression* (2nd ed.), Oxford University Press (2000).

Lewin, B., *Genes VII,* Chapters 18–21, Oxford (2000).

Lodish, H., Berk, A., Zipursky, S.L., Matsudaira, P., Baltimore, D., and Darnell, J., *Molecular Cell Biology* (4th ed.), Chapters 9, 10, 13, and 14, Freeman (2000).

Sumner, A.T., *Chromosomes, Organization and Function,* Blackwell Science (2003).

Wolffe, A., *Chromatin, Structure and Function,* Academic Press (1998).

CHROMOSOME STRUCTURE

Bustin, M., Chromatin unfolding and activation by HMGN chromosomal proteins, *Trends Biochem. Sci.* **26,** 431–437 (2001).

Carey, M. and Smale, S.T., *Transcriptional Regulation in Eukaryotes. Concepts, Strategies, and Techniques,* Cold Spring Harbor Laboratory Press (2000). [A comprehensive guide to the methods used in analyzing transcriptional regulatory mechanisms.]

Cohen, D.E. and Lee, J.T., X-Chromosome inactivation and the search for chromosome-wide silencers, *Curr. Opin. Genet. Dev.* **12,** 219–224 (2002).

Earnshaw, W.C., Large scale chromosome structure and organization, *Curr. Opin. Struct. Biol.* **1,** 237–244 (1991).

Felsenfeld, G. and McGhee, J.D., Structure of the 30 nm chromatin fiber, *Cell* **44,** 375–377 (1986).

Hansen, J.C., Conformation dynamics of the chromatin fiber in solution: Determinants, mechanisms, and functions, *Annu. Rev. Biophys. Biomol. Struct.* **31,** 361–392 (2002).

Harp, J.M., Hanson, B.L., Timm, D.E., and Bunick, G.J., Asymmetries in the nucleosome core particle at 2.5 Å resolution, *Acta Cryst.* **D56,** 1513–1534 (2000).

Kornberg, R.D., Chromatin structure: a repeating unit of histones and DNA, *Science* **184,** 868–871 (1974). [The classic paper first indicating the constitution of nucleosomes.]

Kornberg, R.D. and Lorch, Y., Twenty-five years of the nucleosome, fundamental particle of the eukaryotic chromosome, *Cell* **98,** 285–295 (1999).

Locker, J. (Ed.), *Transcription Factors,* Academic Press (2001).

Luger, K., Mäder, A.W., Richmond, R.K., Sargent, D.F., and Richmond, T.J., Crystal structure of the nucleosome particle at 2.8 Å resolution, *Nature* **389,** 251–260 (1997); Davey, C.A., Sargent, D.F., Luger, K., Maeder, A.W., and Richmond, T.J., Solvent mediated interactions in the structure of the nucleosome core particle at 1.9 Å resolution, *J. Mol. Biol.* **319,** 1097–1113 (2002); *and* Davey, C.A. and Richmond, T.J., The structure of DNA in the nucleosome core, *Nature* **423,** 145–150 (2003).

Luger, K., Structure and dynamic behaviour of nucleosomes, *Curr. Opin. Genet. Dev.* **13,** 127–135 (2003); *and* Akey, C.W. and Luger, K., Histone chaperones and nucleosome assembly, *Curr. Opin. Struct. Biol.* **13,** 6–14 (2003).

Masse, J.E., Wong, B., Yen, Y.-M., Allain, F.H.-T., Johnson, R.C., and Feigon, J., The *S. cerevisiae* architectural HMGB protein NHP6A complexed with DNA: DNA and protein conformational changes upon binding, *J. Mol. Biol.* **323,** 263–284 (2002).

Merika, M. and Thanos, D., Enhanceosomes, *Curr. Opin. Genet. Dev.* **11,** 205–208 (2001).

Ramakrishnan, V., Histone structure, *Curr. Opin. Struct. Biol.* **4,** 44–50 (1994).

Ramakrishnan, V., Finch, J.T., Graziano, V., Lee, P.L., and Sweet, R.M., Crystal structure of globular domain of histone H5 and its implications for nucleosome binding, *Nature* **362,** 219–223 (1993); *and* Ramakrishnan, V., Histone H1 and chromatin higher-order structure, *Crit. Rev. Euk. Gene Exp.* **7,** 215–230 (1997).

Reeves, R. and Beckerbauer, L., HMGI/Y proteins: flexible regulators of transcription and chromatin structure, *Biochim. Biophys. Acta* **1519,** 13–29 (2001).

Thomas, J.O. and Travers, A.A., HMG1 and 2, and related 'architectural' DNA-binding proteins, *Trends Biochem. Sci.* **26,** 167–174 (2001).

van Holde, K. and Zlatanova, J., Chromatin higher order structure: Chasing a mirage? *J. Biol. Chem.* **270,** 8373–8376 (1995). [Presents the arguments that the 30-nm chromatin filament has an irregular structure.]

Weatherall, D.J., Clegg, J.B., Higgs, D.R., and Wood, W.G., The hemoglobinopathies, *in* Scriver, C.R., Beaudet, A.L., Sly, W.S., and Valle, D. (Eds.), *The Metabolic & Molecular Bases of Inherited Disease* (8th ed.), *pp.* 4571–4636, McGraw-Hill (2001). [Contains a discussion of the thalassemias.]

White, C.L., Suto, R.K., and Luger, K., Structure of the yeast nucleosome core particle reveals fundamental changes in internucleosome interactions, *EMBO J.* **20,** 5207–5218 (2001).

Widom, J., Structure, dynamics, and function of chromatin in vitro, *Annu. Rev. Biophys. Biomol. Struct.* **27,** 285–327 (1998).

GENOMIC ORGANIZATION

Berry, M., Grosveld, F., and Dillon, N., A single point mutation is the cause of the Greek form of hereditary persistence of fetal hemoglobin, *Nature* **358,** 499–502 (1992).

Craig, N.L., Craigie, R., Gellert, M., and Lambowitz, A.M., *Mobile DNA II,* Chapters 35, 47, 48, and 49, ASM Press (2002). [Discussions of transposable elements in eukaryotic genomes.]

Deininger, P.L., Batzer, M.A., Hutchinson, C.A., III, and Edgell, M.H., Master genes in mammalian repetitive DNA amplification, *Trends Genet.* **8,** 307–311 (1992).

Hamlin, J.L., Leu, T.-H., Vaughn, J.P., Ma, C., and Dijkwel, P.A., Amplification of DNA sequences in mammalian cells, *Prog. Nucleic Acid Res. Mol. Biol.* **41,** 203–239 (1991).

International Human Genome Sequencing Consortium, Initial sequencing and analysis of the human genome, *Nature* **409,**

860–921 (2001); *and* Venter, J.C., et al., The sequence of the human genome, *Science* **291,** 1304–1351 (2001). [The landmark papers describing the base sequence of the human genome. They contain descriptions of repeating elements and gene distributions in the human genome.]

Kafatos, F.C., Orr, W., and Delidakis, C., Developmentally regulated gene amplification, *Trends Genet.* **1,** 301–306 (1985).

Li, W.-H., Gu, Z., Wang, H., and Nekrutenko, A., Evolutionary analysis of the human genome, *Nature* **409,** 847–849 (2001). [A survey of repetitive elements in the human genome.]

Mandal, R.K., The organization and transcription of eukaryotic ribosomal RNA genes, *Prog. Nucleic Acid Res. Mol. Biol.* **31,** 115–160 (1984).

Maxson, R., Cohn, R., and Kedes, L., Expression and organization of histone genes. *Annu. Rev. Genet.* **17,** 239–277 (1983).

Orgel, L.E. and Crick, F.H.C., Selfish DNA: the ultimate parasite, *Nature* **284,** 604–607 (1980).

Orr-Weaver, T.L., *Drosophila* chorion genes: Cracking the eggshell's secrets, *BioEssays* **13,** 97–105 (1991).

Saccone, C. and Pesole, G., *Handbook of Comparative Genomics. Principles and Methods,* Wiley-Liss (2003).

Schimke, R.T., Gene amplification in cultured cells, *J. Biol. Chem.* **263,** 5989–5992 (1988).

Stamatoyannopoulos, G., Majerus, P.W., Permutter, R.M., and Varmus, H., (Eds.), *The Molecular Basis of Blood Diseases* (3rd ed.), Chapters 2–5, Elsevier (2001). [Discusses hemoglobin genes and their normal and thalassemic expression.]

Südhof, T.C., Goldstein, J.L., Brown, M.S., and Russell, D.W., The LDL receptor gene: a mosaic of exons shared with different proteins, *Science* **228,** 815–828 (1985).

Wainscoat, J.S., Hill, A.V.S., Boyce, A.L., Flint, J., Hernandez, M., Thein, S.L., Old, J.M., Lynch, J.R., Falusi, A.G., Weatherall, D.J., and Clegg, J.B., Evolutionary relationships of human populations from an analysis of nuclear DNA polymorphisms, *Nature* **319,** 491–493 (1986).

CONTROL OF EXPRESSION

Adams, C.C. and Workman, J.L., Nucleosome displacement in transcription, *Cell* **72,** 305–308 (1993).

Andel, F., III, Ladurner, A.G., Inouye, C., Tjian, R., and Nogales, E., Three-dimensional structure of the human TFIID–IIA–IIB complex, *Science* **286,** 2153–2156 (1999).

Andres, A.J. and Thummel, C.S., Hormones, puffs and flies: the molecular control of metamorphosis by ecdysone, *Trends Genet.* **8,** 132–138 (1992).

Ashburner, M., Puffs, genes, and hormones revisited, *Cell* **61,** 1–3 (1990).

Asturias, F.J., Chung, W-H., Kornberg, R.D., and Lorch, Y., Structural analysis of the RSC chromatin-remodeling complex, *Proc. Natl. Acad. Sci.* **99,** 13477–13480 (2002).

Bach, I. and Ostendorff, H.P., Orchestrating nuclear functions: Ubiquitin sets the rhythm, *Trends Biochem. Sci.* **28,** 189–195 (2003); *and* Muratani, M. and Tansey, W.P., How the ubiquitin–proteasome system controls transcription, *Nature Rev. Mol. Cell Biol.* **4,** 192–201 (2003).

Becker, P.B. and Hörz, W., ATP-dependent nucleosome remodeling, *Annu. Rev. Biochem.* **71,** 247–273 (2002); Flaus, A. and Owen-Hughes, T., Mechanisms for ATP-dependent chromatin remodeling, *Curr. Opin. Genet. Dev.* **11,** 148–154 (2001); *and* Fry, C.J. and Peterson, C.L., Chromatin remodeling enzymes: Who's on first? *Curr. Biol.* **11,** R185–R197 (2001).

Bell, A.C., West, A.G., and Felsenfeld, G., Insulators and boundaries: Versatile regulatory elements in the eukaryotic genome, *Science* **291,** 447–450 (2001).

Berger, S., Histone modifications in transcriptional regulation, *Curr. Opin. Genet. Dev.* **12,** 142–148 (2002).

Buratowski, S., The basics of basal transcription by RNA polymerase II, *Cell* **77,** 1–3 (1994).

Burgess-Beusse, B., Farrell, C., Gaszner, M., Litt, M., Mutskov, V., Recillas-Targa, F., Simpson, M., West, A., and Felsenfeld, G., The insulation of genes from external enhancers and silencing chromatin, *Proc. Natl. Acad. Sci.* **99,** 16433–16437 (2002).

Branden, C. and Tooze, J., *Introduction to Protein Structure* (2nd ed.), Chapters 9 and 10, Garland Publishing (1999).

Carey, M. and Smale, S.T., *Transcriptional Regulation in Eukaryotes. Concepts, Strategies, and Techniques,* Cold Spring Harbor Laboratory Press (2000).

Chasman, D.I., Flaherty, K.M., Sharp, P.A., and Kornberg, R.D., Crystal structure of yeast TATA-binding protein and model for interaction with DNA, *Proc. Natl. Acad. Sci.* **90,** 8174–8178 (1993); *and* Nikolov, D.B., Hu, S.-H., Lin, J., Gasch, A., Hoffmann, A., Horikoshi, M., Chua, N.-H., Roeder, R.G., and Burley, S.K., Crystal structure of TFIID TATA-box binding protein, *Nature* **360,** 40–46 (1992).

Chen, F.E. and Ghosh, G., Regulation of DNA binding by Rel/NF-κB transcription factors: structural view, *Oncogene* **18,** 6845–6852 (1999); *and* Berkowitz, B., Huang, D.-B., Chen-Park, F.E., Sigler, P.B., and Ghosh, G., The X-ray crystal structure of the NF-κB p50·p65 heterodimer bound to the interferon β-κB site, *J. Biol. Chem.* **277,** 24694–24700 (2002).

Chen, X., et al., Gene expression patterns in human liver cancers, *Mol. Biol. Cell* **13,** 1929–1939 (2002).

Conway, R.C. and Conway, J.W., General initiation factors for RNA polymerase II, *Annu. Rev. Biochem.* **62,** 161–190 (1993).

Dotson, M.R., Yuan, C.X., Roeder, R.G., Myers, L.C., Gustafsson, C.M., Jiang, Y.W., Li, Y., Kornberg, R.D., and Asturias, F.J., Structural organization of yeast and mammalian mediator complexes, *Proc. Natl. Acad. Sci.* **97,** 14307–14310 (2000); *and* Davis, J.A., Takagi, Y., Kornberg, R.D., and Asturias, F.J., Structure of the yeast RNA polymerse II holoenzyme: Mediator conformation and polymerase interaction, *Mol. Cell* **10,** 409–415 (2002).

Elgin, S.C.R., The formation and function of DNase I hypersensitive sites in the process of gene activation. *J. Biol. Chem.* **263,** 19259–19262 (1988).

Ellenberger, T.E., Getting a grip on DNA recognition: structures of the basic region leucine zipper, and the basic region helix-loop-helix DNA-binding domains, *Curr. Opin. Struct. Biol.* **4,** 12–21 (1994).

Ellenberger, T.E., Brandl, C.J., Struhl, K., and Harrison, S.C., The GCN4 basic region leucine zipper binds DNA as a dimer of uninterrupted α helices: Crystal structure of the protein–DNA complex, *Cell* **71,** 1223–1237 (1992).

Evans, R.M., The steroid and thyroid hormone receptor superfamily, *Science* **240,** 889–895 (1988).

Ferré-d'Amaré, A.R., Prendergast, G.C., Ziff, E.B., and Burley, S.K., Recognition by Max of its cognate DNA through a dimeric b/HLH/Z domain, *Nature* **363,** 38–45 (1993).

Finnin, M.S., Donigian, J.R., and Pavletich, N.P., Structure of the histone deacetylase SIRT2, *Nature Struct. Biol.* **8,** 621–625 (2001).

Funder, J.W., Glucocorticoid and mineralocorticoid receptors: Biology and clinical relevance, *Annu. Rev. Med.* **48,** 231–240 (1997).

Gangloff, Y.-G., Romier, C., Thuault, S., Werten, S., and Davidson, I., The histone fold is a key structural motif of transcription factor TFIID, *Trends Biochem. Sci.* **26,** 250–257 (2001).

Garvie, C.W. and Wolberger, C., Recognition of specific DNA complexes, *Mol. Cell* **8,** 937–946 (2001).

Geiger, J.H., Hahn, S., Lee, S., and Sigler, P.B., Crystal structure of the yeast TFIIA/TBP/DNA complex, *Science* **272,** 830–836 (1996); *and* Tan, S., Hunziker, Y., Sargent, D.F., and Richmond, T.J., Crystal structure of a yeast TFIIA/TBP/DNA complex, *Nature* **381,** 127–134 (1996).

Grewal, S.I.S. and Moazed, D., Heterochromatin and epigenetic control of gene expression, *Science* **301,** 798–802 (2003).

Gross, D.S. and Garrard, W.T., Nuclease hypersensitive sites in chromatin, *Annu. Rev. Biochem.* **57,** 159–197 (1988).

Gustafsson, C.M. and Samuelsson, T., Mediator—a universal complex in transcription regulation, *Mol. Microbiol.* **41,** 1–8 (2001).

Huth, J.R., Bewley, C.A., Nissen, M.S., Evans, J.N.S., Reeves, R., Gronenborn, A.M., and Clore, G.M., The solution structure of an HMG-I(Y)–DNA complex defines a new architectural minor groove binding motif, *Nature Struct. Biol.* **4,** 657–665 (1997).

Iizuka, M. and Smith, M.M., Functional consequences of histone modifications, *Curr. Opin. Genet. Dev.* **13,** 154–160 (2003).

Jacobson, R.H., Ladurner, A.G., King, D.S., and Tjian, R., Structure and function of the human $TAF_{II}250$ double bromodomain module, *Science* **288,** 1422–1425 (2000).

Khorasanizadeh, S. and Rastinejad, F., Nuclear-receptor interactions on DNA-response elements, *Trends Biochem. Sci.* **26,** 384–390 (2001).

Kim, Y., Geiger, J.H., Hahn, S., and Sigler, P.B., Crystal structure of a yeast TBP/TATA-box complex; Kim, J.L., Nikolov, D.B., and Burley, S.K., Co-crystal structure of TBP recognizing the minor groove of a TATA element, *Nature* **365,** 512–520 *and* 520–527 (1993); *and* Nikolov, D.B. and Burley, S.K., 2.1 Å resolution refined structure of TATA box-binding protein (TBP), *Nature Struct. Biol.* **1,** 621–637 (1994).

Klug, A. and Rhodes, D., 'Zinc fingers': a novel protein motif for nucleic acid recognition, *Trends Biochem. Sci.* **12,** 464–469 (1987).

Kouzarides, T., Histone methylation in transcriptional control, *Curr. Opin. Genet. Dev.* **12,** 198–209 (2002); *and* Bannister, A.J., Schneider, R., and Kouzarides, T., Histone methylation: Dynamic or static, *Cell* **109,** 801–806 (2002).

Lee, T.I. and Young, R.A., Transcription of eukaryotic protein-coding genes, *Annu. Rev. Genet.* **34,** 77–137 (2000).

Lemon, B. and Tjian, R., Orchestrated response: A symphony of transcription factors for gene control, *Genes Devel.* **14,** 2551–2569 (2000); *and* Näär, A.M., Lemon, B.D., and Tjian, R., Transcriptional coactivator complexes, *Annu. Rev. Biochem.* **70,** 475–501 (2001).

Li, Q., Peterson, K.R., Fang, X., and Stamatoyannopoulos, G., Locus control regions, *Blood* **100,** 3077–3086 (2002).

Locker, J. (Ed.), *Transcription Factors,* Academic Press (2001).

Luisi, B.F., Xu, W.X., Otwinowski, Z., Freedamn, L.P., Yamamoto, K.R., and Sigler, P.B., Crystallographic analysis of the interaction of the glucocorticoid receptor with DNA, *Nature* **352,** 497–505 (1991).

Malik, S. and Roeder, R.G., Transcriptional regulation through mediator-like coactivators in yeast and metazoan cells, *Trends Biochem. Sci.* **25,** 277–283 (2000).

Marmorstein, R., Structure and function of histone acetyltransferases, *Cell. Mol. Life Sci.* **58,** 693–703 (2001); Marmorstein, R., Protein modules that manipulate histone tails for chromatin regulation, *Nature Rev. Mol. Cell. Biol.* **2,** 422–432 (2001); *and* Marmorstein, R. and Roth, S.Y., Histone acetyltransferases: function, structure, and catalysis, *Curr. Opin. Genet. Dev.* **11,** 155–161 (2001).

Marmorstein, R., Structure of histone deacetylases: Insights into substrate recognition and catalysis, *Structure* **9,** 1127–1133 (2001).

Marmorstein, R., Carey, M., Ptashne, M., and Harrison, S.C., DNA recognition by GAL4: structure of a protein–DNA complex, *Nature* **356,** 408–414 (1992).

Marmorstein, R. and Fitzgerald, M.X., Modulation of DNA-binding domains for sequence-specific DNA recognition, *Gene* **304,** 1–12 (2003).

Martin, G.M., X-Chromosome inactivation in mammals, *Cell* **29,** 721–724 (1982).

McKnight, S.L. and Yamamoto, K.R. (Eds.), *Transcriptional Regulation,* Cold Spring Harbor Laboratory Press (1992). [A two-volume compendium.]

Myers, L.C. and Kornberg, R.D., Mediator of transcriptional regulation, *Annu. Rev. Biochem.* **69,** 729–749 (2000).

Nielsen, P.R., Nietlspach, D., Mott, H.R., Callaghan, J., Bannister, A., Kouzarides, T., Murzin, A.G., Murzin, N.V., and Laue, E.D., Structure of the HP1 chromodomain bound to histone H3 methylated at lysine 9, *Nature* **416,** 103–107 (2002).

Nikolov, D.B., Chen, H., Halay, E.D., Usheva, A.A., Hisatake, K., Lee, D.K., Roeder, R.G., and Burley, S.K., Crystal structure of a TFIIB–TBP–TATA-element ternary complex, *Nature* **377,** 119–128 (1995); *and* Tsai, F.T.F. and Sigler, P.B., Structural basis of preinitiation complex assembly on human Pol II promoters, *EMBO J.* **19,** 25–36 (2000).

Orlando, V., Mapping chromosomal proteins *in vivo* by formaldehyde-crosslinked-chromatin immunoprecipitation, *Trends Biochem. Sci.* **25,** 99–104 (2000).

O'Shea, E.K., Klemm, J.D., Kim, P.S., and Alber, T., X-Ray structure of the GCN4 leucine zipper, a two-stranded, parallel coiled coil, *Science* **254,** 539–544 (1991).

Patikoglou, G. and Burley, S.K., Eukaryotic transcription factor-DNA complexes, *Annu. Rev. Biophys. Biomol. Struct.* **26,** 289–325 (1997); *and* Burley, S.K. and Kamada, K., Transcription factor complexes, *Curr. Opin. Struct. Biol.* **12,** 225–230 (2002).

Pavletich, N.P. and Pabo, C.O., Zinc finger–DNA recognition: Crystal structure of a Zif268-DNA complex at 2.1 Å, *Science* **252,** 809–817 (1991).

Pelz, S.W., Brewer, G., Bernstein, P., Hart, P.A., and Ross, J., Regulation of mRNA turnover in eukaryotic cells, *Crit. Rev. Euk. Gene Express.* **1,** 99–126 (1991); *and* Atwater, J.A., Wisdom, R., and Verma, I.M., Regulated mRNA stability, *Annu. Rev. Genet.* **24,** 519–541 (1990).

Poux, A.N., Cebrat, M., Kim, C.M., Cole, P.A., and Marmorstein, R., Structure of the GCN5 histone acetyltransferase bound to a bisubstrate inhibitor, *Proc. Natl. Acad. Sci.* **99,** 14065–14070 (2002).

Ptashne, M. and Gann, A., *Genes & Signals,* Cold Spring Harbor Laboratory Press (2002). [Discusses mechanisms of genetic regulation.]

Pugh, B.F., Control of gene expression through the regulation of the TATA-binding protein, *Gene* **255,** 1–14 (2000).

Raghow, R., Regulation of messenger RNA turnover in eukaryotes, *Trends Biochem. Sci.* **12,** 358–360 (1987).

Riggs, A.D. and Pfeifer, G.P., X-chromosome inactivation and cell memory, *Trends Genet.* **8,** 169–174 (1992).

Roth, S.Y., Denu, J.M., and Allis, C.D., Histone acetyltransferases, *Annu. Rev. Biochem.* **70,** 81–120 (2001).

Schmiedeskamp, M. and Klevit, R.E., Zinc finger diversity, *Curr. Opin. Struct. Biol.* **4,** 28–35 (1994).

Schwabe, J.W.R., Chapman, L., Finch, J.T., and Rhodes, D., The crystal structure of the estrogen receptor DNA-binding domain bound to DNA: How receptors discriminate between their response elements, *Cell* **75,** 567–578 (1993).

Schwabe, J.W.R. and Klug, A., Zinc mining for protein domains, *Nature Struct. Biol.* **1,** 345–349 (1994). [Discusses the varieties of zinc finger proteins.]

Smale, S.T. and Kadonaga, J.T., The RNA polymerase II core promoter, *Annu. Rev. Biochem.* **72,** 449–479 (2003).

Stamatoyannopoulos, G. and Nienhuis, A.W. (Eds.), *The Regulation of Hemoglobin Switching,* The Johns Hopkins University Press (1991).

Strahl, B.D. and Allis, C.D., The language of covalent histone modifications, *Nature* **403,** 41–45 (2000); *and* Rice, J.C. and Allis, C.D., Histone methylation versus histone acetylation: New insights into epigenetic regulation, *Curr. Opin. Cell Biol.* **13,** 263–273 (2001).

Struhl, K., Duality of TBP, the universal transcription factor, *Science* **263,** 1103–1104 (1994); Rigby, P.W.J., Three in one and one in three: It all depends on TBP, *Cell* **72,** 7–10 (1993); *and* White, R.J. and Jackson, S.P., The TATA-binding protein: a central role in transcription by RNA polymerases I, II, and III, *Trends Genet.* **8,** 284–288 (1992).

Studitsky, V.M., Clark, D.J., and Felsenfeld, G., A histone octamer can step around a transcribing polymerase without leaving the template, *Cell* **76,** 371–382 (1994); Studitsky, V.M., Kassavetis, G.A., Geiduschek, E.P., and Felsenfeld, G., Mechanism of transcription through the nucleosome by eukaryotic RNA polymerase, *Science* **278,** 1960–1965 (1997); *and* Felsenfeld, G., Clark, D., and Studitsky, V., Transcription through nucleosomes, *Biophys. Chem.* **86,** 231–237 (2000).

Tora, L., A unified nomenclature for TATA box binding protein (TBP)-associated factors (TAFs) involved in RNA polymerase II transcription, *Genes Dev.* **16,** 673–675 (2002).

Tsai, M.J. and O'Malley, B., Molecular mechanisms of action of steroid/thyroid receptor superfamily members, *Annu. Rev. Biochem.* **63,** 451–486 (1994).

Turner, B.M., Cellular memory and the histone code, *Cell* **111,** 285–291 (2002).

Veenstra, G.J.C. and Wolffe, A.P., Gene-selective developmental roles of general transcription factors, *Trends Biochem. Sci.* **26,** 665–671 (2001).

Wolffe, S.A., Nekludova, L., and Pabo, C.O., DNA recognition by Cys_2His_2 zinc finger proteins, *Annu Rev. Biophys. Biomol. Struct.* **3,** 183–212 (1999).

Xiao, B., et al., Structure and catalytic mechanism of the human histone methyltransferase SET7/9, *Nature* **421,** 652–656 (2003).

Xie, X., Kokubo, K., Cohen, S.L., Mirza, U.A., Hoffmann, A., Chait, B.T., Roeder, R.G., Nakatani, Y., and Burley, S.K., Structural similarities between TAFs and the heterotetrameric core of the histone octamer, *Nature* **380,** 316–322 (1996); Selleck, W., Howley, R., Fang, Q., Podolny, V., Fried, M.G., Buratowski, S., and Tan, S., A histone fold TAF octamer within the yeast TFIID transcriptional coactivator, *Nature Struct. Biol.* **8,** 695–700 (2001); *and* Werten, S., Mitschler, A., Romier, C., Gangloff, Y.-G., Thuault, S., Davidson, I., and Moras, D., Crystal structure of a subcomplex of human transcription factor TFIID formed by TATA binding protein-associated factors hTAF4 ($hTAF_{II}135$) and hTAF12 ($hTAF_{II}20$), *J. Biol. Chem.* **277,** 45502–45509 (2002).

Yang, X.-J. and Seto, E., Collaborative spirit of histone deacetylases in regulating chromatin structure and gene expression, *Curr. Opin. Genet. Dev.* **13,** 143–153 (2003).

DEVELOPMENT

Bate, M. and Arias, A.M. (Eds.), *The Development of Drosophila melanogaster,* Cold Spring Harbor Laboratory Press (1993).

Blau, H.M., Differentiation requires continuous active control, *Annu. Rev. Biochem.* **61,** 1213–1230 (1992).

Fujioka, M., Emi-Sarker, Y., Yusibova, G.L., Goto, T., and Jaynes, J.B., Analysis of an *even-skipped* rescue transgene reveals both composite and discrete neuronal and early blastoderm enhancers and multi-stripe positioning by gap gene repressor gradients, *Development* **126,** 2527–2538 (1999).

Gehring, W.J., Affolter, M., and Bürglin, T., Homeodomain proteins, *Annu. Rev. Biochem.* **63,** 487–526 (1994); *and* Gehring, W.J., Qian, Y.Q., Billeter, M., Furukobu-Tokunaga, K., Schier, A.F., Resendez-Perez, D., Affolter, M., Otting, G., and Wüthrich, K., Homeodomain–DNA recognition, *Cell* **78,** 211–223 (1994).

Gilbert, S.F., *Developmental Biology* (7th ed.), Sinauer Associates (2003).

Gossler, A. and Balling, R., The molecular and genetic analysis of mouse development, *Eur. J. Biochem.* **204,** 5–11 (1992).

Gurdon, J.B., The generation of diversity and pattern in animal development, *Cell* **68,** 185–199 (1992).

Halder, G., Callaerts, P., and Gehring, W.J., Induction of ectopic eyes by targeted expression of the *eyeless* gene in *Drosophila, Science* **267,** 1788–1792 (1995).

Kenyon, C., If birds can fly, why can't we? Homeotic genes and evolution, *Cell* **78,** 175–180 (1994).

Kissinger, C.R., Liu, B., Martin-Blanco, E., Kornberg, T.B., and Pabo, C.O., Crystal structure of an engrailed homeodomain–DNA complex at 2.8 Å resolution: A framework for understanding homeodomain–DNA interactions, *Cell* **63,** 579–590 (1990).

Krumlauf, R., *Hox* genes in development, *Cell* **78,** 191–201 (1994).

Lawrence, P.A., *The Making of a Fly,* Blackwell Scientific Publications (1992).

Lawrence, P.A. and Morata, G., Homeobox genes: Their function in Drosophila segmentation and pattern formation, *Cell* **78,** 181–191 (1994).

Le Mouellic, H., Lallemand, Y., and Brûlet, P., Homeosis in the mouse induced by a null mutation in the *Hox-3.1* gene, *Cell* **69,** 251–264 (1992).

Mann, R.S. and Morata, G., The developmental and molecular biology of genes that subdivide the body of *Drosophila, Annu. Rev. Cell Dev. Biol.* **16,** 243–271 (2000).

Mavilio, F., Regulation of vertebrate homeobox-containing genes by morphogens, *Eur. J. Biochem.* **212,** 273–288 (1993).

Nüsslein-Volhard, C., Axis determination in the *Drosophila* embryo, *Harvey Lect.* **86,** 129–148 (1992); *and* St. Johnston, D. and Nüsslein-Volhard, C., The origin of pattern and polarity in the *Drosophila* embryo, *Cell* **68,** 201–219 (1992). [Detailed reviews.]

Nüsslein-Volhard, C., The identification of genes controlling development in flies and fishes (Nobel lecture); *and* Wieschaus, E., From molecular patterns to morphogenesis—The lessons from studies on the fruit fly *Drosophila* (Noble lecture), *Angew. Chem. Int. Ed. Engl.* **35,** 2177–2187 *and* 2189–2194 (1996).

Scott, M.P., Development: The natural history of genes, *Cell* **100,** 27–40 (2000).

CANCER AND THE REGULATION OF THE CELL CYCLE

Adams, P.D., Regulation of the retinoblastoma tumor suppressor protein by cyclin/cdks, *Biochim. Biophys. Acta* **1471,** M123–M133 (2001).

Cho, Y., Gorina, S., Jeffrey, P.D., and Pavletich, N.P., Crystal structure of a p53 tumor suppressor–DNA complex: Understanding tumorigenic mutations, *Science* **265,** 346–355 (1994).

Cooper, G.M. and Hausman, R.E., *The Cell. A Molecular Approach* (3rd ed.), Chapter 14, ASM Press (2004).

De Bondt, H.L., Rosenblatt, J., Jancarik, J., Jones, H.D., Morgan, D.O., and Kim, S.-H., Crystal structure of cyclin-dependent kinase 2, *Nature* **363,** 595–602 (1993); Jeffrey, P.D., Russo, A.A., Polyak, K., Gibbs, E., Hurwitz, J., Massagu, J., and Pavletich, N.P., Mechanism of CDK activation revealed by the structure of a cyclin A–CDK2 complex, *Nature* **376,** 313–320 (1995); Russo, A.A., Jeffrey, P.D., Patten, A.K., Massagué, J., and Pavletich, N.P., Crystal structure of the $p27^{Kip1}$ cyclin-dependent-kinase inhibitor bound to the cyclin A–Cdk2 complex, *Nature* **382,** 325–331 (1996); *and* Russo, A.A., Jeffrey, P.D., and Pavletich, N.P., Structural basis of cyclin-dependent kinase activation by phosphorylation, *Nature Struct. Biol.* **3,** 696–700 (1996).

Donehower, L.A., Harvey, M., Slagle, B.L., McArthur, M.J., Montgomery, C.A., Jr., Butel, J.S., and Bradley, A., Mice deficient for p53 are developmentally normal but susceptible to spontaneous tumours, *Nature* **356,** 215–221 (1992).

Haluska, F.G., Tsujimoto, Y., and Croce, C.M., Oncogene activation by chromosome translocation in human malignancy, *Annu. Rev. Genet.* **21,** 321–345 (1987).

Harbour, J.W. and Dean, D.C., The Rb/E2F pathway: expanding roles and emerging paradigms, *Genes Dev.* **14,** 2393–2409 (2000).

Harper, J.W. and Adams, P.D., Cyclin-dependent kinases, *Chem. Rev.* **101,** 2511–2526 (2001).

Hickman, E.S., Moroni, M.C., and Helin, K., The role of p53 and pRB in apoptosis and cancer, *Curr. Opin. Genet. Dev.* **12,** 60–66 (2002).

Hunter, T. and Pines, J., Cyclins and cancer, *Cell* **66,** 1071–1074 (1991).

Jeffrey, P.D., Tong, L., and Pavletich, N.P., Structural basis of inhibition of CDK-cyclin complexes by INK4 inhibitors, *Genes Dev.* **14,** 3115–3125 (2000).

Johnson, D.G. and Walker, C.L., Cyclins and cell cycle checkpoints, *Annu. Rev. Pharmacol Toxicol.* **39,** 295–312 (1999).

Johnstone, R.W., Histone-deacetylase inhibitors: Novel drugs for the treatment of cancer, *Nature Rev. Drug Disc.* **1,** 287–299 (2002).

Lee, E.Y.-H., Chang, C.-Y., Hu, N., Wang, Y.-C.J., Lai, C.-C., Herrup, K., Lee, W.-H., and Bradley, A., Mice deficient for Rb are nonviable and show defects in neurogenesis and haematopoiesis, *Nature* **359,** 288–394 (1992).

Morgan, D.O., Cyclin-dependent kinases: Engines, clocks, and microprocessors, *Annu. Rev. Cell Dev. Biol.* **13,** 261–291 (1997).

Morris, E.J. and Dyson, N.J., Retinoblastoma protein partners, *Adv. Cancer Res.* **82,** 1–54 (2001).

Nigg, E.A., Mitotic kinases as regulators of cell division and its checkpoints, *Nature Rev. Mol. Biol.* **2,** 21–32 (2001).

Norbury, C. and Nurse, P., Animal cell cycles and their control, *Annu. Rev. Biochem.* **61,** 441–470 (1992); *and* Forsburg, S.L. and Nurse, P., Cell cycle regulation in the yeasts *Saccharomyces cerevisiae* and *Schizosaccharomyces pombe, Annu. Rev. Cell Biol.* **7,** 227–256 (1991).

Pavletich, N.P., Mechanisms of cyclin-dependent kinase regulation: Structures of Cdks, their cyclin activators, and Cip and INK4 inhibitors, *J. Mol. Biol.* **287,** 821–828 (1999).

Pollard, T.D. and Earnshaw, W.C., *Cell Biology,* Chapters 43–47 and 49, Saunders (2002).

Russell, P., Checkpoints on the road to mitosis, *Trends Biochem. Sci.* **23,** 399–402 (1998).

Russo, A.A., Jeffrey, P.D., Patten, A.K., Massagué, J., and Pavletich, N.P., Crystal structure of the $p27^{Kip1}$ cyclin-dependent-kinase inhibitor bound to the cyclin A–Cdk2 complex, *Nature* **382,** 325–331 (1996).

Sherr, C.J. and Weber, J.D., The ARF/p53 pathway, *Curr. Opin. Genet. Dev.* **10,** 94–99 (2000).

Vogelstein, B. and Kinzler, K.W., The multistep nature of cancer, *Trends Genet.* **9,** 138–140 (1993).

Volgelstein, B., Lane, D., and Levine, A.J., Surfing the p53 network, *Nature* **408,** 307–310 (2000). [Discusses how p53 integrates the various signals that control cell life and death.]

Xiao, B., Spencer, J., Clements, A., Ali-Khan, N., Mittnacht, S., Broceño, C., Burghammer, M., Parrakis, A., Marmorstein, M., and Gamblin, S.J., Crystal structure of the retinoblastoma tumor suppressor protein bound to E2F and the molecular basis of its regulation, *Proc. Natl. Acad. Sci.* **100,** 2363–2368 (2003); *and* Lee, C., Chang, J.H., Lee, H.S., and Cho, Y., Structural basis for the recognition of the E2F transactivation domain by the retinoblastoma tumor suppressor, *Genes Dev.* **16,** 3199–3212 (2002).

APOPTOSIS

Acehan, D., Jiang, X., Morgan, D.G., Heuser, J.E., Wang, X., and Akey, C.W., Three-dimensional structure of the apoptosome: Implications for assembly, procaspase-9 binding, and activation, *Mol. Cell* **9,** 423–432 (2002).

Chai, J., Wu, Q., Shiozaki, E., Srinivasula, S.M., Alnemri, E.S., and Shi, Y., Crystal structure of a procaspase-7 zymogen: Mechanisms of activation and substrate binding, *Cell* **107,** 399–407 (2001); *and* Riedl, S.J., Fuentes-Prior, P., Renatus, M., Kairies, N., Krapp, S., Huber, R., Salvesen, G.S., and Bode, W., Structural basis for the activation of human procaspase-7, *Proc. Natl. Acad. Sci.* **98,** 14790–14795 (2001).

Desagher, S. and Martinou, J.C., Mitochondria as the central control point of apoptosis, *Trends Cell Biol.* **10,** 369–377 (2000).

Earnshaw, W.C., Martins, L.M., and Kaufmann, S.H., Mammalian caspases: Structure, activation, substrates, and functions during apoptosis, *Annu. Rev. Biochem.* **68,** 383–424 (1999); *and* Grütter, M.G., Caspases: key players in programmed cell death, *Curr. Opin. Struct. Biol.* **10,** 649–655 (2000).

Fesik, S.W., Insights into programmed cell death through structural biology, *Cell* **103,** 272–282 (2000).

Hengartner, M.O., The biochemistry of apoptosis, *Nature* **407,** 770–776 (2000).

Jacobson, M.D. and McCarthy, N. (Eds.), *Apoptosis,* Oxford (2002).

Nagata, S., Fas ligand-induced apoptosis, *Annu. Rev. Genet.* **33,** 29–55 (1999).

Riedl, S.J, Renatus, M., Schwarzenbacher, R., Zhou, Q., Sun, C., Fesik, S.W., Liddington, R.C., and Salvesen, G.S., Structural basis for the inhibition of caspase-3 by XIAP, *Cell* **104,** 791–800 (2001).

Strasser, A., O'Connor, L., and Dixit, V.M., Apoptosis signaling, *Annu. Rev. Biochem.* **69,** 217–245 (2000).

Wei, Y., Fox, T., Chambers, S.P., Sinchak, J., Coll, J.T., Golec, J.M.C., Swenson, L., Wilson, K.P., and Charifson, P., The structures of capsases-1, -3, -7, and -8 reveal the basis for substrate and inhibitor selectivity, *Chem. Biol.* **7,** 423–432 (2000).

Yin, X.M. and Dong, Z. (Eds.), *Essentials of Apoptosis,* Humana Press (2003).

PROBLEMS

1. What is the maximum possible packing ratio of a 10^6-bp segment of DNA; of a 10^9-bp segment of DNA? Assume the DNA is a 20-Å-diameter cylinder with a contour length of 3.4 Å/bp.

2. When an SV40 minichromosome (a closed circular duplex DNA in complex with nucleosomes) is relaxed so that it forms an untwisted circle and is then deproteinized, the consequent closed circular DNA has about −1 superhelical turn for each of the nucleosomes that it originally had. Explain the discrepancy between this observation and the fact that the DNA in each nucleosome is wrapped nearly twice about its histone octamer in a left-handed superhelix.

3. Explain why acidic polypeptides such as polyglutamate facilitate *in vitro* nucleosome assembly.

***4.** Consider a 1 million-bp DNA molecule that has 1500 tandem repeats of a 400-bp sequence with the remainder of the DNA consisting of unique sequences. Sketch the C_0t curve of this DNA when it is sheared into pieces averaging 1000 bp long; when they are 100 bp long.

5. Why do isolated foldback structures, when treated by an endonuclease that cleaves only single-stranded DNA and then denatured, yield complicated C_0t curves?

6. During its 2-month period of maturation, the *Xenopus* oocyte synthesizes ~10^{12} ribosomes. The consequent tremendous rate of rRNA synthesis is only possible because the normal genomic complement of rDNA has been amplified 1500-fold. (a) Why is it unnecessary to likewise amplify the genes encoding the ribosomal proteins? (b) Assuming that rRNA gene amplification occurs in a short time at the beginning of the maturation period, how long would oogenesis require if the rDNA were not amplified?

7. **Hb Kenya** is a β-thalassemia in which the β-globin cluster is deleted between a point in the $^{A}\gamma$-globin gene and the corresponding position in the β-globin gene. Describe the most probable mechanism for the generation of this mutation.

8. Red–green color blindness is caused by an X-linked recessive genetic defect. Hence females rarely exhibit the red–green color-blind phenotype but may be carriers of the defective gene. When a narrow beam of red or green light is projected onto some areas of the retina of such a female carrier, she can readily differentiate the two colors but on other areas she has difficulty in doing so. Explain.

9. Figure 34-53*a* contains a single band just above the bracketed region that increases in density as the Sp1 concentration increases. What is origin of this band?

10. Why do the rare instances of male calico cats all have the abnormal XXY genotype?

11. In *Drosophila*, an *esc*$^-$ homozygote develops normally unless its mother is also an *esc*$^-$ homozygote. Explain.

12. The fusion of cancer cells with normal cells often suppresses the expression of the tumorigenic phenotype. Explain.

NOTES

NOTES

Index

Page references in **bold face** refer to a major discussion of the entry. Positional and configurational designations in chemical names (e.g., 3-, α, *N*-, *p*-, trans, D-, *sn*-) are ignored in alphabetizing. Numbers and Greek letters are otherwise alphabetical as if they were spelled out.